Lecture Notes in Computer Science 7318

Commenced Publication in 1973
Founding and Former Series Editors:
Gerhard Goos, Juris Hartmanis, and Jan van Leeuwen

W0227568

S. Barry Cooper   Anuj Dawar
Benedikt Löwe (Eds.)

How the World Computes

Turing Centenary Conference and
8th Conference on Computability in Europe, CiE 2012
Cambridge, UK, June 18-23, 2012
Proceedings

 Springer

Volume Editors

S. Barry Cooper
University of Leeds, School of Mathematics
Leeds LS2 9JT, UK
E-mail: s.b.cooper@leeds.ac.uk

Anuj Dawar
University of Cambridge, Computer Laboratory
William Gates Building, J.J. Thomson Avenue
Cambridge CB3 0FD, UK
E-mail: anuj.dawar@cl.cam.ac.uk

Benedikt Löwe
University of Amsterdam, Institute for Logic, Language and Computation
P.O. Box 94242, 1090 GE Amsterdam, The Netherlands
E-mail: b.loewe@uva.nl

ISSN 0302-9743   e-ISSN 1611-3349
ISBN 978-3-642-30869-7   e-ISBN 978-3-642-30870-3
DOI 10.1007/978-3-642-30870-3
Springer Heidelberg Dordrecht London New York

Library of Congress Control Number: 2012938860

CR Subject Classification (1998): F.2, F.1, G.2, I.1

LNCS Sublibrary: SL 1 – Theoretical Computer Science and General Issues

Typesetting: Camera-ready by author, data conversion by Scientific Publishing Services, Chennai, India

Printed on acid-free paper

Springer is part of Springer Science+Business Media (www.springer.com)

Preface

CiE 2012: How the World Computes, and the Alan Turing Centenary

In 2003, a group of researchers started their funding proposal for a *Research Training Network* of the European Commission under the unfamiliar acronym CiE with the words:

> We can see now that the world changed in 1936, in a way quite unrelated to the newspaper headlines of that year concerned with such things as the civil war in Spain, economic recession, and the Berlin Olympics. The end of that year saw the publication of a thirty-six page paper by a young mathematician, Alan Turing, claiming to solve a long-standing problem of the distinguished German mathematician David Hilbert.

The proposal (eventually unfunded) went on to describe how "computability as a theory is a specifically twentieth-century development," and how subsequently computability became "both a driving force in our daily lives" and a concept one could talk about with a new precision.

The work and personality of Turing has had a key influence on the development of CiE as an important player in the new multi-disciplinary landscape of twenty-first century computability theory. It has been said that Alan Turing did not so much inhabit different disciplines as investigate one discipline that was fundamental. In recent times, CiE has sought to break down disciplinary barriers, and to sponsor a return to this questioning approach of Turing to "How the World Computes."

Computability in Europe (CiE) is now a hugely successful conference series; an association with close on a thousand members, and growing; has its own journal *Computability*; and edits a high-profile book series *Theory and Applications of Computability*. It was just under five years ago, toward the end of 2007, that the CiE Board set out to acknowledge its debt to Turing by making the Alan Turing Year 2012 a unique and scientifically exciting year. Since then, CiE has formed an international *Turing Centenary Advisory Committee*, which has played an important part in the development of the Turing Centenary programme into a world-wide celebration. The CiE 2012 Turing Centenary Conference will be remembered as a historic event in the continuing development of the powerful explanatory role of computability across a wide spectrum of research areas. We believe that the work presented at CiE 2012 represents the best of current research in the area, and forms a fitting tribute to the short but brilliant trajectory of Alan Mathison Turing (June 23, 1912 – June 7, 1954).

Apart from being a celebration of the Turing Centenary, CiE 2012 was the eighth meeting in our conference series. Both the conference series and the association promote the development of computability-related science, ranging over mathematics, computer science and applications in various natural and engineering sciences such as physics and biology, and also including the promotion of related non-scientific fields such as philosophy and history of computing. This conference was held at the University of Cambridge in England, linking naturally to the semester-long research programme *Semantics & Syntax* at the *Isaac Newton Institute for Mathematical Sciences*.

The first seven of the CiE conferences were held at the University of Amsterdam in 2005, at the University of Wales Swansea in 2006, at the University of Siena in 2007, at the University of Athens in 2008, at the University of Heidelberg in 2009, at the University of the Azores in 2010, and at Sofia University in 2011. The proceedings of these meetings, edited in 2005 by S. Barry Cooper, Benedikt Löwe and Leen Torenvliet, in 2006 by Arnold Beckmann, Ulrich Berger, Benedikt Löwe and John V. Tucker, in 2007 by S. Barry Cooper, Benedikt Löwe and Andrea Sorbi, in 2008 by Arnold Beckmann, Costas Dimitracopoulos and Benedikt Löwe, in 2009 by Klaus Ambos-Spies, Benedikt Löwe and Wolfgang Merkle, in 2010 by Fernando Ferreira, Benedikt Löwe, Elvira Mayordomo and Luís Mendes Gomes, and in 2011 by Benedikt Löwe, Dag Normann, Ivan Soskov, and Alexandra Soskova were published as *Springer Lecture Notes in Computer Science*, Volumes 3526, 3988, 4497, 5028, 5365, 6158, and 6735, respectively.

The annual CiE conference has become a major event, and is the largest international meeting focused on computability theoretic issues. The next meeting in 2013 will be in Milan, Italy, and will be co-located with the conference UCNC 2013 (*Unconventional Computation and Natural Computation*). The series is coordinated by the CiE Conference Series Steering Committee consisting of Luís Antunes (Porto, Secretary), Arnold Beckmann (Swansea), S. Barry Cooper (Leeds), Natasha Jonoska (Tampa FL), Viv Kendon (Leeds), Benedikt Löwe (Amsterdam and Hamburg, Chair), Dag Normann (Oslo), and Peter van Emde Boas (Amsterdam).

The Programme Committee of CiE 2012 was responsible for the selection of the invited speakers and special session organizers and consisted of Samson Abramsky (Oxford), Pieter Adriaans (Amsterdam), Franz Baader (Dresden), Arnold Beckmann (Swansea), Mark Bishop (London), Paola Bonizzoni (Milan), Douglas A. Cenzer (Gainesville FL), S. Barry Cooper (Leeds), Ann Copestake (Cambridge), Anuj Dawar (Cambridge), Solomon Feferman (Stanford CA), Bernold Fiedler (Berlin), Luciano Floridi (Oxford), Marcus Hutter (Canberra), Martin Hyland (Cambridge), Viv Kendon (Leeds), Stephan Kreutzer (Berlin), Ming Li (Waterloo ON), Benedikt Löwe (Amsterdam and Hamburg), Angus Macintyre (Edinburgh), Philip Maini (Oxford), Larry Moss (Bloomington IN), Amitabha Mukerjee (Kanpur), Damian Niwinski (Warsaw), Dag Normann (Oslo), Prakash Panangaden (Montréal QC), Jeff Paris (Manchester), Brigitte Pientka (Montréal QC), Helmut Schwichtenberg (Munich), Wilfried Sieg (Pittsburgh PA), Mariya Soskova (Sofia), Bettina Speckmann (Eindhoven), Christof Teuscher (Portland OR), Peter Van Emde Boas (Amsterdam), Jan Van Leeuwen (Utrecht), Rineke Verbrugge (Groningen).

Structure and Programme of the Conference

The conference had 12 invited plenary lectures, given by Dorit Aharonov (Jerusalem), Verónica Becher (Buenos Aires), Lenore Blum (Pittsburgh PA), Rodney Downey (Wellington), Yuri Gurevich (Redmond WA), Juris Hartmanis (Ithaca NY), Richard Jozsa (Cambridge), Stuart Kauffman (Santa Fe NM), James D. Murray (Princeton NJ), Stuart Shieber (Cambridge MA), Paul Smolensky (Baltimore MD), and Leslie Valiant (Cambridge MA). Six of these plenary lectures have abstracts in this volume. Blum's lecture was the *2012 APAL Lecture* funded by Elsevier, Gurevich's lecture was the *EACSL Invited Lecture* funded by the *European Association for Computer Science Logic*, Murray's lecture was the *Microsoft Research Cambridge Lecture* funded by Microsoft Research, and the lectures by Josza and Valiant were part of a joint event with King's College, Cambridge, on Alan Turing's 100th birthday (June 23, 2012). In addition to the plenary lectures, the conference had two public evening lectures delivered by Andrew Hodges (Oxford) and Ian Stewart (Warwick).

The 2012 conference CiE had six special session on a range of topics. Speakers in the special sessions were invited by the special session organizers and could contribute a paper to this volume. Eighteen of the invited special session speakers made use of this opportunity and their papers are included in these proceedings.

Cryptography, Complexity, and Randomness.
Organizers. Rod Downey (Wellington) and Jack Lutz (Ames IA).
Speakers. Eric Allender, Laurent Bienvenu, Lance Fortnow, Valentine Kabanets, Omer Reingold, and Alexander Shen.

The Turing Test and Thinking Machines.
Organizers. Mark Bishop (London) and Rineke Verbrugge (Groningen).
Speakers. Bruce Edmonds, John Preston, Susan Sterrett, Kevin Warwick, and Jiří Wiedermann.

Computational Models After Turing: The Church-Turing Thesis and Beyond.
Organizers. Martin Davis (Berkeley CA) and Wilfried Sieg (Pittsburg PA).
Speakers. Giuseppe Longo, Péter Németi, Stewart Shapiro, Matthew Szudzik, Philip Welch, and Michiel van Lambalgen.

Morphogenesis/Emergence as a Computability Theoretic Phenomenon.
Organizers. Philip Maini (Oxford) and Peter Sloot (Amsterdam).
Speakers. Jaap Kaandorp, Shigeru Kondo, Nick Monk, John Reinitz, James Sharpe, and Jonathan Sherratt.

Open Problems in the Philosophy of Information.
Organizers. Pieter Adriaans (Amsterdam) and Benedikt Löwe (Amsterdam and Hamburg).
Speakers. Patrick Allo, Luís Antunes, Mark Finlayson, Amos Golan, and Ruth Millikan.

The Universal Turing Machine, and History of the Computer.
Organizers. Jack Copeland (Canterbury) and John Tucker (Swansea).
Speakers. Steven Ericsson-Zenith, Ivor Grattan-Guinness, Mark Priestley, and Robert I. Soare.

CiE 2012 received 178 regular submissions. These were refereed by the Programme Committee and a long list of expert referees without whom the production of this volume would have been impossible. Based on their reviews, 53 of the submissions (29.8%) were accepted for publication in this volume. We should like to thank the subreferees for their excellent work; their names are listed at the end of this preface.

Organization and Acknowledgements

CiE 2012 was organized by Arnold Beckmann (Swansea), Luca Cardelli (Cambridge), S. Barry Cooper (Leeds), Ann Copestake (Cambridge), Anuj Dawar (Cambridge, Chair), Bjarki Holm (Cambridge), Martin Hyland (Cambridge), Benedikt Löwe (Amsterdam), Arno Pauly (Cambridge), Debbie Peterson (Cambridge), Andrew Pitts (Cambridge), and Helen Scarborough (Cambridge).

At the conference, we were able to continue the programme "Women in Computability" funded by the publisher Elsevier and organized by Mariya Soskova

(Sofia). There were five co-located events: the workshop *The Incomputable* organized as part of the Isaac Newton Institute program *Semantics & Syntax* at Chicheley Hall (June 12–15, 2012; organizers: S. Barry Cooper and Mariya Soskova), the conference ACE 2012 at King's College (June 15–16, 2012; organized by Jack Copeland and Mark Sprevak), the workshop *Developments in Computational Models* (DCM 2012) at Corpus Christi College (17 June 2012; organized by Benedikt Löwe and Glynn Winskel), the CiE-IFCoLog student session (June 18–23, 2012; organizers: Sandra Alves and Michael Gabbay), and the conference *Computability and Complexity in Analysis* (CCA 2012) at the Computer Lab of the University of Cambridge (June 24–27, 2012; organizers: Arno Pauly and Klaus Weihrauch).

The organization of CiE 2012 would not have been possible without the financial and/or materials support of our sponsors (in alphabetic order): the Association for Symbolic Logic, Cambridge University Press, Elsevier B.V., the European Association for Computer Science Logic (EACSL), the International Federation for Computational Logic (IFCoLog), IOS Press, the Isaac Newton Institute for Mathematical Sciences, King's College, Cambridge, Microsoft Research Cambridge, Robinson College Cambridge, Science Magazine/AAAS, Springer-Verlag, and the University of Cambridge.

We should also like to acknowledge the support of our non-financial sponsors, the Association Computability in Europe and the European Association for Theoretical Computer Science (EATCS).

We thank Andrej Voronkov for his EasyChair system which facilitated the work of the Programme Committee and the editors considerably. The final preparation of the files for this proceedings volume was done by Bjarki Holm, Steffen Lösch, and Nik Sultana.

April 2012   Anuj Dawar
                                                     S. Barry Cooper
                                                      Benedikt Löwe

Reviewers

Accatoli, Beniamino
Al-Rifaie, Mohammad Majid
Altenkirch, Thorsten
Ambainis, Andris
An, Hyung-Chan
Angione, Claudio
Arai, Toshiyasu
Arvind, Vikraman
Avigad, Jeremy
Bab, Sebastian
Barany, Vince
Barendregt, Henk
Barmpalias, Georgios
Barr, Katie
Beeson, Michael
Beklemishev, Lev
Bernardinello, Luca
Besozzi, Daniela
Bienvenu, Laurent
Binns, Stephen
Blass, Andreas
Bloem, Peter
Blokpoel, Mark
Bodlaender, Hans
Bradfield, Julian
Brattka, Vasco
Braud, Laurent
Braverman, Mark
Bridges, Douglas
Briet, Jop
Carayol, Arnaud
Carlucci, Lorenzo
Cassaigne, Julien
Castiglione, Giuseppa
Cervesato, Iliano
Cook, Stephen
Coquand, Thierry
Daswani, Mayank

Davenport, James
Davis, Martin
Day, Adam
De Mol, Liesbeth
De Weerd, Harmen
Dechesne, Francien
Decker, Hendrik
Delhomme, Christian
Della Vedova, Gianluca
Dennunzio, Alberto
Diener, Hannes
Dimitracopoulos, Costas
Doty, David
Duparc, Jacques
Dziubiński, Marcin
Dębowski, Łukasz
Endriss, Ulle
Evans, Roger
Everitt, Mark
Feige, Uriel
Ferreira, Fernando
Fici, Gabriele
Fokina, Ekaterina
Friis, Nicolai
Galliani, Pietro
Ganchev, Hristo
Ganesalingam, Mohan
Gasarch, William
Gentile, Claudio
Gherardi, Guido
Goncharov, Sergey
Goodman-Strauss, Chaim
Gundersen, Tom
Gutin, Gregory
Hamkins, Joel David
Harizanov, Valentina
Hauser, Marcus
Herbelot, Aurelie

Hertling, Peter
Hindley, Roger
Hinman, Peter
Hinzen, Wolfram
Hirst, Jeff
Holden, Sean
Horsman, Clare
Jockusch, Carl
Johannsen, Jan
Kach, Asher
Kahle, Reinhard
Kapron, Bruce
Kari, Jarkko
Kari, Lila
Khomskii, Yurii
Klin, Bartek
Kobayashi, Satoshi
Kolodziejczyk, Leszek
Kopczynski, Eryk
Kopecki, Steffen
Kracht, Marcus
Krajicek, Jan
Kristiansen, Lars
Kühnberger, Kai-Uwe
Kugel, Peter
Lasota, Sławomir
Lauria, Massimo
Leporati, Alberto
Lesnik, Davorin
Lewis, Andy
León, Carlos
Li, Zhenhao
Lippmann, Marcel
Luttik, Bas
Löding, Christof
Malone, David
Manea, Florin
Marion, Jean-Yves
Martin, Andrew

Meyerovitch, Tom
Michalewski, Henryk
Minari, Pierluigi
Minnes, Mia
Molina-Paris, Carmen
Montalban, Antonio
Moore, Cristopher
Morphett, Anthony
Mumma, John
Mummert, Carl
Murlak, Filip
Murray, Iain
Mycka, Jerzy
Nemoto, Takako
Ng, Selwyn
Novakovic, Novak
Oitavem, Isabel
Pattinson, Dirk
Pauly, Arno
Pin, Jean-Eric
Pradella, Matteo
Primiero, Giuseppe
Radoszewski, Jakub
Rampersad, Narad
Razborov, Alexander
Reimann, Jan
Richomme, Gwénaël
Rimell, Laura
Rubin, Sasha
Ruemmer, Philipp

Salibra, Antonino
Sanders, Sam
Sato, Masahiko
Sattler, Ulrike
Schlage-Puchta,
  Jan-Christoph
Schlimm, Dirk
Schneider, Thomas
Schraudner, Michael
Selivanov, Victor
Semukhin, Pavel
Seyfferth, Benjamin
Shafer, Paul
Shao, Wen
Simonsen, Jakob Grue
Simpson, Stephen
Skrzypczak, Michał
Smets, Sonja
Soloviev, Serguei
Sorbi, Andrea
Soskov, Ivan
Spandl, Christoph
Staiger, Ludwig
Stannett, Mike
Stephan, Frank
Straube, Ronny
Sturm, Monika
Sutskever, Ilya
Szudzik, Matthew
Szymanik, Jakub

Słomczyńska, Katarzyna
Tait, William
Tao, Terence
Tazari, Siamak
Terwijn, Sebastiaan
Tesei, Luca
Teufel, Simone
Thorne, Camilo
Torenvliet, Leen
Toruńczyk, Szymon
Toska, Ferit
Trautteur, Giuseppe
Trifonov, Trifon
Urquhart, Alasdair
van Kreveld, Marc
Vellino, Andre
Vencovska, Alena
Verlan, Sergey
Visser, Albert
Vogt, Paul
Vosgerau, Gottfried
Väänänen, Jouko
Wansing, Heinrich
Welch, Philip
Wiedermann, Jiři
Wiedijk, Freek
Wu, Guohua
Yang, Yue
Zdanowski, Konrad
Zeilberger, Noam

Table of Contents

Ordinal Analysis and the Infinite Ramsey Theorem.................... 1
  Bahareh Afshari and Michael Rathjen

Curiouser and Curiouser: The Link between Incompressibility and
Complexity 11
  Eric Allender

Information and Logical Discrimination 17
  Patrick Allo

Robustness of Logical Depth 29
  Luís Antunes, Andre Souto, and Andreia Teixeira

Turing's Normal Numbers: Towards Randomness...................... 35
  Verónica Becher

Logic of Ruler and Compass Constructions 46
  Michael Beeson

On the Computational Content of the Brouwer Fixed Point Theorem ... 56
  Vasco Brattka, Stéphane Le Roux, and Arno Pauly

Square Roots and Powers in Constructive Banach Algebra Theory 68
  Douglas S. Bridges and Robin S. Havea

The Mate-in-n Problem of Infinite Chess Is Decidable 78
  Dan Brumleve, Joel David Hamkins, and Philipp Schlicht

A Note on Ramsey Theorems and Turing Jumps 89
  Lorenzo Carlucci and Konrad Zdanowski

Automatic Functions, Linear Time and Learning 96
  John Case, Sanjay Jain, Samuel Seah, and Frank Stephan

An Undecidable Nested Recurrence Relation........................ 107
  Marcel Celaya and Frank Ruskey

Hard Instances of Algorithms and Proof Systems.................... 118
  Yijia Chen, Jörg Flum, and Moritz Müller

On Mathias Generic Sets.. 129
  Peter A. Cholak, Damir D. Dzhafarov, and Jeffry L. Hirst

Complexity of Deep Inference via Atomic Flows..................... 139
  Anupam Das

Connecting Partial Words and Regular Languages................... 151
  Jürgen Dassow, Florin Manea, and Robert Mercaş

Randomness, Computation and Mathematics 162
  Rod Downey

Learning, Social Intelligence and the Turing Test: Why an
"Out-of-the-Box" Turing Machine Will Not Pass the Turing Test 182
  Bruce Edmonds and Carlos Gershenson

Confluence in Data Reduction: Bridging Graph Transformation and
Kernelization 193
  *Hartmut Ehrig, Claudia Ermel, Falk Hüffner, Rolf Niedermeier, and
  Olga Runge*

Highness and Local Noncappability................................. 203
  Chengling Fang, Wang Shenling, and Guohua Wu

Turing Progressions and Their Well-Orders 212
  David Fernández Duque and Joost J. Joosten

A Short Note on Spector's Proof of Consistency of Analysis 222
  Fernando Ferreira

Sets of Signals, Information Flow, and Folktales..................... 228
  Mark Alan Finlayson

On the Foundations and Philosophy of Info-metrics.................. 237
  Amos Golan

On Mathematicians Who Liked Logic: The Case of Max Newman 245
  Ivor Grattan-Guinness

Densities and Entropies in Cellular Automata 253
  Pierre Guillon and Charalampos Zinoviadis

Foundational Analyses of Computation 264
  Yuri Gurevich

Turing Machine-Inspired Computer Science Results 276
  Juris Hartmanis

NP-Hardness and Fixed-Parameter Tractability of Realizing Degree
Sequences with Directed Acyclic Graphs 283
  Sepp Hartung and André Nichterlein

A Direct Proof of Wiener's Theorem 293
  Matthew Hendtlass and Peter Schuster

Effective Strong Nullness and Effectively Closed Sets 303
  Kojiro Higuchi and Takayuki Kihara

Word Automaticity of Tree Automatic Scattered Linear Orderings Is
Decidable 313
  Martin Huschenbett

On the Relative Succinctness of Two Extensions by Definitions of
Multimodal Logic 323
  Wiebe van der Hoek, Petar Iliev, and Barteld Kooi

On Immortal Configurations in Turing Machines 334
  Emmanuel Jeandel

A Slime Mold Solver for Linear Programming Problems 344
  Anders Johannson and James Zou

Multi-scale Modeling of Gene Regulation of Morphogenesis 355
  *Jaap A. Kaandorp, Daniel Botman, Carlos Tamulonis, and
  Roland Dries*

Tree-Automatic Well-Founded Trees 363
  Alexander Kartzow, Jiamou Liu, and Markus Lohrey

Infinite Games and Transfinite Recursion of Multiple Inductive
Definitions 374
  Keisuke Yoshii and Kazuyuki Tanaka

A Hierarchy of Immunity and Density for Sets of Reals 384
  Takayuki Kihara

How Much Randomness Is Needed for Statistics? 395
  Bjørn Kjos-Hanssen, Antoine Taveneaux, and Neil Thapen

Towards a Theory of Infinite Time Blum-Shub-Smale Machines 405
  Peter Koepke and Benjamin Seyfferth

Turing Pattern Formation without Diffusion 416
  Shigeru Kondo

Degrees of Total Algorithms versus Degrees of Honest Functions 422
  Lars Kristiansen

A $5n - o(n)$ Lower Bound on the Circuit Size over U_2 of a Linear
Boolean Function 432
  Alexander S. Kulikov, Olga Melanich, and Ivan Mihajlin

Local Induction and Provably Total Computable Functions:
A Case Study 440
  Andrés Cordón–Franco and F. Félix Lara–Martín

What is Turing's Comparison between Mechanism and Writing
Worth? 450
    Jean Lassègue and Giuseppe Longo

Substitutions and Strongly Deterministic Tilesets 462
    Bastien Le Gloannec and Nicolas Ollinger

The Computing Spacetime 472
    Fotini Markopoulou

Unifiability and Admissibility in Finite Algebras 485
    George Metcalfe and Christoph Röthlisberger

Natural Signs 496
    Ruth Garrett Millikan

Characteristics of Minimal Effective Programming Systems 507
    Samuel E. Moelius III

After Turing: Mathematical Modelling in the Biomedical and Social
Sciences: From Animal Coat Patterns to Brain Tumours to Saving
Marriages 517
    James D. Murray

Existence of Faster than Light Signals Implies Hypercomputation
already in Special Relativity 528
    Péter Németi and Gergely Székely

Turing Computable Embeddings and Coding Families of Sets 539
    Víctor A. Ocasio-González

On the Behavior of Tile Assembly System at High Temperatures 549
    Shinnosuke Seki and Yasushi Okuno

Abstract Partial Cylindrical Algebraic Decomposition I: The Lifting
Phase 560
    Grant Olney Passmore and Paul B. Jackson

Multi-valued Functions in Computability Theory 571
    Arno Pauly

Relative Randomness for Martin-Löf Random Sets 581
    *NingNing Peng, Kojiro Higuchi, Takeshi Yamazaki, and
Kazuyuki Tanaka*

On the Tarski-Lindenbaum Algebra of the Class of all Strongly
Constructivizable Prime Models 589
  Mikhail G. Peretyat'kin

Lower Bound on Weights of Large Degree Threshold Functions 599
  Vladimir V. Podolskii

What Are Computers (If They're not Thinking Things)? 609
  John Preston

Compactness and the Effectivity of Uniformization 616
  Robert Rettinger

On the Computability Power of Membrane Systems with Controlled
Mobility 626
  Shankara Narayanan Krishna, Bogdan Aman, and Gabriel Ciobanu

On Shift Spaces with Algebraic Structure 636
  Ville Salo and Ilkka Törmä

Finite State Verifiers with Constant Randomness.................... 646
  A.C. Cem Say and Abuzer Yakaryılmaz

Game Arguments in Computability Theory and Algorithmic
Information Theory 655
  Alexander Shen

Turing Patterns in Deserts 667
  Jonathan A. Sherratt

Subsymbolic Computation Theory for the Human Intuitive Processor ... 675
  Paul Smolensky

A Correspondence Principle for Exact Constructive Dimension......... 686
  Ludwig Staiger

Low_n Boolean Subalgebras 696
  Rebecca M. Steiner

Bringing Up Turing's 'Child-Machine' 703
  Susan G. Sterrett

Is Turing's Thesis the Consequence of a More General Physical
Principle? 714
  Matthew P. Szudzik

Some Natural Zero One Laws for Ordinals Below ε_0 723
  Andreas Weiermann and Alan R. Woods

On the Road to Thinking Machines: Insights and Ideas 733
    Jiří Wiedermann

Making Solomonoff Induction Effective: Or: You Can Learn What You
Can Bound.. 745
    Jörg Zimmermann and Armin B. Cremers

Author Index 755

Ordinal Analysis
and the Infinite Ramsey Theorem

Bahareh Afshari[1] and Michael Rathjen[2]

[1] School of Informatics, University of Edinburgh, Edinburgh,
EH8 9AB, United Kingdom
bafshari@inf.ed.ac.uk
[2] School of Mathematics, University of Leeds, Leeds, LS2 9JT, United Kingdom
rathjen@maths.leeds.ac.uk

Abstract. The infinite Ramsey theorem is known to be equivalent to the statement '*for every set X and natural number n, the n-th Turing jump of X exists*', over RCA_0 due to results of Jockusch [5]. By subjecting the theory RCA_0 augmented by the latter statement to an ordinal analysis, we give a direct proof of the fact that the infinite Ramsey theorem has proof-theoretic strength ε_ω. The upper bound is obtained by means of cut elimination and the lower bound by extending the standard well-ordering proofs for ACA_0. There is a proof of this result due to McAloon [6], using model-theoretic and combinatorial techniques. According to [6], another proof appeared in an unpublished paper by Jäger.

1 Introduction

Ramsey's theorem for infinite sets asserts that for every $k \geq 1$ and colouring of the k-element subsets of $\mathbb{N}$ with finitely many colours, there exists an infinite subset of $\mathbb{N}$ all of whose k-element subsets have the same colour [7]. We shall denote the previous statement, when specialised to a fixed k, by $\mathsf{RT}(k)$. It is well known that ACA_0 is equivalent to $\mathsf{RT}(k)$ for any (outer world) $k \geq 3$ [11]. However, the general assertion of Ramsey's theorem, $\forall x\, \mathsf{RT}(x)$ (abbreviated henceforth by iRT), is stronger than ACA_0.

For $b \geq 1$ we write $F : [A]^n \to b$ to signify that F maps the n-element subsets of A into the set $\{0, \ldots, b-1\}$. $X \subseteq A$ is said to be *monochromatic* for F if F is constant on $[X]^n$.

It is known from work of Jockusch [5, Theorem 5.7] that iRT is not provable in ACA_0. More precisely, for every $n \geq 0$, there is a recursive $F : [\mathbb{N}]^{n+2} \to 2$ such that the n-th Turing jump of $\emptyset$ is recursive in any infinite F-monochromatic $X \subseteq \mathbb{N}$. On the other hand, it also follows from [5, Theorem 5.6] that for every recursive $F : [\mathbb{N}]^n \to b$ and $n \geq 0$ there exists an F-monochromatic X recursive in the n-th Turing jump of $\emptyset$.

For $X, Y \subseteq \mathbb{N}$ and $n < \omega$, let $\mathrm{jump}(n, X, Y)$ abbreviate the formula stating that Y is the n-th Turing jump of X. By relativising the results of [5], we arrive at the following theorem (*cf.* [6]).

S.B. Cooper, A. Dawar, and B. Löwe (Eds.): CiE 2012, LNCS 7318, pp. 1–10, 2012.

Theorem 1. $\mathrm{ACA}_0 + \forall n \forall X \exists Y \mathrm{jump}(n, X, Y)$ *and* $\mathrm{ACA}_0 + \mathsf{iRT}$ *prove the same statements.*

By [6], $\mathrm{ACA}_0 + \mathsf{iRT}$ has the same first order consequences as the theory obtained from PA by iterating the uniform reflection principle arbitrarily often. This will also follow from Theorem 21, in light of the well-known fact that the latter theory has proof-theoretic ordinal ε_ω (see [8]). It is worth mentioning that the paper [12] (whose results actually postdate those in the present paper) contains, among other things, a characterization of the Π_2^0 consequences of ACA_0 augmented by the infinite Ramsey theorem.

We fix a primitive recursive ordering $\preceq$ on ω of order type Γ_0. For $\alpha \preceq \Gamma_0$, let $\mathsf{TI}(\prec \alpha)$ denote the scheme of *transfinite induction on initial segments of* α, i.e.,

$$\forall \xi (\forall \delta \prec \xi A(\delta) \to A(\xi)) \to \forall \xi \prec \bar{\beta} A(\xi),$$

for every $\beta \prec \alpha$ and every arithmetical formula $A(x)$. Here $\bar{\beta}$ denotes the numeral corresponding to the ordinal β. The *proof-theoretic ordinal* of a theory T, is the least α such that T is equivalent to $\mathrm{PA} + \mathsf{TI}(\prec \alpha)$, in the sense that they prove the same statements of arithmetic, and this fact can be established on the basis of PA.

The theory $\mathrm{ACA}_0 + \forall n \forall X \exists Y \mathrm{jump}(n, X, Y)$, commonly denoted by ACA_0', has previously been shown to have proof-theoretic ordinal ε_ω in [6]. The latter paper uses model-theoretic and combinatorial techniques but also draws on proof-theoretic results in order to construct an instance of transfinite induction up to ε_ω that cannot be proven in ACA_0', and thereby indicates that the strength of the theory is bounded by ε_ω. An unpublished proof using proof-theoretic means is attributed to Jäger [6]. However, to the authors' knowledge no proof using cut elimination is available in the published literature. This paper provides a simple proof-theoretic ordinal analysis of the system $\mathrm{ACA}_0 + \forall n \forall X \exists Y \mathrm{jump}(n, X, Y)$. The upper bound is obtained by means of cut elimination and the lower bound by extending the standard well-ordering proofs for ACA_0 as in [9]. For the definitions of systems of comprehension, ordinal notation and other basic definitions we refer the reader to [1,11]. The results presented here form part of [1].

2   The Semi-formal System ACA_∞

Let ACA_∞ be the infinitary calculus corresponding to ACA_0. Informally, the system ACA_∞ is obtained from ACA_0 by replacing the set induction axiom, i.e., $\forall X (0 \in X \wedge \forall x (x \in X \to x + 1 \in X) \to \forall n\, n \in X)$, by the infinitary ω-rule.

The language of ACA_∞ is the same as that of ACA_0 but the notion of a formula comes enriched with predicators. Formulae and *predicators* are defined simultaneously. Literals (atomic or negated atomic formulae) are formulae. Every set variable is a predicator. If $A(x)$ is a formula without set quantifiers, i.e., arithmetical, then $\{x \mid A(x)\}$ is a predicator. If P is a predicator and t is a numerical term, then $t \in P$ and $t \notin P$ are formulae. The other formation rules pertaining to $\wedge, \vee, \forall x, \exists x, \forall X, \exists X$ are as usual.

We shall be working in a Tait-style formalisation of the second order arithmetic. By a sequent Γ we mean a finite set of formulae in the language of second order arithmetic, $\mathcal{L}_2$. Due to the presence of the ω-rule we need only consider formulae without free numerical variables.

The *axioms* of the system ACA$_\infty$ are

- Γ, L where L is a true literal;
- $\Gamma, s \in X, t \notin X$ where s and t are terms of the same value.

The *rules* of ACA$_\infty$ are

$$(\vee_i) \frac{\Gamma, A_i \text{ for } i < 2}{\Gamma, A_0 \vee A_1} \qquad (\wedge) \frac{\Gamma, A_0 \qquad \Gamma, A_1}{\Gamma, A_0 \wedge A_1}$$

$$(\omega) \frac{\Gamma, A(\bar{n}) \text{ for all } n}{\Gamma, \forall x A(x)} \qquad (\exists_1) \frac{\Gamma, A(s)}{\Gamma, \exists x A(x)}$$

$$(\forall_2) \frac{\Gamma, A(X) \ X \text{ not free in } \Gamma}{\Gamma, \forall X A(X)} \qquad (\exists_2) \frac{\Gamma, A(\{x \mid A_0(x)\})}{\Gamma, \exists X A(X)}$$

$$(\mathrm{Pr}_1) \frac{\Gamma, A(t)}{\Gamma, t \in \{x \mid A(x)\}} \qquad (\mathrm{Pr}_2) \frac{\Gamma, \neg A(t)}{\Gamma, t \notin \{x \mid A(x)\}}$$

$$(\mathrm{Cut}) \frac{\Gamma, A \qquad \Gamma, \neg A}{\Gamma}$$

The *rank* of a formula A, denoted $|A|$, is defined as follows.

- $|L| = 0$, if L is a literal.
- $|s \in P| = |s \notin P| = |A(\bar{0})| + 1$, if P is the predicator $\{x \mid A(x)\}$.
- $|A_0 \wedge A_1| = |A_0 \vee A_1| = \max\{|A_0|, |A_1|\} + 1$.
- $|\forall x A(x)| = |\exists x A(x)| = |A(x)| + 1$.
- $|\forall X A(X)| = |\exists X A(X)| = \max\{|A(X)| + 1, \omega\}$.

For ordinals $\alpha, \kappa \prec \Gamma_0$, we write $\mathrm{ACA}_\infty \vdash^\alpha_\kappa \Gamma$ to convey that the sequent Γ is deducible in ACA$_\infty$ via a derivation of length $\preceq \alpha$ containing only cuts on formulae of rank $\prec \kappa$. More precisely, this notion is defined inductively as follows: If Γ is an axiom of ACA$_\infty$ then $\mathrm{ACA}_\infty \vdash^\alpha_\kappa \Gamma$ holds for any α and κ. If $\alpha_i < \alpha$ and $\mathrm{ACA}_\infty \vdash^{\alpha_i}_\kappa \Gamma_i$ hold for all premisses Γ_i of an ACA$_\infty$-inference with conclusion Γ, then $\mathrm{ACA}_\infty \vdash^\alpha_\kappa \Gamma$, provided that in the case of (Cut) the cut-formulae also have ranks $\prec \kappa$.

ACA$_\infty$ corresponds to the system EA* in [9] and can be interpreted into the first level of the semi-formal system of Ramified analysis, RA*. The fact that ACA$_\infty$ enjoys cut elimination is folklore and the proof involves the standard techniques of predicative proof theory. For proofs of the following see, for example, [1,9].

Lemma 2

1. $\mathrm{ACA}_\infty\!\!\mid_{0}^{|A|\cdot 2} \Gamma, A, \neg A$ *for every arithmetical formula A.*
2. *If Γ contains an axiom of ACA_0, then $\mathrm{ACA}_\infty\!\!\mid_{0}^{\omega+k} \Gamma$ for some $k < \omega$.*

Theorem 3 (First Cut Elimination Theorem for ACA_∞). *Let $\alpha, \beta \prec \Gamma_0$ and $k < \omega$. If $\mathrm{ACA}_\infty\!\!\mid_{\beta+k}^{\alpha} \Gamma$, then $\mathrm{ACA}_\infty\!\!\mid_{\beta}^{\omega_k(\alpha)} \Gamma$ where $\omega_0(\alpha) = \alpha$ and $\omega_{k+1}(\alpha) = \omega^{\omega_k(\alpha)}$.*

Theorem 4 (Second Cut Elimination Theorem for ACA_∞). *Let $\omega \preceq \alpha \prec \Gamma_0$. If $\mathrm{ACA}_\infty\!\!\mid_{\omega}^{\alpha} \Gamma$, then $\mathrm{ACA}_\infty\!\!\mid_{0}^{\varepsilon_\alpha} \Gamma$.*

3   An Upper Bound for $\mathrm{ACA}_0 + \mathrm{iRT}$

Let T denote the theory $\mathrm{ACA}_0 + \forall n \forall X \exists Y \mathrm{jump}(n, X, Y)$. We shall obtain an upper bound on the strength of this theory by a combination of embeddings and cut elimination theorems. We first embed T into an intermediate theory T^*. The semi-formal system T^* has the same language as ACA_∞. If A is a formula of T, then we write A^* to refer to any formula obtained from A by substituting all number variables by arbitrary closed terms. The system T^* has as axioms all sequences Γ, A^* such that A is a basic axiom of ACA_0 or the set induction axiom. Moreover, we have the following axioms in T^*.

- $\Gamma, A, \neg A$ if A is arithmetical.
- $\Gamma, \mathrm{jump}(\bar{n}, P, S_n^P)$ for every n and arithmetical predicator P.

In above, S_n^P is the arithmetical predicator which defines the n-th Turing jump of P. The predicator S_n^P is formally defined as follows.

$$S_0^P = \{x \mid \exists u[x = \langle 0, u \rangle \wedge u \in X]\},$$
$$S_{n+1}^P = \{x \mid \exists u[x = \langle \overline{n+1}, u \rangle \wedge \exists v \{u\}^{(S_n^P)_n}(u) = v]\},$$

where $\langle ., . \rangle$ is a primitive recursive pairing function, $\{u\}^Y$ represents the u-th partial recursive function with oracle Y, and $(Y)_n$ denotes the n-slice of the set Y, i.e., $(Y)_n = \{y \mid \langle n, y \rangle \in Y\}$. The logical rules of T^* are the same as in ACA_∞. The rank of a formula and notation $T^*\!\!\mid_{\kappa}^{\alpha} \Gamma$ are defined analogously. From the choice of the axioms of T^* and the fact that the rank of a formula is always strictly less than $\omega + \omega$ we can derive the following.

Theorem 5 (Embedding Theorem). *Suppose $T \vdash B$. Then there exist natural numbers n and m such that $T^*\!\!\mid_{\omega+m}^{n} B^*$ holds for all B^*.*

We now perform partial cut elimination in T^*. The following reduction lemma goes through in the standard way.

Lemma 6 (Reduction Lemma). *Let $n_0, n_1 < \omega$ and $|A| = \kappa \succeq \omega$. If $T^*\!\!\mid_{\kappa}^{n_0} \Gamma, A$ and $T^*\!\!\mid_{\kappa}^{n_1} \Delta, \neg A$, then $T^*\!\!\mid_{\kappa}^{n_0+n_1} \Gamma, \Delta$.*

Proof. By induction on the sum $n_0 + n_1$. We show the interesting case where both A and $\neg A$ are active in the derivations and A is derived via the $(\forall_2)$-rule. Suppose A is of the form $\forall Y A_0(Y)$ and we have

$$\frac{(1) \quad \frac{|m_0}{\kappa} \Gamma, A_0(X), [\forall Y A_0(Y)] \quad m_0 < n_0}{(2) \quad \frac{|n_0}{\kappa} \Gamma, \forall Y A_0(Y)} \ (\forall_2)$$

where X is not free in $\Gamma \cup \{\forall Y A_0(Y)\}$, and

$$\frac{(3) \quad \frac{|m_1}{\kappa} \Delta, \neg A_0(P), [\exists Y \neg A_0(Y)] \quad m_1 < n_1}{(4) \quad \frac{|n_1}{\kappa} \Delta, \exists Y \neg A_0(Y)} \ (\exists_2)$$

where P is an arithmetical predicator. In the above inferences, we write $[B]$ to emphasise that the formula B may or may not have appeared in the premise of the original inference. Applying the induction hypothesis to (2) and (3) yields

$$\frac{|n_0+m_1}{\kappa} \Gamma, \Delta, \neg A_0(P) , \tag{5}$$

and to (1) and (4) yields

$$\frac{|m_0+n_1}{\kappa} \Gamma, \Delta, A_0(X). \tag{6}$$

It is not hard to show that (6) implies $\frac{|m_0+n_1}{\kappa} \Gamma, \Delta, A_0(Q)$ for any arithmetical predicator Q and in particular

$$\frac{|m_0+n_1}{\kappa} \Gamma, \Delta, A_0(P). \tag{7}$$

Since $|A_0(P)| \prec |\forall Y A_0(Y)| = \kappa$, we may perform a cut on (5) and (7) to conclude $\frac{|n_0+n_1}{\kappa} \Gamma, \Delta$ as required. For full details see [1, §3].

Theorem 7 (Cut Elimination Theorem). *If* $T^* \frac{|n}{\omega+m+1} \Gamma$ *for some* $m, n < \omega$, *then* $T^* \frac{|2^n}{\omega+m} \Gamma$.

Proof. By induction on n. If Γ is an axiom, then $\frac{|0}{0} \Gamma$. Otherwise, there are two cases to consider. Suppose Γ is of the form Γ', A and we have

$$\frac{\frac{|n_i}{\omega+m+1} \Gamma', A_i \quad i < \omega, n_i < n}{\frac{|n}{\omega+m+1} \Gamma', A} \ (R)$$

where R is any of the rules of T^* except the cut rule. By applying the induction hypothesis to the premise(s) of the above inference we obtain $\frac{|2^{n_i}}{\omega+m} \Gamma', A_i$. Reapplying the rule (R) yields $\frac{|2^n}{\omega+m} \Gamma$.

If the last inference was a cut, namely,

$$\frac{\frac{|n_0}{\omega+m+1} \Gamma, A \qquad \frac{|n_1}{\omega+m+1} \Gamma, \neg A}{\frac{|n}{\omega+m+1} \Gamma} \ (\text{Cut})$$

where $n_0, n_1 < n$ and $|A| \preceq \omega + m$, by applying the induction hypothesis to the premises of the above cut we obtain

$$\Big|\frac{2^{n_0}}{\omega + m}\ \Gamma, A \quad \text{and} \quad \Big|\frac{2^{n_1}}{\omega + m}\ \Gamma, \neg A\,.$$

If $|A| \prec \omega + m$, by a cut on A we derive $\Big|\frac{2^n}{\omega + m}\ \Gamma$. Otherwise, the Reduction Lemma yields $\Big|\frac{2^{n_0} + 2^{n_1}}{\omega + m}\ \Gamma$, thus by monotonicity we get the desired result.

Corollary 8. *For $n, m < \omega$, if $T^* \Big|\frac{n}{\omega + m}\ \Gamma$, then $T^* \Big|\frac{2_m^n}{\omega}\ \Gamma$ where $2_0^n = n$ and $2_{k+1}^n = 2^{2_k^n}$.*

Finally, to analyse T we shall embed T^* into ACA_∞ so that we can eliminate the remaining cuts and read off an upper bound. First we need the following lemma, which can be verified by induction.

Lemma 9. *There are primitive recursive functions f, g such that for each n, $f(n), g(n) < \omega$ and $\mathrm{ACA}_\infty \Big|\frac{f(n)}{g(n)}\ \Gamma, \mathrm{jump}(\bar n, P, S_n^P)\,.$*

Theorem 10. *Let Γ be a finite set of arithmetical formulae and $k < \omega$. Then $T^* \Big|\frac{k}{\omega}\ \Gamma$ implies $\mathrm{ACA}_\infty \Big|\frac{\varepsilon_k}{0}\ \Gamma$.*

Proof. By induction on k. If Γ is an axiom of T^* we have the following three cases to consider. If Γ is the sequent Γ', A^* where A is a basic axiom of ACA_0 or the set induction axiom, then we have, by Lemma 2, that $\mathrm{ACA}_\infty \Big|\frac{\omega + k}{0}\ \Gamma', A^*$ for some $k < \omega$. If Γ is the sequent $\Gamma', \mathrm{jump}(\bar n, P, S_n^P)$ where P is an arithmetical predicator, Lemma 9 yields $\mathrm{ACA}_\infty \Big|\frac{f(n)}{g(n)}\ \Gamma, \mathrm{jump}(\bar n, P, S_n^P)$. By applying the First Cut Elimination theorem for ACA_∞ we obtain

$$\mathrm{ACA}_\infty \Big|\frac{\omega_{g(n)}(f(n))}{0}\ \Gamma, \mathrm{jump}(\bar n, P, S_n^P)\,,$$

with ω_k being defined in Theorem 3. Since $f(n), g(n) < \omega$, we have $\omega_{g(n)}(f(n)) \prec \varepsilon_0$, and hence may deduce $\mathrm{ACA}_\infty \Big|\frac{\varepsilon_0}{0}\ \Gamma, \mathrm{jump}(\bar n, P, S_n^P)$. Now let Γ be of the form $\Gamma', A, \neg A$ with A being arithmetical. As $\mathrm{ACA}_\infty \Big|\frac{|A| \cdot 2}{0}\ \Gamma, A, \neg A$ holds due to Lemma 2, monotonicity provides the desired result.

Now suppose Γ is derived by an application of a logical rule in T^*. If the last inference is a cut, then we have

$$\frac{T^* \Big|\frac{k_0}{\omega}\ \Gamma, A \qquad T^* \Big|\frac{k_0}{\omega}\ \Gamma, \neg A}{T^* \Big|\frac{k}{\omega}\ \Gamma}\ (\text{Cut})$$

where $k_0 < k$ and A is arithmetical. Applying the induction hypothesis to the premises of the above cut yields $\mathrm{ACA}_\infty \Big|\frac{\varepsilon_{k_0}}{0}\ \Gamma, A$ and $\mathrm{ACA}_\infty \Big|\frac{\varepsilon_{k_0}}{0}\ \Gamma, \neg A$. Applying a cut to A, we conclude that $\mathrm{ACA}_\infty \Big|\frac{\varepsilon_{k_0} + 1}{m}\ \Gamma$ for $m = |A| + 1$. Thus $\mathrm{ACA}_\infty \Big|\frac{\omega_m(\varepsilon_{k_0} + 1)}{0}\ \Gamma$ and so we may deduce $\mathrm{ACA}_\infty \Big|\frac{\varepsilon_k}{0}\ \Gamma$.

If Γ is derived via the $(\forall_2)$-rule, Γ must be of the form $\Gamma', \forall Y A_0(Y)$ and we have

$$\frac{T^*\vert\frac{k_0}{\omega}\, \Gamma', A_0(X) \quad X \text{ is not free in } \Gamma' \quad \text{and} \quad k_0 < k}{T^*\vert\frac{k}{\omega}\, \Gamma', \forall Y\, A_0(Y)} \quad (\forall_2)$$

Applying the induction hypothesis to the premise of the above inference yields $\text{ACA}_\infty\vert\frac{\varepsilon k_0}{0}\, \Gamma', A_0(X)$. Re-applying $(\forall_2)$-rule allows us to deduce $\text{ACA}_\infty\vert\frac{\varepsilon k}{0}\, \Gamma$. The other cases are similar.

Corollary 11. *Every arithmetical theorem of T without set variables is derivable in* $\text{PA} + \text{TI}(\prec \varepsilon_\omega)$.

Proof. Suppose A is an arithmetical sentence and $T \vdash A$. By the Embedding Theorem 5, $T^*\vert\frac{n}{\omega+m}\, A$ holds for some $n, m < \omega$. Cut elimination for T^*, Theorem 7, yields that $T^*\vert\frac{k}{\omega}\, A$ holds for some $k < \omega$. By embedding T^* into ACA_∞ via Theorem 10, we arrive at $\text{ACA}_\infty\vert\frac{\varepsilon k}{0}\, A$. This means that A is derivable in ACA_∞ directly from the axioms and first order rules, and, moreover, if A is of complexity Π_n^0 for some n, then all formulae occurring in this cut-free derivation belong to the same complexity class. By employing a partial truth predicate for Π_n^0-formulae and transfinite induction up to ε_{k+1}, one shows that $\text{PA} + \text{TI}(\prec \varepsilon_\omega) \vdash A$ (cf. [10]).

4   A Lower Bound for $\text{ACA}_0 + \text{iRT}$

To deduce that ε_ω is also a lower bound for the strength of the theory T, we shall show that T can prove the well-foundedness of all ordinals strictly less than ε_ω. Our method is to extend the standard well-ordering proofs for ACA_0 as for instance given in [9, Theorem 23.3]. Let us denote by Sp the operation defined by

$$\text{Sp}(V) = \{\alpha \mid \forall\xi(\xi \subset V \to \xi + \omega^\alpha \subset V)\},$$

where $\xi \subset V$ abbreviates $\forall x(x \prec \xi \to x \in V)$. For sets X and Y we write $X \leq_e Y$ to convey that $\forall x(K_X(x) = \{e\}^Y(x))$, where K_X denotes the characteristic function of the set X. The aim of the next few lemmata is to establish that in the theory T finite iterations of the Sp operator can be coded into a single set. They are easy to verify using the Kleene T-predicate and S-m-n theorem [3]. Detailed proofs can be found in [1, §3]. In the following $X^{(n)}$ denotes the n-th Turing jump of X, i.e., the set Y such that $\text{jump}(n, X, Y)$. We also use X' and X'' respectively for $X^{(1)}$ and $X^{(2)}$.

Lemma 12. *Let $A(x, y, z, U)$ be a Δ_0^0-formula (of the language of ACA_0) with all the free variables exhibited and U being a free set variable. Then there exists a natural number e such that for every $X \subseteq \mathbb{N}$*

$$\{n \mid \forall x \exists y\, A(x, y, n, X)\} \leq_e X''.$$

Proof. Since for every set $X \subseteq \mathbb{N}$ the set $\{\langle x, n \rangle : \exists y A(x, y, n, X)\}$ is recursively enumerable in X (uniformly in X), there exists an index e_0 depending only on the formula $A(x, y, z, U)$ such that for all natural numbers n and sets $X \subseteq \mathbb{N}$,

$$\forall x \exists y A(x, y, n, X) \quad \text{iff} \quad \forall x \langle e_0, \langle x, n \rangle \rangle \in X'.$$

Likewise, there is some d_0 such that for all sets $Y \subseteq \mathbb{N}$, $\{d_0\}^Y$ is total and

$$\forall x \langle e_0, \langle x, n \rangle \rangle \in Y \quad \text{iff} \quad \langle d_0, \langle e_0, n \rangle \rangle \in Y'.$$

It immediately follows from the S-m-n theorem that there is an index e such that $\{n \mid \forall x \exists y A(x, y, n, X)\} \leq_e X''$.

In particular, since $\mathrm{Sp}(X)$ is Π_2^0 in X we can deduce the following.

Corollary 13. ACA_0 *proves the existence of a natural number e that satisfies* $\mathrm{Sp}(X) \leq_e X''$ *for all* $X \subseteq \mathbb{N}$.

Lemma 14. *There is a primitive recursive function* $\circ : \omega \times \omega \to \omega$ *such that for any sets X, Y, Z, if $X \leq_e Y$ and $Y \leq_f Z$, then $X \leq_{eof} Z$.*

Lemma 15. *There is a primitive recursive function N such that $X' \leq_{N(e)} Y'$ whenever $X \leq_e Y$.*

Corollary 16. *There exists a primitive recursive function g such that*

$$\mathrm{Sp}^n(X) \leq_{g(n)} X^{(2n)},$$

where Sp^n is inductively defined as $\mathrm{Sp}^0(X) = X$ and $\mathrm{Sp}^{n+1}(X) = \mathrm{Sp}(\mathrm{Sp}^n(X))$.

Proof. We define g by induction on n. Suppose $n = 0$. Let $g(0)$ be the index of the identity function. Then $\mathrm{Sp}^0(X) \leq_{g(0)} X$ holds trivially. For the induction step suppose $n = k + 1$. By Corollary 13 there is an e (independent of k) such that

$$\mathrm{Sp}^{(k+1)}(X) = \mathrm{Sp}(\mathrm{Sp}^{(k)}(X)) \leq_e (\mathrm{Sp}^{(k)}(X))^{(2)}.$$

Using the induction hypothesis we may assume $\mathrm{Sp}^{(k)}(X) \leq_{g(k)} X^{(2k)}$. Lemma 15 entails

$$(\mathrm{Sp}^{(k)}(X))^{(2)} \leq_{N(N(g(k)))} (X^{(2k)})^{(2)} = X^{(2k+2)},$$

and Lemma 14 yields

$$\mathrm{Sp}^{k+1}(X) \leq_{g(k+1)} X^{(2k+2)},$$

by setting $g(k+1) = e \circ N(N(g(k)))$. Since g is primitive recursive we are done.

Lemma 17. $T \vdash \forall x \forall X \exists Y A(x, X, Y)$, *where A is the formula defined by $(Y)_0 = X \wedge \forall n < x \ (Y)_{n+1} = \mathrm{Sp}((Y)_n)$.*

Proof. We argue informally within T. Given x and X define

$$Y = \{\langle n, z \rangle \mid n \leq x \wedge \{g(n)\}^{X^{(2n)}}(z) \simeq 0\},$$

where g is the primitive recursive function given by Corollary 16. It is easy to see that $(Y)_0 = X$. Moreover, for $n \leq x$,

$$z \in (Y)_n \quad \text{iff} \quad \{g(n)\}^{X^{(2n)}}(z) \simeq 0.$$

By Corollary 16 we have $(Y)_n = \mathrm{Sp}^{(n)}(X)$. Thus $\mathrm{Sp}(\mathrm{Sp}^{(n)}(X)) = \mathrm{Sp}((Y)_n)$, and hence for $n < x$ we can deduce $(Y)_{n+1} = \mathrm{Sp}((Y)_n)$ as required.

Let $\mathsf{Tran}(\prec)$ and $\mathsf{LO}(\prec)$ be abbreviations for formulae stating $\prec$ is transitive and a linear order respectively. $\mathsf{Fund}(\alpha, X)$ is the formula

$$\mathsf{Tran}(\prec) \wedge (\mathsf{Prog}_{\prec}(X) \to \forall \xi \prec \alpha \, (\xi \in X)),$$

where $\mathsf{Prog}_{\prec}(X) = \forall x (\forall y (y \prec x \to y \in X) \to x \in X)$, and $\mathsf{TI}(\alpha, X)$ is the formula

$$\mathsf{LO}(\prec) \wedge \mathsf{Fund}(\alpha, X).$$

The following lemma is well known. For a proof see [9, §21, Lemma 1].

Lemma 18. *For every set X and $\alpha \prec \Gamma_0$,*

$$\mathrm{ACA}_0 \vdash \mathsf{Fund}(\alpha, \mathrm{Sp}(X)) \to \mathsf{Fund}(\omega^\alpha, X).$$

Lemma 19. $T \vdash \mathsf{Fund}(\varepsilon_0, X).$

Proof. We argue informally within T. $\mathsf{Fund}(\alpha, X)$ is progressive in α, therefore it suffices to show $\forall n \mathsf{Fund}(\omega_n(\omega), X)$. By Lemma 17, $A(n+1, X, Y)$ holds for some Y. On the other hand as induction up to ω is available in T for every set, $\mathsf{Fund}(\omega, (Y)_n)$ holds. Since $(Y)_n = \mathrm{Sp}((Y)_{n-1})$, by using Lemma 18 and an internal induction on n we obtain $\mathsf{Fund}(\omega_n(\omega), (Y)_0)$.

We can now show that T proves the well-foundedness of ordinals strictly less than ε_ω.

Theorem 20. *For each $k < \omega$, $T \vdash \mathsf{Fund}(\varepsilon_{\bar{k}}, X)$.*

Proof. Since k is given externally, in the formal theory T it is named by $\bar{k}$, the k-th numeral. Below, for ease of presentation, we shall identify k and $\bar{k}$. The proof proceeds by external induction on k. For a fixed k, by internal recursion on n define $\alpha_0^k = \varepsilon_{k-1} + 1$ and $\alpha_{n+1}^k = \omega^{\alpha_n^k}$. This time start with $\mathsf{Fund}(\alpha_0^k, (Y)_n)$ and proceed as in Lemma 19 to derive $\forall n \mathsf{Fund}(\alpha_n^k, X)$. Since $\sup_n \alpha_n^k = \varepsilon_k$ we can deduce $\mathsf{Fund}(\varepsilon_k, X)$.

5   Conclusion

We have shown that our upper bound for the proof-theoretic ordinal of the theory T is indeed the least one. This allows us to determine the proof-theoretic strength of T, and hence that of the infinite Ramsey theorem.

Theorem 21. *The theory* $\mathrm{ACA}_0 + \mathrm{iRT}$, *i.e.,* ACA_0 *augmented by the infinite Ramsey theorem, proves the same arithmetical statements as* $\mathrm{PA} + \mathrm{TI}(\prec \varepsilon_\omega)$.

Proof. Since $\mathrm{ACA}_0 + \mathrm{iRT}$ is equivalent to $\mathrm{ACA}_0 + \forall n \forall X \exists Y \mathrm{jump}(n, X, Y)$, Theorem 20 implies $\mathrm{ACA}_0 + \mathrm{iRT} \vdash \mathrm{TI}(\varepsilon_{\bar{k}}, X)$ for every $k < \omega$, and thus $\mathrm{ACA}_0 + \mathrm{iRT} \vdash \mathrm{TI}(\prec \varepsilon_\omega)$. Corollary 11 provides the other direction.

Acknowledgements. The first author was supported by the Engineering and Physical Sciences Research Council UK (grant number EP/G012962/1). The second author was supported by a Royal Society International Joint Projects award 2006/R3.

References

1. Afshari, B.: Relative computability and the proof-theoretic strength of some theories, Ph.D. Thesis, University of Leeds, U.K. (2008)
2. Friedman, H., McAloon, K., Simpson, S.: A finite combinatorial principle equivalent to the 1-consistency of predicative analysis. In: Metakides, G. (ed.) Patras Logic Symposion, pp. 197–230. North-Holland (1982)
3. Hinman, P.G.: Recursion-theoretic hierarchies. Springer, Heidelberg (1978)
4. Jäger, G.: Theories for iterated jumps (1980) (unpublished notes)
5. Jockusch, C.G.: Ramsey's theorem and recursion theory. Journal of Symbolic Logic 37, 268–280 (1972)
6. McAloon, K.: Paris-Harrington incompleteness and progressions of theories. Proceedings of Symposia in Pure Mathematics 42, 447–460 (1985)
7. Ramsey, F.P.: On a problem of formal logic. Proceedings of the London Mathematical Society 30(1), 264–286 (1930)
8. Schmerl, U.: A fine structure generated by reflection formulas over primitive recursive arithmetic. In: Boffa, M., McAloon, K., van Dalen, D. (eds.) Studies in Logic and the Foundations of Mathematics, vol. 97, pp. 335–350. Elsevier (1979)
9. Schütte, K.: Proof theory. Springer, Heidelberg (1977)
10. Schwichtenberg, H.: Proof theory: Some applications of cut-elimination. In: Barwise, J. (ed.) Handbook of Mathematical Logic, pp. 867–895. North-Holland, Amsterdam (1977)
11. Simpson, S.G.: Subsystems of second order arithmetic. Springer, Heidelberg (1999)
12. De Smet, M., Weiermann, A.: A Miniaturisation of Ramsey's Theorem. In: Ferreira, F., Löwe, B., Mayordomo, E., Mendes Gomes, L. (eds.) CiE 2010. LNCS, vol. 6158, pp. 118–125. Springer, Heidelberg (2010)

Curiouser and Curiouser: The Link between Incompressibility and Complexity

Eric Allender

Department of Computer Science, Rutgers University, Piscataway, NJ 08855,
United States of America
allender@cs.rutgers.edu

Abstract. This talk centers around some audacious conjectures that attempt to forge firm links between computational complexity classes and the study of Kolmogorov complexity.

More specifically, let R denote the set of Kolmogorov-random strings. Let BPP denote the class of problems that can be solved with negligible error by probabilistic polynomial-time computations, and let NEXP denote the class of problems solvable in nondeterministic exponential time.

Conjecture 1. NEXP = NP^R.

Conjecture 2. BPP is the class of problems non-adaptively polynomial-time reducible to R.

These conjectures are not only audacious; they are obviously false! R is not a decidable set, and thus it is absurd to suggest that the class of problems reducible to it constitutes a complexity class.

The absurdity fades if, for example, we interpret "NP^R" to be "the class of problems that are NP-Turing reducible to R, no matter which universal machine we use in defining Kolmogorov complexity". The lecture will survey the body of work (some of it quite recent) that suggests that, when interpreted properly, the conjectures may actually be true.

1 Introduction

This is a story about mathematical notions that refuse to stay put in their proper domain. Complexity theory is supposed to deal with decidable sets (and preferably with sets that are *very* decidable – primitive recursive at least, and ideally much lower in the complexity hierarchy than that). Undecidable sets inhabit a very different realm, and they look suspiciously out-of-place popping up in a discussion of computational complexity classes.

And yet, the (undecidable) set of Kolmogorov-random strings persists in intruding into complexity-theoretic investigations. It has become much harder to deny that there is a connection, although the precise nature of the relationship remains unclear.

1.1 Cast of Characters

The primary focus of our attention will be a familiar list of deterministic and nondeterministic time- and space-bounded complexity classes: P, NP, PSPACE,

S.B. Cooper, A. Dawar, and B. Löwe (Eds.): CiE 2012, LNCS 7318, pp. 11–16, 2012.
© Springer-Verlag Berlin Heidelberg 2012

NEXP, EXPSPACE, along with the class BPP of languages accepted in polynomial time by probabilistic machines with negligible error, and the class P/poly of problems with polynomial-size circuit complexity. Detailed definitions can be found in a standard text such as [5].

Much of the action in our story revolves around the set of Kolmogorov-random strings. Before this set can be introduced properly, some definitions are required.

Given a Turing machine M, the (plain) Kolmogorov complexity function $C_M(x)$ is defined to be the minimum element of the set $\{|d| : M(d) = x\}$ (and is undefined if this set is empty). A machine U is said to be *universal* for this measure, if

$$\forall M \exists c \forall x C_U(x) \leq C_M(x) + c.$$

As usual in the study of Kolmogorov complexity (see, e.g., [9,8]), we pick one such universal Turing machine and define $C(x)$ to be $C_U(x)$. (There is something rather arbitrary in the selection of U; we shall come back to this later.)

For some applications, a better-behaved Kolmogorov complexity measure is the prefix-free measure $K_M(x)$ that has an identical definition, but where the Turing machine M is restricted to be prefix-free (meaning that if $M(x)$ halts, then M does not halt on input xy for any non-empty string y). It turns out that a universal prefix-free machine U exists such that, for all prefix-free machines M there is a constant c such that for all x $K_U(x) \leq K_M(x) + c$, and we select one such U and define $K(x)$ to be $K_U(x)$. Again, consult [9,8] for details.

A string x is *random* (or *incompressible*) if there is no "description" d with $|d| < |x|$ such that $U(d) = x$. Depending on which notion of Kolmogorov complexity we are using, this gives us two sets of random strings:

- $R_C = \{x : C(x) \geq |x|\}$.
- $R_K = \{x : K(x) \geq |x|\}$.

When it does not make any difference which of these two sets is meant, we shall use the simplified notation "R". (Similarly, if it is necessary to make explicit mention of a universal machine U, we shall refer to R_{K_U} and R_{C_U}.)

2   Some Odd Inclusions

It has long been known [10] that R is Turing-equivalent to the halting problem. However, it is much less clear what can be *efficiently* reduced to R. To date, the only known proof that the halting problem can be Turing-reduced to R via polynomial-size *circuits* relies on the arsenal of derandomization techniques that were developed in the 1990s [2]. For efficient "uniform" reductions (i.e., reductions computed by polynomial-time *machines*), it is not easy to see how to make use of R as an oracle. (To illustrate this, we encourage the reader to spend a minute trying to see how to reduce their favorite NP-complete problem to R.) Thus the following theorem is of some interest.

Theorem 2.1. *The following inclusions hold:*

- BPP $\subseteq$ P$_{tt}^R$ [6].
- PSPACE $\subseteq$ P^R [2].
- NEXP $\subseteq$ NP^R [1].

Here, the notation P$_{tt}^A$ denotes the class of problems that are reducible to A via polynomial-time *truth-table* reductions (also known as "non-adaptive" reductions). These are reductions computed by an oracle Turing machine M that, on input x, computes a list of queries $y_1, \ldots, y_m$, and then asks the oracle about each of these m queries, and then uses the oracle answers to decide whether to accept or reject. (In a more general Turing reduction, the list of queries can depend on the answers that the oracle gives.)

There is indeed something odd about Theorem 2.1. Is it interesting to study *efficient* reductions to an undecidable set? Since R is Turing-equivalent to the halting problem, one ought to wonder whether *every* computably-enumerable set is in P$_{tt}^R$ (in which case, Theorem 2.1 would not be very interesting).

In truth, it *is* still an open question whether the halting problem (and hence every c.e. set) is in P$_{tt}^{R_C}$. In contrast, for the prefix-free measure K, the situation is intriguing, as the next section will relate.

3 An Upper Bound on Complexity

The proofs of the inclusions in Theorem 2.1, such as NEXP $\subseteq$ NP^R, make use of no special properties of the universal Turing machine that defines Kolmogorov complexity. Thus it follows that we actually have the inclusion NEXP $\subseteq$ $\bigcap_U$ NP$^{R_{K_U}}$, where the intersection is taken over all universal prefix-free Turing machines. This might seem to be a trivial observation, but it is actually essential, if we want to obtain an upper bound on the complexity of classes such as NP^R. This is because there exist universal prefix-free Turing machines U such that even P$_{tt}^{R_{K_U}}$ contains arbitrarily complex decidable sets.

However, a paper presented at the 2011 ICALP conference shows that, if we consider only those problems that are reducible to R_K *regardless* of which universal Turing machine is used in defining K-complexity, then we do indeed obtain something that looks very much like a complexity class:

Theorem 3.1. [4]

- BPP $\subseteq$ $\Delta_1^0 \cap \bigcap_U$ P$_{tt}^{R_{K_U}}$ $\subseteq$ PSPACE
- PSPACE $\subseteq$ $\Delta_1^0 \cap \bigcap_U$ P$^{R_{K_U}}$
- NEXP $\subseteq$ $\Delta_1^0 \cap \bigcap_U$ NP$^{R_{K_U}}$ $\subseteq$ EXPSPACE.

Here, as usual, Δ_1^0 denotes the class of decidable sets.

Theorem 3.1 is stated in terms of the prefix-free measure K. It seems reasonable to conjecture that it holds also for the plain measure C, but there does not seem to be an easy way to modify the proof of Theorem 3.1 to deal with the C measure.

The proof of the inclusion $\Delta_1^0 \cap \bigcap_U \mathrm{NP}^{R_{K_U}} \subseteq \mathrm{EXPSPACE}$ proceeds by first observing that an NP-Turing reduction can be simulated by an exponential-time truth-table reduction, and then noting that the PSPACE upper bound on $\Delta_1^0 \cap \bigcap_U \mathrm{P}_{tt}^{R_{K_U}}$ translates into an EXPSPACE upper bound on the class $\Delta_1^0 \cap \bigcap_U \mathrm{EXP}_{tt}^{R_{K_U}}$. Since NP is widely conjectured to be a small subclass of exponential time, it seems likely we are throwing away too much information in the initial step in this argument (replacing an NP-Turing reduction by an exponential-time truth-table reduction). That is, we suspect that the inclusion $\Delta_1^0 \cap \bigcap_U \mathrm{EXP}_{tt}^{R_{K_U}} \subseteq \mathrm{EXPSPACE}$ is not optimal. In fact, we suspect that the inclusion $\mathrm{NEXP} \subseteq \mathrm{NP}^R$ is tight, in the following sense:

Conjecture 3.2. $\mathrm{NEXP} = \Delta_1^0 \cap \bigcap_U \mathrm{NP}^{R_{K_U}}$.

Such a characterization of NEXP in terms of reducibility to R_K would certainly be unusual. Perhaps it would also be useful.

4   Towards a Characterization of BPP

There is more to report, regarding the inclusion $\Delta_1^0 \cap \bigcap_U \mathrm{P}_{tt}^{R_{K_U}} \subseteq \mathrm{PSPACE}$ of Theorem 3.1.

In a still-unpublished paper [3], it is argued that it is likely that the PSPACE upper bound can be improved to $\mathrm{PSPACE} \cap \mathrm{P/poly}$. If it true, then this would imply that $\mathrm{BPP} \subseteq \Delta_1^0 \cap \bigcap_U \mathrm{P}_{tt}^{R_{K_U}} \subseteq \mathrm{PSPACE} \cap \mathrm{P/poly}$. Since there is a dearth of interesting complexity classes between BPP and $\mathrm{PSPACE} \cap \mathrm{P/poly}$, this motivates the following:

Conjecture 4.1. $\mathrm{BPP} = \Delta_1^0 \cap \bigcap_U \mathrm{P}_{tt}^{R_{K_U}}$.

The evidence presented in [3] in support of the P/poly upper bound can be summarized in this way: The authors present a true statement of the form $\forall n \forall j \Psi(n,j)$ (provable in ZF), with the property that if, for each fixed $(\mathbf{n},\mathbf{j})$ there is a proof in Peano Arithmetic of the statement $\psi(\mathbf{n},\mathbf{j})$, then the P/poly upper bound holds. (In fact, under this assumption, for each length n, it suffices to restrict attention to truth-table reductions that make queries only of length $O(\log n)$ and have as oracle a subset of R (possibly a different subset for each input length – which can be encoded as a circuit for inputs of length n).

Motivated largely by the results of [3], Buhrman and Loff [7] have proved a very recent result that can also be seen as supporting the P/poly upper bound. For a polynomial-time reduction from a decidable set A to the undecidable set R, it seems reasonable to hypothesize that the reduction would also work if one used a very high time-complexity approximation to R, such as $R_K^{t(n)}$ for some very rapidly-growing time bound $t(n)$. Buhrman and Loff have shown that, for each decidable set A and polynomial-time truth-table reduction M, it is the case that for *every* large-enough time bound t, if M reduces A to $R_K^{t(n)}$, then $A \in \mathrm{P/poly}$.

Interestingly, the techniques used by Buhrman and Loff also allowed them to show that the sentences $\psi(\mathbf{n,j})$ considered in [3] are, in fact, independent of Peano Arithmetic. Worse, they present a polynomial-time reduction with the property that it can *not* be directly replaced by a reduction that makes queries only of length $O(\log n)$, having as oracle a subset of R. Thus the general approach discussed in [3] will need to be revised substantially, if it is to be used to obtain a P/poly upper bound on this class.

5   Speculations

Since, to date, no interesting characterizations of complexity classes in terms of efficient reductions to R have been obtained, it may be premature to speculate about the usefulness of such a characterization. Nonetheless, it is fun to engage in such speculation. Could it be possible that linking the study of Kolmogorov-random strings to the study of computational complexity classes could enable the application of tools from one domain, to problems where these tools had seemed inapplicable? It is an exciting prospect to contemplate.

The techniques of computability theory usually relativize, which might seem like an impediment to the realization of this program. However, the inclusions $\mathrm{PSPACE} \subseteq \mathrm{P}^R$ and $\mathrm{NEXP} \subseteq \mathrm{NP}^R$ each utilize techniques that do not relativize. Perhaps there is room to explore new combinations of tools and techniques from these fields.

The inclusion $\mathrm{BPP} \subseteq \mathrm{P}_{tt}^R$ does relativize, in the following sense. The argument of [6] shows that, for every decidable set A, every set in BPP^A is P^A-truth-table reducible to R. Thus it is conceivable that a stronger version of Conjecture 4.1 holds, characterizing BPP^A as the class of decidable sets that are P^A-truth-table reducible to R. Thus, any attempt to prove $\mathrm{P} = \mathrm{BPP}$ (as many suspect) by proving that $\mathrm{P} = \Delta_1^0 \cap \bigcap_U \mathrm{P}_{tt}^{R_{K_U}}$ will require some new non-relativizing proof techniques. (Note however that analogous equalities have been proved for some limited classes of truth-table reductions [1].)

At the very least, results such as Theorem 3.1 provide motivation for some questions in computability theory that have not received much attention. For instance, is "$\Delta_1^0\cap$" redundant in each line of Theorem 3.1? That is, if a set is in $\mathrm{NP}^{R_{K_U}}$ for each universal machine U, is it decidable?

Acknowledgments. The research of the author is supported in part by NSF Grants CCF-0830133, CCF-0832787, and CCF-1064785.

References

1. Allender, E., Buhrman, H., Koucký, M.: What can be efficiently reduced to the Kolmogorov-random strings? Annals of Pure and Applied Logic 138, 2–19 (2006)
2. Allender, E., Buhrman, H., Koucký, M., van Melkebeek, D., Ronneburger, D.: Power from random strings. SIAM Journal on Computing 35, 1467–1493 (2006)

3. Allender, E., Davie, G., Friedman, L., Hopkins, S.B., Tzameret, I.: Kolmogorov complexity, circuits, and the strength of formal theories of arithmetic. Tech. Rep. TR12-028, Electronic Colloquium on Computational Complexity (submitted for publication, 2012)
4. Allender, E., Friedman, L., Gasarch, W.: Limits on the Computational Power of Random Strings. In: Aceto, L., Henzinger, M., Sgall, J. (eds.) ICALP 2011. LNCS, vol. 6755, pp. 293–304. Springer, Heidelberg (2011)
5. Arora, S., Barak, B.: Computational Complexity, a modern approach. Cambridge University Press (2009)
6. Buhrman, H., Fortnow, L., Koucký, M., Loff, B.: Derandomizing from random strings. In: 25th IEEE Conference on Computational Complexity (CCC), pp. 58–63. IEEE Computer Society Press (2010)
7. Buhrman, H., Loff, B.: Personal Communication (2012)
8. Downey, R., Hirschfeldt, D.: Algorithmic Randomness and Complexity. Springer (2010)
9. Li, M., Vitanyi, P.: Introduction to Kolmogorov Complexity and its Applications, 3rd edn. Springer (2008)
10. Martin, D.A.: Completeness, the recursion theorem and effectively simple sets. Proceedings of the American Mathematical Society 17, 838–842 (1966)

Information and Logical Discrimination

Patrick Allo

Centre for Logic and Philosophy of Science, Vrije Universiteit Brussel, Pleinlaan 2,
B-1050 Brussels, Belgium
patrick.allo@vub.ac.be

Abstract. Allo & Mares [2] present an "informational" account of logical consequence that is based on the content-nonexpansion platitude.
The core of this proposal is an inversion of the standard direction of explanation: Informational content is not defined relative to a pre-existing
logical space, but it is approached in terms of the level of abstraction at
which information is assessed.

In this paper I focus directly on one of the main ideas introduced
in that paper, namely the contrast between logical discrimination and
deductive strength, and use this contrast to (1) illustrate a number of
open problems for an informational conception of logical consequence,
(2) review its connection with the dynamic turn in logic, and (3) situate
it relative to the research agenda of the philosophy of information.

1 Background and Motivation

1.1 Conceptions of Logical Consequence

By a conception of logic, we primarily mean a conception of its core notion,
namely the concept of logical consequence (what follows from what). Traditionally, logical consequence is analysed along two paths. The first path relies on
the basic platitude that for A to *follow from* Γ, it should be *impossible* for A
to be false while all members of Γ are true. More exactly, according to this first
path, the *prima facie* modal notion of logical consequence can be reduced to the
non-modal notion of truth-preservation. As emphasised by Jc Beall and Greg
Restall [6], this approach is summarised by the *Generalised Tarski Thesis*:

GTT. A conclusion A follows from premises Γ iff every *case* where all the
premises in Γ are true is also a *case* where A is true.

When we provide a formal *precisification* of the GTT, we usually do two things:
First, we give a formal description of what *cases* are; second, we inductively
define what it means for a formula to be true at a case (i.e., we state the truth-
conditions for the different expressions of our language). By following this traditional approach, we give a *model-theoretic* formalisation of a *truth-theoretic*
conception of logical consequence: We use set-theoretical tools to explain what
cases are, and we explain the meaning of our logical vocabulary by stating the
truth-conditions for each type of expression that is in our language (we say, for

S.B. Cooper, A. Dawar, and B. Löwe (Eds.): CiE 2012, LNCS 7318, pp. 17–28, 2012.
© Springer-Verlag Berlin Heidelberg 2012

instance, that $\ulcorner p$ or $q\urcorner$ is true in a case iff $\ulcorner p\urcorner$ is true in that case or $\ulcorner q\urcorner$ is true in that case). In accordance with our use of the GTT-label and its origin in [20], we call this approach Tarskian.

The second path relies on the basic platitude that for A to *follow from* Γ, there must be a derivation of A from Γ, where every step in that derivation is motivated by a primitive rule of inference. Crucially, these primitive rules are often seen as implicit definitions of the meaning of our logical vocabulary. According to this *inferential* account, we rely on proof-theoretical methods to determine the extension of "follows from," and we use proof-conditions to explain the meaning of the logical constants (we say, for instance, that to derive $\ulcorner p$ or $q\urcorner$, we must either derive $\ulcorner p\urcorner$ or derive $\ulcorner q\urcorner$). Given its origin in a famous remark in [12], we call this the Gentzian-approach.

In practice, inferentialism and proof-theoretical accounts of logical consequence (and the meaning of logical constants) are often considered synonymous. Similarly, truth-theoretical conceptions of logical consequence and the model-theoretical means it relies on are traditional allies as well. This received view is attractive when we consider the traditional truth-theoretical account of the consequence relation of classical logic, or when we consider the equally traditional inferentialist account of the consequence relation of intuitionist logic.[1] Yet, it becomes more problematic when we consider a wider variety of logical systems. The standard model-theories of intuitionist or relevant logics,[2] for instance, suggest either that alternative logics change the meaning of "true," or that these models are mere formal tools that, from a philosophical point of view, misconstrue the purpose of developing an alternative to classical logics [1]. As such, the rise of non-classical logics drives a wedge between truth-theoretical accounts of logical consequence and their model-theoretical implementation, and—albeit to a lesser extent—also between proof-theoretical characterisations of logical consequence and their inferentialist philosophical underpinning. It is at this point that informational conceptions of logical consequence become relevant.

Informational conceptions of logic were, perhaps most famously, developed as a response to the charge that the usual model-theory for relevant logics (due to Routley and Meyer)[3] didn't count as a proper semantics. As claimed by Copeland [10, p. 406] the use of (a) possibly incomplete and/or inconsistent cases together with (b) the Routley-star to enforce a non-classical behaviour of its negation-connective, and (c) a ternary relation to enforce a non-classical behaviour of its implication-connective, leads to a model-theory that gets the intended extension of the relevant consequence relation right, but fails to shed a light on the intended meaning of the logical connectives. In other words, we have a pure semantics or model-theory without a corresponding applied semantics.

---

[1] Intuitionist logics reject the classically valid rule of *double-negation elimination*, and hence also the validity of *indirect proof* (to conclude A from the fact that we have derived an absurdity from the negation of A), and of *excluded middle*.

[2] Relevant logics avoid the so-called paradoxes of material and strict implication; mainly: $p \to (q \to p)$, $q \to (p \vee \neg p)$, $(p \wedge \neg p) \to q$, and $q \to (p \to p)$.

[3] The actual details can safely be ignored, but see [3] for an overview.

The informational conceptions of relevant consequence developed by Restall [18] and Mares [15] provide an applied semantics based on Barwise's and Perry's theory of situations [5]. The guiding insights of such approaches are, first, that the possibly incomplete and inconsistent cases should rather be understood in terms of the information that is available in a situation rather than in terms of what is true or false in that situation, second, that for ⌜not−p⌝ to be available in a situation is just for ⌜p⌝ not to be available in all compatible situations, and, third, that for ⌜p implies q⌝ to be available in a situation is just to have access to a regularity or constraint stating that if some situation contains the information that ⌜p⌝ there must also be a related situation that contains the information that ⌜q⌝.

This specific formulation of an informational conception of logic shows that model-theoretic characterisations of logical consequence need not be tied to a narrow truth-conditional interpretation. The close connection between the Routley-Meyer semantics and theories of information that are based on Barwise and Perry's theory of situation, however, does not imply that informational conceptions of logic should ignore proof-theoretical considerations. While it is true that typical information-theoretical conceptions of logic rely on model-theoretical characterisations,[4] Mares [16] explicitly introduces proof-theoretical considerations to motivate the information-conditions for certain expressions, and Paoli [17] suggests an informational reading of sequents.

1.2   Informational Content and Logical Space

In a joint paper with Edwin Mares [2], I suggested that as a mere defence of the Routley-Meyer semantics, the use of situations and information flow between situations falls short of being a full-fledged informational account of logical consequence. Hence, we proposed a more general 'informational' account of logical consequence that is based on the content-nonexpansion platitude.

CN. A follows from Γ iff the content of A does not exceed the combined content of all members of Γ.

We then argued that, while classically there is no difference between an informational and a truth-conditional approach to logic (the information accessible at a world can straightforwardly be analysed as what is true in the set of worlds that cannot be distinguished from the actual world), the contrast between information-conditions and truth-conditions can be exploited in such non-classical logics as relevant and intuitionistic logic.

Because an informational conception leads to an inversion of the standard direction of explanation it allows for theoretical positions that are not readily available to its standard competitors (truth-conditional and inferential conceptions

---

[4] The already mentioned proposals by Restall and Mares [18,15], but also an information-theoretic characterisation of classical logic due to Corcoran [19] and an informational interpretation of constructive substructural logics due to Wansing [22,21].

of logical consequence). It is therefore particularly suited for both non-classical and pluralistically inclined philosophies of logic.

The informational content of a piece of information is almost invariably associated with a certain proportion of a given logical space (or space of possibilities); namely the proportion of the space that is ruled out by that information. On the standard order of explanation, the construction of this logical space is entirely based on the identification of possibility (true somewhere in the logical space) with consistency, and hence of necessity (true everywhere in the logical space) with inconsistency of negation. Concretely, the content associated with a contradiction coincides with the totality of the logical space: Contradictions are false across the whole space, and thus rule out the whole logical space. Analogously, the content associated with a logical truth is empty: Logical truths are true across the whole space (the negation of a logical truth is a contradiction), and thus rule out the null-proportion of the logical space. These considerations yield a logical space that only contains possibilities that are both consistent and complete, and this is entirely in accordance with the standard truth-theoretic principles that either $\ulcorner A \urcorner$ is true or $\ulcorner \text{not} - A \urcorner$ is true, but not both.

As long as we stick to this traditional order of explanation, it is hard to construct a finer logical space (a broader space of possibilities) without rejecting these standard principles. By constructing a logical space that does not depend on truth-theoretic considerations, but only on what can and cannot be distinguished, we keep logical and truth-theoretical considerations separate. The main challenge is to explain how a logical space should be constructed if truth isn't the main criterion.

1.3 Dynamics of Information

The development of an informational conception of logical consequence is just one way to connect logic with information. Another, more recent, trend in the study of logic and information is oriented towards the dynamics of information. Dynamic epistemic and doxastic logics (as well as many of their predecessors) investigate the dynamics of information by looking specifically at, in the traditional sense, how the propositional attitudes of knowledge and belief change in virtue of new information, and, in a more contemporary sense, how the distribution of knowledge and belief in a multi-agent setting evolves in virtue of how these agents exchange information. While extremely fruitful and influential, the dynamic paradigm does not aim to provide an informational conception of logical consequence. This is so because, firstly, the dynamic turn in logic challenges the centrality of the notion of logical consequence [8], and, secondly, because it provides a logical model of informational dynamics, rather than an informational account of logical dynamics (though one may argue that it does both). Yet, even if we grant the difference in focus, it still seems unfortunate that both informational approaches to logic remain unrelated.

Overview

Rather than to reiterate the argumentation of this previous paper, I want to focus directly on one of the main ideas introduced in that paper, namely the contrast between logical discrimination and deductive strength (§2), and use this contrast to review its connection with the dynamic turn in logic (§3.1); to illustrate a number of open problems for an informational conception of logical consequence (§3.2); and to situate it relative to the research agenda of the philosophy of information (§4).

2 Information and Discrimination

The notion of logical discrimination captures what can be "told apart" in a given logical system. We say, for instance, that intuitionist logic can discriminate between p and $\mathtt{not}-\mathtt{not}-p$, whereas classical logic cannot, or that paraconsistent logics can tell different inconsistent theories apart while from a classical viewpoint there's only one such theory, namely the trivial one. Still, even when it is acknowledged that non-classical logics like intuitionist and paraconsistent logic allow for finer distinctions than classical logic, considerations of that kind are not assumed to bear upon core issues in the philosophy of logic. The informational conception of logic we discuss in this paper introduces the notion of logical discrimination as the main criterion for the construction of logical space.

2.1 Two Principles and the Crucial Inversion

As an introduction to how I want to exploit the contrast between logical discrimination and deductive strength, consider the following two principles:[5]

IP. "Whenever there is an increase in available information there is a corresponding decrease in possibilities, and vice versa." [4, p. 491]

DD. "The more a logic proves, the fewer distinctions (or discriminations) it registers" [14, p. 207]

While each of these principles is indifferent with respect to the conceptual order of its relata, (logical) orthodoxy often imposes such an order. The inverse relationship principle **IP** is customarily read as a reduction of the notion of information to that of possibility: What counts as informative depends on a pre-existing space of possibilities. Furthermore, when dealing with logical possibility, the relevant space of possibilities is often further reduced to the logical notions of consistency and inconsistency. In practice, this comes down to a reduction of information to truth-conditions.

---

[5] The first of these principles is usually referred to as the inverse relationship principle. Since the second principle also expresses an inverse relation between two notions, I try to avoid the traditional nomenclature.

Likewise, **DD** merely expresses an inverse relation between the deductive strength and the discriminatory power of a logical system.[6] Yet, the traditional focus on what follows from what in logical theorising suggests that the discriminatory power of a logical system is, if not a mere side-effect of its deductive strength, then at least a consequence of how contents are to be carved out to get the extension of "follows from" right. This is a natural line of thought when one thinks of logical consequence as truth-preservation, for in that case we should make all distinctions that are needed to avoid stepping from truth to falsehood, *but no more than that.*

Barwise [4] already questioned the standard reading of **IP** by proposing a more pragmatic conception of possibilities and impossibilities (which is not restricted to the notions of logical possibility and impossibility):

> My main proposal here is that a good theory of possibility and information should be consistent with [IP]. As I analyze things, impossibilities are those states of the system under investigation that are ruled out by (i.e., incompatible with) the currently available information about the system. States that are not so ruled out are possibilities. Mathematical inquiry, like any other form of successful inquiry decreases possibilities when it increases the available information. (p. 491)
>
> In the case of logical possibility, the available information consists of the laws of logic. Just what the laws amount to, and how they relate the metaphysical and mathematical laws, is a contentious question, however, one we shall not attempt to answer. How one answers it will determine what one counts as a logical possibility. (p. 497)

In [2] we followed a similar path by, in a first move, proposing that how we carve out contents by specifying a logical space is determined by the logical information that is available, and, second, by analysing the notion of available information in terms of how we access and use information in our environment. The emphasis on access and use in the analysis of what counts as available information is rooted in a relational conception of information; it depends on our environment as well as on the kind of agents / reasoners we are. The following quote places this relational conception in the broader context of the philosophy of information:

> The point made was that data are never accessed and elaborated (by an information agent) independently of a *level of abstraction.* Understood as relational entities, data are *constraining affordances*: they allow or invite certain constructs (they are *affordances* for the information agent that can take advantage of them) and resist or impede some others (they are *constraints* for the same agent), depending on the interaction with, and the nature of, the information agent that processes them. [11, p. 87]

The way we access or fail to access certain information is tied to the distributed nature of information (or data). We cannot access all information at once, and

---

[6] As shown by Humberstone, this principle has exceptions, but here we can ignore this complication.

because of that using information often means combining information that is accessed in different situations. By varying the ways in which we access and combine information from different situations, we can conceive of different "reasoning styles" and relate these styles to different logical systems [2, p. 9]. We consider three such styles:

1. When we access information *locally*, and combine information from different situations without thereby always having access to the totality of the information we used to reach our conclusions, the resulting reasoning-style can be codified by a relevant or other substructural logic.
2. When we access information locally, but let the deductive reasoning process be cumulative such as to retain access to the information we used, the resulting reasoning-style can be captured by an intuitionistic logic.
3. When we access information globally (we access all accessible information at once), the resulting reasoning-style is in accordance with classical logic.

This way of relating access and use to logical systems is very natural from the perspective of the frame-semantics for relevant and intuitionist logics [7].

For an informational conception of logic, a logical space, but also a given degree of logical discrimination, or a set of global constraints on a class of models are all abstractions of how we access and use the information in our environment. Global constraints [2, p. 13] are much like the laws of logic mentioned in the quotation by Barwise. Logical spaces, by contrast, can be related to the type of environment we focus on—worlds vs. situations—, or to the perspective on inference and the representation of partial information we adopt—worldly vs. situated—see [2, p. 7], but also [6]. Yet, the notion of a *degree of logical discrimination* is even more closely related to one of Barwise's concerns; namely the fact that a 'given' logical space is often too coarse to account for what we consider genuine increases of available information. By questioning the usual reading of **DD** and using considerations about which distinctions are worth retaining directly for the construction of a logical space, we obtain an alternative perspective on the granularity problem. The main virtue of this type of approach is the connection it establishes between two main concerns in logical (and other types of formal) modelling: the ability to extract information from our model (inference), and the ability to distinguish between relevant properties of the model (discrimination).

2.2 Logic and Granularity

With respect to the notion of granularity, the following view seems fairly uncontroversial: Propositions can be individuated more or less finely. What counts as the correct level of abstraction (henceforth, LoA) or degree of granularity can only be determined once we make clear what we're after. It is, for instance, standard to assume that if we're interested in logical consequence, it suffices to individuate propositions in terms of their truth-conditions to find out whether the content of the conclusion of an argument does not exceed the combined content of its premises. On that account, two propositions should be treated as (logically-) equivalent whenever their truth-conditions coincide. By contrast, truth-conditions are often deemed too coarse to characterise the content of intentional states. Thus—as exemplified by the fact that logic requires us to make fewer distinction than many other theoretical enterprises—different theoretical disciplines require different degrees of granularity.

Yet, orthodoxy has it that while contents can be individuated more or less finely, there's only one "logical" way to discriminate propositional contents; namely in terms of their truth-conditions. When supplemented with the (equally orthodox) view that truth-conditions can only be assigned in accordance with the classical truth-tables, this immediately motivates a monism about logical consequence (there is only one correct consequence relation). An informational conception of logical consequence avoids this second aspect of logical orthodoxy by defending the view that some non-classical ways of carving out contents are as *logical* as the classical way of doing so. This argument is backed by several considerations; some of which are directly related to the functioning of non-classical logics, others are related to how we choose a logical system. To a first approximation, when we settle for a logical system to model or evaluate an argument, the choice is between deductively strong classical consequence relations, and deductively weaker sub-classical systems. When classical logic has unwelcome consequences, we retreat to a weaker logic. This is the received view: Because we end up with a crippled logic, the decision to use a sub-classical logic is a genuine retreat.

With **DD** in mind, an alternative view is readily available: When we opt for a deductively weaker consequence relation, we obtain some additional discriminatory power in return. As a result, when we evaluate a logical system we need to balance the opposite virtues of logical discrimination and deductive strength to decide which logical system is the most appropriate for a given purpose. From that perspective, the standard reading of **DD** as well as the received view on logical revision is mistaken or at least incomplete. While paraconsistent logics are often adopted as a means to avoid triviality in the face of contradiction, such non-explosive logics can equally well be adopted with the intent to discriminate between different, yet classically equivalent, inconsistent theories. In fact, given some minor assumptions, the avoidance of explosion and the ability to tell different contradictions apart are two sides of the same coin.[7] This line of argument is not readily available to the proponent of a truth-theoretical conception of logic.

---

[7] Two expressions A and B are synonymous relative to $\vdash$ ($A \equiv_\vdash B$) iff we have:

$$C_1(A), \ldots, C_n(A) \vdash C_{n+1}(A) \text{ iff } C_1(B), \ldots, C_n(B) \vdash C_{n+1}(B)$$

Where $C_i(B)$ is obtained from $C_i(A)$ by replacing some (but not necessarily all) occurrences of A in $C_i(A)$ by B.

If, following Humberstone, we take *synonymy* as the formal explication of logical discrimination, then (assuming that consequence is *reflexive* and *transitive*, and that *simplification* is valid) "p and not-p" and "q and not-q" are synonymous for a consequence relation iff that consequence relation is explosive.

First, because if our main aim is to avoid stepping from truth to falsehood, it is hard to motivate additional distinctions that do not serve this primary purpose of truth-preservation (Why reject a classically valid argument if it is impossible for the conclusion to be actually false whenever all the premises are actually true?). Second, because finer distinctions require a broader logical space, and this can only be done by including possibly incomplete and/or inconsistent possibilities. This move isn't straightforward if one wishes to preserve a classical conception of truth.

3   Issues and Open Problems

In its standard presentation that is based on the content-nonexpansion platitude, the informational conception of logic is a natural ally of logical pluralism. When put in the words of van Benthem, we should rather say it is an ally of a "reasoning styles" pluralism about logic. Indeed, despite the importance granted to the notion of logical discrimination, the centrality of logical consequence (witness its role in the definition of *synonymy*) still suggests that an informational conception of logic is irrelevant to the concerns of the dynamic turn in logic. This type of objection is implicit in the following fragments from van Benthem [8]:

> In particular, I will argue that logical dynamics sets itself the more ambitious diagnostic goal of explaining why sub-structural phenomena occur, by 'deconstructing' them into classical logic plus an explicit account of the relevant informational events. I see this as a still more challenging departure from traditional logic. (...) The view that logic is really only about consequence relations may have been right at some historical stage of the field. (...) Since the 1930s, modern core logic has been about at least two topics: valid inference, yes—but on a par with that, definability, language and expressive power. (...) And to me, that definability aspect has always been about describing the world, and once we can do that, communicating to others what we know about it. (p. 182–3)

This type of objection raises two important issues for the position I defended in the previous sections. First, it suggests that the informational conception of logic may be too narrow to cover so-called logics for informational dynamics; second, if so-called substructural phenomena can be elucidated in extensions of classical logic, it implies that the sub-classical logics that arise from the need to discriminate propositions more finely may not be the most appropriate or desirable formalism to reach this goal. Below, I discuss each of these objections.

3.1   Information and the Dynamic Turn in Logic

The clue to see how the informational conception really engages with the dynamic turn in logic lies in the already mentioned identification of *inference* with the ability to extract information from a given formal model, and of *discrimination* with the ability to distinguish between different features of that model.

Anyone familiar with the development of dynamic epistemic and doxastic logics will immediately recognise these concerns. Extracting information from an epistemic or doxastic model is what we do when (a1) we assign beliefs or ascribe knowledge to an agent, and (b1) we predict the effect of certain actions on their knowledge or beliefs. Distinguishing relevant features of a model is what we do when (a2) we compare or contrast the epistemic states of different agents, and (b2) compare the effect of different types of actions.

The standard story can thus be extended as follows: If our underlying consequence relation is powerful, our agents will seem highly knowledgeable according to our model; if our underlying consequence relation is more discriminating, our model will not collapse intuitively distinct knowledge or belief-states. While obviously correct, this standard story is also overly reductive. The strength of dynamic epistemic and doxastic logics lies precisely in the wide range of *informational actions* it can incorporate. In dynamic epistemic logic we can contrast public with several types of private and semi-private announcements; in dynamic doxastic logic we can contrast several belief-revision policies. Arguably, this is an increase in discriminatory power that does not obviously correspond to a decrease in deductive power, and thus isn't covered (in a non *ad hoc* way) by the standard story. As such, the dynamic turn in logic clearly poses a challenge for the informational conception of logic. I do not believe this is an insurmountable challenge. The centrality of the content-nonexpansion platitude might have to be reconsidered, and the standard story about how logical discrimination and deductive strength are related will have to be generalised. But this seems feasible. The notion of logical discrimination as well as the double concern of extracting information from a model and the ability to make some distinctions in a model while collapsing others can, therefore, remain unchallenged.

3.2   Guises of Logical Discrimination

Suggesting that logics that revise classical logic may be recaptured as extensions of classical logic isn't specific to the dynamic turn in logic, but is a standard theme in the philosophy of non-classical logic [13,1], and in the discussion of the notion of information in intuitionistic logic [9]. The guiding idea seems to be this: We should not weaken the underlying consequence relation such as to be able to carve out more fine-grained contents, but we should make our languages more expressive such as to both retain the deductive strength of classical logic, and be able to make novel distinctions. The moral of the availability of these two ways of changing our logic is, in my opinion, not that the conservative approach of extending rather than revising our logic is always to be preferred. It only reveals that the study of logical discrimination should not be reduced to that of granularity. Expressivity is equally important! Again, the dynamic turn (and more broadly the development of modal logics) only poses a challenge to the development of an informational conception of logical consequence in the sense that it calls for a further generalisation of its basic concepts (logical discrimination isn't just about the distinctions that are retained by a logic, but also about the distinctions that are already potentially there in a formal language). Such developments do not show that the project of placing logical discrimination at the core of an informational conception of logic is not viable.

4   Informational Semantics and the Philosophy of Information

To conclude, one could ask how this story about logical discrimination relates to the core concerns of the philosophy of information. After all, the concerns about the relation between extensions and revisions of classical logic are already a topic in the philosophy of logic. Likewise, the question of how the content of intentional states should be individuated if doing so on the basis of truth-conditions isn't viable, is a topic in formal semantics, philosophical logic, and the metaphysics of modality. What do we win by regrouping these concerns under the heading of the philosophy of information? There are two complementary answers to that question.

The direct answer is that the inversion in the order of explanation leads to a conception of logical consequence that takes information as its most basic notion. As such, the development of an informational conception of logic falls squarely within the scope of the philosophy of information.

A more indirect answer is related to the method of abstraction (the core method in the philosophy of information [11, III]), and more precisely to the fact that the relational conception of information can be used to motivate the central role of logical discrimination. By adhering to the method of abstraction we accept a form of pluralism that is based on making the relevant LoA explicit, but we do not claim that anything goes. Indeed, because levels of abstraction can be compared, we can require that one level of abstraction refines another level of abstraction without making incompatible claims. This is exactly the situation we have when we compare classical with non-classical logics: both relevant and intuitionist logics provide a refinement of classical logic, whereas classical logic can be seen as the limiting case of each of them.

The interaction between the philosophy of logic and the philosophy of information also works in the opposite direction. Most standard presentations of the method of abstraction suggest that differences in levels of abstraction are best understood as differences in the non-logical vocabulary (e.g., the number of non-logical predicates that are available), but the introduction of non-classical logics in terms of additional logical distinctions suggests that differences in levels of abstraction can equally well be understood as pure differences in logical discrimination. As a result, the development of an informational conception of logic challenges the standard presentation of non-classical logics by according more importance to the notion of logical discrimination, and challenges the standard presentation of the method of abstraction by making clear that even the underlying logic one adheres to is part of the LoA one assumes.

Acknowledgements. The author is a postdoctoral Fellow of the Research Foundation – Flanders (FWO).

References

1. Aberdein, A., Read, S.: The philosophy of alternative logics. In: Haaparanta, L. (ed.) The Development of Modern Logic, pp. 613–723. Oxford University Press, Oxford (2009)
2. Allo, P., Mares, E.: Informational semantics as a third alternative? Erkenntnis, 1–19 (2011), http://dx.doi.org/10.1007/s10670-011-9356-1
3. Anderson, A.R., Belnap, N.D.: Entailment. The Logic of Relevance and Necessity, vol. I. Princeton University Press, Princeton (1975)
4. Barwise, J.: Information and impossibilities. Notre Dame Journal of Formal Logic 38(4), 488–515 (1997)
5. Barwise, J., Perry, J.: Situation and Attitudes. The David Hume Series of Philosophy and Cognitive Science Reissues. CSLI Publications, Stanford (1999)
6. Beall, J.C., Restall, G.: Logical Pluralism. Oxford University Press, Oxford (2006)
7. Beall, J., Brady, R., Dunn, J., Hazen, A., Mares, E., Meyer, R., Priest, G., Restall, G., Ripley, D., Slaney, J., Sylvan, R.: On the ternary relation and conditionality. Journal of Philosophical Logic, 1–18 (May 2011) online First
8. van Benthem, J.: Logical dynamics meets logical pluralism? Australasian Journal of Logic 6, 182–209 (2008)
9. van Benthem, J.: The information in intuitionistic logic. Synthese 167(2), 251–270 (2009)
10. Copeland, B.J.: On when a semantics is not a semantics: Some reasons for disliking the Routley-Meyer semantics for relevance logic. Journal of Philosophical Logic 8(1), 399–413 (1979), http://dx.doi.org/10.1007/BF00258440
11. Floridi, L.: The Philosophy of Information. Oxford University Press, Oxford (2011)
12. Gentzen, G.: Untersuchungen über das logische Schließen. I. Mathematische Zeitschrift 39(1), 176–210 (1935)
13. Haack, S.: Deviant Logic. Some Philosophical Issues. Cambridge University Press, Cambridge (1974)
14. Humberstone, I.L.: Logical discrimination. In: Béziau, J.Y. (ed.) Logica Universalis, pp. 207–228. Birkhäuser Verlag, Basel (2005)
15. Mares, E.: Relevant logic and the theory of information. Synthese 109(3), 345–360 (1997)
16. Mares, E.: General information in relevant logic. Synthese 167(2), 343–362 (2009)
17. Paoli, F.: Substructural logics a primer. Trends in Logic: Studia Logica Library, vol. 13. Kluwer Academic, Dordrecht (2002)
18. Restall, G.: Information flow and relevant logics. In: Seligman, J., Westerståhl, D. (eds.) Logic, Language and Computation: The 1994 Moraga Proceedings, pp. 463–477. CSLI-Press, Stanford (1994)
19. Saguillo, J.M.: Methodological practice and complementary concepts of logical consequence: Tarski's model-theoretic consequence and Corcoran's information-theoretic consequence. History and Philosophy of Logic 30(1), 21–48 (2009)
20. Tarski, A.: On the concept of logical consequence. In: Tarski, A., Corcoran, J. (eds.) Logic, Semantics, Meta-Matematics, 2nd edn., Hackett, Indianapolis (1983)
21. Wansing, H.: The Logic of Information Structures. LNCS (LNAI), vol. 681. Springer, Berlin (1993)
22. Wansing, H.: Informational interpretation of substructural logics. Journal of Logic, Language and Information 2, 285–308 (1993)

Robustness of Logical Depth

Luís Antunes[1], Andre Souto[2], and Andreia Teixeira[1]

[1] Departamento de Ciência de Computadores, Universidade do Porto,
R. Campo Alegre, 1021/1055, 4169-007 Porto, Portugal
[2] Instituto de Telecomunicações, Instituto Superior Técnico, Universidade Técnica de
Lisboa, Torre Norte, Piso 10, Av. Rovisco Pais, 1 1049-001 Lisboa, Portugal

Abstract. Usually one can quantify the subjective notion of useful information in two different perspectives: static resources – measuring the amount of planing required to construct the object; or dynamic resources – measuring the computational effort required to produce the object. We study the robustness of logical depth measuring dynamic resources, proving that small variations in the significance level can cause large changes in the logical depth.

1 Introduction

Philosophy and meaning of information has a long history, however recently interest in this area has increased and researchers have tackled this from different approaches and perspectives, namely its meaning, quantification and measures of information and complexity. In this paper we are interested in measures of meaningful or useful information. In the past there have been several proposals to address this question: sophistication [8,9], logical depth [4], effective complexity [6], meaningful information [12], self-dissimilarity [13], computational depth [3], facticity [1]. Pieter Adriaans [1] divided the several approaches to defined a string as interesting in: i) some amount of computation resources are required to construct the object (Sophistication, Computational Depth). ii) exists a trade-off between the model and the data code under two part code optimization (meaningful information, effective complexity, facticity) and finally iii) it has internal phase transitions (self-dissimilarity).

Solomonoff [11], Kolmogorov [7] and Chaitin [5] independently defined a rigorous measure of the information contained in an individual object x, as the length of the shortest program that produces the object x. This measure is usually called Kolmogorov complexity of x and is denoted by $K(x)$. A randomly generated string, with high probability, has high Kolmogorov complexity, so contains near maximum information. However by being randomly generated, makes it unlikely to have useful information as we can obtain a similar one by flipping fair coins.

Usually one can quantify the subjective notion of useful information, as in item i) previously defined, in two different perspectives: static resources – measuring the amount of planning required to construct the object; or dynamic resources – measuring the computational effort required to produce the object.

S.B. Cooper, A. Dawar, and B. Löwe (Eds.): CiE 2012, LNCS 7318, pp. 29–34, 2012.

Regarding *dynamic resources*, the Kolmogorov complexity of a string x does not take into account the time effort necessary to produce the string from a description of length $K(x)$. Bennett [4] called this effort *logical depth*. Intuitively, a computationally deep string takes a lot of computational effort to be recovered from its shortest description while shallow strings are trivially constructible from their $K(x)$, i.e., the shortest program for x does not require lots of computational power to produce it. After some attempts, Bennett [4] formally defined the s-significant logical depth of an object x as the time required by a standard universal Turing machine to generate x by a program p that is at most s bits longer than its Kolmogorov complexity. Thus, an object is logically deep if it takes a lot of time to be generated from any short description.

Bennett[4] claimed that the significance level in logical depth is due to stability reasons. In this paper we study its robustness, i.e., if small variations in the significance level can cause large changes in the logical depth. In this sense we show that logical depth is not robust.

The rest of the paper is organized as follows: in the next section, we introduce some notation, definitions and basic results needed for the comprehension of the rest of the paper. In Section 3, we prove that logical depth is not a stable measure.

2   Preliminaries

In this paper we use the binary alphabet $\Sigma = \{0,1\}$, $\Sigma^* = \{0,1\}^*$ is the set of all finite binary strings that are normally represented by x, y and z and Σ^n and $\Sigma^{\leq n}$ are the set of strings of length n and the set of strings of length at most n, respectively. We denote the initial segment of length k of a string x with length $|x|$ by $x_{[1:k]}$ and its ith bit by x_i. The function log will always mean the logarithmic function of base 2. $\lfloor k \rfloor$ represents the largest integer smaller or equal than k. All resource-bounds used in this paper are *time constructible*, i.e., there is a Turing machine whose running time is exactly $t(n)$ on every input of size n, for some time t. Given a program p, we denote its running time by $time(p)$. Given two functions f and g, we say that $f \in O(g)$ if there is a constant $c > 0$, such that $f(n) \leq c \cdot g(n)$, for almost all $n \in \mathbb{N}$.

2.1   Kolmogorov Complexity

We refer the reader to the book of Li and Vitányi [10] for a complete study on Kolmogorov complexity.

Definition 1 (Kolmogorov complexity). *Let U be a universal prefix-free Turing machine. The* prefix-free Kolmogorov *complexity of $x \in \Sigma^*$ given $y \in \Sigma^*$ is,*

$$K(x|y) = \min_p \{|p| : U(p, y) = x\}.$$

*The t-*time-bounded prefix-free Kolmogorov complexity *of $x \in \Sigma^*$ given $y \in \Sigma^*$ is,*

$$K^t(x|y) = \min_p \{|p| : U(p, y) = x \text{ in at most } t(|x|) \text{ steps}\}.$$

The default value for the axillary input y for the program p, is the empty string ϵ and to avoid overloaded notation we usually drop this argument in those cases. We choose as a reference universal Turing machine a machine that affects the running time of a program on any other machine by at most a logarithmic factor and the program length by at most a constant number of extra bits.

Definition 2 (c-incompressible). *A string x is c-incompressible if and only if $K(x) \geq |x| - c$.*

A simple counting argument can show the existence of c-incompressible strings. In fact,

Theorem 1. *There are at least $2^n \cdot (1 - 2^{-c}) + 1$ strings $x \in \Sigma^n$ that are c-incompressible.*

Bennett [4] said that a string x is logically deep if it takes a lot of time to be generated from any short description. The c-significant logical depth of an object x is the time that a universal Turing machine needs to generate x by a program that is no more than c bits longer than the shortest description of x.

Definition 3 (Logical depth). *Let x be a string, c a significance level. A string's logical depth at significance level c, is:*

$$ldepth_c(x) = \min_p \{time(p) : |p| \leq K(x) + c \wedge U(p) = x\}.$$

One can however, scale down the running time for program length by using a Busy Beaver function, similar to the notion of Busy Beaver computational depth introduced in [2].

Definition 4 (Busy Beaver function). *The Busy Beaver function, BB, is defined by*

$$BB : \mathbb{N} \to \mathbb{N}$$
$$n \to \max_{p:|p| \leq n} \{running\ time\ of\ U(p)\ when\ defined\}$$

Definition 5 (Busy Beaver logical depth). *The Busy Beaver logical depth, with significance level c, of $x \in \Sigma^n$ is defined as:*

$$ldepth_c^{BB}(x) = \min_l \{\exists p : |p| \leq K(x) + c\ and\ U(p) = x\ in\ time\ \leq BB(l)\}$$

Notice that this is a rescaling of Definition 3, since $BB^{-1}(ldepth_c(x)) = ldepth_c^{BB}(x)$. From this new definition it is clear that $ldepth_c^{BB}(x) \leq K(x) + O(1)$. In fact, one can simulate any computation, keeping track of the amount of computation steps and thus, $K(time(p)) \leq |p| + O(1)$ which implies $ldepth_c^{BB}(x) \leq K(x) + O(1)$.

Notice also that the Busy Beaver logical depth is a static measure based on programs length.

3 Instability of Logical Depth

In this section we prove that even slightly changes of the constant on the significance level c of logical depth determines large variation of this measure.

Theorem 2. *For every sufficiently large n there are constants c, k_1 and k_2 and a string x of length n such that $ldepth_{k_2}(x) \geq 2^n$ and $ldepth_{2c+k_1}(x) \leq O(n \cdot \log n)$.*

Proof. Consider the following set:

$$A = \{x \in \Sigma^n : (\exists p)|p| < n + K(n) - c \wedge U(p) = x \text{ in time } \leq 2^n\}$$

Considering $B = \Sigma^n - A$, we know that B has at least $2^n(1 - 2^{-c})$ elements.
    Let $x \in B$ such that $n + K(n) - c - k_1 \leq K(x) \leq n + K(n) - c - k_2$ for some constants k_1 and k_2. We show that these strings exist in Lemma 3.1 bellow. Thus,

- $ldepth_{k_2}(x) \geq 2^n$.
  Assume that $ldepth_{k_2}(x) < 2^n$, then, by definition of logical depth, there is a program p of size at most $K(x) + k_2 \leq n + K(n) - c - k_2 + k_2 = n + K(n) - c$ such that $U(p) = x$ in time $< 2^n$, which implies that x would be an element of A, which contradicts the choice of x.
- $ldepth_{2c+k_1}(x) \leq O(n)$.
  Since the significance level is $2c + k_1$, then we can consider programs to define its logical depth of length at least $n + K(n) - c - k_1 + 2c + k_1 = n + K(n) + c$ (and of course of length at most $n + K(n) - c - k_2 + 2c + k_1 = n + K(n) + c + k_1 - k_2$). So, if c is sufficiently large to allow a prefix free version of the program `print` to be one of the possible programs, then we conclude that $ldepth_{2c+k_1}(x)$ is at most the running time of `print(x)`, which is at most $O(n \cdot \log n)$.

Lemma 3.1. *Let c be a constant and B be the set described in the last proof. There are constants k_1 and k_2 and strings in B such that $n - c - k_1 \leq K(x) \leq n - c - k_2$.*

Proof. Consider the set $S = \{x \in \Sigma^n : K(x) \geq n + K(n) - c - k_1\}$.
    It is easy to see that every element in S is in B. Let p be the program of size $\leq n + K(n) - c - a$ where a is a constant to be defined later that has the longest running time. Notice that $K(p) \geq n + K(n) - c - a - l$ for some l. In fact, if $K(p) < n + K(n) - c - a - l$ for all l then we could consider the program that runs p^*, the 1st program in the lexicographic order that produces p to obtain p and then run it again and that would be a smaller program that would have longer running time. More formally, consider the program $q = RUN(\cdot)$ where RUN describes the universal Turing machine with some data and run it on U. Since RUN is describable by a constant number of bit, say s, then, if the data is p^*, $|q| = |p^*| + s \leq n + K(n) - a - l + s \leq n + K(n) - a$ for sufficiently large l. Furthermore, $time_U(q) \geq time_U(RUN(p^*)) = time_U(p^*) + time_U(p) \geq time_U(p)$ which contradicts the choice of p.
    Let t be the running time of p and let x be the first string in the lexicographic order such that $K^t(x) \geq n + K(n)$. Thus,

- $K(x) \leq K(p) + b \leq n + K(n) - c - a + b$ for some constant b since from p we can compute t and then compute x.
- $K(x) \geq n + K(n) - c - a$. In fact, if $K(x) < n + K(n) - c - a$ then considering q the prefix-free program that witnesses the Kolmogorov complexity of x we would have that $|q| < n - c - a$ and then $U(q) = x$. Thus, by definition of p we get $time(q) < time(p)$ and hence $K^t(x) \leq n + K(n) - a$ contradicting the choice of x.

Just take $a > b$ and also $k_1 = a$ and $k_2 = a - b$.

Theorem 3. *For every sufficiently large there are constants c, k_1 and k_2 and a string x of length n such that $ldepth_{k_2}^{BB}(x) \geq n$ and $ldepth_{2c+k_1}^{BB}(x) \leq O(\log n)$.*

Proof. The idea is similar to Theorem 2. We rewrite the proof has the reasoning of the conclusions changes a bit.
    Consider the following set:

$$A = \{x \in \Sigma^n : (\exists p)|p| < n + K(n) - c \wedge U(p) = x \text{ in time } \leq BB(n)\}$$

Considering $B = \Sigma^n - A$, we know that B has at least $2^n(1 - 2^{-c})$ elements.
    Let $x \in B$ such that $n + K(n) - c - k_1 \leq K(x) \leq n + K(n) - c - k_2$ for some constants k_1 and k_2. Thus,

- $ldepth_{k_2}^{BB}(x) \geq n$.
    Assume that $ldepth_{k_2}^{BB}(x) < n$, then, by definition of busy beaver logical depth, there is a program p of size at most $K(x) + k_2 \leq n + K(n) - c - k_2 + k_2 = n + K(n) - c$ such that $U(p) = x$ in time $< BB(n)$, which implies that x would be an element of A, contradicting the choice of x.
- $ldepth_{2c+k_1}(x) \leq O(\log n)$.
    Since the significance level is $2c + k_1$, then we can consider programs to define its logical depth of length at least $n + K(n) - c - k_1 + 2c + k_1 = n + K(n) + c$ (and of course of length at most $n + K(n) - c - k_2 + 2c + k_1 = n + K(n) + c + k_1 - k_2$). So, if c is sufficiently large to allow a prefix free version of the program `print` to be one of the possible programs, then we conclude that $ldepth_{2c+k_1}^{BB}(x)$ is at most $BB^{-1}(time(\texttt{print}(x))) = BB^{-1}(n \log n) \leq O(\log n)$ (since the busy beaver function grows faster than any computable function, in particular exponential).

We can adapt the argument presented above to prove that if we allow logarithmic terms on the significance level of logical depth we get a similar result.

Corollary 3.1. *For every n and sufficiently large there are constants c, k_1 and k_2 and a string x of length n such that $ldepth_{k_2 \log n}^{BB}(x) \geq n$ and $ldepth_{(2c+k_1) \log n}^{BB}(x) \leq O(\log n)$.*

Proof. The proof is equal to the previous one with the following adaptations:

$$A = \{x \in \Sigma^n : (\exists p)|p| < n + K(n) - c \log n \wedge U(p) = x \text{ in time } \leq BB(n)\}$$

and with a similar reasoning to Lemma 3.1 we can show the existence of a string in the complement of A satisfying $n + K(n) - c \log n - k_1 \log n \leq K(x) \leq n + K(n) - c \log n - k_2 \log n$.

4 Conclusions

Our major contribution in this paper is the proof that the most commonly used definition of logical depth in the literature is not stable, since small variations in the significance level can cause drastic changes in the value of Logical depth even if we correct it with a Busy Beaver function.

Acknowledgments. We thanks Bruno Bauwens for helpful discussions and comments. This work was supported by FCT projects PEst-OE/EEI/LA0008/2011 and PTDC/EIA-CCO/099951/2008. The authors are also supported by the grants SFRH/BPD/76231/2011 and SFRH/BD/33234/2007 of FCT.

References

1. Adriaans, P.: Facticity as the amount of self-descriptive information in a data set (2012)
2. Antunes, L., Fortnow, L.: Sophistication revisited. Theory of Computing Systems 45(1), 150–161 (2009)
3. Antunes, L., Fortnow, L., van Melkebeek, D., Vinodchandran, N.: Computational depth: concept and applications. Theoretical Computer Science 354(3), 391–404 (2006)
4. Bennett, C.: Logical depth and physical complexity. In: A half-Century Survey on the Universal Turing Machine, pp. 227–257. Oxford University Press, Inc., New York (1988)
5. Chaitin, G.: On the length of programs for computing finite binary sequences. Journal of ACM 13(4), 547–569 (1966)
6. Gell-Mann, M., Lloyd, S.: Information measures, effective complexity, and total information. Complexity 2(1), 44–52 (1996)
7. Kolmogorov, A.: Three approaches to the quantitative definition of information. Problems of Information Transmission 1(1), 1–7 (1965)
8. Koppel, M.: Complexity, depth, and sophistication. Complex Systems 1, 1087–1091 (1987)
9. Koppel, M.: Structure. The Universal Turing Machine: A Half-Century Survey, 2nd edn., pp. 403–419. Springer (1995)
10. Li, M., Vitányi, P.: An Introduction to Kolmogorov Complexity and Its Applications. Springer (2008)
11. Solomonoff, R.: A formal theory of inductive inference, Part I. Information and Control 7(1), 1–22 (1964)
12. Vitányi, P.M.B.: Meaningful information. IEEE Transactions on Information Theory 52(10), 4617–4626 (2006)
13. Wolpert, D.H., Macready, W.: Using self-dissimilarity to quantify complexity. Complexity 12(3), 77–85 (2007)

Turing's Normal Numbers: Towards Randomness

Verónica Becher

Departamento de Computación, Facultad de Ciencias Exactas y Naturales,
Universidad de Buenos Aires, Pabellón I, Ciudad Universitaria, (1428) Buenos Aires,
Argentina
vbecher@dc.uba.ar

Abstract. In a manuscript entitled "A note on normal numbers" and
written presumably in 1938 Alan Turing gave an algorithm that produces
real numbers normal to every integer base. This proves, for the first time,
the existence of computable normal numbers and it is the best solution
to date to Borel's problem on giving examples of normal numbers. Furthermore, Turing's work is pioneering in the theory of randomness that
emerged 30 years after. These achievements of Turing are largely unknown because his manuscript remained unpublished until its inclusion
in his Collected Works in 1992. The present note highlights Turing's ideas
for the construction of normal numbers. Turing's theorems are included
with a reconstruction of the original proofs.

1 On the Problem of Giving Instances of Normal Numbers

The property of *normality* on real numbers, defined by Émile Borel in 1909, is a
form of randomness. A real number is normal to a given integer base if its infinite
expansion is seriously balanced: every block of digits of the same length must
occur with the same limit frequency in the expansion of the number expressed in
that base.[1] For example, if a number is normal to base two, each of the digits '0'
and '1' occur in the limit, half of the times; each of the blocks '00', '01', '10' and '11'
occur one fourth of the times, and so on. A real number that is normal to every
integer base is called *absolutely normal*, or just *normal*. Borel proved that almost
all real numbers are normal (that is, the set of normal numbers has Lebesgue
measure 1), and he asked for an explicit example. Since then it has been easier
to conjecture results on normality than to prove them. In particular, it remains
unproved whether the fundamental mathematical constants such as π, $\sqrt{2}$ and e
are normal to some integer base. Although its has been proved that there exist
numbers that are normal to one base but not to another [9,26], no examples
have been given. There are already many particular constructions of numbers

---

[1] An alternative characterization proves that a real number x is normal to a base b if,
and only if, the sequence $(xb^n)_{n \geq 1}$ is uniformly distributed modulo one [6]. Also, a
real number is normal to a base b if, and only if, its expansion is compressible by no
information lossless finite automaton (injective transducer) [22,18,7].

S.B. Cooper, A. Dawar, and B. Löwe (Eds.): CiE 2012, LNCS 7318, pp. 35–45, 2012.
© Springer-Verlag Berlin Heidelberg 2012

that are normal to a given base, but no explicit instance has been proved normal to two multiplicatively independent bases; see [6] for up to date references.

It is fair to say that Borel's question on providing an example of a normal number (normal to every integer base) is still unresolved because the few known instances are not completely satisfactory: it is desirable to show that a *known* irrational number is normal, or, at least, to exhibit the number explicitly. We would like an example with a simple mathematical definition and such that, in addition of normality, some extra properties are proved. Considering that *computability* is the acceptable notion of constructiveness since the 1930s, we would also like that the number be easily computable. Let us recall that, as defined by Turing [24], the *computable real numbers* are those whose expansion in some integer base can be generated by a mechanical (finitary) method, outputting each of its digits, one after the other.

There is no evident reason for the normal numbers to have a non-empty intersection with the computable numbers. A measure-theoretic argument is not enough to see that these two sets intersect: the set of normal numbers in the unit interval has Lebesgue measure one, but the computable numbers are just countable, hence they form a null set (Lebesgue measure 0). Indeed, there are computable normal numbers, and this result should be attributed to Alan Turing. His manuscript entitled *"A note on normal numbers"*, presumably written in 1938, presents the best answer to date to Borel's question: an algorithm that produces normal numbers. This early proof of existence of computable normal numbers remained largely unknown because Turing's manuscript was only published in 1997 in his Collected Works, edited by J.L.Britton [25]. The editorial notes say that the proof given by Turing is inadequate and speculate that the theorem could be false. In [1] we reconstructed and completed Turing's manuscript, trying to preserve his ideas as accurately as possible and correcting minor errors.

The very first examples of normal numbers were independently given by Henri Lebesgue and Waclaw Sierpiński[2] in 1917 [16,23]. They also lead to computable instances by giving a computable reformulation of the original constructions [2]. Together with Turing's algorithm these are the only known constructions of computable normal numbers. In his manuscript, Turing alerts the reader that the provided examples of normal numbers are not convenient and he explicitly says that one would like that the expansion of such numbers be actually exhibited. From his wording we suppose that he was aware of the problem that the n-th digit in the expansion of a number output by his algorthm is defined by exponentially many operations in n. Actually, a literal reading of Turing's algorithm yields that at most *simple-exponentially* many operations suffice. Our reconstruction worsens this amount to *double-exponentially* many, due to a modification we had to introduce in one expression that Turing wrote without a proof (see Section 2.2). A theorem of Strauss [27] asserts that normal numbers computable in simple exponential time do exits, but this existential result yields no specific instances.

---

[2] Both published their works in the same journal issue, but Lebesgue's dates back to 1909, immediately after Borel's question.

There are two other published constructions of normal numbers, one due to W.M.Schmidt in 1962 [26], the other to M.B.Levin in 1979 [17], but it is still unproved whether they yield computable numbers. Bugeaud in [5] demonstrated the existence of Liouville numbers that are normal. It is an open problem whether there are computable instances. Other non constructive examples of normal numbers follow from the theory of algorithmic randomness (recent reference books are [10,20]; for an overview see [11] in this volume). Since randomness implies normality, the particular real numbers that have been proved random are, therefore, normal. For instance, Chaitin's Omega numbers [8], the halting probabilities of optimal Turing machines with prefix-free domain. But random numbers are not computable, so Omega numbers are not the desired examples.[3]

2   Turing's Construction of Normal Numbers

In his manuscript Turing proves two theorems. Here we discuss the main ideas and include the proofs in accordance to our reconstruction in [1] of the original. We intend our curent presentation to be simpler and more readable. Theorem 1 is a computable version of Borel's fundamental theorem that establishes that almost all real numbers, in the sense of Lebesgue measure, are normal [3]. The theorem gives a construction of a set of real numbers as the limit of computably definable finite approximations. This set has arbitrarily large measure and consists only of normal numbers. This construction is valuable in its own right.

Turing's Theorem 1. *There is a computable function $c(k, n)$ of two integer variables with values consisting of finite sets of pairs of rational numbers such that, for each k and n, if $E_{c(k,n)} = (a_1, b_1) \cup (a_2, b_2) \cup ...(a_m, b_m)$ denotes the finite union of the intervals whose rational endpoints are the pairs given by $c(k, n)$, then $E_{c(k,n)}$ is included in $E_{c(k,n-1)}$ and the measure of $E_{c(k,n)}$ is greater than $1 - 1/k$. And for each k, $E(k) = \bigcap_n E_{c(k,n)}$ has measure $1 - 1/k$ and consists entirely of normal numbers.*

In Theorem 2 Turing gives an algorithm to output the expansion of a normal number in base two. The proof relies on the construction in Theorem 1. The algorithm is a computable functional: it receives an integer value that acts as a parameter to control measure, and an infinite sequence ν in base two to be used as an oracle to possibly determine some digits of the output sequence. When ν is a computable sequence (Turing puts the sequence of all zeros), the algorithm yields a computable normal number. With this result Turing is the first one to prove the existence of computable normal numbers.

Turing's Theorem 2. *There is an algorithm that, given an integer k and an infinite sequence ν of zeros and ones, produces a normal number $\alpha(k, \nu)$ in the unit interval, expressed in base two, such that in order to write down the first n*

---

[3] The family of Omega numbers coincides with the family of random real numbers that can be approximated by a computable non-decreasing sequence of rationals [14].

digits of $\alpha(k, \nu)$ the algorithm requires at most the first n digits of ν. For a fixed k these numbers $\alpha(k, \nu)$ form a set of measure at least $1 - 2/k$.

The algorithm can be adapted to intercalate the bits of the input sequence ν at fixed positions of the output sequence. Thus, one obtains non-computable normal numbers in each Turing degree.

Notation. For an integer base $b \geq 2$, a *digit* in base b is an element in $\{0, ..., b - 1\}$, and a *block* in base b a finite sequence of digits in base b. $|u|$ is the length of a block u, and $u[i..i + r - 1]$ is the inner block of r consecutive digits in a block u starting at position i, for $1 \leq i \leq |u| - r + 1$. A block w *occurs* in a block u at position i if $u[i...i + |w| - 1] = w$. The set of all blocks of length r in base b is denoted by $\{0, ..., (b - 1)\}^r$. For each real number x in the unit interval we consider the unique expansion in base b of the form $x = \sum_{i=1}^{\infty} a_i b^{-i}$, where the integers $0 \leq a_i < b$, and $a_i < b - 1$ infinitely many times. This last condition over a_n is introduced to ensure a unique representation of every rational number. When the base b is fixed, we write $x[i..i + r - 1]$ to denote the inner block of length r in the expansion of x in base b, starting at position i. We write μ to denote Lebesgue measure.

Turing uses the following definition of normality, given by Borel in [4] as a characterising property of normal numbers.

Definition 1 (Normality). *For a real number x and an integer base $b \geq 2$, the number of occurrences of a given block w in the first k digits of the expansion of x in base b is $S(x, b, w, k) = \#\{i : 1 \leq i \leq k - |w| + 1 \text{ and } x[i..i + |w| - 1] = w\}$. The number x is* normal to base b *if for every block w, $\lim_{k \to \infty} \frac{S(x,b,w,k)}{k} = b^{-|w|}$. If x is normal to every base $b \geq 2$ then we say x is* normal.

2.1   Turings's Theorem 1: A Construction via Finite Approximations

The main idea in Turing's Theorem 1 is the construction of a set of normal numbers of arbitrarily large measure, via finite approximations. This is done by pruning the unit interval by stages such that, at the end, one obtains the desired set consisting only of normal numbers. The construction is uniform on a parameter k, whose only purpose is to establish the measure of the constructed set $E(k)$ to be exactly $1 - 1/k$. At each stage n the construction is a finite set of intervals with rational endpoints determined by a computable function $c(k, n)$. At the initial stage 0, the set $E_{c(k,0)}$ is the whole unit interval. At stage n, the set $E_{c(k,n)}$ is the finite approximation to $E(k)$ that results from removing from $E_{c(k,n-1)}$ the points that are *not* candidates to be normal, according to the inspection of an initial segment of their expansions. At the end of this infinite process all rational numbers are discarded, because of their periodic structure. All irrational numbers with an unbalanced expansion are discarded. But also many normal numbers may be discarded, because their initial segments remain unbalanced for too long.

The construction covers all initial segment sizes, all bases, and all blocks by increasing computable functions of the stage n. And it has a decreasing bound on the acceptable discrepancy between the actual number of blocks in the inspected initial segments and the perfect number of blocks expected by the property of normality. These functions (initial segment size, base, block length and discrepancy) are such that, at each stage n, the set of discarded numbers has a small measure. The set $E(k)$, obtained in the limit of the construction, is the countable intersection of the sets $E_{c(k,n)}$ and consists just of normal numbers.

The proof of Theorem 1 depends on a constructive version of the strong law of large numbers: for each base there are a few blocks with too many or too few occurrences of any given shorter block. The expected number of occurrences of a given *digit* in a block of length k is k/b plus or minus a small fraction of k. An upper bound for the number of blocks of length k having the expected occurrences of a given *digit* is proved in Hardy and Wright's book[4] [12], Theorem 148 (also in many books as [6,13,15]).

Definition 2. *The number of blocks of length k in base b where a given block of r digits occurs exactly i times is $p_{b,r}(k,i)$.*

In particular, the number of blocks of length k with exactly i occurrences of a given *digit* is $p_{b,1}(k,i) = \binom{k}{i}(b-1)^{k-i}$.

Lemma 1. *Fix a base $b \geq 2$ and a block length $k > 6b$. For every real number ε such that $6/k \leq \varepsilon \leq 1/b$,*
$$\sum_{i:\ |i-k/b|\geq \varepsilon k} p_{b,1}(k,i)\ <\ 2\,b^k e^{-b\varepsilon^2 k/6}.$$

Turing extends this result to count occurrences of *blocks* instead of *digits*. Lemma 2 corresponds to our reconstruction in [1] where we give the full proof. The upper bound used by Turing in his manuscript is smaller but unproved.

Lemma 2. *Let base $b \geq 2$ and and let k and r be block lengths such that $k > r$. For every real number ε such that $6/\lfloor k/r \rfloor \leq \varepsilon \leq 1/b^r$,*
$$\sum_{i:\ |i-k/b^r|\geq \varepsilon k} p_{b,r}(k,i)\ <\ 2\,b^{k+2r-2} r\ e^{-b^r \varepsilon^2 k/6r}.$$

Lemma 2 provides a lower bound for the measure of the set of real numbers that are candidates to be normal based upon inspection of an initial segment of their expansion in finitely bases. In the following we define $A(\varepsilon, T, L, k)$ as the set of real numbers such that their initial segment of size k in each base up to T has a discrepancy of frequency below ε for each block of length up to L.

Definition 3. *For a real value ε and integer values T, L and k, let*

$$A(\varepsilon, T, L, k) = \bigcap_{2 \leq b \leq T}\ \bigcap_{1 \leq r \leq L}\ \bigcap_{w \in \{0,...,b-1\}^r} \{x \in (0,1)\ :\ |S(x,b,w,k) - k/b^r| < \varepsilon k\}.$$

Observe that $A(\varepsilon, T, L, k)$ is a finite union of intervals with rational endpoints.

---

[4] Since the first edition of *Introduction to the Theory of Numbers* was in 1938 we suppose the material was taught by G.H.Hardy in King's College Cambridge at the time Turing was a student.

Proposition 1. *If* $6/\lfloor k/L \rfloor \leq \varepsilon \leq 1/T^L$, $\mu A(\varepsilon, T, L, k) \geq 1 - 2L\, T^{3L-1} e^{-\varepsilon^2 k/3L}$.

Proof. By Definition 3, the complement of $A(\varepsilon, T, L, k)$ in the unit interval is
$$\overline{A}(\varepsilon, T, L, k) = \bigcup_{2 \leq b \leq T} \bigcup_{1 \leq r \leq L} \bigcup_{w \in \{0,...,b-1\}^r} \overline{B}(\varepsilon, b, w, k), \text{ where the set } \overline{B}(\varepsilon, b, w, k) =$$
$\{x \in (0,1) : |S(x, b, w, k) - k/b^r| \geq \varepsilon k\}$. Observe that if a number x belongs to $\overline{B}(\varepsilon, b, w, k)$ then so does each y such that $x[1..k] = y[1..k]$. Then, the interval $[0.x[1..k]000..., 0.x[1..k](b-1)(b-1)(b-1)...]$, which has measure b^{-k}, is included in $\overline{B}(\varepsilon, b, w, k)$. Recall that $p_{b,r}(k, i)$ (cf. Definition 2) is the number of different blocks of length k in which a given block of length r occurs exactly i times. Letting the block length $r = |w|$ we have $\mu\overline{B}(\varepsilon, b, w, k) \leq b^{-k} \sum_{i:\, |i-k/b^r| \geq \varepsilon k} p_{b,r}(i, k)$.

Applying Lemma 2, $\mu\overline{B}(\varepsilon, b, w, k) < 2\, b^{2r-2} r\, e^{-b^r \varepsilon^2 k/6r}$. Since $1 \leq r \leq L$, $2r/L \leq 2 \leq b^r$. Then, $\varepsilon^2 k/3L \leq b^r \varepsilon^2 k/6r$. This gives a uniform upper bound $\mu\overline{B}(\varepsilon, w, b, k) < 2\, T^{2L-2} L\, e^{-\varepsilon^2 k/3L}$ for all b, r, w such that $2 \leq b \leq T$, $1 \leq r \leq L$ and $w \in \{0, ..., b-1\}^r$. Thus, $\mu\overline{A}(\varepsilon, T, L, k) \geq \sum_{2 \leq b \leq T} \sum_{1 \leq r \leq L} \sum_{w \in \{0,...,b-1\}^r} \mu\overline{B}(\varepsilon, b, w, k)$.
In the third sum there are b^r many blocks w. Using $\sum_{2 \leq b \leq T} \sum_{1 \leq r \leq L} b^r = \sum_{2 \leq b \leq T} \frac{b^{L+1}-1}{b-1} \leq T^{L+1}$, conclude $\mu\overline{A}(\varepsilon, T, L, k) < 2L\, T^{3L-1} e^{-\varepsilon^2 k/3L}$. The proof is completed by taking the complement.

Turing defines the sets A_k as particular instances of the sets $A(\varepsilon, T, L, k)$ where ε, T and L are computable functions of the initial segment size k such that $\varepsilon(k)$ goes to 0 as k increases, and $T(k), L(k)$ are increasing in k. Turing chose the base $T(k)$ to grow sub-linearly in k, and the block length $L(k)$ to grow sub-logarithmically in k, which would yield the maximum discrepancy $\varepsilon(k)$ (according to the bound of Lemma 1). Other assignments are possible.

Definition 4. Let $A_k = A(\varepsilon, T, L, k)$ for $L = \sqrt{\ln k}/4$, $T = e^L$ and $\varepsilon = 1/T^L$.

Proposition 2. *There is k_0 such that for all $k \geq k_0$, $\mu A_k \geq 1 - 1/k(k-1)$.*

Proof. By Definition 4, $L = \sqrt{\ln k}/4$, $T = e^L$ and $\varepsilon = 1/T^L$. Assume $k \geq 2$. Then, $6/\lfloor k/L \rfloor \leq \varepsilon$. By Proposition 1, $\mu A_k \geq 1 - 2L\, T^{3L-1} e^{-\varepsilon^2 k/3L}$. To obtain $\mu A_k \geq 1 - 1/k(k-1)$ it suffices to show $2LT^{3L-1} k^2 \leq e^{\varepsilon^2 k/3L}$, which can be proved to hold for any $k \geq 1$.

From now on let k_0 be the value established in Proposition 2. Turing recursively defines the set $E_{c(k,n)}$ as a subset of A_k with measure *exactly* $1 - 1/k + 1/(k+n)$.

Definition 5. Let $c(k, n)$ be the function of two integer variables with values in finite sets of pairs of rational numbers such that, for each k and n, $E_{c(k,n)} = (a_1, b_1) \cup (a_2, b_2) \cup ...(a_m, b_m)$ denotes the finite union of the intervals whose rational endpoints are given by the pairs in the set $c(k, n)$. For any $k \geq k_0$ let $E_{c(k,0)} = (0,1)$ and $E_{c(k,n+1)} = A_{k+n+1} \cap E_{c(k,n)} \cap (\beta_n, 1)$ where $(\beta_n, 1)$ is an interval such that $\mu E_{c(k,n+1)} = 1 - 1/k + 1/(k+n+1)$.

The β_n above necessarily exists, it is unique, and it is a rational number computable from the two other sets in the definition. Both are a union of finitely many intervals with rational endpoints, so their respective measure are computable, and they are big enough.

Proof of Turing's Theorem 1. We first prove that $\bigcap_{k \geq k_0} A_k$ contains only normal numbers. By way of contradiction assume $x \in \bigcap_{k \geq k_0} A_k$ and x is not normal to base b. Then, $\lim_{k \to \infty} \frac{S(x,b,w,k)}{k} \neq \frac{1}{b^r}$ for some block w of length r. So, there is $\delta > 0$ and there are infinitely many values k such that $|S(x, b, w, k) - k/b^r| > k\delta$. Let $T(k)$, $L(k)$ and $\varepsilon(k)$ be the assignments of Definition 4 and fix $k_1 \geq k_0$ large enough such that $T(k_1) \geq b$, $L(k_1) \geq r$ and $\varepsilon(k_1) \leq \delta$. This is always possible because $T(k)$ and $L(k)$ are increasing in k, and $\varepsilon(k)$ goes to 0 as k increases. Then, for each $k \geq k_1$, $x \in A_k$ and by Definition 3, $|S(x, b, w, k) - k/b^r| < k\,\varepsilon(k) \leq k\delta$, a contradiction. $E(k) \subseteq \bigcap_{i \geq k} A_i$ for $k \geq k_0$; therefore, all real numbers in $E(k)$ are normal. Since $\mu E_{c(k,n)} = 1 - 1/k + 1/(k+n)$, $\mu E(k) = \lim_{n \to \infty} \mu E_{c(k,n)} = 1 - 1/k$. This completes the proof.

2.2   Turing's Theorem 2: An Algorithm to Output Normal Numbers

Turing's algorithm is uniform in the parameter k and it receives as input an infinite sequence ν of zeros and ones. The algorithm works by stages. The main idea is to split the unit interval by halves, successively. It starts with the whole unit interval and at each stage it chooses either the left half or the right half of the current interval. The sequence $\alpha(k, \nu)$ of zeros and ones output by the algorithm is the trace of the left/right selection at each stage. The invariant condition of the algorithm is that the intersection of the current interval with the set $E(k)$ of normal numbers of Theorem 1 has positive measure. Since $E_{c(k,n)}$ is the finite approximation of $E(k)$ at stage n, the algorithm chooses the half of the current interval whose intersection with $E_{c(k,n)}$ reaches a minimum threshold of measure which avoids running out of measure at any later stage. In case both halves reach this minimum, the algorithm uses the n-th symbol of the input sequence ν to decide. The chosen intervals at successive stages are nested and their measures converge to zero; therefore, their intersection contains exactly one number. This is the sequence $\alpha(k, \nu)$ output by the algorithm. The algorithm is correct if the number denoted by $\alpha(k, \nu)$ is normal to base two. This is proved by induction on the stage n, the only non obvious part is the verification of the invariant condition.

Each sequence output by the algorithm has an explicit convergence to normality: in the initial segment of length ℓ in each base up to base $T(\ell)$, all blocks of length up to $L(\ell)$ occur with the expected frequency plus or minus at most $\varepsilon(\ell)$, where $L(\ell) = \sqrt{\ln \ell}/4$, $T(\ell) = e^L$ and $\varepsilon(\ell) = e^{-L^2} = k^{-1/16}$.

The time complexity of the algorithm is the number of needed operations to produce the n-th digit of the output sequence $\alpha(k, \nu)$. This just requires to compute, at each stage n, the measure of the intersection of the current interval with the set $E_{c(k,n)}$. Turing gives no hints on properties of the sets $E_{c(k,n)}$ that could allow for a fast calculation. The naive way does the combinatorial construction of $E_{c(k,n)}$ in a number of operations exponential in n. Turing's algorithm verbatim would have simple-exponential time complexity, but we have been unable to verify its correctness. In our reconstruction in [1] the number of intervals we consider in $E_{c(k,n)}$ is exponentially larger than in Turing's literal formulation, so we end up with *double-exponential* time complexity.

Proof of Turing's Theorem 2. Let k be the integer parameter and ν the input infinite sequence of zeros and ones. We write α to denote the output sequence, $\alpha(i)$ for its digit in position i. Similarly for ν. Redefine the computable function $c(k,n)$ of Theorem 1 as follows. Assuming k is big enough, let $E_{c(k,0)} = (0,1)$ and for $n > 0$, $E_{c(k,n)} = A_{k2^{2n+1}} \cap E_{c(k,n-1)} \cap (\beta_n, 1)$, where $(\beta_n, 1)$ is an interval such that $\mu E_{c(k,n)} = 1 - 1/k + 1/k2^{2n+1}$. Here is the algorithm:

    `Start with` $I_0 = (0,1)$. `At stage` $n > 0$,
    `Split the interval` $I_{n-1} = (a_{n-1}, b_{n-1})$ `into two halves`
        $I_n^0 = (a_{n-1}, \frac{a_{n-1}+b_{n-1}}{2})$ `and` $I_n^1 = (\frac{a_{n-1}+b_{n-1}}{2}, b_{n-1})$.
    `If` $\mu(E_{c(k,n)} \cap I_n^0) > 1/k2^{2n}$ `and` $\mu(E_{c(k,n)} \cap I_n^1) > 1/k2^{2n}$ `then`
        `let` $\alpha(n) = \nu(n)$ `and` $I_n = I_n^{\nu(n)}$.
    `Else if` $\mu(E_{c(k,n)} \cap I_n^1) \le 1/k2^{2n}$ `then`
        `let` $I_n = I_n^0$ `and` $\alpha(n) = 0$.
    `Else, let` $I_n = I_n^1$ `and` $\alpha(n) = 1$.

To show that α is normal, we prove $\alpha \in E(k) = \bigcap_n E_{c(k,n)}$ by induction on n. For $n = 0$, $E_{c(k,0)} = (0,1)$; so, $\mu(E_{c(k,n)} \cap I_0) = 1 > 1/k$. For $n > 0$, assume the inductive hypothesis $\mu(E_{c(k,n)} \cap I_n) > 1/k2^{2n}$. Since the sets $E_{c(k,n)}$ are nested

$$E_{c(k,n+1)} \cap I_n = (E_{c(k,n)} \cap I_n) \setminus ((E_{c(k,n)} \setminus E_{c(k,n+1)}) \cap I_n).$$

So, $\mu(E_{c(k,n+1)} \cap I_n) = \mu(E_{c(k,n)} \cap I_n) - \mu((E_{c(k,n)} \setminus E_{c(k,n+1)}) \cap I_n)$. Then, $\mu(E_{c(k,n+1)} \cap I_n) \ge \mu(E_{c(k,n)} \cap I_n) - \mu(E_{c(k,n)} \setminus E_{c(k,n+1)})$. Using the equality $\mu(E_{c(k,n)} \setminus E_{c(k,n+1)}) = 1/k2^{2n+1} - 1/k2^{2(n+1)+1}$ and the inductive hypothesis, we obtain $\mu(E_{c(k,n+1)} \cap I_n) > 1/k2^{2n} - (1/k2^{2n+1} - 1/k2^{2n+3}) > 2/k2^{2(n+1)}$. It is impossible that both $\mu(E_{c(k,n+1)} \cap I_{n+1}^0)$ and $\mu(E_{c(k,n+1)} \cap I_{n+1}^1)$ be less than or equal to $1/k2^{2(n+1)}$. At least one of the sets $E_{c(k,n+1)} \cap I_{n+1}^i$, for $i \in \{0,1\}$, has measure greater than $1/k2^{2(n+1)}$. The algorithm picks as I_{n+1} the set I_{n+1}^i which fulfills this condition. In case both verify it, the oracle is used to choose left or right. By construction, the expansion of each real number in $E_{c(k,n)} \cap I_n$ starts with $\alpha(0) \, \alpha(1)...\alpha(n)$.

We now prove that for a fixed k, the set of output numbers $\alpha(k,\nu)$ for all possible inputs ν has measure at least $1 - 2/k$. Turing bounds the measure of the unqualified intervals up to stage n, as the n first bits of the sequence ν run through all possibilities. Let $I_m = (\frac{m}{2^{n+1}}, \frac{m+1}{2^{n+1}})$, for $m = 0, 1, ..., 2^{n+1} - 1$. The algorithm discards the interval I_m when $\mu(E_{c(k,n)} \cap I_m) \le 1/k2^{2n}$. The set of intervals that are *not* discarded is recursively defined as follows. Let $M(k,0) = (0,1)$ and for $n > 0$, let $M(k, n+1)$ be the union of the intervals I_m such that $I_m \subseteq M(k,n)$ and $\mu(E_{c(k,n)} \cap I_m) > 1/k2^{2n}$. Then, $\mu(E(k) \cap M(k, n+1))$ equals

$$\mu(E(k) \cap M(k,n)) - \sum_{m=0}^{2^n-1} \mu(E(k) \cap (M(k,n) \setminus M(k,n+1)) \cap \left(\frac{m}{2^n}, \frac{m+1}{2^n}\right)).$$

Each term in the sum is at most $1/k2^{2n}$. Therefore, $\mu(E(k) \cap M(k, n+1)) \geq \mu(E(k) \cap M(k, n)) - 1/k2^n$. Applying this inequality recursively n times, we get $\mu(E(k) \cap M(k, n+1)) \geq \mu(E(k) \cap M(k, 1)) - 1/k \sum_{i=1}^{n} 1/2^n$. Finally, since $E_{c(k,0)} = (0, 1)$ and $k \geq 2$, $M(k, 1) = (0, \frac{1}{2}) \cup (\frac{1}{2}, 1)$; so, $E(k) \cap M(k, 1) = E(k)$. Then, $\mu(E(k) \cap \bigcap_n M(k, n)) > \mu E(k) - 1/k$. Using that $\mu E(k) = 1 - 1/k$, conclude that $E(k) \cap \bigcap_n M(k, n)$ has measure at least $1 - 2/k$.

3   Towards the Theory of Algorithmic Randomness

Turing's manuscript conveys the impression that he had the insight, ahead of his time, that traditional mathematical concepts specified by finitely definable approximations, such as measure or continuity, could be made computational. This point of view has developed under the general name of *effective mathematics*, a part of which is algorithmic randomness. From the modern perspective, Turing's construction of the set of normal numbers in Theorem 1, done via finite approximations, is an instance of a fundamental entity in the theory of algorithmic randomness: a *Martin-Löf test*[5] [19]. Intuitively, a real number is random when when it exhibits the almost-everywhere behavior of all reals, for example its expansion has no predictable regularities. A random real number must pass every test of these properties. Martin-Löf had the idea to focus just in properties definable in terms of computability: a test for randomness is a uniformly computably enumerable sequence of sets whose measure converges to zero. A real number is random if it is covered by no such test. That is to say that it has the almost-everywhere property of avoiding the measure-zero intersection. This definition turned out to be equivalent to the definition of randomness in terms of description complexity [8]. The equivalence between the two been taken as a sign of robustness of the defined notion of randomness.

Definition 6. *1.   A Martin-Löf randomness test, hereafter ML-test, is a uniformly computably enumerable sequence $(V_i)_{i \geq 0}$ of sets of intervals with rational endpoints such that, for each i, $\mu V_i \leq 2^{-i}$.*
2. A real number x is random if for every ML-test $(V_i)_{i \geq 0}$, $x \notin \bigcap_{i \geq 0} V_i$.

Turing's set $E(k)$ of Theorem 1 leads immediately to a ML-test[6]. Hence, it provides a direct proof that randomness implies normality.

Corollary 1. *The sequence $(V_k)_{k \geq 0} = ((0, 1) \setminus E(2^k))_{k \geq 0}$ is a ML-test.*

Proof. By Theorem 1, $E(k) = \bigcap_{n \geq 1} E_{c(k,n)}$, where $c(k, n)$ is computable and for each k and n, $E_{c(k,n)}$ is a finite set of intervals with rational endpoints. So, the complement of each $E_{c(k,n)}$ is also a finite set of intervals with rational

---

[5] Martin-Löf presented the test in terms of sequences of zeros and ones. We give here an alternative formulation in terms of sets of intervals with rational endpoints.
[6] In fact, Theorem 1 yields a *Schnorr test* [21]. This is a ML-test where μV_i is computable uniformly in i. The notion is unchanged if we, instead, let a Schnorr test be a ML-test such that $\mu V_i = 2^{-i}$, for each $i \geq 0$.

endpoints. Then, $(0,1) \setminus E(k) = \bigcup_{n \geq 1}(0,1) \setminus E_{c(k,n)}$ is computably enumerable. Since Turing's construction is uniform in the parameter k, $((0,1) \setminus E(k))_{k \geq 0}$ is uniformly computably enumerable. Finally, since the measure of $E(k)$ is $1 - 1/k$, $\mu((0,1) \setminus E(k)) = 1/k$. Thus, $(V_k)_{k \geq 0} = ((0,1) \setminus E(2^k))_{k \geq 0}$ is a ML-test.

Corollary 2. *Randomness implies normality.*

Proof. If x is not normal then, by Theorem 1, x belongs to no set $E(k)$, for any k. So, $x \in \bigcap_{k \geq 0}(0,1) \setminus E(k)$. By Corollary 1, $(V_k)_{k \geq 0} = ((0,1) \setminus E(2^k))_{k \geq 0}$ is a ML-test. Hence, $x \in \bigcap_{k \geq 0} V_k$; therefore, x is not random.

References

1. Becher, V., Figueira, S., Picchi, R.: Turing's unpublished algorithm for normal numbers. Theoretical Computer Science 377, 126–138 (2007)
2. Becher, V., Figueira, S.: An example of a computable absolutely normal number. Theoretical Computer Science 270, 947–958 (2002)
3. Borel, É.: Les probabilités dénombrables et leurs applications arithmétiques. Rendiconti del Circolo Matematico di Palermo 27, 247–271 (1909)
4. Borel, É.: Leçons sur la thèorie des fonctions, 2nd edn., Gauthier Villars (1914)
5. Bugeaud, Y.: Nombres de Liouville et nombres normaux. Comptes Rendus de l'Académie des Sciences de Paris 335, 117–120 (2002)
6. Bugeaud, Y.: Distribution Modulo One and Diophantine Approximation. Cambridge University Press (2012)
7. Bourke, C., Hitchcock, J., Vinodchandran, N.: Entropy rates and finite-state dimension. Theoretical Computer Science 349(3), 392–406 (2005)
8. Chaitin, G.: A theory of program size formally identical to information theory. Journal ACM 22, 329–340 (1975)
9. Cassels, J.W.S.: On a paper of Niven and Zuckerman. Pacific Journal of Mathematics 2, 555–557 (1952)
10. Downey, R., Hirschfeldt, D.: Algorithmic Randomness and Complexity. Springer (2010)
11. Downey, R.: Randomness, Computation and Mathematics. In: Cooper, S.B., Dawar, A., Löwe, B. (eds.) CiE 2012. LNCS, vol. 7318, pp. 163–182. Springer, Heidelberg (2012)
12. Hardy, G.H., Wright, E.M.: An Introduction to the Theory of Numbers, 1st edn. Oxford University Press (1938)
13. Harman, G.: Metric Number Theory. Oxford University Press (1998)
14. Kučera, A., Slaman, T.: Randomness and recursive enumerability. SIAM Journal on Computing 31(1), 199–211 (2001)
15. Kuipers, L., Niederreiter, H.: Uniform Distribution of Sequences. Dover (2006)
16. Lebesgue, H.: Sur certaines démonstrations d'existence. Bulletin de la Société Mathématique de France 45, 132–144 (1917)
17. Levin, M.B.: On absolutely normal numbers. English translation in Moscow University Mathematics Bulletin 34, 32–39 (1979)
18. Dai, L., Lutz, J., Mayordomo, E.: Finite-state dimension. Theoretical Computer Science 310, 1–33 (2004)
19. Martin-Löf, P.: The Definition of Random Sequences. Information and Control 9(6), 602–619 (1966)

20. Nies, A.: Computability and Randomness. Oxford University Press (2009)
21. Schnorr, C.-P.: Zufälligkeit und Wahrscheinlichkeit. In: Eine algorithmische Begründung der Wahrscheinlichkeitstheorie. Lecture Notes in Mathematics, vol. 218. Springer, Berlin (1971)
22. Schnorr, C.-P., Stimm, H.: Endliche Automaten und Zufallsfolgen. Acta Informatica 1, 345–359 (1972)
23. Sierpiński, W.: Démonstration élémentaire du théorème de M. Borel sur les nombres absolument normaux et détermination effective d'un tel nombre. Bulletin de la Société Mathématique de France 45, 127–132 (1917)
24. Turing, A.M.: On computable numbers, with an application to the Entscheidungsproblem. Proceedings of the London Mathematical Society Series 2 42, 230–265 (1936)
25. Turing, A.M.: A note on normal numbers. In: Britton, J.L. (ed.) Collected Works of A.M. Turing: Pure Mathematics, pp. 263–265. North Holland, Amsterdam (1992); with notes of the editor in 263–265
26. Schmidt, W.M.: On normal numbers. Pacific Journal of Math. 10, 661–672 (1960)
27. Strauss, M.: Normal numbers and sources for BPP. Theoretical Computer Science 178, 155–169 (1997)

Logic of Ruler and Compass Constructions

Michael Beeson

Department of Computer Science, San José State University, 208 MacQuarrie Hall,
San Jose, CA 95192-0249, United States of America

Abstract. We describe a theory **ECG** of "Euclidean constructive geometry". Things that **ECG** proves to exist can be constructed with ruler and compass. **ECG** permits us to make constructive distinctions between different forms of the parallel postulate. We show that Euclid's version, which says that under certain circumstances two lines meet (i.e., a point of intersection exists) is not constructively equivalent to the more modern version, which makes no existence assertion but only says there cannot be two parallels to a given line. Non-constructivity in geometry corresponds to case distinctions requiring different constructions in each case; constructivity requires continuous dependence on parameters. We give continuous constructions where Euclid and Descartes did not supply them, culminating in geometrical definitions of addition and multiplication that do not depend on case distinctions. This enables us to reduce models of geometry to ordered field theory, as is usual in non-constructive geometry. The models of ECG include the set of pairs of Turing's constructible real numbers [7].

1   Introduction

Euclid's geometry, written down about 300 BCE, has been extraordinarily influential in the development of mathematics, and prior to the twentieth century was regarded as a paradigmatic example of pure reasoning.

In this paper, we re-examine Euclidean geometry from the viewpoint of constructive mathematics. The phrase "constructive geometry" suggests, on the one hand, that "constructive" refers to geometrical constructions with straightedge and compass. On the other hand, the word "constructive" may suggest the use of intuitionistic logic. We investigate the connections between these two meanings of the word. Our method is to keep the focus on the body of mathematics in Euclid's *Elements*, and to examine what in Euclid is constructive, in the sense of "constructive mathematics". Our aim in the first phase of this research was to formulate a suitable formal theory that would be faithful to both the ideas of Euclid and the constructive approach of Errett Bishop [4]. We achieved this aim by formulating a theory **ECG** of "Euclidean constructive geometry", first presented in [2], but improved in [3].

In constructive mathematics, if one proves something exists, one has to show how to construct it. In Euclid's geometry, the means of construction are not arbitrary computer programs, but ruler and compass. Therefore it is natural

S.B. Cooper, A. Dawar, and B. Löwe (Eds.): CiE 2012, LNCS 7318, pp. 46–55, 2012.

to look for a theory that has function symbols for the basic ruler-and-compass constructions. The terms of such a theory correspond to ruler-and-compass constructions. Our first main result is that when **ECG** proves that something exists, that something can be constructed with ruler and compass.

In number theory, if one proves an existence theorem, then for a constructive version, one has to show how to compute the desired number as a function of the parameters. In analysis, if one proves an existence theorem, one has to be able to compute approximations to the desired number from approximations to the parameters. In particular, the solution will depend continuously on parameters, at least locally. This feature of constructive analysis depends, in a way, on what we think it means "to be given" a number x. Whatever that may mean, it surely means that we have a way to get a rational approximation to x within any specified limit of accuracy. Geometry is more like analysis than number theory, in the sense that we do not want to assume in advance that points can be given to us all at once in a completely determined location; points are given only approximately, by dots on paper or a computer screen, or in Euclid's case, by indentations in sand (the Greeks drew their diagrams in sand). It might be doubtful whether two such points coincide; in such a case one would have to ask the one who made the diagram to refine it. It follows that in constructive geometry, we should have local continuous dependence of constructions on parameters. We can see that dramatically in computer animations of Euclidean constructions, in which one can select some of the original points and drag them, and the entire construction "follows along." One might formulate a program of "continuous geometry", in which one allows only constructions that depend continuously on parameters. It turns out that this is just another way of viewing constructive geometry, since theorems proved without non-constructive case distinctions will be implemented by continuous ruler-and-compass constructions. One line of research has thus been to identify and repair uses of non-constructive case distinctions. There are several important places where repair is needed, but it is possible. Thus the "C" in **ECG** could just as well be read as "continuous", instead of "constructive."

Once we have a good formal theory of constructive geometry, the possibility opens up to prove independence results. Our most striking results concern the different formulations of Euclid's parallel postulate. Euclid's original version (Euclid 5) is not the same as the version more commonly used today (Playfair's axiom). The difference is that Euclid 5 says that under certain conditions, two lines must meet, while Playfair's axiom says that there cannot be two different parallels to line L through point P. Thus Euclid 5 makes an existence assertion, but Playfair does not. We prove that Playfair does not imply Euclid 5 in **ECG** minus the parallel axiom of **ECG**.

In classical (i.e., nonconstructive) geometry, there are theorems that show that models of geometrical theories all have the form F^2, where F is an ordered field, and the geometrical relations of incidence and betweenness are defined as usual in analytic geometry. Different geometric axioms correspond to different axioms in ordered field theory. When the geometric axioms are those for ruler and compass, we get Euclidean fields (those in which positive elements have square roots). We show that this paradigm extends to constructive geometry as well. This is not trivial, because we need to give geometrical definitions of addition and multiplication (of segments, or of points on a line) that are continuous in parameters, in particular, do not require case distinctions about the sign to construct the sum and product.

Once that is done, we move on to consider the models F^2, and we find that there are three different possible definitions of "constructive Euclidean field". The difference hinges on when the reciprocal $1/x$ is defined: either positive elements have reciprocals, or nonzero elements have reciprocals, or elements without reciprocals are zero. To prove two of these three versions equivalent requires either proof by contradiction or non-constructive case distinctions. Each of these versions corresponds to geometry, with a different version of the parallel axiom.

We obtain our independence proofs by constructing models of one kind of constructive ordered field theory that do not satisfy the next stronger kind. Of course these are not fields in the usual sense, because these field theories are non-constructively equivalent; they are what is known as Kripke models. Their construction involves taking the field elements to be certain real-valued functions, corresponding to points whose location is not yet "pinned down."

This short paper omits proofs, discussion of past axiomatizations of geometry, philosophical discussions, and detailed discussions of the relations between this work and the work of others. All these things can be found in [3]. Heartfelt thanks to Marvin Greenberg and Freek Wiedijk for their comments and suggestions.

2   Is Euclid's Reasoning Constructive?

Euclid's reasoning is generally constructive; indeed the only irreparably non-constructive proposition is Book I, Prop. 2, which shows that the rigid compass can be simulated by a collapsible compass. We just take Euclid I.2 as an axiom, thus requiring a rigid compass in **ECG**. Only one other repair is needed, in the formulation of the parallel axiom, as we shall see below. Euclid did not deal with disjunctions explicitly, and all his theorems are of the form: Given certain points related in certain ways, we can construct (zero or more) other points related to the given points and each other in certain ways. Euclid has been criticized (as far back as Geminus and Proclus) for ignoring case distinctions in a proof, giving a diagram and proof for only one case. Since case distinctions (on whether $ab = cd$ or not) are non-constructive, these omissions are *prima facie* non-constructive. However, these non-constructive proof steps are eliminable, as we shall explain.

An example of such an argument in Euclid is Prop. I.6, whose proof begins

> Let ABC be a triangle having the angle ABC equal to the angle ACB. I say that the side AB is also equal to the side AC. For, if AB is unequal to AC, one of them is greater. Let AB be greater, ...

The same proof also uses an argument by contradiction in the form $\neg x \neq y \rightarrow x = y$. This principle, the "stability of equality", is an axiom of **ECG**, and is universally regarded as constructively acceptable. The conclusion of I.6, however, is negative (has no $\exists$ or $\vee$), so we can simply put double negations in front of every step, and apply the stability of equality once at the end.

Prop. I.26 is another example of the use of the stability of equality: "...DE is not unequal to AB, and is therefore equal to it."

To put the matter more technically, in constructive logic we have $P \rightarrow \neg\neg P$, and although generally we do not have $\neg\neg P \rightarrow P$, we do have it for quantifier-free, disjunction-free P. We can double-negate $A \vee \neg A \rightarrow B$, obtaining $\neg\neg(A \vee \neg A) \rightarrow \neg\neg B$, and then the hypothesis is provable, so we have $\neg\neg B$, and hence B since B is quantifier-free and disjunction-free. The reason why this works throughout Euclid is that the *conclusions* of Euclid's theorems are all quantifier-free and disjunction-free. Euclid never even *thought* of stating a theorem with an "or" in it. The bottom line is that Euclid is constructive as it stands, except for Book I, Prop. 2, and the exact formulation of the parallel postulate. These problems are remedied in **ECG** by taking Book I, Prop. 2 as an axiom and strengthening the parallel postulate as discussed below. We also take as an axiom $\neg\neg \mathbf{B}(x, y, z) \rightarrow \mathbf{B}(x, y, z)$, or "Markov's principle for betweenness", enabling us to drop double negations on atomic sentences.

3   The Elementary Constructions

The Euclidean, or "elementary" constructions, are carried out by constructing lines and circles and marking certain intersection points as newly constructed points. The geometrical theory **ECG** given in [2] has terms to denote the geometrical constructions. These terms can sometimes be "undefined", e.g., if two lines are parallel, their intersection point is undefined. Therefore **ECG** is based on the logic of partial terms LPT [1, p. 97], in which there are atomic formulas $t \downarrow$ expressing that term t has a denotation ("is defined"). A model of such a theory can be regarded as a many-sorted algebra with partial functions representing the basic geometric constructions. Specifically, the sorts are *Point*, *Line*, and *Circle*. We have constants and variables of each sort.

ECG includes function symbols for the basic constructors and accessors, such as $Line(A, B)$ for the line through A and B and $Circle(A, B)$ for the circle through B with center A, and for the "elementary constructions" (each of which has type *Point*):

$$IntersectLines(Line\,K, Line\,L)$$
$$IntersectLineCircle1(Line\,L, Circle\,C)$$
$$IntersectLineCircle2(Line\,L, Circle\,C)$$
$$IntersectCircles1(Circle\,C, Circle\,K)$$
$$IntersectCircles2(Circle\,C, Circle\,K)$$

One can regard circles and lines as mere intermediaries; points are ultimately constructed from other points. (This was proved in [2].)

There is a second constructor for circles, which we can describe for short as "circle from center and radius", as opposed to the first constructor above, "circle from center and point." Specifically $Circle3\,(A,B,C)$ constructs a circle of radius BC and center A, provided $B \neq C$. These two constructors for circles correspond to a "collapsible compass" and a "rigid compass" respectively. The compass of Euclid was a collapsible compass: you cannot use it to "hold" the length BC while you move one point of the compass to A. You can only use it to hold the radius AB while one point of the compass is fixed at A, so in that sense it corresponds to $Circle\,(A,B)$. The second constructor $Circle3$ corresponds to a rigid compass. The theory **ECG** includes $Circle3$, and in [2] we gave reasons why constructive geometry demands a rigid, rather than only a collapsible compass. In short, without a rigid compass, one cannot project a point P onto a line L, without making a case distinction between the case when P lies on L and the case when it does not; and the ability to make such projections is crucial to defining a coordinate system and showing how to perform addition and multiplication on segments.

There are three issues to decide:

- when there are two intersection points, which one is denoted by which term?
- In degenerate situations, such as $Line\,(P,P)$, what do we do?
- When the indicated lines and/or circles do not intersect, what do we do about the term(s) for their intersection point(s)?

Our answers to these questions are as follows. When the indicated lines or circles do not intersect, then the term for their intersection is "undefined". This can best be handled formally using the logic of partial terms, which we do in **ECG**; it can also be handled in other more cumbersome ways without modifying first-order logic. We take $Circle\,(P,P)$ to be defined, i.e., we allow circles of zero radius; that technicality makes the formal development smoother and seems philosophically unobjectionable–we just allow the two points of the compass to coincide. The point here is not so much that circles of zero radius are of interest, but that we do not want to force a case distinction as to whether the two points of the compass are, or are not, coincident. We take the two points of intersection of a line $Line\,(A,B)$ and a circle to occur in the same order as A and B occur on L. That means that lines are treated as having direction. Not only do they have direction, they "come equipped" with two points from which they were constructed. There are function symbols to recover those points from a line. $Line\,(P,P)$ is undefined, since having it defined would destroy continuous dependence of $Line\,(P,Q)$ on P and Q.

The two intersection points $p = IntersectCircles1\,(C,K)$ and $q = IntersectCircles2\,(C,K)$ are to be distinguished as follows: With a the center of C and b the center of K we should have abp a right turn, and abq a left turn. But can "right turn" and "left turn" be defined? What we do is to *define Right* and *Left* using equations involving $IntersectCircles1$ and $IntersectCircles2$; then we give axioms about $Right$ and $Left$, namely that if abc is a left turn, then c and d are on the same side of $Line\,(a,b)$ if and only if abd is a left turn, and c and d are on opposite sides of $Line\,(a,b)$ if and only if abd is a right turn. Note that neither this issue nor its solution has to do with constructivity, but simply with the introduction of function symbols corresponding to the elementary constructions.

4   Models of Ruler-and-Compass Geometry

There are several interesting models of **ECG**, even with classical logic, of which we now mention four. The *standard plane* is $\mathbb{R}^2$, with the usual interpretation of points, lines, and planes. The *Turing plane* has for its points pairs of computable real numbers [7]. The *algebraic plane* has for its points pairs of algebraic numbers. The *Tarski plane* has for points just those points constructible with ruler and compass; this is $\mathbb{K}^2$ where $\mathbb{K}$ is the least subfield of the reals closed under square roots of positive elements. The theory **ECG** uses the same primitive relations as Tarski and Hilbert: betweenness $\mathbf{B}(a, b, c)$ and equidistance $ab = cd$. Hilbert used strict betweenness and Tarski allowed $\mathbf{B}(x, x, x)$; we follow Hilbert.

5   Three Versions of the Parallel Postulate

Let P be a point not on line L. We consider lines through P that do not meet L (i.e., are parallel to L). Playfair's version of the parallel postulate says that two parallels to L through P are equal. Recall that Euclid's postulate 5 is

> *If a straight line falling on two straight lines make the interior angles on the same side less than two right angles, the two straight lines, if produced indefinitely, meet on that side on which are the angles less than the two right angles.*

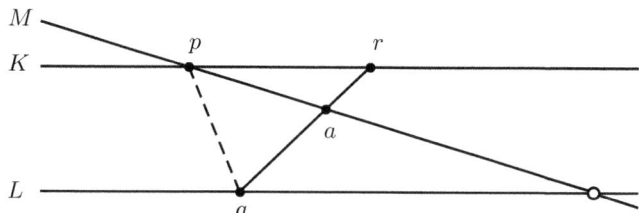

Fig. 1. Euclid 5: M and L must meet on the right side, provided $\mathbf{B}(q, a, r)$ and pq makes alternate interior angles equal with K and L. The point at the open circle is asserted to exist.

But this formulation of Euclid 5 makes use of the notion of "alternate interior angles", while angles are not directly treated in **ECG**, but instead are treated as triples of points. A version of Euclid 5 that does not mention angles is given in Fig. 2.

Although we have finally arrived at a satisfactory formulation of Euclid 5, that formulation is satisfactory only in the sense that it accurately expresses what Euclid said. It turns out that this axiom is not satisfactory as a parallel postulate for **ECG**. The main reason is that it is inadequate to define division geometrically. Here is why: As x gets nearer and nearer to 0, the number $1/x$ requires a line of smaller and smaller slope to meet a certain horizontal line. If x passes through zero, this intersection point "goes to infinity", then is undefined

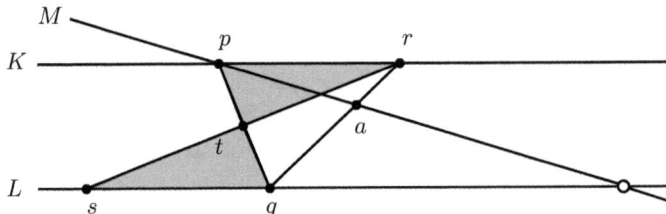

Fig. 2. Euclid 5: M and L must meet on the right side, provided $\mathbf{B}(q, a, r)$ and $pt = qt$ and $rt = st$

when $x = 0$, but then "reappears on the other side", coming in from minus infinity. Without knowing the sign of x, we shall not know on which side of the transversal pq the two adjacent interior angles will make less than two right angles. In other words, with Euclid 5, we shall only be able to divide by a number whose sign we know; and the principle $x \neq 0 \rightarrow x < 0 \vee x > 0$ is not an axiom (or theorem) of **ECG**. The conclusion is that if we want to divide by nonzero numbers, we need to strengthen Euclid's parallel axiom.

We make three changes in Euclid 5 to get the "strong parallel postulate":

(i) We change the hypothesis $\mathbf{B}(q, a, r)$ to $\neg on(a, K)$. In other words, we require that the two adjacent interior angles do not make exactly two right angles, instead of requiring that they make less than two right angles.

(ii) We change the conclusion to state only that M meets L, without specifying on which side of the transversal pq the intersection lies.

(iii) We drop the hypothesis $\neg\, on\,(p, L)$.

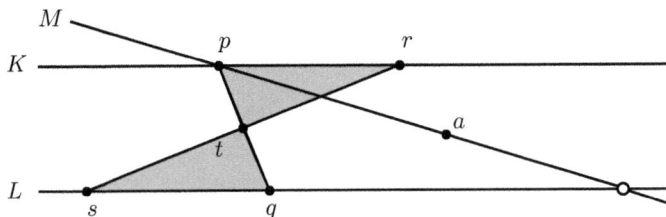

Fig. 3. Strong Parallel Postulate: M and L must meet (somewhere) provided a is not on K and $pt = qt$ and $rt = st$

The strong parallel axiom differs from Euclid's version in that we are not required to know in what *direction* M passes through P; but also the conclusion is weaker, in that it does not specify *where* M must meet L. In other words, the betweenness hypothesis of Euclid 5 is removed, and so is the betweenness conclusion. Since both the hypothesis and conclusion have been changed, it is not immediate whether this new postulate is stronger than Euclid 5, or equivalent, or possibly even weaker, but it turns out to be stronger–hence the name.

6   Constructive Geometry and Euclidean Fields

Classical Euclidean geometry has models $K^2 = K \times K$ where K is a Euclidean field, i.e., an ordered field in which nonnegative elements have square roots. We take that definition also constructively, and we define a *Euclidean* ring to be an ordered ring in which nonnegative elements have square roots. We use a language with symbols $+$ for addition and $\cdot$ for multiplication, and a unary predicate $P(x)$ for "x is positive". A *Euclidean field* is a Euclidean ring in which nonzero elements have reciprocals. Constructively, we also need two weaker notions: Euclidean rings in which positive elements have reciprocals, and Euclidean rings in which elements without reciprocals are zero and if x is greater than a positive invertible element, then x is invertible. These we call *Playfair rings*.

In order to show that the models of some geometrical theory T have the form F^2, one has to define addition and multiplication (of segments or points on a line) within T. This was first done by Descartes in [5], and again (in a different way) by Hilbert in [6]. These constructions, however, involve a non-constructive case distinction on the sign of the numbers being added or multiplied. To repair this problem requires rather more elaborate constructions, and to make those elaborate constructions work, one needs some more elementary case constructions to also work without case distinctions, for example, constructing a line through a point P perpendicular to a line L, without a case distinction as to whether P is or is not on L. These problems are solved, and the (rather lengthy) solutions are presented in detail, in [3].

We can prove that the models of **ECG** are of the form F^2, where F is a Euclidean field. More specifically, given such a field, we can define betweenness, incidence, and equidistance by analytic geometry and verify the axioms of **ECG**. Conversely, and this is the hard part, we can define multiplication, addition, and division of points on a line (having chosen one point as zero), in **ECG**. It turns out that we need the strong parallel axiom to do that. If we replace the parallel axiom of **ECG** by Euclid's parallel postulate, we get instead models of the form F^2, where F is a Euclidean ring in which nonzero elements have reciprocals, but we cannot go the other way by defining multiplication and addition geometrically without the strong parallel axiom. (That is, if we only had Euclid 5, we would need case distinctions, as Hilbert and Descartes did.)

We now work out the field-theoretic version of Playfair's axiom. Playfair says, if P is not on L and K is parallel to L through P, that if line M through P does not meet L then $M = K$. Since $\neg\neg M = K \rightarrow M = K$, Playfair is just the contrapositive of the parallel axiom of **ECG**, which says that if $M \neq K$ then M meets L. Hence it corresponds to the contrapositive of $x \neq 0 \rightarrow 1/x \downarrow$; that contrapositive says that if x has no multiplicative inverse, then $x = 0$. Thus Playfair geometries have models F^2 where F is a Playfair ring (as defined above). (We cannot prove the converse because we need the strong parallel axiom to verify multiplication and addition).

7   What ECG Proves to Exist, Can Be Constructed with Ruler and Compass

In [2], we proved that if **ECG** proves an existential statement $\exists y A(x, y)$, then there is a term t of **ECG** such that **ECG** proves $A(x, t(x))$. In words: things that **ECG** can prove to exist, can be constructed with ruler and compass. Of course, the converse is immediate: things that can be constructed with ruler and compass can be proved to exist in **ECG**. Hence the two meanings of "constructive" coincide for **ECG**: it could mean "proved to exist with intuitionistic logic" or it could mean "constructed with ruler and compass."

The technique of the proof is to apply Gentzen's cut-elimination theorem. What makes it applicable is that the axiomatization of **ECG** has two important properties: it is *quantifier-free*, and it is *disjunction-free*. It was not difficult to axiomatize **ECG** in this way–we just followed Euclid. In [3] we draw on Tarski's approach to achieve a short elegant list of axioms, but that is not essential to the analysis of the parallel axiom. There are many ways to axiomatize geometry.

8   Independence Results for the Parallel Axioms

The reduction of geometry to field theory described above shows that (relative to a base theory), the strong parallel axiom implies Euclid's postulate 5 (since if reciprocals of non-zero elements exist, then of course reciprocals of positive elements exist). (A direct proof is in [3].) And Euclid 5 easily implies Playfair's postulate. Our main theorem is that neither of these two implications can be reversed.

Theorem 1. *Euclid 5 does not imply the strong parallel axiom, and Playfair does not imply Euclid 5, in* **ECG** *minus its parallel axiom.*

Proof sketch of the first claim (A detailed proof can be found in [3]). Since non-constructively, the implications *are* reversible, we cannot hope to give counterexamples. In terms of field theory, we won't be able to construct a Euclidean ring in which positive elements have reciprocals but nonzero elements do not. The proof proceeds by constructing appropriate Kripke models. To show that Euclid 5 does not prove the strong parallel axiom, it suffices to prove the corresponding result in ordered field theory: the axiom that positive elements have reciprocals does not imply that all nonzero elements have reciprocals. That does suffice, in spite of the fact that we have full equivalence between geometry and field theory only for **ECG** and Euclidean fields, for if the weaker geometry proved the strong parallel axiom SP, then the interpretation of SP in field theory would be provable, as *that* direction does work, and the interpretation of SP implies that nonzero elements have reciprocals.

Then we need a Kripke model in which positive elements have reciprocals, but nonzero elements do not necessarily have reciprocals. We construct such a Kripke model whose "points" are functions from $\mathbb{R}$ to $\mathbb{R}$. The function f is *positive semidefinite* if $f(x) \geq 0$ for all real x. Let $\mathbb{K}$ be the least subfield of the reals closed under square roots of positive elements. Let $\mathcal{A}$ be the least ring of real-valued functions containing polynomials with coefficients in $\mathbb{K}$, and closed under reciprocals and square roots of positive semidefinite functions. For example

$$\sqrt{\sqrt{1+t^2}+\sqrt{1+t^4}}+\frac{1}{1+t^2}$$

is in $\mathcal{A}$, but $1/t$ is not in $\mathcal{A}$. We take $\mathcal{A}$ as the root of a Kripke model, interpreting the positivity predicate $P(x)$ to mean x is positive definite. We show (using Pusieux series) that each member of $\mathcal{A}$ has finitely many zeroes and singularities and that there is a countable set Ω including all zeroes and singularities, whose complement is dense in $\mathbb{R}$. For $\alpha \notin \Omega$, we define $\mathcal{A}_\alpha$ by interpreting $P(x)$ to hold if and only if $x(\alpha) > 0$. In our Kripke model, $\mathcal{A}_\alpha$ lies immediately above the root. Now t is a nonzero element without a reciprocal. But if x is positive, then $x(\alpha) > 0$ for all $\alpha \notin \Omega$, and since the complement of Ω is dense and x is continuous, x is positive semidefinite, so $1/x$ exists in $\mathcal{A}$.

9   Conclusions

Euclid needs only two modifications to be completely constructive: we have to postulate a rigid compass, rather than relying on Prop. I.2 to simulate it, and we have to take the strong parallel axiom instead of Euclid 5. With those changes Euclid is entirely constructive, and **ECG** formalizes Euclid nicely. **ECG** has the nice property that things it can prove to exist can be constructed with ruler and compass, and it permits us to distinguish between versions of the parallel axiom with different constructive content, even though non-constructively they are equivalent. The classical constructions used to give geometrical definitions of addition and multiplication involve non-constructive case distinctions, but these can be replaced by more elaborate constructions that are continuous (and constructive), so geometry can still be shown equivalent to the theory of Euclidean fields, and different versions of the parallel axiom correspond to weakenings of the field axiom about reciprocals. We can then use Kripke models whose points are certain real-valued functions to establish formal independence results about the different versions of the parallel axiom.

References

1. Beeson, M.: Foundations of Constructive Mathematics. Springer, Heidelberg (1985)
2. Beeson, M.: Constructive geometry. In: Arai, T. (ed.) Proceedings of the Tenth Asian Logic Colloquium, Kobe, Japan, pp. 19–84. World Scientific, Singapore (2009)
3. Beeson, M.: Foundations of Constructive Geometry (available on the author's website)
4. Bishop, E.: Foundations of Constructive Analysis. McGraw-Hill, New York (1967)
5. Descartes, R.: The Geometry of Rene Descartes as an appendix to Discours de la Methode, 1st edn. Dover, New York (1952); original title, La Geometrie. Facsimile with English translation by Smith, D.E., Latham, M.L.
6. Hilbert, D.: Foundations of Geometry (Grundlagen der Geometrie). Open Court, La Salle (1960); second English edition, translated from the tenth German edition by Unger, L., Original publication date (1899)
7. Turing, A.: On computable numbers, with an application to the Entscheidungsproblem. Proceedings of the London Mathematical Society, Series 2 42 (1936-1937)

On the Computational Content
of the Brouwer Fixed Point Theorem

Vasco Brattka[1], Stéphane Le Roux[2], and Arno Pauly[3]

[1] Department of Mathematics and Applied Mathematics, University of Cape Town,
Private Bag, Rondebosch 7701, South Africa
Vasco.Brattka@uct.ac.za
[2] Department of Mathematics, Technische Universität Darmstadt,
Schlossgartenstraße 7, 64289 Darmstadt, Germany
leroux@mathematik.tu-darmstadt.de
[3] Computer Laboratory, University of Cambridge, William Gates Building, 15 JJ
Thomson Avenue, Cambridge CB3 0FD, United Kingdom
Arno.Pauly@cl.cam.ac.uk

Abstract. We study the computational content of the Brouwer Fixed
Point Theorem in the Weihrauch lattice. One of our main results is that
for any fixed dimension the Brouwer Fixed Point Theorem of that dimension is computably equivalent to connected choice of the Euclidean
unit cube of the same dimension. Connected choice is the operation that
finds a point in a non-empty connected closed set given by negative information. Another main result is that connected choice is complete for
dimension greater or equal to three in the sense that it is computably
equivalent to Weak Kőnig's Lemma. In contrast to this, the connected
choice operations in dimensions zero, one and two form a strictly increasing sequence of Weihrauch degrees, where connected choice of dimension one is known to be equivalent to the Intermediate Value Theorem.
Whether connected choice of dimension two is strictly below connected
choice of dimension three or equivalent to it is unknown, but we conjecture that the reduction is strict. As a side result we also prove that
finding a connectedness component in a closed subset of the Euclidean
unit cube of any dimension greater than or equal to one is equivalent to
Weak Kőnig's Lemma.

1 Introduction

In this paper we continue with the programme to classify the computational content of mathematical theorems in the Weihrauch lattice (see [6,4,3,14,13,5,8]).
This lattice is induced by Weihrauch reducibility, which is a reducibility for partial multi-valued functions $f :\subseteq X \rightrightarrows Y$ on represented spaces X, Y. Intuitively,
$f \leq_{\mathrm{W}} g$ reflects the fact that the function f can be realized with a single application of the function g as an oracle. Hence, if two functions are equivalent in the
sense that they are mutually reducible to each other, then they are equivalent
as computational resources, as far as computability is concerned.

S.B. Cooper, A. Dawar, and B. Löwe (Eds.): CiE 2012, LNCS 7318, pp. 56–67, 2012.
© Springer-Verlag Berlin Heidelberg 2012

Many theorems in mathematics are actually of the logical form

$$(\forall x \in X)(\exists y \in Y)\ P(x, y)$$

and such theorems can straightforwardly be represented by a multi-valued function $f : X \rightrightarrows Y$ with $f(x) := \{y \in Y : P(x,y)\}$ (sometimes partial f are needed, where the domain captures additional requirements that this input x has to satisfy). In some sense the multi-valued function f directly reflects the computational task of the theorem to find some suitable y for any x. Hence, in a very natural way the classification of a theorem can be achieved via a classification of the corresponding multi-valued function that represents the theorem. In this paper we attempt to classify the Brouwer Fixed Point Theorem.

Theorem 1 (Brouwer Fixed Point Theorem 1911). *Every continuous function* $f : [0,1]^n \to [0,1]^n$ *has a fixed point* $x \in [0,1]^n$.

The fact that Brouwer's Fixed Point Theorem cannot be proved constructively has been confirmed in many different ways, most relevant for us is the counterexample in Russian constructive analysis by Orevkov [12], which was transferred into computable analysis by Baigger [1].

Constructions similar to those used for the above counterexamples have been utilized in order to prove that the Brouwer Fixed Point Theorem is equivalent to Weak Kőnig's Lemma in reverse mathematics [17,16] and to analyze computability properties of fixable sets [11], but a careful analysis of these reductions reveals that none of them can be straightforwardly transferred into a *uniform* reduction in the sense that we are seeking here. The results cited above essentially characterize the complexity of fixed points themselves, whereas we want to characterize the complexity of finding the fixed point, given the function. This requires full uniformity.

In the Weihrauch lattice the Brouwer Fixed Point Theorem of dimension n is represented by the multi-valued function $\mathsf{BFT}_n : \mathcal{C}([0,1]^n, [0,1]^n) \rightrightarrows [0,1]^n$ that maps any continuous function $f : [0,1]^n \to [0,1]^n$ to the set of its fixed points $\mathsf{BFT}_n(f) \subseteq [0,1]^n$. The question now is where BFT_n is located in the Weihrauch lattice? It easily follows from a meta theorem presented in [3] that the Brouwer Fixed Point Theorem BFT_n is reducible to Weak Kőnig's Lemma WKL for any dimension n, i.e., $\mathsf{BFT}_n \leq_W \mathsf{WKL}$. However, for which dimensions n do we also obtain the inverse reduction? Clearly not for $n = 0$, since BFT_0 is computable, and clearly not for $n = 1$, since BFT_1 is equivalent to the Intermediate Value Theorem IVT and hence not equivalent to WKL, as proved in [3].[1]

In order to approach this question for a general dimension n, we introduce a choice principle CC_n that we call *connected choice* and which is just the closed choice operation restricted to connected subsets. That is, in the sense discussed above, CC_n is the multi-valued function that represents the following mathematical statement: every non-empty connected closed set $A \subseteq [0,1]^n$ has a point

---

[1] In constructive reverse mathematics the Intermediate Value Theorem is equivalent to Weak Kőnig's Lemma [9], since parallelization is freely available in this framework.

$x \in [0,1]^n$. Since closed sets are represented by negative information (i.e., by an enumeration of open balls that exhaust the complement), the computational task of CC_n consists in finding a point in a closed set $A \subseteq [0,1]^n$ that is promised to be non-empty and connected and that is given by negative information.

One of our main results, presented in Section 4, is that the Brouwer Fixed Point Theorem is equivalent to connected choice for each fixed dimension n, i.e., $\mathsf{BFT}_n \equiv_{\mathrm{W}} \mathsf{CC}_n$. This result allows us to study the Brouwer Fixed Point Theorem in terms of the function CC_n, which is easier to handle since it involves neither function spaces nor fixed points. This is also another instance of the observation that several important theorems are equivalent to certain choice principles (see [3]) and many important classes of computable functions can be calibrated in terms of choice (see [2]). For instance, closed choice on Cantor space $\mathsf{C}_{\{0,1\}^{\mathbb{N}}}$ and on the unit cube $\mathsf{C}_{[0,1]^n}$ are both easily seen to be equivalent to Weak Kőnig's Lemma WKL, i.e., $\mathsf{WKL} \equiv_{\mathrm{W}} \mathsf{C}_{\{0,1\}^{\mathbb{N}}} \equiv_{\mathrm{W}} \mathsf{C}_{[0,1]^n}$ for any $n \geq 1$. Studying the Brouwer Fixed Point Theorem in the form of CC_n now amounts to comparing $\mathsf{C}_{[0,1]^n}$ with its restriction CC_n.

Our second main result, given in Section 5, is that from dimension three onwards connected choice is equivalent to Weak Kőnig's Lemma, i.e., $\mathsf{CC}_n \equiv_{\mathrm{W}} \mathsf{C}_{[0,1]}$ for $n \geq 3$. The backwards reduction is based on a geometrical construction that seems to require at least dimension three in a crucial sense. It is easy to see that connected choice operations for dimensions $0, 1$ and 2 form a strictly increasing sequence of Weihrauch degrees, i.e., $\mathsf{CC}_0 <_{\mathrm{W}} \mathsf{CC}_1 <_{\mathrm{W}} \mathsf{CC}_2 \leq_{\mathrm{W}} \mathsf{CC}_3 \equiv_{\mathrm{W}} \mathsf{C}_{[0,1]}$. The status of connected choice CC_2 of dimension two remains unresolved and we conjecture that it is strictly weaker than choice of dimension three, i.e., $\mathsf{CC}_2 <_{\mathrm{W}} \mathsf{CC}_3$.

In order to prove our results, we use a representation of closed sets by trees of so-called rational complexes, which we introduce in Section 3. It can be seen as a generalization of the well-known representation of co-c.e. closed subsets of Cantor space $\{0,1\}^{\mathbb{N}}$ by trees. As a side result we mention that finding a connectedness component in a closed set for any fixed dimension from one upwards is equivalent to Weak Kőnig's Lemma. This yields conclusions along the line of earlier studies of connected components in [10]. In the following Section 2 we start with a short summary of relevant definitions and results regarding the Weihrauch lattice.

This extended abstract does not contain any proofs.

2   The Weihrauch Lattice

In this section we briefly recall some basic results and definitions regarding the Weihrauch lattice. The original definition of Weihrauch reducibility is due to Weihrauch and has been studied for many years (see [18,19,20,7]). Only recently it has been noticed that a certain variant of this reducibility yields a lattice that is very suitable for the classification of mathematical theorems (see [6,4,3,14,2,13,5]). The basic reference for all notions from computable analysis is [21]. The Weihrauch lattice is a lattice of multi-valued functions on represented spaces. A *representation* δ of a set X is just a surjective partial map

$\delta :\subseteq \mathbb{N}^{\mathbb{N}} \to X$. In this situation we call (X, δ) a *represented space*. In general we use the symbol "$\subseteq$" in order to indicate that a function is potentially partial. Using represented spaces we can define the concept of a realizer. We denote the composition of two (multi-valued) functions f and g either by $f \circ g$ or by fg.

Definition 1 (Realizer). *Let* $f :\subseteq (X, \delta_X) \rightrightarrows (Y, \delta_Y)$ *be a multi-valued function on represented spaces. A function* $F :\subseteq \mathbb{N}^{\mathbb{N}} \to \mathbb{N}^{\mathbb{N}}$ *is called a realizer of* f, *in symbols* $F \vdash f$, *if* $\delta_Y F(p) \in f\delta_X(p)$ *for all* $p \in \mathrm{dom}(f\delta_X)$.

Realizers allow us to transfer the notions of computability and continuity and other notions available for Baire space to any represented space; a function between represented spaces will be called *computable*, if it has a computable realizer, etc. Now we can define Weihrauch reducibility.

Definition 2 (Weihrauch reducibility). *Let* f, g *be multi-valued functions on represented spaces. Then* f *is said to be* Weihrauch reducible *to* g, *in symbols* $f \leq_W g$, *if there are computable functions* $K, H :\subseteq \mathbb{N}^{\mathbb{N}} \to \mathbb{N}^{\mathbb{N}}$ *such that* $K\langle \mathrm{id}, GH \rangle \vdash f$ *for all* $G \vdash g$. *Moreover,* f *is said to be* strongly Weihrauch reducible *to* g, *in symbols* $f \leq_{sW} g$, *if there are computable functions* K, H *such that* $KGH \vdash f$ *for all* $G \vdash g$.

Here $\langle , \rangle$ denotes some standard pairing on Baire space. We note that the relations $\leq_W$, $\leq_{sW}$ and $\vdash$ implicitly refer to the underlying representations, which we mention explicitly only when necessary. It is known that these relations only depend on the underlying equivalence classes of representations, but not on the specific representatives (see Lemma 2.11 in [4]). We use $\equiv_W$ and $\equiv_{sW}$ to denote the respective equivalences regarding $\leq_W$ and $\leq_{sW}$, and by $<_W$ and $<_{sW}$ we denote strict reducibility.

A particularly useful multi-valued function in the Weihrauch lattice is closed choice (see [6,4,3,2]) and it is known that many notions of computability can be calibrated using the right version of choice. We shall focus on closed choice for computable metric spaces, which are separable metric spaces such that the distance function is computable on the given dense subset. We assume that computable metric spaces are represented via their Cauchy representation (see [21] for details).

By $\mathcal{A}_-(X)$ we denote the set of closed subsets of a metric space X, where the index "$-$" indicates that we work with negative information. This information is given by a representation $\psi_- : \mathbb{N}^{\mathbb{N}} \to \mathcal{A}_-(X)$, defined by $\psi_-(p) := X \setminus \bigcup_{i=0}^{\infty} B_{p(i)}$, where B_n is some standard enumeration of the open balls of X with center in the dense subset and rational radius. The computable points in $\mathcal{A}_-(X)$ are called *co-c.e. closed sets*. We now define closed choice for the case of computable metric spaces.

Definition 3 (Closed Choice). *Let* X *be a computable metric space. Then the closed choice operation of this space is defined by* $\mathsf{C}_X :\subseteq \mathcal{A}_-(X) \rightrightarrows X, A \mapsto A$ *with* $\mathrm{dom}(\mathsf{C}_X) := \{A \in \mathcal{A}_-(X) : A \neq \emptyset\}$.

Intuitively, C_X takes as input a non-empty closed set in negative representation (i.e., given by ψ_-) and it produces an arbitrary point of this set as output. Hence,

$A \mapsto A$ means that the multi-valued map C_X maps the input $A \in \mathcal{A}_-(X)$ to the set $A \subseteq X$ as a set of possible outputs.

3   Closed Sets and Trees of Rational Complexes

In this section we want to describe a representation of closed sets $A \subseteq [0,1]^n$ that is useful for the study of connectedness. It is well-known that closed subsets of Cantor space can be characterized exactly as sets of infinite paths of trees. We describe a similar representation of closed subsets of the Euclidean unit cube $[0,1]^n$. While in the case of Cantor space clopen balls are associated to each node of the tree, we now associate finite complexes of rational balls to each node. While infinite paths lead to points of the closed set in case of Cantor space, they now lead to connectedness components.

This representation of closed subsets $A \subseteq [0,1]^n$ of the unit cube will enable us to analyze the relation between connected choice and the Brouwer Fixed Point Theorem in the next section. In this section we shall use this representation in order to show that finding a connectedness component of a closed set A is computably exactly as difficult as Weak Kőnig's Lemma.

We first fix some topological terminology. We work with the *maximum norm* $\| \ \|$ on $\mathbb{R}^n$. By $B(x,r) := \{y \in \mathbb{R}^n : \|x - y\| < r\}$ we denote the *open ball* with center x and radius r and by $B[x,r] := \{y \in \mathbb{R}^n : \|x - y\| \leq r\}$ the corresponding *closed ball*. Since we are using the maximum norm, all these balls are open or closed cubes, respectively (if the radius is positive). By ∂A we denote the topological *boundary*, by $\overline{A}$ the *closure* and by A° the *interior* of a set A. If the underlying space X is clear from the context, then $A^c := X \setminus A$ denotes the *complement* of A. We are now prepared to define rational complexes.

Definition 4 (Rational complex). *We call a set $R := \{B[c_1, r_1], ..., B[c_k, r_k]\}$ of finitely many closed balls $B[c_i, r_i]$ with rational center $c_i \in \mathbb{Q}^n$ and positive rational radius $r_i \in \mathbb{Q}$ an (n–dimensional) rational complex if $\bigcup R$ is connected and $B_1, B_2 \in R$ with $B_1 \neq B_2$ implies $B_1^\circ \cap B_2^\circ = \emptyset$. By $C\mathbb{Q}^n$ we denote the set of n–dimensional rational complexes.*

Each rational complex R can be represented by a list of the corresponding rational numbers $c_1, r_1, ..., c_k, r_k$ and we implicitly assume in the following that this representation is used for the set of rational complexes $C\mathbb{Q}^n$.

In order to organize the rational complexes that are used to approximate sets it is suitable to use trees. We recall that a *tree* is a set $T \subseteq \mathbb{N}^*$ which is closed under prefix, i.e., $u \sqsubseteq v$ and $v \in T$ implies $u \in T$. A function $b : \mathbb{N} \to \mathbb{N}$ is called a *bound* of a tree T if $w \in T$ implies $w(i) \leq b(i)$ for all $i = 0, ..., |w| - 1$, where $|w|$ denotes the *length* of the word w. A tree is called *finitely branching*, if it has a bound. A tree of rational complexes is understood to be a finitely branching tree T (together with a bound) such that to each node of the tree a rational complex is associated with the property that these complexes are compactly included in each other if we proceed along paths of the tree and they are disjoint on any particular level of the tree. We write $A \Subset B$ for two sets $A, B \subseteq \mathbb{R}^n$ if the closure

$\overline{A}$ of A is included in the interior B° of B and we say that A is *compactly included* in B in this case.

Definition 5 (Tree of rational complexes). *We call* (T, f) *a tree of rational complexes if* $T \subseteq \mathbb{N}^*$ *is a finitely branching tree and* $f : T \to C\mathbb{Q}^n$ *is a function such that for all* $u, v \in T$ *with* $u \neq v$

1. $u \sqsubseteq v \Longrightarrow \bigcup f(v) \in \bigcup f(u)$,
2. $|u| = |v| \Longrightarrow \bigcup f(u) \cap \bigcup f(v) = \emptyset$.

In the following we assume that finitely branching trees T are represented as a pair (χ_T, b), where $\chi_T : \mathbb{N}^* \to \{0, 1\}$ is the characteristic function of T and $b : \mathbb{N} \to \mathbb{N}$ is a bound of T. Correspondingly, trees (T, f) of rational complexes are then represented in a canonical way by (χ_T, b, f). We now define which set $A \subseteq [0, 1]^n$ is represented by such a tree (T, f) of rational complexes.

Definition 6 (Closed sets and trees of rational complexes). *We say that a closed set* $A \subseteq \mathbb{R}^n$ *is represented by a tree* (T, f) *of* n*–dimensional rational complexes if one obtains* $A = \bigcap_{i=0}^{\infty} \bigcup_{w \in T \cap \mathbb{N}^i} \bigcup f(w)$.

It is clear that in this way any tree (T, f) of rational complexes actually represents a compact set A. This is because $\bigcup f(w)$ is compact for each $w \in T$ and since T is finitely branching, the set $T \cap \mathbb{N}^i$ is finite for each i, hence $\bigcup_{w \in T \cap \mathbb{N}^i} \bigcup f(w)$ is compact and hence A is compact too. Vice versa, every compact set $A \subseteq \mathbb{R}^n$ can be represented by a tree of n–dimensional rational complexes. For $[0, 1]^n$ we mention the uniform result that even the map $(T, f) \mapsto A$ is computable and has a computable multi-valued right inverse. We assume that trees of rational complexes are represented as specified above and closed sets A are represented as points in $\mathcal{A}_-([0, 1]^n)$.

Proposition 1 (Closed sets and complexes). *Let* $n \geq 1$. *The map* $(T, f) \mapsto A$ *that maps every tree of* n*–dimensional rational complexes* (T, f) *to the closed set* $A \subseteq [0, 1]^n$ *represented by it, is computable and has a multi-valued computable right inverse.*

The representation of closed sets $A \subseteq [0, 1]^n$ by trees of rational complexes also has the advantage that connectedness components of A can easily be expressed in terms of the tree structure. This is made precise by the following lemma. By $[T] := \{p \in \mathbb{N}^{\mathbb{N}} : (\forall i)\, p|_i \in T\}$ we denote the set of *infinite paths* of T, which is also called the *body* of T. Here $p|_i = p(0)...p(i-1) \in \mathbb{N}^*$ denotes the *prefix* of p of length i for each $i \in \mathbb{N}$. We recall that a *connectedness component* of a set A is a connected subset of A that is not included in any larger connected subset of A. Any connectedness component of a subset A is closed in A. According to the following lemma there is bijection between $[T]$ and the set of connectedness components of a non-empty closed set $A \subseteq [0, 1]^n$.

Lemma 1 (Connectedness components). *Let* (T, f) *be a tree of rational complexes and let* $A \subseteq [0, 1]^n$ *be the closed set represented by* (T, f). *Then the sets* $C_p := \bigcap_{i=0}^{\infty} \bigcup f(p|_i)$ *for* $p \in [T]$ *are exactly all connectedness components of* A *(without repetitions).*

As another interesting result we can deduce from Proposition 1 a classification of the operation that determines a connectedness component. We first define this operation. For short we use the notation $\mathcal{A}_n := \{A \in \mathcal{A}_-([0,1]^n) : A \neq \emptyset\}$ for the space of non-empty closed subsets with representation ψ_-.

Definition 7 (Connectedness components). *By* $\mathsf{Con}_n : \mathcal{A}_n \rightrightarrows \mathcal{A}_n$ *we denote the map with* $\mathsf{Con}_n(A) := \{C : C \text{ is a connectedness component of } A\}$ *for every* $n \geq 1$.

The problem Con_n of finding a connectedness component of a closed set has the same strong Weihrauch degree as Weak König's Lemma for every dimension $n \geq 1$.

Theorem 2 (Connectedness components). $\mathsf{Con}_n \equiv_{\mathrm{sW}} \mathsf{WKL}$ *for* $n \geq 1$.

4   Brouwer's Fixed Point Theorem and Connected Choice

In this section we want to show that the Brouwer Fixed Point Theorem is computably equivalent to connected choice for any fixed dimension. We first define these two operations. By $\mathcal{C}(X,Y)$ we denote the *set of continuous functions* $f : X \to Y$ and for short we write $\mathcal{C}_n := \mathcal{C}([0,1]^n, [0,1]^n)$.

Definition 8 (Brouwer Fixed Point Theorem). *By* $\mathsf{BFT}_n : \mathcal{C}_n \rightrightarrows [0,1]^n$ *we denote the operation defined by* $\mathsf{BFT}_n(f) := \{x \in [0,1]^n : f(x) = x\}$ *for* $n \in \mathbb{N}$.

We note that BFT_n is well-defined, i.e., $\mathsf{BFT}_n(f)$ is non-empty for all f, since by the Brouwer Fixed Point Theorem every $f \in \mathcal{C}_n$ admits a fixed point x, i.e., with $f(x) = x$. We now define connected choice.

Definition 9 (Connected choice). *By* $\mathsf{CC}_n :\subseteq \mathcal{A}_n \rightrightarrows [0,1]^n$ *we denote the operation defined by* $\mathsf{CC}_n(A) := A$ *for all non-empty connected closed* $A \subseteq [0,1]^n$ *and* $n \in \mathbb{N}$. *We call* CC_n *connected choice (of dimension* n*).*

Hence, connected choice is just the restriction of closed choice $\mathsf{C}_{[0,1]^n}$ to connected sets. We also use the following notation for the set of fixed points of a functions $f \in \mathcal{C}_n$.

Definition 10 (Set of fixed points). *By* $\mathsf{Fix}_n : \mathcal{C}_n \to \mathcal{A}_n$ *we denote the function with* $\mathsf{Fix}_n(f) := \{x \in [0,1]^n : f(x) = x\}$.

It is easy to see that Fix_n is computable, since $\mathsf{Fix}_n(f) := (f - \mathrm{id})^{-1}\{0\}$ and it is well-known that closed sets in $\mathcal{A}_n$ can also be represented as zero sets of continuous functions (see [21]). We note that the Brouwer Fixed Point Theorem can be decomposed to $\mathsf{BFT}_n = \mathsf{CC}_n \circ \mathsf{Con}_n \circ \mathsf{Fix}_n$.

The main result of this section will be that the Brouwer Fixed Point Theorem and connected choice are (strongly) equivalent for any fixed dimension n (see Theorem 3 below). An important tool for both directions of the proof is the representation of closed sets by trees of rational complexes. The direction

$\mathsf{CC}_n \leq_{\mathrm{sW}} \mathsf{BFT}_n$ can be seen as a uniformization of an earlier construction of Baiger [1] that is in turn built on results of Orevkov [12]. For the other direction $\mathsf{BFT}_n \leq_{\mathrm{sW}} \mathsf{CC}_n$ of the reduction we uniformize ideas of Joseph S. Miller [11] and we use again the representation of closed sets by trees of rational complexes. We also exploit the fact that each rational complex can easily be converted into a simplicial complex. We recall that a *proper n–dimensional rational simplex* is the convex hull of $n + 1$ geometrically independent rational points in $[0, 1]^n$ and a *proper rational simplicial complex* is a set of finitely many proper simplexes such that the interiors of distinct simplexes are disjoint. By SQ^n we denote the set of all such proper rational simplicial complexes and we assume that each such complex is represented by a specification of a list of $n + 1$ geometrically independent rational points for each simplex in the complex. Hence, it is clear that there is a computable $f : \mathsf{CQ}^n \to \mathsf{SQ}^n$ with $\bigcup f(R) = \bigcup R$. That means that we can easily translate each tree of rational complexes into a corresponding tree of rational simplicial complexes (understood in the analogous way). We essentially use Miller's ideas to reduce the Brouwer Fixed Point Theorem uniformly to connected choice. The first observation is that the map $\mathsf{Con}_n \circ \mathsf{Fix}_n$ is computable (which might be surprising in light of Theorem 2).

Proposition 2. $\mathsf{Con}_n \circ \mathsf{Fix}_n : \mathcal{C}_n \rightrightarrows \mathcal{A}_n$ *is computable for all $n \in \mathbb{N}$.*

Since $\mathsf{BFT}_n \supseteq \mathsf{CC}_n \circ \mathsf{Con}_n \circ \mathsf{Fix}_n$ we can directly conclude $\mathsf{BFT}_n \leq_{\mathrm{sW}} \mathsf{CC}_n$ for all n. Together with $\mathsf{CC}_n \leq_{\mathrm{sW}} \mathsf{BFT}_n$ we obtain the following theorem.

Theorem 3 (Brouwer Fixed Point Theorem). $\mathsf{BFT}_n \equiv_{\mathrm{sW}} \mathsf{CC}_n$ *for all n.*

It is easy to see that in general the Brouwer Fixed Point Theorem and connected choice are not independent of the dimension. In case of $n = 0$ the space $[0, 1]^n$ is the one-point space $\{0\}$ and hence $\mathsf{BFT}_0 \equiv_{\mathrm{sW}} \mathsf{CC}_0$ are both computable. In case of $n = 1$ connected choice was already studied in [3] and it was proved that it is equivalent to the Intermediate Value Theorem IVT (see Definition 6.1 and Theorem 6.2 in [3]).

Corollary 1 (Intermediate Value Theorem). $\mathsf{IVT} \equiv_{\mathrm{sW}} \mathsf{BFT}_1 \equiv_{\mathrm{sW}} \mathsf{CC}_1$.

It is also easy to see that the Brouwer Fixed Point Theorem BFT_2 in dimension two is more complicated than in dimension one. For instance, it is known that the Intermediate Value Theorem IVT always offers a computable function value for a computable input, whereas this is not the case for the Brouwer Fixed Point Theorem BFT_2 by Baigger's counterexample [1]. We continue to discuss this topic in Section 5.

Here we point out that Proposition 2 implies that the fixed point set $\mathsf{Fix}_n(f)$ of every computable function $f : [0, 1]^n \to [0, 1]^n$ has a co-c.e. closed connectedness component. Joseph S. Miller observed that also any co-c.e. closed superset of such a set is the fixed point set of some computable function and the following result is a uniform version of this observation. We denote by $(f, g) :\subseteq X \rightrightarrows Y \times Z$ the *juxtaposition* of two functions $f :\subseteq X \rightrightarrows Y$ and $g :\subseteq X \rightrightarrows Z$, defined by $(f, g)(x) = (f(x), g(x))$.

Theorem 4 (Fixability). $(\mathsf{Fix}_n, \mathsf{Con}_n \circ \mathsf{Fix}_n)$ *is computable and has a multivalued computable right inverse for all* $n \in \mathbb{N}$.

Roughly speaking a closed set $A \in \mathcal{A}_n$ together with one of its connectedness components is as good as a continuous function $f \in \mathcal{C}_n$ with A as set of fixed points. As a non-uniform corollary we obtain immediately Miller's original result.

Corollary 2 (Fixable sets, Miller 2002). *A set* $A \subseteq [0,1]^n$ *is the set of fixed points of a computable function* $f : [0,1]^n \to [0,1]^n$ *if and only if it is non-empty and co-c.e. closed and contains a co-c.e. closed connectedness component.*

5   Aspects of Dimension

In this section we want to discuss aspects of dimension of connected choice and the Brouwer Fixed Point Theorem. Our main result is that connected choice is computably universal or complete from dimension three onwards in the sense that it is strongly equivalent to Weak König's Lemma, which is one of the degrees of major importance. In order to prove this result, we use the following geometric construction.

Proposition 3 (Twisted cube). *The function* $T :\subseteq \mathcal{A}_-[0,1] \to \mathcal{A}_3$ *with* $T(A) = (A \times [0,1] \times \{0\}) \cup (A \times A \times [0,1]) \cup ([0,1] \times A \times \{1\})$ *is computable and maps non-empty closed sets* $A \subseteq [0,1]$ *to non-empty connected closed sets* $T(A) \subseteq [0,1]^3$.

Here tuples $(x_1, x_2, x_3) \in T(A)$ have the property that at least one of the first two components provide a solution $x_i \in A$, but the third component provides the additional information which one surely does. If x_3 is close to 1, then surely $x_2 \in A$ and if x_3 is close to 0, then surely $x_1 \in A$. If x_3 is neither close to 0 nor 1, then both $x_1, x_2 \in A$. Hence, there is a computable function H such that $\mathsf{C}_{[0,1]} = H \circ \mathsf{CC}_3 \circ T$, which proves $\mathsf{C}_{[0,1]} \leq_{\mathrm{sW}} \mathsf{CC}_3$. Together with Theorem 3 we obtain the following conclusion.

Theorem 5 (Completeness of three dimensions). *For* $n \geq 3$ *we obtain* $\mathsf{CC}_n \equiv_{\mathrm{sW}} \mathsf{BFT}_n \equiv_{\mathrm{sW}} \mathsf{WKL} \equiv_{\mathrm{sW}} \mathsf{C}_{[0,1]}$.

We note that the reduction $\mathsf{CC}_n \leq_{\mathrm{sW}} \mathsf{C}_{[0,1]^n}$ holds for all $n \in \mathbb{N}$, since connected choice is just a restriction of closed choice and $\mathsf{C}_{[0,1]^n} \equiv_{\mathrm{sW}} \mathsf{C}_{[0,1]} \equiv_{\mathrm{sW}} \mathsf{WKL}$ is known for all $n \geq 1$ (see [2]).

In particular, we get the Baigger counterexample for dimension $n \geq 3$ as a consequence of Theorem 5. A superficial reading of the results of Orevkov [12] and Baigger [1] can lead to the wrong conclusion that they actually provide a reduction of Weak König's Lemma to the Brouwer Fixed Point Theorem BFT_n of any dimension $n \geq 2$. However, this is only correct in a non-uniform way and the corresponding uniform result is still open and does not follow from the known constructions. The Orevkov-Baigger result is built on the following fact.

Proposition 4 (Mixed cube). *The function* $M :\subseteq \mathcal{A}_-[0,1] \to \mathcal{A}_2$ *with* $M(A) = (A \times [0,1]) \cup ([0,1] \times A)$ *is computable and maps non-empty closed sets* $A \subseteq [0,1]$ *to non-empty connected closed sets* $M(A) \subseteq [0,1]^2$.

It follows straightforwardly from the definition that the pairs $(x,y) \in M(A)$ are such that one out of two components x,y is actually in A. In order to express the uniform content of this fact, we introduce the concept of a fraction.

Definition 11 (Fractions). *Let* $f :\subseteq X \rightrightarrows Y$ *be a multi-valued function and* $0 < n \leq m \in \mathbb{N}$. *We define the fraction* $\frac{n}{m}f :\subseteq X \rightrightarrows Y^m$ *such that* $\frac{n}{m}f(x)$ *is the set of all* $(y_1, ..., y_m) \in \mathrm{range}(f)^m$ *with* $|\{i : y_i \in f(x)\}| \geq n$ *for all* $x \in \mathrm{dom}(\frac{n}{m}f) := \mathrm{dom}(f)$.

The idea of a fraction $\frac{n}{m}f$ is that it provides m potential answers for f, at least $n \leq m$ of which have to be correct. The uniform content of the Orevkov-Baigger construction is then summarized in the following result.

Proposition 5 (Dimension two). $\frac{1}{2}\mathsf{C}_{[0,1]} \leq_{\mathrm{sW}} \mathsf{CC}_2 \leq_{\mathrm{sW}} \mathsf{C}_{[0,1]}$.

Proof. With Proposition 4 we obtain $\frac{1}{2}\mathsf{C}_{[0,1]} = \mathsf{CC}_2 \circ M$ and hence $\frac{1}{2}\mathsf{C}_{[0,1]} \leq_{\mathrm{sW}} \mathsf{CC}_2$. The other reduction follows from $\mathsf{CC}_2 \leq_{\mathrm{sW}} \mathsf{C}_{[0,1]^2} \equiv_{\mathrm{sW}} \mathsf{C}_{[0,1]}$.

That is, given a closed set $A \subseteq [0,1]$ we can utilize connected choice CC_2 of dimension 2 in order to find a pair of points (x,y) one of which is in A. This result directly implies the counterexample of Baigger [1] because the fact that there are non-empty co-c.e. closed sets $A \subseteq [0,1]$ without computable point immediately implies that $\frac{1}{2}\mathsf{C}_{[0,1]}$ is not non-uniformly computable (i.e., there are computable inputs without computable outputs) and hence CC_2 is also not non-uniformly computable.

Corollary 3 (Orevkov 1963, Baigger 1985). *There exists a computable function* $f : [0,1]^2 \to [0,1]^2$ *that has no computable fixed point* $x \in [0,1]^2$. *There exists a non-empty connected co-c.e. closed subset* $A \subseteq [0,1]^2$ *without computable point.*

We mention that Proposition 5 does not directly imply $\mathsf{C}_{[0,1]} \equiv_{\mathrm{sW}} \mathsf{CC}_2$, since $\frac{1}{2}\mathsf{C}_{[0,1]} <_{\mathrm{W}} \mathsf{CC}_2$. In the following result we summarize the known relations for connected choice in dependency of the dimension.

Proposition 6. *We obtain* $\mathsf{CC}_0 <_{\mathrm{W}} \mathsf{CC}_1 <_{\mathrm{W}} \mathsf{CC}_2 \leq_{\mathrm{W}} \mathsf{CC}_n \equiv_{\mathrm{W}} \mathsf{C}_{[0,1]}$ *for* $n \geq 3$.

Altogether, we are left with the major open problem whether $\mathsf{C}_{[0,1]} \leq_{\mathrm{W}} \mathsf{CC}_2$ holds or not. We have a conjecture but currently no proof of it.

Conjecture 1 (Brouwer Fixed Point Theorem in dimension two). We conjecture that $\mathsf{CC}_2 <_{\mathrm{W}} \mathsf{C}_{[0,1]}$.

We mention that this conjecture is equivalent to the property that CC_2 is not parallelizable, i.e., to the property that $\widehat{\mathsf{CC}_2} \equiv_{\mathrm{W}} \mathsf{CC}_2$ does not hold. This is because $\widehat{\mathsf{CC}_2} \equiv_{\mathrm{W}} \mathsf{C}_{[0,1]}$ follows from $\mathsf{C}_{\{0,1\}} \leq_{\mathrm{sW}} \mathsf{CC}_2$ and $\widehat{\mathsf{C}_{\{0,1\}}} \equiv_{\mathrm{sW}} \mathsf{C}_{[0,1]}$ and the fact that parallelization is a closure operator, which are known results (see [3]).

6 Conclusions

We have systematically studied the uniform computational content of the
Brouwer Fixed Point Theorem for any fixed dimension and we have obtained
a systematic classification that leaves only the status of the two-dimensional
case unresolved. Besides a solution of this open problem, one can proceed into
several different directions.

For one, one could study generalizations of the Brouwer Fixed Point Theorem, such as the Schauder Fixed Point Theorem or the Kakutani Fixed Point
Theorem. On the other hand, one could study results that are based on the
Brouwer Fixed Point Theorem, such as equilibrium existence theorems in computable economics (see for instance [15]). Nash equilibria existence theorems
have been studied in [13] and they can be seen to be strictly simpler than the
general Brouwer Fixed Point Theorem (in fact they can be considered as linear
version of it). In this context the question arises of how the Brouwer Fixed Point
Theorem can be classified for other subclasses of continuous functions, such as
Lipschitz continuous functions?

Acknowledgements. This project has been supported by the National Research Foundation of South Africa (NRF) and by the German Research Foundation (DFG) through the German-South African project (DFG, 445 SUA-1
13/20/0).

References

1. Baigger, G.: Die Nichtkonstruktivität des Brouwerschen Fixpunktsatzes. Arch.
   Math. Logik Grundlag. 25, 183–188 (1985)
2. Brattka, V., de Brecht, M., Pauly, A.: Closed choice and a Uniform Low Basis
   Theorem. Annals of Pure and Applied Logic 163(8), 986–1008 (2012)
3. Brattka, V., Gherardi, G.: Effective choice and boundedness principles in computable analysis. The Bulletin of Symbolic Logic 17(1), 73–117 (2011)
4. Brattka, V., Gherardi, G.: Weihrauch degrees, omniscience principles and weak
   computability. The Journal of Symbolic Logic 76(1), 143–176 (2011)
5. Brattka, V., Gherardi, G., Marcone, A.: The Bolzano-Weierstrass theorem is the
   jump of weak Kőnig's lemma. Annals of Pure and Applied Logic 163, 623–655
   (2012)
6. Gherardi, G., Marcone, A.: How incomputable is the separable Hahn-Banach theorem? Notre Dame Journal of Formal Logic 50, 393–425 (2009)
7. Hertling, P.: Unstetigkeitsgrade von Funktionen in der effektiven Analysis. Informatik Berichte 208, FernUniversität Hagen, Hagen (November 1996)
8. Hoyrup, M., Rojas, C., Weihrauch, K.: Computability of the Radon-Nikodym
   Derivative. In: Löwe, B., Normann, D., Soskov, I., Soskova, A. (eds.) CiE 2011.
   LNCS, vol. 6735, pp. 132–141. Springer, Heidelberg (2011)
9. Ishihara, H.: Reverse mathematics in Bishop's constructive mathematics.
   Philosophia Scientiae, Cahier special 6, 43–59 (2006)
10. Le Roux, S., Ziegler, M.: Singular coverings and non-uniform notions of closed set
    computability. Mathematical Logic Quarterly 54(5), 545–560 (2008)

11. Miller, J.S.: Pi-0-1 Classes in Computable Analysis and Topology. Ph.D. thesis, Cornell University, Ithaca, USA (2002)
12. Orevkov, V.: A constructive mapping of the square onto itself displacing every constructive point (Russian). Doklady Akademii Nauk 152, 55–58 (1963); translated in: Soviet Math. - Dokl. 4, 1253–1256 (1963)
13. Pauly, A.: How incomputable is finding Nash equilibria? Journal of Universal Computer Science 16(18), 2686–2710 (2010)
14. Pauly, A.: On the (semi)lattices induced by continuous reducibilities. Mathematical Logic Quarterly 56(5), 488–502 (2010)
15. Richter, M.K., Wong, K.C.: Non-computability of competitive equilibrium. Economic Theory 14(1), 1–27 (1999)
16. Shioji, N., Tanaka, K.: Fixed point theory in weak second-order arithmetic. Annals of Pure and Applied Logic 47, 167–188 (1990)
17. Simpson, S.G.: Subsystems of Second Order Arithmetic. Perspectives in Mathematical Logic. Springer, Berlin (1999)
18. von Stein, T.: Vergleich nicht konstruktiv lösbarer Probleme in der Analysis. Diplomarbeit, Fachbereich Informatik, FernUniversität Hagen (1989)
19. Weihrauch, K.: The degrees of discontinuity of some translators between representations of the real numbers. Technical Report TR-92-050, International Computer Science Institute, Berkeley (July 1992)
20. Weihrauch, K.: The TTE-interpretation of three hierarchies of omniscience principles. Informatik Berichte 130, FernUniversität Hagen, Hagen (September 1992)
21. Weihrauch, K.: Computable Analysis. Springer, Berlin (2000)

Square Roots and Powers
in Constructive Banach Algebra Theory

Douglas S. Bridges[1] and Robin S. Havea[2]

[1] Department of Mathematics & Statistics, University of Canterbury, Private Bag
4800, Christchurch, New Zealand
d.bridges@math.canterbury.ac.nz
[2] Department of Mathematics & Computing Science,
University of the South Pacific, Suva, Fiji
robin.havea@usp.ac.fj

Abstract. Several new and improved results about positive integral
powers of hermitian elements, and square roots of positive elements, in
a Banach algebra are proved constructively.

1   Introduction

The purpose of this article is to extend our earlier constructive[1] work on hermitian and positive elements of a separable complex Banach algebra B with identity e [6,5,9]. In particular, we provide conditions—one of which was, unfortunately, lacking in Theorem 4.2 of [6] and the corresponding result in [9]—under which we can prove constructively that positive integral powers of a hermitian element are hermitian; also, we substantially generalise, by a relatively elementary proof, the result in [5] that yields the existence and uniqueness of the square root of a positive element of B.

Although we shall refer to the [6] for much of the background material needed for this paper, for the reader's convenience we here re-present some important notions. First, let B' denote the dual of B. In general, we cannot prove that every $f \in B$ is **normed** in the sense that $\|f\| \equiv \sup \{|f(x)| : x \in B, \|x\| \leqslant 1\}$ exists. However, even when f need not be normed, we adopt the shorthand $\|f\| \leqslant c$ to signify that $|f(x)| \leqslant c$ for all $x \in B$ with $\|x\| \leqslant 1$. An element f of B' is **nonzero** if $|f(x)| > 0$ for some $x \in B$. For each dense sequence $(x_n)_{n \geqslant 1}$ in B we introduce the corresponding **double norm** on B', defined by $|||f||| \equiv \sum_{n=1}^{\infty} 2^{-n} |f(x_n)|$. The topology induced by this norm on the unit ball $B_1' \equiv \{f \in B' : \|f\| \leqslant 1\}$ of B' is independent of the dense sequence relative to which the double norm is defined, and is, in fact, the weak* topology on B_1'.

Now, we may not be able to prove constructively that the **state space** $V_B = \{f \in B' : f(e) = 1 = \|f\|\}$ of B is inhabited (that is, contains an element), let alone weak* compact as it is classically. For this reason we introduce, for each $t > 0$, the approximation

---

[1] We work entirely within the framework of Bishop-style constructive analysis (**BISH**—for more on which, see [2,7,8]).

S.B. Cooper, A. Dawar, and B. Löwe (Eds.): CiE 2012, LNCS 7318, pp. 68–77, 2012.
© Springer-Verlag Berlin Heidelberg 2012

$$V_B^t = \{f \in B' : \|f\| \leqslant 1, \ |1 - f(e)| \leqslant t\}$$

to V_B. The constructive Hahn-Banach theorem ([8], Theorem 5.3.3) is strong enough for us to prove that V_B^t is inhabited; moreover, it is weak* compact for all but countably many $t > 0$. We say that $t > 0$ is **admissible** if V_B^t is weak* compact. We describe V_B as **firm** if (i) it is weak*compact and (ii) for each $\varepsilon > 0$, there exists an admissible $t > 0$ such that for each $f \in V_B^t$, there exists $g \in V_B$ with $|||f - g||| < \varepsilon$. The following result is proved in [6] (Proposition 2.4).

Proposition 1. *If B has firm state space, then so does each Banach subalgebra of B.*

We call an element x of B **hermitian** if for each $\varepsilon > 0$, there exists $t > 0$ such that $|\operatorname{Im} f(x)| < \varepsilon$ for all $f \in V_B^t$; and **positive**—when we write $x \geqslant 0$—if for each $\varepsilon > 0$, there exists $t > 0$ such that $\operatorname{Re} f(x) \geqslant -\varepsilon$ and $|\operatorname{Im} f(x)| < \varepsilon$ for all $f \in V_B^t$. These definitions of *hermitian* and *positive* are classically equivalent to the standard classical ones found in [3], which are constructively too weak to be of much use. The following appears as Lemma 4.1 of [6].

Lemma 1. *Suppose that V_B is firm. Then $a \in B$ is hermitian if and only if $f(a) \in \mathbf{R}$ for each $f \in V_B$, and $a \geqslant 0$ if and only if $f(a) \geqslant 0$ for each $f \in V_B$.*

By a **character** of B we mean a nonzero multiplicative linear functional $u : B \to \mathbf{C}$; such a mapping satisfies $u(e) = 1$ and is normed, with $\|u\| = 1$. The **character space** Σ_B of B comprises all characters of B and is a subset of the unit ball of B'. We cannot prove constructively that the character space of every commutative Banach algebra B is inhabited, let alone that, as classically, it is weak* compact; see page 452 of [2]. However, as we shall see in Proposition 4, we can construct elements of Σ_B under certain conditions on B.

We say that an element x of B is **strongly hermitian** (resp. **strongly positive**) if it is hermitian (resp., positive) and the state space of the closed subalgebra A of B generated by $\{e, x\}$ is the closed convex hull of Σ_A. Classically, the latter condition always holds (see [3], page 206, Lemma 3), so every hermitian (resp. positive) x is strongly hermitian (resp. strongly positive). The main results of this paper are the following.

Theorem 1. *Let B be have firm state space, and let a be a strongly hermitian element of B. Then a^n is hermitian, and a^{2n} is positive, for each positive integer n.*

Theorem 2. *Let B be have firm state space. Let a be a strongly positive element of B, and A the Banach algebra generated by $\{e, a\}$. Then there exists a unique positive element x of A such that $x^2 = a$.*

The first of these is a corrected version of [6] (Theorem 4.2), in which we should have had a hypothesis ensuring that the product of two positive elements of

A is positive. Theorem 2 replaces the restrictive requirement that B be semi-simple, used in [5] (Theorem 3), by the more widely applicable strong positivity hypothesis on a.

2 Products of Hermitian/Positive Elements

When is the product of two hermitian/positive elements of a Banach algebra hermitian/positive?

Proposition 2. *Let B be commutative, with firm state space, and suppose that V_B is the weak*-closed convex hull of Σ_B. Then the product of two hermitian elements of B is hermitian, and the product of two positive elements is positive.*

Proof. Let x and y be hermitian elements of B. Given $f \in V_B$ and $\varepsilon > 0$, pick elements u_k ($1 \leqslant k \leqslant m$) of Σ_B, and corresponding nonnegative numbers λ_k, such that $\sum_{k=1}^{m} \lambda_k = 1$ and $|f(xy) - \sum_{k=1}^{m} \lambda_k u_k(xy)| < \varepsilon$. Since $\Sigma_B \subset V_B$, we have $u_k(x), u_k(y) \in \mathbf{R}$, by Lemma 1; whence

$$|\operatorname{Im} f(xy)| \leqslant \left|\sum_{k=1}^{m} \lambda_k \operatorname{Im} u_k(xy)\right| + \left|f(xy) - \sum_{k=1}^{m} \lambda_k u_k(xy)\right| < 0 + \varepsilon = \varepsilon.$$

But $\varepsilon > 0$ is arbitrary, so $\operatorname{Im} f(xy) = 0$ and therefore $f(xy) \in \mathbf{R}$. Moreover, if $x \geqslant 0$ and $y \geqslant 0$, then

$$\operatorname{Re} f(xy) \geqslant \sum_{k=1}^{m} \lambda_k \operatorname{Re}\left(u_k(x)u_k(y)\right) - \left|f(xy) - \sum_{k=1}^{m} \lambda_k u_k(xy)\right| > 0 - \varepsilon = -\varepsilon.$$

Since $\varepsilon > 0$ is arbitrary, it follows from Lemma 1 that, when x, y are hermitian, $f(xy) \in \mathbf{R}$, and when they are positive, $f(xy) \geqslant 0$.

We are now prepared for the **Proof of Theorem 1.** Under its hypotheses, let A be the closed subalgebra of B generated by $\{e, a\}$. By Proposition 1, V_A is firm; since a is strongly hermitian, the hypotheses of Proposition 2 are satisfied, and the desired conclusions follow almost immediately. ∎

By a **positive linear functional** on B we mean an element f of B' such that $f(x) \geqslant 0$ for each positive element of B; we write $f \geqslant 0$ to signify that f is positive. Every element of the state space V_B is positive (see Section 3 of [6]).

Consider a convex subset K of B_1'. We say that $f \in K$ is a **classical extreme point** of K if

$$\forall_{g,h \in K} \left(f = \tfrac{1}{2}(g+h) \Rightarrow g = h = f\right),$$

and an **extreme point** of K if

$$\forall_{\varepsilon > 0} \exists_{\delta > 0} \forall_{g,h \in K} \left(\left|\left|\left|f - \tfrac{1}{2}(g+h)\right|\right|\right| < \delta \Rightarrow \left|\left|\left|g - h\right|\right|\right| < \varepsilon\right).$$

An extreme point is a classical extreme point, and the converse holds classically if K is also weak* compact. If f is an extreme point of K relative to one double norm on B', then it is an extreme point relative to any other double norm on B'. Proposition 3.1 of [6] states that if the state space V_B is firm, then every extreme point of V_B is an extreme point of the convex set $K^0 = \{f \in B' : f \geqslant 0, f(e) \leqslant 1\}$.

We omit the proof of our next result, which is very close to that on page 38 of [10].

Lemma 2. *Suppose that B is commutative and that the product of two positive elements of B is positive. Let f be a classical extreme point of K^0. Then $f(xy) = f(x)f(y)$ for all $x \in B$ and all positive $y \in B$.*

Proposition 3. *Suppose that B is generated by commuting positive elements, and that the product of two positive elements of B is positive. Then every classical extreme point of K^0 is a multiplicative linear functional on B.*

Proof. Clearly, B is commutative. Let f be a classical extreme point of K^0, and consider first the case where f is nonzero. A simple induction based on Lemma 2 shows that

$$f(xy^n) = f(x)f(y)^n \quad (x \in B, y \in B, y \geqslant 0). \tag{1}$$

Given any $x, y \in B$ and $\varepsilon > 0$, pick positive elements $a_1, \ldots, a_n$ and a complex polynomial $p(\zeta_1, \ldots, \zeta_n)$ such that $\|y - z\| < \varepsilon$, where $z \equiv p(a_1, \ldots, a_n)$. Since finite products of positive elements of B are positive, we see from (1) that $f(xz) = f(x)f(z)$; whence

$$
\begin{aligned}
|f(xy) - f(x)f(y)| &\leqslant |f(xy - xz)| + |f(x)f(z) - f(x)f(y)| \\
&\leqslant \|x(y - z)\| + |f(x)||f(z - y)| \\
&\leqslant 2\|x\|\|y - z\| \leqslant 2\|x\|\varepsilon.
\end{aligned}
$$

Since $\varepsilon > 0$ is arbitrary, we conclude that $f(xy) = f(x)f(y)$. Finally, to remove the condition that f be nonzero, let $x, y \in B$ and suppose that $f(xy) \neq f(x)f(y)$. Then, by the foregoing, f cannot be nonzero; so $f = 0$ and therefore $f(xy) = 0 = f(x)f(y)$, a contradiction. Thus we have $\neg\,(f(xy) \neq f(x)f(y))$ and therefore $f(xy) = f(x)f(y)$.

Proposition 4. *Suppose that B is generated by commuting positive elements and has firm state space, and that the product of two positive elements of B is positive. Let A be a unital Banach subalgebra of B. Then Σ_A is inhabited, and V_A is the double-norm-closed convex hull of Σ_A.*

Proof. By Proposition 1, V_A is firm and hence compact. Since V_A is also convex, it follows from the Krein-Milman theorem ([2], page 363, Theorem (7.5)) that it has extreme points and is the double-norm-closed convex hull of the set of those extreme points. By Proposition 3.1 of [6], every extreme point of V_A is an extreme point, and hence a classical extreme point, of K^0. Since the elements of V_A are nonzero, the result now follows from Proposition 3.

Proposition 4 readily yields a partial converse of Proposition 2:

Corollary 1. *Let a be an element of B all of whose positive integer powers are positive, and let A be the closed subalgebra of B generated by $\{e, a\}$. Then Σ_A is inhabited, and V_A is the double-norm-closed convex hull of Σ_A.*

When—as in the Banach algebra $C(X)$, where X is a compact metric space—is a hermitian element expressible as a difference of positive elements? To answer this, we need to say more about approximations to the character space.

For any dense sequence $(x_n)_{n \geqslant 1}$ in B, we can find a strictly decreasing sequence $(t_n)_{n \geqslant 1}$ of positive numbers converging to 0 such that for each n the set

$$\Sigma_B^{t_n} \equiv \{u \in B_1' : |u(x_j x_k) - u(x_j)u(x_k)| \leqslant t_n \ (1 \leqslant j, k \leqslant n)$$
$$\wedge \ |1 - u(e)| \leqslant t_n\}$$

is (inhabited and) weak* compact ([2], page 460, Proposition (2.7)). The intersection of these sets is the character space Σ_B. For each $x \in B$ we define

$$\|x\|_{\Sigma_B^{t_n}} \equiv \sup \{|u(x)| : u \in \Sigma_b^{t_n}\},$$

which exists since the function $x \rightsquigarrow |u(x)|$ is uniformly continuous on the double-norm compact set $\Sigma_b^{t_n}$.

We recall two important result from constructive Banach algebra theory.

Proposition 5. Sinclair's theorem: *If x is a hermitian element of the Banach algebra B, then $\|x^n\|^{1/n} = \|x\|$ for each positive integer n ([4], pages 293–303).*

Proposition 6. *Let B be commutative, and let $(t_n)_{n \geqslant 1}$ be a decreasing sequence of positive numbers converging to 0 such that Σ^{t_n} is compact for each n. Then the sequences $\left(\|x^n\|^{1/n}\right)_{n \geqslant 1}$ and $(\|x\|_{\Sigma^{t_n}})_{n \geqslant 1}$ are **equiconvergent**: that is, for each term a_m of one sequence and each $\varepsilon > 0$, there exists N such that $b_n \leqslant a_m + \varepsilon$ whenever b_n is a term of the other sequence with $n \geqslant N$ ([2], Chapter 9, Proposition (2.9)).*

Of particular importance for us is the following:

Corollary 2. *Let B be commutative, and let $(t_n)_{n \geqslant 1}$ be a decreasing sequence of positive numbers converging to 0 such that $\Sigma_B^{t_n}$ is compact for each n. Let h be a hermitian element of B. Then $\lim_{n \to \infty} \|h\|_{\Sigma_B^{t_n}} = \|h\|$.*

Proof. By Sinclair's theorem, $\|h^n\|^{1/n} = \|h\|$ for each positive integer n. It follows from Proposition 6 that for each $\varepsilon > 0$, there exists N such that $\|h\|_{\Sigma_B^{t_n}} < \|h\| + \varepsilon$ for all $n \geqslant N$. By that same proposition, for each $n \geqslant N$, there exists m such that $\|h\| = \|h^m\|^{1/m} \leqslant \|h\|_{\Sigma_B^{t_n}} + \varepsilon$. Hence $\left|\|h\| - \|h\|_{\Sigma_B^{t_n}}\right| < \varepsilon$ for all $n \geqslant N$.

Lemma 3. *If x, y are commuting hermitian elements of B, then*

$$\max\left\{\|x\|, \|y\|\right\} \leqslant \|x + iy\|.$$

Proof. Replacing B by the closed subalgebra generated by $\{e, x, y\}$, let $(t_n)_{n \geqslant 1}$ be a decreasing sequence of positive numbers converging to 0 such that Σ^{t_n} is compact for each n. Since $\Sigma_B^{t_n} \subset V_B^{t_n}$ and x, y are hermitian, there exists N such that $\min\left\{|\operatorname{Im} u(x)|, |\operatorname{Im} u(y)|\right\} < \varepsilon$ for each $n \geqslant N$ and each $u \in \Sigma_B^{t_n}$. For such u we have

$$|u(x + iy)| \geqslant |\operatorname{Re} u(x + iy)| = |\operatorname{Re} u(x) - \operatorname{Im} u(y)| > |\operatorname{Re} u(x)| - \varepsilon$$

and therefore $|\operatorname{Re} u(x)| < |u(x + iy)| + \varepsilon \leqslant \|x + iy\| + \varepsilon$. But

$$|u(x)|^2 = (\operatorname{Re} u(x))^2 + (\operatorname{Im} u(x))^2 < |\operatorname{Re} u(x)|^2 + \varepsilon^2,$$

so $|u(x)|^2 \leqslant (\|x + iy\| + \varepsilon)^2 + \varepsilon^2$. Since $u \in \Sigma_B^{t_n}$ is arbitrary, we conclude that $\|x\|_{\Sigma_B^{t_n}}^2 \leqslant (\|x + iy\| + \varepsilon)^2 + \varepsilon^2$. Now, x is hermitian, so by Sinclair's theorem, $\|x^n\|^{1/n} = \|x\|$ for each positive integer n. It follows from Corollary 2 that $\|x\|^2 = \lim_{n \to \infty} \|x\|_{\Sigma_B^{t_n}}^2 \leqslant \|x + iy\|^2$, and hence that $\|x\| \leqslant \|x + iy\|$. Finally, replacing x, y by $-y, x$ in the foregoing, we obtain $\|y\| \leqslant \|-y + ix\| = \|x + iy\|$.

Proposition 7. *Suppose that B is generated by commuting positive elements, and that the product of two positive elements of B is positive. Then for each hermitian element x of B and each $\varepsilon > 0$, there exist positive $a, b \in B$ with $\|x - (a - b)\| < \varepsilon$.*

Proof. There exist commuting positive elements $z_1, \ldots, z_m$ of B with each $\|z_k\| \leqslant 1$, and a polynomial $p(\zeta_1, \ldots, \zeta_m)$ over $\mathbf{C}$, such that

$$\|x - p(z_1, \ldots, z_m)\| < \varepsilon.$$

Write

$$p(\zeta_1, \ldots, \zeta_m) \equiv \sum_{i_1, \ldots, i_m = 1}^{n} \alpha(i_1, \ldots, i_m)\zeta_1^{i_1} \cdots \zeta_m^{i_m}$$

where each $\alpha(i_1, \ldots, i_m) \in \mathbf{C}$. Note that each term $z_1^{i_1} \cdots z_m^{i_m}$ is positive. Perturbing each coefficient $\alpha(i_1, \ldots, i_m)$ by a sufficiently small amount, we can arrange that $\operatorname{Re} \alpha(i_1, \ldots, i_m) \neq 0$ and $\operatorname{Im} \alpha(i_1, \ldots, i_m) \neq 0$ for each tuple $(i_1, \ldots, i_m)$. Let

$$P \equiv \{(i_1, \ldots, i_m) : \operatorname{Re} \alpha(i_1, \ldots, i_m) > 0\},$$

$$Q \equiv \{(i_1, \ldots, i_m) : \operatorname{Re} \alpha(i_1, \ldots, i_m) < 0\},$$

$$a \equiv \sum_{(i_1, \ldots, i_m) \in P} \operatorname{Re} \alpha(i_1, \ldots, i_m) z_1^{i_1} \cdots z_m^{i_m},$$

$$b \equiv -\sum_{(i_1, \ldots, i_m) \in Q} \operatorname{Re} \alpha(i_1, \ldots, i_m) z_1^{i_1} \cdots z_m^{i_m}.$$

74   D.S. Bridges and R.S. Havea

Then $a \geqslant 0$ and $b \geqslant 0$. Moreover,

$$
\varepsilon > \|x - p(z_1, \ldots, z_m)\|
$$

$$
= \left\| x - (a - b) - i \left(\sum_{i_1, \ldots, i_m = 1}^{n} (\operatorname{Im} \alpha(i_1, \ldots, i_m)) z_1^{i_1} \cdots z_m^{i_m} \right) \right\|
$$

where both $x - (a - b)$ and $\sum_{i_1, \ldots, i_m = 1}^{n} (\operatorname{Im} \alpha(i_1, \ldots, i_m)) z_1^{i_1} \cdots z_m^{i_m}$ are hermitian; whence $\|x - (a - b)\| < \varepsilon$, by the preceding lemma.

The approximation of hermitian elements by differences of two positive elements, as in the preceding proposition, is related to classical work on V-algebras and the Vidav-Palmer theorem (see, in particular, Lemma 8 in §38 of [3]). We intend exploring that further in a future paper.

3   The Path to Theorem 2

Our proof of Theorem 2 requires yet more preliminaries, beginning with an estimate that will lead to the continuity of positive square root extraction.

Lemma 4. *Let p be a positive element of the Banach algebra B such that $\|p\| \leqslant 1$, and let A be the Banach algebra generated by $\{e, p\}$. Let $0 < \delta_1, \delta_2 \leqslant 1$, and suppose that there exist positive elements b_1, b_2 of A such that $b_1^2 = e - \delta_1 p$ and $b_2^2 = e - \delta_2 p$. Then*

$$
\|b_1 - b_2\|^2 \leqslant \frac{68}{3} |\delta_1 - \delta_2| (1 + \|p\|).
$$

Proof. Given $\varepsilon > 0$, let

$$
\alpha = \sqrt{\frac{1}{3} (|\delta_1 - \delta_2| (1 + \|p\|) + 2\varepsilon^2)}.
$$

Pick $t_0 > 0$ such that: $V_A^{t_0}$ and $\Sigma_A^{t_0}$ are compact,

$$
\left| u(b_1^2) - u(b_1)^2 \right| < \alpha^2 \text{ and } \left| u(b_2^2) - u(b_2)^2 \right| < \alpha^2 \text{ for each } u \in \Sigma_A^{t_0}, \text{ and}
$$

$$
\min \{\operatorname{Re} f(b_1), \operatorname{Re} f(b_2)\} \geqslant -\alpha \text{ and } \max \{\operatorname{Im} f(b_1), f(b_2)\} \leqslant \alpha \text{ for each } f \in V_A^{t_0}
$$

For each $u \in \Sigma_A^{t_0}$ we have

$$
\begin{aligned}
|u(b_1 - b_2)| \, |u(b_1 + b_2)| &= \left| u(b_1)^2 - u(b_2)^2 \right| \\
&\leqslant \left| u(b_1^2 - b_2^2) \right| + \left| u(b_1^2) - u(b_1)^2 \right| + \left| u(b_2^2) - u(b_2)^2 \right| \\
&< \|b_1^2 - b_2^2\| + 2\varepsilon^2 = |\delta_1 - \delta_2| (1 + \|p\|) + 2\varepsilon^2 = 3\alpha^2.
\end{aligned}
$$

Either $|u(b_1 - b_2)| < 2\alpha$ or $|u(b_1 - b_2)| > \alpha$. In the latter case,

$$
|\operatorname{Re} u(b_1) + \operatorname{Re} u(b_2)| \leqslant |u(b_1 + b_2)| < 3\alpha.
$$

Suppose that $\mathrm{Re}\, u(b_1) > 4\alpha$. Then $\mathrm{Re}\, u(b_2) < 3\alpha - \mathrm{Re}\, u(b_1) < -\alpha$, which, since $u \in V_A^{t_0}$, contradicts our choice of t_0. Hence $-\alpha \leqslant \mathrm{Re}\, u(b_1) \leqslant 4\alpha$ and therefore $|\mathrm{Re}\, u(b_1)| \leqslant 4\alpha$; so

$$\left|u(b_1)^2\right| = (\mathrm{Re}\, u(b_1))^2 + (\mathrm{Im}\, u(b_1))^2 \leqslant 17\alpha^2$$

and therefore $|u(b_1)| \leqslant \sqrt{17}\alpha$. Likewise, $|u(b_2)| \leqslant \sqrt{17}\alpha$, so $|u(b_1 - b_2)| \leqslant 2\sqrt{17}\alpha$, an inequality that also holds in the case $|u\,(b_1 - b_2)| < 2\alpha$. Since $u \in \Sigma_A^{t_0}$ is arbitrary, we now see that $\|b_1 - b_2\|_{\Sigma_A^{t_0}}^2 < 68\alpha^2$. But $b_1 - b_2$ is hermitian, so, by Corollary 2,

$$\|b_1 - b_2\|^2 \leqslant \|b_1 - b_2\|_{\Sigma_A^{t_0}}^2 < \frac{68}{3}\left(|\delta_1 - \delta_2|\,(1 + \|p\|) + 2\varepsilon^2\right).$$

Since $\varepsilon > 0$ is arbitrary, we now obtain the desired conclusion.

Proposition 8. *Let B have firm state space, let a be a strongly positive element of B such that $\|a\| < 1$, and let A be the Banach algebra generated by $\{e, a\}$. Then there exists a positive element s of A such that $s^2 = e - a$.*

Proof. Our proof is based on that of Bonsall and Duncan [3] (page 207, Lemma 7). Those authors use the Gelfand representation theorem and Dini's theorem, the latter lying outside the reach of **BISH** (see [1]). However, we can avoid those two theorems altogether, as follows. First, we note that, by Lemma 5 of [6], $e - a \geqslant 0$ and $\|e - a\| \leqslant 1$. Consider the special case where $\|a\| < 1$. Let $x_0 = 0$ and, for each n,

$$x_{n+1} = \frac{1}{2}(a + x_n^2). \tag{2}$$

A simple induction shows that x_n belongs to A. Noting that $x_1 = \frac{1}{2}a$, suppose that $\|x_n\| < \|a\|$; then

$$\|x_{n+1}\| \leqslant \frac{1}{2}\left(\|a\| + \|x_n\|^2\right) < \frac{1}{2}\left(\|a\| + \|a\|^2\right) = \frac{1 + \|a\|}{2}\,\|a\| < \|a\|.$$

Thus $\|x_n\| < \|a\|$ for each n. Next, observe that, by Proposition 1, V_A is firm; it follows from Proposition 2 that the product of two positive elements of A is positive. Thus if $x_n \geqslant 0$, then $x_n^2 \geqslant 0$, so $a + x_n^2 \geqslant 0$ and therefore $x_{n+1} \geqslant 0$; since $x_0 \geqslant 0$, we conclude that $x_n \geqslant 0$ for each n. In particular, $x_1 - x_0 = x_1 \geqslant 0$. Now suppose that $x_n - x_{n-1} \geqslant 0$. Then since x_n, x_{n-1}, and therefore $x_n + x_{n-1}$ are all positive elements of A,

$$x_{n+1} - x_n = \frac{1}{2}\left(x_n^2 - x_{n-1}^2\right) = \frac{1}{2}\left(x_n + x_{n-1}\right)\left(x_n - x_{n-1}\right) \geqslant 0.$$

Moreover,

$$\|x_{n+1} - x_n\| \leqslant \frac{1}{2}\left(\|x_n\| + \|x_{n-1}\|\right)\|x_n - x_{n-1}\| \leqslant \|a\|\,\|x_n - x_{n-1}\|,$$

so, by another induction, $\|x_{n+1} - x_n\| \leqslant \|a\|^n \|x_1\| = \frac{1}{2} \|a\|^{n+1}$. It follows that if $m > n \geqslant 1$, then

$$\|x_m - x_n\| \leqslant \sum_{k=n}^{m-1} \|x_{k+1} - x_k\| \leqslant \sum_{k=n}^{m-1} \frac{1}{2} \|a\|^{k+1}$$

$$\leqslant \frac{1}{2} \|a\|^{n+1} \sum_{k=0}^{\infty} \|a\|^k = \frac{\|a\|^{n+1}}{2(1 - \|a\|)} \to 0 \text{ as } n \to \infty.$$

Hence $(x_n)_{n \geqslant 1}$ is a Cauchy sequence in the Banach algebra A and therefore converges to a limit $x \in A$. Clearly, x is positive, $\|x\| \leqslant \|a\| < 1$, and x commutes with a. Letting $n \to \infty$ in (2), we obtain $x = \frac{1}{2}(a + x^2)$. Hence $(e - x)^2 = e - 2x + (2x - a) = e - a$. Moreover, $e - x \in A$ and, by Lemma 5 of [6], $e - x \geqslant 0$. Thus $s \equiv e - x$ is a positive square root of $e - a$ in A.

Now consider the general case where $\|a\| \leqslant 1$. For each integer $n \geqslant 2$ set $r_n = 1 - n^{-1}$. Then $r_n a \geqslant 0$ and $\|r_n a\| < 1$. By the foregoing, there exists a positive element s_n of A with $s_n^2 = e - r_n a$. Taking $p = e - a, \delta_1 = \frac{1}{m}$, and $\delta_2 = \frac{1}{n}$ in Lemma 4 now yields

$$\|s_m - s_n\|^2 \leqslant 68 \left| \frac{1}{m} - \frac{1}{n} \right| (1 + \|a\|),$$

from which we see that $(s_n)_{n \geqslant 1}$ is a Cauchy sequence in A. Since A is complete, this sequence has a limit $s \in A$. Clearly, $s \geqslant 0$ and $s^2 = e - a$. Finally, taking $p = e - a$ and $\delta_1 = \delta_2 = 1$ in Lemma 3, we see that s is the unique positive square root of $e - a$ in A.

Finally, we have the **Proof of Theorem 2**. Under the hypotheses of that theorem, if $\|a\| \leqslant 1$, then by Lemma 5 of [5], $e - a \geqslant 0$ and $\|e - a\| \leqslant 1$; whence, by Proposition 8, there exists a unique positive element b of A such that $b^2 = e - (e - a) = a$. In the general case, compute $\delta > 0$ such that $\|\delta a\| \leqslant 1$. There exists a unique positive element p of A such that $p^2 = \delta a$. Then $\delta^{-1/2} p$ is a positive element of A, and $\left(\delta^{-1/2} p \right)^2 = a$. Moreover, if b is a positive square root of a, then $\delta^{1/2} b$ is a positive square root of δa, so $\delta^{1/2} b = p$ and therefore $b = \delta^{-1/2} p$. This establishes the uniqueness of the positive square root of a. ∎

Acknowledgements. The authors thank the Department of Mathematics & Statistics at the University of Canterbury for hosting Havea on several occasions during this work, and the Faculty of Science, Technology and Environment at the University of the South Pacific, Suva, Fiji, for supporting him during those visits.

References

1. Berger, J., Schuster, P.M.: Dini's theorem in the light of reverse mathematics. In: Lindström, S., Palmgren, E., Segerberg, K., Stoltenberg-Hansen, V. (eds.) Logicism, Intuitionism, and Formalism—What has become of them? Synthèse Library, vol. 341, pp. 153–166. Springer, Dordrecht (2009)

2. Bishop, E.A., Bridges, D.S.: Constructive Analysis. Grundlehren der Mathematischen Wissenschaften, vol. 279. Springer, Berlin (1985)

3. Bonsall, F.F., Duncan, J.: Complete Normed Algebras. Ergebnisse der Mathematik
und ihrer Grenzgebiete, vol. 80. Springer, Berlin (1973)

4. Bridges, D.S., Havea, R.S.: Approximating the numerical range in a Banach algebra. In: Crosilla, L., Schuster, P. (eds.) From Sets and Types to Topology and
Analysis. Oxford Logic Guides, pp. 293–303. Clarendon Press, Oxford (2005)

5. Bridges, D.S., Havea, R.S.: Constructing square roots in a Banach algebra. Sci.
Math. Japon. 70(3), 355–366 (2009)

6. Bridges, D.S., Havea, R.S.: Powers of a Hermitian element. New Zealand J.
Math. 36, 1–10 (2007)

7. Bridges, D.S., Richman, F.: Varieties of Constructive Mathematics. London Math.
Soc. Lecture Notes, vol. 97. Cambridge Univ. Press (1987)

8. Bridges, D.S., Vîţă, L.S.: Techniques of Constructive Analysis. Universitext.
Springer, New York (2006)

9. Havea, R.S.: On Firmness of the State Space and Positive Elements of a Banach
Algebra. J. UCS 11(12), 1963–1969 (2005)

10. Holmes, R.B.: Geometric Functional Analysis and its Applications. Graduate Texts
in Mathematics, vol. 24. Springer, New York (1975)

The Mate-in-n Problem
of Infinite Chess Is Decidable

Dan Brumleve[1], Joel David Hamkins[2,3,4], and Philipp Schlicht[5]

[1] Topsy Labs, Inc., 140 Second Street, 6th Floor, San Francisco, CA 94105,
United States of America
[2] Department of Philosophy, New York University, 5 Washington Place, New York,
NY 10003, United States of America
[3] Mathematics, CUNY Graduate Center, The City University of New York, 365 Fifth
Avenue, New York, NY 10016, United States of America
[4] Mathematics, College of Staten Island of CUNY, Staten Island, NY 10314,
United States of America
jhamkins@gc.cuny.edu
[5] Mathematisches Institut, Rheinische Friedrich-Wilhelms-Universität Bonn,
Endenicher Allee 60, 53115 Bonn, Germany
schlicht@math.uni-bonn.de

Abstract. The mate-in-n problem of infinite chess—chess played on an
infinite edgeless board—is the problem of determining whether a designated player can force a win from a given finite position in at most
n moves. Although a straightforward formulation of this problem leads
to assertions of high arithmetic complexity, with $2n$ alternating quantifiers, the main theorem of this article nevertheless confirms a conjecture
of the second author and C. D. A. Evans by establishing that it is computably decidable, uniformly in the position and in n. Furthermore, there
is a computable strategy for optimal play from such mate-in-n positions.
The proof proceeds by showing that the mate-in-n problem is expressible
in what we call the first-order structure of chess $\mathfrak{Ch}$, which we prove (in
the relevant fragment) is an automatic structure, whose theory is therefore decidable. The structure is also definable in Presburger arithmetic.
Unfortunately, this resolution of the mate-in-n problem does not appear
to settle the decidability of the more general winning-position problem,
the problem of determining whether a designated player has a winning
strategy from a given position, since a position may admit a winning
strategy without any bound on the number of moves required. This issue
is connected with transfinite game values in infinite chess, and the exact
value of the omega one of chess ω_1^{chess} is not known.

Infinite chess is chess played on an infinite edgeless chess board, arranged like the
integer lattice $\mathbb{Z} \times \mathbb{Z}$. The familiar chess pieces—kings, queens, bishops, knights,
rooks and pawns—move about according to their usual chess rules, with bishops
on diagonals, rooks on ranks and files and so on, with each player striving to place
the opposing king into checkmate. There is no standard starting configuration in
infinite chess, but rather a game proceeds by setting up a particular position on

S.B. Cooper, A. Dawar, and B. Löwe (Eds.): CiE 2012, LNCS 7318, pp. 78–88, 2012.
© Springer-Verlag Berlin Heidelberg 2012

the board and then playing from that position. In this article, we shall consider
only finite positions, with finitely many pieces; nevertheless, the game is sensible for positions with infinitely many pieces. We came to the problem through
Richard Stanley's question on mathoverflow.net [6].

The *mate-in-n* problem of infinite chess is the problem of determining for
a given finite position whether a designated player can force a win from that
position in at most n moves, counting the moves only of the designated player.
For example, figure 1 exhibits an instance of the mate-in-12 problem, adapted
from a position appearing in [2]. We provide a solution at the article's end.

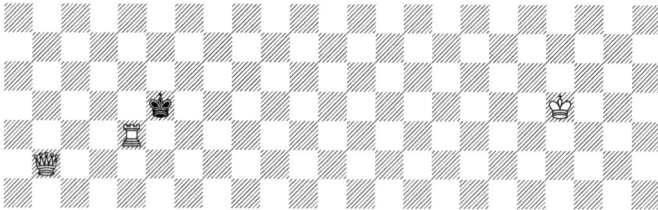

Fig. 1. White to move on an infinite, edgeless board. Can white force mate in 12 moves?

A naive formulation of the mate-in-n problem yields arithmetical assertions
of high arithmetic complexity, with $2n$ alternating quantifiers: *there is a white
move, such that for any black reply, there is a counter-play by white*, and so on.
In such a formulation, the problem does not appear to be decidable. One cannot
expect to search an infinitely branching game tree even to finite depth. Nevertheless, the second author of this paper and C. D. A. Evans conjectured that
it should be decidable anyway, after observing that this was the case for small
values of n, and the main theorem of this paper is to confirm this conjecture.

Before proceeding, let us clarify a few of the rules as they relate to infinite
chess. A position should have at most one king of each color. There is no rule
for pawn promotion, as there is no boundary for the pawns to attain. In infinite
chess, we abandon as limiting the 50-move rule (asserting that 50 moves without
a capture or pawn movement is a draw). A play of the game with infinitely
many moves is a draw. We may therefore abandon the three-fold repetition rule,
since any repetition could be repeated endlessly and thus attain a draw, if both
players desire this, and if not, then it needn't have been repeated. This does
not affect the answer to any instance of the mate-in-n problem, because if the
opposing player can avoid mate-in-n only because of a repetition, then this is
optimal play anyway. Since we have no official starting rank of pawns, we also
abandon the usual initial two-step pawn movement rule and the accompanying
en passant rule. Similarly, there is no castling in infinite chess.

Chess on arbitrarily large finite boards has been studied, for example in [3],
but the winning positions of infinite chess are not simply those that win on
all sufficiently large finite boards. For example, two connected white rooks can
checkmate a black king on any large finite board, but these pieces have no

checkmate position on an infinite board. The lack of edges in infinite chess seems to make it fundamentally different from large finite chess, although there may be a link between large finite chess and infinite chess on a quarter-infinite board.

Since infinite chess proceeds from an arbitrary position, there are a few additional weird boundary cases that do not arise in ordinary chess. For example, we may consider positions in which one of the players has no king; such a player could never herself be checkmated, but may still hope to checkmate her opponent, if her opponent should have a king. We shall not consider positions in which a player has two or more kings, although one might adopt rules to cover this case. In ordinary chess, one may construe the checkmate winning condition in terms of the necessary possibility of capturing the king. That is, we may imagine a version of chess in which the goal is simply to capture the opposing king, with the additional rules that one must do so if possible and one must prevent this possibility for the opponent on the next move, if this is possible. Such a conception of the game explains why we regard it as illegal to leave one's own king in check and why one may still checkmate one's opponent with pieces that are pinned to one's own king,[1] and gives rise to exactly the same outcomes for the positions of ordinary chess, with the extra king-capture move simply omitted. This conception directs the resolution of certain other initial positions: for example, a position with white to move and black in check is already won for white, even if white is in checkmate.

The *stalemate-in-n* problem is the problem of determining, for a given finite position whether a designated player can force a stalemate (or a win) in at most n moves, meaning a position from which the player to move has no legal moves. The *draw-in-n-by-k-repetition* problem is the problem of determining for a given finite position whether a designated player can in at most n moves force a win or force the opponent into a family of k positions, inside of which he may force his opponent perpetually to remain. The *winning-position* problem is the problem of determining whether a designated player has a winning strategy from a given finite position. The *drawn-position* problem is the problem of determining whether a designated player may force a draw or win from a given finite position.

Main Theorem 1. *The mate-in-n problem of infinite chess is decidable.*

1. *Moreover, there is a computable strategy for optimal play from a mate-in-n position to achieve the win in the fewest number of moves.*
2. *Similarly, there is a computable strategy for optimal opposing play from a mate-in-n position, to delay checkmate as long as possible. Indeed, there is a computable strategy to enable any player to avoid checkmate for k moves from a given position, if this is possible.*
3. *In addition, the stalemate-in-n and draw-in-n-by-k-repetition problems are also decidable, with computable strategies giving optimal play.*

---

[1] Even in ordinary finite chess one might argue that a checkmate position with pinned pieces should count merely as a draw, because once the opposing king is captured, one's own king will be killed immediately after; but we do not advance this argument.

Allow us briefly to outline our proof method. We shall describe a certain first order structure $\mathfrak{Ch}$, the structure of chess, whose objects consist of all the various legal finite positions of infinite chess, with various predicates to indicate whether a position shows a king in check or whether one position is reachable from another by a legal move. The mate-in-n problem is easily expressed as a $\Sigma_{2n} \vee \Pi_{2n}$ assertion in the first-order language of this structure, the language of chess. In general, of course, there is little reason to expect such complicated assertions in a first-order structure to be decidable, as generally even the Σ_1 assertions about an infinite computable structure are merely computably enumerable and not necessarily decidable. To surmount this fundamental difficulty, the key idea of our argument is to prove that for any fixed finite collection A of pieces, the reduct substructure $\mathfrak{Ch}_A$ of positions using only at most the pieces of A is not only computable, but in the restricted language of chess is an *automatic* structure, meaning that the domain of objects can be represented as strings forming a regular language, with the predicates of the restricted language of chess also being regular. Furthermore, the mate-in-n problem for a given position has the same truth value in $\mathfrak{Ch}$ as it does in $\mathfrak{Ch}_A$ and is expressible in the restricted language of chess. We may then appeal to the decidability theorem of automatic structures [5,1], which asserts that the first order theory of any automatic structure is decidable, to conclude that the mate-in-n problem is decidable. An alternative proof proceeds by arguing that $\mathfrak{Ch}_A$ is interpretable in Presburger arithmetic $\langle \mathbb{N}, + \rangle$, and this also implies that the theory of chess with pieces from A is decidable. The same argument methods apply to many other decision problems of infinite chess. Despite this, the method does not seem to generalize to positions of infinite game value—positions from which white can force a win, but which have no uniform bound on the number of moves it will take—and as a result, it remains open whether the general winning-position problem of infinite chess is decidable. Indeed, we do not know even whether the winning-position problem is arithmetic or even whether it is hyperarithmetic.

One might begin to see that the mate-in-n problem is decidable by observing that for mate-in-1, one needn't search through all possible moves, because if a distant move by a long-range piece gives checkmate, then all sufficiently distant similar moves by that piece also give checkmate. Ultimately, the key aspects of chess on which our argument relies are: (i) no new pieces enter the board during play, and (ii) the distance pieces—bishops, rooks and queens—move on straight lines whose equations can be expressed using only addition. Thus, the structure of chess is closer to Presburger than to Peano arithmetic, and this ultimately is what allows the theory to be decidable.

Let us now describe the first-order structure of chess $\mathfrak{Ch}$. Informally, the objects of this structure are all the various finite positions of infinite chess, each containing all the information necessary to set-up and commence play with an instance of infinite chess, namely, a finite listing of pieces, their types, their locations and whether they are still in play or captured, and an indication of whose turn it is. Specifically, define that a *piece designation* is a tuple $\langle i, j, x, y \rangle$, where $i \in \{\mathrm{K}, \mathrm{k}, \mathrm{Q}, \mathrm{q}, \mathrm{B}, \mathrm{b}, \mathrm{N}, \mathrm{n}, \mathrm{R}, \mathrm{r}, \mathrm{P}, \mathrm{p}\}$ is the piece type, standing for king,

queen, bishop, knight, rook or pawn, with upper case for white and lower case
for black; j is a binary indicator of whether the piece is in play or captured; and
$(x, y) \in \mathbb{Z} \times \mathbb{Z}$ are the coordinates of the location of the piece on the chessboard
(or a default value when the piece is captured). A finite *position* is simply a
finite sequence of piece designations, containing at most one king of each color
and with no two live pieces occupying the same square on the board, together
with a binary indicator of whose turn it is to play next. One could easily define
an equivalence relation on the positions for when they correspond to the same
set-up on the board—for example, extra captured bishops do not matter, and
neither does permuting the piece designations—but we shall actually have no
need for that quotient. Let us denote by Ch the set of all finite positions of
infinite chess. This is the domain of what we call the structure of chess $\mathfrak{Ch}$.

We shall next place predicates and relations on this structure in order to indicate various chess features of the positions. For example, WhiteToPlay(p) holds
when position p indicates that it is white's turn, and similarly for BlackToPlay(p).
The relation OneMove(p, q) holds when there is a legal move transforming position p into position q. We adopt the pedantic formalism for this relation that the
representation of the pieces in p and q is respected: the order of listing the pieces
is preserved and captured pieces do not disappear, but are marked as captured.
The relation BlackInCheck(p) holds when p shows the black king to be in check,
and similarly for WhiteInCheck(p). We define BlackMated(p) to hold when it is
black's turn to play in p, black is in check, but black has no legal move; the dual
relation WhiteMated(p) when it is white to play, white is in check, but has no
legal move. Similarly, BlackStalemated(p) holds when it is black's turn to play,
black is not in check, but black has no legal move; and similarly for the dual
WhiteStalemated(p). The *structure of chess* $\mathfrak{Ch}$ is the first-order structure with
domain Ch and with all the relations we have mentioned here. The language is
partly redundant, in that several of the predicates are definable from the others,
so let us refer to the language with only the relations WhiteToPlay, OneMove,
BlackInCheck and WhiteInCheck as the restricted language of chess. Later, we
shall also consider expansions of the language.

Since the OneMove(p, q) relation respects the order in which the pieces are
enumerated, the structure of chess $\mathfrak{Ch}$ naturally breaks into the disjoint components $\mathfrak{Ch}_A$, consisting of those positions whose pieces come from a *piece specification* type A, that is, a finite list of chess-piece types. For example, $\mathfrak{Ch}_{KQQkb}$
consists of the chess positions corresponding to the white king and two queens
versus black king and one bishop problems, enumerated in the KQQkb order,
with perhaps some of these pieces already captured. Since there is no pawn promotion in infinite chess or any other way to introduce new pieces to the board
during play, any game of infinite chess beginning from a position with piece specification A continues to have piece specification A, and so chess questions about
a position p with type A are answerable in the substructure $\mathfrak{Ch}_A$. We consider
the structure $\mathfrak{Ch}_A$ to have only the restricted language of chess, and so it is a
reduct substructure rather than a substructure of $\mathfrak{Ch}$.

We claim that the mate-in-n problem of infinite chess is expressible in the structure of chess $\mathfrak{Ch}$. Specifically, for any finite list A of chess-piece types, there are assertions $\mathrm{WhiteWins}_n(p)$ and $\mathrm{BlackWins}_n(p)$ in the restricted language of chess, such that for any position p of type A, the assertions are true in $\mathfrak{Ch}_A$ if and only if that player has a strategy to force a win from position p in at most n moves. This can be seen by a simple inductive argument. For the $n = 0$ case, and using the boundary-case conventions we mentioned earlier, we define $\mathrm{WhiteWins}_0(p)$ if it is black to play, black is in checkmate and white is not in check, or it is white to play and black is in check. Next, we recursively define $\mathrm{WhiteWins}_{n+1}(p)$ if either white can win in n moves, or it is white's turn to play and white can play to a position from which white can win in at most n moves, or it is black's turn to play and black indeed has a move (so it is not stalemate), but no matter how black plays, white can play to a position from which white can win in at most n moves. It is clear by induction that these assertions exactly express the required winning conditions and have complexity $\Sigma_{2n} \vee \Pi_{2n}$ in the language of chess (since perhaps the position has the opposing player going first), or complexity $\Sigma_{2n+1} \vee \Pi_{2n+1}$ in the restricted language of chess, since to define the checkmate condition from the in-check relation adds an additional quantifier.

Lemma 1. *For any finite list A of chess-piece types, the structure $\mathfrak{Ch}_A$ is automatic.*

Proof. This crucial lemma is the heart of our argument. What it means for a structure to be automatic is that it can be represented as a collection of strings forming a regular language, and with the predicates also forming regular languages. The functions are replaced with their graphs and handled as predicates. We refer the reader to [4] for further information about automatic structures. We shall use the characterization of regular languages as those consisting of the set of strings that are acceptable to a read-only Turing machine. We shall represent integers x and y with their signed binary representation, consisting of a sign bit, followed by the binary representation of the absolute value of the integer. Addition and subtraction are recognized by read-only Turing machines.

Let us discuss how we shall represent the positions as strings of symbols. Fix the finite list A of chess-piece types, and consider all positions p of type A. Let N be the length of A, that is, the number of pieces appearing in such positions p. We shall represent the position p using $3N + 1$ many strings, with end-of-string markers and padding to make them have equal length. One of the strings shall be used for the turn indicator. The remaining $3N$ strings shall represent each of the pieces, using three strings for each piece. Recall that each piece designation of p consists of a tuple $\langle i, j, x, y \rangle$, where i is the piece type, j is a binary indicator for whether the piece is in play or captured and (x, y) are the coordinates of the location of the piece in $\mathbb{Z} \times \mathbb{Z}$. Since we have fixed the list A of piece types, we no longer need the indicator i, since the kth piece will have the type of the kth symbol of A. We represent j with a string consisting of a single bit, and we represent the integers x and y with their signed binary representation. Thus, the position altogether consists of $3N+1$ many strings. Officially, one pads the strings to the same length and interleaves them together into one long string, although for regularity one may skip this interleaving step and simply work with the strings on separate tapes of a multi-tape machine. Thus, every position $p \in \mathrm{Ch}_A$ is represented by a unique sequence of $3N + 1$ many strings. We now argue that the collection of such sequences of strings arising from positions is regular, by recognizing with a read-only multi-tape Turing machine that they obey the necessary requirements: the turn indicator and the alive/captured indicators are just one bit long; the binary representation of the locations has the proper form, with no leading zeros (except for the number 0 itself); the captured pieces display the correct default location information; and no two live pieces occupy the same square. If all these tests are passed, then the input does indeed represent a position in $\mathfrak{Ch}$.

Next, we argue that the various relations in the language of chess are regular. For this, it will be helpful to introduce a few more predicates on the collection of strings. From the string representing a position p, we can directly extract from it the string representing the location of the ith piece in that position. Similarly, the coding of whose turn it is to play is explicitly given in the representation, so the relations $\mathrm{WhiteToPlay}(p)$ and $\mathrm{BlackToPlay}(p)$ are regular languages.

Consider the relation $\mathrm{Attack}_i(p, x, y)$, which holds when the ith piece of the position represented by p is attacking the square located at (x, y), represented with signed binary notation, where by attack we mean that piece i could move so as to capture an opposing piece on that square, irrespective of whether there is currently a friendly piece occupying that square or whether piece i could not in fact legally move to that square because of a pin condition. We claim that the Attack relation is a regular language. It suffices to consider the various piece types in turn, once we know the piece is alive. If the ith piece is a king, we check that its location in p is adjacent to (x, y), which is a regular language because $\{ (c, d) \mid |c - d| = 1 \}$, where c and d are binary sequences representing signed integers, is regular. Similarly, the attack relation in the case of pawns and knights is also easily recognizable by a read-only Turing machine. Note that for bishops, rooks and queens, other pieces may obstruct an attack. Bishops move on diagonals, and two locations (x_0, y_0) and (x_1, y_1) lie on the same diagonal if and only if they have the same sum $x_0 + y_0 = x_1 + y_1$ or the same difference $x_0 - y_0 = x_1 - y_1$, which is equivalent to $x_0 + y_1 = x_1 + y_0$, and since signed binary addition is regular, this is a regular condition. When two locations (x_0, y_0) and (x_1, y_1) lie on the same diagonal, then a third location (a, b) obstructs the line connecting them if and only if it also has that sum $a + b = x_0 + y_0 = x_1 + y_1$ or difference $a - b = x_0 - y_0 = x_1 - y_1$, and also has $x_0 < a < x_1$ or $x_1 < a < x_0$. This is a regular condition on the six variables $(x_0, y_0, x_1, y_1, a, b)$, because it can be verified by a read-only Turing machine. The order relation $x < y$ is a regular requirement on pairs (x, y), because one can check it by looking at the signs and the first bit of difference. So the attack relation for bishops is regular. Rooks move parallel to the coordinate axes, and so a rook at (x_0, y_0) attacks the square at (x_1, y_1), if it is alive and these two squares are different but lie either on the same rank $y_0 = y_1$, or on the same file $x_0 = x_1$, and there is no obstructing piece. This is clearly a condition that can be checked by a read-only Turing machine, and so it is a regular requirement. Finally, the attack relation for queens is regular, since it reduces to the bishop and rook attack relations.

It follows now that the relation WhiteInCheck(p), which holds when the position shows the white king in check, is also regular. With a read-only Turing machine we can clearly recognize whether the white king is indicated as alive in p, and then we simply take the disjunction over each of the black pieces listed in A, as to whether that piece attacks the location square of the white king. Since the regular languages are closed under finite unions, it follows that this relation is regular. Similarly the dual relation BlackInCheck(p) is regular.

Consider now the Move$_i(p, x, y)$ relation, similar to the attack relation, but which holds when the ith piece in position p is alive and may legally move to the square (x, y). It should be the correct player's turn; there should be no obstructing pieces; the square at (x, y) should not be occupied by any friendly piece; and the resulting position should not leave the moving player in check. Each of these conditions can be verified as above. A minor complication is presented by the case of pawn movement, since pawns move differently when capturing than when not capturing, and so for pawns one must check whether there is an opposing piece at (x, y) if the pawn should move via capture.

Consider next the relation OneMove$_i(p, q)$, which holds when position p is transformed to position q by a legal move of piece i. With a read-only Turing machine, we can easily check that position p indicates that it is the correct player's turn, and by the relations above, that piece i may legally move to its location in q, that any opposing piece occupying that square in p is marked as captured in q, and that none of the other pieces of p are moved or have their capture status modified in q. Thus, this relation is a regular language.

Finally, we consider the relation OneMove(p, q) in $\mathfrak{Ch}_A$, which holds when position p is transformed to position q by a single legal move. This is simply the disjunction of OneMove$_i(p, q)$ for each piece i in A, and is therefore regular, since the regular languages are closed under finite unions.

Thus, we have established that the domain of $\mathfrak{Ch}_A$ is regular and all the predicates in the restricted language of chess are regular, and so the structure $\mathfrak{Ch}_A$ is automatic, establishing the lemma.   ∎

We now complete the proof of the main theorem. Since the structure $\mathfrak{Ch}_A$ is automatic, it follows by the decidability theorem of automatic structures [5,1] that the first-order theory of this structure is uniformly decidable. In particular, since the mate-in-n question is expressible in this structure—and we may freely add constant parameters—it follows that the mate-in-n question is uniformly decidable: there is a computable algorithm, which given a position p and a natural number n, determines yes-or-no whether a designated player can force a win in at most n moves from position p. Furthermore, in the case that the position p is mate-in-n for the designated player, then there is a computable strategy providing optimal play: the designated player need only first find the smallest value of n for which the position is mate-in-n, and then search for any move leading to a mate-in-$(n-1)$ position. This value-reducing strategy achieves victory in the fewest possible number of moves from any finite-value position. Conversely, there is a computable strategy for the losing player from a mate-in-n position to avoid inadvertantly reducing the value on a given move, and thereby delay the checkmate as long as possible. Indeed, if a given position p is not mate-in-n, then we may computably find moves for the opposing player that do not inadvertantly result in a mate-in-n position. Finally, we observe that the stalemate-in-n and draw-in-n-by-k-repetition problems are similarly expressible in the structure $\mathfrak{Ch}_A$, and so these are also decidable and admit computable strategies to carry them out. This completes the proof of the main theorem.   ∎

An essentially similar argument shows that the structure of chess $\mathfrak{Ch}_A$, for any piece specification A, is definable in Presburger arithmetic $\langle \mathbb{N}, + \rangle$. Specifically, one codes a position with a fixed length sequence of natural numbers, where each piece is represented by a sequence of numbers indicating its type, whether it is still in play, and its location (using extra numbers for the sign bits). The point is that the details of our arguments in the proof of the main theorem show that the attack relation and the other relations of the structure of chess are each definable from this information in Presburger arithmetic. Since Presburger arithmetic is decidable, it follows that the theory of $\mathfrak{Ch}_A$ is also decidable.

We should like to emphasize that our main theorem does not appear to settle the decidability of the winning-position problem, the problem of determining whether a designated player has a winning strategy from a given position. The point is that a player may have a winning strategy from a position, without there being any finite bound on the number of moves required. Black might be able to delay checkmate any desired finite amount, even if every play ends in his loss, and there are positions known to be winning but not mate-in-n for any n. These are precisely the positions with infinite game value in the recursive assignment of ordinal values to winning positions: already-won positions for white have value 0; if a position is white-to-play, the value is the minimum of the values of the positions to which white may play, plus one; if it is black-to-play, and all legal plays have a value, then the value is the supremum of these values. The winning positions are precisely those with an ordinal value, and this value is a measure of the distance to a win. A mate-in-n position, with n minimal, has value n. A position with value ω has black to play, but any move by black will be mate-in-n for white for some n, and these are unbounded. The omega one of chess, denoted ω_1^{chess}, is defined in [2] to be the supremum of the values of the finite positions of infinite chess. The exact value of this ordinal is an open question, although an argument of Blass appearing in [2] establishes that $\omega_1^{\text{chess}} \leq \omega_1^{ck}$, as well as the accompanying fact that if a player can win from position p, then there is a winning strategy of hyperarithmetic complexity. Although we have proved that the mate-in-n problem is decidable, we conjecture that the general winning-position problem is undecidable and indeed, not even arithmetic.

Consider briefly the case of three-dimensional infinite chess, as well as higher-dimensional infinite chess. Variants of three-dimensional (finite) chess arose in the late nineteenth century and have natural infinitary analogues. Without elaborating on the details—there are various reasonable but incompatible rules for piece movement—we remark that the method of proof of our main theorem works equally well in higher dimensions.

Corollary 1. *The mate-in-n problem for k-dimensional infinite chess is decidable.*

Results in [2] establish that the omega one of infinite positions in three-dimensional infinite chess is exactly true ω_1; that is, every countable ordinal arises as the game value of an infinite position of three-dimensional infinite chess.

We conclude the article with a solution to the chess problem we posed in figure 1. With white Qe5+, the black king is forced to the right kg4, and then white proceeds in combination Rg3+ kh4 Qg5+ ki4, rolling the black king towards the white king, where checkmate is delivered on white's 13$^\text{th}$ move. Alternatively, after Qe5+ kg4, white may instead play Ks4, moving his king to the left, forcing the two kings together from that side, and this also leads to checkmate by the queen on the 13th move. It is not possible for white to checkmate in 12 moves, because if the two kings do not share an adjacent square, there is no checkmate position with a king, queen and rook versus a king. Thus, white must force the two kings together, and this will take at least 12 moves, after which the checkmate move can now be delivered, meaning at least 13 moves.[2] However, white can force a stalemate in 12 moves, by moving Qe5+, and then afterwards moving only the white king towards the black king, achieving stalemate on the 12th move, as the black king is trapped on f4 with no legal move. White can force a draw by repetition in 3 moves, by trapping the black king in a 4×4 box with the white queen and rook at opposite corners, via Qe5+ kg4 Ri3 kh4 Qf6+, which then constrains the black king to two squares.

Acknowledgements. The research of the second author has been supported in part by grants from the National Science Foundation, the Simons Foundation and the CUNY Research Foundation.

References

1. Blumensath, A., Grädel, E.: Automatic structures. In: 15th Annual IEEE Symposium on Logic in Computer Science, Santa Barbara, CA, pp. 51–62. IEEE Comput. Soc. Press, Los Alamitos (2000), http://dx.doi.org/10.1109/LICS.2000.855755
2. Evans, C.D., Hamkins, J.D., Woodin, W.H.: Transfinite game values in infinite chess (in preparation)
3. Fraenkel, A.S., Lichtenstein, D.: Computing a perfect strategy for $n \times n$ chess requires time exponential in n. J. Combin. Theory Ser. A 31(2), 199–214 (1981), http://dx.doi.org/10.1016/0097-3165(81)90016-9

---

[2] C. D. A. Evans (US national master) confirms this mate-in-13-but-not-12 analysis.

4. Khoussainov, B., Minnes, M.: Three lectures on automatic structures. In: Logic Colloquium 2007. Lect. Notes Log, vol. 35, pp. 132–176. Assoc. Symbol. Logic, La Jolla (2010)
5. Khoussainov, B., Nerode, A.: Automatic Presentations of Structures. In: Leivant, D. (ed.) LCC 1994. LNCS, vol. 960, pp. 367–392. Springer, Heidelberg (1995)
6. Stanley (mathoverflow.net/users/2807), R.: Decidability of chess on an infinite board. MathOverflow, `http://mathoverflow.net/questions/27967` (version: July 20, 2010)

A Note on Ramsey Theorems and Turing Jumps

Lorenzo Carlucci[1] and Konrad Zdanowski[2]

[1] Dipartimento di Informatica, Università di Roma Sapienza, Via Salaria 113,
00198 Roma, Italy
carlucci@di.uniroma1.it

[2] Uniwersytet Kardynała Stefana Wyszyńskiego w Warszawie, ul. Dewajtis 5,
01-815 Warszawa, Poland
k.zdanowski@uksw.edu.pl

Abstract. We give a new treatment of the relations between Ramsey's Theorem, $\mathbf{ACA}_0$ and $\mathbf{ACA}_0'$. First we combine a result by Girard with a colouring used by Loebl and Nešetril for the analysis of the Paris-Harrington principle to obtain a short combinatorial proof of $\mathbf{ACA}_0$ from Ramsey Theorem for triples. We then extend this approach to $\mathbf{ACA}_0'$ using a characterization of this system in terms of preservation of well-orderings due to Marcone and Montalbán. We finally discuss how to apply this method to $\mathbf{ACA}_0^+$ using an extension of Ramsey's Theorem for colouring relatively large sets due to Pudlàk and Rödl and independently to Farmaki.

1 Introduction

The proof-theoretic and computability-theoretic strength of Ramsey Theorem have been intensively studied. By Ramsey's Theorem for n-tuples and c colours we here mean the assertion that every colouring of the n-tuples of $\mathbf{N}$ in c colours admits an infinite monochromatic set. We refer to this principle by $\mathbf{RT}_c^n$. In the context of Reverse Mathematics [12], a full characterization is known for the Infinite Ramsey Theorem for n-tuples with $n \geq 3$. In particular, it is known that for every $n \geq 3$, $\mathbf{RT}_2^3$ is equivalent to $\mathbf{ACA}_0$ (i.e., to $\mathbf{RCA}_0$ augmented by the assertion that the Turing jump of any set exists) and that $\forall n \mathbf{RT}_2^n$ is equivalent to $\mathbf{ACA}_0'$ (i.e., $\mathbf{RCA}_0$ augmented by the assertion that for all n the n-th Turing jump of any set exists).

By a *well-ordering preservation principle* we mean an assertion of the form $\forall \mathcal{X}(\mathsf{WO}(\mathcal{X}) \to \mathsf{WO}(\mathcal{F}(\mathcal{X})))$, where $\mathcal{X}$ is a linear order, $\mathsf{WO}(\mathcal{F})$ is the Π_1^1 sentence expressing that $\mathcal{X}$ is a well-ordering, and $\mathcal{F}$ is an operator from linear orders to linear orders. An old result of Girard [4] shows that $\mathbf{ACA}_0$ is equivalent to the well-ordering preservation principle $\forall \mathcal{X}(\mathsf{WO}(\mathcal{X}) \to \mathsf{WO}(2^{\mathcal{X}}))$, or, equivalently, to $\forall \mathcal{X}(\mathsf{WO}(\mathcal{X}) \to \mathsf{WO}(\omega^{\mathcal{X}}))$, where $2^{\mathcal{X}}$ and $\omega^{\mathcal{X}}$ have the expected meaning. Well-ordering preservation principles have recently attracted new interest. Analogues of Girard's result for systems stronger that $\mathbf{ACA}_0$ have been recently obtained by Afshari and Rathjen [1] and independently by Marcone and Montalbán [9] using proof-theoretic and computability-theoretic methods, respectively. The systems $\mathbf{ACA}_0'$, $\mathbf{ACA}_0^+$, $\Pi_{\omega^\alpha}^0\text{-}\mathbf{CA}_0$ and $\mathbf{ATR}_0$ have been proved equivalent to well-ordering preservation principles of increasing strength.

S.B. Cooper, A. Dawar, and B. Löwe (Eds.): CiE 2012, LNCS 7318, pp. 89–95, 2012.

In this note we outline a combinatorial method for obtaining implications of the form "This type of Ramsey Theorem implies closure under this type of Turing jump over $\mathbf{RCA}_0$". In particular we combine Girard's characterization of $\mathbf{ACA}_0$ and Marcone-Montalbàn's characterization of $\mathbf{ACA}_0'$ with some partitions defined by Loebl and Nešetril in their beautiful combinatorial proof of the independence of Paris-Harrington Theorem from Peano Arithmetic [8] to obtain new proofs of the following results.

$$\mathbf{RCA}_0 \vdash \mathbf{RT}_2^4 \to \mathbf{ACA}_0, \text{ and } \mathbf{RCA}_0 \vdash \forall n \mathbf{RT}_2^n \to \mathbf{ACA}_0'.$$

Furthermore, we strongly conjecture that this approach can be extended to prove that a Ramsey-type theorem for bicolourings of exactly large sets due to Pudlàk-Rödl [11] and independently to Farmaki (see, e.g., [3]) implies $\mathbf{ACA}_0^+$ (i.e., equivalently, $\mathbf{RCA}_0$ augmented by the assertion that the ω Turing jump of any set exists). The effective and proof-theoretic content of this theorem have been recently fully analyzed by the authors in [2]. We conjecture that the method can be extended to relate the general version of the latter theorem [3] to the systems $\Pi_{\omega^\alpha}^0\text{-}\mathbf{CA}_0$, for $\alpha \in \omega_1^{\mathrm{CK}}$, using their characterization in [9] in terms of the well-ordering preservation principles $\forall \mathcal{X}(\mathsf{WO}(\mathcal{X}) \to \mathsf{WO}(\varphi(\alpha, \mathcal{X})))$.

2 $\mathbf{RT}^3$, $\mathbf{ACA}_0$, and ω-Exponentiation

We give a new proof of the fact that Ramsey Theorem for triples implies Arithmetical Comprehension (over $\mathbf{RCA}_0$). We use Girard's result [4,6,9] that $\forall \mathcal{X}(\mathsf{WO}(\mathcal{X}) \to \mathsf{WO}(\omega^\mathcal{X}))$ implies $\mathbf{ACA}_0$ over $\mathbf{RCA}_0$. We give a direct combinatorial argument showing that Ramsey Theorem for triples implies $\forall \mathcal{X}(\mathsf{WO}(\mathcal{X}) \to \mathsf{WO}(\omega^\mathcal{X}))$. Surprisingly, the needed colourings come from Loebl and Nešetril's [8] combinatorial proof of the unprovability of the Paris-Harrington principle from Peano Arithmetic.

For linear orders we use the same notations as in [9]. In particular, we define the following operator $\omega^\mathcal{X}$ from linear orders to linear orders. Given a linear order $\mathcal{X}$, $\omega^\mathcal{X}$ is the set of finite strings $\langle x_0, x_1, \ldots, x_{k-1} \rangle$ of elements of $\mathcal{X}$ where $x_0 \geq_\mathcal{X} x_1 \geq_\mathcal{X} \cdots \geq_\mathcal{X} x_{k-1}$. The intended meaning of $\langle x_0, x_1, \ldots, x_{k-1} \rangle$ is $\omega^{x_0} + \cdots + \omega^{x_{k-1}}$. The order $\leq_{\omega^\mathcal{X}}$ on $\omega^\mathcal{X}$ is the lexicographic order. If $\alpha = \omega^{\alpha_0} n_0 + \cdots + \omega^{\alpha_k} n_k$ in Cantor Normal Form, we denote n_i by $c_i(\alpha)$ and α_i by $e_i(\alpha)$. If $\alpha > \beta$ we denote by $\Delta(\alpha, \beta)$ the least index at which the Cantor Normal Forms of α and β differ.

Proposition 1. *Over* $\mathbf{RCA}_0$, $\mathbf{RT}_3^3$ *implies* $\forall \mathcal{X}(\mathsf{WO}(\mathcal{X}) \to \mathsf{WO}(\omega^\mathcal{X}))$.

Proof. Assume $\mathbf{RT}_3^3$. Suppose $\neg\mathsf{WO}(\omega^\mathcal{X})$. We show $\neg\mathsf{WO}(\mathcal{X})$. We define a colouring $C^{(\alpha)} : [\mathbf{N}]^3 \to 3$ with an explicit sequence parameter α of intended type $\alpha : \mathbf{N} \to field(\omega^\mathcal{X})$. $C^{(\alpha)}$ (with parameter α) is defined by the following case distinction.

$$C^{(\alpha)}(i,j,k) = \begin{cases} 1 & \text{if } \Delta(\alpha_i, \alpha_j) > \Delta(\alpha_j, \alpha_k), \\ 2 & \text{if } \Delta(\alpha_i, \alpha_j) \leq \Delta(\alpha_j, \alpha_k) \wedge c_{\Delta(\alpha_i,\alpha_j)}(\alpha_i) > c_{\Delta(\alpha_j,\alpha_k)}(\alpha_j), \\ 3 & \text{if } \Delta(\alpha_i, \alpha_j) \leq \Delta(\alpha_j, \alpha_k) \wedge c_{\Delta(\alpha_i,\alpha_j)}(\alpha_i) \leq c_{\Delta(\alpha_j,\alpha_k)}(\alpha_j). \end{cases}$$

Let $\alpha : \mathbf{N} \to field(\omega^{\mathcal{X}})$ be an infinite descending sequence in $\omega^{\mathcal{X}}$. Let H be an infinite $C^{(\alpha)}$-homogeneous set. Consider $(\beta_i)_{i \in \mathbf{N}}$, where $\beta_i = \alpha_{h_i}$ and $H = \{h_1, h_2, \dots\}$ in increasing order.

Case 1. The colour of $C^{(\alpha)}$ on $[H]^3$ is 1. Then

$$\Delta(\beta_1, \beta_2) > \Delta(\beta_2, \beta_3) > \dots$$

Contradiction, since $\Delta(\beta_i, \beta_{i+1}) \in \mathbf{N}$.

Case 2. The colour of $C^{(\alpha)}$ on $[H]^3$ is 2. Then

$$c_{\Delta(\beta_1, \beta_2)} > c_{\Delta(\beta_2, \beta_3)} > \dots$$

Contradiction, since $c_{\Delta(\beta_i, \beta_{i+1})} \in \mathbf{N}$.

Case 3. The colour of $C^{(\alpha)}$ on $[H]^3$ is 3. If $\Delta(\beta_i, \beta_j) = \Delta(\beta_j, \beta_k)$, since $c_{\Delta(\beta_i, \beta_j)}(\beta_i) \leq c_{\Delta(\beta_j, \beta_k)}(\beta_j)$, and $\beta_i > \beta_j$, it must be the case that $e_{\Delta(\beta_i, \beta_j)}(\beta_i) > e_{\Delta(\beta_j, \beta_k)}(\beta_j)$.

If $\Delta(\beta_i, \beta_j) < \Delta(\beta_j, \beta_k)$ then $e_{\Delta(\beta_i, \beta_j)}(\beta_i) > e_{\Delta(\beta_j, \beta_k)}(\beta_j)$, since $\beta_i > \beta_j > \beta_k$. Thus, in any case

$$e_{\Delta(\beta_1, \beta_2)}(\beta_1) > e_{\Delta(\beta_2, \beta_3)}(\beta_2) > \dots.$$

In other words, $\alpha' : \mathbf{N} \to \mathcal{X}$ defined by

$$i \mapsto e_{\Delta(\alpha_{h_i}, \alpha_{h_{i+1}})}(\alpha_{h_i}),$$

is the desired infinite descending sequence in $\mathcal{X}$. □

Corollary 1. *Over* $\mathbf{RCA}_0$, $\mathbf{RT}_3^3$ *implies* $\mathbf{ACA}_0$.

Recall the following result of Jockusch's (a similar observation is made at the end of [8]).

Proposition 2 (Jockusch [7]). *For every* n, c, *for every recursive* $C : [\mathbf{N}]^n \to c$ *there exists a recursive* $C' : [\mathbf{N}]^{n+1} \to 2$ *such that the* C'-*homogeneous sets are just the* C-*homogeneous sets.*

Proof. For the proof it is sufficient to define C' as follows. For $S \in [\mathbf{N}]^{n+1}$, $C'(S) = 0$ if $[S]^n$ is C-homogeneous and $C'(S) = 1$ otherwise. □

Thus, restriction to two colours is inessential for the study of the strength of Ramsey Theorem.

Corollary 2. *Over* $\mathbf{RCA}_0$, $\mathbf{RT}_2^4$ *implies* $\mathbf{ACA}_0$.

3 $\forall n\mathbf{RT}^n$, $\mathbf{ACA}_0'$, and Iterated ω-Exponentiation

We give a new combinatorial proof of the fact that the full Ramsey Theorem $\forall n\mathbf{RT}^n$ implies closure under all finite Turing jumps (over $\mathbf{RCA}_0$). This result is originally due to McAloon [10].

We use Marcone and Montalbàn's [9] result that $\forall n \forall \mathcal{X}(\text{WO}(\mathcal{X}) \rightarrow \text{WO}(\omega^{\langle n, \mathcal{X} \rangle}))$ is equivalent to $\textbf{ACA}_0'$ over $\textbf{RCA}_0$. We give a direct combinatorial argument showing that Ramsey Theorem implies the latter well-ordering preservation principle. Again, the needed colourings come from [8].

We define, as in [9], $\omega^{\langle 0, \mathcal{X} \rangle} = \mathcal{X}$ and $\omega^{\langle n+1, \mathcal{X} \rangle} = \omega^{\omega^{\langle n, \mathcal{X} \rangle}}$.

Proposition 3. $(\textbf{RCA}_0)$ $\forall n \forall c \textbf{RT}_c^n$ *implies* $\forall n \forall \mathcal{X}(\text{WO}(\mathcal{X}) \rightarrow \text{WO}(\omega^{\langle n, \mathcal{X} \rangle})))$.

Proof. We define a family of colourings $C_h^{(\alpha)} : [\textbf{N}]^{h+2} \rightarrow d(h)$ for $h \geq 1$, where d is a primitive recursive function to be read off from the proof. The definitions of the colourings $C_h^{(\alpha)}$ feature an explicit infinite sequence parameter α of intended type $\alpha : \textbf{N} \rightarrow field(\omega^{\langle h, \mathcal{X} \rangle})$. $C_1^{(\alpha)}$ is the colouring $C^{(\alpha)} : [\textbf{N}]^3 \rightarrow 3$ defined in the proof of Proposition 1. We define $C_h^{(\alpha)}$ for $h \geq 2$ as follows. Given $i_1, \ldots, i_{h+2}$ let

$$v_1 = (C_1^{(\alpha)}(i_1, i_2, i_3), C_1^{(\alpha)}(i_2, i_3, i_4), \ldots, C_1^{(\alpha)}(i_{h-1}, i_h, i_{h+1})),$$

and

$$v_2 = (C_1^{(\alpha)}(i_2, i_3, i_4), C_1^{(\alpha)}(i_3, i_4, i_5), \ldots, C_1^{(\alpha)}(i_h, i_{h+1}, i_{h+2})).$$

$$C_h^{(\alpha)}(i_1, \ldots, i_{h+2}) = \begin{cases} (v_1, v_2) & \text{if } \neg(v_1 = v_2 = (3, \ldots, 3)), \\ C_{h-1}^{(\alpha[i_1, \ldots, i_{h+2}])}(i_1, \ldots, i_{h+1}) & \text{otherwise.} \end{cases}$$

where $\alpha[i_1, \ldots, i_{h+2}]$ is the sequence

$$\alpha_1, \ldots, \alpha_{i_1-1}, e_{\Delta(\alpha_{i_1}, \alpha_{i_2})}(\alpha_{i_1}), \ldots, e_{\Delta(\alpha_{i_{h+1}}, \alpha_{i_{h+2}})}(\alpha_{i_{h+1}}), \alpha_{i_{h+2}}, \alpha_{i_{h+2}+1}, \ldots$$

For well-definedness, we observe that in the second case the arguments are strictly decreasing in $\omega^{\langle h-1, \mathcal{X} \rangle}$.

When considering $C_h^{(\alpha)}$ for a fixed sequence α we sometimes write $C_h(\alpha_{i_1}, \ldots, \alpha_{i_{h+2}})$ for $C_h^{(\alpha)}(i_1, \ldots, i_{h+2})$.

Let h and $\alpha : \textbf{N} \rightarrow field(\omega^{\langle h, \mathcal{X} \rangle})$ be given such that α is strictly descending. Let H be an infinite $C_h^{(\alpha)}$-homogeneous set. For $h = 1$ we have already proved the theorem. Let $h \geq 2$. We show how to compute an $\mathcal{X}$-descending sequence given α and H. Let $\{s_1, s_2, \ldots, \}$ be an enumeration of H in increasing order.

We first argue that for every $j_1 < \cdots < j_{h+2}$, the corresponding values v_1 and v_2 relative to $\alpha_{s_{j_1}}, \ldots, \alpha_{s_{j_{h+2}}}$ satisfy $v_1 = v_2 = (3, \ldots, 3)$.

To exclude the other cases we argue as follows. Let $\beta_1, \ldots, \beta_{h+2}$ in the rest of the argument be arbitrary $\alpha_{i_1}, \ldots, \alpha_{i_{h+2}}$ with $\{i_1, \ldots, i_{h+2}\}_< \subset H$.

Case 1. The colour is (v_1, v_2) with $v_1 \neq v_2$. This is easily seen to be impossible. Let s be minimum such that v_1 and v_2 differ at position s for the first time. Then $C_1(\beta_i, \beta_{i+1}, \beta_{i+2}) = C_1(\beta_{i+1}, \beta_{i+2}, \beta_{i+3})$, for $i < s$ but $C_1(\beta_s, \beta_{s+1}, \beta_{s+2}) \neq C_1(\beta_{s+1}, \beta_{s+2}, \beta_{s+3})$. Now consider the $(h+2)$-tuple $(\beta_2, \beta_3, \ldots, \beta_{h+2}, \beta^*)$ where β^* is any choice of an element smaller than β_{h+2} in the infinite homogeneous set H. The first component of the vector associated to this tuple is v_2. But then $v_1 = v_2$ should hold by homogeneity.

Case 2. The colour is (v, v) for some v and $v \neq (3, 3, \ldots, 3)$.

Case 2.1. $v \neq (i, i, \ldots, i)$ for $i \in \{1, 2\}$. This is easily seen to be impossible. Let $s \in [1, h-1]$ be minimum such that $i = C_1(\beta_s, \beta_{s+1}, \beta_{s+2}) \neq C_1(\beta_{s+1}, \beta_{s+2}, \beta_{s+3}) = j$. Then $v_1 = (i, i, \ldots, i, j, \ldots)$. On the other hand $i = C_1(\beta_2, \beta_3, \beta_4) = \cdots = C_1(\beta_s, \beta_{s+1}, \beta_{s+2})$ and $v_2 = v_1$ imposes $C_1(\beta_{s+1}, \beta_{s+2}, \beta_{s+3}) = i$. Contradiction.

Case 2.2. If $v = (1, \ldots, 1)$ then the sequence $(\Delta(\beta_i, \beta_{i+1}))_{i \in \mathbf{N}}$ is strictly descending in $\mathbf{N}$.

Case 2.3. If $v = (2, \ldots, 2)$ then the sequence $(c_{\Delta(\beta_i, \beta_{i+1})})_{i \in \mathbf{N}}$ is strictly descending in $\mathbf{N}$.

Since Cases 1 and 2 cannot hold, we have the following.

$$(\exists b)(\forall j_1, \ldots, j_{h+2})[C_{h-1}^{\alpha[s_{j_1}, \ldots, s_{j_{h+2}}]}(s_{j_1}, \ldots, s_{j_{h+1}}) = b].$$

Therefore

$$(\forall j_1 < \cdots < j_{h+2})[e_{\Delta(\alpha_{s_{j_1}}, \alpha_{s_{j_2}})}(\alpha_{s_{j_1}}) > \cdots > e_{\Delta(\alpha_{s_{j_{h+1}}}, \alpha_{s_{j_{h+2}}})}(\alpha_{s_{j_{h+1}}})].$$

From this we can conclude that the sequence α' defined as follows

$$n \mapsto e_{\Delta(\alpha_{s_n}, \alpha_{s_{n+1}})}(\alpha_{s_n})$$

is infinite descending in $\omega^{\langle h-1, \mathcal{X}\rangle}$. Furthermore we claim that $C_{h-1}^{(\alpha')}$ is constant. Let $j_1 < \cdots < j_{h+1}$ and $i_1 < \cdots < i_{h+1}$ be arbitrary. We show that

$$C_{h-1}^{(\alpha')}(s_{j_1}, \ldots, s_{j_{h+1}}) = C_{h-1}^{(\alpha')}(s_{i_1}, \ldots, s_{i_{h+1}}).$$

The following chain of identities holds by $C_h^{(\alpha)}$-homogeneity of H and by definition of $C_h^{(\alpha)}$ and of α'.

$$
\begin{aligned}
C_{h-1}^{\alpha'}(j_1, \ldots, j_{h+1}) =\ & C_{h-1}(e_{\Delta(\alpha_{s_{j_1}}, \alpha_{s_{j_2}})}(\alpha_{s_{j_1}}), \ldots, e_{\Delta(\alpha_{s_{j_{h+1}}}, \alpha_{s_{j_{h+1}+1}})}(\alpha_{s_{j_{h+1}}})) \\
=\ & C_{h-1}^{\alpha[s_{j_1}, \ldots, s_{j_{h+1}}, s_{j_{h+1}+1}]}(s_{j_1}, \ldots, s_{j_{h+1}}) \\
=\ & C_h^{(\alpha)}(s_{j_1}, \ldots, s_{j_{h+1}}, s_{j_{h+1}+1}) \\
=\ & C_h^{(\alpha)}(s_{i_1}, \ldots, s_{i_{h+1}}, s_{i_{h+1}+1}) \\
=\ & C_{h-1}^{\alpha[s_{i_1}, \ldots, s_{i_{h+1}}, s_{i_{h+1}+1}]}(s_{i_1}, \ldots, s_{i_{h+1}}) \\
=\ & C_{h-1}(e_{\Delta(\alpha_{s_{i_1}}, \alpha_{s_{i_2}})}(\alpha_{s_{i_1}}), \ldots, e_{\Delta(\alpha_{s_{i_{h+1}}}, \alpha_{s_{i_{h+1}+1}})}(\alpha_{s_{i_{h+1}}})) \\
=\ & C_{h-1}^{(\alpha')}(s_{i_1}, \ldots, s_{i_{h+1}})
\end{aligned}
$$

Then, by iterating the above argument $(h-1)$ times we obtain the desired descending sequence in $\mathcal{X}$, computably in α and H. □

Corollary 3. *Over* $\mathbf{RCA}_0$ $\forall n\mathbf{RT}_2^n$ *implies* $\mathbf{ACA}_0'$.

Proof. We just need to replace each colouring $C_h^{(\alpha)} : [\mathbf{N}]^{h+2} \to d(h)$ used in the proof of Proposition 3 by a colouring $D_h^{(\alpha)} : [\mathbf{N}]^{h+3} \to 2$ such that $C_h^{(\alpha)}$ and $D_h^{(\alpha)}$ have the same homogeneous sets. This is possible by Proposition 2. □

4   Large Sets, $\mathbf{ACA_0^+}$, and the ε Function

We discuss how to apply the proof-technique from the previous sections to a natural extension of Ramsey Theorem for colouring relatively large sets (in the sense of Paris and Harrington [5]). A finite set $X \subseteq \mathbf{N}$ is *large* if $\mathrm{card}(X) > \min(X)$ and is *exactly large* if $\mathrm{card}(X) = \min(X) + 1$. The principle of interest is the following.

Theorem 1 (Pudlàk-Rödl [11] and Farmaki [3]). *For every infinite subset M of $\mathbf{N}$, for every colouring C of the exactly large subsets of $\mathbf{N}$ in two colours, there exists an infinite set $L \subseteq M$ such that every exactly large subset of L gets the same colour by C.*

We refer to the statement of the above Theorem as $\mathbf{RT}(!\omega)$. The effective and proof-theoretic content of $\mathbf{RT}(!\omega)$ has been recently characterized in [2]. In that paper it is shown that $\mathbf{RT}(!\omega)$ is equivalent to $\mathbf{ACA_0^+}$ over $\mathbf{RCA_0}$. We strongly conjecture that the technique from the previous sections can be applied to give a completely different proof of the implication $\mathbf{RCA_0} \vdash \mathbf{RT}(!\omega) \to \mathbf{ACA_0^+}$, using Marcone-Montalbàn's result that $\forall \mathcal{X}(\mathrm{WO}(\mathcal{X}) \to \mathrm{WO}(\varepsilon_{\mathcal{X}}))$ implies $\mathbf{ACA_0^+}$ over $\mathbf{RCA_0}$ (Theorem 5.23 in [9]).

We use the following notation from [9]. An ordering $\varepsilon_{\mathcal{X}}$ is defined from a linear order $\mathcal{X}$ as follows. The elements of the new ordering are the formal terms defined inductively as follows. (1) 0, and ε_x for $x \in \mathcal{X}$, (2) If t_1, t_2 are formal terms then $t_1 + t_2$ is a formal term. (3) If t is a formal term then ω^t is a formal term. By induction on terms we define the order relation $\leq_{\varepsilon_{\mathcal{X}}}$ and a normal form simultaneously as follow. A term $t = t_0 + \cdots + t_k$ is in normal form if either $t = 0$ (i.e., $k = 0$ and $t_0 = 0$) or the following points (A-B) hold. (A) $t_0 \geq_{\varepsilon_{\mathcal{X}}} t_1 \geq_{\varepsilon_{\mathcal{X}}} \cdots \geq_{\varepsilon_{\mathcal{X}}} t_k$, and (B) each t_i is either 0, ε_x with $x \in \mathcal{X}$ or ω^{s_i} where s_i is in normal form and $s_i \neq \varepsilon_x$ for any x. Every term t can be written in normal form using the following points (i-iv). (i) + is associative, (ii) $s + 0 = 0 + s = s$, (iii) If $s <_{\varepsilon_{\mathcal{X}}} r$ then $\omega^s + \omega^r = \omega^r$, (iv) $\omega^{\varepsilon_x} = \varepsilon_x$. Given $t = t_0 + \cdots + t_n$ and $s = s_0 + \cdots + s_m$ in normal form, $t \leq_{\varepsilon_{\mathcal{X}}} s$ if and only if either (a) $t = 0$, or (b) $t = \varepsilon_x$ and ε_y occurs in s for some $y \geq_{\mathcal{X}} x$, or (c) $t = \omega^{t'}$, $s_0 = \varepsilon_y$ and $t' \leq_{\varepsilon_{\mathcal{X}}} \varepsilon_y$, or (d) $t = \omega^{t'}$, $s_0 = \omega^{s'}$ and $t' \leq_{\varepsilon_{\mathcal{X}}} s'$, or (e) $k > 0$ and $t_0 <_{\varepsilon_{\mathcal{X}}} s_0$, or (f) $k > 0$ and $t_0 = s_0$, $m > 0$ and $t_1 + \cdots + t_n \leq_{\varepsilon_{\mathcal{X}}} s_1 + \cdots + s_m$.

The next proposition follows from Theorem 9 in [2] and Theorem 3.4 in [9]. We believe that an alternative proof can be obtained using the techniques of the previous sections, by taking the proof of Theorem 3.4 in [9] as a model. We know how to reduce towers of exponentiation of arbitrary height starting with homogeneous sets for the colourings from Proposition 3. The extraction procedure in Proposition 3 uses the following computable operation on notations: from a sequence $(\alpha_i)_{i \in I}$ the sequence $(e_{\Delta(\alpha_{i_n}, \alpha_{i_{n+1}})}(\alpha_{i_n}))_{n \in \mathbf{N}}$ is extracted, where $\{i_1, i_2, \dots\}$ is an enumeration of I in increasing order.

Proposition 4. ($\mathbf{RCA_0}$) $\mathbf{RT}(!\omega)$ *implies* $\forall \mathcal{X}(\mathrm{WO}(\mathcal{X}) \to \mathrm{WO}(\varepsilon_{\mathcal{X}}))$.

References

1. Afshari, B., Rathjen, M.: Reverse Mathematics and well-ordering principles: a pilot study. Ann. Pure Appl. Log. 160(3), 231–237 (2009)
2. Carlucci, L., Zdanowski, K.: The strength of Ramsey Theorem for colouring relatively large sets, http://arxiv.org/abs/1204.1134
3. Farmaki, V., Negrepontis, S.: Schreier Sets in Ramsey Theory. Trans. Am. Math. Soc. 360(2), 849–880 (2008)
4. Girard, J.-Y.: Proof Theory and Logical Complexity. Biblipolis, Naples (1987)
5. Harrington, L., Paris, J.: A mathematical incompleteness in Peano Arithmetic. In: Barwise, J. (ed.) Handbook of Mathematical Logic, pp. 1133–1142. North-Holland (1977)
6. Hirst, J.: Reverse Mathematics and ordinal exponentiation. Ann. Pure App. Log. 66(1), 1–18 (1994)
7. Jockusch Jr., C.G.: Ramsey's Theorem and Recursion Theory. J. Symb. Log. 37(2), 268–280 (1972)
8. Loebl, M., Nešetřil, J.: An unprovable Ramsey-type theorem. Proc. Am. Math. Soc. 116(3), 819–824 (1992)
9. Marcone, A., Montalbàn, A.: The Veblen function for computability theorists. J. Symb. Log. 76(2), 575–602 (2011)
10. McAloon, K.: Paris-Harrington incompleteness and transfinite progressions of theories. In: Nerode, A., Shore, R.A. (eds.) Recursion Theory. Proceedings of Symposia in Pure Mathematics, vol. 42, pp. 447–460. American Mathematical Society (1985)
11. Pudlàk, P., Rödl, V.: Partition theorems for systems of finite subsets of integers. Discret. Math. 39(1), 67–73 (1982)
12. Simpson, S.G.: Subsystems of Second Order Arithmetic. Springer (1999)

Automatic Functions, Linear Time and Learning

John Case[1], Sanjay Jain[2], Samuel Seah[3], and Frank Stephan[2,3]

[1] Department of Computer and Information Sciences, University of Delaware,
Newark, DE 19716-2586, United States of America
case@cis.udel.edu
[2] Department of Computer Science, National University of Singapore, Singapore
117417, Republic of Singapore.
sanjay@comp.nus.edu.sg
[3] Department of Mathematics, National University of Singapore, Singapore 119076,
Republic of Singapore
a0030907@nus.edu.sg, fstephan@comp.nus.edu.sg

Abstract. The present work determines the exact nature of *linear time computable* notions which characterise automatic functions (those whose graphs are recognised by a finite automaton). The paper also determines which type of linear time notions permit full learnability for learning in the limit of automatic classes (families of languages which are uniformly recognised by a finite automaton). In particular it is shown that a function is automatic iff there is a one-tape Turing machine with a left end which computes the function in linear time where the input before the computation and the output after the computation both start at the left end. It is known that learners realised as automatic update functions are restrictive for learning. In the present work it is shown that one can overcome the problem by providing work-tapes additional to a resource-bounded base tape while keeping the update-time to be linear in the length of the largest datum seen so far. In this model, one additional such worktape provides additional learning power over the automatic learner model and the two-work-tape model gives full learning power.

1 Introduction

In inductive inference, automatic learners and linear time learners have played an important role, as both are considered as valid notions to model severely resource-bounded learners. On one hand, Pitt [18] observed that recursive learners can be made to be linear time learners by delaying; on the other hand, when learners are formalised by using automata updating a memory in each cycle with an automatic function, the corresponding learners are not as powerful as non-automatic learners [10] and cannot overcome their weakness by delaying. The relation between these two models is that automatic learners are indeed linear time learners [4] but not vice versa. This motivates to study the connection between linear time and automaticity on a deeper level.

S.B. Cooper, A. Dawar, and B. Löwe (Eds.): CiE 2012, LNCS 7318, pp. 96–106, 2012.

It is well known that a finite automaton recognises a regular language in linear time. One can generalise the notion of automaticity from sets to relations and functions [3,9,13] and say that a relation or a function is automatic iff an automaton recognises its graph, see Section 2. For automatic functions it is not directly clear that they are in linear time, as recognising a graph and *computing the output of a string from the input* are two different tasks. Interestingly, in Section 2 below, it is shown that automatic functions coincide with those computed by linear time one-tape Turing machines which have the input and output both starting at the left end of the tape. In other words, a function is automatic iff it is linear-time computable with respect to the most restrictive variant of this notion; increasing the number of tapes or not restricting the position of the output on the tape results in a larger complexity class.

Section 3 is dedicated to the question on how powerful a linear time notion must be in order to capture full learning power in inductive inference. In respect to the automatic learners [4,10,11], it has been the practice to consider hypotheses spaces whose membership relation is automatic (that is, uniformly regular) and which are called automatic families. It turned out that certain automatic families which are learnable by a recursive learner cannot be learnt by an automatic learner. One can simulate automatic learners by using one tape; in the present work this tape (called base-tape) is restricted in length by the length of the longest datum seen so far — this results in longest word size memory limited automatic learners as studied in [10]. In each cycle, the learner reads one datum about the set to be learnt and revises its memory and conjecture. The question considered is how much extra power is added to the learner by permitting additional work-tapes which do not have length-restrictions; in each learning cycle the learner can, however, only work on these tapes in time linear in the length of the longest example seen so far. It can be shown that by a clever archivation technique, two additional work tapes can store all the data observed in a way that a successful learner can be simulated. When having only one additional work-tape, the current results are partial: One can simulate a learner when the time-constraint of the computation is super-linear and all languages in the class to be learnt are infinite.

2   Automatic Functions and Linear Time

Informally, an automatic function is a function from strings to strings whose graph is recognised by a finite automaton. More formally, this is based on the notion of convolution. The convolution of two strings $x = x_1 x_2 \ldots x_m$ and $y = y_1 y_2 \ldots y_n$ is defined as follows: Let $x' = x_1' x_2' \ldots x_r'$ and $y' = y_1' y_2' \ldots y_r'$, where (i) $r = \max(\{m, n\})$; (ii) $x_i' = x_i$, if $i \le m$, $x_i' = \#$ otherwise; (iii) $y_i' = y_i$, if $i \le n$, $y_i' = \#$ otherwise. Now, $\mathrm{conv}(x, y) = (x_1', y_1')(x_2', y_2') \ldots (x_r', y_r')$. One can define the convolution of a fixed number of strings similarly in order to define functions which have a fixed number of inputs instead of one. One can use convolutions also to define functions with several inputs computed by one-tape Turing machines; therefore the exposition in this section just follows the basic case of mapping strings to strings.

Now a function f is called automatic iff there is a finite automaton which recognises the convoluted input-output pairs; that is, given $\text{conv}(x, y)$, the automaton accepts iff x is in the domain of f and $f(x) = y$. The importance of the concept of automatic functions and automatic relations is that every function or relation, which is first-order definable from finite number of automatic functions and relations, is automatic again and the corresponding automaton can be computed effectively from the other automata. This gives the second nice fact that structures consisting of automatic functions and relations have a decidable first-order theory [9,13]. The main result of this section is that the following three models are equivalent:

- automatic functions;
- functions computed in deterministic linear time by a one-tape Turing machine where input and output start at the same position;
- functions computed in non-deterministic linear time by a one-tape Turing machine where input and output start at the same position.

This equivalence is shown in the following two results, where the first one generalises prior work [4, Remark 2].

Theorem 1. *Let f be an automatic function. Then there is a deterministic linear time one-tape Turing machine which replaces any legal input x on the tape by $f(x)$ starting at the same position as x before.*

Proof. Assume that a deterministic automaton with c states (numbered 1 to c, where 1 is the starting state) accepts a word of the form $\text{conv}(x, y) \cdot (\#, \#)$ iff x is in the domain of f and $y = f(x)$; the automaton rejects any other sequence.

Suppose input is $x = x_1 x_2 \ldots x_r$. Now one considers the following additional work-tape symbols consisting of all tuples $(a, s_1, s_2, \ldots, s_c)$: where a is $\#$ or one of x_k's, and for $d \in \{1, 2, \ldots, c\}$, s_d takes the values $-$, $+$ or $*$. Now consider the k-th cell: $s_d = -$ iff there is no word of the form $y_1 y_2 \ldots y_{k-1}$ such that the automaton on input $(x_1, y_1)(x_2, y_2) \ldots (x_{k-1}, y_{k-1})$ reaches the state d; $s_d = +$ iff there is exactly one such word; $s_d = *$ iff there are at least two such words. Here the x_i and y_i can also be $\#$ when a word has been exhausted.

Now the Turing machine simulating the automaton replaces the cell to the left of the input by o, the cell containing x_1 by $(x_1, +, -, \ldots, -)$. Then, for each new cell with entry x_k (from the input or $\#$ if that has been exhausted) the automaton replaces x_k by $(x_k, s_1, s_2, \ldots, s_c)$ under the following conditions, (where the entry in the previous cell was $(x_{k-1}, s_1', s_2', \ldots, s_c')$):

- s_d is $+$ iff there is exactly one (y_{k-1}, d') such that $s'_{d'}$ is $+$ and the automaton transfers on (x_{k-1}, y_{k-1}) from state d' to d and there is no pair (y_{k-1}, d') such that $s'_{d'}$ is $*$ and the automaton transfers on (x_{k-1}, y_{k-1}) from d' to d;
- s_d is $*$ iff there are at least two pairs (y_{k-1}, d') such that $s'_{d'}$ is $+$ and the automaton transfers on (x_{k-1}, y_{k-1}) from state d' to d or there is at least one pair (y_{k-1}, d') such that $s'_{d'}$ is $*$ and the automaton transfers on (x_{k-1}, y_{k-1}) from d' to d;
- s_d is $-$ iff for all pairs (y_{k-1}, d') such that the automaton transfers on (x_{k-1}, y_{k-1}) from d' to d, it holds that $s'_{d'}$ is $-$.

Note that the third case applies iff the first two do not apply. The automaton replaces each symbol in the input as above until it reaches the cell where the intended symbol $(a, s_1, s_2, \ldots, s_c)$ has $s_d = +$ for some accepting state d. If this happens, the Turing machine turns around, memorises the state d, erases this cell and goes backward.

When the Turing machine comes backward from the cell $k + 1$ to the cell k, where the state memorised for the cell $k + 1$ is d', then it determines the unique (d, y_k) such that $s_d = +$ (as stored in cell k) and the automaton transfers from d to d' on (x_k, y_k); now the Turing machine replaces the symbol on cell k by y_k (if $y_k \neq \#$) and by the blank symbol (if $y_k = \#$). Then the automaton keeps the state d in the memory and goes to the left and repeats this process until it reaches the cell which has the symbol o on it. Once the Turing machine reaches there, it replaces this symbol by the blank and terminates.

For the verification, note that the output $y = y_1 y_2 \ldots$ (with $\#$ appended) satisfies that the automaton, after reading $(x_1, y_1)(x_2, y_2) \ldots (x_k, y_k)$, is always in a state d with $s_d = +$ (as written in cell $k + 1$ in the algorithm above), as the function value y is unique in x; thus, whenever the automaton ends up in an accepting state d with $s_d = +$ then the input-output-pair $conv(x, y) \cdot (\#, \#)$ has been completely processed and $x \in dom(f) \wedge f(x) = y$ has been verified. Therefore, the Turing machine can turn and follow the unique path, marked by $+$ symbols, backwards in order to reconstruct the output from the input and the markings. All superfluous symbols and markings are removed from the tape in this process. As y depends uniquely on x, the automaton accepting $conv(x, y)$ can accept at most c symbols after the word x; hence the runtime of the Turing machine is bounded by $2 \cdot (|x| + c + 2)$, that is, the runtime is linear.

Note that this Turing machine makes two passes, one from the origin to the end of the word and one back. These two passes are needed as the function $f(x_1 x_2 \ldots x_{k-1} x_k) = x_k x_2 \ldots x_{k-1} x_1$ shows, where the first and last symbol are exchanged and the others remain unchanged.

For the converse direction, assume that a function is computed by a nondeterministic one-tape Turing machine in linear time in a way that input and output starts at the same position on the tape. For an input x, any two nondeterministic accepting computations have to produce the same output $f(x)$.

Furthermore, the runtime of each computation has to follow the same linear bound $c \cdot (|x| + 1)$, independent of whether the computation ends up in an accepting state or not.

Theorem 2. *Let f be a function computed by a non-deterministic one-tape Turing machine in linear time, with the input and output starting at the same position. Then f is automatic.*

Proof. The proof is based on crossing-sequence methods, see [8] and [16, Section VIII.1]. Without loss of generality one can assume that there is a special symbol o left of the input occurring only there and that the automaton each time turns when it reaches this position. Furthermore, it starts there and returns to that position at the end; a computation accepts only when the full computation has been accomplished and the automaton has returned to its origin o. By a result of Hartmanis [7] and Trakhtenbrot [19], there is a constant c' such that an accepting computation visits each cell of the Turing tape at most c' times; otherwise the function f would not be linear time computable. This permits to represent the computation locally by storing for each visit to a cell — the direction from which the Turing machine entered the cell, in which state it was, what activity it did and in which direction it left the cell. This gives, for each cell, only a constant amount of information — which can be stored in the cell using a sufficiently large alphabet.

Now a non-deterministic automaton can recognise the set

$$\{\text{conv}(x, y) : x \in dom(f) \wedge y = f(x)\}$$

by guessing on each cell, the local information of the visits of the Turing machine, and comparing it with the information from the previous cell and checking whether it is consistent; furthermore, the automaton checks whether, on the k-th cell, y_k is written after all the guessed activity of the Turing machine and whether this activity is consistent with the initial value x_k. The automaton passes over the full word and accepts $\text{conv}(x, y)$ iff the non-deterministic computation transforms some input of the form $ox\#^*$ into some output of the form $oy\#^*$. These techniques are standard and the final verification is left to the reader.

Remark 3. One might ask whether the condition on the input and output starting at the same position is really needed. The answer is "yes". Assume by way of contradiction that it would not be needed and that all functions linear time computable by a one-tape Turing machine without any restrictions on output positions are automatic. Then one could consider the free monoid over $\{0, 1\}$. For this monoid, the following function could be computed from $\text{conv}(x, y)$: The output is $z = f(x, y)$ if $y = xz$; the output is $\#$ if such a z does not exist. For this, the machine just compares x_1 with y_1 and erases (x_1, y_1), x_2 with y_2 and erases (x_2, y_2) and so on, until it reaches (a) a pair of the form (x_m, y_m) with $x_m \neq y_m$ or (b) a pair of the form $(x_m, \#)$ or (c) a pair of the form $(\#, y_m)$ or (d) the end of the input. In cases (a) and (b) the output has to be

and the machine just erases all remining input symbols and puts the special symbol # to denote the special case; in case (c) the value z is just obtained by changing all remaining input symbols $(\#, y_k)$ to y_k and the Turing machine terminates. In case (d) the valid output is the empty string and the Turing machine codes it adequately on the tape. Hence f would be automatic. But now one could first-order define concatenation g by letting $g(x, z)$ be that y for which $f(x, y) = z$; this would give that the concatenation is automatic, which is known to be false. Hence the condition on the starting-positions cannot be dropped.

3   Linear Time Learners

In order to evaluate the complexity of a learner, the following assumptions are made.

Definition 4. A learner M is a machine which maintains some memory and in each cycle receives as input one word to be learnt, updates its memory and then outputs an hypothesis. The machine is organised as follows:

- Tape 0 (base tape) contains convolution of the input, the output and some information (memory) which is not longer than the longest word seen so far (plus a constant). Input and output on tape 0 always start at a fixed position o.
- Tapes $1, 2, \ldots, k$ are normal tapes, whose contents and head position are not modified during change of cycle, which the Turing machine can use for archiving information and doing calculations.
- The machine has in each cycle a time allowance linear in the length of the largest example seen so far. Without loss of generality, tape 0 stores this bound.

The learner is said to have k additional work-tapes iff it has in addition to tape 0 also the tapes $1, 2, \ldots, k$.

Note that if only tape 0 is present, the model is equivalent to an automatic learner with the memory bounded by the size of the longest datum seen so far (plus a constant) [4,10].

The class of languages to be learnt is represented by an automatic family $\{L_e : e \in I\}$; automatic families [10,11] are the automata-theoretic counterpart of indexed families [1,15] which were widely used in inductive inference to represent the class to be learnt. The basic model of inductive inference [1,2,6,12,17] is that the learner M reads cycle by cycle a list $w_0, w_1, \ldots$ of all the words in a language L_e and at the same time M outputs a sequence $e_0, e_1, \ldots$ of indices, in each cycle one of them, such that $L_{e_k} = L_e$ for almost all k. As the equivalence of indices is automatic, one can take the hypothesis space I to be one-one and therefore the criterion would indeed have that $e_k = e$ for almost all k. In a one-one hypothesis space, the index e of a finite language L_e has, up to an additive constant, the same length as the longest word in L_e; this follows from [11, Theorem 3.5] using that $d_{run}(R)$ is the longest word in R, for a finite set R.

This observation is crucial as otherwise the time-constraint on the learner would prevent the learner from eventually outputting the right index; for infinite languages this is not a problem as the language must contain arbitrary long words.

Angluin [1] gave a criterion when a class is learnable in general. This criterion, adjusted to automatic families, says that a class is learnable iff for every $e \in I$ there exists a finite set D such that there is no $d \in I$ with $D \subseteq L_d \subset L_e$. The main question of this section is which learnable classes can also be learnt by a linear-time learner with k additional work tapes. For $k = 0$ this is in general not possible, as automatic learners fail to learn various learnable classes [10], for example the class of all sets $\{0,1\}^* - \{x\}$ and the class of all sets $L_e = \{x \in \{0,1\}^{|e|} : x \neq e\}$.

Freivalds, Kinber and Smith [5] introduced limitations on the long term memory into inductive inference, Kinber and Stephan [14] transferred it to the field of language learning. Automatic learners have similar limitations and are therefore not able to learn all learnable automatic classes [4,10]. The usage of additional work-tapes for linear time learners permits to overcome these limitations, the next results specify how many additional work-tapes are needed. Recall from above that work-tapes are said to be *additional* iff they are in addition to the base tape.

Theorem 5. *The automatic family consisting of $L_\varepsilon = \{0,1\}^*$, $L_{x0} = \{0,1\}^* \cup \{x2\} - \{x\}$ and $L_{x1} = \{0,1\}^* \cup \{x2\}$ does not have an automatic learner but has a linear-time learner using one additional work-tape.*

Proof. An automatic learner cannot memorise all the data from $\{0,1\}^*$ it sees; therefore one can show, see [10], that there are two finite sequences of words, one containing x and one not containing x, such that the learner has the same long term memory after having seen both sequences. If one presents to the learner after one of these sequences all the elements of L_{x0}, then the automatic learner has actually no way to find out whether the overall language presented is L_{x0} or L_{x1}, therefore it cannot learn the class.

A linear time learner with one additional work-tape (called tape 1) initially conjectures L_ε and uses tape 1 to archive all the examples seen at the current end of the written part of the tape. When the learner sees a word of the form $x2$, it maintains a copy of it in the memory part of tape 0. In each subsequent cycle, the learner scrolls back tape 1 by one word and compares the word there as well as the current input with $x2$; if one of these two is x then the learner changes its conjecture to L_{x1}, else it keeps its conjecture as L_{x0}. In the case that the origin of tape 1 is reached, the learner from then onwards ignores tape 1 and only compares the incoming input with $x2$.

Theorem 6. *Every learnable automatic family has a linear-time learner using two additional work-tapes.*

Proof. Jain, Luo and Stephan [10] showed that for every learnable automatic family $\{L_e : e \in I\}$ there is an automatic learner with a memory bound of the length of the longest example seen so far (plus a constant) which learns the class from every fat text (a text in which every element of the language appears infinitely often). So the main idea is to use the two additional tapes in order to simulate and feed the learner M on tape 0 with a fat text. In each cycle, tape 0 is updated from a pair $\text{conv}(mem_k, w_k)$ to $\text{conv}(mem_{k+1}, e_k)$, where mem_k is the long-term memory of M before the k-th cycle and w_k is the k-th input word and e_k is the conjecture issued in this cycle.

The update is now done in a way such that instead of one learning cycle, two are done by first mapping $\text{conv}(mem_k, w_k)$ to $\text{conv}(mem', e')$ and then mapping $\text{conv}(mem', t)$ to $\text{conv}(mem_{k+1}, e_k)$, where t is the word on tape 1 at the current position (which can be accessed by scrolling the tape accordingly). Furthermore, after reading the word t, tape 1 is scrolled to the starting position of the previous word; when the tape is at the beginning, the direction of taking out the words is altered until the end of the tape is reached. In each cycle, the current datum w_k is also appended at the end of tape 2. If at the end of a cycle, words are taken out in a forward manner from tape 1 and the end of tape 1 is reached, then the roles of tape 1 and tape 2 are exchanged, so that each word on tape 2 is then given to M for learning, while tape 1 is used to archive the new words.

It is easy to see that in each cycle, the time spent is proportional to $|mem_k| + |x_k| + |t|$ and thus linear in the length of the longest word seen so far (plus a constant); note that mem', e', e_k are also bounded by that length (plus a constant). Furthermore, each input word is infinitely often put through M as each of the words observed gets archived on one of the two tapes. Hence M learns the language. It follows that the given automatic family $\{L_e : e \in I\}$ is learnt by a linear time Turing machine with two additional work-tapes.

Open Problem 7. *It is unknown whether one can learn every in principal learnable automatic class using an automatic learner augmented by only one work-tape.*

The next result shows that, if one allows a bit more than just linear time, then one can learn, using one work-tape, all learnable automatic classes of infinite languages. The result could even be transferred to families of arbitrary r.e. sets as the simulated learner is an arbitrary recursive learner.

Theorem 8. *Assume that $\{L_e : e \in I\}$ is an automatic family where every L_e is infinite and M is a recursive learner which learns this family. Furthermore, assume that f, g are recursive functions with the property that $f(n) \geq m$ whenever $n \geq g(m)$ (so g is some type of inverse of f). Then there is a learner N which learns the above family, using only one additional work tape, and satisfies the following constraint: if n is the length of the longest example seen so far, then only the cells $1, 2, \ldots, n$ of tape 0 can be non-empty and the update time of N in the current cycle is $O(n \cdot f(n))$.*

Further investigations deal with the question what happens if one does not add further worktapes to the learner but uses other methods to store memory. Indeed, the organisation in a tape is a bit arkward and using a queue solves some problems. A queue is a tape where one reads at one end and writes at the opposite end, both the reading and writing heads are unidirectional and cannot overtake each other. Tape 0 satisfies the same constraints as in the model of additional work tapes and one also has the constraint that in each cycle only linearly many symbols (measured in the length of the longest datum seen so far) is stored in the queue and retrieved from it.

Theorem 9. *Every learnable automatic family has a linear-time learner using one additional queue as a data structure.*

Proof. The learner simulates an automatic learner M using fat text, similarly as in Theorem 6. Let M in the k-th step map (mem_k, w_k) to (mem_{k+1}, e_k) for M's memory mem_k. At the beginning of a cycle the learner has $\text{conv}(v_k, -, mem_k, -)$ on Tape 0 where v_k is the current datum, mem_k the archived memory of M and "$-$" refers to irrelevant or empty content. In the k-th cycle, the learner scans four times over Tape 0 from beginning to the end and each time afterwards returns to the beginning of the tape:

1. Copy v_k from Tape 0 to the write-end of the queue;
2. Read w_k from the read-end of the queue and update Tape 0 to $\text{conv}(v_k, w_k, mem_k, -)$;
3. Copy w_k from Tape 0 to the write-end of the queue;
4. Simulate M on Tape 0 in order to map (mem_k, w_k) to (mem_{k+1}, e_k) and update Tape 0 to $\text{conv}(v_k, w_k, mem_{k+1}, e_k)$.

It can easily be verified that this algorithm permits to simulate M using the data type of a queue and that each cycle takes only time linear in the length of the longest datum seen so far.

A further data structure investigated is to the provision of additional stacks. Tape 0 remains a tape in this model and has still to obey to the resource-bound of not being longer than the longest word seen so far. Theorems 5 and 6 work also with one and two stacks, respectively, as the additional work-tapes are actually used like stacks.

Theorem 10. *There are some automatic classes which can be learnt with one additional stack but not by an automatic learner. Furthermore, every in principle learnable automatic class can be learnt by a learner using two additional stacks.*

Furthermore, the next result shows that in general one stack is not enough; so one additional stack gives only intermediate learning power while two or more additional stacks give the full learning power.

Theorem 11. *The class of all $L_e = \{x \in \{0,1\}^{|e|} : x \neq e\}$ with $e \in \{0,1\}^* \cup \{2\}^*$ cannot be learnt by a learner using one additional stack.*

4   Conclusion

Automatic functions are shown to be the same as functions computed in linear time by one-tape Turing machines with input and output starting at the left end of the machine. Furthermore, linear time learner can be modelled by having a base tape of the length of the longest datum seen so far plus additional structures which can either be additional Turing machine work tapes, queues or stacks. In each cycle the learner runs in time linear in the longest example seen so far, updates the base tape and accesses the additional storage devices also only to retrieve or store a linear number of symbols. It is shown that two additional work tapes, two additional stacks or one additional queue give full learning power; furthermore, the learning power of one additional stack is properly intermediate and the learning power of one additional work tape is better than no additional work tape. It is an open problem whether there is a difference in the learning power of one and two additional work tapes.

Acknowledgements. The second author was supported in part by NUS grants C252-000-087-001 and R252-000-420-112. The fourth author was supported in part by NUS grant R252-000-420-112.

References

1. Angluin, D.: Inductive inference of formal languages from positive data. Information and Control 45, 117–135 (1980)
2. Blum, L., Blum, M.: Toward a mathematical theory of inductive inference. Information and Control 28, 125–155 (1975)
3. Blumensath, A., Grädel, E.: Automatic structures. In: 15th Annual IEEE Symposium on Logic in Computer Science (LICS), pp. 51–62 (2000)
4. Case, J., Jain, S., Le, T.D., Ong, Y.S., Semukhin, P., Stephan, F.: Automatic Learning of Subclasses of Pattern Languages. In: Dediu, A.-H., Inenaga, S., Martín-Vide, C. (eds.) LATA 2011. LNCS, vol. 6638, pp. 192–203. Springer, Heidelberg (2011)
5. Freivalds, R., Kinber, E., Smith, C.H.: On the impact of forgetting on learning machines. Journal of the ACM 42, 1146–1168 (1995)
6. Mark Gold, E.: Language identification in the limit. Information and Control 10, 447–474 (1967)
7. Hartmanis, J.: Computational complexity of one-tape Turing machine computations. Journal of the Association of Computing Machinery 15, 411–418 (1968)
8. Hennie, F.C.: Crossing sequences and off-line Turing machine computations. In: Sixth Annual Symposium on Switching Circuit Theory and Logical Design, pp. 168–172 (1965)
9. Hodgson, B.R.: Décidabilité par automate fini. Annales des sciences mathématiques du Québec 7(1), 39–57 (1983)
10. Jain, S., Luo, Q., Stephan, F.: Learnability of Automatic Classes. In: Dediu, A.-H., Fernau, H., Martín-Vide, C. (eds.) LATA 2010. LNCS, vol. 6031, pp. 321–332. Springer, Heidelberg (2010)
11. Jain, S., Ong, Y.S., Pu, S., Stephan, F.: On automatic families. In: Proceedings of the eleventh Asian Logic Conference in Honour of Professor Chong Chitat on his Sixtieth Birthday, pp. 94–113. World Scientific (2012)
12. Jain, S., Osherson, D.N., Royer, J.S., Sharma, A.: Systems That Learn, 2nd edn. MIT Press (1999)

13. Khoussainov, B., Nerode, A.: Automatic Presentations of Structures. In: Leivant, D. (ed.) LCC 1994. LNCS, vol. 960, pp. 367–392. Springer, Heidelberg (1995)
14. Kinber, E., Stephan, F.: Language learning from texts: mind changes, limited memory and monotonicity. Information and Computation 123, 224–241 (1995)
15. Lange, S., Zeugmann, T., Zilles, S.: Learning indexed families of recursive languages from positive data: a survey. Theoretical Computer Science 397, 194–232 (2008)
16. Odifreddi, P.: Classical Recursion Theory. Studies in Logic and the Foundations of Mathematics, vol. II, 143. Elsevier (1999)
17. Osherson, D., Stob, M., Weinstein, S.: Systems That Learn, An Introduction to Learning Theory for Cognitive and Computer Scientists. Bradford — The MIT Press, Cambridge, Massachusetts (1986)
18. Pitt, L.: Inductive inference, DFAs, and Computational Complexity. In: Jantke, K.P. (ed.) AII 1989. LNCS (LNAI), vol. 397, pp. 18–44. Springer, Heidelberg (1989)
19. Trakhtenbrot, B.A.: Turing computations with logarithmic delay. Algebra i Logika 3, 33–48 (1964)

An Undecidable Nested Recurrence Relation

Marcel Celaya and Frank Ruskey

Department of Computer Science, University of Victoria,
Victoria, BC, V8W 3P6, Canada

Abstract. Roughly speaking, a recurrence relation is *nested* if it contains a subexpression of the form $\ldots A(\ldots A(\ldots)\ldots)$. Many nested recurrence relations occur in the literature, and determining their behavior seems to be quite difficult and highly dependent on their initial conditions. A nested recurrence relation $A(n)$ is said to be *undecidable* if the following problem is undecidable: given a finite set of initial conditions for $A(n)$, is the recurrence relation calculable? Here *calculable* means that for every $n \geq 0$, either $A(n)$ is an initial condition or the calculation of $A(n)$ involves only invocations of A on arguments in $\{0, 1, \ldots, n-1\}$. We show that the recurrence relation

$$A(n) = A(n - 4 - A(A(n-4))) + 4A(A(n-4))$$
$$+ A(2A(n - 4 - A(n-2)) + A(n-2))$$

is undecidable by showing how it can be used, together with carefully chosen initial conditions, to simulate Post 2-tag systems, a known Turing complete problem.

1 Introduction

In the defining expression of a recurrence relation $R(n)$, one finds at least one application of R to some function of n. The Fibonacci numbers, for example, satisfy the recurrence $F(n) = F(n-1) + F(n-2)$ for $n \geq 2$. A recurrence relation $R(n)$ is called *nested* when the defining expression of R contains at least two applications of R, one of which is contained in the argument of the other.

Many sequences defined in terms of nested recurrences have been studied over the years. One famous example is Hofstadter's Q sequence, which is defined by the recurrence

$$Q(n) = Q(n - Q(n-1)) + Q(n - Q(n-2)), \tag{1}$$

with initial conditions $Q(1) = Q(2) = 1$. This sequence is very chaotic, and a plot of the sequence demonstrates seemingly unpredictable fluctuation about the line $y = x/2$. It remains an open question whether Q is defined on all positive integers, despite its introduction in [8] over 30 years ago. Indeed, if it happens

S.B. Cooper, A. Dawar, and B. Löwe (Eds.): CiE 2012, LNCS 7318, pp. 107–117, 2012.
© Springer-Verlag Berlin Heidelberg 2012

that there exists some m such that $m < Q(m-1)$ or $m < Q(m-2)$, then the calculation of $Q(m)$ would require an application of Q to a negative integer outside its domain. While little is known about the Q sequence, other initial conditions that give rise to much better behaved sequences that also satisfy the Q recurrence have been discovered [7], [12].

Another sequence defined in terms of a nested recurrence is the Conway-Hofstadter sequence

$$C(n) = C(C(n-1)) + C(n - C(n-1)),\qquad (2)$$

with initial conditions $C(1) = C(2) = 1$. Unlike the Q sequence, this sequence is known to be well-defined for $n \geq 1$, and in fact Conway proved that $\lim_{n\to\infty} C(n)/n = 1/2$. Plotting the function $C(n) - n/2$ reveals a suprising, fractal-like structure. This sequence is analyzed in depth in [9].

Another sequence whose structure is mainly understood, but is extraordinarily complex, is the Hofstadter-Huber V sequence, defined by

$$V(n) = V(n-V(n-1)) + V(n-V(n-4)), \text{ with } V(1) = V(2) = V(3) = V(4) = 1.\qquad (3)$$

It was first analyzed by Balamoham, Kuznetsov and Tanny [2] and recently Allouche and Shallit showed that it is 2-automatic [1].

Some of these nested recurrences are well-behaved enough to have closed forms. Hofstadter's G sequence, for example, is defined by

$$G(n) = n - G(G(n-1)), \text{ with } G(0) = 0.\qquad (4)$$

This sequence has closed form $G(n) = \lfloor (n+1)/\phi \rfloor$, where ϕ is the golden ratio [6]. A sequence due to Golomb [7], defined by $G(1) = 1$ and $G(n) = G(n - G(n-1)) + 1$ when $n > 1$, is the unique increasing sequence in which every $n \geq 1$ appears n times, and has closed form $G(n) = \lfloor (1 + \sqrt{8n})/2 \rfloor$.

Despite their wide variation in behaviour, all of these recursions are defined in terms of only three simple operations: addition, subtraction, and recurrence application. Of these, the latter operation makes them reminiscent of certain discrete systems—particularly the cellular automaton. Consider, for instance, the Q sequence defined above. It is computed at any point by looking at the two values immediately preceeding that point, and using them as "keys" for a pair of "lookups" on a list of previously computed values, the results of which are summed together as the next value. It is well-known that many cellular automata are Turing complete; an explanation of how to simulate any Turing machine in a suitably-defined one-dimensional cellular automaton is given in [13]. With respect to nested recurrences, therefore, two questions naturally arise. First, does there exist, in some sense, a computationally universal nested recurrence defined only in terms of the aforementioned three operations? Second, given a nested recurrence, is it capable of universal computation? In this paper we aim to clarify the first question and answer it in the positive. A related approach was taken by Conway to show that a generalization of the Collatz problem is undecidable [4].

2   Tag Systems

The *tag system,* introduced by Emil Post in [11], is a very simple model of computation. It has been used in many instances to prove that some mathematical object is Turing complete. This was done, for example, with the one-dimensional cellular automaton known as Rule 110; the proof uses a variant of the tag system [5].

Such a system consists of a finite alphabet of symbols Σ, a set of production rules $\Delta : \Sigma \to \Sigma^*$, and an initial word W_0 from Σ^*. Computation begins with the initial word W_0, and at each step of the computation the running word $w_1 \ldots w_k$ is transformed by the operation

$$w_1 \ldots w_k \vdash w_3 \ldots w_k \Delta \left(w_1 \right).$$

In other words, at each step, the word is truncated by two symbols on the left but extended on the right according to the production rule of the first truncated symbol. In this paper, we adopt the convention that lowercase letters represent individual symbols while uppercase letters represent words.

If the computation at some point yields a word of length 1, truncation of the first two symbols cannot occur; it is at this point the system is said to *halt.* The *halting problem for tag systems* asks: given a tag system, does it halt? As is the case for Turing machines, the halting problem for tag systems is undecidable [10].

Although this definition of tag systems has *two* symbols deleted at each step, there's no reason why this number need be fixed at two. In general, an *m-tag system* is a tag system where at each step m symbols are removed from the beginning of the running word. The number m is called the *deletion number* of the tag system. It is known that $m = 2$ is the smallest number for which m-tag systems are universal [3]. Thus, only 2-tag systems are considered in the remainder of these notes.

Example 1. Two tag systems are depicted below, both of which share alphabet $\Sigma = \{a, b, c\}$ and production rules Δ given by $a \to abb$, $b \to c$, and $c \to a$. The initial word of the left tag system is *abcb,* while the initial word of the right tag system is *abab.* Observe that one is periodic while the other halts.

```
        abcb              abab
         cbabb             ababb
          abba              abbabb
           baabb             babbabb
            abbc              bbabbc
             bcabb             abbcc
              abbc              bccabb
               ...              cabbc
                                 bbca
                                  cac
                                   ca
                                    a
```

3   A Modified Tag System

The goal of this paper is to show that the recurrence given in the abstract can simulate some universal model of computation. In particular, we wish to show that if we encode the specification of some abstract machine as initial conditions for our recurrence, then the resulting sequence produced by the recurrence will somehow encode every step of that machine's computation. The tag system model seems like a good candidate for this purpose, since the entire run of a tag system can be represented by a single, possibly infinite string we'll call the *computation string*. For example, the string corresponding to the tag system above and to the right is *ababbabbabbccabbcacaa*. A specially-constructed nested recurrence A would need only generate such a string on $\mathbb{N} = \{0, 1, 2, \ldots\}$ to simulate a tag system; each symbol would be suitably encoded as an integer.

Ideally, the sequence defined by the nested recurrence can be calculated one integer at a time using previously computed values. It would therefore make sense to find some tag-system-like model of computation capable of generating these strings one symbol at a time. That way, the computation of the nth symbol of a string in this new model can correspond to the calculation of $A(n)$ (or, more likely, some argument linear in n). With this motivation in mind, we now introduce a modification of the tag system model.

A *reverse tag system* consists of a finite set of symbols Σ, a set of production rules $\delta : \Sigma^2 \to \Sigma$, a function $d : \Sigma \to \mathbb{N}$, and an initial word $W_0 \in \Sigma^*$. While an ordinary tag system modifies a word by removing a *fixed* number of symbols from the beginning and adding a *variable* number of symbols to the end, the situation is reversed in a reverse tag system.

A single computation step of a reverse tag system is described by the operation

$$w_1 \ldots w_k \vdash w_{d(y)+1} \ldots w_k y,$$

where $y = \delta(w_1, w_k)$. Given a word that starts with w_1 and ends with w_k, the production rule for the pair (w_1, w_k) yields a symbol y which is appended to the end of the word. Then, the first $d(y)$ symbols are removed from the beginning of the word. The number $d(s)$ we'll call the *deletion number* of the symbol s. If at some point the deletion number of y exceeds k, then the reverse tag system *halts*.

Example 2. Let $\Sigma = \{a, b\}$, $d(a) = 0$, and $d(b) = 2$. Define δ by

$$(a, a) \to b$$
$$(a, b) \to b$$
$$(b, a) \to b$$
$$(b, b) \to a.$$

It takes 12 steps before this reverse tag system with initial word $W_0 = baaab$ becomes periodic.

```
baaab
baaaba
 aabab
  babb
  babba
   bbab
   bbaba
    abab
     abb
      bb
      bba
       ab
        b
       ba
        b
        ...
```

4 Simulating a Tag System with a Reverse Tag System

Consider a tag system $T = (\Sigma, \Delta, W_0)$ such that each production rule of Δ yields a nonempty string. The goal of this section is to construct a reverse tag system $R = (\Sigma', \delta, d, W_0')$ which simulates T.

This construction begins with Σ'. Some notation will be useful to represent the elements that are to appear in Σ'. Let $[s]_j$ denote the symbol "s_j", where s is a symbol in Σ and j is an integer.

For each $s_i \in \Sigma$, write $\Delta(s_i)$ as $s_{i,\ell_i} \ldots s_{i,2} s_{i,1}$. For each symbol $s_{i,j}$ in this word, the symbol $[s_{i,j}]_j$ shall appear in Σ'. For example, if $a \to abc$ is a production rule of Δ, then Σ' contains the three symbols $[a]_3$, $[b]_2$, and $[c]_1$. If $W_0 = q_1 q_2 \ldots q_m$, the symbols $[q_1]_1, [q_2]_1, \ldots, [q_m]_1$ are also included in Σ'. Constructed this way, Σ' contains no more symbols than the sum of the lengths of the words in $\Delta(\Sigma)$ and the word W_0.

The production rules of δ include the rules

$$\delta([s_i]_*, [*]_1) = [s_{i,\ell_i}]_{\ell_i}$$
$$\delta([s_i]_*, [s_{i,j}]_j) = [s_{i,j-1}]_{j-1}$$

taken over all $s_i \in \Sigma$, all $j \in \{2, 3, \ldots, \ell_i\}$, and all possibilites for the $*$'s. Note that this specification of δ doesn't necessarily exhaust all possible pairs of $(\Sigma')^2$, however, any remaining pairs can be arbitrarily specified because they are never used during the computation of R.

Finally, the deletion numbers are specified by

$$d\left([s]_j\right) = \begin{cases} 0, & j > 1 \\ 2, & j = 1 \end{cases}$$

for all $[s]_j \in \Sigma'$, and if $W_0 = q_1 q_2 \ldots q_m$, then $W_0' = [q_1]_1 [q_2]_1 \ldots [q_m]_1$.

Example 3. This example demonstrates a simulation of the tag system T on the left in Example 1 using a reverse tag system $R = (\Sigma', \delta, d, W_0')$.

The production rules in Example 1 are

$$a \to abb \qquad b \to c \qquad c \to a.$$

To properly simulate T, the three symbols a_3, b_2, b_1 are needed for the first rule, the symbol c_1 is needed for the second, and the symbol a_1 is needed for the third. The initial word for R is $W_0' = a_1 b_1 c_1 b_1$. Taking all these symbols together, we have $\Sigma' = \{a_1, b_1, c_1, b_2, a_3\}$.

If we take "$*$" to mean "any symbol or subscript, as appropriate," the production rules δ can be written as follows:

$$(a_*, *_1) \to a_3 \ (b_*, *_1) \to c_1 \ (c_*, *_1) \to a_1$$
$$(a_*, a_3) \to b_2$$
$$(a_*, b_2) \to b_1$$

Finally, every symbol with a subscript of 1 gets a deletion number of two, and zero otherwise:

$$d(a_1) = d(b_1) = d(c_1) = 2$$
$$d(b_2) = d(a_3) = 0.$$

The output of R is depicted to the right. Compare the marked rows with the output of T in Example 1.

$a_1 b_1 c_1 b_1 \leftarrow$
$a_1 b_1 c_1 b_1 a_3$
$a_1 b_1 c_1 b_1 a_3 b_2$
$c_1 b_1 a_3 b_2 b_1 \leftarrow$
$a_3 b_2 b_1 a_1 \leftarrow$
$a_3 b_2 b_1 a_1 a_3$
$a_3 b_2 b_1 a_1 a_3 b_2$
$b_1 a_1 a_3 b_2 b_1 \leftarrow$
$a_3 b_2 b_1 c_1 \leftarrow$
$a_3 b_2 b_1 c_1 a_3$
$a_3 b_2 b_1 c_1 a_3 b_2$
$b_1 c_1 a_3 b_2 b_1 \leftarrow$
$a_3 b_2 b_1 c_1 \leftarrow$
$\ldots$

One point worth mentioning is that if a reverse tag system halts while simulating an ordinary tag system, then the simulated tag system must halt also. However, the converse is not true! A reverse tag system might keep rolling once it has completed the simulation of a halting tag system. The reverse tag system in Example 2 is a good example of this; it can survive even when there's only one symbol, while ordinary tag systems always require at least two.

Theorem 1. *Let* $T = (\Sigma, \Delta, W_0)$ *be a tag system such that each production rule of* Δ *yields a nonempty string, and let* R *be a reverse tag system constructed as above in terms of* T. *Suppose* $k > 0$, $w_1 \ldots w_k \in \Sigma^*$, *and* $\Delta(w_1) = z_\ell z_{\ell-1} \ldots z_1$. *If* $i_1, i_2, \ldots, i_{k-1}$ *are such that* $[w_j]_{i_j} \in \Sigma'$, *then*

$$[w_1]_{i_1} \cdots [w_{k-1}]_{i_{k-1}} [w_k]_1 \vdash^* [w_3]_{i_3} \cdots [w_k]_1 [z_\ell]_\ell \cdots [z_1]_1$$

in R.

Proof. We have by construction of R that

$$[w_1]_{i_1} \cdots [w_{k-1}]_{i_{k-1}} [w_k]_1 \vdash [w_1]_{i_1} \cdots [w_{k-1}]_{i_{k-1}} [w_k]_1 [z_\ell]_\ell$$
$$\vdash [w_1]_{i_1} \cdots [w_{k-1}]_{i_{k-1}} [w_k]_1 [z_\ell]_\ell [z_{\ell-1}]_{\ell-1}$$
$$\vdots \qquad \vdots$$
$$\vdash [w_1]_{i_1} \cdots [w_{k-1}]_{i_{k-1}} [w_k]_1 [z_\ell]_\ell [z_{\ell-1}]_{\ell-1} \cdots [z_2]_2$$
$$\vdash [w_3]_{i_3} \cdots [w_{k-1}]_{i_{k-1}} [w_k]_1 [z_\ell]_\ell [z_{\ell-1}]_{\ell-1} \cdots [z_2]_2 [z_1]_1.$$

$\square$

5 Simulating a Reverse Tag System with a Recurrence

While it's possible to describe how the recurrence A simulates a reverse tag system, a better approach is to introduce another, simpler recurrence B which does this simulation, then show how A reduces to B. The simpler recurrence, without initial conditions, is:

$$B\left(n\right) = \begin{cases} B\left(n-2\right) + 2B\left(B\left(n-1\right)\right), & \text{if } n \text{ is even} \\ B\left(2B\left(n-2-B\left(n-1\right)\right) + B\left(n-2\right)\right), & \text{if } n \text{ is odd.} \end{cases}$$

Consider a reverse tag system $R = \left(\Sigma, \delta, d, W_0\right)$. The simulation of R by B necessitates encoding δ and d as initial conditions of B. In order to do this, every symbol in Σ and every possible pair in $\Sigma^2 = \Sigma \times \Sigma$ must be represented by a unique integer. Then, invoking B on such an integer would correspond to evaluating δ or d, whatever the case may be. In order to avoid conflicts doing this, any integer representation of symbols and symbol pairs $\alpha : \Sigma \cup \Sigma^2 \to \mathbb{N}$ must be injective.

Assuming $\Sigma = \{s_1, s_2, \ldots, s_t\}$, one such injection is defined as follows:

$$\alpha\left(s_i\right) = 4^{i+1} + 2 = 2^{2i+2} + 2, \text{ and}$$
$$\alpha\left(s_i, s_j\right) = 2\alpha\left(s_i\right) + \alpha\left(s_j\right) = 2^{2i+3} + 2^{2j+2} + 6.$$

The fact that α is injective can be seen by considering the binary representation of such numbers. Each of the bitstrings of $\alpha\left(s_1\right), \ldots, \alpha\left(s_t\right)$ are clearly distinct from one another, and the bitstring of $\alpha\left(s_i, s_j\right)$ for any $i, j \in \{1, 2, \ldots, t\}$ "interleaves" the bitstrings of $\alpha\left(s_i\right)$ and $\alpha\left(s_j\right)$. The constant 2 term in the definition of α is important in the next section, when the A recurrence is considered.

The initial conditions of B are constructed so that the encoding of d occurs on $\alpha\left(\Sigma\right)$, and the encoding of δ occurs on $\alpha\left(\Sigma^2\right)$. For $i, j \in \{1, 2, \ldots, t\}$, the encoding for d and δ is done respectively as follows:

$$B\left(\alpha\left(s_i\right)\right) = 1 - d\left(s_i\right) \tag{5}$$
$$B\left(\alpha\left(s_i, s_j\right)\right) = \alpha\left(\delta\left(s_i, s_j\right)\right).$$

Is it worth noting that because of (5), $B(n)$ can take on negative values.

The largest value attained by α is

$$\alpha\left(s_t, s_t\right) = 3\alpha\left(s_t\right) = 3 \cdot 4^{t+1} + 6.$$

Let $c_0 = \alpha\left(s_t, s_t\right) + 2$. For the remainder of initial conditions that appear before c_0 and don't represent a symbol or symbol pair under α, B is assigned zero. One observes that even though the number of initial conditions specified is exponential in the size of Σ, only a polynomial number of these are actually nonzero.

The way the B recurrence simulates R is that R's computation string, as represented under α, is recorded on the odd integers, while the length of the running word is recorded on the even integers. Thus, for large enough n, the pair $\left(B\left(2n+1\right), B\left(2n+2\right)\right)$ represents exactly one step of R's computation. The simulation begins with the initial word $W_0 = q_1 q_2 \ldots q_m$. Specifically, the m integers $\alpha\left(q_1\right), \ldots, \alpha\left(q_m\right)$ are placed on the first m odd integers that come after

Table 1. Illustration of initial conditions

$B(c_0+k)$	$\alpha(q_1)$	0	$\alpha(q_2)$	0	...	$\alpha(q_{m-1})$	0	$\alpha(q_m)$	$2m-2$
k	1	2	3	4	...	$2m-3$	$2m-2$	$2m-1$	$2m$

c_0. The value $2m-2$ is then immediately placed after the last symbol of W_0; it is the last initial condition of B and signifies the length of the initial word. Beyond this point, the recurrence of B takes effect. An illustration of these initial conditions is given in Table 5.

We now formalize what is meant by "B simulates R." As mentioned previously, B will alternatingly output symbols and word lengths. We encode the symbols and word lengths produced by B in the following manner: any symbol $s \in \Sigma$ is encoded as the integer $\alpha(s)$, while the length k of some computed word is recorded in the output of B as the value $2k-2$.

Suppose that at the ith computation step of R, the word $W = w_1 w_2 \ldots w_k$ is produced. We shall say that B *computes the ith step of R at n* if the following equalities hold:

$$(B(n-2k+1), \ldots, B(n-3), B(n-1)) = (\alpha(w_1), \alpha(w_2), \ldots, \alpha(w_k))$$
$$B(n) = 2k-2.$$

This terminology is justified, since if B computes the ith step of R at n, then these equalities allow W to be reconstructed from the output of B near n. If there exist constants r, s such that for all $i \in \mathbb{N}$, B computes the ith step of R at $ri + s$ whenever step i exists, then we shall say that B *simulates R*.

Theorem 2. *With the above initial conditions, B simulates $R = (\Sigma, \delta, d, W_0)$.*

Proof. If we suppose that the 0th step of R yields the initial word W_0, then by Table 5 it is clear that B computes the 0th step of R at $c_0 + 2m$.

Assume that B computes the ith step of R at $2n$, where, again, we assume the word produced at step i is $w_1 w_2 \ldots w_k$. We would like to show that B computes the $(i+1)^{\text{th}}$ step of R at $2n+2$. Showing this, by induction, would prove the theorem.

If $y = \delta(w_1, w_k)$, then the word produced by R at step $i+1$ is $w_{d(y)+1} \ldots w_k y$. The last symbol of this word is y and length of this word is $k+1-d(y)$. Therefore, to prove the theorem, we need only show that

$$B(2n+1) = \alpha(y)$$
$$B(2n+2) = 2(k+1-d(y)) - 2$$
$$= 2(k-d(y)).$$

We first consider the point $2n + 1$. Since this point is odd, we have

$$
\begin{aligned}
B(2n+1) &= B(2B(2n-1-B(2n)) + B(2n-1)) \\
&= B(2B(2n-1-2(k-1)) + B(2n-1-2(k-k))) \\
&= B(2\alpha(w_1) + \alpha(w_k)) \\
&= B(\alpha(w_1, w_k)) \\
&= \alpha(\delta(w_1, w_k)) \\
&= \alpha(y).
\end{aligned}
$$

The point $2n + 2$ is even, thus

$$
\begin{aligned}
B(2n+2) &= B(2n) + 2B(B(2n+1)) \\
&= 2k - 2 + 2B(\alpha(y)) \\
&= 2k - 2 + 2(1 - d(y)) \\
&= 2(k - d(y)).
\end{aligned}
$$

$\square$

The above theorem describes the behaviour of B when R does not halt. If R halts at any point, then there exists some even n such that $B(n) = -2$. Then, $B(n+1) = B(2B(n+1) + B(n-1))$, and so B is not calculable. Thus, B with the prescribed initial conditions is calculable if and only if R does not halt.

6   Reducing A to B

It remains to show that the output of B is effectively the same as the output of the recurrence

$$
\begin{aligned}
A(n) =& A(n-4-A(A(n-4))) + 4A(A(n-2)) \\
&+ A(2A(n-4-A(n-2)) + A(n-4)),
\end{aligned} \tag{6}
$$

given the right initial conditions.

Once more, suppose we have a reverse tag system $R = (\Sigma, \delta, d, W_0)$. One restriction that will be made on R is that $d(\Sigma) = \{0, 2\}$. Section 3 demonstrated how, despite this restriction, R can still simulate an ordinary tag system. The goal at the beginning of these notes, to show that A is Turing complete, is therefore still in reach.

Assume there are t symbols in Σ, and m symbols in the initial word W_0. Let $c_0 = \alpha(s_t, s_t) + 2$, as before. We now specify the initial conditions of A. For $n = 0, 1, \ldots, c_0$, A and B will share the same initial conditions. Immediately after, we'll have, for $0 \le n < m$ and $0 \le j < 4$,

$$
A(c_0 + 4n + j) = \begin{cases} 0, & j = 0, 2 \\ B(c_0 + 2n + 1), & j = 1 \\ 2B(c_0 + 2n + 2), & j = 3 \end{cases} \tag{7}
$$

The next theorem, stated without proof for space reasons, demonstrates how to obtain the sequence B from A. A proof of this theorem can be viewed online at http://arxiv.org/abs/1203.0586.

Theorem 3. *Using the given initial conditions for A and B, A is calculable if and only if B is calculable. If B is calculable, then* (7) *holds for all $n \geq 0$.*

7 Concluding Remarks

In this paper, we have shown the existence of an undecidable nested recurrence relation. Furthermore, like its more well known cousins (1), (2), (3) and (4), our recurrence relation (6) is formed only from the operations of addition, subtraction, and recursion. Thus the result lends support to the idea that, in general, it will be difficult to prove broad results about nested recurrence relations. It will be interesting to try to determine whether other nested recurrence relations, such as (1), are decidable or not. If it is undecidable then it will certainly involve extending the techniques that are presented here, since the form of the recursion seems to prevent lookups in the manner we used.

Acknowledgements. The research was supported in part by an NSERC Discovery Grant. The authors would like to thank the anonymous referees for their suggestions and for providing reference [4].

References

1. Allouche, J.P., Shallit, J.: A variant of Hofstadter's sequence and finite automata. arXiv:1103.1133v2 (2011)
2. Balamohan, B., Kuznetsov, A., Tanny, S.: On the behavior of a variant of Hofstadter's Q-sequence. J. Integer Sequences 10, 29 pages (2007)
3. Cocke, J., Minsky, M.: Universality of tag systems with $p = 2$. J. ACM 11(1), 15–20 (1964)
4. Conway, J.H.: Unpredictable iterations. In: Proceedings of the 1972 Number Theory Conference, pp. 49–52 (August 1972)
5. Cook, M.: Universality in elementary cellular automata. Complex Systems 15(1), 1–40 (2004)
6. Downey, P.J., Griswold, R.E.: On a family of nested recurrences. Fibonacci Quarterly 22(4), 310–317 (1984)
7. Golomb, S.: Discrete chaos: sequences satisfying "strange" recursions (1991) (preprint)
8. Hofstadter, D.R.: Gödel, Escher, Bach: An Eternal Golden Braid. Basic Books (1979)
9. Kubo, T., Vakil, R.: On Conway's recursive sequence. Discrete Mathematics 152(1), 225–252 (1996)

10. Minsky, M.L.: Recursive unsolvability of Post's problem of "Tag" and other topics in theory of Turing machines. The Annals of Mathematics 74(3), 437–455 (1961)
11. Post, E.L.: Formal reductions of the general combinatorial decision problem. American Journal of Mathematics 65(2), 197–215 (1943)
12. Ruskey, F.: Fibonacci meets Hofstadter. Fibonacci Quarterly 49(3), 227–230 (2011)
13. Smith, A.R.: Simple computation-universal cellular spaces and self-reproduction. In: IEEE Conference Record of the 9th Annual Symposium on Switching and Automata Theory 1968, pp. 269–277 (October 1968)

Hard Instances of Algorithms and Proof Systems

Yijia Chen[1], Jörg Flum[2], and Moritz Müller[3]

[1] Department of Computer Science and Engineering, Shanghai Jiao Tong University,
Dongchuan Road, No. 800, 200240 Shanghai, China
`yijia.chen@cs.sjtu.edu.cn`
[2] Abteilung für mathematische Logik, Albert-Ludwigs-Universität Freiburg,
Eckerstraße 1, 79104 Freiburg, Germany
`joerg.flum@math.uni-freiburg.de`
[3] Kurt Gödel Research Center for Mathematical Logic,
Währinger Straße 25, 1090 Wien, Austria
`moritz.mueller@univie.ac.at`

Abstract. If the class TAUT of tautologies of propositional logic has no almost optimal algorithm, then every algorithm $\mathbb{A}$ deciding TAUT has a polynomial time computable sequence witnessing that $\mathbb{A}$ is not almost optimal. We show that this result extends to every Π_t^p-complete problem with $t \geq 1$; however, assuming the Measure Hypothesis, there is a problem which has no almost optimal algorithm but is decided by an algorithm without such a hard sequence. Assuming that a problem Q has an almost optimal algorithm, we analyze whether every algorithm deciding Q, which is not almost optimal algorithm, has a hard sequence.

1 Introduction

Let $\mathbb{A}$ be an algorithm deciding a problem Q. A sequence $(x_s)_{s \in \mathbb{N}}$ of strings in Q is *hard for* $\mathbb{A}$ if it is computable in polynomial time and the sequence $(t_{\mathbb{A}}(x_s)_{s \in \mathbb{N}})$ is not polynomially bounded in s.[1] Here, $t_{\mathbb{A}}(x)$ denotes the number of steps the algorithm $\mathbb{A}$ takes on input x. Clearly, if $\mathbb{A}$ is polynomial time, then $\mathbb{A}$ has no hard sequences. Furthermore, an almost optimal algorithm for Q has no hard sequences either. Recall that an algorithm $\mathbb{A}$ is *almost optimal for Q* if for any other algorithm $\mathbb{B}$ deciding Q and all $x \in Q$ the running time $t_{\mathbb{A}}(x)$ is polynomially bounded in $t_{\mathbb{B}}(x)$. In fact, if $(x_s)_{s \in \mathbb{N}}$ is a hard sequence for an algorithm, then one can superpolynomially speed up it on $\{x_s \mid s \in \mathbb{N}\}$, so it cannot be almost optimal.

Central to this paper is the question: To what extent can we show that algorithms which are not almost optimal have hard sequences? Our starting point is the following result (more or less explicit in [3,11]):

> If TAUT, *the class of tautologies of propositional logic, has no almost optimal algorithm, then every algorithm deciding* TAUT *has hard sequences.*

First we generalize this result from the Π_1^p-complete problem TAUT to all problems which are Π_t^p-complete for some $t \geq 1$:

---

[1] All notions will be defined in a precise manner later.

S.B. Cooper, A. Dawar, and B. Löwe (Eds.): CiE 2012, LNCS 7318, pp. 118–128, 2012.
© Springer-Verlag Berlin Heidelberg 2012

(∗) *If a Π_t^p-complete problem Q has no almost optimal algorithm, then every algorithm deciding Q has hard sequences.*

Apparently there are some limitations when trying to show (∗) for all problems Q as we prove:

(+) *If the Measure Hypothesis holds, then there is a problem which has no almost optimal algorithm but is decided by an algorithm without hard sequences.*

Perhaps one would expect that one can strengthen (∗) and show that even if a Π_t^p-complete problem Q has an almost optimal algorithm, then every algorithm, which is not almost optimal and decides Q, has a hard sequence. However, we show:

> *If the Measure Hypothesis holds, then every problem with padding and with an almost optimal algorithm is decided by an algorithm which is not almost optimal but has no hard sequences.*

As an algorithm deciding a problem Q which is not almost optimal can be polynomially speeded up on an infinite subset of Q, by (+) we see that, at least under the Measure Hypothesis, this notion of speeding up (e.g., considered in [13]) is weaker than our notion of the existence of a hard sequence.

Assume that $Q :=$ TAUT (or any Π_t^p-complete Q) has no almost optimal algorithm; thus, by (∗), every algorithm deciding Q has a hard sequence. Can we even effectively assign to every algorithm deciding Q a hard sequence? We believe that under reasonable complexity-theoretic assumptions one should be able to show that such an effective procedure or at least a polynomial time procedure does not exist, but we were not able to show it. However, recall that by a result due to Stockmeyer [13] and rediscovered by Messner [10] we know:

> *For every* EXP-*hard problem Q there is a polynomial time effective procedure assigning to every algorithm solving Q a sequence hard for it.*

Hence, if EXP $= \Pi_t^p$, then for every Π_t^p-hard problem Q there is a polynomial time effective procedure assigning a hard sequence to every algorithm deciding Q.

Our proof of (∗) generalizes to nondeterministic algorithms. This "nondeterministic statement" yields a version for Π_t^p-complete problems of a result that Krajíček derived for non-optimal propositional proof systems: If TAUT has no optimal proof system, then for every propositional proof system $\mathbb{P}$ there is a polynomial time computable sequence $(\alpha_s)_{s \in \mathbb{N}}$ of propositional tautologies α_s which only have superpolynomial $\mathbb{P}$-proofs; moreover, he showed that the α_s can be chosen with $s \leq |\alpha_s|$. While it is well-known that for any problem Q nondeterministic algorithms deciding Q and proof systems for Q are more or less the same, the relationship between deterministic algorithms and propositional proof systems is more subtle. Nevertheless, we are able to use (∗) to derive a statement on hard sequences for Π_t^p-complete problems Q without a *polynomially* optimal proof system.

As a byproduct, we obtain results in "classical terms" (that is, not referring to hard sequences). For example, we get for $t \geq 1$:

> *Let Q be Π_t^p-complete. Then, Q has an almost optimal algorithm if and only if Q has a polynomially optimal proof system.*
>
> *If some Π_t^p-complete has no almost optimal algorithm, then every Π_t^p-hard problem has no almost optimal algorithm.*

It is still open whether there exist problems outside of NP with optimal proof systems. We show their existence (in NE) assuming the Measure Hypothesis. Krajíček and Pudlák [7] proved that NE = coNE implies that TAUT has an optimal proof system, a result later strengthened by [8,1].

If for an algorithm $\mathbb{A}$ deciding a problem Q we have a hard sequence $(x_s)_{s\in\mathbb{N}}$ satisfying $s \le |x_s|$, then $\{x_s \mid s \in \mathbb{N}\}$ is a *hard set for* $\mathbb{A}$, that is, a polynomial time decidable subset of Q on which $\mathbb{A}$ is not polynomial time. Messner [10] has shown for any Q with padding that all algorithms deciding Q have hard sets if and only if Q has no polynomially optimal proof system. We show for arbitrary Q that the existence of hard sets for all algorithms is equivalent to the existence of an effective enumeration of all polynomial time decidable subsets of Q, a property which has turned out to be useful in various contexts (cf. [12,3,4]). We analyze what Messner's result means for proof systems.

The content of the sections is the following. In Section 2 we recall some concepts. We deal with hard sequences for algorithms in Section 3 and for proof systems in Section 4. Section 5 is devoted to hard sets and Section 6 contains the results and the examples of problems with special properties obtained assuming that the Measure Hypothesis holds. Finally Section 7 gives an effective procedure yielding hard sequences for nondeterministic algorithms for coNEXP-hard problems. Due to space limitations we defer almost all proofs to the full version of this extended abstract.

2   Preliminaries

By $n^{O(1)}$ we denote the class of polynomially bounded functions on the natural numbers. We let Σ be the alphabet $\{0, 1\}$ and $|x|$ the length of a string $x \in \Sigma^*$. We identify problems with subsets of Σ^*. *In this paper we always assume that Q denotes a decidable and nonempty problem.*

We assume familiarity with the classes P (polynomial time), NP (nondeterministic polynomial time) and the classes Π_t^p for $t \ge 1$ (the "universal" class of the tth level of the polynomial hierarchy). In particular, $\Pi_1^p = $ coNP.

The *Measure Hypothesis* [5] is the assumption "NP does not have measure 0 in E."' For the corresponding notion of measure we refer to [9]. This hypothesis is sometimes used in the theory of resource bounded measures.

A problem $Q \subseteq \Sigma^*$ has *padding* if there is a function $pad : \Sigma^* \times \Sigma^* \to \Sigma^*$ computable in logarithmic space having the following properties:

- For any $x, y \in \Sigma^*$, $|pad(x,y)| > |x| + |y|$ and $\big(pad(x,y) \in Q \iff x \in Q\big)$.
- There is a logspace algorithm which, given $pad(x,y)$ recovers y.

By $\langle \ldots, \ldots \rangle$ we denote some standard logspace computable tupling function with logspace computable inverses.

If $\mathbb{A}$ is a deterministic or nondeterministic algorithm and $\mathbb{A}$ accepts the string x, then we denote by $t_{\mathbb{A}}(x)$ the minimum number of steps of an accepting run of $\mathbb{A}$ on x; if $\mathbb{A}$ does not accept x, then $t_{\mathbb{A}}(x)$ is not defined. By $L(\mathbb{A})$ we denote the language accepted by $\mathbb{A}$. We use deterministic and nondeterministic Turing machines with Σ as alphabet as our basic computational model for algorithms (and we often use the notions "algorithm" and "Turing machine" synonymously). If necessary we shall not distinguish between a Turing machine and its code, a string in Σ^*. *By default, algorithms are deterministic.* If an algorithm $\mathbb{A}$ on input x eventually halts and outputs a value, we denote it by $\mathbb{A}(x)$.

3   Hard Sequences for Algorithms

In this section we derive the results concerning the existence of hard sequences for Π_t^p-complete problems.

Let $Q \subseteq \Sigma^*$. A deterministic (nondeterministic) algorithm $\mathbb{A}$ deciding (accepting) Q is *almost optimal* if for every deterministic (nondeterministic) algorithm $\mathbb{B}$ deciding (accepting) Q we have

$$t_{\mathbb{A}}(x) \leq \big(t_{\mathbb{B}}(x) + |x|\big)^{O(1)}$$

for all $x \in Q$. Note that nothing is required for $x \notin Q$.

Clearly, every problem in P (NP) has an almost optimal (nondeterministic) algorithm. There are problems outside P with an almost optimal algorithm (see Messner[10, Corollary 3.33]; we slightly improve his result in Theorem 22 of Section 6). However, it is not known whether there are problems outside NP having an almost optimal nondeterministic algorithm and it is not known whether there are problems with padding outside P having an almost optimal algorithm. We show in Theorem 23 of Section 6 that the former is true if the Measure Hypothesis holds.

Definition 1. Let $Q \subseteq \Sigma^*$.
(1) Let $\mathbb{A}$ be a deterministic (nondeterministic) algorithm deciding (accepting) Q. A sequence $(x_s)_{s \in \mathbb{N}}$ is *hard for* $\mathbb{A}$ if $\{x_s \mid s \in \mathbb{N}\} \subseteq Q$, the function $1^s \mapsto x_s$ is computable in polynomial time, and $t_{\mathbb{A}}(x_s)$ is not polynomially bounded in s.
(2) The problem Q *has hard sequences for algorithms (for nondeterministic algorithms)* if every (nondeterministic) algorithm deciding Q has a hard sequence.

In the proof of the following lemma we show that an algorithm $\mathbb{A}$ can be superpolynomially speeded up on $\{x_s \mid s \in \mathbb{N}\}$ if $(x_s)_{s \in \mathbb{N}}$ is hard for $\mathbb{A}$.

Lemma 2. *Let $\mathbb{A}$ be a deterministic (nondeterministic) algorithm deciding (accepting) Q. If $\mathbb{A}$ has a hard sequence, then $\mathbb{A}$ is not almost optimal.*

As already remarked in the Introduction, part (b) of the next theorem, the main
result of this section, generalizes the corresponding result for $Q = \text{TAUT}$ due to
Krajíček.

Theorem 3. *Let Q be a Π_t^p-complete problem for some $t \geq 1$. Then:*
(a) *Q has no almost optimal algorithm $\iff$ Q has hard sequences for algorithms.*
(b) *Q has no almost optimal nondeterministic algorithm $\iff$ Q has hard
sequences for nondeterministic algorithms.*

Remark 4. For $Q = \text{TAUT}$, part (a) is implicit in [3,11]. In fact, there it is shown
that a halting problem polynomially isomorphic to TAUT has hard sequences for
algorithms if it has no almost optimal algorithm. In Remark 14 we show how this
can be extended to every coNP-complete problem using known results relating
almost optimal algorithms and proof systems.

Lemma 2 yields the implications from right to left in Theorem 3. The following
considerations will yield a proof of the converse direction. For a nondeterministic
algorithm $\mathbb{A}$ and $s \in \mathbb{N}$ let $\mathbb{A}^s$ be the algorithm that rejects all $x \in \Sigma^*$ with
$|x| > s$. If $|x| \leq s$, then it simulates s steps of $\mathbb{A}$ on input x; if this simulation
halts and accepts, then $\mathbb{A}^s$ accepts; otherwise it rejects.

For $Q \subseteq \Sigma^*$ we consider the *deterministic (nondeterministic) algorithm subset
problem* $\text{DAS}(Q)$ $(\text{NAS}(Q))$

$\text{DAS}(Q)$ $(\text{NAS}(Q))$
  Instance: A (nondeterministic) algorithm $\mathbb{A}$ and 1^s with
    $s \in \mathbb{N}$.
  Question: $L(\mathbb{A}^s) \subseteq Q$?

The following two lemmas relate the equivalent statements in Theorem 3 (a)
(in Theorem 3 (b)) to a statement concerning the complexity of $\text{DAS}(Q)$ (of
$\text{NAS}(Q)$).

Lemma 5. (a) *If $\langle \mathbb{A}, 1^s \rangle \in \text{DAS}(Q)$ is solvable in time $s^{f(\mathbb{A})}$ for some function
f, then Q has an almost optimal algorithm.*
(b) *If there is a nondeterministic algorithm $\mathbb{V}$ accepting $\text{NAS}(Q)$ such that for
all $\langle \mathbb{A}, 1^s \rangle \in \text{NAS}(Q)$ we have $t_{\mathbb{V}}(\langle \mathbb{A}, 1^s \rangle) \leq s^{f(\mathbb{A})}$ for some function f, then
Q has an almost optimal nondeterministic algorithm.*

If Q is Π_t^p-complete, then $\text{NAS}(Q)$ and hence $\text{DAS}(Q)$ are in Π_t^p, too (this is the
reason why 1^s and not just s is part of the input of $\text{NAS}(Q)$ and of $\text{DAS}(Q)$).
Thus, together with Lemma 5 the following lemma yields the remaining claims
of Theorem 3.

Lemma 6. (a) *Assume that $\text{DAS}(Q) \leq_p Q$, that is, that $\text{DAS}(Q)$ is polynomial
time reducible to Q. If $\langle \mathbb{A}, 1^s \rangle \in \text{DAS}(Q)$ is not solvable in time $s^{f(\mathbb{A})}$ for
some function f, then Q has hard sequences for algorithms.*

(b) *Assume that* $\mathrm{NAS}(Q) \leq_p Q$. *If there is no nondeterministic algorithm* $\mathbb{V}$ *accepting* $\mathrm{NAS}(Q)$ *such that for all* $\langle \mathbb{A}, 1^s \rangle \in \mathrm{NAS}(Q)$ *we have* $t_{\mathbb{V}}(\langle \mathbb{A}, 1^s \rangle) \leq s^{f(\mathbb{A})}$ *for some function* f, *then* Q *has hard sequences for nondeterministic algorithms.*

Remark 7. In the proof of Theorem 3 we use the assumption that Q is Π_t^p-complete only to ensure that $\mathrm{NAS}(Q) \leq_p Q$ (cf. Lemma 6). This condition is also fulfilled for every Q complete, say, in one of the classes E or PSPACE. Thus the statements of Theorem 3 hold for such a Q.

Remark 8. Assume that Q is Π_t^p-complete and has padding (for $t = 1$, the set TAUT is an example of such a Q). If Q has no almost optimal algorithm, then every algorithm $\mathbb{B}$ deciding Q has a hard sequence $(x_s)_{s \in \mathbb{N}}$ with $s \leq |x_s|$. Then, in particular

$$\{x_s \mid s \in \mathbb{N}\} \in \mathrm{P} \qquad \text{and} \qquad \mathbb{B} \text{ is not polynomial time on } \{x_s \mid s \in \mathbb{N}\}.$$

In fact, it is well-known that for Q with padding we can replace any polynomial time reduction to Q by a length-increasing one. An analysis of the proof of Lemma 6 shows that then we can get hard sequences $(x_s)_{s \in \mathbb{N}}$ with $s \leq |x_s|$.

In contrast to the last remark, for the validity of the next lemma it is important that we do not require $s \leq |x_s|$ in our definition of hard sequence.

Lemma 9. *Assume that* $\mathbb{S}$ *is a polynomial time reduction from* Q *to* Q' *and let* $\mathbb{B}$ *be a (nondeterministic) algorithm deciding (accepting)* Q'. *If* $(x_s)_{s \in \mathbb{N}}$ *is a hard sequence for* $\mathbb{B} \circ \mathbb{S}$, *then* $(\mathbb{S}(x_s))_{s \in \mathbb{N}}$ *is a hard sequence for* $\mathbb{B}$.

Therefore, if $Q \leq_p Q'$ *and* Q *has hard sequences for (nondeterministic) algorithms then so does* Q'.

We derive two consequences of our results:

Corollary 10. *Assume* $t \geq 1$ *and let* Q *and* Q' *be* Π_t^p-*complete. Then,* Q *has an almost optimal algorithm if and only if* Q' *has an almost optimal algorithm.*

Corollary 11. *Let* $t \geq 1$ *and assume that the some* Π_t^p-*complete problem has no almost optimal algorithm. Then every* Π_t^p-*hard problem has no almost optimal algorithm.*

4   Hard Sequences for Proof Systems

In this section we translate the results on hard sequences from algorithms to proof systems. We first recall some basic definitions.

A *proof system for* Q is a polynomial time algorithm $\mathbb{P}$ computing a function from Σ^* onto Q. If $\mathbb{P}(w) = x$, we say that w is a $\mathbb{P}$-*proof* of x.

Let $\mathbb{P}$ and $\mathbb{P}'$ be proof systems for Q. An algorithm $\mathbb{T}$ is a *translation from* $\mathbb{P}'$ *into* $\mathbb{P}$ if $\mathbb{P}(\mathbb{T}(w')) = \mathbb{P}'(w')$ for every $w' \in \Sigma^*$. Note that translations always exist. A translation is *polynomial* if it runs in polynomial time.

A proof system $\mathbb{P}$ for Q is *p-optimal* or *polynomially optimal* if for every proof system $\mathbb{P}'$ for Q there is a polynomial translation from $\mathbb{P}'$ into $\mathbb{P}$. A proof system $\mathbb{P}$ for Q is *optimal* if for every proof system $\mathbb{P}'$ for Q and all $w' \in \Sigma^*$ there is a $w \in \Sigma^*$ such that $\mathbb{P}(w) = \mathbb{P}'(w')$ and $|w| \leq |w'|^{O(1)}$. Clearly, any p-optimal proof system is optimal.

We often will make use of the following relationship between the optimality notions for algorithms and that for proof systems (see [7,10]).

Theorem 12. *(1) For every Q we have (a) $\Rightarrow$ (b) and (b) $\Rightarrow$ (c); moreover (a), (b), and (c) are all equivalent if Q has padding. Here*
  (a) Q has a p-optimal proof system.
  (b) Q has an almost optimal algorithm.
  (c) There is an algorithm that decides Q and runs in polynomial time on every subset X of Q with $X \in \mathrm{P}$.
(2) For every Q we have (a) $\Longleftrightarrow$ (b), (b) $\Rightarrow$ (c), and (c) $\Rightarrow$ (d); moreover (a)–(d) are all equivalent if Q has padding. Here
  (a) Q has an optimal proof system.
  (b) Q has an almost optimal nondeterministic algorithm.
  (c) There is a nondeterministic algorithm that accepts Q and runs in polynomial time on every subset X of Q with $X \in \mathrm{NP}$.
  (d) There is a nondeterministic algorithm that accepts Q and runs in polynomial time on every subset X of Q with $X \in \mathrm{P}$.

We use our results of Section 3 to extend the equivalence between (a) and (b) of part (1) of Theorem 12 to arbitrary Π_t^p-complete problems:

Theorem 13. *Let Q be a Π_t^p-complete problem for some $t \geq 1$. Then:*

Q *has a p-optimal proof system* $\Longleftrightarrow$ Q *has an almost optimal algorithm.*

Remark 14. Using Theorem 12, for every coNP-complete Q we get a simple, direct proof of
  if Q has no almost optimal algorithm, then Q has hard sequences for algorithms
using the result for $Q = \textsc{Taut}$ (that we already knew by Remark 4). In fact, assume that Q has no almost optimal algorithm. Then $\textsc{Taut}$ has no almost optimal algorithm; otherwise, $\textsc{Taut}$ has a p-optimal proof system by the equivalence of (a) and (b) in part (1) of Theorem 12 ($\textsc{Taut}$ has padding!). As $Q \leq_p \textsc{Taut}$, then Q has a p-optimal proof system too (cf. [8, Lemma 1]) and hence, again by Theorem 12, an almost optimal algorithm, a contradiction. Thus, $\textsc{Taut}$ has hard sequences for algorithms. As $\textsc{Taut} \leq_p Q$, by Lemma 9 the problem Q has hard sequences for algorithms, too.

We already mentioned that for every $Q \subseteq \Sigma^*$ there is a well-known and straightforward correspondence between proof systems and nondeterministic algorithms preserving the optimality notions, so that the proof of the equivalence between (a) and (b) in Theorem 12 (2) is immediate. Thus the translation of our results for nondeterministic algorithms to proof systems is easy and we omit it here. Moreover, the corresponding results are due to Krajíček [6] who proved them by quite different means.

Definition 15. (1) Let $\mathbb{P}$ be a proof systems for Q. A sequence $(x_s)_{s\in\mathbb{N}}$ is *hard for* $\mathbb{P}$ if $\{x_s \mid s \in \mathbb{N}\} \subseteq Q$, the function $1^s \mapsto x_s$ is computable in polynomial time, and there is no polynomial time algorithm $\mathbb{W}$ with $\mathbb{P}(\mathbb{W}(1^s)) = x_s$ for all $s \in \mathbb{N}$.

(2) The problem Q *has hard sequences for proof systems* if every proof system for Q has a hard sequence.

For $Q = \text{TAUT}$ the following result is already known (cf., e.g., the survey [2, Section 11]). We give a new proof that works for any, not necessarily paddable Π_t^p-complete problem Q.

Theorem 16. *Let Q be a Π_t^p-complete problem for some $t \geq 1$. Then:*

Q has no p-optimal proof system iff Q has hard sequences for proof systems.

Again an analysis of the proof of this theorem shows that for Q with padding, we can require that the claimed hard sequence $(x_s)_{s\in\mathbb{N}}$ satisfies $s \leq |x_s|$.

5   Hard Subsets

If for an algorithm $\mathbb{A}$ deciding a problem Q we have a hard sequence $(x_s)_{s\in\mathbb{N}}$ satisfying $s \leq |x_s|$, then $\{x_s \mid s \in \mathbb{N}\}$ is a polynomial time decidable subset of Q on which $\mathbb{A}$ is not polynomial time. We then speak of a hard set for $\mathbb{A}$ even if its elements cannot be generated in polynomial time. More precisely:

Definition 17. Let $Q \subseteq \Sigma^*$.
(1) Let $\mathbb{A}$ be a deterministic or nondeterministic algorithm accepting Q. A subset X of Q is *hard for* $\mathbb{A}$ if $X \in \text{P}$ and $\mathbb{A}$ is not polynomial time on X.
(2) The problem Q *has hard sets for (nondeterministic) algorithms* if every (nondeterministic) algorithm deciding Q has a hard set.

Using these notions the equivalences (a) $\Leftrightarrow$ (c) and (a) $\Leftrightarrow$ (d) in Theorem 12 (1) and (2), respectively, can be expressed in the following way:

Assume that Q has padding. Then
(1) Q has no almost optimal algorithm $\iff Q$ has hard sets for algorithms.
(2) Q has no almost optimal nondeterministic algorithm $\iff Q$ has hard sets for nondeterministic algorithms.

Hence, we get (we leave the nondeterministic variant to the reader):

Corollary 18. *Assume Q has padding.*
(a) If Q has hard sequences for algorithms, then Q has hard sets for algorithms.
(b) If in addition Q is Π_t^p-complete, then Q has hard sequences for algorithms if and only if Q has hard sets for algorithms.

Assume that Q has an almost optimal algorithm. Then, in general, one cannot show that every algorithm deciding Q, which is not almost optimal, has a hard set. In fact, Messner [10, Corollary 3.33] has presented a P-immune Q_0 with an almost optimal algorithm. Of course, no algorithm deciding Q_0 has a hard set.

For an arbitrary problem Q the existence of hard subsets is equivalent to a (non-)listing property. We introduce this property.

Let C be the complexity class P or NP. A set X is a C-subset of Q if $X \subseteq Q$ and $X \in C$. We write List(C, Q) and say that there is a *listing of the* C-*subsets of Q by* C-*machines* if there is an algorithm that, once having been started, lists Turing machines $\mathbb{M}_1, \mathbb{M}_2, \ldots$ of type C such that $\{L(\mathbb{M}_i) \mid i \geq 1\} = \{X \subseteq Q \mid X \in C\}$.

For Q with padding the equivalences in the following proposition were known [12].

Theorem 19. *(1) Q has hard sets for algorithms $\iff$ not List(P, Q).*
(2) Every nondeterministic algorithm $\mathbb{A}$ accepting Q is not polynomial on at least one subset X of Q with $X \in$ NP $\iff$ not List(NP, Q).

We close this section by introducing hard subsets for proof systems and stating the corresponding result.

Definition 20. (1) Let $\mathbb{P}$ be a proof system for Q. A subset X of Q is *hard for $\mathbb{P}$* if $X \in$ P and there is no polynomial time algorithm $\mathbb{W}$ such that $\mathbb{P}(\mathbb{W}(x)) = x$ for all $x \in X$.
(2) Q *has hard sets for proof systems* if every proof system for Q has a hard set.

The following result can be obtained along the lines of the proof of Theorem 16.

Theorem 21. *Let Q be a problem with padding. Then:*

Q has no p-optimal proof system if and only if Q has hard sets for proof systems.

6   Assuming the Measure Hypothesis

In this section we present some examples of problems with special properties, some yield limitations to possible extensions of results mentioned in this paper. Most are proven assuming the Measure Hypothesis.

Recall that an algorithm $\mathbb{A}$ deciding Q is *optimal* if for every algorithm $\mathbb{B}$ deciding Q we have

$$t_\mathbb{A}(x) \leq (t_\mathbb{B}(x) + |x|)^{O(1)}$$

for *all* $x \in \Sigma^*$. Clearly, every problem in P has an optimal algorithm.

Theorem 22. *(1) There exist problems in* $E \setminus P$ *with optimal algorithms.*
(2) If the Measure Hypothesis holds, then there exist problems in $NP \setminus P$ *with optimal algorithms.*

Here, $E := \bigcup_{d \in \mathbb{N}} \text{DTIME}(2^{d \cdot n})$. Messner [10] showed the existence of problems in $E \setminus P$ with *almost* optimal algorithms. The question whether there are sets outside of NP with optimal proof systems was stated by Krajíček and Pudlák [7] and is still open. As already mentioned, they proved that TAUT has an optimal proof system if $E = NE$ $(:= \bigcup_{d \in \mathbb{N}} \text{NTIME}(2^{d \cdot n}))$. We are able to show:

Theorem 23. *If the Measure Hypothesis holds, then there exist problems in* $NE \setminus NP$ *with optimal proof systems (or, equivalently, with almost optimal non-deterministic algorithms).*

Concerning algorithms which are not almost optimal but do not have hard sequences we derive the following results.

Theorem 24. *Let Q be a problem with padding and with an almost optimal algorithm. If the Measure Hypothesis holds, then there is an algorithm deciding Q, which is not almost optimal and has hard sets but does not have hard sequences.*

The following example shows that the padding hypothesis is necessary in Theorem 24.

Example. Let $Q := \{1^n \mid n \in \mathbb{N}\}$. As $Q \in P$, it has an almost optimal algorithm. However, the set Q itself is a hard set and $(1^s)_{s \in \mathbb{N}}$ a hard sequence for every non-optimal (that is, for every superpolynomial) algorithm deciding Q.

Corollary 25. *If the Measure Hypothesis holds, then the following are equivalent for $t \geq 1$:*
(i) No Π_t^p-complete problem has an almost optimal algorithm.
(ii) Every non-almost optimal algorithm deciding a Π_t^p-complete problem has hard sequences.

Theorem 26. *If the Measure Hypothesis holds, there is a problem which has hard sets for algorithms (and hence has no almost optimal algorithm) but has algorithms without hard sequences.*

7 Getting Hard Sequences in an Effective Way

We have mentioned in the Introduction that Stockmeyer [13] has shown that for every EXP-hard problem Q there is a polynomial time procedure assigning to every algorithm deciding Q a hard sequence. Based on his proof we derive a "nondeterministic" version.

Theorem 27. *Let Q be a coNEXP-hard problem. Then there is a polynomial time computable function $g : \Sigma^* \times \{1\}^* \to \Sigma^*$ such that for every nondeterministic algorithm $\mathbb{A}$ accepting Q the sequence $\big(g(\mathbb{A}, 1^s)\big)_{s \in \mathbb{N}}$ is hard for $\mathbb{A}$.*

Acknowledgments. The authors thank the John Templeton Foundation for its support through Grant #13152. Yijia Chen is affiliated with BASICS and MOE-MS Key Laboratory for Intelligent Computing and Intelligent Systems which is supported by National Nature Science Foundation of China (61033002). Moritz Müller thanks the FWF (Austrian Research Fund) for its support through Grant number P 23989 - N13.

References

1. Ben-David, S., Gringauze, A.: On the existence of optimal propositional proof systems and oracle-relativized propositional logic. Electronic Colloquium on Computational Complexity (ECCC), Technical Report TR98-021 (1998)
2. Beyersdorff, O.: On the correspondence between arithmetic theories and propositional proof systems - a survey. Mathematical Logic Quarterly 55(2), 116–137 (2009)
3. Chen, Y., Flum, J.: On p-Optimal Proof Systems and Logics for PTIME. In: Abramsky, S., Gavoille, C., Kirchner, C., Meyer auf der Heide, F., Spirakis, P.G. (eds.) ICALP 2010, Part II. LNCS, vol. 6199, pp. 321–332. Springer, Heidelberg (2010)
4. Chen, Y., Flum, J.: Listings and logics. In: Proceedings of the 26th Annual IEEE Symposium on Logic in Computer Science (LICS 2011), pp. 165–174. IEEE Computer Society (2011)
5. Hitchcock, J.M., Pavan, A.: Hardness hypotheses, derandomization, and circuit complexity. In: Lodaya, K., Mahajan, M. (eds.) FSTTCS 2004. LNCS, vol. 3328, pp. 336–347. Springer, Heidelberg (2004)
6. Krajíček, J.: Bounded arithmetic, propositional logic, and complexity theory. Cambridge University Press (1995)
7. Krajíček, J., Pudlák, P.: Propositional proof systems, the consistency of first order theories and the complexity of computations. The Journal of Symbolic Logic 54, 1063–1088 (1989)
8. Köbler, J., Messner, J.: Complete problems for promise classes by optimal proof systems for test sets. In: Proceedings of the 13th IEEE Conference on Computational Complexity (CCC 1998), pp. 132–140 (1998)
9. Mayordomo, E.: Almost every set in exponential time is P-bi-immune. Theoretical Computer Science 136(2), 487–506 (1994)
10. Messner, J.: On the Simulation Order of Proof Systems. PhD Thesis, Univ. Erlangen (2000)
11. Monroe, H.: Speedup for natural problems and noncomputability. Theoretical Computer Science 412(4-5), 478–481 (2011)
12. Sadowski, Z.: On an optimal propositional proof system and the structure of easy subsets of TAUT. Theoretical Computer Science 288(1), 181–193 (2002)
13. Stockmeyer, L.: The Complexity of Decision Problems in Automata Theory. PhD. Thesis, MIT (1974)

On Mathias Generic Sets

Peter A. Cholak[1], Damir D. Dzhafarov[1], and Jeffry L. Hirst[2]

[1] Department of Mathematics, 255 Hurley Building, University of Notre Dame,
Notre Dame, IN 46556-4618, United States of America
cholak@nd.edu, ddzhafar@nd.edu
[2] Department of Mathematical Sciences, 322 Walker Hall, Appalachian State
University, Boone, NC 28608-2091, United States of America
jlh@math.appstate.edu

Abstract. We present some results about generics for computable
Mathias forcing. The n-generics and weak n-generics in this setting form
a strict hierarchy as in the case of Cohen forcing. We analyze the complexity of the Mathias forcing relation, and show that if G is any n-generic
with $n \geq 3$ then it satisfies the jump property $G^{(n-1)} = G' \oplus \varnothing^{(n)}$. We
prove that every such G has generalized high degree, and so cannot have
even Cohen 1-generic degree. On the other hand, we show that G, together with any bi-immune $A \leq_T \varnothing^{(n-1)}$, computes a Cohen n-generic.

1 Introduction

Forcing has been a central technique in computability theory since it was introduced (in the form we now call Cohen forcing) by Kleene and Post to exhibit a
degree strictly between $\mathbf{0}$ and $\mathbf{0}'$. The study of the algorithmic properties of Cohen generic sets, and of the structure of their degrees, has long been a rich source
of problems and results. In the present paper, we propose to undertake a similar
investigation of generic sets for (computable) Mathias forcing, and present some
of our initial results in this direction.

Mathias forcing was perhaps first used in computability theory by Soare in
[11] to build an infinite set with no subset of strictly higher degree. Subsequently,
it became a prominent tool for constructing infinite homogeneous sets for computable colorings of pairs of integers, as in Seetapun and Slaman [9], Cholak,
Jockusch, and Slaman [2], and Dzhafarov and Jockusch [4]. It has also found
applications in algorithmic randomness, in Binns, Kjos-Hanssen, Lerman, and
Solomon [1].

We show below that a number of results for Cohen generics hold also for Mathias generics, and that a number of others do not. The main point of distinction
is that neither the set of conditions, nor the forcing relation is computable, so
many usual techniques do not carry over. We begin with background in Section
2, and present some preliminary results in Section 3. In Section 4 we characterize
the complexity of the forcing relation, and in Section 5 we prove a number of
results about the degrees of Mathias generic sets, and about their relationship
to Cohen generic degrees. We indicate questions along the way we hope will be
addressed in future work.

S.B. Cooper, A. Dawar, and B. Löwe (Eds.): CiE 2012, LNCS 7318, pp. 129–138, 2012.
© Springer-Verlag Berlin Heidelberg 2012

2   Definitions

We assume familiarity with the terminology particular to Cohen forcing in computability theory. (For background on computability theory, see [10]. For background on Cohen generic sets, see Section 1.24 of [3].) The definition of the Mathias forcing partial order is standard, but its formalization in the setting of computability theory requires some care. A slightly different presentation is given in [1, Section 6], over which ours has the benefit of reducing the complexity of the set of conditions from Σ_3^0 to Π_2^0.

Definition 1

1. A (computable Mathias) pre-condition *is a pair* (D, E) *where* D *is a finite set*, E *is a computable set, and* $\max D < \min E$.
2. A (computable Mathias) condition *is a pre-condition* (D, E), *such that* E *is infinite*.
3. A pre-condition (D^*, E^*) *extends a pre-condition* (D, E), *written* $(D^*, E^*) \leq (D, E)$, *if* $D \subseteq D^* \subseteq D \cup E$ *and* $E^* \subseteq E$.
4. A set A *satisfies a pre-condition* (D, E) *if* $D \subseteq A \subseteq D \cup E$.

By an *index* for a pre-condition (D, E) we shall mean a pair (d, e) such that d is the canonical index of D and $E = \{x : \Phi_e(x) \downarrow = 1\}$. By adopting the convention that for all x, if $\Phi_e(x) \downarrow$ then $\Phi_e(y) \downarrow \in \{0, 1\}$ for all $y \leq x$, it follows that Φ_e is total if E is infinite, i.e., if (D, E) is a condition. Of course, if E is finite then Φ_e may only be defined on a proper initial segment of ω.

The definition makes the set of all indices Π_1^0. However, we can pass to a computable subset containing an index for every pre-condition. Namely, define a strictly increasing computable function g by

$$\Phi_{g(d,e)}(x) = \begin{cases} 0 & \text{if } x \leq \max D_d, \\ \Phi_e(x) & \text{otherwise.} \end{cases}$$

Then the set of pairs of the form $(d, g(d, e))$ is computable, and each is an index for a pre-condition. Moreover, if (d, e) is an index as well, then it and $(d, g(d, e))$ index the same pre-condition. Formally, all references to pre-conditions in the sequel will be to indices from this set, and we shall treat D and E as numbers when convenient.

Note that whether one pre-condition extends another is a Π_2^0 question. By our convention about partial computable functions, the same question for conditions is seen to be Π_1^0.

In what follows, a Σ_n^0 *set of conditions* refers to a Σ_n^0-definable set of pre-conditions, each of which is a condition. (Note that this is not the same as the set of all conditions satisfying a given Σ_n^0 definition, as discussed further in the next section.) We call such a set *dense* if it contains an extension of every condition, and define what it means to meet or avoid such a set as usual.

Definition 2. *Fix $n \in \omega$.*

1. *A set G is* Mathias n-generic *if it meets or avoids every Σ_n^0 set of conditions.*
2. *A set G is* weakly Mathias n-generic *if it meets every dense such set.*

We call a degree *generic* if it contains a set that is n-generic for all n.

It is easy to see that for every $n \geq 2$, there exists a Mathias n-generic $G \leq_T \varnothing^{(n)}$ (indeed, even $G' \leq_T \varnothing^{(n)}$). This is done just as in Cohen forcing (see [6, Lemma 2.6]), but as there is no computable listing of Σ_n^0 sets of conditions, one goes through the Σ_n^0 sets of pre-conditions and checks which of these consist of conditions alone. We pass to some other basic properties of generics. We refer to Mathias n-generics below simply as n-generics when no confusion is possible.

3   Basic Results

Note that the set of all conditions is Π_2^0. Thus, the set of conditions satisfying a given Σ_n^0 definition is Σ_n^0 if $n \geq 3$, and Σ_3^0 otherwise. For $n < 3$, we may thus wish to consider the following stronger form of genericity, which has no analogue in the case of Cohen forcing.

Definition 3. *A set G is* strongly n-generic *if, for every Σ_n^0-definable set of pre-conditions $\mathcal{P}$, either G satisfies some condition in $\mathcal{P}$ or G meets the set of conditions not extended by any condition in $\mathcal{P}$.*

Proposition 1. *For $n \geq 3$, a set is strongly n-generic if and only if it is n-generic. For $n \leq 2$, a set is strongly n-generic if and only if it is 3-generic.*

Proof. Evidently, every strongly n-generic set is n-generic. Now suppose $\mathcal{P}$ is a Σ_n^0 set of pre-conditions, and let $\mathcal{C}$ consist of all the conditions in $\mathcal{P}$. An infinite set meets or avoids $\mathcal{P}$ if and only if it meets or avoids $\mathcal{C}$, so every $\max\{n,3\}$-generic set meets or avoids $\mathcal{P}$. For $n \geq 3$, this means that every n-generic set is strongly n-generic, and for $n \leq 2$ that every 3-generic set is strongly n-generic.

It remains to show that every strongly 0-generic set is 3-generic. Let $\mathcal{C}$ be a given Σ_3^0 set of conditions, and let R be a computable relation such that (D, E) belongs to $\mathcal{C}$ if and only if $(\exists a)(\forall x)(\exists y)R(D, E, a, x, y)$. Define a strictly increasing computable function g by

$$\Phi_{g(D,E,a)}(x) = \begin{cases} \Phi_E(x) & \text{if } (\exists y)R(D, E, a, x, y) \text{ and } \Phi_E(x) \downarrow, \\ \uparrow & \text{otherwise,} \end{cases}$$

and let $\mathcal{P}$ be the computable set of all pre-conditions of the form $(D, g(D, E, a))$. If $(D, E) \in \mathcal{C}$ then Φ_E is total and so there is an a such that $\Phi_{g(D,E,a)} = \Phi_E$. If, on the other hand, (D, E) is a pre-condition not in $\mathcal{C}$ then for each a there is an x such that $\Phi_{g(D,E,a)}(x) \uparrow$. Thus, the members of $\mathcal{C}$ are precisely the conditions in $\mathcal{P}$, so an infinite set meets or avoids $\mathcal{C}$ if and only if it meets or avoids $\mathcal{P}$. In particular, every strongly 0-generic set meets or avoids $\mathcal{C}$.

As a consequence, we shall restrict ourselves to 3-genericity or higher from now on, or at most weak 2-genericity. Without further qualification, n below will always be a number ≥ 3.

Proposition 2. *Every n-generic set is weakly n-generic, and every weakly n-generic set is $(n-1)$-generic.*

Proof. The first implication is clear. For the second, let a Σ^0_{n-1} set $\mathcal{C}$ of conditions be given. Let $\mathcal{D}$ be the class of all conditions that are either in $\mathcal{C}$ or else have no extension in $\mathcal{C}$, which is clearly dense. If $n \geq 4$, then $\mathcal{D}$ is easily seen to be Σ^0_n (actually Π^0_{n-1}) as saying a condition (D, E) has no extension in $\mathcal{C}$ is written

$$\forall (D^*, E^*)[[(D^*, E^*) \text{ is a condition } \wedge (D^*, E^*) \leq (D, E)] \implies (D^*, E^*) \notin \mathcal{C}].$$

If $n = 3$, this makes $\mathcal{D}$ appear to be Σ^0_4 but since $\mathcal{C}$ is a set of conditions only, we can re-write the antecedent of the above implication as

$$D \subseteq D^* \subset D \cup E \wedge (\forall x)[\Phi_{E^*}(x) \downarrow = 1 \wedge \Phi_E(x) \downarrow \implies \Phi_E(x) = 1]$$

to obtain an equivalent Σ^0_3 definition. In either case, then, a weakly n-generic set must meet $\mathcal{D}$, and hence must either meet or avoid $\mathcal{C}$.

The proof of the following proposition is straightforward. (The first half is proved much like its analogue in the Cohen case. See, e.g., [8, Corollary 2.7].)

Proposition 3. *Every weakly n-generic set G is hyperimmune relative to $\varnothing^{(n-1)}$. If G is n-generic, then its degree forms a minimal pair with $\mathbf{0}^{(n-1)}$.*

Corollary 1. *Not every n-generic set is weakly $(n+1)$-generic.*

Proof. Take any n-generic $G \leq_T \varnothing^{(n)}$. Then G is not hyperimmune relative to $\varnothing^{(n+1)}$, and so cannot be weakly $(n+1)$-generic.

We shall separate weakly n-generic sets from n-generic sets in Section 5, thereby obtaining a strictly increasing sequence of genericity notions

$$\text{weakly 3-generic} \Longleftarrow \text{3-generic} \Longleftarrow \text{weakly 4-generic} \Longleftarrow \cdots$$

as in the case of Cohen forcing. In many other respects, however, the two types of genericity are very different. For instance, as noted in [2, Section 4.1], every Mathias generic G is cohesive, i.e., satisfies $G \subseteq^* W$ or $G \subseteq^* \overline{W}$ for every computably enumerable set W. In particular, if we write $G = G_0 \oplus G_1$ then one of G_0 or G_1 is finite. This is false for Cohen generics, which, by an analogue of van Lambalgen's theorem due to Yu [12, Proposition 2.2], have relatively n-generic halves. Thus, no Mathias generic can be even Cohen 1-generic.

Question 1. What form of van Lambalgen's theorem holds for Mathias forcing?

Another basic fact is that every Mathias n-generic G is high, i.e., satisfies $G' \geq_T \varnothing''$. (See [1, Corollary 6.7], or [2, Section 5.1] for a proof.) By contrast, it is a well-known result of Jockusch [6, Lemma 2.6] that every Cohen n-generic set G satisfies $G^{(n)} \equiv_T G \oplus \varnothing^{(n)}$. As no high G can satisfy $G'' \leq_T G \oplus \varnothing''$, it follows that no Mathias generic can have even Cohen 2-generic degree. This does not prevent a Mathias n-generic from having Cohen 1-generic degree, as there are high 1-generic sets, but we show this does not happen either in Corollary 4.

4   The Forcing Relation

Much of the discrepancy between Mathias and Cohen genericity stems from the fact that the complexity of forcing a formula, defined below, does not agree with the complexity of the formula. Our forcing language here is the typical one of formal first-order arithmetic plus a set variable, X, and the epsilon relation, $\in$.

We regard every Σ_0^0 (i.e., bounded quantifier) formula φ with no free number variables as being written in disjunctive normal form according to some fixed effective procedure for doing so. Call a disjunct *valid* if the conjunction of all the equalities and inequalities in it is true, which can be checked computably. For each i (ranging over the number of valid disjuncts), let $P_{\varphi,i}$ be the set of all n such that $n \in X$ is a conjunct of the ith valid disjunct, and $N_{\varphi,i}$ the set of all n such that $n \notin X$ is a conjunct of the ith valid disjunct. Canonical indices for these sets can be determined uniformly effectively from an index for φ.

Definition 4. *Let (D, E) be a condition and let $\varphi(X)$ be a formula with only the set variable X free. If φ is Σ_0^0, say (D, E) forces $\varphi(G)$, written $(D, E) \Vdash \varphi(G)$, if for some i, $P_{\varphi,i} \subseteq D$ and $N_{\varphi,i} \subseteq \overline{D \cup E}$. From here, extend the definition of $(D, E) \Vdash \varphi(G)$ to arbitrary φ inductively according to the standard definition of strong forcing (e.g., as in [3, p. 100, footnote 22, items (iii)–(v)]).*

Remark 1. If $\varphi(X)$ is Σ_0^0 with only the set variable X free and A is a set then $\varphi(A)$ holds if and only if there is an i such that $P_{\varphi,i} \subseteq A$ and $N_{\varphi,i} \subseteq \overline{A}$. Hence, $(D, E) \Vdash \varphi(G)$ if and only if $\varphi(D \cup F)$ holds for all finite $F \subset E$.

Lemma 1. *Let (D, E) be a condition and let $\varphi(X)$ be a formula in exactly one free set variable.*

1. *If φ is Σ_0^0 with no free number variables then the relation $(D, E) \Vdash \varphi(G)$ is computable.*
2. *If φ is Π_1^0, Σ_1^0, or Σ_2^0, then so is the relation $(D, E) \Vdash \varphi(G)$.*
3. *For $n \geq 2$, if φ is Π_n^0 then the relation of $(D, E) \Vdash \varphi(G)$ is Π_{n+1}^0.*
4. *For $n \geq 3$, if φ is Σ_n^0 then the relation $(D, E) \Vdash \varphi(G)$ is Σ_{n+1}^0.*

Proof. We first prove 1. If φ is as hypothesized and $\varphi(D \cup F)$ does not hold for some finite $F \subset E$, then neither does $\varphi(D \cup (F \cap (\bigcup_i P_{\varphi,i} \cup N_{\varphi,i})))$. So by Remark 1, we have that $(D, E) \Vdash \varphi(G)$ if and only if $\varphi(D \cup F)$ holds for all finite $F \subset E \cap (\bigcup_i P_{\varphi,i} \cup N_{\varphi,i})$, which can be checked computably.

For 2, suppose that $\varphi(X) \equiv (\forall x)\theta(x, X)$, where θ is Σ_0^0. We claim that (D, E) forces $\varphi(G)$ if and only if $\theta(a, D \cup F)$ holds for all a and all finite $F \subset E$, which makes the forcing relation Π_1^0. The right to left implication is clear. For the other, suppose there is an a and a finite $F \subset E$ such that $\theta(a, D \cup F)$ does not hold. Writing $\theta_a(X)$ for the formula $\theta(a, X)$, let $D^* = D \cup F$ and

$$E^* = \{x \in E : x > \max D \cup F \cup \bigcup_i P_{\theta_a,i} \cup N_{\theta_a,i}\},$$

so that (D^*, E^*) is a condition extending (D, E). Then if (D^{**}, E^{**}) is any extension of (D^*, E^*), we have that

$$D^{**} \cap (\bigcup_i P_{\theta_a,i} \cup N_{\theta_a,i})) = (D \cup F) \cap (\bigcup_i P_{\theta_a,i} \cup N_{\theta_a,i})),$$

and so $\theta(a, D^{**})$ cannot force $\theta(a, G)$. Thus (D, E) does not force $\varphi(G)$. The rest of 2 follows immediately, since forcing a formula that is Σ_1^0 over another formula is Σ_1^0 over the complexity of forcing that formula.

We next prove 3 for $n = 2$. Suppose that $\varphi(G) \equiv (\forall x)(\exists y)\theta(x, y, X)$ where θ is Σ_0^0. Our claim is that $(D, E) \Vdash \varphi(G)$ if and only if, for every a and every condition (D^*, E^*) extending (D, E), there is a finite $F \subset E^*$ and a number $k > \max F$ such that

$$(D^* \cup F, \{x \in E^* : x > k\}) \Vdash (\exists y)\theta(a, y, G), \tag{1}$$

which is a Π_3^0 definition. Since the condition on the left side of (1) extends (D^*, E^*), this definition clearly implies forcing. For the opposite direction, suppose $(D, E) \Vdash \varphi(G)$ and fix any a and $(D^*, E^*) \le (D, E)$. Then by definition, there is a b and a condition (D^{**}, E^{**}) extending (D^*, E^*) that forces $\theta(a, b, G)$. Write $\theta_{a,b}(X) = \theta(a, b, X)$, and let $F \subset E^*$ be such that $D^{**} = D^* \cup F$. Since $\theta_{a,b}(D^* \cup F)$ holds, we must have $P_{\theta_{a,b},i} \subseteq D^* \cup F$ and $N_{\theta_{a,b},i} \cap (D^* \cup F) = \varnothing$ for some i. Thus, if we let $k = \max N_{\theta_{a,b},i}$, we obtain (1).

To complete the proof, we prove 3 and 4 for $n \ge 3$ by simultaneous induction on n. Clearly, 3 for $n - 1$ implies 4 for n, so we already have 4 for $n = 3$. Now assume 4 for some $n \ge 3$. The definition of forcing a Π_{n+1}^0 statement is easily seen to be Π_2^0 over the relation of forcing a Σ_n^0 statement, and hence Π_{n+2}^0 by hypothesis. Thus, 3 holds for $n + 1$.

We shall see in Corollary 2 in the next section that the complexity bounds in parts 3 and 4 of the lemma cannot be lowered to Σ_n^0 and Π_n^0, respectively. As a consequence, n-generics only decide all Σ_{n-1}^0 formulas, and not necessarily all Σ_n^0 formulas.

Proposition 4. *Let G be n-generic, and for $m \le n$ let $\varphi(X)$ be a Σ_m^0 or Π_m^0 formula in exactly one free set variable. If (D, E) is any condition satisfied by G that forces $\varphi(G)$, then $\varphi(G)$ holds.*

Proof. If $m = 0$, then φ holds of any set satisfying (D, E), whether it is generic or not. If $m > 0$ and the result holds for Π_{m-1}^0 formulas, it also clearly holds for Σ_m^0 formulas. Thus, we only need to show that if $m > 0$ and the result holds for Σ_{m-1}^0 formulas then it also holds for Π_m^0 formulas. To this end, suppose $\varphi(X) \equiv (\forall x)\theta(x, X)$, where θ is Σ_{m-1}^0. For each a, let $\mathcal{C}_a$ be the set of all conditions forcing $\theta(a, X)$, which has complexity at most Σ_n^0 by Lemma 1. Hence, G meets or avoids each $\mathcal{C}_a$. But if G were to avoid some $\mathcal{C}_a$, say via a condition (D^*, E^*), then (D^*, E^*) would force $\neg\theta(a, G)$, and then (D, E) and (D^*, E^*) would have a common extension forcing $\theta(a, G)$ and $\neg\theta(a, G)$. Thus, G meets every $\mathcal{C}_a$, so $\theta(a, G)$ holds for all a by hypothesis, meaning $\varphi(G)$ holds.

Remark 2. It is not difficult to see that if $\varphi(G)$ is the negation of a Σ_m^0 formula then any condition (D, E) forcing $\varphi(G)$ forces an equivalent Π_m^0 formula. Thus, if G is n-generic and satisfies such a condition, then $\varphi(G)$ holds.

5 Degrees of Mathias Generics

We begin here with a jump property for Mathias generics similar to that of Jockusch for Cohen generics. It follows that the degrees $\mathbf{d}$ satisfying $\mathbf{d}^{(n-1)} = \mathbf{d}' \cup \mathbf{0}^{(n-1)}$ yield a strict hierarchy of subclasses of the high degrees.

Theorem 1. *For all $n \geq 2$, if G is n-generic then $G^{(n-1)} \equiv_T G' \oplus \varnothing^{(n)}$.*

Proof. That $G^{(n-1)} \geq_T G' \oplus \varnothing^{(n)}$ follows from the fact that G is high, as discussed above. That $G^{(n-1)} \leq_T G' \oplus \varnothing^{(n)}$ is trivial for $n = 2$. To show it for $n \geq 3$, we wish to decide every $\Sigma_{n-1}^{0,G}$ sentence using $G' \oplus \varnothing^{(n)}$. Let $\varphi_0(X), \varphi_1(X), \ldots$, be a computable enumeration of all Σ_{n-1}^0 sentences in exactly one free set variable, and for each i let $\mathcal{C}_i$ be the set of conditions forcing $\varphi_i(G)$, and $\mathcal{D}_i$ the set of conditions forcing $\neg\varphi_i(G)$. Then $\mathcal{D}_i$ is the set of conditions with no extension in $\mathcal{C}_i$, so if G meets $\mathcal{C}_i$ it cannot also meet $\mathcal{D}_i$. On the other hand, if G avoids $\mathcal{C}_i$ then it meets $\mathcal{D}_i$ by definition. Now by Lemma 1, each $\mathcal{C}_i$ is Σ_n^0 since $n \geq 3$, and so it is met or avoided by G. Thus, for each i, either G meets $\mathcal{C}_i$, in which case $\varphi_i(G)$ holds by Proposition 4, or else G meets $\mathcal{D}_i$, in which case $\neg\varphi_i(G)$ holds by Remark 2. To conclude the proof, we observe that $G' \oplus \varnothing^{(n)}$ can decide, uniformly in i, whether G meets $\mathcal{C}_i$ or $\mathcal{D}_i$. Indeed, from a given i, indices for $\mathcal{C}_i$ and $\mathcal{D}_i$ (as a Σ_n^0 set and a Π_n^0 set, respectively) can be found uniformly computably, and then $\varnothing^{(n)}$ has only to produce these sets until a condition in one is found that is satisfied by G, which can in turn be determined by G'.

Corollary 2. *For every $n \geq 2$ there is a Π_n^0 formula in exactly one free set variable, the relation of forcing which is not Π_n^0. For every $n \geq 3$ there is a Σ_n^0 formula in exactly one free set variable, the relation of forcing which is not Σ_n^0.*

Proof. It suffices to prove the second part, as it implies the first by the proof of Lemma 1. For consistency with Theorem 1, we fix $n \geq 4$ and prove the result for $n - 1$. If forcing every Σ_{n-1}^0 formula were Σ_{n-1}^0, then the proof of the theorem could be carried out computably in $G' \oplus \varnothing^{(n-1)}$ instead of $G' \oplus \varnothing^{(n)}$. Hence, we would have $G^{(n-1)} \equiv_T G' \oplus \varnothing^{(n-1)}$, contradicting that G must be high.

The following result is the analogue of Theorem 2.3 of Kurtz [8] that every $A >_T \varnothing^{(n-1)}$ hyperimmune relative to $\varnothing^{(n-1)}$ is Turing equivalent to the $(n-1)$st jump of a weakly Cohen n-generic set. The proof, although mostly similar, requires a few important modifications. The main problem is in coding A into $G^{(n-2)}$, which, in the case of Cohen forcing, is done by appending long blocks of 1s to the strings under construction. As the infinite part of a Mathias condition can be made very sparse, we cannot use the same idea here. We highlight the changes below, and only sketch the rest of the details. Recall that a set is *co-immune* if its complement has no infinite computable subset.

Proposition 5. *If $A >_T \varnothing^{(n-1)}$ is hyperimmune relative to $\varnothing^{(n-1)}$, then $A \equiv_T$ $G^{(n-2)}$ for some weakly n-generic set G.*

Proof. Computably in A, we build a sequence $(D_0, E_0) \geq (D_1, E_1) \geq \cdots$ of conditions, beginning with $(D_0, E_0) = (\varnothing, \omega)$. Let $\mathcal{C}_0, \mathcal{C}_1, \ldots$ be a listing of all Σ_n^0 sets of pre-conditions, and fixing a $\varnothing^{(n-1)}$-computable enumeration of each $\mathcal{C}_i$, let $\mathcal{C}_{i,s}$ be the set of all pre-conditions enumerated into $\mathcal{C}_i$ by stage $p_A(s)$. We may assume that $\langle D, E \rangle \leq s$ for all $(D, E) \in \mathcal{C}_{i,s}$. Let $B_0, B_1, \ldots$ be a uniformly $\varnothing^{(n-1)}$-computable sequence of pairwise disjoint co-immune sets. Say $\mathcal{C}_i$ *requires attention* at stage s if there exists $b \leq p_A(s)$ in $B_i \cap E_s$ and a condition (D, E) in $\mathcal{C}_{i,s}$ extending $(D_s \cup \{b\}, \{x \in E_s : x > b\})$.

At stage s, assume (D_s, E_s) is given. If there is no $i \leq s$ such that $\mathcal{C}_i$ requires attention at stage s, set $(D_{s+1}, E_{s+1}) = (D_s, E_s)$. Otherwise, fix the least such i. Choose the least corresponding b and earliest enumerated extension (D, E) in $\mathcal{C}_{i,s}$, and let $(D^*, E^*) = (D, E)$. Then obtain (D^{**}, E^{**}) from (D^*, E^*) by forcing the jump, in the usual manner. Finally, let k be the number of stages $t < s$ such that $(D_t, E_t) \neq (D_{t+1}, E_{t+1})$, and let $(D^{***}, E^{***}) = (D^{**} \cup \{b\}, \{x \in E^{**} : x > b\})$, where b is the least element of $B_{A(k)} \cap E^{**}$. If $\langle D^{***}, E^{***} \rangle \leq s+1$, set $(D_{s+1}, E_{s+1}) = (D^{***}, E^{***})$, and otherwise set $(D_{s+1}, E_{s+1}) = (D_s, E_s)$.

By definition, the B_i must intersect every computable set infinitely often, and so the entire construction is A-computable. That $G = \bigcup_s D_s$ is weakly n-generic can be verified much like in Kurtz's proof, but using the $\varnothing^{(n-1)}$-computable function h where $h(s)$ is the least t so that for each (D, E) with $\langle D, E \rangle \leq s$ there exists $b \leq t$ in $B_i \cap E$ and $(D^*, E^*) \in \mathcal{C}_{i,t}$ extending $(D \cup \{b\}, \{x \in E : x > b\})$. That $G^{(n-2)} \leq_T A$ follows by Theorem 1 from G' being forced during the construction and thus being A-computable. Finally, to show $A \leq_T G^{(n-2)}$, let $s_0 < s_1 < \cdots$ be all the stages $s > 0$ such that $(D_{s-1}, E_{s-1}) \neq (D_s, E_s)$. The sequence $(D_{s_0}, E_{s_0}) > (D_{s_1}, E_{s_1}) \cdots$ can be computed by $G^{(n-2)}$ as follows. Given (D_{s_k}, E_{s_k}), the least $b \in G - D_{s_k}$ must belong to some B_i, and since $G^{(n-2)}$ computes $\varnothing^{(n-1)}$ it can tell which B_i. Then $G^{(n-2)}$ can produce $\mathcal{C}_i$ until the first (D^*, E^*) extending $(D_{s_k} \cup \{b\}, \{x \in E_{s_k} : x > b\})$, and then obtain (D^{**}, E^{**}) from (D^*, E^*) by forcing the jump. By construction, G satisfies (D^{**}, E^{**}) and $(D_{s_{k+1}}, E_{s_{k+1}}) = (D^{**} \cup \{b\}, \{x \in E^{**} : x > b\})$ for the least $b \in G - D_{s_{k+1}}$. And this b is in B_1 or B_0 depending as k is or is not in B.

Corollary 3. *Not every weakly n-generic set is n-generic.*

Proof. By the previous proposition, $\varnothing^{(n)} \equiv_T G^{(n-2)}$ for some weakly n-generic set G. By Theorem 1, if G were n-generic we would have $\varnothing^{(n+1)} \equiv_T G^{(n-1)} \equiv_T G' \oplus \varnothing^{(n)} \equiv_T \varnothing^{(n)}$, which cannot be.

In spite of Theorem 1, we are still left with the possibility that some Mathias n-generic set has Cohen 1-generic degree. We now show that this cannot happen.

Theorem 2. *If G is n-generic then it has $\mathbf{GH}_1$ degree, i.e., $G' \equiv_T (G \oplus \varnothing')'$.*

Proof. A condition (D, E) forces $i \in (G \oplus \varnothing')'$ if there is a $\sigma \in 2^{<\omega}$ such that that $\Phi_i^\sigma(i) \downarrow$ and for all $x < |\sigma|$,

$$\sigma(x) = 1 \implies (D, E) \Vdash x \in G \oplus \varnothing';$$
$$\sigma(x) = 0 \implies (D, E) \Vdash x \notin G \oplus \varnothing'.$$

This is thus a Σ_2^0 relation, as forcing $x \in G \oplus \varnothing'$ and $x \notin G \oplus \varnothing'$ are Σ_1^0 and Π_1^0, respectively. We claim that (D, E) forcing $i \notin (G \oplus \varnothing')'$, i.e., $\neg(i \in (G \oplus \varnothing')')$, is equivalent to (D, E) having no finite extension that forces $i \in (G \oplus \varnothing')'$, and hence is Π_2^0. That forcing implies this fact is clear. In the other direction, suppose (D, E) does not force $i \notin (G \oplus \varnothing')'$, and so has an extension (D^*, E^*) that forces $i \notin (G \oplus \varnothing')'$. Let σ witness this fact, as above. Then if P and N consist of the $x < |\sigma|$ such that $\sigma(2x) = 1$ and $\sigma(2x) = 0$, respectively, σ witnesses that $(D \cup P, \{x \in E : x > \max P \cup N\})$ also forces $i \in (G \oplus \varnothing')'$.

We now show that $G' \geq_T (G \oplus \varnothing')'$. Let $\mathcal{C}_i$ be the set of conditions that force $i \in (G \oplus \varnothing')'$, and $\mathcal{D}_i$ the set of conditions that force $i \notin (G \oplus \varnothing')'$. Then $\mathcal{C}_i$ is Σ_3^0 and $\mathcal{D}_i$ is Π_2^0, and indices for them as such can be found uniformly from i. Each $\mathcal{C}_i$ must be either met or avoided by G, and as in the proof of Theorem 1, G meets $\mathcal{C}_i$ if and only if it does not meet $\mathcal{D}_i$. Which of the two is the case can be determined by G' since $G' \geq_T \varnothing''$ and $\mathcal{C}_i$ and $\mathcal{D}_i$ are both c.e. in $\varnothing''$. By Proposition 4, G' can thus determine whether $i \in (G \oplus \varnothing')'$, as desired.

Recall that a degree $\mathbf{d}$ is $\mathbf{GL}_n$ if $\mathbf{d}^{(n)} = (\mathbf{d} \cup \mathbf{0}')^{(n-1)}$, and that no such degree can be $\mathbf{GH}_1$. It was shown by Jockusch and Posner [7, Corollary 7] that every $\overline{\mathbf{GL}}_2$ degree computes a Cohen 1-generic set. Hence, we obtain the following:

Corollary 4. *Every Mathias n-generic set has $\overline{\mathbf{GL}}_m$ degree for all $m \geq 1$. Hence, it is not of Cohen 1-generic degree, but does compute a Cohen 1-generic.*

We leave open the following question, which we have so far been unable to answer. Partial answers are given in the subsequent results.

Question 2. Does every Mathias n-generic set compute a Cohen n-generic set?

Theorem 3. *If G is Mathias n-generic, and $A \leq_T \varnothing^{(n-1)}$ is bi-immune (i.e., A and $\overline{A}$ are each co-immune), then $G \oplus A$ computes a Cohen n-generic.*

Proof. For every set $S = \{s_0 < s_1 < \cdots\}$, define $S_A = A(s_0)A(s_1)\cdots$, which is a string in $2^{<\omega}$ if S is finite, and a sequence in 2^ω otherwise. Now let $\mathcal{C}_0, \mathcal{C}_1, \ldots$ be a listing of all Σ_n^0 subsets of $2^{<\omega}$, together with fixed $\varnothing^{(n-1)}$-computable enumerations. For each i, let $\mathcal{D}_i$ be the set of all conditions (D, E) such that the string D_A belongs to $\mathcal{C}_i$. Then $\mathcal{D}_i$ is a Σ_n^0 set of conditions, and as such must be met or avoided by G. If G meets $\mathcal{D}_i$ then G_A, viewed as an element of 2^ω, meets $\mathcal{C}_i$. If G avoids $\mathcal{D}_i$, we claim that G_A must avoid $\mathcal{C}_i$. Indeed, suppose G avoids $\mathcal{D}_i$ via (D, E). Since A and $\overline{A}$ are co-immune, they intersect E infinitely often, and so if D_A had an extension τ in $\mathcal{C}_i$, we could make a finite extension (D^*, E^*) of (D, E) so that $D_A^* = \tau$. This extension would belong to $\mathcal{D}_i$, a contradiction.

Thus, for example, the join of G with any non-computable $A \leq_T \varnothing'$ computes a Cohen n-generic, as every such A is bi-immune [5, Corollary 5 (iii)].

Proposition 6. *If G is Mathias n-generic and H is Cohen n-generic then H is not many-one reducible to G.*

Proof. Seeking a contradiction, suppose f is a computable function such that $f(H) \subseteq G$ and $f(\overline{H}) \subseteq \overline{G}$. The set of conditions (D, E) with $E \subseteq \mathrm{ran}(f)$ is Σ_3^0-definable, and must be met by G else $G \cap \mathrm{ran}(f)$ would be finite and H would be computable. So fix a condition (D, E) in this set satisfied by G. For all $a > \max D$, we then have that $a \in G$ if and only if $a \in E$ and $f^{-1}(a) \subseteq H$. It follows that $G \leq_T H$, and hence that $G \equiv_T H$, contradicting our observation at the end of Section 3 that no Mathias n-generic can have Cohen n-generic degree.

Acknowledgements. The first author was partially supported by NSF grant DMS-0800198; the second author was partially supported by an NSF Postdoctoral Fellowship; the third author was partially supported by grant ID#20800 from the John Templeton Foundation. The opinions expressed in this publication are those of the authors and do not necessarily reflect the views of the John Templeton Foundation. The authors are grateful to Christopher P. Porter and the anonymous referees for helpful comments.

References

1. Binns, S., Kjos-Hanssen, B., Lerman, M., Solomon, R.: On a conjecture of Dobrinen and Simpson concerning almost everywhere domination. J. Symbolic Logic 71, 119–136 (2006)
2. Cholak, P.A., Jockusch Jr., C.G., Slaman, T.A.: On the strength of Ramsey's theorem for pairs. J. Symbolic Logic 66, 1–55 (2001)
3. Downey, R.G., Hirschfeldt, D.R.: Algorithmic randomness and complexity. Theory and Applications of Computability. Springer, New York (2010)
4. Dzhafarov, D.D., Jockusch Jr., C.G.: Ramsey's theorem and cone avoidance. J. Symbolic Logic 74, 557–578 (2009)
5. Jockusch Jr., C.G.: The degrees of bi-immune sets. Z. Math. Logik Grundlagen Math. 15, 135–140 (1969)
6. Jockusch Jr., C.G.: Degrees of generic sets. In: Drake, F.R., Wainer, S.S. (eds.) Recursion Theory: its Generalisation and Applications. London Math. Soc. Lecture Note Ser., vol. 45, pp. 110–139. Cambridge University Press, Cambridge (1980)
7. Jockusch Jr., C.G., Posner, D.B.: Double jumps of minimal degrees. J. Symbolic Logic 43, 715–724 (1978)
8. Kurtz, S.A.: Notions of weak genericity. J. Symbolic Logic 48, 764–770 (1983)
9. Seetapun, D., Slaman, T.A.: On the strength of Ramsey's theorem. Special Issue: Models of arithmetic. Notre Dame J. Formal Logic 36, 570–582 (1995)
10. Soare, R.I.: Computability theory and applications. Theory and Applications of Computability. Springer, New York (2012)
11. Soare, R.I.: Sets with no subset of higher degree. J. Symbolic Logic 34, 53–56 (1969)
12. Yu, L.: Lowness for genericity. Arch. Math. Logic 45, 233–238 (2006)

Complexity of Deep Inference via Atomic Flows

Anupam Das

Department of Computer Science, University of Bath, Claverton Down,
BA2 7AY, United Kingdom
a.das@bath.ac.uk

Abstract. We consider the fragment of deep inference free of compression mechanisms and compare its proof complexity to other systems, utilising 'atomic flows' to examine size of proofs. Results include a simulation of Resolution and dag-like cut-free Gentzen, as well as a separation from bounded-depth Frege.

1 Introduction

Deep inference differs from other proof formalisms by allowing derivations themselves to be composed by logical connectives. There has recently been a lot of activity in the proof complexity of deep inference [2], most notably that a cut-free system, KS^+, quasipolynomially simulates Frege systems [12] [3]. It is conjectured that this can be improved to a polynomial simulation, so finding lower bounds for KS^+ is probably as hard as finding one for Frege, which has escaped proof complexity theorists for years.

However this quasipolynomial simulation relies crucially on the presence of dag-like behaviour, manifested in deep inference by a particular rule, *cocontraction*: $\dfrac{A}{A \wedge A}$. Without it we have a minimal complete system closed under deep inference, KS. This system is free of compression mechanisms, in that a proof of a conjunction can be 'partitioned' into proofs of each conjunct, unlike proofs that are dag-like or contain cut.

It is conjectured that KS is unable to polynomially simulate KS^+ [2], raising the question of exactly where it fits in the hierarchy of proof systems.

In this paper we focus on upper bounds and simulations to demonstrate the relative strength of KS. Our arguments employ *atomic flows* [10], diagrams that track structural changes in a proof but forget logical information, to show that cocontraction, and certain other steps, can be sometimes eliminated from a proof in polynomial time. A comprehensive introduction to atomic flows can be found in [11].

Our main result is a polynomial simulation of dag-like cut-free Gentzen systems (dag-Gen^-) in KS, improving on the simulation of the tree-like system in [2]. This also places KS in the gap between dag-Gen^- and a variation augmented with elimination rules ($Gen^\star$), shown in [7] to simulate KS^+, thereby quasipolynomially simulating Frege by the aforementioned result. This is discussed further in conclusion 7.2.

Fig. 1 summarises our results, and full proofs of results can be found in [8].

S.B. Cooper, A. Dawar, and B. Löwe (Eds.): CiE 2012, LNCS 7318, pp. 139–150, 2012.
© Springer-Verlag Berlin Heidelberg 2012

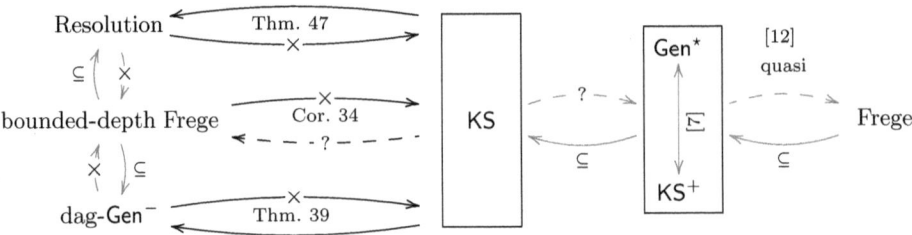

Fig. 1. Relative complexity of systems after results in this paper

2   Deep Inference

We work in propositional logic over the basis $\{\bar{\ }, \wedge, \vee\}$ with formulae in negation normal form. Syntactic equivalence of formulae is denoted $\equiv$.

Definition 1 (Rules and Systems). *An* inference rule *is a binary relation on formulae decidable in polynomial time, and a* system *is a set of rules. We define the rules we use below, and the systems* $\mathsf{KS} = \{\mathsf{ai}\!\downarrow, \mathsf{aw}\!\downarrow, \mathsf{ac}\!\downarrow, \mathsf{s}, \mathsf{m}\}$, $\mathsf{KS}^+ = \mathsf{KS} \cup \{\mathsf{aw}\!\uparrow, \mathsf{ac}\!\uparrow\}$, $\mathsf{SKS} = \mathsf{KS}^+ \cup \{\mathsf{ai}\!\uparrow\}$ *and* $\overline{\mathsf{KS}} = \{\mathsf{ai}\!\uparrow, \mathsf{aw}\!\uparrow, \mathsf{ac}\!\uparrow, \mathsf{s}, \mathsf{m}\}$.*
We also have a logical rule $=$ *which allows us to apply laws of associativity, commutativity and basic equations with units [2].*

Atomic structural rules   Linear logical rules

$\mathsf{ai}\!\downarrow \dfrac{\mathsf{t}}{a \vee \bar{a}}$ $\mathsf{aw}\!\downarrow \dfrac{\mathsf{f}}{a}$ $\mathsf{ac}\!\downarrow \dfrac{a \vee a}{a}$ $\mathsf{s} \dfrac{A \wedge [B \vee C]}{(A \wedge B) \vee C}$

identity   *weakening*   *contraction*   *switch*

$\mathsf{ai}\!\uparrow \dfrac{a \wedge \bar{a}}{\mathsf{f}}$ $\mathsf{aw}\!\uparrow \dfrac{a}{\mathsf{t}}$ $\mathsf{ac}\!\uparrow \dfrac{a}{a \wedge a}$ $\mathsf{m} \dfrac{(A \wedge B) \vee (C \wedge D)}{[A \vee C] \wedge [B \vee D]}$

cut   *coweakening*   *cocontraction*   *medial*

Definition 2 (Proofs and Derivations). *We define derivations, and premiss and conclusion functions* (pr, cn *resp.*), *inductively. Every formula A is a derivation with* $\mathsf{pr}(A) \equiv A \equiv \mathsf{cn}(A)$. *For derivations* Φ, Ψ: *if* $\star \in \{\wedge, \vee\}$ *then* $\Phi \star \Psi$ *is a derivation with premiss* $\mathsf{pr}(\Phi) \star \mathsf{pr}(\Psi)$ *and conclusion* $\mathsf{cn}(\Phi) \star \mathsf{cn}(\Psi)$; *if* $\rho \dfrac{\mathsf{cn}(\Phi)}{\mathsf{pr}(\Psi)}$ *is an instance of a rule* ρ, $\rho \dfrac{\Phi}{\Psi}$ *is a derivation with premiss* $\mathsf{pr}(\Phi)$ *and conclusion* $\mathsf{cn}(\Psi)$.

If $\mathrm{pr}(\Phi) \equiv \mathsf{t}$ *then we call* Φ *a* proof. *If* Φ *is a derivation in a system* $\mathcal{S}$ *with premiss* A*, conclusion* B*, we write* $\Phi \Vert \mathcal{S}\,\genfrac{}{}{0pt}{}{A}{B}$*. If* $A \equiv \mathsf{t}$*, i.e.,* Φ *is a proof, we write* $\Phi \Vert \mathcal{S}\,\genfrac{}{}{0pt}{}{}{B}$*.*

Proposition 3 **([1])**. *Each rule* ρ *below is derivable in* $\{\mathsf{s}, \mathsf{m}, \mathsf{a}\rho\}$*:*

$$\mathsf{i}{\downarrow}\,\frac{\mathsf{t}}{A \vee \bar{A}} \qquad \mathsf{w}{\downarrow}\,\frac{\mathsf{f}}{A} \qquad \mathsf{c}{\downarrow}\,\frac{A \vee A}{A} \qquad \mathsf{i}{\uparrow}\,\frac{A \wedge \bar{A}}{\mathsf{f}} \qquad \mathsf{w}{\uparrow}\,\frac{A}{\mathsf{t}} \qquad \mathsf{c}{\uparrow}\,\frac{A}{A \wedge A}$$

We shall use the above 'generic' rules as abbreviations for their derivations.

Definition 4 (Complexity). *We define the* size $|\Phi|$ *of a derivation* Φ *to be the number of atom occurrences in* Φ*. A system* $\mathcal{S}$ p-simulates *a system* $\mathcal{T}$ *if each* $\mathcal{T}$*-proof can be polynomially transformed into an* $\mathcal{S}$*-proof of the same conclusion.*

3   Atomic Flows

Definition 5 (Atomic Flows). *For an* SKS *derivation* Φ *we define its* atomic flow*,* $fl(\Phi)$*, to be the diagram obtained by tracing the path of each atom through the derivation, designating structural rules by the following corresponding nodes:*

$$\mathsf{ai}{\downarrow}\,\frac{\mathsf{t}}{a \vee \bar{a}} \;\longmapsto\; \quad \mathsf{aw}{\downarrow}\,\frac{\mathsf{f}}{a} \;\longmapsto\; \quad \mathsf{ac}{\downarrow}\,\frac{a \vee a}{a} \;\longmapsto\;$$

$$\mathsf{ai}{\uparrow}\,\frac{a \wedge \bar{a}}{\mathsf{f}} \;\longmapsto\; \quad \mathsf{aw}{\uparrow}\,\frac{a}{\mathsf{t}} \;\longmapsto\; \quad \mathsf{ac}{\uparrow}\,\frac{a}{a \wedge a} \;\longmapsto\;$$

We consider flows as graphs embedded in the plane with the six types of nodes above. Note that edges may be pending *at either end*.

We define the size of a flow ϕ, denoted $|\phi|$, to be its number of edges.

Definition 6. *We define a rewriting system* norm *on flows in Fig. 2.*

Proposition 7 ([10]). norm *is terminating and confluent.*

Notation 8. *If a flow* ψ *is the normal form of a flow* ϕ *under a terminating, confluent rewriting system* r*, then we write* $\phi \underset{\mathsf{r}}{\to} \psi$*.*

Definition 9. *If* $\mathcal{R}$ *is a relation on atomic flows we say that* $\mathcal{R}$ *lifts polynomially to* SKS *if, whenever* $(fl(\Phi), \psi) \in \mathcal{R}$*, we can construct a derivation* Ψ *in time polynomial in* $|\Phi| + |\psi|$ *with the same premiss and conclusion as* Φ *and atomic flow* ψ*.*

Theorem 10 ([10]). $\underset{\mathsf{norm}}{\longrightarrow}$ *lifts polynomially to* SKS*.*

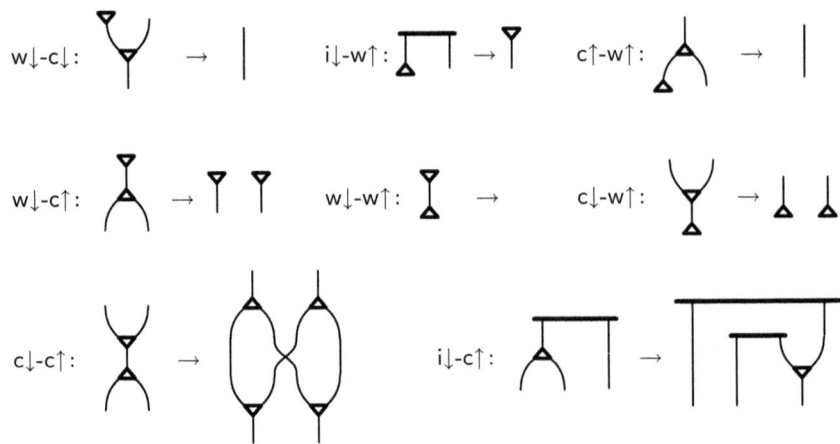

Fig. 2. Local rewriting rules for the system norm

Corollary 11. *If ϕ is the flow of a KS^+ proof, $\phi \xrightarrow[\text{norm}]{} \psi$ then ψ is the flow of a KS proof.*

Example 12. In Fig. 3 we give a derivation, its flow and a reduction under norm.

We consider atoms to be positive or negative, under some valid assignment of polarity. We use the terms 'upwards' and 'downwards' with regard to derivations and flows, interpreted in the natural way, independently from the notion of direction defined below.

Definition 13 (Paths). *To each edge we assign a direction: downwards if the atom associated with it is positive, upwards if it is negative.*
    We define a path *in a flow to be a directed path between pending edges.*

Example 14. We give the following flow and all its paths:

$$23679, 23678,$$
$$4578, 4579,$$
$$1.$$

where $+$, $-$ indicate the polarity of the atom associated with an edge, under some assignment, and $\star$ indicates that either polarity may be correctly assigned.
    Notice that the number of paths is invariant under valid assignments of polarity.

The following results allow us to estimate the size of the normal form of a flow, under norm, without actually constructing it.

Observation 15. *Reducing under norm conserves the number of paths in a flow.*

Recall that, in a proof, there are no assumptions, and so the flow of a proof can have no edge with upper end pending; it must be attached to an identity or weakening node.

Let $\#(\rho, \phi)$ be the number of ρ-nodes in a flow ϕ, and $\ulcorner\phi\urcorner$ be its number of paths.

Observation 16. *If ϕ is the flow of a* KS *proof, then* $\ulcorner\phi\urcorner = \#(\mathsf{ai}{\downarrow}, \phi)$.

Theorem 17. *If ϕ is the flow of a* KS^+ *proof, $\phi \xrightarrow[\text{norm}]{} \psi$, then* $|\psi| = O(|\phi| + \ulcorner\phi\urcorner)$.

Proof. Decompose ψ into its identity fragment ψ_1 and weakening fragment ψ_2. Note that each rule involving $\mathsf{w}{\downarrow}$ or $\mathsf{w}{\uparrow}$ reduces the size of the flow, so $|\psi_2| \leq |\phi|$.

Notice that $|\psi_1| = 2 \cdot \#(\mathsf{ai}{\downarrow}, \psi_1) + \#(\mathsf{ac}{\downarrow}, \psi_1)$. However, clearly, the number of contractions cannot outnumber the number of edges emanating from identity steps, so we have $|\psi_1| \leq 4 \cdot \#(\mathsf{ai}{\downarrow}, \psi_1)$. By Obs. 16 we then have $|\psi_1| \leq 4 \cdot \ulcorner\psi\urcorner$, and by Obs. 15 that $|\psi_1| \leq 4 \cdot \ulcorner\phi\urcorner$, whence $|\psi| = |\psi_1| + |\psi_2| \leq |\phi| + 4 \cdot \ulcorner\phi\urcorner$.

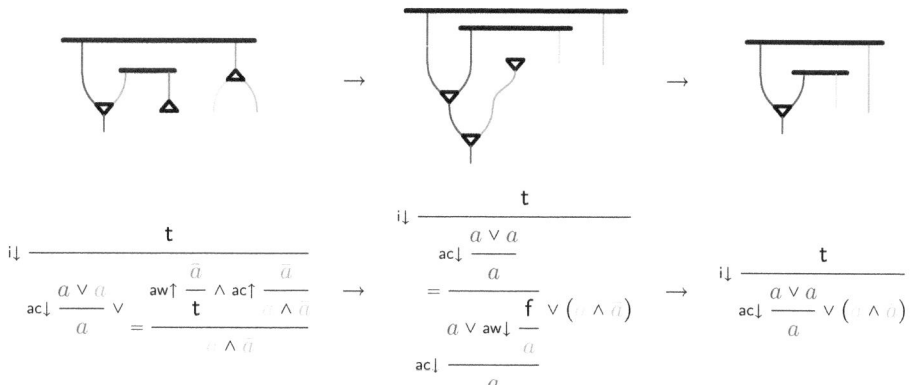

Fig. 3. An example of a derivation, its flow and a reduction under norm

Remark 18. The main contributor to an increase of flow size reducing under norm is the rule $\mathsf{c}{\downarrow}$-$\mathsf{c}{\uparrow}$. It can sometimes cause an exponential blowup [10].

The following proposition estimates the increase in size caused by $\mathsf{c}{\downarrow}$-$\mathsf{c}{\uparrow}$.

Proposition 19. *If in every directed path of a flow ϕ there are at most k alternations of $\mathsf{ac}{\uparrow}$ and $\mathsf{ac}{\downarrow}$ nodes then* $\ulcorner\phi\urcorner = |\phi|^{O(k)}$.

4   Truth Tables and Tableaux

Bruscoli and Guglielmi have proved that tree-like cut-free Gentzen cannot p-simulate KS, by way of the Statman tautologies [2]. We offer a new proof here, via truth tables.

Observation 20. *A truth table proof for a formula A has size $|A| \cdot 2^{\#A}$, where $\#A$ is the number of distinct propositional variables in A.*

Lemma 21. KS^+ *p-simulates truth tables.*

Proof. Let τ be a tautology. For each partial assignment $\mathcal{A}$, defined on just those atoms appearing in τ construct a derivation $\Phi_{\mathcal{A}}(\tau)$ by structural induction on τ:

$$\Phi_{\mathcal{A}}(a) \equiv a \quad , \quad \Phi_{\mathcal{A}}(A \wedge B) \equiv \Phi_{\mathcal{A}}(A) \wedge \Phi_{\mathcal{A}}(B) \quad , \quad \Phi_{\mathcal{A}}(A \vee B) \equiv \dfrac{= \dfrac{\Phi_{\mathcal{A}}(B)}{\mathsf{w}\downarrow \dfrac{\mathsf{f}}{A} \vee B}}{}$$

where, in the last case, when τ is a disjunction, choose a disjunct B that is true under $\mathcal{A}$. It is clear that each $\Phi_{\mathcal{A}}(\tau)$ has conclusion τ and premiss a conjunction of literals; moreover this conjunction of literals is satisfied by $\mathcal{A}$.

Let $\gamma_{\mathcal{A}}$ be the conjunction of all literals satisfied by $\mathcal{A}$ such that each literal appears at most once. Then there is a derivation of the form: $\begin{array}{c} \gamma_{\mathcal{A}} \\ \| \{\mathsf{aw}\uparrow, \mathsf{ac}\uparrow\}. \\ \mathsf{pr}(\Phi_{\mathcal{A}}(\tau)) \end{array}$

By distributivity, derived on the left, we can construct a proof Ψ of $\bigvee_{\mathcal{A}} \gamma_{\mathcal{A}}$ in $\{\mathsf{ai}\downarrow, \mathsf{ac}\uparrow, \mathsf{s}, \mathsf{m}\}$, and then apply contractions to obtain the proof, on the right:

$$\text{Distributivity:} \quad \dfrac{= \dfrac{\mathsf{c}\uparrow \dfrac{A}{A \wedge A} \wedge [B \vee C]}{\mathsf{s}\dfrac{[B \vee C] \wedge A}{A \wedge \dfrac{[B \vee C] \wedge A}{B \vee (C \wedge A)}}}}{\mathsf{s}\dfrac{A \wedge [B \vee (A \wedge C)]}{(A \wedge B) \vee (A \wedge C)}} \quad , \quad \bigvee_{\mathcal{A}} \left[\begin{array}{c} \Psi \| \{\mathsf{ai}\downarrow, \mathsf{ac}\uparrow, \mathsf{s}, \mathsf{m}\} \\ \gamma_{\mathcal{A}} \\ \| \{\mathsf{aw}\uparrow, \mathsf{ac}\uparrow\} \\ \mathsf{pr}(\Phi_{\mathcal{A}}) \\ \Phi_{\mathcal{A}} \| \{\mathsf{w}\downarrow\} \\ \tau \end{array} \right] \begin{array}{c} \\ \| \{\mathsf{c}\downarrow\} \\ \tau \end{array}$$

Theorem 22. KS *p-simulates truth tables.*

Proof. In the above proofs all $\mathsf{c}\uparrow$ steps are above all $\mathsf{c}\downarrow$ steps, so by Prop. 19 the number of paths is polynomial in the size of the flow. The result follows by Thms. 17 and 10.

Notation 23. *Let tree/dag-Gen^- denote cut-free Gentzen with tree/dag proofs resp.*

Proposition 24 ([6]). *Tree-Gen^- cannot p-simulate truth tables.*

Corollary 25. *Tree-Gen^- cannot p-simulate KS.*

Proof. Immediate from Prop. 24 and Thm. 22.

5   Separations via the Functional Pigeonhole Principle

We show Gen^-, Resolution and bounded-depth Frege systems cannot p-simulate KS, by reducing under norm Jeřábek's KS^+ proofs of the functional pigeonhole principle.

Conversely we give a simulation of Resolution, and some extensions, in KS.

5.1   Polynomial-Size Proofs of the Functional Pigeonhole Principle

The functional pigeonhole principle is a class of propositional tautologies asserting that there is no injective function from a set of size $n + 1$ to a set of size n.

Definition 26 (Functional Pigeonhole Principle)

$$\mathsf{FPHP}_n \equiv \bigvee_{i=0}^{n} \bigwedge_{j=1}^{n} \bar{a}_{ij} \vee \bigvee_{i=0}^{n} \bigvee_{j=1}^{n-1} \bigvee_{j'=j+1}^{n} (a_{ij} \wedge a_{ij'}) \vee \bigvee_{j=1}^{n} \bigvee_{i=0}^{n-1} \bigvee_{i'=i+1}^{n} (a_{ij} \wedge a_{i'j})$$

Theorem 27 ([15][14]). *Bounded-depth Frege has only exponential-size proofs of* FPHP_n.

Corollary 28. *Resolution and* Gen^- *have only exponential-size proofs of* FPHP_n.

Theorem 29 ([4]). *There are polynomial-size Frege proofs of* FPHP_n.

Proposition 30 ([2]). SKS *is polynomially equivalent to Frege systems.*

Lemma 31. *Every* SKS *proof* Φ *of a formula A can be polynomially transformed to a* KS *proof of* $A \vee \bigvee_i (a_i \wedge \bar{a}_i)$, *where a_i are the distinct propositional variables in A.*

Lemma 32 ([12]). *There are polynomial-size proofs* Θ_n *in* KS^+ *of* FPHP_n.

Proof (Jeřábek). By Thm. 29, Prop. 30 and Lemma 31 we can build KS proofs Φ_n of $\mathsf{FPHP}_n \vee \bigvee_{i,j}(a_{ij} \wedge \bar{a}_{ij})$ that have size polynomial in n. For each atom a_{st} we construct a derivation $\Psi_n^{a_{st}}$ in $\mathsf{KS}^+ \setminus \{\mathsf{ac}\!\downarrow\}$ from $a_{st} \wedge \bar{a}_{st}$ to FPHP_n, on the left below, and apply contractions to obtain the proof, on the right:

$$
2 \cdot s \cfrac{
=\cfrac{a_{st} \wedge \bar{a}_{st}}{
a_{st} \wedge \bar{a}_{st} \wedge \mathsf{i}\!\downarrow \cfrac{\mathsf{t}}{\bigwedge_{j \neq t} \bar{a}_{sj} \vee \bigvee_{j \neq t} a_{sj}}
}
}{
\mathsf{w}\!\downarrow \cfrac{
\bigwedge_j \bar{a}_{sj} \vee \left(n \cdot \mathsf{ac}\!\uparrow \cfrac{\cfrac{a_{st}}{\bigwedge_{j \neq 0} a_{st}} \wedge \bigvee_{j \neq t} a_{sj}}{\bigvee_{j \neq t} a_{st} \wedge a_{sj}} \Big\| \{\mathsf{s}\} \right)
}{
\mathsf{FPHP}_n
}
}
\quad , \quad
\left[\cfrac{\Phi_n \Big\| \mathsf{KS}}{\mathsf{FPHP}_n \vee \bigvee_{i,j} \left(\cfrac{a_{ij} \wedge \bar{a}_{ij}}{\Psi_n^{a_{ij}} \Big\| \mathsf{KS}^+ \setminus \{\mathsf{ac}\!\downarrow\}}{\mathsf{FPHP}_n} \right)} \Big\| \{\mathsf{c}\!\downarrow\} \atop \mathsf{FPHP}_n \right]
$$

Theorem 33. *There are polynomial-size proofs in* KS *of* FPHP_n.

Proof. In Θ_n above, there are 2 alternations between $\mathsf{c}\!\downarrow$ and $\mathsf{c}\!\uparrow$ nodes, so $\ulcorner fl(\Theta_n) \urcorner = |fl(\Theta_n)|^{O(2)}$ by Prop. 19. The result follows by Thms. 17 and 10.

Corollary 34. Gen^-, *Resolution, bounded-depth Frege cannot p-simulate* KS.

Proof. Immediate from Thm. 27, Cor. 28 and Thm. 33.

146   A. Das

5.2   A Polynomial Simulation of Resolution and Some Extensions

We give a p-simulation of resolution systems in KS, first noticed in [9].

Definition 35. *We define* Resolution *by the following CNF rewriting rules:*

$$
\mathsf{RES} = \left\{ \mathsf{res}\,\frac{A \wedge [B \vee a] \wedge [\bar{a} \vee C]}{A \wedge [B \vee C]}, \mathsf{ac}{\downarrow}\,\frac{A \wedge [B \vee a \vee a]}{A \wedge [B \vee a]}, \mathsf{dag}\,\frac{A \wedge B}{A \wedge B \wedge B}, \mathsf{aw}{\downarrow}\,\frac{A \wedge B}{A \wedge [B \vee b]} \atop A \right\}
$$

modulo associativity and commutativity. A RES *refutation is a derivation* $\|\underset{f}{\mathsf{RES}}$.

Proposition 36. *Define* $\mathsf{w}{\downarrow}\text{-}\mathsf{i}{\uparrow}: \underset{}{\underline{\top}\,\underline{\bot}} \to \underline{\bot}$ *and* $\mathsf{w} = \{\mathsf{w}{\downarrow}\text{-}\mathsf{c}{\uparrow}, \mathsf{w}{\downarrow}\text{-}\mathsf{w}{\uparrow}, \mathsf{w}{\downarrow}\text{-}\mathsf{i}{\uparrow}, \mathsf{w}{\downarrow}\text{-}\mathsf{c}{\downarrow}\}$.

Then w *is terminating, confluent and* $\underset{\mathsf{w}}{\to}$ *lifts polynomially to* SKS. *[10]*

Lemma 37. RES *refutations can be polynomially transformed to ones in* $\overline{\mathsf{KS}}$.

Proof. We derive a generalisation res′ of res on the left, and eliminate ac↓-steps by the translation on the right. Finally, aw↓ steps are eliminated by Prop. 36.

$$
2 \cdot \mathsf{s}\,\frac{\dfrac{[B \vee C] \wedge [\bar{C} \vee D]}{C \wedge \bar{C}}}{B \vee \mathsf{i}{\uparrow}\,\dfrac{}{\mathsf{f}} \vee D} \quad , \quad \mathsf{res}\,\frac{\left[B \vee \mathsf{ac}{\downarrow}\,\dfrac{a \vee a}{a}\right] \wedge [\bar{a} \vee D]}{B \vee D} \to \mathsf{res}'\,\frac{[B \vee a \vee a] \wedge \left[\mathsf{ac}{\uparrow}\,\dfrac{\bar{a}}{\bar{a} \wedge \bar{a}} \vee D\right]}{B \vee D}
$$
$$
=\,\frac{}{B \vee D}
$$

Lemma 38. *We can transform a* $\overline{\mathsf{KS}}$ *refutation of* $\bar{A}$ *in linear time to a* KS *proof of* A.

Proof (Sketch). 'Flip' the refutation upside-down, replace every formula with its negation and every 'up' rule with its corresponding 'down' rule.

Theorem 39. KS *p-simulates Resolution systems.*

Proof. Immediate from Lemmata 38 and 37.

Finally, the simulation above can be extended to some basic extensions of Resolution, introduced by Krajíček [13], where literals are replaced by conjunctions of literals.

Definition 40. RES(f) *consists of the rules of* RES, *with atomic variables varying over conjunctions of literals, and the rule* $\wedge\,\dfrac{A \wedge [B \vee \bigwedge_i a_i] \wedge [\bigwedge_j b_j \vee C]}{A \wedge [B \vee (\bigwedge_i a_i \wedge \bigwedge_j b_j) \vee C]}$.

Additionally, in a derivation Φ, *no conjunction of literals may be larger than* $f(|\Phi|)$.

Proposition 41. KS *p-simulates* RES(f) *for any function* f.

Proof. $\wedge$ is derivable in $\{\mathsf{s}\}$, and the old rules can be dealt with as before.

6   A Simulation of Dag-Like Cut-Free Gentzen

We consider Gen in its one-sided variation, e.g., GS1p in [16], and identify sequents with the disjunction of their formulae, as an abuse of notation.

We now give a translation of dag-like cut-free Gentzen proofs to KS^+, and then KS. Naively we could just apply a generic cocontraction to simulate each dag-step, duplicating the entire sequent, but this may lead to an exponential blowup reducing under norm by Rmk. 18.

Instead we notice that, when two branches of a dag step are brought together by a $\wedge$ step, we only need to cocontract the formulae which are common ancestors to the conjuncts of the $\wedge$ step. For example:

$$
\mathsf{dag}\frac{\Gamma, A, B, C}{\wedge\frac{\Gamma, A, B, C \quad \Gamma, A, B, C}{\Gamma, \Gamma, [A \vee B] \wedge [A \vee C], B, C}} \quad \rightarrow \quad \mathsf{w}{\downarrow}\frac{\Gamma}{\Gamma \vee \Gamma} \vee \mathsf{c}{\uparrow}\frac{\mathsf{w}{\downarrow}\frac{A}{A \vee B} \wedge \mathsf{w}{\downarrow}\frac{A}{A \vee C}}{} \vee B \vee C
$$

When there are other rules between the dag and $\wedge$ steps, we can translate them into deep steps, inside the conjunction $[A \vee B] \wedge [A \vee C]$ above, for example.

Definition 42. *For a sequent Γ and formula A occurring in a Gen^- derivation π, let $\mathsf{Anc}_\Gamma(A)$ denote the set of ancestors of A occurring in Γ.*

Definition 43. *A* contraction loop *in a flow ϕ is a $(\mathsf{c}{\uparrow}, \mathsf{c}{\downarrow})$ pair of nodes (ν_1, ν_2) in ϕ, with ν_1 above ν_2, where there are two (or more) disjoint paths between ν_1 and ν_2.*

Lemma 44. *There is a polynomial transformation T from dag-Gen^- derivations to KS^+ satisfying the following properties:*

$$
T: \ {\overset{\Gamma}{\underset{\Delta}{\|\mathrm{tree\text{-}Gen}^-}}} \ \rightarrow \ {\overset{\Gamma}{\underset{\Delta}{\|\mathsf{KS}}}}
\qquad
T: \ {\overset{\Gamma}{\underset{\Sigma}{\overset{\pi_1\|}{\Delta}}_{\pi_2\|}}} \ \rightarrow \ {\overset{\Gamma}{\underset{\Sigma}{\overset{T\pi_1\|}{\Delta}}_{T\pi_2\|}}}
\ ,\qquad
T: \ \mathsf{dag}\frac{\Gamma}{\wedge\frac{{\overset{\Gamma}{\underset{\Delta,A}{\|\mathrm{dag\text{-}Gen}^-}}} \quad {\overset{\Gamma}{\underset{B,\Sigma}{\|\mathrm{dag\text{-}Gen}^-}}}}{\Delta, A \wedge B, \Sigma}} \ \rightarrow
$$

$$
\begin{aligned}
&= \frac{\Gamma}{\Gamma' \vee \mathsf{c}{\uparrow}\frac{X}{X \wedge X}}\\
&\quad {\overset{}{\underset{R \vee (A \wedge B)}{\star\|\mathsf{KS}^+}}}\\
&= \frac{}{\Gamma' \vee R}\\
&\quad {\underset{\Delta \vee \Sigma}{\|\mathsf{KS} \vee (A \wedge B)}}
\end{aligned}
$$

where $X = \mathsf{Anc}_\Gamma(A) \cap \mathsf{Anc}_\Gamma(B)$ and $\Gamma' = \Gamma \setminus X$, R is some formula, and there are no contraction loops in $\star$.

Observation 45. *If π is a dag-Gen^- derivation then, by the subformula property and the properties in Lemma 44, there are no contraction loops in $fl(T\pi)$.*

Lemma 46. *If there are no contraction loops in a flow ϕ then $\ulcorner\phi\urcorner$ is polynomial in $|\phi|$.*

Proof. Define $\mathsf{c}\colon$ $\to$. c is terminating, confluent and if $\phi \underset{\mathsf{c}}{\to} \psi$ then $|\phi| = |\psi|$ and $\ulcorner\phi\urcorner \le \ulcorner\psi\urcorner$. If ϕ has no contraction loops then, in ψ, all $\mathsf{c}{\downarrow}$-nodes are above all $\mathsf{c}{\uparrow}$-nodes. So $\ulcorner\phi\urcorner \le \ulcorner\psi\urcorner = |\psi|^{O(3)} = |\phi|^{O(3)}$ by Prop. 19.

Theorem 47. KS *p-simulates dag-like cut-free Gentzen systems.*

Proof. Immediate from Obs. 45, Lemmata 46, 44 and Thms. 17, 10.

7   Conclusions

We have seen that KS is a surprisingly powerful system, despite lacking any mechanism to compress proofs. As well as the simulations of Resolution and dag-like cut-free Gentzen, it cannot be p-simulated by bounded-depth Frege, one of the strongest 'weak' systems, and also has polynomial-size proofs of the functional pigeonhole principle.

7.1   Atomic Flows as a Tool for Complexity Analysis

Atomic flows are a useful tool to analyse and manipulate derivations; often we can avoid the possibly exponential blowup in eliminating cocontractions. Further work could investigate whether we can always avoid this blowup, via a local or global flow reduction.

7.2   Dag-Like Cut-Free Gentzen Systems and Variations

Results in [7] show that the addition of elimination rules (below) to dag-Gen⁻ result in a system Gen* that is p-equivalent to KS⁺, and so quasipolynomially simulates Frege.

$$\vee\text{-Elim}\ \frac{\Gamma, A \vee B}{\Gamma, A, B} \quad , \quad \wedge\text{-Elim-L}\ \frac{\Gamma, A \wedge B}{\Gamma, A} \quad , \quad \wedge\text{-Elim-R}\ \frac{\Gamma, A \wedge B}{\Gamma, B}$$

On the other hand we showed that, without these modifications, Gen⁻ cannot even p-simulate KS, and in fact that KS fits neatly between these two variations:

$$\text{dag-Gen}^- < \text{KS} \le_? \text{KS}^+ = \text{Gen}^* \le_? \text{Frege}$$

The restriction on proofs caused by the subformula property seems to be critical; it would be interesting to investigate its effects on proof complexity in general.

We regard KS to be an uncompressed system: every proof of a conjunction $A \wedge B$ can be partitioned into a proof of A and a proof of B, with no sharing between them, by substituting t for one of the conjuncts and reducing every line in the proof by $=$.

Consequently, for any dag-Gen$^-$ proof of a conjunction $A \wedge B$ there are KS proofs of A and B whose sizes sum to the size of the initial proof, for some notion of size globally accurate up to a polynomial. We thus argue that the sharing effect of dagness in cut-free Gentzen systems serves solely to do some of the work of deep inference, but not all of it due to the strict separation between the two systems.

We notice that the separation of KS and tree-Gen$^-$ in [2] is in fact just a special case of Thm. 47, since dag-like cut-free Gentzen has polynomial-size proofs of the Statman tautologies [5].

7.3   Stronger Systems

We showed that bounded-depth Frege cannot p-simulate KS but did not consider the other direction. We conjecture that they are incomparable, due to the dissimilar ways they are defined. Similar questions persist for other 'stronger' systems, e.g., Cutting Planes, although ongoing research suggests we may be able to obtain a separation of KS from Cutting Planes.

Acknowledgements. The author would like to thank Alessio Guglielmi and Tom Gundersen for all their time and effort in discussing this paper, and the anonymous referees for their useful comments.

References

1. Brünnler, K., Tiu, A.F.: A local system for classical logic (2001); preprint (WV-01-02) (2001)
2. Bruscoli, P., Guglielmi, A.: On the proof complexity of deep inference. ACM Transactions on Computational Logic 10(2), article 14, 1–34 (2009)
3. Bruscoli, P., Guglielmi, A., Gundersen, T., Parigot, M.: Quasipolynomial normalisation in deep inference via atomic flows and threshold formulae (2009) (submitted)
4. Buss, S.R.: Polynomial size proofs of the propositional pigeonhole principle. Journal of Symbolic Logic 52(4), 916–927 (1987)
5. Clote, P., Kranakis, E.: Boolean Functions and Computation Models. Springer (2002)
6. D'Agostino, M.: Are tableaux an improvement on truth-tables? Journal of Logic, Language and Information 1, 235–252 (1992)
7. Das, A.: On the Proof Complexity of Cut-Free Bounded Deep Inference. In: Brünnler, K., Metcalfe, G. (eds.) TABLEAUX 2011. LNCS, vol. 6793, pp. 134–148. Springer, Heidelberg (2011)
8. Das, A.: Complexity of deep inference via atomic flows (2012)(preprint)
9. Guglielmi, A.: Resolution in the calculus of structures (2003) (preprint)
10. Guglielmi, A., Gundersen, T.: Normalisation control in deep inference via atomic flows II (2008) (preprint)
11. Gundersen, T.: A General View of Normalisation Through Atomic Flows. Ph.D. thesis, University of Bath (2009)

12. Jeřábek, E.: Proof complexity of the cut-free calculus of structures. Journal of Logic and Computation 19(2), 323–339 (2009)
13. Krajíček, J.: On the weak pigeonhole principle (2001)
14. Krajíček, J., Pudlák, P., Woods, A.: An exponential lower bound to the size of bounded depth frege proofs of the pigeonhole principle. Random Structures & Algorithms 7(1), 15–39 (1995)
15. Pitassi, T., Beame, P., Impagliazzo, R.: Exponential lower bounds for the pigeonhole principle. Computational Complexity 3, 97–140 (1993)
16. Troelstra, A., Schwichtenberg, H.: Basic Proof Theory. Cambridge Tracts in Theoretical Computer Science, vol. 43. Cambridge University Press (1996)

Connecting Partial Words
and Regular Languages

Jürgen Dassow[1], Florin Manea[2], and Robert Mercaş[1]

[1] Fakultät für Informatik, Otto-von-Guericke-Universität Magdeburg, Postfach 4120,
39016 Magdeburg, Germany
dassow@iws.cs.uni-magdeburg.de, robertmercas@gmail.com
[2] Institut für Informatik, Christian-Albrechts-Universität zu Kiel, 24098 Kiel,
Germany
flm@informatik.uni-kiel.de

Abstract. We initiate a study of languages of partial words related to
regular languages of full words. First, we study the possibility of expressing a regular language of full words as the image of a partial-words-
language through a substitution that only replaces the hole symbols of
the partial words with a finite set of letters. Results regarding the structure, uniqueness and succinctness of such a representation, as well as
a series of related decidability and computational-hardness results, are
presented. Finally, we define a hierarchy of classes of languages of partial
words, by grouping together languages that can be connected in strong
ways to regular languages, and derive their closure properties.

1 Introduction

Two DNA strands attach one to the other, normally, in a complementary way
according to their nucleotides. That is, each purine, A or G, creates a hydrogen
bond with one complementary pyrimidine, T or C, respectively. But, sometimes,
it is the case that this process goes wrong, allowing G-T bonds. Starting from
this situation, and motivated by the need of having a way to recover (as much
as possible) and work with a correct DNA sequence, Berstel and Boasson [1]
suggested the usage of partial words as a suitable mathematical model. Partial
words are words that beside regular letters contain an extra "joker" symbol, also
called "hole" or "do-not-know" symbol, that matches all symbols of the original
alphabet, which were investigated already in the 1970s [3]. Going back to the
initial example, one could recover the information regarding a DNA sequence
from the badly bonded pair of DNA strands by associating actual letters to the
positions where the bonds were correct and holes to the positions where the
bonds were not correctly formed. Besides the above motivation, partial words
may find applications in other fields, as well; for instance, they can be seen
as data sequences that are corrupted either by white noise or other external
factor, or even incomplete or insufficient information, that one needs in some
process. In the last decade a lot of combinatorial and algorithmic properties
of partial words have been investigated (see the survey [2], and the references

S.B. Cooper, A. Dawar, and B. Löwe (Eds.): CiE 2012, LNCS 7318, pp. 151–161, 2012.
© Springer-Verlag Berlin Heidelberg 2012

therein). Surprisingly, so far, no study of classes of languages of partial words (or sets of partial words that have common features) was performed. The only work on a connected topic, that we are aware of, is [4]. There, the concept of restoration of punctured languages and several similarity measures between full-words-languages, related to this concept, were investigated. More precisely, puncturing a word means replacing some of its letters with holes; from a language of punctured words, its restoration was obtained by taking all the languages that can be punctured to obtain the respective language. The results of [4] regarded classes of full-words-languages defined by applying successively puncturing and restoration operations to classes of languages from the Chomsky hierarchy.

The study of the class of regular languages, the most restrictive class of the Chomsky-hierarchy, has been one of the central topics in theoretical computer science. This class of languages, defined usually either as the class of languages accepted by finite automata or as the class of languages described by regular expressions, was extensively studied throughout the last seventy years (starting from the early 1940s) and, besides its impact in theory (mostly in language theory, but also in complexity theory, for instance), it was shown to have a wide range of applications. Regular languages, and the various mechanisms used to specify them, were used, for instance, in compilers theory, circuit design, text editing, pattern matching, formal verification, DNA computing, or natural language processing (see [7]).

In this work, we aim to establish a stronger connection between the attractive notions mentioned above: partial words, on one side, and regular languages, on the other side. First, we show how we can (non-trivially) represent every regular language as the image of a regular language of partial words through a substitution that defines the letters that may replace the hole (called $\diamond$-substitution, in the following). Moreover, we show that such a representation can be useful: for some regular languages, there exist deterministic finite automata accepting languages of partial words that represent the full-word-language and are exponentially more succinct than the minimal deterministic finite automaton accepting that language. Unfortunately, it may also be the case when the minimal non-deterministic finite automaton accepting a language is exponentially more succinct than any deterministic automaton accepting a language of partial words representing the same language. Generally, automata accepting languages of partial words representing a given full-words language can be seen as intermediate between the deterministic finite automata and the non-deterministic automata accepting that language. We also present a series of algorithmic and complexity results regarding the possibility of representing a regular language as the image of a language of partial words through a $\diamond$-substitution.

Motivated by the above results, that connect in a meaningful way languages of partial words to regular languages of full words, and by the theoretical interest of studying systematically such languages, we define a series of classes of languages of partial words. Each of these classes contains languages that can be placed in a particular strong relation with the regular languages. Further, we investigate these classes from a language theoretic point of view, show that they form a hierarchy, and establish their closure properties.[1]

We begin the paper with a series of basic definitions. Let V be a non-empty finite set of symbols called an *alphabet*. Each element $a \in V$ is called a *letter*. A *full word* (or, simply, word) over V is a finite sequence of letters from V while a *partial word* over V is a finite sequence of letters from $V \cup \{\diamond\}$, the alphabet V extended with the distinguished hole symbol $\diamond$. The *length* of a (partial) word u is denoted by $|u|$ and represents the total number of symbols in u; the number of occurrences of a symbol a in a (partial) word w is denoted $|w|_a$. The *empty (partial) word* is the sequence of length zero and is denoted by λ. We denote by V^* (respectively, $(V \cup \{\diamond\})^*$) the set of words (respectively, partial words) over the alphabet V and by V^+ (respectively, $(V \cup \{\diamond\})^+$) the set of non-empty words (respectively, non-empty partial words) over V. The catenation of two (partial) words u and v is defined as the (partial) word uv. Recall that V^* (where the alphabet V may include the $\diamond$ symbol) is the free monoid generated by V, under the operation of catenation of words; the unit element in this monoid is represented by the empty word λ. A language L of full words over an alphabet V is a subset of V^*; a language of partial words L over an alphabet V (that does not contain the $\diamond$ symbol) is a subset of $(V \cup \{\diamond\})^*$. Given a language L we denote by $\mathbf{alph}(L)$ (the alphabet of L) the set of all the letters that occur in the words of L; for the precision of the exposure, we say that a language L of full (respectively, partial) words is over V, with $\diamond \notin V$, if and only if $\mathbf{alph}(L) = V$ (respectively, $\mathbf{alph}(L) = V \cup \{\diamond\}$). For instance, $L = \{abb, ab\diamond\}$ has $\mathbf{alph}(L) = \{a, b, \diamond\}$, thus, is a language of partial words over $\{a, b\}$. Note that the catenation operation can be extended to languages; more precisely, if L_1 and L_2 are languages over V, we define their catenation $L_1 L_2 = \{w_1 w_2 \mid w_1 \in L_1, w_2 \in L_2\}$.

Let u and v be two partial words of equal length. We say that u is *contained* in v, denoted $u \sqsubset v$, if $u[i] = v[i]$ for all $u[i] \in V$; moreover, u and v are *compatible*, denoted $u \uparrow v$, if there exists a word w such that $u \sqsubset w$ and $v \sqsubset w$. These notions can be extended to languages. Let L and L' be two languages of partial words with $\mathbf{alph}(L) \cup \mathbf{alph}(L') = V \cup \{\diamond\}$ and $\diamond \notin V$. We say that L is *contained* in L', denoted $L \sqsubset L'$, if, for every word $w \in L$, there exists a word $w' \in L'$ such that $w \sqsubset w'$. We say that L is *compatible* to L', denoted $L \uparrow L'$, if, for each $w \in L$, there exists $w' \in L'$ such that $w \uparrow w'$ and, for each $v' \in L'$, there exists $v \in L$ such that $v' \uparrow v$.

A substitution is a mapping $h : V^* \to 2^{U^*}$ with $h(xy) = h(x)h(y)$, for $x, y \in V^*$, and $h(\lambda) = \{\lambda\}$; h is completely defined by the values $h(a)$ for all $a \in V$. A morphism is a particular type of a substitution for which $h(a)$ contains exactly one element for all $a \in V$; i.e., a morphism is a function $h : V^* \to U^*$ with $h(xy) = h(x)h(y)$ for $x, y \in V^*$. A $\diamond$-substitution over V is a substitution with $h(a) = \{a\}$, for $a \in V$, and $h(\diamond) \subseteq V$. Here we assume that $\diamond$ can replace any symbol of V.

---

[1] A technical appendix containing full proofs of our results can be found at the webpage: https://www.informatik.uni-kiel.de/zs/pwords

In this paper, DFA stands for deterministic finite automaton and NFA for non-deterministic finite automaton; the language accepted by a finite automaton M is denoted $L(M)$. Also, the set of all the regular languages is denoted by **REG**; by $\mathbf{REG_{full}}$, we denote the set of all the regular languages of full words. Further definitions regarding finite automata and regular languages can be found in [7], while partial words are surveyed in [2].

2   Definability by Substitutions

Let us begin our investigation by presenting several results regarding the way regular languages can be expressed as the image of a language of partial words through a substitution.

Lemma 1. *Let $L \subseteq (V \cup \{\diamond\})^*\{\diamond\}(V \cup \{\diamond\})^*$ be a language of partial words and let σ be a $\diamond$-substitution over V. There exists a language L' such that $\sigma(L) = \sigma(L')$ and $|w|_\diamond = 1$ for all $w \in L'$.*

Lemma 2. *Let L be a regular language over V and let σ be a $\diamond$-substitution over V. Then there exists a maximal (with respect to set inclusion) language $L' \subseteq L$ which can be written as $\sigma(L'')$, where L'' is a language of partial words such that any word in L'' has exactly one hole. Moreover, L' and L'' are regular languages and, provided that L is given by a finite automaton accepting it, one can algorithmically construct a finite automaton accepting L' and L''.*

Proof. Let $\sigma(\diamond) = V'$.

We start by noting that a word w belongs to $\sigma(L'')$ for a language of partial words L'', whose elements contain at least one hole each, if and only if there exist the words x and y such that $w = xay$, for some $a \in V'$ and $\{x\}V'\{y\} \subseteq L$.

Now let $M = (Q, V, q_0, F, \delta)$ be a DFA accepting L.

Let $q \in Q$. We define the language R_q as follows. A word w of L is in R_q if and only if there exists the partial word $x\diamond y$, compatible with w, with $\delta(q_0, x) = q$, $S = \delta(q, V') \subseteq Q$, $\delta(S, y) \subseteq F$. Basically, R_q is the set of the words for which there exists a compatible partial word $x\diamond y$ with exactly one hole, such that x labels a path from q_0 to q in A and any word from $\{x\}V'\{y\}$ is in L.

Clearly, $R_q = \{x \mid x \in V^*, \delta(q_0, x) = q\}V'\{y \mid y \in V^*, \delta(q', y) \in F$ for all $q' \in \delta(q, V)\}$. It follows that R_q is regular and an automaton accepting this language can be constructed starting from M. Moreover, $R_q = \sigma(H_q)$, for $H_q = \{x \mid x \in V^*, \delta(q_0, x) = q\}\{\diamond\}\{y \mid y \in V^*, \delta(q', y) \in F$ for all $q' \in \delta(q, V)\}$.

Take now $L' = \cup_{q \in Q} R_q$ and $L'' = \cup_{q \in Q} H_q$. Then L' is regular, as all the languages R_q are regular, and $L' = \sigma(L'')$. We only have to show that L' is maximal. If there exists $L_1 \subset L$ and a language of partial words L_2, whose elements contain exactly one hole each, such that $\sigma(L_2) = L_1$, then for every word w of L_1 there exist the words x and y such that $x\diamond y \in L_2$ and $w \in \sigma(x\diamond y) = \{x\}V'\{y\} \subseteq \sigma(L_2) = L_1$. Thus, $w \in R_q$ for $q = \delta(q_0, x)$. Therefore, $L_1 \subseteq L'$.   □

Note that the sets R_q with $q \in Q$ are not a partition of L, as they are not necessarily mutually disjoint.

Next we introduce two relations connecting partial-words-languages and full-words-languages.

Definition 1. *Let $L \subseteq V^*$ be a language and σ be a $\diamond$-substitution over V. We say that L is σ-defined by the language L', where $L' \subseteq (V \cup \{\diamond\})^*$ is a partial-words-language, if $L = \sigma(L')$. Moreover, we say that L is essentially σ-defined by L', where $L' \subseteq (V' \cup \{\diamond\})^*$, if $L = \sigma(L')$ and every word in L' contains at least a $\diamond$-symbol.*

Obviously, for any regular language L over V, there is a regular language L' of partial words and a $\diamond$-substitution σ over V such that $\sigma(L') = L$, i.e., L is σ-defined by L'. For instance, take the set L' of the words obtained by replacing in the words of L some occurrences of a symbol $a \in V$ by $\diamond$, and the $\diamond$-substitution σ over V that maps $\diamond$ to $\{a\}$. More relevant ways of defining a regular language, in the sense of Definition 1, are presented in this section. We begin by characterizing the essentially definable languages.

Assume that the regular language $L \subseteq V^*$ is essentially σ-definable for some $\diamond$-substitution σ over V. Then $\sigma(L') = L$ for some appropriate language L' such that any word of L' contains at least a hole. By Lemma 1, we get that there is a regular language L'' of partial words such that $\sigma(L'') = \sigma(L')$ and each word of L'' contains exactly one hole. Now by Lemma 2 and its proof we get immediately the following characterisation of σ-definable languages.

Theorem 1. *Let L be a regular language of full words over V and σ a $\diamond$-substitution over V. Then L is essentially σ-definable if and only if $L = \bigcup_{q \in Q} R_q$ (where R_q is given in the proof of Lemma 2).* $\square$

We also easily get the following decidability results.

Theorem 2. *i) Given a regular language L over V and a $\diamond$-substitution σ over V, it is decidable whether L is essentially σ-definable.*
ii) Given a regular language L over V, one can algorithmically identify all $\diamond$-substitutions σ for which L is essentially σ-definable.

Proof. By the previous results, testing whether L is essentially σ-definable is equivalent to testing whether L and $L' = \bigcup_{q \in Q} R_q$ are equal. Because the equality of two regular languages is decidable, the first statement follows. The second statement follows by an exhaustive search in the (finite) set of all $\diamond$-substitutions σ over V for those that essentially define L. $\square$

The following consequence of Lemma 2 is worth noting, as it provides a canonical non-trivial representation of regular languages.

Theorem 3. *Given a regular language $L \subseteq V^*$ and a $\diamond$-substitution σ over V, there exists a unique regular language $L_\diamond$ of partial words that fulfils the following three conditions: (i) $L = \sigma(L_\diamond)$, (ii) for any language L_1 with $\sigma(L_1) = L$ we have $\{w \mid w \in L_1, |w|_\diamond \geq 1\} \sqsubset \{w \mid w \in L_\diamond, |w|_\diamond \geq 1\}$, and (iii) $(L_\diamond \cap V^*) \cap \sigma(\{w \mid w \in L_\diamond, |w|_\diamond \geq 1\}) = \emptyset$.*

Proof. Using the sets defined in Lemma 2, take $L_\diamond = L'' \cup (L \setminus L')$. The conclusion follows easily.   □

Motivated by this last result, we now turn to the descriptional complexity of representing a regular language of full words by regular languages of partial words. We are interested in the question whether there is a regular language $L \subseteq V^*$, a regular language $L' \subseteq (V \cup \{\diamond\})^*$, and a $\diamond$-substitution σ over V with $\sigma(L') = L$ such that the minimal DFA accepting L' has a (strictly) lower number of states than the minimal DFA accepting L? In other words, are there cases when one can describe in a more succinct way a regular language via a language of partial words and a substitution that define it? Moreover, can we decide algorithmically whether for a given regular language L there exist a language of partial words and a substitution providing a more succinct description of L?

Let L be a regular language of full words over V. We denote by $\min_{\mathrm{DFA}}(L)$ the number of states of the (complete) minimal DFA accepting L. Furthermore, let $\min_{\mathrm{NFA}}(L)$ denote the number of states of a minimal NFA accepting L. Moreover, for a regular language L let $\min_{\mathrm{DFA}}^\diamond(L)$ denote the minimum number of states of a (complete) DFA accepting a regular language $L' \subseteq (V \cup \{\diamond\})^*$ (where $\diamond$ is considered as an input symbol) for which there exists a $\diamond$-substitution σ over V such that $\sigma(L') = L$.

We have the following relation between the defined measures.

Theorem 4. *i) For every regular language L we have*

$$\min_{\mathrm{DFA}}(L) \geq \min_{\mathrm{DFA}}^\diamond(L) \geq \min_{\mathrm{NFA}}(L).$$

ii) There exist regular languages L such that

$$\min_{\mathrm{DFA}}(L) > \min_{\mathrm{DFA}}^\diamond(L) > \min_{\mathrm{NFA}}(L).$$

By the previous result one can see that, for certain substitutions σ, minimal DFAs accepting languages of partial words that σ-define a given full-words-regular language can be seen as intermediate between the minimal DFA and the minimal NFA accepting that language: they provide a succinct representation of that language, while having a limited non-determinism.

In fact, one can show that the differences $\min_{\mathrm{DFA}}(L) - \min_{\mathrm{DFA}}^\diamond(L)$ and $\min_{\mathrm{DFA}}^\diamond(L) - \min_{\mathrm{NFA}}(L)$ can be arbitrarily large; more precisely, we may have an exponential blow-up with respect to both relations.

Theorem 5. *Let n be a natural number, $n \geq 3$. There exist regular languages L and L' such that $\min_{\mathrm{DFA}}^\diamond(L) \leq n + 1$ and $\min_{\mathrm{DFA}}(L) = 2^n - 2^{n-2}$ and $\min_{\mathrm{NFA}}(L') \leq 2n + 1$ and $\min_{\mathrm{DFA}}^\diamond(L') \geq 2^n - 2^{n-2}$.*

The following remark provides an algorithmic side of the results stated above.

Remark 1. Given a DFA accepting a regular language L we can construct algorithmically a DFA with $\min_{\mathrm{DFA}}^\diamond(L)$ states, accepting a regular language of partial words L', and a $\diamond$-substitution σ over $\mathbf{alph}(L)$, such that L is σ-defined by L'. By exhaustive search, we take a DFA M with at most $\min_{\mathrm{DFA}}(L)$ states, whose transitions are labelled with letters from an alphabet included in $\mathbf{alph}(L) \cup \{\diamond\}$, and a ◇-substitution σ over **alph**(L). We transform M into an NFA accepting $\sigma(L(M))$ by replacing the transitions labelled with ◇ by $|\sigma(◇)|$ transitions labelled with the letters of $\sigma(◇)$, respectively. Next, we construct the DFA equivalent to this NFA, and check whether it accepts L or not (that is, $\sigma(L(M)) = L$). From all the initial DFAs we keep those with minimal number of states, since they provide the answer to our question. It is an open problem whether such a DFA can be obtained by a polynomial time deterministic algorithm; however, we conjecture that the problem is computationally hard.

We conclude by showing the hardness of a problem related to definability.

Theorem 6. *Consider the problem P: "Given a DFA accepting a language L of full words, a DFA accepting a language L' of partial words, and a ◇-substitution σ over **alph**(L), decide whether $\sigma(L') \neq L$." This problem is NP-hard.*

Proof. In [6], the following problem was shown to be NP-complete:
P': "Given a list of partial words $S = \{w_1, w_2, \ldots, w_k\}$ over the alphabet V with $|V| \geq 2$, each partial word having the same length L, decide whether there exists a word $v \in V^L$ such that v is not compatible with any of the partial words in S."

We show here how problem P' can be reduced in polynomial time by a many-one reduction to problem P. Indeed, take an instance of P': a list of partial words $S = \{w_1, w_2, \ldots, w_k\}$ over the alphabet V with $|V| \geq 2$, each having the same length L. We can construct in polynomial time a DFA M accepting exactly the language of partial words $\{w_1, w_2, \ldots, w_k\}$. Also, we can construct in linear time a DFA M' accepting the language of full words V^L. It is clear that for $L(M)$ and the substitution σ, mapping the letters of V to themselves and ◇ to V, we have $\sigma(L(M)) \neq V^L$ (that is, the answer to the input M, M' and σ of problem P is positive) if and only if the answer to the given instance of P' is also positive. Since solving P' is not easier than solving P, we conclude our proof. □

Theorem 6 provides a simple way to show the following well known result.

Corollary 1. *The problem of deciding whether a DFA M and an NFA M' accept different languages is NP-hard.*

3 Languages of Partial Words

While the results of the last section study the possibility and efficiency of defining a regular language as the image of a (regular) language of partial words, it seems interesting to us to take an opposite point of view, and investigate the languages of partial words whose images through a substitution (or all possible substitutions) are regular. Also, languages of partial words compatible with at least one regular language (or only with regular languages) seem worth investigating.

The definitions of the first three classes considered in this section follow the main lines of the previous section. We basically look at languages of partial words that can be transformed, via substitutions, into regular languages.

Definition 2. *Let L be a language of partial words over V.*
1. *We say that L is $(\forall\sigma)$-regular if $\sigma(L)$ is regular for all the $\diamond$-substitutions σ over alphabets that contain V and do not contain $\diamond$.*
2. *We say that L is* **max**-*regular if $\sigma(L)$ is regular, where σ is a $\diamond$-substitution over V' with $\sigma(\diamond) = V'$, and $V' = V$ if $V \neq \emptyset$, and V' is a singleton with $\diamond \notin V'$, otherwise.*
3. *We say that L is $(\exists\sigma)$-regular if there exists a $\diamond$-substitution σ over a nonempty alphabet V', that contains V and does not contain $\diamond$, such that $\sigma(L)$ is regular.*
The classes of all $(\forall\sigma)$-regular, **max**-*regular, and $(\exists\sigma)$-regular languages are denoted by* $\mathbf{REG}_{(\forall\sigma)}$, $\mathbf{REG}_{\mathbf{max}}$, *and, respectively,* $\mathbf{REG}_{(\exists\sigma)}$.

We consider, in the following, two classes of languages of partial words that are defined starting from the concept of compatibility.

Definition 3. *Let L be a language of partial words over V.*
4. *We say that L is $(\exists)$-regular if exists a regular language L' of full words such that $L \uparrow L'$.*
5. *We say that L is $(\forall)$-regular if every language L' of full words such that $L \uparrow L'$ is regular.*
The class of all the $(\exists)$-regular languages is denoted $\mathbf{REG}_{(\exists)}$, *while that of $(\forall)$-regular languages by* $\mathbf{REG}_{(\forall)}$.

According to the definitions from [4], the $(\exists)$-regular languages are those whose restoration contains at least a regular language, while $(\forall)$-regular languages are those whose restoration contains only regular languages.

We start with the following result.

Theorem 7. *For every non-empty alphabet V with $\diamond \notin V$ there exist an undecidable language L of partial words over V, such that:*
i) $\sigma(L) \in \mathbf{REG}$ for all substitutions σ over V, and $\sigma'(L) \notin \mathbf{REG}$ for the $\diamond$-substitution σ' with $\sigma'(\diamond) = V \cup \{c\}$, where $c \notin V$.
ii) every language $L' \subseteq V^*$ of full words, which is compatible with L, is regular and there is an undecidable language $L'' \subseteq (V')^*$, where V' strictly extends V, which is compatible with L.

Proof. Let $L_1 \subseteq V^*$ be an undecidable language (for instance, L_1 can be constructed by the classical diagonalization technique $L_1 = \{a^n \mid$ the nth Turing machine in an enumeration of the Turing machines with binary input does not accept the binary representation of $n\}$, where $a \in V$) and $L = V^* \cup \{w \mid w \in L_1\}$. Clearly, for any $\diamond$-substitution σ over V, we have $\sigma(L) = V^*$. However, if we take a letter $c \notin V$ and the $\diamond$-substitution σ' which replaces $\diamond$ by $V \cup \{c\}$ we obtain an undecidable language $\sigma'(L)$. This concludes the proof of (i). To show (ii) we just have to note that the only language contained in V^* compatible with L is V^*, and, if we take a letter $c \notin V$ and replace $\diamond$ by c (or, in other words, if we see $\diamond$ as the conventional symbol c), we obtain an undecidable language compatible with L. $\qquad\square$

We can now show a first result regarding the classes previously defined.

Theorem 8. REG = REG$_{(\forall\sigma)}$ $\subset$ REG$_{\mathbf{max}}$.

Proof. It is rather clear that **REG**$_{(\forall\sigma)}$ $\subseteq$ **REG**$_{\mathbf{max}}$.

Since **REG** is closed to substitutions it follows that **REG** $\subseteq$ **REG**$_{(\forall\sigma)}$.

It is also not hard to see that **REG**$_{(\forall\sigma)}$ $\subseteq$ **REG** (given a language L in **REG**$_{(\forall\sigma)}$, one can take the special substitution that replaces $\diamond$ with a symbol that does not occur in **alph**(L) and obtain a regular language; therefore L is a regular language if $\diamond$ is seen as a normal symbol).

By Theorem 7, **REG**$_{\mathbf{max}}$ contains an undecidable language; indeed, given an non-empty alphabet V, the language L defined in its proof for V is in **REG**$_{\mathbf{max}}$ according to (i). The strictness of the inclusion **REG** $\subsetneq$ **REG**$_{\mathbf{max}}$ follows. $\square$

The next result gives some insight on the structure of the class **REG**$_{\mathbf{max}}$.

Theorem 9. *Let $L \in$ **REG**$_{\mathbf{max}}$ be a language of partial words over $V \neq \emptyset$ and σ the $\diamond$-substitution used in the definition of **REG**$_{\mathbf{max}}$. Then there exists a maximal language (with respect to set inclusion) $L_0 \in$ **REG**$_{\mathbf{max}}$ of partial words over V such that $\sigma(L_0) = \sigma(L)$. Moreover, given an automaton accepting L, an automaton accepting L_0 can be constructed.*

It is also not hard to see that any language from **REG**$_{\mathbf{max}}$ whose words contain only holes is regular.

The following relation also holds:

Theorem 10. REG$_{\mathbf{max}}$ $\subset$ REG$_{(\exists\sigma)}$ $\subset$ REG$_{(\exists)}$.

Proof. The non-strict inclusions **REG**$_{\mathbf{max}}$ $\subseteq$ **REG**$_{(\exists\sigma)}$ $\subseteq$ **REG**$_{(\exists)}$ are immediate. We show now that each of the previous inclusions is strict.

Take $L = \{(ab)^n \diamond b(ab)^n \mid n \in \mathbf{N}\}$. Considering σ a $\diamond$-substitution as in the definition of **REG**$_{\mathbf{max}}$, we have $\sigma(L) \cap \{w \mid w \in \{a, b\}^*, w$ contains $bb\}$ is the language $\{(ab)^n bb(ab)^n\}$ which is not regular. Thus, $\sigma(L)$ is not regular, and L is not in **REG**$_{\mathbf{max}}$. However, it is clearly in **REG**$_{(\exists\sigma)}$, as when we take the substitution $\sigma(a) = \{a\}$, $\sigma(b) = \{b\}$, and $\sigma(\diamond) = \{a\}$, we have $\sigma(L) = \{(ab)^{2n+1} \mid n \in \mathbf{N}\}$, which is a regular language. This shows that **REG**$_{\mathbf{max}}$ $\subset$ **REG**$_{(\exists\sigma)}$.

Now, take $L = \{(ab)^n \diamond b(ab)^n \mid n \in \mathbf{N}\} \cup \{(ab)^n a\diamond(ab)^n \mid n \in \mathbf{N}\}$. This language is not in **REG**$_{(\exists\sigma)}$ by arguments similar to above, but it is in **REG**$_{(\exists)}$ as it is compatible with $\{(ab)^{2n+1} \mid n \in \mathbf{N}\}$. $\square$

As already shown, all the languages in **REG**$_{(\forall)}$ are in **REG** = **REG**$_{(\forall\sigma)}$; however, not all the languages in **REG** are in **REG**$_{(\forall)}$. The following statement characterizes exactly the regular languages that are in **REG**$_{(\forall)}$.

Theorem 11. *Let L be a regular partial-words-language over V. Then $L \in$ **REG**$_{(\forall)}$ if and only if the set $\{w \mid |w|_\diamond \geq 1, w \in L\}$ is finite.*

The previous result provides a simple procedure for deciding whether a regular partial-words-language is in **REG**$_{(\forall)}$ or not. We simply check (taking as input a DFA for that language) whether there are finitely many words that contain $\diamond$ or not. If yes, we accept the input and confirm that the given language is in **REG**$_{(\forall)}$; otherwise, we reject the input.

Theorem 11 has also the following consequence.

Theorem 12. $\mathbf{REG}_{(\forall)} \subset \mathbf{REG}$.

In many previous works (surveyed in [2]), partial words were defined by replacing specific symbols of full words by $\diamond$, in a procedure that resembles the puncturing of [4]. Similarly, in [5], partial words were defined by applying the finite transduction defined by a deterministic generalised sequential machine (DGSM) to full words, such that $\diamond$ appears in the output word. Accordingly, we can define a new class of partial-words-languages, $\mathbf{REG_{gsm}}$, using this automata-theoretic approach. Let L be a language of partial words over V, with $\diamond \in \mathbf{alph}(L)$; L is **gsm-regular**, and is in $\mathbf{REG_{gsm}}$, if there exists a DGSM M and a regular language L' such that L is obtained by applying the finite transduction defined by M to L'. It is not hard to show that $\mathbf{REG_{gsm}} = \mathbf{REG} \setminus \mathbf{REG_{full}}$.

By the Theorems 8,10,11, and 12 we get the following hierarchies:

$$\mathbf{REG_{full}} \subset \mathbf{REG}_{(\forall)} \subset \mathbf{REG} = \mathbf{REG}_{(\forall\sigma)} \subset \mathbf{REG_{max}} \subset \mathbf{REG}_{(\exists\sigma)} \subset \mathbf{REG}_{(\exists)}$$

$$\mathbf{REG} \setminus \mathbf{REG_{full}} = \mathbf{REG_{gsm}} \subset \mathbf{REG} = \mathbf{REG}_{(\forall\sigma)}$$

Finally, the closure properties of the defined classes are summarized in the following table. Note that y (respectively, n) at the intersection of the row associated with the class $\mathcal{C}$ and the column associated with the operation $\circ$ means that $\mathcal{C}$ is closed (respectively, not closed) under operation $\circ$. A special case is the closure of $\mathbf{REG_{max}}$ under union and concatenation: in general this class is not closed under these operations, but when we apply them to languages of $\mathbf{REG_{max}}$ over the same alphabet we get a language from the same class.

Class	$\cup$	$\cap$	$\cap\mathbf{REG}$	$\mathbf{alph}(L)^* \setminus L$	$*$	$\cdot$	ϕ	ϕ^{-1}	σ
$\mathbf{REG}_{(\forall)}$	y	y	y	n	n	n	n	n	n
$\mathbf{REG} = \mathbf{REG}_{(\forall\sigma)}$	y	y	y	y	y	y	y	y	y
$\mathbf{REG_{max}}$	n/y	n	n	n	y	n/y	n	n	n
$\mathbf{REG}_{(\exists\sigma)}$	n	n	n	n	y	n	n	n	n
$\mathbf{REG}_{(\exists)}$	y	n	n	y	y	y	n	n	n

Acknowledgements. The work of Florin Manea is supported by the DFG grant 582014. The work of Robert Mercaş is supported by the Alexander von Humboldt Foundation.

References

1. Berstel, J., Boasson, L.: Partial words and a theorem of Fine and Wilf. Theoretical Computer Science 218, 135–141 (1999)
2. Blanchet-Sadri, F.: Algorithmic Combinatorics on Partial Words. Chapman & Hall/CRC Press (2008)
3. Fischer, M.J., Paterson, M.S.: String matching and other products. In: Complexity of Computation, Proceedings of SIAM-AMS, vol. 7, pp. 113–125 (1974)
4. Lischke, G.: Restoration of punctured languages and similarity of languages. Mathematical Logic Quarterly 52(1), 20–28 (2006)

5. Manea, F., Mercaş, R.: Freeness of partial words. Theoretical Computer Science 389(1-2), 265–277 (2007)
6. Manea, F., Tiseanu, C.: Hard Counting Problems for Partial Words. In: Dediu, A.-H., Fernau, H., Martín-Vide, C. (eds.) LATA 2010. LNCS, vol. 6031, pp. 426–438. Springer, Heidelberg (2010)
7. Rozenberg, G., Salomaa, A.: Handbook of Formal Languages. Springer-Verlag New York, Inc., Secaucus (1997)

Randomness, Computation and Mathematics

Rod Downey

[1] School of Mathematics, Statistics and Operations Research, Victoria University,
P. O. Box 600, Wellington, New Zealand
[2] Isaac Newton Institute for Mathematical Sciences, 20 Clarkson Road,
Cambridge CB3 0EH, United Kingdom
rod.downey@vuw.ac.nz

Abstract. This article examines some of the recent advances in our understanding of algorithmic randomness. It also discusses connections with various areas of mathematics, computer science and other areas of science. Some questions and speculations will be discussed.

1 Introduction

The Copenhagen interpretation of quantum physics suggests to us that randomness is essential to our understanding of the universe. Mathematics has developed many tools to utilize randomness in the development of algorithms and in combinatorial (and other) techniques. For instance, these include Markov Chain Monte Carlo and the metropolis algorithms, methods central to modern science, the probabilistic method is central to combinatorics. Computer science has its own love affair with randomness such as its uses in cryptography, fast algorithms and proof techniques.

Nonetheless, it is not clear what each community means by "randomness". Moreover, even when we agree to try one of the formalizations of the notion of randomness based on computation there is also no clear understanding on how this should be interpreted and the extent to which the applications in the disparate arenas can be reconciled.

In this article I will look at the long term programme of understanding the meaning of randomness via an important part of Turing's legacy, the theory of algorithmic computation: *algorithmic randomness*. The last decade has seen some quite dramatic advances in our understanding of algorithmic randomness. In particular, we have seen significant clarification as to the mathematical relationship between algorithmic computational power of infinite random sources and algorithmic randomness. Much of this material has been reported in the short surveys Downey [27], Nies [53] and long surveys [26,30] and long monographs Downey and Hirschfeldt [29] and Nies [52]. Also the book edited by Hector Zenil [78] has a lot of discussion of randomness of varying levels of technicality, many aimed at the general audience.

S.B. Cooper, A. Dawar, and B. Löwe (Eds.): CiE 2012, LNCS 7318, pp. 162–181, 2012.

To my knowledge, Turing himself though that randomness was a physical phenomenon, and certainly recognized the noncomputable nature of generating random strings. For example, from Turing [71], we have the following quote[1]:

> " An interesting variant on the idea of a digital computer is a "digital computer with a random element." These have instructions involving the throwing of a die or some equivalent electronic process; one such instruction might for instance be, "Throw the die and put the-resulting number into store 1000." Sometimes such a machine is described as having free will (though I would not use this phrase myself)."

John von Neumann (e.g., [75]) also recognized the noncomputable nature of generating randomness, and both seem to believe that physical procedures would be necessary. Von Neumann's quote is famous:

> "Any one who considers arithmetical methods of producing random digits is, of course, in a state of sin."

Arguably this idea well predated any notion of computation, but the germ of this can be seen in the following quotation of Joseph Bertrand [14] in 1889.

> "How dare we speak of the laws of chance?
> Is not chance the antithesis of all law?"

There has been a developing body of work seeking to understand not just the theory of randomness but how it arizes in mathematics.

For example, we have also seen an initiative (whose roots go back to work of Demuth [25]) towards using these ideas in the understanding of almost everywhere behaviour and differentiation in analysis (such as Brattka, Miller, Nies [15]). Also halting probabilities are natural and turn up in places apparently removed from such considerations. For instance they turned up naturally in the study of subshifts of finite type (Hochman and Meyerovitch [39], Simpson [66,68]), fractals (Braverman and Yampolsky [16,17]) (as we see later), we see randomness giving insight into Ergodic theory such as Avigad [6], Bienvenu et al. [13] and Franklin et al. [33].

Randomness has long been intertwined with computer science, (although some regard this as a matter of debate such as Gregorieff and Ferbus [38]) being central to things like polynomial identity testing, proofs like all known proofs of Toda's Theorem and the PCP theorem, as well as cryptographic security. A nice programme of Allender and his co-workers (e.g., [4,3]) suggests that perhaps complexity classes can be understood by understanding how they relate to the collections of strings which are algorithmically random according to various measures.

In this article I will try to give a brief outline of theses topics, and make some tentative suggestions for lines of investigation.

---

[1] I am indebted to Veronica Becher for discussions of Turing's and von Neumann's thoughts on randomness.

My assumption of the reader of this paper is that they are not well versed in the theory of algorithmic randomness. I will assume that they have a basic training in computability theory to the level of a first course in logic. If you are at all excited by what you read I urge you to look at the surveys or the books suggested above for fuller accounts.

2   Basics

I will refer to members of $\{0,1\}^* = 2^{<\omega}$ as *strings*, and infinite binary sequences (members of 2^ω, Cantor space) as *reals*. 2^ω is endowed with the tree topology, which has as basic clopen sets

$$[\sigma] := \{X \in 2^\omega : \sigma \prec X\},$$

where $\sigma \in 2^{<\omega}$. The *uniform* or *Lebesgue measure* on 2^ω is induced by giving each basic open set $[\sigma]$ measure $\mu([\sigma]) := 2^{-|\sigma|}$.

We identify an element X of 2^ω with the set $\{n : X(n) = 1\}$. The space 2^ω is measure-theoretically identical with the real interval $[0,1]$, although the two are not homeomorphic as topological spaces, so we can also think of elements of 2^ω as elements of $[0,1]$. We shall let $X \restriction n$ denote the first n bits of X.

The earliest work trying to give meaning to the randomness of an individual source was that of von Mises who argued as follows. The real should certainly have to obey the frequency laws like the law of large numbers. Thus

$$\lim_{n \to \infty} \frac{\{m \mid m \le n \wedge X(m) = 1\}}{n} = \frac{1}{2}.$$

This property is called *normality* and was studied by Borel and others. In fact, any random real clearly should be *absolutely normal*, normal to any basis. It is easy to construct such numbers computably (an interest of Turing discussed in Veronica Becher's article in this volume [8]). In fact any polynomial time random real (in any reasonable sense) is absolutely normal.

von Mises' idea was to consider any possible *selection* of a subsequence and ask that it was normal: Let $f : \omega \to \omega$ be an increasing injection, a selection function. Then a random X should satisfy the following.

$$\lim_{n \to \infty} \frac{\{m \mid m \le n \wedge X(f(m)) = 1\}}{n} = \frac{1}{2}.$$

von Mises work predated the work in the 30's,culminating in the classic paper of Turing [70], clarifying the notion of computable function. Thus von Mises had no canonical choice for "acceptable selection rules". However, Wald [76,77] showed that for any *countable* collection of selection functions, there is a sequence that is random in the sense of von Mises. Church [21] proposed restricting f to (partial) computable increasing functions. This gives rise to notions now called *computable stochasticity*, and *partial computable stochasticity*.

This was how matters stood until the work of Ville. [73] In the following, $S(\alpha, n)$ is the number of 1's in the first n bits of α and similarly S_f for the selected places.

Theorem 1 (Ville's Theorem [73]). *Let E be any countable collection of selection functions. Then there is a sequence $\alpha = \alpha_0 \alpha_1 \ldots$ such that the following hold.*

1. *$\lim_n \frac{S(\alpha,n)}{n} = \frac{1}{2}$.*
2. *For every $f \in E$ that selects infinitely many bits of α, we have $\lim_n \frac{S_f(\alpha,n)}{n} = \frac{1}{2}$.*
3. *For all n, we have $\frac{S(\alpha,n)}{n} \le \frac{1}{2}$.*

The killer is item 3 which says that there are never situations with more 1's than 0's in the first n bits of α. That is plainly non-random. Ville suggested adding a further statistical law, the law of iterated logarithms, to von Mises' definition. However, we might well ask "How we can be sure that adding this law would be enough?". Why should we expect there not to be a further result like Ville's (which there is, see [29]) exhibiting a sequence that satisfies both the law of large numbers and the law of iterated logarithms, yet clearly fails to have some other basic property that we would naturally associate with randomness?

We could add more and more statistical laws to our collection of desiderata for random sequences, but there is no reason to believe we would ever be done, and we certainly do not want a definition of randomness that changes with time, if we can avoid it. Martin-Löf's fundamental idea in [55] was to define an abstract notion of a performable statistical test for randomness, and require that a random sequence pass *all* such tests. He did so by effectivizing the notion of a set of measure 0. The way to think about Martin-Löf's definition below is that as we effectively shrink the measure of the open sets we regard as "tests", we are specifying reals satisfying them more and more.

In the below a Σ_1^0 class is a computably enumerable collection $\{[\sigma] \mid \sigma \in W\}$ for some c.e. set W of strings. Alternatively think of this as a c.e. set of intervals in the interval $[0,1]$.

Definition 1 (Martin-Löf [55])

1. *A Martin-Löf test is a sequence $\{U_n\}_{n \in \omega}$ of uniformly Σ_1^0 classes such that $\mu(U_n) \le 2^{-n}$ for all n.*
2. *A class $C \subset 2^\omega$ is Martin-Löf null if there is a Martin-Löf test $\{U_n\}_{n \in \omega}$ such that $C \subseteq \bigcap_n U_n$.*
3. *A set $A \in 2^\omega$ is Martin-Löf random if $\{A\}$ is not Martin-Löf null.*

Now there are three main views of algorithmic randomness. The above is called the *measure-theoretical paradigm*.

We briefly discuss the two other main paradigms in algorithmic randomness as they are crucial to our story. The first is the *computational paradigm*: Random sequences are those whose initial segments are all hard to describe, or, equivalently, hard to compress.

We think of Turing machines U with input τ giving a string σ. We regard τ as a description of σ and the shortest such is regarded as the intrinsic information in σ. The plain U-Kolmogorov complexity $C_U(\sigma)$ of σ is the *length* of the shortest

τ with $U(\tau) = \sigma$. Turing machines can be enumerated $U_0, U_1, \ldots$ and hence we can remove the machine dependence by defining a new (universal) machine

$$U(0^e 1 \tau) = U_e(\tau),$$

so that we can define for this machine M, $C(\sigma) = C_M(\sigma)$ and for all e, $C(\sigma) \le C_{U_e}(\sigma) + e + 1$. We shall use the notation $\le^+$ for constants and shall write $C(\sigma) \le^+ C_{U_e}(\sigma)$.

A simple counting argument due to Kolmogorov [44] shows that as $C(\sigma) \le^+ |\sigma|$ (using the identity machine), there must be strings of length n with $C(\sigma) \ge n$. We call such strings *C-random*.

We would like to define a real to be random by saying for all n, $C(\alpha \upharpoonright n) \ge^+ n$. Unfortunately, there are no such random reals due to a phenomenon called complexity oscillations, which (in a quantitative way) say that in very long strings σ there must segments with $C(\sigma \upharpoonright n) < n$. This oscillation really due to the fact that on input τ, we don't just get the *bits* of τ as information but the *length* of τ as well. Thus we are losing the intentional meaning that the bits of τ are processed by U to produce σ. To get around this first Levin [48,49] and later Chaitin [19] suggested using *prefix-free machines* to capture this intentional meaning.

Prefix free machines work like telephone numbers. If $U(\tau) \downarrow$ (i.e., halts) then for all $\hat{\tau}$ comparable with τ, $U(\hat{\tau}) \uparrow$.

Already we see a theme that there is not one but perhaps *many* notions of computational compressibility of relevance to understanding randomness. In the case of prefix free complexity, in some sense we know we are on the correct track, due to the following theorem which can be interpreted as saying (for discrete spaces) that Occam's razor and Baye's theorem give the same result (in that the shortest description is essentially the probability that the string is output).

Theorem 2 (Coding Theorem-Levin [48,49], Chaitin [19]). *For all σ,* $K(\sigma) =^+ -\log(Q(\sigma))$ *where $Q(\sigma)$ is $\mu(\{\tau \mid U(\tau) = \sigma\})$.*

Using this notion, and noticing that the universal machine above would be prefix-free if all the U_e were prefix free, we can define the prefix-free Kolmogorov complexity $K(\sigma)$.

Definition 2 (Levin [49], Chaitin [19]). *A set A is 1-random if $K(A \upharpoonright n) \ge^+ n$).*

Theorem 3 (Schnorr). *A real A is Martin-Löf random iff it is 1-random.*

Hence the two paradigms converge on a common intuition. It is easy to see that since there are only countably many machines, a real is random with probability 1. The classic example of a 1-random real is Chaitin's *halting probability* (for a universal prefix-free machine U):

$$\Omega = \sum_{\{\sigma \mid U(\sigma) \downarrow\}} 2^{-|\sigma|},$$

the measure of the domain of U (which has meaning as the domain of U is a prefix free set of strings).

It would seem that the definition of Ω is thoroughly machine independent but in the same spirit as Myhill's theorem, we can define a reducibility we call Solovay reducibility, and show that there is only one Ω in this sense. To wit, we observe that $\Omega = \lim_s \Omega_s$ where $\Omega_s = \sum_{\{\sigma | U(\sigma)[s]\downarrow\}} 2^{-|\sigma|}$, (i.e., s steps of computation), and hence Ω is what is called a left c.e.-real. We can define a notion of reducibility on left c.e.-reals $\alpha \leq_S \beta$ iff there is a partial computable function f and a constant c, such that for all rationals q (we assume all reals are nonrational for uniformity), if $q < \alpha$ then $f(q) \downarrow$ and $|\alpha - q| \leq c|\beta - f(q)|$. The culmination of a series of papers is the Kučera-Slaman theorem which states that there is really only one left-c.e. random real.

Theorem 4 (Kučera-Slaman Theorem [46]). α is 1-random and left-c.e. iff for all left c.e.-reals β, $\beta \leq_S \alpha$.

The final randomness paradigm is the one based on prediction. The *unpredictability paradigm* is that we should not be able to predict the next bit of a random sequences even if we know all preceding bits, in the same way that a coin toss is unpredictable even given the results of previous coin tosses.

Definition 3 (Levy [50]). *A function $d : 2^{<\omega} \to \mathbb{R}^{\geq 0}$ is a martingale[2] if for all σ,*

$$d(\sigma) = \frac{d(\sigma 0) + d(\sigma 1)}{2}.$$

d *is a* supermartingale *if for all σ,*

$$d(\sigma) \geq \frac{d(\sigma 0) + d(\sigma 1)}{2}.$$

A (super)martingale d succeeds on a set A if $\limsup_n d(A \upharpoonright n) = \infty$. The collection of all sets on which d succeeds is called the success set *of d, and is denoted by $S[d]$.*

The idea is that a martingale $d(\sigma)$ represents the capital that we have after betting on the bits of σ while following a particular betting strategy ($d(\lambda)$ being our starting capital). The *martingale condition* $d(\sigma) = \frac{d(\sigma 0) + d(\sigma 1)}{2}$ is a fairness condition, ensuring that the expected value of our capital after a bet is equal to our capital before the bet. Ville [73] proved that the success sets of (super)martingales correspond precisely to the sets of measure 0.

Now again we shall need a notion of effective betting strategy. We shall say that the martingale is computable if d is a computable function (with range $\mathbb{Q}$, without loss of generality), and we shall say that d is c.e. iff d is given by an effective approximation $d(\sigma) = \lim_s d_s(\sigma)$ where $d_{s+1}(\sigma) \geq d_s(\sigma)$. This means

---

[2] A more complex notion of martingale is used in probability theory. We shall discuss this notion, and the connection between it and ours, in [29], where it is discussed how computable martingale processes can be used to characterize 1-random reals.

that we are allowed to bet more as we become more confident of the fact that σ is the more likely outcome in the betting, as time goes on.

Theorem 5 (Schnorr [64,65]). *A set is* 1-*random iff no c.e. (super)martingale succeeds on it.*

These all seem basic theorems from long ago, but there remain a lot of things we don't understand even around these basic theorems. For example, here are three questions around these theorems.

First, it seems strange that to define randomness we use c.e. martingales and not computable ones. Based on this possible defect, Schnorr defined two other notions of randomness, *computable randomness* (where the martingales are all computable) and Schnorr randomness (where we use the Martin-Löf definition but insist that $\mu(U_k) = 2^{-k}$ rather than $\leq 2^{-k}$ so we know precisely the $[\sigma]$ in U_k uniformly) meaning in each case that the randomness notion is a computable rather and computably enumerable one. We know that Martin-Löf randomness implies computable randomness which implies Schnorr randomness, and none of these implications are reversible. The first question is:"Can we use some kind of computable randomness to define 1-randomness?". The suggested method to do this is to use a computable *but nonmonotonic* notion of randomness, where we have a betting strategy which bets on bits one at a time, but instead of being increasing can bet in some arbitrary order, and may not bet on all bits. The order is determined by what has happened so far. This gives a notion called *Kolmogorov-Loveland* (or nonmonotonic) randomness and the following question has been open for quite a while.

Question 1 (Muchnik, Semenov, and Uspensky [59]). Is every nonmonotonically random sequence 1-random?

A discussion of known results can be found in [29].

The second and third questions actually stem from the proof where we show that there is a translation of Martin-Löf tests into c.e. supermartingales. There, we start with a uniformly c.e. sequence $R_0, R_1, \ldots$ of prefix-free generators for a Martin-Löf test. We build a c.e. supermartingale d that bets evenly on $\sigma 0$ and $\sigma 1$ until it finds that, say, $\sigma 0 \in R_n$, at which point it starts to favor $\sigma 0$, to an extent determined by n. If later d finds that $\sigma 1 \in R_m$, then what it does is determined by the relationship between m and n. If $m < n$ then d still favors $\sigma 0$, though to a lesser extent than before. If $m = n$ then d again bets evenly on $\sigma 0$ and $\sigma 1$. If $m > n$ then d switches allegiance and favors $\sigma 1$. This can happen several times, as we find more R_i to which $\sigma 0$ or $\sigma 1$ belong.

The computable enumerability of d is essential in the above. A computable supermartingale (which we have seen we may assume is rational-valued without loss of generality) has to decide which side to favor, if any, immediately. Hitchcock has asked whether an intermediate notion, where we allow our super-martingale to be c.e. but do not allow it to switch allegiances in the way described above, is still powerful enough to capture 1-randomness. The purest version of this question was suggested by Kastermans. A *Kastergale* is a pair consisting of a partial computable function $g : 2^{<\omega} \to \{0,1\}$ and a c.e. supermartingale d such that $g(\sigma) \downarrow= i$ iff $\exists s\,(d_s(\sigma i) > d_s(\sigma(1-i)))$ iff $d(\sigma i) > d(\sigma(1-i))$. A set is *Kastermans random* if no Kastergale succeeds on it. A *Hitchgale* is the same as a Kastergale, except that in addition the proportion $\frac{d_s(\tau j)}{d_s(\tau)}$ (where we regard $\frac{0}{0}$ as being 0) is a Σ_1^0 function, so that if we ever decide to bet some percentage of our capital on τj, then we are committed to betting at least that percentage from that point on, even if our total capital on τ increases later. A set is *Hitchcock random* if no Hitchgale succeeds on it. It is unknown if these notions differ from 1-randomness and the import is that *is any bias allowed in the definition of 1-randomness?*

The message also is that there are many kinds of randomness and they each give insight. Variations of the notion of randomness include Kurtz or weak randomness, Demuth randomness, finite randomness, resource bounded randomness (for analyzing complexity classes), etc. For instance, weak randomness asks that X belongs to all Σ_1^0 classes of measure 1. We refer mostly to [29,52] for more. There are similarly many kinds of Kolmogorov complexities such as process and monotone complexities (which solve the "C-" problem by asking that the action of machines be continuous rather than prefix free). To wit, if $U(\sigma) \downarrow$ and $U(\nu) \downarrow$, and $\sigma \preceq \nu$, then $U(\sigma) \preceq U(\nu)$. There are various interpretations of this idea, such as U being a multifunction (so that U is really a c.e. collection of pairs of strings) called Km, monotone complexity, but for all of them, an analog of Schnorr's Theorem holds so that α is 1-random iff $K(\alpha \upharpoonright n) \geq^+ n$ for all n. In most cases, $K(\alpha \upharpoonright n) =^+ n$ since the identity machine is monotone.

These ideas and associated probability measures have seen applications into geometric measure theory such as Jan Reimann's new proof of (classical) Frostmann's Lemma using methods from effective randomness ([62])[3]. These continuous Kolmogorov complexities tend to be less well understood. Work of Adam Day (see [29]) gives new methods for building machines. One hallmark is Day's remarkable improvement [23] of Gác's Theorem [35] that the Coding Theorem fails for continuous spaces.

For the remainder of this paper we shall need some further (stronger) notions of randomness. We can strengthen the idea of randomness by giving the computational devices more compression power via oracles. Then if $\emptyset^{(n)}$ denotes the n-th iterate of the halting problem, we say that X is $n + 1$-random iff $K^{\emptyset^{(n)}}(X \upharpoonright n) \geq^+ n$ for all n.

It is a surprising fact that for all n, n-randomness can be defined purely in terms of K with no oracle. This follows by the next result.

Theorem 6 (Bienvenu, Muchnik, Shen, and Vereschagin [12]). $K^{\emptyset'}(\sigma) = \limsup_m K(\sigma \mid m) \pm O(1)$.

Hence A would be 2-random iff for all n, $\limsup_m K(A \upharpoonright n \mid m) \geq^+ n$. In some cases, we know of natural definitions of n-randomness. For instance, we have

---

[3] As we soon see, Simpson ([68]) has similar applications of effective measure to derive classical results in Hausdorff dimension.

seen that it is impossible for a real to have $C(X \upharpoonright n) \geq^+ n$ for *all* n, but Martin-Löf showed in his original paper that there are reals X with $C(X \upharpoonright n) \geq^+ n$ for *infinitely many* n. Joe Miller and later Nies, Stephan and Terwijn showed that such randoms are precisely the 2-randoms, and later Miller showed that the 2-randoms are exactly those that achieve maximal prefix-free complexity $(n + K(n))$ infinitely often. Also Becher and Gregorieff [9] have a kind of index set characterizations of higher notions of randomness. I know of no other natural definitions, such as for the 3-randoms.

Another subtext in these investigations is to dispense with Kolmogorov complexity altogether. The idea is to redo algorithmic randomness using total machines.

Definition 4 (Bienvenu and Merkle [11]). *A computable function f is a Solovay function if $\sum_n 2^{-f(n)} < \infty$ and $\liminf_n f(n) - K(n) < \infty$ (in other words, there is a c such that $f(n) \leq K(n) + c$ for infinitely many n).*

Solovay functions were first constructed by Solovay, but any reasonable time bounded version of prefix-free Kolmogorov complexity give rise to one. (An observation of Hölzl, Kräling, and Merkle [40].) Building on earlier work of Gács, and of Miller and Yu, recently Merkle, Miller and Nies have proven that a set A is 1-random iff $C(A \upharpoonright n) \geq n - g(n) - O(1)$ for any Solovay function g. In fact by themselves, Solovay functions characterize 1-randomness.

Theorem 7 (Bienvenu and Downey [10]). *Let f be a computable function. The following are equivalent.*

1. *f is a Solovay function.*
2. *$\sum_n 2^{-f(n)}$ is a 1-random real.*

Further extensions on this theme, generalizations and relationships with randomness have been found. See [29], and the later material on K-triviality.

We have left out the vast amount of work on effective dimensions. In the same way as we effectivize measure, we can effectivize fractional measure. Theorems include characterizations due to Mayordomo [56] that effective Hausdorff dimension of X is equal to $\liminf_{n \to \infty} \frac{K(X \upharpoonright n)}{n}$ and the characterization of effective packing dimension by Athreya, Hitchcock, Lutz, and Mayordomo [5] as $\limsup_{n \to \infty} \frac{K(X \upharpoonright n)}{n}$ (C can replace K in both cases). These concepts have been shown to have fascinating interactions with computability, such as characterizing degree classes, and as we discuss later, have been used to give new proofs of classical theorems. I don't have space to discuss further, but see [29].

3   Randomness and Classical Computability

Interactions of measure, randomness and computability go way back to the early years of the subject such as the paper de Leeuw et al. [24] where, amongst other things, it is proven that a set X is enumerable from a set of oracles of positive measure iff X is computably enumerable. As a consequence, we get a result later rediscovered by Sacks that if a real X is computable from a collection of sources of positive measure, then X must be computable. Nevertheless, a classical result is the following saying that random sources can have computational power.

Theorem 8 (Kučera [45], Gács [36]). *For every set X, there is a random Y such that $X \leq_{wtt} Y$, where $\leq_{wtt}$ is Turing reducibility with use bounded by a computable function.*

The above argues that 1-random reals are not random enough to correlate to the thesis that random reals should have no computational power. This intuition was clarified by Stephan who proved the following[4].

Theorem 9 (Stephan [69]). *Suppose a random real is powerful enough to compute a $\{0,1\}$-valued function f such that for all n, $f(n) \neq \varphi_n(n)$ (i.e., a PA degree). Then $\emptyset' \leq_T X$, so that it is a "false random."*

Thus we can wash away lots of computational power by raising the level of randomness. For example, it can be shown that X is weakly 2-random (i.e., in every Σ_2^0 class of measure 1) iff X is 1-random and its degree forms a minimal pair with $\emptyset'$. Hence no (weakly) 2-random real can bound a PA degree. A remarkable theorem here is the following, demonstrating a deep relationship between PA degrees and randomness.

Theorem 10 (Barmpalias, Lewis, and Ng [7]). *Every PA degree is the join of two 1-random degrees.*

There has been a huge amount of work concerning the interplay between things like PA degrees and weakenings of the notion of fixed point free functions ($f(n) \neq \varphi_n(n)$). For example, you can show that this ability corresponds to traceing, and the speed of growth of the initial segment complexity of a real. As an illustration, A is h-complex if $C(A \restriction n) \geq h(n)$ for all n. A is *autocomplex* if there is an A-computable order h such that A is h-complex.

Theorem 11 (Kjos-Hanssen, Merkle, and Stephan [41]). *A set is autocomplex iff it is of DNC degree.*

Another illustration of the interplay of notions of randomness and Turing degrees is the theorem.

Theorem 12 (Nies, Stephan, and Terwijn [54]). *If a nonhigh set (i.e., $A' \not\geq_T \emptyset^{(2)}$.) is Schnorr random then it is 1-random.*

In fact it is possible to show that within the high degrees the separations between computable, Schnorr, and Martin-Löf randomness always occur. In the

---

[4] Interpreted by Hirschfeldt as saying that there are two methods of passing a stupidity test. One is the be the genuine article. The other is to be like Ω is be so smart that you know what a stupid person would say.

hyperimmune-free (or computably dominated $\mathbf{a}$, meaning that for every $f \leq \mathbf{a}$ there is a computable g with $f(n) < g(n)$ for all n) degrees, weak randomness coincides with all of these as well as weak 2-randomness.

There is a delicate interweaving of randomness notions and properties of Turing degrees. For example, Kurtz and Kautz long ago showed us that every 2-random degree is hyperimmune (i.e., $\exists f \leq \mathbf{a}(\forall g)(g$ computable $\rightarrow \exists^{\infty} n(f(n) > g(n)))$.) Moreover the "almost all" theory of degrees is decidable by another old result of Stillwell. We refer to [29] for a lot more on this, and similar things concerning effective dimensions.

I cannot leave this part of the survey without mentioning the long sequences of results about lowness notions. For any reasonable property P we say that X is *low for* P if $P^X = P$. For example, being low for the Turing jump means that $X' \equiv_T \emptyset'$. A set A is low for 1-randomness iff A does not make any 1-randoms nonrandom. You can also have a notion of lowness for tests, meaning that every (effective nullset)^A can be covered by an effective nullset. In all cases the lowness notion for randomness and for tests have turned out to coincide with a single recent exception of "difference randomness" found by Diamondstone and Fanklin (paper in preparation).

Now it is not altogether clear that noncomputable sets low for 1-randomness should exist. But they do and form a remarkable class called the K-trivials. That is, they coincide with the class of reals A such that forall n, $K(A \upharpoonright n) \leq^+ K(n)$. (In fact Bienvenu and Downey [10] showed that it is enough to put a Solovay function on the right side.) Many properties of this class have been shown, and particularly Andre Nies proved the deep result that A is K-trivial iff A is low for Martin-Löf randomness iff A is useless as a compressor, $K^A =^+ K$. (Nies [51]). A good account of this material can be found in Nies [52,53], but things are constantly changing, with maybe 17 characterizations of this class. We also refer to [29] for the situation up to mid-2010.

Other randomness notions give quite different lowness notions. For example, there are no noncomputable reals which are low for C and none low for computable randomness. On the other hand, lowness for Schnorr and Kurtz randomness give interesting subclasses of the hyperimmune-free degrees characterized by notions of being computably dominated, and fixed point free functions in the case of Kurtz. Work here is still ongoing and many results proven, but the pattern remains very opaque. Even for a fixed real like Ω (i.e., when does Ω^X remain random?) results are quite interesting. In the case of Ω, X is low for Ω and X is computable from the halting problem, then X is K-trivial, whereas if X is random, then it is 2-random. (Results of Joe Miller, see [29].)

These classes again relate to various refinements of the jump and to "tracing" which means giving an effective collection of *possibilities* for (partial) functions computable from the degree at hand. Again this has taken on a life of its own, and such methods have been used to solve questions from classical computability theory. For instance, Downey and Greenberg's [28] used "strong jump traceability" to solve a longstanding question of Jockusch and Shore on pseudo-jump operators and cone avoidance. Strongly jump traceable reals have their own agenda and form a fascinating class, cf., e.g., [20].

The final material for this section is the deep results of Reimann and Slaman who were looking at the question (first discussed by Levin): given $X \not\equiv_T \emptyset$, is there a measure relative to which X is random?

Clearly we can concentrate a measure on a real, but assuming that we are not allowed to do this the answer is still that every noncomputable real can be made random. On the other hand, if we ask that there are no atoms in the measure, the situation is very different. We get a class of *never continuously n-random* reals. For each n this class is countable, but the proof of this requires magical things like big fragments of Borel determinacy, *provably*. The reader should look at Reimann and Slaman [63].

4   (Some) Applications

4.1   Left Out

I apologize to the workers who are using approximations to C like commercial compression packages to apply Kolmogorov complexity to measure things like common information[5]. As an illustration, I refer to the work of Vitanyi and his co-workers who do phylogenetic analysis (in biology and music evolution, etc) by replacing metrics like maximum parsimony by common information defined via Kolmogorov complexity. (Cf., e.g., [22,72].)

4.2   Ergodic Theory

A very important part of classical mathematics is concerned with recurrent actions of some process. For example, A d-dimensional *shift* of finite type is a collection of colourings of $\mathbb{Z}^d$ defined by local rules and a shift action (basically saying certain colourings are illegal). Its (Shannon) *entropy* is the asymptotic growth in the number of legal colourings. More formally, consider $G = (\mathbb{N}^d, +)$ or $(\mathbb{Z}^d, +)$, and A a finite set of symbols. We give A the discrete topology and A^G the product topology. The *shift action* of G on A^G is

$$(S^g x)(h) = x(h + g), \text{ for } g, h \in G \wedge x \in A^G.$$

A *subshift* is $X \subseteq A^G$ such that $x \in X$ implies $S^g x \in X$ (i.e., shift invariant). The classical area of *Symbolic Dynamics* studies subshifts usually of "finite type." Such subshifts are well known to be connected to number theory, Ramsey theory etc.

The following is a recent theorem showing that Ω occurs naturally in this setting.

---

[5] The earliest calssical application of Kolgorogov compexity I know of is an old one by Schnorr and Fuchs [34] sharpening aspects of Markov Chain Monte Carlo.

Theorem 13 (Hochman and Meyerovitch, [39]). *The values of entropies of subshifts of finite type over $\mathbb{Z}^d$ for $d \geq 2$ are exactly the complements of halting probabilities.*

I remark that in the same way that Reimann proved Frostman's Lemma using effective methods, Simpson [68] has proven classical results using our effective methods. Simpson studies topological entropy for subshifts X and the relationship with Hausdorff dimension. If $X \subset A^G$ use the standard metric $\rho(x,y) = 2^{-|F_n|}$ where n is as large as possible with $x \upharpoonright F_n = u \upharpoonright F_n$ and $F_n = \{-n, \ldots, n\}^d$. In discussions with co-workers, Simpson proved that the classical dimension equals the entropy (generalizing a difficult result of Furstenburg 1967) using effective methods, which were much simpler.

Theorem 14 (Simpson [68]). *If X is a subshift (closed and shift invariant), then the effective Hausdorff dimension of X is equal to the classical Hausdorff dimension of X is equal to the entropy, moreover there are calculable relationships between the effective and classical quantities. (See Simpson's home page for his recent talks and more precise details.)*

Other types of Ergodic behaviour have been studied.

The general setting is the following. Let (X, μ) be a probability space, and $T : X \to X$ measure preserving so that for measurable $A \subseteq X$, $\mu(T^{-1}A) = \mu(A)$. Such a map is *invariant* if $T^{-1}A = A$ except on a measure 0 set. Finally the map is *ergodic* if every T-invariant subset is either null or co-null. The shift operator above (say, on Cantor space so that $T(a_0 a_1 \ldots) = a_1 a_2 \ldots$) is an ergodic action with the Bernoulli product measure.

A classic theorem of Poincaré is that if T is ergodic on (X, μ), then for all $E \subseteq X$ of positive measure and *almost all* $x \in X$, $T^n(x) \in E$ for infinitely many n. For a set of measurable subsets E of X, we call an x a *Poincaré point* if $T^n(x) \in Q$ for all $Q \in E$ of positive measure. Long ago Kučera [45] showed that X is 1-random iff X is a Poincaré point for the shift operator with respect to the collection of effectively closed subsets of 2^ω.

Bienvenu et al. proved the following extension of this result.

Theorem 15 (Bienvenu, et al. [13]). *Let T be computable ergodic on a computable probability space (X, μ). Then $x \in X$ is 1-random iff x is a Poincaré point for all effectively closed subsets of X.*

We remark that the notion of a computable probability space is natural and along the lines of the Pour-El Richards [61] version of computable metric space. There are again a lot of results here. Franklin et al. [33] looked at the classic Birkhoff ergodic theorem for $f \in L^1(X)$ (namely $\lim_{n\to\infty} \frac{1}{n} \sum_{i<n} f(T^i(x)) = \int f d\mu$.) and showed that 1-random points satisfy Birkhoff's ergodic theorem. For other interpretations and stronger hypotheses (that the measure of the closed sets is computable), Gács, Hoyrup and Rojas [37], showed that the Birkhoff points are precisely the Schnorr randoms. This is currently an area of intense activity, and many of the classical ergodic theorems remain to be studied. For example,

Furstenburg's one with its applications to arithmetical progressions would seem a natural candidate.

This is also related to metamathematical studies, and here we refer the reader to Avigad [6].

Another interesting application of the ideas from algorithmic randomness is to the area of Julia sets. Recall that this is described by $z \mapsto z^2 + \alpha z$, where $\alpha = e^{2\pi i\theta}$. Braverman and Yampolsky [16,17] showed that in general even for computable θ, Julia sets can coincide with complements of Ω.

4.3 Differentiability Is the Same as Randomness

In his blog, Terry Tao remarks that Ergodic theorems and classical theorems from analysis such as the Lebesgue theorem that functions of bounded variation are differentiable almost everywhere are closely related. In Bishop's book, they have almost the same proof. It is thus not surprising that we see such theorems giving rise to randomness notions. This is an idea going way back to the work of Oswald Demuth, a constructivist from Prague. It is being actively pursued by Brattka, Nies, Miller and others.

Recall that the Denjoy upper and lower derivatives for a function f are defined as follows.

$$\overline{D}f(x) = \limsup_{h \to 0} \frac{f(x) - f(x+h)}{h} \text{ and } \underline{D}f(x) = \liminf_{h \to 0} \frac{f(x) - f(x+h)}{h}.$$

The Denjoy derivate exists iff both of the above quantities exist and are finite. The idea in this is that slopes like those in the definitions can be considered to be martingales.

Using this for one direction, various notions of randomness can be characterized by (i) varying the strength of the notion of computable real valued function (e.g., Markov computable, type 2 computable etc) (ii) varying the theorem.

For an illustration, we have the following.

Theorem 16 (Brattka, Miller and Nies [15]). *z is computably random iff every computable (in the type two sense) increasing function $f[0,1] \to \mathbb{R}$ is differentiable at x.*

There are similar results relating 1-randomness of x to its differentiability of functions of bounded variation. There is still a lot of activity here, and class like Lipschitz functions and many other classical almost everywhere behaviour in analysis are found to correlate to various notions of randomness. The paper [15] is an excellent introduction to this material.

We might speculate that this could also be related to the general purpose analog computer studied by Shannon, Martin-Pour-El, Ruebel and others last century.

5   Relationships between Random Strings and Complexity Classes

A very interesting programme is due to Allender and his co-workers (and others such as Day). At first glance it seems rather strange, but the idea is to look at resource bounded reductions to highly noncomputable objects like clean versions of $R_C = \{\sigma : C(\sigma) \geq \frac{|\sigma|}{2}\}$, and similarly R_Q for any other Kolmogorov complexity Q. Long ago, Kummer [47] showed that R_C is tt-complete. This is by no means an obvious fact and the proof uses $\mathbf{0}''$ nonuniformity to build the reduction. It is not necessarily true for R_K and depends on the choice of universal machine, a fact established by Muchnik (in [58]), using a fascinating game-theoretical argument (see [29] for details).

Kummer's reduction was double exponential length increasing and one might ask what does P^{R_C} look like. Clearly P^{R_C} has noncomputable sets of strings in it, but the idea that this is an artifact of the choice of universal machine. The correct class to look at is

$$\cap_U P^{R_{C_U}}.$$

Sometimes it is suggested that this should be intersected with the computable sets, but Allender conjectures that this makes no difference, $\cap_U P^{R_{C_U}} \cap \text{COMP} = \cap_U P^{R_{C_U}}$.

In [4] it is proven that $P = \cap_U \{A : A \leq^p_{dtt} R_{C_U}\} \cap \text{COMP}$ where the reductions are restricted to polynomial time disjunctive truth table ones. Some of the results so far, for any variants of the Kolmogorov complexity (so we drop the subscript) are BPP $\subseteq \{A : A \leq^p_{tt} R\} \cap \text{COMP}$, PSPACE $\subseteq P^R \cap \text{COMP}$, and NEXP $\subseteq \text{NP}^R \cap \text{COMP}$. Specifically it is open for these containments if we can drop the $\cap$COMP. The containments might actually be equality, and these are important open questions. Recently, Allender, Friedman and Gasarch [2] have tightened two of these for prefix-free complexity to BPP $\subseteq \{A : A \leq^p_{tt} R_K\} \cap \text{COMP} \subseteq$ PSPACE and NEXP $\subseteq \text{NP}^{R_K} \cap \text{COMP} \subseteq \text{EXPSPACE}$. Interestingly, these proofs come from sharpening Muchnik's game method, along with the fact that the natural home for strategies is PSPACE.

The methods for some of these results use extractors. These are methods of taking weak sources of randomness and producing pseudo-randomness from them, and are particularly successful if you either have independent sources, or some "true" randomness like a physical source assuming quantum assumptions. those have found other uses in algorithmic randomness, such as Zimand's proof [79] that two sources of nonzero effective Hausdorff dimension can together compute a degree which has Hausdorff dimension 1. It is known that one source is not enough as Miller [57] has shown that there is a Turing degree of fractional effective Hausdorff dimension. (See [29]. It is still open if a Turing degree can be minimal and have effective Hausdorff dimension 1.)

6   Physics

In this last section I will mention a few things of relevance. First, it is possible to look at various natural phenomena which are regarded as random, such as, say,

Brownian motion. Fouche [31,32], Kjos-Hanssen and Nerode [42] and B. Kjos-Hanssen and T. Szabados [46] have a nice body of work here, showing that, for instance, 1-randomness can be used to understand Brownian motion.

Another major area of randomness is quantum physics under the Copenhagen interpretation. Some physicists claim that this produces *true randomness*. In the same way that we don't know if the universe can produce any incomputability, it seems that we don't know if it can even produce 1-randomness, say. In spite of this, it seems that we can buy true randomness by Internet, via companies using semi-transparent mirrors. One such company is *Quantis*: quantum mechanical random number generator produced and sold by id Quantique of the University of Geneva. They seem to pass reasonable practical statistical tests.

It seems that this is a hypothesis that might be analyzed. Assuming that the universe is a (computable) manifold and assuming the Copenhagen interpretation, we could ask if we could produce initial segments of random reals. Calude, Svozil and others are looking at this idea, e.g., [1,18]

7   Conclusion

This is my interpretation of a few themes and high points for the exciting area of algorithmic randomness. Space considerations preclude me including more. I do hope I have at least wetted your interest in this fascinating subject.

Acknowledgements. Research supported by the Marsden Fund of New Zealand. This paper was written whilst the author was a visiting fellow at the Isaac Newton Institute for Mathematical Science, Cambridge, UK, as part of the Alan Turing "Semantics and Syntax" programme, in 2012.

References

1. Abbott, A.A., Calude, C.S., Svozil, K.: Incomputability of quantum physics (in preparation)
2. Allender, E., Friedman, L., Gasarch, W.: Limits on the Computational Power of Random Strings. In: Aceto, L., Henzinger, M., Sgall, J. (eds.) ICALP 2011. LNCS, vol. 6755, pp. 293–304. Springer, Heidelberg (2011)
3. Allender, E., Buhrman, H., Koucký, M., van Melkebeek, D., Ronneburger, D.: Power from Random Strings. SIAM J. Comp. 35, 1467–1493 (2006)
4. Allender, E., Buhrman, H., Koucký, M.: What Can be Efficiently Reduced to the Kolmogorov-Random Strings? Annals of Pure and Applied Logic 138, 2–19 (2006)
5. Athreya, K., Hitchcock, J., Lutz, J., Mayordomo, E.: Effective strong dimension in algorithmic information and computational complexity. SIAM Jour. Comput. 37, 671–705 (2007)
6. Avigad, J.: The metamathematics of ergodic theory. Annals of Pure and Applied Logic 157, 64–76 (2009)

7. Barmpalias, G., Lewis, A., Ng, K.M.: The importance of Π^0_1 classes in effective randomness. JSL 75(1), 387–400 (2010)
8. Becher, V.: Turing's Normal Numbers: Towards Randomness. In: Cooper, S.B., Dawar, A., Löwe, B. (eds.) CiE 2012. LNCS, vol. 7318, pp. 35–45. Springer, Heidelberg (2012)
9. Becher, V., Grigorieff, S.: From index sets to randomness in $\emptyset^n$, Random reals and possibly infinite computations. Journal of Symbolic Logic 74(1), 124–156 (2009)
10. Bienvenu, L., Downey, R.: Kolmogorov complexity and Solovay functions. In: STACS 2009, pp. 147–158 (2009)
11. Bienvenu, L., Merkle, W.: Reconciling Data Compression and Kolmogorov Complexity. In: Arge, L., Cachin, C., Jurdziński, T., Tarlecki, A. (eds.) ICALP 2007. LNCS, vol. 4596, pp. 643–654. Springer, Heidelberg (2007)
12. Bienvenu, L., Muchnik, A., Shen, A., Vereshchagin, N.: Limit complexities revisited. In: STACS 2008 (2008)
13. Bienvenu, L., Day, A., Hoyrup, M., Mezhirov, I., Shen, A.: Ergodic-type characterizations of algorithmic randomness. To appear in Information and Computation
14. Bertrand, J.: Calcul des Probabilités (1889)
15. Brattka, V., Miller, J., Nies, A.: Randomness and differentiability (to appear)
16. Braverman, M., Yampolsky, M.: Non-Computable Julia Sets. Journ. Amer. Math. Soc. 19(3) (2006)
17. Braverman, M., Yampolsky, M.: Computability of Julia Sets. Springer (2008)
18. Calude, C., Svozil, K.: Quantum randomness and value indefiniteness. Advanced Science Letters 1, 165–168 (2008)
19. Chaitin, G.: A theory of program size formally identical to information theory. Journal of the ACM 22, 329–340 (1975)
20. Cholak, P., Downey, R., Greenberg, N.: Strong-jump traceablilty. I. The computably enumerable case. Advances in Mathematics 217, 2045–2074 (2008)
21. Church, A.: On the concept of a random sequence. Bulletin of the American Mathematical Society 46, 130–135 (1940)
22. Cilibrasi, R., Vitanyi, P.M.B., de Wolf, R.: Algorithmic clustering of music based on string compression. Computer Music J. 28(4), 49–67 (2004)
23. Day, A.: Increasing the gap between descriptional complexity and algorithmic probability. Transactions of the American Mathematical Society 363, 5577–5604 (2011)
24. de Leeuw, K., Moore, E.F., Shannon, C.E., Shapiro, N.: Computability by probabilistic machines. In: Shannon, C.E., McCarthy, J. (eds.) Automata studies. Annals of Mathematics Studies, vol. 34, pp. 183–212. Princeton University Press, Princeton (1956)
25. Demuth, O.: The differentiability of constructive functions of weakly bounded variation on pseude-numbers. Comment. Math. Univ. Carolina. 16, 583–599 (1975)
26. Downey, R.: Five Lectures on Algorithmic Randomness. In: Chong, C., Feng, Q., Slaman, T.A., Woodin, W.H., Yang, Y. (eds.) Computational Prospects of Infinity, Part I: Tutorials. Lecture Notes Series, Institute for Mathematical Sciences, National University of Singapore, vol. 14, pp. 3–82. World Scientific, Singapore (2008)
27. Downey, R.: Algorithmic randomness and computability. In: Proceedings of the 2006 International Congress of Mathematicians, vol. 2, pp. 1–26. European Mathematical Society (2006)
28. Downey, R., Greenberg, N.: Pseudo-jump operators and SJTHard sets (submitted)
29. Downey, R., Hirschfeldt, D.: Algorithmic Randomness and Complexity. Springer (2010)

30. Downey, R., Hirschfeldt, D., Nies, A., Terwijn, S.: Calibrating randomness. Bulletin Symbolic Logic 12, 411–491 (2006)
31. Fouche, W.: The descriptive complexity of Brownian motion. Advances in Mathematics 155, 317–343 (2000)
32. Fouche, W.: Dynamics of a generic Brownian motion: Recursive aspects. Theoretical Computer Science 394, 175–186 (2008)
33. Franklin, J., Greenberg, N., Miller, J., Ng, K.M.: Martin-Loef random points satisfy Birkhoff's ergodic theorem for effectively closed sets. Proc. Amer. Math. Soc. (to appear)
34. Fuchs, H., Schnorr, C.: Monte Carlo methods and patternless sequences. In: Operations Research Verfahren, Symp., Heidelberg, vol. XXV, pp. 443–450 (1977)
35. Gács, P.: On the relation between descriptional complexity and algorithmic probability. Theoretical Computer Science 22, 71–93 (1983)
36. Gács, P.: Every set is reducible to a random one. Information and Control 70, 186–192 (1986)
37. Gács, P., Hoyrup, M., Rojas, C.: Randomness on computable probability spaces, a dynamical point of view. Theory of Computing Systems 48(3), 465–485 (2011)
38. Gregorieff, S., Ferbus, M.: Is Randomness native to Computer Science? Ten years after. In: [78], pp. 243–263 (2011)
39. Hochman, M., Meyerovitch, T.: A characterization of the entropies of multidimensional shifts of finite type. Annals of Mathematics 171(3), 2011–2038 (2010)
40. Hölzl, R., Kräling, T., Merkle, W.: Time-Bounded Kolmogorov Complexity and Solovay Functions. In: Královič, R., Niwiński, D. (eds.) MFCS 2009. LNCS, vol. 5734, pp. 392–402. Springer, Heidelberg (2009)
41. Fortnow, L., Lee, T., Vereshchagin, N.K.: Kolmogorov Complexity with Error. In: Durand, B., Thomas, W. (eds.) STACS 2006. LNCS, vol. 3884, pp. 137–148. Springer, Heidelberg (2006)
42. Kjos-Hanssen, B., Nerode, A.: Effective dimension of points visited by Brownian motion. Theoretical Computer Science 410(4-5), 347–354 (2009)
43. Kjos-Hanssen, B., Szabados, T.: Kolmogorov complexity and strong approximation of Brownian motion. Proc. Amer. Math. Soc. 139(9), 3307–3316 (2011)
44. Kolmogorov, A.N.: Three approaches to the quantitative definition of information. Problems of Information Transmission 1, 1–7 (1965)
45. Kučera, A.: Measure, Π_1^0 classes, and complete extensions of PA. In: Recursion Theory Week, Oberwolfach. Lecture Notes in Mathematics, vol. 1141, pp. 245–259. Springer, Berlin (1984-1985)
46. Kučera, A., Slaman, T.: Randomness and recursive enumerability. SIAM J. on Comp. 31, 199–211 (2001)
47. Kummer, M.: On the Complexity of Random Strings(Extended abstract). In: Puech, C., Reischuk, R. (eds.) STACS 1996. LNCS, vol. 1046, pp. 25–36. Springer, Heidelberg (1996)
48. Levin, L.: Some theorems on the algorithmic approach to probability theory and information theory. Dissertation in Mathematics Moscow University (1971)
49. Levin, L.: Laws of information conservation (non-growth) and aspects of the foundation of probability theory. Problems of Information Transmission 10, 206–210 (1974)
50. Lévy, P.: Théorie de l'Addition des Variables Aléatoires. Gauthier-Villars (1937)
51. Nies, A.: Lowness properties and randomness. Advances in Mathematics 197(1), 274–305 (2005)

52. Nies, A.: Computability and Randomness. Oxford University Press (2009)
53. Nies, A.: Interactions of computability and randomness. In: Ragunathan, S. (ed.) Proceedings of the International Congress of Mathematicians, pp. 30–57 (2010)
54. Nies, A., Stephan, F., Terwijn, S.A.: Randomness, relativization, and Turing degrees. JSL 70(2), 515–535 (2005)
55. Martin-Löf, P.: The definition of random sequences. Information and Control 9, 602–619 (1966)
56. Mayordomo, E.: A Kolmogorov complexity characterization of constructive Hausdorff dimension. Infor. Proc. Lett. 84, 1–3 (2002)
57. Miller, J.: Extracting information is hard: a Turing degree of non-integral effective Hausdorff dimension. Advances in Mathematics 226(1), 373–384 (2011)
58. Muchnik, A.A., Positselsky, S.P.: Kolmogorov entropy in the context of computability theory. Theor. Comp. Sci. 271, 15–35 (2002)
59. Muchnik, A.A., Semenov, A., Uspensky, V.: Mathematical metaphysics of randomness. Theor. Comp. Sci. 207(2), 263–317 (1998)
60. Nies, A., Miller, J.: Randomness and computability: Open questions. Bull. Symb. Logic. 12(3), 390–410 (2006)
61. Poul-El, M., Richards, I.: Computability in Analysis and Physics. Springer (1989)
62. Reimann, J.: Effectively closed classes of measures and randomness. Annals of Pure and Applied Logic 156(1), 170–182 (2008)
63. Reimann, J., Slaman, T.: Randomness for continuous measures (to appear), draft available from Reimann's web site
64. Schnorr, C.P.: A unified approach to the definition of a random sequence. Mathematical Systems Theory 5, 246–258 (1971)
65. Schnorr, C.P.: Zufälligkeit und Wahrscheinlichkeit. Eine algorithmische Begründung der Wahrscheinlichkeitstheorie. Lecture Notes in Mathematics, vol. 218. Springer, Berlin (1971)
66. Simpson, S.: Medvedev Degrees of 2-Dimensional Subshifts of Finite Type. Ergodic Theory and Dynamical Systems (to appear)
67. Simpson, S.: Mass Problems Associated with Effectively Closed Sets. Tohoku Mathematical Journal 63(4), 489–517 (2011)
68. Simpson, S.: Symbolic Dynamics: Entropy = Dimension = Complexity (2011) (to appear)
69. Stephan, F.: Martin-Löf random sets and PA-complete sets. In: Logic Colloquium 2002. Lecture Notes in Logic, vol. 27, pp. 342–348. Association for Symbolic Logic (2006)
70. Turing, A.: On computable numbers with an application to the Entscheidungsproblem. Proceedings of the London Mathematical Society 42, 230–265 (1936); Correction in Proceedings of the London Mathematical Society 43, 544–546 (1937)
71. Turing, A.: Computing machinery and intelligence. Mind 59, 433–460 (1950)
72. Vitanyi, P.: Information distance in multiples. IEEE Trans. Inform. 57(4), 2451–2456 (2011)
73. Ville, J.: Étude Critique de la Notion de Collectif. Gauthier-Villars (1939)
74. von Mises, R.: Grundlagen der Wahrscheinlichkeitsrechnung. Math. Z 5, 52–99 (1919)
75. von Neumann, J.: Various techniques used in connection with random digits. In: Householder, A.S., Forsythe, G.E., Germond, H.H. (eds.) Monte Carlo Method. National Bureau of Standards Applied Mathematics Series, vol. 12, pp. 36–38 (1951)

76. Wald, A.: Sur le notion de collectif dans la calcul des probabilitiés. Comptes Rendes des Seances de l'Académie des Sciences 202, 1080–1083 (1936)
77. Wald, A.: Die Weiderspruchsfreiheit des Kollektivbegriffes der Wahrscheinlichkeitsrechnung. Ergebnisse eines mathematischen Kolloquiums 8, 38–72 (1937)
78. Zenil, H. (ed.): Randomness Through Computation: Some Answers, More Questions. World Scientific, Singapore (2011)
79. Zimand, M.: Two Sources Are Better Than One for Increasing the Kolmogorov Complexity of Infinite Sequences. In: Hirsch, E.A., Razborov, A.A., Semenov, A., Slissenko, A. (eds.) CSR 2008. LNCS, vol. 5010, pp. 326–338. Springer, Heidelberg (2008)

Learning, Social Intelligence and the Turing Test
Why an "Out-of-the-Box" Turing Machine Will Not Pass the Turing Test

Bruce Edmonds[1] and Carlos Gershenson[2]

[1] Centre for Policy Modelling, Manchester Metropolitan University, Aytoun Building, Aytoun Street, Manchester M1 3GH, United Kingdom
bruce@edmonds.name
[2] Departmento de Ciencias de la Computación, Instituto de Investigaciones en Matemáticas Aplicadas y en Sistemas, Universidad Nacional Autónoma de México, Ciudad Universitaria, A.P. 20-726, 01000 México D.F., México
cgg@unam.mx

Abstract. The Turing Test checks for human intelligence, rather than any putative general intelligence. It involves repeated interaction requiring learning in the form of adaption to the human conversation partner. It is a macro-level post-hoc test in contrast to the definition of a Turing machine, which is a prior micro-level definition. This raises the question of whether learning is just another computational process, i.e., can be implemented as a Turing machine. Here we argue that learning or adaption is fundamentally different from computation, though it does involve processes that can be seen as computations. To illustrate this difference we compare (a) designing a Turing machine and (b) learning a Turing machine, defining them for the purpose of the argument. We show that there is a well-defined sequence of problems which are not effectively designable but are learnable, in the form of the bounded halting problem. Some characteristics of human intelligence are reviewed including it's: interactive nature, learning abilities, imitative tendencies, linguistic ability and context-dependency. A story that explains some of these is the Social Intelligence Hypothesis. If this is broadly correct, this points to the necessity of a considerable period of acculturation (social learning in context) if an artificial intelligence is to pass the Turing Test. Whilst it is always possible to 'compile' the results of learning into a Turing machine, this would not be a designed Turing machine and would not be able to continually adapt (pass future Turing Tests). We conclude three things, namely that: a purely "designed" Turing machine will never pass the Turing Test; that there is no such thing as a general intelligence since it necessarily involves learning; and that learning/adaption and computation should be clearly distinguished.

1 Introduction

The approaches in Turing's two most famous papers contrast markedly. The definition of a computation, in the form of a Turing Machine (Turing machine),

S.B. Cooper, A. Dawar, and B. Löwe (Eds.): CiE 2012, LNCS 7318, pp. 182–192, 2012.
© Springer-Verlag Berlin Heidelberg 2012

is a micro-level specification of a device (its design and rules for operation) from which computable functions can be defined [24]. It specifies what happens when a Turing machine that has been built is set going. It is a formal definition that defines the set of computable functions. The Turing Test (Turing Test) is a macro-level test that is applied to an existing entity that is "running" [25]. It is not formally defined but a practical test, intended to be feasible to implement. Here intelligence is not something to be proved but demonstrated.

As pointed out by French [10], the Turing Test is not a test of a putative "general intelligence" but a test of a specific kind of intelligence – normal human intelligence. There may well be intelligent entities that might not pass the Turing Test, for example a human suffering from influenza or an alien whose language we do not know. The point of the Turing Test is that if some entity passes it, then it is hard to deny that this entity is intelligent – it short-cuts possible quibbling, and thus opens up the possibility that an artificial entity could be judged as intelligent.

The Turing Test consists of a conversation over a period of time between a tester and the entity being tested. This requires an ability to learn or adapt to what the tester has said, including: the topic of conversation, the style, the detected context that the tester is coming from, and the importance given to particular issues. Clearly the Turing Test is harder the longer it goes on for. It is far easier to fool someone if one only talks to them for a limited period of time (using, for example, rote learned scripts) than if there is time for topics to be revisited with more testing questions, checking consistency with what went before as well as common knowledge and assumptions. It is the interactive and adaptive nature of the Turing Test that makes it so challenging, revealing shallow strategies (such as simplistic syntactic approaches) as inadequate.[1]

The arguments in this paper rest upon the difference between computation (defined by a halting Turing machine) and adaption, which is an essential part of the intelligence that is tested for by the Turing Test. Thus in section 2 we argue that computation and adaptivity are different kinds of things, giving an example where they can be shown to differ. We then briefly review some of the characteristics of human intelligence in section 3 and look at some of the consequences in terms of passing the Turing Test using purely designed Turing machines in section 4. We then consider the broader nature of intelligence and its role in section 5 before concluding.

2   Learning vs. Computation

The relationship between adaptive processes, i.e., learning in the broadest sense, and computation is not straightforward. Clearly learning involves processes that can sensibly be characterized as a computation, as the field of Machine Learning

---

[1] Here we assume that the Turing Test is conducted over a suitably extended period of time, since this tests intelligence as we know it, what is called the "Long Term Turing Test" in [8].

amply demonstrates. Similarly, some computations – when acting upon some internal data structure and outputting an updated data structure in response to new information – might be sensibly thought of as a learning process. The physical device of "a computer" can clearly be set up to do both learning and computation.[2] Thus the question arises whether they do, fundamentally, differ. In particular it is sometimes assumed that learning processes are just a particular kind of computation. The core of our argument here is that the Turing machine (or equivalent) is not an *adequate* model of learning processes, and hence misses out a crucial aspect of intelligence, its adaptivity.

Many different ways of defining the computable functions turned out to be the same, resulting in the "Church-Turing" thesis that these *were* the class of effective computations, including observed processes in the real world [5]. This meant that the details of the computation processes were not deemed as crucial, but rather what was, and was not, finitely computable. Turing also showed that there was a Universal Turing machine, one that could be given a program number and then effectively execute that program. Thus, in a deep sense there is a universal characterization of computation. An observed, physical process is meaningfully characterized as a computation if its inputs compared to outputs can be effectively predicted (at least at the micro, step-by-step level) by a computable function (via a suitable mapping).

There is no such agreement on the definition of a learning process. Rather there are many different kinds of learning process, each with different properties and assumptions. Indeed the "No Free Lunch" [28] theorems from Machine Learning show (in an ultimate, abstract sense) that no learning algorithm is better than any other. In other words, to find a better learning algorithm you need to use some prior knowledge about the class of situations within which the adaption is taking place (in Machine Learning terms this is characterized as the search problem), which means that the class of situations being learned about is a proper subset of all possible situations. This indicates that there is no universally best learning algorithm, however clever the learning strategy is (e.g., using meta-reasoning/learning, or lots of special cases).

Thus, for our argument, we shall define a kind of learning process, an *adequate incremental learner*, as follows:

- As a computational process (e.g., a Turing machine) plus a model, which is a set of data;
- Where the data can be judged as to its truth or adequacy about something exterior, which we shall call the learning "target" (for example via another Turing machine that produces a prediction, or output, from the model with respect to the target);

---

[2] It is interesting that the first use of the word "computers" referred to people – for example the women that did calculations as part of the effort to break codes during WWII at Bletchley Park. The word seems to have been transferred to the devices that took over this task when they were constructed [1].

- Where the computation is iteratively provided with information from the target such that during each iteration the data is modified by the Turing machine using the information so that the updated model is at least as adequate as before;[3]
- After a finite number of iterations the model becomes maximally adequate.

This is clearly not a universal definition of learning, since it excludes known learning processes (e.g., ones that sometimes degrade the model). However, it is also clearly *a* learning process and will thus do for our purposes. Here we use it to show that there is something that can be learned (in the above sense) that cannot be computed (by a Turing machine). The particular problem we shall use for this demonstration is that of finding the lookup table (or Turing machine) that solves the "Limited Halting Problem".

The "Limited Halting Problem" is a sequence of increasingly difficult problems, indexed by integer, n, and defined as follows. Let $\{P_1, P_2, P_3...\}$ be an effective enumeration of Turing machine's. The limited halting problem is that of "given n, does $P_i(j)$ eventually halt where $i, j \leq n$". Let us call this sequence of problems $\{H_1, H_2, H_3, ...\}$, where $H_n(i, j)$ is 1 if $i, j \leq n$ and $P_i(j)$ eventually halts, 0 if $i, j \leq n$ and $P_i(j)$ does not halt, and undefined otherwise. Each problem H_i is computationally decidable, since it can be implemented as a simple 2D look-up table with the rows $\{1, ...n\}$ for possible program indices, P_i, and columns $\{0, ...n1\}$ for possible inputs, j, and entries 0 or 1 depending on whether $P_i(j)$ eventually halts. The problem is not the existence of this table but of finding the right entries for it. The definition of computability is not constructive, it is sufficient that there *exist* a program to compute a function, not that we can find or implement this program.

The point *is* that there is an adequate incremental learner that can learn the lookup table for H_n given any n but there is *no* Turing machine that can implement this lookup table (or equivalent Turing machine) given any n. We now give an informal proof of each part.

The following non-terminating algorithm establishes that, given an integer n the lookup table that solves H_i can be learnt by an adequate incremental learner.

```
Build a nxn table with all entries 0
s := 1
Repeat for ever
  For i from 1 to n
    For j from 0 to n-1
      Calculate Pi(j) for s steps
      If it has terminated
        then change entry at (I,j) to 1
    Next j
  Next i
```

---

[3] One can see the Turing machine as presented with an index representing the set of information: new information from the target, and the present model and outputting an index representing the updated model (which would replace the old version of the model).

Any $P_i(j)$ that halts will do so after a finite period of time, so eventually the table being adapted by the above algorithm will have the correct entries for solving H_n although one might well not know when one gets to this point and there is no effective method for knowing whether one has (otherwise we could solve the general halting problem). This algorithm fulfills the definition for an adequate incremental learner defined above.

Now to show that there is no effective method for finding the Turing machine that implements the solution to H_n, given n. Suppose there were a computable function $f(x)$ such that $P_{f(n)}$ computes H_n. In other words, of effectively finding the program index that implements the table for H_n, (which we know exists). This is equivalent to having a general and effective procedure for constructing the Turing machine for H_n given n.

If there were such a computable function, $f(x)$, then to decide whether $P_i(j)$ halts, we can calculate $P_{f(max(i,j))}(i,j)$. This is defined since $i,j \leq max(i,j)$. See if the answer is 1 or 0. Thus $P_{f(n)}$ is computable via the universal Turing machine [5]. Thus if $f(x)$ were computable we could effectively solve the general halting problem, which we know is impossible [24]. Thus $f(x)$ is not computable, that is to say there is no Turing machine that computes it.

Thus there is a learning process whose "resulting" model is not computable by a Turing machine. Learning is different from computation. Of course, in a way, this is obvious since a computation is, as defined, not a process but a formally defined function (albeit possibly defined using a process), whilst learning is an on-going process. Learning can be seen as a kind of non-predefined change in a computational process [13]. A Turing machine cannot implement this since the change is not predefined.

Formalists may well be dissatisfied with the above demonstration since the Turing machine has to halt whilst the adequate incremental learner does not. However this difference is at the crux of the matter. Computation, as defined by a halting Turing machine, is not a process but the *result* of a process – the Turing machine is merely a means of defining which functions are computable.[4] A possible confusion might arise if people conflate what we call a "computer" (the physical object we use) and what is formally defined as "computable" (using a halting Turing machine or other definition). It is true that the intermediate state of any process that implements a complete computation could, itself, be seen as a computation of that intermediate state, but that intermediate state is not part of the definition of the complete computation.[5] This difference becomes important when we are considering what sort of entity could pass the Turing Test.

---

[4] There are non-procedural ways of defining computable functions, e.g., using lambda calculus.

[5] Each intermediate state is the result of a different computation, but this is not the same as the one an intermediate state is part of unless it happens to be the final state.

3   Human Intelligence

We briefly consider some of the characteristics of human intelligence, since the Turing Test tests for an ability that results from human intelligence: the ability to hold a recognisably "normal" conversation. These include being able to:

- continually react to social signals,
- imitate and learn from others,
- detect what is the appropriate social context,
- react in ways appropriate to the detected social context,
- imagine what it is like to be other people,
- use other people for filtering relevant information from the environment,
- use other people to act on behalf of ourselves,
- make alliances and friendships, maintained by frequent interaction,
- acquire and effectively use a large body of knowledge that is shared with others,
- reason in ways which might be accepted by others using a shared, but implicit, common knowledge,
- talk to others and make them understand our intention and meaning.

All of these characteristics (and many more) are explained by the "Social Intelligence Hypothesis" (SIH) [19].[6] The SIH seeks to explain the evolutionary advantage provided by human brains. The explanation can be summarised as follows: our brains give us the social ability to coordinate and develop a commonly held set of knowledge and behaviours; this allows groups of individuals to inhabit specialized ecological niches (e.g., to live in the Tundra or Kalahari [22]); this ability to collectively adapt to and successfully inhabit a variety of niches gives us selective advantage. Under this view our brain is an adaptive organ to give us social abilities, what might be called *social intelligence*, including the characteristics listed above. However it also implies that these social abilities are primary and classic displays of intelligence (including reasoning, and problem solving) are by-products. Being able to solve a Rubrick's Cube or play chess has no selective advantage,[7] but being part of a society that invents such puzzles and games and talks about them is!

4   Design and the Turing Test

It is interesting that Turing chose a *social* test for intelligence years before the social roots of intelligence were widely appreciated. However, these social roots

---

[6] The related idea of "Machiavellian intelligence" had been around in primatology (e.g., [7]) before it was taken up in anthropology, focussing on the cognitive 'arms race' that might have occurred in terms of the competitive evolution in the social ability to make alliances [2]. The SIH is more general in its formulation covering a broad range of social abilities.

[7] It is not even effective as a display for attracting a potential mate, as those on the back row of my undergraduate mathematics lectures demonstrated.

have implications for passing the Turing Test. In particular it indicates that being part of the appropriate human culture is a key part of social intelligence, and not just an 'extra' that needs to be added once individual intelligence is sorted out. If this is the case, a considerable period of acculturation within a human society is necessary to pass the Turing Test.[8] A purely individual intelligence without such training would not suffice.

If a significant amount of learning is necessary to obtain a social intelligence that might pass the Turing Test, then to a large extent, this intelligence is not *designed* but developed in context. In other words, what one intentionally puts into a Turing machine, i.e., by design, is not sufficient to pass the Turing Test, but rather the huge body of context-specific information that is usually accumulated by humans as they grow into a culture. This is what we mean when we say that an "out-of-the-box" Turing machine will not pass the Turing Test.

Of course, once cultural information has been learnt by some entity, this could be 'compiled' into a complex Turing machine, but we would not have *designed* this in any meaningful sense. Such a compiled entity would be a static entity that has stopped learning. Such a Turing machine might pass muster in a one-shot and short Turing Test. However, if the test were extended in time then the conversation itself would form part of the cultural information that needs to be learnt about by the entity, so that in subsequent interactions it can appropriately refer back to what has been said. The difference is made clear if one has a conversation with a human with an impaired ability to lay down new memories, but who retains all the long-term memories before a certain date.[9] For the first period of time such people seem normal, but over any period of time their lack of memory becomes apparent.[10]

5   Intelligence in General

It should be obvious that the sort of intelligence that could pass the Turing Test requires reasoning *and* learning, it cannot be reasoning alone.[11] One consequence of this, along with the "No Free Lunch" theorems mentioned above [28], is that there is no such thing as *general* intelligence. That is to say that (unlike computation) intelligence cannot be general, rather that different intelligences are each suited to particular problems and/or environments. Clearly, if the SIH is true, our intelligence is particularly suited to living in groups that develop a collective body of knowledge, habits, norms, skills, stories etc. Whatever "clever" algorithms we invent, using meta-reasoning, meta-learning, special cases, etc. will

---

[8] Indeed, anecdotal accounts have it that Turing joked that such machines might be teased whilst they attended human schools.

[9] That such conditions exist and can be brought on by many different causes can be seen in many clinical accounts, e.g., [10].

[10] Of course, it might be that such a person might pass the Turing Test if people with such conditions were expected as a possible participant. However, they would be clearly distinguishable from a healthy adult human.

[11] Reasoning is roughly coincident to computation, since any formal reasoning can be implemented as a computation and any computation as formal reasoning.

have "blind spots" where its algorithms are not as effective (w.r.t. its goals) as another (and possibly simpler) algorithm. Any algorithmic elaboration will have downsides as well as new abilities. In other words, intelligence is relative to the goals and environment of an entity, it is not an ordinal quality, with humans having the most.

Once one accepts that intelligence cannot have an ideal, we can distinguish different types of intelligence. For example, rats can be smarter than humans at navigating mazes. Does this imply that rats are smarter than humans? Well, it is better to clarify what is being tested. In the Turing Test where human social interaction is being tested humans tend to do pretty well. But different types of tests allow us to explore different types of intelligence, in animals and machines. For example, social insects can be pretty effective at collective decision making [11]. Plants could also be said to be intelligent [23], as well as bacteria [2].[12] We could take a narrow definition of intelligence and apply it to humans, or embrace a broad diversity of intelligences and understand human intelligence better by relating it to other types. For example, Randall Beer defines intelligence as "the ability to display adaptive behaviour" [1].

The same applies for artificial intelligence: we can take the narrow path of attempting to program human intelligence, which we argue is not achievable without learning/adaption; or we can take the broad path of exploiting all types of biological and artificial intelligences for solving problems in the most diverse areas. This broad approach seems more suitable for the parts of our complex world that are not structured around human sociability. Given the fact that problems are changing constantly in unpredictable ways due to their interactions [13], systems we build to solve these problems must adapt and learn constantly, matching the timescale at which problems change. We can predict that, as the complexity of problems increase, artificial systems will focus more and more on learning and adaption rather than on directly programming static solutions.

None of this invalidates the Turing Test that, if passed, would unequivocally establish that a substantial intelligence, as impressive as human intelligence, had been brought into being. We would be forced to recognise its legitimacy since we impute intelligence in our fellow humans on a similar basis.

6   Understanding Ourselves

Understanding human intelligence is obviously hard, due to its complexity. However, there is another difficulty as well, a "touchy" subject for us humans. We seek to understand ourselves using whatever cognitive means are at our disposal and what we use for this purpose effects our self-image.

The first and obvious way of understanding ourselves is by observing and understanding others around us. Whilst we clearly understand others by imagining how we would feel or think in their situation [6], it seems to also be true that,

---

[12] Though it might be more accurate to attribute the intelligence in many cases to the process of evolution (the underlying *learning* process) rather than the entity that results.

during development, we use our observations of how others behave to help us understand ourselves [15]. The circular bootstrapping of our own identity produces a very strong association between our perception of our own intelligence and that of others. It is perhaps this that makes the Turing Test so compelling: we cannot but consider the entities with whom we converse as similar to ourselves.

A second way of understanding ourselves is by using analogies with devices around us. Whilst the Victorians might have conceived of the world and persons in terms of intricate clockwork, we tend to use the computer. Thinking of ourselves as computers is natural, since we have many things in common with modern computational devices: they are interactive, responsive and can be programmed to behave in human-like ways, for example by learning. Intensive interaction with computers can lead to a strong association with such devices, to the extent that we see ourselves as a kind of computer and impute many human characteristics upon them [26]. Since the Turing machine is a formal model of a computation, and being able to be predicted by such a model is what makes a device a "computer",[13] this leads to an association of Turing machines and our intelligence. However, as the above arguments show, this analogy captures only some of the complete picture.

A third way of understanding ourselves, is as the pinnacle of evolution, as possessing (essentially) a completely general intelligence. Somehow it is assumed that the human brain, together with its creations: paper, maths, computers etc., is not limited in what it can understand. This is not a view of any individual, of course, but rather that eventually humankind will be able to work anything out. Surely, if people do indeed think this, this is nothing more than sheer hubris. However forms of this thinking seem to be implicit in some of the thinking about the Turing Test, e.g., French's [10] denigration of 'subcognitive' aspects of human intelligence and his criticism of the Turing Test as "only" being a test of human intelligence.

We (along with others) wish to work towards a different understanding of intelligence, by looking to other parts of life, their ecology and the process of evolution. This tries to relate the nature of intelligence to why it evolved. It is a more interactive view that wishes to place individual entities within a broader web of interactions, so that the interactions within an entity are just part of this broader web. It is admitted that, currently, this view does not have formal models of the strength of the Turing machine, but rather a plethora of simulations and approaches.

7 Conclusions

There are good and general models of computation, going back to Turing's original paper but there is no equivalent for the interactive and continual process

---

[13] It is sensible to think of these devices as "computers", since one can understand what they are doing since, with practice, we can perform simple calculations and work out what small bits of computer code are doing. The micro-level model as a kind of Turing machine predicts the device's behaviour accurately.

that we call learning, let alone the more general phenomena of adaptive interaction. The Turing Test nicely shows the difference between the two, being a
test that requires the latter. Perhaps Turing's later paper will have the effect of
moving on thinking about ways of formally representing intelligence, and free it
from the limited analogy of the Turing machine.

Acknowledgements. BE was partially supported by the Engineering and
Physical Sciences Research Council, grant number EP/H02171X/1. CG was partially supported by SNI membership 47907 of CONACyT, Mexico. We thank the
referees for their further suggested reading, including [16], [21] and [27] which
were interesting but we did not feel clarified the arguments here.

References

1. Beer, R.D.: Intelligence as adaptive behavior: an experiment in computational neu-
   roethology. Academic Press (1990)
2. Ben-Jacob, E., Becker, I., Shapira, Y., Levine, H.: Bacterial linguistic communication and social intelligence. Trends in Microbiology 12(8), 366–372 (2004)
3. Byrne, Whiten, A.: Machiavellian intelligence. Oxford University Press (1988)
4. Copeland, B.J.: The Modern History of Computing. In: Edward, N. (ed.) The
   Stanford Encyclopedia of Philosophy (fall 2008 edn.) (2008),
   http://plato.stanford.edu/archives/fall2008/entries/computing-history
5. Cutland, N.: Computability, an introduction to recursive function theory.
   Cambridge University Press (1980)
6. Tomasello, M.: The cultural origins of human cognition. Harvard University Press
   (1999)
7. De Waal, F.: Chimpanzee politics: power and sex among apes. John Hopkins University Press (1989)
8. Edmonds, B.: The constructability of artificial intelligence (as defined by the Turing
   Test). Journal of Logic Language and Information 9, 419–424 (2000)
9. Edmonds, B.: The social embedding of intelligence: how to build a machine that
   could pass the Turing Test. In: Epstein, R., Roberts, G., Beber, G. (eds.) Parsing
   the Turing Test, pp. 211–235. Springer (2008)
10. French, R.M.: Subcognition and the limits of the Turing Test. Mind 99, 53–64
    (1989)
11. Garnier, S., Gautrais, J., Theraulaz, G.: The biological principles of swarm intelli-
    gence. Swarm Intelligence 1(1), 3–31 (2007)
12. Gershenson, C.: Cognitive paradigms: Which one is the best? Cognitive Systems
    Research 5(2), 135–156 (2004)
13. Gershenson, C.: Computing Networks: A General Framework to Contrast Neural
    and Swarm Cognitions, Paladyn. Journal of Behavioral Robotics 1(2), 147–153
    (2010)
14. Gershenson, C.: The implications of interactions for science and philosophy. Tech-
    nical Report, 04, C3, UNAM, Mexico (2011)
15. Guimond, S., et al.: Social comparison, self-stereotyping, and gender differences in
    self-construals. Journal of Personality and Social Psychology 90(2), 221–242 (2006)
16. Glymour, C.: Learning, prediction and causal Bayes nets. Trends in Cognitive
    Sciences (1), 43–48 (2003)

17. Humphrey, N.K.: The social function of the intellect. In: Bateson, P.P.G., Hinde, R.A. (eds.) Growing Points in Ethology, Cambridge University Press, Cambridge (1976)
18. Kirshner, H.S.: Approaches to intellectual and memory impairments. In: Gradley, W.G., et al. (eds.) Neurology in Clinical Practice, 5th edn., ch. 6, Butterworth-Heinemann (2008)
19. Kummer, H., Daston, L., Gigerenzer, G., Silk, J.: The social intelligence hypothesis. In: Weingart, et al. (ed.) Human by Nature: Between Biology and the Social Sciences, pp. 157–179. Lawrence Erlbaum Associates (1997)
20. Lane, H.: The Wild Boy of Aveyron. Harvard University Press (1976)
21. Marr, D.: Vision: A Computational Approach. Freeman & Co., San Francisco (1982)
22. Reader, J.: Man on Earth. Penguin Books (1990)
23. Trewavas, A.: Aspects of plant intelligence. Annals of Botany 92(1), 1–20 (2003)
24. Turing, A.M.: On computable numbers, with an application to the Entscheidungsproblem. Proc. of the London Mathematical Society 2 42, 230–265 (1936)
25. Turing, A.M.: Computing machinery and intelligence. Mind 59, 433–460 (1950)
26. Turkle, S.: The second self: computers and the human spirit. Simon and Schuster (1984)
27. van Rooij, I.: Tractable Cognition: Complexity Theory in Cognitive Psychology. PhD Thesis, Katholieke Universiteit Nijmegen, Netherlands (1998), http://www.nici.ru.nl/~irisvr/PhDthesis.pdf
28. Wolpert, D.H.: The lack of a priori distinctions between learning algorithms. Neural Computation 8(7), 1341–1390 (1996)

Confluence in Data Reduction: Bridging Graph Transformation and Kernelization

Hartmut Ehrig, Claudia Ermel, Falk Hüffner, Rolf Niedermeier, and Olga Runge

Institut für Softwaretechnik und Theoretische Informatik, Technische Universität Berlin, Ernst-Reuter-Platz 7, 10587 Berlin, Germany
{hartmut.ehrig,claudia.ermel,falk.hueffner}@tu-berlin.de,
{rolf.niedermeier,olga.runge}@tu-berlin.de

Abstract Kernelization is a core tool of parameterized algorithmics for coping with computationally intractable problems. A *kernelization* reduces in polynomial time an input instance to an equivalent instance whose size is bounded by a function only depending on some problem-specific parameter k; this new instance is called problem kernel. Typically, problem kernels are achieved by performing efficient data reduction rules. So far, there was little study in the literature concerning the mutual interaction of data reduction rules, in particular whether data reduction rules for a specific problem always lead to the same reduced instance, no matter in which order the rules are applied. This corresponds to the concept of confluence from the theory of rewriting systems. We argue that it is valuable to study whether a kernelization is confluent, using the NP-hard graph problems (EDGE) CLIQUE COVER and PARTIAL CLIQUE COVER as running examples. We apply the concept of critical pair analysis from graph transformation theory, supported by the AGG software tool. These results support the main goal of our work, namely, to establish a fruitful link between (parameterized) algorithmics and graph transformation theory, two so far unrelated fields.

1 Introduction

Theoretical Computer Science is usually divided into algorithm-oriented research and description-oriented research (as witnessed by the two volumes "Algorithms and Complexity" and "Formal Methods and Semantics" of the Handbook of Theoretical Computer Science [15]). Unfortunately, the corresponding research communities typically work in two "parallel worlds" with relatively little interaction. In this work, we propose a new link between algorithmics and formal methods that may lead to a fruitful "interdisciplinary" field of research. More specifically, we develop a connection between efficient preprocessing of NP-hard (graph) problems by kernelization [2,11] and the theory of graph transformations [6,20]: We employ the concept of confluence of rewriting systems to show "uniqueness results" for problem kernels. This leads to the natural concept of confluent data reduction rules, having a number of both theoretical and practical benefits as discussed in the following.

S.B. Cooper, A. Dawar, and B. Löwe (Eds.): CiE 2012, LNCS 7318, pp. 193–202, 2012.

Confluence in kernelization. Data reduction, also known as polynomial-time preprocessing, is a classic approach for dealing with NP-hard combinatorial optimization problems (see [2,11] for surveys). The idea is to remove redundant parts of the input, thereby obtaining a hard "core" of the instance. Costly algorithms need then only be applied to this core. Data reduction is thus useful in virtually any approach to solving computationally hard problems, whether heuristic, approximative, or exact. Formally, we consider only decision problems, and a (*data*) *reduction rule* replaces in polynomial time a given problem instance I by an instance I' with $|I'| < |I|$. We say that the rule is *correct* when I is a yes-instance iff I' is a yes-instance. An instance to which none of a given set of reduction rules applies is called *reduced* with respect to these rules.

While they are a standard technique for practitioners, only fairly recently have data reduction rules been the subject of wider theoretical analyses, using the concept of a *problem kernel* [2,11]. This notion comes from the field of *parameterized complexity* [5,18,9], where performance of algorithms is analyzed not just in terms of the problem size n, but also in terms of a parameter k, for example the solution size. A *kernelization* is a data reduction that creates an equivalent instance whose size depends only on the parameter k, and not on the original input size n anymore (see Section 2 for a more formal definition).

We call a terminating set of data reduction rules *confluent* if any order of application of the rules yields a unique reduced instance, up to isomorphism. Confluence is a standard concept from graph transformation theory (see below). There are a number of reasons why it seems useful to investigate whether data reduction rules are confluent: If they are, then the rules are robust in a sense; we obtain a unique starting point for further processing after the data reduction has been performed. In an implementation of the rules, we can apply the rules in any order without worrying about the result, and can thus optimize for the speed of their application. If the rules are not confluent, this might indicate some "slack" in the rules: some orders of application might lead to worse results, that is, larger kernels. Investigating all this might lead to improved reduction rules. Further, insight on the interaction of data reduction rules can lead to faster kernelizations. Confluence was also exploited in [14], who showed that for their problem, one order of application of data reduction yields some desired property of the reduced instance, and another order yields a different desired property. A proof of confluence now shows that a reduced instance has both properties. Finally, proving confluence is also a good way to check for possible conflicts between data reduction rules, since all possible interactions need to be taken into account. It might also give an incentive to create "minimal" kernelization rules in order to make confluence proofs easier, which could give a sharper picture of what exactly is needed to achieve a kernel.

If we allowed data reduction in kernelization with restriction of the rule execution order, we can force confluent kernelization in a trivial way by allowing only one execution order. In this paper, we avoid this trivial case by allowing any execution order.

Confluence of graph transformation systems. The theory of graph grammars and graph transformation systems has been started in the early 1970s [8] as a generalization of Chomsky grammars and term rewriting systems, which are based on strings and trees, respectively. The main idea is the rule-based modification of graphs. Graph transformations are most suitable to model the operational semantics of visual languages and also to define model transformations between different kinds of models. Several approaches for graph transformations are known [20], including logical and algebraic approaches. A graph transformation system consists of a set of graph rules, which are applied in a non-deterministic way, leading to graph transformation steps $G \Longrightarrow H$ and sequences $G \stackrel{*}{\Longrightarrow} H$. A single rule consists of a left-hand side graph *LHS*, a right-hand side graph *RHS*, and their intersection graph. To apply a rule, a *match* is sought, that is, a subgraph in the input graph that is isomorphic to *LHS*. This subgraph without the intersection graph is then deleted, resulting in a context graph, which is glued together with *RHS* at the nodes and edges of the intersection graph. A graph transformation system is called *confluent* if for each pair of graph transformation sequences $G \stackrel{*}{\Longrightarrow} G_1$, $G \stackrel{*}{\Longrightarrow} G_2$, there is a graph G_3 together with sequences $G_1 \stackrel{*}{\Longrightarrow} G_3$ and $G_2 \stackrel{*}{\Longrightarrow} G_3$.

There are numerous applications in software engineering, concurrency, and distributed systems [7], where confluence of graph transformations plays an important role. Confluence together with termination, that is, non-existence of infinite transformation sequences, implies that any order of applying the rules as long as possible yields a unique graph, up to isomorphism. Moreover, we obtain for isomorphic input graphs isomorphic reduced graphs [6].

In order to show confluence it is sufficient to show local confluence and termination [13,17], where local confluence means confluence for the special case that the given sequences from G to G_1 and G_2 are transformation steps $G \Longrightarrow G_1$ and $G \Longrightarrow G_2$, where in each step only one transformation rule is applied. Data reduction rules for kernelization for graph problems define graph transformation systems based on undirected graphs, such that the general concepts of (local) confluence and termination are applicable.

Structure of the Paper. After presenting basic concepts and definitions about kernelization and critical pair analysis in Section 2, we present our two case studies *Clique Cover* and *Partial Clique Cover* in Section 3 and 4, respectively, by showing that the corresponding data reduction rule sets yield problem kernels and are confluent. Section 5 concludes with an outlook to future work. Due to limited space, we defer some proofs and details to the full version of this paper.

2   Basic Concepts of Kernelization and Critical Pair Analysis

Kernelizations. A *parameterized problem* can be defined by a set of instances (x, k), where k is called the *parameter* [5,9,18]. Let L be a parameterized problem. A *reduction to a problem kernel* or *kernelization* is a transformation via data reduction rules of an instance (x, k) to an instance (x', k') (the *problem kernel* of instance (x, k)), such that

- $(x, k) \in L \iff (x', k') \in L$,
- $|x'| \le g(k)$ for some arbitrary computable function g depending only on k,
- $k' \le k$, and
- the transformation runs in polynomial time.

We call $g(k)$ the problem kernel of the parameterized problem L.

Critical Pair Analysis in Graph Transformation Theory. The algebraic theory of graph transformations [6] provides a specific technique known from term rewriting systems [13], called *critical pair analysis*, which has been generalized to graph transformation systems in [19]. Critical pair analysis supports the verification of local confluence using the software system AGG [1]. The main idea is to show local confluence not for all pairs of (a possibly infinite number of) transformation steps $G \implies G_1$ and $G \implies G_2$ via rules r_1 resp. r_2, but only for all *critical pairs*. A pair of transformation steps is called a *critical pair* if it is *conflicting in a minimal context* in the following sense: The pair $G \implies G_1$, $G \implies G_2$ via r_1, r_2 is called *parallel independent* if there are transformation steps $G_1 \implies G_3$, $G_2 \implies G_3$ via r_2, r_1 leading to the same G_3. A pair is called *conflicting* if it is not parallel independent, and it has *minimal context* if each vertex and edge in G belongs to the match of r_1 or r_2 in G. For a graph transformation system with a finite number of rules based on finite graphs, there is a finite number of critical pairs. All of them can be computed automatically by the graph transformation analysis tool AGG [1]. The Local Confluence Theorem for algebraic graph transformations [6] implies local confluence of a graph transformation system provided that all critical pairs are *strictly confluent*, where "strictness" is an additional technical condition for the transformations. The verification of strict confluence for critical pairs can also be supported by AGG and is applied to data reduction in Sections 3 and 4.

The application of critical pair analysis to data reduction rules, however, is not yet fully automated. The first reason is that the Local Confluence Theorem [6] based on critical pairs is valid for directed graphs (with parallel edges and loops) and several other kinds of graphs, but not yet proved for undirected graphs as considered for data reduction in this paper. The second reason is that data reduction rules in general are rule schemes in the sense of graph transformation theory, where rule schemes can be applied to an unbounded number of vertices, and rules are applied to a constant-size subgraph. Each rule schema corresponds to a—possibly infinite—set of rules in the sense of [6]. For these reasons, we prove confluence directly in Sections 3 and 4; in the case of PARTIAL CLIQUE COVER, the proof is quite complex, based on a large number of case distinctions. These proofs depend strictly on the specific rules. It is an interesting challenge for future work to extend the theory of graph transformations [6]—and the corresponding tool AGG—to handle also data reduction in a more general way.

3   Case Study Clique Cover

We use the well-known NP-hard CLIQUE COVER problem for our first case study.

CLIQUE COVER
Instance: An undirected graph $G = (V, E)$ and an integer $k \geq 0$.
Question: Is there a set of at most k cliques in G such that each edge
in E has both its endpoints in at least one of the selected cliques?

For an instance (G, k), we call a set of at most k cliques that covers all edges
a *solution*. Choosing CLIQUE COVER[1] has several reasons: It is a conceptually
simple graph problem, and the best known (theoretical) data reduction rules so
far are easy to understand and also applied in practice [10]. Moreover, CLIQUE
COVER has a kernelization with a size bound of 2^k vertices [10,12], and it was recently shown that under standard complexity-theoretic assumptions, this cannot
be improved to a polynomial bound [4].

Kernelization for Clique Cover. For the currently only known kernelization
for CLIQUE COVER with parameter k, the following data reduction rules are
used [10,12].[2]

Rule 1. *Remove isolated vertices, that is, vertices with no neighbors.*

Rule 2. *If there is an isolated edge, then delete it and decrease k by one.*

Two vertices $u, v \in V$ are called *twins* if $\{u, v\} \in E$ and u and v have exactly
the same neighbors (except for v and u, respectively).

Rule 3. *If $\{u, v\}$ are twins and $\{u, v\}$ is not an isolated edge, then delete u (that
is, remove it from the vertex set and all incident edges from the edge set).*

Theorem 1 ([10,12]). *Rules 1 to 3 are correct and yield a problem kernel for*
CLIQUE COVER *with at most 2^k vertices.*

Note that for technical correctness of the kernel (as defined in Section 2), we
need to check whether after exhaustive application of Rules 1 to 3 there are
more than 2^k vertices left, and if so, the instance is replaced with a small "no"-
instance (for instance, $k + 1$ disjoint edges). We omit such trivial checks in the
following.

---

[1] Note that in the literature sometimes also covering vertices instead of edges by
cliques is called CLIQUE COVER.
[2] We note that [10] uses different rules involving "covered edges", which are equivalent
to the rules presented here if the initial instance does not have covered edges (except
that Rule 3' from [10] does not treat isolated edges correctly; as already noted in
[12], they require a special case.)

Confluence of Data Reduction for Clique Cover. We now show that the kernelization rules from Theorem 1 are confluent.

Theorem 2. *The set of Rules 1 to 3 for* CLIQUE COVER *is confluent.*

Proof. Clearly, the order of application for Rule 1 and Rule 2 with respect to any of the three rules is not relevant, since their application does not affect the applicability of other rules. It remains to show that the relative order of applications of Rule 3 does not matter.

If we consider two vertices as equivalent when they are twins, we obtain an equivalence relation on the vertex set. Thus, we can partition the vertex set into the equivalence classes of this relation, called *twin classes*. Note that every twin class forms a clique in the graph. Let the *twin graph*[3] of a graph be a graph with the twin classes as vertices and an edge between two twin classes if there is an edge between one vertex from one class and one vertex from the other class.

The twin graph does not change (up to isomorphism) when Rule 3 is applied, since u and v must be from the same twin class and the rule thus always leaves at least one vertex in any twin class. Further, Rule 3 is applicable until a twin class contains exactly one vertex (if it is connected to vertices outside the twin class) or two vertices (if it is an isolated clique). Since the twin graph and the number of vertices per twin class uniquely represent a graph up to isomorphism, we obtain confluence.   □

This proof also yields a shortcut to calculate the result of the kernelization, whose naive calculation would require $O(|E| \cdot |V|^2)$ time (The authors of [10] only state the running time of $O(|V|^4)$ for Rules 1 to 3 plus another rule).

Corollary 1. *A 2^k-vertex kernel for* CLIQUE COVER *can be found in linear time.*

Proof. From the proof of Theorem 2, we can see that it is sufficient to calculate the twin graph, contract each twin class to a single vertex, and then delete isolated vertices and edges. Finding the twin graph can be done in linear time [16, Corollary 7.4], so the kernelization can be done in linear time, too.   □

Confluence via Critical Pair Analysis. As pointed out in Section 1, the standard way to show confluence of a rule set in graph transformation theory [6] is to construct all critical pairs and to show for each critical pair that it is strictly confluent. The approach has been shown for directed graphs [6], and we are confident that it can also be extended to undirected graphs as considered in this paper, in particular to data reduction for CLIQUE COVER and PARTIAL CLIQUE COVER. Note that data reduction rules, like Rule 2, may also change the parameter k, but this is not essential for confluence and will be disregarded in this section.

Actually, Rule 3 is a rule scheme in the sense of graph transformation theory, which can be represented by the following family of rules $R3.m$ for $m \geq 1$:

---

[3] Twin classes and the twin graph have been used before for data reduction under the names *critical cliques* and *critical clique graph* (see e.g. [11]).

LHS u $R3.m(u,v)$ RHS

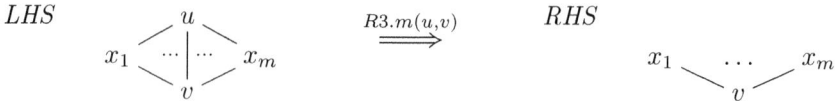

The rule describes the deletion of u. Applying the rule to a graph G means to find an occurrence of the left-hand side LHS in G satisfying $N[u] = N[v] = \{u, v, x_1, \ldots, x_m\}$, and to replace this occurrence by the right-hand side RHS.

For graphs with n vertices, we only have to consider rules Rule 1, Rule 2, Rule 3.1, $\ldots$, Rule 3.r with $r = n - 2$, because rules with $r > n - 2$ cannot be applied. Fig. 1 shows the table computed by the AGG tool [1] giving the number of critical pairs (CP) for each pair of rules and $r = 3$. Clicking on an entry in the CP table table (e.g. the highlighted field showing 12 minimal conflicts for Rule 3.2 and Rule 3.3 where rule Rule 3.2 is applied first), the 12 conflicting situations of these two rules are shown in detailed graphical views. Vertices and edges in the rules (in the bottom of Fig. 1) are numbered to define their conflicting overlapping situation. We can see one of the 12 conflicts in the overlapping graph P in the upper right part of Fig. 1, where vertex 1 and edges 5, 6 and 8 shall be deleted by Rule 3.2, but vertex 1 and edges 6 and 8 are also needed for the application of Rule 3.3 which is supposed to delete vertex 4 and its incident edges.

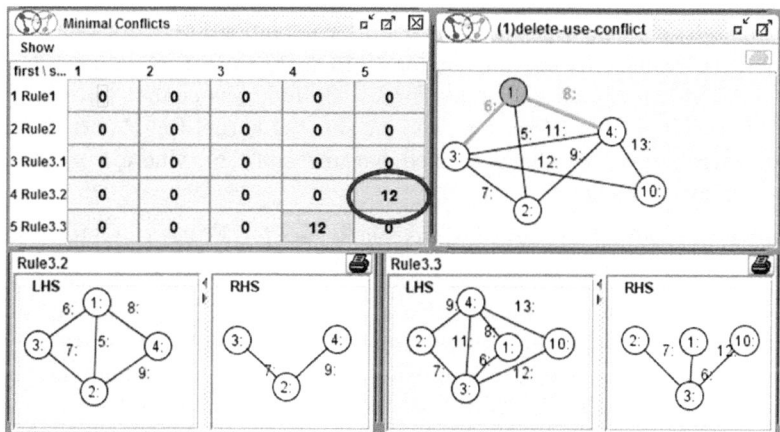

Fig. 1. CP table for CLIQUE COVER Rules 1 to 3.3, and one critical pair in detail

For each critical pair $P_1 \xLeftarrow{r_1} P \xRightarrow{r_2} P_2$ of the rule set in Fig. 1, we have shown strict confluence using AGG, essentially by applying the rules from the rule set as long as possible to P_1 and to P_2, leading to reduced graphs $\bar{P}_1$ and $\bar{P}_2$, and showing that they are isomorphic, as indicated in the diagram to the right.

The critical pairs can be computed automatically, and the reduction sequences $P_1 \overset{*}{\Longrightarrow} \bar{P}_1, P_2 \overset{*}{\Longrightarrow} \bar{P}_2$, and the isomorphism for $\bar{P}_1$ and $\bar{P}_2$ can be checked interactively using the tool AGG.

4 Case Study Partial Clique Cover

We provide a second, more demanding case study: PARTIAL CLIQUE COVER, a generalization of CLIQUE COVER where some edges C are annotated as already covered, and only uncovered edges need to be covered by cliques. Due to space constraints, we can only sketch our results.

We generalize Rules 1 and 2 in a canonical way.

Rule 4 ([10, Rule 1]). *Remove isolated vertices and vertices that are only incident to covered edges.*

Rule 5. *If there is an isolated edge, then delete it and, if the edge was not covered, decrease the parameter by one.*

We then adapt Rule 3 as follows:

Rule 6. *Let u, v be twins. Mark all edges incident to u as covered if the following covering conditions hold:*[4]

$$\forall x \in V \setminus \{u,v\} : \{u,x\} \in C \iff \{v,x\} \in C \tag{1}$$

$$\{u,v\} \notin C \Rightarrow \exists x \in V \setminus \{u,v\} : \{v,x\} \notin C. \tag{2}$$

Unfortunately, the new rules do not yield a problem kernel for PARTIAL CLIQUE COVER with respect to the parameter k. In fact, we can show that PARTIAL CLIQUE COVER is already NP-hard for $k = 3$, and thus cannot have a problem kernel unless P = NP. However, we can show a kernel for PARTIAL CLIQUE COVER with respect to the combined parameter (k, c), where $c = |C|$ is the number of covered edges.

Theorem 3. *Rules 4 to 6 yield a problem kernel for PARTIAL CLIQUE COVER with at most 2^{k+c} vertices.*

The idea of the proof is to show that if a PARTIAL CLIQUE COVER instance has more than 2^{k+c} vertices, then we can construct a CLIQUE COVER instance, find a data reduction opportunity there using Rules 1 to 3, and using this also find a data reduction for the PARTIAL CLIQUE COVER instance; in this way, we can use the bounds from Theorem 1.

Next, we claim that the new rules are confluent; the omitted proof uses local confluence and a somewhat involved case distinction.

Theorem 4. *Rules 4 to 6 for PARTIAL CLIQUE COVER are confluent.*

The challenge in proving Theorem 4 is that we cannot use the twin graph anymore as in Theorem 2, since it might be required for an optimal solution to cover twins with different cliques. In addition to a direct proof, for graphs of bounded size Theorem 4 can be shown using critical pair analysis by AGG [1].

---
[4] Note that if we drop either (1) or (2), then the rule is not correct.

5   Discussion and Future Work

Seemingly for the first time, our work establishes a fruitful link between graph transformation theory and the theory of kernelization from parameterized algorithmics. While considering (comparatively simple) kernelizations for edge clique covering, already several theoretical and technical challenges popped up when proving confluent kernelizations. We believe that to analyze whether a set of data reduction rules is confluent is a well-motivated and natural theoretical question of practical relevance with the potential for numerous opportunities for (interdisciplinary) future research between so far unrelated research communities.

As to research questions that are more rooted in graph transformation theory, it is first important to extend the theory of critical pair analysis to undirected graphs. We are confident that this works not only for the examples in this paper. Moreover, it is an important challenge to extend critical pair analysis from rules considering a constant-size subgraph to so-called rule schemes with unbounded number of vertices, that is, to transfer the "amalgamation" [3] of rewriting rules to this new context.

As to research on confluent kernelization rooted more in algorithmics, it appears to be of general interest to investigate how confluent problem kernels may help in deriving both upper and lower bounds for problem kernel sizes. In addition, it remains to study how confluence may contribute to speeding up kernelization algorithms and how the knowledge of having a uniquely determined problem kernel can help subsequent solution strategies that build on top of the kernel. Finally, studying confluence of data reduction and kernelization beyond graph problems, for example for string or set problems, remains a future task.

Acknowledgments. We thank Jiong Guo (Universität des Saarlandes) for pointing to an NP-hardness proof for PARTIAL CLIQUE COVER already for $k \geq 3$ covering cliques (see Section 4). The third author was supported by DFG project PABI (NI 369/7-2).

References

1. AGG: Attributed Graph Grammar Tool. TU Berlin (2011),
   http://tfs.cs.tu-berlin.de/agg
2. Bodlaender, H.L.: Kernelization: New Upper and Lower Bound Techniques. In: Chen, J., Fomin, F.V. (eds.) IWPEC 2009. LNCS, vol. 5917, pp. 17–37. Springer, Heidelberg (2009)
3. Böhm, P., Fonio, H.R., Habel, A.: Amalgamation of graph transformations: a synchronization mechanism. Journal of Computer and System Sciences 34, 377–408 (1987)
4. Cygan, M., Kratsch, S., Pilipczuk, M., Pilipczuk, M., Wahlström, M.: Clique cover and graph separation: New incompressibility results. Tech. Rep. arXiv:1111.0570v1 [cs.DS], arXiv (2011)
5. Downey, R.G., Fellows, M.R.: Parameterized Complexity. Springer (1999)

6. Ehrig, H., Ehrig, K., Prange, U., Taentzer, G.: Fundamentals of Algebraic Graph Transformation. EATCS Monographs in Theoretical Computer Science. Springer (2006)
7. Ehrig, H., Kreowski, H.J., Montanari, U., Rozenberg, G. (eds.): Handbook of Graph Grammars and Computing by Graph Transformation. Concurrency, Parallelism and Distribution, vol. 3. World Scientific (1999)
8. Ehrig, H., Pfender, M., Schneider, H.: Graph grammars: an algebraic approach. In: Proc. IEEE Symposium on Switching and Automata Theory, pp. 167–180. IEEE (1973)
9. Flum, J., Grohe, M.: Parameterized Complexity Theory. Springer (2006)
10. Gramm, J., Guo, J., Hüffner, F., Niedermeier, R.: Data reduction and exact algorithms for clique cover. ACM Journal of Experimental Algorithmics 13, 2.2:1–2.2:15 (2008)
11. Guo, J., Niedermeier, R.: Invitation to data reduction and problem kernelization. ACM SIGACT News 38(1), 31–45 (2007)
12. Gyárfás, A.: A simple lower bound on edge coverings by cliques. Discrete Mathematics 85(1), 103–104 (1990)
13. Huet, G.: Confluent reductions: Abstract properties and applications to term rewriting systems. Journal of the ACM 27(4), 797–821 (1980)
14. Kneis, J., Mölle, D., Richter, S., Rossmanith, P.: A bound on the pathwidth of sparse graphs with applications to exact algorithms. SIAM Journal on Discrete Mathematics 23(1), 407–427 (2009)
15. van Leeuwen, J. (ed.): Handbook of Theoretical Computer Science. MIT Press (1990)
16. McConnell, R.M.: Linear-time recognition of circular-arc graphs. Algorithmica 37(2), 93–147 (2003)
17. Newman, M.H.A.: On theories with a combinatorial definition of equivalence. Annals of Mathematics 43(2), 223–242 (1942)
18. Niedermeier, R.: Invitation to Fixed-Parameter Algorithms. Oxford Lecture Series in Mathematics and Its Applications, vol. 31. Oxford University Press (2006)
19. Plump, D.: Confluence of Graph Transformation Revisited. In: Middeldorp, A., van Oostrom, V., van Raamsdonk, F., de Vrijer, R. (eds.) Processes, Terms and Cycles: Steps on the Road to Infinity. LNCS, vol. 3838, pp. 280–308. Springer, Heidelberg (2005)
20. Rozenberg, G.: Handbook of Graph Grammars and Computing by Graph Transformations: Foundations, vol. 1. World Scientific (1997)

Highness and Local Noncappability

Chengling Fang[1], Wang Shenling[1,2] and Guohua Wu[1]

[1] Division of Mathematical Sciences, School of Physical and Mathematical Sciences
Nanyang Technological University, 21 Nanyang Link, Singapore 637371
`fang0032@e.ntu.edu.sg, guohua@ntu.edu.sg`
[2] College of Information Science and Technology, Beijing Normal University
Haidian District, Beijing 100875, P.R. China
`wang0362@e.ntu.edu.sg`

Abstract. Let $\mathbf{a}$ be a nonzero incomplete c.e. degree. Say that $\mathbf{a}$ is locally noncappable if there is a c.e. degree $\mathbf{c}$ above $\mathbf{a}$ such that no nonzero c.e. degree below $\mathbf{c}$ can form a minimal pair with $\mathbf{a}$, and $\mathbf{c}$ is a witness of such a property of $\mathbf{a}$. Seetapun proved that every nonzero incomplete c.e. degree is locally noncappable, and Stephan and Wu proved recently that such witnesses can always be chosen as high$_2$ degrees. This latter result is optimal as certain classes of c.e. degrees, such as nonbounding degrees, plus-cupping degrees, etc., cannot have high witnesses. Here, a c.e. degree is nonbounding if it bounds no minimal pairs, and is plus-cupping if every nonzero c.e. degree below it is cuppable.

In this paper, we prove that for any nonzero incomplete c.e. degree $\mathbf{a}$, there exist two incomparable c.e. degrees $\mathbf{c}, \mathbf{d} > \mathbf{a}$ witnessing that $\mathbf{a}$ is locally noncappable, and that $\mathbf{c} \vee \mathbf{d}$, the joint of $\mathbf{c}$ and $\mathbf{d}$, is high. This result implies that both classes of the plus-cuppping degrees and the nonbounding c.e. degrees do not form an ideal, which was proved by Li and Zhao by two separate constructions.

1 Introduction

Lachlan [4] and Yates [9] proved independently the existence of minimal pairs, refuting a conjecture of Shoenfield. A c.e. degree is *cappable* if it is either $\mathbf{0}$ or a part of a minimal pair, and a c.e. degree is noncappable, if it is not cappable. The existence of noncappable degrees was proved by Yates in [9]. In [1], Ambos-Spies, Jockusch, Shore and Soare proved that all the cappable degrees form an ideal of the class of c.e. degrees, while all the noncappable degrees form a strong filter of the class of c.e. degrees. They also proved that a c.e. degree is noncappable if and only if it is low-cuppable.

As an variant of noncappable degrees, Seetapun proposed in his thesis [7] the notion of *locally noncappable degrees*. Here a c.e. degree $\mathbf{a}$ is *locally noncappable* if there is a c.e. degree $\mathbf{c}$ above $\mathbf{a}$ such that no nonzero c.e. degree $\mathbf{w}$ below $\mathbf{c}$ forms a minimal pair with $\mathbf{a}$. We say that $\mathbf{c}$ witnesses that $\mathbf{a}$ is locally noncappable. Seetapun proved in his thesis [7] that every nonzero incomplete c.e. degree is locally noncappable. This result was later published in [3] by Giorgi, but with one Σ_3 outcome missing, so Giorgi's construction is not complete. Recently,

S.B. Cooper, A. Dawar, and B. Löwe (Eds.): CiE 2012, LNCS 7318, pp. 203–211, 2012.

Stephan and Wu [2] proved Seetapun's result by showing that such witnesses can always be chosen as high$_2$ degrees. This is optimal as certain classes of c.e. degrees, such as nonbounding degrees, plus-cupping degrees, etc., cannot have high witnesses. In this paper, we prove that any nonzero incomplete c.e. degree can have two incomparable c.e. degrees above it witnessing that it is locally noncappable and the joint of these two degrees is high.

Theorem 1. *For any nonzero incomplete c.e. degree* $\mathbf{a}$, *there exist two incomparable c.e. degrees* $\mathbf{c}$, $\mathbf{d} > \mathbf{a}$ *witnessing that* $\mathbf{a}$ *is locally noncappable, and* $\mathbf{c} \vee \mathbf{d}$, *the joint of* $\mathbf{c}$ *and* $\mathbf{d}$, *is high.*

Obviously, Theorem 1 implies both classes of the plus-cuppping degrees and the nonbounding c.e. degrees do not form ideals, as $\mathbf{c} \vee \mathbf{d}$ is high, and hence bounds a minimal pair, and also noncuppable degrees. These two consequences were proved by Li and Zhao in [6], by two separate constructions.

Theorem 1 also implies the the class of those degrees bounding no bases of Slaman triples do not form an ideal, even though these degrees are downwards closed. In [5], Leonhardi proved the existence of a high$_2$ c.e. degree bounding no bases of Slaman triples. here a nonzero c.e. degree $\mathbf{a}$ is called a base of a Slaman triple if there are c.e. degrees $\mathbf{b}$ and $\mathbf{c}$ with $\mathbf{c} \not\leq \mathbf{b}$ such that for any $\mathbf{w} \leq \mathbf{a}$, if $\mathbf{w} \neq \mathbf{0}$, then $\mathbf{c} \leq \mathbf{b} \vee \mathbf{w}$. Shore and Slaman proved in [8] that each high c.e. degree bounds a Slaman triple, and hence a base of such a triple.

Our notation and terminology are quite standard and generally follow Soare [10]. A parameter p is defined to be fresh at a stage s if $p > s$ and p is the least number not mentioned so far in the construction.

2   Requirements

Let $\mathbf{a}$ be a nonzero and incomplete c.e. degree and A is a c.e. set in $\mathbf{a}$. To prove Theorem 1, we shall construct two c.e. sets C, D and a p.c. functional Λ satisfying the following requirements:

$\mathcal{H}_e$: $\mathrm{Tot}(e) = \lim_x \Lambda^{A \oplus C \oplus D}(e, x)$,

$\mathcal{Q}_e^C$: $C \neq \Phi_e^{A \oplus D}$,

$\mathcal{Q}_e^D$: $D \neq \Phi_e^{A \oplus C}$,

$\mathcal{R}_e^C$: $W_e = \Phi_e^{A \oplus C} \Rightarrow W_e \leq_T A$ or $\exists X_e \leq_T A, W_e(X_e$ c.e. and incomputable),

$\mathcal{R}_e^D$: $V_e = \Psi_e^{A \oplus D} \Rightarrow V_e \leq_T A$ or $\exists Y_e \leq_T A, V_e(Y_e$ c.e. and incomputable).

Here $\{(W_e, V_e, \Phi_e, \Psi_e) : e \in \omega\}$ is an effective enumeration of all quadruples (W, V, Φ, Ψ) of c.e. sets W, V and p.c. functionals Φ, Ψ. The set $\mathrm{Tot} = \{i : \varphi_i$ is total$\}$ is a Π_2^0-complete set. Let $\mathbf{c}$ be the degree of $A \oplus C$, and $\mathbf{d}$ be the degree of $A \oplus D$. By the $\mathcal{Q}$-requirements, $\mathbf{c}$ and $\mathbf{d}$ are two incomparable degrees above $\mathbf{a}$. The $\mathcal{H}$-requirements ensure that $\mathbf{c} \vee \mathbf{d}$ is high. By the $\mathcal{R}$-requirements, both $\mathbf{c}$ and $\mathbf{d}$ witness that $\mathbf{a}$ is locally noncappable.

2.1   A $\mathcal{Q}_e^C$ Strategy

A $\mathcal{Q}_e^C$-strategy α is a standard Sacks coding strategy (here, a $\mathcal{Q}_e^D$-strategy is similar). That is, even though A is not in our control, we can still satisfy the $\mathcal{Q}_e^C$-requirement by the assumption that A is incomplete. α will run cycles j for $j \in \omega$, and all cycles of α define a functional Ξ_α jointly. Each cycle j tries to find a number x_j such that $C(x_j) \neq \Phi_e^{A \oplus D}(x_j)$, and if cycle j fails to make it, then this cycle will define $\Xi_\alpha^A(j) = K(j)$ successfully. As A is incomplete, some cycle, j say, will find an x_j such that $C(x_j) \neq \Phi_e^{A \oplus D}(x_j)$. Cycle j proceeds as follows:

(1) Choose x_j as a fresh number.
(2) Wait for $\Phi_e^{A \oplus D}(x_j) \downarrow = 0$.
(3) Preserve $D \lceil \varphi_e(x_j)$ from other strategies, and define $\Xi_\alpha^A(j) = K(j)$ with use $\xi_\alpha(j) = \varphi_e(x_j)$. Start cycle $j + 1$ simultaneously.
(4) Wait for $A \lceil \varphi_e(x_j)$ or $K(j)$ to change.
  (a) If $A \lceil \varphi_e(x_j)$ changes first, then cancel all cycles $j' > j$ and drop the D-restraint of cycle j to 0. Go back to step 2.
  (b) If $K(j)$ changes first, then stop cycles $j' > j$, and go to step 5.
(5) Put x_j into C. Wait for $A \lceil \varphi_e(x_j)$ to change.
(6) Define $\Xi_\alpha^A(j) = K(j) = 1$ with use 0, and start cycle $j + 1$.

Cycle j has two outcomes:

(j, f) : Cycle j waits forever at step 2 or 5.
  (*The $\mathcal{Q}_e^C$-strategy is satisfied via witness x_j in an obvious way.*)
(j, ∞) : Cycle j runs infinitely often.
  (*Cycle j returns from step 4 to 2 infinitely often, and hence $\Phi_e^{A \oplus D}(x_j)$ diverges, the $\mathcal{Q}_e^C$-strategy is satisfied.*)

Note that as A is assumed to be incomplete, it is impossible for α to run infinitely many cycles, with each cycle runs only finitely often, since otherwise, Ξ_α^A is defined as a total function and $\Xi_\alpha^A = K$, a contradiction.

α has outcomes (j, ∞) and (j, f) with $(j, \infty) < (j, f)$ and for any $j_1 < j_2$, $(j_1, *) < (j_2, \dagger)$, where $*, \dagger \in \{\infty, f\}$.

We introduce some notions for convenience when we talk about such cycles. Say that *cycle j acts* if it chooses a fresh number x_j as its attacker at step 1, or it enumerates x_j into C at step 5. Say that *cycle j is active* at stage s if at this stage, when α is visited, α is running cycle j, except the situation that cycle j is just started at stage s.

2.2   An $\mathcal{R}_e^C$ Strategy

Assume that a strategy β works on $\mathcal{R}_e^C$, we define the length of agreement function $l(\beta, s)$ at stage s to be

$$l(\beta, s) = \max\{x < s : (\forall y < x)[W_e(y)[s] = \Phi_e^{A \oplus C}(y)[s]]\},$$

and the maximum length of agreement function at stage s is defined to be

$$m(\beta, s) = \max\{l(\beta, t) : t < s \text{ and } t \text{ is a } \beta\text{-stage}\}.$$

Say that a stage s is a β-expansionary stage if $s = 0$ or s is a β-stage with $l(\beta, s) > m(\beta, s)$.

At β-expansionary stages, we shall construct a c.e. set X_e and two p.c. functionals Γ_β and Δ_β such that if $W_e = \Phi_e^{A \oplus C}$, then either $W_e \leq_T A$ or X_e is incomputable with $X_e = \Gamma_\beta^A = \Delta_\beta^{W_e}$.

β has two outcomes $\infty <_L f$. ∞ denotes that β has infinitely many expansionary stages, and f for finitely many. Below outcome ∞, to ensure that X_e is incomputable, we need to satisfy the following subrequirements

$$\mathcal{S}_{e,i}^C : \qquad X_e \neq \varphi_i$$

for all i.

In the construction, we always define the use $\delta_\beta(x) = x$, which will ensure that $\Delta_\beta^{W_e}$ is totally defined. When $\Gamma_\beta^A(x)$ is defined at a β-expansionary stage s, we shall define $\gamma_\beta(x)[s] > \varphi_e(x)[s]$, and as a consequence, when $\varphi_e(x)$ diverges, $\gamma_\beta(x)$ also diverges.

2.3   An $\mathcal{S}_{e,i}^C$ Strategy

Assume that η is an $\mathcal{S}_{e,i}^C$-strategy below the ∞ outcome of an $\mathcal{R}_e^C$-strategy β, and η has two parameters x_η and z_η. The basic η-strategy is almost the same as the one given in [2]:

(1) Pick x_η as a fresh number.
(2) Wait for $\varphi_i(x_\eta) \downarrow = 0$.
(3) Assign $z_\eta = x_\eta$ and let x_η be undefined. Go back to step 1, and wait for A to change below z_η. (If such a change occurs, then $\Gamma_\beta^A(z_\eta)$ is undefined.)
(4) (Open a gap)
    Create a link between β and η, wait for W_e to change below z_η.
(5a) (Close a gap successfully)
    If W_e has a change below z_η, then η performs the diagonalization by putting z_η into X_e. Cancel the link to close the gap. In this case, η is satisfied forever.
(5b) (Close a gap unsuccessfully)
    If W_e does not have a change below z_η, then cancel the link to close the gap. Impose a restraint on C. Request that $\gamma_\beta(z_\eta)$ be defined greater than $\varphi_e(z_\eta)$.

We shall define a p.c. functional Θ_η such that if η opens infinitely many different gaps (each gap associated with different value of z_η), then Θ_η^A is totally defined and computes W_e correctly, and hence η provides a global win for β.

In the construction, we define Θ_η in step (5b), i.e., when a gap for z_η is closed unsuccessfully, for all $x \leq z_\eta$, if $\Theta_\eta^A(x) \uparrow$, then we define $\Theta_\eta^A(x) = W_e(x)$ with $\theta_\eta(x) = \varphi_e(x)$.

The period between a stage at which a gap is closed unsuccessfully and the stage at which the next gap is open is called a cogap. Note that, when a gap is closed unsuccessfully, a restraint is imposed on C to prevent W_e from changing below z_η. But, as A is given, A can change below $\varphi_e(z_\eta)$ during the cogaps, which may change W_e below z_η. In other words, during a cogap for z_η, if W_e changes on some number $x \leq z_\eta$, then $Delta_\beta^{W_e}(z_\eta)$ and $\Gamma_\beta^A(z_\eta)$ are both undefined and η is allowed to enumerate z_η into X_e to satisfy $\mathcal{S}_{e,i}^C$. In case that W_e does not have such a change, then at the next β-expansionary stage, we shall define $\Gamma_\beta^A(z_\eta)$ again, with a big γ-use. Here comes a special feature of this strategy: a gap can be closed and reopen, by changes of A, infinitely many times, and this feature corresponds to a divergence outcome, as $\Phi_e^{A \oplus C}$ diverges on some argument below z_η.

An η-strategy has four outcomes $s <_L g <_L d <_L w$, the outcome s denotes the case that a gap is closed successfully; the outcome g denotes the case that η opens infinitely many different gaps (each gap associated with different value of z_η, and is closed eventually), we can show that if $W_e = \Phi_e^{A \oplus C}$ then $W_e = \Theta_\eta^A$, a global win for β; the outcome d denotes the case that η opens only finitely many different gaps and for the last gap, A changes below the corresponding use $\varphi_e(z_\eta)$ infinitely often, and hence $\Phi_e^{A \oplus C}(z_\eta) \uparrow$, another global win for β; the outcome w denotes the case that η waits for $\varphi_i(x_\eta) \downarrow= 0$ forever for some x_η. We shall reconsider $\mathcal{S}_{e,i}^C$-strategies after the $\mathcal{H}$-strategies are introduced.

An $\mathcal{R}_e^D$-strategy is similar to an $\mathcal{R}_e^C$-strategy and hence we have the following subrequirements

$$\mathcal{S}_{e,i}^D : Y_e \neq \varphi_i$$

for all i.

An $\mathcal{S}_{e,i}^D$-strategy is similar to an $\mathcal{S}_{e,i}^C$-strategy.

2.4 An $\mathcal{H}_e$ Strategy

Assume τ works on $\mathcal{H}_e$. We define the length of convergence function $l(\tau, s)$ at stage s to be

$$l(\tau, s) = \max\{x < s : (\forall y < x)[\varphi_e(y)[s] \downarrow]\},$$

and the maximum length of convergence function at stage s is defined to be

$$m(\tau, s) = \max\{l(\tau, t) : t < s \text{ and } t \text{ is a } \tau\text{-stage}\}.$$

Say that a stage s is a τ-expansionary stage if $s = 0$ or s is a τ-stage with $l(\tau, s) > m(\tau, s)$. τ has two outcomes $\infty <_L f$. If τ has ∞ outcome, i.e., φ_e is total, then we shall define $\Lambda^{A \oplus C \oplus D}(e, x) = 1$ for almost all x; if τ has f outcome, i.e., φ_e is not total, then we shall define $\Lambda^{A \oplus C \oplus D}(e, x) = 0$ for almost all x.

Note that Λ is a global p.c. functional built by us through the whole construction. $\Lambda^{A\oplus C\oplus D}(e,x)$ is undefined automatically if some number $\leq \lambda(e,x)$ is enumerated into C or D. As a consequence, $\lambda(e,x)$ may be lifted when $\Lambda^{A\oplus C\oplus D}(e,x)$ is redefined later.

If we defined $\Lambda^{A\oplus C\oplus D}(e,x) = 0$ under the f outcome at a previous stage, and now we see φ_e converges on more arguments, i.e., τ changes its outcome from f to ∞, then we want to (re)define $\Lambda^{A\oplus C\oplus D}(e,x) = 1$, but first we need to undefine all the previous $\Lambda^{A\oplus C\oplus D}(e,x) = 0$. So generally, at a τ-expansionary stage, we shall put the $\lambda(e,x)$ use into C to undefine $\Lambda^{A\oplus C\oplus D}(e,x) = 0$, and we (re)define $\Lambda^{A\oplus C\oplus D}(e,x) = 1$ with use $\lambda(e,x)$ as -1 at τ-expansionary stages since we never want to undefine it later. This means that we only care about the $\lambda(e,x)$ use defined under the f outcome.

Actually, in the construction, we need to consider the restraint imposed on τ, so we shall define *boundary* $bd(\tau)$ of τ (playing a role of the restraint) as follows: when τ is visited for the first time, we define $bd(\tau)$ to be a fresh number, whenever τ is initialized, we shall redefine $bd(\tau)$ as a fresh number. At a τ-expansionary stage s, we shall put the $\lambda(e,x)$ use with $bd(\tau) < \lambda(e,x) \leq s$ into C to undefine $\Lambda^{A\oplus C\oplus D}(e,x) = 0$.

One may ask whether it is true that the $\mathcal{H}_e$-strategy τ only enumerates the λ-uses only into C when it needs to undefine $\Lambda^{A\oplus C\oplus D}(e,x) = 0$. The answer is "no", as, for example, when A is of nonbounding degree or of plus-cupping degree, then $A\oplus C$ cannot have high degree. Here we fix C just for simplicity, and in the actual construction, we also need to put the $\lambda(e,x)$ use into D sometimes, based on which gap is open. We shall explain below the idea of putting these λ-uses into C and D alternatively, to make $A \oplus C \oplus D$ of high degree.

2.5   Interaction between Strategies

Now we consider the interaction between the highness strategies and the gap-cogap argument used in $\mathcal{S}$-strategies.

If β is an $\mathcal{R}_e^C$-strategy working below the ∞ outcome of an $\mathcal{H}$-strategy τ, i.e., $\tau^\frown\langle\infty\rangle \subset \beta$, then, in the definition of β-expansionary stages, we only use β-believable computations, where a computation $\Phi_e^{A\oplus C}(x)[s]$ is β-believable, if for each n with $bd(\tau) < n \leq s$, the $\lambda(e',n)[s]$ is already defined as -1, where we assume that τ is an $\mathcal{H}_{e'}$-strategy.

A nontrivial case is when an $\mathcal{H}_{e_0}$-strategy τ works between β and a substrategy η with

$$\beta^\frown\langle\infty\rangle \subset \tau \subseteq \tau^\frown\langle\infty\rangle \subset \eta.$$

Suppose that η opens a gap and a link between β and η is created at some β-expansionary stage, s_0 say. At the next β-expansionary stage s_1 (β-believable computations are considered), and suppose this gap is closed unsuccessfully, then the link is cancelled, and during the cogap for η after stage s_1, no number less than s_1 can be put into C (to prevent W_e from changing). It is this restraint that forces us to enumerate λ-uses into D. That is, when β closes the gap for η at stage s_1, for all the $\mathcal{H}$-strategies τ with $\beta^\frown\langle\infty\rangle \subset \tau \subseteq \tau^\frown\langle\infty\rangle \subset \eta$, we shall put $\lambda(e(\tau),x)$ into D

to lift the λ-uses if needed. Thus, after stage s_1, all the numbers enumerated into C in the future by such $\mathcal{H}$-strategies are greater than s_1.

We now consider details of such interactions. Recall that we a gap is closed, we do extend the definitions of Γ_β^A and $\Delta_\beta^{W_e}$, and if the gap is closed unsuccessfully, W_e does not change below z_η, but A changes below $\varphi_e(z_\eta)$. Without loss of generality, we only consider the case when a gap is closed (unsuccessfully) and reopen infinitely often. Than is, during each cogap, A changes on small numbers and reopens the same gap. In this case, in [2], when a gap is reopen, η is visited immediately with outcome d. It is this feature that is consistent with construction of a high$_2$ degree. However, for our construction, we need to make some modifications, as we want to make $A \oplus C \oplus D$ high.

Here is the idea: After a gap for z_η is closed unsuccessfully by β, at the next β-expansionary stage, if A changes below $\varphi_e(z_\eta)$, but W_e does not change below z_η, we shall not let β reopen this gap immediately. Instead, if η is visited at this stage, then we let η reopen this gap. That is, in the construction, at an η-stage, if there is a gap which was closed unsuccessfully by β at a previous stage, and if A changes below the corresponding use $\varphi_e(z_\eta)$ at this η-stage, then we shall let η reopen this gap. Note that, in the construction, when a gap is reopened, it must be reopened by η, not by β. This will avoid the problem above, since if a gap can be closed unsuccessfully and reopened infinitely often, then the $\mathcal{H}$-strategies between β and η must have outcome the same as the one seen at η.

Note that, in a cogap for η, it may happen that we see W_e change below z_η at a β-expansionary stage, but $\Gamma_\beta^A(z_\eta) \downarrow$ at this stage. We now describe whey this can happen in detail. Suppose that a gap for z_η is closed unsuccessfully by β at a β-expansionary stage s_1, and we define $\Delta_\beta^{W_e}(z_\eta)$ and $\Gamma_\beta^A(z_\eta)$ at the end of stage s_1, with $\gamma_\beta(z_\eta)[s_1] > \varphi_e(z_\eta)[s_1]$. At the next β-expansionary stage, $\Gamma_\beta^A(z_\eta) \uparrow$, but $\Delta_\beta^{W_e}(z_\eta) \downarrow$. Assume that η is not visited at stage s_2 (and hence η cannot reopen this gap at stage s_2), then we define $\Gamma_\beta^A(z_\eta)$ at the end of stage s_2, with $\gamma_\beta(z_\eta)[s_2] > \varphi_e(z_\eta)[s_2]$. Assume that $\varphi_e(z_\eta)[s_2] > \varphi_e(z_\eta)[s_1]$, and hence the $\mathcal{H}$-strategies between β and η may enumerate some small λ-uses into C, changing the computation $\Phi_e^{A \oplus C}(z_\eta)[s_2]$. Thus, at the next β-expansionary stage s_3, it may happen that $\Delta_\beta^{W_e}(z_\eta) \uparrow$, but $\Gamma_\beta^A(z_\eta) \downarrow$, i.e., W_e changes below z_η, but A does not change below $\varphi_e(z_\eta)$. In this case, we can not let η perform the diagonalization at stage s_3 even though W_e changes below z_η.

On the other hand, at an η-stage, if a gap (which was closed unsuccessfully by β at a previous stage) can not be reopened, i.e., A does not change below $\varphi_e(z_\eta)$, then we shall consider a new value of z_η (if any) for η and check whether we can open a new gap. If we can open a new gap, then we say that the existing gap is closed completely.

The same argument applies when β is an $\mathcal{Q}^D$-strategy.

Finally, we consider the interaction between an $\mathcal{R}^C$-strategy and an $\mathcal{R}^D$-strategy. Suppose that an $\mathcal{R}_e^D$-strategy β' works between an $\mathcal{R}_{e'}^C$-strategy β and an $\mathcal{S}_{e',i'}^C$-strategy η with

$$\beta^\frown \langle \infty \rangle \subset \beta' \subset \beta'^\frown \langle \infty \rangle \subset \eta \subset \eta^\frown \langle \mathcal{O} \rangle \subset \tau \subset \tau^\frown \langle \infty \rangle \subset \eta',$$

where $\mathcal{O} \in \{g, d\}$, τ is an $\mathcal{H}$-strategy and η' is an $\mathcal{S}_{e,i}^{D}$-strategy and $0 \leq e' \leq e$. Then we may have that a gap is open (or reopened) for η and creates a link between β and η and η' reopens a gap and creates a link between β' and η' at the same stage, s_0 say. That is, we have two crossed links (β, η) and (β', η') at stage s_0. At the next β-expansionary stage s_1, suppose that β closes the gap for η unsuccessfully and cancels the link (β, η), it requires that all the $\mathcal{H}$-strategies with ∞ outcome between β and η enumerate the λ-uses into set D to lift λ-uses if needed to protect C. And, η imposes a C-restraint after stage s_1 till the stage, s_2 say, at which the next gap for η is open. But, during this cogap for η, we shall travel the link (β', η') created at stage s_0. Suppose that we travel the link (β', η') at stage t ($s_1 < t < s_2$) say, and β' closes the gap for η' unsuccessfully, thus it requires that all the $\mathcal{H}$-strategies with ∞ outcome between β' and η' enumerate the λ-uses into set C to lift λ-uses if needed to protect D. In particular, suppose that the $\mathcal{H}$-strategy τ is required to enumerate the λ-uses into set C to lift λ-uses at stage t, these λ-uses may be less than s_1. But, η does not allow small numbers ($\leq s_1$) to be enumerated into C before stage s_2. So η (i.e., β) injures the satisfaction of β'. Thus, such crossed links (β, η) and (β', η') should be avoided.

In the construction, we shall use a backup strategy to deal with this. That is, we shall put a backup strategy $\hat{\beta}$ below $\eta^{\frown}\langle\mathcal{O}\rangle$ to try to satisfy the $\mathcal{R}_e^D$-requirement. Therefore, in the construction, there will never be two crossed links (β, η) and (β', η') as above at the same stage.

Actually, in a gap-cogap argument, the crossed links need to be avoided generally. So, in our construction, suppose that we have $\beta^{\frown}\langle\infty\rangle \subset \beta' \subset \beta'^{\frown}\langle\infty\rangle \subset \eta$, where β can be an $\mathcal{R}_e^C$-strategy or $\mathcal{R}_e^D$-strategy, β' can be an $\mathcal{R}_{e'}^C$-strategy or $\mathcal{R}_{e'}^D$-strategy ($e \leq e'$), and η is a substrategy of β. Then, when η has outcome g or d, we shall say that β' becomes inactive or injured, and we need to arrange a back-up strategy for β' under this outcome. Moreover, it's not hard to see that, for a fixed e, there are at most finitely many backup $\mathcal{R}_e^C$-strategies ($\mathcal{R}_e^D$-strategy is similar) on any path of the construction tree and the longest $\mathcal{R}_e^C$-strategy is responsible to satisfy the $\mathcal{R}_e^C$ requirement.

Note that, we may have two nested links (β, η) and (β', η') at the same stage in the construction, where β is an $\mathcal{R}_e^C$-strategy and β' is an $\mathcal{R}_e^D$-strategy, and η is an $\mathcal{S}_{e,i}^C$-strategy and η' is an $\mathcal{S}_{e',i'}^D$-strategy with $\beta^{\frown}\langle\infty\rangle \subset \beta' \subset \beta'^{\frown}\langle\infty\rangle \subset \eta' \subset \eta'^{\frown}\langle\mathcal{O}\rangle \subset \eta$, where $\mathcal{O} \in \{g, d\}$. That is, we may have that a gap is open (or reopened) for η' and creates a link between β' and η' and η reopens a gap and creates a link between β and η at the same stage, s_0 say. At the next β-expansionary stage s_1, suppose β closes the gap for η unsuccessfully and cancels the link (β, η), so it requires that the $\mathcal{H}$-strategies with ∞ outcome between β and η enumerate the λ-uses into set D to lift λ-uses if needed. (The enumeration of numbers into D at stage s_1 may have effect on the computation involved in the gap for η', but this is allowed since the gap for η' is open at stage s_1.) η will impose a C-restraint below s_1 after stage s_1 till the stage, s_2 say, at which the next gap for η is open. But, before stage s_2, i.e., during this cogap for η, we shall travel the link (β', η') created at stage s_0 and β' may close the gap for η' unsuccessfully and so it requires that the $\mathcal{H}$-strategies with ∞ outcome between

β' and η' enumerate the λ-uses into set C to lift λ-uses if needed. Note that such λ-uses are already lifted at stage s_1 and hence must be greater than s_1, so we can enumerate them into C and there is no conflict.

This completes the description of our strategies for Theorem 1, and the possible interactions between them. A full construction can proceed on a priority tree, and Theorem 1 is proved.

Acknowledgements. G. Wu was partially supported by NTU grant RG37/09, M52110101 and grant MOE2011-T2-1-071 from MOE of Singapore.

References

1. Ambos-Spies, K., Jockusch Jr., C.G., Shore, R.A., Soare, R.I.: An algebraic decomposition of the recursively enumerable degrees and the coincidence of several degree classes with the promptly simple degrees. Trans. Amer. Math. Soc. 281, 109–128 (1984)
2. Stephan, F., Wu, G.: Highness, nonboundings and presentations (to appear)
3. Giorgi, M.B.: The computably enumerable degees are locally non-cappable. Arch. Math. Logic 43, 121–139 (2004)
4. Lachlan, A.H.: Lower bounds for pairs of recursively enumerable degrees. Proc. London Math. Soc. 16, 537–569 (1966)
5. Leonhardi, S.D.: Nonbounding and Slaman triples. Ann. Pure Appl. Logic 79, 139–163 (1996)
6. Li, A., Zhao, Y.: Plus cupping degrees do not form an ideal. Science in China, Ser. F. Information Sciences 47, 635–654 (2004)
7. Seetapun, D.: Contributions to Recursion Theory, Ph.D. thesis, Trinity College (1991)
8. Shore, R.A., Slaman, T.A.: Working below a high recursively enumerable degree. J. Symb. Logic 58, 824–859 (1993)
9. Yates, C.E.M.: A minimal pair of recursively enumerable degrees. J. Symbolic Logic 31, 159–168 (1966)
10. Soare, R.I.: Recursively Enumerable Sets and Degrees. Perspectives in Mathematical Logic. Springer (1987)

Turing Progressions and Their Well-Orders

David Fernández Duque[1] and Joost J. Joosten[2]

[1] Departamento de Filosofía y Lógica y Filosofía de la Ciencia, Universidad de
Sevilla, Calle Camilo José Cela, sin número, Sevilla 41018, Spain
`davidstofeles@gmail.com`
[2] Departamento de Lógica, História i Filosofia de la Ciéncia, Universitat de
Barcelona, Montalegre, 6, 08001 Barcelona, Catalonia, Spain
`jjoosten@ub.edu`

Abstract. We see how Turing progressions are closely related to the
closed fragment of GLP, polymodal provability logic. Next we study natural
well-orders in GLP that characterize certain aspects of these Turing
progressions.

1 Turing Progressions and Modal Logic

Gödel's Second Incompleteness Theorem tells us that any sound recursive theory that is strong enough to code syntax will not prove its own consistency. Thus, adding $\mathsf{Con}(T)$ to such a theory T will yield a strictly stronger theory. Turing took up this idea in his seminal paper [6] to consider recursive ordinal progressions of some recursive sound base theory T:

$$
\begin{aligned}
T_0 &:= T; \\
T_{\alpha+1} &:= T_\alpha + \mathsf{Con}(T_\alpha); \\
T_\lambda &:= \bigcup_{\alpha<\lambda} T_\alpha \qquad \text{for limit } \lambda.
\end{aligned}
$$

As we shall see, poly-modal provability logics turn out to be suitably well equipped to talk about Turing progressions. These logics have modalities $[n]$ that shall be interpreted as "provable in EA using all true Π_n sentences" abbreviated $[n]_{\mathrm{EA}}$. By EA we denote *Elementary Arithmetic* which is a formal arithmetic theory axiomatized by the recursive equations for successor, addition and multiplication, by open induction together with an axiom stating the totality of exponentiation. Often we shall not distinguish a modal formula from its arithmetical interpretation.

We recall ([5]) that the provability logic of any Σ_1-sound theory extending EA is Gödel Löb's provability logic **GL** as defined below. Various mathematical statements can be expressed within **GL** like Gödel's Second Incompleteness Theorem: $\Diamond\top \to \Diamond\Box\top$. It is also not hard to see that finite Turing progressions are definable in **GL** as T_n is provably equivalent to $T + \Diamond^n_T\top$. Transfinite progressions are not expressible in the modal language with just one modal operator. However, using stronger provability predicates provides a way out (see [3]):

Proposition 1. $T + \langle n+1 \rangle_T \top$ *is a* Π_{n+1} *conservative extension of*
$T + \{\langle n \rangle^k_T \top \mid k \in \omega\}$.

S.B. Cooper, A. Dawar, and B. Löwe (Eds.): CiE 2012, LNCS 7318, pp. 212–221, 2012.
© Springer-Verlag Berlin Heidelberg 2012

The provability behavior of these mixed modalities is fully described by what we call GLP_ω. We give a more general definition GLP_Λ for any ordinal Λ.

Definition 1. *For Λ an ordinal, the logic GLP_Λ is the propositional normal modal logic that has for each $\alpha < \Lambda$ a modality $[\alpha]$ and is axiomatized by the following schemata:*

$$[\alpha](\chi \to \psi) \to ([\alpha]\chi \to [\alpha]\psi)$$
$$[\alpha]([\alpha]\chi \to \chi) \to [\alpha]\chi$$
$$\langle\alpha\rangle\psi \to [\beta]\langle\alpha\rangle\psi \qquad \textit{for } \alpha < \beta,$$
$$[\alpha]\psi \to [\beta]\psi \qquad \textit{for } \alpha \leq \beta.$$

The rules of inference are Modus Ponens and necessitation for each modality: $\frac{\psi}{[\alpha]\psi}$. By GLP we denote the class-size logic that has a modality $[\alpha]$ for each ordinal α and all the corresponding axioms and rules. **GL** *refers to* GLP_1.

As suggested by Proposition 1, for the sake of Turing progressions a particular interest lies in GLP_Λ^0, the closed fragment – that is, the modal formulas that have no propositional variables but rather just $\bot$ and $\top$ – of these GLP_Λ. We shall call iterated consistency statements in this closed fragments *worms* in reference to the heroic worm-battle, a variant of the Hydra battle (see [1]).

Definition 2 (Worms, S, S_α). *By S we denote the set of* worms *of GLP which is inductively defined as $\top \in S$ and $A \in S \Rightarrow \langle\alpha\rangle A \in S$. Similarly, we inductively define for each ordinal α the set of worms S_α where all ordinals are at least α as $\top \in S_\alpha$ and $A \in S_\alpha \wedge \beta \geq \alpha \Rightarrow \langle\beta\rangle A \in S$.*

We shall denote worms by roman uppercase letters like $A, B, \ldots$ and often we associate worms with the strings of ordinals occurring in them whence we can concatenate worms. In denoting worms we might use any hybrid combination between the formal definition and its associated string. For example, we might equally well write $\omega 0 \omega$, as $\langle\omega\rangle 0 \omega$, or $\langle\omega\rangle\langle 0\rangle\langle\omega\rangle\top$.

The next easy lemma ([2], [4]) is the basis of most of our reasoning and we shall use it often in the remainder of this paper without explicit mention.

Lemma 1

1. $\mathsf{GLP} \vdash AB \to A$
2. *For a closed formula ϕ and a worm B, if $\beta < \alpha$, then*
   $\mathsf{GLP} \vdash (\langle\alpha\rangle\phi \wedge \langle\beta\rangle B) \leftrightarrow \langle\alpha\rangle(\phi \wedge \langle\beta\rangle B)$;
3. *For a closed formula ϕ and a worm B, if $\beta < \alpha$, then*
   $\mathsf{GLP} \vdash (\langle\alpha\rangle\varphi \wedge [\beta]B) \leftrightarrow \langle\alpha\rangle(\varphi \wedge [\beta]B)$;
4. *If $A \in S_{\alpha+1}$, then $\mathsf{GLP} \vdash A \wedge \langle\alpha\rangle B \leftrightarrow A\alpha B$;*
5. *If $A, B \in S_\alpha$ and $\mathsf{GLP} \vdash A \leftrightarrow B$, then*
   $\mathsf{GLP} \vdash A\alpha C \leftrightarrow B\alpha C$.

Worms can be conceived as the backbone of GLP^0. It is known that each closed formula of GLP is equivalent a Boolean combination of worms. Moreover, a decision procedure for GLP^0 proceeds via a reduction to worms. Also, there are various important generalizations of Proposition 1 in terms of worms. In particular (as in Prop. 1, there are some technical conditions on T, see [3]):

Proposition 2. *For each ordinal $\alpha < \epsilon_0$ there is some GLP_ω-worm A such that $T + A$ is Π_1 equivalent to T_α.*

To get generalizations of this lemma beyond ϵ_0 one should consider more than ω modalities. In particular $[\alpha]$ should be read as "provable in EA together with all true hyperarithmetical sentences of level α". This paper focusses on the modal calculus involved in such generalizations.

2   Omega Sequences

We define an order $<_\alpha$ on S_α by $A <_\alpha B :\Leftrightarrow \mathsf{GLP} \vdash B \to \langle\alpha\rangle A$. In [2] and [4] it is shown that $<_\alpha$ defines a well-order on S_α modulo provably equivalence. We can consider the ordering $<_\alpha$ also on the full class S. As we shall see $<_\alpha$ is no longer linear on S. However, we shall see that it is still well-founded. Anticipating this, we can define $\Omega_\alpha(A)$, the generalized $<_\alpha$ order-type of a worm A.

Definition 3. *Given an ordinal ξ and a worm A, we define a new ordinal $\Omega_\xi(A)$ by induction on $<_\xi$ by $\Omega_\xi(A) = \sup_{B <_\xi A}(\Omega_\xi(B)+)$. Likewise, we define $o_\xi(A) = \sup\{o_\xi(B) + 1 \mid B \in S_\xi \wedge B <_\xi A\}$.*

With this, we can assign to each worm A a sequence of order-types $\Omega(A)$ for the sequence $(\Omega_\xi(A))_{\xi\in\mathsf{On}}$.

To link back to the Turing progressions we mention here that the $\Omega_0(A)$ corresponds to the α from Proposition 2 and the $\Omega_\xi(A)$ for $\xi > 0$ correspond to further generalizations of Proposition 2. For worms that contain only natural numbers this has been established as stated in Proposition 2 and it is conjectured to hold also for worms containing larger ordinals. Thus, in the remainder of this paper it is good to bear in mind that the omega sequences have important proof-theoretic content.

In further sections we shall show how to calculate $\Omega_\xi(A)$ for given ξ and A. In the current section we shall see how questions about Ω_ξ can be recursively reduced to questions about o_ζ. For this reduction we need the syntactical definitions of *head* and *remainder*.

Definition 4. *Let A be a word. By $h_\xi(A)$ we denote the ξ-head of A. Recursively: $h_\xi(\epsilon) = \epsilon$, and $h_\xi(\zeta \star A) = \zeta \star h_\xi(A)$ if $\zeta \geq \xi$ and $h_\xi(\zeta \star A) = \epsilon$ if $\zeta < \xi$. Likewise, by $r_\xi(A)$ we denote the ξ-remainder of A: $r_\xi(\epsilon) = \epsilon$, and $r_\xi(\zeta \star A) = r_\xi(A)$ if $\zeta \geq \xi$ and $r_\xi(\zeta \star A) = \zeta \star A$ if $\zeta < \xi$.*

We obviously have $A = h_\xi(A) \star r_\xi(A)$ for all ξ and A and, clearly, over GLP, $h_\xi(A) \star r_\xi(A)$ and $h_\xi(A) \wedge r_\xi(A)$ are equivalent as the first symbol of $r_\xi(A)$ is less than ξ and $h_\xi(A) \in S_\xi$ (see Lemma 1).

Lemma 2. *$(A \to \langle\xi\rangle B) \Leftrightarrow [(h_\xi(A) \to \langle\xi\rangle h_\xi(B)) \wedge (A \to r_\xi(B))].$*

Proof. "$\Rightarrow$" Clearly, $B \leftrightarrow h_\xi(B) \wedge r_\xi(B)$ whence $A \to r_\xi(B)$. Likewise $A \leftrightarrow h_\xi(A) \wedge r_\xi(A)$. As $h_\xi(A), h_\xi(B) \in S_\xi$ we know that either $h_\xi(A) = h_\xi(B)$,

$h_\xi(B) \to \langle \xi \rangle\, h_\xi(A)$, or $h_\xi(A) \to \langle \xi \rangle\, h_\xi(B)$. By assumption $A \to \langle \xi \rangle\, B$ whence $A \to \langle \xi \rangle\, h_\xi(B) \wedge r_\xi(B)$.

Suppose now $h_\xi(A) = h_\xi(B)$. Then, $h_\xi(A) \wedge r_\xi(A) \to \langle \xi \rangle\, h_\xi(A) \wedge r_\xi(A)$ whence also

$$h_\xi(A) \wedge r_\xi(A) \to \langle \xi \rangle\, (h_\xi(A) \wedge r_\xi(A)).$$

The latter is equivalent to $A \to \langle \xi \rangle\, A$ which contradicts the ireflexivity of $<_\xi$.

Likewise, the assumption that $h_\xi(B) \to \langle \xi \rangle\, h_\xi(A)$ contradicts the reflexivity of $<_\xi$ and we conclude that $h_\xi(A) \to \langle \xi \rangle\, h_\xi(B)$.

"$\Leftarrow$" This is the easier direction.

$$\begin{aligned} A &\leftrightarrow h_\xi(A) \wedge r_\xi(A) \\ &\to \langle \xi \rangle\, h_\xi(B) \wedge r_\xi(B) \\ &\to \langle \xi \rangle\, (h_\xi(B) \wedge r_\xi(B)) \\ &\to \langle \xi \rangle\, B. \end{aligned}$$

Note that this lemma indeed recursively reduces the general $<_\xi$ question between to words, to the $<_\xi$ questions between words in S_ξ. Note that $<_\xi$ is not tree-like. For example, we see that both $011 <_1 10111 <_1 1111$ and $011 <_1 11011 <_1 1111$ while 10111 and 11011 are $<_1$ incomparable.

3   Basic Properties of Omega Sequences

In these final sections we give a full characterization of the sequences $\Omega(A)$. That is, we shall determine for given A each of the values $\Omega_\xi(A)$ and shall classify at what coordinates ξ the $\Omega(A)$ sequence changes value. Clearly, $\Omega(A)$ defines a weakly decreasing sequence of ordinals.

Lemma 3. *For $\xi < \zeta$ we have that $\Omega_\xi(A) \geq \Omega_\zeta(A)$.*

Proof. In general we have for $\xi < \zeta$ that $A \to \langle \zeta \rangle B$ implies $A \to \langle \xi \rangle B$. Thus, any $<_\zeta$ sequence is automatically also a $<_\xi$ sequence.

Lemma 4. $\Omega_\xi(A) = o_\xi(h_\xi(A))$

Proof. Suppose $A_0 <_\xi A_1 <_\xi \ldots <_\xi A$, then

$$h_\xi(A_0) <_\xi h_\xi(A_1) <_\xi \ldots <_\xi h_\xi(A)$$

by Lemma 2 whence $\Omega_\xi(A) \leq o_\xi h_\xi(A)$.

On the other hand, if $B <_\xi h_\xi(A)$, then $h_\xi(A) \to \langle \xi \rangle B$. But as $A \leftrightarrow h_\xi(A) \wedge r_\xi(A)$ we also have $A \to \langle \xi \rangle B$. Consequently $o_\xi h_\xi(A) \leq \Omega_\xi(A)$.

Corollary 1. *The $\Omega(A)$ sequence has a maximal non-zero coordinate. In particular, the maximal non-zero coordinate is given by $\Omega_{\mathsf{First}(A)}(A)$, where $\mathsf{First}(A)$ is the first element of the non-empty word A.*

Proof. Clearly, $h_{\mathsf{First}(A)}(A) \neq \epsilon$ whence by Lemma 4, $\Omega_{\mathsf{First}(A)}(A) \neq 0$. On the other hand, for $\xi > \mathsf{First}(A)$, clearly $h_\xi(A) = \epsilon$ whence $\Omega_\xi(A) = 0$.

It is good to have reduced $\Omega_\xi(A)$ to $o_\xi(A)$ as in [2], and [4] a full calculus for it is given. Let $e_0\alpha := -1 + \omega^\alpha$ and let e_α denote the Veblen progression based on e_0. That is, each e_α enumerates the ordinals which are simultaneous fixed points of all the e_β for $\beta < \alpha$. We define $e^0\alpha := \alpha$ and $e^{\omega^\xi + \zeta} = e_\xi \circ e^\zeta$ whenever $\zeta < \omega^\xi + \zeta$. Further, we define $\xi\uparrow\zeta := \xi + \zeta$ and for $\xi < \zeta$ we define $\xi\downarrow\zeta$ to be the unique ordinal such that $\xi\uparrow(\xi\downarrow\zeta) = \zeta$. These operations are naturally extended to worms by simultaneously applying them to all occurrences of ordinals.

Lemma 5

1. $o(0^n) = n$;
2. If $A = A_1 0 \ldots 0 A_n$, then $o(A) = \omega^{o_1(A_n)} + \ldots + \omega^{o_1(A_1)}$;
3. For $A \in S_\xi$ we have $o_\xi(A) = o(\xi \downarrow A)$;
4. $o(\xi \uparrow A) = e^\xi o(A)$.

4   Successor Coordinates

First, let us compute $\Omega_{\xi+1}(A)$ in terms of $\Omega_\xi(A)$. By $\ell\alpha$ we denote the unique β such that $\alpha = \alpha' + \omega^\beta$ for $\alpha > 0$. We define $\ell 0 := 0$.

Lemma 6. *Given an ordinal ξ and a worm A,*

$$o_{\xi+1}h_{\xi+1}h_\xi(A) = \ell o_\xi h_\xi(A).$$

Proof. We write $h_\xi(A)$ as $A_0\xi \ldots \xi A_n$. Clearly, $h_{\xi+1}(h_\xi(A)) = A_0$. We shall now see that $\ell(o_\xi(h_\xi(A))) = o_{\xi+1}(A_0)$.

To this end, we observe that

$$\begin{aligned}
o_\xi h_\xi(A) &= o_\xi(A_0\xi \ldots \xi A_n) \\
&= o\Big((\xi\downarrow A_0)0\ldots0(\xi\downarrow A_n)\Big) \\
&= \omega^{o_1(\xi\downarrow A_n)} + \ldots + \omega^{o_1(\xi\downarrow A_0)} \\
&= \omega^{o_{\xi+1}(A_n)} + \ldots + \omega^{o_{\xi+1}(A_0)}
\end{aligned}$$

Consequently $\ell o_\xi h_\xi(A) = o_{\xi+1}(A_0)$, as desired.

Now we are ready to describe the relation between successor coordinates of the $\Omega(A)$ sequence.

Theorem 1. $\Omega_{\xi+1}(A) = \ell\Omega_\xi(A)$

Proof.

$$\begin{aligned}
\Omega_{\xi+1}(A) &= o_{\xi+1}h_{\xi+1}(A) & \text{by Lemma 2} \\
&= o_{\xi+1}h_{\xi+1}h_\xi(A) & \text{by Lemma 6} \\
&= \ell o_\xi h_\xi(A) & \\
&= \ell\Omega_\xi(A) & \text{by Lemma 2.}
\end{aligned}$$

Theorem 1 tells us what the relation between successor coordinates of $\Omega(A)$ is. However, it does not directly tell us when successor coordinates are different. If $\Omega_\xi(A)$ is a fixed point of $\zeta \mapsto \omega^\zeta$ then $\Omega_\xi(A) = \Omega_{\xi+1}(A)$.

5   Equal Coordinates

Theorem 2 below gives us a characterization of when different coordinates attain different or equal values. Before we can state and prove this theorem we need some notation and back-ground on Cantor Normal Forms (CNFs).

For $\alpha \in \mathsf{On}$ we define N_α and the syntactic operation $\mathsf{CNF}(\alpha) := \sum_{i=1}^{N_\alpha} \omega^{\xi_i}$ to be the unique CNF expression of α. Next, we define for an ordinal α the set of its *Cantor Normal Form Approximation* as the set of partial sums of $\mathsf{CNF}(\alpha)$, that is, if $\mathsf{CNF}(\alpha) = \sum_{i=1}^{N_\alpha} \omega^{\xi_i}$, then

$$\mathsf{CNA}(\alpha) \; := \; \left\{ \sum_{i=1}^{k} \omega^{\xi_i} : 0 \le k \le N_\alpha \right\}.$$

We also define the *Cantor Normal Form Projection* of some ordinal ζ on another ordinal ξ as follows:

$$\mathsf{CNP}(\zeta, \xi) \; := \; \max\{\xi' \in \mathsf{CNA}(\xi) \mid \xi' \le \zeta\}.$$

Note that $\mathsf{CNP}(\zeta, \xi)$ is defined for all $\zeta, \xi \in \mathsf{On}$ by setting $\max \varnothing = 0$.

For $\alpha, \beta, \gamma \in \mathsf{On}$ we define

$$\alpha \sim_\gamma \beta \quad :\Leftrightarrow \quad \mathsf{CNP}(\alpha, \gamma) = \mathsf{CNP}(\beta, \gamma).$$

In words, $\alpha \sim_\gamma \beta$ whenever there is no partial sum of the CNF of γ that falls in between α and β.

The just-defined notions of $\mathsf{CNA}(\xi)$, $\mathsf{CNP}(\zeta, \xi)$ and $\alpha \sim_\gamma \beta$ are needed to characterize the $\xi \downarrow \zeta$ operation.

Lemma 7

1. $\forall \zeta \le \xi \; \zeta \downarrow \xi = \mathsf{CNP}(\zeta, \xi) \downarrow \xi$;

2. $\forall \zeta \le \xi \exists ! \eta \in \mathsf{CNA}(\xi) \; \zeta \downarrow \xi = \eta \downarrow \xi$;

3. *For* $\xi, \zeta \le \eta$, *we have*
$$\xi \downarrow \eta = \zeta \downarrow \eta \; \Leftrightarrow \; \xi \sim_\eta \zeta.$$

Proof. (1.) We consider $\zeta \le \xi$. Now let η be shorthand for $\max\{\eta' \in \mathsf{CNA}(\xi) \mid \eta' \le \zeta\} = \mathsf{CNP}(\zeta, \xi)$. The claim is that $\zeta \downarrow \xi = \eta \downarrow \xi$. Let $\mathsf{CNF}(\xi) = \sum_{i=1}^{N_\xi} \omega^{\xi_i}$.

As $\eta = \sum_{i=1}^{k} \omega^{\xi_i}$ for some $k \le N_\xi$, we see that

$$\eta \downarrow \xi = \sum_{i=k+1}^{N_\xi} \omega^{\xi_i}$$

for $k < N_\xi$ and $\eta \downarrow \xi = 0$ for $k = N_\xi$. We now claim that $\zeta + (\eta \downarrow \xi) = \xi$ so that $\zeta \downarrow \xi = \eta \downarrow \xi$ follows from the fact that

$$\forall \zeta < \xi \; \exists ! \delta \; \zeta + \delta = \xi.$$

We may assume $\zeta > \eta$ otherwise $\zeta + (\eta{\downarrow}\xi) = \xi$ is trivial.

Thus,

$$\eta = \sum_{i=1}^{k} \omega^{\xi_i} < \zeta \le \sum_{i=1}^{k+1} \omega^{\xi_i}.$$

As by the definition of η we see that $\zeta \le \sum_{i=1}^{k+1} \omega^{\xi_i}$ cannot be an equality whence

$$\eta = \sum_{i=1}^{k} \omega^{\xi_i} < \zeta < \sum_{i=1}^{k+1} \omega^{\xi_i}.$$

Thus, $\eta \in CNA(\zeta)$ and $\zeta + \sum_{i=k+1}^{N_\xi} \omega^{\xi_i} = \xi$, whence

$$\sum_{i=k+1}^{N_\xi} \omega^{\xi_i} = \zeta{\downarrow}\xi = \sum_{i=1}^{k} \omega^{\xi_i} = \eta{\downarrow}\xi.$$

(2.) Follows from (1.) once we realize that for different η and η' both in $\mathsf{CNA}(\xi)$ we have $\eta{\downarrow}\xi \ne \eta'{\downarrow}\xi$.

(3). From the proof of (1.) we see that

$$\xi{\downarrow}\eta = \zeta{\downarrow}\eta \;\Leftrightarrow\; \max\{\eta'{\in}\mathsf{CNA}(\eta) \mid \eta' \le \xi\} = \max\{\eta'{\in}\mathsf{CNA}(\eta) \mid \eta' \le \zeta\}$$

where the latter is precisely the definition of $\xi \sim_\eta \zeta$.

Once we have this lemma to characterize the $\xi{\downarrow}\zeta$ operation, we are armed to prove a characterization for when two coordinates in $\Omega(A)$ are equal. But first we need a definition of when a worm A is in Beklemishev Normal Form (BNF). Recursively we say that the empty worm $\epsilon \in$ BNF, and if $A_i \in S_{\xi+1}{\cap}$BNF, with $A_i \ge_{\xi+1} A_{i+1}$, then $A_n \alpha \dots \alpha A_1 \in$BNF. It is easy to see that if a worm is in BNF, then so are its head and remainder. From [2], and [4] we know that the set $S{\cap}$BNF is well-ordered by $<_0$ and that o_0 provides an isomorphism between $\langle S, <_0 \rangle$ and $\langle \mathsf{On}, < \rangle$.

Theorem 2. *For $A \in BNF$, the following five conditions are equivalent.*

1. $\Omega_\xi(A) = \Omega_\zeta(A)$
2. $o_\xi h_\xi(A) = o_\zeta h_\zeta(A)$
3. $\xi{\downarrow}h_\xi(A) = \zeta{\downarrow}h_\zeta(A)$
4. $h_\xi(A) = h_\zeta(A)$ and $\xi{\downarrow}h_\xi(A) = \zeta{\downarrow}h_\zeta(A)$
5. $h_\xi(A) = h_\zeta(A)$ and $\forall \eta \in h_\xi(A)$, $\xi \sim_\eta \zeta$

Proof. $(1.) \Leftrightarrow (2.)$ is just Lemma 4.

$(2.) \Leftrightarrow (3.)$: Observe that $o_\xi(h_\xi(A)) = o(\xi{\downarrow}h_\xi(A))$ and $o_\zeta(h_\zeta(A)) = o(\zeta{\downarrow}h_\zeta(A))$. As o defines an isomorphism between S and On, we obtain

$$o_\xi h_\xi(A) = o_\zeta h_\zeta(A) \;\Leftrightarrow\; \xi{\downarrow}h_\xi(A) = \zeta{\downarrow}h_\zeta(A).$$

(3.) $\Leftrightarrow$ (4.): Suppose $h_\xi(A) \neq h_\zeta(A)$. W.l.o.g. we may assume that $\zeta < \xi$ whence

$$\mathsf{Length}(h_\xi(A)) < \mathsf{Length}(h_\zeta(A))$$

and also

$$\mathsf{Length}(h_\xi(A)) = \mathsf{Length}(\xi{\downarrow}h_\xi(A)) < \mathsf{Length}(\zeta{\downarrow}h_\zeta(A)) = \mathsf{Length}(h_\zeta(A)).$$

As $A \in \mathrm{BNF}$, also $h_\xi(A)$ and $h_\zeta(A)$ are in BNF, whence also $\xi{\downarrow}h_\xi(A)$, $\zeta{\downarrow}h_\zeta(A)$ are in BNF. We know that normal forms are graphically unique so that $\xi{\downarrow}h_\xi(A) \neq \zeta{\downarrow}h_\zeta(A)$ whence $o(\xi{\downarrow}h_\xi(A)) \neq o(\zeta{\downarrow}h_\zeta(A))$.

(4.) $\Leftrightarrow$ (5.):

$$\begin{aligned}
&h_\xi(A) = h_\zeta(A) \wedge \xi{\downarrow}h_\xi(A) = \zeta{\downarrow}h_\zeta(A) &&\Leftrightarrow \\
&h_\xi(A) = h_\zeta(A) \wedge \forall\, \eta{\in}h_\xi(A)\ \xi{\downarrow}\eta = \zeta{\downarrow}\eta \Leftrightarrow && \text{by Lemma 7.3} \\
&h_\xi(A) = h_\zeta(A) \wedge \forall\, \eta{\in}h_\xi(A)\ \xi \sim_\eta \zeta
\end{aligned}$$

6   Limit Coordinates

The results so far have already provided us with quite some insight about what the sequences $\varOmega(A)$ look like. By Lemma 3 we know that the set of values that occur in $\varOmega(A)$ is finite. Moreover, by Theorem 1 we know exactly the values at successor coordinates. In particular, we know that if the value of $\varOmega(A)$ at ξ is the same as at the successor coordinate, then it remains the same for all further successors.

The question remains what happens at limit ordinals coordinates. In this section we shall determine at what limit ordinals a new value can be attained and how the new value relates to previous values. Let us start out the analysis by formulating a negative version of Theorem 2.

Lemma 8. *For $\xi < \zeta$ we have that*

$$\varOmega_\xi(A) > \varOmega_\zeta(A) \;\Leftrightarrow\; (\exists\, \eta{\in}h_\xi(A)\ \xi{\leq}\eta{<}\zeta) \;\vee\; (\exists\, \eta{\in}h_\xi(A)\ \mathsf{CNP}(\xi,\eta){<}\mathsf{CNP}(\zeta,\eta)).$$

Proof. By contraposing equivalence (1.) $\Leftrightarrow$ (5.) of Theorem 2 we get

$$\varOmega_\zeta(A) \neq \varOmega_\xi(A) \;\Leftrightarrow\; h_\xi(A){\neq}h_\zeta(A) \;\vee\; \exists\, \eta{\in}h_\zeta(A)\ \xi \nsim_\eta \zeta.$$

But, as $\zeta < \xi$ we see

$$h_\xi(A){\neq}h_\zeta(A) \;\Leftrightarrow\; \exists\, \eta{\in}h_\zeta(A)\ \zeta \leq \eta < \xi.$$

Likewise,

$$\exists\, \eta{\in}h_\zeta(A)\ \xi \nsim_\eta \zeta \;\Leftrightarrow\; \exists\, \eta{\in}h_\zeta(A)\ \mathsf{CNP}(\zeta,\eta){\neq}\mathsf{CNP}(\xi,\eta).$$

As $\zeta < \xi$ we have

$$\mathsf{CNP}(\zeta,\eta){\neq}\mathsf{CNP}(\xi,\eta) \;\Leftrightarrow\; \mathsf{CNP}(\zeta,\eta){<}\mathsf{CNP}(\xi,\eta).$$

The first question to ask is at which limit coordinates the sequence $\Omega(A)$ can change. Let us first write precisely what it means for the sequence $\Omega(A)$ to change at some coordinate ζ. We express this by the formula

$$\mathsf{Change}(\zeta, A) \;\; := \;\; \exists \xi{<}\zeta \; (\Omega_\xi(A){>}\Omega_\zeta(A) \; \wedge \; \forall \eta \; (\xi{\leq}\eta{<}\zeta \to \Omega_\xi(A){=}\Omega_\zeta(A))).$$

The next lemma gives an alternative characterization of $\mathsf{Change}(\zeta, A)$.

Lemma 9. $\mathsf{Change}(\zeta, A) \;\; \Leftrightarrow \;\; \forall \xi{<}\zeta \; \Omega_\xi(A){>}\Omega_\zeta(A)$

Proof. For $\zeta \in \mathsf{Succ}$ this is clear. If $\zeta \in \mathsf{Lim}$, then $\{\Omega_\xi(A) \mid \xi < \zeta\}$ is a finite set as all the $\Omega_\xi(A) \in \mathsf{On}$ and these are weakly decreasing. Thus, at some point below ζ the sequence must stabilize.

We can now characterize at what limit ordinals the sequence $\Omega(A)$ can change.

Theorem 3. *For $\zeta \in \mathsf{Lim}$:* $\mathsf{Change}(\zeta, A) \;\; \Leftrightarrow \;\; \exists \xi{\in}h_\zeta(A) \; \zeta{\in}\mathsf{CNA}(\xi)$

Proof. For $\zeta \in \mathsf{Lim}$ we reason:

$$\begin{aligned}
\mathsf{Change}(\zeta, A) &\;\; \Leftrightarrow \text{By Lemma 9} \\
\forall \xi{<}\zeta \; \Omega_\xi(A){>}\Omega_\zeta(A) &\;\; \Leftrightarrow \text{By Lemma 8} \\
\forall \xi{<}\zeta \; (\exists \eta{\in}h_\xi(A) \; \xi{\leq}\eta < \zeta \;\vee\; \exists \eta{\in}h_\xi(A) \; \mathsf{CNP}(\xi, \eta){<}\mathsf{CNP}(\zeta, \eta)) &\;\; \Leftrightarrow \\
\forall \xi \; (\xi_0{<}\xi{<}\zeta \to \exists \eta{\in}h_\zeta(A) \; \mathsf{CNP}(\xi, \eta){<}\mathsf{CNP}(\zeta, \eta))
\end{aligned}$$

where $\xi_0 := \sup\{\xi' \in A \mid \xi' < \zeta\}$. Note that for these ξ, indeed, we have $h_\xi(A) = h_\zeta(A)$. We now claim that the latter is equivalent to $\exists \eta{\in}h_\zeta(A) \; \zeta{\in}\mathsf{CNA}(\eta)$. Clearly, if $\zeta \in \mathsf{CNA}(\eta)$ for some $\eta \in h_\zeta(A)$, then $\xi{\downarrow}\eta < \zeta{\downarrow}\eta$ for each $\xi < \zeta$.

For the converse direction, suppose $\zeta \notin \mathsf{CNA}(\eta)$ for all $\eta \in h_\zeta(A)$. Then, for all ξ' with

$$\sup\{\xi \mid \exists \eta{\in}h_\zeta(A) \; (\xi{\in}\mathsf{CNA}(\eta) \;\wedge\; \xi < \zeta)\} < \xi' < \zeta$$

we have $\xi' \sim_\eta \zeta$ for all $z \in h_\zeta(A)$, whence by Theorem 2 $\Omega_{\xi'}(A) = \Omega_\zeta(A)$.

Now that we have fully determined at which limit coordinates a change can occur the only thing left to establish is the size of the change. In other words, if $\mathsf{Change}(\zeta, A)$ for some $\zeta \in \mathsf{Lim}$, how does $\Omega_\zeta(A)$ relate to $\Omega_\xi(A)$ for $\xi < \zeta$. In order to answer this question, we need a generalization of Lemma 5.4.

Lemma 10. $o_\xi(\zeta{\uparrow}A) = e^\zeta o_\xi(A)$ *for $A \in S_\xi$.*

Proof. We claim that for $B \in S_\xi$ we have that $\xi{\downarrow}(\zeta{\uparrow}B) = \zeta{\uparrow}(\xi{\downarrow}B)$. From this claim the statement follows easily from Lemma 5.4.

$$o_\xi(\zeta{\uparrow}A) = o(\xi{\downarrow}(\zeta{\uparrow}A)) = o(\zeta{\uparrow}(\xi{\downarrow}A)) = e^\zeta o(\xi{\downarrow}A) = e^\zeta o_\xi(A).$$

Thus, we only need to prove our claim. Clearly, it suffices to show the claim for any ordinal $\eta \geq \xi$ instead of for any word in S_ξ. By definition, $\xi{\downarrow}(\zeta{\uparrow}z){=}\delta \;\Leftrightarrow\; \xi + \delta = \eta + \zeta$. Likewise, $\xi{\downarrow}\eta = \delta' \;\Leftrightarrow\; \xi + \delta' = \eta$. As $\zeta{\uparrow}(\xi{\downarrow}\eta) = \xi{\downarrow}\eta + \zeta = \delta' + \zeta$ we obtain $\xi + \delta' = \eta \;\Rightarrow\; (\xi + \delta') + \zeta = \eta + \zeta$ and by associativity of ordinal addition also $\xi + (\delta' + \zeta) = \eta + \zeta$. We conclude that $\delta' + \zeta = \delta$ which translates exactly to $\zeta{\uparrow}(\xi{\downarrow}\eta) = \xi{\downarrow}(\zeta{\uparrow}\eta)$, quod erat demostrandum.

With this technical lemma at hand we are ready to prove the concluding theorem of this section.

Theorem 4. *Let $\zeta \in \mathsf{Lim}$, and let $\xi < \zeta$ be such that, whenever $\xi' \in [\xi, \zeta)$, it follows that $\Omega_\xi(A) = \Omega_{\xi'}(A)$. Then,*

$$\Omega_\xi(A) = e^{-\xi + \zeta} \Omega_\zeta(A) = e_{\ell\zeta} \Omega_\zeta(A).$$

Proof. As the values of $\Omega_{\xi'}(A)$ do not change for $\xi \leq \xi' < \zeta$ we know in particular by Theorem 2 that

$$h_\xi(A) \;=\; h_\zeta(A). \tag{1}$$

Likewise by Theorem 2 we see that $-\xi + \zeta = \omega^{\ell\zeta}$. Let $\delta = -\xi + \zeta$.
   Then,

$$\Omega_\zeta(A) = o_\zeta h_\zeta(A) = o_{\xi + (\xi \downarrow \zeta)}(h_\zeta(A)) = o_{\xi + \delta}(h_\zeta(A)) = o_\xi(\delta \downarrow h_\zeta(A)). \tag{2}$$

Thus,

$$
\begin{array}{lll}
\Omega_\xi(A) & = & \\
o_\xi(h_\xi(A)) & = & \text{By (1)} \\
o_\xi(h_\zeta(A)) & = & \\
o_\xi(\delta\uparrow(\delta\downarrow h_\zeta(A))) & = & \text{By Lemma 10} \\
e^\delta o_\xi(\delta\downarrow h_\zeta(A)) & = & \text{By (2)} \\
e^\delta \Omega_\zeta(A) & = & \text{By definition of } e^\alpha \\
e_{\ell\zeta} \Omega_\zeta(A). & &
\end{array}
$$

Note that this theorem establishes the size of limit coordinates both in case a change does occur and in case no change occurs. The latter case can only be so when $\Omega_\zeta(A)$ is a fixed point of $e_{\ell\zeta}$.

References

1. Beklemishev, L.D.: Provability algebras and proof-theoretic ordinals, I. Annals of Pure and Applied Logic 128, 103–124 (2004)
2. Beklemishev, L.D.: Veblen hierarchy in the context of provability algebras. In: Proceedings of the Twelfth International Congress on Logic, Methodology and Philosophy of Science. Kings College Publications (2005)
3. Beklemishev, L.D.: Reflection principles and provability algebras in formal arithmetic. Russian Mathematical Surveys 60(2), 197–268 (2005)
4. Beklemishev, L.D., Fernández Duque, D., Joosten, J.J.: On transfinite provability logic (under preparation, 2012)
5. Solovay, R.M.: Provability interpretations of modal logic. Israel Journal of Mathematics 28, 33–71 (1976)
6. Turing, A.M.: Systems of logics based on ordinals. Proc. London Math. Soc. Series 2 45, 161–228 (1939)

A Short Note on Spector's Proof
of Consistency of Analysis

Fernando Ferreira

Universidade de Lisboa, Campo Grande, Ed. C6, 1749-016 Lisboa, Portugal
fjferreira@fc.ul.pt

Abstract. In 1962, Clifford Spector gave a consistency proof of analysis using so-called bar recursors. His paper extends an interpretation of arithmetic given by Kurt Gödel in 1958. Spector's proof relies crucially on the interpretation of the so-called numerical double negation shift principle. The argument for the interpretation is ad hoc. On the other hand, William Howard gave in 1968 a very natural interpretation of bar induction by bar recursion. We show directly that, within the framework of Gödel's interpretation, numerical double negation shift is a consequence of bar induction.

The 1958 paper [4] of Kurt Gödel presented an interpretation (now known as the *dialectica* interpretation) of Heyting arithmetic HA into a quantifier-free calculus T of finite-type functionals. The terms of T denote certain *computable functionals of finite type* (a primitive notion in Gödel's paper, as it were): the so-called primitive recursive functionals in the sense of Gödel. These terms can be rigorously defined and they include as primitives the *combinators* (a bureaucracy of terms for dealing with the "logical" part of the calculus) and the arithmetical constants: 0, the successor constant and, importantly, the *recursors*.[1] The *dialectica* interpretation assigns to each formula A of the language of first-order arithmetic a (quantifier-free) formula $A_D(x, y)$ of the language of T, and Gödel showed that if HA $\vdash A$, then there is a term t (in which y does not occur free) of the language of T such that T $\vdash A_D(t, y)$.[2] The combinators play a central role in showing the preservation of the interpretation under (intuitionistic) logic and, unsurprisingly, the recursors play an essential role in interpreting the induction axioms.

It is convenient to extend the *dialectica* interpretation to Heyting arithmetic in all finite types HA$^\omega$.[3] Within the language of this theory, one can formulate the *characteristic principles* of the interpretation:

---

[1] The reader can consult [11], [1] or [7] for a precise description of the calculus T and of its terms in particular. These are good references for details concerning the *dialectica* interpretation.

[2] We are taking some liberties here (and will take in the sequel). Rigorously, either one should speak of tuple of variables $x := x_1, \ldots, x_n$ and $y := y_1, \ldots, y_m$ or allow convenient product types.

[3] Some delicate issues concerning equality in higher types arise at this point (if not before). See [1] for a discussion of these matters.

S.B. Cooper, A. Dawar, and B. Löwe (Eds.): CiE 2012, LNCS 7318, pp. 222–227, 2012.
© Springer-Verlag Berlin Heidelberg 2012

AC^ω: $\forall x \exists y B(x, y) \to \exists f \forall x B(x, f(x))$,

MP^ω: $\neg \forall x A(x) \to \exists x \neg A(x)$,

$\mathsf{IP}^\omega_\forall$: $(\forall x A(x) \to \exists y C(y)) \to \exists y (\forall x A(x) \to C(y))$,

where x and y may be of any finite type, A is quantifier-free and B, C are arbitrary. The first principle is a form of choice, the second is a finite-type form of Markov's principle, and the third is an independence of premises statement. These principles arise in virtue of the very definition of the assignment $A \rightsquigarrow A_D$ and, accordingly, have trivial interpretations. They are sufficient to prove the equivalence between a given formula A of the language of HA^ω and its *dialectica* translation $A^D := \exists x \forall y A_D(x, y)$.[4]

The *dialectica* interpretation extends to (finite-type) Peano arithmetic PA^ω by composing it with a negative translation $A \rightsquigarrow A^g$. Therefore, if A is a consequence of PA^ω then A^g is provable in HA^ω and, hence, there is a term t of the language of T such that $T \vdash (A^g)_D(t, y)$.

In the last paragraph of his 1958 paper, Gödel suggests the construction of systems stronger than T. Presumably, calculi with more terms can interpret (via the *dialectica* blueprint) stronger systems of arithmetic. In this note, we are concerned with the system obtained from PA^ω by adding full numerical comprehension (obviously, this theory contains full second-order arithmetic, a.k.a. *analysis*). The formulation of the numerical comprehension principle CA in finite type arithmetic takes the form

$$\exists f \forall n (fn = 0 \leftrightarrow A(n)),$$

for arbitrary formulae A (n is a number variable). Clifford Spector isolated in [10] the so-called *principle of numerical double negation shift* (principle F in Spector's own paper):

$$\forall n \neg\neg P(n) \to \neg\neg \forall n P(n),$$

where n is a number variable and P is arbitrary. This principle is important because of the following result:[5]

Theorem 1 (Kreisel). *The negative translation of* CA *is provable in the theory* $\mathsf{HA}^\omega + \mathsf{AC}^\omega$ *together with the principle of numerical double negation shift.*

Proof. The theory HA^ω proves $\forall n \neg\neg (A^g(n) \lor \neg A^g(n))$. By numerical double negation shift, $\neg\neg \forall n (A^g(n) \lor \neg A^g(n))$. Equivalently, $\neg\neg \forall n \exists k((k = 0 \to A^g(n)) \land (k \neq 0 \to \neg A^g(n)))$. By AC^ω, $\neg\neg \exists f \forall n ((fn = 0 \to A^g(n)) \land (fn \neq 0 \to \neg A^g(n)))$. This formula is intuitionistically equivalent to the negative translation of CA. □

---

[4] This result is due to Mariko Yasugi in [14]. There is a brief discussion in [3] clarifying the role of the characteristic principles.

[5] See [8] and note 4 of [10].

Spector's paper [10] was published posthumously.[6] The paper was submitted by Kreisel on Spector's behalf. The first footnote (written by Kreisel, as were all footnotes) and a postscript by Gödel explain the origin of the paper. In the paper, Spector generalizes Brouwer's principle of bar induction to higher types.[7] We work with the following version of bar induction: For any given formulas A and B, if

(1) $\forall f \exists k \, A(\overline{f}(k))$

(2) $\forall s, s'(A(s') \wedge s'$ is a initial subsequence of $s \rightarrow A(s))$

(3) $\forall s(A(s) \rightarrow B(s))$

(4) $\forall s \, (\forall x B(\hat{s}x) \rightarrow B(s))$

then $B(\langle\rangle)$. In the above, x is of an arbitrary given type, f is an infinite sequence of elements of the type of x, and s, s' are finite sequences of elements of the type of x (also, $\overline{f}(k)$ is the finite sequence $\langle f0, \ldots, f(k-1)\rangle$ and $\hat{s}x$ is the sequence s concatenated with the element x).[8],[9] The great novelty is that Spector also introduces definitions by so-called bar recursion.[10] There are now new terms, arising from the bar recursors: an extension of Gödel's T calculus has been produced. Spector points that "since our immediate objective is to obtain a quantifier-free interpretation of analysis, bar recursion rather than bar induction is appropriate." The goal is, of course, to witness (with the help of the new terms) the *dialectica* translations of the instances of the numerical double negation shift principle. By Kreisel's result above, this would solve the problem of providing an

---

[6] Spector died at the age of thirty in the summer of 1961 of acute leukemia, after spending the academic year in the Institute for Advanced Study in Princeton, New Jersey.

[7] A statement of bar induction (or bar theorem) by Brouwer himself can be found in [2]. Van Atten has a discussion of this principle in his book [12] on Brouwer.

[8] The usual presentation of finite type arithmetic has no primitive types for finite sequences (of a given fixed type). However, there are manners of representing finite sequences in finite type arithmetic. We will not worry about these issues here.

[9] Bar induction in the sense of Brouwer demands that x be a number variable. Brouwerian bar induction is already sufficient to interpret Σ_1^0 comprehension (and so, by a well-known fact, arithmetical comprehension). This can be seen by analyzing carefully the proofs of the Kreisel result above and of the main result below. Spector introduced bar induction for x of any given type.

[10] We do not define bar recursion in this paper (for modern references and discussions, see note 1). In contrast to recursion, bar recursion is not well defined in the full set theoretic structure of finite type functionals (it is not a classical principle). It is nevertheless defined in the structure of continuous functionals and in the strong majorizability model. Bar induction is, on the other hand, a classical principle.

interpretation for CA. In the crucial Section 10 of his paper, Spector produces an ad hoc witness solution.[11]

Why does Spector introduce and discuss bar induction? After all, his interpretation of analysis does not formally require it. He explains that "bar induction (is discussed) primarily to point out the relationship between bar recursion and the bar theorem" and candidly observes that "bar recursion is a principle of definition and bar induction a corresponding principle of proof." This is exactly right. The situation is analogous with that of induction/recursion (*pace* the remark in note 10). In effect, William Howard proved in [6] that the principle of bar induction is interpretable in the extended calculus, with bar recursors. The proof is very natural and, if I may add, has a certain character of inexorability about it.[12]

So, there is this very satisfying picture:

$$
\begin{array}{ccc}
\text{induction} & \text{\rule{3cm}{0.4pt}} & \text{recursion} \\
\text{bar induction} & \text{\rule{3cm}{0.4pt}} & \text{bar recursion}
\end{array}
$$

The following result shows that the principle of double negation shift follows from the characteristic principles and bar induction and, therefore (by Howard), has a *dialectica* interpretation in the extended calculus.

Theorem 2. *In the theory* HA^ω *together with the characteristic principles, the principle of bar induction implies numerical double negation shift.*

Proof. Under the hypothesis of the theorem, we must show that $\forall n \neg\neg P(n) \to \neg\neg \forall n P(n)$ is a consequence of bar induction (P arbitrary). By the characteristic principles, the formula $P(n)$ is equivalent to a formula of the form $\exists x \forall y Q(x, y, n)$, with Q quantifier-free. Let us assume $\forall n \neg\neg P(n)$ and $\neg \forall n P(n)$ in order to derive a contradiction. Consider $A(s) :\equiv \exists i < |s| \exists y \neg Q(s_i, y, i)$ and $B(s) :\equiv A(s)$, where s is of the type of finite sequences of elements of type x ($|s|$ denotes the length of s). It is clear that $\neg B(\langle\rangle)$. Therefore, if one proves the hypothesis (1) to (4) of the principle of bar induction, we get a contradiction. Hypothesis (2) and (3) hold trivially. Hypothesis (1) is a consequence of the assumption that

---

[11] The heart of the matter lies in the solution of a system of equations arising from the *dialectica* translation of the principle of double negation shift. Due the presence of many negations, this system is rather cryptic. It is a kind of brute fact that, somehow, one is able to solve it with bar recursive functionals. (This is not to say that the proof in unmotivated, as Spector describes in his paper the motivation for his proof.) The intuitionistic laws $\neg\neg A_1 \wedge \ldots \wedge \neg\neg A_n \to \neg\neg(A_1 \wedge \ldots \wedge A_n)$ are "miniaturizations" of the principle of double negation shift and, by Gödel's paper, their *dialectica* translations have witnessing solutions in the T calculus. It is a non trivial exercise to find such solutions, even for $n = 2$. Paulo Oliva in [9] discusses these solutions in detail and tries to motivate the use of bar recursion for the *dialectica* interpretation of double negation shift in terms of a limit process when the number of conjuncts goes to infinity.

[12] Avigad and Feferman say in [1] that "while the proof requires some effort, the underlying idea is straightforward."

$\neg\forall n P(n)$. In effect, this assumption says that $\neg\forall n\exists x\forall y Q(x,y,n)$. By AC^ω and intuitionistic logic, we get $\forall f\neg\forall n, y Q(fn,y,n)$. Markov's principle MP^ω, on the other hand, entails $\forall f\exists n, y\neg Q(fn,y,n)$. Therefore, $\forall f\exists k\exists i < k\exists y\neg Q(fi,y,i)$, i.e., $\forall f\exists k A(\overline{f}(k))$.

Let us argue (4). Given a finite sequence s, suppose $\forall x B(\hat{s}x)$, i.e.:

$$\forall x(\exists i < |s|\exists y\neg Q(s_i,y,i)\vee\exists y\neg Q(x,y,|s|)).$$

By the assumption $\forall n\neg\neg P(n)$, we have $\neg\neg P(|s|)$. Using intuitionistic logic and Markov's principle MP^ω, we get $\neg\forall x\exists y\neg Q(x,y,|s|)$. We now appeal to the intuitionistic law

$$\forall x(\phi\vee\psi(x)),\neg\forall x\,\psi(x)\ \Rightarrow\ \neg\neg\phi$$

(where x does not occur free in ϕ) to infer $\neg\neg\exists i < |s|\exists y\neg Q(s_i,y,i)$. By Markov's principle MP^ω once again, we finally conclude $B(s)$.   □

Spector's paper has an appendix in which he aims to "indicate how bar induction can be used to obtain an interpretation (of analysis) in a system with quantifiers" (cf. p. 8 of [10]). I read this statement as proposing to show that the translations A^D of instances A of (negative translations of) numerical comprehension somehow follow from bar induction. Rather than presenting an x-witnessing term for $A_D(x,y)$, Spector sets himself the goal of proving the statement $\exists x\forall y A_D(x,y)$. Section 12.1 of the appendix describes an informal proof of this result for the particular case of comprehension for Σ^0_1 predicates.[13] Our result above can be viewed as providing a formal argument for the general case.[14,15]

Acknowledgements. This work was partially financed by Fundação para a Ciência e a Tecnologia, project reference PTDC/MAT/104716/2008.

References

1. Avigad, J., Feferman, S.: Gödel's functional ("Dialectica") interpretation. In: Buss, S.R. (ed.) Handbook of Proof Theory. Studies in Logic and the Foundations of Mathematics, vol. 137, pp. 337–405. North Holland, Amsterdam (1998)
2. Brouwer, L.E.J.: Über Definitionsbereiche von funktionen. Mathematische Annalen 93, 60–75 (1927); English translation in [13], pp. 457–463
3. Ferreira, F.: A most artistic package of a jumble of ideas. Dialectica 62, 205–222 (2008); Special Issue: Gödel's *dialectica* Interpretation. Guest editor: Thomas Strahm

---

[13] It is Spector himself who says that the proof is informal. It uses a Brouwerian continuity principle and it is best seen as an argument in Brouwerian (intuitionistic) analysis. Note that the proof only uses bar induction of numerical type (cf. note 9).

[14] According to a letter of Spector to Kreisel mentioned in the first footnote of [10], this seemed to be the intent of Spector were he able to complete his paper.

[15] The main theorem of this note can also be obtained from the work of Howard in [6]. Our proof has the advantage of being very direct.

4. Gödel, K.: Über eine bisher noch nicht benützte Erweiterung des finiten Stand-punktes. Dialectica 12, 280–287 (1958); Reprinted with an English translation in [5], pp. 240–251
5. Gödel, K., Feferman, S., et al. (eds.): Collected Works, vol. II. Oxford University Press, Oxford (1990)
6. Howard, W.A.: Functional interpretation of bar induction by bar recursion. Compositio Mathematica 20, 107–124 (1968)
7. Kohlenbach, U.: Applied Proof Theory: Proof Interpretations and their Use in Mathematics. Springer Monographs in Mathematics. Springer, Berlin (2008)
8. Kreisel, G.: Interpretation of analysis by means of constructive functionals of finite types. In: Heyting, A. (ed.) Constructivity in Mathematics, pp. 101–128. North Holland, Amsterdam (1959)
9. Oliva, P.: Understanding and Using Spector's Bar Recursive Interpretation of Classical Analysis. In: Beckmann, A., Berger, U., Löwe, B., Tucker, J.V. (eds.) CiE 2006. LNCS, vol. 3988, pp. 423–434. Springer, Heidelberg (2006)
10. Spector, C.: Provably recursive functionals of analysis: a consistency proof of analysis by an extension of principles in current intuitionistic mathematics. In: Dekker, F.D.E. (ed.) Recursive Function Theory: Proceedings of Symposia in Pure Mathematics, vol. 5, pp. 1–27. American Mathematical Society, Providence (1962)
11. Troelstra, A.S. (ed.): Metamathematical Investigation of Intuitionistic Arithmetic and Analysis. Lecture Notes in Mathematics, vol. 344. Springer, Berlin (1973)
12. van Atten, M.: On Brouwer. Wadsworth (2004)
13. van Heijenoort, J. (ed.): From Frege to Gödel. Harvard University Press (1967)
14. Yasugi, M.: Intuitionistic analysis and Gödel's interpretation. Journal of the Mathematical Society of Japan 15, 101–112 (1963)

Sets of Signals, Information Flow, and Folktales

Mark Alan Finlayson

Computer Science and Artificial Intelligence Laboratory, Massachusetts Institute of
Technology, 32 Vassar Street, Cambridge MA 02139, United States of America
markaf@mit.edu

Abstract. I apply Barwise and Seligman's theory of information flow
to understand how sets of signals can carry information. More precisely I
focus on the case where the information of interest is not present in any
individual signal, but rather is carried by correlations between signals.
This focus has the virtue of highlighting an oft-neglected process, viz., the
different methods that apply categories to raw signals. Different methods
result in different information, and the set of available methods provides
a way of characterizing relative degrees of intensionality. I illustrate my
points with the case of folktales and how they transmit cultural information. Certain sorts of cultural information, such as a grammar of hero
stories, are not found in any individual tale but rather in a set of tales.
Taken together, these considerations lead to some comments regarding
the "information unit" of narratives and other complex signals.

1   A Theory of Information Flow

In their book "Information Flow: The Logic of Distributed Systems," Barwise
and Seligman [1] present a mathematically sophisticated theory of *how things
can carry information about other things.* Barwise and Seligman started from
Dreske's seminal work on information flow [2], and expanded and formalized
his observations, integrating his approach with related approaches, resulting in
a more general formulation. (From here on out I will refer to this general formulation as the "DBS" theory of information flow, short for Dreske-Barwise-
Seligman). I observe that the DBS theory is, in fact, even more general than it
at first appears, and it is my aim to illustrate how it can be used to frame and
describe several important facets of information flow, knowledge, and belief that
were left unelaborated in both Barwise and Seligman's and Dreske's work. In
particular, I will show how the DBS theory, without modification, can be used
to conceptualize two important items which Dreske touched upon only tantaliz-
ingly: learning and intensionality. I show how this conceptualization brings into
relief a part of information channels that is often taken for granted in philosophical analyses, namely, the process by which categories are applied to raw signals.
I will then apply these insights to make some comments on the information
content of cultural narratives (folktales).

I set the stage by reviewing in brief the relevant parts of the DBS theory.
The theory involves, at its core, *classifications* and *infomorphisms.* These two

S.B. Cooper, A. Dawar, and B. Löwe (Eds.): CiE 2012, LNCS 7318, pp. 228–236, 2012.

objects are used to model how information flows across *distributed systems*, which are systems that can be analyzed in terms of both a whole and constituent parts. In Barwise and Seligman's terminology, an *information channel* brings classifications and infomorphisms together into a full model of the information flow of a particular system.

We shall lose no generality if we restrict ourselves to discussing a distributed system W comprising only two parts, a proximal part P, to which we have direct access and be thought of as the "receiving" end, and a distal part D, from which information is flowing. There are infomorphisms that map properties of the classifications of the distal and proximal parts to the whole; call these d and p, respectively. To provide a concrete example to discuss, let us take Barwise and Seligman's example of a nuclear reactor: in this case W is the whole reactor, D will be the reactor core, and P will be a gauge in the reactor control room, and d and p are the regularities that connect the core to the reactor to the gauge.

A *classification* is similar to what one thinks of when considering the standard classification task in cognitive psychology: it is a set of labels or classes that may be applied to some object or phenomenon. Classifications can be, for example, mutually exclusive (e.g., {SQUARE,CIRCLE}), exhaustive (e.g., {TRUE,FALSE}), or overlapping (e.g., {TALL,FAT}). They can also be none of those things. Importantly, though, each part, as well as the whole, receives a classification. For our reactor example we might consider the reactor core D to be classified by the exclusive types NORMAL and OVERHEATING, the reactor status gauge can show one of GREEN or RED, while the reactor overall can be in one of the four states achieved by the cross product of these two classifications.

An *infomorphism* relates classifications on a part to classifications on the whole. It is a way of describing how classifications are transformed as the information they carry moves through the distributed system, from one part to another: they are models of the regularities that allow information flow. In such a system one infomorphism d may be applied to the distal part's classification to obtain a classification on the whole, and then another infomorphism p may be applied in reverse to the classification on the whole to obtain a classification on the proximal part. We need not say too much about infomorphisms except that, as they are applied in the forward and reverse directions, the resulting classifications loose some of their guarantees and internal relationships and are downgraded to what Barwise and Seligman call *local logics*. In the reactor example, the combined infomorphisms from reactor core to reactor whole, and then from reactor whole to control room gauge, given a reactor in working order, results in a display of GREEN on the gauge when the core is NORMAL, and a display of RED on the gauge when the core is OVERHEATING. Thus information flows from the distal part of the system (the reactor core) to the proximal part of the system (the control room gauge).

The details are not critical to my argument, but there are two essential points to take away from this description. First is that regularities across the system, modeled by chains of infomorphisms, are what allow information to flow from one part of the system to another. (It is often helpful to think of these regularities, like Dretske did, as lawful relationships, but one must remember that not all regularities are lawful.) Second, classifications are the language by which the information is communicated and information flow is relative to the classifications of the whole and its parts.

2   Information Flow via Sets of Signals

Information flow in the DBS theory is intimately connected, whether explicitly or implicitly, with *signals*. The examples in Dretske's and Barwise and Seligman's work are all concerned with individual signals. A signal is not defined precisely in either work, but one gathers it follows the natural intuitions: signals carry information and they are relatively localized in time and space. Signals flow across a distributed system from the distal part to the proximal part. They are the messages that contain the information. A light flashing SOS, reading the symbols off a map, a speech act: these are all signals.

I turn to an interesting and important case, that where the information of interest is not present in any individual signal, but rather is carried by correlations between signals. Regularities in a distributed system can result not only in information carried in a single signal; certain types of regularities can also result in correlations between signals.

How can this be so? Here is an example. Let us consider the reactor, and ask a simple question: Is the reactor more often NORMAL or OVERHEATING? Perhaps not a very interesting question, and one whose answer is obvious to anyone who knows much about nuclear reactors and how they are designed and run. But imagine that you know very little at all about nuclear reactors. Then, certainly, you would agree that if you were to learn the answer to this question about a particular reactor, then you would be the recipient of information. How we answer the question is straightforward: we simple observe the gauge periodically, noting whether it the gauge is NORMAL or OVERHEATING. Eventually, we stop and count up our observations, and whichever type outnumbers the other, that is our answer for this particular reactor.

How often we observe the gauge, for how long, doesn't matter much for my argument. What is important is that we cannot know, by observing any individual signal from gauge, whether the reactor is more often NORMAL or not. The information is not contained in any one signal, it is only contained in a collection of those signals. Now, one may object that this question is contrived and uninteresting, and does not represent the sort of information we are interested in studying. But, in fact, this is a common scientific question: "Is it more likely that X or Y for some type of signal?" Doctors, for example, ask the question of whether or not patients are more likely to die or be cured (or something in between) when they use or don't use a particular drug. Engineers ask whether a building is more likely to fall down in an earthquake when designed this way or that. Astronomers ask whether it is more likely for a type of star to go supernova sooner rather than later, or not all, if it has this or that characteristic. All of these examples are more complex than the reactor example, in that answering these questions usually involves correlating signals from multiple parts of the system at many different times, using much more complex methods, but the basic principle is the same: one cannot obtain the information from any one signal. The set of signals itself becomes an information channel.

How may this be analyzed within the DBS theory? The distributed system becomes, not a single instance at a particular time of the system under consideration, but a set of distributed systems, each one at different times. Each instance might be a specific time instance of a particular distributed system, they might be different instances of different distributed systems (all similar in some relevant way), or some mixture in between. The infomorphisms still reflect the regularities that underlie the system, they now just describe regularities spread across distributed system instances, and thus, time and space. The classification may be thought of as all the possible answers to the question — what those who do statistical analyses might call the *hypothesis space* of the problem. When we finally determine what is the actual answer, we have pinned a particular type on the receiving end of our set-of-signals distributed system, and we are the recipient of information.

This focus on the set of signals and the observation that scientists use sets of signals to answer scientific questions highlights an important fact: the way one correlates the signals in the set is key to the extraction of the information. Different methods result in different information flowing across the system. This choice of method contains much of the contribution of science: how do we process the raw data so as to uncover the information that we seek?

Naturally, the differences in information between methods may result from different classifications used or implied by each method. This is exactly the same as in the individual signal case where different information flows when we have a different classification for a part or the whole. But, there is an important distinction I would like to highlight, namely, that different methods might produce different answers for the *same classification*. For example, some correlation techniques might give a wider or narrower range for an answer (on an ordinal scale); on the other hand, a different technique might give a completely different answer. Thus in the scientific literature much effort is spent on justifying one's technique on principled grounds, and much is made of two different techniques converging on the same answer.

3   Learning and Intensionality

The above treatment shows that the DBS theory may be applied beyond examples containing a single signal. This allows us to frame two phenomena that are left unelaborated in the DBS analysis.

The first phenomenon is learning. Dretske noted that "Learning, the acquisition of concepts, is a process whereby we acquire the ability to extract ... information from the sensory representation." [3, p. 61] Learning can be described in the DBS theory by framing it as a set-of-signals information channel. We begin with individual signals that are unclassified. Moving up to the set of signals level, we apply a correlation method for identifying the type that applies to particular signals in particular circumstances. Having learned this classification we may return to the single signal case, and apply the newly learned classification. In the reactor example, suppose we learn, via observations at multiple times, and application of a particular correlation method, that certain gauges on the reactor always move in synchrony. This is a classification. When next presented with an individual signal, where perhaps we can observe only a single gauge, we can infer the state of the other gauges from a single observation. Similarly with what presumably happens when a child learns a new word. Daddy says "airplane!" and points. This happens several times. Perhaps there are some near-misses that aid learning ("No, honey, that's a butterfly.") Eventually, by correlations between all these signals, the child learns the category, and now can say "airplane" herself when seeing only a single signal. Learning has occurred.

The second phenomenon is degree of intensionality. Dretske said: "Our experience of the world is rich in information in a way that our consequent beliefs are not. ... The child's experience of the world is (I rashly conjecture) as rich and as variegated as that of the most knowledgable adult. What is lacking is a capacity to exploit these experiences in the generation of reliable beliefs (knowledge) about what the child sees. I, my daughter, and my dog can all see the daisy. I see it as a daisy. My daughter sees it simple as a flower. And who knows about my dog?" [3, p. 60] Dreske describes these differences in the perceptions as differences in *intensionality*. We can characterize this degree of intensionality by equating it with the method for extracting the information from the signal. The more sophisticated the correlation method, the more complex and varied the proximal classification, the more *intensionality* we assign to the agent in question. (This observation might lead us to hope that we can provide a full or partial order over intensionalities. Unfortunately this is not to be — see the next section.)

There are a number additional observations that may be made on this topic. For example, if we talk about information carried by sets of signals, why not talk about information carried by sets of sets of signals? Or sets of sets of sets? This, perhaps, is the same as talking about learning about learning, and so forth. We might also explore how the scientific method in general, or specific fields of inquiry, such as machine learning, are illuminated by these observations. We could examine in more detail how the learning method intervenes between signal and classification. But rather than explore these interesting lines of inquiry, I turn my attention to an application of these observations to a domain of particular interest to me: cultural information as carried by sets of folktales.

4   Folktales and Narrative Structure

My switchings gears to the topic of cultural information as carried by sets of folktales may seem like a *non sequitur*. I assert several reasons for the attention. First, narratives are an excellent example of a complex signal which contains myriad subtle sorts of information. Everyone is familiar with folktales, and so they will serve as an effective proxy for all sorts of complex signals with multiple possible interpretations without the overhead of detailed setting of the ground. I would like to use what I know about them to explore more this idea of varying degrees of intensionality, and for this purpose they have the advantage of us not yet knowing, scientifically speaking, what exactly is the information contained in them, and therefore we need not suspend our disbelief to imagine that there may be several ways of interpreting the information contained in folktales: we have several proposals (I will consider two) and we don't know which one (or ones) is right. Second, these observations allow me to pose, and explore a bit, some interesting questions about the nature of information carried in narratives. Third, narratives and culture are of central importance my work, and I am the one writing this paper. So bear with me.

Folktales specifically, and narratives in general, clearly communicate information aside from any considerations of their properties across a set. They are like any other text: they communicate information as individual objects. In a folktale in particular, and narratives in general, we can learn things such as who the characters are, what their plans and goals are, and what they are doing and when. (Although, usually being about a fictional world, it is an interesting question whether this information translates into knowledge.) This sort of information, the sort contained in an individual tale, is not the information we are interested in here. I am interested, rather, in information that is communicated across a set of tales.

There are numerous types of information communicated by a set of folktales. My work so far has focused on a particular sort, that of narrative structure of the plot [4]. This information corresponds to a grammar for plots, specific to the culture in question. Much like a natural language, a folktale grammar provides a set of symbols (plot pieces) and rules of combination that allow us to build folktales in that culture. Much like the grammar of a natural language is not captured in a single sentence, the grammar of the folktales is not captured in any single tale. There have been many proposals for the form of these grammars, proposals that span the range from universal theories across all stories, to highly culturally-specific theories for certain genres of folktales. I will contrast two examples, the first being Vladimir Propp's theory of the morphology of the folktale [6].

Propp's theory lays out a grammar in three levels, where the middle level, that of the *function*, has 31 pieces that describe the major constituents of Russian fairy tales. These pieces include such plot fragments as *Villainy*, *Struggle* (between the Hero and Villain), and *Reward* (of the Hero for defeating the Villain). I demonstrated that these plot pieces and rules of combination, rather than being figments of Propp's imagination, can be learned by a computerized correlation method from the actual tales [4]. I call the method *Analogical Story Merging*, which is modification of a machine learning technique called *Bayesian Model Merging* that relies on correlations uncovered by a statistical process leveraging Bayes' rule. Key to the method is a bias function called the *prior* that tells the method what similarities it should consider important when considering what parts of the folktales may be patterns. In the case of learning Propp's theory, the prior focuses the method on three important features: the semantic character of events; which characters are involved in those events; and where the events occur in the timeline of the tale.

This is all well and good. We have a method that extracts higher-level plot patterns from sets of folktales, where the patterns themselves cannot be seen by examining just a single tale. We have identified information flow from a set of tales, and in contrast to other information in an individual folktale, there is a fair chance that this information actually reflects something in reality (rather than a fictional world) — it likely reflects the ideas of participants in the culture under examination, such as the sorts of bad things that can happen to people, the proper conduct of a heroic person, and the rewards for heroic behavior. But is this the only information transmitted by sets of folktales?

Consider a competing proposal for narrative structure, that of Lévi-Strauss [5]. In his structural analysis of myth, he identified units of analysis that he called *mythemes*, where each mytheme was a set of semantic units unified by their treatment of a common theme, such as *death* or *familial relations*. In contrast to Propp's so-called *diachronic* analysis of the tales, where each function is laid out in the order it is encountered in the tale, Lévi-Strauss organized his analysis *synchronically*, where mythemes are organized by theme regardless of their position in the tale. Moreover, Lévi-Strauss's 'grammar' (if it may be called that) is highly constrained, consisting of two paired binary oppositions arranged in a specific relationship. While I don't have an algorithm that demonstrates learning Lévi-Strauss's theory from stories, it is clear that the method I used for learning Proppian structures would not be sufficient, merely from theoretical limitations of grammatical inference.

Given both Propp's and Lévi-Strauss's analyses, what are we to say about the information they contain, relative to one another? Lévi-Strauss's theory is not a specialization of Propp's, or vice-versa — they are completely orthogonal. One needs a completely different method to learn Lévi-Strauss's theory from the stories. So which is the actual information carried across the set? The answer is clear, in that it depends on the method one uses. Both theories, if underwritten by regularities in the distributed system (of people, culture and folktales) describe equally valid information carried by the set. They are two completely different interpretations of what is going on.

This observation points the way toward understanding what is going on with different degrees of intensionality. Indeed, these examples show that intensionality is not a matter of degrees at all. We have an intuitive ranking of the dog's, child's, and Dretske's classifications of the daisy, but these are all relative to an implicit value judgement about merely one aspect of the classification method. Dretske's classification is a refinement of the child's, and the child's a refinement of the dog's (we suppose), and we have implicitly associated a more refined classification with a higher degree of intensionality. But in the case of narrative structure, there is no such refinement relationship. The two theories attend to quite different patterns in the texts at hand. Thus we see clearly that intensionality, in the general case, does not come with a clear intuitive ranking. Intensionality is relative to some measure on the method we are using to extract our information. We may find, for a particular situation, that complexity of the method, or complexity of classification, or utility for some purpose, are the appropriate way to rank and order intensionalities. We have refined the question from *what makes this processor of information more intensional than that one?* to *what characteristics of the classification method explain our intuitions of degree of intensionality?*

5   Information in Individual Narratives

Information can flow from a set of narratives. What can we do with it? We can of course go back and apply that knowledge to a single narrative. For example, in a Proppian-style analysis, suppose we have learned from our set of narratives that there is something we decide to call a *Villainy* in the culture in question, and it takes certain specific forms. The method shows us what to pay attention to for when looking for villainies, and so now we can (usually) pick out a villainy in an individual story. This does not mean, of course, that the individual narrative contains the information about the nature of villainies — we learned that from the set of narratives. Our concept of villainy depended on analyzing the whole set. It contains the information that there is (or is not) a villainy in that particular narrative. This is just another way of emphasizing, as the DBS theory does, that information flow is relative to the receiver. Interestingly, for cultural narratives, what information flows at the narrative structure level is a function not only of the method used (e.g., a Proppian-style analysis, or Lévi-Straussian analysis, or something else), but also a function of the contents of the set itself. Change the set of tales, to folktales from another culture, and you get different functions [4, §6.1.5]. This naturally leads to the questions of how we decide what set of narratives to analyze? What principles should guide that selection? In my work, and Propp's, the principle was a representative selection of a particular genre of folktale from a particular culture. For other purposes the principle could be quite different.

There is a second point of interest. Naturally, even if one keeps the selection principle the same, the nature of the information extracted from the set of folktales varies with the number of tales in the set [4, Fig. 5-3]. For smaller sets, we learn fewer Proppian, and they are learned with less fidelity. Thus, in a sense, when fewer tales support the higher-level analysis, the information carried by the individual tale is coarser, and the information "chunk size" is larger. One would imagine, when doing a Proppian analysis on a set that properly contains the set of tales analyzed by Propp, that one might find more than 31 functions. (Indeed, I noted a possible missing function of this sort [4, §5.5.4].) In this case, with more tales, the information chunk is smaller, and the information carried by the tale is finer.

In a Proppian-style analysis, if Propp's functions can be considered the "information unit" or "chunk size" of the plot of the narrative, relative to some particular set of folktales, how do we know when we have the right sized chunk? I cannot think of any philosophical reason that the chunks, in this particular case, will be of one size rather than another. I imagine it will boil down to an experimental question, where one may find, upon adding more and more folktales to the original set, that the chunk size does not get any smaller. For another style of analysis one may find that there is no stable point, and the chunk size always depends on the number of narratives added. This could potentially be a discriminator between effective and ineffective theories of narrative structure.

Acknowledgements. This work was funded by the Office of Naval Research under award number N00014-09-1-0597. Any opinions, findings, and conclusions or recommendations expressed here are those of the author and do not necessarily reflect the views of the Office of Naval Research.

References

1. Barwise, J., Seligman, J.: Information Flow: The Logic of Distributed Systems. Cambridge University Press, Cambridge (1997)
2. Dretske, F.I.: Knowledge and the Flow of Information. MIT Press, Cambridge (1981)
3. Dretske, F.I.: Précis of knowledge and the flow of information. The Behavioral and Brain Sciences 6(1), 55–90 (1983)
4. Finlayson, M.A.: Learning Narrative Structure from Annotated Folktales. Ph.D. thesis, Department of Electrical Engineering and Computer Science. Massachusetts Institute of Technology (2011)
5. Lévi-Strauss, C.: The structural study of myth. The Journal of American Folklore 68(270), 428–444 (1955)
6. Propp, V.: Morphology of the Folktale. University of Texas Press, Austin (1968)

On the Foundations and Philosophy of Info-metrics

Amos Golan

Info-Metrics Institute and Department of Economics, American University, 4400
Massachusetts Avenue, Washington, DC 20016, United States of America
agolan@american.edu

1 Background, Objective and Motivation

Among the set of open questions in philosophy of information posed by Floridi
[5,6] is the question of "What Is the Dynamics of Information?" For recent
discussion see Crnkovic and Hofkirchner [2]) and a complimentary summary of
open questions in the interconnection of philosophy of information and computation (Adriaans [1]). The broad definition of "dynamics of information" includes
the concept of "information processing." This article concentrates on that concept, redefines it as "info-metrics," and discusses some open questions within
info-metrics.

Inference and processing of limited information is one of the most fascinating
universal problems. We live in the information age. Information is all around
us. But be it much or little information, perfect or blurry, complementary or
contradicting, the main task is always how to process this information such that
the inference – derivation of conclusions from given information or premises – is
optimal.

The emerging field of info-metrics is the science and art of inference and quantitatively processing information. It crosses the boundaries of all sciences and
provides a mathematical and philosophical foundation for inference with finite,
noisy or incomplete information. Info-metrics is at the intersection of information theory, statistical methods of inference, applied mathematics, statistics and
econometrics, complexity theory, decision analysis, modeling and the philosophy
of science. From mystery solving to the formulation of all theories – we must infer
with limited and blurry observable information. The study of info-metrics helps
in resolving a major challenge for all scientists and all decision makers of how
to reason under conditions of incomplete information. Though optimal inference
and efficient information processing are at the heart of info-metrics, these issues
cannot be developed and studied without understanding information, entropy,
statistical inference, probability theory, information and complexity theory as
well as the meaning and value of information, data analysis and other related
concepts from across the sciences. Info-metrics is based on the notions of information, probabilities and relative entropy. It provides a unified framework for
reasoning under conditions of incomplete information.

Though much progress has been made, there are still many deep philosophical and conceptual open questions in info-metrics: What is a correct inference

S.B. Cooper, A. Dawar, and B. Löwe (Eds.): CiE 2012, LNCS 7318, pp. 237–244, 2012.

method? How should a new theory be developed? Is a unified approach to inference, learning and modeling necessary? If so, does info-metrics provide that unified framework? Or even simpler questions related to real data analyses and correct processing of different types and sizes of blurry data fall at the heart of info-metrics. Simply stated, modeling flaws and inference with imperfect information is a major challenge. Inconsistencies between theories and empirical predictions are observed across all scientific disciplines. Info-metrics deals with the study of that challenge. These issues – the fundamental problem, current state, thoughts on a framework for potential solutions and open questions – are the focus of that article.

The answer to the above questions demands a better understanding of both information and information processing. That includes understanding the types of information observed, connecting it to the fundamental – often unobserved – entities of interest and then meshing it all together and processing it in a consistent way that yields the best theories and the best predictions. Info-metrics provides that mathematical and philosophical framework. It generalizes the Maximum Entropy (ME) principle (Jaynes [10,11]) which builds on the principle of insufficient reason. The ME principle states that in any inference problem, the probabilities should be assigned by maximizing Shannon's information measure called entropy (Shannon [13]) subject to all available information. Under this principle only the relevant information is used. All information enters as constraints in an optimization process: constraints on the probability distribution representing our state of uncertainty. Maximizing the entropy subject to no constraints yields the uniform distribution representing a state of complete uncertainty. Introducing meaningful information as constraints in the optimization takes the distribution away from uniformity. The more information there is, the further away the resulting distribution is from uniformity or from a state of complete uncertainty. (For detailed discussion on the philosophy of information, as well as dynamic information, across the sciences see van Benthem, and Adriaans [4]; the resent text of van Benthem [3]; the proceedings of the Info-Metrics Workshop on the Philosophy of Information [12].)

In this article, I will provide a summary of the state of info-metrics, the universality of the problem and a solution framework. I will discuss some of the open questions and provide a number of examples from across the scientific spectrum throughout this article.

2   Examples and Framework

Much has been written on information. Much has been written on inference and prediction. But the interdisciplinary combined study of information and efficient information processing is just starting. It is based on very sound foundations and parts of it have long history within specific disciplines. It provides the needed link to connect all sciences and decision processes. To demonstrate the basic issues in info-metrics and the generality of the problem, consider the following very simple examples.

Consider "betting" on the state of the economy as is conveyed by the data. This is expressed nicely in a recent Washington Post editorial (August 9, 2011): "Top officials of the Federal Reserve, and their staff, assembled around a gigantic conference table to decide what, if anything, they should do to help the flagging U.S. economy. Just about every member of this group, headed by Chairman Ben S. Bernanke, was an expert in economics, banking, finance or law. Each had access to the same data. And yet, after hours of discussion, they could not agree...." What would you do? How would you interpret the data? Would you follow the Chairman or would you construct your own analysis? There is not enough information to ensure one solution. There are many stories that are fully consistent with the available information and with the data in front of each one of the Fed members. The Chairman's solution is only one such story. The Fed members face blurry and imperfect data. Simply phrased, the question is what is the solution to X+Y = more-or-less 10. Unlike the previous case, even if you have more information (say, X = more-or-less 3), there are still infinitely many solutions to this two blurry equations problem. Using information theory and the tools of info-metrics yields the most conservative solution.

Now, rather than "betting" on the story the "data" represent, consider a judge "betting" on the District Attorney's story based on the observable evidence, or a Justice "betting" on the correct interpretation of the constitution as is nicely expressed by Justice Souter's remarks at the 2011 Harvard Commencement: "...The reasons that constitutional judging is not a mere combination of fair reading and simple facts extend way beyond the recognition that constitutions have to have a lot of general... . Another reason is that the Constitution contains values that may well exist in tension with each other, not in harmony. Yet another reason is that the facts that determine whether a constitutional provision applies may be very different from facts like a person's age or the amount of the grocery bill; constitutional facts may require judges to understand the meaning that the facts may bear before the judges can figure out what to make of them. And this can be tricky..." The justices are faced with the same evidence (information). But in addition to the hard evidence, the justices incorporate in their decision process other information such as their own values, their own subjective information, and their own interpretation of the written word. They end up disagreeing even though they face the same hard evidence. Is there a trivial way of solving it? No. But, there is a way to consistently incorporate the hard evidence and the softer information such as prior beliefs, value judgment, and imprecise meanings of words within the information-theoretic and info-metrics framework mentioned earlier. Building on these examples, I discuss the universal problem of information processing and a framework for inference with the available information.

3   Universality

Generally speaking, what is common to decisions made by Supreme Court justices, communication among individuals, scientific theories, literary reviews, art critiques, image reconstruction, data analyses, and decision making? In each case, inference is made and information is processed. This information may be complete or incomplete, perfect or blurry, objective or subjective at the decision time or at the moment of analysis. The justices are faced with the same evidence (information). But in addition to the evidence, each justice decides the case using her/his own values and her/his own subjective information and interpretation of the written word. They end up disagreeing even though they face the same hard evidence. When communicating with one another, individuals use their own interpretation and understanding of each word, phrase, gesture and facial expression. Again, the subjective information is incorporated with the hard evidence – the words. Scientific theories are based on observed information (and observable data) together with sets of axioms and assumptions. Some of these axioms and assumptions are unobservable and cannot be verified. They reflect the very basic beliefs together with the minimally needed information the scientist needs to deduce and infer a new theory. They are the fundamentals necessary for building a theory but they are not always observed. Again, the subjective information (beliefs, assumptions, axioms) is incorporated with the hard evidence – the observed data. Literary and art reviewers are also faced with the same evidence: the painting, the text, the dance, the poem. But, in addition to evidence, they evaluate it based on their own subjective information. Image and data processing involve observable, unobservable and subjective information. Since all data processing involve noisy information, it is impossible to do inference without incorporating subjective and other structural information. To complicate things further, some of the observed or unobserved information may incorporate contradicting pieces of information making it practically impossible to reconcile it within a single, unified theory. Even worse: if the different pieces of information are complementary, how much value should each one receive? The common thread across these examples is the need for coherent and efficient information processing and inference. The problem is universal across all disciplines and across all decision makers.

4 The Starting Premise – The Observer

If one views the scientist as an observer (not a common view within the social sciences) then the universal problem of inferring with limited information can be resolved. That solution framework serves as a common thread across all sciences. The idea is very simple. Regardless of the system or question studied, the researcher observes only a certain amount of information or evidence. The observer needs to process that information in a scientifically honest way. That means that she has to process the observed information in conjunction with the unobserved information coming from theoretical considerations, intuition or subjective beliefs. That processing should be done within an optimal inference framework while taking into account the relationship between the observable and the unobservable; the relationship of the fundamental (often unobserved) entity of interest and its relationship to the observable quantities; the connection between the micro state (usually unobserved) and the global macro state of the system (usually observed); the value of the observed information; specification of the theoretical and other subjective information; and finally even if one has all the answers and inference is made, it needs to be verified.

As an observer the scientist and the decision maker become detectives that must combine all the types of information, process it all and make their inference. When possible they validate their theory or decision and update their inference accordingly. With that view we are back at the universal problem of inference with limited information. But now we already acknowledge the advantage of tackling that challenge as external observers. Since we deal with information, it seems natural that the study of this problem takes us back to information theory and the earlier work of Jaynes on the Maximum Entropy formalism.

5   Open Questions

Keeping in mind that all information is finite and that the observed information is often very limited and blurry, all inferential problems are inherently underdetermined (as discussed above). Further, the information observed is of different types: "hard" data, "soft" data and prior information. The "hard" data are just the traditional observed quantities. The "soft" data represent our own subjective beliefs and interpretation as well as axioms, assumptions and possible unverifiable theories. The prior information represents our prior understanding of the process being studied (often it is based on some verifiable data or on information coming from previous studies or experiments).

I will start with a few examples of small and large data and then I will provide a short list of open questions.

A number of simple representative examples capture some of the basic problems raised so far. Consider for example analyzing a game between two – or more – individuals. Traditionally, one starts by assuming the players are rational or maximize a certain objective function. Other structure (on the information set each one of the players has) is also assumed. But in reality the researcher does not observe the individual's preferences (or objectives) but rather observes the actions taken by these players. But similar to previous examples, this means that the problem is inherently underdetermined: there are many stories/games that are consistent with the observed actions. The info-metrics framework described here allows one to analyze such problems. Bono and Golan (2011) formulate that problem. They formulate a solution concept without making assumptions about expected utility maximization, common knowledge or beliefs. Beliefs, strategies and the degree to which players maximize expected utility are endogenously determined as part of the solution. To achieve this, rather than solving the game from the players' point of view, they analyze the game as an "observer" who is not engaged in the process of the game but observes the actions taken by the players. They use an information-theoretic approach which is based on the Maximum Entropy principle. They also compare their solution concept with Bayesian Nash equilibrium and provide a way to test and compare different models and different modeling assumptions. They show that alternative uses of the observer's information lead to alternative interpretations of rationality. These alternative interpretations of rationality may prove useful in the context of ex post arbitration or the interpretation of experimental data because they indicate who is motivating whom.

This example brings out some of the fundamental questions a researcher has to deal with when constructing a theory and processing information in the behavioral or social sciences. For example, how can one connect the unobserved preferences (and rationality) with the observed actions? Or, how much observed information one needs in order to solve such a problem? Or what inference method is the "correct" method to use? Or how can the theory be validated with the observed information? All of these questions arise naturally when one deals with small and noisy data within the social sciences.

Consider now a different example dealing with decomposing mass spectra of gas mixtures from noisy measurements (Toussaint, Golan and Dose [14]). Given noisy observations the objective here is estimate the unknown cracking patterns and the concentrations of the contributing molecules. Again, unless more structure is imposed, the problem is underdetermined. The authors present an information-theoretic inversion method called generalized maximum entropy (GME) for decomposing mass spectra of gas mixtures from noisy measurements. In this GME approach to the noisy, underdetermined inverse problem, the joint entropies of concentration, cracking, and noise probabilities are maximized subject to the measured data. This provides a robust estimation for the unknown cracking patterns and the concentrations of the contributing molecules. The method is applied to mass spectroscopic data of hydrocarbons, and the estimates are compared with those received from a Bayesian approach. The authors also show that the GME method is efficient and is computationally fast. Like the previous example, an information-theoretic constrained optimization framework is developed. Here as well some fundamental questions arise. For example, how should one handle the noise if the exact underlying distribution is unknown? How can one connect the observed noisy data with the basic entities of interest (concentration and cracking in this case)?

Similar inferential problems exist also with big data. One trivial example is image reconstruction or a balancing of a very large matrix. The first problem is how to reduce the data (or the dimensionality of the problem) so the reconstruction will be computationally efficient. Within the information-theoretic constrained optimization (or inverse) framework discussed here one can solve that problem as well. Generally speaking, the inverse problem is transformed into a generalized moment problem, which is then solved by an information theoretic method. This estimation approach is robust for a whole class of distributions and allows the use of prior information. The resulting method builds on the foundations of information-theoretic methods, uses minimal distributional assumptions, performs well and uses efficiently all the available information (hard and soft data). This method is computationally efficient. For more image reconstruction examples see Golan, Bhati and Buyuksahin [8]. For other examples within the natural sciences see Golan and Volker [9]. For more derivations and examples (small and large data, mostly within the social sciences) see Golan [7].

Below I provide a partial list of the fundamental open questions in info-metrics. Naturally, some of these questions are not independent of one another, but it is helpful to include each one of these questions separately.

1. What is information?
2. What information do we observe?
3. How can we connect the observed information to the basic unit (or entity) of interest?
4. How should we process the information we observe while connecting it to the basic unit of interest?
5. Can we quantify all types of information?
6. How can we handle contradicting evidence (or information)?
7. How can we handle complimentary evidence (or information)?
8. Can we define a concept of "useful" information?
9. Is there a way to assign value to information? If so is it an absolute or a relative value?
10. How is the macro level information connected to the basic micro level information?
11. How to do inference with finite information?
12. Is there a way for modelling (and developing theories) with finite/limited information?
13. How can we validate our theories?
14. Is there a unique relationship between information and complexity?
15. What is a correct inference method? Is it universal to all inferential problems? (What are the mathematical foundations of that method?)
16. Is the same information processing and inference framework applies across all sciences?
17. Is a unified approach to inference and learning necessary and useful?

The above list is not complete but it provides a window toward some pressing issues within info-metrics that need more research. A more detailed discussion and potential answers to some of these questions is part of a current and future research agenda.

6   Summary

In this paper, I summarized some of the fundamental and philosophical principles of info-metrics. In addition to presenting the basic ideas of info-metrics (including its precise definition), I demonstrated some of the basic problems via examples taken from across the disciplines. A short discussion of the different types of information is provided as well. These include the "hard" information (data), "soft" information (possible theories, intuition, beliefs, conjectures, etc), and priors. A generic framework for an information-theoretic inference is discussed. The basic premise is that since all information is finite/limited (or since we need to process information in finite time) we can construct all information processing as a general constrained optimization problem (information-theoretic inversion procedure). This framework builds on Shannon's entropy (Shannon [13]), on Bernoulli's principle of insufficient reason (published eight years after his death in 1713) and on Jaynes principle of maximum Entropy (Jaynes [10,11]) and further generalizations (e.g., Golan [7]).

The paper concludes with a partial list of open questions in info-metrics. These questions are related directly to quantitative information processing and inference and to the meaning of information. In future research these questions will be tackled one at a time.

References

1. Adriaans, P.: Some open problems in the study of information and computation (2011), http://staff.science.uva.nl/~pietera/open_problems.html
2. Crnkovic, G.D., Hofkirchner, W.: Floridi's: Open Problems in Philosophy of Information. Ten Years Later. Information 2, 327–359 (2011)
3. van Benthem, J.: Logical Dynamics of Information and Interaction, pp. 1–384. Cambridge University Press, Cambridge (2011)
4. van Benthem, J., Adriaans, P.: Philosophy of Information. North Holland, Amsterdam (2008)
5. Floridi, L.: Open problems in the Philosophy of Information. Metaphilosophy 35, 554–582 (2004)
6. Floridi, L.: The Philosophy of Information, pp. 1–432. Oxford University Press, Oxford (2011)
7. Golan, A.: Information and Entropy Econometrics – A Review and Synthesis. Foundations and Trends in Econometric 2(1-2), 1–145 (2008); Also appeared as a book - paperback
8. Golan, Bhati, A., Buyuksahin, B.: An Information-Theoretic Approach for Image Reconstruction: The Black and White Case. In: Knuth, K. (ed.) 25th International Workshop on Bayesian Inference and Maximum Entropy Methods in Science and Engineering, MAXENT (2005)
9. Golan, A., Volker, D.: A Generalized Information Theoretical Approach to Tomographic Reconstruction. J. of Physics A: Mathematical and General, 1271-1283 (2001)
10. Jaynes, E.T.: Information Theory and Statistical Mechanics. Physics Review 106, 620–630 (1957a)
11. Jaynes, E.T.: Information Theory and Statistical Mechanics II. Physics Review 108, 171–190 (1957b)
12. Proceedings of Info-Metrics Institute Workshop on the Philosophy of Information. American University, Washington, DC (2011)
13. Shannon, C.E.: A Mathematical Theory of Communication. Bell System Technical Journal 27, 379–423 (1948)
14. Toussaint, U.V., Golan, A., Dose, V.: Maximum Entropy Decomposition of Quadruple Mass Spectra. Journal of Vacuum Science and Technology A 22(2), 401–406 (2004)

On Mathematicians Who Liked Logic
The Case of Max Newman

Ivor Grattan-Guinness

Middlesex University Business School, The Burroughs,
Hendon, London NW4 4BT, England

Abstract. The interaction between mathematicians and (formal) logicians has always been much slighter than one might imagine. After a brief review of examples of very partial contact in the period 1850-1930, the case of Max Newman is reviewed in some detail. The rather surprising origins and development of his interest in logic are recorded; they included a lecture course at Cambridge University, which was attended in 1935 by Alan Turing.

1   Cleft

One might expect that the importance to many mathematicians of means of proving theorems, and their desire in many contexts to improve the level of rigour of proof, would motivate them to examine and refine the logic that they were using. However, inattention has long been common.

A very important source of maintaining the cleft during the 19th century is the founding from the late 1810s onwards of the 'mathematical analysis' of real variables, grounded upon an articulated theory of limits, by the French mathematician A.-L. Cauchy. He and his followers extolled rigour, especially careful nominal definitions of major concepts and detailed proofs of theorems. From the 1850s onwards this aim was enriched by the German mathematician Karl Weierstrass and his many followers, who brought in, for example, multiple limit theory, definitions of irrational numbers, and an increasing use of symbols – and from the early 1870s, Georg Cantor and his set theory. However, absent from all these developments was explicit attention to any kind of logic.

This silence continued among the many set theorists who participated in the inauguration of measure theory, functional analysis and integral equations[1]. Artur Schoenflies and Felix Hausdorff were particularly hostile to logic, targeting Bertrand Russell. Even the extensive dispute over the axiom of choice focussed mostly on its legitimacy as an assumption in set theory and mathematics and use of higher-order quantification [Moore 1982]: its ability to state an infinitude of independent choices within *finitary* logic constituted a special difficulty for logicists such as Russell.

---

[1] The history of mathematical analysis is well covered; cf. especially [Rosenthal 1923; Bottazzini 1986; Medvedev 1991; Jahnke 2003]. A similar story obtains for complex-variable analysis.

S.B. Cooper, A. Dawar, and B. Löwe (Eds.): CiE 2012, LNCS 7318, pp. 245–252, 2012.

The creators of symbolic logics were exceptional among mathematicians in attending to logic, but also they made little impact on their colleagues. The algebraic tradition with George Boole, C. S. Peirce, Ernst Schröder and others from the mid 19th century was just a curiosity to most of their contemporaries. Similarly, when mathematical logic developed from the late 1870s, especially with Giuseppe Peano's 'logistic' programme at Turin from around 1890, he gained many followers there [Roero and Luciano 2010] but few elsewhere. However, followers in the 1900s included the Britons Russell and A. N. Whitehead, who adopted logistic (include Cantor's set theory) and converted it into their 'logicistic' thesis that all the "objects" of mathematics could be obtained; G. H. Hardy but not many other mathematicians responded [Grattan-Guinness 2000; chs. 8-9]. From 1903 onwards Russell publicised the mathematical logic and arithmetic logicism put forward from the late 1870s by Gottlob Frege, which had gained little attention hitherto even from students of foundations and did not gain much more in the following decades. In the late 1910s David Hilbert started the definitive phase of his programme of metamathematics and attracted several followers at Göttingen University and a few elsewhere; however, its impact among mathematicians was limited even in Germany[2].

The next generations of mathematicians include a few distinguished students of foundations. For example, in the USA from around 1900 E. H. Moore studied Peano and Hilbert and passed on an interest in logic and model theory to his student Oswald Veblen, then to Veblen's student Alonzo Church, and then to his students Stephen Kleene and Barkley Rosser [Aspray 1991]. At Harvard Peirce showed multiset theory to the Harvard philosopher Josiah Royce, who was led on to study logic, and especially to supervise around 1910 C. I. Lewis, Henry Sheffer, Norbert Wiener, Morris Cohen and C. J. Ducasse, the last the main founder of the Association of Symbolic Logic in the mid 1930s [Grattan-Guinness 2002]. In central Europe Johann von Neumann included metamathematics and axiomatic set theory among his concerns [Hallett 1984; ch. 8], while in Poland a distinguished group of logicians did not mesh with a distinguished group of mathematicians even though made both made much use of set theory [Kuratowski 1980]; for example, their joint journal *Fundamenta mathematicae* (1920+) rarely carried logical articles.

But the normal attitude of mathematicians remained indifference. For instance, around 1930 Alfred Tarski and others proved the 'deduction theorem' [Tarski 1941; 125-130]; [Kleene 1952; 90-98]; it gained the apathy of the mathematical community, although it came to be noted by the Bourbaki group, who normally were hostile to logics. (Maybe the reason was that their compatriot Jacques Herbrand had proved versions of it; if so, it was his sole impact on

---

[2] There does not seem to be an integrated social history of metamathematics, but one can be pieced together from the temporally ordered trio [Peckhaus 1992; Sieg 1999; Menzler-Trott 2001]. In the early stages Hilbert based logic on the existence of a thought-source, a rather peculiar notion already found in Dedekind; Zermelo worked with truth-functions, but the place of logic in his set theory is modest. Compare [Peckhaus 1994].

French mathematics.) Also, Kurt Gödel's theorems [Gödel 1931] on the incompletability of first-order arithmetic were appreciated fairly quickly by students of foundations, but the community did not become widely aware of them until the mid 1950s [Grattan-Guinness 2011].

2   Newman's Course at Cambridge

Turing's own career provides a good example of the cleft; for when he submitted his paper [Turing 1936] on computability to the London Mathematical Society they could not referee it properly, because Max Newman was the only other expert in Britain and he had been involved in its preparation [Hodges 1983, 109-114]; they seem to have accepted it on Newman's word. But this detail prompts an historical question that has not been examined: why was the mathematician Newman also a logician? An answer is given in the rest of this article; more details are given in [Grattan-Guinness 2012a]

Although he did not publish much on logic, it is clear that Newman was familiar with the technicalities of several parts of it. In particular, during his wartime period at Bletchley Park he published three technical papers, one written jointly with Turing [Newman 1942, 1943; Newman and Turing 1943]. At Cambridge in the 1930s he had taught a course on 'Foundations of mathematics', which, as is well known, was crucial for Turing when he attend it in 1935; for from it learnt about recursion theory and Gödel numbering from the only Briton who was familiar with it.

Ready for the academic year 1933-1934, Newman ran the course only for the two succeeding years before it was closed down, perhaps because of disaffection among staff as well as among students. In particular, Hardy, despite his familiarity with foundations, opined to Newman in 1937: 'though "Foundations" is now a highly respectable subject, and everybody ought to know something about it, it is, (like dancing or "groups") slightly dangerous for a bright young mathematician!'[3]. Somehow Newman continued to set questions for five of the six years that he was to remain at Cambridge before moving to Bletchley Park in 1942;[4] The questions for 1939 may have been set by Turing, who, presumably in resistance against Hardy's coolness, was invited to give a lecture course on foundations in the Lent term of 1939. He was asked to repeat it in 1940; but by then he also was at Bletchley Park[5]. How had Newman got involved in logic in the first place?

---

[3] Newman's archive, Saint John's College Cambridge; thanks to David Anderson much of it is available in digital form at http://www.cdpa.co.uk/Newman/. Individual items are cited in the style 'NA, [box] a- [folder] b- [document] c'; here 2-12-3.

[4] A Mathematics Tripos course in 'logic' was launched in 1944 by S. W. P. Steen. The Moral Sciences Tripos continued to offer its long-running course on the more traditional parts of 'Logic'.

[5] On Turing's teaching, cf. [Hodges 1983; 153,177] and the Faculty Board minutes for 29 May 1939.

3   Newman's Way in to Logic

Maxwell Hermann Alexander Neumann (1897-1984) was born in London to a German father and an English mother. He gained a scholarship to St John's College Cambridge in 1915 and took Part I of the Mathematics Tripos in the following year. But carrying the surname 'Neumann' in Britain in the Great War was not a good idea; the family changed its surname to 'Newman', and Max had to leave his college until 1919, when he returned and completed Part II of the Tripos very well in 1921.

Then, very unusually, he spent much of the academic year 1922-1923 at Vienna University. He went with two other members of his college.

One was Lionel Penrose; as a schoolboy he had been interested in Russell's mathematical logic, and he specialised in traditional versions of logic as taught in the Moral Sciences Tripos. But he also examined mathematical logic, and may well have at least have alerted his friend Newman to the subject, which was absent from the Mathematics Tripos. He became interested also in psychology (well represented in his own Tripos syllabus), especially its bearing upon logic, and he wanted to meet Sigmund Freud and Karl Bühler and other psychologists in Vienna. He seems to have initiated the visit to Vienna; his family was wealthy enough to sustain it, especially as at that inflationary time British money would last a long time in Vienna. His friendship with Newman was multi-faceted and deep.

The other was Rolf Gardiner, who was later active in organic farming and folk dancing, enthused for the Nazis [Moore-Colyer 2001], and was to be the father of the conductor Sir John Eliot Gardiner. His younger sister Margaret came along; she became an artist and a companion to the biologist Desmond Bernal. She recalled 'the still deeply impoverished town' of Vienna, where Penrose and Newman would walk side-by-side down the street playing a chess game in their heads [Gardiner 1988; 61-68].

Of Newman's contacts with the mathematicians in Vienna we have only a welcoming letter of July 1922 from ordinary professor Wilhelm Wirtinger;[6] but it seems clear that his experience of Viennese mathematics was decisive in changing the direction of his researches. His principal research interest was to become topology, which was *not* a speciality of British mathematics. By contrast, in Vienna some of Wirtinger's own work related to the topology of surfaces; in 1922 the University recruited Kurt Reidemeister, who was to become a specialist in combinatorial topology, like Newman himself; Leopold Vietoris was a junior staff member; and a student was Karl Menger (though rather ill at the time and away from Vienna).

Most notably, ordinary professor Hans Hahn was not only a specialist in the topology of curves, and in real-variable mathematical analysis; he also regarded formal logic as both a research and as a teaching topic. In particular, while Newman was there he ran a preparatory seminar on 'algebra and logic', and in later years held two full seminars on *Principia mathematica*. In addition,

---

[6] The Wirtinger letter is at NA, 2-1-2.

he supervised doctoral student Kurt Gödel working on the completeness of the first-order functional calculus with identity, and as editor of the *Monatshefte für Mathematik und Physik* published both that thesis and the sequel paper [Gödel 1931] on the incompletability of first-order arithmetic (which was to be registered as Gödel's higher doctorate).

Hahn also engaged in philosophical debates. When he had studied at Vienna University from the mid 1890s to his higher doctorate in 1905 he had participated in some of the discussion groups that surrounded certain chairs in the university. After teaching elsewhere for several years, he returned to Vienna University as a full professor of mathematics in 1921. During 1922 he led the move to appoint to the chair of natural philosophy the German physicist and philosopher Moritz Schlick; after arriving in 1923 Schlick created what was to be known as the 'Vienna Circle', with Hahn as a leading member[7]. Further, while the Circle had no agreed philosophy among all its members, Schlick, Hahn and later Carnap strongly advocated positivism and empiricism, acknowledging major influences from Ernst Mach (who had held that chair in the 1890s) and Russell.

4   After Vienna

After his return Newman developed as a (pioneer) British topologist, with a serious interest in logic and logic education and (as we shall soon see) a readiness to engage with Russell's philosophy; surely one sees heavy Viennese influences here, especially from Hahn.

In 1923 Newman applied for a college fellowship. He submitted a paper [Newman 1923a] on the avoiding the axioms of choice in developing the theory of functions of a real variable that was published that year[8], some unidentifiable discussion of solutions of Laplace's equation, and a long unpublished essay [Newman 1923b] in the philosophy of science that was completed in August. Its title, 'The foundations of mathematics from the standpoint of physics', could well have originated in a Viennese chat. Maybe he wrote some of it there; unfortunately the 161 folios do not contain any watermarks.

In this essay Newman took the world of idealised objects that was customary adopted in applied mathematics (smooth bodies, light strings, and so on) as 'certain ideals, or abstractions [...] not applicable to those of real physical objects', and contrasted it with the world of real physical that one encounters and on which he wished to philosophise. He distinguished between these two kinds of philosophising by *the different logics that they used*. The idealisers would draw on the two-valued logic, for which he cited a recent metamathematical paper by Hilbert [1922] as a source; but those interested in real life would go to constructive logic, on which he cited papers by Brouwer [1918-1919] and Hermann Weyl [1921].

We see here Newman's notable readiness to admit logical pluralism, and to put logics at the centre of the analysis of a philosophical problem; most unusual for

---

[7] On the Vienna Circle, cf. [Stadler 2001]; on Hahn, cf. [Sigmund 1995].

[8] On the context, cf. *passim* in [Rosenthal 1923] and [Medvedev 1991].

a mathematician, and far more Viennese than Cantabrigian. His college referees, Ebenezer Cunningham and H. F. Baker, were not impressed by the essay but recommended the award of the Fellowship. He neither revised the essay nor seemed to seek its publication, although occasionally he alluded to its concerns; and it must be at least a major source of his recognition of the importance of logics.

This essay built upon the awareness of logic that he must have gained at Cambridge from Penrose. That contact will have continued, for after Vienna Penrose wrote several manuscripts on mathematical logic, especially the psychological aspects, in which he was influenced by Russell and also by Ludwig Wittgenstein's notion of tautology given in the recent *Tractatus logico-philosophicus* (1922). He worked on a doctoral dissertation on the psychology of mathematics, but then abandoned it[9]. From 1925 he studied for a degree in medicine at Cambridge and London, and became a distinguished geneticist, psychiatrist and statistician, and also father of the mathematicians Oliver and Sir Roger Penrose [Harris 1973].

An occasion for Newman to exercise his logical and philosophical talents arose when he attended a set of philosophical lectures that Russell gave in Trinity College Cambridge in 1926. They went into a book on 'the analysis of matter' [Russell 1927]. Newman helped Russell to write two chapters, and when the book appeared he criticised its philosophical basis most acutely in [Newman 1928]; Russell accepted the criticisms, which stimulated Newman to write Russell two long letters on logic and on topology [Grattan-Guinness 2012b].

Newman continued to pioneer both topology and logic at Cambridge. Doubtless with topology in mind, in 1936, a year before sinking the foundations course, Hardy had proposed Newman as a Fellow of the Royal Society, with J. E. Littlewood as seconder, although Newman was no Hardy-Littlewood analyst; the election was made in 1939. Newman used the Society to support his logical cause. In 1950 he proposed Turing as a Fellow, seconded by Russell, the election being accepted in 1951; five years later he wrote the obituary [Newman 1955] of Turing. In 1966 Newman proposed and Russell seconded Gödel as Foreign Member, duly gained two years later[10]. In 1970 he agreed to be the obituarist of Russell, to be helped by the philosopher A. J. Ayer, but he was not well enough to oblige. He died in 1984.

Among mathematicians who came to like logic, Newman is a very unusual case. The (sparse) evidence suggests two sources: Penrose's early interest; and the unusual mixture of mathematics, logic and philosophy in Vienna, which drew him also to topology. Thus he changed directions; had he stayed in Cambridge in 1922-1923, he would have surely continued in the direction indicated by the paper on avoiding the axioms of choice, namely, Hardy-Littlewood mathematical analysis. But then his interest in logic could have waned (and in topology never have flowered), so that maybe no foundations course would have existed for budding Hardy-Littlewood mathematical analyst Turing to take and thereby to

---

[9] In the Penrose Papers, University College London Archives, cf. especially boxes 20-21 and 26-28.

[10] Information comes from Royal Society Archives, and NA, 2-15-10 to -13.

learn of the subjects of recursive functions and undecidability. Then the story of Bletchley Park and afterwards could have been very different; neither he nor this alternative Newman would have been the obvious choices to go there, nor would they have been as effective as they actually were. The way that things turned out for Newman and Turing contained some strokes of luck!

References

Aspray, W.: Oswald Veblen and the Origins of Mathematical Logic at Princeton. In: Drucker, T. (ed.) Perspectives on the History Of Mathematical Logic, pp. 54–70. Birkhäuser, Boston (1991)

Bottazzini, U.: The Higher Calculus. A History of Real and Complex Analysis from Euler to Weierstrass. Springer, New York (1986)

Gardiner, M.: A Scatter of Memories. Free Association Books, London (1988)

Gödel, K.: Über formal unentscheidbare Sätze der Principia Mathematica und verwandter Systeme. Monatshefte für Mathematik und Physik 38, 173–198 (1931); Many reprs. and transs.

Grattan-Guinness, I.: The Search For Mathematical Roots, 1870-1940. Logics, Set Theories and the Foundations of Mathematics from Cantor through Russell to Gödel. Princeton University Press, Princeton (2000)

Grattan-Guinness, I.: Re-interpreting "λ": Kempe on Multisets and Peirce on Graphs, 1886-1905. Transactions of the C. S. Peirce Society 38, 327–350 (2002)

Grattan-Guinness, I.: The Reception of Gödel's 1931 Incompletability Theorems by Mathematicians, and Some Logicians, up to the Early 1960s. In: Baaz, M., Papadimitriou, C.H., Putnam, H.W., Scott, D.S., Harper, C.L. (eds.) Kurt Gödel and the Foundations of Mathematics. Horizons of Truth, pp. 55–74. Cambridge University Press, Cambridge (2011)

Grattan-Guinness, I.: Discovering the Logician Max Newman (in preparation, 2012a)

Grattan-Guinness, I.: Logic, Topology and Physics: Max Newman to Bertrand Russell (1928) (in preparation, 2012b)

Hallett, M.: Cantorian Set Theory and Limitation of Size. Clarendon Press, Oxford (1984)

Harris, H.: Lionel Sharples Penrose. Biographical Memoirs of Fellows of the Royal Society 19, 521–561 (1973); Repr. in Journal of Medical Genetics 11, 1–24 (1974)

Hilbert, D.: Die logischen Grundlagen der Mathematik. Mathematische Annalen 88, 151–165 (1922); Repr. in Gesammelte Abhandlungen, vol. 3, pp. 178-191. Springer, Berlin (1935)

Hodges, A.: Alan Turing: the Enigma. Burnett Books and Hutchinson, London (1983)

Jahnke, N.H.: A History of Analysis. American Mathematical Society, Providence (2003)

Kleene, S.C.: Introduction to Metamathematics. van Nostrand, Amsterdam (1952)

Kuratowski, K.: A Half Century of Polish Mathematics. Polish Scientific Publishers, Oxford (1980)

Medvedev, F.A.: Scenes from the History of Real Functions. Birkhäuser, Basel (1991); translated by R. Cooke

Menzler-Trott, E.: Gentzens Problem. Birkhäuser, Basel (2001); English ed.: Logic's Lost Genius: the Life of Gerhard Gentzen. American Mathematical Society and London Mathematical Society, Providence (2007)

Moore, G.H.: Zermelo's Axiom of Choice. Springer, New York (1982)

Moore-Colyer, R.J.: Rolf Gardiner, English Patriot and the Council for the Church and Countryside. The Agricultural History Review 49, 187–209 (2001)

Newman, M.H.A.: On Approximate Continuity. Transactions of the Cambridge Philosophical Society 23, 1–18 (1923a)

Newman, M.H.A.: The Foundations of Mathematics from the Standpoint of Physics (1923b) manuscript, Saint John College Archives, item F 33.1

Newman, M.H.A.: Mr. Russell's "Causal Theory of Perception". Mind 37, 137–148 (1928)

Newman, M.H.A.: On Theories with a Combinatorial Definition of "Equivalence". Annals of Mathematics 43, 223–243 (1942)

Newman, M.H.A.: Stratified Systems of Logic. Proceedings of the Cambridge Philosophical Society 39, 69–83 (1943)

Newman, M.H.A.: Alan Mathison Turing. Biographical Memoirs of Fellows of the Royal Society 1, 253–263 (1955)

Newman, M.H.A., Turing, A.: A Formal Theorem in Church's Theory of Types. Journal of Symbolic Logic 7, 28–33 (1943)

Peckhaus, V.: Hilbert, Zermelo und die Institutionalisierung der mathematischen Logik. Deutschland. Berichte zur Wissenschaftsgeschichte 15, 27–38 (1992)

Peckhaus, V.: Logic in Transition: the Logical Calculi of Hilbert (1905) and Zermelo (1908). In: Prawitz, D., Westerståhl, D. (eds.) Logic and Philosophy of Science in Uppsala, pp. 311–323. Kluwer, Dordrecht (1994)

Roero, C.S., Luciano, E.: La scuola di Giuseppe Peano. In: Roero (ed.) Peano e la sua scuola, Fra matematica, logica e interlingua, Atti del Congresso internazionale di studi, Torino, October 6-7, 2008, vol. xi–xviii, pp. 1–212. Deputazione Subalpina di Storia Patria (2010)

Rosenthal, A.: Neuere Untersuchungen über Funktionen reeller Veränderlichen. In: Encyklopädie der mathematischen Wissenschaften, vol. 2, pt. C, (article IIC9), pp. 851–1187. Teubner, Leipzig (1923)

Russell, B.A.W.: The Analysis of Matter. Kegan Paul, London (1927)

Sieg, W.: Hilbert Programs: 1917-1922. Bulletin of Symbolic Logic 5, 1–44 (1999)

Sigmund, K.: A Philosopher's Mathematician: Hans Hahn and the Vienna Circle. The Mathematical Intelligencer 17(4), 16–19 (1995)

Stadler, F.: The Vienna Circle. Springer, Vienna (2001)

Tarski, A.: Introduction to Logic and to the Methodology of the Deductive Sciences. Oxford University Press, New York (1941); (1st edn., translated by O. Helmer)

Turing, A.M.: On Computable Numbers, with an Application to the Entscheidungsproblem. Proceedings of the London Mathematical Society 42(2), 230–265 (1936)

Weyl, C.H.H.: Über die neue Grundlagenkrise der Mathematik. Mathematische Zeitschrift 10, 39–79 (1921); Repr. in Gesammelte Abhandlungen, vol. 2, pp. 143-180. Springer, Berlin (1968)

Densities and Entropies in Cellular Automata

Pierre Guillon[1,2] and Charalampos Zinoviadis[1]

[1] Department of Mathematics, University of Turku, 20014 Turku, Finland
chzino@utu.fi
[2] CNRS & Institut de Mathématiques de Luminy, Campus de Luminy, Case 907,
13288 Marseille cedex 9, France
pguillon@math.cnrs.fr

Abstract. Following work by Hochman and Meyerovitch on multi-dimensional SFT, we give computability-theoretic characterizations of the real numbers that can appear as the topological entropies of one-dimensional and two-dimensional cellular automata.

1 Introduction

Cellular automata are a widely-used model for complex systems or computation, consisting in a network of cells each of whose is in one among a finite number of states, that is updated synchronously in parallel as a function of the sates of its neighbors. Their entropy is a measure of how complex or random the local long-term behavior can look like. The entropy of cellular automata has been proven uncomputable in [1] (see also [13] for subshifts), but the question remained whether the entropy of a single given cellular automaton could be an uncomputable number. Recently, M. Hochman and T. Meyerovitch have characterized the entropies of 2-dimensional SFT [7] and 3-dimensional CA [5] as, respectively, the right-computable numbers and the limits of computable increasing sequences of such numbers. We prove here that these two classes still characterize the possible entropies of, respectively, 1-dimensional and 2-dimensional CA. To do so, we adapt their homogeneous encoding [7], J. Kari's determinization signals [8] and P. Gács's self-similar construction [3]. The result brings new equivalences between classes that are equally natural in computability theory and dynamical systems; we also believe that the construction in itself is promising, and could help understand the real computational power of these natural models.

In Section 2 we introduce the notions and a brief state of the art. In Section 3 we state our main results, and the following sections are devoted to sketching their proofs. The algorithmic part, as well as entropy proofs and a sketch of the main construction can be found in [4].

2 Preliminaries

2.1 Configurations

$\mathbb{N}$ will denote the set of natural numbers, $\mathbb{N}_1$ the set $\mathbb{N} \setminus \{0\}$ of positive natural numbers and $[\![i, j]\!]$ the integer interval $\{i, \ldots j\}$, for $0 \leq i \leq j$. $\mathbb{R}_+$ is the set of nonnegative real numbers.

S.B. Cooper, A. Dawar, and B. Löwe (Eds.): CiE 2012, LNCS 7318, pp. 253–263, 2012.

Let A be a finite set called the **alphabet** and $d \in \mathbb{N}_1$ the **dimension**. Any element x of $A^{\mathbb{Z}^d}$ is called a **configuration**, and x_i is called the **state** of **cell** i. The set of configurations forms a compact topological space when endowed with the product of the discrete topology.

For any $q \in A$, $^\infty q^\infty$ denotes the q-**uniform** configuration of $A^{\mathbb{Z}^d}$, all of whose cells are in state q. If $U \subset \mathbb{Z}^d$, $x_{|U}$ is the **pattern** representing the restriction of x to U. For instance, we can define the central pattern $x_{|B(r)}$ of width r, where $B(r) = [\![-r, r]\!]^d$.

2.2 Symbolic Dynamics

$\mathbb{Z}^d$ acts on $A^{\mathbb{Z}^d}$ by the **shift**: to any $k \in \mathbb{Z}^d$ we associate the homeomorphism $\sigma^k : A^{\mathbb{Z}^d} \to A^{\mathbb{Z}^d}$ defined by $\forall x \in A^{\mathbb{Z}^d}, \forall i \in \mathbb{Z}^d, \sigma^k(x)_i = x_{i+k}$. A ($d$-dimensional, or dD) **subshift** is the set $X = \left\{ x \in A^{\mathbb{Z}^d} \,\middle|\, \forall U \underset{\text{finite}}{\subset} \mathbb{Z}^d, k \in \mathbb{Z}^d, \sigma^k(x)_{|U} \notin \mathcal{F} \right\}$ of configurations that avoid some particular set $\mathcal{F}$ of finite patterns. Equivalently, a subshift is a subset which is invariant by σ^k for any $k \in \mathbb{Z}^d$ and topologically closed. It is of **finite type** (SFT) if $\mathcal{F}$ can be chosen finite.

Let $X \subset A^{\mathbb{Z}^d}$ be a subshift. The **language** of support $U \subset \mathbb{Z}^d$ of X is $\mathcal{L}_U(X) = \left\{ x_{|U} \,\middle|\, x \in X \right\}$. Its **complexity** of support U is $\mathcal{K}_U(X) = |\mathcal{L}_U(X)|$. The (topological) **entropy** of X is $\mathcal{H}(X) = \lim_{r \to \infty} \frac{\log \mathcal{K}_{B(r)}(X)}{|B(r)|}$. This is always a limit, but may be infinite. Note that if $Y \subset B^{\mathbb{Z}^d}$ is another subshift, then $X \times Y$ can be essentially seen as a subshift of $(A \times B)^{\mathbb{Z}^d}$, and its entropy is the sum of those of X and of Y.

A subshift $Y \subset B^{\mathbb{Z}^d}$ is a **letter factor** of $X \subset A^{\mathbb{Z}^d}$ if there exists some **letter projection** $\pi : A \to B$ such that the corresponding global map $\Pi : X \to Y$, defined by the parallel application of π, is onto (we say that X **letter-factors** onto Y). A subshift is called **sofic** if it is a letter factor of some SFT.

The same definitions hold for (one-sided) subshifts over $A^{\mathbb{N}_1}$.

The **trace** of X according to vector $\boldsymbol{v} \in \mathbb{Z}^d$ and width k is the $(d-1)$D subshift $\tau_{\boldsymbol{v}}^k(X) = \left\{ (x_{|[\![0,k[\![\times \{0\} + n\boldsymbol{v}})_{n \in \mathbb{Z}} \,\middle|\, x \in X \right\}$ over alphabet A^k. The **directional entropy** according to vector $\boldsymbol{v}$ is the limit $\mathcal{H}_{\boldsymbol{v}}(X)$ of the entropies of $\tau_{\boldsymbol{v}}^k(X)$, when k goes to infinity (see [9]). One can see that $\mathcal{H}_{\boldsymbol{e_2}}(X) = \lim_{k \to \infty} \lim_{r \to \infty} \frac{\log \mathcal{N}_{k,r}(X)}{r}$, where $\mathcal{N}_{k,r}(X) = \mathcal{K}_{[\![0,k[\![\times [\![0,r[\![}(X)$.

Let X and Y be 2D subshifts. We say that X **simulates** Y with parameters B, T if there exists $Z \subset X$ such that $X = \bigcup_{0 \le i < B, 0 \le j < T} \sigma^{(i,j)}(Z)$ and that $Z^{<B \times T>} = \left\{ (x_{|[\![kB,(k+1)B[\![\times [\![lT,(l+1)T[\![})_{k,l \in \mathbb{Z}} \,\middle|\, (x_{i,j})_{i,j \in \mathbb{Z}} \in Z \right\}$ is a subshift that letter-factors onto Y, i.e., any configuration of X can be divided into $B \times T$ rectangles that project onto letters of Y. A simulation is an r-**simulation** if the letters onto which an array of $(2r+1)$ horizontally consecutive rectangles of size $B \times T$ project uniquely determine the central rectangle.

The following lemma will be useful in the sequel. $(\boldsymbol{e_1}, \boldsymbol{e_2})$ denotes the canonical base for $\mathbb{Z}^2$.

Lemma 1 ([4]). *Let X and Y be 2D subshifts such that X l-simulates Y with parameters B, T. Then, $\mathcal{H}_{e_1}(X) \leq \mathcal{H}_{e_1}(Y)/B$ and $\mathcal{H}_{e_2}(X) \leq \mathcal{H}_{e_2}(Y)/T$.*

2.3 Cellular Automata and Determinism

A **cellular automaton** (CA) is a system $F : A^{\mathbb{Z}^d} \to A^{\mathbb{Z}^d}$ such that $F\sigma^k = \sigma^k F$; equivalently there is a **radius** $r \in \mathbb{N}_1$ and a **local rule** $f : A^{\mathcal{B}(r)} \to A$ such that $\forall x \in A^{\mathbb{Z}^d}, \forall i \in \mathbb{Z}^d, F(x)_i = f(x_{|i+V})$. The **entropy** $\mathcal{H}(F)$ of F is the limit, when r goes to infinity, of the entropy of the subshift $\left\{ (F^t(x)_{|\mathcal{B}(r)})_{t \in \mathbb{N}} \,\middle|\, x \in A^{\mathbb{Z}^d} \right\}$.

We say that an SFT $X \subset A^{\mathbb{Z}^2}$ is **south-deterministic** if there is a map $F : \tau^1_{e_1} \to \tau^1_{e_1}$ that maps any line of a valid tiling to a unique line that can appear above, *i.e.*, $\forall x \in X, j \in \mathbb{Z}, F(x_{\mathbb{Z} \times \{j\}}) = x_{\mathbb{Z} \times \{j+1\}}$. It is known that F can actually be taken to be the restriction of a CA over alphabet $A \sqcup \{\bot\}$, where $\bot$ must be understood as "extension not defined"; and the entropy of F is equal to $\mathcal{H}_{e_1}(X)$ (intuitively, this comes from the fact that state $\bot$ will remain forever and not contribute to the entropy). X is **south-west-deterministic** if there is the same kind of CA on the diagonal, *i.e.*, $\forall x \in X, j \in \mathbb{Z}, F((x_{i,j})_{i=-j}) = (x_{i,j})_{i=1-j}$.

Let us say that a 2D subshift is **S0-sofic** if it is a letter-factor of some south-deterministic SFT with null entropies, *i.e.*, directional entropy 0 according to any vector.

2.4 Effectiveness

In $A^{\mathbb{Z}^d}$, it is easy to enumerate computationally a base of open sets (consider the sets of configurations sharing a given pattern as a central pattern). That way, we can define an **effectively closed** subset $S \subset A^{\mathbb{Z}^d}$ as the complement of the union of a computable sequence of open sets. It is an **effective subshift** if, besides, it is a subshift. For instance, trace of SFT are effective subshifts. Effectively closed sets can also be defined in other Cantor sets; in $A^{\mathbb{N}_1}$ they correspond to sets of configurations that are not ultimately rejected when scanned by some given TM. An **effective system** is an effectively closed subset $S \subset (A^{\mathbb{N}_1})^{\mathbb{Z}^d}$ which is invariant by the $\mathbb{Z}^d$-shift. Intuitively, it is a dynamical system where the preimages of open sets can be computed.

A Π_1 (or right-computable) number is the limit of a decreasing computable sequence of rational numbers. A Σ_2 number is the limit of an increasing computable sequence of Π_1 numbers. Equivalently, there exists an algorithm that on input k outputs the code of another algorithm $\mathcal{M}_k$ such that $\mathcal{M}_k$ enumerates the approximations of a Π_1 number h_k, the sequence h_k is increasing and converges to h. The set of Σ_2 is strictly larger than the set of Π_1 numbers, which, in turn, is strictly larger than the set of computable (Δ_1) numbers. We refer to [14] for more on these classes of numbers (and many more).

Remark 1. The binary representations of real numbers from an interval $[0, \alpha]$ form an effectively closed subset of $2^{\mathbb{N}}$ if and only if α is Π_1.

3 Results

Some evidence of the *computing power* of a given model can be given by studying the class of numbers that can be realized as entropy. Elegant characterizations have recently been achieved for multidimensional SFT.

Theorem 1 ([7,5]). *For $d \geq 2$, the class of entropies of d-dimensional SFT (resp. d-dimensional sofic subshifts, effective subshifts) is $\mathbb{R}_+ \cap \Pi_1$.*

In the broader case of effective systems (and as a consequence for high-dimension CA), the class of entropies that can be realized is larger.

Theorem 2 ([5]). *For $d \geq 3$, the class of entropies of d-dimensional CA (resp. effective systems) is $\mathbb{R}_+ \cap \Sigma_2 \cup \{\infty\}$.*

The last two theorems have left open the case of entropies realized by 1D and 2D CA, that are both included in Σ_2. The main purpose of the present article is to solve these two remaining cases. The first step of the answer is given by the following result:

Theorem 3 ([11]). *The entropy of a 1D CA is equal to the entropy of some trace of the corresponding 2D SFT.*

From the theorem above, the entropy of a 1D CA is thus Π_1. We will actually prove that the converse is also true.

Theorem 4. *The class of entropies of 1D CA is $\mathbb{R}_+ \cap \Pi_1$.*

This class of numbers is thus strictly weaker than the possible entropies of 3D CA, characterized in [5]. However, this is not true for the 2D case.

Theorem 5. *The class of entropies of 2D CA is $\mathbb{R}_+ \cap \Sigma_2 \cup \{\infty\}$.*

4 Construction

4.1 Density Encoding

This subsection is devoted to encoding data in the density of the configurations. The most relevant is actually the binary case, which follows the construction in [7]. A **1-net** is a family $(2^n \mathbb{Z} + k_n)_{n \in \mathbb{N}_1}$ of pairwise disjoint subsets of $\mathbb{Z}$ called **levels**, where $(k_n)_{n \in \mathbb{N}_1} \in \mathbb{Z}^{\mathbb{N}_1}$. It can be seen that for any 1-net, there is at most one cell $i \in \mathbb{Z}$ which does not belong to any level.

Let us denote $|u|_a$ the number of occurrences of letter a in word u. The **frequency** of a letter $a \in A$ in some one-dimensional configuration $x \in A^{\mathbb{Z}}$ is, if ever it exists, the limit $\delta_a(x) = \lim_{r \to \infty} |x_{|\mathcal{B}(r)}|_a / |\mathcal{B}(r)|$.

If $\alpha, \beta \in A^{\mathbb{N}_1}$, we note $\alpha \sim \beta$ if $\alpha = \beta$ or there exists $i \in \mathbb{N}_1$ such that $\forall j < i, \alpha_j = \beta_j$, and $\forall j > i, \alpha_i = \beta_j$ and $\alpha_j = \beta_i$. This is an equivalence relation, for which all the classes have cardinal one or two. As an example, two binary sequences are equivalent for $\sim$ if and only if they represent binary expansions of the same real number in $[0, 1[$. Let $\widetilde{A^{\mathbb{N}_1}}$ be the quotient of $A^{\mathbb{N}_1}$ by this equivalence relation. It can be endowed with the induced topology from the product topology. We will often confuse a sequence x and its equivalence class.

If $\alpha \in A^{\mathbb{N}_1}$, we note $\mathcal{D}_\alpha \subset A^{\mathbb{Z}}$ the set of *Tœplitz* configurations which are constantly equal to α_n on level $2^n\mathbb{Z} + k_n$ for some 1-net $(2^n\mathbb{Z} + k_n)_{n\in\mathbb{N}}$. If $S \subset A^{\mathbb{N}_1}$, we note $\mathcal{D}_S = \bigcup_{\alpha\in S} \mathcal{D}_\alpha$. These sets have interesting properties.

Remark 2.

1. For any nonempty closed set $S \subset A^{\mathbb{N}_1}$, $\mathcal{D}_S$ is a nonempty subshift.
2. The frequency of any letter $a \in A$ in any configuration $x \in \mathcal{D}_\alpha$ is $\sum_{\alpha_i = a} 2^{-i}$. In particular if α is binary, then it is a binary expansion of $\delta_1(\alpha)$.
3. If $\alpha \sim \beta$, then $\mathcal{D}_\alpha = \mathcal{D}_\beta$; otherwise, $\mathcal{D}_\alpha \cap \mathcal{D}_\beta = \emptyset$.
4. Let $x \in \mathcal{D}_\alpha$, $j \in \mathbb{Z}$, and i be an odd number. Then $x_{|i\mathbb{Z}+j}$ is still in $\mathcal{D}_\alpha$.

Point 3 of the previous remark suggests that it is relevant to talk about $\mathcal{D}_\alpha$ (resp. $\mathcal{D}_S$) for an equivalence class $\alpha \in \widetilde{A^{\mathbb{N}_1}}$, or for a real number $\alpha \in [0, 1]$ (resp. a set $S \subset \widetilde{A^{\mathbb{N}_1}}$ of classes).

Moreover, the sequence α encoded in the densities of the subshift can actually (up to equivalence) be effectively approximated by reading finite patterns.

Lemma 2 ([4]). *There exists a TM $\mathcal{M}_\triangleright$ which, given a word u over alphabet A, outputs a word v such that, if $u = x_{|[0,2^n[}$ for some $x \in \mathcal{D}_\alpha$ and some $n \in \mathbb{N}$, then $v = \beta_{[1,n]}$ for some $\beta \sim \alpha'$ and $\alpha'_{[1,n]} = \alpha_{[1,n]}$.*

We say that a TM has input in $\widetilde{A^{\mathbb{N}_1}}$ if it reads sequences of $A^{\mathbb{N}_1}$ as input, and gives the same result for sequences in the same equivalence class. We can also assume that, if $\alpha \sim \alpha'$, then this TM stops after the same number of steps for α and α'.

Lemma 3 ([4]). *For any TM $\tilde{\mathcal{M}}$ with input in $\widetilde{A^{\mathbb{N}_1}}$, there exists a TM $\mathcal{M}$ with input in $A^{\mathbb{N}_1}$ such that:*

- *If $\tilde{\mathcal{M}}$ halts over input $\alpha \in A^{\mathbb{N}_1}$, then there exists $k \in \mathbb{N}$ such that for any configuration $x \in \mathcal{D}_\alpha$, $\mathcal{M}$ halts over input $x_{|[0,k[}$ before time k;*
- *otherwise, $\mathcal{M}$ does not halt over any input $x \in \mathcal{D}_\alpha$.*

The following corollary is a direct application of Lemma 3 with a machine rejecting configurations outside some effectively closed set.

Corollary 1. *If $S \subset \widetilde{A^{\mathbb{N}_1}}$ is effectively closed, then $\mathcal{D}_S$ is an effective subshift.*

4.2   Checking Homogeneity

Our proof involves a deterministic SFT that is built layer by layer: the state of each cell is in a product of alphabets that we define one after the other, each layer having to respect some local constraints in how it can be superimposed with the previous ones. For $\alpha \in A^{\mathbb{N}_1}$ (resp. $S \subset A^{\mathbb{N}_1}$), let us note $\mathcal{D}_\alpha^*$ (resp. $\mathcal{D}_S^*$)

the set of configurations $x \subset A^{\mathbb{Z}^2}$ which are constant vertically, and where each row $(x_{i,k})_{i\in\mathbb{Z}}$ is in $\mathcal{D}_S$, for $k \in \mathbb{Z}$.

The purpose of this subsection is to build an SFT which checks that some layer is well homogeneous, in the sense of the following lemma; this follows [7, Section 6], but contrary to this, keeping determinism and null entropies forces us to go back to the actual SFT construction rather than directly invoke Mozes's theorem for 2×2-substitutions.

Lemma 4. $\mathcal{D}^*_{A^{\mathbb{N}_1}}$ *is S0-sofic.*

We will only give a sketch of the proof. A **2-net** is a family $(I_n \times J_n)_{n\in\mathbb{N}_1}$ of products of levels of two 1-nets $(I_n)_{n\in\mathbb{N}_1}$ and $(J_n)_{n\in\mathbb{N}_1}$. Each $I_n \times J_n$ itself is called the *level n* of the net. The I_n (resp. J_n) being pairwise disjoint, it follows that a horizontal (resp. vertical) line can intersect at most one level of the 2-net. If $i \in I_n$, then $\{i\} \times \mathbb{Z}$ is called a *column of level n*. By definition, columns of level n appear with horizontal period 2^n.

In [12], Robinson constructed an SFT R in which every configuration is divided regularly into squares of size 2^n for every n. In particular, he mentions, in other terms, the following property about the good repartition of a particular state called a cross.

Lemma 5 ([12]). *For every $x \in R$, the set $\{i \in \mathbb{Z}^2 \mid x_i$ is a cross$\}$ is a 2-net.*

Now, this SFT has been made deterministic in [8], by adding to it a layer with signals that forbid some configurations that would share the same bottom-left half as another one. The result can be restated as follows.

Lemma 6 ([8]). *There exists a south-west-deterministic SFT $\overrightarrow{R}$ that letter-factors onto some nonempty subsystem of R.*

Proof (of Lemma 4). Let us first define a south-west-deterministic SFT $\tilde{R}'$, in which configurations are vertically constant and correspond horizontally to $\mathcal{D}^*_{A^{\mathbb{N}_1}}$. $\tilde{R}' \subset R \times A^{\mathbb{Z}^2} \times A^{\mathbb{Z}^2}$ is defined with three layers: the first one contains the deterministic Robinson SFT $\overrightarrow{R}$; the second one is constant horizontally; the third one is constant vertically. We additionally require that if the first layer is a cross, then the other two must coincide. $\tilde{R}'$ is south-west-deterministic, since all three of its layers are. Now it is not difficult to turn this SFT into a south-deterministic one, by simply considering $\tilde{R} = \left\{ (x_{(i,j-i)})_{(i,j)\in\mathbb{Z}^2} \mid (x_{i,j})_{(i,j)\in\mathbb{Z}^2} \in \tilde{R}' \right\}$, whose columns correspond to columns of $\tilde{R}'$, but lines correspond to north-west-to-south-east diagonals of $\tilde{R}'$.

Null entropies come from the substitutive nature of R, which is transmitted to $\tilde{R}$. More details about this can be found in [4]. □

4.3   Checking the Density

In this section, we construct a south-deterministic SFT with null entropies which letter-factors onto $\mathcal{D}^*_S$. In the SFT, there is a special layer which consists exactly in $\mathcal{D}_S^*$: from Lemma 4, we can a priori assume that all configurations of this layers are in $\mathcal{D}_{A^{\mathbb{N}_1}}^*$, by implicitly having a layer in $\tilde{R}$. We will now add a layer whose purpose is to check that if $x \in \mathcal{D}_\alpha^*$ is read from this layer, with $\alpha \in \widetilde{A^{\mathbb{N}_1}}$, then α is really in the wanted set S, by simulating the application of a machine $\mathcal{M}$ corresponding to the machine $\tilde{\mathcal{M}}$ that rejects any configuration that is not in S (see Lemma 3).

A naive simulation of the machine for an infinite time would create invalid limit configurations. A solution to this problem is to build the additional layer in a self-similar way, in the fashion of [3,2,6]: we build a family of south-deterministic SFT (Y_n) such that Y_n simulates the TM for n steps, and also simulates Y_{n+1} with some parameters B_n, T_n. That way, if n was not enough to figure out that the input had to be rejected, then a higher level will notice it. More precisely, Y_n will be able to apply the TM over the input $x_{|B_{n+1}[0, B_n[+j}$ for some $j \in [0, B_{n+1}[$. The simulation of Y_{n+1}, as defined previously, consists in dividing naturally every valid configuration of Y_n into rectangles of size $B_n \times T_n$ called the Y_n-**macrotiles**. An important feature is that this family admits a uniform description: one single SFT is actually described. Each configuration is conscious of the level Y_n it belongs to, and will check that it simulates a configuration of the next one. The details of the construction ensuring these conditions can be found in [4].

The following lemma applies machine $\mathcal{M}$ from Lemma 3 to finite configurations composed of some arithmetic progressions in lines of the SFT, that are still in $\mathcal{D}_\alpha$. Null entropies come from the self-simulation.

Lemma 7 ([4]). *If $S \subset \widetilde{A^{\mathbb{N}_1}}$ is an effectively closed set, then $\mathcal{D}_S^*$ is S0-sofic.*

4.4   From Density to Entropy

Finally, let us see how Lemma 7 can be used to prove Theorem 4: it simply independently splits each letter 1 into two letters, so that its density is transformed into entropy.

Proof (of Theorem 4). One direction corresponds to Theorem 3. Let us prove the converse. Should we make the product with the shift over $2^{\lfloor \alpha \rfloor}$ symbols, whose entropy is $\lfloor \alpha \rfloor$, we can assume that $\alpha \in [0, 1[$.

Let F be the shift composed with the CA corresponding to the deterministic SFT given by Lemma 7 for the effectively closed set S consisting of binary representations of real numbers from the interval $[0, \alpha]$, A its alphabet, and $\pi : A \to 2$ be the corresponding letter projection. Let $\tilde{F}$ be the CA over alphabet $(A \times \{0\}) \sqcup (\pi^{-1}(1) \times \{1\})$ such that the first component performs F and the second one performs the shift. , *i.e.*, in the first component we can see the 0-entropy F and, in the second one the one-dimensional subshift:

$$\mathcal{D}_S^\nabla = \left\{ (y_i)_{i \in \mathbb{Z}} \in 2^{\mathbb{Z}} \mid \exists (x_i)_{i \in \mathbb{Z}} \in \mathcal{D}_S, \forall i \in \mathbb{Z}, \text{ if } x_i = 0, \text{ then } y_i = 0 \right\}.$$

It is known that the entropy of a product is the sum of the entropies, hence the entropy of $\tilde{F}$ is that of $\mathcal{D}_S^\nabla$.

$\mathcal{K}_U(\mathcal{D}_S^\nabla) = \sum_{u \in \mathcal{L}_U(\mathcal{D}_S)} 2^{|u|_1}$ can be bounded by $\mathcal{K}_U(\mathcal{D}_S) 2^{\sup_{u \in \mathcal{L}_U(\mathcal{D}_S)} |u|_1}$. Hence, the entropy $\mathcal{H}(\mathcal{D}_S^\nabla)$ is:

$$\mathcal{H}(\mathcal{D}_S^\nabla) = \lim_{r \to \infty} \frac{\log \mathcal{K}_{\mathcal{B}(r)}(\mathcal{D}_S^\nabla)}{|\mathcal{B}(r)|} \le \mathcal{H}(\mathcal{D}_S) + \lim_{r \to \infty} \sup_{u \in \mathcal{L}_{\mathcal{B}(r)}(\mathcal{D}_S)} \frac{|u|_1}{|\mathcal{B}(r)|}.$$

However, since $\mathcal{H}(\mathcal{D}_S) = 0$, $\mathcal{H}(\mathcal{D}_S^\nabla)$ is not more than the maximal density α of configurations of $\mathcal{D}_S$. Conversely, if $x \in \mathcal{D}_\alpha \subset \mathcal{D}_S$, then $\mathcal{L}_{\mathcal{B}(r)}(\mathcal{D}_S^\nabla) \supset \{(x_i, y_i)_{|i|<r} | \forall i \in \mathbb{Z}, y_i \in \{x_i, 2x_i\}\}$; hence $\mathcal{K}_{\mathcal{B}(r)}(\mathcal{D}_S) \ge 2^{|x_{\mathcal{B}(r)}|_1}$ and $\mathcal{H}(\mathcal{D}_S^\nabla) \ge \limsup_{r \to \infty} |x_{\mathcal{B}(r)}|_1 = \alpha$. Therefore, $\mathcal{H}(\tilde{F}) = \mathcal{H}(\mathcal{D}_S^\nabla) = \alpha$.   □

5   The Second Dimension

Let us now prove Theorem 5, dealing with 2D CA. The first inclusion is direct from Theorem 2. The idea here will be to realize, in each horizontal slice, some right-computable number, as in the previous section. These slices will actually be parameterized by some index encoded in its density, that is increased by one between consecutive slices, and that will give a sequence approximating the wanted Σ_2. The trick is that the encoding has to be spare in order to prevent limit configurations to achieve too much entropy; this has to be compensated by having actual groups of consecutive slices hold the same parameter.

Let us denote by $(w)_4$ the 4-ary representation of a natural number $w \in \mathbb{N}$ over $\{0', 1', 2', 3'\}$. Let $(S_k)_{k \in \mathbb{N}_1}$ be a computable sequence of effectively closed subsets of $\widetilde{A^{\mathbb{N}_1}}$, and $S = \{*^k (w)_4 y \,|\, k \in \mathbb{N}, 0 \le w \le 4^k - 1, y \in \mathcal{D}_{S_k}\} \cup \{*^\infty, \sharp^\infty\}$ a set of sequences over alphabet $A = \{*, 0', 1', 2', 3', 0, 1, \sharp\}$. Consider $S' = S_1 \cup S_2 \cup S_3$, where:

$$S_1 = \{(z, z') \in S^2 \,|\, \exists k, y, w \in [\![0, 4^k - 1]\![, z = *^k(w)_4 y \text{ and } z' = *^k(w+1)_4 y\};$$
$$S_2 = \{(z, z') \in S^2 \,|\, \exists k, y, z = *^k(4^k - 1)_4 y \text{ and } z' = \sharp^\infty\};$$
$$S_3 = \{(z, z') \in S^2 \,|\, z = *^\infty \text{ and } z' = *^\infty \text{ or } \exists k, y, z' = *^k(0)_4 y\}.$$

Lemma 8 ([4]). *S' is an effectively closed subset of $\widetilde{(A \times A)^{\mathbb{N}_1}}$.*

We are now ready to characterize the entropies of 2D CA. Similarly to the one-dimensional case, a 2D CA corresponds to a south-deterministic 3D SFT, up to adding a spreading state, and its entropy can be seen as the directional one for the south-to-north unitary vector.

Proof (of Theorem 5). Let α_k be a computable sequence of Π_1 numbers, $S_k = [0, \alpha_k]$, $\mathcal{M}$ the TM given by Lemma 8, Y the 2D SFT given by Lemma 7.

Consider now the following 3D SFT Y': each horizontal slice must satisfy the conditions of Y. The only vertical local constraint we add is the following: the second letter (in A) of the pair held by a tile is equal to the first letter of the tile on top of it. Intuitively, the way to think about this is that when a horizontal

Densities and Entropies in Cellular Automata   261

slice is considering whether it should accept or reject its input (the first sequence it holds), it can also read as input the sequence of the slice above it (the second sequence).

Y' is south-deterministic. Indeed, every horizontal slice is an element of Y, which is a 2D south-deterministic SFT. Hence, if we know a slice $x_{|\mathbb{Z} \times \{n\} \times \mathbb{Z}}$, we can uniquely determine $x_{|\mathbb{Z} \times \{n+1\} \times \mathbb{Z}}$. Moreover, Y' has null entropies, as a subshift of an infinite product of 2D SFT with null entropies.

Let us now modify the SFT in order to get the wanted entropy. We need to understand the structure of the configurations. From now on, we forget the second sequence encoded in every horizontal slice and we work only with the first one. If z_k is the sequence encoded in the kth horizontal slice, then the sequence $(z_k)_{k \in \mathbb{Z}}$ can only have one of the following forms:

- $z_k = *^\infty$, for all $k \in \mathbb{Z}$;
- there exist $m \in \mathbb{Z}, k \in \mathbb{N}_1$ and $y \in \mathcal{D}_{[0,\alpha_k]}$ such that $z_i = *^\infty$ for $i < m$, $z_i = *^k(i - m)_4 y$ for $m \leq i < m + 4^k$, and $z_i = \sharp^\infty$ for $i \geq m + 4^k$.
- $z_k = \sharp^\infty$, for all $k \in \mathbb{Z}$;

This follows directly from the definition of S'. For $k \in \mathbb{N}_1$, let $Y'(k) \subseteq Y'$ consist of those configurations whose horizontal slices are either $*^\infty$, $\sharp^\infty$, or contain $*^k(0)_4 y$ for some $y \in \mathcal{D}_{[0,\alpha_k]}$. It is a subshift.

Let us allow splitting of the letter 1 into two (by adding a second, binary, layer, as in the proof of Theorem 4), independently in every horizontal slice. Then, in configurations of the subsystem $Y'(k)$ there are 4^k slices where splitting is done and each one contributes up to $4^{-k}\beta$ to the entropy, where $\beta \in [0, \alpha_k]$ is such that $y \in \beta$. This happens because in every slice, y is encoded in 2-net starting from level $2k$. Since splitting is done independently in 4^k slices, the entropy of the subsystem $Y'(k)$ is β. By the variational principle, and since the nonwandering system of the CA is included in the disjoint union of the $Y'(k)$ and the trivial subsystems, we have that the entropy of F in the vertical direction is:

$$\mathcal{H}(F) = \sup_{k \in \mathbb{N}} \sup_{0 \leq \beta \leq \alpha_k} \beta = \sup_{k \in \mathbb{N}} \alpha_k,$$

which is the wanted Σ_2 number.   □

6   Conclusion

We have reached a characterization of the entropies of CA in terms of computability classes. This is inspired by what had been done over multidimensional SFT, but the construction presents some intrinsically interesting points, such as determinization widgets, self-similar construction, or a generalized encoding of configurations into densities.

This problem helps us understand what kind of results on tilings could be adapted to CA, that is when one of the dimensions of the system actually represents a deterministic temporal evolution. It could be interesting to try to adapt some more results from multidimensional symbolic dynamics, such as the substitutions of [10], or the characterization of subactions in [5,2]. Nevertheless, when translating into cellular automata, we will in general have to deal with wandering points, which could be omitted here in the study of entropy but may sometimes alter significantly the results.

Among open problems, we could try to characterize the entropies of restricted classes of CA: requiring transitivity constraints, or reversibility. The latter case might be achieved by adapting our proof while requiring two-way determinism in the underlying tilings (but again extending it to a full set of configurations may be difficult). We could also study the entropies of other computationally-inspired dynamical systems, such as Turing machines with moving tapes.

Acknowledgements. This project was supported by the Academy of Finland Grant 131558. The second author was also supported by the Finnish Academy of Science and Letters and the Turku Center for Computer Science. The first author was supported by the ANR Projet Blanc "EMC". Special thanks to Alexis Ballier, Timo Jolivet, Jarkko Kari and Pascal Vanier for interesting discussion around that matter.

References

1. Čulik II, K., Hurd, L.P., Kari, J.: The topological entropy of cellular automata is uncomputable. Ergodic Theory & Dynamical Systems 12(2), 255–265 (1992)
2. Durand, B., Romashchenko, A., Shen, A.: Fixed-point tile sets and their applications (September 2010), draft
3. Gács, P.: Reliable cellular automata with self-organization. Journal of Statistical Physics 102(1-2), 45–267 (2001), http://www.cs.bu.edu/fac/gacs/recent-publ.html
4. Guillon, P., Zinoviadis, C.: Densities and entropies in cellular automata (2012), arXiv:1204.0949
5. Hochman, M.: On the dynamics and recursive properties of multidimensional symbolic systems. Inventiones Mathematicæ 176(1), 131–167 (2009), http://www.springerlink.com/content/h664281759545081
6. Hochman, M.: Expansive directions for $\mathbb{z}^2$ actions. Ergodic Theory & Dynamical Systems 31(1), 91–112 (2011)
7. Hochman, M., Meyerovitch, T.: A characterization of the entropies of multidimensional shifts of finite type. Annals of Mathematics 171(3), 2011–2038 (2010), http://pjm.math.berkeley.edu/annals/ta/080814-Hochman/080814-Hochman-v1.pdf
8. Kari, J.: The nilpotency problem of one-dimensional cellular automata. SIAM Journal on Computing 21(3), 571–586 (1992)
9. Milnor, J.: On the entropy geometry of cellular automata. Complex Systems 2(3), 357–385 (1988)
10. Mozes, S.: Tilings, substitution systems and dynamical systems generated by them. Journal d'analyse mathématique 53, 139–186 (1988)

11. Park, K.K.: Entropy of a skew product with a F^2 -action. Pacific Journal of Mathematics 172(1), 227–241 (1996), `http://projecteuclid.org/euclid.pjm/1102366193`
12. Robinson, R.M.: Undecidability and nonperiodicity for tilings of the plane. Inventiones Mathematicæ 12(3) (1971)
13. Simonsen, J.G.: On the computability of the topological entropy of subshifts. Discrete Mathematics & Theoretical Computer Science 8, 83–96 (2006), `www.dmtcs.org/dmtcs-ojs/index.php/dmtcs/article/download/456/1602`
14. Zheng, X., Weihrauch, K.: The Arithmetical Hierarchy of Real Numbers. In: Kutyłowski, M., Wierzbicki, T., Pacholski, L. (eds.) MFCS 1999. LNCS, vol. 1672, pp. 23–33. Springer, Heidelberg (1999)

Foundational Analyses of Computation

Yuri Gurevich

Microsoft Research, One Microsoft Way, Redmond, WA 98052-6399,
United States of America

Give me a fulcrum, and I shall move the world.

—Archimedes

Abstract. How can one possibly analyze computation in general? The task seems daunting if not impossible. There are too many different kinds of computation, and the notion of general computation seems too amorphous. As in quicksand, one needs a rescue point, a fulcrum. In computation analysis, a fulcrum is a particular viewpoint on computation that clarifies and simplifies things to the point that analysis becomes possible.

We review from that point of view the few foundational analyses of general computation in the literature: Turing's analysis of human computations, Gandy's analysis of mechanical computations, Kolmogorov's analysis of bit-level computation, and our own analysis of computation on the arbitrary abstraction level.

1 Introduction

Algorithms and computations are closely related concepts. Syntactically algorithms are programs (or recipes) but semantically they specify computations. And the only computations that we consider here are algorithmic (also known as mechanical). In this paper, we abstract from the syntax of algorithms, so that analysis of algorithms and analysis of computation are one and the same.

Turing's analysis of algorithms was provoked by the Entscheidungsproblem, the problem whether the validity of first-order formulas is computable. Logicians have been interested in what functions are computable, and Turing's analysis is often seen from that point of view. But there may be much more to an algorithm than its input-output behavior. In general algorithms perform tasks, and computing functions is a rather special class of tasks.

Here we concentrate on foundational analyses of algorithms/computations, not on what functions are computable.

2 Turing

Alan Turing analyzed computation in his 1936 paper "On Computable Numbers, with an Application to the Entscheidungsproblem" [21]. The validity relation on first-order formulas can be naturally represented as a real number,

S.B. Cooper, A. Dawar, and B. Löwe (Eds.): CiE 2012, LNCS 7318, pp. 264–275, 2012.

and the Entscheidungsproblem becomes whether this particular real number is computable. "Although the subject of this paper is ostensibly the computable numbers, it is almost equally easy to define and investigate computable functions of an integral variable or a real or computable variable, computable predicates, and so forth. The fundamental problems involved are, however, the same in each case, and I have chosen the computable numbers for explicit treatment as involving the least cumbrous technique" [21, p. 230].

How could Turing analyze computation in such generality? The world of algorithms is large and diverse. Explicitly or implicitly, he imposed some constraints on the computations in consideration. And he found a fulcrum. We start with the fulcrum. There were no computers in Turing's time[1] but that does not seem to make Turing's task much simpler. Humans are hard to analyze. Amazingly Turing found a way to do just that: Ignore how the algorithm is given, ignore what human computers have in their minds, and concentrate on what the computers do, what their observable behavior is. That is his fulcrum.

One may argue that Turing did not ignore the mind. He speaks about the state of mind of the human computer explicitly and repeatedly. For example, he says that "[t]he behaviour of the computer at any moment is determined by the symbols which he is observing, and his 'state of mind' at that moment" [21, p. 250]. But Turing postulates that "the number of states of mind which need be taken into account is finite." The computer just remembers the current state of mind, and even that is not necessary: "we avoid introducing the 'state of mind' by considering a more physical and definite counterpart of it. It is always possible for the computer to break off from his work, to go away and forget all about it, and later to come back and go on with it. If he does this he must leave a note of instructions (written in some standard form) explaining how the work is to be continued. This note is the counterpart of the 'state of mind'."

Turing introduced abstract computing machines that became known as Turing machines (and constructed a universal Turing machine). He defined a real number to be computable "if its decimal can be written down by a [Turing] machine" [21, p. 230]. His thesis was that Turing computable numbers "include all numbers which could naturally be regarded as computable" (Turing [21, p. 230]). He used the thesis to prove the undecidability of the Entscheidungsproblem. To convince the reader of his thesis, Turing used three arguments.

Reasonableness: He gave examples of large classes of real numbers which are [Turing] computable.

Robustness: He gave another explicit definition of computability and proved it is equivalent to the original one "in case the new definition has a greater intuitive appeal." The robustness argument was strengthened in the appendix where, after learning about Church's explicit definition of computability [6], he proved the equivalence of their definitions.

Appeal to Intuition: He analyzed computation appealing directly to intuition.

---

[1] "Numerical calculation in 1936 was carried out by human beings; they used mechanical aids for performing standard arithmetical operations, but these aids were not programmable" (Gandy [8, p. 12]).

The first two arguments are important but insufficient. There are other reasonable and robust classes of computable real numbers, e.g., the class of primitive recursive real numbers. The direct appeal to intuition is crucial.

While Turing's analysis is very general, his algorithms are subject to some constraints. Here are some of them.

Symbolic: Computation is symbolic (or digital, symbol-pushing).
Sequential Time: Computation splits into a sequence of steps.
Bounded Work: Only bounded work is performed at any one step.
Isolated Computation is self-contained. No oracle is consulted, and nobody interferes with the computation either during a computation step or in between steps. The whole computation of the algorithm is determined by the initial state.

2.1   Discussion

Q[2] Did Turing really impose the symbolic constraint?

A : Yes, he did. "Computing is normally done by writing certain symbols on paper," writes Turing [21, p. 249], and he analyses only such computations.

Q : Is this really a constraint?

A : These days we are so accustomed to digital computations that the symbolic constraint may not look like a constraint. But it is. Non-symbolic computations have been performed by humans from ancient times [13, §3].

Q : I came across a surprising remark of Gödel that Turing's argument "is supposed to show that mental procedures cannot go beyond mechanical procedures" [9]. I believe that Turing's goal was to analyze mechanical procedures. Since such procedures were executed by humans in his time, he had to analyze human execution of mechanical procedures; there was no other way.

A : We may never know what goal was in Turing's head; let's hear Gödel's argument.

Q : "What Turing disregards completely is the fact that mind, in its use, is not static, but constantly developing, i.e., that we understand abstract terms more and more precisely as we go on using them, and that more and more abstract terms enter the sphere of our understanding. There may exist systematic methods of actualizing this development, which could form part of the procedure" (Gödel, [9]).

A : Gödel raises a possibility that there exists a sophisticated decision procedure for the Entscheidungsproblem that can be executed by gifted mathematicians.

Q : Hmm, if gifted mathematicians can reliably execute a procedure, they should be able to figure out how to program it, and then the procedure is mechanical.

A : Well, it is hard to delimit human creativity. Certainly Turing did not do that.

Q : And didn't intend to, I am sure. But let me change the topic. You said nothing about Church's arguments in favor of his definition of computability.

---

[2] Q is my inquisitive friend Quisani, and A is the author.

A: Church had strong arguments that his definition of reasonable and robust. In particular, he and his student Kleene proved that a numerical function is expressible in Church's λ-calculus if and only if it is expressible in Gödel's recursive calculus. Church's thesis was that [Gödel's] recursive functions include all numerical functions that are "effectively calculable".

Q: Here is another quote. "For the actual development of the (abstract) theory of computation, where one must build up a stock of particular functions and establish various closure conditions, both Church's and Turing's definition are equally awkward and unwieldy. In this respect, general recursiveness is superior" (Sol Feferman, [8, p. 6]). Do you buy that?

A: Indeed, the recursive approach has been dominant in mathematical logic, but Turing's approach dominates in computer science and it influenced the early design of digital computers.

3   Kolmogorov

Andrei Kolmogorov analyzed computation in abstraction from the computer. Kolmogorov's fulcrum seems to be the idea that computations, independently from the computer, satisfy nontrivial constraints. In a 1953 talk to the Moscow Mathematical Society [14], he stipulated that every algorithmic process satisfies the following constraints.

Sequentiality: An algorithmic process splits into steps whose complexity is bounded in advance.

Elementary Steps: Each step consists of a direct and unmediated transformation of the current state S to the next state S^*.

Locality: Each state S has an active part of size bounded in advance. The direct and unmediated transformation of S to S^* is based only on the information about the active part of S and applies only to the active part.

These ideas gave rise to a new computation model developed by Kolmogorov and his student Vladimir Uspensky [15]. Instead of a linear tape, a Kolmogorov machine has a graph of bounded degree (so that there is a bound on the number of edges attached to any vertex), with a fixed number of the types of vertices and a fixed number of the types of edges. We speculated in [10] that "the thesis of Kolmogorov and Uspensky is that every computation, performing only one restricted local action at a time, can be viewed as (not only being simulated by, but actually being) the computation of an appropriate KU machine." Uspensky agreed [22, p. 396].

We do not know what analysis, if any, allowed Kolmogorov and Uspensky to arrive from the constraints above at the particular architecture of Kolmogorov machines. "As Kolmogorov believed," wrote Uspensky [22, p. 395], "each state of every algorithmic process ... is an entity of the following structure. This entity consists of elements and connections; the total number of them is finite. Each connection has a fixed number of elements connected. Each element belongs to some type; each connection also belongs to some type. For every given algorithm the total number of element types and the total number of connection types are bounded." In that approach, the number of nonisomorphic active zones is finite (because of a bound on the size of the active zones), so that the state transition can be described by a finite program.

Leonid Levin told us that Kolmogorov thought of computation as a physical process developing in space and time, that the edges of Kolmogorov machine reflect physical closeness of computation elements [16]. But then, as we mentioned in [12], the dimensionality of the space may grow with the input size.

Kolmogorov's analysis has not been well known. In this connection, let us point out these references: [1,10,22,23].

4   Gandy

Gandy analyzed computation in his 1980 paper "Church's Thesis and Principles for Mechanisms" [7]. In this section, by default, quotations are from that paper.

> Turing's analysis of computation by a human being does not apply directly to mechanical devices ... Our chief purpose is to analyze mechanical processes and so to provide arguments for ...
>
> **Thesis M.** *What can be calculated by a machine is computable.*

Contrary to human computers, a machine can perform parallel actions. Kolmogorov machines are fine "but at each step only a bounded portion of the whole state [of a Kolmogorov machine] is changed." Thesis M "must take parallel working into account." A question arises what machines are.

> (1) In the first place I exclude from consideration devices which are *essentially* analogue machines. ...I shall distinguish between "mechanical devices" and "physical devices" and consider only the former. The only physical presuppositions made about mechanical devices ... are that there is a lower bound on the linear dimensions of every atomic part of the device and that there is an upper bound (the velocity of light) on the speed of propagation of changes.
> (2) Secondly we suppose that the progress of calculation by a mechanical device may be described in discrete terms, so that the devices considered are, in a loose sense, digital computers.
> (3) Lastly we suppose that the device is deterministic; that is, the subsequent behaviour of the device is uniquely determined once a complete description of its initial state is given.
>
> After these clarifications we can summarize our argument for a more definite version of Thesis M in the following way.
>
> **Thesis P.** *A discrete deterministic mechanical device satisfies principles I–IV below.*

Later, discussing how to describe computation states, Gandy says that he wants "the form of description to be sufficiently abstract to apply uniformly to mechanical, electrical or merely notional devices." After all the clarifications, it is not clear what Gandy's notion of machine is; see [18] in this connection. Gandy does presume that machine computations are sequential-time and isolated; these are two of the four constraints in §2. Sequential time parallelism is known as synchronous.

Principles I-IV are precise though require too many definitions to be stated precisely here. The four principles entail Gandy's main theorem: "What can be calculated by a machine is computable."

Principle I asserts in particular that, for any machine, the states can be described by hereditarily finite sets[3] and there is a transition function F such that, if x describes an initial state, then $Fx, F(Fx), \ldots$ describe the subsequent states. Principles II are III are technical restrictions on the state descriptions and the transition function respectively. Principle IV generalizes Kolmogorov's locality constraint to parallel computations.

> We now come to the most important of our principles. In Turing's analysis the requirement that the action depend only on a bounded portion of the record was based on a human limitation. We replace this by a physical limitation [Principle IV] which we call the *principle of local causation*. Its justification lies in the finite velocity of propagation of effects and signals: contemporary physics rejects the possibility of instantaneous action at a distance.

A preliminary version of Principle IV gives a good idea about the intentions behind the principle.

> **Principle IV** (Preliminary version). The next state, Fx, of a machine can be reassembled from its restrictions to overlapping "regions" s and these restrictions are locally caused. That is, for each region s of Fx there is a causal neighborhood $t \subseteq \mathrm{TC}(x)$ of bounded size such that $Fx \upharpoonright s$ [the restriction of Fx to s] depends only on $x \upharpoonright t$ [the restriction of x to t].

4.1 Comments

It isn't clear to us what Gandy's fulcrum was and even whether he had a fruitful viewpoint on machine computations. We recently [13] criticized Gandy's approach. Here we add just a few remarks.

The only parallelism that Gandy considers is synchronous. That is restrictive. Nowadays asynchronous machine computations are common.

The principle of local causality does not apply to all synchronous parallel algorithms. Gandy himself mentions one counterexample, namely Markov's normal algorithms [17]. The principle fails in the circuit model of parallel computation, the oldest model of parallel computation in computer theory. The reason is that the model allows gates to have unbounded fan-in. We illustrate this on the example of a first-order formula $\forall x R(x)$ where $R(x)$ is atomic. The formula gives rise

---

[3] A set x is *hereditarily finite* if its transitive closure $\mathrm{TC}(x)$ is finite. Here $\mathrm{TC}(x)$ is the least set t such that $x \in t$ and such that $z \in y \in t$ implies $z \in t$.

to a collection of circuits C_n of depth 1. Circuit C_n has n input gates, and any unary relation R on $\{1, \ldots, n\}$ provides an input for C_n. Circuit C_n computes the truth value of the formula $\forall x R(x)$ in one step, and the value depends on the whole input. Ironically, it is easy to construct a Turing machine that simulates sequentially these parallel computations.

Hereditarily finite sets are finite. The finiteness constraint is understandable taking into account that Gandy's goal was to confirm Church's thesis. However, taking into account that Gandy's machines are isolated (and thus non-interactive), the finiteness constraint excludes some useful algorithms. For example it excludes a simple algorithm that consumes a stream of numbers keeping track of the maximum of the numbers seen so far. The finiteness constraint is not necessarily satisfied by Turing machines. In particular, a Turing machine can execute the stream algorithm above if the whole stream is written on its initial tape.

We accept that computation states can be described in set theoretic terms. A problem arises how to make the transition function work with such a description. Many of Gandy's technical problems are related to this problem, and indeed describing algorithms in Gandy's terms is rather challenging.

This said, let us emphasize that Gandy pioneered the axiomatic approach in foundational analysis of algorithms. He bravely attacked the hard problem of a general analysis of machine computations. Wilfried Sieg adopted Gandy's approach and reworked Gandy's axioms, see [23] and references there, but he did not clarify or justify Gandy's fulcrum. The problem of a general analysis of machine computations is wide open. In our view, the notion of machine computation is evolving and will be evolving for the foreseeable future; think of quantum computers for example. The notion has not matured enough to lend itself to formal analysis.

5   Analyzing Computations on Their Native Levels of Abstraction

5.1   Motivation

By the 1980s, there were plenty of computers and software. A problem arose how to specify software. The most popular approaches to this problem were denotational semantics and algebraic specifications. Both approaches were proudly declarative. The declarative character of specifications was supposed to be an advantage. Indeed, declarative specifications tend to be more comprehensible, higher-level (that is of higher level of abstraction) and cleaner than operational, executable specifications, which is great. But executable specifications have their own advantages. You can "play" with them: run them, test, debug. In principle, you can verify properties of a declarative or executable spec mathematically, and sometimes you have to, and there are better and better tools to do that. In practice though, mathematical verification is out of the question in an overwhelming majority of cases, and the possibility to test specs is indispensable. Declarative specifications are static while software evolves. In most cases, it is virtually impossible to keep a declarative spec and an implementation in sync. In the case of an executable spec, you can test whether the implementation conforms to the spec (or, if the spec was reverse-engineered from an implementation, whether the spec is consistent with the implementation).

A question arises whether an executable specification have to be low-level and detailed? This leads to a theoretical, even foundational problem. Is there an executable specification of any algorithm A on the level of abstraction of A itself? For example, imagine that you conceived a wonderful algorithm. How would you specify it succinctly in an executable way? A natural-language explanation would not do as it is not executable. Besides, such an explanation may introduce ambiguities and misunderstanding. You can program your algorithm in a conventional programming language but this will surely introduce lower-level details.

Turing and Kolmogorov machines are executable but low-level. Consider for example Turing-machine implementations of these two versions of Euclid's algorithm for the greatest common divisor of two natural numbers: the original version where you advance by means of differences, and a faster (and higher-level) version where you advance by means of divisions. The chances are that divisions were reduced to differences in the Turing machine implementation, and the distinction of the abstraction levels disappeared.

Can we generalize Turing and Kolmogorov machines in order to solve the foundational problem in question? The answer turns out to be positive, at least for sequential algorithms [12], synchronous parallel algorithms [2], and interactive algorithms [3,4].

Following Kolmogorov, we consider computation in abstraction from the computer. Following Gandy, we use an axiomatic approach. The fulcrum for the sequential case is this. Every algorithm A has its native level of abstraction. On that level, the states can be faithfully represented by first-order structures of a fixed vocabulary in such a way that state transitions become just sets of assignments. The fulcrums for the parallel and interactive cases are built on this fulcrum. Here we restrict attention to sequential algorithms and do not cover parallel and interactive ones. Sequential algorithms are also known as classical as they had been virtually the only algorithms from time immemorial to the 1950s. The three stipulations of Kolmogorov in §3 give a great informal description of sequential algorithms. In the rest of this section, algorithms are by default sequential.

5.2   Constraints

Sequential Time: *Any algorithm A is associated with a nonempty collection $\mathcal{S}(A)$ of states, a subcollection $\mathcal{I}(A) \subseteq \mathcal{S}(A)$ of initial states and a (possibly partial) state transition map $\tau_A : \mathcal{S}(A) \longrightarrow \mathcal{S}(A)$.*

Q: Your algorithm A is deterministic: $\tau_A(X)$ is determined by state X. Why not to make τ_A multi-valued?

A : In our view, this would involve intra-step (within a single step) inter-action with the environment [12, §9]. Intra-step interactive algorithms are analyzed in [3,4]. Note in this connection that we do not rule out inter-step interaction with the environment. In other words, the environment can intervene between the steps of the algorithm A. If the intervention results in a legitimate state of A, the algorithm A continues to run. So, in general, the steps of A are interleaved with those of the environment, and thus the behavior of A is not necessarily determined by the initial step.

Recall that a first-order structure X is a nonempty set (the base set of X) with relations and operations; the vocabulary of X consists of the names of those relations and operations. For example, if the vocabulary of X consists of one binary relation then X is a directed graph.

Abstract State: *The states of an algorithm A can be faithfully represented by first-order structures of the same finite vocabulary, which we call the vocabulary of A, in such a way that*

- *τ_A does not change the base set of a state,*
- *collections $\mathcal{S}(A)$ and $\mathcal{I}(A)$ are closed under isomorphisms, and*
- *any isomorphism from a state X to a state Y is also an isomorphism from $\tau_A(X)$ to $\tau_A(Y)$.*

Q : You claim that first-order structures are sufficiently general to faith-fully represent the states of any algorithm?

A : I have been making that claim from the 1980s. The collective experience of computer science seems to corroborate the claim.

Q : But maybe the notion of first-order structure is too broad. Consider for example, natural numbers with the usual arithmetic relations and operations plus the unary relation $T(n)$ that is true if and only if the Turing machine number n halts on the empty tape. Starting with such a structure, a simple algorithm solves the halting problem.

A : Suppose, more generally, that T is produced by some process, not necessarily algorithmic. For example, T is the result of some measure-ment or coin flipping. How would you rule out your particular version of T?

Q : OK, but I have another question about the postulate. That base-set preservation sounds restrictive. A graph algorithm may extend the graph with new nodes.

A : And where will the algorithm take those nodes? From some reserve? Make that reserve a part of your initial state.

Q : Now, why should the collection of states be closed under isomor-phisms, and why should the state transition respect isomorphisms?

A : Every algorithm works at its native level of abstraction. Irrelevant details should not matter. Consider a graph algorithm for example. In an implementation, nodes may be integer numbers, but the algorithm cannot examine whether a node is even or odd or which of the nodes is greater. These are implementation details irrelevant to the graph algorithm. And if the algorithm does take advantage of the integer representation of nodes then its vocabulary should reflect the relevant part of arithmetic.

According to Kolmogorov's informal definition of sequential algorithms, there is a bound on the amount of work done during any one step. But how to measure step complexity or the work done during one step? Fortunately the abstract state constraint helps.

Note that, according to the sequential-time constraint, the next state $\tau_A(X)$ of an algorithm A depends only on the current state X of A. The executor does not need to remember any history (even the current position in the program); all that is reflected in the state. If the executor is human and writes something on scratch paper, that paper should be a part of the computation state.

In order to change the given state X into $\tau_A(X)$, the algorithm A explores a portion of X and then performs the necessary changes of the values of the predicates and operations of X. According to Kolmogorov's informal definition, the explored portion, the "active zone", is bounded. And the change from X to $\tau_A(X)$, let us call it $\Delta_A(X)$, depends only on the results of exploration. Formally, $\Delta_A(X)$ can be defined as the collection of equations $F(\bar{a}) = b$ where b is a new value of a vocabulary function F at point $\bar{a}$.

But how does the algorithm know what to explore and what to change? That information is normally supplied by the program, and it should be applicable to all the states. In the light of the abstract state constraint, it should be given symbolically, in terms of the vocabulary of A.

Bounded Exploration: *There exists a finite set T of terms (or expressions) in the vocabulary of algorithm A such that $\Delta_A(X) = \Delta_A(Y)$ whenever states X, Y of A coincide over T.*

5.3 Definition and the Representation Theorem

Now think of the sequential-time constraint as a postulate where $\mathcal{S}(A)$ is just a nonempty collection of things. Think of the abstract-state and bounded-exploration constraints as postulates that clarify/restrict what those things are and how the map τ_A works.

Definition 1. A (sequential) algorithm is any entity that satisfies the sequential-time, abstract-state and bounded-exploration postulate.

Abstract state machines (ASMs) were defined in [11]. Here we restrict attention to sequential ASMs, which are undoubtedly algorithms.

Theorem 1 ([12]). *For every algorithm A, there exists a sequential ASM with the same states and the same state transition function.*

5.4   Deriving Church's Thesis

Our postulates do not entail Church's thesis. The reason is that initial states of sequential algorithms may be uncomputable. The halting problem for Turing machines may be encoded in an initial state. Think also of ruler-and-compass algorithms or the Gauss elimination procedure; they satisfy our postulates but cannot be simulated by Turing machines. An arithmetical-state postulate of [5] asserts that only undeniably-computable operations are available in initial states; see details in [5].

Theorem 2 ([5]). *Church's thesis follows from the sequential-time, abstract-state, bounded-exploration and arithmetical-state postulates.*

Acknowledgements. Many thanks to Andreas Blass and Oron Shagrir for useful comments.

References

1. Blass, A., Gurevich, Y.: Algorithms: A quest for absolute definitions. In: Paun, G., et al. (eds.) Current Trends in Theoretical Computer Science, pp. 283–311. World Scientific (2004); also in: Olszewski, A. (ed.): Church's Thesis After 70 Years Ontos Verlag, pp. 24–57. Ontos Verlag (2006)
2. Blass, A., Gurevich, Y.: Abstract state machines capture parallel algorithms. ACM Trans. on Computational Logic 4(4), 578–651 (2003); Correction and Extension, Same Journal 9(3), article 19 (2008)
3. Blass, A., Gurevich, Y.: Ordinary interactive small-step algorithms. ACM Trans. Computational Logic 7(2), Part I, 363–419 (2006); plus 8:3 , articles 15 and 16 (Parts II and III) (2007)
4. Blass, A., Gurevich, Y., Rosenzweig, D., Rossman, B.: Interactive small-step algorithms. Logical Methods in Computer Science 3(4) (2007); papers 3 and 4 (Part I and Part II)
5. Dershowitz, N., Gurevich, Y.: A natural axiomatization of computability and proof of Church's thesis. Bull. of Symbolic Logic 14(3), 299–350 (2008)
6. Church, A.: An unsolvable problem of elementary number theory. American Journal of Mathematics 58, 345–363 (1936)
7. Gandy, R.: Church's thesis and principles for mechanisms. In: Barwise, J., et al. (eds.) The Kleene Symposium, pp. 123–148. North-Holland (1980)
8. Gandy, R.O., Yates, C.E.M. (eds.): Collected works of A.M. Turing: Mathematical logic. Elsevier (2001)
9. Göedel, K.: A philosophical error in Turing's work. In: Feferman, S., et al. (eds.) Kurt Gödel: Collected Works, vol. II, p. 306. Oxford University Press (1990)
10. Gurevich, Y.: On Kolmogorov machines and related issues. Bull. of Euro. Assoc. for Theor. Computer Science 35, 71–82 (1988)
11. Gurevich, Y.: Evolving algebra 1993: Lipari guide. In: Börger, E. (ed.) Specification and Validation Methods, pp. 9–36. Oxford Univ. Press (1995)
12. Gurevich, Y.: Sequential abstract state machines capture sequential algorithms. ACM Trans. on Computational Logic 1(1), 77–111 (2000)
13. Gurevich, Y.: What Is an Algorithm? In: Bielikova, M., et al. (eds.) SOFSEM 2012. LNCS, vol. 7147, pp. 31–42. Springer, Heidelberg (2012)

14. Kolmogorov, A.N.: On the concept of algorithm. Uspekhi Mat. Nauk 8(4), 175–176 (1953) (Russian)
15. Kolmogorov, A.N., Uspensky, V.A.: On the definition of algorithm. Uspekhi Mat. Nauk 13(4), 3–28 (1958) (Russian); English translation in AMS Translations 29, 217–245 (1963)
16. Levin, L.A.: Private communication (2003)
17. Markov, A.A.: Theory of algorithms. Trans. of the Steklov Institute of Mathematics 42 (1954) (Russian); English translation by the Israel Program for Scientific Translations, 1962; also by Kluwer (2010)
18. Shagrir, O.: Effective computation by humans and machines. Minds and Machines 12, 221–240 (2002)
19. Shagrir, O.: Göedel on Turing on computability. In: Olszewski, A., et al. (eds.) Church's Thesis After 70 Years, pp. 393–419. Ontos-Verlag (2006)
20. Sieg, W.: On computability. In: Irvine, A. (ed.) Handbook of the Philosophy of Mathematics, pp. 535–630. Elsevier (2009)
21. Turing, A.M.: On computable numbers, with an application to the Entscheidungsproblem. Proceedings of London Mathematical Society 2(42), 230–265 (1936)
22. Uspensky, V.A.: Kolmogorov and mathematical logic. Journal of Symbolic Logic 57(2), 385–412 (1992)
23. Uspensky, V.A., Semenov, A.L.: Theory of algorithms: main discoveries and applications, Nauka (1987) (Russian), Kluwer (2010) (English)

Turing Machine-Inspired Computer Science Results

Juris Hartmanis

Department of Computer Science,
Cornell University, Ithaca, NY 14850, United States of America
jh@cs.cornell.edu

Abstract. This paper discusses how the Turing machine model directly inspired and guided developments in theoretical computer science. In particular, the Turing machine model was ideal for the creation of computational complexity theory, which has grown into an essential part of theoretical computer science and has found application in other disciplines. The machine operation count was used to define time-bounded computations and the tape squares used defined the tape or memory-bounded computations. The definition and exploration of the corresponding asymptotic complexity classes followed naturally.

I received my PhD from California Institute of Technology in 1955 with a dissertation in lattice theory. After two delightful years at Cornell I spent a summer in the newly formed Information Studies Section of the General Electric Research Lab in Schenectady, New York. This Section's task was to build up a research effort to lay the scientific foundations for the emerging information and computing technologies. For me this was an exciting introduction to a potentially great new research area. By the end of the summer I had finished my first paper on linear coding networks and was fully committed to return to the Lab and dedicate myself to computer science research. I did so a year later in 1958, after fulfilling my earlier commitment to Ohio State University.

For computer science these were interesting and exciting times as academic programs in computer science were created and research areas were defined and established. We got to know personally many of the scientists who explored computer science. My early research at the GE Research Lab was concentrated on finite automata and their decomposition into smaller automata from which they could be realized. I was joined in this effort by Dick Stearns who came to the Information Studies Section after a summer job at the Lab and completion of his PhD dissertation in game theory at Princeton University. We collaborated intensively and explored whatever material we could find on computer science. I personally was very impressed by Shannon's work on information theory. I was surprised and very impressed that Shannon had captured precisely such vague concepts as amount of information of a source and capacity of a noisy information channel and could derive quantitative results about how much information can be transmitted over such channels [14]. I wondered if an equivalent quantitative theory could be developed to measure the difficulty of computational

S.B. Cooper, A. Dawar, and B. Löwe (Eds.): CiE 2012, LNCS 7318, pp. 276–282, 2012.

problems. My efforts in this direction based on entropy did not succeed. I lacked the right concepts or models for a quantitative theory of computing. My mathematical education had not exposed me to concepts of computability neither via recursive function theory nor Turing's work. I also recall that Dick Stearns was not familiar with Turing's work. Our exposure to Turing's ideas was a dramatic event for us. We studied Turing's paper [17] with excitement and were delighted in the simplicity of the Turing machine model and the beauty of the capture of computability via this model. The Turing machine was indeed a very powerful intellectual tool. We read more and played intellectually with the new concepts and models. We also read and were influenced by a tech report by Hisao Yamada, later published as [18], which studied a special time-bounded class of computations. All this led Dick and me to explore the Turing machine model as a foundation for a quantitative theory of computing.

We quickly convinced ourselves that the Turing machine model was robust under "reasonable" changes of the model. Turing had already observed that adding tapes to the model did not change what it could compute. Our task was to show that reasonable changes did not dramatically change the computation time and to quantify the changes. We explored multi-tape machines, machines with two and higher dimensional tapes as well as other modifications. In most of these cases we could show that these modifications could not speed up the computation by more than the square root of the one-tape machine time. We defined time-bounded computational complexity classes as consisting of all problems solvable in a number of steps bounded by some function of the input length n. In other words, the computational complexity classes were defined by how fast the difficulty of the computation grew with the length of the input. We derived hierarchy results by time-bounded diagonalization that showed that there are computations with very sharp time bounds; a slight increase in the computation time would yield a larger complexity class. Though these first results about computations with sharp time bounds were obtained by diagonalization and were not about specific practical problems, they guaranteed the existence of such problems and contributed to the interest in computational complexity [7]. As we have learned later, proving lower computational bounds for specific problems may be a very difficult indeed. Think of the $P \overset{?}{=} NP$ problem [6].

It is interesting to note that the machine model naturally suggested how to bound the diagonalization time by adding a "clock" running in parallel (on separate tapes) with the diagonalzation process and shutting it off in the desired time. Note that these clocks which bounded the complexity class runtime to $T(n)$ steps had to compute $T(n)$ in $T(n)$ steps to shut off the diagonalization in $T(n)$ steps. Originally we thought that these fast computable "clocks" were not an essential part of the result and conjectured that they could be eliminated. A short time later, my first PhD student Alan Borodin proved the gap theorem in his PhD dissertation [4]. The gap theorem showed that there exist arbitrarily large gaps in the hierarchy of complexity classes. In essence, for any monotonically growing recursive function, $g(n) > n$, there exists a recursive function $T(n) > n$ such that the complexity classes bounded by $T(n)$ and $g(T(n))$ are identical.

Clearly, $T(n)$ had to be a function that required much more than $T(n)$ steps to compute it. Thus some sort of an "easy to compute clock" was essential for the diagonalization to obtain computations with sharp time bounds.

Borodin's gap theorem was proved in Blum's axiomatic complexity theory, independent of any concrete machine models, and thus holding for all complexity measures satisfying the Blum axioms, including time- and tape-bounded computations. The gap theorem was discovered independently by B. Trakhtenbrot in 1964 behind the "Iron Curtain" and by Borodin in 1969 and is known as the Borodin-Trakhtenbrot theorem. We shall return to Blum axioms a bit later.

In the study of time bounded computations Dick and I also explored the computation of real numbers. We showed that all algebraic numbers were computable in n^2 time. We also found transcendental numbers computable in real-time (the successive digits of the number can each be printed by a Turing machine in a fixed amount of time). Our failure to show that there are real-time computable irrational algebraic numbers led us to the following conjecture: The real-time computable numbers are either rational or transcendental. That is, there are no real-time computable irrational algebraic numbers. Should this be so, we would have a proof for a large set of easily computable numbers that they are transcendental. This would establish a very interesting contrast between the computational complexity of irrational algebraic numbers and transcendental numbers. So far, we know of no real progress on this conjecture.

Dick Stearns and I met Manuel Blum at MIT while he was working on his dissertation and we exchanged ideas about computational complexity and compared our approaches. Dick and I were very impressed by the elegance Manuel's abstract axiomatic approach and the sweeping generality of his and the subsequent results formulated in terms of the Blum axioms [2]. We were particularly impressed by Blum's speedup theorem, which showed that for any complexity measure satisfying the Blum axioms, there exist computations for which any given algorithm can be "speeded up" by any prescribed arbitrarily large amount [3]. For example, there exist computations such that for any given algorithm for them there exists another algorithm which runs in loglog-time of the given algorithm. These highly abstract axiomatic results nicely complemented our concrete complexity results.

It is interesting to observe that Manuel, steeped in recursive function theory at MIT, chose the abstract, axiomatic approach to complexity theory and that Dick and I, at the GE Lab recently exposed to Turing's work, chose the Turing machine model more directly to initiate the study of concrete computation complexity theory.

After the early exploration of time bounded computations, Dick, I and Phil Lewis turned to the investigation of tape- or memory-bounded computations [15]. This work progressed very rapidly and we derived sharp hierarchy theorems and related results. A very fruitful modification of the tape-bounded Turing machine was the separation of a read-only input tape from the read-write work tape. This was indeed a model-suggested modification and it allowed us to investigate the rich set of interesting complexity classes requiring little tape for their solution. We showed that there existed non-regular languages recognizable on $\log\log n$ tape and that no non-regular language could be recognized with less tape. Thus there was a complexity gap between a constant amount of tape and $\log\log n$ tape. We had a similar gap for time-bounded computation of one-tape machines between linear time (n) and $n\log n$ time. We also showed that there existed non-regular context-free languages that could be recognized on $\log\log n$ tape (in essence, without the ability to count up to the length of the input). One of our nicest results was that all context-free languages can be recognized on $(\log n)^2$ tape. This was one of the first computational complexity results about a practical computer science problem. At the same time, our colleague Dan Younger at the GE Lab showed that context-free languages can be recognized in n^3 time. This result is now known as the CYK algorithm, for Cocke-Younger-Kasami, testifying to the rapid growth of work in complexity theory [5,11,19]. These early explorations were followed by a stream of interesting results by us and other scientists, including Savitch's result [13] that nondeterministic tape-bounded computations can be performed on deterministic machines not using more than the square of the amount of tape used by the nondeterministic computation, and the Immerman–Szelepcsényi result that nondeterministic space computations are closed under complementation [10,16].

Some of the unsolved concrete complexity class separation problems involve tape bounded computations:

$$LOGTAPE \overset{?}{=} P, \ P \overset{?}{=} NP, \ NP \overset{?}{=} PTAPE, \ PTAPE \overset{?}{=} EXPTIME.$$

We know by diagonalization arguments that there is an inequality in the sequence

$$LOGTAPE \overset{?}{=} P, \ P \overset{?}{=} NP, \ NP \overset{?}{=} PTAPE,$$

we just do not know where the break occurs! The same holds for the sequence

$$P \overset{?}{=} NP, \ NP \overset{?}{=} PTAPE, \ PTAPE \overset{?}{=} EXPTIME.$$

Another interesting use of machine model interaction yielded a set of undecidability results about context-free languages [8]. Context-free languages, *CFL*s, are the languages accepted by nondeterministic push-down automata, *PDA*s, or generated by context-free grammars, *CFG*s. For any Turing machine M, let *VAL-M* designate the set of sequences of instantaneous descriptions of valid computations by M in which every second instantaneous description is reversed. It is easily seen that a nondeterministic pushdown automaton can recognize the set of all invalid computations of M, *INVAL-M*, by nondeterministically selecting a presumably valid instantaneous description of M's computation and pushing it onto the stack and pulling it up and comparing it to the following alleged instantaneous description of M's computation. If an error or mismatch between the two allegedly consecutive instantaneous descriptions is found, the input is accepted. This links context-free languages to Turing machine computations and we see that the context-free language *INVAL-M* accepts all inputs if and only if M accepts the empty set. Thus it is recursively undecidable if a *PDA* accepts all strings over the input alphabet or, equivalently, if a context-free grammar generates all possible strings of its terminals. With a bit of cleverness it follows that for context-free languages or context-free grammars the following problems are recursively undecidable:

Is the complement of L empty?
Is $L = L'$?
Is L contained in L'?
Is complement of L finite?
Is complement of L regular?
Is complement of L a CFL?
Is the set of prefixes of the complement of L recursive?
Is the CFL L ambiguous?

Finally, these methods easily prove that no nontrivial property on the recursively enumerable sets can be recursively decided for the quotients of two $CFLs$. Some of these results were already known and proved by other methods [1].

Since the early explorations of computational complexity using Turing machine models, other models more directly modeling random access computers have been proposed and explored yielding some interesting results.

A random access machine model, RAM, consists of an set of operations and an set of registers $R_0, R_1, R_2, \ldots$ capable of storing a nonnegative integer in binary representation. The set of operations consist of assignment between registers, indirect addressing, addition and proper subtraction of two register contents, Boolean bitwise operations between two register contents, and conditional jump. The computation is started with the input in R_0 and all other registers set to zero. The instructions of the program are executed in sequence until a conditional jump is executed and the computation continues. The computation time is the count of operations executed [9,12].

For this RAM model we get reassuring results:

$$RAM\text{-}POLYTIME = P, \ NRAM\text{-}POLYTIME = NP.$$

The surprise comes when we add multiplication of register contents as a one-step operation to the RAM operation set; we call it an $MRAM$. The $MRAM$ can build, in n operations, exponentially long register contents on which it can perform parallel logical bit operations in one step. This power leads to the following results:

$$MRAM\text{-}POLYTIME = PSPACE = NMRAM\text{-}POLYTIME.$$

Since we believe that P is not equal to NP nor NP to $PSPACE$, we must conclude the multiplication made the RAM into an unrealistic model for computational complexity theory. At the same time, it is reassuring that the polynomial time RAM computations (with registers of unbounded length) yield the classical polynomial time complexity classes, P and NP, respectively. These results give additional evidence for the adequacy of the Turing machine model for computational complexity theory. At the same time, the more powerful $MRAM$

model wipes out the distinctions between complexity classes of direct interest in computer science. In general, the separation of complexity classes defined by different concrete complexity measures are among the many unsolved problems in complexity theory, including the famous $P \overset{?}{=} NP$ problem.

Since these early days of computational complexity research, much of it inspired and guided by the Turing machine model, computational complexity has grown into an essential and vital part of theoretical computer science with many beautiful and important results as well as applications in computer science and other disciplines.

References

1. Bar-Hillel, Y., Perlis, M., Shamir, E.: On formal properties of simple phrase structure grammars. Z. Phonetik, Sprachwissen. Komm. 14, 143–172 (1961)
2. Blum, M.: A machine-independent theory of the complexity of recursive functions. J. Assoc. Comput. Mach. 14(2), 322–336 (1967)
3. Blum, M.: On effective procedures for speeding up algorithms. J. Assoc. Comput. Mach. 18(2), 290–305 (1971)
4. Borodin, A.B.: Computational complexity and the existence of complexity gaps. J. Assoc. Comput. Mach. 19(1), 158–174 (1972)
5. Cocke, J., Schwartz, J.T.: Programming languages and their compilers: Preliminary notes. Tech. rep., Courant Institute (1970)
6. Cook, S.A.: The complexity of theorem proving procedures. In: Proc. 3rd Symp. Theory of Computing, pp. 151–158. ACM, New York (1971)
7. Hartmanis, J., Stearns, R.E.: On the computational complexity of algorithms. Trans. Amer. Math. Soc. 117, 285–306 (1965)
8. Hartmanis, J.: Context-free languages and Turing machine computations. J. Symbolic Logic 37(4), 759 (1972)
9. Hartmanis, J., Simon, J.: On the power of multiplciation in random-access machines. In: Proc. 15th Annu. IEEE Sympos. Switching Automata Theory, pp. 13–23 (1974)
10. Immerman, N.: Nondeterministic space is closed under complement. SIAM J. Comput. 17, 935–938 (1988)
11. Kasami, T.: An efficient recognition and syntax algorithm for context-free languages. Tech. Rep. AFCRL-65-758, Air Force Cambridge Research Lab (1965)
12. Pratt, V.R., Rabin, M.O., Stockmeyer, L.J.: A characterization of the power of vector machines. In: Proc. ACM Symp. Theory of Computation (STOC 1974), pp. 122–134 (1974)
13. Savitch, W.: Relationship between nondeterministic and deterministic tape complexities. J. Comput. Syst. Sci. 4(2), 177–192 (1970)
14. Shannon, C.E.: A mathematical theory of communication. Bell System Technical Journal 27, 379–423, 623–656 (1948)
15. Stearns, R., Hartmanis, J., Lewis, R.: Hierarchies of memory limited computations. In: Proc. IEEE Conf. Switching Circuit Theory and Logical Design, pp. 179–190 (1965)

16. Szelepcsényi, R.: The method of forcing for nondeterministic automata. Bull. EATCS 33, 96–100 (1987)
17. Turing, A.M.: On computable numbers with an application to the Entscheidungsproblem. Proc. London Math. Soc. 42, 230–265 (1936); erratum: Ibid. 43, 544–546 (1937)
18. Yamada, H.: Real-time computation and recursive functions not real-time computable. IEEE Trans. Electronic Computers EC-11(6), 753–760 (1962)
19. Younger, D.H.: Recognition and parsing of context-free languages in time n^3. Information and Control 10(2), 189–208 (1967)

NP-Hardness and Fixed-Parameter Tractability of Realizing Degree Sequences with Directed Acyclic Graphs

Sepp Hartung and André Nichterlein

Institut für Softwaretechnik und Theoretische Informatik, Technische Universität Berlin, TEL 12-4, Ernst-Reuter-Platz 7, 10587 Berlin, Germany
{sepp.hartung,andre.nichterlein}@tu-berlin.de

Abstract. In graph realization problems one is given a degree sequence and the task is to decide whether there is a graph whose vertex degrees match the given sequence. This realization problem is known to be polynomial-time solvable when the graph is directed or undirected. In contrast, we show NP-completeness for the problem of realizing a given sequence of pairs of positive integers (representing indegrees and outdegrees) with a *directed acyclic graph*, answering an open question of Berger and Müller-Hannemann [FCT 2011]. Furthermore, we classify the problem as fixed-parameter tractable with respect to the parameter "maximum degree".

Keywords: graph realization problems, combinatorial algorithms, parameterized complexity, realizing topological orderings.

1 Introduction

Berger and Müller-Hannemann introduced the following problem [1]:

DAG REALIZATION
Input: A multiset $\mathcal{S} = \left\{ \binom{a_1}{b_1}, \ldots, \binom{a_n}{b_n} \right\}$ of integer pairs with $a_i, b_i \geq 0$.
Question: Is there a directed acyclic graph (without parallel arcs and self-loops) that admits a labeling of its vertex set $\{v_1, \ldots, v_n\}$ such that for all $v_i \in V$ the indegree is a_i and the outdegree is b_i?

If the *degree sequence* $\mathcal{S}$ is a yes-instance, then $\mathcal{S}$ is called *realizable* and the corresponding directed acyclic graph (DAG for short) D is called a *realizing DAG* for $\mathcal{S}$. Berger and Müller-Hannemann showed that this problem is polynomial-time solvable for special types of degree sequences, but left the complexity of the general problem as their main open question [1]. We answer this question by showing that DAG REALIZATION is NP-complete. Moreover, on the positive side we classify DAG REALIZATION as fixed-parameter tractable with respect to the parameter maximum degree $\Delta := \max\{a_1, b_1, \ldots, a_n, b_n\}$. The corresponding algorithm actually constructs for yes-instances a realizing DAG.

S.B. Cooper, A. Dawar, and B. Löwe (Eds.): CiE 2012, LNCS 7318, pp. 283–292, 2012.

Related Work. It has been known for a long time that deciding whether a given degree sequence (a multiset of positive integers) is realizable with an *undirected graph* is polynomial-time solvable. There are characterizations for realizable degree sequences due to [5] and algorithms by Havel and Hakimi [11,10]. In the case, where one asks whether there is a *directed graph* realizing the given degree sequence (a multiset of positive integer pairs), has also been intensively studied: Cf. [3,8,7,17] for characterizations of digraph realizations and [13] for polynomial-time algorithms. The problem of realizing degree sequences has also been studied in context of (loop-less) multigraphs, where the aim is to minimize or maximize the number of multi-edges [12].

Preliminaries. We set $\mathbb{N} := \{0, 1, 2, \ldots\}$. A parameterized problem is *fixed-parameter tractable* if any instance (I, k), consisting of the "classical" problem instance I and the parameter $k \in \mathbb{N}$, can be solved in $f(k) \cdot n^c$ time. Thereby, f is a computable function solely depending on k and $c \in \mathbb{N}$ is a constant. For a more detailed introduction to parameterized algorithmics and complexity we refer to the monographs [4,6,15].

We denote directed graphs by $D = (V, A)$ with vertex set V and arc set $A \subseteq V \times V$. The indegree of $v \in V$ is denoted by $d^-(v)$ and the outdegree by $d^+(v)$. Correspondingly, for a degree sequence $\mathcal{S}$ and an *element* $s \in \mathcal{S}$ with $s = \binom{a}{b}$, we set $d^-(s) := a$ and $d^+(s) := b$. A directed graph $D = (V, A)$ is a *DAG* if it does not contain a cycle. A cycle is a vertex sequence $v_1, \ldots, v_l$ such that for all $1 \leq i < l : (v_i, v_{i+1}) \in A$ and $(v_l, v_1) \in A$. Each DAG D admits a *topological ordering* , that is, an ordering of all its vertices $v_1, \ldots, v_n$ such that for all arcs $(v_i, v_j) \in A$ it holds that $i < j$. Consequently, for a realizing DAG we call a corresponding topological ordering a *realizing topological ordering* .

We use the *opposed order* $\leq_{\mathrm{opp}}$ for the elements of a degree sequence $\mathcal{S}$, as introduced in [1]:

Definition 1. $\binom{a_1}{b_1} \leq_{\mathrm{opp}} \binom{a_2}{b_2} \iff (a_1 \leq a_2 \wedge b_1 \geq b_2)$

Note that there might be elements in the degree sequence $\mathcal{S}$ that are not comparable with respect to the opposed order. However, we can always assume that a realization does not collide with the opposed order and thus DAG REALIZATION is polynomial-time solvable in case of all elements of $\mathcal{S}$ are comparable.

Lemma 1 ([1, Corollary 3]). *Let* $\mathcal{S} = \left\{ \binom{a_1}{b_1}, \ldots, \binom{a_n}{b_n} \right\}$ *be a realizable degree sequence. Then, there exists a realizing topological ordering* ϕ *such that for all* $1 \leq i, j \leq n$ *with* $s_i = \binom{a_i}{b_i} \leq_{\mathrm{opp}} \binom{a_j}{b_j} = s_j$ *and* $s_i \neq s_j$, *it holds that in* ϕ *the position of the vertex that corresponds to* s_i *is smaller than the position of the vertex that corresponds to* s_j.

Our paper is organized as follows: Section 2 contains the proof of the NP-completeness and in Section 3 we show that DAG REALIZATION is fixed-parameter tractable with respect to the parameter maximum degree Δ.

Due to the space constraints we omit most of the proofs in Section 2 and the detailed correctness proof of the algorithm that we describe in

Section 3. For the general public a full version of the paper is available from http://arxiv.org/abs/1110.1510v2.

2   NP-Completeness

In this section we show the NP-hardness of DAG REALIZATION by giving a polynomial-time many-to-one reduction from the strongly NP-hard problem 3-PARTITION [9, SP15]:

3-PARTITION

Input: A multiset $\mathcal{A} = \{a_1, \ldots, a_{3m}\}$ of $3m$ positive integers and an integer B with $\sum_{i=1}^{3m} a_i = mB$ and $\forall i : B/4 < a_i < B/2$.

Question: Is there a partition of the $3m$ integers from $\mathcal{A}$ into m disjoint triples such that in every triple the three elements add up to B?

This section is organized as follows: First we describe the construction of our reduction and explain the idea of how it works. Then, we prove the correctness in the remainder of the section.

Construction. Given an instance $(\mathcal{A}, B)$ of 3-PARTITION, we construct an equivalent instance $\mathcal{S}$ of DAG REALIZATION as follows:

$$\mathcal{S} := X_0, X_1, \ldots, X_m, \alpha_1, \alpha_2, \ldots, \alpha_{3m}$$

where $\alpha_i := \binom{a_i}{a_i}$, $1 \leq i \leq 3m$. The X_i, $0 \leq i \leq m$, are subsequences which we formally define after giving the idea of the construction. We call an element from a subsequence X_i an *x-element* and the α_j are called *a-elements*. In a realizing DAG D the vertices realizing x-elements are called *x-vertices* and the vertices realizing a-elements are called *a-vertices*.

The intuition of the construction is that a DAG D realizing $\mathcal{S}$ (if it exists) looks as follows: The vertices realizing elements of a subsequence X_i, $0 \leq i \leq m$, form a "block" in a realizing topological ordering ϕ. These blocks are a skeletal structure in any realizing topological ordering and there are m "gaps" between them. The construction ensures that these gaps are filled with a-vertices and, moreover, the indegree and outdegree of all the a-vertices in a gap sum up to B. Hence, these m gaps require to partition the a-vertices into m sets where each set has in total in- and outdegree B and, thus, correspond to a solution for the 3-PARTITION instance where we reduce from. In the reverse direction, for each triple in a solution of a 3-PARTITION instance the corresponding a-vertices will be used to fill up one gap. See Figure 1 for an example of the construction.

To achieve the mentioned skeletal structure of the subsequences $X_0, \ldots, X_m$, we require the corresponding x-vertices to form a *transitive tournament*. That is a DAG with n vertices and $\binom{n}{2}$ arcs that realizes the degree sequence $\{\binom{0}{n-1}, \binom{1}{n-2}, \ldots, \binom{n-1}{0}\}$. Observe that there is only one DAG realizing such a sequence and, furthermore, such a transitive tournament admits only one topological ordering.

Now, we complete the reduction by defining the subsequences $X_0, \ldots, X_m$. As indicated in Figure 1, X_0 and X_m contain B elements and the

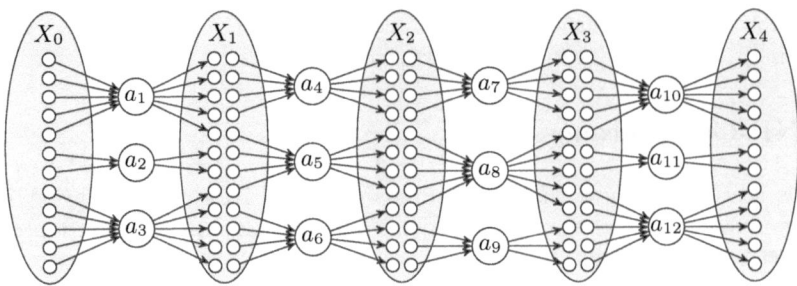

Fig. 1. A schematic representation of a DAG that realizes a degree sequence $\mathcal{S}$ that is constructed from a 3-PARTITION instance with $B = 12$ and $m = 4$. There are five blocks marked by the gray ellipses and four gaps between them. In each gap there are three a-vertices, altogether having in- and outdegree B. The sets X_i, $1 \leq i \leq 3$, are partitioned into two parts of size B. The vertices in the left part have B incoming arcs from the a-vertices that fill the gap between X_{i-1} and X_i. Correspondingly, the vertices in the right part have B outgoing arcs to the a-vertices that fill the gap between X_i and X_{i+1}. Consequently, the first and the last block X_0 and X_4 are of size B. The in- and outdegree of the a-vertices in each triple sum up to B.

other subsequences contain $2B$ elements. The subsequence X_0 consists of the elements $x_0^0, x_0^1, \ldots, x_0^{B-1}$. This subsequence corresponds to the x-vertices $v_0^0, \ldots, v_0^{B-1}$ forming the first block in a realizing DAG for $\mathcal{S}$. Remember that the x-vertices are supposed to form a transitive tournament. To achieve this, v_0^j has $(B-1-j)$ outgoing arcs to $v_0^{j+1}, \ldots, v_0^{B-1}$ and $(m-1)2B+B = (2m-1)B$ outgoing arcs to the x-vertices in the subsequent blocks. Furthermore, v_0^j has j incoming arcs from the x-vertices $v_0^0, \ldots, v_0^{j-1}$. Finally, each x-vertex in $v_0^0, \ldots, v_0^{B-1}$ has one outgoing arc to one of the three subsequent a-vertices. Hence, the corresponding x-element of v_0^j is as follows:

$$x_0^j := \binom{j}{(B-1-j) + (2m-1)B + 1} = \binom{j}{2mB - j}.$$

Analogously, the subsequence X_m consists of B elements $x_m^0, x_m^1, \ldots, x_m^{B-1}$ with

$$x_m^j := \binom{(2m-1)B + j + 1}{B - 1 - j}.$$

For $0 < i < m$, the subsequence X_i consists of $2B$ elements $x_i^0, x_i^1, \ldots, x_i^{2B-1}$. Let $v_i^0, \ldots, v_i^{2B-1}$ denote the corresponding x-vertices. Then, v_i^j has $(i-1)2B + B = (2i-1)B$ incoming arcs from the x-vertices in the preceding blocks and j incoming arcs from $v_i^0, \ldots, v_i^{j-1}$. Furthermore, v_i^j has $(m-i-1)2B + B = (2m-2i-1)B$ outgoing arcs to the subsequent blocks and $2B - 1 - j$ outgoing arcs to $v_i^{j+1}, \ldots, v_i^{B-1}$. Finally, if $j < B$, then v_i^j has an incoming arc from one of the three preceding a-vertices. Otherwise, if $j \geq B$, then v_i^j has an outgoing arc to one of the three subsequent a-vertices. Hence, the corresponding x-element of v_i^j is as follows:

$$x_i^j := \begin{pmatrix} (2i-1)B + j + 1 \\ (2m - 2i + 1)B - 1 - j \end{pmatrix} \qquad \text{if } j < B,$$

$$x_i^j := \begin{pmatrix} (2i-1)B + j \\ (2m - 2i + 1)B - j \end{pmatrix} \qquad \text{if } j \geq B.$$

Observe that the strong NP-hardness of 3-PARTITION is essential to prove the polynomial running time of the reduction: The size of the constructed DAG REALIZATION instance is upper-bounded by a polynomial in the values of the integers in $\mathcal{A}$. Since 3-PARTITION is strongly NP-hard, it remains NP-hard when the values of the integers in $\mathcal{A}$ are bounded by a polynomial in the input size. Hence, the size of the DAG REALIZATION instance is polynomially bounded in the size of the 3-PARTITION instance. Clearly, the construction can be computed in polynomial time.

Correctness. Next, we prove the correctness of the construction given above. Therefore, throughout this subsection let $(\mathcal{A}, B)$ be an instance of 3-PARTITION and let $\mathcal{S}$ be the constructed degree sequence.

Lemma 2. *If $(\mathcal{A}, B)$ is a yes-instance of 3-PARTITION, then $\mathcal{S}$ is a yes-instance of DAG REALIZATION.*

To show the reverse direction, we first need two observations.

- Observation 1: In any DAG D realizing $\mathcal{S}$, the a-vertices form an independent set and the x-vertices form a transitive tournament.
- Observation 2: If $\mathcal{S}$ is realizable, then there is a realizing topological ordering where the x-vertices are ordered as follows:

$$x_0^0, x_0^1, \ldots, x_0^{B-1}, x_1^0, \ldots, x_1^{2B-1}, x_2^0, \ldots, x_2^{2B-1}, x_3^0, \ldots, x_{m-1}^{2B-1}, x_m^0, \ldots, x_m^{B-1}.$$

Lemma 3. *If $\mathcal{S}$ is a yes-instance of DAG REALIZATION, then $(\mathcal{A}, B)$ is a yes-instance of 3-PARTITION.*

Proof. Let $D = (V, A)$ be the realization of $\mathcal{S}$ with a topological ordering ϕ. Let v_i^j be the x-vertex realizing x_i^j and let u_i be the a-vertex realizing a_i. Furthermore, $\text{pos}_\phi(v)$ denotes the position of v in the topological ordering ϕ. Since $\mathcal{S}$ is a yes-instance, we can assume by Observation 2 that $\text{pos}_\phi(v_i^j) < \text{pos}_\phi(v_i^\ell)$ for $j < \ell$ and that $\text{pos}_\phi(v_i^j) < \text{pos}_\phi(v_k^\ell)$ for $i < k$.

From Observation 1 it follows that none of the x-vertices $v_0^0, v_0^1, \ldots, v_0^{B-1}$ has an incoming arc from an a-vertex, but each has one outgoing arc to an a-vertex. Hence, we can assume that $\text{pos}_\phi(u_i) > \Phi(v_0^{B-1})$ for all $1 \leq i \leq 3m$. Observe that each x-vertex $v_1^0, v_1^1, \ldots, v_1^{B-1}$ has one incoming arc from an a-vertex and no outgoing arc to an a-vertex. Hence, we can assume that there are a-vertices $u_{i_1}, u_{i_2}, \ldots, u_{i_\ell}$ with $\mathrm{pos}_\phi(v_0^{B-1}) < \mathrm{pos}_\phi(u_{i_j}) < \mathrm{pos}_\phi(v_1^0)$ and $\sum_{j=1}^{\ell} a_{i_j} = B$. Since $B/4 < a_j < B/2$ for all $1 \leq j \leq 3m$, it follows that $\ell = 3$.

The vertices $v_1^B, \ldots, v_1^{2B-1}$ also have no incoming arc from an a-vertex but each of them has an outgoing arc to an a-vertex. Also, each of the vertices $v_2^0, \ldots, v_2^{B-1}$ needs one incoming arc from an a-vertex. Thus, we can assume that in the topological ordering ϕ of D there are three a-vertices between v_1^{2B-1} and v_2^0 such that their indegrees and also their outdegrees sum up to B. Analogously, it follows for all $1 \leq i < m$ that there are three a-vertices $u_{j_1^i}, u_{j_2^i}, u_{j_3^i}$ with $\mathrm{pos}_\phi(v_i^{2B-1}) < \mathrm{pos}_\phi(u_{j_1^i}) < \mathrm{pos}_\phi(u_{j_2^i}) < \mathrm{pos}_\phi(u_{j_3^i}) < \mathrm{pos}_\phi(v_{i+1}^0)$ and $\sum_{\ell=1}^{3} a_{j_\ell^i} = B$. Thus, $(\mathcal{A}, B)$ is a yes-instance of 3-PARTITION.   □

Our construction together with Lemma 2 and Lemma 3 yields the NP-hardness of DAG REALIZATION. Containment in NP is easy to see: Guessing a n-vertex DAG and checking whether or not it is a realization for $\mathcal{S}$ is clearly doable in polynomial time. Hence, we arrive at the following theorem.

Theorem 1. DAG REALIZATION *is NP-complete.*

Berger and Müller-Hannemann gave a polynomial-time algorithm for DAG RE-ALIZATION if the degree sequence can be ordered with respect to the opposed order [1]. Hence, one may search for other polynomial-time solvable special cases. One way to identify such special cases is to have a closer look on NP-hardness proofs and to check whether certain "quantities" need to be unbounded in order to make the proof (many-to-one reduction) work [14,16]. In our NP-hardness proof the maximum degree Δ is unbounded. We show in the next section that DAG REALIZATION is polynomial-time solvable for constant maximum degree. Indeed, we can even show fixed-parameter tractability with respect to the parameter Δ.

3   Fixed-Parameter Tractability

Denoting the maximum degree in a degree sequence by Δ, in this section we show that DAG REALIZATION is fixed-parameter tractable with respect to the parameter Δ. To describe the basic idea that our fixed-parameter algorithm is based on, we need the following definition.

Definition 2. *Let $\phi = v_1, v_2, \ldots, v_n$ be a topological ordering for a DAG D. For all $1 \leq i \leq n$, the potential at position i is a vector $p_i^\phi \in \mathbb{N}^\Delta$ where $p_i^\phi[l]$ for $1 \leq l \leq \Delta$ is the number of vertices in the subsequence $v_1, \ldots, v_i$ that have in D at least l neighbors in the subsequence $v_{i+1}, \ldots, v_n$. The value of the potential p_i^ϕ is $\omega(p_i^\phi) := \sum_{l=1}^{\Delta} p_i^\phi[l]$.*

If the topological ordering ϕ is clear from the context, then we write p instead of p^ϕ. Observe that, for any potential $p_i \in \mathbb{N}^\Delta$, it holds that $p_i[j] \geq p_i[j+1]$ for all $1 \leq j < \Delta$.

Algorithm Outline. Our algorithm consists of two parts. First, if the degree sequence of a DAG REALIZATION instance admits a DAG realization where at any position the value of the potential is at least Δ^2, then we shall find such a "high-potential" realization with the algorithm that is described in Subsection 3.2. Otherwise, by exploiting the fact that the value of all potentials is upper-bounded, we shall find a "low potential" realization with the algorithm described in Subsection 3.3.

3.1 General Terms and Observations

In this section we introduce some general notations and observations that will be used in the algorithms to find high potential as well as low potential realizations.

Notation: For a topological ordering $\phi = v_1, \ldots, v_n$ and two indices $1 \leq i \leq j \leq n$, set $\phi[i, j] := v_i, v_{i+1}, \ldots, v_j$. The set $\{v_i, \ldots, v_j\}$ is also denoted by $\phi[i, j]$.

Definition 3. *Let* $\mathcal{S} = \left\{ \binom{a_1}{b_1}, \ldots, \binom{a_n}{b_n} \right\}$ *be a degree sequence. Two tuples* $\binom{a_i}{b_i}$ *and* $\binom{a_j}{b_j}$ *are of the same* type *if* $a_i = a_j$ *and* $b_i = b_j$. *Furthermore, a type* $\binom{a_i}{b_i}$ *is a* good type *if* $a_i \leq b_i$ *and otherwise it is a* bad type.

Note that there are at most $(\Delta + 1)^2$ different types.

Well-Connected DAGs. Berger and Müller-Hannemann already observed that, given a degree sequence $\mathcal{S} = \left\{ \binom{a_1}{b_1}, \ldots, \binom{a_n}{b_n} \right\}$, one can check in polynomial time whether $\mathcal{S}$ is realizable by a DAG with a corresponding topological ordering $v_1, \ldots, v_n$ where $d^-(v_i) = a_i$ and $d^+(v_i) = b_i$ [2]. This implies that it is sufficient to compute the correct ordering of the elements in $\mathcal{S}$ as they appear in a topological ordering of a realizing DAG. To prove this, the main observation is that for any topological ordering one can construct at least one corresponding DAG by *well-connecting* consecutive vertices.

Definition 4. *Let D be a DAG with a corresponding topological ordering $\phi = v_1, \ldots, v_n$. The* remaining outdegree *at position j of vertex v_i, $1 \leq i \leq j \leq n$, is the number of v_i's neighbors in the subsequence $\phi[j, n]$. Furthermore, D is* well-connected *if for all vertices $v_i \in \phi$ it holds that v_i is connected to the $d^-(v_i)$ vertices in $\phi[1, i-1]$ that have the highest remaining outdegree at position $i - 1$.*

It is not hard to show that in a well-connected DAG the potential at position i can be easily determined from the potential at position $i - 1$.

Cut-Out Subsequences. The following lemma shows that if in a topological ordering $\phi[1, n]$ there are two indices $1 \leq i < j \leq n$ with equal potential, then we can cut out $\phi[i + 1, j]$ resulting in a topological ordering $\phi[1, i][j + 1, n]$. One can show that one can reinsert $\phi[i + 1, j]$ at any position that fulfills some reasonable conditions. Thereby, the corresponding degree of each vertex do not change. Cutting out subsequences and reinserting them appropriately is the main operation that we perform in order to "restructure" a topological ordering such that we can exploit the resulting regular structure in our algorithms.

Lemma 4. *Let $\phi = v_1, \ldots, v_n$ be a realizing topological ordering for the degree sequence $\mathcal{S} = \left\{ \binom{a_1}{b_1}, \ldots, \binom{a_n}{b_n} \right\}$. If there are two indices $1 \leq i < j \leq n$ such that $p_i^\phi = p_j^\phi$, then the sequence $\phi' = \phi[1, i]\phi[j+1, n]$ is a realizing topological ordering for the degree sequence that results from $\mathcal{S}$ by deleting the degrees of the vertices in $\phi[i+1, j]$. Moreover, the potential $p_{i+l}^{\phi'}$ is equal to p_{j+l}^ϕ for all $1 \leq l \leq n-j$.*

Lemma 4 shows that from a topological ordering ϕ we can cut out a subsequence $\phi[i+1, j]$ whenever $p_i^\phi = p_j^\phi$. Informally speaking, one can show that the subsequence $\phi[i+1, j]$ can be inserted into any topological ordering ϕ' at position b whenever there is "enough potential" from the left part $\phi'[1, b]$ to satisfy the indegrees of $\phi[i+1, j]$. Then, the structure of $\phi[i+1, j]$ guarantees that the "remaining potential" of $\phi'[1, b]\phi[i+1, j]$ is sufficient to satisfy the indegree of $\phi'[b+1, n]$ and thus $\phi'[1, b]\phi[i+1, j]\phi'[b+1, n]$ is a topological ordering.

3.2   High Potential Sequences

In this subsection we show that if a realizable sequence admits a realizing topological ordering where at some position the value of the potential is at least Δ^2, a so-called *high potential realizing topological ordering* , then there is also a realizing topological ordering ϕ that is of the following "pattern": The ordering ϕ can be partitioned into four sub-sequences $I \circ G \circ B \circ E$ (where $\circ$ is the concatenation). The sequence I is an initializing sequence that "establishes" a potential of value at least Δ^2, a so-called *high potential*. Correspondingly, at the end there is a sequence E that reduces the value of the potential from a value that is greater than Δ^2 to zero. Furthermore, I and E are of length at most $\Delta^{2\Delta}$ and thus can be guessed in $O((\Delta + 1)^2)^{\Delta^{2\Delta}}) = O(\Delta^{2\Delta^{2\Delta}})$ time. The subsequence G, which is of arbitrary length, only consists of good types and, correspondingly, B is of arbitrary length but only consists of bad types.

Our strategy to prove that there is a high potential realizing topological ordering with the pattern $I \circ G \circ B \circ E$ is as follows. Let $\phi = v_1, \ldots, v_n$ be an arbitrary high potential realizing topological ordering and let i be the minimum position with high potential and, symmetrically, let j be the maximum position with high potential. First, one can show that we can assume that $i \leq \Delta^{2\Delta}$ and $j \geq n - \Delta^{2\Delta}$. Towards this the main argument is that if $i > \Delta^{2\Delta}$, since there are $O(\Delta^{2\Delta})$ potentials with value less than Δ^2, there have to be two positions $1 \leq l_1 < l_2 < i$ with $p_{l_1} = p_{l_2}$. Then, by Lemma 4, we can cut out $\phi[l_1 + 1, l_2]$ from ϕ and one can show that it can be reinserted right behind i, resulting in a realizing topological ordering $\phi[1, l_1]\phi[l_2 + 1, i]\phi[i + 1, l_2]\phi[i + 1, n]$. By iteratively applying this operation, we end up with a realizing topological ordering where the minimum position with high potential is at most $\Delta^{2\Delta}$. A symmetric argument holds for the maximum position j with high potential. Second, one can show that the vertices in $\phi[i + 1, j]$ can be arbitrarily sorted under the constraint that at first vertices of good type occur, and then they are followed by the bad type vertices.

Altogether, this shows that in order to check whether there is a high potential realizing topological ordering it is sufficient to branch into all possibilities to choose I and E, insert the remaining vertices sorted by good and bad types in between, and, finally, check whether this ordering is a topological ordering.

Theorem 2. *If a* DAG REALIZATION *instance admits a high potential realizing topological ordering, then it can be solved in* $O(\Delta^{4\Delta^{2\Delta}} \cdot n)$ *time.*

3.3 Low Potential Sequences

In this section, we shall provide an algorithm that finds a *low potential realization* (if one exists) for a DAG REALIZATION instance. That is, a realization such that in the corresponding topological ordering the value of all potentials is strictly less than Δ^2.

As in the high potential case, the main idea is to restrict the length of the parts in a realizing topological ordering that have to be guessed by brute-force. In the low potential case, we can exploit that there are at most $\Delta^{2\Delta}$ potentials with value less than Δ^2 and thus if the length of a realizing topological ordering is greater than $\Delta^{2\Delta}$, then there have to be two positions with equal potential. We call subsequences with equal potential at their beginning and end *super-types*. Then, by cutting-out super-types and reinserting them appropriately we can bound the distance between two subsequent positions with the same potential, implying a restricted number of different super-types. Hence, if one removes in a realizing topological ordering all repetitions of a super-type, then the length of the resulting *non-repeating ordering* can be upper-bounded in a function of Δ.

The algorithm works as follows: In the first step it branches into all possibilities to choose a non-repeating ordering. In the second step, we check via solving an ILP (integer linear program) whether this non-repeating ordering can be extended by inserting repeating super-types to a realizing topological ordering of the input degree sequence.

Theorem 3. *If a degree sequence admits a low potential realizing topological ordering, then it can be found in* $\Delta^{\Delta^{\Delta^{O(\Delta)}}} \cdot n$ *time.*

Theorems 2 and 3 together lead to the main theorem of this section.

Theorem 4. DAG REALIZATION *is fixed-parameter tractable with respect to the parameter maximum degree* Δ.

Note that Theorem 4 is a mere classification result: The corresponding running time is $\Delta^{\Delta^{\Delta^{O(\Delta)}}} \cdot n$. It is dominated by the low potential case.

4   Conclusion and Open Questions

Answering an open question by Berger and Müller-Hannemann [1] we proved the NP-completeness of DAG REALIZATION. Following the spirit of deconstructing intractability we figured out the necessity of large degrees in the NP-hardness proof by showing fixed-parameter tractability for DAG REALIZATION with respect to the maximum degree Δ. The natural questions whether DAG REAL-IZATION is solvable in single-exponential time and whether it admits a polynomial-size problem kernel with respect to the parameter Δ arises. In our NP-hardness reduction other parameters occur with unbounded values, for instance, the number of types. Investigating this parameter is an interesting task for future work.

Acknowledgements.   We thank Annabell Berger and Matthias Müller-Hannemann for fruitful discussions about DAG REALIZATION. Moreover, we are grateful to Rolf Niedermeier for helpful comments improving the presentation.

References

1. Berger, A., Müller-Hannemann, M.: Dag Realizations of Directed Degree Sequences. In: Owe, O., Steffen, M., Telle, J.A. (eds.) FCT 2011. LNCS, vol. 6914, pp. 264–275. Springer, Heidelberg (2011)
2. Berger, A., Müller-Hannemann, M.: How to Attack the NP-Complete Dag Realization Problem in Practice. In: Klasing, R. (ed.) SEA 2012. LNCS, vol. 7276, pp. 51–62. Springer, Heidelberg (2012)
3. Chen, W.K.: On the realization of a (p, s)-digraph with prescribed degrees. J. Franklin Inst. 281(5), 406–422 (1966)
4. Downey, R.G., Fellows, M.R.: Parameterized Complexity. Springer (1999)
5. Erdős, P., Gallai, T.: Graphs with prescribed degrees of vertices. Math. Lapok 11, 264–274 (1960) (in Hungarian)
6. Flum, J., Grohe, M.: Parameterized Complexity Theory. Springer (2006)
7. Fulkerson, D.: Zero-one matrices with zero trace. Pacific J. Math. 10(3), 831–836 (1960)
8. Gale, D.: A theorem on flows in networks. Pacific J. Math. 7, 1073–1082 (1957)
9. Garey, M.R., Johnson, D.S.: Computers and Intractability: A Guide to the Theory of NP-Completeness. Freeman (1979)
10. Hakimi, S.: On realizability of a set of integers as degrees of the vertices of a linear graph. i. J. SIAM 10(3), 496–506 (1962)
11. Havel, V.: A remark on the existence of finite graphs. Casopis Pest. Mat. 80, 477–480 (1955)
12. Hulett, H., Will, T.G., Woeginger, G.J.: Multigraph realizations of degree sequences: Maximization is easy, minimization is hard. Oper. Res. Lett. 36(5), 594–596 (2008)
13. Kleitman, D., Wang, D.: Algorithms for constructing graphs and digraphs with given valences and factors. SIAM J. Discrete Math. 6(1), 79–88 (1973)
14. Komusiewicz, C., Niedermeier, R., Uhlmann, J.: Deconstructing intractability—a multivariate complexity analysis of interval constrained coloring. J. Discrete Algorithms 9(1), 137–151 (2011)
15. Niedermeier, R.: Invitation to Fixed-Parameter Algorithms. Oxford University Press (2006)
16. Niedermeier, R.: Reflections on multivariate algorithmics and problem parameterization. In: Proc. 27th STACS. LIPIcs, vol. 5, pp. 17–32. IBFI Dagstuhl, Germany (2010)
17. Ryser, H.: Combinatorial properties of matrices of zeros and ones. Canadian J. Math. 9, 371–377 (1957)

A Direct Proof of Wiener's Theorem

Matthew Hendtlass and Peter Schuster

Pure Mathematics
University of Leeds
Leeds LS2 9JT, England
{mmmrh,pmtpsc}@leeds.ac.uk

Abstract. In functional analysis it is not uncommon for a proof to proceed by contradiction coupled with an invocation of Zorn's lemma. Any object produced by such an application of Zorn's lemma does not in fact exist, and it is likely that the use of Zorn's lemma is artificial. It has turned out that many proofs of this sort can be simplified, both in form and complexity, with the principle of open induction isolated by Raoult as a substitute for Zorn's lemma. If moreover the theorem under consideration is sufficiently concrete, then a far weaker instance of induction suffices and, with some massaging, one may obtain a fully constructive proof. In the present note we apply this method to Gelfand's proof of Wiener's theorem, producing first a simple direct proof of Wiener's theorem, and then an even simpler constructive proof. With this example in mind we look toward developing a more generally applicable technique.

1 Introduction

This paper is the beginning of a programme to systematically constructivise theorems from functional analysis which typically, in the classical setting, rely on a proof by contradiction coupled with an invocation of Zorn's lemma. Since the ideal objects produced by an application of Zorn's lemma in a proof by contradiction do not exist (hence the contradiction), proofs of this form can often be simplified by giving a direct proof using the principle of open induction isolated by Raoult [12]. As pointed out in [13] for the setting of abstract algebra, if a theorem proved with open induction has finite input data, then a finite partial order carries the required instance of induction, giving an elementary proof. In analysis we do not deal with finite inputs, but if a result is sufficiently concrete, then we can often make do with standard mathematical induction.

Proving concrete theorems directly with open induction removes unneccesary appeal to the ideal objects of Zorn's lemma, thus allowing us to work with objects of concrete existence. This reduces the logical complexity, modulo open induction, of our proofs and removes a major hurdle in constructivising proofs of this form. Ultimately, we are interested in uniform techniques to translate proofs which use Zorn's lemma and contradiction into proofs via induction.

In this paper we consider Gelfand's proof of Wiener's theorem on the invertibility of absolutely convergent Fourier series. We first use open induction for a

S.B. Cooper, A. Dawar, and B. Löwe (Eds.): CiE 2012, LNCS 7318, pp. 293–302, 2012.

simple direct variant of this proof , and then give an even simpler constructive proof—motivated by the first—which only requires a weak form of countable choice.

Our aim is not to give either an elementary or a constructive proof of Wiener's theorem, tasks which have been performed by others. The purpose of the present paper is to demonstrate the general approach this programme will take.

Functional analysis has already received plenty of attention from a constructive point of view [4,2,7,8,6,3], including constructive proofs of Wiener's theorem [7].[1]

We have in mind the following explicit statement of Wiener's $1/f$ theorem.

Wiener's Theorem. If $(a_n)_{n \in \mathbb{Z}}$ is a sequence of complex numbers with

$$\sum_{n=-\infty}^{\infty} |a_n| < \infty$$

such that the function $f : [0, 2\pi] \to \mathbb{C}$ never vanishes which is defined by

$$f(t) = \sum_{n=-\infty}^{\infty} a_n e^{int},$$

then there exists a sequence $(b_n)_{n \in \mathbb{Z}}$ of complex numbers with

$$\sum_{n=-\infty}^{\infty} |b_n| < \infty$$

such that $fg(t) = 1$ for all $t \in [0, 2\pi]$ where

$$g(t) = \sum_{n=-\infty}^{\infty} b_n e^{int}.$$

A constructive proof of this statement will embody an algorithm which takes as input an absolutely summable sequence of complex numbers $(a_n)_{n \in \mathbb{Z}}$ and outputs another absolutely summable sequence of complex numbers $(b_n)_{n \in \mathbb{Z}}$ such that $g(t) = 1/f(t)$ for all t where f, g are as above.

We take as our base theory a suitable fragment of Aczel's constructive Zermelo-Fraenkel set theory (**CZF**) [1]. We indicate in brackets when our proofs require more than that, e.g. the classical Zermelo-Fraenkel set theory with choice (**ZFC**).

2   A Direct Proof of Wiener's Theorem

In this section we give a simple proof, using open induction and classical logic, of Wiener's theorem. This however requires some preparation.

---

[1] Coquand and Spitters are preparing a treatise on constructive Banach algebra theory.

Open induction [12] is a classical reformulation of Zorn's lemma. Let P be a chain complete partial order. A subset A of P is *open* if A intersects a chain C in P whenever the supremum $\bigvee C$ of C is in A; and A is *progressive* if

$$\forall_{x \in P} \left(\forall_{y > x} y \in A \to x \in A \right).$$

The most general form of *open induction* is:

OI. If A is an open progressive subset of a chain complete poset P, then $A = P$.

It is easy to show that **OI** is equivalent to Zorn's lemma in **ZF**.

Our proof of Wiener's theorem is a recasting of Gelfand's short and elegant proof using Gelfand theory. That such a proof is possible is not surprising since Wiener's theorem is a concrete statement about relatively simple objects; there also have been elementary proofs of Wiener's theorem before [11].

Let $\mathcal{A}$ be the set of all functions of the form

$$f(t) = \sum_{n=-\infty}^{\infty} a_n e^{int}$$

where $(a_n)_{n \in \mathbb{Z}}$ is a sequence of complex numbers such that

$$\sum_{n=-\infty}^{\infty} |a_n| < \infty \, ;$$

each $f \in \mathcal{A}$ will be uniformly continuous with $f(0) = f(2\pi)$. With addition and multiplication on $\mathcal{A}$ given by the pointwise operations, and with the norm

$$\|f\| = \sum_{n=-\infty}^{\infty} |a_n|$$

where f is as above, $\mathcal{A}$ forms a commutative Banach algebra with unit $1 : x \mapsto 1$.

With this notation, Wiener's theorem can be succinctly stated as

Theorem 1 (ZFC). *If $f \in \mathcal{A}$ is nonzero everywhere, then $1/f \in \mathcal{A}$.*

Gelfand's proof uses the fact that the maximal ideals of $\mathcal{A}$ are of the form

$$M_t = \{f \in \mathcal{A} : f(t) = 0\}$$

with $t \in [0, 2\pi]$ (see e.g. [9, page 26-27]) and proceeds as follows. Let $f \in \mathcal{A}$ be such that f is nonzero everywhere and suppose that $1/f \notin \mathcal{A}$. Then the ideal $\langle f \rangle$ generated by f is a proper ideal; by Zorn's lemma, applied to the poset of proper ideals of $\mathcal{A}$ ordered by inclusion, there exists a maximal ideal M which contains $\langle f \rangle$. But by the above characterisation of maximal ideals, $M = M_t$ for some $t \in [0, 2\pi]$, which is absurd; whence $1/f \in \mathcal{A}$.

Our proof, which apart from Gelfand's method is motivated by the interplay between varieties and ideals characteristic of algebraic geometry, applies open induction on the—a fortiori, proper—ideals of $\mathcal{A}$ of the form

$$I(S) = \{f \in \mathcal{A} : f(s) = 0 \text{ for all } s \in S\} = \bigcap_{t \in S} M_t ,$$

where S is a nonempty subset of $[0, 2\pi]$. Our first task is to show that the collection P of these ideals, partially ordered by inclusion, is chain complete. Before so doing we notice that I is antimonotone, and transforms a union of subsets into the intersection of their ideals.

Lemma 1 (ZF). *Let S, S' be subsets of $[0, 2\pi]$. If $I(S) \subseteq I(S')$, then $S' \subseteq \overline{S}$. In particular, $I(S) = I(S')$ if and only if $\overline{S} = \overline{S'}$.*

Proof. Suppose $x \in S'$ and $d = \rho(x, S) > 0$. Let $f \in \mathcal{A}$ be a bump function on $[0, 2\pi]$ which is zero outside $(x - d, x + d) \cap [0, 2\pi]$ and nonzero at x. Then f is in $I(S)$ but is not in $I(S')$—a contradiction. Hence $x \in \overline{S}$.

We pause here to examine the structure of P; in particular to show that P is a lattice in **ZF**. We first observe that $I([0, 2\pi]) = \{0\}$ is the least element of P, and that the maximal elements of P are those of the form M_t with $t \in [0, 2\pi]$.

We now prove that P is a lattice. It is clear that the meet $I(S) \wedge I(S')$ of $I(S)$ and $I(S')$ is $I(S) \cap I(S') = I(S \cup S')$. The join $I(S) \vee I(S')$ of $I(S)$ and $I(S')$ is $I(\overline{S} \cap \overline{S'})$: it is clear that $I(\overline{S} \cap \overline{S'})$ is an upper bound for $I(S)$ and $I(S')$. Let $T \subseteq [0, 2\pi]$ be such that $I(T)$ is also an upper bound for $I(S)$ and $I(S')$. Then, by Lemma 1, $T \subseteq \overline{S}$ and $T \subseteq \overline{S'}$. Hence $T \subseteq \overline{S} \cap \overline{S'}$, so $I(\overline{S} \cap \overline{S'}) \subseteq I(T)$.

Although we have focused our attention on ideals of $\mathcal{A}$, Lemma 1 tells us that the poset P is anti-isomorphic to the poset Q of nonempty closed subsets of $[0, 2\pi]$; whence the dual poset $P \cong Q^{op}$ is isomorphic to the poset of proper open subsets of $[0, 2\pi]$, of course with the standard topology. The latter is chain complete in **ZF** (in a compact space, the union of a chain of proper open subsets is proper); whence the following is immediate.[2]

Proposition 1 (ZF). *The poset P is chain complete.*

We need to be more precise on this. The *zero set* of an ideal H of $\mathcal{A}$ is

$$Z(H) = \{t \in [0, 2\pi] : h(t) = 0 \text{ for all } h \in H\} = \{t \in [0, 2\pi] : M_t \supseteq H\} .$$

This Z is antimonotone, and converts a sum of ideals into the intersection of their zero sets. By Lemma 1, $ZI(S) = \overline{S}$ for every $S \subseteq [0, 2\pi]$; in particular, $ZI(S) = S$ for every $S \in Q$ and $IZ(F) = F$ for every $F \in P$. So the aforementioned anti-isomorphism between P and Q assigns $F \in P$ to $Z(F) \in Q$ and $S \in Q$ to $I(S) \in P$. Hence if C is a chain in P, then $\bigvee C = IZ(H)$ where $H = \bigcup C$.

---

[2] This has kindly been pointed out to us be one of the referees.

We say that two sets D and E meet, denoted $D \between E$, if $D \cap E$ is nonempty.[3] The *Jacobson radical* of an ideal H of $\mathcal{A}$ is

$$\operatorname{Jac}(H) = \{r \in \mathcal{A} : H \between 1 + \langle 1 - rs \rangle \text{ for every } s \in \mathcal{A}\}.$$

This definition is classically equivalent to the more common one:

Lemma 2 (ZFC). *For every ideal H of $\mathcal{A}$ we have*

$$\operatorname{Jac}(H) = \bigcap_{M \supseteq H} M$$

where M ranges over the maximal ideals of $\mathcal{A}$.

While this lemma—which of course holds for an arbitrary commutative ring—is usually proved by means of Zorn's lemma (see for example [10, IX.1.1], also for more discussion), a proof by open induction is possible. [4]

Now, if H is an ideal of $\mathcal{A}$, then $IZ(H) = \operatorname{Jac}(H)$, because $t \in Z(H)$ precisely when $M_t \supseteq H$. In all, if C is a chain in P, then $\bigvee C = \operatorname{Jac}(H)$ where $H = \bigcup C$.

Here finally is our **proof of Theorem 1.** To show that $1/f \in \mathcal{A}$ is tantamount to show that $1 \in \langle f \rangle$ or, equivalently, that $1 + \langle f \rangle \between \{0\}$. We will show this last condition by an open induction in P. To this end let

$$A = \{F \in P : 1 + \langle f \rangle \between F\}.$$

We first show that A is open. Let C be a chain in P, and $f \in A$. If $\operatorname{Jac}(H) \between 1 + \langle f \rangle$, say $h = 1 - fg$ belongs to $\operatorname{Jac}(H)$, then $H \between 1 + \langle 1 - h \rangle = 1 + \langle fg \rangle$. Hence $H \between 1 + \langle f \rangle$ as required.

To see that A is progressive, let $F \in P$ with $F = I(S)$ where $S \subseteq [0, 2\pi]$. We have two cases, a base case and the induction case, depending on the size of S.

If $|S| = 1$, that is, $S = \{t\}$ for some $t \in [0, 2\pi]$, then $F = M_t$ is a maximal ideal of $\mathcal{A}$. Setting $g = -f(t)^{-1}$, which actually belongs to $\mathbb{C}$, we have that

$$(1 + fg)(t) \;=\; 1 + f(t)g(t) \;=\; 0,$$

so $1 + fg \in F$ and $1 + \langle f \rangle \between F$.

If $|S| \geqslant 2$, pick distinct elements t_1, t_2 of S and let $\delta = |t_1 - t_2|/2$. By Lemma 1 and the induction hypothesis,

$$1 + \langle f \rangle \between I(S_i) \qquad (i = 1, 2)$$

where $S_i = S \setminus (t_i - \delta, t_i + \delta)$. Hence the multiplicative set $1 + \langle f \rangle$ meets

$$I(S_1)I(S_2) \;\subseteq\; I(S_1) \cap I(S_2) \;=\; I(S_1 \cup S_2) \;=\; F.$$

Thus A is progressive. We can now apply open induction to conclude that $A = P$. In particular, $\{0\} \in P$, which is to say that $1 + \langle f \rangle \between \{0\}$ or, equivalently, $1/f \in \mathcal{A}$.

---

[3] To our knowledge this notation has been coined by Giovanni Sambin.
[4] Simon Huber has kindly communicated to us a proof of this kind.

3   Constructing the Inverse

We are concerned not with giving a constructive proof of Theorem 1, this has been done [7], but with the following question: Can we adapt the proof of Theorem 1 to actually construct the element $1/f$ of $\mathcal{A}$?

Let us look more closely at the induction in the proof of Theorem 1. We suppose that $1 + \langle f \rangle \between I(S_1)$ and $1 + \langle f \rangle \between I(S_2)$; let $-g_1, -g_2 \in \mathcal{A}$ be witnesses of these statements: $1 - fg_i \in I(S_i)$ $(i = 1, 2)$. Then

$$(1 - fg_1)(1 - fg_2) = 1 - fg_1 - fg_2 + fg_1fg_2$$
$$= 1 - f(g_1 + g_2 - g_1fg_2)$$

is in

$$I(S_1)I(S_2) \subseteq I(S_1) \cap I(S_2) = I(S_1 \cup S_2).$$

Given a well ordering $(x_\alpha)_{\alpha < 2^{\aleph_0}}$ of $[0, 2\pi]$ we can recast our proof of Wiener's theorem via open induction as a proof by transfinite induction. Using the above, we can construct a sequence of elements $(g_\alpha)_{\alpha \leq 2^{\aleph_0}}$ of $\mathcal{A}$ such that g_α is zero on $\{x_\beta : \beta \leq \alpha\}$. Then $1/f = g_{2^{\aleph_0}}$.

We want to restrict the induction in our proof of Wiener's theorem to mathematical induction. Suppose we are given a sequence $(q_n)_{n \in \mathbb{N}}$ of elements of $[0, 2\pi]$. Using the argument in the proof of Theorem 1 we construct, by primitive recursion, functions $g_n \in \mathcal{A}$ $(n \in \mathbb{N})$ such that $fg_n(q_m) = 1$ for each $m \leq n$. To do so we set $g_1 = f(q_1)^{-1}$, which again belongs to $\mathbb{C}$, and

$$g_n = f(q_n)^{-1} + g_{n-1} - f(q_n)^{-1}g_{n-1}f$$

for $n > 1$; moreover, by mathematical induction,

$$1 - g_n f = \prod_{k=1}^{n} (1 - f/f(q_k)).$$

We would now like to pick a sequence $(q_n)_{n \in \mathbb{N}}$ such that

(i) $(q_n)_{n \in \mathbb{N}}$ is dense in $[0, 2\pi]$ and
(ii) the sequence $(g_n)_{n \in \mathbb{N}}$ in $\mathcal{A}$ converges.

For then this limit must, by continuity, be the inverse of f. This, however, does not work: for example, let $f : t \mapsto e^{i\pi t}$. By construction, each g_n is a polynomial in f, but $1/f, t \mapsto e^{-i\pi t}$, cannot be approximated by a polynomial in $e^{i\pi t}$. This process, however, proves very straightforward when f is real valued, and with some elementary geometry we can extend to the general case.

In the following we use, in addition to some small fragment of **CZF**, the *weak countable choice principle* from [5]:

WCC. Let $(A_n)_{n \geqslant 1}$ be a sequence of nonempty sets such that if $n \neq m$, then either A_n is a singleton or A_m is a singleton. Then there exists a sequence $(a_n)_{n \geqslant 1}$ with $a_n \in A_n$ for each n.

The principle WCC is valid in classical **ZF** but independent of Friedman's **IZF**.[5]
    We denote by $\| \cdot \|_\infty$ the supremum norm on $\mathcal{A}$, and reserve $\| \cdot \|$ for the Wiener norm

$$\|f\| = \sum_{n=-\infty}^{\infty} |a_n|;$$

hereafter we use $\mathcal{A}$ exclusively for $\mathcal{A}$ with the Wiener norm. We need the following two results from elementary calculus, which we do not prove.

Lemma 3. If $f \in \mathcal{A}$, then

$$\|f\| \leqslant \|f\|_\infty + \left\| \frac{d}{dx} f \right\|_\infty.$$

Lemma 4. Let f be a real valued function on $[0, 2\pi]$. If $\|f\|_\infty < 1/4$, then

$$\left\| \frac{d}{dx} f^2 \right\|_\infty < \frac{1}{2} \left\| \frac{d}{dx} f \right\|_\infty.$$

Theorem 2 (WCC). If $f \in \mathcal{A}$ with $\inf |f| > 0$, then $1/f \in \mathcal{A}$.

Proof. For $S \subseteq \mathbb{C}$, define

$$\mathcal{A}(S) = \{ f \in \mathcal{A} : \forall_{t \in [0, 2\pi]} f(t) \in S \}.$$

We first consider the special case where $f \in \mathcal{A}(0, \infty)$. We will construct a sequence $(q_n)_{n \in \mathbb{N}}$ in $[0, 2\pi]$ such that the sequence $(g_n)_{n \in \mathbb{N}}$, defined above, converges to the inverse of f. By construction

$$\|g_n f - 1\|_\infty = \left\| \prod_{k=1}^{n} (1 - f/f(q_k)) \right\|_\infty \leqslant \prod_{k=1}^{n} \|1 - f/f(q_k)\|_\infty.$$

Let

$$r = 1 - \frac{\inf f}{\sup f},$$

and suppose that $\inf f < \sup f$. Pick $q \in [0, 2\pi]$ such that

$$f(q) \geqslant \sup f/(1+r).$$

If we now set $q_n = q$ for each $n \in \mathbb{N}$, then

$$\|g_n f - 1\|_\infty = \left\| \prod_{k=1}^{n} (1 - f/f(q_k)) \right\|_\infty \leqslant \|1 - f/f(q)\|_\infty^n \leqslant r^n$$

---

[5] The latter can be shown by way of a sheaf model as provided e.g. in [14].

and thus $\|g_n f - 1\|_\infty \to \infty$ as $n \to \infty$. Now let $n \in \mathbf{N}$ be such that

$$\|g_n f - 1\|_\infty < 1/4.$$

Then, by Lemma 4,

$$\left\|\frac{d}{dx}(1 - g_{2^k n} f)\right\|_\infty = \left\|\frac{d}{dx}(1 - f/f(q))^{2^k n}\right\|_\infty < \frac{1}{2^k}\left\|\frac{d}{dx}(1 - f/f(q))^n\right\|_\infty ,$$

and hence $\left\|\frac{d}{dx}(1 - g_{2^k n} f)\right\|_\infty \to 0$ as $k \to \infty$. It now follows from Lemma 3 that $1 - g_{2^k n} f \to 0$ in $\mathcal{A}$ as $k \to \infty$. Since $\mathcal{A}$ is complete, $1/f = \lim_{k\to\infty} g_{2^k n}$ is in $\mathcal{A}$.

Now consider an arbitrary $f \in \mathcal{A}(0, \infty)$. As in [5], by WCC we can choose an increasing binary sequence $(\lambda_n)_{n\geqslant 1}$ such that

$$\lambda_n = 0 \Rightarrow \sup f - \inf f < 1/n;$$
$$\lambda_n = 1 \Rightarrow \sup f - \inf f > 1/(n+1).$$

Without loss of generality we may assume that $\lambda_1 = 0$. Pick any $p \in [0, 2\pi]$, and let $n \geqslant 1$. If $\lambda_n = 0$, then we set $g_n = 1/f(p)$; in this case

$$\|g_n f - 1\|_\infty = \|f/f(p) - 1\|_\infty \leqslant \max\left\{1 - \frac{\inf f}{f(p)}, \frac{\sup f}{f(p)} - 1\right\}$$
$$\leqslant \left(1 - \frac{\inf f}{f(p)}\right) + \left(\frac{\sup f}{f(p)} - 1\right) = \frac{\sup f - \inf f}{f(p)} < \frac{1}{n \inf f}.$$

If however $\lambda_k = 1 - \lambda_{k-1}$, then we follow the first part of this proof with $f' = f/f(p)$ in place of f, construct a sequence $(g'_n)_{n\geqslant 1}$ in $\mathcal{A}$ such that $\|g'_n f' - 1\|_\infty \to 0$ as $n \to \infty$, and set $g_m = g'_{m-k+1}/f(p)$ for every $m \geqslant k$. In this case,

$$\|g_m f - 1\|_\infty = \|g'_{m-k+1} f' - 1\|_\infty \leqslant r^{m-k+1}$$

for all $m \geqslant k$, simply because $\inf f'/\sup f' = \inf f/\sup f$.

In all, $\|g_n f - 1\|_\infty \to 0$ as $n \to \infty$, from which we conclude as in the first part.

Now let f be an arbitrary element of $\mathcal{A}$ satisfying the hypothesis, and set

$$f^*(t) = \sum_{n\in\mathbb{Z}} a_n e^{-int}.$$

Note that f^* is in $\mathcal{A}$. Since the map $t \mapsto -t$ corresponds to a reflection in the real axis, ff^* is real valued; in fact, $ff^* = |f|^2 \geqslant \inf |f|^2 > 0$. By the first part of this proof, there is $g \in \mathcal{A}$ such that $ff^*g = 1$; then $1/f = f^*g$ is also in $\mathcal{A}$.

4   A More General Result

In our classical proof of Theorem 1 we use not much that is specific to the algebra of continuous periodic functions with absolutely summable Fourier coefficients. This observation allows us to abstract from that particular context, as follows.

Let X be a topological space, and let $\mathcal{A}$ be a subalgebra of $\mathcal{C}(X)$. Consider the partial order $P(\mathcal{A})$ of—a fortiori, proper—ideals of the form

$$I(S) = \{f \in \mathcal{A} : f(s) = 0 \text{ for all } s \in S\},$$

where S is a nonempty subset of X.

We say that $\mathcal{A}$ *isolates the points of* X if for each $x \in X$ and each open neighbourhood W of x there exists an f such that $f(x)$ is nonzero and f is zero on $-W$. For any such $\mathcal{A}$ the proof of Lemma 1 works equally well; whence in particular the following counterpart of Proposition 1 holds:

Proposition 2 (ZF). *Let X be a compact topological space. If a subalgebra $\mathcal{A}$ of $\mathcal{C}(X)$ isolates the points of X, then $P(\mathcal{A})$ is chain complete.*

Also, the computation of the supremum of a chain can be carried over to $P(\mathcal{A})$.

Theorem 3 (ZFC). *Let $\mathcal{A}$ be a subalgebra of $\mathcal{C}(X)$ that isolates the points of the compact Hausdorff space X. If $f \in \mathcal{A}$ is nonzero everywhere, then $1/f \in \mathcal{A}$.*

We will prove this theorem as an example of the proof principle from [13]. This proof principle was in fact motivated by the proof of Theorem 1 given above.

Let U be a unary predicate on a partial order P with finite meets. An element $x \in P$ is *reducible* if it can be written as the meet of two strictly larger elements of P, i.e. there exist $y, z \in P$ such that $y > x$, $z > x$, and $x = y \wedge z$. The predicate U is said to be *good* if for every $x \in P$, either x is reducible or $U(x)$. Further, U is *meet closed* if $U(x)$ holds whenever $x = y \wedge z$, $U(y)$, and $U(z)$.

Now the proof principle from [13] gives the following.

Theorem 4 (OI). *Let U be an open predicate on a chain complete poset P. If U is meet closed and good, then $\forall_{x \in P} U(x)$.*

To prove Theorem 3, it suffices to show that the predicate

$$U(S) \equiv 1 + \langle f \rangle \between I(S)$$

on the chain complete poset $P(\mathcal{A})$ is open, good, and meet closed, where the meet operation is intersection. To do so we follow the proof of Theorem 1 given above; in particular, the proof that U is open is completely analogous.

Suppose $|S| = 1$; then $S = \{t\}$ for some $t \in X$. Setting $g = -f(t)^{-1}$ we have

$$(1 + fg)(t) \;=\; 1 + f(t)g(t) \;=\; 0,$$

so $1 + fg \in I(S)$ and $U(S)$. On the other hand, if $|S| \geqslant 2$, pick distinct elements t_1, t_2 of S and let W_1, W_2 be disjoint open subsets of X such that $t_i \in W_i$. Then

$$(S \setminus W_1) \cup (S \setminus W_2) = S;$$

whence $I(S)$ is reducible in view of the counterpart of Lemma 1. In all, we have proved that U is good.

It remains to show that U is meet closed. To this end, let $I(S) = I(S_1) \cap I(S_2)$, and suppose that $U(S_1), U(S_2)$; that is,

$$1 + \langle f \rangle \between I(S_i) \qquad (i = 1, 2).$$

Hence $1 + \langle f \rangle$ meets

$$I(S_1)I(S_2) \subseteq I(S_1) \cap I(S_2) = I(S),$$

so $U(S)$. Theorem 3 now follows from Theorem 4.

References

1. Aczel, P., Rathjen, M.: Notes on Constructive Set Theory. Report No. 40, Institut Mittag-Leffler, Royal Swedish Academy of Sciences (2001)
2. de Bruijn, N.G., van der Meiden, W.: Notes on Gelfand's theory. Indag. Math. 29, 467–474 (1967)
3. Bridges, D.S.: Constructive Functional Analysis. Research Notes in Mathematics, vol. 28. Pitman, London (1979)
4. Bishop, E.A., Bridges, D.S.: Constructive Analysis. Grundlehren der Math. Wiss, vol. 279. Springer, Heidelberg (1985)
5. Bridges, D., Richman, F., Schuster, P.: A weak countable choice principle. Proc. Amer. Math. Soc. 128, 2749–2752 (2000)
6. Cohen, P.: A note on constructive methods in Banach algebras. Proc. Amer. Math. Soc. 12, 159–163 (1961)
7. Coquand, T., Spitters, B.: Constructive theory of Banach algebras. Journal of Logic and Analysis 2(11), 1–15 (2010)
8. Coquand, T., Stolzenberg, G.: The Wiener lemma and certain of its generalizations. Bull. Amer. Math. Soc. (N.S.) 24, 1–9 (1991)
9. Gelfand, I.M., Raikov, D.A., Chilov, G.E.: Les anneaux normés commutatifs. Gauthier-Villars, Paris (1964)
10. Lombardi, H., Quitté, C.: Algèbre commutative. Méthodes constructives. Modules projectifs de type fini, Mathématiques en devenir, vol. 107. Calvage & Mounet, Paris (2012)
11. Newman, D.: A simple proof of Wiener's $1/f$ theorem. Proc. Amer. Math. Soc. 48, 264–265 (1975)
12. Raoult, J.: Proving open properties by induction. Inform. Process. Letters 29, 19–23 (1988)
13. Schuster, P.: Induction in algebra: a first case study. In: Twenty-Seventh Annual ACM/IEEE Symposium on Logic in Computer Science, LICS 2012 (2012)
14. Troelstra, A.S., van Dalen, D.: Constructivism in mathematics: An introduction, Volume II. Studies in Logic and the Foundations of Mathematics, vol. 123. North-Holland (1988)

Effective Strong Nullness and Effectively Closed Sets

Kojiro Higuchi and Takayuki Kihara

Mathematical Institute, Tohoku University, Sendai 980-8578, Japan
sa7m24@math.tohoku.ac.jp

Abstract. The strongly null sets of reals have been widely studied in the context of set theory of the real line. We introduce an effectivization of strong nullness. A set of reals is said to be *effectively strongly null* if, for any computable sequence $\{\varepsilon_n\}_{n \in \omega}$ of positive rationals, a sequence of intervals I_n of diameter ε_n covers the set. We show that for Π_1^0 subsets of 2^ω effective strong nullness is equivalent to another well studied notion called *diminutiveness*: the property of not having a computably perfect subset. In addition, we also investigate the Muchnik degrees of effectively strongly null Π_1^0 subsets of 2^ω. Let MLR and DNC be the sets of all Martin-Löf random reals and diagonally noncomputable functions, respectively. We prove that neither the Muchnik degree of MLR nor that of DNC is comparable with the Muchnik degree of a nonempty effectively strongly null Π_1^0 subsets of 2^ω with no computable element.

1 Introduction

1.1 Background

Miniaturization of set-theoretic notions is sometimes useful in computability theory. For example, set-theoretic forcing is transformed into a notion called arithmetical forcing and n-generic reals, and it has become a fundamental tool in computability theory. There is another set-theoretical notion that we expect its miniaturization to play an important role. The notion is known as *strong nullness* which is introduced by Borel in 1919. Careful consideration on measure theoretic behavior of sets of reals has profound significance in the study on algorithmic randomness [8,13]. Binns [2,3,4] made a deep study on notions stronger than being measure zero/Hausdorff dimension zero, and clarified an interesting connection among such measure theoretic smallness, Muchnik degrees, and Kolmogorov complexity.

Inspired by these previous works, we introduce an effectivization of strong nullness. In contrast to Laver's model [12] of ZFC in which all strongly measure zero sets are countable, we always have an effectively strongly null set of reals which is uncountable and Π_1^0 definable. Indeed, we show that, for closed subsets of reals, effective strong nullness is equivalent to another well studied notion called *diminutiveness* [4]. In addition, we also investigate the Muchnik degrees of effectively strongly null Π_1^0 subsets of 2^ω. Let MLR and DNC be the

S.B. Cooper, A. Dawar, and B. Löwe (Eds.): CiE 2012, LNCS 7318, pp. 303–312, 2012.
© Springer-Verlag Berlin Heidelberg 2012

sets of all Martin-Löf random reals and diagonally noncomputable functions, respectively. We prove that neither the Muchnik degree of MLR nor that of DNC
is comparable with the Muchnik degree of a nonempty effectively strongly null
Π_1^0 subsets of 2^ω with no computable element. Finally, we see some interactions
between measure theoretic smallness and Kolmogorov complexity. A real $x \in 2^\omega$
is *K-trivial* if $K(x \restriction n) \leq K(n) + O(1)$, and x is *complex* if $K(x \restriction f(n)) \geq n$
for some computable function f, where K denotes the prefix-free Kolmogorov
complexity. By using a non-basis result for small Π_1^0 sets, we construct a perfect
Π_1^0 subset of 2^ω whose elements are neither K-trivial nor complex.

1.2   Notation

Let $\omega = \{0,1,2,\cdots\}$ denote the set of all natural numbers; $\omega^\omega = \{f \mid f : \omega \to \omega\}$, Baire space; $2^\omega = \{f \mid f : \omega \to \{0,1\}\}$, Cantor space; $\omega^{<\omega}$, the set of all
finite strings of natural numbers; and $2^{<\omega}$, the set of all finite binary strings.
We define $\omega^{\leq\omega} = \omega^{<\omega} \cup \omega^\omega$ and $2^{\leq\omega} = 2^{<\omega} \cup 2^\omega$. We use $\emptyset$ to denote the empty
string or the empty set. For a set A, we use $\#A$ to denote the cardinal number
of A. For $\sigma, \tau \in \omega^{<\omega}$ and $\rho, \rho' \in \omega^{\leq\omega}$, we use $\sigma \subset \rho$ to mean that σ is an initial
segment of ρ, i.e., ρ extends σ; $\sigma \mid \tau$ to mean that σ and τ are incomparable,
i.e., neither $\sigma \subset \tau$ nor $\sigma \supset \tau$; $\sigma\rho$ or $\sigma^\frown\rho$ to denote the concatenation of σ and
ρ, i.e., the string σ followed by ρ; $\mathrm{lh}(\rho)$ and $|\rho|$ to denote the length of ρ, i.e.,
the cardinal number of the domain of ρ; $\rho \restriction n$ to denote the initial segment of
ρ of the length n for any $n \leq \mathrm{lh}(\rho)$; $\rho \cap \rho'$ to denote the longest common initial
segment of ρ and ρ'; $\rho \oplus \rho'$ to denote the string ρ'' with $\rho''(2n) = \rho(n)$ and
$\rho''(2n+1) = \rho'(n)$, when $\mathrm{lh}(\rho) = \mathrm{lh}(\rho')$; $[\![\sigma]\!]$ to denote the set $\{f \in \omega^\omega \mid \sigma \subset f\}$
or the set $\{f \in 2^\omega \mid \sigma \subset f\}$ depending on the context. We often identify a natural
number n with the string $\langle n \rangle$ of the length 1. Let $A \subset \omega^{<\omega}$ and $P, Q \subset \omega^\omega$. $[\![A]\!]$
denotes the set $\bigcup_{\sigma \in A}[\![\sigma]\!]$; $[A]$, the set $\{f \in \omega^\omega \mid (\forall n \in \omega)[f \restriction n \in A]\}$; $\mathrm{Ext}(P)$,
the set $\{\sigma \mid [\![\sigma]\!] \cap P \neq \emptyset\}$; $\mathrm{Br}(P)$, the set $\{\sigma \cap \tau \mid \sigma, \tau \in \mathrm{Ext}(P) \ \& \ \sigma \mid \tau\}$; $\mathrm{Brl}(P)$,
the set $\{\mathrm{lh}(\sigma) \mid \sigma \in \mathrm{Br}(P)\}$; $\mathrm{Ext}(A)$, $\mathrm{Br}(A)$ and $\mathrm{Brl}(A)$ denote the sets $\mathrm{Ext}([A])$,
$\mathrm{Br}([A])$ and $\mathrm{Brl}([A])$, respectively; $P \times Q$ denotes the set $\{f \oplus g \mid f \in P \ \& \ g \in Q\}$;
and $P + Q$, the set $\{0f, 1g \mid f \in P \ \& \ g \in Q\}$. A set $T \subset \omega^{<\omega}$ is called a *tree* if T
is closed under taking initial segments, i.e., $\tau \in T$ if $\tau \subset \sigma$ for some $\sigma \in T$. For a
tree T, $\sigma \in T$ is an *immediate successor* of τ in T if $\tau \subset \sigma$ and $\mathrm{lh}(\sigma) = \mathrm{lh}(\tau)+1$;
T is *finitely branching* if every element in T has at most finitely many immediate
successors.

Let $P, Q \subset \omega^\omega$. P is *strongly reducible* to Q, denoted by $P \leq_s Q$, if there is a
computable function $\Phi : Q \to P$; P is *strongly comparable* with Q if $P \leq_s Q$ or
$P \geq_s Q$; otherwise, *strongly incomparable*; P is *strongly equivalent* to Q, denoted
by $P \equiv_s Q$, if $P \leq_s Q$ and $P \geq_s Q$. The *strong degree* of P is the equivalence
class of P under the equivalence relation $\equiv_s$. P is *weakly reducible* to Q, denoted
by $P \leq_w Q$, if $P \leq_s \{g\}$ for all $g \in Q$. *Weak comparability*, *weak incomparability*,
weak equivalence and *weak degree* are defined in the same way. The arithmetical
hierarchy is introduced in the usual way.

2 Effective Strong Nullness

2.1 Combinatorial Theorem

We first show that a combinatorial theorem. While we use the theorem to characterize effective strong nullness we define later, the theorem itself is interesting.

Theorem 1. *Let $T \subset \omega^{<\omega}$ be a finitely branching tree. Suppose that $[T] \setminus [\![A]\!] \neq \emptyset$ holds for any $A \subset T \setminus \{\emptyset\}$ such that*

$$(\forall n \in \omega)[\#\{\sigma \in A \mid \mathrm{lh}(\sigma) = n + 1\} \leq \#\{\tau \in T \mid \mathrm{lh}(\tau) = n\}].$$

Then there exists a length-preserving embedding from $2^{<\omega}$ into T.

Proof. Let $\varphi(T')$ denote the condition that $[T'] \setminus [\![A]\!] \neq \emptyset$ holds for any $A \subset T' \setminus \{\emptyset\}$ such that

$$(\forall n \in \omega)[\#\{\sigma \in A \mid \mathrm{lh}(\sigma) = n + 1\} \leq \#\{\tau \in T' \mid \mathrm{lh}(\tau) = n\}].$$

For $\sigma \in T$, we define $T(\sigma) = \{\tau \in \omega^{<\omega} \mid \sigma\tau \in T\}$. It suffices to show that for any $\sigma \in T$ with $\varphi(T(\sigma))$ there exist at least two immediate successors σi, σj of σ in T with $\varphi(T(\sigma i))$ and $\varphi(T(\sigma j))$.

Fix $\sigma \in T$ with $\varphi(T(\sigma))$. Let $\sigma k_0, \sigma k_1, \cdots, \sigma k_n \in T$ be all immediate successors of σ in T with $k_0 < k_1 < \cdots < k_n$. Suppose that there is at most one immediate successor of σ in T with the property φ. Let $i \leq n$ satisfy $\varphi(\sigma k_i)$ if there is such a natural number. For any $j \in \{0, 1, \cdots, n\} \setminus \{i\}$, choose A_j witnessing $\neg\varphi(T(\sigma k_j))$. It is easy to see that $\{k_i\} \cup \bigcup_{j \neq i}\{k_j\tau \mid \tau \in A_j\}$ witnesses that $\neg\varphi(T(\sigma))$. We have a contradiction. Thus there exist at least two immediate successors of σ in T with the property φ.   □

2.2 Effective Strong Nullness and Diminutiveness

Now we turn to define effective strong nullness and diminutiveness.

Definition 1. *A set $M \subset 2^{\omega}$ is called effectively strongly null if for any computable sequence $\{k_i\}_{i \in \omega}$ of natural numbers there exist finite strings σ_i, $i \in \omega$, of length $\geq k_i$ such that $\{[\![\sigma_i]\!]\}_{i \in \omega}$ is an open cover of M, i.e., $M \subset \bigcup_{i \in \omega}[\![\sigma_i]\!]$ holds.*

Proposition 1. *Let C be an effectively strongly null closed subset of 2^{ω}. Given any computable sequence $\{k_i\}_{i \in \omega}$ of naturals, there exist finitely many finite strings σ_i, $i \leq n$, of the length k_i such that $\{[\![\sigma_i]\!]\}_{i \leq n}$ is an open cover of C.*

Proof. It is trivial by the effective strong nullness and the compactness of C.   □

Recall that a *perfect subset* of 2^{ω} means a nonempty closed set with no isolated point.

Definition 2. *A perfect set $P \subset 2^\omega$ is said to be computably perfect if there exists a computable function $F : \omega \to \omega$ such that*

$$(\forall n \in \omega)(\forall f \in P)(\exists g \in P)[n \le \mathrm{lh}(f \cap g) \le F(n)].$$

A subset of 2^ω is said to be diminutive if it contains no computably perfect subset.

Proposition 2 (Binns [4, Lemma 2.4]). *If a Π_1^0 subset of 2^ω contains a computably perfect subset, then it contains a computably perfect Π_1^0 subset.* □

Observe that a perfect set P is computably perfect if and only if there exists a computable function $F : \omega \to \omega$ such that for any $n \in \omega$ and any $f \in P$ there exists an element $g \in P$ such that $F(n) \le \mathrm{lh}(f \cap g) < F(n+1)$ holds. Using this equivalence, the following proposition is easily proved.

Proposition 3. *Every computably perfect subset of 2^ω is not effectively strongly null.*

Proof. Let P be a computably perfect subset of 2^ω. Choose a computable function $F : \omega \to \omega$ such that

$$(\forall n \in \omega)(\forall f \in P)(\exists g \in P)[F(n) \le \mathrm{lh}(f \cap g) < F(n+1)].$$

Now it is clear that $\{F(n+1)\}_{n \in \omega}$ witnesses P is not effectively strongly null. □

Corollary 1. *Every effectively strongly null subset of 2^ω is diminutive.* □

Using Theorem 1, we show the converse also holds for closed subsets of 2^ω.

Theorem 2. *A closed subset of 2^ω is diminutive if and only if it is effectively strongly null.*

Proof. Fix a closed set $C \subset 2^\omega$. Suppose that a computable sequence $\{k_i\}_{i \in \omega}$ witnesses C is not effectively strongly null. We show that C contains a computably perfect subset. We may safely assume that $k_i < k_{i+1}$ for all $i \in \omega$. Define $F : \omega \to \omega$ recursively by $F(0) = 0$ and $F(n+1) = F(n) + 2^{k_{F(n)}}$ for all $n \in \omega$. Note that $k_{F(n)} \ge n$ and $k_{F(n+1)} - k_{F(n)} \ge 2^{k_{F(n)}}$ for all $n \in \omega$. Define $\{T_n\}_{n \in \omega}$ by $T_0 = \{\emptyset\}$ and

$$T_{n+1} = \{\sigma \in \mathrm{Ext}(C) \mid \mathrm{lh}(\sigma) = k_{F(n+1)}\}.$$

Let $T = \bigcup_{n \in \omega} T_n$. Since C is closed, note that $C = \bigcap_{n \in \omega} [\![T_n]\!]$. The effective strong nullness of C implies that $C \setminus [\![A]\!] \ne \emptyset$ holds if $A \subset T \setminus \{\emptyset\}$ satisfies that

$$\#\{\sigma \in A \mid \mathrm{lh}(\sigma) = k_{F(n+1)}\} \le \#\{\tau \in \mathrm{Ext}(C) \mid \mathrm{lh}(\tau) = k_{F(n)}\} \le 2^{k_{F(n)}}$$

for all $n \in \omega$. Naturally, $(T, \subset)$ can be seen as a graph of a finitely branching tree and can be embedding into $\omega^{<\omega}$ so that the image form a finitely branching tree on ω with the assumption of Theorem 1. Thus $2^{<\omega}$ has a length-preserving embedding into $(T, \subset)$ by Theorem 1. This implies that C has a computably perfect subset witnessed via $n \mapsto k_{F(n+1)}$. □

2.3   Lattice Operators

Theorem 2 is very useful for investigating the properties of effective strong nullness. We first see relation between effective strong nullness and lattice operators.

Proposition 4. *Let P and Q be closed subsets of 2^ω. Then $P + Q$ is effectively strongly null if and only if so are P and Q.*

Proof. This is because $P + Q$ contains a computably perfect subset if and only if one of P and Q contains a computably perfect subset by the definition of $+$.   □

Proposition 5. *Let P and Q be nonempty closed subsets of 2^ω. If $P \times Q$ is effectively strongly null, then so are P and Q.*

Proof. It is clear that if one of P and Q contains a computably perfect subset, then so $P \times Q$ does.   □

To show the converse of the above proposition, we need the following lemma.

Lemma 1. *Let P be an effectively strongly null Π_1^0 subset of 2^ω. Given a computable sequence $\{a_i\}_{i \in \omega}$ of naturals, we can (uniformly) find a computable increasing function $F : \omega \to \omega$ and a computable sequence of finite strings σ_i, $i \in \omega$, of the length a_i such that $[\![\sigma_{F(n)}]\!], [\![\sigma_{F(n)+1}]\!], \cdots , [\![\sigma_{F(n+1)-1}]\!]$ are an open cover of P.*

Proof. By Proposition 1, we know that for any $k \in \omega$ there are finitely many finite strings σ_{k+i}, $i \leq m$, of the length a_{k+i} such that $\{[\![\sigma_{k+i}]\!]\}_{i \leq m}$ is an open cover of P. Moreover, we can find such sequences uniformly in a given $k \in \omega$ since P is Π_1^0. From this, it is clear that the lemma holds.   □

Theorem 3. *If P and Q are effectively strongly null Π_1^0 subsets of 2^ω, then $P \times Q$ is also effectively strongly null.*

Proof. To show that $P \times Q$ is effectively strongly null, fix a computable sequence $\{b_i\}_{i \in \omega}$ of naturals. Let $\{a_i\}_{i \in \omega}$ be a strictly increasing computable sequence of natural numbers such that $b_i \leq 2a_i$ for all $i \in \omega$ and, applying Lemma 1 to P and $\{a_i\}_{i \in \omega}$, take a computable function F and a computable sequence $\{\sigma_i\}_{i \in \omega}$ as in Lemma 1. Here we can safely assume that $F(0) = 0$. Since Q is also effectively strongly null, there exist finite strings τ_n, $n \in \omega$, of the length $a_{F(n+1)}$ which generate an open cover of Q. For each $i \in \omega$, define $\rho_i = \sigma_i \oplus (\tau_{n_i} \upharpoonright \mathrm{lh}(\sigma_i))$, where n_i is the unique natural number such that $F(n_i) \leq i < F(n_i + 1)$. Since $\mathrm{lh}(\sigma) = a_i$, we have $\mathrm{lh}(\rho_i) = 2a_i$ for all $i \in \omega$.

It suffices to show that $\{\rho_i\}_{i \in \omega}$ generates an open cover of $P \times Q$. Fix $f \oplus g \in P \times Q$. Since $g \in Q$, there exists $n \in \omega$ such that $\tau_n \subset g$. By the choice of finite strings σ_i, $F(n) \leq i < F(n+1)$, there exists m with $F(n) \leq m < F(n+1)$ such that $\sigma_m \subset f$. We have $\rho_m \subset f \oplus g$ and, thus, $P \times Q \subset \bigcup_{i \in \omega} [\![\rho_i]\!]$.   □

Corollary 2. *For nonempty Π_1^0 subsets of 2^ω, the following three statements are pairwise equivalent:*

1. *P and Q are effectively strongly null.*
2. *$P + Q$ is effectively strongly null.*
3. *$P \times Q$ is effectively strongly null.*

    □

2.4   Closure Properties

We shall see that effective strong nullness is closed under taking subsets and is closed under taking the images of computable functions for Π_1^0 subsets of 2^ω.

Proposition 6. *Every subset of an effectively strongly null subset of 2^ω is again effectively strongly null.* □

For a partial computable function Φ on 2^ω, a finite binary string σ and a natural number n, we use $\Phi(\sigma; n)$ to denote the computation of Φ with an oracle σ, an input n and step $\mathrm{lh}(\sigma)$ and we use $\Phi(\sigma)$ to denote the finite string τ of the maximum length such that $\Phi(\sigma; n) = \tau(n)$ for all $n < \mathrm{lh}(\tau)$.

Theorem 4. *The image $\Phi(P)$ of an effectively strongly null Π_1^0 subset P of 2^ω under a computable function $\Phi : P \to 2^\omega$ is again effectively strongly null.*

Proof. Fix an effectively strongly null Π_1^0 subset P of 2^ω and a computable function $\Phi : P \to 2^\omega$, and assume, contrary to our theorem, $\Phi(P)$ is not effectively strongly null. Let a computable sequence $\{k_i\}_{i\in\omega}$ of naturals be a witness of this assumption. Since P is Π_1^0, we can find a computable sequence $\{k_i'\}_{i\in\omega}$ of naturals such that $\mathrm{lh}(\sigma) \geq k_i'$ implies $\mathrm{lh}(\Phi(\sigma)) \geq k_i$ for all $\sigma \in \mathrm{Ext}(P)$. Using the effective strong nullness of P, choose a sequence of finite strings σ_i of length k_i' such that $P \subset \bigcup_{i\in\omega} [\![\sigma_i]\!]$. We have an open cover $\{[\![\Phi(\sigma_i)]\!]\}_{i\in\omega}$ of $\Phi(P)$, contradicting our assumption that $\{k_i\}_{i\in\omega}$ witnesses that $\Phi(P)$ is not effectively strongly null. Thus $\Phi(P)$ is effectively strongly null. □

A set M of ω^ω is called *computably bounded (c.b.)* if there is a computable function $f : \omega \to \omega$ with $g(n) < f(n)$ for any $g \in M$ and $n \in \omega$. It is well-known that every c.b. Π_1^0 set is computably homeomorphic to a Π_1^0 subset of 2^ω.

Remark 1. We can extend Definition 1 and Definition 2 to c.b. closed subsets of ω^ω in the straightforward way. Also, Theorem 2, Corollary 2, Proposition 6 and Theorem 4 can be easily extended to c.b. closed subsets of ω^ω. We shall use these extended theorems later.

Theorem 5 (Simpson [14, Theorem 4.7]). *Let $P \subset \omega^\omega$ be a nonempty c.b. Π_1^0 set and let $\Phi : P \to \omega^\omega$ be a computable function. Then $\Phi(P)$ is a nonempty c.b. Π_1^0 subset of ω^ω.* □

Theorem 6 (Simpson [14, Lemma 6.9]). *Let $M \subset \omega^\omega$ and let $P \subset \omega^\omega$ be a nonempty c.b. Π_1^0 set. If $M \leq_{\mathrm{w}} P$, then P contains a nonempty c.b. Π_1^0 subset Q with $M \leq_{\mathrm{s}} Q$.* □

We prove the following theorem using the technique of the proof of Corollary 2.16 in Binns [2].

Theorem 7. *Let $\mathfrak{A}$ be a set of nonempty c.b. Π_1^0 subsets of ω^ω which is closed under taking nonempty Π_1^0 subset and taking the images of computable functions. Let $P \subset \omega^\omega$ and let $Q \in \mathfrak{A}$. If $P \leq_{\mathrm{w}} Q$, then some subset of P is in $\mathfrak{A}$.*

Proof. Suppose that $P \leq_w Q$. By Theorem 6, there exists a computable function $\Phi : Q' \to P$ for some nonempty Π_1^0 subset Q' of Q. The image $\Phi(Q')$ is a nonempty c.b. Π_1^0 subset of P by Theorem 5. We have $\Phi(Q') \in \mathfrak{A}$ by the closure properties of $\mathfrak{A}$.   □

Applying the theorem to $\mathfrak{A}$ as the set of all nonempty Π_1^0 effectively strongly null subsets of 2^ω, we have the following corollary.

Corollary 3. *If a subset P of 2^ω is weakly reducible to a nonempty effectively strongly null Π_1^0 subset of 2^ω, then P contains a nonempty effectively strongly null Π_1^0 subset.*   □

2.5 MLR and DNC

We denote MLR the set of all Martin-Löf random elements of 2^ω and denote DNC the set of all diagonally non-computable elements of ω^ω. We refer to the texts by Downey and Hirschfeldt [8], Nies [13] and Soare [15] for the definitions of these notions. Simpson [14] proved that MLR is weakly incomparable with any perfect thin Π_1^0 subset of 2^ω. We use the technique of his proof to show that MLR, DNC are incomparable with any nonempty effectively strongly null Π_1^0 subset of 2^ω with no computable element. We use the facts that every nonempty Π_1^0 subset of MLR is weakly equivalent to MLR and that MLR contains a nonempty Π_1^0 subset. See [14].

Theorem 8. *Let $P \subset \mathrm{MLR}$ be a nonempty Π_1^0 set and let $\Phi : P \to \omega^\omega$ be a computable function. If $\Phi(P)$ contains no computable element, then $\Phi(P) \equiv_w P$.*

Proof. By Simpson [14, Corollary 4.9], we know that $\Phi(f) \leq_{tt} f$ for all $f \in P$, where $\leq_{tt}$ refers to the *truth-table reducibility*. Additionally, by Demuth [6, Lemma 30], we know that, for any $f \in P$, $\Phi(f)$ is Turing equivalent to an element of MLR. Thus $\mathrm{MLR} \leq_w \Phi(P) \leq_w P \equiv_w \mathrm{MLR}$ holds. Hence $\Phi(P) \equiv_w P$.   □

Theorem 9. *Let $\mathfrak{A}$ be a set of nonempty c.b. Π_1^0 subsets of ω^ω which is closed under taking nonempty Π_1^0 subset and taking the images of computable functions. Let $P \subset \omega^\omega$ be weakly reducible to MLR and let $Q \in \mathfrak{A}$ contain no computable element. Suppose that every c.b. Π_1^0 subset of P is not in $\mathfrak{A}$. Then P and Q are weakly incomparable.*

Proof. Since $P \leq_w Q$ implies that P contains a nonempty Π_1^0 subset in $\mathfrak{A}$ by Theorem 7, we have $P \not\leq_w Q$. Suppose that $Q \leq_w P$. Since $P \leq_w \mathrm{MLR}$, we have $Q \leq_w \mathrm{MLR}$. Choose a nonempty Π_1^0 set $R \subset \mathrm{MLR}$ and a computable function $\Phi : R \to Q$. By Theorem 6 and Theorem 8, we have $R \equiv_w \Phi(R)$. By the closure properties of $\mathfrak{A}$, we have $\Phi(R) \in \mathfrak{A}$. On the other hand, we have $P \leq_w R \equiv_w \Phi(R)$. A contradiction. Thus $Q \not\leq_w P$.   □

Proposition 7. *Every nonempty Π_1^0 subset of MLR contains a computably perfect subset.*

Proof. By Simpson [14, Lemma 8.9], every nonempty Π_1^0 subset of MLR is of positive measure. By Hertling [10, Proposition 8], we know that any closed subset of 2^ω of positive measure contains a computably perfect subset. Thus the proposition holds. $\square$

Applying Theorem to P = MLR and $\mathfrak{A}$ as the set of all nonempty c.b. Π_1^0 effectively strongly null subsets of ω^ω, we have the following corollary.

Corollary 4. *Let Q be an effectively strongly null Π_1^0 subset of 2^ω with no computable element. Then Q is weakly incomparable with MLR.* $\square$

Proposition 8. *Every nonempty c.b. Π_1^0 subset of DNC is computably perfect.*

Proof. Let $P \subset$ DNC be a nonempty c.b. Π_1^0 set. Using Recursion Theorem, given a $\sigma \in 2^{<\omega}$, we can effectively find n_σ such that $\{n_\sigma\}(n_\sigma)$ is the unique value of $f(n_\sigma)$ for some $s \in \omega$ and some $f \in P_s \cap [\![\sigma]\!]$ (if exist) such that, for any $f, g \in P_s \cap [\![\sigma]\!]$, $f(n_\sigma) = g(n_\sigma)$, where P_s is a clopen set which is the s-th approximation of P. Then the computable function $m \mapsto \max\{n_\sigma \mid \sigma \in 2^{<\omega}\ \&\ \mathrm{lh}(\sigma) = m\}$ witnesses that P is computably perfect. $\square$

It is well-known that DNC $\leq_\mathrm{w}$ MLR. See, for instance, Giusto/Simpson [9, Lemma 6.18]. Thus applying the theorem to P = DNC and $\mathfrak{A}$ as the set of all nonempty c.b. Π_1^0 effectively strongly null subsets of ω^ω, we have the following corollary.

Corollary 5. *Let Q be an effectively strongly null Π_1^0 subset of 2^ω with no computable element. Then Q is weakly incomparable with DNC.* $\square$

Remark 2. Indeed, one direction of Corollary 5, i.e., DNC is weakly reducible to no diminutive Π_1^0 subset of 2^ω, can be obtained easily using Theorem 2.12 and Corollary 2.14 of Binns [4].

Remark 3. By Binns [4, Theorem 3.8, Theorem 3.9] and Binns [3], we have that thinness or smallness imply diminutiveness. Thus Corollary 4 and Corollary 5 hold even when we replace "effectively strongly null" with "thin" or "small". Here, Simpson [14, Theorem 9.15] showed that MLR is weakly incomarable with any nonempty thin perfect Π_1^0 subset of 2^ω.

3   Kolmogorov Complexity, and K-Triviality

A real $x \in 2^\omega$ is *complex* [11] if there is a computable function $f : \omega \to \omega$ such that $K(x \restriction f(n)) \geq n$ for any $n \in \omega$, where $K : 2^{<\omega} \to \omega$ denotes the prefix-free Kolmogorov complexity. By combining Theorem 2 and a result from Binns [4, Theorem 2.13], we have the following characterization:

Corollary 6. *The following are pairwise equivalent for any Π_1^0 set $P \subseteq 2^\omega$:*

1. *P is effectively strongly null;*
2. *No element of P is complex.*
3. *There exists a real $x \in 2^\omega$ such that no element of P wtt-computes x.*

The previous works on measure theoretic smallness of Π_1^0 sets implies the following non-basis theorem:

Theorem 10 (Small Non-Basis Theorem). *For any Π_1^0 set $P \subseteq 2^\omega$, if $\mathrm{Brl}(P)$ is not dominated by any computable function, then every element of P is neither complex nor computable in any 1-generic real.*

Proof. Any such Π_1^0 set is said to be *small* ([2]). By Binns [3,4], every small Π_1^0 set is diminutive. By Cenzer/Kihara/Weber/Wu [5], every small Π_1^0 set is *immune*, i.e., there exists no infinite c.e. set of proper initial segments of elements of the set. By Binns [4, Theorem 2.13], every element of a diminutive Π_1^0 set is non-complex. By Demuth/Kučera [7], every 1-generic real computes no element of an immune Π_1^0 set. □

The Small Non-Basis Theorem 10 may have some applications. A real $x \in 2^\omega$ is *K-trivial* if there is a constant $c \in \omega$ such that $K(x \upharpoonright n) \le K(n) + c$ for any $n \in \omega$; A real $x \in 2^\omega$ is *i.o. K-trivial* [1] if there is a constant $c \in \omega$ such that $K(x \upharpoonright n) \le K(n) + c$ for infinitely many $n \in \omega$. Barmpalias/Vlek [1] showed that, if a real is computable in a 1-generic, then it is i.o. K-trivial; and there is a Π_1^0 subset of 2^ω consisting of i.o. K-trivial reals but which does not contain any K-trivial reals.

Theorem 11. *There is a Π_1^0 set $P \subseteq 2^\omega$ which satisfies the following:*

1. *Every element of P is i.o. K-trivial,*
2. *P contains infinitely many complex reals,*
3. *No element of P is K-trivial.*
4. *No element of P is computable in a 1-generic real.*

Proof (Sketch). Our idea is to add a strategy to the construction in Barmpalias-Vlek [1, Theorem 2.10] to ensure immunity for Π_1^0 sets. □

Theorem 12. *There is a nonempty (perfect) Π_1^0 subset of 2^ω which satisfies the following conditions:*

1. *No element of P is complex,*
2. *No element of P is K-trivial,*
3. *No element of P is computable in a 1-generic real.*

Proof (Sketch). Our idea is to add a strategy to the construction in Barmpalias-Vlek [1, Theorem 2.10] to ensure smallness for Π_1^0 sets, via finite injury.

Requirements. We need to construct a nonempty Π_1^0 set $P \subseteq 2^\omega$ satisfying the following K-trivial avoiding requirements $\{\mathcal{K}_c\}_{c\in\omega}$ and smallness requirements $\{\mathcal{S}_e\}_{e\in\omega}$:

$$\mathcal{K}_c : (\exists h^\star \in \omega)(\forall f \in P)\ K(f \upharpoonright h^\star) > 2\log h^\star + c,$$
$$\mathcal{S}_e : \Phi_e \text{ total unbounded} \implies (\exists n)\ [\Phi_e(n), \Phi_e(n+1)] \cap \mathrm{Brl}(P) = \emptyset.$$

Here, $\{\Phi_e\}_{e\in\omega}$ is an effective enumeration of all partial computable function, and $[l, r]$ denotes the interval $\{m : l \le m \le r\}$.

The $\mathcal{K}$-strategies ensure that no element of P is K-trivial; The $\mathcal{S}$-strategies ensure that P is small, by Binns [3, Theorem 2.10]. By Small Non-Basis Theorem 10, every element of P is neither complex nor computable in a 1-generic real, as desired. □

Acknowledgements. The authors were supported by JSPS Research Fellowships.

References

1. Barmpalias, G., Vlek, C.S.: Kolmogorov complexity of initial segments of sequences and arithmetical definability. Theor. Comput. Sci. 412(41), 5656–5667 (2011)
2. Binns, S.: Small Π_1^0 classes. Arch. Math. Log. 45(4), 393–410 (2006)
3. Binns, S.: Hyperimmunity in $2^\mathbb{N}$. Notre Dame Journal of Formal Logic 48(2), 293–316 (2007)
4. Binns, S.: Π_1^0 classes with complex elements. J. Symb. Log. 73(4), 1341–1353 (2008)
5. Cenzer, D., Kihara, T., Weber, R., Wu, G.: Immunity and non-cupping for closed sets. Tbilisi Math. J. 2, 77–94 (2009)
6. Demuth, O.: A notion of semigenericity. Comment. Math. Univ. Carolinae 28, 71–84 (1987)
7. Demuth, O., Kučera, A.: Remarks on 1-genericity, semigenericity and related concepts. Comment. Math. Univ. Carolinae 28, 85–94 (1987)
8. Downey, R.G., Hirschfeldt, D.R.: Algorithmic randomness and complexity. Theory and Applications of Computability, 883 pages. Springer (2010)
9. Giusto, M., Simpson, S.G.: Located sets and reverse mathematics. J. Symb. Log. 65(3), 1451–1480 (2000)
10. Hertling, P.: Surjective functions on computably growing Cantor sets. J. UCS 3(11), 1226–1240 (1997)
11. Kjos-Hanssen, B., Merkle, W., Stephan, F.: Kolmogorov complexity and the recursion theorem. Trans. Amer. Math. Soc. 363(10), 5465–5480 (2011)
12. Laver, R.: On the consistency of Borel's conjecture. Acta Math. 137(1), 151–169 (1976)
13. Nies, A.: Computability and Randomness. Oxford Logic Guides. Oxford University Press (2009)
14. Simpson, S.G.: Mass problems and randomness. Bull. Symb. Log. 11(1), 1–27 (2005)
15. Soare, R.I.: Recursively Enumerable Sets and Degrees. Perspectives in Mathematical Logic. Springer, Heidelberg (1987)

Word Automaticity of Tree Automatic Scattered Linear Orderings Is Decidable

Martin Huschenbett

Fakultät Informatik und Automatisierung, Fachgebiet Theoretische Informatik,
Technische Universität Ilmenau, Postfach 100565, 98684 Ilmenau, Germany
martin.huschenbett@tu-ilmenau.de

Abstract A tree automatic structure is a structure whose domain can be encoded by a regular tree language such that each relation is recognisable by a finite automaton processing tuples of trees synchronously. Words can be regarded as specific simple trees and a structure is word automatic if it is encodable using only these trees. The question naturally arises whether a given tree automatic structure is already word automatic. We prove that this problem is decidable for tree automatic scattered linear orderings. Moreover, we show that in case of a positive answer a word automatic presentation is computable from the tree automatic presentation.

1   Introduction

The fundamental idea of automatic structures can be traced back to the 1960s when Büchi, Elgot, Rabin, and others used finite automata to provide decision procedures for the first-order theory of Presburger arithmetic $(\mathbb{N}; +)$ and several other logical problems. Hodgson generalised this idea to the concept of *automaton decidable* first-order theories. Independently of Hodgson and inspired by the successful employment of finite automata and their methods in group theory, Khoussainov and Nerode [4] initiated the systematic investigation of *automatic structures*. Recalling the efforts from the 1960s, Blumensath [2] extended this concept notion beyond finite automata to finite automaton models recognising infinite words, finite trees, or infinite trees.

Basically, a countable relational structure is *tree automatic* or *tree automatically presentable* if its elements can be encoded by finite trees in such a way that its domain and its relations are recognisable by finite automata processing either single trees or tuples of trees synchronously. A structure is *word automatic* if its elements can be encoded using only specific simple trees which effectively represent words. In contrast to the more general concept of *computable structures* and based on the strong closure properties of recognisability, automatic structures provide pleasant algorithmic features. In particular, they possess decidable first-order theories.

Due to this latter fact, the concept of automatic structures gained a lot attention which led to noticeable progress (cf. [1,6]). Automatic presentations were found for many structures, some structures where shown to be tree but not word

S.B. Cooper, A. Dawar, and B. Löwe (Eds.): CiE 2012, LNCS 7318, pp. 313–322, 2012.

automatic, for instance Skolem arithmetic $(\mathbb{N}; \times)$, whereas other structures, like the random graph, were proven to be neither word nor tree automatic. For some classes of structures it was even possible to characterise its automatic members, for example an ordinal is word automatic respectively tree automatic precisely if it is less than ω^ω respectively ω^{ω^ω}. Certain extensions of first-order logic were shown to preserve decidability of the corresponding theory. The question whether two automatic structures are isomorphic turned out to be highly undecidable in general as well as for some restricted classes of structures. At the same time, the isomorphism problem for word automatic ordinals was proven to be decidable. Last but not least, the different classes of automatic structures was characterised by means of interpretations in universal structures.

Due to the fact that word automaticity is a special case of tree automaticity, the question naturally arises whether a given tree automatic structure is already word automatic. As far as we know, this problem was neither solved in general nor for any restricted class of structures. For that reason, we investigate the respective question for scattered linear orderings in this paper. Actually, we prove the corresponding problem to be decidable and our main result is as follows:

Theorem 1.1. *Given a tree automatic presentation $\mathcal{P}$ of a scattered linear ordering $\mathfrak{L}$, it is decidable whether $\mathfrak{L}$ is word automatic. In case $\mathfrak{L}$ is word automatic, one can compute a word automatic presentation of $\mathfrak{L}$ from $\mathcal{P}$.*

Since every well-ordering is scattered, this result still holds if $\mathfrak{L}$ is assumed to be an ordinal. The proof of Theorem 1.1 splits into three parts. First, we introduce the notion of *slim* tree languages and prove this property to be decidable (Theorem 3.2). Second, we show that a slim domain is sufficient for a tree automatic structure to be word automatic (Theorem 4.1). Last, we demonstrate that this condition is also necessary in case of scattered linear orderings (Theorem 5.1). Altogether, Theorem 1.1 follows from the three mentioned theorems.

2   Background

In this section we recall the necessary notions of logic, automatic structures (cf. [1,6]), tree automata (cf. [3]), and linear orderings. We agree that the natural numbers $\mathbb{N}$ include 0 and that $[m, n] = \{m, m + 1, \ldots, n\} \subseteq \mathbb{N}$ for all $m, n \in \mathbb{N}$.

Logic. A *(relational) signature* $\tau = (\mathcal{R}, \mathrm{ar})$ is a finite set $\mathcal{R}$ of *relation symbols* together with a map $\mathrm{ar} \colon \mathcal{R} \to \mathbb{N}$ assigning to each $R \in \mathcal{R}$ its *arity* $\mathrm{ar}(R) \geq 1$. A τ-structure $\mathfrak{A} = \big(A; (R^{\mathfrak{A}})_{R \in \mathcal{R}}\big)$ consists of a set $A = \mathrm{dom}(\mathfrak{A})$, its *domain*, and an $\mathrm{ar}(R)$-ary relation $R^{\mathfrak{A}} \subseteq A^{\mathrm{ar}(R)}$ for each $R \in \mathcal{R}$. *First order logic* FO_τ over τ is defined as usual, including an equality predicate. A *sentence* is a formula without free variables. Writing $\varphi(\bar{x})$ means that all free variables of the formula φ are among the entries of the tuple $\bar{x} = (x_1, \ldots, x_n)$. The set $\varphi^{\mathfrak{A}}$ consists of all $\bar{a} \in A^n$ satisfying $\mathfrak{A} \models \varphi(\bar{a})$, where the latter is defined as usual.

Automatic Structures. The set of all *(finite) words* over an alphabet Σ is $\Sigma^\star$, the *empty word* is ε, and the *length* of w is $|w|$. Subsets of $\Sigma^\star$ are called *languages* and $L \subseteq \Sigma^\star$ is *regular* if it can be *recognised* by some (non-deterministic) finite automaton.

Let $\square \notin \Sigma$ be a new symbol and $\Sigma_\square = \Sigma \cup \{\square\}$. For $n \geq 1$ consider an n-tuple $\bar{w} = (w_1, \ldots, w_n) \in (\Sigma^\star)^n$ of words with $w_i = a_{i,1} a_{i,2} \ldots a_{i,m_i}$ for all $i \in [1, n]$. Let $m = \max\{m_1, \ldots, m_n\}$ and $a_{i,j} = \square$ for $j \in [m_i + 1, m]$. The *convolution* of $\bar{w}$ is the word $\otimes \bar{w} = \bar{a}_1 \ldots \bar{a}_m \in (\Sigma_\square^n)^\star$ with $\bar{a}_j = (a_{1,j}, \ldots, a_{n,j}) \in \Sigma_\square^n$ for all $j \in [1, m]$. An n-ary relation $R \subseteq (\Sigma^\star)^n$ is *automatic* if the language $\otimes R \subseteq (\Sigma_\square^n)^\star$, which consists of all $\otimes \bar{w}$ with $\bar{w} \in R$, is regular.

A τ-structure $\mathfrak{A}$ with $\operatorname{dom}(\mathfrak{A}) \subseteq \Sigma^\star$ is *(word) automatic* if $\operatorname{dom}(\mathfrak{A})$ is regular and $R^{\mathfrak{A}}$ is automatic for all $R \in \mathcal{R}$. A *(word) automatic presentation* of $\mathfrak{A}$ is a tuple $(\mathcal{A}_{\operatorname{dom}}; (\mathcal{A}_R)_{R \in \mathcal{R}})$ of finite automata such that $\mathcal{A}_{\operatorname{dom}}$ recognises $\operatorname{dom}(\mathfrak{A})$ and $\mathcal{A}_R$ recognises $\otimes R^{\mathfrak{A}}$. Abusing notation, we call any structure $\mathfrak{B}$ which is isomorphic to some word automatic structure $\mathfrak{A}$ also *(word) automatic*.

Tree Automata. A *tree domain* is a non-empty, finite, and prefix-closed subset $D \subseteq \{0, 1\}^\star$ satisfying $u0 \in D$ iff $u1 \in D$ for all $u \in D$. A *tree* over Σ is a map $t: D \to \Sigma$ where $\operatorname{dom}(t) = D$ is a tree domain. The set of all trees is denoted by T_Σ and its subsets are called *(tree) languages*. For some $t \in T_\Sigma$ and $u \in \operatorname{dom}(t)$ the *subtree* of t *rooted* at u is the tree $t{\restriction}u \in T_\Sigma$ defined by

$$\operatorname{dom}(t{\restriction}u) = \{\, v \in \{0, 1\}^\star \mid uv \in \operatorname{dom}(t) \,\} \quad \text{and} \quad (t{\restriction}u)(v) = t(uv).$$

A *(deterministic bottom-up) tree automaton* $\mathcal{A} = (Q, \iota, \delta, F)$ over Σ consists of a finite set Q of *states*, a *start state function* $\iota: \Sigma \to Q$, a *transition function* $\delta: \Sigma \times Q \times Q \to Q$, and a set $F \subseteq Q$ of *accepting states*. For each $t \in T_\Sigma$ a state $\mathcal{A}(t) \in Q$ is defined recursively by $\mathcal{A}(t) = \iota(t(\varepsilon))$ if $\operatorname{dom}(t) = \{\varepsilon\}$ and $\mathcal{A}(t) = \delta\big(t(\varepsilon), \mathcal{A}(t{\restriction}0), \mathcal{A}(t{\restriction}1)\big)$ otherwise. The language *recognised* by $\mathcal{A}$ is the set of all $t \in T_\Sigma$ with $\mathcal{A}(t) \in F$. A language $L \subseteq T_\Sigma$ is *regular* if it can be recognised by some tree automaton.

The *convolution* of $\bar{t} = (t_1, \ldots, t_n) \in (T_\Sigma)^n$ is the tree $\otimes \bar{t} \in T_{\Sigma_\square^n}$ defined by $\operatorname{dom}(\otimes \bar{t}) = \operatorname{dom}(t_1) \cup \cdots \cup \operatorname{dom}(t_n)$ and $(\otimes \bar{t})(u) = \big(t_1'(u), \ldots, t_n'(u)\big)$, where $t_i'(u) = t_i(u)$ if $u \in \operatorname{dom}(t_i)$ and $t_i'(u) = \square$ otherwise. A relation $R \subseteq (T_\Sigma)^n$ is *automatic* if the language $\otimes R \subseteq T_{\Sigma_\square^n}$ is regular.

Tree automatic structures and *tree automatic presentations* are defined like in the word automatic case, but based on trees and tree automata.

Linear Orderings. A *linear ordering* is a structure $\mathfrak{A} = (A; <^{\mathfrak{A}})$ where $<^{\mathfrak{A}}$ is a *strict* linear order relation on A. The ordering $\mathfrak{A}$ is *scattered* if $(\mathbb{Q}; <)$ cannot be embedded into $\mathfrak{A}$. Obviously, every well-ordering is scattered. For any two linear orderings $\mathfrak{A}$ and $\mathfrak{B}$ we define another linear ordering $\mathfrak{A} \cdot \mathfrak{B}$ by $\operatorname{dom}(\mathfrak{A} \cdot \mathfrak{B}) = \operatorname{dom}(\mathfrak{A}) \times \operatorname{dom}(\mathfrak{B})$ and $(a_1, b_1) <^{\mathfrak{A} \cdot \mathfrak{B}} (a_2, b_2)$ iff either $a_1 <^{\mathfrak{A}} a_2$ or $a_1 = a_2$ and $b_1 <^{\mathfrak{B}} b_2$. Finally, if $\mathfrak{A}_1$ can be embedded into $\mathfrak{B}_1$ and $\mathfrak{A}_2$ into $\mathfrak{B}_2$, then $\mathfrak{A}_1 \cdot \mathfrak{A}_2$ can be embedded into $\mathfrak{B}_1 \cdot \mathfrak{B}_2$.

3   Slim and Fat Tree Languages

In this section, we introduce the notion of slim tree languages and show that it is decidable whether the language recognised by a given tree automaton is slim.

Definition 3.1. *The* thickness $\varnothing(t)$ *of a tree* $t \in T_\Sigma$ *is the maximal number of nodes on any level, i.e.,*

$$\varnothing(t) = \max \{ \, |\mathrm{dom}(t) \cap \{0,1\}^\ell| \mid \ell \geq 0 \, \} \in \mathbb{N}.$$

For every $K \geq 1$ *the set of all* $t \in T_\Sigma$ *with* $\varnothing(t) \leq K$ *is denoted by* $T_{\Sigma,K}$. *A tree language* $L \subseteq T_\Sigma$ *is* slim *if there exists some* $K \geq 1$ *such that* $L \subseteq T_{\Sigma,K}$, *otherwise* L *is* fat.

A tree automaton $\mathcal{A}$ is *reduced* if for every state q of $\mathcal{A}$ there is a tree $t \in T_\Sigma$ with $\mathcal{A}(t) = q$. For every tree automaton $\mathcal{A}$ one can compute a reduced tree automaton which recognises the same language and has no more states than $\mathcal{A}$.

Theorem 3.2. *Given a reduced tree automaton* $\mathcal{A}$, *it is decidable whether the tree language* L *recognised by* $\mathcal{A}$ *is slim or fat. If* L *is slim, then* $L \subseteq T_{\Sigma,2^{n-1}}$, *where* n *is the number of states of* $\mathcal{A}$.

For the rest of this section we fix a reduced tree automaton $\mathcal{A} = (Q, \iota, \delta, F)$. The proof of Theorem 3.2 essentially depends on an inspection of the directed graph $G_\mathcal{A} = (Q, E_\mathcal{A})$ with

$$(p, q) \in E_\mathcal{A} \quad \text{iff} \quad \exists a \in \Sigma, r \in Q \colon \delta(a, p, r) = q \text{ or } \delta(a, r, p) = q. \tag{1}$$

Clearly, this graph is computable from $\mathcal{A}$. The lemma below is shown by applying the idea of pumping to tree automata. Therein, the *height* $h(t)$ of a tree $t \in T_\Sigma$ is the number

$$h(t) = \max \{ \, |u| \mid u \in \mathrm{dom}(t) \, \} \in \mathbb{N}.$$

Lemma 3.3. *For every* $q \in Q$ *the following are equivalent:*

(1) there are infinitely many $t \in T_\Sigma$ *satisfying* $\mathcal{A}(t) = q$,
(2) there is a tree $t \in T_\Sigma$ *satisfying* $h(t) \geq n$ *and* $\mathcal{A}(t) = q$, *where* $n = |Q|$,
(3) $G_\mathcal{A}$ *contains a cycle from which* q *is reachable.*

An edge $(p, q) \in E_\mathcal{A}$ is *special* if in the definition of $E_\mathcal{A}$ in Eq. (1) the state $r \in Q$ can be chosen such that it satisfies the conditions of Lemma 3.3 (for r in place of q). Since condition (3) is decidable, it is decidable whether an edge is special. The key idea for proving Theorem 3.2 is stated by the following lemma:

Lemma 3.4. *The following are equivalent:*

(1) the tree language L *recognised by* $\mathcal{A}$ *is fat,*
(2) there is a tree $t \in L$ *satisfying* $\varnothing(t) > 2^{n-1}$, *where* $n = |Q|$,
(3) $G_\mathcal{A}$ *contains a cycle including a special edge and from which* F *is reachable.*

The proof of this lemma works similar to the one of Lemma 3.3. Since condition (3) is decidable given $\mathcal{A}$ as input, Theorem 3.2 follows.

4 Slim Tree Automatic Structures Are Word Automatic

This section is devoted to the proof of the following theorem:

Theorem 4.1. *Let $\mathfrak{A}$ be a tree automatic structure such that $\mathrm{dom}(\mathfrak{A})$ is slim. Then, $\mathfrak{A}$ is already word automatic and one can compute a word automatic presentation of $\mathfrak{A}$ from a tree automatic presentation of $\mathfrak{A}$.*

The idea of the proof is the following. Let $K \geq 1$ be such that $\mathrm{dom}(\mathfrak{A}) \subseteq T_{\Sigma,K}$. We give an alphabet $\widehat{\Sigma}$ and an injective map $C \colon T_{\Sigma,K} \to \widehat{\Sigma}^{\star}$, the *encoding*, such that $C(L)$ is regular for all regular $L \subseteq T_{\Sigma,K}$ (Proposition 4.5) and $C(R)$ is automatic for all automatic relations $R \subseteq (T_{\Sigma,K})^n$ (Proposition 4.6). Thus, the structure $C(\mathfrak{A})$ is word automatic. A word automatic presentation of $C(\mathfrak{A})$ is computable since both propositions are effective and Theorem 3.2 allows for computing a suitable K. Although it is possible to show both propositions using automata, it is much more convenient to accomplish this by means of logic.

4.1 Monadic Second Order Logic

Monadic second order logic MSO_{τ} extends FO_{τ} by *set variables*, which range over subsets of the domain and are denoted by capital letters, quantifiers for these variables, and the formula "$x \in X$" (cf. [7]). Let $\tau = (\mathcal{R}, \mathrm{ar})$ and τ' be two signatures. An *(MSO-)interpretation* of a τ-structure $\mathfrak{A}$ in a τ'-structure $\mathfrak{B}$ is a pair $\langle f, \mathcal{I} \rangle$ consisting of an injective map $f \colon \mathrm{dom}(\mathfrak{A}) \to \mathrm{dom}(\mathfrak{B})$ and a tuple $\mathcal{I} = (\Delta; (\Phi_R)_{R \in \mathcal{R}})$ of $\mathsf{MSO}_{\tau'}$-formulae with free FO-variables only such that $f(\mathrm{dom}(\mathfrak{A})) = \Delta^{\mathfrak{B}}$ and $f(R^{\mathfrak{A}}) = \Phi_R^{\mathfrak{B}}$ for each $R \in \mathcal{R}$. In fact, f induces an isomorphism between $\mathfrak{A}$ and $\mathcal{I}(\mathfrak{B}) = (\Delta^{\mathfrak{B}}; (\Phi_R^{\mathfrak{B}})_{R \in \mathcal{R}})$. Replacing in an MSO_{τ}-formula $\varphi(\bar{x})$ all symbols $R \in \mathcal{R}$ with Φ_R and relativising quantifiers to Δ yields an $\mathsf{MSO}_{\tau'}$-formula $\varphi^{\mathcal{I}}(\bar{x})$ satisfying $\mathfrak{A} \models \varphi(\bar{a})$ iff $\mathfrak{B} \models \varphi^{\mathcal{I}}(f(\bar{a}))$ for all $\bar{a} \in A^n$.

For an alphabet Σ the signature $\mathsf{W}\Sigma$ consists of one binary relation symbol $\leq$ and a unary symbol P_a for each $a \in \Sigma$. Every word $w = a_1 a_2 \ldots a_{|w|} \in \Sigma^{\star}$ is regarded as a $\mathsf{W}\Sigma$-structure with domain $\mathrm{dom}(w) = \{1, \ldots, |w|\}$, $\leq^w$ being the natural order on $\mathrm{dom}(w)$, and $i \in P_a^w$ iff $a_i = a$. For fixed numbers $m, r \in \mathbb{N}$, relations like $x = y + m$ and $x \equiv r \,(\mathrm{mod}\, m)$ are expressible in $\mathsf{MSO}_{\mathsf{W}\Sigma}$. The language *defined* by an $\mathsf{MSO}_{\mathsf{W}\Sigma}$-sentence Φ is the set of all $w \in \Sigma^{\star}$ with $w \models \Phi$.

The signature $\mathsf{T}\Sigma$ is similar to $\mathsf{W}\Sigma$ but contains two binary symbols S_0 and S_1 instead of $\leq$. Each tree $t \in T_{\Sigma}$ is considered as a $\mathsf{T}\Sigma$-structure with domain $\mathrm{dom}(t)$, $(u, v) \in S_d^t$ iff $ud = v$ $(d = 0, 1)$, and $u \in P_a^t$ iff $t(u) = a$. The language *defined* by some $\mathsf{MSO}_{\mathsf{T}\Sigma}$-sentence Φ is the set of all $t \in T_{\Sigma}$ with $t \models \Phi$.

The following theorem holds for word languages as well as for tree languages:

Theorem 4.2 (cf. [7]). *A language L is regular iff it is definable in MSO, and both conversions, from automata to formulae and vice versa, are effective.*

4.2   The Encoding and Preservation of Regularity

For the rest of this section fix the $K \geq 1$ from above. The
first objective is to give the encoding $C \colon T_{\Sigma,K} \to \widehat{\Sigma}^{\star}$,
where \$ is a new symbol and $\widehat{\Sigma} = \Sigma \times \{0,1\} \cup \{\$\}$. For
a tree $t \in T_{\Sigma,K}$ of height $m = h(t)$ its *encoding* $C(t) =$
$\sigma_0 \sigma_1 \ldots \sigma_m$ is made up of $m+1$ blocks $\sigma_0, \ldots, \sigma_m \in \widehat{\Sigma}^K$
describing the individual levels of t. More specifically, σ_ℓ
consists of the labels of the ℓ-th level from left to right,
each enriched by a bit stating whether the corresponding

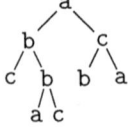

Fig. 1. The tree t_{ex}

node possesses children, and is padded up to length K by \$ symbols. For ex-
ample, the tree $t_{ex} \in T_{\{a,b,c\}}$ in Figure 1 on the right satisfies $\varnothing(t_{ex}) = 4$ and is,
under the assumption $K = 5$, encoded by the word

$$C(t_{ex}) = \langle a,1 \rangle \$\$\$\$ \langle b,1 \rangle \langle c,1 \rangle \$\$\$ \langle c,0 \rangle \langle b,1 \rangle \langle b,0 \rangle \langle a,0 \rangle \$ \langle a,0 \rangle \langle c,0 \rangle \$\$\$ \,.$$

Formally, for each $\ell \in [0,m]$ let $u_{\ell,1}, \ldots, u_{\ell,s_\ell}$ be the lexicographic enumeration
(w.r.t. $0 < 1$) of $\mathrm{dom}(t) \cap \{0,1\}^\ell$. For $r \in [1, s_\ell]$ we let $c_{\ell,r} = 1$ if $u_{\ell,r}$ is an inner
node, i.e., $u_{\ell,r}\{0,1\} \subseteq \mathrm{dom}(t)$, and $c_{\ell,r} = 0$ if $u_{\ell,r}$ is a leaf. Finally, we put

$$\sigma_\ell = \langle t(u_{\ell,1}), c_{\ell,1} \rangle \langle t(u_{\ell,2}), c_{\ell,2} \rangle \ldots \langle t(u_{\ell,s_\ell}), c_{\ell,s_\ell} \rangle \$^{K-s_\ell} \,.$$

The main tool for studying the map $C \colon T_{\Sigma,K} \to \widehat{\Sigma}^{\star}$ is the following lemma:

Lemma 4.3. *For all $t \in T_{\Sigma,K}$ there is an MSO-interpretation $\langle f_C, \mathcal{I}_C \rangle$ of t
in $C(t)$ such that $\mathcal{I}_C$ does not depend on t.*

Proof. Observe that for each inner node u of t the children of u are the $(2s-1)$-th
and $2s$-th node on the next level, where s is the number of inner nodes from left
up to u on its level. Formally, for an inner node $u_{\ell,r}$ we have $u_{\ell,r}d = u_{\ell+1,2s-1+d}$,
where $d \in \{0,1\}$ and $s = c_{\ell,1} + \cdots + c_{\ell,r}$. Based on this observation, one can
give an interpretation $\langle f_C, \mathcal{I}_C \rangle$ of t in $C(t)$ such that $f_C(u_{\ell,r}) = \ell \cdot K + r$. □

As a first consequence, we obtain $t \cong \mathcal{I}_C(t) = \mathcal{I}_C(t') \cong t'$, and hence $t = t'$, for
all $t, t' \in T_{\Sigma,K}$ with $C(t) = C(t')$, i.e., C is injective. The proof of the fact that
C preserves regularity is mainly based on Lemma 4.3 and Lemma 4.4 below.

Lemma 4.4. *Let $\sigma \in \widehat{\Sigma}^{\star}$. There exists a tree $t \in T_{\Sigma,K}$ with $C(t) = \sigma$ iff
$\sigma = \sigma_0 \sigma_1 \ldots \sigma_n$ for some $n \geq 0$ and $\sigma_0, \ldots, \sigma_n \in \widehat{\Sigma}^K$ satisfying (a) and (b):*

*(a) $\sigma_\ell = \alpha_{\ell,1} \ldots \alpha_{\ell,s_\ell} \$^{K-s_\ell}$ for some $s_\ell \geq 1$ and $\alpha_{\ell,1}, \ldots, \alpha_{\ell,s_\ell} \in \Sigma \times \{0,1\}$ and
for each $\ell \in [0,n]$.*
*(b) $s_0 = 1$, $s_{\ell+1} = 2 \cdot (c_{\ell,1} + \cdots + c_{\ell,s_\ell})$ for $0 \leq \ell < n$, and $c_{m,1} + \cdots + c_{m,s_m} = 0$,
where $\alpha_{\ell,r} = \langle a_{\ell,r}, c_{\ell,r} \rangle$.*

Proof. To see that $C(t)$ has the required shape, notice that (b) mainly reflects the
relationship between the numbers of nodes on two adjacent levels. Conversely, if
$\sigma \in \widehat{\Sigma}^{\star}$ is of the required shape, then there is a tree $t \in T_\Sigma$ with $t \cong \mathcal{I}_C(\sigma)$ and
it turns out that $\varnothing(t) \leq K$ and $C(t) = \sigma$.

Proposition 4.5. *Let $L \subseteq T_{\Sigma,K}$ be a regular language. Then, the language $C(L) \subseteq \widehat{\Sigma}^\star$ is also regular and one can compute a finite automaton recognising $C(L)$ from a tree automaton recognising L.*

Proof. Let Γ_C be an $\mathsf{MSO}_{\mathsf{W}\widehat{\Sigma}}$-sentence which expresses the requirement on the shape of σ from Lemma 4.4. By Theorem 4.2, there is an $\mathsf{MSO}_{\mathsf{T}\Sigma}$-sentence Φ defining $L \subseteq T_{\Sigma,K}$. Then, the $\mathsf{MSO}_{\mathsf{W}\widehat{\Sigma}}$-sentence $\Gamma_C \wedge \Phi^{\mathcal{I}_C}$ defines $C(L)$ and, again by Theorem 4.2, this language is regular. Finally, all employed constructions are effective. ☐

4.3 Preservation of Automaticity

The purpose of this subsection is to complete the proof of Theorem 4.1.

Proposition 4.6. *Let $R \subseteq (T_{\Sigma,K})^n$ be an automatic relation. Then, the relation $C(R) \subseteq (\widehat{\Sigma}^\star)^n$ is also automatic and one can compute a finite automaton recognising $\otimes C(R)$ from a tree automaton recognising $\otimes R$.*

Basically, the key idea behind the proof is the same as for Proposition 4.5 though it is more involved. Let $\bar{t} = (t_1, \ldots, t_n) \in (T_{\Sigma,K})^n$. Due to cardinality reasons, $\otimes \bar{t}$ is commonly not directly interpretable in $\otimes C(\bar{t})$ but only in an n-fold copy of $\otimes C(\bar{t})$. This is formalised by means of the injective monoid morphism

$$H \colon (\widehat{\Sigma}_\square^n)^\star \to (\widehat{\Sigma}_\square^n)^\star, \bar{\alpha}_1 \ldots \bar{\alpha}_m \mapsto \bar{\alpha}_1^n \ldots \bar{\alpha}_m^n.$$

The interpretation of $\otimes \bar{t}$ in $H(\otimes C(\bar{t}))$ embraces two aspects which are better considered separately. Thus, we define an intermediate structure $\mathrm{II}\bar{t}$ which extends the disjoint union of the t_i's on domain $\mathrm{dom}(\mathrm{II}\bar{t}) = \bigcup_{i \in [1,n]} \{i\} \times \mathrm{dom}(t_i)$ by a binary relation $L^{\mathrm{II}\bar{t}}$, relating all (i, u) and (j, v) with $|u| = |v|$, and unary relations $Q_i^{\mathrm{II}\bar{t}} = \{i\} \times \mathrm{dom}(t_i)$ for each $i \in [1, n]$. Altogether, we give several interpretations whose formulae naturally do not depend on the specific choice of $\bar{t}$. An overview of the whole setting is depicted in Figure 2.

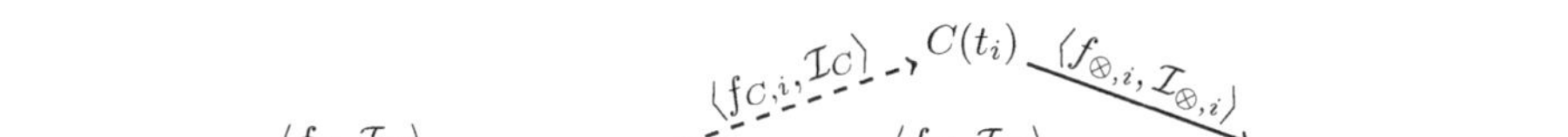

Fig. 2. Interpretations involved in proving Proposition 4.6

The Interpretation $\langle f_{\mathrm{II}}, \mathcal{I}_{\mathrm{II}} \rangle$. Standard techniques as the MSO-definability of the transitive closure in combination with the relation L yield an MSO-formula $E(x, y)$ with $\mathrm{II}\bar{t} \models E((i, u), (j, v))$ iff $u = v$. Using this formula, one can construct an interpretation $\langle f_{\mathrm{II}}, \mathcal{I}_{\mathrm{II}} \rangle$ of $\otimes \bar{t}$ in $\mathrm{II}\bar{t}$ such that $f_{\mathrm{II}}(u) = (i, u)$, where i is minimal with $u \in \mathrm{dom}(t_i)$.

The Interpretations $\langle f_{\otimes,i}, \mathcal{I}_{\otimes,i} \rangle$. For all $i \in [1, n]$ and $\bar{w} \in (\widehat{\Sigma}^\star)^n$ one can easily give an interpretation $\langle f_{\otimes,i}, \mathcal{I}_{\otimes,i} \rangle$ of w_i in $H(\otimes \bar{w})$ such that $f_{\otimes,i}(p) = (p-1) \cdot n + i$.

The Interpretation $\langle f_H, \mathcal{I}_H \rangle$. For $i \in [1,n]$ let $\langle f_{C,i}, \mathcal{I}_C \rangle$ be the interpretation of t_i in $C(t_i)$ from Lemma 4.3. Then the map $f_H \colon \mathrm{dom}(\mathrm{II}\bar{t}) \to \mathrm{dom}\big(H(\otimes C(\bar{t}))\big)$ with $f_H(i,u) = f_{\otimes,i}(f_{C,i}(u))$ is injective and one can construct formulae $\mathcal{I}_H$ such that $\langle f_H, \mathcal{I}_H \rangle$ is an interpretation of $\mathrm{II}\bar{t}$ in $H\big(\otimes C(\bar{t})\big)$.

Proof (of Proposition 4.6). Let Γ_H be an $\mathsf{MSO}_{\mathsf{w}\widehat{\Sigma}_{\square}^n}$-sentence defining the language $H\big(\otimes(\widehat{\Sigma}^\star)^n\big) \subseteq (\widehat{\Sigma}_{\square}^n)^\star$. If Φ defines $\otimes R$, then

$$\Gamma_H \wedge \bigwedge_{i \in [1,n]} \Gamma_C^{\mathcal{I}_{\otimes,i}} \wedge (\Phi^{\mathcal{I}_{\mathrm{II}}})^{\mathcal{I}_H}$$

defines $H\big(\otimes C(R)\big)$. Since H is an injective monoid morphism, $\otimes C(R)$ is regular as well. Finally, all employed constructions are effective.   □

5 Fat Tree Automatic Ordinals Are Not Word Automatic

The goal of this section is to give the last missing piece for the proof of Theorem 1.1, namely the following theorem:

Theorem 5.1. *Let $\mathfrak{L}$ be a tree automatic scattered linear ordering such that* $\mathrm{dom}(\mathfrak{L})$ *is fat. Then, $\mathfrak{L}$ is not word automatic.*[1]

The theorem below states the necessary condition on word automatic linear orderings we use to show non-automaticity:

Theorem 5.2 (Khoussainov, Rubin, Stephan [5]). *If $\mathfrak{L}$ is a word automatic linear ordering, then its FC-rank is finite.*

Actually, we do not need any details on the *FC-rank* (finite condensation rank) besides the fact that every scattered linear ordering $\mathfrak{L}$, having the property that for each $r \geq 1$ at least one linear ordering from

$$\mathcal{N}_r = \big\{\, \mathfrak{A}_1 \cdot \mathfrak{A}_2 \cdots \mathfrak{A}_r \mid \mathfrak{A}_1, \ldots, \mathfrak{A}_r \in \{(\mathbb{N}; <), (\mathbb{N}; >)\} \,\big\}$$

can be embedded into $\mathfrak{L}$, has infinite *FC-rank*. For any linear ordering $\mathfrak{A}$ and all $a_1, a_2 \in \mathrm{dom}(\mathfrak{A})$ we define $\mathrm{cmp}_{\mathfrak{A}}(a_1, a_2) \in \{-1, 0, 1\}$ to be -1 if $a_1 <^{\mathfrak{A}} a_2$, 0 if $a_1 = a_2$, and 1 if $a_2 <^{\mathfrak{A}} a_1$. The main idea of the proof of Theorem 5.1 is given by the following two lemmas:

Lemma 5.3. *Let $\mathfrak{A}$ and $\mathfrak{A}_1, \ldots, \mathfrak{A}_r$ be infinite linear orderings with $\mathrm{dom}(\mathfrak{A}) = \mathrm{dom}(\mathfrak{A}_1) \times \cdots \times \mathrm{dom}(\mathfrak{A}_r)$ and satisfying the following two conditions:*

(1) $\mathrm{cmp}_{\mathfrak{A}}(\bar{a}, \bar{b})$ is determined by $\mathrm{cmp}_{\mathfrak{A}_1}(a_1, b_1), \ldots, \mathrm{cmp}_{\mathfrak{A}_r}(a_r, b_r)$ for all $\bar{a}, \bar{b} \in A$,
(2) if $\bar{a}, \bar{b} \in A$ differ only in the i-th component, then $\mathrm{cmp}_{\mathfrak{A}}(\bar{a}, \bar{b}) = \mathrm{cmp}_{\mathfrak{A}_i}(a_i, b_i)$.

Then, there exists a permutation π of $\{1, \ldots, r\}$ such that $\mathfrak{A}$ is isomorphic to $\mathfrak{A}_{\pi(1)} \cdot \mathfrak{A}_{\pi(2)} \cdots \mathfrak{A}_{\pi(r)}$.

---

[1] This claim fails if we permit non-injective tree automatic presentations.

Lemma 5.4. *Let $\mathfrak{L} = (L; <)$ be a tree automatic scattered linear ordering, $(\mathcal{A}; \mathcal{A}_<)$ an automatic presentation of $\mathfrak{L}$, n the number of states of $\mathcal{A}$, and $r \geq 1$. If there exists some tree $t \in L$ with $\varnothing(t) \geq r \cdot 2^n$, then there are infinite linear orderings $\mathfrak{A}_1, \ldots, \mathfrak{A}_r$ such that $\mathfrak{A}_1 \cdot \mathfrak{A}_2 \cdots \mathfrak{A}_r$ can be embedded into $\mathfrak{L}$.*

Proof (of Lemma 5.4). To simplify notation, we put $[\![s, t]\!]_< = \mathcal{A}_<\big(\otimes(s, t)\big)$ for all $s, t \in T_\Sigma$. Moreover, we assume w.l.o.g. that from $[\![s, t]\!]_<$ one can deduce whether $s = t$ holds true. Then, $\mathrm{cmp}_{\mathfrak{L}}(s, t)$ *is determined by* $[\![s, t]\!]_<$ for all $s, t \in L$, i.e., there is a map f from the state set of $\mathcal{A}_<$ to $\{-1, 0, 1\}$ such that $\mathrm{cmp}_{\mathfrak{L}}(s, t) = f\big([\![s, t]\!]_<\big)$ for all $s, t \in L$.

Let $\mathfrak{T} \in L$ be a tree and $\ell \geq n$ such that $\big|\mathrm{dom}(\mathfrak{T}) \cap \{0, 1\}^\ell\big| \geq r \cdot 2^n$. Thus, there exist at least r mutually distinct $u \in \mathrm{dom}(\mathfrak{T}) \cap \{0, 1\}^{\ell-n}$ for which there is a $v \in \{0, 1\}^n$ with $uv \in \mathrm{dom}(\mathfrak{T})$, say $u_1, \ldots, u_r$. For $\bar{t} = (t_1, \ldots, t_r) \in (T_\Sigma)^r$ let $\mathfrak{T}[\bar{t}] \in T_\Sigma$ be the tree obtained from $\mathfrak{T}$ by replacing for each $i \in [1, r]$ the subtree rooted at u_i with t_i. Then, $\mathcal{A}(\mathfrak{T}[\bar{t}])$ is determined by the r states $\mathcal{A}(t_1), \ldots, \mathcal{A}(t_r)$ for all $\bar{t} \in (T_\Sigma)^r$. Moreover, for $\bar{s} \in (T_\Sigma)^r$ the tree $\otimes\big(\mathfrak{T}[\bar{s}], \mathfrak{T}[\bar{t}]\big)$ is obtained from $\otimes(\mathfrak{T}, \mathfrak{T})$ by replacing for each $i \in [1, r]$ the subtree rooted at u_i with $\otimes(s_i, t_i)$. Consequently, $\big[\![\mathfrak{T}[\bar{s}], \mathfrak{T}[\bar{t}]\big]\!]_<$ is determined by the r states $[\![s_1, t_1]\!]_<, \ldots, [\![s_r, t_r]\!]_<$ for all $\bar{s}, \bar{t} \in (T_\Sigma)^r$.

Observe that the height $h(\mathfrak{T} \lceil u_i)$ is at least n for each $i \in [1, r]$. Therefore, by Lemma 3.3 and Ramsey's theorem for infinite, undirected, finitely coloured graphs, there exists an infinite set $A_i \subseteq T_\Sigma$ of trees $t \in T_\Sigma$ with $\mathcal{A}(t) = \mathcal{A}(\mathfrak{T} \lceil u_i)$ such that

$$c(s, t) = \big\{[\![s, s]\!]_<, [\![t, t]\!]_<, [\![s, t]\!]_<, [\![t, s]\!]_<\big\}$$

is the same set Q_i for all distinct $s, t \in A_i$. It turns out that Q_i has exactly three elements and $[\![s, s]\!]_< = [\![t, t]\!]_<$ for all $s, t \in A_i$.

Now, put $A = A_1 \times \cdots \times A_r$. For each $\bar{t} \in A$ we have $\mathcal{A}(\mathfrak{T}[\bar{t}]) = \mathcal{A}(\mathfrak{T})$ and hence $\mathfrak{T}[\bar{t}] \in L$. We define a linear ordering $\mathfrak{A} = \big(A; <^{\mathfrak{A}}\big)$ by $\bar{s} <^{\mathfrak{A}} \bar{t}$ iff $\mathfrak{T}[\bar{s}] < \mathfrak{T}[\bar{t}]$. By definition, $\mathfrak{A}$ can be embedded into $\mathfrak{L}$.

For $i \in [1, r]$, $\bar{a} \in A$, and $t \in A_i$ we let $\bar{a}_{i/t} \in A$ be the tuple $\bar{a}$ with the i-th component replaced by t. Then, for all $\bar{a}, \bar{b}$ and $s, t \in A_i$ we obtain $\big[\![\mathfrak{T}[\bar{a}_{i/s}], \mathfrak{T}[\bar{a}_{i/t}]\big]\!]_< = \big[\![\mathfrak{T}[\bar{b}_{i/s}], \mathfrak{T}[\bar{b}_{i/t}]\big]\!]_<$ and hence $a_{i/s} <^{\mathfrak{A}} a_{i/t}$ iff $b_{i/s} <^{\mathfrak{A}} b_{i/t}$. Thus, defining a linear ordering $\mathfrak{A}_i = \big(A_i; <^{\mathfrak{A}_i}\big)$ by $s <^{\mathfrak{A}_i} t$ iff $\bar{a}_{i/s} <^{\mathfrak{A}} \bar{a}_{i/t}$ is independent from the specific choice of $\bar{a} \in A$. Clearly, $\mathrm{cmp}_{\mathfrak{A}_i}(s, t)$ is determined by $[\![s, t]\!]_<$ for all $s, t \in A_i$. Since Q_i contains exactly three elements, $[\![s, t]\!]_<$ is determined by $\mathrm{cmp}_{\mathfrak{A}_i}(s, t)$ for all $s, t \in A_i$ as well. Hence, the linear orderings $\mathfrak{A}$ and $\mathfrak{A}_1, \ldots, \mathfrak{A}_r$ satisfy the condition of Lemma 5.3 below and consequently $\mathfrak{A}_{\pi(1)} \cdots \mathfrak{A}_{\pi(r)}$ can be embedded into $\mathfrak{L}$. $\square$

Finally, we are in a position to prove Theorem 5.1.

Proof (of Theorem 5.1). Let $(\mathcal{A}; \mathcal{A}_<)$ be an automatic presentation of $\mathfrak{L}$ and n the number of states of $\mathcal{A}$. Since $\mathrm{dom}(\mathfrak{L})$ is fat, for any $r \geq 1$ there is a $t \in \mathrm{dom}(\mathfrak{L})$ with $\varnothing(t) \geq r \cdot 2^n$. Let $\mathfrak{A}_1, \ldots, \mathfrak{A}_r$ be the infinite linear orderings from Lemma 5.4. For each $i \in [1, r]$ some $\mathfrak{B}_i \in \big\{(\mathbb{N}; <), (\mathbb{N}; >)\big\}$ can be embedded into $\mathfrak{A}_i$. Then, $\mathfrak{B}_1 \cdot \mathfrak{B}_2 \cdots \mathfrak{B}_r \in \mathcal{N}_r$ can be embedded into $\mathfrak{A}_1 \cdot \mathfrak{A}_2 \cdots \mathfrak{A}_r$ and consequently into $\mathfrak{L}$. Hence, $\mathfrak{L}$ has infinite *FC*-rank and is, by Theorem 5.2, *not* word automatic. $\square$

6 Conclusions

Altogether, we proved that is decidable whether a given tree automatic scattered linear ordering is already word automatic. Taking a closer look at the proof reveals that the problem is solvable nondeterministically in logarithmic space, provided the tree automaton recognising the domain is reduced.

The restriction to scattered linear orderings naturally rises the question whether this result holds true for general linear orderings. Unfortunately, this problem cannot be solved by means of our technique since the ordering $(\mathbb{Q}; <)$ of the rationals admits a word automatic as well as a fat tree automatic presentation. As the Boolean algebra of finite and co-finite subsets of $\mathbb{N}$ shares this feature, the same pertains to an analogue of Theorem 1.1 for Boolean algebras. In spite of that, we suggest trying to apply the technique to other classes of structures, such as groups or trees, for which a necessary condition on its automatic members is known.

As a first step, suppose that $\mathfrak{A}$ is a tree automatic structure and φ a first order formula which defines a scattered linear order on $\mathfrak{A}$. If $\mathfrak{A}$ is already word automatic so is $\big(\mathrm{dom}(\mathfrak{A}); \varphi^{\mathfrak{A}}\big)$ and hence $\mathrm{dom}(\mathfrak{A})$ is slim. Thus, $\mathfrak{A}$ is word automatic if, and only if, $\mathrm{dom}(\mathfrak{A})$ is slim. Consequently, word automaticity is uniformly decidable for the class of tree automatic structures which admit a first order definable scattered linear order. In particular, the decision procedure needs no knowledge of the formula defining the order.

Using similar arguments in combination with a technique and a result from [5], one can further show that it is decidable whether a given tree automatic finitely branching tree is already word automatic. However, for arbitrary, not necessarily finitely branching trees this problem remains open.

Finally, Theorem 1.1 provides a decidable characterisation of all tree automatic ordinals $\alpha \geq \omega^{\omega}$. Finding such a characterisation for each $\omega^{\omega^{k}}$ with $k \in \mathbb{N}$ possibly turns out to be the main ingredient for showing that the isomorphism problem for tree automatic ordinals is decidable.

References

1. Bárány, V., Grädel, E., Rubin, S.: Automata-based presentations of infinite structures. In: Esparza, J., Michaux, C., Steinhorn, C. (eds.) Finite and Algorithmic Model Theory, pp. 1–76. Cambridge University Press (2011)
2. Blumensath, A.: Automatic structures. Diploma thesis, RWTH Aachen (1999)
3. Gécseg, F., Steinby, M.: Tree languages. In: Rozenberg, G., Salomaa, A. (eds.) Handbook of Formal Languages, vol. 3, pp. 1–68. Springer, Heidelberg (1997)
4. Khoussainov, B., Nerode, A.: Automatic Presentations of Structures. In: Leivant, D. (ed.) LCC 1994. LNCS, vol. 960, pp. 367–392. Springer, Heidelberg (1995)
5. Khoussainov, B., Rubin, S., Stephan, F.: Automatic linear orders and trees. ACM Transactions on Computational Logic 6(4), 675–700 (2005)
6. Rubin, S.: Automata presenting structures: A survey of the finite string case. Bulletin of Symbolic Logic 14(2), 169–209 (2008)
7. Thomas, W.: Languages, automata, and logic. In: Rozenberg, G., Salomaa, A. (eds.) Handbook of Formal Languages, vol. 3, pp. 384–455. Springer, Heidelberg (1997)

On the Relative Succinctness of Two Extensions by Definitions of Multimodal Logic

Wiebe van der Hoek[1], Petar Iliev[1], and Barteld Kooi[2]

[1] Department of Computer Science, University of Liverpool, Liverpool L69 7ZF,
United Kingdom
{wiebe,pvi}@liverpool.ac.uk
[2] Department of Theoretical Philosophy, University of Groningen, Oude
Boteringestraat 52, 9712 GL Groningen, The Netherlands
B.P.Kooi@rug.nl

Abstract. The growing number of logics has lead to the question: How do we compare two formalisms? A natural answer is: We can compare their expressive power and computational properties. There is, however, another way of comparing logics that has attracted attention recently, namely in terms of representational succinctness, i.e., we can ask whether one of the logics allows for a more "economical" encoding of information than the other. Using extended-syntax trees that correspond to game trees for the Addler-Immerman games, we prove that two well-known abbreviations in multimodal logic lead to an exponential increase in succinctness.

1 Introduction

Consider the following general question. Let L_1 and L_2 be two formalisms that express the same class of "properties". Is one of the formalisms representationally more succinct (allowing for a more "economical" representation of information) than the other and by how much? A famous instance of this general question is whether there is a family of Boolean functions for which boolean circuits can be exponentially more succinct than Boolean formulas. It is believed that the answer to this question is beyond our current mathematical knowledge and techniques. Indeed, it is convincingly argued in [5] that we are far from understanding representational succinctness and we need to first master some very basic problems in order to develop our intuition and mathematical toolbox.

Here we prove that the multimodal logic $[\wedge\Gamma]ML$ is exponentially more succinct than the logic $[\vee\Gamma]ML$ and vice versa. Hence, there are semantic properties that are more economically expressed by using the $[\wedge\Gamma]$ operator; similarly, there are semantic properties that are more economically expressed by using the $[\vee\Gamma]$ operator.

A formula of the form $[\wedge\Gamma]\varphi$ is an abbreviation of the multimodal logic (ML) formula $\bigwedge_{a\in\Gamma}[a]\varphi$, where Γ is a finite set of relation indices. In the same way, a formula of the form $[\vee\Gamma]\varphi$ is an abbreviation of the ML formula $\bigvee_{a\in\Gamma}[a]\varphi$. Readers familiar with epistemic logic will notice that the $[\wedge\Gamma]$ modality corresponds

S.B. Cooper, A. Dawar, and B. Löwe (Eds.): CiE 2012, LNCS 7318, pp. 323–333, 2012.

to the "everybody knows" operator, while $[\vee\Gamma]$ corresponds to the "somebody knows" operator. It is obvious that adding formulas of the form $[\wedge\Gamma]\varphi$ to ML and, thus, obtaining the logic $[\wedge\Gamma]ML$ does not lead to an increase in expressive power; what is more, the computational complexity of the satisfiability problem for $[\wedge\Gamma]ML$ is the same as that for ML [6]. In the same way, $[\vee\Gamma]ML$ is not more expressive than ML. However, it is known that both $[\wedge\Gamma]ML$ and $[\vee\Gamma]ML$ are exponentially more succinct than ML [3]. Proving that $[\wedge\Gamma]ML$ is exponentially more succinct than $[\vee\Gamma]ML$ and vice versa, is a further continuation of the line of work started in [3]. Furthermore, it shows that formulas of the form $[\vee\Gamma]\varphi$ and $[\wedge\Gamma]\varphi$ introduce two different types of "information" compression.

2 Preliminaries

2.1 Multimodal Logic

Definition 1 (Multimodal $[\vee\Gamma][\wedge\Gamma]ML_n^m$ logic). *The signature of the multimodal logic $[\vee\Gamma][\wedge\Gamma]ML_n^m$ is a pair $S = \{P, I\}$, where $P = \{p_0, p_1 \ldots p_n\}$ is a finite set of propositional symbols and $I = \{a_1, \ldots, a_m\}$ is a finite set of indices. Let $\mathcal{P}I$ be the set of nonempty subsets of I. The formulas of $[\vee\Gamma][\wedge\Gamma]ML_n^m$ are built according to the rule:*

$$\psi := \bot \mid \top \mid p \in P \mid \neg\psi \mid \psi \vee \psi \mid \psi \wedge \psi \mid [a]\psi \mid \langle a\rangle\psi \mid [\vee\Gamma]\psi \mid [\wedge\Gamma]\psi,$$

where $a \in I$ and $\Gamma \in \mathcal{P}I$.

Definition 2 ($[\wedge\Gamma]ML_n^m$, $[\vee\Gamma]ML_n^m$, and ML_n^m). *The formulas of the logic $[\wedge\Gamma]ML_n^m$ are the formulas of the logic $[\vee\Gamma][\wedge\Gamma]ML_n^m$ with the exception of all formulas of the form $[\vee\Gamma]\psi$. Similarly, the formulas of the logic $[\vee\Gamma]ML_n^m$ are the formulas of the logic $[\vee\Gamma][\wedge\Gamma]ML_n^m$ with the exception of all formulas of the form $[\wedge\Gamma]\psi$. Finally, the formulas of multimodal logic ML_n^m are the formulas of $[\vee\Gamma][\wedge\Gamma]ML_n^m$ logic with no formulas of the forms $[\wedge\Gamma]\psi$ and $[\vee\Gamma]\psi$ allowed.*

The semantics of $[\vee\Gamma][\wedge\Gamma]ML_n^m$ is given as follows.

Definition 3 (Model). *A model for the signature $S = \{P, I\}$ is a triple $M = \langle W, R, V \rangle$, where*

- *W is a set of points;*
- *$R : I \to 2^{W \times W}$ is a function that assigns a binary relation $R(a)$ on W to every $a \in I$. We write $wR_a v$ for $(w, v) \in R(a)$ and say that v can be reached from w in one a-step.*
- *$V : P \to 2^W$ is a function that assigns a subset $V(p) \subseteq W$ to every $p \in P$.*

*A model $M = \langle W, R, V \rangle$ is said to be **finite** if W is finite. Given a model $M = \langle W, R, V \rangle$, a **pointed model** is a pair (M, w), where $w \in W$. Sets of pointed models are denoted $\mathbb{M}, \mathbb{N}, \mathbb{M}_1, \mathbb{N}_1, \mathbb{M}_2, \mathbb{N}_2$, etc.*

We define the notion "formula φ is true in a pointed model (M, w)" in the usual way (see for example [2]). In particular:

$(M, w) \models \langle a \rangle \psi$ iff there is a $v \in W$ such that $wR_a v$ and $(M, v) \models \psi$;
$(M, w) \models [a]\psi$ iff $(M, v) \models \psi$ for all $v \in W$ such that $wR_a v$;
$(M, w) \models [\wedge\Gamma]\psi$ iff $(M, w) \models \bigwedge_{a \in \Gamma}[a]\psi$;
$(M, w) \models [\vee\Gamma]\psi$ iff $(M, w) \models \bigvee_{a \in \Gamma}[a]\psi$.

Given this semantics, it is obvious that $[\wedge\Gamma]ML_n^m$, $[\vee\Gamma]ML_n^m$, and $[\vee\Gamma][\wedge\Gamma]ML_n^m$ are just extensions by definition of the logic ML_n^m. In addition, if $\Gamma = \{a\}$, then both $[\wedge\Gamma]\psi$ and $[\vee\Gamma]\psi$ are the formula $[a]\psi$. Hence, in what follows, we assume that the set Γ contains at least two indices. As usual, for every pointed model (M, w) and every formula $\langle a \rangle \psi$,

$$(M, w) \models \langle a \rangle \psi \leftrightarrow \neg[a]\neg\psi.$$

From now on, if (M, w) is a pointed model, where $M = \langle W, R, V \rangle$, we write $v \in M$ instead of $v \in W$; all models and all sets of pointed models are finite. We write $\mathbb{M} \models \varphi$ to mean that for all $(M, w) \in \mathbb{M}$, $(M, w) \models \varphi$. Note that if $\mathbb{M} = \varnothing$, then for every $[\vee\Gamma][\wedge\Gamma]ML_n^m$ formula φ, it is trivially true that $\mathbb{M} \models \varphi$. We are going to use the well-known fact that if two pointed models (M, w) and (N, v) are bisimilar, then for every $[\vee\Gamma][\wedge\Gamma]ML_n^m$ formula φ, $(M, w) \models \varphi$ if and only if $(N, v) \models \varphi$ (see [2]).

We define the following operations on pointed models and sets of pointed models:

Definition 4. *Let (M, w) and $\mathbb{M} = \{(M_1, w_1), \ldots, (M_k, w_k)\}$ be a pointed model and a set of pointed models for the signature $S = \{P, I\}$. Let $a \in I$ and $\Gamma = \{a_i, \ldots, a_j\}$ be a subset of I. Then*

- $[a](M, w) = \{(M, v) \mid v \in M \text{ and } wR_a v\}$. *Intuitively, $[a](M, w)$ is the set of all pointed models that can be reached from w by making one a-step. Note that if there is no point $v \in M$ such that $wR_a v$, then $[a](M, w) = \varnothing$.*
- $[a]\mathbb{M} = [a](M_1, w_1) \cup \ldots \cup [a](M_k, w_k)$.
- $[\wedge\Gamma]\mathbb{M} = [a_i]\mathbb{M} \cup \ldots \cup [a_j]\mathbb{M}$, *i.e., $[\wedge\Gamma]\mathbb{M}$ is the union of $[a]\mathbb{M}$ for all $a \in \Gamma$.*
- *Suppose that $\mathbb{M} \models [\vee\Gamma]\psi$. Hence, for every $(M_i, w_i) \in \mathbb{M}$, there is a subset $\Gamma_i \subseteq \Gamma$ such that for every $a \in \Gamma_i$, $(M_i, w_i) \models [a]\psi$. Therefore, we can construct the set*
  $[\vee\Gamma(\psi)]\mathbb{M} = \bigcup_{a \in \Gamma_1}[a](M_1, w_1) \cup \ldots \cup \bigcup_{a \in \Gamma_k}[a](M_k, w_k)$.
- *Suppose that $(M, w) \models \langle a \rangle \psi$. Hence, there is at least one $v \in M$ such that $wR_a v$ and $(M, v) \models \psi$. We construct the nonempty set of all such pointed models, i.e., $\langle a(\psi) \rangle (M, w) = \{(M, v) \mid v \in M \text{ and } wR_a v, \text{and} (M, v) \models \psi\}$.*
- *Suppose $\mathbb{M} \models \langle a \rangle \psi$. Therefore, we can form the nonempty set of pointed models*
  $\langle a(\psi) \rangle \mathbb{M} = \langle a(\psi) \rangle (M_1, w_1) \cup \ldots \cup \langle a(\psi) \rangle (M_k, w_k)$.
- *Suppose that $\mathbb{M} \models \neg[\vee\Gamma]\psi$. Hence, for every $(M, w) \in \mathbb{M}$ and every $a \in \Gamma$, there is at least one $v \in M$ such that $wR_a v$ and $(M, v) \models \neg\psi$. We form the set $\neg[\vee\Gamma(\psi)]\mathbb{M} = \bigcup_{a \in \Gamma}\langle a(\neg\psi) \rangle (M_1, w_1) \cup \ldots \cup \bigcup_{a \in \Gamma}\langle a(\neg\psi) \rangle (M_k, w_k)$.*

- *Suppose that* $\mathbb{M} \models \neg[\wedge\Gamma]\psi$. *Hence, for every* $(M_i, w_i) \in \mathbb{M}$, *there is a nonempty subset* $\Gamma_i \subseteq \Gamma$ *such that for every* $a \in \Gamma_i$ *there is at least one* $v \in M_i$ *for which* $w_i R_a v$ *and* $(M_i, v) \models \neg\psi$. *Therefore, we can construct the nonempty set*
$$\neg[\wedge\Gamma(\psi)]\mathbb{M} = \bigcup_{a\in\Gamma_1}\langle a(\neg\psi)\rangle(M_1, w_1) \cup \ldots \cup \bigcup_{a\in\Gamma_k}\langle a(\neg\psi)\rangle(M_k, w_k).$$

2.2 Extended Syntax Trees

Every $[\vee\Gamma][\wedge\Gamma]ML_n^m$ formula φ can be represented by a syntax tree in the usual way, i.e., every leaf of the tree is labeled with an atomic symbol occurring in φ and inner nodes are labeled with Boolean connectives or modal operators occurring in φ.

The notion of extended syntax tree of a first-order formula was introduced in [5]. Extended syntax trees correspond to game trees for the Adler-Immerman games defined in [1]. The reader can think about these trees as normal syntax trees in which every node has an additional semantic label that consists of two sets of pointed models.

Definition 5 (Extended Syntax Tree). *Let* φ *be a* $[\vee\Gamma][\wedge\Gamma]ML_n^m$ *formula and let* $\mathbb{M}$ *and* $\mathbb{N}$ *be two sets of pointed models such that* $\mathbb{M} \models \varphi$ *and* $\mathbb{N} \models \neg\varphi$. *The extended syntax tree* $T_\varphi^{\langle\mathbb{M}\circ\mathbb{N}\rangle}$ *is defined inductively on the structure of* φ *as follows:*

(φ **is a propositional symbol** $p \in P$): $T_p^{\langle\mathbb{M}\circ\mathbb{N}\rangle}$ *consists of a single node* t *that has a* **syntax label** $synl(t) := p$ *and a* **semantic label** $seml(t) := \langle\mathbb{M}\circ\mathbb{N}\rangle$.

(φ **is** $\top$): $T_\top^{\langle\mathbb{M}\circ\mathbb{N}\rangle}$ *consists of a single node* t *with* $synl(t) := \top$ *and* $seml(t) := \langle\mathbb{M}\circ\mathbb{N}\rangle$.
Note that in this case $\mathbb{N} = \varnothing$.

(φ **is** $\bot$): $T_\bot^{\langle\mathbb{M}\circ\mathbb{N}\rangle}$ *consists of a single node* t *with* $synl(t) := \bot$ *and* $seml(t) := \langle\mathbb{M}\circ\mathbb{N}\rangle$.
Note that in this case $\mathbb{M} = \varnothing$.

(φ **is** $\neg\psi$): $T_{\neg\psi}^{\langle\mathbb{M}\circ\mathbb{N}\rangle}$ *has a root node* t *with* $synl(t) := \neg$ *and* $seml(t) := \langle\mathbb{M}\circ\mathbb{N}\rangle$.
The unique child of t *is the root of* $T_\psi^{\langle\mathbb{N}\circ\mathbb{M}\rangle}$. *Note that* $\mathbb{N} \models \psi$ *and* $\mathbb{M} \models \neg\psi$.

(φ **is** $\psi_1 \wedge \psi_2$): $T_{\psi_1\wedge\psi_2}^{\langle\mathbb{M}\circ\mathbb{N}\rangle}$ *has a root node* t *with* $synl(t) := \wedge$ *and* $seml(t) := \langle\mathbb{M}\circ\mathbb{N}\rangle$. *The first child of* t *is the root of* $T_{\psi_1}^{\langle\mathbb{M}\circ\mathbb{N}_1\rangle}$. *The second child of* t *is the root of* $T_{\psi_2}^{\langle\mathbb{M}\circ\mathbb{N}_2\rangle}$, *where, for* $i \in \{1,2\}$, $\mathbb{N}_i = \{(N, v) \in \mathbb{N} \mid (N, v) \models \neg\psi_i\}$.
Note that $\mathbb{N} = \mathbb{N}_1 \cup \mathbb{N}_2$, $\mathbb{M} \models \psi_1 \wedge \psi_2$, *and* $\mathbb{N}_1 \models \neg\psi_1$, $\mathbb{N}_2 \models \neg\psi_2$.

(φ **is** $\psi_1 \vee \psi_2$): $T_{\psi_1\vee\psi_2}^{\langle\mathbb{M}\circ\mathbb{N}\rangle}$ *has a root node* t *with* $synl(t) := \vee$ *and* $seml(t) := \langle\mathbb{M}\circ\mathbb{N}\rangle$.
The first child of t *is the root of* $T_{\psi_1}^{\langle\mathbb{M}_1\circ\mathbb{N}\rangle}$. *The second child of* t *is the root of* $T_{\psi_2}^{\langle\mathbb{M}_2\circ\mathbb{N}\rangle}$, *where, for* $i \in \{1,2\}$, $\mathbb{M}_i = \{(M, v) \in \mathbb{M} \mid (M, v) \models \psi_i\}$.
Note that $\mathbb{M} = \mathbb{M}_1 \cup \mathbb{M}_2$, $\mathbb{M}_1 \models \psi_1$, *and* $\mathbb{M}_2 \models \psi_2$, $\mathbb{N} \models \neg(\psi_1 \vee \psi_2)$.

(φ **is** $[a]\psi$): $T_{[a]\psi}^{\langle\mathbb{M}\circ\mathbb{N}\rangle}$ *has a root node* t *with* $synl(t) := [a]$ *and* $seml(t) := \langle\mathbb{M}\circ\mathbb{N}\rangle$.

The unique child of t is the root of $T_\psi^{\langle[a]\mathbb{M}\circ\langle a(\neg\psi)\rangle\mathbb{N}\rangle}$. Note that $[a]\mathbb{M} \models \psi$ and $\langle a(\neg\psi)\rangle\mathbb{N} \models \neg\psi$.

(φ is $\langle a\rangle\psi$): $T_{\langle a\rangle\psi}^{\langle\mathbb{M}\circ\mathbb{N}\rangle}$ has a root node t with $synl(t) := \langle a\rangle$ and $seml(t) :=$ $\langle\mathbb{M}\circ\mathbb{N}\rangle$.

The unique child of t is the root of $T_\psi^{\langle\langle a(\psi)\rangle\mathbb{M}\circ[a]\mathbb{N}\rangle}$. Note that $\langle a(\psi)\rangle\mathbb{M} \models \psi$ and $[a]\mathbb{N} \models \neg\psi$.

(φ is $[\wedge\Gamma]\psi$): $T_{[\wedge\Gamma]\psi}^{\langle\mathbb{M}\circ\mathbb{N}\rangle}$ has a root node t with $synl(t) := [\wedge\Gamma]$ and $seml(t) :=$ $\langle\mathbb{M}\circ\mathbb{N}\rangle$. The unique child of t is the root of $T_\psi^{\langle[\wedge\Gamma]\mathbb{M}\circ\neg[\wedge\Gamma(\psi)]\mathbb{N}\rangle}$. Note that $[\wedge\Gamma]\mathbb{M} \models \psi$ and $\neg[\wedge\Gamma(\psi)]\mathbb{N} \models \neg\psi$.

(φ is $[\vee\Gamma]\psi$): $T_{[\vee\Gamma]\psi}^{\langle\mathbb{M}\circ\mathbb{N}\rangle}$ has a root node t with $synl(t) := [\vee\Gamma]$ and $seml(t) :=$ $\langle\mathbb{M}\circ\mathbb{N}\rangle$.

The unique child of t is the root of $T_\psi^{\langle[\vee\Gamma]\mathbb{M}\circ\neg[\vee\Gamma(\psi)]\mathbb{N}\rangle}$. Note that $[\vee\Gamma]\mathbb{M} \models \psi$ and $\neg[\vee\Gamma(\psi)]\mathbb{N} \models \neg\psi$.

Then next proposition follows immediately from Definition 5.

Proposition 1. *For any $[\vee\Gamma][\wedge\Gamma]ML_n^m$ formula φ and any sets of pointed models $\mathbb{M}$, $\mathbb{N}$ such that $\mathbb{M} \models \varphi$ and $\mathbb{N} \models \neg\varphi$, if the root of the extended syntax tree $T_\varphi^{\langle\mathbb{M}\circ\mathbb{N}\rangle}$ has a child-node that is the root of the syntax tree $T_\psi^{\langle\mathbb{O}\circ\mathbb{P}\rangle}$, then $\mathbb{O} \models \psi$ and $\mathbb{P} \models \neg\psi$.*

Figure 1 shows the extended syntax tree $T_{[\vee\Gamma]p\wedge\langle a\rangle\langle b\rangle\top}^{\langle\mathbb{M}\circ\mathbb{N}\rangle}$ of the formula $[\vee\Gamma]p \wedge \langle a\rangle\langle b\rangle\top$, where $\mathbb{M} = \{(M_1, s_1), (M_2, s_2)\}$ and $\mathbb{N} = \{(M_3, s_3), (M_4, s_4)\}$. The lowermost black square in a model M_i that is in the semantic label of the root of the tree is the point s_i.

Definition 6 (Branches). *A branch B in an extended-syntax tree is any path leading from the root of the tree to a leaf. We denote by $I(B)$ the word $i_1 \ldots i_k$, formed by the indices of all the nodes with syntactic labels of the form $[i]$ or $\langle i\rangle$ occurring along a branch B when traversing the branch from the root to its leaf.*

For instance, the branch B (right) of the game tree of Figure 1 satisfy $I(B) = ab$. Note that given an extended syntax tree $T_\varphi^{\langle\mathbb{M}\circ\mathbb{N}\rangle}$, the "shape" of the tree depends solely on φ, e.g., if we disregard the semantic labels of the nodes, the three extended syntax trees $T_{\langle a\rangle\langle b\rangle\top\wedge[\vee\Gamma]p}^{\langle\mathbb{M}\circ\mathbb{N}\rangle}$, $T_{\langle a\rangle\langle b\rangle\top\wedge[\vee\Gamma]p}^{\langle\varnothing\circ\mathbb{N}\rangle}$, and $T_{\langle a\rangle\langle b\rangle\top\wedge[\vee\Gamma]p}^{\langle\varnothing\circ\varnothing\rangle}$ are actually the usual syntax tree of the formula $\langle a\rangle\langle b\rangle\top \wedge [\vee\Gamma]p$. Hence, the following definition is unambiguous.

Definition 7 (Formula Size). *Let φ be a $[\vee\Gamma][\wedge\Gamma]ML_n^m$ formula and let $\mathbb{M}$ and $\mathbb{N}$ be two arbitrary sets of pointed models such that $\mathbb{M} \models \varphi$ and $\mathbb{N} \models \neg\varphi$. The size of φ (denoted $||\varphi||$) is the number of verticies occurring in $T_\varphi^{\langle\mathbb{M}\circ\mathbb{N}\rangle}$.*

We enumerate the nodes of an extended syntax tree $T_\varphi^{\langle\mathbb{M}\circ\mathbb{N}\rangle}$ in increasing order starting from the root of the tree; it's children are enumerated from left to right, etc., i.e., the nodes are enumerated as in a breadth-first search algorithm.

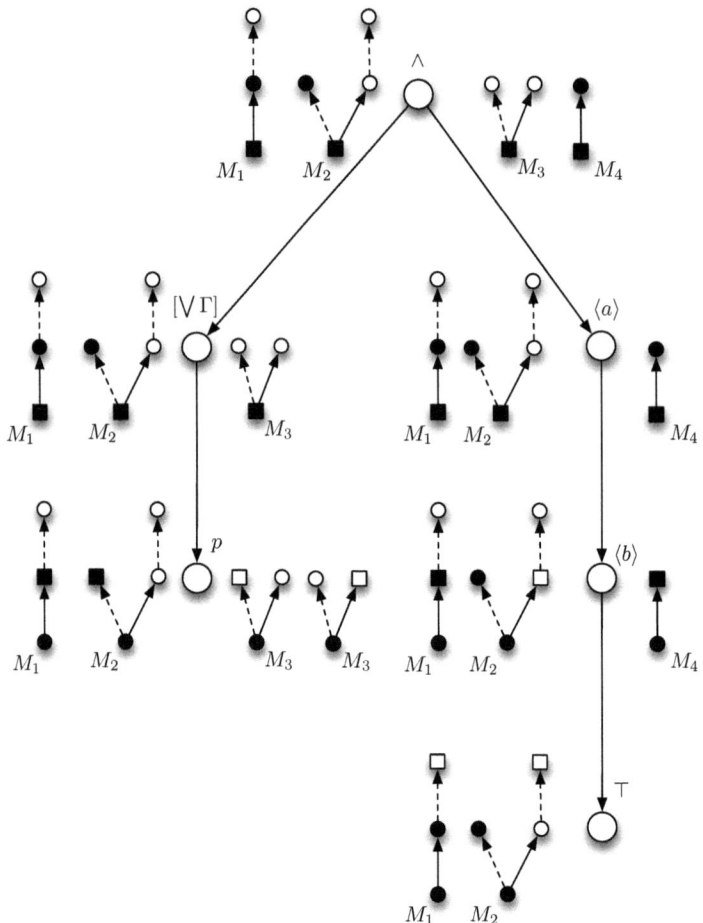

Fig. 1. The extended syntax tree $T_{[\vee\Gamma]p\wedge\langle a\rangle\langle b\rangle\top}^{\langle \mathbb{M}\circ\mathbb{N}\rangle}$, where $\Gamma = \{a, b\}$. The current points in the pointed models are denoted by a square, the rest of the points are denoted by circles. Black squares and circles denote the points where the atom p is true. The solid arrows in the models denote relation steps indexed with a, the dashed arrows are indexed with b. The big white circles denote nodes in the syntax tree.

2.3   Succinctness

Definition 8. *Let $\mathbb{C}$ be a non-empty set of models and let L_1 and L_2 be two logics. We say that L_1 is at least as expressive as L_2 on $\mathbb{C}$, written $L_2 \leq_{\mathbb{C}}^{\text{expr}} L_1$, if for every $\varphi_2 \in L_2$, there is a formula φ_1 in L_1 such that $\mathbb{C} \models \varphi_1 \leftrightarrow \varphi_2$.*

Following [5], our formal definition of succinctness is:

Definition 9 (Succinctness). *Let L_1, L_2 be two logics. Let $\mathbb{C}$ be a class of models such that $L_1 \leq_{\mathbb{C}}^{\text{expr}} L_2$. Let $\mathcal{F}$ be a class of functions $f : \mathbb{N} \to \mathbb{R}$. We say that L_1 is $\mathcal{F}$-succinct in L_2 on $\mathbb{C}$, and write $L_1 \leq_{\mathbb{C}}^{\mathcal{F}} L_2$, iff there is a function $f \in \mathcal{F}$ such that for every L_1-formula φ_1, there is an L_2-formula φ_2 for which the following is true:*

- $\mathbb{C} \models \varphi_1 \leftrightarrow \varphi_2$;
- $\|\varphi_2\| \leq f(\|\varphi_1\|)$.

Intuitively, this means that $\mathcal{F}$ gives an upper bound on the size of L_2 formulas needed to express all of L_1 on $\mathbb{C}$. Therefore, by saying that L_1 *is exponentially more succinct than L_2*, we mean $L_1 \leq_{\mathbb{C}}^{\text{expr}} L_2$ and $L_1 \not\leq_{\mathbb{C}}^{\text{SUBEXP}} L_2$, i.e., the size of formulas of L_2 expressing all of L_1 on $\mathbb{C}$ cannot be bounded from above by a sub-exponential function.

In order to prove that $L_1 \not\leq_{\mathbb{C}}^{\text{SUBEXP}} L_2$ for two logics L_1 and L_2, it is sufficient to show that there are two infinite sequences of formulas $\varphi_1, \varphi_2, \ldots$ in L_1 and $\chi_1, \chi_2, \ldots$ in L_2, and rational numbers k and t such that

1. $\|\varphi_n\| = kn + t$;
2. $\|\chi_n\| \geq 2^i$;
3. χ_n is the shortest formula in L_2 such that $\mathbb{C} \models \varphi_n \leftrightarrow \chi_n$.

3   Main Results

For a natural number $n \geq 1$, let $[\wedge \Gamma]^n$ stand for $\overbrace{[\wedge \Gamma] \ldots [\wedge \Gamma]}^{n \text{ times } [\wedge \Gamma]}$ and similarly for $[\vee \Gamma]^n$.

Definition 10. *Let $S = \{P, I\}$ be a signature where P contains at least one propositional symbol p and I contains at least two indices a and b and let $\Gamma = \{a, b\}$. For every $n \geq 1$, the $[\wedge \Gamma]ML_n^m$ formulas φ_n, the $[\vee \Gamma]ML_n^m$ formulas θ_n, and the ML_n^m formulas ψ_n and χ_n are defined as follows.*

$$\varphi_n := \neg[\wedge \Gamma]^n \neg p, \quad \psi_1 := \langle a \rangle p \vee \langle b \rangle p;$$
$$\psi_n := \langle a \rangle \psi_{n-1} \vee \langle b \rangle \psi_{n-1}, n > 1.$$
$$\theta_n := [\vee \Gamma]^n p, \quad \chi_1 := [a]p \vee [b]p;$$
$$\chi_n := [a]\chi_{n-1} \vee [b]\chi_{n-1}, n > 1.$$

It is easily seen that the formulas in the right column are equivalent to the formulas in the left column. It is obvious that the length of the latter is linear in n while the length of the former is exponential in n.

Firstly, we would like to prove that there is no sequence of $[\vee \Gamma]ML_n^m$ formulas δ_n such that φ_n is equivalent to δ_n and at the same time the length of δ_n is subexponential in n. To this end, for every $n \geq 1$, we define a set of models $\mathbb{M}^n$ such that $\mathbb{M}^n \models \varphi_n$. Then we find a set of models $\mathbb{N}^n$ such that $\mathbb{N}^n \models \neg \varphi_n$.

Finally, we prove that for every $[\vee\Gamma]ML_n^m$ formula δ such that $\mathbb{M}^n \cup \mathbb{N}^n \models \varphi_n \leftrightarrow \delta$, the extended syntax tree $T_\delta^{\langle \mathbb{M}^n \circ \mathbb{N}^n \rangle}$ has at least 2^n nodes. Intuitively, it is clear that the main difficulty in such a proof stems from the "power" of the $[\vee\Gamma]$ operator. While $(M, w) \models [a]\varphi$ means that all points reachable from w in one a-step satisfy the formula φ, $(M, w) \models [\vee\Gamma]\varphi$ means that *there is at least one index $i \in \Gamma$* such that *all* points reachable from w in one i-step satisfy φ. Therefore, if we manage to define the models in $\mathbb{M}^n$ and $\mathbb{N}^n$ in such a way as to make the $[\vee\Gamma]$ operator "useless", i.e., the possibility offered by $[\vee\Gamma]$ to non-deterministically chose relation steps is eliminated, our task will be easier.

Secondly, we would like to prove that there is no $[\wedge\Gamma]ML_n^m$ formula γ_n of subexponential length such that θ_n is equivalent to γ_n. Again, the main problem is the power of the $[\wedge\Gamma]$ modality. Guided by the same intuition, for every $n \geq 1$, we define sets of models $\mathbb{O}^n$ and $\mathbb{P}^n$ where $\mathbb{O}^n \models \theta_n$, $\mathbb{P}^n \models \neg\theta_n$, and for every $[\wedge\Gamma]ML_n^m$ formula γ such that $\mathbb{O}^n \cup \mathbb{P}^n \models \theta_n \leftrightarrow \gamma$, the syntax tree $T_\gamma^{\langle \mathbb{O}^n \circ \mathbb{P}^n \rangle}$ has at least 2^n nodes. As for $[\vee\Gamma]ML_n^m$, we define the models in $\mathbb{O}^n$ and $\mathbb{P}^n$ so that the operator $[\wedge\Gamma]$ is of no "use".

Definition 11 and items (b) and (c) from Lemma 1 are the formalization of this idea.

Definition 11 (Tree models). *Let the signature $S = \{P, I\}$ be as in Definition 10. Figure 2 shows the sets of models $\mathbb{M}^n, \mathbb{N}^n, \mathbb{O}^n$, and $\mathbb{P}^n$.*

The tree-like models in $\mathbb{M}^1, \mathbb{N}^1, \mathbb{O}^1, \mathbb{P}^1$ are built as shown. The models in $\mathbb{O}^{n+1}$ and $\mathbb{P}^{n+1}$, for example, are defined recursively by taking a model from $\mathbb{O}^1$, erasing the propositional symbols, and then using the leaves of the tree as roots for the models from $\mathbb{O}^n$ and $\mathbb{P}^n$ as shown[1]. The same strategy is employed in the construction of the models in $\mathbb{M}^{n+1}$ and $\mathbb{N}^{n+1}$.

We denote the root of any of the models in $\mathbb{M}^n$, $\mathbb{N}^n, \mathbb{O}^n$, $\mathbb{P}^n$ by r. For any model M_{xk}^{n+1}, such that $x \in \{a, b\}$, the pair (M_{xk}^{n+1}, M_k^n) stands for the pointed model (M_{xk}^{n+1}, r), where r is the root of M_k^n. Similarly for (M_{xk}^{n+1}, N^n) and (N^{n+1}, N^n).

Lemma 1. *Let the sequences of formulas ϕ_n, θ_n, ψ_n, and χ_n be defined as in Definition 10 and let $\mathbb{M}^n$, $\mathbb{N}^n$, $\mathbb{O}^n$, and $\mathbb{P}^n$ be as in Definition 11. For every n, and all sets of pointed models $\$$ and $\mathbb{T}$:*

(a) $\mathbb{M}^n \models \varphi_n$, $\mathbb{N}^n \models \neg\varphi_n$, $\mathbb{O}^n \models \theta_n$, $\mathbb{P}^n \models \neg\theta_n$;

(b) *For any (M_w^n, r), there is neither a formula $[\vee\Gamma]\psi$ such that $(M_w^n, r) \models [\vee\Gamma]\psi$ and $(N^n, r) \models \neg[\vee\Gamma]\psi$ nor a formula $[\vee\Gamma]\psi$ such that $(N^n, r) \models [\vee\Gamma]\psi$ and $(M_w^n, r) \models \neg[\vee\Gamma]\psi$.*

(c) *For any (O_w^n, r), there is neither a formula $[\wedge\Gamma]\psi$ such that $(O_w^n, r) \models [\wedge\Gamma]\psi$ and $(P_w^n, r) \models \neg[\vee\Gamma]\psi$ nor a formula $[\vee\Gamma]\psi$ such that $(P_w^n, r) \models [\wedge\Gamma]\psi$ and $(O_w^n, r) \models \neg[\vee\Gamma]\psi$.*

---

[1] Intuitively, the subscript in the name of the model O_{ak}^{n+1} encodes a path (starting with an a-step) leading from the root of the tree to a leaf satisfying the proposition p. The same path in the model P_{ak}^{n+1} leads to a point that does not satisfy p. Apart from this difference, the models O_{ak}^{n+1} and P_{ak}^{n+1} look the same.

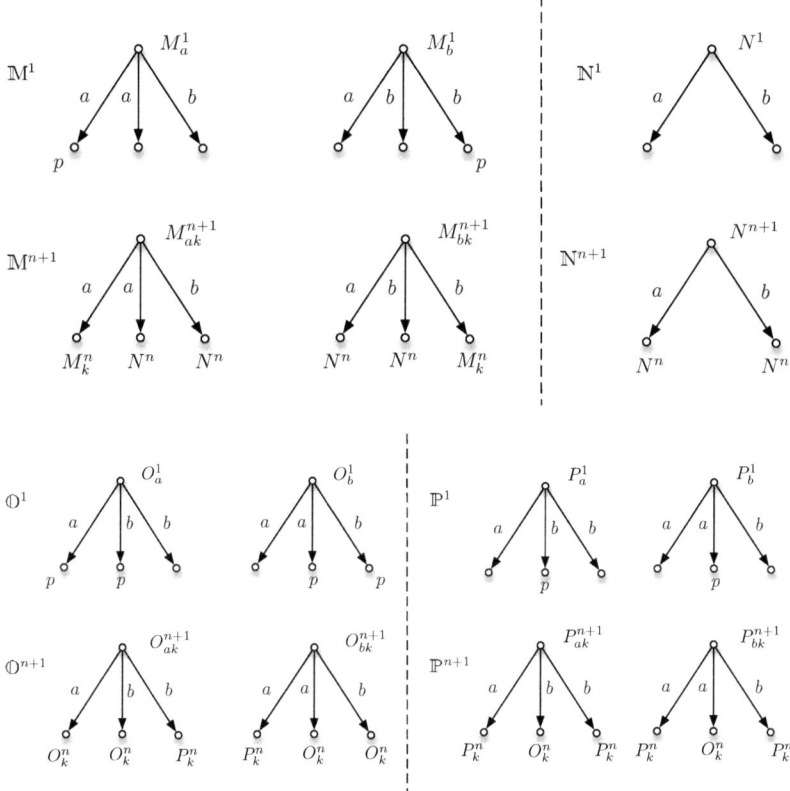

Fig. 2. The sets of models $\mathbb{M}^n$, $\mathbb{N}^n$, $\mathbb{O}^n$, and $\mathbb{P}^n$

(d) *For any* $[\vee\Gamma]ML_n^m$ *formula* φ *such that* $\$ \cup \{(M_w^n, r)\} \models \varphi$ *and* $\mathbb{T} \cup \{(N^n, r)\} \models \neg\varphi$, *the extended syntax tree* $T_\varphi^{\langle \$ \cup \{(M_w^n, r)\} \circ \{(N^n, r)\} \cup \mathbb{T} \rangle}$ *has a branch* B, *where* $I(B) = w$.

(e) *For any* $[\wedge\Gamma]ML_n^m$ *formula* φ *such that* $\$ \cup \{(O_w^n, r)\} \models \varphi$ *and* $\{(P_w^n, r)\} \cup \mathbb{T} \models \neg\varphi$, *the extended syntax tree* $T_\varphi^{\langle \$ \cup \{(O_w^n, r)\} \circ \{(P_w^n, r)\} \cup \mathbb{T} \rangle}$ *has a branch* B, *where* $I(B) = w$.

Proof

(a) It is easily seen that $\mathbb{M}^n \models \psi_n$, $\mathbb{N}^n \models \neg\psi_n$, $\mathbb{O}^n \models \chi_n$, $\mathbb{P}^n \models \neg\chi_n$.

(b) We prove only the case $n > 1$ and $w = ak$. The case $w = bk$ or $n = 1$ is analogous. Assume there is a formula $[\vee\Gamma]\psi$ such that $(M_{ak}^n, r) \models [\vee\Gamma]\psi$ and $(N^n, r) \models \neg[\vee\Gamma]\psi$. Consider the extended syntax tree $T_{[\vee\Gamma]\psi}^{\langle \{(M_{ak}^n, r)\} \circ \{(N^n, r)\} \rangle}$. Using the last item from Definition 5, we see that there must be two points s and t in N^n such that $rR_a s$ and $(N^n, s) \models \neg\psi$ and $rR_b t$ and $(N^n, t) \models \neg\psi$. There are only two such points and they are the roots of the models N^{n-1}.

On the other hand, we must have at least one of the following $(M_{ak}^n, r) \models [a]\psi$
or $(M_{ak}^n, r) \models [b]\psi$. This however is impossible, since in the first case there is
a $z \in M_{ak}^n$ such that $rR_a z$ and z is the root of the tree N^{n-1}; whereas, in the
second case, there is a u such that $rR_b u$ and u is the root of the model N^{n-1}.
In either case, the root of $T_{[\vee\Gamma]\psi}^{\langle\{(M_{ak}^n,r)\}\circ\{(N^n,r)\}\rangle}$ must have a child of the form
$T_\psi^{\langle\$\cup\{(M_{ak}^n,N^{n-1})\}\circ\{(N^n,N^{n-1})\}\cup\mathbb{T}\rangle}$ which is impossible since, using Proposition
1, we have two bisimilar models, namely (M_{ak}^n, N^{n-1}) and (N^n, N^{n-1}), for
which $(M_{ak}^n, N^{n-1}) \models \psi$ and $(N^n, N^{n-1}) \models \neg\psi$. The assumption that there
is $[\vee\Gamma]\psi$ such that $(N^n, r) \models [\vee\Gamma]\psi$ and $(M_w^n, r) \models \neg[\vee\Gamma]\psi$ leads to a
contradiction in a similar way.

(c) The assumption that there is such a formula plus the rule for formulas of the
form $[\wedge\Gamma]\psi$ from Definition 5, leads to a contradiction as in item(b) above.

(d) Let φ, $\$$, and $\mathbb{T}$ be as described and let $T_\varphi^{\langle\$\cup\{(M_w^n,r)\}\circ\{(N^n,r)\}\cup\mathbb{T}\rangle}$ be the extended syntax tree. Suppose $n > 1^2$. We consider only the case $w = ak$. The
case $w = bk$ is similar. Let rt denote the root of the tree. It is obvious that
the syntax label of rt cannot be one of the following "$\top$", "$\bot$", "p", and item
(b) implies that $synl(rt)$ cannot be "$[\vee\Gamma]$". Therefore, $synl(rt)$ is either "$\neg$",
"$\vee$", "$\wedge$", "$[i]$" or "$\langle i\rangle$" for $i \in \{a, b\}$. If $synl(rt)$ is one of the Boolean connectives, then rt will have as a child the root of a tree $T_\psi^{\langle\$_1\cup\{(M_{ak}^n,r)\}\circ\{(N^n,r)\}\cup\mathbb{T}_1\rangle}$
or $T_\psi^{\langle\mathbb{T}_1\cup\{(N^n,r)\}\circ\{(M_{ak}^n,r)\}\cup\$_1\rangle}$ for some sub-formula ψ of φ. The reasoning
above shows that the root node rt_1 of this tree cannot have a syntactic label
of the form "$\top$", "$\bot$", "p", or "$[\vee\Gamma]$". Hence, $T_\varphi^{\langle\$\cup\{(M_{ak}^n,r)\}\circ\{(N^n,r)\}\cup\mathbb{T}\rangle}$ must
contain at least one node that is the root of a tree $T_\psi^{\langle\$_l\cup\{(M_{ak}^n,r)\}\circ\{(N^n,r)\}\cup\mathbb{T}_l\rangle}$
or $T_\psi^{\langle\mathbb{T}_l\cup\{(N^n,r)\}\circ\{(M_{ak}^n,r)\}\cup\$_l\rangle}$ for some sub-formula ψ of φ, that has one of
the forms $[a]\theta$, $\langle a\rangle\theta$, $\langle b\rangle\theta$ or $[b]\theta$. Let n be the first such node in the enumeration of $T_\varphi^{\langle\$\cup\{(M_{ak}^n,r)\}\circ\{(N^n,r)\}\cup\mathbb{T}\rangle}$. It is easy to see that for all formulas $\langle b\rangle\theta$
and $[b]\theta$, $(M_{ak}^n, r) \models \langle b\rangle\theta$ if and only if $(N^n, r) \models \langle b\rangle\theta$, and $(M_{ak}^n, r) \models [b]\theta$
if and only if $(N^n, r) \models [b]\theta$. Hence, ψ is either of the form $[a]\theta$ or $\langle a\rangle\theta$.
Consider the tree $T_\psi^{\langle\$_l\cup\{(M_{ak}^n,r)\}\circ\{(N^n,r)\}\cup\mathbb{T}_l\rangle}$. In this case, ψ is not of the
form $[a]\theta$ since the root of the tree would contain a child with semantic label
$\$_l \cup \{(M_{ak}^n, N^{n-1})\} \circ \{(N^n, N^{n-1})\} \cup \mathbb{T}_l\rangle$. Given the fact that (M_{ak}^n, N^{n-1})
and (N^n, N^{n-1}) are bisimilar, we arrive at a contradiction using Proposi-
tion 1. Hence ψ has the form $\langle a\rangle\theta$; moreover, the bisimilarity of (M_{ak}^n, N^{n-1})
and (N^n, N^{n-1}) implies that the root of the tree can only have a child with
semantic label $\langle\$_{l+1} \cup \{(M_{ak}^n, M_k^{n-1})\} \circ \{(N^n, N^{n-1})\} \cup \mathbb{T}_{l+1}\rangle$. It is obvious
that this reasoning can be repeated until the branch B such that $I(B) = ak$
is constructed. In the case of $T_\psi^{\langle\mathbb{T}_l\cup\{(N^n,r)\}\circ\{(M_{ak}^n,r)\}\cup\$_l\rangle}$, similar considera-
tions show that ψ has the form $[a]\theta$ and that there is a branch B such that
$I(B) = ak$.

(e) The proof is analogous to the proof of item (d).

---

2 The proof of the case $n = 1$ will become clear from the proof for $n > 1$.

Theorem 1. *Let* $\mathbb{M}$ *denote the union of all* $\mathbb{M}_n$ *and* $\mathbb{N}_n$ *models. Let* $\mathbb{O}$ *denote the union of all* $\mathbb{O}_n$ *and* $\mathbb{P}_n$ *models. Then*

(1) $[\wedge\Gamma]ML_n^m \not\leq_{\mathbb{M}}^{\mathrm{SUBEXP}} [\vee\Gamma]ML_n^m$;

(2) $[\vee\Gamma]ML_n^m \not\leq_{\mathbb{O}}^{\mathrm{SUBEXP}} [\wedge\Gamma]ML_n^m$.

Proof. We prove (2). The proof of (1) is similar. Item (e) from Lemma 1 implies that the syntax tree $T_\delta^{\langle \mathbb{O}^n \circ \mathbb{P}^n \rangle}$ of any $[\wedge\Gamma]ML_n^m$ formula δ such that $\mathbb{O}^n \models \delta$ and $\mathbb{P}^n \models \neg\delta$ contains 2^n different branches. Hence, any $[\wedge\Gamma]ML_n^m$ formula δ such that $\mathbb{O} \models [\vee\Gamma]^n p \leftrightarrow \delta$ has size at least 2^n. Since $[\vee\Gamma]^n p$ is equivalent to the formula χ_n from Definition 10, we see that $[\vee\Gamma]ML_n^m$ is exactly exponentially more succinct than $[\wedge\Gamma]ML_n^m$.

4   Conclusion and Open Questions

As mentioned in the introduction, the results presented here are relevant to epistemic logic. One disadvantage of such an interpretation is that all relations in our models are neither reflexive, nor symmetric, nor transitive, whereas the semantics of epistemic logic is usually given via models in which all relations are relations of equivalence. We believe that our results can be extended to this class of models as well, although we conjecture that the proofs will be technically more challenging than ours.

On a more general note. The study of representational succinctness is a vastly underdeveloped subject and many of the known results rely on unproven computational complexity conjectures (see for example [4]). Improving the proofs of at least some these results so that they are not dependent on such conjectures will greatly increase our understanding of this phenomenon.

References

1. Adler, M., Immerman, N.: An n! lower bound on formula size. ACM Transactions on Computational Logic 4(3), 296–314 (2003)
2. Blackburn, P., de Rijke, M., Venema, Y.: Modal logic. Cambridge University Press (2001)
3. French, T., van der Hoek, W., Iliev, P., Kooi, B.: Succinctness of epistemic languages. In: Proceedings of the 22nd International Joint Conference on Artificial Intelligence (IJCAI), pp. 881–886 (2011)
4. Gogic, G., Papadimitriou, C., Selman, B., Kautz, H.: The comparative linguistics of knowledge representation. In: Proceedings of the 14th International Joint Conference on Artificial Intelligence (IJCAI), pp. 862–869 (1995)
5. Grohe, M., Schweikardt, N.: The succinctness of first-order logic on linear orders. In: Proceedings of the 19th Annual IEEE Symposium on Logic in Computer Science (LICS), pp. 438–447 (2004)
6. Lutz, C.: Complexity and succinctness of public announcement logic. In: Proceedings of the 5th International Joint Conference on Autonomous Agents and Multiagent Systems (AAMAS), pp. 137–143 (2006)

On Immortal Configurations in Turing Machines

Emmanuel Jeandel

Laboratoire d'Informatique, de Robotique et de Microélectronique de Montpellier,
UMR 5506—CC 477, 161 rue Ada, 34095 Montpellier, Cedex 5, France
jeandel@lirmm.fr

Abstract. We investigate the immortality problem for Turing machines and prove that there exists a Turing Machine that is immortal but halts on every recursive configuration. The result is obtained by combining a new proof of Hooper's theorem [11] with recent results on effective symbolic dynamics.

In this paper we investigate the behaviour of Turing machines seen as dynamical systems. In this context, a Turing machine does not start from a specified initial state and tape, but from any configuration. In this context, we call a configuration *immortal* if the machine runs forever starting from it. The seminal article of Hooper [7] proves that we cannot decide whether a Turing machine has an immortal configuration (is immortal). However, the result does not say anything about what the immortal configurations look like. In fact, in the construction by Hooper, if the Turing machine has an immortal configuration, then it has one where the tape is almost completely empty.

A few results give some structure on the set of immortal configurations: Kurka [12] asked whether there always exists a (temporally) *periodic* configuration, and was refuted by Blondel et al. [1]. Delvenne and Blondel [6] proved that the set of immortal configurations, if nonempty, always contains temporally quasi-periodic configurations.

In this article, we go further, and give some news results on the set of immortal configurations. We prove in particular in Theorem 2 that there exists a Turing machine for which the set of immortal configurations is nonempty and contains no computable configurations.

The main ingredient is a new proof of Hooper's theorem by Kari and Ollinger [11] combined with an encoding into subshifts of an effective set with no recursive points. We first define all relevant vocabulary in the next section, then proceed to the proof.

1 The Immortality Problem

1.1 Definitions

We first give some definitions and properties of Turing machines. Unless specified, our Turing machines will have only one biinfinite tape. We use the

S.B. Cooper, A. Dawar, and B. Löwe (Eds.): CiE 2012, LNCS 7318, pp. 334–343, 2012.
© Springer-Verlag Berlin Heidelberg 2012

moving-tape convention: the head is always at position 0 and the tape, rather than the head, is moving. In this context, the state of the system can be described only by the state of the Turing machine and the content of the tape.

Let Σ be the alphabet of the tape, and S the set of states of the Turing machine M. A *configuration* (also called instantaneous description) is an element of $\mathcal{C} = S \times \Sigma^{\mathbb{Z}}$. The Turing machine can now be seen as a partial function T (as an halting configuration has no image) on $\mathcal{C}$.

A configuration $c \in \mathcal{C}$ is *immortal* for M if M runs forever starting from the configuration c. We denote by $I(M)$ the immortal configurations (if they exist) of M. $I(M)$ is therefore the largest set on which the Turing machine map T is total, and $(I(M), T)$ is therefore a dynamical system.

Note in particular that a configuration is not only a tape, but the combination of a tape and a state. As a consequence, one cannot restrict itself to the dynamics of the Turing machine starting from some specific state, but must take into account computations starting from *any* state.

The first known result is given by Hooper:

Theorem 1 ([7]). *There is no algorithm that decides, given a Turing Machine M, whether $I(M)$ is empty.*

It is important to note that the proof of Hooper and subsequent proofs [11] do not give any insights into the structure of $I(M)$. Indeed, in these proofs, $I(M)$, if not empty, always contains a *finite* configuration (ie every cell of the tape, except a finite number, contains the same symbol). We shall prove here that $I(M)$ can be more intricate, and in particular might contain only nonrecursive configurations.

1.2   Effective Sets

To understand the computational properties of $I(M)$, we need the following notion:

Definition 1. *A subset $S \subseteq \{0,1\}^{\mathbb{N}}$ is an* effective set *(also called a Π_1^0-set) if there exists an oracle Turing machine M so that S is exactly the set of oracles on which the Turing machine, starting from the initial state and the empty tape, runs forever.*

There exist many equivalent definitions. For example, S is effective if there exists a recursive set of finite words L so that S is exactly the set of infinite words containing no prefix in L. There exists an extensive literature on effective sets, and we refer the reader to [3,4].

Now, up to a slight (recursive) encoding of the set of configurations $\mathcal{C}$, $I(M)$ can be seen as a subset of $\{0,1\}^{\mathbb{N}}$, and it is quite clear it is effective: it is easy to build a Turing machine that, using $c \in \mathcal{C}$ as on oracle, simulates M on input c.

From the fact that $I(M)$ is effective, we already obtain many properties about its structure, cf., e.g., the basis theorems in [4]. We also know [9] that there exist (nonempty) effective sets with no recursive points (A point $w \in \{0, 1\}^{\mathbb{N}}$ is recursive if there is a Turing machine that outputs w_n given n).

So now, the question is as follows: Are the immortal sets as rich as the effective sets? To answer this question, we shall look at the Turing degree of points of an immortal set. If x and y are two infinite words (or configurations), we say that $x \leq_T y$ if there exists a Turing machine that outputs x given y as an oracle. The Turing degree of a word is then its equivalence class for the relation $\leq_T \cap \geq_T$. The degree of recursive points is usually denoted $\mathbf{0}$.

Our first observation is that there is no way to encode any effective sets into immortal sets, preserving, e.g., the Turing degrees, due to the following lemma:

Lemma 1 ([12]). *If $I(M)$ is nonempty, then one of the following is true:*

- *it contains a configuration $c \in I(M)$ so that M, starting from c, never reads the symbols in any position $i < 0$ of the tape of c.*
- *it contains a configuration $c \in I(M)$ so that M, starting from c, never reads the symbols in any position $i > 0$ of the tape of c.*

A reformulation is that a Turing machine with moving tape is never positively expansive, see [12]. As a consequence, take the configuration $c \in I(M)$ given by the lemma for which the Turing Machine, wlog, never reads any cell in any position $i < 0$ of c. Then all configurations $c' \in \mathcal{C}$ identical with c on position $i \geq 0$ are also immortal configurations. Choosing the other bits of c' wisely proves we can find in $I(M)$ configurations of any Turing degree greater than the degree of (the right part of) c.

Proposition 1. *If $I(M)$ is nonempty, there exists a Turing degree d so that $I(M)$ contains configurations of any Turing degree above d.*

As there exist nonempty effective sets where any two different points have incomparable Turing degrees [9], there exist effective sets that cannot be encoded as immortal sets. So we need a weaker definition. The good notion is Muchnik equivalence [17]:

Definition 2. *Two subsets S_1 and S_2 of $\{0, 1\}^{\mathbb{N}}$ are Muchnik equivalent if for every $x_1 \in S_1$, there is a point $x_2 \in S_2$ computable with oracle x_1, and conversely.*

Intuitively, two sets S_1 and S_2 are Muchnik equivalent if they contain the same "minimal" Turing degrees. In particular, S_1 contains a recursive point if and only if S_2 contains a recursive point. It is important to note that the transform from x_1 into x_2 need not to be uniform in x_1, that is the Turing machine transforming x_1 into x_2 may depend on x_1. When the transform is uniform, we speak of *Medvedev* equivalence. See [17] for more information on mass problems and Muchnik and Medvedev equivalence.

We now can state our result:

Theorem 2 (Main Result). *For every effective set S, there exists a Turing machine M so that $I(M)$, the set of immortal configurations of M, is Muchnik equivalent to S.*

Corollary 1. *There exists a Turing machine M so that $I(M)$ is nonempty and contains only nonrecursive configurations.*

1.3 Effective Subshifts

Before going to the proof, we need another ingredient, coming from symbolic dynamics. To introduce this notion, we look at *traces* of Turing machines: For a configuration c, the *trace* of c is the word $u \in (\Sigma \times S)^{\mathbb{N}}$ where u_i contains the letter in position 0 of the tape and the state at the i-th step during the execution of M on input c. The trace is well defined only on configurations $c \in I(M)$ (otherwise u would be finite). Let $T(M)$ be the set of traces of M.

Now we look at the map from c to its trace $u(c)$. First of all, it is clear that $u(c)$ is computable from c. Furthermore, we can reconstruct the cells of the tape of c read by the Turing machine from $u(c)$ (which, depending of c, might or not be all the cells). In particular, there exists a configuration d so that $u(d) = u(c)$ that is computable in $u(c)$: Let $I = [a, b] \subseteq \mathbb{N}$ (possibly with $a = -\infty$ or $b = +\infty$) be the set of cells of c that are visited during the computation of M, and reconstruct the tape of d on I using the trace $u(c)$ and take all other cells to be s for some arbitrary symbol $s \in \Sigma$. In particular, we just have proven that $I(M)$ and $T(M)$ are Muchnik-equivalent. (It is important to note that the interval I is not always computable given $u(c)$: therefore the transformation from $u(c)$ to d is not uniform in $u(c)$. (In technical terms, we obtain here only a Muchnik equivalence, and not a Medvedev-equivalence).

$T(M)$ has interesting properties. It is an effective set: by a compactness argument, $u \in T(M)$ if for every length $n > 0$, there exists a configuration $c \in \mathcal{C}$ so that u coincides with the (possibly finite) trace of c on positions $i < n$. $T(M)$ is also closed under shift: If $u \in T(M)$ then $\sigma(u)$ defined by $\sigma(u)_i = u_{i+1}$ is also in $T(M)$. This means $T(M)$ is what is called an *effective subshift*:

Definition 3. *A subset $S \subseteq \Sigma^{\mathbb{N}}$ is an effective (right-sided) subshift if it is effective and closed under shift.*

An equivalent definition is as follows:

Definition 4. *Let $L \subseteq \Sigma^\star$. We denote by $S(L), S^+(L), S^-(L)$ respectively the set of biinfinite , right infinite, left infinite words over Σ that contain no word in L as a factor (substring).*

Then $S \subseteq \Sigma^{\mathbb{Z}}$ (resp. $\Sigma^{\mathbb{N}}, \Sigma^{\mathbb{Z}^-}$) is an effective twosided (resp. right-sided, left-sided) subshift if $S = S(L)$ (resp. $S^+(L), S^-(L)$) for some recursive language L.

Subshifts are an important tool to understand dynamical systems, cf., e.g., [13]. *Effective* subshifts is the computable counterpart of subshifts, and has received increasing attention in recent years [14,18,2]. In particular we obtained a result similar to Proposition 1 for subshifts in [8].

Based on these properties, it is reasonable to try to encode an effective subshift, rather than an effective set, as an immortal set. To be able to do this, we need the following:

Theorem 3 ([14]). *For every effective set S, there exists a language $L \subseteq \{0,1\}^\star$ so that $S(L)$ (resp. $S^+(L), S^-(L)$) is Muchnik-equivalent to S.*

(Proposition 3.1 in [14] is given only for $S^+(L)$ but it is not hard to see it works for left-sided and two-sided subshifts as well).

2   Proof of the Main Result

Now we are able to explain how the proof will work. We shall start from an effective subshift $S(L)$ given by a recursive set L, and code it into a Turing machine. The machine will have two tracks: The second one will be read-only and contain a biinfinite word w, and the machine will try to prove that $w \in S(L)$. However due to Lemma 1, the Turing machine might on some inputs prove only that some infinite prefix (resp. suffix) of w is in $S^-(L)$ (resp. $S^+(L)$).

To do this, we shall start from the proof of the undecidability of the immortality problem by Kari and Ollinger [11], and explain how to modify it for our purposes. In particular, a thorough examination of [11] by the reader is encouraged. We shall try as much as possible to use the same notations.

2.1   Counter Machines

Usual proofs of the undecidability of the immortality problem usually start with a counter machine. There will be no difference here. However note that we need these machines to accept tapes, i.e., infinite words. For this to make sense, we shall consider oracle counter machines. To simplify the definition, we shall suppose that the oracle is over the binary alphabet $\{0,1\}$.

In an oracle counter machine, one of the counters (the first here, for reasons soon to be apparent) is used to represent the position inside the oracle. Informally, an oracle counter machine thus contains three types of instructions: lookup instructions (test if a counter is nonzero), oracle lookup (test if the letter of the oracle is nonzero) and modifying instructions (increase/decrease a counter).

We now define it formally. Let $\Phi = \{-1,0,1\}$ and $\Upsilon = \{0,1\}$.

An oracle k-counter machine is given by a tuple (S, k, T) where S is a finite set of states, $k \in \mathbb{N}$ is the number of counters, and $T \subseteq S \times \{1,\ldots,k,o\} \times \Upsilon \times \Phi \times S$ the transition relation.

Let w be an infinite word over $\{0,1\}$ that will be used as oracle for the machine. A configuration of a k-counter machine is an element of $S \times \mathbb{N}^k$. An instruction (s,i,u,v,s') can be applied only on configurations of the form (s,c) and will lead to (s',c') (denoted by $(s,c) \vdash (s',c')$) with respect to the following rules:

- If $i \neq o$, the instruction can only be applied if $u = \min(1, c_i)$. That is, u is the result of the test whether the i-th counter is empty. If $i = o$, the instruction can only be applied if $w_{c_1} = u$, that is if the letter in position c_1 of w is u.
- The instruction will then update the counter c_i and the state depending on v. If $i = o$, it does nothing ($c' = c$). Otherwise $c'_j = c_j$ for $j \neq i$ and $c'_i = c_i + v$. Note that it is not allowed to decrease the value of a counter that is already zero.

Note that the first counter plays a special role. A counter machine computes by applying instructions. If at some point there is no instruction that can be applied, then the machine halts. We shall only be interested in deterministic counter machines (DCM), for which for every state s, there is at most two tuples $(s, i, u, v, s') \in T$, and these tuples share the same i and differ on u.

Let s_0 be a special (initial) state of the counter machine. We say that a word $w \in \{0, 1\}^{\mathbb{N}}$ is accepted by a counter machine if, starting from $(s_0, 0^k)$, the computation of the counter machine never halts. If we follow the usual encoding of Turing machines into 3-counter machines [15], we can prove easily:

Lemma 2. *For every effective subset $S \subseteq \{0, 1\}^{\mathbb{N}}$, there exists an oracle 4-counter machine that accepts exactly S.*

We shall need this machine to be *reversible*. Informally, a DCM M is reversible (RCM) if there exists a DCM M' so that for every oracle w, $(s, c) \vdash (s', c')$ by M iff $(s', c') \vdash (s, c)$ by M', see [11] for a syntactical definition. Any DCM can be extended into a RCM by using two additional counters to store the previous instructions, so that we have

Lemma 3. *For every effective subset $S \subseteq \{0, 1\}^{\mathbb{N}}$, there exists a oracle reversible 6-counter machine that accepts exactly S.*

Finally we add to this machine a new counter, that increases every two instructions. This ensures that every infinite computation of the counter machine, regardless of the first configuration, will never be periodic. We thus obtain an oracle reversible 7(!)-counter machine.

2.2   Turing Machines

We now explain how the clever construction of Kari and Ollinger works, starting from an ordinary 2-counter machine. We shall in the next section explain how to extend it for our purpose.

We begin by a generic simulation of a counter machine by Turing machines.

Let M be a 2-counter machine. Let $\Gamma = \{@, 0, x, y\}$ be a set of 4 different symbols.

A naive but effective way to encode these machines into Turing machines can be described as below: The state of the Turing machine will contain the state of the counter machine, and the tape will contain the word $@0^{c_1}x0^{c_2}y$ where c_i is the value of the i-th counter.

@0000000000000x0000000000000y
^

Now the behaviour of the Turing machine is quite clear: When it is at the position of @, it will scan the tape until x/y, deducing whether the i-th counter is zero, and changing it if necessary (which might need to shift the counters of indices greater than i), then coming back to @.

The main problem of this simulation is that it always has immortal uninteresting configurations, corresponding to unbounded searches for one of the delimitor symbols: If the symbol is not present in the configuration, the search will be infinite.

To prevent this problem, the idea, originally from Hooper [7], is to use recursive calls. Suppose we are searching for the symbol a:

@000000000000x0000000000000y
^

  ?a

If the symbol a is not found in the next 3 cells of the tape, then we write @xy over the next 3 symbols of the tape, then *recursively* call the Turing machine.

@@xy000000000x0000000000000y
^

  s0

When (if) this nested simulation stops by reading the symbol a, the entire nested simulation is erased by doing it in *reverse*. We then can continue the search for a starting from the next symbol[1]

@000000000000x0000000000000y
^

  ?s

We refer the reader to [11] for more details. In particular we need to change the first symbol @ into many different symbols to be able to keep into memory the state of the Turing machine before the recursive call.

An important feature of this construction is the following. If the counter machine has no periodic configuration (which happens as soon as one counter keeps increasing during the computation), then any computation of the Turing machine will contain computations of the counter machine starting from (almost) *anywhere*, in the following sense:

Proposition 2. *Let c be an immortal configuration of M. Then there exists an infinite interval I with the following property: for every $i \in I$, and every n so that $i + n \in I$, there exists a time during the computation of the Turing machine when the cells from i to $i + n$ contains the word @$xy0^{n-2}$, the head is in position i, and a recursive computation is started.*

---

[1] The construction in [11] actually starts the search from three symbols to the right (as we already know there is no symbol s in the next two symbols). It is clear that this does not change anything for their proof, but make Proposition 2 below true.

2.3   The Main Construction

We now explain how the construction can be extended to work in our context. As explained above, we may see the Turing machine as having two tracks (but only one head): the first one has the original construction, and the other one a biinfinite word w over $\{0,1\}$. When the counter machine starts a computation from position i, it will try to accept the infinite word $u \in \{0,1\}^{\mathbb{N}}$ defined by $u_j = w_{i+j}$. Note that reading the letter of the oracle is quite easy: the position of the letter x is always where we want to read the oracle (that's why we chose the first counter to contain the position of the oracle)

Dealing with 7 counters instead of 2 requires no additional machinery, so we are nearly done.

However one problem remains. Suppose we start from a RCM recognizing the language $S \subseteq \{0,1\}^{\mathbb{N}}$. We now look at an infinite run of the Turing machine and we look at the interval I defined by Proposition 2. There are two cases for I:

- $I = [a, \infty[$ (possibly with $a = -\infty$) In this case, for all $i \geq a$, there exist arbitrary large computation starting from the cell i. This means that for all $i \geq a$, the word $(u_j)_{j \geq i} \in S$.
- $I = [-\infty, a[$. In this case, we can say nothing: we cannot find any position i for which we are certain that the word $(u_j)_{j \geq i}$ is in S.

We thus have to use a last additional trick to make the whole thing work: during the recursive call of the left searches, instead of using the same machinery, we shall use another reversible counter machine, and run it *in the opposite direction*. That is, for the 2-counter machine, we shall start, e.g., from $Y X \mathbf{c}$ and the simulation will extend to the left, rather than the right.

Now this new construction starts from two reversible counter machines R_1 and R_2 and has the following property (the lemma is given for a oracle 2-counter machine. The change for a 7-counter machine is obvious):

Lemma 4. *Let c be an immortal configuration for the Turing Machine M. Then there exists an infinite interval $I = [a, b]$ (with $a = -\infty$ or $b = \infty$) with the following property*

- *for every $i \in I$, and every $n \geq 0$ so that $i + n \in I$, there exists a time during the computation of the Turing machine when the cells of the first track from i to $i + n$ contain the word $\mathbf{c} x y 0^{n-2}$, and a computation from R_1 is started from i.*
- *for every $i \in I$, and every $n \geq 0$ so that $i - n \in I$, there exists a time during the computation of the Turing machine when the cells from $i - n$ to i contain the word $0^{n-2} Y X \mathbf{c}$, and a computation from R_2 is started from i.*

We can now use all this refined construction to finally prove our main result:

Theorem 2 (Main Result). *For every effective set S, there exists a Turing machine M so that $I(M)$, the set of immortal configurations of M, is Muchnik equivalent to S.*

Proof. We start from S, and use theorem 3 to obtain a recursive language L so that the three subshifts defined by L are Muchnik equivalent to S. Now let R_1 and R_2 be two RCM that recognize respectively words with no factor in L and words with no factor in the mirror of L, and M be the Turing machine simulating R_1 and R_2 as above.

We now look at the immortal configurations for M.

- Let w be an biinfinite word avoiding L. Now it is clear that the configuration containing w on the second track and $@xy$ on the first track, with the head aligned with $@$, is an immortal configuration. Indeed, all factors of w (resp. of its mirror) are not in L (resp. its mirror), so that all computations of R_1 and R_2 will not halt. Hence, for every $w \in S(L)$, there exists an immortal configuration $c \in I(M)$ so that c is computable in w. In particular for every $u \in S$, there exists $c \in I(M)$ so that c is computable from u.
- Let x be a immortal configuration of the Turing machine, and I given by Lemma 4. Let u be the second track of x
  - If $I = [a, \infty[$, then the word w defined by $w_j = u_{a+j}$ avoids L, hence there exists a right-infinite word avoiding L that is Turing reducible to x. By Muchnik equivalence of $S^+(L)$ and S, there exists $u \in S$ that is computable from x.
  - If $I =] - \infty, b[$, then the word w defined by $w_j = u_{b-j}$ avoids the mirror of L, hence there exists a left-infinite word avoiding L that is Turing reducible to x. By Muchnik equivalence of $S^-(L)$ and S, there exists $u \in S$ that is computable from x.

  In both cases, we have found a word $u \in S$ that is computable from x.

  □

It is important to note that we do not know, given a configuration x, whether the set I is left or right infinite, as we would need to simulate the Turing machine to do this. This means that the reduction from x to a right-infinite word is not uniform in x. As a side note, this means we have only proven a Muchnik equivalence and not a Medvedev equivalence.

3   Conclusion

We finish with an application. There exists a well-known encoding of Turing machines into affine maps [5]. Using this encoding, we can prove:

Corollary 2. *There exists a piecewise affine map with rational coefficients and endpoints from $[0, 1]^2$ to itself so that all computable points converge in finite time to $(0, 0)$. However, there exists a noncomputable point that does not converge in finite time to $(0, 0)$.*

Using the encoding of piecewise affine rational maps into tilings [10], we have in particular an alternate proof of Myers' theorem [16]: There exists a tiling system that produces tilings, but no recursive tilings.

Acknowledgements. The author was partially supported by ANR-09-BLAN-0164.

References

1. Blondel, V.D., Cassaigne, J., Nichitiu, C.: On the presence of periodic configurations in Turing machines and in counter machines. Theoretical Computer Science 289(1), 573–590 (2002)
2. Cenzer, D., Dashti, A., King, J.L.F.: Computable symbolic dynamics. Mathematical Logic Quarterly 54(5), 460–469 (2008)
3. Cenzer, D., Remmel, J.: Π_1^0 classes in mathematics. In: Handbook of Recursive Mathematics - Volume 2: Recursive Algebra, Analysis and Combinatorics. Studies in Logic and the Foundations of Mathematics, vol. 2, 139, ch. 13, pp. 623–821. Elsevier (1998)
4. Cenzer, D., Remmel, J.: Effectively Closed Sets. ASL Lecture Notes in Logic (2011) (in preparation)
5. Collins, P., van Schuppen, J.H.: Observability of Hybrid Systems and Turing Machines. In: 43rd IEEE conference on Decision and Control, pp. 7–12 (2004)
6. Delvenne, J.C., Blondel, V.D.: Quasi-periodic configurations and undecidable dynamics for tilings, infinite words and Turing machines. Theoretical Computer Science 319, 127–143 (2004)
7. Hooper, P.K.: The Undecidability of the Turing Machine Immortality Problem. Journal of Symbolic Logic 31(2), 219–234 (1966)
8. Jeandel, E., Vanier, P.: Turing degrees of multidimensional SFTs. submitted to Theoretical Computer Science, arXiv:1108.1012v2
9. Jockusch, C.G., Soare, R.I.: Degrees of members of Π_1^0 classes. Pacific J. Math. 40(3), 605–616 (1972)
10. Kari, J.: A small aperiodic set of Wang tiles. Discrete Mathematics 160, 259–264 (1996)
11. Kari, J., Ollinger, N.: Periodicity and Immortality in Reversible Computing. In: Ochmański, E., Tyszkiewicz, J. (eds.) MFCS 2008. LNCS, vol. 5162, pp. 419–430. Springer, Heidelberg (2008)
12. Kurka, P.: On topological dynamics of Turing machines. Theoretical Computer Science 174, 203–216 (1997)
13. Lind, D.A., Marcus, B.: An Introduction to Symbolic Dynamics and Coding. Cambridge University Press, New York (1995)
14. Miller, J.S.: Two Notes on subshifts. Proceedings of the American Mathematical Society 140(5), 1617–1622 (2012)
15. Minsky, M.L.: Computation: Finite and Infinite Machines. Prentice-Hall (1967)
16. Myers, D.: Non Recursive Tilings of the Plane II. Journal of Symbolic Logic 39(2), 286–294 (1974)
17. Simpson, S.G.: Mass problems associated with effectively closed sets. Tohoku Mathematical Journal 63(4), 489–517 (2011)
18. Simpson, S.G.: Medvedev Degrees of 2-Dimensional Subshifts of Finite Type. Ergodic Theory and Dynamical Systems (2011)

A Slime Mold Solver
for Linear Programming Problems

Anders Johannson[1] and James Zou[2]

[1] Department of Mathematics Uppsala University, P.O. Box 480 751 06 Uppsala,
Sweden
[2] School of Engineering and Applied Sciences, Harvard University, Cambridge MA
02138, United States of America

Abstract. *Physarum polycephalum* (true slime mold) has recently
emerged as a fascinating example of biological computation through morphogenesis. Despite being a single cell organism, experiments have observed that through its growth process, the Physarum is able to solve
various minimum cost flow problems. This paper analyzes a mathematical model of the Physarum growth dynamics. We show how to encode
general linear programming (LP) problems as instances of the Physarum.
We prove that under the growth dynamics, the Physarum is guaranteed
to converge to the optimal solution of the LP. We further derive an
efficient discrete algorithm based on the Physarum model, and experimentally verify its performance on assignment problems.

1 Introduction

How do biological systems process information and solve optimization problems?
There has been a growing interest to understand these questions within a computation framework. The agenda is two fold: first, can we understand and analyze
biological systems in terms of algorithms; and second, can we design new algorithms inspired by biology. Prominent examples of such "natural algorithms"
include flocking algorithms motivated by the collective behavior of birds and
fish, and swarming algorithms motived by ant foraging behavior [1–3].

In this paper, we analyze the morphogenesis of *Physarum polycephalum* (true
slime mold) as a natural algorithm. The Physarum is a single cell organism with
multiple nuclei. Recent experiments have shown that it exhibits a surprising
ability to solve complex optimization problems [4, 5]. In one set of experiments,
researchers place food at various places akin to cities on a map [5]. As it grows,
the Physarum is able to process the input (food) in a de-centralized manner,
and converge to the distance minimizing network connecting these food sources[1].
Mathematicians and biologists have proposed a dynamical systems model that
captures essential behaviors of Physarum growth [6]. Simulation and analysis of
this model give mechanistic insight of how such a simple organism can solve the
Shortest Path Problem [6–9].

---

[1] An experiment from [4] is http://www.youtube.com/watch?v=BZUQQmcR5-g

S.B. Cooper, A. Dawar, and B. Löwe (Eds.): CiE 2012, LNCS 7318, pp. 344–354, 2012.
© Springer-Verlag Berlin Heidelberg 2012

Here we show how a general Linear Programming Problem can be encoded as an instance of a Physarum. We prove that under the Physarum growth model, the appropriate quantity is guaranteed to converge to the optimal solution of the LP. Our result draws on new analysis techniques inspired from the negative cost cycle algorithm. In addition, we derive a discrete algorithm for solving LPs from the dynamical systems. The algorithm is efficient to implement. We apply it to solve instances of the Assignment Problem, and report simulation results.

2   The Generalized Physarum Solver

A Model of Physarum Growth. A Physarum contains a network of veins that it uses to transport nutrients across its body. The veins are modeled as a graph $\mathbf{G}$ [6]. The vertices $\mathbf{V}$ are where multiple veins meet, and an edge $e \in \mathbf{E}$ is a segment of the vein. Each edge is a tube described by length c_e and cross-sectional area σ_e. Vertex i is associated with pressure p_i. We assume that $\mathbf{G}$ is directed. So edge ij goes from vertex i to j. The flow on ij is $x_{ij} = \sigma_{ij} \frac{p_i - p_j}{c_{ij}}$. $\mathbf{G}$ may contain multi-edges. In particular, a bidirectional edge can be modeled by two edges: ij and ji, with distinct σ_{ij} and σ_{ji}.

The Physarum growth model describes the time evolution of σ_{ij} and p_i on $\mathbf{G}$. The expansion and contraction of a vein segment is governed by:

$$\frac{d}{dt}\sigma_{ij}(t) = \sigma_{ij}(t)\frac{p_i - p_j}{c_{ij}} - \sigma_{ij}(t). \tag{1}$$

The first term on the left captures the positive feedback: the greater the flow on ij, the more the vein segment will expand. If $p_i < p_j$ implying that $x_{ij} < 0$, then σ_{ij} shrinks. The second term reflect the natural decay of the Physarum.

The pressure p_i is obtained from flow conservation: the flow into a vertex (except for source or sink) equals the flow out from it. Let $\mathbf{A}$ be the $|\mathbf{V}|\mathrm{x}|\mathbf{E}|$ incidence matrix of $\mathbf{G}$. The conservation equation is $\mathbf{Ax}(\mathbf{t}) = \mathbf{b}$, where $b_i = 0$ for all i except for $b_s = 1$ and $b_u = -1$. Vertices s and u are the source and sink. The conservation equation can be rewriten as $\mathbf{AW}(\mathbf{t})\mathbf{A}^{\mathbf{T}}\mathbf{p}(\mathbf{t}) = \mathbf{b}$, where $\mathbf{W}(t) = \mathrm{diag}(\sigma_e(t)/c_e)$. If $\mathbf{A}$ has full row rank, then $\mathbf{p}(\mathbf{t})$ can be uniquely solved for.

The Physarum growth dynamic thus has two coupled processes:

1. The vein sizes $\sigma_{ij}(t)$ grow according to Eqn. 1, which depends on $\mathbf{p}(\mathbf{t})$.
2. For a given set of $\sigma_{ij}(t)$, flow conservation uniquely determines $\mathbf{p}(\mathbf{t})$.

Under this growth model, $\{x_{ij}(t)\}$ evolves on $\mathbf{G}$. As $t \to \infty$, the flow $\mathbf{x}(\mathbf{t})$ converges to 1 on edges in the shortest s-u path, and to 0 on all other edges [8].

LP Problems. A general linear programming (LP) problem is to $\min \mathbf{c}^{\mathbf{T}}\mathbf{x}$ subject to the contraints $\mathbf{Ax} = \mathbf{b}$, $\mathbf{x} \geq 0$. Let $\mathbf{\Phi} = \{\mathbf{x} : \mathbf{Ax} = \mathbf{b}\}$ and $\mathbf{\Phi}^+ = \{\mathbf{x} : \mathbf{Ax} = \mathbf{b}, \mathbf{x} \geq 0\}$. We work in cases where $\mathbf{\Phi}^+$ forms a bounded polytope and A has full row rank, which are common assumptions in LP applications. We assume that the LP has a unique optimal solution. Any small perturbation to $\mathbf{c}$ leads to unique optimum.

Keeping the graph notations, we label columns of $\mathbf{A}$ by $e \in \mathbf{E}$. Each column $\mathbf{A_e}$ is analogous to an edge and is associated with a "conductance" $\sigma_e \geq 0$, which is an auxiliary variable. Every constraint equation in $\mathbf{Ax} = \mathbf{b}$ corresponds to a vertex and is associated with a pressure p_i. We define the generalized *Physarum Solver* by the dynamics

$$\frac{d}{dt}\sigma_e = \sigma_e(\psi_e - 1) \tag{2}$$

where $\psi_e = (\mathbf{A_e^T p})/\mathbf{c_e}$ is the "pressure gradient"and $\mathbf{p}$ satisfies Kirchhoff's Law

$$\mathbf{AWA^T p} = \mathbf{b} \text{ with } \mathbf{W} = \mathrm{diag}(\sigma_e/c_e). \tag{3}$$

As before, the "flow" is given by $x_e(t) = \sigma_e(t)\psi_e(t)$. Note that $\mathbf{Ax} = \mathbf{AWA^T p} = \mathbf{b}$, though $x_e(t)$ could be negative if $\psi_e < 0$. So $\mathbf{x}(t) \in \mathbf{\Phi}$.

Duality. The pressure, p_i, is the dual variable associated with each constraint. If the dynamical system reaches a stationary point, then for all e, $0 = \frac{d}{dt}\sigma_e = \sigma_e(\frac{\mathbf{A_e^T P}}{c_e} - 1)$. If $x_e > 0$, then $\sigma_e > 0$ and stationarity implies $\mathbf{A_e^T P} = c_e$. This is precisely the complementary slackness condition. The stationary point of the Physarum Solver is the optimal solution to the LP.

The main result of the paper is the following theorem.

Theorem 1. *Suppose the LP problem has a unique optimal solution. Then from any positive initial configuration $\sigma(\mathbf{0}) > \mathbf{0}$, the flow of the Physarum Solver $\mathbf{x(t)}$ converges exponentially fast to the optimal solution.*

3   Proof of Convergence

3.1   LP with Positive Costs

We first assume that $c_e > 0 \; \forall e$. In the next section we shall show how to extend the proof to general $\mathbf{c}$. Here we give the outline of the proof; the technical details are in the Appendix.

We multiply Eqn. 2 by e^t and integrate to obtain

$$\sigma(t) = (1 - e^{-t})\tilde{\mathbf{x}}(t) + e^{-t}\sigma(0) \tag{4}$$

with

$$\tilde{\mathbf{x}}(t) \doteq \frac{1}{1 - e^{-t}} \int_0^t \mathbf{x}(s)e^{-(t-s)}ds. \tag{5}$$

Since $\tilde{\mathbf{x}}(t)$ is a weighted average of $\mathbf{x}(s)$ for $s \leq t$ and each $\mathbf{x}(s)$ satisfies $\mathbf{Ax}(s) = \mathbf{b}$, we also have $\mathbf{A\tilde{x}}(t) = \mathbf{b}$.

Definition 1. *A circulation is a $|\mathbf{E}|$ dimensional vector γ such that $\mathbf{A\gamma} = \mathbf{0}$.*

There must exists e such that $\gamma_e < 0$, for otherwise the feasible region $\mathbf{\Phi}^+$ would be unbounded. Let $\gamma_- \doteq \{e : \gamma_e < 0\}$ denote the negative edges of γ. Similarly define $\gamma_+ \doteq \{e : \gamma_e > 0\}$. We say a circulation has negative cost if $\mathbf{c^T}\gamma < 0$.

Lemma 1. *In a bounded LP problem, $\mathbf{x}$ is an optimal solution if and only if there is no negative cost circulation γ such that $\gamma_- \subseteq supp(\mathbf{x})$.*

Our strategy is to prove that under Eqn. 2, the flow on γ_- becomes exponentially small for any negative circulation γ. We start with the following Lemma.

Lemma 2. *Let γ be a negative cost circulation. Then for all $\epsilon > 0$, $\exists e$ with $\gamma_e < 0$ and t such that $\sigma_e(t) < \epsilon$.*

Proof. Let $Z_\gamma \doteq e^{-\sum_{e \in \mathbf{E}} \gamma_e c_e \log \sigma_e}$. Since $\sum_{e \in \mathbf{E}} \gamma_e c_e \psi_e = \mathbf{p}^\mathbf{T} \mathbf{A} \gamma = \mathbf{0}$,

$$\frac{d}{dt} Z_\gamma = Z_\gamma \sum_{e \in \mathbf{E}} \gamma_e c_e (1 - \psi_e) = Z_\gamma \mathbf{c}^\mathbf{T} \gamma < \mathbf{0}. \tag{6}$$

Therefore Z_γ decays exponentially. On the other hand,

$$Z_\gamma(t) = \frac{\Pi_{e \in \gamma_-} \sigma_e(t)^{|\gamma_e| c_e}}{\Pi_{e \in \gamma_+} \sigma_e(t)^{\gamma_e c_e}}, \tag{7}$$

and by Lemma 4 (Appendix), all σ_e are bounded above for large t and hence the denominator is bounded above as t increases. As the ratio decays exponentially, the numerator must decay exponentially and at least one of the terms in its product must be exponentially small.

The next lemma shows that we can clean up a solution $\mathbf{x}$ of $\mathbf{Ax} = \mathbf{b}$ by essentially setting x_e to 0 if x_e is sufficiently small.

Lemma 3. *Given $\mathbf{x} \in \mathbf{\Phi}$, and $\mathbf{S} \subseteq \mathbf{E}$. There are constants M_1 and K such that if $\forall e \in \mathbf{S}$, $|x_e| < \epsilon^{|\mathbf{S}|+1} < M_1/2m$ for some $\epsilon < \min\{1, \frac{M_1}{2}, \frac{1}{2K}\}$, then there exists $\mathbf{y} \in \mathbf{\Phi}$ satisfying:*

- $\|\mathbf{x} - \mathbf{y}\|_\infty < K\epsilon$,
- $supp(\mathbf{y}) \subseteq supp(\mathbf{x}) \backslash \mathbf{S}$,
- $x_e y_e \geq 0 \ \forall e$.

Combining the previous two lemmas, we prove that $\mathbf{x}(t)$ converges to the unique optimal solution under the Physarum dynamics.

Proof of Theorem 1 for LP with Positive Costs
Consider the time-averaged solution $\tilde{\mathbf{x}}(t)$ defined above. Let $\mathbf{S} = \{e : \tilde{x}_e(t) < 0\}$. From Eqn. 4 and the fact that $\sigma(t)$ is bounded for large t, it follows that

$$\tilde{\mathbf{x}}(t) = \sigma(t) + O(e^{-k_1 t}) \tag{8}$$

for some constant k_1. Since $\sigma(t) > 0$, we have $|\tilde{x}_e(t)| \sim O(e^{-k_1 t})$, $\forall e \in \mathbf{S}$. For sufficiently large t so that the inequality requirements of Lemma 3 are satifistied, we obtain a $\mathbf{y}(t)$ such that

$$\mathbf{y}(t) = \tilde{\mathbf{x}}(t) + O(e^{-k_2 t}) \tag{9}$$

for constant k_2. Moreover $\mathbf{y}(t) \geq 0$ is feasible.

Let $\hat{\mathbf{y}}$ denote the optimal feasible solution of the LP. Let $e \in \mathrm{supp}(\mathbf{y})/\mathrm{supp}(\hat{\mathbf{y}})$. Then by Lemma 6 (Appendix), there is a basic feasible solution τ such that

$$y_e \leq k_3 \min\{y_{e'} : e' \in \mathrm{supp}(\tau)\} \tag{10}$$

for constant k_3. Consider $\gamma = \hat{\mathbf{y}} - \tau$. This is a circulation since $\mathbf{A}\gamma = \mathbf{0}$, and it has negative cost. By Lemma 2, $\exists\, e' \in \mathrm{supp}(\tau)$ such that $\sigma_{e'}(t)$ decays exponentially. Eqns. 8 and 9 implies that $y_{e'}(t) \sim O(e^{-k_4 t})$ for some constant k_4. Combined with Eqn. 10, this implies that $y_e \sim O(e^{-k_4 t})$ for all $e \in \mathrm{supp}(\mathbf{y})/\mathrm{supp}(\hat{\mathbf{y}})$. Applying Lemma 2 again to remove all such e, we find $\tilde{\mathbf{y}} \in \Phi^+$ such that

$$\tilde{\mathbf{y}} = \mathbf{y}(\mathbf{t}) + O(e^{-k_5 t}) = \tilde{\mathbf{x}}(t) + O(e^{-k_2 t}) + O(e^{-k_5 t}). \tag{11}$$

But $\tilde{\mathbf{y}}$ has support contained in $\mathrm{supp}(\hat{\mathbf{y}})$, implies $\tilde{\mathbf{y}} = \hat{\mathbf{y}}$ since the optimal solution is a unique basic feasible solution. This concludes that $\tilde{\mathbf{x}}(t) = \hat{\mathbf{y}}(t) + O(e^{-k_6 t})$.

As $\mathbf{x}(\mathbf{t})$ is continuous, the weighted average $\tilde{\mathbf{x}}(t)$ converges to $\hat{\mathbf{y}}$ exponentially implies $\mathbf{x}(t)$ also converges to $\hat{\mathbf{y}}$ exponentially.   □

3.2   LP with General Costs

First we show how to deal with when $c_e = 0$ for some e. The assumption that Φ^+ is a bounded polytope implies $\exists\, M$ such that $x_e < M$ for all $e \in \mathbf{E}$ and $\mathbf{x} \in \Phi^+$. Given $\{\mathbf{A}, \mathbf{b}, \mathbf{c}\}$, consider the LP $\{\mathbf{A}, \mathbf{b}, \hat{\mathbf{c}}\}$, where $\hat{c}_e = c_e$ if $c_e > 0$ and $\hat{c}_e = \epsilon$ if $c_e = 0$. This problem can be solved by the Physarum Solver since $\hat{\mathbf{c}} > \mathbf{0}$. Let $\mathbf{x_1}$ be the optimal solution of $\{\mathbf{A}, \mathbf{b}, \mathbf{c}\}$ and let $\mathbf{x_2}$ be its second lowest cost solution. Define the gap $\delta = \mathbf{c}^T\mathbf{x_2} - \mathbf{c}^T\mathbf{x_1}$ and set $\epsilon < \frac{\delta}{|E|M}$. It's easy to see that $\mathbf{x_1}$ is also the optimal solution of $\{\mathbf{A}, \mathbf{b}, \hat{\mathbf{c}}\}$. Hence it suffices to apply the Physarum Solver to solve $\{\mathbf{A}, \mathbf{b}, \hat{\mathbf{c}}\}$.

Now consider if $c_e < 0$ for some $e \in \mathbf{U} \subseteq \mathbf{E}$. For each such e add a new variable $\hat{e}$ and the constraint equation $x_e + x_{\hat{e}} = M$. Modify the costs by setting $c'_e = 0$ and $c'_{\hat{e}} = |c_e|$ for $e \in \mathbf{U}$. The new LP problem $\{\mathbf{A}', \mathbf{b}', \mathbf{c}'\}$ can by solved by the Physarum Solver. There is a one-to-one mapping between the feasible sets of $\{\mathbf{A}, \mathbf{b}, \mathbf{c}\}$ and $\{\mathbf{A}', \mathbf{b}', \mathbf{c}'\}$ that maps the optimal solutions to each other.

3.3   Relations to Other LP Solvers

The Physarum dynamics does not maintain primal feasibility. Even though $\mathbf{Ax}(t) = \mathbf{b}$ always holds, the gradient $\frac{A_e^T p}{c_e}$ could be negative, resulting in $x_e(t) < 0$ for some t. Though our proof shows that $\mathbf{x}(t)$ stays exponentially close to the feasible polytope Φ^+. Similarly, the potentials may violate dual feasibility: $\mathbf{A}_e^T \mathbf{p}(t) > c_e$ for some e.

In the Physarum solver, $\mathbf{x}(t)$ does not live on the boundary of Φ^+ and in this sense is more related to interior point methods than to the Simplex Algorithm. Unlike interior point methods, the Physarum is not guaranteed to monotonically reduce the cost $\mathbf{c}^T\mathbf{x}(t)$ or increase the dual objective $\mathbf{p}^T\mathbf{b}$, and is not restricted to be inside the polytope [10].

The Physarum is conceptually most similar to the negative cost cycle algorithm. In this algorithm, a negative cost cycle is identified and a flow is pushed on this cycle until one of the directed edges has flow 0 [10]. As our proof shows, this is also the Physarum's strategy. Think of the graph case for intuition. If there is a negative cost oriented cycle γ, such that all edges ij of γ are unsaturated: $p_i - p_j < c_{ij}$. Then σ_{ij} increase for all $ij \in \gamma$, and an additional negative cost flow is pushed $\{x_{ij}(t + \triangle) - x_{ij}(t) : ij \in \gamma\}$.

4   The Discrete Physarum Solver

4.1   Finite Difference Approximation

The continuous dynamics of the generalized Physarum Solver makes it difficult to be implemented and analyzed as an algorithm. We use finite difference method to discretize Eqn. 1

$$\sigma_e(t + \triangle) \approx \sigma_e(t) + \left(\frac{\mathbf{A_e^T p}}{c_e} - 1\right)\sigma_e(t)\triangle. \tag{12}$$

By setting $\triangle = 1$, we obtain a simple discrete algorithm.

Algorithm 1. Discrete Physarum Solver

INPUT: LP problem instance $(\mathbf{A}, \mathbf{b}, \mathbf{c})$
1: Initialize $\{\sigma_e(0) > 0, e \in \mathbf{E}\}$
2: **while** $\sigma(t)$ not converged **do**
3:    Compute $\mathbf{p}(t)$ by solving $\mathbf{AWA^T p} = \mathbf{b}$, where $\mathbf{W} = \mathrm{diag}(\frac{\sigma_e(t)}{c_e})$.
4:    **if** $\mathbf{A^T p}(t) > 0$ **then**
5:       $\sigma_e(t + 1) = \sigma_e(t)\frac{\mathbf{A_e^T p}(t)}{c_e} \doteq x(t)$
6:    **else**
7:       $\sigma_e(t) = \epsilon$
8:    **end if**
9:    $t = t + 1$
10: **end while**

In the algorithm, $\epsilon \ll 1$ is a positive constant. We say σ has converged if $\sigma(t + 1) - \sigma(t)$ is small. We check for whether the algorithm has converged to the optimal solution by testing for approximate complementary slackness: $|\frac{A_e^T p}{c_e} - 1| < \delta, \forall e$ for some small δ. The most computationally intensive step is to solve Kirchhoff's Equation for $\mathbf{p}(t)$. Recent progress shows this can be done in near-linear time in $|\mathbf{E}|$ [11]. Because of the large finite difference approximation we have taken, the candidate solution $\mathbf{x}(t)$ is not guaranteed to converge to the optimal. We show via simulation that the Discrete Physarum Solver does find the optimum for the special class of Assignment problems.

4.2   The Assignment Problem

We test the discrete Physarum algorithm on a standard class of LP problems: the Assignment Problem [10]. In the Assignment Problem, there are N workers and N tasks. Every worker and task pair has an associated cost. The goal is to assign each worker to a unique task such that the total cost incurred is minimal.

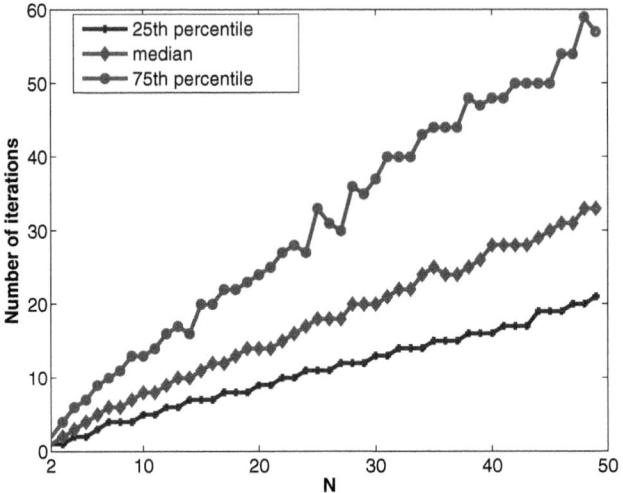

Fig. 1. Number of iterations to find the optimal solution versus the size of the Assignment Problem. N is the number of tasks. For each N, plotted are the 25th, median, and 75th percentile of the number of iterations.

This can be represented by a complete bipartite graph $\mathbf{G}$ with worker nodes and task nodes. Edges are directed from worker nodes to task nodes. Edge ij connecting worker i and task j is associated a cost c_{ij}. $\mathbf{A}$ is the incidence matrix, and $b_i = 1$ if i is a worker vertex; $b_j = -1$ for task vertex. The update is $\sigma_{ij}(t+1) = \sigma_{ij}(t)\frac{p_i-p_j}{c_{ij}} = x_{ij}(t)$ if $p_i > p_j$; else $\sigma_{ij}(t+1) = \epsilon$. To check for optimality, we derive a hard assignment from $\mathbf{x}(t)$. For each i, identify $j_i = \arg\max_j\{x_{ij}(t)\}$. Set $x_{ij_i} = 1$ and to 0 on all other edges. Then we test for complementary slackness.

For each N from 2 to 50, we randomly generate 1000 weighted complete bipartite graphs. The costs are iid samples from the uniform distribution. We applied the Discrete Physarum Algorithm to each of these assignment problems. After every iteration of the algorithm, we check for optimality. If the optimal solution is found, the number of iterations so far is recorded; otherwise the algorithm continues.

In all the assignment problem instances tested, the Discrete Physarum Solver always converged to the optimal solution. For each N, we plot the 25th, median, and 75th percentile of the number of iterations the Discrete Physarum Solver took before reaching the optimal solution. The median number of iterations grows roughly linear in the size of the problem.

Acknowledgments. The authors would like to thank David Parkes, Devavrat Shah, and David Sumpter for helpful discussions and feedback on the paper.

References

1. Chazelle, B.: Natural Algorithms. In: Proceedings of the 20th Symposium of Discrete Algorithms (2009)
2. Sumpter, D.: Collective Animal Behavior. Princeton University Press (2010)
3. Camazine, S., Deneubourg, J., Franks, N., Sneyd, J., Theraulaz, G., Bonabeau, E.: Self-Organization in Biological Systems. Princeton University Press (2003)
4. Tero, A., Takagi, S., Saigusa, T., Ito, K., Bebber, D., Fricker, M., Yumiki, K., Kobayashi, R., Nakagaki, T.: Rules for biological inspired adaptive network design. Science 327(5964), 439–442 (2010)
5. Nakagaki, T., Yamada, H., Toth, A.: Maze-solving by an amoeboid organism. Nature 407, 470 (2000)
6. Nakagaki, T., Tero, A., Kobayashi, R.: A mathematical model for adaptive transport network in path finding by true slime mold. Journal of Theoretical Biology 244(4), 553–564 (2007)
7. Miyaji, T., Ohnishi, I.: Physarum can solve the shortest path problem on Riemannian surface mathematically rigorously. International Journal of Pure and Applied Mathematics 47(3), 353–369 (2008)
8. Ito, K., Johansson, A., Nakagaki, T., Tero, A.: Convergence properties for the Physarum solver. arXiv:1101.5249v1 (2011)
9. Bonifaci, V., Mehlhorn, K., Varma, G.: Physarum can compute shortest paths. In: Proceedings of the 23th Symposium of Discrete Algorithms (2012)
10. Bertsimas, D., Tsitsiklis, J.: Introduction to Linear Optimization. Athena Scientific (1997)
11. Christiano, P., Kelner, J., Madry, A., Spielman, D., Teng, S.: Electrical flows, Laplacian systems, and faster approximation of maximum flow in undirected graphs. In: Proceedings of the 43rd ACM Symposium on Theory of Computing (2011)

A   Proofs of Technical Lemmas

Proof of Lemma 1

Suppose there exists a negative cost circulation γ. Let $\epsilon = max_e\{\frac{x_e}{\lceil\gamma_e\rceil}\}$. Then $\mathbf{y} = \mathbf{x} + \epsilon\gamma$ satisfies $\mathbf{y} \geq 0$, $\mathbf{Ay} = \mathbf{b}$, and $\mathbf{y}$ has lower cost than $\mathbf{x}$. This implies $\mathbf{x}$ is not an optimal solution. Conversely, suppose $\mathbf{x}$ is not optimal. Let $\mathbf{y}$ be the optimal solution and $\gamma = \mathbf{y} - \mathbf{x}$. Then γ is a negative cost circulation. □

Lemma 4. *For all e, $\sigma_e(t)$ is bounded for large t.*

352   A. Johannson and J. Zou

Proof. Rearranging Eqn. 4, $\tilde{\mathbf{x}}(t) > -e^{-t}\sigma(0)$. Let $\mathbf{S}(t) = \{e : \tilde{x}_e(t) < 0\}$. Then for t sufficiently large, the condition for Lemma 3 is satisfied and there exists $\mathbf{y}(t)$ such that $\mathbf{y}(t) \geq 0$, $\mathbf{A}\mathbf{y}(t) = \mathbf{b}$, and $||\tilde{\mathbf{x}}(t) - \mathbf{y}(t)||_\infty < Ce^{-t}$ for some constant C. The vector $\mathbf{y}(t)$ lies in the feasible region of the LP, which we assume to be a bounded polytopy, implies $||\mathbf{y}(t)||_\infty$ is bounded. This implies that $||\tilde{\mathbf{x}}(t)||$ is bounded and hence $\sigma(t)$ is bounded.

Definition 2. *Given* $\mathbf{A}$ *and* $\mathbf{x}$, *a* ***positive re-orientation*** *of* $\mathbf{A}$ *with respect to* $\mathbf{x}$ *is* $\mathbf{A}'$ *and* $\mathbf{x}'$ *such that*

- $x'_e = |x_e|$ *for all* e
- $\mathbf{A}'_e = \mathbf{A}_e$ *for* e *where* $x_e \geq 0$
- $\mathbf{A}'_e = -\mathbf{A}_e$ *for* e *where* $x_e < 0$

Definition 3. *A* ***positive cycle*** *is a vector* γ *indexed by* $e \in \mathbf{E}$ *such that* $\mathbf{A}\gamma = 0$, $\gamma_e \geq 0, \forall e$, *and* $supp(\gamma)$ *is minimal. Since* γ *is ambigous up to a constant, we set* $\min_{e \in supp(\gamma)}\{\gamma_e\} = 1$. *The set of positive cycles of* A *is denoted by* $pcyc(A)$.

Let $bfs(A)$ denote the set of basic feasible solutions of the LP $\mathbf{A}\mathbf{x} = \mathbf{b}$, $\mathbf{x} \geq 0$.

Lemma 5. *Given* $\mathbf{A}$, $\mathbf{x} \in \mathbf{\Phi}$ *and* $e \in \mathbf{E}$ *such that* $|x_e| < \epsilon < M_1/2$. *Then* $\exists$ $\mathbf{y} \in \mathbf{\Phi}$ *such that:*

- $||\mathbf{x} - \mathbf{y}||_\infty < 3\epsilon\frac{M_2}{M_1}$,
- $supp(\mathbf{y}) \subseteq supp(\mathbf{x})\backslash e$,
- $x_e y_e \geq 0$ *for all* e,

Here

$$M_1 = \min_{A'} \min_{\tau \in bfs(A') \cup pcyc(A')} \{\tau_e : e \in \mathbf{E} \text{ and } \tau_e > 0\} \tag{13}$$

$$M_2 = \max_{A'} \max_{\tau \in bfs(A') \cup pcyc(A')} \{\tau_e : e \in \mathbf{E}\} \tag{14}$$

where min and max are over all re-orientation A' *of* A.

Proof. After re-orientation, we can assume wlog that $\mathbf{x} \geq 0$. Then $\mathbf{x}$ lies in the polyhedron defined by $\mathbf{A}'\mathbf{x} = \mathbf{b}$ and $\mathbf{x} \geq \mathbf{0}$. We can express $\mathbf{x}$ as a convex linear combination of basic feasible solutions and positive cycles [10]: $\mathbf{x} = \sum_{\tau \in bfs(A') \cup pcyc(A')} c_\tau \tau$, such that $c_\tau > 0$ and $\sum_{\tau \in bfs(A')} c_\tau = 1$. We can split the sum into two groups of τ depending on if $\tau_e = 0$ or $\tau_e > 0$:

$$\mathbf{x} = \sum_{\tau:\tau_e=0} c_\tau \tau + \sum_{\tau:\tau_e>0} c_\tau \tau. \tag{15}$$

Let $\alpha = \sum_{\tau:\tau_e>0} c_\tau$. Then

$$\alpha M_1 \leq \sum_{\tau:\tau_e>0} c_\tau \tau_e = x_e < \epsilon < \frac{M_1}{2} \tag{16}$$

implying $\alpha < \frac{\epsilon}{M_1} < \frac{1}{2}$. Therefore there is at least one bfs τ such that $c_\tau > 0$ and $\tau_e = 0$. Define

$$\mathbf{y} = \frac{1}{1-\alpha} \sum_{\tau \in \text{bfs}(A'):\tau_e=0} c_\tau \tau + \sum_{\tau \in \text{pcyc}(A'):\tau_e=0} c_\tau \tau. \tag{17}$$

Then $\mathbf{y}$ is the convex combination of at least one bfs and positive cycles, and is also a feasible solution. In particular $x_e y_e \geq 0$ for all e. It's also clear that $\text{supp}(\mathbf{y}) \subseteq \text{supp}(\mathbf{x})$. Finally,

$$\|\mathbf{x} - \mathbf{y}\|_\infty \leq \frac{\alpha}{1-\alpha} \| \sum_{\tau \in bfs(A'):\tau_e=0} c_\tau \tau \|_\infty + \| \sum_{\tau \in \text{pcyc}:\tau_e>0} c_\tau \tau \|_\infty. \tag{18}$$

Since $\alpha < \frac{1}{2}$, $\frac{\alpha}{1-\alpha} < 2\alpha$. Moreover,

$$\| \sum_{\tau \in bfs(A'):\tau_e=0} c_\tau \tau \|_\infty < M_2 \sum_{\tau \in bfs(A'):\tau_e=0} c_\tau < M_2 \tag{19}$$

, and $\| \sum_{\tau:\tau_e>0} c_\tau \tau \|_\infty < M_2 \alpha$. We have

$$\|x - y\|_\infty \leq 3\alpha M_2 < 3\epsilon \frac{M_2}{M_1}. \tag{20}$$

We use induction to remove multiple small edges.

Proof of Lemma 3

Let M_1, M_2 be the same as in the previous lemma, and set $K = 3\frac{M_2}{M_1}$. Let $S = \{e_1, e_2, ..., e_{|S|}\}$. It's clear that properties 2 and 3 are satisfied during induction, so we focus on 1. In the base case, we apply the previous lemma to e_1 to construct $\mathbf{y_1}$, such that $\|\mathbf{x} - \mathbf{y_1}\|_\infty < \epsilon^{|S|+1} K$. The induction claim that we shall prove is: if $\exists \mathbf{u}$ that removes $\{e_1, e_2, ...e_{l-1}\}$ from $\mathbf{x}$, and $\|\mathbf{x} - \mathbf{u}\|_\infty < K\epsilon^n$, $n \leq |S|+1$, then we can construct $\mathbf{y}$ such that $\mathbf{y}$ removes $\{e_1, ..., e_l\}$ from $\mathbf{x}$, and $\|\mathbf{x} - \mathbf{y}\|_\infty < K\epsilon^{n-1}$. We can then iterate $|S|$ times to construct $\mathbf{y}$ such that $\|\mathbf{x} - \mathbf{y}\|_\infty < K\epsilon$.

To prove the induction claim, observe that $\|\mathbf{x} - \mathbf{u}\|_\infty < K\epsilon^n < \frac{\epsilon^{n-1}}{2}$. Since $n \leq |S|+1$, $x_{e_l} < \epsilon^{|S|+1} < \epsilon^n$ and

$$u_{e_l} < x_{e_l} + \frac{\epsilon^{n-1}}{2} < \epsilon^{n-1} < \frac{M_1}{2}. \tag{21}$$

Apply Lemma to remove e_l from $\mathbf{u}$ and construct $\mathbf{y}$ such that $\|\mathbf{y} - \mathbf{u}\|_\infty < K\epsilon^{n-1}$. Then

$$\|\mathbf{x} - \mathbf{y}\|_\infty < \|\mathbf{x} - \mathbf{u}\|_\infty + \|\mathbf{y} - \mathbf{u}\|_\infty < K\epsilon^n + K\epsilon^{n-1} < K\epsilon^{n-1}. \tag{22}$$

□

Let $\beta_{\min} = \min_{\tau \in \text{bfs}}\{\tau_e : e \in \text{supp}(\tau)\}$ and $\beta_{\max} = \max_{\tau \in \text{bfs}}\{\tau_e : e \in \text{supp}(\tau)\}$.

Lemma 6. *Let* $\mathbf{x} \in \mathbf{\Phi}^+$ *and* $e \in \mathbf{E}$ *be in the support of* $\mathbf{x}$. *Then there is a basic feasible solution* $\hat{\tau} \in \mathbf{\Phi}^+$ *such that*

$$\min \mathbf{x}(supp(\hat{\tau})) \geq \frac{\beta_{min}}{\beta_{max}} \frac{x_e}{|supp(\mathbf{x})|} \tag{23}$$

Proof. Since $\mathbf{\Phi}^+$ is a bounded polytope, $\mathbf{x}$ is a convex linear combination of basic feasible solutions. In particular, there exists $\tau \in$ bfs such that $\mathrm{supp}(\tau) \subseteq \mathrm{supp}(\mathbf{x})$. Let $c_\tau = min\{\frac{x_e}{\tau_e} : e \in \mathrm{supp}(\tau)\}$. Then $\mathbf{x} - c_\tau \tau \in \mathbf{\Phi}^+$ and has support strictly contained in the support of $\mathbf{x}$. Iterating this shows that we can write $\mathbf{x} = \sum c_\tau \tau$, where the number of positive c_τ is at most $|\mathrm{supp}(\mathbf{x})|$. Let $\hat{\tau} = \mathrm{argmax}_\tau \{c_\tau \tau_e\}$, then

$$c_{\hat{\tau}} \beta_{max} |\mathrm{supp}(\mathbf{x})| \geq c_{\hat{\tau}} \hat{\tau}_e |\mathrm{supp}(\mathbf{x})| \geq x_e. \tag{24}$$

Furthermore

$$\min \mathbf{x}(\mathrm{supp}(\hat{\tau})) \geq c_{\hat{\tau}} \min \hat{\tau}(\mathrm{supp}(\hat{\tau})) \geq c_{\hat{\tau}} b_{min} \tag{25}$$

and the desired inequality follows.

Multi-scale Modeling
of Gene Regulation of Morphogenesis

Jaap A. Kaandorp, Daniel Botman, Carlos Tamulonis, and Roland Dries

Faculty of Sciences, Universiteit van Amsterdam, Science Park 904, 1098 XH
Amsterdam, The Netherlands

Abstract. In this paper we demonstrate a spatio-temporal gene regulatory network for early gastrulation in the sea anemone *Nematostella vectensis*. We measure gene expression during early gastrulation using a gene expression quantification tool. We measure gene expression during early gastrulation when the embryo is more or less radial symmetrical and where there is only one body axis (the aboral - oral body axis in the case of *Nematostella*), allowing us, to use a one-dimensional model. The shape changes are induced by mechanical forces that are defined as rules or explicitly coupled to genetic regulation. The morphological evolution of the *Nematostella* embryo during gastrulation is described by a cell-based model, which can potentially be coupled with the one-dimensional gene regulatory network.

1 Introduction

A spectacular achievement in modern biology is the discovery of gene regulatory networks involved in pattern formation during embryo development [4]. An important next step is to understand the complex regulatory structure and dynamics of these networks, which in turn are closely coupled with cell dynamics. The network dynamics vary across space and time and are influenced by biomechanical events (e.g., formation of cell layers, cell migration, cell death, and cell division). To provide insight into the dynamical behavior of such complex regulatory systems, detailed models are required, which can be tested against available experimental observations. Several modelling formalisms have been proposed to simulate the dynamics of developmental gene regulatory networks [5, 13, 22]. Here, we focus on spatio-temporal models of gene regulation, which capture quantitative aspects of pattern formation during embryo development [12, 20]. However, obtaining such models is not straightforward. First of all, it is often not clear which network-modelling formalism should be used. Due to the biochemical complexity of eukaryotic transcription, it is currently impossible to derive network models from first principles [20]. Instead, many phenomenological formalisms have been used, from Boolean models to differential equation models and stochastic formalisms [5]. There are currently no widely applicable guidelines for the choice of model, which depends on the specific problem under study, as well as the availability and quality of gene expression data.

S.B. Cooper, A. Dawar, and B. Löwe (Eds.): CiE 2012, LNCS 7318, pp. 355–362, 2012.
© Springer-Verlag Berlin Heidelberg 2012

Another important research problem is the determination of model parameters. Even models of moderately sized networks contain a large number of parameters, which determine production, diffusion, decay rates, regulatory interaction rates, as well as the regulatory topology of a gene network. These parameters are often difficult (if not impossible) to measure. Instead, they have to be inferred by fitting models to data using global, non-linear optimization. This type of reverse-engineering approach poses a number of significant challenges. One major issue for parameter inference is that the observed dynamical behaviour of the system can often be explained by multiple distinct regulatory mechanisms. This generally happens because the optimization problem is ill-posed or insufficiently constrained by data [2]. Alternatively, parameters can be difficult to determine due to correlations between them [2, 10]. Model validation based on additional experimental evidence is required to decide which of the alternative mechanisms is applicable to the real biological system. This is often time-consuming and technically challenging. Therefore, it is essential to decrease the number of alternative predictions that need to be tested experimentally. One way of achieving this is to take additional criteria into account in the identification of the best explanatory model. Usually, the accuracy with which a model reproduces observed expression patterns is measured by a cost function based on the sum of squared differences between model and data (single-objective optimisation). Multi-objective optimization, however, can take other desirable properties into account, such as model robustness and known information about the underlying genetic regulatory network, to constrain the optimization procedure and produce fewer but more realistic solutions.

The *Drosophila* (fruit fly) gap gene interaction network has been derived in quite some detail, partly through the study of mRNA hybridizations during early embryogenesis [9, 12, 20]. The *Drosophila* case study is particularly convenient for spatial gene expression analysis, because the embryos remain in the same shape during their early development and gene expression is measured along a nearly straight axis. Early embryogenesis in Drosophila occurs in a cigar-shaped embryo-sac with quasi-radial symmetry that does not change shape and in which there are no moving cells. Gene expression can therefore be quantified along the main axis of symmetry (the anterior posterior axis) and the whole system can be described with a one-dimensional spatio-temporal regulatory network. However, nearly all other animals rapidly change their morphology by cell division, cell movement, blastula formation and gastrulation. In order to quantify gene expression for comparison and analysis gene expression patterns should be mapped to their tissue shape.

In this paper we demonstrate a spatio-temporal gene regulatory network for early gastrulation in the sea anemone *Nematostella vectensis*. We measure gene expression during early gastrulation using a gene expression quantification tool developed by [6]. We shall model formation of gene expression patterns in static or changing geometries. Depending on the stage and symmetry of the gene expression patterns of interest this can be done in 1D, 2D or 3D. In this paper we measure gene expression during early gastrulation when embryo is more or less radial symmetrical and where there is only one body axis (the aboral-oral body axis in the case of *Nematostella*), allowing us, again, to use a one-dimensional model. The shape changes are induced by mechanical forces that are defined as rules or explicitly coupled to genetic regulation. The morphological evolution of the *Nematostella* embryo during gastrulation is described by a cell-based model [24], which can potentially be coupled with the one-dimensional gene regulatory network.

2   Methods

Gene Expression Patterns in Nematostella Vectensis

For quantifying gene expression during gastrulation we have used a number of published in situ hybridisations of *Nematostella* embryos (Fig. 1). Fig. 1A shows Anthox1 expression [8] in the blastula stage. Anthox1 expression sets up the first body axis (the aboral- oral axis) in the system and breaks the symmetry of the blastula stage. Fig 1B shows snail expression [16], snail is expressed exactly opposite of Anthox1 expression around the future oral opening. In both pictures the shape of the embryo is approximated with a spline [6] and the aboral site is located at the left and the oral site at the right.

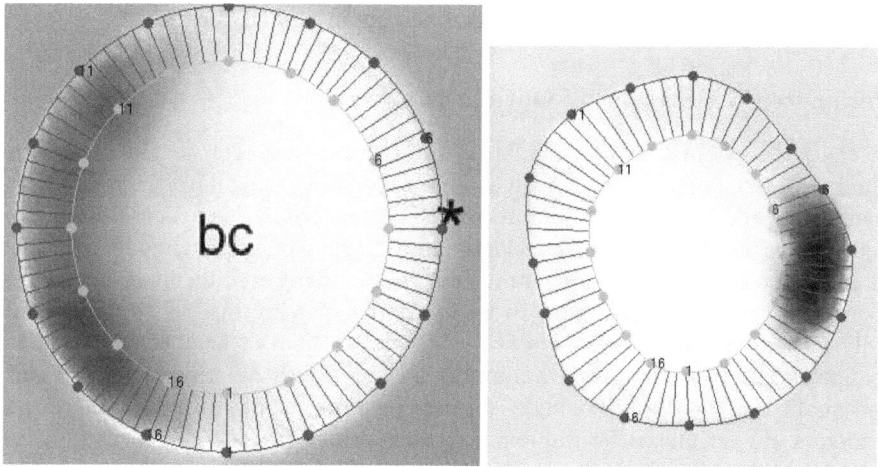

Fig. 1. Two examples of quantified in situ hybridisations of *Nematostella vectensis*. The left picture shows Anthox1 expression (the original picture can be found in [8]. The right picture shows snail expression (the original picture is described in [16]. In both pictures the shape of the embryo is approximated with a spline and in both pictures the aboral site is located at the left and the oral site at the right.

Modelling Gene Regulatory Networks

A number of different modelling formalisms have been proposed to simulate the dynamics of developmental gene regulatory networks [5]. Boolean network models [13] or generalised logical models (e.g., [7, 22] have been applied to model pattern formation and cell-cycle regulation in a qualitative manner. Here we focus on spatio-temporal models of gene regulation, which capture quantitative aspects of pattern formation during embryo development. Here we use the connectionist gene circuit model introduced by Reinitz and colleagues [12,17,19,23]:

$$\frac{dg_i{}^a}{dt} = R^a \phi \{ \sum_{b=1}^{N_g} W^{ab} g_i^b + h^a \} - \lambda g_i^a + D^a \nabla^2 g_i^a \qquad (1)$$

where a and b denote gene products (targets and regulators respectively) and i the cell (or nucleus) number (i.e., position in space). In Eq. 1 gene product concentrations depend on three main factors: (1) The first term describes the regulation of gene product synthesis in each cell. The set of genes is assumed to be identical in each cell. This is represented by the weight matrix W whose elements W^{ab} characterize the regulatory effect of gene b on gene a. N_g denotes the total number of genes in the model. ϕ is a sigmoid function with range [0,1], R^a is the production rate, h^a is a threshold parameter, which describes the expression of gene product a in absence of any regulators b. (2)The second term represents the decay of the gene products and (3)the third term represents the exchange of diffusible products between neighbouring cells.

Parameter Estimation by Optimisation

In the connectionist model regulatory interactions are represented by a large number of model parameters, such as kinetic constants, regulatory weights, Hill-coefficients and/or threshold parameters. The precise mathematical formulation depends on the level of detail considered in each gene regulatory model, but they all have in common that these parameter values are often difficult or impossible to measure. Instead, they have to be inferred by fitting models to experimentally measured gene expression data using global, non-linear optimization. The standard optimisation approach minimizes a single cost function that represents the quality of the data-fit (single-objective optimisation). In most studies, parameters are estimated by minimizing the sum of squared differences between model and data in which the residual function is given by:

$$E(\theta) = \sum_{i,a,t} (g_i^a(t, \theta)_{\text{model}} - g_i^a(t)_{\text{data}})^2 \qquad (2)$$

where the summation is performed over all cell, genes and time points for which we have data. Box constraints and penalty functions are used to reduce the search space and also to ensure that the parameter values are within biophysical limits [9, 12].

Cell-Based Modelling

The model for gene regulation can be coupled with a physically based model of cells. A great variety of cell-based modelling approaches have been published with levels of detail varying from point cells to complex polygons [1, 3, 11, 15, 18, 21].

Important aspects in early gastrulation are cell shape and cell membrane properties that mediate cell-cell interactions in a complex manner where adhesion strength may vary spatially. This cell polarity can also be utilized to direct cell division and cell movement. To each cell or region a gene regulatory network can be attributed and also signalling between cells, regions and the environment can be modelled. At early stages the number of cells is limited and cell migration mainly occurs during gastrulation. For this stage, we use a cell-based model, in which each cell is represented individually by a polygon (2D, [24]) or a polyhedron (3D, unpublished data).

3   Results and Discussion

The in situ hybridisations (Fig. 1) are decomposed into segments that are located along an approximately circular region. We have quantified gene expression in every segment in Fig 1. If we start measuring gene expression at the aboral site, we can transform the expression patterns shown in Fig 1 into the one-dimensional plot shown in Fig. 2. The x-axis represents a spatial axis in the system and the y-axis the quantification of gene expression in every segment along the circular decomposition constructed in the blastula stage. Fig. 2 summarizes the information from several published in situ hybridizations during the early gastrulation of *Nematostella* [8, 14, 16, 25].

The quantified gene expression patterns shown in Fig. 2 can be fit efficiently to the connectionist gene expression model with a genetic algorithm-based optimisation method [9] in which the error function shown in Eq. 2 is minimized. Fig 3 shows one (preliminary) approximation of the gene expression pattern in Fig. 2. It is a dynamical solution: first there is a maternal gradient beta-catenin, followed by the formation of an Anthox1 gradient and finally a gradient of snail expression around the oral pore.

Currently we are working on coupling the model of gene regulation with the cell-based model of gastrulation. Fig. 4 shows an example of the 2D cell-based model of gastrulation in *Nematostella vectensis* [24]. The simulation starts with a series of wedge-shaped cells in a circular configuration (top left picture in Fig. 4). In the blue cells there is no snail expression, while in the red region snail is being expressed. Snail expression radically changes the biomechanics of the cells: in the red cells, cell adhesion decreases while the red cells remain attached at the apical ends. This different biomechanical behaviour results into the formation of bottle cells and apical constriction [24]. Furthermore the red cells are becoming motile and form filipodia which zipper with the rigid layer of blue cells. These biomechanical events - the apical constriction and the zipping up by the filopodia with the blue cells - are sufficient to obtain a fully gastrulated structure (right bottom

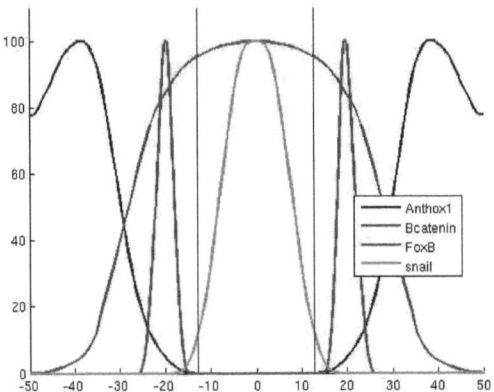

Fig. 2. Quantified gene expression of Anthox1, beta-catenin, FoxB and snail in in situ hybridisations of the blastula stage of *Nematostella vectensis* [8, 14, 16, 25]. Gene expression was measured in every segment along a circular decomposition at the blastula stage (Fig. 1). The x-axis represents the spatial axis in the system and the measurements start at the aboral segment in Fig. 1. The y-axis represents the normalized quantification of the gene expression in every segment along the decomposition.

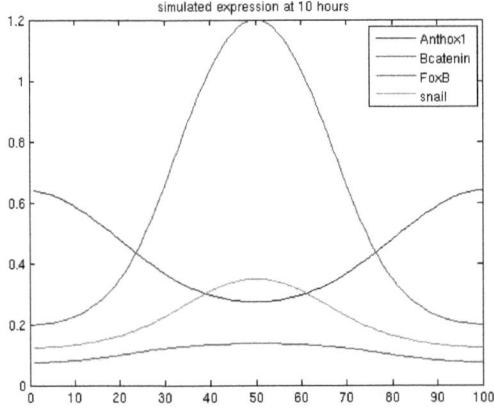

Fig. 3. Simulated gene expression using the connectionist model (Eq. 1). This solution is based on the data shown in Fig. 2. The picture shows simulated gene expression after 10 hours. The x-axis is a spatial axis in the system (similar like in Fig. 2).

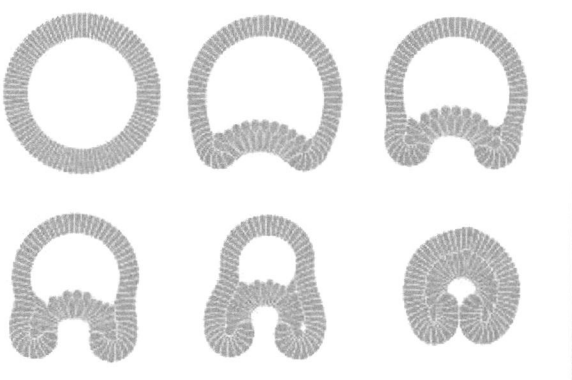

Fig. 4. 2D cell-based model [24] of gastrulation in *Nematostella vectensis*. The colours indicate regions with different simulated gene expression.

picture in Fig. 4) where two cell layers consisting of endodermal (red cells) and ectodermal cells are formed. In future work, we shall combine both the connectionist gene regulatory model with the mechanical model of gastrulation to obtain a full genetically regulated cell-based model of the developing *Nematostella* embryo.

Acknowledgements. Carlos Tamulonis was funded by the *Fundação para a Ciência e a Tecnologia* Portugal (grant SFRH/BD/29117/2006. Daniel Botman was funded by the FP7 project BioPreDyn.

References

1. Armstrong, N.J., Painter, K.J., Sheratt, J.A.: A continuum approach to modelling cell-cell adhesion. J. Theor. Biol. 243, 98–113 (2006)
2. Ashyraliyev, M., Fomekong Nanfack, Y., Kaandorp, J.A., Blom, J.G.: Systems biology: Parameter estimation for biochemical models. FEBS Journal 276, 886–902 (2009)
3. Dallon, J.C., Othmer, H.G.: How cellular movement determines the collective force generated by the Dictyostelium discoideum slug. J. Theor. Biol. 231, 203–222 (2004)
4. Davidson, E.H.: The regulatory genome: gene regulatory networks in development and evolution. Academic Press, London (2006)
5. de Jong, H.: Modeling and simulation of genetic regulatory systems: A literature review. J. Comp. Biol. 9, 67–103 (2002)
6. de Jong, J.: Quantitative analysis of gene expression in *Nematostella vectensis*. Master's thesis, Faculty of Science, University of Amsterdam (2009)
7. Fauré, A., Naldi, A., Chaouiya, C., Thieffry, D.: Dynamical analysis of a generic Boolean model for the control of the mammalian cell cycle. Bioinformatics 22(14), e124–e131 (2006)

362 J.A. Kaandorp et al.

8. Finnerty, J.R., Pang, K., Burton, P., Paulson, D., Martindale, M.Q.: Origins of Bilateral Symmetry: Hox and Dpp expression in a Sea Anemone Science, vol. 304, pp. 1335–1336 (2004)
9. Fomekong Nanfack, Y., Kaandorp, J.A., Blom, J.G.: Efficient parameter estimation for spatio-temporal models of pattern formation: Case study of Drosophila melanogaster. Bioinformatics 23, 3356–3363 (2007)
10. Gutenkunst, R.N., Waterfall, J.J., Casey, F.P., Brown, K.S., Myers, C.R., et al.: Universally Sloppy Parameter Sensitivities in Systems Biology Models. PLoS Comput. Biol. 3(10) (2007)
11. Honda, H., Tanemura, M., Nagai, T.: A three-dimensional vertex dynamics cell model of space-filling polyhedra simulating cell behavior in a cell aggregate. J. Theor. Biol. 226, 439–453 (2004)
12. Jaeger, J., Surkova, S., Blagov, M., Janssens, H., Kosman, D., Kozlov, K.N., Manu, M.E., Vanario-Alonso, C.E., Samsonova, M., Sharp, D.H., Reinitz, J.: Dynamic control of positional information in the early Drosophila blastoderm. Nature 430, 368–371 (2004)
13. Kauffman, S.A.: The Origins of Order: Self Organization and Selection in Evolution. Oxford University Press, Oxford (1993)
14. Magie, C.R., Pang, K., Martindale, M.Q.: Martindale Genomic inventory and expression of Sox and Fox genes in the cnidarian Nematostella vectensis. Dev. Genes Evol. 215, 618–630 (2005)
15. Maree, A.F., Hogeweg, P.: Modelling Dictyostelium discoideum morphogenesis: the culmination. Bull. Math. Biol. 64(2), 327–353 (2002)
16. Martindale, M.Q., Pang, K., Finnerty, J.R.: Investigating the origins of triploblasty: mesodermal gene expression in a diploblastic animal, the sea anemone Nematostella vectensis (phylum, Cnidaria; class, Anthozoa). Development 131, 2463–2474 (2004)
17. Mjolsness, E., Sharp, D.H., Reinitz, J.: A connectionist model of development. J. Theor. Biol. 152, 429–453 (1991)
18. Odell, G., Oster, G., Burnside, B., Alberch, P.: A mechanical model for epithelial morphogenesis. J. Math. Biol. 9, 291–295 (1980)
19. Reinitz, J., Sharp, D.H.: Mechanisms of eve stripe formation. Mechanisms of Development 49, 133–158 (1995)
20. Reinitz, J., Hou, S., Sharp, D.H.: Transcriptional Control in Drosophila. ComPlexUs 1, 54–64 (2003)
21. Schaller, G., Meyer-Hermann, M.: Multicellular tumor spheroid in an off-lattice Voronoi-Delaunay cell model. Phys. Rev. E. Stat. Nonlin. Soft. Matter Phys. 71(5 pt. 1), 51910 (2005)
22. Sánchez, L., Thieffry, D.: A logical analysis of the Drosophila gap-gene system. J. Theor. Biol. 211, 115–141 (2001)
23. Surkova, S., Kosman, D., Kozlov, K., Manu Myasnikova, E., et al.: Characterization of the Drosophila segment determination morphome. Dev. Biol. 313, 844–862 (2008)
24. Tamulonis, C., Postma, M., Marlow, H., Magie, C., de Jong, J., Kaandorp, J.A.: Morphometrics and Modeling of Nematostella vectensis Gastrulation. Developmental Biology 351, 217–228 (2011)
25. Wikramanayake, A.H., Hong, M., Lee, P.N., Pang, K., Byrum, C.A., Bince, J.M., Xu, R., Martindale, M.Q.: An ancient role for beta-catenin in the evolution of axial polarity and germ layer segregation. Nature 426, 446–450 (2003)

Tree-Automatic Well-Founded Trees

Alexander Kartzow[1,*], Jiamou Liu[2], and Markus Lohrey[1,*]

[1] Universität Leipzig, Germany, Institut für Informatik, Johannisgasse 26, 04103
Leipzig, Germany
{kartzow,lohrey}@informatik.uni-leipzig.de
[2] Department of Computer Science, University of Auckland, Private Bag 92019
Auckland, New Zealand
jiamou.liu@aut.ac.nz

Abstract. We investigate tree-automatic well-founded trees. For this, we introduce a new ordinal measure for well-founded trees, called ∞-rank. The ∞-rank of a well-founded tree is always bounded from above by the ordinary (ordinal) rank of a tree. We also show that the ordinal rank of a well-founded tree of ∞-rank α is smaller than $\omega \cdot (\alpha + 1)$. For string-automatic well-founded trees, it follows from [16] that the ∞-rank is always finite. Here, using Delhommé's decomposition technique for tree-automatic structures, we show that the ∞-rank of a tree-automatic well-founded tree is strictly below ω^ω. As a corollary, we obtain that the ordinal rank of a string-automatic (resp., tree-automatic) well-founded tree is strictly below ω^2 (resp., ω^ω). The result for the string-automatic case nicely contrasts a result of Delhommé, saying that the ranks of string-automatic well-founded partial orders reach all ordinals below ω^ω. As a second application of the ∞-rank we show that the isomorphism problem for tree-automatic well-founded trees is complete for level $\Delta^0_{\omega^\omega}$ of the hyperarithmetical hierarchy (under Turing-reductions). Full proofs can be found in the arXiv-version [11] of this paper.

1 Introduction

Various classes of infinite but finitely presented structures received a lot of attention in algorithmic model theory [2]. Among the most important such classes of structures is the class of *string-automatic structures* [13]. A (relational) structure is string-automatic if its universe is a regular set of words and all relations can be recognized by synchronous multi-tape automata. During the past 15 years a theory of string-automatic structures has emerged. This theory was developed along two interrelated branches. The first is a structural branch, which leads to (partial) characterizations of particular classes of string-automatic structures [6,12,14,15,18]. The second is an algorithmic branch, which leads to numerous decidability and undecidability, as well as complexity results for important algorithmic problems for string-automatic structures [4,14,17]. One of the most fundamental results for string-automatic structures states that their first-order theories are uniformly decidable [13].

---

* The first and third author are supported by the DFG research project GELO.

S.B. Cooper, A. Dawar, and B. Löwe (Eds.): CiE 2012, LNCS 7318, pp. 363–373, 2012.
© Springer-Verlag Berlin Heidelberg 2012

By replacing strings and string automata by trees and tree automata, Blumensath [3] generalized string-automatic structures to *tree-automatic structures* and proved that their first-order theories are still uniformly decidable. However compared to string-automatic structures, the theory of tree-automatic structures is less developed. The only non-trivial characterization of a class of tree-automatic structures we are aware of concerns ordinals. Delhommé proved in [6] that an ordinal is tree-automatic if and only if it is strictly below ω^{ω^ω}. Some complexity results for first-order theories of tree-automatic structures are shown in [17]. Recently, Huschenbett proved that it is decidable whether a given tree-automatic scattered linear order is string-automatic [9].

In this paper, we study tree-automatic well-founded trees.[1] Our main tool is a an ordinal measure for well-founded trees called ∞-rank, which is related to the classical (ordinal) rank of a well-founded tree. Consider a well-founded tree $\mathfrak{T}$ with root r. The rank of $\mathfrak{T}$ is the smallest ordinal, which is strictly larger than the ranks of the subtrees rooted in the children of r. In contrast to this, we only require the ∞-rank of $\mathfrak{T}$ to be (i) strictly larger than the ∞-ranks of those subtrees that are rooted (up to isomorphism) in *infinitely* many children of r and (ii) to be at least as large as the ∞-ranks of those subtrees that are rooted (up to isomorphism) in *finitely* many children of r. For instance, if a tree $\mathfrak{T}$ has finite depth, then ∞-rank$(\mathfrak{T})$ is the largest number $i \in \mathbb{N}$ such that the tree $\mathbb{N}^{\leq i}$ can be embedded into $\mathfrak{T}$.

Clearly, the ∞-rank of a well-founded tree is bounded from above by the classical (ordinal) rank of a tree. We also show that the rank of a well-founded tree of ∞-rank α is strictly bounded by $\omega \cdot (\alpha + 1)$. For string-automatic well-founded trees, it follows from [16] that the ∞-rank is always finite. Here, using a refinement of Delhommé's decomposition technique for tree-automatic structures [6], we show that the ∞-rank of a tree-automatic well-founded tree is strictly below ω^ω. As a corollary, we obtain that the rank of a string-automatic (resp., tree-automatic) well-founded tree is strictly below ω^2 (resp., ω^ω). The result for the string-automatic case nicely contrasts a result from [6,12], saying that the ranks of string-automatic well-founded partial orders reach exactly all ordinals below ω^ω.

Our second application of the ∞-rank concerns the isomorphism problem for tree-automatic well-founded trees. In [16], it was shown that the isomorphism problem for string-automatic well-founded trees is complete for level Δ^0_ω of the hyperarithmetical hierarchy. In other words, the isomorphism problem for string-automatic well-founded trees is recursively equivalent to true arithmetic. We show that the ∞-rank of well-founded computable trees determines the complexity of the isomorphism problem in the following sense: The isomorphism problem for well-founded computable trees of ∞-rank at most $\lambda + k$ (where $k \in \mathbb{N}$ and λ is a computable limit ordinal) belongs to level $\Sigma^0_{\lambda+3(k+1)}$ of the hyperarithmetical hierarchy. Since we know that the ∞-rank of a tree-automatic well-founded tree is strictly below ω^ω, we can use this fact and show that the isomorphism

---

[1] In this paper *tree* always refers to an order tree $\mathfrak{T} = (T, \leq)$ as opposed to a successor tree, i.e., a tree is a partial order (without successor relation).

problem for tree-automatic well-founded trees belongs to level $\Delta^0_{\omega^\omega} = \Sigma^0_{\omega^\omega} \cap \Pi^0_{\omega^\omega}$ of the hyperarithmetical hierarchy. We also provide a corresponding lower bound w.r.t. Turing-reductions. Thus, the isomorphism problem for tree-automatic well-founded trees is $\Delta^0_{\omega^\omega}$-complete under Turing-reductions.

Let us remark that for non-well-founded order trees, the isomorphism problem is complete for Σ^1_1 (the first existential level of the analytical hierarchy) already in the string-automatic case [16], and this complexity is in a certain sense maximal, since the isomorphism problem for the class of all computable structures is Σ^1_1-complete as well [5,7]. Let us also emphasize that all our results only hold for order trees, i.e., trees are seen as particular partial orders.

2   Preliminaries

A *relational structure* $\mathfrak{S}$ consists of a *domain* D and atomic relations on the set D. In this paper we shall only consider structures with countable domains. Let $\mathfrak{A} = (A, \leq)$ be a partial order. A subset $B \subseteq A$ is a *chain* if for all $a, b \in B$, $a \leq b$ or $b \leq a$. A subset $B \subseteq A$ is an *antichain* if for all pairs of distinct $a, b \in B$, neither $a \leq b$ nor $b \leq a$.

Trees and Forests. A *forest* is a partial order $\mathfrak{F} = (F, \leq)$ where for every $a \in F$ the set $\{b \in F \mid b < a\}$ is a finite chain. A *tree* is a forest which has a smallest element, which is called the *root* of the tree. Thus, a forest is a disjoint union of (an arbitrary number of) trees. For a given forest $\mathfrak{F}$, we denote by $\langle \mathfrak{F} \rangle$ the tree that results from adding a new root, i.e., a new smallest element, to $\mathfrak{F}$. If F is the domain of $\mathfrak{F}$ we denote by $\langle F \rangle$ the domain of $\langle \mathfrak{F} \rangle$. For a node u in $\mathfrak{F}$, let $\mathfrak{F}(u)$ be the subtree of $\mathfrak{F}$ at u, i.e., $\mathfrak{F}(u)$ is the restriction of $\mathfrak{F}$ to the set $\{v \in F \mid v \geq u\}$. We define the *successor relation* of $\mathfrak{F}$ as

$$E_{\mathfrak{F}} = \{(x, y) \in F \times F \mid x < y, \neg \exists z : x < z < y\}.$$

For $x \in F$ the set of *children* of x in $\mathfrak{F}$ is $E_{\mathfrak{F}}(x) = \{y \in F \mid (x, y) \in E_{\mathfrak{F}}\}$. A forest $\mathfrak{F} = (F, \leq)$ is *well-founded* if it does not contain an infinite ascending chain $a_1 < a_2 < a_3 < \cdots$.

Let us now define inductively the classical (ordinal) rank of a well-founded tree as well as the new notion of ∞-rank. We use standard terminology concerning ordinals; cf., e.g., [20]. For a set of ordinals M, let $\sup(M)$ be its supremum, where $\sup(\emptyset) = 0$. Let $\mathfrak{T}$ be a well-founded tree with root r. Thus, $C = E_{\mathfrak{T}}(r)$ is the set of children of the root. We define the *rank* of $\mathfrak{T}$ inductively as the ordinal

$$\mathrm{rank}(\mathfrak{T}) = \sup\{\mathrm{rank}(\mathfrak{T}(a)) + 1 \mid a \in C\}.$$

We define the ordinal ∞-rank$(\mathfrak{T})$ inductively using $\alpha = \sup\{\infty\text{-rank}(\mathfrak{T}(a)) \mid a \in C\}$:

$$\infty\text{-rank}(\mathfrak{T}) = \begin{cases} \alpha & \text{if } \{a \in C \mid \infty\text{-rank}(\mathfrak{T}(a)) = \alpha\} \text{ is finite,} \\ \alpha + 1 & \text{otherwise.} \end{cases}$$

The ∞-rank of a forest $\mathfrak{F}$ without smallest element is ∞-rank$(\mathfrak{F}) = \infty$-rank$(\langle \mathfrak{F} \rangle)$. A simple application of König's lemma shows that a well-founded forest $\mathfrak{F}$ has ∞-rank 0 if and only if $\mathfrak{F}$ is finite; in contrast, the rank of a finite tree can reach any finite ordinal. More generally, ∞-rank$(\mathfrak{F}) = n < \omega$ if and only if there is an embedding of the tree $\mathbb{N}^{\leq n}$ (the tree of height n where every non-leaf has $\aleph_0$ many children) into $\langle \mathfrak{F} \rangle$ but no embedding of $\mathbb{N}^{\leq n+1}$ into $\langle \mathfrak{F} \rangle$. The following lemma is crucial for studying the ∞-rank:

Lemma 1. *Let $\mathfrak{F} = (F, \leq)$ be a well-founded forest. There are only finitely many $a \in F$ with ∞-rank$(\mathfrak{F}(a)) = \infty$-rank$(\mathfrak{F})$.*

Proof. Let $\alpha = \infty$-rank$(\mathfrak{F})$. We show that $D = \{ a \in \langle F \rangle \mid \infty$-rank$(\langle \mathfrak{F} \rangle (a)) = \alpha \}$ is finite. Note that D is a downward-closed subset of the tree $\langle \mathfrak{F} \rangle$. Assume that this set is infinite. Since $\langle \mathfrak{F} \rangle$ is well-founded, König's lemma implies that D contains a node a which has infinitely many children a_i $(i \in \mathbb{N})$ that all belong to D. But then $\alpha = \infty$-rank$(\langle \mathfrak{F} \rangle (a)) \geq \alpha + 1$, which is a contradiction. $\square$

It is obvious that ∞-rank$(\mathfrak{T}) \leq$ rank$(\mathfrak{T})$ for every well-founded tree $\mathfrak{T}$. On the other hand, we can also bound rank$(\mathfrak{T})$ in terms of ∞-rank$(\mathfrak{T})$ as follows.

Lemma 2. *For a well-founded tree $\mathfrak{T}$ we have* rank$(\mathfrak{T}) < \omega \cdot (\infty$-rank$(\mathfrak{T}) + 1)$.

Proof. Let $\mathfrak{T} = (T, \leq)$ We proceed by induction on ∞-rank$(\mathfrak{T})$. If T is finite, then ∞-rank$(\mathfrak{T}) = 0$ and rank$(\mathfrak{T}) \leq |T| < \omega$. Now assume that ∞-rank$(\mathfrak{T}) = \alpha$ for some ordinal $\alpha > 0$ such that the theorem holds for all trees of ∞-rank strictly below α. By Lemma 1, $T_\alpha = \{ a \in T \mid \infty$-rank$(\mathfrak{T}(a)) = \alpha \}$ is a finite and downward-closed subset of T. Let $M_\alpha \subseteq T_\alpha$ be the set of $\leq$-maximal elements of T_α and consider a tree $\mathfrak{T}(a)$ for $a \in M_\alpha$. The definition of M_α implies the following. If $b \in T$ with $b > a$, then ∞-rank$(\mathfrak{T}(b)) = \beta$ for some ordinal $\beta < \alpha$. By the induction hypothesis it follows that rank$(\mathfrak{T}(b)) < \omega \cdot (\beta + 1) \leq \omega \cdot \alpha$. In particular, rank$(\mathfrak{T}(b)) < \omega \cdot \alpha$ for all children b of a. Thus, rank$(\mathfrak{T}(a)) \leq \omega \cdot \alpha$. Finally, since T_α is a finite set, we have

$$\text{rank}(\mathfrak{T}) \leq \sup\{\text{rank}(\mathfrak{T}(a)) \mid a \in M_\alpha\} + |T_\alpha| \leq \omega \cdot \alpha + |T_\alpha| < \omega \cdot (\alpha + 1). \quad \square$$

Note that the upper bound of $\omega \cdot (\infty$-rank$(\mathfrak{T}) + 1) = \omega \cdot \infty$-rank$(\mathfrak{T}) + \omega$ is optimal as for any $n < \omega$, the linear order of size n has ∞-rank 0 but rank $\omega \cdot 0 + n$.

Finite Labeled Trees. A *finite binary tree* is a prefix-closed finite subset $T \subseteq \{0, 1\}^*$, i.e., $uv \in T$ implies $u \in T$. We denote the set of all finite binary trees by $\mathcal{T}_2^{\text{fin}}$. Let $\preceq$ be the prefix relation on $\{0, 1\}^*$. Clearly $(T, \preceq)$ is a tree in the above sense.

Let Σ be a finite alphabet. A finite Σ-*labeled* binary tree is a pair (T, λ), where $T \in \mathcal{T}_2^{\text{fin}}$ and $\lambda : T \to \Sigma$ is a labeling function. By $\mathcal{T}_{2,\Sigma}^{\text{fin}}$ we denote the set of all finite Σ-labeled binary trees. Elements of $\mathcal{T}_{2,\Sigma}^{\text{fin}}$ are denoted by lower case letters $(s, t, \dots)$.

Next, we define the *convolution* $t_1 \otimes \cdots \otimes t_n$ of $t_1, \dots, t_n \in \mathcal{T}_{2,\Sigma}^{\text{fin}}$ as follows: Let $t_i = (T_i, \lambda_i)$ where $\lambda_i : T_i \to \Sigma$ and $\diamond \notin \Sigma$. Let $T = \bigcup_{i=1}^n T_i$ and define

$\lambda_i' : T \to \Sigma \cup \{\diamond\}$ by $\lambda_i'(u) = \lambda_i(u)$ for $u \in T_i$ and $\lambda_i'(u) = \diamond$ for $u \in T \setminus T_i$. Then $t_1 \otimes \cdots \otimes t_n$ is the finite $((\Sigma \cup \{\diamond\})^n \setminus \{\diamond\}^n)$-labeled tree (T, λ) where λ is defined by $\lambda(u) = (\lambda_1'(u), \ldots, \lambda_n'(u))$ for each $u \in T$.

Tree Automata and Tree-Automatic Structures. For $T \in \mathcal{T}_2^{\text{fin}}$ let

$$\text{cl}(T) = T \cup \{ui \mid u \in T, i \in \{0,1\}\}$$

be its closure, which is again prefix-closed. Let Σ be a finite alphabet. A *tree automaton over* Σ is a tuple $\mathcal{A} = (Q, \Delta, Q_I, Q_F)$, where Q is the finite set of states, $Q_I \subseteq Q$ is the set of initial states, $Q_F \subseteq Q$ is the set of final states, and $\Delta \subseteq (Q \setminus Q_F) \times \Sigma \times Q \times Q$ is the transition relation. Given $t = (T, \lambda) \in \mathcal{T}_{2,\Sigma}^{\text{fin}}$, a *successful run* of $\mathcal{A}$ on t is a mapping $\rho : \text{cl}(T) \to Q$ such that (i) $\rho(\varepsilon) \in Q_I$, (ii) $\rho(\text{cl}(T) \setminus T) \subseteq Q_F$, and (iii) for every $d \in T$, $(\rho(d), \lambda(d), \rho(d0), \rho(d1)) \in \Delta$. By $L(\mathcal{A})$ we denote the set of all $t \in \mathcal{T}_{2,\Sigma}^{\text{fin}}$ on which $\mathcal{A}$ has a successful run. A set $L \subseteq \mathcal{T}_{2,\Sigma}^{\text{fin}}$ is called *regular* if there is a tree automaton $\mathcal{A}$ over Σ with $L = L(\mathcal{A})$.

An n-ary relation $R \subseteq (\mathcal{T}_{2,\Sigma}^{\text{fin}})^n$ is called *tree-automatic* if there is a tree automaton $\mathcal{A}_R$ over $(\Sigma \cup \{\diamond\})^n \setminus \{\diamond\}^n$ such that $L(\mathcal{A}_R) = \{t_1 \otimes \cdots \otimes t_n \mid (t_1, \ldots, t_n) \in R\}$. A relational structure $\mathfrak{S}$ is called *tree-automatic* over Σ if its domain is a regular subset of $\mathcal{T}_{2,\Sigma}^{\text{fin}}$ and each of its atomic relations is tree-automatic; any tuple $\mathbb{P}$ of automata that accepts the domain and the relations of $\mathfrak{S}$ is called a *tree-automatic presentation* of $\mathfrak{S}$. In this case, we write $\mathfrak{S}(\mathbb{P})$ for $\mathfrak{S}$. If a tree-automatic structure $\mathfrak{S}$ is isomorphic to a structure $\mathfrak{S}'$, then $\mathfrak{S}$ is called a *tree-automatic copy* of $\mathfrak{S}'$ and $\mathfrak{S}'$ is *tree-automatically presentable*. In this paper we sometimes abuse the terminology referring to $\mathfrak{S}'$ as simply tree-automatic and calling a tree-automatic presentation of $\mathfrak{S}$ also a tree-automatic presentation of $\mathfrak{S}'$. We also simplify our statements by saying "given/compute a tree-automatic structure $\mathfrak{S}$" for "given/compute a tree-automatic presentation $\mathbb{P}$ of a structure $\mathfrak{S}(\mathbb{P})$". The structures $(\mathbb{N}, +)$ and $(\mathbb{N}, \times)$ are examples of tree-automatic structures. We shall make use of the following simple lemma.

Lemma 3. *For every tree-automatic structure there is an isomorphic tree-automatic structure $\mathfrak{S}$ over the alphabet $\{a\}$, i.e., the domain can be seen as a subset of $\mathcal{T}_2^{\text{fin}}$.*

Consider $\text{FO} + \exists^\infty + \exists^{n,m} + \exists^{\text{chain}}$, i.e., first-order logic extended by the quantifiers $\exists^\infty$ (there exists infinitely many), $\exists^{n,m}$ (there exists finitely many and the exact number is congruent n modulo m, where $m, n \in \mathbb{N}$) and the chain-quantifier $\exists^{\text{chain}}$ (if $\varphi(x, y)$ is some formula, then $\exists^{\text{chain}}\varphi(x, y)$ asserts that $\varphi(x, y)$ defines a partial order that contains an infinite ascending chain). Results from [3,10,21] show that the $\text{FO} + \exists^\infty + \exists^{n,m} + \exists^{\text{chain}}$ theory of any tree-automatic structure $\mathfrak{S}$ is (uniformly) decidable. Note that the property of being a tree is expressible in $\text{FO} + \exists^\infty$ and well-foundedness of a tree is expressible in $\text{FO} + \exists^{\text{chain}}$. Hence, we get:

Theorem 4. *It is decidable whether a given tree-automatic structure is a well-founded tree.*

Let $\mathcal{K}$ be a class of tree-automatic presentations. The *isomorphism problem* $\mathrm{Iso}(\mathcal{K})$ is the set of pairs $(\mathbb{P}_1, \mathbb{P}_2) \in \mathcal{K} \times \mathcal{K}$ of tree-automatic presentations with $\mathfrak{S}(\mathbb{P}_1) \cong \mathfrak{S}(\mathbb{P}_2)$. If $\mathcal{K}$ is the class of tree-automatic presentations for a class $\mathcal{C}$ of relational structures (e.g., trees), then we shall briefly speak of the isomorphism problem for (tree-automatic members of) $\mathcal{C}$. The isomorphism problem for the class of all tree-automatic structures is complete for Σ_1^1, the first level of the analytical hierarchy; this holds already for (non well-founded) string-automatic trees [14,16].

Hyperarithmetical Sets. We use standard terminology concerning recursion theory; cf., e.g., [19]. We use the definition of the hyperarithmetical hierarchy from Ash and Knight [1] (cf. [8]). We first define inductively a set of *ordinal notations* $O \subseteq \mathbb{N}$. Simultaneously we define a mapping $a \mapsto |a|_O$ from O into ordinals and a strict partial order $<_O$ on O. The set O is the smallest subset of $\mathbb{N}$ satisfying the following conditions:

- $1 \in O$ and $|1|_O = 0$, i.e., 1 is a notation for the ordinal 0.
- If $a \in O$, then also $2^a \in O$. We set $|2^a|_O = |a|_O + 1$ and let $b <_O 2^a$ if and only if $b = a$ or $b <_O a$.
- If $e \in \mathbb{N}$ is such that Φ_e (the eth partial computable function) is total, $\Phi_e(n) \in O$ for all $n \in \mathbb{N}$, and $\Phi_e(0) <_O \Phi_e(1) <_O \Phi_e(2) <_O \cdots$, then also $3 \cdot 5^e \in O$. We set $|3 \cdot 5^e|_O = \sup\{|\Phi_e(n)|_O \mid n \in \mathbb{N}\}$ and let $b <_O 3 \cdot 5^e$ if and only if there is an $n \in \mathbb{N}$ with $b <_O \Phi_e(n)$.

An ordinal α is *computable* if there is an $a \in O$ with $|a|_O = \alpha$. The smallest non-computable ordinal is the Church-Kleene ordinal ω_1^{ck}. If $a \in O$ then the restriction of the partial order $(O, <_O)$ to $O_a = \{b \in O \mid b <_O a\}$ is isomorphic to the ordinal $|a|_O$ [1, Proposition 4.9]. Based on ordinal notations we define the *hyperarithmetical hierarchy*. For this we define sets $H(a)$ for each $a \in O$ as follows:

- $H(1) = \emptyset$,
- $H(2^b) = H(b)'$ (the Turing jump of $H(b)$; cf., e.g., [19]),
- $H(3 \cdot 5^e) = \{\langle b, n \rangle \mid b <_O 3 \cdot 5^e, n \in H(b)\}$; here $\langle \cdot, \cdot \rangle$ denotes some computable pairing function.

Spector has shown that $|a|_O = |b|_O$ implies that $H(a)$ and $H(b)$ are Turing equivalent. The levels of the hyperarithmetical hierarchy can be defined as follows, where α is a computable ordinal.

- Σ_α^0 is the set of all subsets $A \subseteq \mathbb{N}$ that are recursively enumerable in some $H(a)$ with $|a|_O = \alpha$ (by Spector's theorem, the concrete choice of a is irrelevant).
- Π_α^0 is the set of all complements of Σ_α^0 sets.
- $\Delta_\alpha^0 = \Sigma_\alpha^0 \cap \Pi_\alpha^0$, i.e., Δ_α^0 is the set of all subsets $A \subseteq \mathbb{N}$ that are Turing-reducible to some $H(a)$ with $|a|_O = \alpha$.

For any two computable ordinals α and β, $\alpha < \beta$ implies $\Sigma_\alpha \cup \Pi_\alpha \subsetneq \Delta_\beta$. The union of all classes Σ_α^0 where $\alpha < \omega_1^{ck}$ yields the class of all *hyperarithmetical sets.* By a classical result of Kleene, the hyperarithmetical sets are exactly the sets in $\Delta_1^1 = \Sigma_1^1 \cap \Pi_1^1$, where Σ_1^1 is the first existential level of the analytical hierarchy, and Π_1^1 is the set of all complements of Σ_1^1-sets.

3   Bounding the ∞-rank of Tree-Automatic Well-Founded Trees

The first main result of this paper is:

Theorem 5. *If $\mathfrak{T}$ is a tree-automatic well-founded tree, then $\infty\text{-rank}(\mathfrak{T}) < \omega^\omega$.*

Before we sketch a proof of this result, let us first deduce a corollary:

Corollary 6. *For $\mathfrak{T} = (T, \leq)$ a string-automatic (tree-automatic, respectively) well-founded tree we have $\mathsf{rank}(\mathfrak{T}) < \omega^2$ ($\mathsf{rank}(\mathfrak{T}) < \omega^\omega$, respectively).*

Proof. For a string-automatic well-founded tree $\mathfrak{T}$, $\infty\text{-rank}(\mathfrak{T})$ is finite by [16][2]. With Lemma 2 we get $\mathsf{rank}(\mathfrak{T}) \leq \omega \cdot i < \omega^2$ for some $i \in \mathbb{N}$. For a tree-automatic well-founded tree $\mathfrak{T}$ we have $\infty\text{-rank}(\mathfrak{T}) < \omega^\omega$ by Theorem 5. Thus, there is some $i \in \mathbb{N}$ such that $\infty\text{-rank}(\mathfrak{T}) \leq \omega^i$. With Lemma 2 we get $\mathsf{rank}(\mathfrak{T}) < \omega \cdot \omega^i + \omega < \omega^\omega$.   □

Note that Corollary 6 contrasts with results on the ranks of string-automatic well-founded partial orders.[3] By [6,12], the ordinal ranks of string-automatic well-founded partial orders are the ordinals strictly below ω^ω. In fact, the result still holds for partial orders without infinite chains [11]. Moreover, Delhommé's characterization of tree-automatic ordinals yields tree-automatic well-founded partial orders of rank α for each $\alpha < \omega^{\omega^\omega}$ [6].

Let us sketch the proof of Theorem 5. The first part relies on Delhommé's decomposition technique for tree-automatic structures from [6] where he proved that the ordinal ω^{ω^ω} is not tree-automatic. Let us explain his decomposition technique for a tree-automatic graph $\mathfrak{G} = (V, E)$. Because of Lemma 3 we can assume that $V \subseteq \mathcal{T}_2^{\mathsf{fin}}$. Consider a tree automaton $\mathcal{A}$ that accepts a subset of $V \otimes \mathcal{T}_2^{\mathsf{fin}}$, and for each $s \in \mathcal{T}_2^{\mathsf{fin}}$ let $\mathfrak{G}_s$ be the subgraph of $\mathfrak{G}$ induced by the set $\{t \in V \mid t \otimes s \in L(\mathcal{A})\}$.

Delhommé's main proposition from [6] shows that every subgraph $\mathfrak{G}_s$ can be obtained from a finite set of subgraphs $\mathcal{C}$ by using the operations of *box-augmentation* and *sum-augmentation*. Roughly speaking, a graph $\mathfrak{G}$ is a sum-augmentation of subgraphs $\mathfrak{G}_1, \ldots, \mathfrak{G}_n$ if it is the disjoint union of $\mathfrak{G}_1, \ldots, \mathfrak{G}_n$ where we may add edges between different $\mathfrak{G}_i$ (but not within a single $\mathfrak{G}_i$). $\mathfrak{G}$ is a box-augmentation of the graphs $\mathfrak{G}_1, \ldots, \mathfrak{G}_n$ with node sets $V_1, \ldots, V_n$

---

[2] In [16], the notion of *embedding rank* of an arbitrary tree is defined. Comparison of the definitions shows that the embedding rank of a well-founded tree is finite iff its ∞-rank is finite.

[3] The rank generalizes naturally to all well-founded partial orders (considering roots as maximal elements of trees).

if the node set of $\mathfrak{G}$ is the product $\prod_{i=1}^n V_i$ and for every $1 \le i \le n$ and all $v_1 \in V_1, \ldots, v_{i-1} \in V_{i-1}, v_{i+1} \in V_{i+1}, \ldots, v_n \in V_n$, the subgraph of $\mathfrak{G}$ induced by the set $\{(v_1, \ldots, v_{i-1}, v, v_{i+1}, \ldots, v_n) \mid v \in V_i\}$ is isomorphic to $\mathfrak{G}_i$.

Now, let ν be a function that maps graphs to some set M such that isomorphic graphs are mapped to the same element. We say that $m \in M$ is ν-sum-indecomposable (ν-box-indecomposable, resp.) if for all graphs $\mathfrak{G}, \mathfrak{G}_1, \ldots, \mathfrak{G}_n$ such that $\mathfrak{G}$ is a sum-augmentation (box-augmentation, resp.) of $\mathfrak{G}_1, \ldots, \mathfrak{G}_n$ the following implication holds: If $\nu(\mathfrak{G}) = m$ then $\nu(\mathfrak{G}_i) = m$ for some $1 \le i \le n$. Delhommé's decomposition result implies that the set $\{\nu(\mathfrak{G}_s) \mid s \in \mathcal{T}_2^{\mathrm{fin}}\}$ contains only finitely many values that are both ν-sum-indecomposable and ν-box-indecomposable.

In order to show that ω^{ω^ω} is not tree-automatic, Delhommé takes a tree-automatic copy $\mathfrak{G} = (V, \le)$ of some ordinal and a tree automaton $\mathcal{A}$ for the first-order formula $y < x$. Hence, the substructures $\mathfrak{G}_s$ are the initial segments of $\mathfrak{G}$. Moreover let ν_0 be the function that maps an initial segment of $\mathfrak{G}$ to the corresponding ordinal. Delhommé proves that every ordinal of the form ω^{ω^α} is both ν_0-sum-indecomposable and ν_0-box-indecomposable. Hence, $\mathfrak{G} = (V, \le)$ can contain only finitely many initial segments of the form ω^{ω^α}, which is not the case for ω^{ω^ω}.

We follow a similar strategy. Heading for a contradiction to Theorem 5, take a well-founded tree-automatic forest $\mathfrak{F} = (F, \le)$ with $\infty\text{-rank}(\mathfrak{F}) = \omega^\omega$. Let $\mathcal{A}$ be a tree automaton for the first-order formula $x \le y$. Hence, the substructures $\mathfrak{F}_s$ are the subtrees $\mathfrak{F}(v)$ of $\mathfrak{F}$ for $v \in F$. It is not difficult to show that for every ordinal $\alpha < \omega^\omega$, $\mathfrak{F}$ must contain a subtree of ∞-rank α. In particular, $\mathfrak{F}$ contains a subtree of ∞-rank ω^i for every $i \in \mathbb{N}$. Now, let ν_1 be the function that maps a subtree $\mathfrak{F}(v)$ to its ∞-rank. We would obtain a contradiction by proving that every ordinal of the form ω^α is both ν_1-sum-indecomposable and ν_1-box-indecomposable. Indeed, we can prove that every ordinal of the form ω^α is ν_1-sum-indecomposable. But there is a problem with ν_1-box-indecomposability: The ordinals 0 and 1 are the only ν_1-box-indecomposable ordinals. The problem is that any forest can be embedded into the box-augmentation of two infinite antichains. Hence, box-augmentations of two infinite antichains may have arbitrarily high ∞-rank. To solve this problem, we observe that the box-augmentations that are used for building up the subtrees $\mathfrak{F}_s$ of $\mathfrak{F}$ ($s \in \mathcal{T}_2^{\mathrm{fin}}$) have a particular property that we call tame colorability (this is joint work with Martin Huschenbett). If a graph $\mathfrak{G} = (V, E)$ is a box-augmentation of subgraphs $\mathfrak{G}_i = (V_i, E_i)$ $(1 \le i \le n)$, then this box-augmentation is *tamely colorable* if for each $1 \le i \le n$ there is a finite coloring c_i of $V_i \times V_i$ such that whether $((v_1, \ldots v_n), (v'_1, \ldots, v'_n)) \in E$ only depends on the colors $c_i(v_i, v'_i)$ for $1 \le i \le n$. A careful analysis of box-augmentations of forests shows that ω^α is also ν_1-tamely-colorable-box-indecomposable (tamely-colorable-box-indecomposability is defined as box-indecomposability, but only considering tamely colorable box-augmentations). As in Delhommé's argument, we conclude that a well-founded tree-automatic forest $\mathfrak{F}$ only contains finitely many subtrees of pairwise distinct ∞-ranks of the form ω^i. Hence, $\infty\text{-rank}(\mathfrak{F}) < \omega^\omega$ and Theorem 5 follows.

4   The Isomorphism Problem for Well-Founded Tree-Automatic Trees

It turns out that the ∞-rank for well-founded computable trees yields an upper bound on the recursion-theoretic complexity of the isomorphism problem. Recall that we defined trees and forests as particular partial orders. For the isomorphism problem, it is useful to assume that also the direct successor relation is computable. When speaking of a *computable forest* in the following theorem, we mean a forest $\mathfrak{F} = (F, \leq)$ such that $F \subseteq \mathbb{N}$, $\leq \, \subseteq \mathbb{N} \times \mathbb{N}$, and the direct successor relation $E_{\mathfrak{F}}$ are all computable sets.[4] Note that the direct successor relation of a tree-automatic forest is still tree-automatic (and hence computable) because it is first-order definable.

Lemma 7. *Let α be a computable ordinal and assume that $\alpha = \lambda + k$, where $k \in \mathbb{N}$ and either $\lambda = 0$ or λ is a limit ordinal. The isomorphism problem for well-founded computable trees of ∞-rank at most α belongs to level $\Sigma^0_{\lambda+2(k+1)}$ of the hyperarithmetical hierarchy.*

For the proof of Lemma 7 we use a characterization of the hyperarithmetical levels by *computable infinitary formulas* (cf. [1]). These are first-order formulas over the structure $(\mathbb{N}, +, \times)$, where countably infinite conjunctions and disjunctions are allowed. Computability of such an infinitary formula means that for an infinite conjunction $\bigvee_{n \in \mathbb{N}} \varphi_n$ there is a computable function that maps n to a representation of φ_n (note that φ_n may again contain infinite conjunctions and disjunctions), and similarly for infinite disjunctions. Roughly speaking, such an infinitary formula can be encoded by a computable tree (the syntax tree of the formula) and if that tree has rank α (a computable ordinal) then the relation defined by the formula belongs to level α of the hyperarithmetical hierarchy.

In order to prove Lemma 7, we construct for a computable ordinal α a computable infinitary formula $\mathsf{iso}_\alpha(x, y)$ that is satisfied in a well-founded computable forest $\mathfrak{F}$ if and only if $\mathfrak{F}(x)$ and $\mathfrak{F}(y)$ have ∞-rank at most α and $\mathfrak{F}(x) \cong \mathfrak{F}(y)$. The construction is carried out inductively along the ordinal α, similarly to the proof of Lemma 25 in [16].

From Theorem 5 and Lemma 7 it follows that the isomorphism problem for well-founded tree-automatic trees belongs to $\Pi^0_{\omega^\omega}$. Using similar formulas as those constructed in our proof of Lemma 7, we can also show that the isomorphism problem for well-founded tree-automatic trees belongs to $\Sigma^0_{\omega^\omega}$. Hence, we get:

Corollary 8. *The isomorphism problem for well-founded tree-automatic trees belongs to $\Delta^0_{\omega^\omega}$.*

---

[4] On the other hand, if we would omit the requirement of a computable direct successor relation in Lemma 7, then we would only have to replace the constant 2 in the lemma by a larger value.

Let us now turn to lower bounds. Our main technical result is:

Lemma 9. *From a given $i \in \mathbb{N}$, one can compute a well-founded tree-automatic tree $\mathfrak{V}_i$ such that the following holds: From a given $\Pi^0_{\omega^i}$-set $P \subseteq \mathbb{N}$ (represented, e.g., by a computable infinitary formula) and $n \in \mathbb{N}$ one can compute a well-founded tree-automatic tree $\mathfrak{W}_{P,n}$ such that $n \in P$ if and only if $\mathfrak{V}_i \cong \mathfrak{W}_{P,n}$.*

We prove Lemma 9 by a reduction from the isomorphism problem for well-founded *computable trees*. We use a construction from [8]. Basically, [8, Proposition 3.2] states Lemma 9 for well-founded computable trees (instead of well-founded tree-automatic trees) and all computable ordinals (instead of ordinals ω^i). It turns out that the trees constructed in [8] for a certain ordinal α consist of computable subtrees of a "universal" well-founded computable tree $\mathfrak{S}_\alpha$. In case $\alpha = \omega^i$ for $i \in \mathbb{N}$ we can moreover show that the tree $\mathfrak{S}_{\omega^i}$ is tree-automatic. Roughly speaking this yields a weaker version of Lemma 9, where instead of well-founded tree-automatic trees we have the (tree-automatic) trees $\mathfrak{S}_{\omega^i}$ enriched by a computable unary predicate K on the node set of $\mathfrak{S}_{\omega^i}$. In fact K can be assumed to be a subset of the leaves of $\mathfrak{S}_{\omega^i}$; it yields a computable subtree of the universal tree $\mathfrak{S}_{\omega^i}$ by removing all leaves from K. Finally, to get rid of K we use a technique from [16]. In Lemma 41 from [16] it was shown that there are non-isomorphic (string-automatic) trees U_0 and U_1 with the following property: from an index of a computable set of strings $L \subseteq \{0,1\}^*$ one can compute a string-automatic forest $\mathfrak{F}_L$ of height 3 such that: (i) the set of roots is $\{0,1\}^*$, (ii) if $x \in L$ then $\mathfrak{F}_L(x) \cong U_0$, and (iii) if $x \notin L$ then $\mathfrak{F}_L(x) \cong U_1$. In [16], this statement was used in order to reduce the isomorphism problem for (non-well-founded) computable trees to the isomorphism problem for (non-well-founded) string-automatic trees. Hence, the latter problem is Σ^1_1-complete. In our situation, we first prove a tree version of [16, Lemma 41], where $\{0,1\}^*$ is replaced by $\mathcal{T}^{\text{fin}}_2$. Then, one can eliminate the additional computable unary predicate K on the leaves of $\mathfrak{S}_{\omega^i}$. For this, every leaf from K is replaced by the height-3 tree U_0, whereas all other leaves are replaced by the height-3 tree U_1. This yields a well-founded tree automatic tree that encodes the pair $(\mathfrak{S}_{\omega^i}, K)$.

From Lemma 9 we can deduce that the isomorphism problem for well-founded tree-automatic trees is $\Delta^0_{\omega^\omega}$-hard under Turing-reductions. With Corollary 8, we finally obtain our main result for the isomorphism problem.

Theorem 10. *The isomorphism problem for well-founded tree-automatic trees is $\Delta^0_{\omega^\omega}$-complete under Turing-reductions.*

Acknowledgment. We thank Martin Huschenbett for helpful discussions concerning Delhommé's decomposition result.

References

1. Ash, C.J., Knight, J.F.: Computable structures and the hyperarithmetical hierarchy. Studies in Logic and the Foundations of Mathematics, vol. 144. North-Holland Publishing Co., Amsterdam (2000)
2. Bárány, V., Grädel, E., Rubin, S.: Automata-based presentations of infinite structures. In: Finite and Algorithmic Model Theory. London Mathematical Society Lecture Notes Series, vol. 379, pp. 1–76. Cambridge University Press (2011)
3. Blumensath, A.: Automatic structures. Diploma thesis, RWTH Aachen (1999)
4. Blumensath, A., Grädel, E.: Finite presentations of infinite structures: Automata and interpretations. Theory Comput. Syst. 37, 642–674 (2004)
5. Calvert, W., Knight, J.F.: Classification from a computable viewpoint. Bull. Symbolic Logic 12(2), 191–218 (2006)
6. Delhommé, C.: Automaticité des ordinaux et des graphes homogènes. C.R. Acad. Sci. Paris Ser. I 339, 5–10 (2004)
7. Goncharov, S.S., Knight, J.F.: Computable structure and antistructure theorems. Algebra Logika 41(6), 639–681 (2002)
8. Hirschfeldt, D.R., White, W.M.: Realizing levels of the hyperarithmetic hierarchy as degree spectra of relations on computable structures. Notre Dame J. Form. Log. 43(1), 51–64 (2002)
9. Huschenbett, M.: Word automaticity of tree automatic scattered linear orderings is decidable. Technical report, arXiv.org (2012), http://arxiv.org/abs/1201.5070
10. Kartzow, A.: First-Order Model Checking On Generalisations of Pushdown Graphs. PhD thesis, TU Darmstadt (2011)
11. Kartzow, A., Lohrey, M., Liu, J.: Tree-automatic well-founded trees. Technical report, arXiv.org (2012), http://arxiv.org/abs/1201.5495
12. Khoussainov, B., Minnes, M.: Model theoretic complexity of automatic structures. Ann. Pure Appl. Logic 161(3), 416–426 (2009)
13. Khoussainov, B., Nerode, A.: Automatic Presentations of Structures. In: Leivant, D. (ed.) LCC 1994. LNCS, vol. 960, pp. 367–392. Springer, Heidelberg (1995)
14. Khoussainov, B., Nies, A., Rubin, S., Stephan, F.: Automatic structures: richness and limitations. Log. Methods Comput. Sci. 3(2):2:2, 18 (2007)
15. Khoussainov, B., Rubin, S., Stephan, F.: Automatic linear orders and trees. ACM Trans. Comput. Log. 6(4), 675–700 (2005)
16. Kuske, D., Liu, J., Lohrey, M.: The isomorphism problem on classes of automatic structures with transitive relations. To appear in Trans. Amer. Math. Soc. (2012)
17. Kuske, D., Lohrey, M.: Automatic structures of bounded degree revisited. J. Symbolic Logic 76(4), 1352–1380 (2011)
18. Oliver, G.P., Thomas, R.M.: Automatic Presentations for Finitely Generated Groups. In: Diekert, V., Durand, B. (eds.) STACS 2005. LNCS, vol. 3404, pp. 693–704. Springer, Heidelberg (2005)
19. Rogers, H.: Theory of Recursive Functions and Effective Computability. McGraw-Hill (1968)
20. Rosenstein, J.: Linear Ordering. Academic Press (1982)
21. To, A.W., Libkin, L.: Recurrent Reachability Analysis in Regular Model Checking. In: Cervesato, I., Veith, H., Voronkov, A. (eds.) LPAR 2008. LNCS (LNAI), vol. 5330, pp. 198–213. Springer, Heidelberg (2008)

Infinite Games and Transfinite Recursion of Multiple Inductive Definitions

Keisuke Yoshii and Kazuyuki Tanaka

Mathematical Institute, Tohoku University, Sendai 980-8578, Japan
keisuke.yoshii@gmail.com, tanaka@math.tohoku.ac.jp

Abstract. The purpose of this research is to investigate the logical strength of weak determinacy of Gale-Stewart games from the standpoint of reverse mathematics. It is known that the determinacy of Σ_1^0 sets (open sets) is equivalent to system ATR$_0$ and that of Σ_2^0 corresponds to the axiom of Σ_1^1 inductive definitions. Recently, much effort has been made to characterize the determinacy of game classes above Σ_2^0 within second order arithmetic. In this paper, we show that for any $k \in \omega$, the determinacy of $\Delta((\Sigma_2^0)_{k+1})$ sets is equivalent to the axiom of transfinite recursion of Σ_1^1 inductive definitions with k operators, denote $[\Sigma_1^1]^k$-IDTR. Here, $(\Sigma_2^0)_{k+1}$ is the difference class of $k + 1$ Σ_2^0 sets and $\Delta((\Sigma_2^0)_{k+1})$ is the conjunction of $(\Sigma_2^0)_{k+1}$ and co-$(\Sigma_2^0)_{k+1}$.

1 Introduction

In 1953, Gale and Stewart introduced the following infinite games. Two players I and II alternatively choose an element from $\mathbb{N}$ and construct an infinite sequence $f \in \mathbb{N}^{\mathbb{N}}$. Player I is said to win the game if the resulting sequence f belongs to a given set A, and player II wins otherwise. We denote this game G_A, and call A the winning set. If A is an open set (a closed set, or a Borel set), G_A is called an open game (a closed game, or a Borel game, respectively). A strategy for player I (for player II) is a function σ from $\mathbb{N}^{\text{even}}$ to $\mathbb{N}$ (a function τ from $\mathbb{N}^{\text{odd}}$ to $\mathbb{N}$), where $\mathbb{N}^{\text{even}}$ ($\mathbb{N}^{\text{odd}}$) is a set of finite sequences of natural numbers with even (odd) length. A game G_A is said to be determinate if one of the players has a winning strategy in G_A, i.e., a strategy with which the player always wins.

Researches on the determinacy of infinite games have been conducted in descriptive set theory. It is provable in ZFC that a Borel game is determinate, but the same thing does not hold for an analytic game. These facts simply represent that the strength of the determinacy of G_A varies depending on the complexity of the winning set A.

Aside from descriptive set theory, researches classifying the strength of determinacy within second order arithmetic have been started by J. Steel and Tanaka along the Reverse Mathematics program. The goal of the program is to answer the following questions: *What set existence axioms are necessary and sufficient to prove the theorems of ordinary mathematics?* See Simpson [11] for the major results.

S.B. Cooper, A. Dawar, and B. Löwe (Eds.): CiE 2012, LNCS 7318, pp. 374–383, 2012.
© Springer-Verlag Berlin Heidelberg 2012

In [12], it is already shown that Δ_2^0-determinacy and Δ_2^1-CA are incomparable, which means that the determinacy strength can not be characterized by a standard comprehension axiom. In fact, even Δ_1^1-determinacy can not imply Δ_2^1-CA ([5]). Thus, different kinds of set existence axioms such as inductive definitions and transfinite recursions are needed to characterize the determinacy statements.

In [13], it is shown that Σ_2^0-determinacy is equivalent to the axiom Σ_1^1-ID of Σ_1^1 inductive definitions. Then, in [6], to pin down Δ_3^0-determinacy, we introduced a new axiom of multiple inductive definitions as well as the different hierarchy $(\Sigma_2^0)_\alpha, 0 < \alpha < \omega_1$. (see also [5]). In this paper, we treat the determinacy of the game in the Wadge classes $\Delta((\Sigma_2^0)_{k+1})$ which is between $(\Sigma_2^0)_k$ and $(\Sigma_2^0)_{k+1}$, and prove that it is equivalent to the axiom of iterated inductive definition with multiple operators, $[\Sigma_1^1]^k$-IDTR. This more or less answers Open Question 29 in A. Montalbán's survey paper [8]. See section 4.

2   Preliminaries

In this section, we recall some basic definitions and facts about second order arithmetic. The language $\mathcal{L}_2$ of second-order arithmetic is a two-sorted language with number variables $x, y, z, \ldots$ and unary function variables $f, g, h, \ldots$, consisting of constant symbols $0, 1, +, \cdot, =, <$. We also use set variables $X, Y, Z, \ldots$, intending to range over the $\{0, 1\}$-valued functions, that is, the characteristic functions of sets.

The formulas can be classified as follows:

- φ is *bounded* (Π_0^0) if it is built up from atomic formulas by using propositional connectives and bounded number quantifiers $(\forall x < t), (\exists x < t)$, where t does not contain x.
- φ is Π_0^1 if it does not contain any function quantifier. Π_0^1 formulas are called *arithmetical* formulas.
- $\neg\varphi$ is Σ_n^i if φ is a Π_n^i-formula $(i \in \{0, 1\}, n \in \omega)$.
- $\forall x_1 \cdots \forall x_k \varphi$ is Π_{n+1}^0 if φ is a Σ_n^0-formula $(n \in \omega)$,
- $\forall f_1 \cdots \forall f_k \varphi$ is Π_{n+1}^1 if φ is a Σ_n^1-formula $(n \in \omega)$.

We loosely say that a formula is Σ_n^i (resp. Π_n^i) if it is equivalent over a base theory (such as ACA$_0$) to a $\psi \in \Sigma_n^i$ (resp. Π_n^i).

We now define some popular axiom schemata of second order arithmetic.

Definition 1. *Let $\mathcal{C}$ be a set of $\mathcal{L}_2$ formulas.*

(1) $\mathcal{C}$-IND: $(\varphi(0) \wedge \forall x(\varphi(x) \to \varphi(x+1))) \to \forall x \varphi(x)$,
    where $\varphi(x)$ belongs to $\mathcal{C}$.
(2) $\mathcal{C}$-TI: *for any well-ordering $<_X$, $(\forall x(\forall y <_X x \; \varphi(y) \to \varphi(x))) \to \forall x \varphi(x)$,*
    where $\varphi(x)$ belongs to $\mathcal{C}$.
(3) $\mathcal{C}$-CA : $\exists X \forall x(x \in X \leftrightarrow \varphi(x))$,
    where $\varphi(x)$ belongs to $\mathcal{C}$ and X does not occur freely in $\varphi(x)$.

(4) $\mathcal{C} \cap \mathcal{C}^-$-**CA** : $\forall x(\varphi(x) \leftrightarrow \psi(x)) \rightarrow \exists X \forall x(x \in X \leftrightarrow \varphi(x))$,
    where $\varphi(x)$ and $\neg\psi(x)$ belong to $\mathcal{C}$ and X does not occur freely in $\varphi(x)$.
(5) $\mathcal{C}$-**AC** : $\forall x \exists X \varphi(x, X) \rightarrow \exists X \forall x \varphi(x, X_x)$,
    where $\varphi(x, X)$ belongs to $\mathcal{C}$ and $X_x = \{y : (x, y) \in X\}$.

The system ACA_0 consists of the ordered semiring axioms for $(\omega, +, \cdot, 0, 1, <)$, Σ_1^0-CA and Σ_1^0-IND. For a set Λ of sentences, Λ_0 denotes the system consisting of ACA_0 plus Λ.

By Δ_n^i-CA, we denote $\Sigma_n^i \cap (\Sigma_n^i)^-$-CA. We can easily show that for any $k \geq 0$,

$$\Delta_k^1\text{-}\mathsf{CA}_0 \subset \Sigma_k^1\text{-}\mathsf{AC}_0.$$

Moreover, if $k = 2$, the above two axioms are known to be equivalent to each other.

For a formula φ with a distinct variable f ranging over $\mathbb{N}^{\mathbb{N}}$, we associate a two-person *game* G_φ (or simply denote φ) as follows: player I and player II alternately choose natural numbers (starting with I) to form an infinite sequence $f \in \mathbb{N}^{\mathbb{N}}$ and I (resp. II) wins iff $\varphi(f)$ (resp. $\neg\varphi(f)$). We say that φ is *determinate* if one of the players has a *winning strategy* $\sigma : \mathbb{N}^{<\mathbb{N}} \rightarrow \mathbb{N}$ in the game φ. For a class $\mathcal{C}$ of formulas, $\mathcal{C}$-**Det** is the axiom which states that any game in $\mathcal{C}$ is determinate.

3   Weak Determinacy of Games and Inductive Definitions

3.1   Inductive Definitions

We start by formalizing the axiom of inductive definition. An operator $\Gamma : P(\mathbb{N}) \rightarrow P(\mathbb{N})$ belongs to a class $\mathcal{C}$ of formulas iff its graph $\{(x, X) : x \in \Gamma(X)\}$ belongs to $\mathcal{C}$. Γ is said to be *monotone* iff $\Gamma(X) \subset \Gamma(Y)$ whenever $X \subset Y$. By mon-$\mathcal{C}$, we shall denote the class of monotone operators in $\mathcal{C}$.

A relation W is a *pre-ordering* iff it is reflexive, connected and transitive. W is a *pre-wellordering* iff it is a well-founded pre-ordering. The *field* of W is the set $F = \{x : \exists y \, (x, y) \in W \vee (y, x) \in W\}$. An axiom of inductive definition asserts the existence of a pre-wellordering constructed by iterative application of a given operator.

Definition 2. *Let $\mathcal{C}$ be a set of $\mathcal{L}_2$ formulas. $\mathcal{C}$-ID asserts that for any operator $\Gamma \in \mathcal{C}$, there exists a set $W \subset \mathbb{N} \times \mathbb{N}$ such that*

1. *W is a pre-wellordering on its field F,*
2. *$\forall x \in F \quad W_x = \Gamma(W_{<x}) \cup W_{<x}$,*
3. *$\Gamma(F) \subset F$,*

where $W_x = \{y \in F : (y, x) \in W\}$ and $W_{<x} = \{y \in F : (y, x) \in W$ and $(x, y) \notin W\}$.

We write $\mathcal{C}$-MI to denote mon-$\mathcal{C}$-ID. We note that for a monotone operator Γ, the second condition of the above definition can be replaced by

$$\forall x \in F \quad W_x = \Gamma(W_{<x}).$$

It is also easy to see that for any class $\mathcal{C}$, $\mathcal{C}$-MI$_0$ implies $\mathcal{C}$-CA$_0$.

Finally, we have

Theorem 3 (Tanaka [13]). *Over* RCA$_0$, Σ_2^0-Det, Σ_1^1-MI, *and* Σ_1^1-ID *are equivalent.*

3.2   Difference Sets

Now, we consider more complex games than Σ_2^0.

Definition 4. *Let* $\mathcal{C}$ *and* $\mathcal{C}'$ *be classes of formulas. We denote the classes of formulas in the form* $\phi \wedge \psi$ ($\phi \in \mathcal{C}, \psi \in \mathcal{C}'$) *as* $\mathcal{C} \wedge \mathcal{C}'$, *and* $\neg\psi$ ($\psi \in \mathcal{C}$) *as* $\neg\mathcal{C}$.

Definition 5. *For all* $n, k \geq 1$, *we define the classes of the formulas* $(\Sigma_n^0)_k$, $(\Pi_n^0)_k$ *as follows.*

- $(\Sigma_n^0)_1 = \Sigma_n^0$, $\quad (\Sigma_n^0)_k = \Sigma_n^0 \wedge (\Pi_n^0)_{k-1}$ *if* $k > 1$,
- $(\Pi_n^0)_k = \neg(\Sigma_n^0)_k$.

Lemma 6. *For any* $n, k \geq 1$, *the following hold.*

- $(\Sigma_n^0)_k = \Pi_n^0 \wedge (\Sigma_n^0)_{k-1}$ *if* k *is even.*
- $(\Sigma_n^0)_k = \Sigma_n^0 \vee (\Sigma_n^0)_{k-1}$ *if* k *is odd.*

Definition 7. *Let* $\mathcal{C}$ *and* $\mathcal{C}'$ *be a class of formulas. If there exists a* $\mathcal{C}$-*formula* ψ, *a* $\neg\mathcal{C}'$-*formula* η, *and a* $\mathcal{C}'$-*formula* η' *such that for all* $f \in X^{\mathbb{N}}(\varphi(f) \leftrightarrow ((\psi(f) \wedge \eta(f)) \vee (\neg\psi(f) \wedge \eta'(f))))$, *then we call a formula* φ *a* Sep$(\mathcal{C}, \mathcal{C}')$-*formula.*

The above theorem can be easily extended as follows.

Theorem 8 (Mashiko et al. [4]). *Over* RCA$_0$, Σ_1^1-ID$_0$, Sep(Δ_1^0, Σ_2^0)-Det, *and* Sep(Σ_1^0, Σ_2^0)-Det *are equivalent.*

Definition 9. *Let* $\mathcal{C}$ *be a class of formulas. If there exists a* $\neg\mathcal{C}$-*formula* φ' *such that for all* $f \in \mathbb{N}^{\mathbb{N}}(\varphi(f) \leftrightarrow \varphi'(f))$, *we call a* $\mathcal{C}$-*formula* φ *a* $\Delta(\mathcal{C})$-*formula.*

Our objective of this paper is to pin down the strength of $\Delta((\Sigma_2^0)_k)$-determinacy. For this purpose, we use the following fact.

Theorem 10. *Over* RCA$_0$, *the class* $\Delta((\Sigma_n^0)_{k+1})$ *is equivalent to the class* Sep$(\Delta_n^0, (\Sigma_n^0)_k)$.

The following lemma can be easily shown by induction.

Lemma 11. *For any $n, k \geq 1$, any $(\Sigma_n^0)_k$ formula φ can be expressed as follows:*

$$\varphi \leftrightarrow \begin{cases} (\varphi_1 \wedge \neg\varphi_2) \vee \cdots \vee (\varphi_{k-1} \wedge \neg\varphi_k) & \text{if } k \text{ even} \\ (\varphi_1 \wedge \neg\varphi_2) \vee \cdots \vee (\varphi_{k-1} \wedge \neg\varphi_k) \vee \varphi_k & \text{if } k \text{ odd} \end{cases} \tag{1}$$

where φ_i's are Σ_n^0 formulas and for each $i < k$, $\varphi_{i+1} \to \varphi_i$ holds. Conversely, if a formula φ can be expressed as above, it is $(\Sigma_n^0)_k$. Moreover, a $(\Pi_n^0)_k$ formula can be expressed analogously.

Then the theorem can be proved as follows. First, to show $\mathsf{Sep}(\Delta_n^0, (\Sigma_n^0)_k) \subseteq \Delta((\Sigma_n^0)_{k+1})$, we need to prove the next lemma.

Lemma 12. *Fix any $k, n \geq 1$. For a Σ_n^0 formula $\psi_0(f)$, a Π_n^0 formula $\psi_1(f)$, a $(\Pi_n^0)_k$ formula $\eta(f)$, and a $(\Sigma_n^0)_k$ formula $\eta'(f)$, there exist a $(\Sigma_n^0)_{k+1}$ formula $\zeta_0(f)$ and a $(\Pi_n^0)_{k+1}$ formula $\zeta_1(f)$ such that*

$$\forall f(\psi_0(f) \leftrightarrow \psi_1(f))$$
$$\to \forall f\Big((\zeta_0(f) \leftrightarrow \zeta_1(f)) \wedge \Big(\zeta_0(f) \leftrightarrow ((\psi_0(f) \wedge \eta(f)) \vee (\neg\psi_0(f) \wedge \eta'(f)))\Big)\Big).$$

Proof. Assume that k is even. Take a Σ_n^0 formula $\psi_0(f)$, a Π_n^0 formula $\psi_1(f)$, a $(\Pi_n^0)_k$ formula $\eta(f)$ and a $(\Sigma_n^0)_k$ formula $\eta'(f)$ such that $\forall f(\psi_0(f) \leftrightarrow \psi_1(f))$ holds. Since $\psi(f) \wedge \eta(f)$ is $(\Sigma_n^0)_{k+1}$ formula, by the lemma 11, there exist Σ_n^0 formulas $\eta_1(f), \cdots, \eta_{k+1}(f)$ such that the following hold: $\eta_{i+1}(f) \to \eta_i(f)$ $(1 \leq i < k)$ and $(\psi(f) \wedge \eta(f)) \leftrightarrow ((\eta_1(f) \wedge \neg\eta_2(f)) \vee (\eta_3(f) \wedge \neg\eta_4(f)) \vee \cdots \vee (\eta_{k-1}(f) \wedge \neg\eta_k(f)) \vee \eta_{k+1}(f))$. Also, since $\neg\psi(f) \wedge \eta'(f)$ is a $(\Sigma_n^0)_k$ formula, by lemma 11, there exist Σ_n^0 formulas $\eta_1'(f), \cdots, \eta_k'(f)$ such that the following hold: $\eta_{i+1}'(f) \to \eta_i'(f)$ $(1 \leq i < k)$ and $(\neg\psi(f) \wedge \eta'(f)) \leftrightarrow ((\eta_1'(f) \wedge \neg\eta_2'(f)) \vee (\eta_3'(f) \wedge \neg\eta_4'(f)) \vee \cdots \vee (\eta_{k-1}'(f) \wedge \neg\eta_k'(f)))$.

Now, for $1 \leq i \leq k$, if we let $\phi_i(f) \equiv \eta_i(f) \vee \eta_i'(f)$, $\phi_{k+1}(f) \equiv \eta_{k+1}(f)$, we have $\phi_{i+1}(f) \to \phi_i(f)$ and $((\psi(f) \wedge \eta(f)) \vee (\neg\psi(f) \wedge \eta'(f))) \leftrightarrow ((\phi_1(f) \wedge \neg\phi_2(f)) \vee \cdots \vee (\phi_{k-1}(f) \wedge \neg\phi_k(f)) \vee \phi_{k+1}(f))$. Thus, by lemma 11, there exists a $(\Sigma_n^0)_{k+1}$ formula $\zeta_0(f)$ such that $\zeta_0(f) \leftrightarrow ((\psi(f) \wedge \eta(f)) \vee (\neg\psi(f) \wedge \eta'(f)))$ holds. By the same argument, there exists a $(\Sigma_n^0)_{k+1}$ formula $\zeta_1(f)$ such that $\zeta_1(f) \leftrightarrow ((\psi(f) \wedge \neg\eta(f)) \vee (\neg\psi(f) \wedge \neg\eta'(f)))$ holds.

Since $\zeta_0(f) \leftrightarrow \neg\zeta_1(f)$ and $\neg\zeta_1(f)$ is a $(\Pi_n^0)_{k+1}$ formula, the statement holds. Clearly, we can prove it in the same way when k is odd. $\qquad\square$

The reverse inclusion $\mathsf{Sep}(\Delta_n^0, (\Sigma_n^0)_k) \supseteq \Delta((\Sigma_n^0)_{k+1})$ can be shown by a similar argument. Thus, we obtain the theorem 10.

3.3 Tranfinite Recursion of Inductive Definitions

Following [4], we introduce an axiom $\mathcal{C}$-IDTR which asserts the existence of sets constructed by transfinite recursion of the inductive definitions.

Definition 13. *The formal system* C-IDTR$_0$ *consists of* ACA$_0$ *and the following axiom scheme (C-IDTR): for any well-ordering* $\preceq$ *and* C-*operator* Γ, *there exists a sequence* $\langle V^r : r \in \text{field}(\preceq)\rangle$ *satisfying the following conditions.*

1. V^r *is pre-well ordering on its field* $F^r = \text{field}(V^r)$.
2. $\forall x \in F^r(V_x^r = \Gamma^{F^{\prec r}}(V_{<x}^r) \cup V_{<x}^r)$.
3. $\Gamma^{F^{\prec r}}(F^r) \subset F^r$.

where $V_x^r = \{y \in F^r : y \leq_{V^r} x\}$, $V_{<x}^r = \{y \in F^r : y <_{V^r} x\}$, $F^{\prec r} = \bigcup \{F^{r'} : r' \prec r\}$.

Especially, [mon-C]-IDTR$_0$ *is also written as* C-MITR$_0$. *For* C-MITR$_0$, *the second condition of the definition may be replaced by* $\forall x \in F^r$ $(V_x^r = \Gamma^{F^{\prec r}}(V_{<x}^r))$.

As mentioned above, in C-IDTR$_0$ the inductive definitions with C operator Γ are iterated transfinitely many times. More precisely, we apply the inductive definition and obtain a pre-well ordering V^{r_0} on its field F^{r_0}. Then, by taking F^{r_0} as an oracle, we again apply the inductive definition with operator $\Gamma^{F^{r_0}}$. Then, we obtain a pre-well ordering V^{r_1} on its field F^{r_1}. We iterate this procedure transfinitely many times along the given well-ordering $\preceq$, and then we obtain the sequence of pre-well orderings $\langle V^r : r \in \text{field}(\preceq)\rangle$.

In [4], it is already stated that Σ_1^1-IDTR$_0$ and $\Delta((\Sigma_2^0)_2)$-determinacy are equivalent. We here sketch the proof.

Theorem 14. *Over* RCA$_0$, *the following are equivalent.*

(i) $\Delta((\Sigma_2^0)_2)$-Det.
(ii) Σ_1^1-IDTR$_0$.
(iii) Sep(Δ_2^0, Σ_2^0)-Det.

Proof. By theorem 10, it is sufficient to prove the equivalence of (ii) and (iii).

(ii) $\rightarrow$ (iii). First, we recall that Δ_2^0-Det $\leftrightarrow$ Π_1^1-TR$_0$, and Σ_2^0-Det $\leftrightarrow$ Σ_1^1-ID ([12], [13]). Then, to construct a winning strategy of a Sep(Δ_2^0, Σ_2^0) game, we may match up winning strategies for Δ_2^0 games and Σ_2^0 games. Thus, it is not so hard to see that Sep(Δ_2^0, Σ_2^0)-Det is proved from Σ_1^1-IDTR$_0$ by a similar way as in the proof of theorem 5.1 of [12].

(iii)$\rightarrow$(ii). We construct a following Sep(Δ_2^0, Σ_2^0)-game. The proof technique is a refinement of previous researches. In particular, we count on the proof of theorem 3.1. in [13]. A play of the game starts with player I's choosing a pair (y^*, r^*) and arising a question "$y^* \in F^{r^*}$?".

For this question, player II answers 1, which means "Yes", if II thinks $y^* \in F^{r^*}$. Then, II constructs a sequece of pre-well orderings $\langle V^r : r \preceq r^*\rangle$ so that $y^* \in F_1^{r^*}$. In this game the player who constructs the pre-well orderings is called Pro. Conversely, the other player, who tries to point out mistakes in the construction of Pro, is called Con.

Pro wins the game if Pro can construct the sequence of pre-wellorderings satisfying the all conditions of Σ_1^1-IDTR$_0$ with $y^* \in F^{r^*}$. Con wins otherwise.

Now we assume that Con has a winning strategy in this game. (This means that Con knows "the correct construction" of the sequence of pre-wellorderings.)

As Pro constructs the pre-well orderings, since Con knows what the right elements are for those sets, Con can points out (if any) unsuitable elements in the Pro's construction. (We define a term *challenge* to Pro's construction as the same meaning of *pointing out*.) There are two crucial techniques of this proof in the ways of challenges, which are not used in the proof of [13]. The first one is that the *switches* of the players' roles, Pro and Con, occur during the game. For example, assume that when Pro insists that $x \notin F^r$ in the game, Con thinks $x \in F^r$. Then, Con will make a challenge to that Pro's assertion $x \notin F^r$, and Con becomes Pro and constructs V_x^r on the field(F^r) such that $x \in F^r$.

The other technique is that we let Con point out unsuitable elements until Con exhausts all challenges in the Pro's construction. We consider the following example. In the proof, we define a rule of challenge which requires Con to make challenge in descending order. For examples, once Con made a challenge to $x \not\leq_{V^r} y$ then Con can not do it in $V^{r'}$ for any $r' \succ r$ or $F^r \setminus V_y^r$ after that challenge.

From this rule, if Con challenges without violating the rules, Con wins in the following two cases:

1. Con can point out the mistakes in Pro's construction infinitely many times.
2. Con can show that the last challenged element is actually not the suitable one, if Con points out only finitely many elements.

For the first case, if Con can challenge infinitely many times, it means there is an infinite descending sequence in Pro's construction. The second case is the crucial point of the proof. If Con points out only finitely many times it means that Con agrees that there is no inappropriate elements any more in the Pro's construction. In other words, Pro and Con agree on the initial segments of V^{r^*}. For example, suppose that the last challenge occurs to Pro's assertion $x \not\leq_{V^r} y$ for any x, y, r. Then, Con and Pro agree on the parts $V_{<y}^r$ and $\langle V^{r'} : r' \prec r \rangle$.

Con will use that agreed-part of the two players to show that $y \leq_{V^r} x$. $\square$

3.4   Multiple Inductive Definitions and Determinacy of Games

In [6], inductive definitions with multiple operators are introduced. By combining transfinite recursion with such inductive definitions, we introduce a new axiom scheme $[\Sigma_1^1]^k$-IDTR$_0$, and we prove that it is equivalent to $\Delta((\Sigma_2^0)_{k+1})$-determinacy. For the simplicity, we here consider only the case $k = 2$.

Definition 15. *The formal definition of $[S_0, S_1]$-IDTR$_0$ consists of* ACA$_0$ *and the following axiom scheme: Let S_0 and S_1 are collections of operators. The axiom scheme $[S_0, S_1]$-IDTR$_0$ asserts the following. For any well-ordering $\preceq$ and any $\Gamma_0 \in S_0, \Gamma_1 \in S_1$, there exist $\langle W^r : r \in \text{field}(\preceq) \rangle$, $\langle V^{r,x} : r \in \text{field}(\preceq), x \in F_1^r \rangle$ and $\langle V^{r,\infty} : r \in \text{field}(\preceq) \rangle$ such that the following are all satisfied.*

1. *W^r is pre-wellordering on its field F_1^r.*
2. *$\forall x \in F_1^r \cup \{\infty\}$*
   – $V^{r,x}$ is pre-wellordering on its field $F_0^{r,x}$.

$$- V_y^{r,x} = \Gamma_0^{F_1^{\prec r} \oplus W_{<x}^r}(V_{<y}^{r,x}) \cup V_{<y}^{r,x} \text{ for all } y \in F_0^{r,x}.$$

$$- W_x^r = \Gamma_1^{F_1^{\prec r}}(F_0^{r,x}) \cup W_{<x}^r.$$

$$- \Gamma_0^{F_1^{\prec r} \oplus W_{<x}^r}(F_0^{r,x}) \subset F_0^{r,x}.$$

3. $W_\infty^r = W_{<\infty}^r = F_1^r$.

where $F_1^{\prec r} = \oplus\{F_1^{r_i} : r_i \prec r\}$. Note also that $X \oplus Y = \{2x : x \in X\} \cup \{2y + 1 : y \in Y\}$.

We see how the sequence of pre-well orderings are constructed. We assume that $W^{\prec r}$ has been constructed and consider the rth construction of $[\mathcal{C}]^2\text{-IDTR}_0$. $\Gamma_0^{F_1^{\prec r}}$-operator is applied until we get the fixed point F_0^{r,x_0}. Then, S_1-operator with oracle $F_1^{\prec r}$, which is $\Gamma_1^{F_1^{\prec r}}$, is applied to F_0^{r,x_0}, and $\Gamma_1^{F_1^{\prec r}}(F_0^{r,x_0}) = W_{x_0}^r$ is constructed. Next, this $W_{x_0}^r$ is joined to $F_1^{\prec r}$, written as $F_1^{\prec r} \oplus W_{x_0}^r$, and then S_0-operator becomes $\Gamma_0^{F_1^{\prec r} \oplus W_{x_0}^r}$. This procedure is repeated until F_1^r becomes the common fixed point of those two operators. Note that the pre-well orderings constructed by $[\mathcal{C}]^2\text{-ID}_0$ is equal to the r_0th (i.e., the first) construction by $[\mathcal{C}]^2\text{-IDTR}_0$, where r_0 is the $\preceq$-least element of field($\preceq$).

Then, we obtain the following theorem.

Theorem 16. *The following is provable over* RCA_0. *For any* $k \geq 1$,

$$\Delta((\Sigma_2^0)_{k+1})\text{-Det} \leftrightarrow [\Sigma_1^1]^k\text{-IDTR}_0.$$

Proof. The proof becomes complicated, but we can use the similar ideas which are shown in the sketch of the proof of theorem 14 to prove this theorem.   □

4   Conclusion and Future Studies

In this paper, we introduced the axiom of transfinite recursion of Σ_1^1 inductive definitions with k operators, denote $[\Sigma_1^1]^k\text{-IDTR}_0$, and showed that it is equivalent to the determinacy of $\Delta((\Sigma_2^0)_{k+1})$ sets. A key fact used in the proof is that a $\Delta((\Sigma_2^0)_{k+1})$ set is expressed as a $\mathsf{Sep}(\Delta_2^0, (\Sigma_2^0)_k)$ set, namely a Δ_2^0-separated union of a $(\Sigma_2^0)_k$ set and $(\Pi_2^0)_k$ set. By virtue of this fact, we can utilize a difference hierarchy for a Δ_2^0 set (cf. [12], [6]) to construct a winning strategy for a $\Delta((\Sigma_2^0)_{k+1})$ game.

In [6], the exact determinacy strength of Δ_3^0 sets has been pinned down in terms of transfinte combinations of Σ_1^1 inductive definitions. We should notice that their axiom for transfinte combinations of Σ_1^1 inductive definitions is much stronger than our $[\Sigma_1^1]^k\text{-IDTR}$ here. However, it is worth studying such an axiom as $[\Sigma_1^1]^\alpha\text{-IDTR}$, where α is an ordinal, to refine their result on Δ_3^0 games.

Montalbán ans Shore [7] show that for any $m \geq 1$, $\Pi_{m+2}^1\text{-CA}$ proves the determinacy of $(\Sigma_3^0)_m$ sets, but $\Delta_{m+2}^1\text{-CA}$ does not. Thus, $(\Sigma_3^0)_\omega$-determinacy is not provable over Z_2. Then, Montalbán [8] raises Question 28 to classify the precise strength of $(\Sigma_3^0)_m$-determinacy.

In [1], Bradfield has shown that the sets of Player I's winning positions of a $(\Sigma_2^0)_k$-game are exactly the same as the $(k+1)$-level of μ-calculus alternation hierarchy Σ_{k+1}^μ. Then, Bradfield [2] claims that the hierarchy $\langle \Sigma_n^\mu, n \in \omega \rangle$ is strict, that is, for any k in ω, we have $\Sigma_k^\mu \subsetneq \Sigma_{k+1}^\mu$. This result easily follows from the previous result on multiple inductive definitions ([6]) together with observation that for any k in ω, Π_2^1-CA$_0$ proves the consistency of Δ_2^1-CA$_0 + (\Sigma_2^0)_k$-determinacy, while it does not prove the consistency of $(\Sigma_2^0)_{<\omega}$-determinacy. (cf. Heinatsch and Möllerfeld [3])

From the main result of this paper, we shall also obtain the following refinement. First of all, the hierarchy $\langle \Pi_n^\mu, n \in \omega \rangle$ is naturally defined and so is $\langle \Delta_n^\mu, n \in \omega \rangle$. Then, by the argument of this paper, we can associate a Δ_{n+1}^μ formula with transfinite recursion of a Σ_k^μ formula. Moreover, for any k in ω, we have $\Sigma_k^\mu \subsetneq \Delta_{k+1}^\mu \subsetneq \Sigma_{k+1}^\mu$ by a similar observation as above. Details will appear in the future literature.

The research to characterize the determinacy strength in Cantor space has been conducted by Nemoto, MedSalem, and Tanaka. See [10] and table 2 of [9]. See also Question 29 in A. Montalbán's survey paper [8]. To investigate the relationships between the determinacies strength in Cantor space and Baire space, a technique which translates the games in Baire space into the games in Cantor space was introduced in [10]. *Translate* here means that, for a game $\phi(f)$ in Baire space, the game $\phi^*(f)$ in Cantor space is constructed so that the same player with the winning strategy in $\phi(f)$ has a winning strategy. By constructing this king of game $\phi^*(f)$, if we assume that $\phi \in \mathcal{C}$ and $\phi^* \in \mathcal{C}'$, then we can prove over appropriate system that $\mathcal{C}'$-Det* is provable from $\mathcal{C}$-Det. By applying this method, we obtain that, for any $k > 0$, $[\Sigma_1^1]^k$-IDTR$_0$ is equivalent to $\Delta((\Sigma_2^0)_{k+2})$-Det*.

Subsystem of SOA	Determinacy in Baire space	Determinacy in Cantor Space
ATR$_0$	Δ_1^0, Σ_1^0	Δ_2^0, Σ_2^0
Π_1^1-CA$_0$	$\Delta((\Sigma_1^0)_2), (\Sigma_1^0)_2$	Sep(Σ_1^0, Σ_2^0)
Π_1^1-TR$_0$	Δ_2^0	$\Delta((\Sigma_2^0)_2)$
Σ_1^1-ID$_0$	Σ_2^0 / Sep(Σ_1^0, Σ_2^0)	$(\Sigma_2^0)_2$
Σ_1^1-IDTR$_0$	$\Delta((\Sigma_2^0)_2)$	$\Delta((\Sigma_2^0)_3)$
$\vdots$	$\vdots$	$\vdots$
$[\Sigma_1^1]^k$-ID$_0$	$(\Sigma_2^0)_k$	$(\Sigma_2^0)_{k+1}$
$[\Sigma_1^1]^k$-IDTR$_0$	$\Delta((\Sigma_2^0)_{k+1})$	$\Delta((\Sigma_2^0)_{k+2})$
$\vdots$	$\vdots$	$\vdots$
$[\Sigma_1^1]^\omega$-ID$_0$	$(\Sigma_2^0)_\omega$	$(\Sigma_2^0)_\omega$
$[\Sigma_1^1]^{TR}$-ID$_0$	Δ_3^0	Δ_3^0

The above diagram shows the results on determinacy strength of infinite games in second order arithmetic. The left column contains subsystems of second order arithmetic from weaker to stronger. The center and the right most columns contain classes of the games in Baire space and Cantor space, respectively. Each row represents that a certain axiom is equivalent to the determinacy of the corresponding games over appropriate systems (RCA_0, but with Π_3^1-TI for the last row).

Acknowledgements. The first author was partially supported by Global COE program. This research was partially supported by KAKENHI 23340020.

References

1. Bradfield, J.C.: Fixpoints, games and the difference hierarchy. Theor. Inform. Appl. 37, 1–15 (2003)
2. Bradfield, J.C.: The modal μ-calculus alternation hierarchy is strict. Theor. Comput. Sci. 195, 133–153 (1998)
3. Heinatsch, C., Möllerfeld, M.: The determinacy strength of Π_2^1-comprehension. Ann. Pure Appl. Logic 161, 1462–1470 (2010)
4. Mashiko, K., Tanaka, K., Yoshii, K.: Determinacy of the Infinite Games and Inductive Definition in Second Order Arithmetic. In: RIMS Kokyuroku, vol. 1729, pp. 167–177 (2011)
5. MedSalem, M.O., Tanaka, K.: Δ_3^0-determinacy, comprehension and induction. Journal of Symbolic Logic 72, 452–462 (2007)
6. MedSalem, M.O., Tanaka, K.: Weak determinacy and iterations of inductive definitions. Lect. Notes Ser. Inst. Math. Sci. Natl. Univ. Singap., vol. 15. World Sci. Publ., Hackensack (2008)
7. Montalbán, A., Shore, R.A.: The Limits of determinacy in second order arithmetic (preprint)
8. Montalbán, A.: Open Questions in Reverse Mathematics. Bulletin of Symbolic Logic 17, 431–454 (2011)
9. Nemoto, T.: Determinacy of Wadge classses and subsystems of second order arithmetic. Math. Log. Quart. 55(2), 154–176 (2009)
10. Nemoto, T., MedSalem, M.O., Tanaka, K.: Infinite games in the Cantor space and subsystems of second order arithmetic. Math. Log. Quart. 53, 226–236 (2007)
11. Simpson, S.G.: Subsystems of Second Order Arithmetic. Springer (1999)
12. Tanaka, K.: Weak axioms of determinacy and subsystems of analysis I (Δ_2^0 games). Z. Math. Logik Grundlag. Math. 36, 481–491 (1990)
13. Tanaka, K.: Weak axioms of determinacy and subsystems of analysis II (Σ_2^0 games). Ann. Pure Appl. Logic 52, 181–193 (1991)

A Hierarchy of Immunity and Density
for Sets of Reals

Takayuki Kihara

Mathematical Institute, Tohoku University, Sendai 980-8578, Japan
kihara.takayuki.logic@gmail.com

Abstract. The notion of immunity is useful to classify degrees of non-computability. Meanwhile, the notion of immunity for topological spaces can be thought of as an opposite notion of density. Based on this viewpoint, we introduce a new degree-theoretic invariant called *layer density* which assigns a value n to each subset of Cantor space. Armed with this invariant, we shed light on an interaction between a hierarchy of density/immunity and a mechanism of type-two computability.

1 Introduction

1.1 Summary

The study of *immunity* was initiated essentially by Post in 1944. Demuth-Kučera [5] studied the notion of immunity for closed sets in Baire space. Immunity for a closed set indicates that it is *"far from dense"*. They showed that any 1-generic real computes no element of any immune co-c.e. closed set, and hence no 1-generic real computes a Martin-Löf random real. Binns [1] introduced many notions of *hyperimmunity* for closed sets to classify degrees of difficulty of co-c.e. closed sets. Cenzer-Kihara-Weber-Wu [4] started the systematic study on immunity for closed sets. Higuchi-Kihara [6] clarified that such notions indicating being *"nearly/far from dense"* are extremely useful to study a hierarchy of *nonuniform computability on sets of reals*. We investigate a hierarchy of properties that are *"nearly dense"*, by introducing a new degree-theoretic invariant called *layer density* which assigns a value n to each subset of any computable metric space. In this way, we shed light on an interaction between a hierarchy of density and a mechanism of type-two computability. We also continue the work [6] on the structure *inside the Turing upward closure* of any co-c.e. closed set.

1.2 Notation and Convention

Much of our notation in this paper follows that in [6]. For basic terminology on Computability Theory and Computable Analysis, see [3,8,9]. For any sets X and Y, f is said to be a *function from X to Y* if $\text{dom}(f) \supseteq X$ and $\text{range}(f) \subseteq Y$ hold. We use the symbol $\frown$ for concatenation. For $\sigma \in \omega^{<\omega}$, we let $|\sigma|$ denote the length of σ. Moreover, $f \restriction n$ denotes the unique initial segment of f of length

S.B. Cooper, A. Dawar, and B. Löwe (Eds.): CiE 2012, LNCS 7318, pp. 384–394, 2012.

n. We also define $[\sigma] = \{f \in \omega^\omega : f \supset \sigma\}$. For a tree $T \subseteq \omega^{<\omega}$, let $[T]$ denote the set of all infinite paths through T. For a subset A of a space X, $\mathrm{cl}(A)$, and $\mathrm{ext}(A)$ denote the closure, and the exterior of A, respectively. A *representation* ρ of a space X is a surjection $\rho :\subseteq \omega^\omega \to X$. Let $\mathcal{A}_-(X)$ denote the hyperspace consisting of closed subsets of X represented by $\psi_- : \alpha \mapsto X \setminus \bigcup_n B_{\alpha(n)}$. Here, $\{B_n\}_{n\in\omega}$ is a fixed countable base of X. A computable element of $\mathcal{A}_-(X)$ (i.e., $\psi_-(\alpha)$ for some computable $\alpha \in \omega^\omega$) is called a *co-c.e. closed set* or a Π_1^0 *class*.

2 Computability with Layers

2.1 Density and Immunity

Let X be a topological space, and $\mathcal{B}$ be a collection of open sets in X. A subset $S \subseteq X$ is said to be $\mathcal{B}$-*dense* if it intersects with all nonempty open sets contained in $\mathcal{B}$. By restricting $\mathcal{B}$, one may introduce various "*pre-dense*" properties. For instance, *immunity* [4] and *hyperimmunity* [1] can be introduced in this way. A variety of interactions are known between density/immunity and degrees of difficulty [4,5,7]. To introduce nice $\mathcal{B}$-density notion, we consider the following effective notion for open sets: An open set $S \subseteq X$ is *bi-c.e. open* if both S and $\mathrm{ext}(S)$ are c.e. open. We fix $X = 2^\omega$. A sequence $\{B_n\}$ of open rational balls is *nontrivial* if it contains no empty set, and $\liminf_n \mathrm{diam}(B_n) = 0$; *computable* if it is uniformly computable (hence, $\bigcup_n B_n$ is c.e. open); and *decidable* if it is computable, and $\bigcup_n B_n$ is bi-c.e. open. Let $P \subseteq 2^\omega$ be a closed set, and let T_P^{ext} denote the tree $\{\sigma \in 2^{<\omega} : P \cap [\sigma] \neq \emptyset\}$. Cenzer et al. [4] introduced the following notion: P is *immune* if T_P^{ext} contains no infinite computable subset. P is *tree-immune* if T_P^{ext} contains no infinite computable subtree.

Proposition 1. *Let $P \subseteq 2^\omega$ be a closed set with no computable element. Then, P is not immune if and only if it is $\mathcal{B}$-dense for some nontrivial computable sequence $\mathcal{B}$ of open balls; P is not tree-immune if and only if it is $\mathcal{B}$-dense for some nontrivial decidable sequence $\mathcal{B}$ of pairwise disjoint open balls.*

Proof. Assume that P is $\mathcal{B}$-dense via an infinite computable sequence $\mathcal{B}$ of open balls. For each $B \in \mathcal{B}$, we choose the smallest clopen set $[\sigma]$ including B, and enumerate $[\sigma]$ into another sequence $\mathcal{B}^*$. As $\liminf_{B\in\mathcal{B}} \mathrm{diam}(B) = 0$, the sequence $\mathcal{B}^*$ is infinite. It is easy to see that P is also $\mathcal{B}^*$-dense. Therefore, P is not immune. Another direction is obvious.

Assume that P is not tree-immune via an infinite computable tree $V \subseteq T_P^{ext}$. As P has no computable element, V has infinitely many leaves, i.e., $L = \{\sigma \in V : (\forall i < 2)\ \sigma^\frown i \notin V\}$ is infinite. Then, we define $\mathcal{B} = \{[\sigma] : \sigma \in L\}$. To enumerate the exterior of $\bigcup\mathcal{B}$, for each $\sigma \in 2^{<\omega}$, we define $(\sigma^\frown i)^* = \sigma^\frown(1 - i)$ for each $i < 2$. Then, the exterior of $\bigcup\mathcal{B}$ is generated by the computable set $\{\sigma \in 2^{<\omega} \setminus V : \sigma^* \in V\}$, since $[V]$ has no interior. Hence, $\bigcup\mathcal{B}$ is bi-c.e. open.

Conversely, assume that P is $\mathcal{B}$-dense for a decidable sequence $\mathcal{B} = \{[\sigma_n]\}_{n\in\omega}$ of open balls. Then, there is a computable enumeration of all strings σ that are comparable with σ_n for some $n \in \omega$, since $\mathcal{B} = \{\sigma_n\}_{n\in\omega}$ is computable. Moreover,

$[\sigma] \subseteq \text{ext}(\bigcup \mathcal{B})$ if and only if there is no $n \in \omega$ such that σ is comparable with σ_n. Hence, the set U consisting of all strings $\sigma \in 2^{<\omega}$ which are comparable with some σ_n is computable, since $\text{ext}(\bigcup \mathcal{B})$ is c.e. open. Then, we can compute the tree $V = \{\sigma \in 2^{<\omega} : (\exists n \in \omega) \, \sigma \subseteq \sigma_n\}$ as follows: If $\sigma \notin U$, then declare $\sigma \notin V$. If $\sigma \in U$, then σ must be comparable with some σ_n. Wait for the least such $n \in \omega$, and if $\sigma \subseteq \sigma_n$, then declare $\sigma \in V$. Otherwise, declare $\sigma \notin V$. This algorithm correctly computes V, since the sequence $\{\sigma_n\}_{n \in \omega}$ is pairwise incomparable. Then, for each $\sigma \subseteq \sigma_n$, the open ball $[\sigma] \supseteq [\sigma_n]$ intersects with P, by $\mathcal{B}$-density of P. $\qquad \square$

By considering layers $\{B_j\}_{j \in \omega}, \{B_{j,k}\}_{j,k \in \omega}, \{B_{j,k,l}\}_{j,k,l \in \omega}, \dots$ of open balls hitting a set $P \subseteq 2^\omega$, we may strengthen the notion of $\mathcal{B}$-density. Here, it is required that P is $\{B_j\}_{j \in \omega}$-dense; $P \cap B_j$ is $\{B_{j,k}\}_{k \in \omega}$-dense for each $j \in \omega$; $P \cap B_j \cap B_{j,k}$ is $\{B_{j,k,l}\}_{l \in \omega}$-dense for each $j, k \in \omega$, $\dots$

Definition 1. *Let Y be a subset of $X = 2^\omega$.*

1. *A sequence $\{B_{n,m}\}_{(n,m) \in I \times J}$ of open balls is an J-refinement of $\{A_n\}_{n \in I}$ in Y if it is pairwise disjoint, and $B_{n,m} \subseteq A_n$ for any $(n,m) \in I \times J$.*
2. *A sequence $\{\mathcal{B}_k\}_{k<n}$ (resp. $\{\mathcal{B}_k\}_{k \in \omega}$) of decidable sequences of nonempty open rational balls is an n-layer in Y (resp. an ∞-layer) if $\mathcal{B}_{k+1} = \{B_{i,j}^{k+1}\}_{i,j}$ is an ω-refinement of $\mathcal{B}_k = \{B_i^k\}_i$ in Y, and $\{B_{i,j}^{k+1}\}_{j \in \omega}$ is decidable uniformly in i, for any $k < n - 1$ (resp. for any $k \in \omega$).*
3. *For $n \in \omega \cup \{\infty\}$, a set $P \subseteq X$ is n-layered if there is an n-layer $\mathfrak{B} = \{\mathcal{B}_k\}$ in P such that P is $\bigcup \mathfrak{B}$-dense, where $\mathcal{B}_0 = \{X\}$.*
4. *The layer density of a set $P \subseteq X$ is defined as follows:*

$$\text{density}(P) = \sup\{n \in \omega \cup \{\infty\} : P \text{ is } n\text{-layered }\}.$$

Here, the ordering on $\omega \cup \{\omega, \infty\}$ is defined as $n < \omega < \infty$ for any $n \in \omega$.

Proposition 2. *Let P be a subset of 2^ω. Then, P is empty if and only if $\text{density}(P) = 0$; If $Q \subseteq P$, then $\text{density}(Q) \leq \text{density}(P)$; If P is dense, then P is ∞-layered.* $\qquad \square$

Proposition 3. *Let $P \subseteq 2^\omega$ be a closed set with no computable element. Then, $P \subseteq 2^\omega$ is n-layered if and only if there is a sequence $\{T_i\}_{i<n}$ of infinite computable trees such that $[T_n] \subseteq P$ for any $i < n$, and $T_i \subseteq T_{i+1}^{ext}$ for any $i < n - 1$.*

Proof. Assume that $P \subseteq 2^\omega$ has such a sequence $\{T_i\}_{i \leq n}$ of infinite computable trees. We effectively enumerate all leaves $\{\sigma_k^i\}_{k \in \omega}$ of the tree T_i, for each $i < n$. Then, as Proposition 1, $\{2^\omega, \{[\sigma_k^0]\}_{k \in \omega}, \dots, \{[\sigma_k^{n-1}]\}_{k \in \omega}\}$ forms an n-layer of P.

Conversely, assume that $P \subseteq 2^\omega$ is n-layered via $\{\mathcal{B}_i\}_{i \leq n}$. As in the proof of Proposition 1, without loss of generality, we may assume $\mathcal{B}_i$ is of the form $\{[\sigma_k^i]\}_{k \in \omega}$, for each $i \leq n$. Then, we define $T_i = \{\sigma \in 2^{<\omega} : (\exists k \in \omega) \, \sigma \subseteq \sigma_k^{i+1}\}$. We can see that T_i is computable for each $i < n$, as Proposition 1. Then, $\{T_0, T_1, \dots, T_{n-1}, T_P\}$ is the desired sequence. $\qquad \square$

Example 1. Let P be a co-c.e. closed subset of 2^ω. Then, for a fixed computable tree T_P with $P = [T_P]$, we have the computable set $\{\rho_n\}_{n \in \omega}$ of all leaves of T_P. The *concatenation* $P^\frown P$ is defined by $\bigcup_n \rho_n{}^\frown P$. Consider $P^{(1)} = P$; $P^{(n+1)} = P^\frown P^{(n)}$; $P^{(\omega)} = \bigcup_n \rho_n{}^\frown P^{(n)}$; and $P^{(\infty)} = \bigcup_n P^{(n)}$. Then, density$(P^{(n)}) \geq n$; density$(P^{(\omega)}) \geq \omega$; and density$(P^{(\infty)}) = \infty$. See also Higuchi-Kihara [6].

2.2 Learnability on Topological Spaces

When we try to extract effective content in classical mathematics, we sometimes encounter the notion of nonuniform computability [2,10]. The deep structures of subnotions of nonuniformly computability have been studied [6].

Definition 2 (Learnability). *Let X be a topological space with a representation $\theta :\subseteq \omega^\omega \to X$, and fix a new symbol $? \notin X$.*

1. *The* representation $\theta_?$ *of the space $X_? = X \cup \{?\}$ is defined as $\theta_?(\langle 0 \rangle^\frown \alpha) = \theta(\alpha)$, and $\theta_?(\langle 1 \rangle^\frown \alpha) = ?$, for any $\alpha \in \omega^\omega$.*
2. *A sequence $\{f_n\}_{n \in \omega}$ of partial functions $f_n :\subseteq Y \to X_?$ is ?-good if $? \in \{f_n(\alpha), f_{n+1}(\alpha)\}$ whenever $f_n(\alpha) \neq f_{n+1}(\alpha)$.*
3. *The* discrete limit *of a ?-good sequence $\{f_n\}_{n \in \omega}$ of partial functions $f_n :\subseteq Y \to X_?$ is a partial function $\lim_n f_n :\subseteq Y \to X$ defined as follows.*

$$\lim_n f_n(\alpha) = \begin{cases} f_t(\alpha), & \text{if } (\forall s \geq t) \, f_s(\alpha) \neq ?, \\ \text{undefined}, & \text{if } (\exists^\infty s) \, f_s(\alpha) = ?. \end{cases}$$

4. *A function $f :\subseteq Y \to X$ is* learnable *if it is the discrete limit of a computable ?-good sequence $\{f_n\}_{n \in \omega}$ of partial functions $f_n :\subseteq Y \to X_?$.*
5. *An* anti-Popperian point *of a ?-good sequence $\{f_n\}_{n \in \omega}$ is a point $\alpha \in \omega^\omega$ such that $f_n(\alpha) = ?$ at most finitely many $n \in \omega$, but $\lim_n f_n(\alpha)$ is undefined.*
6. *A function $f : Y \to X$ is* eventually Popperian learnable *(abbreviated as e.P. learnable) if it is the discrete limit of a computable ?-good sequence $\{f_n\}_{n \in \omega}$ of partial functions $f_n :\subseteq Y \to X_?$ with no anti-Popperian points.*

Lemma 1 (Blum-Blum Locking). *Let (X, d) be a Polish space with a representation, and Q be a closed set in X. For every learnable function $\Gamma : Q \to P$, there is an open set $U \subseteq X$ such that $Q \cap U \neq \emptyset$, and the restriction $\Gamma|_U : Q \cap U \to P$ is computable.*

Proof. Suppose not. Fix a learnable function $\Gamma = \lim_s \Gamma_s : Q \to P$ witnessing the falsity of the assertion. Then, for any open set U_0^* and every $s_0 \in \omega$, there is $s_1 \geq s_0$ such that the open set $U_1 = \Gamma_{s_1}^{-1}\{?\}$ has a nonempty intersection with Q. Then U_1 contains an open ball $\{p \in X : d(p, q) < \varepsilon\}$ with $q \in Q$ and $\varepsilon > 0$. Pick $U_1^* = \{p \in X : d(p, q) < \min\{\varepsilon/2, 2^{-n}\}\} \subseteq U_1$. By iterating this procedure, we can get a decreasing sequence $\{U_n^*\}_{n \in \omega}$. Choose $x_n \in U_n^* \cap Q$. Then, $\{x_n\}_{n \in \omega}$ converges to an element $x \in Q \cap \bigcap_n \text{cl}(U_n^*)$. By our choice of $\{U_n^*\}_{n \in \omega}$, we see that $\Gamma_s(x) = ?$ for infinitely many $s \in \omega$. Consequently, $\Gamma(x) = \lim_s \Gamma_s(x)$ is undefined, i.e., $\text{dom}(\Gamma) \not\supseteq Q$. □

3 Degrees of Difficulty

3.1 Layer Density as a Degree-Theoretic Invariant

Theorem 1. *Let $P, Q \subseteq 2^\omega$ be co-c.e. closed sets with no computable element. If a computable function exists from P to Q, then $\mathrm{density}(P) \leq \mathrm{density}(Q)$.*

Proof. A sequence $\{T_m\}_{m<n}$ of infinite computable trees is said to be an n-*layer* if $T_m^{ext} \subseteq T_{m+1}$ for each $m < n - 1$. This definition is essentially equivalent to the definition of n-layers of open balls, by Proposition 3. Let P be an n-layered co-c.e. closed set with an n-layer $\{T_m\}_{m<n}$, and Q be a co-c.e. closed set. Let Φ be a computable function from P to Q. As P is co-c.e. closed, we may safely assume that Φ is total. It suffices to show that the sequence $\{\Phi(T_m)\}_{m<n}$ of images of T_m's under Φ forms an n-layer of Q. Note that $\Phi(T_m)$ is computable for any $m \leq n$, by totality of Φ. Fix $m < n - 1$. For each leaf ρ of $\Phi(T_m)$, we must have a leaf ρ^* of T_m with $\Phi(\rho^*) = \rho$. As $T_m \subseteq T_{m+1}^{ext}$, there are infinitely many nodes of T_{m+1} extending ρ^*. By weak König's lemma, T_{m+1} has an infinite path g extending ρ^*, and then g belongs to P, since $[T_{m+1}] \subseteq P$. Therefore, $\Phi(g) \in Q$ by our assumption that $\mathrm{dom}(\Phi)$ includes P. Then, $\Phi(T_{m+1})$ has a path $\Phi(g) \in Q$ extending $\Phi(\rho^*) = \rho$, i.e., $\rho \in \Phi(T_m)$ is extendible in $\Phi(T_{m+1})$. Hence, we have $\Phi(T_m) \subseteq (\Phi(T_{m+1}))^{ext}$, as desired. $\square$

Definition 3. *Fix $P \subseteq X$. The* layer density *of a point $\alpha \in X$ on P is defined as $\mathrm{density}_P(\alpha) = \inf\{\mathrm{density}(P \cap O) : \alpha \in O \in \Sigma_1^0(X)\}$. For $n \in \omega \cup \{\omega, \infty\}$ a point $\alpha \in X$ is an n-layered accumulation point of P if $\mathrm{density}_P(\alpha) \geq n$.*

Theorem 2. *Let $P, Q \subseteq 2^\omega$ be co-c.e. closed sets with no computable element. If a learnable function exists from P to Q, then $\mathrm{density}(P) \leq \max\{\omega, \mathrm{density}(Q)\}$.*

Proof. Fix an ∞-layered co-c.e. closed set $P \subseteq 2^\omega$ and a computable function $F : P \to 2^\omega$. By Blum-Blum Locking Lemma 1, there is a string σ extendible in $P^\heartsuit = \{\alpha \in P : \mathrm{density}_P(\alpha) = \mathrm{density}(P)\}$ such that $F \restriction [\sigma]$ is computable, since $P^\heartsuit$ is nonempty and closed. Moreover, $\mathrm{density}(P^\heartsuit) = \mathrm{density}(P) = \infty$. The image of an ∞-layer by a computable function is again an ∞-layer. Therefore, $F(P)$ is ∞-layered. $\square$

For elements a, b of a lattice L, we say that a *cups to* b if a is one-half of a witness of join-reducibility of b. For a bounded lattice L and $a \in L$, we also say that a *is cuppable in* L if a cups to $\max L$. We define preorders $\leq_1^1$ and $\leq_\omega^1$ on $\mathcal{P}(\omega^\omega)$ as follows: $P \leq_1^1 Q$ (resp. $P \leq_\omega^1 Q$) if there is a partial computable (resp. learnable) function F on ω^ω such that $\mathrm{dom}(F) \supseteq P$ and $F(P) \subseteq Q$. The structures $\mathcal{P}(\omega^\omega)/\equiv_1^1$ and $\mathcal{P}(\omega^\omega)/\equiv_\omega^1$ form lattices, where the supremum in these lattices are given by $P \otimes Q = \{p \oplus q : (p, q) \in P \times Q\}$. The former lattice is called the *Medvedev lattice*, and the latter lattice is said to be the *degrees of nonlearnability* [6].

Theorem 3. *For each $n \in \omega \cup \{\infty\}$, let LD_n denote the set of all Medvedev degrees of n-layered co-c.e. closed sets in 2^ω. Then, the set LD_n is a principal prime ideal in LD_1, and every element of LD_{n+1} is noncuppable in LD_n.*

Moreover, LD_∞ is a principal prime ideal in the degrees of nonlearnability of nonempty co-c.e. closed sets.

Proof. See Cenzer et al. [4, Corollary 4.13]. Indeed, the top element of LD_n is the Medvedev degree of $\mathsf{PA}^{(n)}$, where PA denotes the set of all consistent complete theories extending Peano Arithmetic. For principality, by Higuchi-Kihara [6], $\mathsf{PA}^{(n+1)}$ is noncuppable in LD_n, i.e., $\mathsf{PA}^{(n+1)}$ does not cup to $\mathsf{PA}^{(n)}$. □

Fix a countable base $\mathfrak{D}$ of Cantor space 2^ω. A set $P \subseteq 2^\omega$ is *totally ∞-layered* if it is ∞-layered, and there exists a computable function $\mathfrak{B} : \mathfrak{D} \times \omega \to (\mathfrak{D}^\omega)^{<\omega}$ such that $\mathfrak{B}(U, n)$ forms an n-layer of $P \cap U$, whenever $P \cap U$ is ∞-layered.

Example 2. Fix a co-c.e. closed set $P = [T_P] \subseteq 2^\omega$. Then $P^{\blacktriangledown}$ denotes the set of all infinite paths through the tree consisting of strings of the form $\rho_0{}^\frown\tau(0){}^\frown\rho_1{}^\frown\tau(1)$ ${}^\frown\rho_2{}^\frown \dots {}^\frown\rho_{|\tau|-1}{}^\frown\tau(|\tau|-1){}^\frown\sigma$, where $\sigma, \tau \in T_P$ and each ρ_i is a leaf of T_P. Then, $P^{\blacktriangledown}$ is totally ∞-layered, and $(P^{\blacktriangledown})^{\heartsuit} = \{\alpha \in P^{\blacktriangledown} : \mathrm{density}_{P^{\blacktriangledown}}(\alpha) = \mathrm{density}(P^{\blacktriangledown})\}$ is co-c.e. closed.

Theorem 4. *If a totally ∞-layered set P has a co-c.e. closed subset $P^\star$ consisting of ∞-layered accumulation points, then P is noncuppable in the degrees of nonlearnability of co-c.e. closed subsets of 2^ω.*

Lemma 2. *Let $C(X)$ denote the space of all continuous functions on X. There exists a computable function $\Xi : C(\omega^\omega) \times \mathcal{A}_-(2)^\omega \times (2^{<\omega})^\omega \times \omega^\omega \to \omega^\omega$ such that, for any $(f, H, (\sigma_i)_{i\in\omega}, \alpha) \in C(\omega^\omega) \times \mathcal{A}_-(2)^\omega \times (2^{<\omega})^\omega \times \omega^\omega$, if the image of $f|_{[\sigma_i]\otimes\{\alpha\}}$ intersects with the product set $H \subseteq 2^\omega$ for every $i \in \omega$, then $\Xi(f, H, (\sigma_i)_{i\in\omega}, \alpha)$ is contained in H.*

Proof. Indeed, the proof of Cenzer et al. [4, Theorem 5.2] is uniform, where their theorem states that, if a co-c.e. closed set P is $\mathcal{B}$-dense for some infinite computable sequence $\mathcal{B} = \{[\sigma_i]\}_{i\in\omega}$ of intervals (i.e., P is not immune), then it does not cup to any separating class $H \in \mathcal{A}_-(2)^\omega$. In other words, if a computable function $f : P \otimes R \to H$ exists, then we have a computable function $\Xi : \omega^\omega \to \omega^\omega$ such that $\Xi(\alpha) \in H$ for any $\alpha \in R$. □

Proof (Theorem 4). Fix a learnable function $F = \lim_s F_s : P \otimes R \to \mathsf{PA}$. Note that $P^\star \otimes \{g\}$ is closed for any $g \in R$. Therefore, by Blum-Blum Locking Lemma 1, there must exist an extendible string ρ in $P^\star$ such that $G_\rho = F|_{(P^\star \cap [\rho])\otimes\{g\}}$ is computable. Then, we can find a sequence $\{\sigma_i^\rho\}_{i\in\omega}$ extending ρ such that $P^\star \cap [\sigma_i^\rho] \neq \emptyset$, since P is totally ∞-layered. Therefore, $\Xi(G_\rho, \mathsf{PA}, (\sigma_i^\rho)_{i\in\omega}, g)$ is contained in PA, where Ξ is a computable function in Lemma 2. From an input $g \in R$, one can learn a ρ^g such that $\rho^g \in P^\star$ and $\Gamma_s|_{\rho^g\otimes\{g\}} = \Gamma_{|\rho^g|}|_{\rho^g\otimes\{g\}}$ for any $s \geq |\rho^g|$, since the assertion $\Gamma_s|_Y = \Gamma_t|_Y$ is equivalent to the following: for any clopen set $[\sigma]$ and any $u \in [t, s]$, such that $\Gamma_t^{-1}(\{?\}) \cap Y \neq \emptyset$. Here recall that $\{?\}$ is a clopen set in $(\omega^\omega)_?$, and hence, $\Gamma_t^{-1}(\{?\})$ is c.e. open. Therefore, there is a $\Pi_1^0(g)$ statement characterizing ρ^g, uniformly in $g \in R$. Then, we have a learnable function $h = \lim_s h_s : R \to 2^\omega$ which maps g to such ρ^g. Define $\Delta_s(g) =?$ if $h_s(g) =?$, and $\Delta_s(g) = \Xi(G_{h_s(g)}, \mathsf{PA}, (\sigma_i^{h_s(g)})_{i\in\omega}, g)$ otherwise. It is easy to see that the learnable function $\Delta = \lim_s \Delta_s$ maps R into PA. □

3.2   Topological Games and Popperian Learnability

By Lewis-Shore-Sorbi [7], the initial segment $(\mathbf{0}, \mathbf{d}]$ below the Medvedev degree $\mathbf{d}$ of a dense set in ω^ω has no co-c.e. closed set. There are other density-like properties making *co-c.e.-free initial segments*:

For a set $S \subseteq X$, the two-players game $\mathfrak{G}_S$ is defined as follows: Each *play* is a decreasing sequence $\{U_n\}_{n \in \omega}$ of open sets with $S \cap U_n \neq \emptyset$. For a play $p = \{U_n\}_{n \in \omega}$, Player II *wins* on p if $S \cap \bigcap_n U_n \neq \emptyset$. Otherwise, Player I wins. If Player II has a winning strategy for the game $\mathfrak{G}_S$, then S is called *Choquet*.

$$
\begin{array}{lcccccccc}
\text{Player I:} & U_0 & & & U_2 & & & U_4 & & \cdots \\
 & & \supsetneq & & & \supsetneq & & & \supsetneq & & \supsetneq \\
\text{Player II:} & & & U_1 & & & U_3 & & & U_5 & & \cdots
\end{array}
$$

Theorem 5. *Assume that a set $P \subseteq 2^\omega$ contains a Choquet subset $C \subseteq P$ whose closure has a dense subset of computable points. For any co-c.e. closed set $Q \subseteq 2^\omega$, if an e.P. learnable function exists from P to Q, then Q contains a computable element.*

Proof. Let $F :\subseteq \omega^\omega \to \omega^\omega$ be a partial learnable function. A partial computable function $f :\subseteq \omega^{<\omega} \to \omega^{<\omega} \cup \{?\}$ is said to be an *approximation of F* if:

- (?-goodness) $f(\sigma^-) \not\subseteq f(\sigma)$ occurs only when $? \in \{f(\sigma^-), f(\sigma)\}$;
- (Convergence) $F(x) = \lim_s f(x \upharpoonright s)$, for any $x \in \mathrm{dom}(F)$.

Fix a winning strategy ψ_{II} for Player II on the Choquet game $\mathfrak{G}_C$, a co-c.e. closed set $Q \subseteq 2^\omega$ with no computable element, and suppose that an e.P. learnable function $F : P \to Q$ exists. Fix also an approximation $f : \omega^{<\omega} \to \omega^{<\omega} \cup \{?\}$ of F. Choose any string τ_i with $[\tau_i] \cap C \neq \emptyset$. Since $\mathrm{cl}(C)$ has a dense subset of computable points, C is dense at a computable point $\beta_i \supset \tau_i$. Note that, if $f(\beta_i \upharpoonright n) \neq ?$ for any $n \geq |\tau_i|$, then $[f(\sigma)] \cap Q = \emptyset$ for some $\sigma \subset \beta_i$. Otherwise, since F is e.P., we have $\lim_n f(\beta_i \upharpoonright n) \in Q$. However, monotonicity of $\{f(\beta_i \upharpoonright n)\}_{|\tau_i| \leq n \in \omega}$ implies that $\lim_n f(\beta_i \upharpoonright n)$ is computable. This contradicts our assumption that Q contains no computable element. If $f(\sigma) \notin T_Q$ happens for some $\sigma \subset \beta_i$ extending τ_i, for any $\alpha \in C$ extending σ, we have $f(\sigma) \not\subseteq \lim_s f(\alpha \upharpoonright s) \in Q$, since $F(C) \subseteq Q$. Therefore, $f(\sigma^*) = ?$ must occur for some σ^* with $\tau_i \subset \sigma^* \subset \alpha$. Then, define $\psi_{\mathrm{I}}(\tau_i) = \sigma^*$. Player II extend it to $\tau_{i+1} = \psi_{\mathrm{II}}(\psi_{\mathrm{I}}(\tau_i))$. Eventually an infinite increasing sequence $\{\tau_i\}_{i \in \omega}$ is constructed, and then $h = \lim_i \tau_i \in C$ by the property of ψ_{II}. However, $\lim_n f(h \upharpoonright n)$ does not converge. Therefore, $h \notin \mathrm{dom}(F)$. □

Definition 4. *Fix a set $S \subseteq X$, and we consider the Choquet game $\mathfrak{G}_S$.*

1. *A function ψ is a strategy if, for a given open set b_n in X, $\psi(b_n)$ is an open subset of b_n, and $S \cap \psi(b_n) \neq \emptyset$ whenever $S \cap b_n \neq \emptyset$.*
2. *A function ψ is a prestrategy if, for a given previous move a_θ, $\psi(a_\theta)$ is a pair $\langle b_{\theta 0}, b_{\theta 1} \rangle$ of open sets with $b_{\theta 0} \cup b_{\theta 1} \subseteq a_\theta$, or $\psi(a_\theta) = \mathrm{RESIGN}$, where we declare that $S \cap \mathrm{RESIGN} = \emptyset$.*
3. *For a strategy ψ_{I} and a prestrategy ψ_{II}, the preplay $\psi_{\mathrm{I}} \otimes \psi_{\mathrm{II}}$ produced by ψ_{I} and ψ_{II} is a collection $\langle a_{\langle\rangle}, b_{\theta j}, a_{\theta j} \rangle_{\theta \in 2^{<\omega}, j<2}$, where $a_{\langle\rangle} = \psi_{\mathrm{I}}(\langle\rangle)$, $\psi_{\mathrm{II}}(a_\theta) = \langle b_{\theta 0}, b_{\theta 1} \rangle$, and $a_{\theta j} = \psi_{\mathrm{I}}(b_{\theta j})$ for any $\theta \in 2^{<\omega}$, and $j < 2$.*

I II I II I II I II I ...

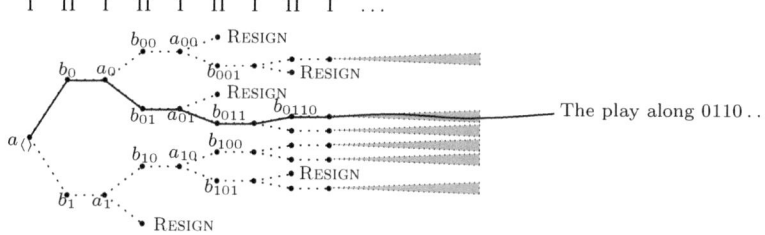

Fig. 1. A preplay on a given Choquet game

4. For a preplay $p = \langle a_{\langle\rangle}, b_{\theta j}, a_{\theta j}\rangle_{\theta \in 2^{<\omega}, j<2}$, the play of p along $h \in 2^\omega$ is defined by the infinite sequence $p|_h = \langle a_{\langle\rangle}, b_\theta, a_\theta\rangle_{\theta \subset h}$.
5. The play tree $\mathrm{Play}(\psi_\mathrm{I} \otimes \psi_\mathrm{II})$ of a preplay $\psi_\mathrm{I} \otimes \psi_\mathrm{II} = \langle a_{\langle\rangle}, b_{\theta j}, a_{\theta j}\rangle_{\theta \in 2^{<\omega}, j<2}$ is defined by $\mathrm{Play}(\psi_\mathrm{I} \otimes \psi_\mathrm{II}) = \{\theta : (\forall \eta \subseteq \theta)\, b_\eta \neq \mathrm{RESIGN}\}$. For a partial preplay $\pi \subset \psi_\mathrm{I} \otimes \psi_\mathrm{II}$, the play tree $\mathrm{Play}(\pi)$ is also defined in the same manner.
6. A prestrategy ψ_II for Player II is winning if, for every strategy ψ_I for Player I, Player II wins on the play of $\psi_\mathrm{I} \otimes \psi_\mathrm{II}$ along any infinite path h through $\mathrm{Play}(\psi_\mathrm{I} \otimes \psi_\mathrm{II})$, i.e., $S \cap \bigcap_n (\psi_\mathrm{I} \otimes \psi_\mathrm{II}|_h)(n) \neq \emptyset$ for any $h \in [\mathrm{Play}(\psi_\mathrm{I} \otimes \psi_\mathrm{II})]$.
7. A function ψ is a playful strategy if it is a prestrategy, and the play tree $\mathrm{Play}(\phi \otimes \psi)$ has an infinite path for any strategy ϕ.
8. If Player II has a computable winning playful strategy for the game $\mathfrak{G}_S$, then S is called **PA-Choquet**.

A partial computable function $\beta : \omega^{<\omega} \to \omega^\omega$ is a *dense choice of computable points* in C if $C \cap [\sigma]$ is dense at the point $\beta(\sigma)$, whenever $C \cap [\sigma]$ is nonempty.

Theorem 6. *Assume that a set $P \subseteq 2^\omega$ contains a PA-Choquet subset $C \subseteq P$ whose closure has a dense choice of computable points. For any co-c.e. closed set $Q \subseteq 2^\omega$ and any $R \subseteq \omega^\omega$, if an e.P. learnable function exists from $P \otimes R$ to Q, then an e.P. learnable function exists from R to Q.*

Proof. Fix a computable winning playful strategy ψ_II for the player II on the Choquet game $\mathfrak{G}_C$, a co-c.e. closed set $Q \subseteq 2^\omega$, and an e.P. learnable function $F : P \otimes R \to Q$ with an approximation $f : \omega^{<\omega} \to \omega^{<\omega} \cup \{?\}$. Let β be a dense choice of computable points in C. Fix $g \in R$.

Strategies S_θ^g. We introduce a strategy S_θ^g for each $\theta \in 2^{<\omega}$. There are four states for strategies, ACTIVE, CHANGED, REFUTED, and RESIGNED. First we declare the root strategy $S_{\langle\rangle}^g$ to be ACTIVE. Assume that, on a partial play on the Choquet game $\mathfrak{G}_C$, the θ-th move τ_θ^g of ψ_II is given, S_θ^g is ACTIVE, and there is no ACTIVE strategy S_κ^g for $\kappa \subsetneq \theta$. We determine the state of the θ-th strategy S_θ^g as follows:

- S_θ^g is CHANGED if $f(\sigma \oplus g) =?$ for some $\tau_\theta^g \subset \sigma \subset \beta(\tau_\theta^g)$.
- S_θ^g is REFUTED if $f(\sigma \oplus g) \notin T_Q$ for some $\tau_\theta^g \subset \sigma \subset \beta(\tau_\theta^g)$.
- S_θ^g is RESIGNED when we find that θ does not extend to an infinite path through the play tree $\mathrm{Play}(\psi_\mathrm{I} \otimes \psi_\mathrm{II})$ of the winning strategy ψ_II.

If S_θ^g is declared to be CHANGED, or RESIGNED, then we withdraw the previous declaration that S_θ^g is ACTIVE, and close the strategy S_θ^g.

Play on Choquet Game $\mathfrak{G}_C$. Now we determine the next move of Player I, i.e., define $\psi_I(\tau_\theta^g)$. If S_θ^g is REFUTED or RESIGNED, then Player I takes no action. If S_θ^g is CHANGED, then Player I chooses the least σ such that S_θ^g is refuted at σ, and put $\psi_I(\tau_\theta^g) = \sigma$. Then, by using the winning strategy ψ_{II}, Player II chooses the $(\theta 0)$-th move $\tau_{\theta 0}^g$ and the $(\theta 1)$-th move $\tau_{\theta 1}^g$, from the partial play $\psi_I(\tau_\theta^g)$, i.e., $\psi_{II}(\psi_I(\tau_\theta^g)) = \langle \tau_{\theta 0}^g, \tau_{\theta 1}^g \rangle$, and declare that the strategies $S_{\theta 0}^g$ and $S_{\theta 1}^g$ are ACTIVE. Note that τ_θ^g and the state of S_θ^g at each stage are partial computable uniformly in θ and g, since ψ_{II} and β are computable.

Observation. For any $g \in R$, consider the following binary tree V^g consisting of all binary strings $\theta \in 2^{<\omega}$ such that S_θ^g is declared to be ACTIVE at some stage. Claim that V^g has no infinite path. If V^g has an infinite path h, then f must outputs ? infinitely often along $p_h = \bigcup_{\theta \subset h} \tau_\theta$. However, p_h is constructed along the winning strategy ψ_{II}, and p_h is an infinite path through the play tree $\text{Play}(\psi_I \otimes \psi_{II})$, since no substring of h is RESIGNED. As ψ_{II} is winning, p_h must belong to $C \subseteq P$. It implies that $F(p_h \oplus g) = \lim_s f(p_h \oplus g \restriction s)$ does not converges, and note that $p_h \oplus g \in C \otimes \{g\} \subseteq P \otimes R$, This contradicts our assumption that the domain of F includes $P \otimes R$.

Thus, at some stage, all declarations of strategies on V^g are determined. Moreover, each leaf of V^g which is not assigned RESIGN by ψ_{II} must be declared to be ACTIVE at almost all stages. Because C is dense at $\beta(\tau_\rho^g)$ for each leaf $\rho \in V^g$ which is not declared to be RESIGNED, and then $\lim_s f(\beta(\tau_\rho^g) \oplus g \restriction s)$ is total, since $F = \lim f$ is e.P., and each leaf $\rho \in V^g$ must not be declared to be CHANGED. In particular, $\lim_s f(\beta(\tau_\rho^g) \oplus g \restriction s) \in Q$.

Learning Procedure. We construct an e.P. learnable function $G : R \to Q$. The learner $G(g)$ tries to find an ACTIVE leaf ρ of V^g at each stage s, and set $G(g) = F(\beta(\tau_\theta^g) \oplus g)$. Each time his guess on an eventually ACTIVE leaf of V^g is changed, an approximation of G returns ?. If g is contained in R, then by finiteness of V^g, an approximation of $G(g)$ eventually finds an ACTIVE leaf of V^g. If $g \notin R$, then $G(g)$ may yet fail to find an ACTIVE leaf of V^g. But then its approximation returns ? infinitely often. Otherwise, $G(g)$ is defined to be $F(\beta(\tau_\theta^g) \oplus g)$, and then it is e.P., since F is e.P. By the previous observation, the e.P. learnable function G maps R into Q as desired.   □

Definition 5 (Higuchi-Kihara [6]). *Fix* $\sigma \in \omega^{<\omega}$, *and* $i \in \omega$. *Then the* i-th *projection of* σ *is inductively defined as follows.*

$$\text{pr}_i(\langle \rangle) = \langle \rangle, \qquad \text{pr}_i(\sigma) = \begin{cases} \text{pr}_i(\sigma^-)^\frown n, & \text{if } \sigma = \sigma^{-\frown}\langle i, n \rangle, \\ \text{pr}_i(\sigma^-), & \text{otherwise.} \end{cases}$$

Furthermore, the projection of $x \in \omega^\omega$ *is defined to be* $\text{pr}_i(x) = \lim_n \text{pr}_i(x \restriction n)$.

Theorem 7. *For every co-c.e. closed set* $P \subseteq 2^\omega$, *for each* $k \geq 2$, *the set* $\text{TEAM}_k\text{LEARNING}(P) = \{x \in \omega^\omega : (\exists i < k)\ \text{pr}_i(x) \in P^{(\infty)}\}$ *is a* Σ_3^0 *subset*

*of 2^ω which has the same Turing upward closure as P, and has a PA-Choquet
subset whose closure has a dense choice of computable points.*

Proof. Set $S = \mathrm{TEAM}_2\mathrm{LEARNING}(P)$. Straightforwardly, we can check that S is
Σ_3^0, and it has the same Turing upward closure as P. Consider the following set:

$$C = \{x \in \omega^\omega : \mathbf{pr}_0(x) \in P^{(\infty)} \ \& \ (\forall n \in \omega) \, \mathbf{pr}_1(x \upharpoonright n) \in T_P^{ext}\}.$$

Clearly, C is a subset of S. To construct a dense choice β of computable points
in the closure of C, we fix a leaf of T_P. Given σ, if it has a nonempty intersection
with C, then $\mathbf{pr}_0(\sigma)$ must be of the form $\rho_0{}^\frown\rho_1{}^\frown \ldots {}^\frown\rho_n{}^\frown\tau$, where ρ_i is a leaf of
T_P for each $i \leq n$, and τ is a node of T_P. By a uniformly computable way, we
can calculate the position of a leaf $\tau{}^\frown\eta$ of T_P. Then, define $\beta(\sigma)$ as follows:

$$\beta(\sigma) = \sigma{}^\frown(0^{|\eta|} \oplus \eta){}^\frown(0^{|\rho|} \oplus \rho){}^\frown(0^{|\rho|} \oplus \rho){}^\frown \ldots {}^\frown(0^{|\rho|} \oplus \rho){}^\frown(0^{|\rho|} \oplus \rho){}^\frown \ldots$$

Here, $0^{|\alpha|} \oplus \alpha$ denotes the string $\langle 0, \alpha(0), 0, \alpha(1), \ldots, 0, \alpha(|\alpha|-1)\rangle$. Clearly, $\beta(\sigma)$
is contained in the closure of C.

Now we construct a strategy ψ for Player II on Choquet game $\mathfrak{G}_C$ as follows:
Given $a_\theta \in \omega^{<\omega}$, the θ-th move of Player I, first check whether $\mathbf{pr}_1(a_\theta)$ has an
extension in T_P of length $\max\{|\mathbf{pr}_1(a_\theta)|, |\theta|\}$ or not. If not (it is possible because
of the past moves by Player II), Player II resigns the game $\mathfrak{G}_C$, i.e., $\psi_{\mathrm{II}}(a_\theta) =$
RESIGN. Otherwise, when $|\mathbf{pr}_1(a_\theta)| > |\theta|$, Player II does not act, i.e., $\psi_{\mathrm{II}}(a_\theta) =$
$\langle a_\theta, a_\theta \rangle$. If $\mathbf{pr}_1(a_\theta) \leq |\theta|$, then Player II returns $\psi_{\mathrm{II}}(a_\theta) = \langle a_\theta{}^\frown\langle 1, 0\rangle, a_\theta{}^\frown\langle 1, 1\rangle\rangle$.
By our construction of the strategy ψ_{II}, for every $\psi_{\mathrm{I}} \otimes \psi_{\mathrm{II}}|_h$ along any infinite
path h through the play tree $\mathrm{Play}(\psi_{\mathrm{I}} \otimes \psi_{\mathrm{II}})$, the 1-st projection of $\bigcap_n \psi_{\mathrm{I}} \otimes \psi_{\mathrm{II}}|_h$
must be contained in P. Therefore, $\bigcap_n \psi_{\mathrm{I}} \otimes \psi_{\mathrm{II}}|_h$ is contained in C. Moreover,
P is equal to the set of all infinite paths through $\mathrm{Play}(\psi_{\mathrm{I}} \otimes \psi_{\mathrm{II}})$. Consequently,
ψ_{II} is a winning playful strategy of Player II.   □

Acknowledgements. This work was supported by a Grant-in-Aid for JSPS
fellows.

References

1. Binns, S.: Hyperimmunity in $2^{\mathbb{N}}$. Notre Dame Journal of Formal Logic 48(2), 293–316 (2007)
2. Brattka, V., Gherardi, G.: Effective choice and boundedness principles in computable analysis. Bulletin of Symbolic Logic 17(1), 73–117 (2011)
3. Brattka, V., Presser, G.: Computability on subsets of metric spaces. Theor. Comput. Sci. 305(1-3), 43–76 (2003)
4. Cenzer, D., Kihara, T., Weber, R., Wu, G.: Immunity and non-cupping for closed sets. Tbilisi Math. J. 2, 77–94 (2009)
5. Demuth, O., Kučera, A.: Remarks on 1-genericity, semigenericity and related concepts. Comment. Math. Univ. Carolinae 28, 85–94 (1987)
6. Higuchi, K., Kihara, T.: Inside the Muchnik degrees: Discontinuity, learnability, and constructivism (preprint)

7. Lewis, A.E.M., Shore, R.A., Sorbi, A.: Topological aspects of the Medvedev lattice. Arch. Math. Log. 50(3-4), 319–340 (2011)
8. Soare, R.I.: Recursively Enumerable Sets and Degrees. Perspectives in Mathematical Logic, xVIII+437 pages. Springer, Heidelberg (1987)
9. Weihrauch, K.: Computable Analysis: An Introduction. Texts in Theoretical Computer Science, 285 pages. Springer (2000)
10. Ziegler, M.: Real computation with least discrete advice: A complexity theory of nonuniform computability with applications to effective linear algebra. Annals of Pure and Applied Logic 163(8), 1108–1139 (2012),
    http://www.sciencedirect.com/science/article/pii/S016800721100203X

How Much Randomness Is Needed for Statistics?

Bjørn Kjos-Hanssen[1], Antoine Taveneaux[2], and Neil Thapen[3,4]

[1] University of Hawai'i at Mānoa, Honolulu, HI 96822, United States of America
`bjoern@math.hawaii.edu`
[2] Laboratoire d'Informatique Algorithmique: Fondements et Applications (LIAFA),
Université Paris Diderot—Paris 7, 75205 Paris Cedex 13, France
`taveneaux@liafa.jussieu.fr`,
[3] Academy of Sciences of the Czech Republic, 115 67 Praha 1, Czech Republic
`thapen@math.cas.cz`
[4] Isaac Newton Institute for Mathematical Sciences, 20 Clarkson Road, Cambridge
CB3 0EH, United Kingdom

Abstract. In algorithmic randomness, when one wants to define a randomness notion with respect to some non-computable measure λ, a choice needs to be made. One approach is to allow randomness tests to access the measure λ as an oracle (which we call the "classical approach"). The other approach is the opposite one, where the randomness tests are completely effective and do not have access to the information contained in λ (we call this approach "Hippocratic"). While the Hippocratic approach is in general much more restrictive, there are cases where the two coincide. The first author showed in 2010 that in the particular case where the notion of randomness considered is Martin-Löf randomness and the measure λ is a Bernoulli measure, classical randomness and Hippocratic randomness coincide. In this paper, we prove that this result no longer holds for other notions of randomness, namely computable randomness and stochasticity.

1 Introduction

In algorithmic randomness theory we are interested in which almost sure properties of an infinite sequence of bits are effective or computable in some sense. Martin-Löf defined randomness with respect to the uniform fair-coin measure μ on 2^ω as follows.

A sequence $X \in 2^\omega$ is *Martin-Löf random* if we have $X \notin \bigcap_{n \in \mathbb{N}} \mathcal{U}_n$ for every sequence of uniformly Σ_1^0 (or effectively open) subsets of 2^ω such that $\mu(\mathcal{U}_n) \leq 2^{-n}$.

Now if we wish to consider Martin-Löf randomness for a Bernoulli measure μ_p (i.e. a measure such that the ith bit is the result of a Bernoulli trial with parameter p) we have have two ways to extend the previous definition.

The first option is to consider p as an oracle (with an oracle p we can compute μ_p) and relativize everything to this oracle. Then X is μ_p-Martin-Löf random if for every sequence $(\mathcal{U}_n)_{n \in \mathbb{N}}$ of uniformly $\Sigma_1^0[p]$ sets such that

S.B. Cooper, A. Dawar, and B. Löwe (Eds.): CiE 2012, LNCS 7318, pp. 395–404, 2012.
© Springer-Verlag Berlin Heidelberg 2012

$\mu_p(\mathcal{U}_n) \leq 2^{-n}$ we have $X \notin \bigcap_{n \in \mathbb{N}} \mathcal{U}_n$. We will call this approach the *classical* notion of Martin-Löf randomness relative to μ_p.

Another option is to keep the measure μ_p "hidden" from the process which describes the sequence $(\mathcal{U}_n)$. One can merely replace μ by μ_p in Martin-Löf's definition but still require $(\mathcal{U}_n)$ to be uniformly Σ_1^0 in the unrelativized sense. This notion of randomness was introduced by Kjos-Hanssen [4] who called it *Hippocratic randomness*; Bienvenu, Doty and Stephan [1] used the term *blind randomness*.

Kjos-Hanssen showed that for Bernoulli measures, Hippocratic and classical randomness coincide in the case of Martin-Löf randomness. Bienvenu, Gács, Hoyrup, Rojas and Shen [2] extended Kjos-Hanssen's result to other classes of measures. Here we go in a different direction and consider weaker randomness notions, such as computable randomness and stochasticity. We discover the contours of a dividing line for the type of betting strategy that is needed in order to render the probability distribution superfluous as a computational resource.

We view *statistics* as the discipline concerned with determining the underlying probability distribution μ_p by looking at the bits of a random sequence. In the case of Martin-Löf randomness it is possible to determine p ([4]), and therefore Hippocratic randomness and classical randomness coincide. In this sense, Martin-Löf randomness is sufficient for statistics to be possible, and it is natural to ask whether smaller amounts of randomness, such as computable randomness, are also sufficient.

Notation. Our notation generally follows Nies' monograph [5]. We write 2^n for $\{0, 1\}^n$, and for sequences $\sigma \in 2^{\leq \omega}$ we will also use σ to denote the real with binary expansion $0.\sigma$, that is, the real $\sum_{i=1}^{\infty} \sigma(i)2^{-i}$. We use ε to denote the empty word, $\sigma(n)$ for the nth element of a sequence and $\sigma \upharpoonright n$ for the sequence formed by the first n elements. For sequences ρ, σ we write $\sigma \prec \rho$ if σ is a proper prefix of ρ and denote the concatenation of σ and ρ by $\sigma.\rho$ or simply $\sigma\rho$. Throughout the paper we set $n' = n(n-1)/2$.

1.1 Hippocratic Martingales

Formally a martingale is a function $\mathcal{M} : 2^{<\omega} \to \mathbb{R}^{\geq 0}$ satisfying

$$\mathcal{M}(\sigma) = \frac{\mathcal{M}(\sigma 0) + \mathcal{M}(\sigma 1)}{2}.$$

Intuitively, such a function arises from a betting strategy for a fair game played with an unbiased coin (a sequence of Bernoulli trials with parameter $1/2$). In each round of the game we can choose our stake, that is, how much of our capital we will bet, and whether we bet on heads (1) or tails (0). A coin is tossed, and if we bet correctly we win back twice our stake.

Suppose that our betting strategy is given by some fixed function S of the history σ of the game up to that point. Then it is easy to see that the function $\mathcal{M}(\sigma)$ giving our capital after a play σ satisfies the above equation. On the other hand, from any $\mathcal{M}$ satisfying the equation we can recover a corresponding strategy S.

More generally, consider a biased coin which comes up heads with probability $p \in (0, 1)$. In a fair game played with this coin, we would expect to win back $1/p$ times our stake if we bet correctly on heads, and $1/(1-p)$ times our stake if we bet correctly on tails. Hence we define a p-martingale to be a function satisfying

$$\mathcal{M}(\sigma) = p\mathcal{M}(\sigma 1) + (1-p)\mathcal{M}(\sigma 0).$$

We can generalize this further, and for any probability measure μ on 2^ω define a μ-martingale to be a function satisfying

$$\mu(\sigma)\mathcal{M}(\sigma) = \mu(\sigma 1)\mathcal{M}(\sigma 1) + \mu(\sigma 0)\mathcal{M}(\sigma 0).$$

For the Bernoulli measure with parameter p, we say that a sequence $X \in 2^\omega$ is p-computably random if for every total, p-computable p-martingale $\mathcal{M}$, the sequence $(\mathcal{M}(X \restriction n))_n$ is bounded.

This is the classical approach to p-computable randomness. Under the Hippocratic approach, the bits of the parameter p should not be available as a computational resource. The obvious change to the definition would be to restrict to p-martingales $\mathcal{M}$ that are computable without an oracle for p. However this does not give a useful definition, as p can easily be recovered from any non-trivial p-martingale. Instead we will define p-Hippocratic computable martingales in terms of their stake function (or strategy) S.

We formalize S as a function $2^{<\omega} \to [-1, 1] \cap \mathbb{Q}$. The absolute value $|S(\sigma)|$ gives the fraction of our capital we put up as our stake, and we bet on 1 if $S(\sigma) \geq 0$ and on 0 if $S(\sigma) < 0$. Given $\alpha \in (0, 1)$, the α-martingale $\mathcal{M}^\alpha$ arising from S is then defined inductively by

$$\mathcal{M}^\alpha(\varepsilon) = 1$$

$$\mathcal{M}^\alpha(\sigma 1) = \mathcal{M}^\alpha(\sigma)\left(1 - |S(\sigma)| + \frac{|S(\sigma)|}{\alpha}1_{\{S(\sigma) \geq 0\}}\right)$$

$$\mathcal{M}^\alpha(\sigma 0) = \mathcal{M}^\alpha(\sigma)\left(1 - |S(\sigma)| + \frac{|S(\sigma)|}{1-\alpha}1_{\{S(\sigma) < 0\}}\right)$$

where, for a formula T, we use the notation $1_{\{T\}}$ to mean the function which takes the value 1 if T is true and 0 if T is false.

We define a p-Hippocratic computable martingale to be a p-martingale $\mathcal{M}^p$ arising from some total computable (without access to p) stake function S. We say that a sequence $X \in 2^\omega$ is p-Hippocratic computably random if for every p-Hippocratic computable martingale $\mathcal{M}$, the sequence $(\mathcal{M}(X \restriction n))_n$ is bounded.

In Section 2 below we show that for all $p \in$ MLR the set of p-Hippocratic computably random sequences is strictly bigger than the set of p-computable random sequences. More precisely, we show that we can compute a sequence $Q \in 2^\omega$ from p such that Q is p-Hippocratic computably random. In a nutshell, the proof works as follows. We use the number p in two ways. To compute the ith bit of Q, the first i bits of p are treated as a parameter $r = 0.p_0 \ldots p_i$, and we pick the ith bit of Q to look like it has been chosen at random in a Bernoulli trial with bias r. To do this, we use some fresh bits of p (which have not been used so far in the construction of Q) and compare them to r, to simulate the trial. Since these bits of p were never used before, if we know only the first $i - 1$ bits of Q they appear random, and thus the ith bit of Q indeed appears to be chosen at random with bias r. Since $r = 0.p_1p_2\ldots p_i$ converges quickly to p (this convergence is faster than the deviations created by statistical noise in a real sequence of Bernoulli trials with parameter p), we are able to show that Q overall looks p-random as long as we do not have access to p, in other words, that Q is p-Hippocratic computably random.

1.2   Hippocratic Stochasticity and KL Randomness

In Section 3 we consider another approach to algorithmic randomness, known as stochasticity. It is reasonable to require that a random sequence satisfies the law of large numbers, that is, that the proportion of 1s in the sequence converges to the bias p. But, for an unbiased coin, the string

$$010101010\ldots$$

satisfies this law but is clearly not random. Following this idea, we say that a sequence X is p-Kolmogorov-Loveland (or KL) stochastic if there is no p-computable way to select infinitely many bits from X, where we are not allowed to know the value of a bit before we select it, without the selected sequence satisfying the law of large numbers (see Definition 1 for a formal approach).

For this paradigm the Hippocratic approach is clear: we consider only selection functions which are computable without an oracle for p. We show in Theorem 2 that for $p \in \Delta_2^0 \cap \text{MLR}$ there exists a sequence Q which is p-Hippocratic KL stochastic but not p-KL stochastic. Again we use p as a random bit generator and create a sequence Q that appears random for a sequence of Bernoulli trials, where the bias of the ith trial is q_i for a certain sequence $(q_i)_i$ converging to p. Intuitively, the convergence is so slow that it is impossible to do (computable) statistics with Q to recover p, and we are able to show that without access to p the sequence Q is p-KL stochastic.

At the end of Section 3 we consider another notion, Kolmogorov-Loveland randomness. We give a simple argument to show that if we can compute p from every p-Hippocratic KL random sequence, then the p-Hippocratic KL random sequences and the p-KL random sequences are the same (and vice versa).

Some technical definitions and lemmas follow in the appendix.

2   Computable Randomness

In this section we show that for any Martin-Löf random bias p, p-computable randomness is a stronger notion than p-Hippocratic computable randomness.

Theorem 1. *Let $\alpha \in \text{MLR}$. There exists a sequence $Q \in 2^\omega$, computable in polynomial time from α, such that Q is α-Hippocratic computably random.*

Before giving the proof, we remark that a stronger version of the theorem is true: the sequence Q is in fact α-Hippocratic partial computably random (meaning that we allow the martingale to be a partial computable function, see Definition 7.4.3 in [3]).

Also, a sceptic could (reasonably) complain that it is not really natural for us to make bets without any idea about our current capital. However if we add an oracle to give the integer part of our capital at each step (or even an approximation with accuracy 2^{-n} when we bet on the nth bit), Theorem 1 remains true and the proof is the same. In the same spirit we could object that it is more natural to have a stake function giving the amount of our bet (to be placed only if we have a capital large enough) and not the proportion of our capital. For this definition of a Hippocratic computable martingale, similarly the theorem remains true and the proof is the same.

Proof. Let $\alpha \in \text{MLR}$. Then α is not rational and cannot be represented by a finite binary sequence and we can suppose that $0 < \alpha < 1/2$. Recall that $n' = n(n-1)/2$ and that we freely identify a sequence X (finite or infinite) with the real number with the binary expansion $0.X$.

The proof has the following structure. First, we describe an algorithm to compute a sequence Q from α. To compute each bit Q_n of Q we will use a finite initial segment of α as an approximation of α, and we will compare this with some other fresh bits of α which we treat as though they are produced by a random bit generator. In this way Q_n will approximate the outcome of a Bernoulli trial with bias α.

Second, we will suppose for a contradiction that there is an α-Hippocratic computable martingale (that is, a martingale that arises from a stake function computable without α) such that the capital of this martingale is not bounded on Q. We will show that we can use this stake function to construct a Martin Löf test $(U_n)_n$ such that α does not pass this test.

So let $Q = Q_1 Q_2 \ldots$ be defined by the condition that:

$$Q_n = \begin{cases} 0 & \text{if } 0.\alpha_{n'+1} \ldots \alpha_{n'+n} \geq 0.\alpha_1 \ldots \alpha_n, \\ 1 & \text{otherwise.} \end{cases}$$

We can compute Q in polynomial time from α, as we can compute each bit Q_n in time $O(n^2)$.

Now let $S : 2^{<\omega} \to \mathbb{Q} \cap [-1,1]$ be a computable stake function. We will write $\mathcal{M}^X$ for the X-martingale arising from S. Suppose for a contradiction that

$$\limsup_{n \to \infty} \mathcal{M}^\alpha(Q \upharpoonright n) = \infty.$$

Our goal is to use this to define a Martin-Löf test which α fails. The classical argument (see Theorem 6.3.4 in [3]) would be to consider the sequence of sets

$$V_j = \{X \in 2^\omega | \exists n \ \mathcal{M}^\alpha(X \upharpoonright n) > 2^j\},$$

but without oracle access to α this is not Σ_1^0, and does not define a Martin-Löf test. However it turns out that we can use a similar sequence of sets, based on the idea that, although we cannot compute $\mathcal{M}^\alpha$ precisely, we can approximate it using the approximation $\alpha_1 \ldots \alpha_{n'}$ of α. For this we will use the following lemma, proved in the appendix.

Lemma 1. *For $\alpha \in MLR$, there exists $m \in \mathbb{N}$ such that if $\sigma \succcurlyeq (\alpha \upharpoonright m')$ and $\tau \succcurlyeq (\alpha \upharpoonright m')$ then for all $\eta \in 2^{<\omega}$ and all $n \geq m$ we have:*

$$\text{if } 0 < \tau - \sigma < 2^{-n'} \text{ and } |\eta| \leq n+1 \text{ then } |\mathcal{M}^\sigma(\eta) - \mathcal{M}^\tau(\eta)| \leq 2^{-n}.$$

Let m be given by Lemma 1 and let ρ be $\alpha \upharpoonright m'$. Without loss of generality we may assume $2^{-m} < \rho$. Let $\Gamma : 2^{\leq \omega} \to 2^{\leq \omega}$ be the operator which converts $\alpha_1 \ldots \alpha_{n'}$ into $Q_1 \ldots Q_n$. That is, $\Gamma(\alpha_1 \ldots \alpha_k) = Q_1 \ldots Q_n$ where n is the biggest integer such that $n' \leq k$. This notation naturally extends to infinite sequences so we may write $\Gamma(\alpha) = Q$. We consider the uniform sequence of Σ_1^0 sets

$$U_j' = \{X_1 \ldots X_{k'} | \rho \preccurlyeq X_1 \ldots X_{k'} \text{ and } \mathcal{M}^{X_1 \ldots X_{k'}}(\Gamma(X_1 \ldots X_{k'})) > 2^j\}.$$

We let U_j denote the set of infinite sequences with a prefix in U_j'. By Lemma 1,

$$|\mathcal{M}^\alpha(\Gamma(\alpha_1 \ldots \alpha_{k'})) - \mathcal{M}^{\alpha_1 \ldots \alpha_{k'}}(\Gamma(\alpha_1 \ldots \alpha_{k'}))| < 2^{-k} \leq 1$$

for all sufficiently large k. Since $\mathcal{M}^\alpha$ increases unboundedly on $Q = \Gamma(\alpha)$ it follows that $\alpha \in U_j$ for all j.

To show that (U_j) is a Martin-Löf test, it remains to show that the measure of U_j is small. Since $\sigma \mapsto \mathcal{M}^\sigma(\sigma)$ is almost a α-martingale, where σ runs over the prefixes of α, we will use a lemma similar to the Kolmogorov inequality (see Theorem 6.3.3 in [3]). We postpone the proof to the appendix.

Lemma 2. *For any number $n \geq m$, any extension $\sigma \succcurlyeq \rho$ of length n' and any prefix-free set $Z \subseteq \bigcup_{k \in \mathbb{N}} \{0,1\}^{k'}$ of extensions of σ, we have*

$$\sum_{\tau \in Z} 2^{-|\tau|} \mathcal{M}^\tau(\Gamma(\tau)) \leq 2^{-|\sigma|} e^2 \left[1 + \mathcal{M}^\sigma(\Gamma(\sigma))\right].$$

Now fix j and let W_j be a prefix-free subset of U_j' with the property that the set of infinite sequences with a prefix in W_j is exactly U_j. Then by the definition of U_j', if $\tau \in W_j$ then $\mathcal{M}^\tau(\Gamma(\tau)) \geq 2^j$. Hence by Lemma 2 we have:

$$\mu(U_j) = \sum_{\tau \in W_j} 2^{-|\tau|} \leq \sum_{\tau \in W_j} \frac{\mathcal{M}^\tau(\Gamma(\tau))}{2^j} 2^{-|\tau|} \leq \frac{2^{-|\rho|} e^2 \left(1 + \mathcal{M}^\rho(\Gamma(\rho))\right)}{2^j}.$$

Since $2^{-|\rho|}(1 + \mathcal{M}^\rho(\Gamma(\rho)))$ is constant, this shows that (U_j) is a Martin-Löf test. As $\alpha \in \bigcap_j U_j$ it follows that $\alpha \notin MLR$. This is a contradiction.

Notice that this proof makes use of the fact that in our betting strategy we have to proceed monotonically from left to right through the string, making a decision for each bit in turn as we come to it. This is why our construction is able to use α as a random bit generator, because at each step it can use bits that were not used to compute the previous bits of Q. Following this idea the question naturally arises: if we are allowed to use a non-monotone strategy, then are the classical and Hippocratic random sequences the same? We explore this question in the next section.

3   Kolmogorov-Loveland Stochasticity and Randomness

We define Kolmogorov-Loveland stochasticity and show that, in this setting, the Hippocratic and classical approaches give different sets. We also consider whether this is true for Kolmogorov-Loveland randomness, and relate this to a statistical question.

3.1   Definitions

For a finite string $\sigma \in \{0,1\}^n$, we write $\#0(\sigma)$ for $|\{k < n | \sigma(k) = 0\}|$ and $\#1(\sigma)$ for $|\{k < n | \sigma(k) = 1\}|$. We write $\Phi(\sigma)$ for $\#1(\sigma)/n$, the frequency of 1s in σ.

Definition 1 (Selection Function). *A KL selection function is a partial function*

$$f : 2^{<\omega} \to \{scan, select\} \times \mathbb{N}.$$

We write $f(\sigma)$ as a pair $(s(\sigma), n(\sigma))$ and in this paper we insist that for all σ and $\rho \succ \sigma$ we have $n(\rho) \neq n(\sigma)$, so that each bit is read at most once.

Given input X, we write (V_f^X) for the sequence of strings seen (with bits either scanned or selected) by f, so that

$$V_f^X(0) = X(n(\varepsilon))$$
$$V_f^X(k+1) = V_f^X(k).X(n(V_f^X(k))).$$

We write U_f^X for the subsequence of bits selected by f. Formally U_f^X is the limit of the monotone sequence of strings (T_f^X) where

$$T_f^X(0) = \varepsilon$$
$$T_f^X(k+1) = \begin{cases} T_f^X(k) & \text{if } s(V_f^X(k)) = scan \\ T_f^X(k).n(V_f^X(k)) & \text{if } s(V_f^X(k)) = select. \end{cases}$$

Informally, the function is used to select bits from X in a non-monotone way. If V is the string of bits we have read so far, $n(V)$ gives the location of the next bit of X to be read. Then "$s(V) = scan$" means that we will just read this bit, whereas "$s(V) = select$" means that we will add it to our string T of selected bits.

Definition 2 (p-KL stochastic sequence). *A sequence X is p-KL stochastic if for all p-computable KL selection functions f (notice that f can be a partial function) such that the limit U_f^X of (T_f^X) is infinite, we have*

$$\lim_{k \to \infty} \Phi(T_f^X(k)) = p.$$

A sequence X is p-Hippocratic KL stochastic if for all KL selection functions f, computable without an oracle p, such that U_f^X is infinite, we have

$$\lim_{k \to \infty} \Phi(T_f^X(k)) = p.$$

For technical reasons we informally introduce two more definitions (formal definitions are given in section C.1 in the Appendix). A generalized Bernoulli measure
λ is a probability measure such that each bit is seen as an independent Bernoulli
trial with parameter $b_i^\lambda \in (0,1)$. A KL martingale is equipped with a KL style
selection function which allows it to bet on bits in a non-monotone way. A sequence X is KL random for a measure if every KL martingale using the measure
is bounded on X.

3.2   Hippocratic Stochasticity Is Not Stochasticity

We will show that, despite the fact that we now allow non-monotone strategies,
once again there exist sequences computable from α which are α-Hippocratic KL
stochastic, for $\alpha \in \mathrm{MLR} \cap \Delta_2^0$ (recall that Chaitin's constant Ω is the prototypical
example of such an α).

We remark that our proof shows also that for $\alpha \in \mathrm{MLR} \cap \Delta_2^0$ the Hippocratic
and classical versions of MWC-stochasticity are different (see Definition 7.4.1 in
[3] for a formal definition).

Theorem 2. *Let $\alpha \in \mathrm{MLR} \cap \Delta_2^0$. There exists a sequence $Q \in 2^\omega$, computable
from α, such that Q is α-Hippocratic KL stochastic.*

Proof. We will first define the sequence Q, and then show that Q is λ-KL random
for a certain generalized Bernoulli measure λ for which the parameters (b_i^λ) converge to α. Finally we will show that it follows that Q is actually α-Hippocratic
KL stochastic.

Since $\alpha \in \Delta_0^2$, by Shoenfield's limit lemma α is the limit of a computable
sequence of real numbers (although the convergence must be extremely slow,
since α is not computable). In particular there exists a computable sequence of
finite strings (β^k) such that $\beta^k \in \{0,1\}^k$ and

$$\lim_{k \to \infty} \beta^k = \alpha.$$

We define Q_k by

$$Q_k = \begin{cases} 1 & \text{if } 0.\beta^k \geq 0.\alpha_{k'+1} \ldots \alpha_{k'+k} \\ 0 & \text{otherwise.} \end{cases}$$

We set $Q = Q_1 Q_2 \ldots$. Intuitively, as in the proof of Theorem 1, we are using
α as a random bit generator to simulate a sequence of Bernoulli trials with
parameter β^k.

Notice that the transformation mapping α to Q is a total computable function.
We know that in general if g is total computable, and X is (Martin-Löf) random
for the uniform measure μ, then $g(X)$ is random for the measure $\mu \circ g^{-1}$ (see
[8] for a proof of this fact). Since $\alpha \in \mathrm{MLR}$, in our case this tell us that Q is
random for exactly the generalized Bernoulli measure λ given by $b_i^\lambda = \beta^i$.

Lemma 3 ([7] and in [3] p.311). *If X is Martin-Löf random for a computable
generalized Bernoulli measure λ, then X is λ-KL random.*

The proof is essentially the same as for the uniform measure on 2^ω. We postpone it to the appendix. It follows that Q is λ-KL random and by the next lemma, also proved in the appendix, we can conclude that Q is α-Hippocratic KL stochastic, completing the argument.

Lemma 4. *Let λ be a computable generalized Bernoulli measure and suppose*

$$\lim_{i \to \infty} b_i^\lambda = p.$$

Then every λ-KL random sequence is p-Hippocratic KL stochastic.

3.3   Kolmogorov-Loveland Randomness

We have shown that for computability randomness and non-monotone stochasticity, whether a string is random can depend on whether or not we have access to the actual bias of the coin. It is natural to ask if this remains true for Kolmogorov-Loveland randomness.

Lemma 5 ([6]). *For sequences $X, Y \in 2^\omega$, $X \oplus Y$ is p-KL random if and only if both X is p-KL^Y-random and Y is p-KL^X-random, and this remains true in the Hippocratic setting (that is, where the KL martingales do not have oracle access to p).*

The proof is a straightforward adaptation of the proof given in [6] (Proposition 11). Using this lemma we can show an equivalence between our question and a statistical question.

Theorem 3. *The following two sentences are equivalent.*

1. *The p-Hippocratic KL random and p-KL random sequences are the same.*
2. *From every p-Hippocratic KL random sequence X, we can compute p.*

Proof. For (1) $\Rightarrow$ (2), we know that if X is p-KL random then it must satisfy the law of the iterated logarithm (see [9] for this result). Hence we know how quickly $\Phi(X \upharpoonright k)$ converges to p and using this we can (non-uniformly) compute p from X.

For (2) $\Rightarrow$ (1), suppose that X is p-Hippocratic KL random but not p-KL random. Let $X = Y \oplus Z$. Then by Lemma 5, Y (say) is not p-KL^Z random, meaning that there is a KL martingale $(\mathcal{M}, f)$ which, given oracle access to p and Z, wins on Y. On the other hand, both Y and Z remain p-Hippocratic KL random, so in particular by (2) if we have oracle access to Z then we can compute p. But this means that we can easily convert $(\mathcal{M}, f)$ into a p-Hippocratic KL martingale which wins on X, since to answer oracle queries to either Z or p it is enough to scan Z and do some computation.

Acknowledgements. The authors would like to thank Samuel Buss for his invitation to University of California, San Diego. Without his help and the university's support this paper would never exist. Taveneaux's research has been helped by a travel grant of the "Fondation Sciences Mathématiques de Paris". Kjos-Hanssen's research was partially supported by NSF (USA) grant no. DMS-0901020. Thapen's research was partially supported by grant IAA100190902 of GA AV ČR and by Center of Excellence CE-ITI under grant P202/12/G061 of GA CR and RVO: 67985840.

References

1. Bienvenu, L., Doty, D., Stephan, F.: Constructive dimension and Turing degrees. Theory of Computing Systems 45(4), 740–755 (2009)
2. Bienvenu, L., Gács, P., Hoyrup, M., Rojas, C., Shen, A.: Algorithmic tests and randomness with respect to a class of measures. Proceedings of the Steklov Institute of Mathematics 271, 41–102 (2011)
3. Downey, R.G., Hirschfeldt, D.: Algorithmic Randomness and Complexity. Springer (2010)
4. Kjos-Hanssen, B.: The probability distribution as a computational resource for randomness testing. J. Log. Anal. 2 (2010)
5. Nies, A.: Computability and Randomness. Oxford University Press (2009)
6. Merkle, W., Miller, J.S., Nies, A., Reimann, J., Stephan, F.: Kolmogorov-Loveland randomness and stochasticity. Ann. Pure Appl. Logic 138, 183–210 (2006)
7. Muchnik, A.A., Semenov, A.L., Uspensky, V.A.: Mathematical metaphysics of randomness. Theoret. Comput. Sci. 207, 263–317 (1998)
8. Shen, A.: One more definition of random sequence with respect to computable measure. In: First World Congress of the Bernoulli Society on Math. Statistics and Probability Theory, Tashkent (1986)
9. Wang, Y.: Resource bounded randomness and computational complexity. Theoretical Computer Science 237, 33–55 (2000)

Towards a Theory of Infinite Time Blum-Shub-Smale Machines

Peter Koepke and Benjamin Seyfferth

Mathematisches Institut, Rheinische Friedrich-Wilhelms-Universität Bonn,
Endenicher Allee 60, 53115 Bonn, Germany
{koepke,seyfferth}@math.uni-bonn.de

Abstract. We introduce a generalization of Blum-Shub-Smale machines on the standard real numbers $\mathbb{R}$ that is allowed to run for a transfinite ordinal number of steps before terminating. At limit times, register contents are set to the ordinary limit of previous register contents in $\mathbb{R}$. It is shown that each such machine halts before time ω^ω or diverges. We undertake first steps towards estimating the computational strength of these new machines.

1 Introduction

In the spirit of ordinal computability — the study of classical models of computations generalized to transfinite ordinal numbers — we study a variation of the Blum-Shub-Smale (BSS) machine introduced in [1]. In contrast to established models of ordinal computability, such as Hamkins' and Lewis' infinite time Turing machines (ITTMs) [4] and Koepke's ordinal Turing machines (OTMs) [5], these machines employ real numbers in the classical continuum $\mathbb{R}$ as opposed to elements of Baire space ${}^\omega\omega$ or Cantor space ${}^\omega 2$. The topological differences matter as soon as we consider limits (see below). Variations thereof, be it in allowing infinitely many registers or changes in the limit behavior, might very well change the computational strength. In this paper, we aim for the "weakest" possible generalization of BSS machines into ordinal time. We believe that already this restricted model shows interesting properties.

Our machines have a finite number n of *registers*, each containing a real number. Generalizations to other fields and rings are possible but shall not be of concern to this paper. The computation is steered by a finite program $P \subseteq \omega \times \{f \mid f : \mathbb{R}^n \xrightarrow{\text{rational}} \mathbb{R}^n\} \times \{0,1\} \times \omega \times \omega$, containing commands of the form (i, ϕ, j, k, l), where i is the index of the command at hand, ϕ is a rational map (with rational coefficients), and j tells us if the command represents a *computation node* or a *branch node*. In case $j = 0$, we are at a computation node, the register content $x \in \mathbb{R}^n$ is replaced by $\phi(x)$ and the next command (index $i+1$) is carried out next. The values of k and l are ignored in this case. Otherwise, $j = 1$ and we are at a branching node. This means that the register content is left unchanged and, depending on whether $\phi(x) > 0$, the next command will be the one with index k. If on the other hand $\phi(x) \leq 0$, command number l is

S.B. Cooper, A. Dawar, and B. Löwe (Eds.): CiE 2012, LNCS 7318, pp. 405–415, 2012.
© Springer-Verlag Berlin Heidelberg 2012

carried out next. We can assume the indices of a given program's commands to form an initial segment of the natural numbers and that no index appears twice. In case a command index is called for which no command in the program exists, the computation halts.

Note 1. As a minor technical detail we would like to note that, as in the original paper [1], we avoid discontinuity points of rational functions by putting decision nodes before each computation or decision node to check whether the denominator of the rational function to be evaluated is 0. If not, we continue as planned, if yes, an infinite loop is entered and the computation diverges.

So far, we have outlined a standard BSS machine with the additional restriction that the rational functions present in computation and branching nodes do not allow for arbitrary real coefficients. We add irrational coefficients in form of *parameters* later on. We now make our machines access the transfinite: In order for the machine to run for infinitely many steps, we have to *define* the register content at limit times. In the established theories of infinite time or ordinal Turing and register machines, often an inferior or superior limit is used for this purpose. Instead, we want to restrict ourselves here to ordinary limits of sequences of real numbers. This immediately implies that there will be situations where an infinite time BSS machine will, e.g., be properly defined at any finite time but not at the first limit time ω because the register contents do not converge. We can imagine the machine to "crash" in such a case and say that for such a combination of program and input no valid computations exists. In case of converging register contents we also have to come up with a command that is carried out at a limit time. For this we shall indeed use the inferior limit, i.e., the command with the least index that was used cofinally often below the limit. Note that we do not introduce a dedicated limit state.

Let us put things together in the following definition.

Definition 1. *Let $n \in \omega$ be a number of registers. Let $k < n$ and let P be a $n + k$-register BSS program. The* infinite time BSS machine computation *(ITBM computation) with parameters $p_1, p_2, \ldots, p_k \in \mathbb{R}$ by P on some input $x \in \mathbb{R}^n$ is defined as the transfinite sequence $(C_t)_{t \in \theta} = (R(t), I(t))_{t \in \theta} \in {}^{\theta}(\mathbb{R}^{n+k} \times \omega)$ where:*

(a) *$\theta \in \mathrm{Ord}$ or $\theta = \mathrm{Ord}$*
(b) *$R(0) = (x, p_1, p_2, \ldots, p_k)$ and $I(0) = 0$*
(c) *(computation node) If $t < \theta$ and $I(t) = i$ let $(i, \phi, 0, k, l)$ be the command of P with index i. Then $R(t+1) = \phi(R(t))$ and $I(t+1) = i+1$.*
(d) *(branching node) If $t < \theta$ and $I(t) = i$ let $(i, \phi, 1, k, l)$ be the command of P with index i. Then $R(t+1) = R(t)$ and if furthermore $\phi(R(t)) > 0$ then $I(t+1) = k$; if on the other hand $\phi(R(t)) \leq 0$, then $I(t+1) = l$.*
(e) *If $t < \theta$ is a limit and $y = \lim_{s \to t} R(s)$, then $R(t) = y$ and $I(t) = \liminf_{s \to t} I(s)$.*
(f) *If $\theta < \mathrm{Ord}$, then θ is a successor ordinal and $I(\theta - 1)$ calls a command index that is not in P (the machine halts (in θ-many steps)).*

We define ITBM computable functions on the reals:

Definition 2. *A function $f : \mathrm{dom} f \to Y$ where $\mathrm{dom} f, Y \subseteq \mathbb{R}$ is called* ITBM computable *in parameters $p_1, p_2, \ldots, p_k$ if there is an at least $k+1$-register ITBM program P s.t. for every $x \in \mathrm{dom} f$ the computation by P on input $(x, 0, 0, \ldots, 0, p_1, p_2, \ldots, p_k)$ halts and the final register content is of the form $(f(x), \cdot, \cdot, \ldots, \cdot)$. On input $x \notin \mathrm{dom} f$ the computation is required to diverge. We call such a function* ITBM computable *if $k = 0$, i.e., no real parameters are necessary.*

The use of one limit step enables us to compute the classical elementary functions that are defined by power series as illustrated by the following examples. While such functions as the exponential function can be computed in classical recursive analysis, they are not computable in the standard BSS model [2].

Example 1. The exponential function $e : \mathbb{R} \to \mathbb{R}, x \mapsto e^x = \sum_{k=0}^{\omega} \frac{x^k}{k!}$ is ITBM computable: Define a 5-register program that computes the desired function if $|x| < 1$. Later we shall describe the modifications necessary to work for any x.

Algorithm 1

```
input R₁ = x;
set R₂ := R₃ := R₄ := R₅ := 1; (initialize)
call loop;
loop:
if R₂ = 0 then set R₁ := R₅ and halt, else continue;
set R₄ := R₄/(R₄+1); (store 1/i, where i is the current iteration)
set R₃ := R₃ * R₄; (store 1/i!)
set R₂ := R₂ * R₁; (store xⁱ)
set R₅ := R₅ + (1/R₂)/R₃; (store i!/xⁱ)
call loop;
```

If $|x| < 1$, all register contents converge and R_5 contains the desired output at time ω, which is correctly recognized when $R_2 = 0$. We can adapt the algorithm for $|x| \geq 1$ by adding a case distinction in the beginning and, in case $|x| \geq 1$, save $\frac{1}{x^i}$ in register three. Then register three converges also at limit times. Of course, the command that updates R_5 inside the loop has to be changed accordingly.

Example 2. The sine function $sin : \mathbb{R} \to \mathbb{R}, x \mapsto \sum_{k=0}^{\omega} (-1)^k \frac{x^{(2k+1)}}{(2k+1)!}$ and the cosine function $cos : \mathbb{R} \to \mathbb{R}, x \mapsto \sum_{k=0}^{\omega} (-1)^k \frac{x^{2k}}{(2k)!}$ are ITBM computable. This is proven by the previous example and the fact that $(-1)^k$ can be recovered from $(-\frac{1}{2})^k$ and $(\frac{1}{2})^k$, both of the latter which can be convergently stored and updated in separate registers.

Note 2. Common to this type of examples is that some tricks are necessary to make all registers converge at limits. Auxiliary registers used for scratch work often do not contain converging content. If their content is bounded, however,

408   P. Koepke and B. Seyfferth

one can simply divide the register content cofinally often by a fixed number and keep track of how often this division has occurred. Unbounded contents are best stored as their multiplicative inverse. Both approaches can be imagined as pushing the relevant data contained in a register into increasingly later places in their decimal/binary expansion. Compound limits like $\omega \cdot \omega$ are an additional problem, as scratch registers cannot be set to arbitrary values after limit times without sacrificing convergence at the compound limit. However, in every limit stage towards a compound limit the register content will be bounded if treated like above. So, in order to ensure convergence at the compound limit, these bounds themselves need to converge. See the next chapter for an example.

2   Clockable Ordinals

Since Hamkins' and Lewis' paper on ITTMs [4], determining those ordinals that appear as halting times on empty inputs has proved to be important for the study of machine models. Since our machines do not halt at limit times, we are interested in machines that run for some limit number α many steps and halt in the next step:

Definition 3. *An ordinal α is called* ITBM clockable *if there is an $n \in \omega$ and an n-register ITBM program that halts on input $0 \in \mathbb{R}^n$ in exactly $\alpha + 1$ many steps.*

The algorithms above prove that ω is clockable, but let us establish this anew with an algorithm that uses only one register.

Lemma 1. *The first limit ordinal ω is ITBM clockable.*

Proof (by algorithm).

Algorithm 2

```
set  R₁ := 1;
call loop;
loop :
if  R₁ = 0 then halt else continue;
set  R₁ := R₁/2 ;
call loop;
```

We can clock $\omega \cdot n$ by having n loops in separate lines of code, where loop1 calls loop2 and loop i calls loop $i + 1$ instead of the halting command, each time resetting R_1 to 1. Instead, we could also use a separate register to perform a countdown from n. When trying to extend this approach trivially to clock $\omega \cdot \omega$, we run into a problem that is connected to the fact that $\omega \cdot \omega$ is a compound limit, i.e., a limit of limits: At time $\omega \cdot \omega$, R_1 will have cofinally often been set from 0 to 1, so convergence or R_1 fails. While this is easily fixed as seen below, it hints at the limitations imposed by the strict limit rule and why the supremum of ITBM clockable ordinals might be quite low in the ordinals.

Lemma 2. *The first compound limit ordinal $\omega \cdot \omega$ is ITBM clockable.*

Proof (by algorithm)

Algorithm 3

```
set  R₁ := 1;
set  R₂ := 1;
call loop;
loop:
if  R₂ = 0 then halt else continue;
if  R₁ = 0 then set R₂ := R₂/2 and set R₁ := R₂ and continue,
else continue;
set  R₁ := R₁/2 ;
call loop;
```

In this algorithm, R_1 is halved repeatedly to detect limits. Once a limit time is reached ($R_1 = 0$), R_2 is halved and R_1 is reset not to 1 but to the current value of R_2. Once R_2 hits 0, we have found the compound limit $\omega \cdot \omega$. At every limit, every register content converges.

As before, it is easy to clock finite multiples of the form $\omega \cdot \omega \cdot n$. Also, if we extend the algorithm to use extra registers $R_3, R_4, \ldots, R_n$ in the same manner as we extended the ω-algorithm with R_2, we can in fact clock any finite power ω^n. However, this is as far as we can get, as ω^ω turns out to be the supremum of the ITBM clockable ordinals.

First observe that at any limit time, the register contents are a fixed point for every command that has been carried out cofinally often:

Lemma 3. *Let $(C_t)_{t<\theta}$ be an ITBM computation by some program P and let $\alpha < \theta$ be a limit ordinal. Then the first command in the computation that alters the register contents after time α has not been carried out cofinally often below α.*

Proof. Let $(c, \phi, 0, \cdot, \cdot) \in P$ be a computation node which is called cofinally often below α. Let c be called at some time $\beta > \alpha$, where for all $\alpha < \gamma < \beta$ the register content has not been changed yet, i.e., $R(\alpha) = R(\gamma) = R(\beta)$. Since ϕ may be assumed as locally continuous (cf. Note 1), we get:

$$
\begin{aligned}
R(\beta + 1) &= \phi(R(\beta)) \\
&= \phi(R(\alpha)) \\
&= \phi(\lim_{t \to \alpha} R(t)) \\
&= \lim_{t \to \alpha \wedge I(t)=c} \phi(R(t)) \\
&= \lim_{t \to \alpha \wedge I(t)=c} R(t + 1) \\
&= R(\alpha)
\end{aligned}
$$

So the first computation node that changes the register content after time α cannot have been called cofinally below α.

Theorem 1. *Let P be a program with k computation nodes. Then in any computation $(C_t)_{t<\theta}$ according to P, the register contents stabilize before ω^{k+1}.*

Proof. If $k = 0$ then the computation halts after finitely many steps or diverges since the program contains only finitely many branch nodes: The computation may run through these nodes in a finite sequence or in an infinite loop. Every branch node may trigger halting depending on its rational function evaluated on the unchanged input. After finitely many steps, every node in the sequence or loop has been visited once. If the computation didn't halt up to this point, the program will go on forever as the register content is never changed.

So let the hypothesis hold for k. Let P be a program with $(k + 1)$-many computation nodes. Suppose the register contents change after ω^{k+2}. The first computation node c responsible for a new register content is not used cofinally often below ω^{k+2}. Let α be the supremum over the times $< \omega^{k+2}$ when c was carried out nontrivially. We can view $(C_t)_{\alpha \leq t < \omega^k + 2}$ as the computation by $P \setminus \{c\}$ on input $R(\alpha)$. Inductively, the register content of this computation stabilizes in ω^{k+1}-many steps. Since $\alpha + \omega^{k+1} < \omega^{k+2}$ this means that the original computation stabilizes before ω^{k+2}. But a computation that stabilizes before a limit can never change its register content again.

Once the register contents have stabilized, an ITBM computation diverges or may run for an additional finitely many steps before halting. So we get:

Theorem 2. *The supremum of ITBM clockable ordinals is ω^ω.*

The above argument is in fact independent of the input:

Corollary 1. *Every ITBM computation halts before ω^ω-many steps or diverges.*

Corollary 2. *If an ITBM computation diverges, the register content at time ω^ω is not changed any more and can be considered the* pseudo-output *of the diverging computation.*

3   Connections to Other Models of Computation

From a computability perspective, reals provide ample possibilities as codes for complex objects. We can code and decode into reals with our ITBMs by interpreting the binary expansion of a real in the interval $[0, 1]$ as an element of the Cantor space ${}^\omega 2$, i.e., an ω-long sequence of 0's and 1's. The binary expansion of a real is not necessarily unique, so two binary strings representing the same real will appear to our machines as equivalent.

Let us give an algorithm to retrieve the n-th binary digit b_n of an $x = 0.b_0 b_1 b_2 \cdots \in [0, 1)$.

Algorithm 4

```
input  R₁ = x;
input  R₂ = n;
set  R₃ = ½;  ( = 2⁰)
set  R₄ = 0;  (the current approximation to x)
call loop;
loop:
if  R₂ := 0 then call lastloop else continue; if  R₄ + R₃ > R₁
then continue (do not add  R₃ to the approximation  R₄ if the
result would exceed x)
else set  R₄    :=    R₄  +  R₃ and continue; (add  R₃ to the
approximation)
set  R₃ := R₃/2;
set  R₂ := R₂ - 1;  (= #remaining_iterations)
call loop;
lastloop:
if  R₄ + R₃ > x then set  R₁ := 0 and halt else set  R₁ := 1 and
halt;
```

With this algorithm we can also do local changes to the binary bits of x in the fashion of a Turing machine. So finite Turing computations can obviously be implemented on BSS/ITBMs. Also, the halting problem for Turing machines becomes ITBM computable:

Lemma 4. *The classical halting problem is ITBM computable.*

Proof. Since standard Turing computations are ITBM computable, we can generate Turing programs successively. So, in iteration n carry out the first n Turing programs for n many steps on empty input, using a dedicated simulation register. In step n, use only the n-th and later binary digits for the Turing simulation, so at time ω, this register will have converged to 0 (cf. Note 2). Once the algorithm finds that some program (say, the i-th) halts on its input, change the i-th binary digit of an initially zero output register to 1. Since there is a finite time when all halting computations of programs with index $< n$ will have halted, this register converges to the halting problem.

So our machines have more computing strength than Turing machines or classical BSS machines. Also, the type-2 Turing machines of computable analysis (see [8]) can easily be simulated by ITBMs: The output tape of a type-2 Turing machine, when modeled as an ITBM register, converges by definition. Input tapes do not change their content and the finitely many work tapes can be made convergent like in Note 2.

Using the above coding, ITBMs can also operate on functions. A continuous function $f : \mathbb{R} \to \mathbb{R}$ may be input to an ITBM as its restriction to the rationals $f^{\mathbb{Q}} = f \upharpoonright \mathbb{Q} : \mathbb{Q} \to \mathbb{R}$ coded into a real $p_f \in [0,1]$. This requires a fixed enumeration of rational numbers $q : \omega \to \mathbb{Q}$ which may be chosen as ITBM computable and an ITBM computable bijective pairing function $\langle \cdot, \cdot \rangle : \omega \times \omega \rightarrow \omega$. Then $f^{\mathbb{Q}}(x) = y$ may be expressed as "the $\langle q^{-1}(x), i \rangle$-th binary digit of p_f is exactly the i-th digit of y". By computing a convergent sequence of rationals for a given real (nested intervals) we can compute the function value at this real. This takes ω-many steps: Produce, for all $n < \omega$, the approximations of x up to n binary digits in a similar way to Algorithm 4 as a sequence of rationals converging to x. For every such approximation, decode from p_f the function value up to n digits. This defines a sequence of rationals that converge to $f(x)$.

Example 3. The derivative of a differentiable function is ITBM computable.

Proof. Given a point x, have an ITBM evaluate the differential quotient in x using only $f(x)$ itself and rational approximations to $f(x)$.

An upper bound on strength of ITBMs is given by the strength of ITTMs:

Lemma 5. *Let P be an n-register BSS program. There is an ITTM program Q and a map $f : \mathbb{R}^n \rightarrow {}^\omega 2$ s.t. for every $x \in \mathbb{R}$ the ITTM computation by Q on input $f(x)$ halts and returns the information that either no computation by P on x exists, or that P halts on x with output $y \in {}^\omega 2$, or that the P diverges on x with pseudo-output $y \in {}^\omega 2$, where $f^{-1}(y)$ is the final register content of the ITBM computation by P on x.*

Proof. We assume that the ITTM we are working with has a finite number of read-write tapes, which can be accomplished of interlacing these tapes onto the single scratch tape in the definition of ITTMs in [4]. Due to Corollary 2, we know that after time ω^ω also a diverging ITBM computation does not change its register content anymore. Since it is easy for an ITTM to construct a well order of ω of length ω^ω on one tape, it is possible for an ITTM to code the complete ITBM computation $(R_t, I_t)_{t < \omega^\omega}$ up to time ω^ω on the output tape.

So let us begin with defining a map $f' : \mathbb{R} \rightarrow {}^\omega 2$. First, imagine the value $f'(x)$ to be contained in three elements of ${}^\omega 2$ (i.e., three Turing tapes) where the first tape contains only a 0 or 1 in the first cell to specify the sign of x, the second contains the maximal exponent n s.t. $2^n \leq x$, and the third contains the binary expansion of $\frac{x}{2^n}$, ignoring the decimal (binary) point and normalized in the following way: Binary expansions that end on an infinite trail of 1s are replaced by the unique binary expansion of the same number that ends on a trail of 0s. Then, use an ITTM computable pairing function to interlace these three tapes into one. The function f' can easily be extended to a function $f : \mathbb{R}^n \rightarrow {}^\omega 2$.

It is easy to see that an ITTM is perfectly capable — albeit with an enormous time consumption — of computing addition, subtraction, multiplication and division on elements of ${}^\omega 2$ and of normalizing the result in the way described above. So, given the input, the program Q will start carrying out the sequence of ITBM commands in P and writing the results (and the program instructions used) one after another in the respective cells of the output tape. At limit times, the output tape contains all the information of the previous times, so it is easy for the Q to check whether everything converges and compute the limit if one exists. If not, output that there is no valid computation by P on x. If yes, Q will continue its simulation of P up to time ω^ω. At time ω^ω, it can replace the output tape content with $f(y)$ where y is P's output or pseudo-output.

It turns out that ITTMs are much stronger than needed, as will be clear from the following. Gödel's constructible model L of set theory is closely related to infinite time computations. We shall use Ronald Jensen's J_α-hierarchy to study definability in L. The $J_\alpha[\overrightarrow{x}]$-hierarchy relativized to the real parameters $\overrightarrow{x}$ is defined by the following recursion on the ordinals: $J_0[\overrightarrow{x}] = \emptyset$; $J_{\alpha+1}[\overrightarrow{x}]$ is the closure of $J_\alpha[\overrightarrow{x}] \cup \{J_\alpha[\overrightarrow{x}]\}$ under all rudimentary functions, using also the parameters $\overrightarrow{x}$. The rudimentary functions are simple set theoretic functions which include the formation of ordered pairs. Also first-order definitions over structures can be computed rudimentarily. Note that the ordinal height of $J_\alpha[\overrightarrow{x}]$ is $J_\alpha[\overrightarrow{x}] \cap \mathrm{Ord} = \omega \cdot \alpha$. In the sequel, the level $J_{\omega^\omega}[\overrightarrow{x}]$ will play a special role. It is also a member of the standard L-hierarchy, and indeed $J_{\omega^\omega}[\overrightarrow{x}] = L_{\omega^\omega}[\overrightarrow{x}]$ is the least level beyond ω where the two hierarchies coincide.

In the following we assume that real numbers are coded by their binary expansions. Then a real number a will be a function from ω into the set $\{0, 1, ., -\}$, where . denotes the binary dot and $-$ is the minus sign. The real a can be considered a subset of $H_\omega = J_1[\overrightarrow{x}]$. We show that an ITBM computation can be uniformly defined along the $J_\alpha[\overrightarrow{x}]$-hierarchy.

Lemma 6. *Let $(C_t)_{t\in\theta}$ be an ITBM computation according to a program P on input $\overrightarrow{x} \in \mathbb{R}^n$. Then for all $\alpha > 0$ the following hold:*

(i) If $t < \omega \cdot \alpha$ then $C \upharpoonright t+1 \in J_{1+\alpha}[\overrightarrow{x}]$.
(ii) $C \upharpoonright \omega \cdot \alpha$ is uniformly $\Sigma_1(J_{1+\alpha}[\overrightarrow{x}])$ definable in the parameters $\overrightarrow{x}$ and H_ω.

Proof. By the set theoretic recursion theorem, Definition 1 yields a definition of $(C_t)_{t\in\theta}$ of the form

$$\begin{aligned}
y = C_t &\leftrightarrow \exists f : t+1 \to V[f(0) = G_0(\overrightarrow{x}) \\
&\wedge \forall u < t(f(u+1) = G_1(f(u), H_\omega)) \\
&\wedge \forall u < t(limit(u) \to f(u) = G_2(f \upharpoonright u, H_\omega)) \\
&\wedge y = f(t)].
\end{aligned} \tag{1}$$

The functions G_0, G_1, G_2 are rudimentary: G_0 produces the initial configuration C_0 from the input $\overrightarrow{x}$. This amounts to assembling $(R(0), I(0))$ from the components of $\overrightarrow{x}$. This operation is certainly rudimentary.

The function G_1 is the transition function from the configuration at time u to the configuration at time $u+1$. The transition involves some case distinctions and the application of rational functions to register contents. The complexity of real arithmetic is indeed arithmetical in the arguments: for reals a and b the relations $a = b$, $a < b$ and the reals $a + b$, $a \cdot b$, $a - b$, and $\frac{a}{b}$ (in case $b \neq 0$) are first-order definable over the structure (H_ω, a, b). Since $H_\omega = J_1[\overrightarrow{x}] \in J_2[\overrightarrow{x}]$, real arithmetic is rudimentary, using the extra parameter H_ω.

The function G_2 performs the ITBM limit operation at limit times $u < \theta$. Given $f \upharpoonright u$, the ordinary limit in the binary reals can be obtained by first-order quantification over the multisorted structure $(H_\omega, u, f \upharpoonright u)$. So G_2 is rudimentary.

This means that definition (1) is a Σ_1-definition of $y = C_t$ whose kernel is rudimentary. Rudimentary predicates are absolute with respect to any level of the $J_\alpha[.]$-hierarchy. Note that the unique witness for the existential quantifier in (1) is $C \upharpoonright (t+1)$.

We now show the lemma by simultaneous induction on $\alpha > 0$. By our considerations so far, (i) for α implies (ii) for α.

Case $\alpha = 1$. If $t < \omega \cdot \alpha = \omega$ then $C \upharpoonright t+1$ is built from $\vec{x}$ and H_ω by finitely many applications of the rudimentary functions G_0 and G_1. Since $J_{1+1}[\vec{x}]$ is closed under rudimentary functions, $C \upharpoonright t + 1 \in J_{1+1}[\vec{x}]$.

Case $\alpha = \beta + 1$, where the lemma holds for β. Then $C \upharpoonright \omega \cdot \beta$ is $\Sigma_1(J_{1+\beta}[\vec{x}])$ definable in parameters, and hence $C \upharpoonright \beta \in J_{1+\alpha}[\vec{x}]$. For $t \in [\omega \cdot \beta, \omega \cdot \alpha)$, $C \upharpoonright t+1$ can be built from $C \upharpoonright \beta$ by finitely many applications of the rudimentary functions G_0 and G_1. Hence $C \upharpoonright t + 1 \in J_{1+\alpha}[\vec{x}]$.

For α being a limit ordinal, property (i) for all $\beta < \alpha$ immediately implies property (i) at α.

Corollary 3. *Every ITBM computable real is an element of $L_{\omega^\omega} = J_{\omega^\omega}$.*

A set $A \subseteq \mathbb{R}$ is ITBM *decidable* if there is an ITBM program that outputs 0 on inputs $x \in A$ and 1 otherwise. A set $B \subset \mathbb{R}$ is ITBM *semi-decidable* if there is a program P that halts only on the $x \in B$.

Corollary 4. *All ITBM (semi-)decidable sets of reals lie in $L_{\omega^\omega}[\mathbb{R}]$.*

4   Remarks and Open Questions

As with other models of ordinal computability, we have established a connection between computability and Gödel's constructibility. A natural conjecture is that Corollary 3 can be reversed. Can the constructible hierarchy up to L_{ω^n} be "simulated" by an ITBM? This will require ITBMs to be able to do simple syntactic operations and inductions up to ω^n, for every natural number n.

ITBMs form "pointwise" limits at every limit time. This should lead to connections with the Baire hierarchy of functions.

Relaxing the limit rules by going to lim inf's, e.g., will allow computations to go on beyond time ω^ω and will lead to stronger notions of computability. In [6] resp. [7], [3] we studied machines with finitely many registers that contain natural numbers. At limit times register contents are the lim inf's of previous register contents. The weaker machines in [6] crash when one of these lim inf's is ∞ whereas in [7], [3] that register is reset to 0. One can use similar limit rules for infinite time generalizations of Blum-Shub-Smale machines. Those generalizations obviously are able to simulate the machines of [6] resp. [7], [3] and should thus allow to compute all hyperarithmetic reals resp. finitely iterated hyperjumps.

References

1. Blum, L., Shub, M., Smale, S.: On a theory of computation and complexity over the real numbers: NP-completeness, recursive functions and universal machines. Bulletin of the American Mathematical Society 21(1), 1–46 (1989)
2. Brattka, V.: The emperors new recursiveness: The epigraph of the exponential funciton in two models of computability. In: Ito, M., Imaoka, T. (eds.) Words, Languages and Combinatorics III, pp. 63–72. World Scientific Publishing, Singapore (2003)
3. Carl, M., Fischbach, T., Koepke, P., Miller, R., Nasfi, M., Weckbecker, G.: The basic theory of infinite time register machines. Archive for Mathematical Logic 49(2), 249–273 (2010)
4. Hamkins, J.D., Lewis, A.: Infinite time Turing machines. The Journal of Symbolic Logic 65(2), 567–604 (2000)
5. Koepke, P.: Turing computations on ordinals. The Bulletin of Symbolic Logic 11, 377–397 (2005)
6. Koepke, P.: Infinite Time Register Machines. In: Beckmann, A., Berger, U., Löwe, B., Tucker, J.V. (eds.) CiE 2006. LNCS, vol. 3988, pp. 257–266. Springer, Heidelberg (2006)
7. Koepke, P., Miller, R.: An Enhanced Theory of Infinite Time Register Machines. In: Beckmann, A., Dimitracopoulos, C., Löwe, B. (eds.) CiE 2008. LNCS, vol. 5028, pp. 306–315. Springer, Heidelberg (2008)
8. Weihrauch, K.: Computable analysis. An Introduction. Springer, Heidelberg (2000)

Turing Pattern Formation without Diffusion

Shigeru Kondo

Graduate School of Frontier Biosciences, Osaka University, 1-3 Yamadaoka, Suita,
Osaka 565-0871, Japan
skondo@fbs.osaka-u.ac.jp

Abstract. Using the pigmentation pattern of zebrafish as the experimental system, we have been studying the mechanism of skin pattern formation. Recent findings of the cellular interactions among the two types of pigment cells, melanophores and xanthophores, are uncovering the cellular and molecular mechanisms. With these data, we now can answer the crucial question, "Is this a Turing mechanism or not?" We have identified the molecular basis of three interactions between the pigment cells. All of them are transferred at the tip of the dendrites of cells. In spite of the expectation of many theoretical biologists, there is no diffusion of the chemicals involved in the patterning mechanism. However, we also found that the lengths of the dendrites are different among the interactions, which makes it possible to generate the conditions of Turing pattern formation, "local positive feedback and long range negative feedback". Although it does not contain "diffusion", it may be appropriate to be called as a Turing mechanism.

1 Introduction

Over the past three decades, studies at the molecular level have revealed that a wide range of physiological phenomena are regulated by complex networks of cellular or molecular interactions [1]. The complexity of such networks gives rise to new problems, however, as the behavior of such systems often defies immediate or intuitive understanding. Mathematical approaches can help to facilitate the understanding of complex systems, and to date these have taken two primary forms. The first of these involves analyzing every element of a network quantitatively and simulating all interactions by computation [1]. This strategy is effective in relatively simple systems, for example, the metabolic pathway in a single cell, and is extensively explored in the field of systems biology. However, for more complex systems in which spatiotemporal parameters take on importance, it becomes almost impossible to make a meaningful prediction. In such cases, a second strategy involving simple mathematical modeling from which the details of the system are omitted can be more effective in extracting the nature of the complex system [4]. The reaction-diffusion (RD) model [6] proposed by Alan Turing is a masterpiece of this sort of mathematical modeling, one that is capable of explaining how spatial patterns develop autonomously.

In the RD model, Turing used a simple system of "two diffusible substances interacting with each other" to represent patterning mechanisms in the embryo, and found that such systems have the ability to generate spatial patterns

S.B. Cooper, A. Dawar, and B. Löwe (Eds.): CiE 2012, LNCS 7318, pp. 416–421, 2012.
© Springer-Verlag Berlin Heidelberg 2012

autonomously. Unfortunately Turing died soon after publishing this legendary paper, but simulation studies of the model have shown that this system can replicate most biological spatial patterns [3,5,2]. Later, a number of mathematical models [3] have been proposed, but in most of them, Turings basic idea that "the mutual interaction of elements results in spontaneous pattern formation" is followed. The RD model is now recognized as a standard among mathematical theories that deal with biological pattern formation.

However, this model has yet to gain widespread acceptance among experimental biologists. One of the major causes for this is the gap between the mathematical simplicity of the model and the complexity of the real world. The hypothetical molecules in the original RD model have been so idealized for the purposes of mathematical analysis that it seems nearly impossible to adapt the model directly to the complexity of real biological systems. However this is a misunderstanding to which experimental researchers tend to succumb. We can understand the logic of pattern formation using even simple models, and by adapting this logic to very complex biological phenomena, it becomes easier to extract the essence of the underlying mechanisms. Genomic data and new analytic technologies have caused a shift in the target of research in developmental biology from the identification of molecules to understanding the behavior of complex networks, making the RD model more important as a tool for theoretical analysis.

2   Finding Turing Patterns in Real Systems

During embryogenesis, a great variety of periodic structures develop from various non-periodic cell or tissue sources, suggesting that waves of the sort generated by Turing or related mechanisms may underlie a wide range of developmental processes. Using modern genetic and molecular techniques, it is entirely possible to identify putative elements of interactive networks that fulfill the criteria of short-range positive feedback and long-range negative feedback, but finding the network alone is not enough. Skeptics rightly point out that just because there is water, it does not mean there are waves. No matter how vividly or faithfully a mathematical simulation might replicate an actual biological pattern, this alone does not constitute proof that the simulated state reflects the reality. This has certainly been another major hurdle in identifying compelling examples of Turing patterns in living systems. The solution, however, is not so complicated; in order to show that a wave exists, we need to identify the dynamic properties of the pattern that is predicted by the computer simulation. The first experimental demonstrations have focused on pattern formation in the skin, because the specific characteristics of Turing patterns are more evident in two dimensions than in one.

3   Turing Patterns in Vertebrate Skin

The first observation of the dynamic properties of Turing patterns in nature was made by Kondo and Asai in a study of horizontal stripes in the tropical fish,

Pomacanthus imperator. Recently their work has shown that this dynamic nature is shared by many fish species, including the well-established model organism, zebrafish. While zebrafish stripes may appear to be stationary, experimental perturbation of the pattern triggers slow changes. As can be seen in Figure 1, following laser ablation of pigment cells in a pair of black horizontal stripes, the lower line shifts upward before stabilizing in a Bell-like curve. As a result, the spatial interval between the lines is maintained, even when their direction changes. This striking behavior is entirely predicted by simulation of the Turing mechanism.

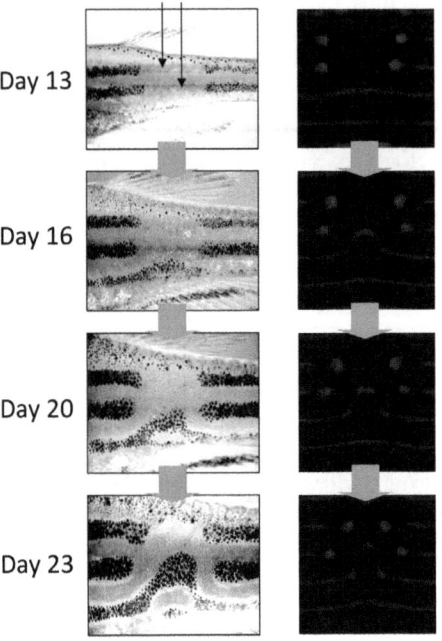

Day 13

Day 16

Day 20

Day 23

Fig. 1.

4   Zebrafish

Fortunately, the zebrafish is amenable to a variety of experimental approaches, and it is hoped that this will lead to the identification of the circuit of interactions that generates these patterns. Work to date has shown that the skin patterns of this fish are set up and maintained by interactions between pigmented cells. Nakamasu [5] worked out the interaction network among the pigment cells (cf. Figure 2). Although the shape of the network is different from that of the original Turing model, it fits the short-range positive, long-range negative feedback description.

long-range
negative feedback

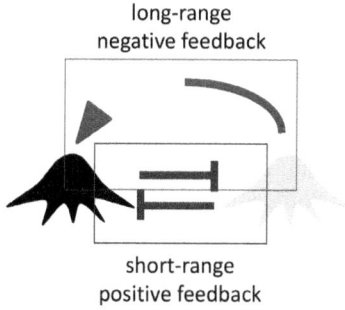

short-range
positive feedback

Fig. 2.

5   Identification of the Network at the Cellular and Molecular Level

The network among the pigment cells that Nakamasu [5] had determined was a very rough estimation of the whole system deduced from cell survival and migration. To clarify the mechanism of pattern formation, we needed to identify each interaction at the level of genes and molecules. To complete this project, we have isolated the mutant genes that cause the abnormal skin pattern, and analyzed the function of the genes with a newly developed procedure to maintain the pigment cells in vitro. With this in vitro system, we were able to observe the cell-cell interaction directly without any disturbance of other cells.

6   Short Range Repulsion of the Pigment Cells Is Correspond to the Local Positive Feedback [2]

The in vitro observation of cells has shown that melanophores and xanthophores interact directly with each other via the tip of xanthophore dendrites. Xanthophores extend the dendrites toward melanophores actively. Melanophores touched by xanthophores move to escape from the touch, but the xanthophore moves to follow the melanophores. It is obvious that mutual signal transduction occurs at the xanthophore dendrites. This repulsive behavior of the pigment cells is theoretically possible to induce the complete segregation of the two types of pigment cells that are seen in the real zebrafish. As this effect enhances the local dominance of one cell type, it can be recognized as local positive feedback. One of the genes involved in the repulsive behavior is connexin418, a gap junction gene. Artificially changing the activity of this gene gives rise to various skin patterns in zebrafish.

7   Long Range Survival Help by the Locally Competing Cell Is Correspond to the Long Negative Feedback (Unpublished Data)

In vivo experiments have suggested that the existence of xanthophores is necessary for the survival of melanophores, and the functional distance of this interaction is assumed to be "long range". After a systematic search with modern molecular-genetic technologies, we recently identified a membrane bound signaling molecule as being responsible for this effect. Although these molecules are bound to the membrane, melanophores spread long "legs" (pseudopodia) that are as long as a half width of the zebrafish stripe. Therefore, the membrane molecule can transfer the signal to cells located far apart. As melanophores are competing with xanthophores locally, this relay (melanophore > (negative) > xanthophore > (positive) > melanophore) of the signal acts as the "long range negative feedback loop".

8   No Diffusion in the Identified Network

The discovered network satisfies the conditions for Turing pattern formation, "local positive feedback and long range negative feedback", and computer simulation of this mechanism can generate the stripe patterns. Currently we are continuing experiments to clarify the molecular details of pigment pattern formation. One of the interesting outcomes of the study is that it is now possible to generate a wide variety (nearly all kinds of skin pattern seen in the animal world) of patterns on the zebrafish skin by modulating the nucleotide sequence of a single gene, connexin418. Hence we are sure that the network of interaction we discovered is the core of the pattern forming mechanism in zebrafish, and provably in many other animals. One thing that we, and most other researchers did not expected is that "there is no diffusion" in the system. Therefore, the identified mechanism is not a RD-mechanism. However, the specific functional distance of each signal is determined by the length of dendrites and pseudopodia.

In the classic RD mechanism, the width of the stripes is determined by the diffusion constants and the reaction (production and decay) ratios of the hypothetical molecules. All such values are determined, in the real body of animals, by the activity of the cells and tissues. The problem is that such cellular activity is not stable. Production and release of the hypothetical "activator and inhibitor" might change drastically depending on environmental conditions. This should make the resulting pattern extremely unstable. Another problem of diffusion is "3D diffusion". Skin pigmentation occurs in a 2D field. But if the signal were transferred by diffusion, we should account for diffusion towards the inside and outside of the body. Diffusion of the signaling molecule to the outside of the plane should destroy the Turing pattern. These difficulties have been noticed by experimental biologists, making them skeptical of the Turing mechanism. However, we now know that the signals are transferred by the dendrites and pseudopodia. These cellular structures are far more stable than diffusion, and the directionality is obvious. So, we expect that the skeptical feeling among experimental biologists we disappear soon. We expect they will accept more actively the theoretical models of biological pattern formation.

9   Does the Discovered Mechanism Belong to "Turing Mechanism"?

As the identified mechanism does not contain "diffusion", it is not a RD mechanism. If the definition of a Turing mechanism is identical to an RD mechanism, then the identified mechanism cannot be called a Turing mechanism. If I have some right to name it as the person who found it, I still would like to call it as a Turing mechanism. One of the reasons is that the essence of the pattern forming mechanism does not depend on the device of signal transfer. The most important and innovative idea of Turing was that the difference of functional distance can give rise to the periodic pattern. So any kind of signaling mechanism can give rise to similar pattern. The reason why Turing used "diffusion" might be that the details of cell biology were not yet clear at that stage. As the discovered mechanism still retains the principle of the idea of Turing, I would like to call it as a Turing mechanism.

10   Future Directions

Interestingly, most of the genes involved in pigment pattern formation are very common genes that function during morphogenesis of the animal embryo. This fact implies a possibility that the identified molecular mechanism is not only governing the pigment pattern formation in the skin but also controlling the structure of the animal body. To test this interesting possibility, we are now examining the effect of these genes on various morphological phenomena. It has been questioned for long whether a Turing mechanism governs many of the pattern formation of animals or not. The answer may come up in a few years.

References

1. Alon, U.: An introduction to systems biology. CRC Press, Taylor and Francis Group, London (2006)
2. Inaba, M., Yamanaka, H., Kondo, S.: Pigment pattern formation by contact-dependent depolarization. Science 335(677) (2012)
3. Kondo, S., Asai, R.: A reaction-diffusion wave on the skin of the marine angelfish Pomacanthus. Nature 376(765) (1995)
4. May, R.M.: Simple mathematical models with very complicated dynamics. Nature 261(459) (1976)
5. Nakamasu, A., Takahashi, M., Kondo, S.: Interactions between zebrafish pigment cells responsible for the generation of Turing patterns. Proc. Natl. Acad. Sci. U.S.A. 106(8429) (2009)
6. Turing, A.M.: The chemical basis of morphogenesis. Bull. Math. Biol. 52(153) (1990)

Degrees of Total Algorithms versus Degrees of Honest Functions

Lars Kristiansen

Department of Mathematics, P.O. Box 1053, Blindern, 0316 Oslo, Norway
`larsk@math.uio.no`

Abstract. We prove a few theorems elucidating the relationship between to different approaches to subrecursive degree theory. One approach has its roots in the theory of algorithms and Turing degrees. The other approach has its roots in subrecursive hierarchies of fast-growing functions.

1 Introduction

Two obviously different, but also apparently related, degree structures have recently been studied in the literature.

Cai [3,4] studies degrees of *algorithms* computing total functions. The algorithms are given by an enumeration of, e.g., Turing machines. The reducibility relation $i \leq_p j$ holds if $T + \text{tot}(\Phi_j) \vdash \text{tot}(\Phi_i)$ where T is a sufficiently strong first-order number theory and $\text{tot}(\Phi_e)$ is a Π_2^0-sentence stating that the algorithm e computes a total function. In this paper we shall work with Peano Arithmetic (PA) and we define $i \leq_p j$ by $\text{PA} + \text{tot}(\Phi_j) \vdash \text{tot}(\Phi_i)$. The PA-provability degrees are the equivalence classes induced on the algorithms computing total functions by the reducibility relation $\leq_p$.

Kristiansen et al. [12] studies degrees of *honest functions*. Honest functions are number-theoretic functions with certain properties. Typical examples of such functions are the fast-growing functions found in a spine sequence for a subrecursive hierarchy. The reducibility relation $f \leq_{\alpha E} g$ holds if the function f can be defined from the function g by (Kalmar) elementary operations and ordinal recursion up to the ordinal α. The honest α-elementary degrees are the equivalence classes induced on the honest functions by $\leq_{\alpha E}$. Now, ϵ_0 is the proof-theoretic ordinal of PA, and the zero degree in the structure of honest ϵ_0-elementary degrees consists of all honest functions being provably total in PA.

So, we are dealing with two degree structures: The PA-provability degrees are degrees of algorithms, and two algorithms that compute the same function will not necessarily be of the same PA-provability degree – indeed, there will be an algorithm computing a trivial function like 2^x in any PA-provability degree. The honest ϵ_0-elementary degrees are degrees of functions, and a trivial function like 2^x belongs to the zero degree. The aim of this paper is to state a few theorems elucidating relationship between these two structures.

S.B. Cooper, A. Dawar, and B. Löwe (Eds.): CiE 2012, LNCS 7318, pp. 422–431, 2012.

2   Preliminaries

A function ϕ is *elementary* if ϕ can be generated from the initial functions 2^x, max, 0, $\mathcal{S}$ (successor), $\mathcal{I}_i^n$ (projections) by composition and bounded primitive recursion. (Recall that f is generated by *bounded primitive recursion* over g, h and j when $f(\vec{x}, 0) = g(\vec{x})$ and $f(\vec{x}, y + 1) = h(\vec{x}, y, f(\vec{x}, y))$ and $f(\vec{x}, y) \leq j(\vec{x}, y)$.) It is easy to see that for any elementary function ϕ, we have $\phi(\vec{x}) \leq 2_k^{\max(\vec{x})}$ where k is a fixed number and $2_0^x = x$ and $2_{k+1}^x = 2^{2_k^x}$. Moreover, the elementary predicates are closed under propositional operations and bounded quantifiers, and uniform systems for coding finite sequences of natural numbers are available inside the class of elementary functions. We are quite informal and indicate the use of coding functions with the notations $\langle \ldots \rangle$ and $(x)_i$ where $(\langle x_0, \ldots, x_i, \ldots, x_n \rangle)_i = x_i$. For more on the elementary functions, cf., e.g., [20].

The reader should have some knowledge of degree theory, and we assume familiarity with the *computable functions*, *indexes* for computable functions, *computation trees* and other well-known concepts of computability theory. For more on basic computability theory, cf., e.g., [18] or [19].

Furthermore, we assume the reader is familiar with formalised first-order arithmetic, in particular, we assume familiarity with PA. We shall work with a version of PA where we have function symbols and defining equations for all the elementary functions, and we use $\mathfrak{N}$ to denote the standard model for this first-order theory. If we simply say that a PA-statement is true, we mean that the statement is true in $\mathfrak{N}$. We may display the free variables of a PA-statement A by using the standard notation $A(x_1, \ldots, x_n)$. If we use this notation, then we shall display all the free variables in the statement. For more on PA, cf., e.g., [13] or [5].

The reader is also required to know a little bit about ordinals and proof theory.

3   Cai's Degree Structure

Let $\mathcal{U}$ be a function such that $\mathcal{U}(\langle x_1, \ldots, x_m \rangle) = x_m$, that is, a function giving the last coordinate of a sequence number. Let $\mathcal{T}$ be the Kleene predicate, that is, the predicate $\mathcal{T}(e, \langle x_1, \ldots, x_n \rangle, t)$ holds iff t is a computation tree for the eth computable function on the inputs $x_1, \ldots, x_n$. The predicate $\mathcal{T}$ and the function $\mathcal{U}$ are elementary, and for each total computable function ϕ we have $\phi(x_1, \ldots, x_n) = \mathcal{U}(\mu z[\mathcal{T}(e, \langle x_1, \ldots, x_n \rangle, z)])$ when e is a computable index for ϕ.

Let $\phi_i(x) = \mathcal{U}(\mu t[\mathcal{T}(i, x, t)])$, and let $\Phi_i(x_1, x_2, x_3)$ be a Δ_0-statement (in the language of PA) such that

$$\mathfrak{N} \models \Phi_i(x_1, x_2, x_3) \left[s_{a,b,t}^{x_1, x_2, x_3} \right] \quad \Leftrightarrow \quad \mathcal{T}(i, a, t) \wedge \mathcal{U}(t) = b \,.$$

(It is trivial that such a Δ_0-statement exists since we are working with a version of PA that contains function symbols for all the elementary functions.) Now, $\phi_0, \phi_1, \phi_2, \ldots$ is an enumeration of (all) the unary computable functions, whereas

$\Phi_0, \Phi_1, \Phi_2, \ldots$ is an enumeration of (some) Δ_0-statements of PA. Moreover, we have

$$\mathfrak{N} \models \exists z \Phi_i(x_1, x_2, z) \left[s_{a,b}^{x_1,x_2} \right] \quad \Leftrightarrow \quad \phi_i(a) = b \ .$$

Let $\mathrm{tot}(\Phi_i) \equiv \forall x \exists y z \Phi_i(x, y, z)$. The index i of the statement Φ_i describes an algorithm (that computes the function ϕ_i). If the statement $\mathrm{tot}(\Phi_i)$ is true, the algorithm given by i is total (the algorithm yields an output for any input). We define the set of *total algorithms*, written $\mathcal{A}_{\mathrm{tot}}$, by $\mathcal{A}_{\mathrm{tot}} = \{ i \mid \mathfrak{N} \models \mathrm{tot}(\Phi_i) \}$ and we define the reducibility relation $\leq_{\mathrm{p}}$ over the total algorithms by

$$i \leq_{\mathrm{p}} j \quad \Leftrightarrow \quad \mathsf{PA} + \mathrm{tot}(\Phi_j) \vdash \mathrm{tot}(\Phi_i) \ .$$

The PA-*provability degrees of (total) algorithms* are the equivalence classes induced on $\mathcal{A}_{\mathrm{tot}}$ by $\leq_{\mathrm{p}}$.

Cai [3,4] investigates the PA-provability degrees of total algorithms. He proves that the structure is a distributive lattice; that any degree strictly above the zero degree, has an incomparable degree; that a minimal pair exists. He also studies a jump operator. Cai's proofs are based on classical computability-theoretic constructions. For more results and details, see [3,4].

4   The Honest Elementary Degrees

A function $f : \mathbb{N} \to \mathbb{N}$ is *honest* if it is monotone, that is, $f(x) \leq f(x+1)$; dominates 2^x, that is, $f(x) \geq 2^x$; and has elementary graph, that is, the relation $f(x) = y$ is elementary. A function ϕ is *elementary in* a function ψ, written $\phi \leq_E \psi$, if ϕ can be generated from the initial functions ψ, 2^x, max, 0, $\mathcal{S}$ (successor), $\mathcal{I}_i^n$ (projections) by composition and bounded primitive recursion. The *honest elementary degrees* are the equivalence classes induced on the honest functions by the reducibility relation $\leq_E$. We use $f \leq g$ to denote $\forall x[f(x) \leq g(x)]$, and f^k denotes k iterations of the function f, that is, $f^0(x) = x$ and $f^{k+1}(x) = f(f^k(x))$.

Honest degree theory has its roots in subrecursion theory developed in the 1970s. Some relevant papers are those by Meyer & Ritchie [17] and Machtey [14,15,16]. Influenced by this work, Kristiansen introduced the honest elementary degrees about 20 years later. The first published proof of the next theorem can be found in Kristiansen [6].

Theorem 1 (Growth Theorem). *Let f and g be honest functions. Then, we have*

$$g \leq_E f \quad \Leftrightarrow \quad g \leq f^k \text{ for some fixed } k \ .$$

In the 1990s, Kristiansen used The Growth Theorem to investigate honest elementary degrees and related subjects. This research is published in a thesis [9] and four papers [6,7,8,10]. The structure of honest elementary degrees is a comparable to a classical computability-theoretic degree structure, e.g., the structure of Turing degrees, but the Growth Theorem makes it possible to abandon classical computability-theoretic proof methods and investigate this structure by asymptotic analysis and methods of number-theoretic nature. To prove that $g \leq_E f$, it is sufficient to provide a fixed k such that $g \leq f^k$; to prove that $g \not\leq_E f$, it is sufficient to prove that such a k does not exist. Thus, there is no need for the standard computability-theoretic machinery involving enumerations, diagonalisations and constructions with requirements to be satisfied. This makes the proofs concise and transparent.

We know a lot about the structure of honest elementary degrees: The structure is a distributive lattice with very strong density proprieties. We have capability and cupability results. We have studied a jump operator, and low_n and high_n degrees exist (for any $n \in \mathbb{N}$). For more results and details, see [12,11].

5   The Honest ϵ_0-Elementary Degrees

The honest α-elementary degrees, where α is some ordinal less than or equal to ϵ_0, were recently introduced by Kristiansen, Weiermann & Schlage-Puchta [12]. An α-elementary degree is a generalisation of an elementary degree.

Our approach to ordinals and ordinal recursion is based on work by Weiermann, and others, cf., e.g., [2]. The reader should recall that any ordinal strictly less than ϵ_0 can be written in Cantor normal form, that is, in the form $\omega^{\alpha_1} + \ldots + \omega^{\alpha_n} + 0$ where $\alpha_1 \geq \ldots \geq \alpha_n$ and each α_i is in Cantor normal form. In particular, any $k \in \mathbb{N}$ can be written in the form $k = \underbrace{\omega^0 + \ldots + \omega^0}_{k \text{ copies}} + 0$.

For any ordinal $\alpha < \epsilon_0$, we define the *norm of* α, written $N(\alpha)$, by induction over the structure of α's Cantor normal form: $N(0) = 0$; $N(\beta+\gamma) = N(\beta)+N(\gamma)$; and $N(\omega^\beta) = 1 + N(\beta)$. We define the *$\alpha$-iterate of the unary function* ϕ, written ϕ_α, by $\phi_0(x) = \phi(x)$ and

$$\phi_\alpha(x) = \max\{\phi_\beta\phi_\beta(x) \mid \beta < \alpha \ \wedge \ N(\beta) \leq N(\alpha) + x\}$$

for any α such that $0 < \alpha < \epsilon_0$. (The set $\{\beta \mid \beta < \alpha \ \wedge \ N(\beta) \leq N(\alpha) + x\}$ is finite.)

Let $\alpha \leq \epsilon_0$. A function ϕ is *α-elementary in* a function ψ, written $\phi \leq_{\alpha E} \psi$, if ϕ can be generated from the initial functions ψ, 2^x, max, 0, S (successor), $\mathcal{I}_i^n$ (projections) by composition, bounded primitive recursion and β-iteration where $\beta < \alpha$. Let SLim denote the class of all infinite additive principal numbers $\leq \epsilon_0$, that is, SLim $= \{\omega^\beta \mid \epsilon_0 \geq \beta > 0\}$.

Fix $\alpha \in$ SLim. It is proved in [12] that the reducibility relation $\leq_{\alpha E}$ induce a degree structure on the honest functions. The honest α-elementary degrees are the equivalence classes induced on the honest functions by $\leq_{\alpha E}$. The next theorem is also proved in [12].

Theorem 2 (Generalised Growth Theorem). *Let f and g be honest functions, and let $\alpha \in$ SLim. Then, we have*

$$g \leq_{\alpha E} f \ \Leftrightarrow \ g \leq f_\beta \text{ for some fixed } \beta < \alpha .$$

For any $\alpha \in$ SLim, the structure of honest α-elementary degrees seems to be very similar to the structure of honest elementary degrees. When the Generalised Growth Theorem is available, we can easily transform proofs of theorems on elementary degrees into proofs of corresponding theorems on α-elementary degrees. Hence, structure of honest α-elementary degrees is a distributive lattice with strong density properties; we have a jump operator; for any n, there exists low_n and high_n degrees; and so on. It is an open problem whether the structure of honest α-elementary degrees is isomorphic to the structure of honest β-elementary degrees for any $\alpha, \beta \leq \epsilon_0$.

We need a few more definitions before we can state our next theorem: Δ_0-statements and Σ_1-statements of the form $A(x_1, \ldots, x_n, y)$ will be called *representations*. A representation $A(x_1, \ldots, x_n, y)$ is a *representation of* a function $\phi(x_1, \ldots, x_n)$ when

$$\mathfrak{N} \models A \left[s_{a_1, \ldots, a_n, b}^{x_1, \ldots, x_n, y} \right] \quad \Leftrightarrow \quad \phi(a_1, \ldots, a_n) = b \qquad (*)$$

holds for any assignment s. If (*) holds and A is a Δ_0-statement, we shall say that A is an *honest* representation of ϕ. (Any honest function has an honest representation.) For any representation $A(\vec{x}, y)$, let $\text{tot}(A)$ denote the statement $\forall \vec{x} \exists y A$.

Now we are ready to state a theorem relating $\leq_{\epsilon_0 E}$-reducibility to PA-provability of Π_2-statements in extensions of PA. A proof of this theorem can be found in [12].

Theorem 3. *Let f and g be honest functions. Then, we have $f \leq_{\epsilon_0 E} g$ if, and only if, for any representation A of g there exists a representation B of f such that $\mathsf{PA} + \text{tot}(A) \vdash \text{tot}(B)$.*

Algorithms and representations are two sides of the same coin. When we have an algorithm i (computing the function ϕ_i), then we also have a representation of the function ϕ_i: the statement $\exists t \Phi_i(x, y, t)$ is a representation of ϕ_i. When we have a representation $\exists z C(x, y, z)$ of a function ξ, then we also have an algorithm i computing ξ such that $\mathsf{PA} + \text{tot}(C) \vdash \text{tot}(\Phi_i)$. (Let i be the algorithm that (i) takes input x; (ii) searches for the least number $\langle y, z \rangle$ such that $C(x, y, z)$ holds; and (iii) gives output y.) Hence, the next theorem relating $\leq_p$-reducibility and $\leq_{\epsilon_0 E}$-reducibility, is a straightforward consequence of Theorem 3.

Theorem 4. *Let f and g be honest functions. Then, we have $f \leq_{\epsilon_0 E} g$ if, and only if, for any algorithm i computing g there exists an algorithm j computing f such that $j \leq_p i$.*

6   Honest Associates and PA^+ Degrees

The first-order theory PA^+ is PA extended with all true Π_1-statements. We define the reducibility relation $\leq_{p+}$ over the total algorithms by

$$i \leq_{p+} j \quad \Leftrightarrow \quad \mathsf{PA}^+ + \text{tot}(\Phi_j) \vdash \text{tot}(\Phi_i) .$$

The PA^+-*provability degrees of (total) algorithms* are the equivalence classes induced on $\mathcal{A}_{\text{tot}}$ by $\leq_{p+}$.

Now, $i \leq_p j$ implies $i \leq_{p+} j$, but $i \leq_{p+} j$ does not imply $i \leq_p j$. Hence the structure of PA^+-provability degrees is coarser than the structure of PA-provability degrees, and inside each PA^+-provability degrees there is an infinite structure of PA-provability degrees.

Next we shall define a function ψ_i called the *honest associate* of the algorithm i. Intuitively, $\psi_i(x)$ yields the computation trees for the algorithm i on all inputs $\leq x$. After defining ψ_i, we shall prove the one of the main theorems of this paper:

Theorem 5 (Growth Theorem for Algorithms). *For any $i, j \in \mathcal{A}_{\mathrm{tot}}$, we have*

$$i \leq_{p+} j \quad \Leftrightarrow \quad \psi_i \leq (\psi_j)_\alpha \text{ for some } \alpha < \epsilon_0 .$$

We define *honest associate* of the algorithm e, written ψ_e, by

$$\psi_e(x) = \langle t_0, t_1, \ldots, t_x \rangle$$

where $t_i = \mu z [\mathcal{T}(e, i, z)]$ (for $i = 0, \ldots, x$). Let $|t| = x$ be a Δ_0-statement of PA being true iff t encodes a sequence of length x; let $(t)_i = z$ be a Δ_0-statement of PA being true iff the ith component in the sequence encoded by t equals z; let $T(e, i, z)$ be a Δ_0-statement of PA defining the Kleene predicate $\mathcal{T}(e, i, z)$; and finally, let

$$\Psi_e(x, t) \equiv |t| = x \wedge \forall i \leq x \exists z \leq t [(t)_i = z \wedge T(e, i, z)] .$$

Note the following:

- $\psi_0, \psi_1, \psi_2 \ldots$ is an enumeration of computable functions
- if ψ_i is a total function, then ψ_i is an honest functions
- ψ_i is total iff ϕ_i is total
- $\Psi_0, \Psi_1, \Psi_2 \ldots$ is an enumeration of Δ_0-statements, and we have

$$\mathfrak{N} \models \Psi_i(x_1, x_2) \left[s_{a,b}^{x_1, x_2} \right] \quad \Leftrightarrow \quad \psi_i(a) = b$$

and thus, Ψ_i is an honest representation of ψ_i.

Lemma 1. *For any $i, j \in \mathcal{A}_{\mathrm{tot}}$, we have*

$$i \leq_{p+} j \quad \Leftrightarrow \quad PA^+ + \mathrm{tot}(\Psi_j) \vdash \mathrm{tot}(\Psi_i) .$$

Proof. It should be obvious that we have $PA \vdash \mathrm{tot}(\Psi_k) \leftrightarrow \mathrm{tot}(\Phi_k)$ (for any $k \in \mathbb{N}$). Since PA^+ is an extension of PA, we also have

$$PA^+ \vdash \mathrm{tot}(\Psi_k) \leftrightarrow \mathrm{tot}(\Phi_k) . \tag{*}$$

By the definition of $\leq_{p+}$ and (*), we have $i \leq_{p+} j$ iff $PA^+ + \mathrm{tot}(\Phi_j) \vdash \mathrm{tot}(\Phi_i)$ iff $PA^+ + \mathrm{tot}(\Psi_j) \vdash \mathrm{tot}(\Psi_i)$. □

Lemma 2. *Let $\alpha < \epsilon_0$, and let f be an honest function. Then, f_α is an honest function. Moreover, for any honest representation F of f, there exists an honest representation F_α of f_α such that $PA + \mathrm{tot}(F) \vdash \mathrm{tot}(F_\alpha)$.*

Lemma 2 holds since transfinite induction up to ϵ_0 can be carried out in PA. The proof is long and technical, but there should be no surprise that the lemma holds. (Proofs of related results can be found in [12]; see Lemma 17 and Theorem 16.) The next lemma is the crucial lemma needed for the proof of the Growth Theorem for Algorithms. Lemma 2 is used to prove the right-left implication of the next lemma.

Lemma 3. *For any $i, j \in \mathcal{A}_{\text{tot}}$, we have*

$$\mathsf{PA}^+ + \text{tot}(\Psi_j) \vdash \text{tot}(\Psi_i) \quad \Leftrightarrow \quad \psi_i(x) \le (\psi_j)_\alpha(x) \text{ for some } \alpha < \epsilon_0 \; .$$

Proof. The left-right implication is a consequence of the following claim.

> **(Claim)** Let h be an honest function, and let A_h be any honest representation of h. Furthermore, let B_ϕ be a representation of the function ϕ. If $\mathsf{PA}^+ + \text{tot}(A_h) \vdash \text{tot}(B_\phi)$, then we have $\phi \le h_\gamma$ for some $\gamma < \epsilon_0$.

This claim can be verified by making minor modifications and extensions to the proofs found in Blankertz & Weiermann [1]. A derivation in a Tait-style calculus of a set of PA-formulas can be embedded in a calculus based on so-called F-controlled derivations. F-controlled derivations are similar to the usual derivations in a first-order Tait-style calculus, but an infinitary ω-rule replaces the standard $\forall$-rule, and moreover, the F-controlled derivations embody some explicit information about $\exists$-witnesses, derivation lengths and complexity of cut formulas. Let Γ denote a set of PA-formulas. The Embedding Lemma in [1] roughly states that if $\mathsf{PA} \vdash \Gamma$, then we have a derivation $F_\gamma \vdash^\alpha_r \Gamma$ in the F-controlled calculus where $\gamma, \alpha < \epsilon_0$; $r < \omega$; and F is an elementary honest function. The ordinal α gives information about the height of the derivation; the natural number r gives an upper bound on the complexity of the cut formulas involved in the derivation; and the function F_γ yields upper bounds on $\exists$-witnesses. The Embedding Lemma can be extended to the following lemma.

> **(Extended Embedding Lemma)** Let h be an honest function, and let A_h be an honest representation of h. If $\mathsf{PA}^+ + \text{tot}(A_h) \vdash \Gamma$, then there exists $\gamma, \alpha < \epsilon_0$ and $r < \omega$ such that $h_\gamma \vdash^\alpha_r \Gamma$.

The proof of this lemma is very similar to the proof of the Embedding Lemma in [1] except that we now also have to assure that

(i) everything works well when we are working with PA^+ in place of PA
(ii) we have $h_\gamma \vdash^\alpha_0 \text{tot}(A_h)$ for some $\gamma, \alpha < \epsilon_0$.

It is easy to prove that both (i) and (ii) go through. (Indeed, the very proof given in [1] works for the theory PA^+.) When the Extended Embedding Lemma is proved, we can proceed as in [1] and carry out cut-elimination in the F-controlled calculus, and then, we can use the existence of cut-free proofs to prove that $\mathsf{PA}^+ + \text{tot}(A_h) \vdash \text{tot}(B_\phi)$ entails that there exists $\gamma < \epsilon_0$ such that $\phi \le h_\gamma$. This completes a sketch of a possible way to prove (Claim). It should also be possible to prove this claim by interpreting derivations (in the theory $\mathsf{PA}^+ + \mathrm{tot}(A_h)$) in Gödel's system T.

We turn to the proof of the right-left implication. The function ψ_j is honest, and Ψ_j is an honest representation of ψ_j. By Lemma 2, we have an honest representation $(\Psi_j)_\alpha$ of $(\psi_j)_\alpha$ such that $\mathsf{PA} + \mathrm{tot}(\Psi_j) \vdash \mathrm{tot}((\Psi_j)_\alpha)$. Now, assume $\psi_i(x) \leq (\psi_j)_\alpha(x)$. Then

$$\forall xy \left[(\Psi_j)_\alpha(x,y) \ \rightarrow \ \exists z \leq y \Psi_i(x,z) \right]$$

is a true Π_1-statement, and thus, this statement is an axiom of PA^+. (The statement $(\Psi_j)_\alpha$ is a Δ_0-statement as $(\Psi_j)_\alpha$ is an honest representation.) Hence, we have

- $\mathsf{PA} + \mathrm{tot}(\Psi_j) \vdash \mathrm{tot}((\Psi_j)_\alpha)$
- $\mathsf{PA}^+ \vdash \forall xy \left[(\Psi_j)_\alpha(x,y) \ \rightarrow \ \exists z \leq y \Psi_i(x,z) \right]$

and then we also have $\mathsf{PA}^+ + \mathrm{tot}(\Psi_j) \vdash \mathrm{tot}(\Psi_i)$. $\qquad\square$

The Growth Theorem for Algorithms (Theorem 5) follows straightforwardly from Lemma 1 and Lemma 3.

Corollary 1. *Let $i, j \in \mathcal{A}_{\mathrm{tot}}$ and recall that ψ_i and ψ_j are the honest associates of respectively i and j. The following statements are equivalent: (i) $i \leq_{\mathrm{p+}} j$. (ii) $\psi_i \leq (\psi_j)_\alpha$ for some $\alpha < \epsilon_0$. (iii) $\psi_i \leq_{\epsilon_0 \mathrm{E}} \psi_j$. (iv) For any representation A of ψ_j there exists a representation B of ψ_i such that $\mathsf{PA} + \mathrm{tot}(A) \vdash \mathrm{tot}(B)$. (v) For any algorithm j' computing ψ_j there exists an algorithm i' computing ψ_i such that $i' \leq_{\mathrm{p}} j'$.*

Proof. (i) and (ii) are equivalent by the Growth Theorem for Algorithms; (ii) and (iii) are equivalent by Theorem 2; (iii) and (iv) are equivalent by Theorem 3; and (iii) and (v) are equivalent by Theorem 4. $\qquad\square$

7   The Isomorphism Theorem

We use $\equiv_r$ to denote the equivalence relation induced by the reducibility relation $\leq_r$.

Lemma 4. *Let f be an honest function. Then, there exists ψ_i such that $f \equiv_E \psi_i$ (and thus $f \equiv_{\epsilon_0 \mathrm{E}} \psi_i$).*

Proof. We need the following claim.

(**Claim**) Let ξ be an elementary function. Then there exists an index i and a fixed $k \in \mathbb{N}$ such that $\xi(x_1, \ldots, x_n) = \phi_i(\langle x_1, \ldots, x_n \rangle)$ and

$$\mathcal{T}(i, \langle x_1, \ldots, x_n \rangle, t) \ \Rightarrow \ t \leq 2_k^{\langle x_1, \ldots, x_n \rangle} .$$

This claim is proved by induction over the build-up of ξ from the initial functions 2^x, max, 0, $\mathcal{S}$, $\mathcal{I}_i^n$ by composition and bounded primitive recursion.

Let f be an arbitrary honest function. We shall prove that there exists $i \in \mathcal{A}_{\text{tot}}$ such that $f \equiv_E \psi_i$.

Let $\chi_f(x,y) = 0$ if $f(x) = y$; otherwise, let $\chi_f(x,y) = 1$. Now, χ_f is elementary as f is honest. Thus, by the claim, there exists $j \in \mathcal{A}_{\text{tot}}$ and a fixed $k \in \mathbb{N}$ such that
$$\phi_j(\langle x,y \rangle) = \chi_f(x,y) \quad \text{and} \quad \mathcal{T}(j, \langle x,y \rangle, t) \Rightarrow t \leq 2_k^{\langle x,y \rangle} .$$
This entails that there exists $i \in \mathcal{A}_{\text{tot}}$ and a fixed $\ell \in \mathbb{N}$ such that
$$\phi_i(x) = \mu y[\chi_f(x,y) = 0] = f(x) \quad \text{and} \quad \mathcal{T}(i,x,t) \Rightarrow t \leq 2_\ell^{f(x)} .$$
We shall now consider ψ_i, that is, the honest associate of ϕ_i. We have $\psi_i(x) = \langle t_0, \ldots, t_x \rangle$ where $t_r = \mu z \leq 2_\ell^{f(r)}[\mathcal{T}(i,r,z)]$ (for $r = 0, \ldots x$). It follows that there exists a fixed $m \in \mathbb{N}$ such that $\psi_i(x) \leq 2_m^{f(x)}$, and thus, we also have $\psi_i \leq f^{m+1}$, and thus, by the Growth Theorem, $\psi_i \leq_E f$. Moreover, as $f \leq \psi_i$, the Growth Theorem also yields $f \leq_E \psi_i$.   □

Theorem 6 (Isomorphism). *The structure of* PA$^+$*-degrees of algorithms is isomorphic to the structure of honest ϵ_0-elementary degrees.*

Proof. Any function in the sequence $\psi_0, \psi_1, \psi_2, \ldots$ is honest, and Lemma 4 assures that any honest function is $\equiv_{\epsilon_0 E}$-equivalent to some function in the sequence $\psi_0, \psi_1, \psi_2, \ldots$. The Generalised Growth Theorem assures that we have
$$g \leq_{\epsilon_0 E} f \quad \Leftrightarrow \quad g \leq f_\alpha \text{ for some fixed } \alpha < \epsilon_0$$
for any honest functions f and g. The Growth Theorem for Algorithms (Theorem 5 above) assures that we have
$$i \leq_{\text{p}+} j \quad \Leftrightarrow \quad \psi_i \leq (\psi_j)_\alpha \text{ for some fixed } \alpha < \epsilon_0$$
for any $i, j \in \mathcal{A}_{\text{tot}}$. It follows that the degree structure induced on the honest function by $\leq_{\epsilon_0 E}$ is isomorphic to the structure induced on $\mathcal{A}_{\text{tot}}$ by $\leq_{\text{p}+}$.   □

Proof methods based on computability-theoretic constructions work for PA-provability degrees, but they do not work for PA$^+$-provability degrees. Proof methods based on growth of honest functions work for PA$^+$-provability degrees, but they do not for PA-provability degrees. The proof methods developed by Cai are needed to investigate the structure of PA-provability degrees. Still, some results on PA$^+$-provability degrees have corollaries for PA-provability degrees as $i \not\leq_{\text{p}+} j$ implies $i \not\leq_{\text{p}} j$. One example of such a corollary is that there exists an infinite anti-chain of PA-provability degrees. This follows from Theorem 6 since there exists infinite anti-chains in the structure of honest ϵ_0-elementary degrees. (It is proved in [9] that any countable partial ordering can be embedded in the structure of elementary degrees. A similar proof will yield the same result for the structure of honest ϵ_0-elementary degrees. Thus, we can find infinite anti-chains in the structure of honest ϵ_0-elementary degrees.)

Acknowledgements. The author wants to thank Andreas Weiermann, Mingzhong Cai and many others for enlightening discussions and advice. In particular, he wants to thank Albert Visser for making him aware of the nice properties of the theory PA^+. An anonymous referee also deserves some thanks.

References

1. Blankertz, B., Weiermann, A.: How to Characterize Provably Total Functions. In: Hajek (ed.) Gödel 1996. LNL, vol. 6, pp. 205–213. Springer, Heidelberg (1996)
2. Buchholz, W., Cichon, A., Weiermann, A.: A Uniform Approach to Fundamental Sequences and Hierarchies. Mathematical Logic Quarterly 40, 273–286 (1994)
3. Cai, M.: Degrees of Relative Provability. Accepted for publication in the Notre Dame Journal of Formal Logic (manuscript)
4. Cai, M.: Elements of Classical Recursion Theory: Degree-Theoretic Properties and Combinatorial Properties. PhD Thesis, Department of Mathematics, Cornell University (2011)
5. Kaye, R.: Models of Peano Arithmetic. Clarendon Press, Oxford (1991)
6. Kristiansen, L.: Information Content and Computational Complexity of Recursive Sets. In: Hajek (ed.) Gödel 1996. LNL, vol. 6, pp. 235–246. Springer, Heidelberg (1996)
7. Kristiansen, L.: A Jump Operator on Honest Subrecursive Degrees. Archive for Mathematical Logic 37, 105–125 (1998)
8. Kristiansen, L.: Subrecursive Degrees and Fragments of Peano Arithmetic. Archive for Mathematical Logic 40, 365–397 (2001)
9. Kristiansen, L.: Papers on Subrecursion Theory. Dr Scient Thesis, Department of Informatics, University of Oslo (1996) ISBN 82-7368-130-0
10. Kristiansen, L.: Low_n, High_n, and Intermediate Subrecursive Degrees. In: Calude, Dinneen (eds.) Combinatorics, Computation and Logic, pp. 286–300. Springer, Singapore (1999)
11. Kristiansen, L., Lubarsky, R.S., Weiermann, A., Schlage-Puchta, J.-C.: On the Structure of Honest Elementary Degrees. In: Friedman, Koerwien, Müller (eds.) Accepted for Publication in the Proceedings of the Infinity Project
12. Kristiansen, L., Weiermann, A., Schlage-Puchta, J.-C.: Streamlined Subrecursive Degree Theory. Annals of Pure and Applied Logic 163, 698–716 (2012)
13. Lindström, P.: Aspects of Incompleteness. LNL, vol. 10. Springer, Berlin (1997)
14. Machtey, M.: Augmented Loop Languages and Classes of Computable Functions. Journal of Computer and System Sciences 6, 603–624 (1972)
15. Machtey, M.: The Honest Subrecursive Classes are a Lattice. Information and Control 24, 247–263 (1974)
16. Machtey, M.: On the Density of Honest Subrecursive Classes. Journal of Computer and System Sciences 10, 183–199 (1975)
17. Meyer, A.R., Ritchie, D.M.: A Classification of the Recursive Functions. Zeitschr. f. Math. Logik und Grundlagen d. Math. Bd. 18, 71–82 (1972)
18. Odifreddi, P.: Classical Recursion Theory. North-Holland (1989)
19. Rogers, H.: Theory of Recursive Functions and Effective Computability. McGraw Hill (1967)
20. Rose, H.E.: Subrecursion. Functions and Hierarchies. Clarendon Press, Oxford (1984)

A $5n - o(n)$ Lower Bound on the Circuit Size over U_2 of a Linear Boolean Function

Alexander S. Kulikov[1,2], Olga Melanich[1], and Ivan Mihajlin[3]

[1] Laboratory of Mathematical Logic, Steklov Institute of Mathematics
at St. Petersburg, 27 Fontanka, St. Petersburg 191023, Russia
[2] Algorithmic Biology Laboratory, St. Petersburg Academic University, Khlopina
8/3, St. Petersburg, 194021, Russia
[3] St. Petersburg State Polytechnical University, Polytechnicheskaya, 29,
St. Petersburg, 195251, Russia

Abstract. We give a simple proof of a $5n - o(n)$ lower bound on the circuit size over U_2 of a linear function $f(x) = Ax$ where $A \in \{0,1\}^{\log n \times n}$ (here, U_2 is the set of all Boolean binary functions except for parity and its complement).

1 Introduction

Proving lower bounds on the circuit complexity of explicit Boolean functions is a central problem of theoretical computer science. Despite of many efforts currently we can only prove small linear lower bounds: $3n - o(n)$ for circuits over the full binary basis B_2 [1,2] and $5n - o(n)$ for the basis $U_2 = B_2 \setminus \{\oplus, \equiv\}$ [3]. These lower bounds are proved for single-output Boolean functions. However, even less is known for multi-output functions. Though intuitively it seems that computing several functions must be harder than computing just one of them no stronger lower bounds are known for multi-output functions. E.g., if instead of one output we consider $o(n)$ outputs then the strongest lower bounds over B_2 and U_2 are still $3n - o(n)$ and $5n - o(n)$, respectively.

In this note, we prove a $5n - o(n)$ lower bound on the circuit size over U_2 of a linear Boolean function $f(x) = Ax$ where all the columns of the matrix $A \in \{0,1\}^{\log n \times n}$ are pairwise different and non-zero. In fact, we prove a wider result:
$$C_{U_2}(Ax \oplus b) \geq 5(n - m),$$
where $A \in \{0,1\}^{m \times n}$ is a matrix consisting of n different columns and $b \in \{0,1\}^m$ is any vector.

The advantage of the proof is that it contains almost no case analysis though is based on the standard gate elimination method. First, we show that an optimal circuit for such a function does not contain out-degree one variables. For out-degree two variables, we show that either the considered circuit is not optimal or by appropriate substitution at least five gates can be eliminated. Finally, if a circuit contains a variable of degree three then it is straightforward to eliminate five gates.

S.B. Cooper, A. Dawar, and B. Löwe (Eds.): CiE 2012, LNCS 7318, pp. 432–439, 2012.
© Springer-Verlag Berlin Heidelberg 2012

2   Definitions

By $B_{n,m}$ we denote the set of all Boolean functions $f \colon \{0,1\}^n \to \{0,1\}^m$. $B_{n,1}$ is denoted just by B_n. For $f \in B_{n,m}$, by f_i we denote the i-th component of f (thus, $f_i \in B_n$).

A circuit over a basis $\Omega \subseteq B_2$ is a DAG where each vertex has in-degree either 0 or 2 and unbounded out-degree. Vertices of in-degree 0 are called input gates and are labeled by variables $x_1, \dots, x_n$. All other vertices are called gates and are labeled by binary functions from Ω. Some of the gates are also marked as output gates. Edges connecting gates are usually called wires. The out-degree of a variable is the number of gates fed by this variable.

Such a circuit computes in a natural way a function from $B_{n,m}$ where m is the number of output gates. The size of a circuit is its number of gates not including the input gates. The two widely studied bases Ω are B_2 and $U_2 = B_2 \setminus \{\oplus, \equiv\}$.

The 16 binary functions $f(x_1, x_2)$ from B_2 are usually classified as follows:

- two constants: $0, 1$;
- four degenerate functions: x_1, $x_1 \oplus 1$, x_2, $x_2 \oplus 1$;
- eight AND-type functions: $(x_1 \oplus a)(x_2 \oplus b) \oplus c$, where $a, b, c \in \{0,1\}$;
- two XOR-type functions: $x_1 \oplus x_2 \oplus a$, where $a \in \{0,1\}$.

It is easy to see that gates labeled by functions from the first two classes can be easily removed from a circuit. Also, circuits over U_2 do not contain gates computing XOR-functions. In the following, we assume without loss of generality that a circuit over U_2 consists of AND-type gates only.

For a function $f \in B_{n,m}$, by $C_\Omega(f)$ we denote the minimal size of a circuit over Ω computing f.

3   Known Lower Bounds

3.1   Single-Output Functions

By a counting argument, Shannon [4] showed that the circuit complexity (over both B_2 and U_2) of almost all functions from B_n is $\Theta(2^n / n)$. For explicit functions, the following lower bounds are known.

For the basis B_2, Schnorr [5] proved a $2n - \Theta(1)$ lower bound for a wide class of functions satisfying some natural property. Paul [6] proved a $2n - o(n)$ lower bound for the storage access function and a $2.5n - o(n)$ lower bound for a modification of the storage access function. Stockmeyer [7] proved a $2.5n - \Theta(1)$ lower bound for many symmetric functions. Blum [1] generalized Paul's proof to get a $3n - o(n)$ lower bound. Kojevnikov and Kulikov [8] proved a $7n/3 - \Theta(1)$ lower bound for functions of high degree. Demenkov and Kulikov [2] proved a $3n - o(n)$ lower bound for affine dispersers.

For the basis U_2, Schnorr [5] proved that the circuit complexity of parity is $3n - 3$. Zwick [9] proved a $4n - \Theta(1)$ lower bound for certain symmetric functions. Lachish and Raz [10] proved a $4.5n - o(n)$ lower bound for strongly two-dependent functions. Iwama and Morizumi [3] improved the bound to $5n - o(n)$.

3.2   Multi-output Functions

By counting one can show that almost all functions from $B_{n,n}$ have circuit complexity $\Theta(2^n)$.

Several lower bounds on complexity of multi-output functions are discussed in Hiltgen's thesis [11] (section 4.3, "Lower Bounds on the Complexity of Vector Functions").

Lamagne and Savage [12] proved the following result: if, for $f \in B_{n,m}$, all f_i's are different, then

$$C_\Omega(f) \geq \min_{1 \leq i \leq m} C_\Omega(f_i) + m - 1.$$

In other words, the result says that if one needs to compute m different functions instead of just one then at least $m - 1$ additional gates are needed. Using this simple method one can prove $4n - o(n)$ and $6n - o(n)$ lower bounds on the circuit complexity over B_2 and U_2, respectively, for a function from $B_{n,n}$. E.g., we can take a function $g \in B_n$ with $C_{B_2}(g) \geq 3n - o(n)$ and define $f = (f_1, \ldots, f_n)$ where $f_i(x_1, \ldots, x_n) = g(x_1, \ldots, x_n) \oplus x_i$.

Blum and Seysen [13] proved that $C_{B_2}(\mathrm{AND}, \mathrm{NAND}) = 2n - 2$ and moreover any optimal circuit for (AND, NAND) consists of two independent trees. Red'kin [14] proved that the circuit complexity over B_2 of a binary adder is $2.5n - 3$. The binary adder is a function from $B_{n,n/2+1}$ that outputs the sum of two input $n/2$-bit numbers. Interlando et al. [15] studied the circuit complexity over $\{\oplus\}$ of the same function studied in this note. They proved that $C_{\{\oplus\}}(Ax) = 2n - o(n)$. Chashkin [16] proved that $C_{B_2}(Ax) = 2n - o(n)$ and also constructed a function from $B_{n,n}$ with circuit complexity (over B_2) $3n - o(n)$. Hiltgen [11] studied so-called feebly one-way functions, i.e., permutations from $B_{n,n}$ such that $C(f) < C(f^{-1})$. He constructed examples of such f's with $C_{B_2}(f) = n - \Theta(1)$ and $C_{B_2}(f^{-1}) = 2n - \Theta(1)$.

4   A $5n - o(n)$ Lower Bound

In this section, we consider functions $f \in B_{n,m}$ of the form $f(x) = Ax \oplus b$, where $A \in \{0,1\}^{m \times n}$ is a matrix with n different non-zero columns and $b \in \{0,1\}^m$ is any vector. First, note that after assigning a Boolean value to any variable one gets a function of the same type (the corresponding column of A is eliminated and some bits of b are changed; all the columns of A are still different and non-zero). This allows us to prove a lower bound by induction. Next, we prove a few elementary lemmas for circuits computing such functions.

Recall that in this section we consider only circuits over U_2. One can assume without loss of generality that such a circuit consists of AND-type gates only. The main property of such a gate is the following: if a variable feeds an AND-type gate then one can assign a constant to this variable so that the gate becomes constant.

Lemma 1. *If some input variable x_i of a circuit computing f feeds exactly one gate then this circuit is not optimal.*

Proof. Assume for the sake of contradiction that x_i has out-degree 1.

If all outputs either depend only on x_i or does not depend on x_i at all then x_i does not need to feed any gate. In this case we can feed a constant instead of x_i into the gate that is fed by x_i.

Otherwise at least one of the outputs depends on x_i and at least one other variable. Let G be the other input of the gate that is fed by x_i. Note that G does not depend on x_i. The considered gate computes a function of the form $(x_i \oplus a)(g \oplus b) \oplus c$. Assume that there is an assignment to all variables but x_i under which G is equal to b. Then the considered gate trivializes to the constant c and the resulting circuit does not depend on x_i. This contradicts the fact that even under this substitution the considered output still depends on x_i. This means that under all possible substitutions G is equal to $b \oplus 1$, i.e., G is constant. This, in turn, means that the circuit is not optimal.

Lemma 2. *Let P and Q be gates fed by a variable x_i and let Q be a direct successor of P. Then one can reconstruct the circuit without increasing its size so that P and Q are not connected by a wire.*

Proof. Let G be the other successor of P. Then Q depends on G and x_i. Note that Q cannot compute a XOR-type function of x_i and G since to compute the XOR of two variables in U_2 one needs three gates. Thus, we can change the binary function computed in Q and put a wire to Q not from P, but directly from G. The transformation is illustrated in Fig. 1.

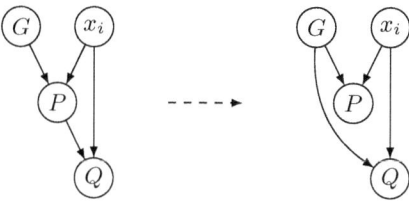

Fig. 1. A transformation described in Lemma 2

Theorem 1

$$C_{U_2}(f) \geq 5(n - m).$$

Proof. The proof is by induction on $n - m$. Without loss of generality we may assume that A does not contain all-zero columns.

There are two cases when the statement holds by trivial reasons: $n \leq m$ and $m = 1$ (if $m = 1$ then $n = 1$ since all the columns must be different).

Assume now that $n > m$ and $m > 1$. If there is a row containing exactly one 1 (say, at i-th column), then the corresponding output depends only on x_i. We may then assign $x_i = 0$. This removes the i-th column from the matrix as well as

the corresponding row. It is easy to see that the resulting circuit still computes
the resulting function. It is also easy to see that all the columns in the resulting
matrix are different.

Let now all the rows A contain at least two 1's and consider an optimal
circuit computing f. None of the input variables is an output gate. Note also the
following property of the function f: even if we assign all variables but x_i there
is at least one output that still depends on x_i. Also, by Lemma 1 all variables
feed at least two gates.

Consider a top gate P, i.e., a gate fed by two variables x_i and x_j. As the
circuit is optimal these variables are different and each of them feeds at least
one other gate.

We now consider several cases. In each of the cases we show that it is possible
to assign some variable a constant so that at least 5 gates are eliminated. Note
that all the gates that become constant are not output gates as all output gates
depend on at least two variables. This means that each such gate has at least
one successor. Note also that Lemma 2 allows us to assume that if two gates are
fed by the same variable then neither of them is fed by the other one.

Case 1. One of x_i and x_j (say, x_i) feeds at least three gates (call them P, Q, R).
Then there exists a constant $c \in \{0, 1\}$ such that the substitution $x_i = c$ makes
at least two of these gates constant (say, P and Q). Hence, this substitution
eliminates P, Q, R and all the successors of P and Q (Fig. 2).

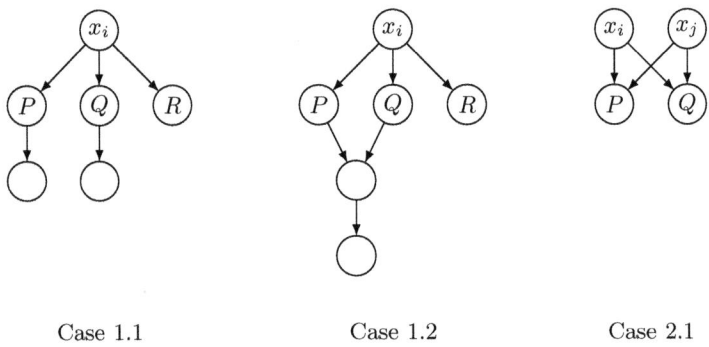

Case 1.1   Case 1.2   Case 2.1

Fig. 2. Cases 1.1, 1.2, and 2.1

Case 1.1. If the total number of successors of P and Q is at least 2, then at
least five gates are eliminated.

Case 1.2. If P and Q have a single successor then it also becomes constant and
its successor is also eliminated (and if R happens to be this last successor then
R becomes constant and also its successor is eliminated).

Case 2. Both x_i and x_j feed exactly two gates. Denote the other successors of
x_i and x_j by R and Q, respectively.

Case 2.1. $Q = R$ (Fig. 2). We show that this case is just impossible. Indeed, since P and Q are AND-type gates,

$$P = (x_i \oplus a_1)(x_j \oplus b_1) \oplus c_1 \,,$$
$$Q = (x_i \oplus a_2)(x_j \oplus b_2) \oplus c_2 \,,$$

for some constants $a_1, b_1, c_1, a_2, b_2, c_2 \in \{0, 1\}$. Note that $a_1 \neq a_2$. Otherwise by a substitution $x_i = a_1$ we would make the circuit independent of x_j. By exactly the same reason $b_1 \neq b_2$. But then the circuit does not distinguish between the assignments $\{x_i = a_1, x_j = b_1 \oplus 1\}$ and $\{x_i = a_1 \oplus 1, x_j = b_1\}$ (under both these substitutions $P = c_1$ and $Q = c_2$). This cannot be the case since all the columns of A are different and hence there is at least one output that either depends on x_i and does not depend on x_j or vice versa.

Case 2.2. $Q \neq R$. Since the circuit is a DAG the gates can be topologically sorted. Fix some topological order and assume that R precedes Q in it. This, in particular, means that R does not depend on Q.

We would like to show that there is a constant $c \in \{0, 1\}$ such that the substitution $x_j = c$ makes both gates P and R constant. Let G be the other input of R (Fig. 3). As P and R are AND-type gates there exist constants $a_1, b_1, c_1, a_2, b_2, c_2 \in \{0, 1\}$ such that

$$P = (x_j \oplus a_1)(x_i \oplus b_1) \oplus c_1 \,,$$
$$R = (x_j \oplus a_2)(g \oplus b_2) \oplus c_2 \,.$$

Note that under the substitution $x_i = b_1$ the gate P trivializes to the constant c_1. Then if for some substitution to all other variables except for x_j the gate G has value b_2, then the gate R also trivializes to the constant c_2 and the circuit becomes independent of the variable x_j while this cannot be the case. Thus,

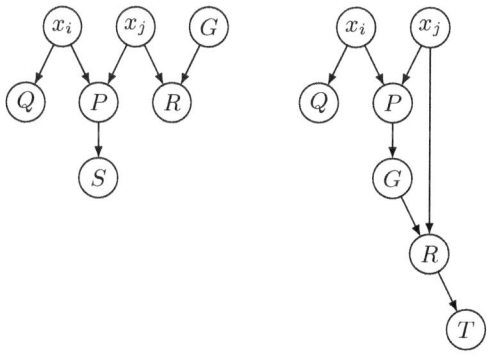

Case 2.2.1   Case 2.2.2

Fig. 3. All subcases of case 2.2

438 A.S. Kulikov, O. Melanich, and I. Mihajlin

$x_i = b_1$ implies $G = b_2 \oplus 1$. A crucial observation is that then $x_j = a_1$ also implies $G = b_2 \oplus 1$. Indeed there is no path in the circuit from Q to G (by the assumption that R precedes Q), hence G can only depend on x_i through P. And we know that if $P = c_1$ then $G = b_2 \oplus 1$.

Altogether, if we assign $x_j = a_1$ then P and G trivialize to constant, R also becomes a constant as it is now fed by two constants. All the successors of P, G, R are also eliminated. Also, in the resulting circuit x_i has out-degree 1 hence by Lemma 1 at least one additional gate can be eliminated. To show that in all possible cases at least 5 gates are eliminated we consider two subcases depending on the successors of P (the two subcases are shown at Fig. 3). Denote by S some successor of P. Note that Lemma 2 guarantees that $S \neq Q$ and $S \neq R$.

Case 2.2.1. $S \neq G$. Then the substitution $x_j = a_1$ kills the gates P, R, G, S. After that at least one additional gate can be eliminated due to Lemma 1.

Case 2.2.2. $S = G$. Note that if any of P, G, R has out-degree more than 1 we again remove at leats 5 gates. Assume now that all of them have out-degree exactly 1. Denote the only successor of R by T and consider the other input of T. It is not a constant and it does not depend on x_j. Thus, by an appropriate substitution to all variables but x_j we can trivialize the gate T and make the circuit independent of x_j, a contradiction.

Acknowledgements. Research is partially supported by Russian Foundation for Basic Research (11-01-00760-a and 11-01-12135-ofi-m-2011), RAS Program for Fundamental Research, Grant of the President of Russian Federation (NSh-3229.2012.1), and Computer Science Club scholarship.

References

1. Blum, N.: A Boolean function requiring $3n$ network size. Theoretical Computer Science 28, 337–345 (1984)
2. Demenkov, E., Kulikov, A.S.: An Elementary Proof of a $3n - o(n)$ Lower Bound on the Circuit Complexity of Affine Dispersers. In: Murlak, F., Sankowski, P. (eds.) MFCS 2011. LNCS, vol. 6907, pp. 256–265. Springer, Heidelberg (2011)
3. Iwama, K., Morizumi, H.: An Explicit Lower Bound of $5n-o(n)$ for Boolean Circuits. In: Diks, K., Rytter, W. (eds.) MFCS 2002. LNCS, vol. 2420, pp. 353–364. Springer, Heidelberg (2002)
4. Shannon, C.E.: The synthesis of two-terminal switching circuits. Bell System Technical Journal 28, 59–98 (1949)
5. Schnorr, C.P.: Zwei lineare untere Schranken für die Komplexität Boolescher Funktionen. Computing 13, 155–171 (1974)
6. Paul, W.J.: A $2.5n$-lower bound on the combinational complexity of Boolean functions. SIAM Journal of Computing 6(3), 427–433 (1977)
7. Stockmeyer, L.J.: On the combinational complexity of certain symmetric Boolean functions. Mathematical Systems Theory 10, 323–336 (1977)
8. Kojevnikov, A., Kulikov, A.S.: Circuit Complexity and Multiplicative Complexity of Boolean Functions. In: Ferreira, F., Löwe, B., Mayordomo, E., Mendes Gomes, L. (eds.) CiE 2010. LNCS, vol. 6158, pp. 239–245. Springer, Heidelberg (2010)

9. Zwick, U.: A $4n$ lower bound on the combinational complexity of certain symmetric boolean functions over the basis of unate dyadic Boolean functions. SIAM Journal on Computing 20, 499–505 (1991)
10. Lachish, O., Raz, R.: Explicit lower bound of $4.5n - o(n)$ for boolean circuits. In: Proceedings of the Annual Symposium on Theory of Computing (STOC), pp. 399–408 (2001)
11. Hiltgen, A.P.L.: Cryptographically Relevant Contributions to Combinational Complexity Theory. ETH Series in Information Processing, vol. 3 (1994)
12. Lamagna, E.A., Savage, J.E.: On the logical complexity of symmeric switching functions in monotone and and complete bases. Technical report, Brown University (1973)
13. Blum, N., Seysen, M.: Characterization of all optimal networks for a simultaneous computation of AND and NOR. Acta Informatica 21(2), 171–181 (1984)
14. Red'kin, N.P.: Minimal realization of a binary adder. Problemy kibernetiki 38, 181–216 (1981) (in Russian)
15. Interlando, J.C., Byrne, E., Rosenthal, J.: The Gate Complexity of Syndrome Decoding of Hamming Codes. Applications of Computer Algebra (ACA), 1–5 (2004)
16. Chashkin, A.V.: On complexity of Boolean matrices, graphs and corresponding Boolean matrices. Diskretnaya Matematika 6(2), 43–73 (1994) (in Russian)

Local Induction and Provably Total Computable Functions: A Case Study

Andrés Cordón–Franco and F. Félix Lara–Martín

Departamento Ciencias de la Computación e Inteligencia Artificial, Facultad de
Matemáticas. Universidad de Sevilla, C/ Tarfia, s/n, 41012 Sevilla, Spain
{acordon,fflara}@us.es

Abstract. Let $I\Pi_2^-$ denote the fragment of Peano Arithmetic obtained
by restricting the induction scheme to parameter free Π_2 formulas. Answering a question of R. Kaye, L. Beklemishev showed that the provably
total computable functions (p.t.c.f.) of $I\Pi_2^-$ are, precisely, the primitive
recursive ones. In this work we give a new proof of this fact through
an analysis of the p.t.c.f. of certain *local versions* of induction principles
closely related to $I\Pi_2^-$. This analysis is essentially based on the equivalence between local induction rules and restricted forms of iteration. In
this way, we obtain a more direct answer to Kaye's question, avoiding the
metamathematical machinery (reflection principles, provability logic,...)
needed for Beklemishev's original proof.

1   Introduction

An important notion in studying the computational content of a fragment of
Arithmetic is that of its *provably total computable functions*. A number–theoretic
computable function $f : \mathbb{N}^k \to \mathbb{N}$ is said to be a provably total computable
function (p.t.c.f.) of a theory T, written $f \in \mathcal{R}(T)$, if there is a Σ_1 formula
$\varphi(\vec{x}, y)$ such that:

1. φ defines the graph of f in the standard model of Arithmetic $\mathbb{N}$; and
2. $T \vdash \forall \vec{x}\, \exists! y\, \varphi(\vec{x}, y)$.

Observe that condition 1. amounts to the computability of f, whereas condition
2. yields an implicit measure of the complexity of f attending to the logical
principles needed to prove that a Σ_1–definition of f defines a total function. Since
it was introduced by G. Kreisel in the 1950s this notion has been widely studied,
and nice recursion–theoretic and computational complexity characterizations of
the sets $\mathcal{R}(T)$ have been obtained for a good number of theories T. For instance,
by a classical result due independently to G. Mints, C. Parsons and G. Takeuti,
the class of p.t.c.f. of the scheme of induction for Σ_1–formulas $I\Sigma_1$ equals to the
class of the primitive recursive functions PR. Indeed, all classes $\mathcal{R}(I\Sigma_n)$, $n \geq 1$,
can be characterized in terms of the Fast Growing Hierarchy up to the ordinal
ε_0. As for weak fragments below $I\Sigma_1$, their p.t.c.f. have been characterized in
terms of subrecursive operators (bounded recursion, bounded minimization, ...)

S.B. Cooper, A. Dawar, and B. Löwe (Eds.): CiE 2012, LNCS 7318, pp. 440–449, 2012.

as well as in terms of computational complexity classes. In fact, their classes
of p.t.c.f. have been intensively investigated in connection with important open
problems in Complexity Theory, mainly in the context of Bounded Arithmetic.

In spite of the wide range of the theories considered, a number of uniform
methods for characterizing the p.t.c.f. of an arithmetic theory are available. E.g.
Herbrand analyses as developed by W. Sieg in [9], S. Buss' witnessing method [5]
or, in general, proof–theoretic techniques using Cut elimination theorem. However, for some particular fragments of Peano Arithmetic none of these standard
methods seems to be applicable. Of special interest is the case of the scheme of
parameter free Π_2–induction, $I\Pi_2^-$, given by the induction scheme

$$I_\varphi: \quad \varphi(0) \wedge \forall x \, (\varphi(x) \rightarrow \varphi(x+1)) \ \rightarrow \ \forall x \, \varphi(x)\,,$$

restricted to $\varphi(x) \in \Pi_2^-$ (as usual, we write $\varphi(x) \in \Gamma^-$ to mean that φ is in Γ
and contains no other free variables than x). Since $I\Sigma_1^- \subseteq I\Pi_2^-$ and $I\Sigma_1$ is Σ_3–
conservative over $I\Sigma_1^-$ [8], it follows that every primitive recursive function is
provably total in $I\Pi_2^-$; and R. Kaye asked whether the p.t.c.f. of $I\Pi_2^-$ are exactly
the primitive recursive ones. This question remained elusive until [4], where L.
Beklemishev gave a positive answer using modal provability logic techniques.
Although quite elegant, Beklemishev's answer only provides an indirect solution.
Firstly, he reformulated $I\Pi_2^-$ in terms of local reflection principles (reflection
principles in Arithmetic are axiom schemes expressing the statement that "if a
formula φ is provable in a theory T then φ is valid"). Secondly, he derived the
result as an application of a conservation theorem for local reflection principles
whose proof leans upon properties of Gödel–Löb provability logic **GL**.

In this work we obtain a more direct answer to Kaye's question, avoiding the
metamathematical machinery needed for Beklemishev's proof. In fact, our proof
that $\mathcal{R}(I\Pi_2^-) = PR$ will follow the lines of standard arguments for characterizing
classes $\mathcal{R}(T)$. Let us consider, for instance, a proof that $\mathcal{R}(I\Sigma_1) = PR$. Such a
proof typically proceeds in two steps.

- Step 1: $I\Sigma_1$ is Π_2–conservative over the *inference rule* version of the principle
  of Σ_1–induction Σ_1–IR. So, $\mathcal{R}(I\Sigma_1) = \mathcal{R}(\Sigma_1\text{–IR})$.
- Step 2: Applications of Σ_1–IR correspond to applications of the primitive
  recursion operator.

The main obstacle to apply this argument to $I\Pi_2^-$ is that there is no simple,
direct argument to reduce $I\Pi_2^-$ to an inference rule version of it. Here we solve
this problem by showing that $I\Pi_2^-$ is equivalent to $I(\Sigma_2^-, \mathcal{K}_2)$, a certain version
of the parameter free Σ_2–induction scheme where the elements x for which the
induction axiom claims $\varphi(x)$ to hold are restricted to be Σ_2–definable elements.
Equipped with this result, it is easy to obtain that $I\Pi_2^-$ is Π_2 (in fact, Π_3)
conservative over the corresponding inference rule version $(\Sigma_2, \mathcal{K}_2)$–IR. Then,
we show that applications of $(\Sigma_2, \mathcal{K}_2)$–IR correspond to (restricted forms) of the
iteration operator and thus all functions in $\mathcal{R}(I\Pi_2^-)$ are primitive recursive.

Our analysis also yields a new conservation theorem for fragments of Peano
Arithmetic, which is of independent interest. Namely, we prove that $I\Pi_2^-$ is Π_3–
conservative over $I\Sigma_1$. This improves on a previous result by Beklemishev in [4],

where conservativity between these theories with respect to boolean combinations of Σ_2–sentences was established.

We close this section by giving a precise definition of the auxiliary scheme that will be central in our analysis of the class of p.t.c.f. of $I\Pi_2^-$.

Let $\mathcal{L} = \{0, S, +, \cdot, <\}$ denote the language of first order Arithmetic. If Γ is a set of formulas of $\mathcal{L}$, then $I\Gamma$ is the theory axiomatized over Robinson's Q by the induction scheme, I_φ, restricted to formulas $\varphi(x) \in \Gamma$. If free variables other that x are not allowed, we write $\varphi(x) \in \Gamma^-$ and, accordingly, $I\Gamma^-$ denotes the theory axiomatized over Q by the axioms I_φ, for $\varphi(x) \in \Gamma^-$.

Definition 1. $I(\Sigma_2, \mathcal{K}_2)$ *is the theory given by* $I\Sigma_1^-$ *together with the scheme*

$$\varphi(0) \wedge \forall x \,(\varphi(x) \to \varphi(x+1)) \to$$
$$\to \forall x_1, x_2 \,(\delta(x_1) \wedge \delta(x_2) \to x_1 = x_2) \to \forall x \,(\delta(x) \to \varphi(x))$$

where $\varphi(x) \in \Sigma_2$ *and* $\delta(x) \in \Sigma_2^-$. *The natural inference rule associated to this scheme, denoted* $(\Sigma_2, \mathcal{K}_2)$*–IR, is given by:*

$$\frac{\varphi(0) \wedge \forall x \,(\varphi(x) \to \varphi(x+1))}{\forall x_1, x_2 \,(\delta(x_1) \wedge \delta(x_2) \to x_1 = x_2) \to \forall x \,(\delta(x) \to \varphi(x))}$$

where $\delta(x) \in \Sigma_2^-$ *and* $\varphi(x) \in \Sigma_2$. *Finally, if we restrict the scheme to* $\varphi(x) \in \Sigma_2^-$, *we obtain the parameter free counterpart of* $I(\Sigma_2, \mathcal{K}_2)$, *denoted* $I(\Sigma_2^-, \mathcal{K}_2)$.

Remark 1. Firstly, let us recall that, given a model $\mathfrak{A}$, $\mathcal{K}_2(\mathfrak{A})$ denotes the set of elements of $\mathfrak{A}$ that are definable in $\mathfrak{A}$ by a formula $\delta(x) \in \Sigma_2$. This explains why $\mathcal{K}_2$ appears in our notation for these theories. Secondly, if $\mathfrak{A} \models I\Sigma_1^-$, then $\mathcal{K}_2(\mathfrak{A}) \prec_2 \mathfrak{A}$ (i.e., $\mathcal{K}_2(\mathfrak{A})$ is a Π_2–elementary substructure of $\mathfrak{A}$). This property plays an important role in what follows and it is because of it that $I(\Sigma_2, \mathcal{K}_2)$ is axiomatized over $I\Sigma_1^-$ instead of over a weaker system (such as Q or $I\Delta_0$).

A key fact is that $I(\Sigma_2^-, \mathcal{K}_2)$ provides an alternative formulation of $I\Pi_2^-$:

Lemma 2. $I\Pi_2^- \equiv I(\Sigma_2^-, \mathcal{K}_2)$.

Proof. We only prove that $I(\Sigma_2^-, \mathcal{K}_2)$ extends $I\Pi_2^-$. The converse is similar. Let $\mathfrak{A} \models I(\Sigma_2^-, \mathcal{K}_2)$ and $\varphi(x) \in \Pi_2^-$ such that $\mathfrak{A} \models \varphi(0) \wedge \forall x \,(\varphi(x) \to \varphi(x+1))$. Assume $\mathfrak{A} \models \exists x \,\neg\varphi(x)$. Since $\mathfrak{A} \models I\Sigma_1^-$, $\mathcal{K}_2(\mathfrak{A}) \prec_2 \mathfrak{A}$ and there is $a \in \mathcal{K}_2(\mathfrak{A})$ such that $\mathfrak{A} \models \neg\varphi(a)$. Let $\delta(v)$ be a Σ_2 formula defining the element a and let $\theta(x)$ be $\exists v \,(\delta(v) \wedge \neg\varphi(v - x))$. Clearly, $\mathfrak{A} \models \theta(0) \wedge \forall x \,(\theta(x) \to \theta(x+1))$. By $I(\Sigma_2^-, \mathcal{K}_2)$, $\mathfrak{A} \models \theta(a)$ and so $\mathfrak{A} \models \neg\varphi(0)$, which is a contradiction.

Given a theory T and an inference rule R, we denote by $[T, R]$ the closure of T under first order logic and unnested applications of R. We denote by $T + R$ the closure of T under first order logic and (nested) applications of R. Therefore, $T + R = \bigcup_{k \in \omega} [T, R]_k$, where $[T, R]_0 = T$ and $[T, R]_{k+1} = [[T, R]_k, R]$.

The first step in the analysis of $I\Pi_2^-$ is a suitable reduction of $I(\Sigma_2, \mathcal{K}_2)$ to a fragment defined by the rule $(\Sigma_2, \mathcal{K}_2)$–IR. Indeed, we have:

Proposition 3. $I(\Sigma_2, \mathcal{K}_2)$ *is* Π_3*–conservative over* $I\Sigma_1^- + (\Sigma_2, \mathcal{K}_2)$*–IR.*

Very conveniently, this reduction can be carried out by the same tools used to derive the reduction of $I\Sigma_1$ to Σ_1–IR (e.g., by adapting the cut–elimination argument used in [3] to derive a similar reduction for the Collection scheme). Alternatively, in [6], lemma 3.6, we gave a model–theoretic proof of this result using the notion of a Σ_{n+1}–closed model, following the methods introduced by Avigad in [1].

2   Local Induction and Restricted Iteration

Next step in our analysis is to show that applications of $(\Sigma_2, \mathcal{K}_2)$–IR correspond to (a restricted form of) the iteration operator. To this end, we shall consider extensions of $\mathcal{L}$ obtained by adding a finite set of unary function symbols, $\mathcal{F} = \{f_1, \ldots, f_n\}$, and a (finite or countable) set of new constant symbols, C. Through this section we consider a fixed set of constants, C, and we shall denote by $\mathcal{L}_\mathcal{F}$ the language $\mathcal{L} + \{f_1, \ldots, f_n\} + C$. If g is a new unary function symbol then $\mathcal{L}_{\mathcal{F},g}$ will denote the language $\mathcal{L}_{\{f_1, \ldots, f_n, g\}}$.

Definition 4. *Let* $f \in \mathcal{F}$ *a unary function symbol and* T *an* $\mathcal{L}_\mathcal{F}$*–theory. We say that* f *is an* iterable non decreasing function *over* T *if the theory* T *proves:*

$$\forall x_1, x_2 \, (x_1 \le x_2 \to f(x_1) \le f(x_2)), \ \ and \ \forall x \, (x^2 < f(x))$$

Let $\Sigma_0^\mathcal{F}$ be the class of bounded formulas of $\mathcal{L}_\mathcal{F}$. Classes $\Sigma_{n+1}^\mathcal{F}$ and $\Pi_{n+1}^\mathcal{F}$ are defined as usual. The theory $I\Sigma_0^\mathcal{F}$ is the $\mathcal{L}_\mathcal{F}$–theory axiomatized over Q by

- The induction axiom I_φ for each formula $\varphi \in \Sigma_0^\mathcal{F}$, and
- Axioms for each $f \in \mathcal{F}$:
  $$\forall x_1, x_2 \, (x_1 \le x_2 \to f(x_1) \le f(x_2)), \ and \ \forall x \, (x^2 < f(x))$$

This is a basic theory to deal with the *iteration* of f and to guarantee the usual properties of the iteration of a nondecreasing function with a $\Pi_0^\mathcal{F}$–definable graph. The basic facts provable in this theory were stated in [6]. Next result collects together the facts that we shall need in the present context.

Proposition 5. *For each* $f \in \mathcal{F}$ *there exists a formula* $IT_f(z, x, y) \in \Sigma_0^\mathcal{F}$ *such that the following formulas are theorems of* $I\Sigma_0^\mathcal{F}$:

1. $IT_f(z, x, y_1) \wedge IT_f(z, x, y_2) \to y_1 = y_2.$
2. $(IT_f(0, x, y) \leftrightarrow x = y) \wedge (IT_f(1, x, y) \leftrightarrow f(x) = y).$
3. $IT_f(z + 1, x, y) \leftrightarrow \exists y_0 \le y \, (IT_f(z, x, y_0) \wedge f(y_0) = y).$
4. $IT_f(z, x, y) \to \forall z_0 < z \, \exists y_0 < y \, IT_f(z_0, x, y_0).$
5. $z \ge 1 \wedge IT_f(z, x, y) \to x^2 < y \wedge z \le y.$
6. $z \ge 1 \wedge x_1 \le x_2 \wedge IT_f(z, x_1, y_1) \wedge IT_f(z, x_2, y_2) \to y_1 \le y_2.$
7. $IT_f(z_1, x, y_0) \wedge IT_f(z_2, y_0, y) \to IT_f(z_1 + z_2, x, y).$

In what follows we use a more suggestive notation and write $f^z(x) = y$ instead of $IT_f(z, x, y)$.

Definition 6. *We say that* $f \in \mathcal{F}$ *is a* dominating function *over* T *if, for any term* $t(x)$ *of* $\mathcal{L}_{\mathcal{F}}$, *there exists* $k \in \omega$ *such that* T *proves*

$$\forall x \, (t(x) \leq f^k(x + \sigma(t)))$$

where $\sigma(t) = c_1 + \cdots + c_m$ *and* $c_1, \ldots, c_m$ *are all the constants occurring in* $t(x)$.

Lemma 7. *Let* T *be an extension of* $I\Sigma_0^{\mathcal{F}}$ *and let* $f \in \mathcal{F}$ *a (iterable nondecreasing) dominating function over* T. *Then, for each term* $t(x_1, \ldots, x_m)$ *of* $\mathcal{L}_{\mathcal{F}}$ *whose variables are among* $x_1, \ldots, x_m$, *there exists* $k \in \omega$ *such that*

$$T \vdash t(x_1, \ldots, x_m) < f^k(x_1 + \cdots + x_m + \sigma(t)).$$

Remark 2. Languages $\mathcal{L}_{\mathcal{F}}$ and the notion of a dominating function are tailored to deal with the following situation. Assume $\Gamma = \{\theta_1, \ldots, \theta_m\}$ is a finite set of Σ_0–formulas with only two free variables, say x and y, and for each $j = 1, \ldots, m$, $\bar{\theta}_j(x, y)$ denotes the formula $\forall u \leq x \exists v \leq y \, \theta_j(u, v)$. Let $\mathcal{F} = \{f_1, \ldots, f_m, f\}$ be a set of unary function symbols and let T be the extension of $I\Sigma_0^{\mathcal{F}}$ with the following additional axioms:

- For each $j = 1, \ldots, m$, $\forall x \, (f_j(x) = y \leftrightarrow \exists y_0 \leq y \, (\bar{\theta}_j(x, y_0) \wedge y = (x+1)^2 + y_0)$.
- $\forall x \, (f(x) = (x+1)^2 + f_1(x) + \cdots + f_m(x))$.

Then, every $h \in \mathcal{F}$ is an iterable nondecreasing function over T and f is a dominating function over T. This last fact can be proved by induction on terms. The most interesting case occurs when $t(x)$ is a product of two terms, $t_1(x) \cdot t_2(x)$. By induction hypothesis, $t_1(x) \leq f^k(x + \sigma(t_1))$ and $t_2(x) \leq f^l(x + \sigma(t_2))$, for some $k \geq \max(l, 2)$ (so, for every u, $f^k(u) \geq k \geq 2$.) Then,

$$t(x) \leq (t_1(x) + t_2(x))^2 \leq f(t_1(x) + t_2(x)) \leq f(f^k(x + \sigma(t_1)) + f^l(x + \sigma(t_2)))$$
$$\leq f(2 \cdot f^k(x + \sigma(t))) \leq f((f^k(x + \sigma(t)))^2) \leq f^{k+2}(x + \sigma(t))$$

and we conclude that $t(x) \leq f^{k+2}(x + \sigma(t))$. The remaining cases are similar.

As a final step in the analysis of $(\Sigma_2, \mathcal{K}_2)$–IR and due to technical reasons, it will be convenient to denote the Σ_2–definable elements by closed terms of an extended language. This motivates the introduction of the following *local induction rules*.

Definition 8. *For each set of formulas* Γ *and each set of closed terms* Λ_0 *of* $\mathcal{L}_{\mathcal{F}}$ *we consider the rules (where* $\varphi(x) \in \Gamma$ *and* $t \in \Lambda_0$):

$$(\Gamma, \Lambda_0)\text{–}IR : \quad \frac{\varphi(0) \wedge \forall x \, (\varphi(x) \rightarrow \varphi(x+1))}{\varphi(t)}$$

$$(\Gamma, \Lambda_0)\text{–}IR_0 : \quad \frac{\forall x \, (\varphi(x) \rightarrow \varphi(x+1))}{\varphi(0) \rightarrow \varphi(t)}$$

Definition 9. *We say that* Λ_0 *is* exponentially closed *over* T *if for every* $t, s \in \Lambda_0$ *there exists* $t' \in \Lambda_0$ *such that* $[T, (\Sigma_1^{\mathcal{F}}, \Lambda_0)\text{–}IR] \vdash \exists y \leq t' \, (s^t = y)$.

These rules were intensively studied in [6], where the following results were obtained. From now on, we assume that T is a fixed extension of $I\Sigma_0^{\mathcal{F}}$ obtained by adding a set of $\Pi_1^{\mathcal{F}}$ sentences, and Λ_0 denotes the set of all closed terms of a sublanguage of $\mathcal{L}_{\mathcal{F}}$ extending $\mathcal{L}$ and containing the set of constants C (and so Λ_0 is closed under sum and product).

Remark 3. Let us note that under these assumptions T satisfies a natural version of Parikh's theorem (see [7], chapter 5, theorem 1.4). This fact will be used extensively without further comments.

In addition, we assume that there is $f \in \mathcal{L}_{\mathcal{F}}$ a dominating function over T and Λ_0 is exponentially closed over T. Then, we have (see lemma 4.8, lemma 4.10 and theorem 4.14 of [6]):

Proposition 10. $T + (\Pi_2^{\mathcal{F}}, \Lambda_0)\text{–IR} \equiv T + \{\forall x \, \exists y \, (f^t(x) = y) : \ t \in \Lambda_0\}.$

Theorem 11. $T + (\Pi_2^{\mathcal{F}}, \Lambda_0)\text{–IR}_0$ is $\Pi_2^{\mathcal{F}}$–conservative over $T + (\Pi_2^{\mathcal{F}}, \Lambda_0)\text{–IR}.$

Here we extend our work in [6] and obtain a new theorem on these local induction systems that will be crucial to derive our main result. The ideas involved are similar to the ones used in [6] to obtain Proposition 10 and Theorem 11.

Theorem 12. $T + I\Sigma_1^{\mathcal{F}}$ extends $T + (\Sigma_2^{\mathcal{F}}, \Lambda_0)\text{–IR}.$

Proof. The arguments used in [2], proposition 2.1, can be easily adapted to yield that for every $k \in \omega$, $[T, (\Sigma_2^{\mathcal{F}}, \Lambda_0)\text{–IR}]_k \equiv [T, (\Pi_2^{\mathcal{F}}, \Lambda_0)\text{–IR}_0]_k$. So it is enough to prove that for every $k \in \omega$, $T + I\Sigma_1^{\mathcal{F}}$ extends $[T, (\Pi_2^{\mathcal{F}}, \Lambda_0)\text{–IR}_0]_k$. We proceed by induction on $k \in \omega$:

Case $k = 0$ is trivial; so, let us assume that $T + I\Sigma_1^{\mathcal{F}}$ extends $[T, (\Pi_2^{\mathcal{F}}, \Lambda_0)\text{–IR}_0]_k$. Let $t \in \Lambda_0$ and $\varphi(u, v) \in \Pi_2^{\mathcal{F}}$ such that

$$(\dagger) \qquad [T, (\Pi_2^{\mathcal{F}}, \Lambda_0)\text{–IR}]_k \vdash \forall u \, (\varphi(u, v) \to \varphi(u + 1, v)).$$

We must prove that $T + I\Sigma_1^{\mathcal{F}} \vdash \varphi(0, v) \to \varphi(t, v)$.

Without loss of generality, we can assume that $\varphi(u, v) \equiv \forall x \, \exists y \, \varphi_0(u, x, y, v)$, with $\varphi_0(u, x, y, v) \in \Sigma_0^{\mathcal{F}}$. Let g be a new unary function symbol and T^g the extension of $T + I\Sigma_0^{\mathcal{F}, g}$ obtained by adding the sentences:

$$\forall x_1, x_2 \, (x_1 \leq x_2 \to g(x_1) \leq g(x_2)), \quad \forall x \, (x^2 < g(x)) \quad \text{and} \quad \forall x \, (f(x) \leq g(x)).$$

Thus, g is a dominating (iterable nondecreasing) function over T^g. By $(\dagger)$, it follows that $[T^g, (\Pi_2^{\mathcal{F}, g}, \Lambda_0)\text{–IR}]_k \vdash \varphi^g$, where φ^g is the following sentence:

$$\forall u \, (\forall x \, \exists y \leq g(x + u + v) \, \varphi_0(u, x, y, v) \to \forall x \, \exists y \, \varphi_0(u + 1, x, y, v)).$$

Claim. There is a closed term $\tau_0 \in \Lambda_0$ such that $T^g + \forall x \, \exists y \, (g^{\tau_0}(x) = y)$ proves

$$\forall u \, (\forall x \, \exists y \leq g(x + u + v) \, \varphi_0(u, x, y, v) \to \forall x \, \exists y \leq g^{\tau_0}(u + x + v) \, \varphi_0(u + 1, x, y, v))$$

Proof of Claim: We distinguish two cases:

Case 1: $k = 0$. Then $T^g \vdash \varphi^g$. Hence, by Parikh's theorem, there exists a term $s(u, x, v)$ of $\mathcal{L}_{\mathcal{F},g}$ such that

$$T^g \vdash \forall u \, (\forall x \, \exists y \leq g(x+u+v) \, \varphi_0(u, x, y, v) \to \forall x \, \exists y \leq s(u, x, v) \, \varphi_0(u+1, x, y, v))$$

By Lemma 7, there is $m \in \omega$ such that $T^g \vdash s(u, x, v) < g^m(u + x + v + \sigma(s))$. By induction on z it can be proved that

$$T^g \vdash g^u(x + z) = y_1 \wedge g^{u+z}(x) = y_2 \to y_1 \leq y_2$$

and, thus, if $\tau_0 = m + \sigma(s)$ then $\tau_0 \in \Lambda_0$ and the result follows.

Case 2: $k \geq 1$. Since $[T^g, (\Pi_2^{\mathcal{F},g}, \Lambda_0)\text{--IR}]_k \vdash \varphi^g$ and φ^g is a $\Pi_2^{\mathcal{F},g}$–formula, by Theorem 11, $T^g + (\Pi_2^{\mathcal{F},g}, \Lambda_0)\text{--IR}$ also proves φ^g. It follows from Proposition 10 that there exist $t_1, \ldots, t_n \in \Lambda_0$ such that

$$T^g + \{\forall x \, \exists y \, (g^{t_j}(x) = y) : \; j = 1, \ldots, n\} \vdash \varphi^g.$$

Let $r = t_1 + \cdots + t_n$. Then, by part (4) of Proposition 5, $T^g + \forall x \, \exists y \, (g^r(x) = y)$ extends $T^g + \{g^{t_j} \text{ is total} : \; j = 1, \ldots, n\}$. Let h be a new unary function symbol and let T^h be the extension of T^g obtained by adding to T^g the axiom $\forall x \, (g^r(x) = h(x))$. Then $T^h \vdash \varphi^g$ and T^h is conservative over T^g.

By Proposition 5, h is an iterable nondecreasing function over T^h and $T^h \vdash \forall x \, (g(x) \leq h(x))$. Therefore, h is a dominating function over T^h and T^h extends $I\Sigma_0^{\mathcal{F},g,h}$. By Parikh's theorem, there is a term $s(u, x, v)$ of $\mathcal{L}_{\mathcal{F},g,h}$ such that

$$T^h \vdash \forall u \, (\forall x \, \exists y \leq g(x+u+v) \, \varphi_0(u, x, y, v) \to \forall x \, \exists y \leq s(u, x, v) \, \varphi_0(u+1, x, y, v))$$

and, by Lemma 7, there is $m \in \omega$ such that $T^h \vdash s(u, x, v) < h^m(u+x+v+\sigma(s))$. Recall that $T^h \vdash h^u(x + z) = y_1 \wedge h^{u+z}(x) = y_2 \to y_1 \leq y_2$ and, thus, if $\sigma_0 = m + \sigma(s)$ then $\sigma_0 \in \Lambda_0$ and $T^h + \forall x \, \exists y \, (h^{\sigma_0}(x) = y)$ proves

$$\forall u \, (\forall x \, \exists y \leq g(x+u+v) \, \varphi_0(u, x, y, v) \to \forall x \, \exists y \leq h^{\tau_0}(u+x+v) \, \varphi_0(u+1, x, y, v))$$

Using part (7) of Proposition 5, we can prove, by $\Sigma_0^{\mathcal{F},g,h}$–induction, that

$$T^h \vdash h^z(x) = y \leftrightarrow g^{r \cdot z}(x) = y$$

As a consequence, $T^h + \forall x \, \exists y \, (h^{\sigma_0}(x) = y)$ proves

$$\forall u \, (\forall x \, \exists y \leq g(x+u+v) \, \varphi_0(u, x, y, v) \to \forall x \, \exists y \leq g^{r \cdot \sigma_0}(u+x+v) \, \varphi_0(u+1, x, y, v))$$

Hence, putting $\tau_0 = r \cdot \sigma_0 \in \Lambda_0$, the result follows concluding the proof of Claim.

Let $\mathfrak{A} \models T + I\Sigma_1^{\mathcal{F}}$ and $c \in \mathfrak{A}$ such that $\mathfrak{A} \models \varphi(0, c)$. We shall show that $\mathfrak{A} \models \varphi(t, c)$. Let $\psi(x, y, c) \in \Sigma_0^{\mathcal{F}}$ the formula

$$\forall z \leq x \exists w \leq y \, (\varphi_0(0, z, w, c) \wedge y = w + f(x)).$$

Then $\mathfrak{A} \models \forall x \exists y \psi(x, y, c)$ and the formula $\psi(x, y, c) \wedge \forall z < y \neg \psi(x, z, c)$ defines a total nondecreasing function $H : \mathfrak{A} \to \mathfrak{A}$. There is a $\Sigma_0^{\mathcal{F}}$ formula, that we denote by $H^z(x) = y$, defining the iteration of H and, since $\mathfrak{A} \models I\Sigma_1^{\mathcal{F}}$, we have

$$\mathfrak{A} \models \forall x \, \forall z \, \exists y \, (H^z(x) = y).$$

Let $\theta(u, v)$ be the following $\Pi_1^{\mathcal{F}}$ formula:

$$u > t \vee \forall x \, \forall y_1 \, \left[H^{\tau_0^u}(x + u + v) = y_1 \to \exists y \le y_1 \, \varphi_0(u, x, y, v) \right].$$

Since $\mathfrak{A} \models \forall x \, \exists y \, (H(x) = y)$, by definition of $\theta(u, v)$ we have $\mathfrak{A} \models \theta(0, v)$. Let us show that $\mathfrak{A} \models \forall u \, (\theta(u, v) \to \theta(u + 1, v))$.

Pick $a, b \in \mathfrak{A}$ such that $\mathfrak{A} \models a \le t \wedge \theta(a, b)$. Then, the formula $H^{\tau_0^a}(x) = y$ defines a total nondecreasing function in $\mathfrak{A}$ and we can use it to get an expansion of $\mathfrak{A}$ to a model $\mathfrak{A}^g$ of T^g such that $\mathfrak{A}^g \models \forall x \, \exists y \le g(x + a + b) \, \varphi_0(a, x, y, b)$. By part (7) of Proposition 5, we can prove by $\Sigma_0^{\mathcal{F},g}$–induction on z that

$$\mathfrak{A}^g \models \forall z \le \tau_0 \, [g^z(x + a + b) = H^{\tau_0^a \cdot z}(x + a + b)]$$

In particular, $\mathfrak{A}^g \models \forall x \, (g^{\tau_0}(x+a+b) = H^{\tau_0^a \cdot \tau_0}(x+a+b))$ and, as a consequence, $\mathfrak{A}^g \models T^g + \forall x \, \exists y \, (g^{\tau_0}(x) = y)$. Hence, by the Claim, we conclude that $\mathfrak{A}^g \models \forall x \, \exists y \le g^{\tau_0}(x + a + b) \, \varphi_0(a + 1, x, y, b)$ and, therefore, $\mathfrak{A} \models \theta(a + 1, b)$.

We have shown that $\mathfrak{A} \models \theta(0, v) \wedge \forall u \, (\theta(u, v) \to \theta(u + 1, v))$, and we know that $\mathfrak{A} \models I\Pi_1^{\mathcal{F}}$ (because $I\Sigma_1^{\mathcal{F}} \equiv I\Pi_1^{\mathcal{F}}$), so, $\mathfrak{A} \models \forall u \, \theta(u, b)$. In particular, since

$$\mathfrak{A} \models \theta(t, v) \to \forall x \, \exists y \le H^{\tau_0^t}(t + x + v) \, \varphi_0(t, x, y, v),$$

we conclude $\mathfrak{A} \models \varphi(t, v)$.

3   Main Result

We are now ready to obtain the main results. Firstly, we need a version of Theorem 12 in the language of first–order Arithmetic.

Lemma 13. $I\Sigma_1$ *extends* $I\Delta_0 + (\Sigma_2, \mathcal{K}_2)$–*IR.*

Proof. Let $\mathfrak{A} \models I\Sigma_1$ and $\varphi(x) \in \Sigma_2$ such that

$$(\bullet) \quad I\Delta_0 + (\Sigma_2, \mathcal{K}_2)\text{–IR} \vdash \varphi(0) \wedge \forall x \, (\varphi(x) \to \varphi(x + 1)).$$

We must show that for every $\delta(u) \in \Sigma_2^-$,

$$(\star) \quad \mathfrak{A} \models \forall x_1 \, \forall x_2 (\delta(x_1) \wedge \delta(x_2) \to x_1 = x_2) \to \forall x \, (\delta(x) \to \varphi(x)).$$

By $(\bullet)$ there exist formulas $\varphi_1(x), \ldots, \varphi_r(x) \in \Sigma_2$ and $\delta_1(x), \ldots, \delta_r(x) \in \Sigma_2^-$ such that $I\Delta_0$ plus the sentences

$$\alpha_j : \quad \forall x_1 \, \forall x_2 (\delta_j(x_1) \wedge \delta_j(x_2) \to x_1 = x_2) \to \forall x \, (\delta_j(x) \to \varphi_j(x))$$

$(j = 1 \ldots, r)$ proves $\varphi(0) \wedge \forall x \, (\varphi(x) \to \varphi(x+1))$. More precisely for each $j \leq r$,

$$I\Delta_0 + \bigwedge_{1 \leq i < j} \alpha_i \vdash \varphi_j(0) \wedge \forall x \, (\varphi_j(x) \to \varphi_j(x+1)),$$

and $I\Delta_0 + \bigwedge_{i=1}^{r} \alpha_i \vdash \varphi(0) \wedge \forall x \, (\varphi(x) \to \varphi(x+1))$.

Let $E = \{j : 1 \leq j \leq r, \, \mathfrak{A} \models \neg \exists x \delta_j(x)\}$ and, for each $j \in E$, let $\theta_j(x,y) \in \Pi_0$ such that $\neg \exists x \, \delta_j(x)$ is equivalent to $\forall x \exists y \, \theta_j(x,y)$. Let m the cardinal of E and let $\mathcal{F} = \{f_1, \ldots, f_m, f\}$ a set of new unary function symbols. From the set of Σ_0 formulas $\Gamma = \{\theta_j(x,y) : \, j \in E\}$, we define a theory T as in Remark 2. Let $\mathcal{L}(\mathfrak{A})$ denote the language obtained by adding to $\mathcal{L}$ a constant symbol $\underline{a}$, for each $a \in \mathfrak{A}$. Put $T' = T + D_{\Pi_1}(\mathfrak{A})$, where $D_{\Pi_1}(\mathfrak{A})$ is the Π_1–diagram of $\mathfrak{A}$. Let Λ_0 be the set of closed terms of $\mathcal{L}(\mathfrak{A})$ containing only constants of the form $\underline{a}$ for $a \in \mathcal{K}_2(\mathfrak{A})$. Then $\mathfrak{A}$ has a natural expansion $\mathfrak{A}_\mathcal{F}$ to the language $\mathcal{L}_\mathcal{F} \cup \mathcal{L}(\mathfrak{A})$ such that $\mathfrak{A}_\mathcal{F} \models T' + I\Sigma_1^\mathcal{F}$. By Proposition 12, $\mathfrak{A}_\mathcal{F} \models T' + (\Sigma_2^\mathcal{F}, \Lambda_0)$–IR. Given $\delta(x) \in \Sigma_2^-$, we can distinguish several cases:

If $\mathfrak{A} \models \neg \exists x \, \delta(x)$ then $(\star)$ obviously holds. On the other hand, if $\mathfrak{A} \models \neg \forall x_1 \, \forall x_2 (\delta(x_1) \wedge \delta(x_2) \to x_1 = x_2)$, since this is a Σ_2–sentence and T' extends $D_{\Pi_1}(\mathfrak{A})$, we have that $T' \vdash \neg \forall x_1 \, \forall x_2 (\delta(x_1) \wedge \delta(x_2) \to x_1 = x_2)$. So,

$$T' \vdash \forall x_1 \, \forall x_2 (\delta(x_1) \wedge \delta(x_2) \to x_1 = x_2) \to \forall x \, (\delta(x) \to \varphi(x)).$$

In that way $(\star)$ holds again. We must deal with a last case: $\mathfrak{A} \models \exists! x \, \delta(x)$.

Then there exists $d \in \mathcal{K}_2(\mathfrak{A})$ such that $\mathfrak{A} \models \delta(d)$ and $\underline{d} \in \Lambda_0$. In order to verify $(\star)$ it is enough to show that $T' + (\Sigma_2^\mathcal{F}, \Lambda_0)$–IR $\vdash \varphi(\underline{d})$.

We prove, by induction on j, that for all $j = 1, \ldots, r$, $T' + (\Sigma_2^\mathcal{F}, \Lambda_0)$–IR $\vdash \alpha_j$. Let $j \leq r$, and assume that $T' + (\Sigma_2^\mathcal{F}, \Lambda_0)$–IR $\vdash \bigwedge_{1 \leq i < j} \alpha_i$. Then

$$(\bullet)_j \quad T' + (\Sigma_2^\mathcal{F}, \Lambda_0)\text{–IR} \vdash \varphi_j(0) \wedge \forall x \, (\varphi_j(x) \to \varphi_j(x+1)).$$

If $j \in E$ or $\mathfrak{A} \models \neg \forall x_1 \, \forall x_2 (\delta_j(x_1) \wedge \delta_j(x_2) \to x_1 = x_2)$ then, reasoning as in previous cases, we conclude that $T' \vdash \alpha_j$. If $\mathfrak{A} \models \exists! x \, \delta_j(x)$, then there exists $b \in \mathcal{K}_2(\mathfrak{A})$ such that $\mathfrak{A} \models \delta_j(b)$ and $\underline{b} \in \Lambda_0$. Using $(\bullet)_j$ we get $T' + (\Sigma_2^\mathcal{F}, \Lambda_0)$–IR $\vdash \varphi_j(\underline{b})$. As a consequence, $T' + (\Sigma_2^\mathcal{F}, \Lambda_0)$–IR $\vdash \exists x \, (\delta(x) \wedge \varphi_j(x))$, and it follows that $T' + (\Sigma_2^\mathcal{F}, \Lambda_0)$–IR $\vdash \alpha_j$, as required.

We have proved that $T' + (\Sigma_2^\mathcal{F}, \Lambda_0)$–IR $\vdash \bigwedge_{j=1}^{r} \alpha_j$; hence

$$T' + (\Sigma_2^\mathcal{F}, \Lambda_0)\text{–IR} \vdash \varphi(0) \wedge \forall x \, (\varphi(x) \to \varphi(x+1))$$

It follows that $T' + (\Sigma_2^\mathcal{F}, \Lambda_0)$–IR $\vdash \varphi(\underline{d})$ and, as a consequence, $(\star)$ holds.

Our last theorem extends a previous conservation result obtained in [4] and, as a direct corollary, yields the characterization of the p.t.c.f. of $I\Pi_2^-$.

Theorem 14. *$I\Pi_2^-$ is Π_3–conservative over $I\Sigma_1$.*

Proof. Let θ be a Π_3 sentence provable in $I\Pi_2^-$. Then $I(\Sigma_2, \mathcal{K}_2) \vdash \theta$ by Lemma 2 and $I\Sigma_1^- + (\Sigma_2, \mathcal{K}_2)$–IR $\vdash \theta$ by Proposition 3. We need the following fact:

Claim. $I\Sigma_1^- + (\Sigma_2, \mathcal{K}_2)\text{–IR} \equiv I\Sigma_1^- + (I\Delta_0 + (\Sigma_2, \mathcal{K}_2)\text{–IR})$

Proof of Claim: Each axiom of $I\Sigma_1^-$ is a Σ_3 sentence, so it is enough to prove that for every $\sigma_0(u) \in \Pi_2$,

$$[I\Delta_0, (\Sigma_2, \mathcal{K}_2)\text{–IR}] + \exists u\, \sigma_0(u) \text{ extends } [I\Delta_0 + \exists u\, \sigma_0(u), (\Sigma_2, \mathcal{K}_2)\text{–IR}].$$

Assume $I\Delta_0 + \exists u\, \sigma_0(u) \vdash \varphi(0) \wedge \forall x\, (\varphi(x) \to \varphi(x+1))$, with $\varphi(x) \in \Sigma_2$, and let $\psi(x, u) \in \Sigma_2$ be $\sigma_0(u) \to \varphi(x)$. Then, $I\Delta_0 \vdash \psi(0, u) \wedge \forall x\, (\psi(x, u) \to \psi(x+1, u))$, and, therefore, $[I\Delta_0, (\Sigma_2, \mathcal{K}_2)\text{–IR}] \vdash U_\delta \to \forall x\, (\delta(x) \to \psi(x, u))$, where $\delta(x)$ is in Σ_2^- and U_δ denotes the sentence $\forall x_1 \forall x_2(\delta(x_1) \wedge \delta(x_2) \to x_1 = x_2)$. Then it holds that $[I\Delta_0, (\Sigma_2, \mathcal{K}_2)\text{–IR}]$ also proves

$$\exists u\, \sigma_0(u) \to (U_\delta \to \forall x\, (\delta(x) \to \varphi(x)))$$

and so $[I\Delta_0, (\Sigma_2, \mathcal{K}_2)\text{–IR}] + \exists u\, \sigma_0(u) \vdash U_\delta \to \forall x\, (\delta(x) \to \varphi(x))$, as required.

It follows from this Claim and Lemma 13 that $I\Sigma_1$ extends $I\Sigma_1^- + (\Sigma_2, \mathcal{K}_2)\text{–IR}$ and, therefore, $I\Sigma_1 \vdash \theta$.

Corollary 15. *The class of provably total computable functions of $I\Pi_2^-$ is the class of primitive recursive functions.*

Acknowledgements. This work was partially supported by grant MTM2008-06435 of Ministerio de Ciencia e Innovación, Spain.

References

1. Avigad, J.: Saturated models of universal theories. Annals of Pure and Applied Logic 118, 219–234 (2002)
2. Beklemishev, L.D.: Induction rules, reflection principles and provably recursive functions. Annals of Pure and Applied Logic 85(3), 193–242 (1997)
3. Beklemishev, L.D.: A proof–theoretic analysis of collection. Archive for Mathematical Logic 37(5-6), 275–296 (1998)
4. Beklemishev, L.D.: Parameter free induction and provably total computable functions. Theoretical Computer Science 224, 13–33 (1999)
5. Buss, S.: The Witness Function Method and Provably Recursive Functions of Peano Arithmetic. In: Westertahl, D., Prawitz, D., Skyrms, B. (eds.) Proceedings of the 9th International Congress on Logic, Methodology and Philosophy of Science, pp. 29–68. Elsevier, North–Holland, Amsterdam (1994)
6. Cordón–Franco, A., Fernández–Margarit, A., Lara–Martín, F.F.: On conservation result for parameter–free Π_n–induction. In: Cégielski, P. (ed.) Studies in Weak Arithmetics, pp. 49–97. CSLI Publications, Stanford (2010)
7. Hájek, P., Pudlák, P.: Metamathematics of First–Order Arithmetic. Perspectives in Mathematical Logic. Springer (1993)
8. Kaye, R., Paris, J., Dimitracopoulos, C.: On parameter free induction schemas. The Journal of Symbolic Logic 53(4), 1082–1097 (1988)
9. Sieg, W.: Herbrand Analyses. Archive for Mathematical Logic 30, 409–441 (1991)

What is Turing's Comparison between Mechanism and Writing Worth?

Jean Lassègue[1] and Giuseppe Longo[2]

[1] Centre de Recherche en Épistémologie Appliquée (CREA), École Polytechnique,
32, boulevard Victor, 75015 Paris, France
{jean.lassegue,giuseppe.longo}@polytechnique.edu
[2] Centre International de Recherches en Philosophie, Lettres, Savoirs (CIRPHLES),
Département de philosophie, École Normale Supérieure, 45 rue dUlm,
75005 Paris, France
longo@ens.fr

Abstract. In one of the many and fundamental side-remarks made by Turing in his 1950 paper (The Imitation Game paper), an analogy is made between Mechanism and Writing. Turing is aware that his Machine is a writing/re-writing mechanism, but he doesn't go deeper into the comparison. Striding along the history of writing, we shall hint here at the nature and the role of alphabetic writing in the invention of Turing's (and today's) notion of computability. We shall stress that computing is a matter of alphabetic sequence checking and replacement, far away from the physical world, yet related to it once the role of physical measurement is taken into account. Turing Morphogenesis paper, 1952, provides the guidelines for the modern analysis of "continuous dynamics" at the core Turing's late and innovative approach to bio-physical processes.

1 Introduction

In his 1950 philosophical article, Turing rather offhandedly used a comparison between mechanism and writing he didn't take time to develop: "(Mechanism and writing are from our point of view almost synonymous)" [28, p. 456]. The relationship between the two terms is far from being straightforward though. Mechanism seems to be understood by Turing as a branch of physics and, in a broader perspective, as the ideal deterministic world-view natural sciences should pursue. On the other hand, even if writing could be stretched to fit into linguistics it is neither a type of knowledge nor a scientific paradigm but rather a technology used for recording data in many different areas. Therefore how could mechanism and writing be compared? Turing's remark could of course be considered as a simple digression in the course of his article and the sentence just quoted is indeed put into brackets. But it is worth trying to stick to this comparison and follow it as far as it can lead us to. This is the goal of the next few pages. First of all, we should refine the comparison between the two terms to make it clearer.

S.B. Cooper, A. Dawar, and B. Löwe (Eds.): CiE 2012, LNCS 7318, pp. 450–461, 2012.
© Springer-Verlag Berlin Heidelberg 2012

2   Mechanism as a Scientific Paradigm

Resting upon the intellectual development of Turing himself, we know firstly that he extended Hilbert's formalist program as far as to come up against one of its inner limitations: the negative result of the halting problem in 1936 is a paradigmatic example of a limitation in decidability as well as the birth certificate of computer science. Now, as hinted below, there is a deep and non-trivial relationship between (computational) undecidability and (deterministic) unpredictability in mathematical physics. Secondly, Turing managed to work out a new kind of non-predictive determinism in his 1952 article by studying symmetry breaking and non-linear processes in morphogenesis. In this article, Turing puts forward the notion of an "exponential drift" - what is nowadays called "sensitivity to initial conditions". Because of this "drift", a fluctuation / perturbation below measurement, may be amplified over time and may induce an *unpredictable*, yet major (i.e., *observable*) change in the genesis of forms. As is well known since Poincaré, but not much studied since then (with a few exceptions: Hadamard, Pontyagrin, Birkhoff...), unpredictability pops out as soon as non-linearity expresses dynamical interactions between components of a system, such as in the action/reaction/diffusion system Turing studies in 1952 (of course, technically, he works at the solutions given by the linear approximation of the system, yet he is aware of, and greatly interested by, the key properties of non-linearity). Of course, unpredictability is at the *interface* between the mathematical determination, e.g., equations, evolution functions..., and the physical processes which is being modeled by these written equations, functions etc..

In reference to this unpredictability, Turing makes the following observation in his 1950 paper, concerning his Discrete State Machine (DSM, as he then calls his Logical Computing Machine), "...[in a DSM], it is always possible to predict all future states... This is reminiscent of Laplace's view... The prediction which we are considering is, however, rather nearer to practicability than that considered by Laplace" [28, p. 440].

In fact, he explains, the Universe and its processes are subject to the exponential drift and gives the following example: "The displacement of a single electron by a billionth of a centimetre at one moment might make the difference between a man being killed by an avalanche a year later, or escaping.". On the contrary, and here lies the greatest effectiveness of his approach, "...It is an essential property of... [DSMs] that this phenomenon does not occur. Even when we consider the actual physical machines instead of the idealized machines...", prediction is possible, [28, p. 440]. Of course, stresses Turing, there may be a program which is so long that it is hard (practically impossible) to predict its behavior; yet, this is a practical issue, a very different one from the core theoretical property, that is deterministic unpredictability in non-linear dynamical systems, due in particular to the exponential drift.

In view of its developments, which originated from Turing's DSM, computer science appears to be predictive determinism made real, and no one should be allowed to belittle what an incredible feat it is to have made it possible since many (and perhaps most) physical processes precisely do not belong to predictive determinism but to the non-predictive one Turing explored in his last years. From this point of view, computer science came not so much as a surprise but as a kind of miracle, even if it became more and more clear that mechanism cannot be extended to most physical processes in nature (the "frictionless" simple pendulum and pulsars are some of the known exceptions), as was established as early as Poincaré's famous example in celestial mechanics concerning The Three Body Problem: the sun and one planet give a predictable Keplerian orbit, one more planet and the non-linear interactions let the system go, at equilibrium, along unstable orbits (almost always). Therefore, what should be kept in mind in Turing's comparison is not "mechanism" as such but the opposition between two paradigms in science: predictive (mechanistic) determinism on the one hand and nonpredictive determinism on the other, the two of which Turing explored with close scrutiny and in which he produced fundamental results.

In the case of writing, a similar remark should be done. It is not writing proper that the attention should be focused on, but the opposition between explicit and implicit processes in the recording of data. It will therefore be argued below that computer science inherits a very specific writing structure which gives a compelling rationale to the comparison put forward by Turing: *explicit processes in mechanism and writing are of the same nature.* But in order to reach this conclusion, we have to hark back to the history of writing.

3   The Long History of Writing

3.1   Writing Numbers and Languages

Turing does not specify which kind of writing should be compared to mechanism and we are left only with conjectures. Let us therefore start with the most primitive written system in the West as described in [24]. If we go back to this primitive use of writing, we find that it was first used in Mesopotamia around -3100 B.C. to *count* specific goods to be stored and that the recording of natural languages came only in second, at a later stage. The counting system itself which was utilized was much older.

Prehistoric Counting System. In order to record quantities of goods which were presumably stocked for conservation and redistribution, Mesopotamian accountants made use of a pre-historical system of tokens that can be traced as far back as 7500 B.C. This system of tokens had three remarkable features. Firstly, tokens were geometrical forms, that bore no resemblance to the good which they referred to. Secondly, these tokens were used to refer to specific goods: each kind of geometrical token would refer to a single kind of good. Therefore, the system implied a great variety of tokens, in fact as many as the kinds of goods to be accounted for. Thirdly, tally was made possible through sheer repetition: using as many tokens as the number of pieces would indicate the quantity of the good in stock. By its geometrical form, each token was bestowed a qualitative meaning referring to the type of good and by its repetition, a quantitative one, referring to the number of pieces under consideration. It should be emphasized that this system is purely graphical and does not presuppose any use of natural language, even if it was presumably accompanied by verbal expression.

Involvement of Language. Later on, the tally system evolved and verbal language became necessary to keep track of the name of the donators. The written recording of words in natural language capable of referring to names would follow the numerical system which was purely graphical. But the transformation of the writing system had a consequence on the writing of numbers in return. It was not thanks to repetition that the tokens were now capable of referring to numbers: it was through their *names* in natural language that they could play the role of predicates[1]. This meant a fantastic economy of means since from then on, no repetition of tokens was needed to express plurality and a single token interpreted as a quantity could now avoid the tedious and often mistaken repetition of the same token. But in return, it would also induce the use of natural language as the foundation for the whole counting system: writing was not entirely graphical anymore and would then oscillate between a visible, graphical part and an invisible, verbal one (for example the graphical token "ÅA" would *mean* five units), the relationship between the two being subject to technical innovations all along the history of writing.

3.2 Three Technical Innovations in the History of Verbal Writing

Three of these innovations are worth mentioning in order to specify in what sense mechanism and writing can be considered as synonyms.

Phonograms. As we just noticed, as soon as it became necessary to go beyond graphical subitizing in order to keep a numerical track of what was in stock, *speech* was used in the up to then seemingly silent and only graphical process of recording. Phonograms (addition of syllables to form a new word like in a rebus) were devised and the way tokens standing for numbers and those standing for goods was therefore reorganized.

A phonogram uses the *first* syllable of a word chosen in advance (let us call is a "master-word") as a part of its own composition.[2] In order to compose a new phonogram from "master-words", the *first* syllable must therefore be recognized as such and be used at some place (the first or any other one fixed in advance) in the new word. This process follows therefore a numerical (ordinal) pattern in which the places (first, second and so on) taken by the syllables are crucial. The disassociation of a syllable (recognized as the first one in the master-word used as

---

[1] Cf. [25, pp. 162–167]: "Remarkably, no new signs were created to express abstract numbers. Instead, the signs referring to measures of grain simply acquired another meaning: the wedge standing for a small unit of grain became '1', and the circle representing a large unit of grain became '10'."

[2] E.g., [25, pp. 162–167] uses as an example the composition of the modern phonogram 'Lucas' by the composition of the token for "Lu" (retrieved from the first syllable of the Sumerian word for mouth "Lu") and of the token for "Cas" (retrieved from the first syllable of the Sumerian word for man "Ka").

blueprint) plays the same role as a numerical predicate which is first dissociated and then placed next to a word to compose a sentence (like in the expression '3 jars'). This writing system implies that one has to know in advance which are the "master-words" new phonograms can be composed from: any phonographic composition presupposes that the speaker has a knowledge of the set of syllables used in the "master-words" of the language.

Alphabets. Contrary to phonograms the construction of which depends on "master-words" used as blueprints for syllables, alphabets seem at first sight to form a finite list of tokens that record the sounds used for the formation of all the words of a given lexicon. Graphically speaking though, two different types of alphabets must be distinguished for they do not record the same phonological realities: a consonantal alphabet or "abjad"[3] tends to record the *syllables* of a given language and therefore carries the semantic value attached to its syllables whereas a vocalic-consonantal alphabet or "alphabet"[4]" in the strict sense of the term tends to record the *phonemes* of a given language, apart from any semantic consideration. Let us make this difference clear.

Consonantal Alphabet. A consonantal alphabet or "abjad", is a type of writing system where each symbol stands for a consonant, leaving the reader to supply the appropriate vowel and complete the syllable. It is particularly fit for languages (like Hebrew or Arabic) that have only a few vowels, for the semantic indetermination regarding the completion of the syllable is rather limited and easily supplemented by a reader, provided that he or she is knowledgeable in the language which is spoken. In this type of writing, the knowledge of the language is therefore a mandatory prerequisite, for no word can be read if the reader is not able to supply the vowels.

Vocalic-consonantal Alphabet. Alphabets in the strict sense of the word are systems of writing that do not record syllables but phonemes, i.e., sounds that are considered as having the same *function* in speech but do not have any meaning in themselves. From this point of view, abjads and alphabets, even though they are sometimes both called alphabets, are not based on the same principle (cf. [10]) and they differ on at least three features. Firstly, alphabets are not restricted to languages the structure of which possesses only a few vowels: any language can be written with an alphabet because what is recorded is just phonemes, that is phonological realities and not syllables. Secondly, no previous knowledge of the language is required: anybody who can read alphabetically can read a foreign language which is alphabetically written, even without understanding a word of what is written (try with Turk or Vietnamese if you are not familiar with these languages or just read something when you are tired until you eventually realize you have been reading without understanding a word of what was written... This is just impossible in Arabic or Hebrew). Thirdly, and as a consequence, no intervention is required from the reader, which means that *the process of alphabetical*

---

[3] Standing for the first four letters in Arabic writing: 'Alif, Bā', Ǧīm, Dāl.

[4] Standing for the first two letters in Greek writing: Alpha, Bêta.

reading can be transferred to a machine. This is the true reason for the comparison Turing put forward in his 1950 article: mechanism and alphabetic writing only involve entirely explicit processes of pattern recognition of characters that have an objective value in the sense that they do not depend either on the type of language (natural or artificial language as long as it is written) or on the type of reader (human or machine).

The oldest example of an alphabet is very well known: it is the Greek one, which appeared in the 8th century B.C. and from which all other alphabets derive. Its linguistic originality has been commented upon through and through but what is most striking from our point of view is just one fact: *alphabetical reading is potentially mechanisable.* That is why the Hilbertian formalist program as well as the works by Schönfinkel and Curry (see below) and all formalisms in the 30's *inherit a Greek alphabetical-mechanistic structure* which is the core of computability, after a long history we cannot fully describe in details here. This alphabetical-mechanistic structure is not, so to say, the alpha and omega of deterministic science, as Turing was well aware in his 1952 article, both in mathematical physics and in linguistics. For it is true that explicit processes deprived of any semantic structure lack what is most fundamental in spoken languages as well as in mathematical (and biological) structures: their plasticity, their versatility and, nonetheless, their incredible stability. But one has to make clear what this deterministic structure exactly is to be able to go beyond.

4 The Alphabetic Combinators

In the 1920's, Schönfinkel and Curry independently proposed an algebra of signs for expressing and computing logical predicates (cf. [8,9]) for a technical introduction and an historical account). Curry's formalism, which is still a reference, is based on two "combinators", K and S, that *act on (combine) signs* according to the following rules:

$$(KX)Y > X \quad \text{read: ``}KXY \text{ reduces to } X\text{''}$$
$$SXYZ > XZ(YZ) \text{ read: ``}SXYZ \text{ reduces to } XZ(YZ)\text{''}$$

By convention, association is intended to the left: $XYZ \equiv (XY)Z$ (the first sequence is *identical* to the second).

Surprisingly enough, by combining these "manipulators of signs" in a type-free way (there are no restrictions on what can be applied to what), one can compute a large class of functions from sequences of signs to sequences of signs. For example, $I \equiv SKK$ computes the identity function:

$$IX \equiv SKKX > KX(KX) > X,$$

while $SIIX > XX$, or $(S(K(SI))(S(KK)I))XY > YX$, where, of course, X, Y, $Z \dots$ are arbitrary signs or *sequences of signs* (within a parenthesis!). Truth values may also be encoded: $T \equiv K$, $F \equiv KI$. And, by a smart coding of numbers as combinators (take $0 \equiv KI$ and... keep going in a non-obvious way, see the references or below), one can compute all the Turing computable functions. Let's stress that Curry's Combinatory Logic, at the origin of the formal computability of the '30s, is a pure *calculus of alphabetic signs*. Uninterpreted, or meaningless (and this is crucial), sequences of alphabetic signs are formally copied, erased, iterated in a ... *potentially mechanisable way*, following Hilbert's request for formal systems. There is even no intended mathematical meaning, as there was no "interpretation" of this calculus on algebraic or geometric structures at the time. The point of course is to understand what "interpretation" means. Yet, this calculus is where the conception of the modern computing machine began: by a "meaningless" action of signs over signs, formally codified by axioms and/or rules (the rules for K and S above). All what is needed is a "pattern matching of signs" (one should better say: sequence matching, i.e., checking that two sequences are identical) and *sequence replacement* (replace a sequence by another, according to the rules).

Shortly later, Church, also in Princeton, invented the λ-calculus [6]. This calculus has a more "mathematical flavor", since it provides a purely formal theory of functions, based on "functional abstraction" over variables ($\lambda y.X$ below is the functional analogous of $\{y/X\}$ in Set Theory). Consider a string of signs, X, possibly containing a (free) variable y, then write $\lambda y.X$ as the function that operationally behaves as follows, by "reduction" again, ">":

$$(\lambda y.X)Z > [Z/y]X$$

where $[Z/y]$ is short for Z replacing y in all free occurrences of y in X (a variable y is bound or "not free" when it occurs in the field of a $\lambda y.(...)$, cf. [3] for details). No more needs to be said: by borrowing for functions the notion of "abstraction" used for sets, one obtains an amazingly powerful calculus with just one axiom (and a few obvious rules for reduction, ">"). Integer numbers, for example, are easily coded by $0 \equiv \lambda x.\lambda y.y$ and $n \equiv \lambda x.\lambda y.x(x...(xy)...)$, where x occurs n times. Curry's and Church's alphabetic calculi are equivalent modulo a simple translation (yet, some subtleties are required, cf. [12,3]).

When these calculi and Turing's computable functions, jointly with Kleene's system [13], were proved equivalent as for their number-theoretic expressiveness (they compute the same class of functions), λ-calculus played a pivotal role: the proofs were given under the form of "equivalent λ-expressiveness". The Turing-Church Thesis was then justified by these rather surprising equivalences: it was fair to assume that all formal systems a la Hilbert could at most define the same class of functions, the Turing computable ones. The meaningless manipulation of alphabetic strings is at the core of these formalisms, following Hilbert's notion of a formal system. Type Theory [7,11] made the link between *propositions* of Formal Logic and *types* that one can associate to typable terms of Curry-Church calculi.

Are these purely formal-alphabetic systems "in the (physical) world"? They are not. As we said earlier on, they are grounded on one of the most fantastic (and earliest) abstraction invented by humans: alphabetic writing. They stress the core idea of it: signs have no and must not have meaning. In the alphabetic systems of writing, meaning pops out phonetically: the sound of voice gives meaning to the sequences of letters.

However, let's make a distinction. In the formal systems for computation, a form of meaning is possible: the *operational* one given by the rules and axioms for manipulating them. Consider, say, the $KXY > X$ rule mentioned above. One may say that it gives a "meaning" to K as an *operator* on signs. No meaning yet, in no ordinary, nor mathematical sense, just an operational definition. The non obvious mathematical structures "interpreting" these calculi were invented in the late '60, [26] and were later given a general set-theoretic frame, see ([18,3]) for the type free calculus, and a categorical one, for both the type-free and the typed calculi [27,19,14,1].

The difference of these forms of "meaning" should be clear. The first is the purely mechanical processing of uninterpreted signs: (re-)writing as a mechanism. As explained above, it is the direct result of the amazing abstraction proposed by the Greek invention of an uninterpreted alphabet, at the origin of our modern form of writing. These signs are so meaning independent that they may be used for a game of signs, a combinatorial mechanics which is at the true origin of modern computing: a machine can operate according to the rules for the K and S operators. "Mechanism and writing are from our point of view almost synonymous", says Turing. The so-called operational semantics is just the specification of rules to operate mechanically on signs. Indeed, once Turing made explicit the distinction between software and hardware, a vast area of programming styles were directly based on Lambda Calculus and Combinatory Logic: the functional languages (LISP, Edinburg ML...), the Haskell style's languages (from Haskell B. Curry), which are also functional, but with a more combinatorial flavor. Theory of Programming is largely indebted to "term-rewriting", [5], the discipline that generalized Curry's and Church's calculi by the general analysis of alphabetic "sequence checking and sequence replacement systems of mechanical rules", at the core of Turing's intuition.

The mathematical semantics instead, deals with meaning in a rather different way. One has to interpret those signs and their operations over independently established mathematical structures, deriving their "meaning", in principle, from radically different conceptual constructions. It is like a "translation" into a different language that *per se* bears meaning by different or totally independent conceptual experiences and contexts. The challenge is not obvious. Let's see why. Any sequence of signs can be applied to any other. So XX is well formed, say. A way to understand that X may operate on X is to consider X as a function acting on ... X. Typically, in λ-calculus, $(\lambda x.M)N$ works even when $N \equiv (\lambda x.M)$, whose *mathematical meaning* is that of a function applied to itself. How to construct a non-trivial mathematical universe where this is possible? The universe must be non-trivial as it must contain all integer numbers and, as we know from Turing's equivalences, all computable functions on them. Some non obvious categorical properties allow this, in particular the existence of "reflexive objects", i.e., spaces D such that all the morphisms (functions) on D, call it $D \to D$, may be isomorphically embedded into D. This is set-theoretically impossible, unless

$D = \{1\}$, since $D \to D$ is always provably larger than D. The construction of suitable categories is given in the references: they must have enough morphisms to interpret all computable functions on integers, but must not contain all set-theoretic functions. Here, we just want to stress that that categorical interpretation may be really considered as providing "mathematical meaning": one has to construct, by totally independent tools from the operatorial signs, a mathematical structure (a category, in its precise technical sense), where, typically, this challenging self-application is understood by an embedding of a space of morphisms (functions), $D \to D$, into its own domain of definition, D.

But why should these categories "add meaning"? First, as we said, a translation in another language does add meaning, per se. Second, categories are defined in terms of diagrams, that is of visual objects. Categorical diagrams may be always described by sets of equations, yet only their visual display makes the symmetries manifest, while they are just implicit in the equations. Dualities (like adjunctions, a deep concept in Category Theory) are symmetries and play a major role in the mathematics of categories. So, in the end, meaning is added to the mechanical signs' pushing, by their interpretation into geometric structures, including diagrams and their gestaltic power to make us understand. Symmetries are meaningful to us: they organize space, correlate structures and visualized concepts.

Of course, by adding structural meaning we lose effectiveness. In Combinatory logic and λ-calculus, the reduction operation, ">", is decidable and its transitive and symmetric closure, "=", is semi-decidable. That is, a machine can perform them and, given two terms, M and N, it may decide whether $M > N$ or semi-decide whether $M = N$. Instead, in all proper set-theoretic and categorical models, the interpretation of ">" and "=" are identified and are not semi-decidable. No machine can (semi-)decide the equality of infinite objects, like the functions or morphisms in the categories needed for the semantics of these type-free calculi. In a sense, one goes from the Laplacian universe where dynamics and computations are predictable (one may "decide" the future), as observed by Turing, to richer universes where meaning is grounded on manifolded conceptual experiences (mathematical structures, gestalts...), far away from the purely mechanical notion of "decidability" by an algorithm.

5   The Undecidable

There exist deep and non-obvious correlations between Gödel-Turing undecidability and Poincaré's deterministic unpredictability, at the core of modern non-linear dynamics, cf. [21]. One thing should be clear though: our systems of equations generally yield computable solutions, if any. Writing and solving the most complex systems of (non-)linear equations is a matter of writing and rewriting, that is one has to write them and apply an *algorithm* that we invented to solve them, if any. This is entirely effective, including the result (see below for some exceptions), unless one is crazy enough as to put a non-computable real, Chaitin's Ω, say, as a coefficient or an exponent in the equations. Some computationalists derive from this that "the world is computable" (against Turing's claim: "The displacement of a single electron by a billionth of a centimetre... " may induce major unpredictable/incomputable changes, [28, p. 440]. As he saw perfectly well, it is the question of physical measurement that changes everything, i.e., the relation between mathematics and the physical world by way of a measurement (the "small error... about the size", [28, p. 451]. It is our alphabetic writing of equations and solutions that is computable. And most of the time mathematics "misses the world": unpredictability means that the solution of the intended equational/functional system for a physical process diverges from that process, i.e., very soon it does not describe it anymore, or "prediction becomes impossible", as observed since Poincaré. The key issue is "how to associate a number to a process", whether it is an integer or a computable real, that is, the problem is *measurement*, the only way by which we access to the physical processes.

Take a physical double pendulum: the two equations determining it, if implemented on a digital computer (they have computable solutions, of course), do not allow to predict this chaotic physical process, in view of the inevitable choice of an input number *and* of (Turing's) "exponential drift": two (computable real) numbers within the best measurement interval *may very soon yield radically different trajectories*. In [16], unpredictability is proved for the planetary system (many more equations), at a reasonable astronomical time scale (about 1 million years). There is no way to compute and predict with arbitrary precision processes that are enough sensitive to initial conditions. The approximation in (classical!) measurement has lower bounds of various sorts: thermal fluctuation, gravitational effects, elasticity... Thus most *physical* processes, far from "computing" non-computable *functions*, just do not *(well) define* input-output functions, [20,21]: starting on the same input as an interval provided by measurement, a physical process the *mathematical* representation of which is chaotic, will soon lead to diverging trajectories, thus to very different measurable outputs. As for actual implementations, the Shadowing Lemmas, [22], at most guarantee (and only for hyperbolic non-linear systems), that, *for any* discrete space-time trajectory, *there exists* a continuous one approximating it (and not the converse!). So far, we dealt with the unpredictability of *physical* processes modeled by *computable functions*.

There exist two remarkable exceptions to the computability of the mathematical solutions of interesting mathematical systems. One is given in [4], the other in [23]. These effectively written systems yield *only* non-computable solutions. Yet, by looking closely at their proofs, one sees that this doesn't say much, per se, about the physical world, but only about the "mechanisms of writing" used.

The first proof reduces, by a complex argument, to Turing's halting problem, thus to a purely diagonal argument within the formalism. As for the second, the issue is more delicate and it has been extensively discussed, see [31]. The very elegant system used for computability over real numbers, [30], (there are many and they are not equivalent, a challenge for the extension of Church Thesis to computable "continua"), uses as input values the computable reals. One then shows that *computability implies continuity* (over the Cantor-set topology). Then the proof goes by proving the non-continuity of the unique solution, under computable initial conditions. This non-continuity is clearly a relevant physico-mathematical fact, to be discussed first. But... what is exactly its *physical* sense? Poincaré, from a purely mathematical analysis to the non-analyticity (almost everywhere) of the solution of the Three (planetary) Body Problem, gave a physico-mathematical meaning to the existence of diverging coefficients in the approximating series: minor fluctuations in the initial conditions (below measurement) could mathematically yield divergent trajectories. Predicting is a matter of "pre-dicere" (to say or, better, to write in advance), that is to *measure* firstly, then *compute* by (re-)writing and compare written numerical results and measurement obtained at the end of the process.

We lack, so far, a close analysis of the physical relevance of the incomputability (due to a discontinuity) of the wave equations (by the way, are these as sound, for actual waves, as Newton-Laplace equations for planets?). This should be discussed in reference to actual physical measurement as the only way we have to "extract" numbers from a natural process. We can only say, so far, that the computations lead to a "mathematical singularity" (a discontinuity), an issue *within* the mechanism of term (re-)writing and solving equations.

And so we are back to our founding father. Alphabetic writing and its formal re-writing are effective (they are "synonymous to mechanism"): incomputability is their internal challenge. Each time, the relations between mathematical writing and the world must be closely analyzed, as Poincaré did in reference to physical measurement: there is no way we can deduce the existence of physical super-computers from internal (yet very deep) mathematical games of signs. The fact that the world does not compute does not need to mean that it "super-computes", i.e., it computes non-computable functions, as some wrongly claim. As we said, it may simply mean that it doesn't even (well) define a function. This is so for a double pendulum, the planets or any system to be mathematically described as sensitive to initial conditions. It may also be the case that our mathematical system badly models the world; typically, Navier-Stokes and wave equations badly apply close to borders - what is the relevance of this fact, as for measurement? More generally, what is the relationship between (effective) mathematical writing and physical processes which may be only related by means of physical measurement?

References

1. Asperti, A., Longo, G.: Categories, Types and Structures; Category Theory for the working computer scientist. M.I.T. Press (1991)
2. Bailly G, Longo G.: Mathematics and Natural Sciences: the Physical Singularity of Life. Imperial Coll. Press, London (2011)
3. Barendregt, H.: The Lambda Calculus: its syntax and semantics. North Holland, Amsterdam (1984)
4. Braverman, M., Yampolski, M.: Non-computable Julia sets. Journ. Amer. Math. Soc. 19, 551–578 (2006)
5. Bezem, M., Klop, J.W., Roel de Vrijer, R.: Term Rewriting Systems. Cambridge Univ. Press, Cambridge (2003)

6. Church, A.: A set of postulates for the foundation of Logic. Annals of Math. 33(2), 346–366, 34, 37–54 (1932-1933)
7. Church, A.: A formalisation of the simple theory of types. JSL 5, 56–58 (1940)
8. Curry, H.B., Feys, E.: Combinatory Logic, vol. I. North-Holland, Amsterdam (1968)
9. Curry, H.B., Hindley, J.R., Seldin, J.: Combinatory Logic, vol. II. North-Holland, Amsterdam (1972)
10. Herrenschmidt, C.: Les trois écritures; langue, nombre, code. Gallimard, Paris (2007)
11. Hindley, R.: Basic Simple Type Theory. Cambridge University Press, Cambridge (1997)
12. Hindley, R., Longo, G.: Lambda-calculus models and extensionality. Zeit. für Mathemathische Logik und Grundlagen der Mathematik 26(2), 289–310 (1980)
13. Kleene, S.C.: Lambda definability and recursiveness. Duke Math. J. 2, 340–353 (1936)
14. Lambek, J., Scott, P.J.: Introduction to higher order Categorial Logic. Cambridge Univ. Press, Cambridge (1986)
15. Lassègue's page (downloadable articles), http://www.lassegue.net
16. Laskar, J.: Large scale chaos in the Solar System. Astron. Astrophysics 287, L9–L12 (1994)
17. Longo's page (downloadable articles), http://www.di.ens.fr/users/longo
18. Longo, G.: Set-theoretical models of lambda-calculus: Theories, expansions, isomorphisms. Annals of Pure and Applied Logic 24, 153–188 (1983)
19. Longo, G., Moggi, E.: A category-theoretic characterization of functional completeness. Theoretical Computer Science 70(2), 193–211 (1990)
20. Longo, G.: Critique of Computational Reason in the Natural Sciences. In: Gelenbe, E., et al. (eds.) Fundamental Concepts in Computer Science. Imperial College Press, London (2008)
21. Longo G.: Incomputability in Physics and Biology. To appear in MSCS (2012) (see Longo's web page); Preliminary version: Ferreira, F., Löwe, B., Mayordomo, E., Mendes Gomes, L. (eds.): CiE 2010. LNCS, vol. 6158. Springer, Heidelberg (2010)
22. Yu, P.S.: Shadowing in dynamical systems. Springer (1999)
23. Pour-El, M.B., Richards, J.I.: Computability in analysis and physics. Perspectives in Mathematical Logic. Springer, Berlin (1989)
24. Schmandt-Besserat, D.: How Writing Came About. University of Texas Press, Austin (1992)
25. Schmandt-Besserat, D.: From Tokens to Writing: the Pursuit of Abstraction. Bull. Georg. Natl. Acad. Sci. 175(3), 162–167 (2007)
26. Scott, D.: Continuous lattices. In: Lawvere (ed.) Toposes, Algebraic Geometry and Logic. SLNM, vol. 274, pp. 97–136. Springer, Berlin (1972)
27. Scott, D.: Data types as lattices. SIAM Journal of Computing 5, 522–587 (1976)
28. Turing, A.M.: Computing Machinery and Intelligence. Mind LIX, 433–460 (1950)
29. Turing, A.M.: The Chemical Basis of Morphogenesis. Philo. Trans. Royal Soc. B 237, 37–72 (1952)
30. Weihrauch, K.: Computable analysis. Springer, Berlin (2000)
31. Weihrauch, K., Zhong, N.: Is wave equation computable or computers can beat the Turing Machine? Proc. London Math. Soc. 85(3), 312–332 (2002)

Substitutions and Strongly Deterministic Tilesets

Bastien Le Gloannec and Nicolas Ollinger

Laboratoire d'Informatique Fondamentale d'Orléans (LIFO), Université d'Orléans,
BP 6759, 45067 Orléans Cedex 2, France
{Bastien.Le-Gloannec,Nicolas.Ollinger}@univ-orleans.fr

Abstract. Substitutions generate hierarchical colorings of the plane. Despite the non-locality of substitution rules, one can extend them by adding compatible local matching rules to obtain locally checkable colorings as the set of tilings of finite tileset. We show that for 2 × 2 substitutions the resulting tileset can furthermore be chosen strongly deterministic, a tile being uniquely determined by any two adjacent edges. A tiling by a strongly deterministic tileset can be locally reconstructed starting from any infinite path that cross every line and column of the tiling.

A strongly deterministic tileset is a finite set of Wang tiles, square tiles with colored edges, having the property that, for any two adjacent edges, no two tiles share the same pair of colors. This generalization of Kari's NW-deterministic tilesets [1], introduced to study dynamical properties of cellular automata, was introduced by Kari and Papasoglu [2] who constructed a strongly deterministic aperiodic tileset by enrichment of Robinson's aperiodic tileset [3]. The result was more recently extended by Lukkarila [4] who proved that the Domino Problem remains undecidable for strongly deterministic tilesets. Strong determinism adds to the locally checkable property of tilings the ability to recover uniquely from finite holes, the complete description of the whole tiling being encoded into every 8-connected infinite path that cross every line and column of the tiling. Moreover, the tiling can be reconstructed, that is recomputed, by iteratively applying local rules. Notice that this notion is different from the robustness to error introduced by Durand et al. [5], albeit strong determinism can be used as a tool to built robust tilesets. See Lukkarila [4] for links with self-healing in self-assembly.

A substitution rule associates a finite pattern of letters to every letter. Bigger and bigger patterns are obtained by iterating the rule, leading to a notion of limit set: colorings of the plane generated by the substitution. Substitutions provide a convenient tool to organize areas in space, for example to built the skeleton of a computation scheme. Indeed, a large amount of constructions on tilings, for example to construct aperiodic tilings [3, 5–7], involve the enforcement of the limit set of some substitution using local matching rules. General constructions have been provided in the literature [7–10] to enforce limit sets of different kinds of substitutions by encoding locally checkable encodings inside tilings.

In this paper, we provide an effective method to associate to every two-by-two substitution s a strongly deterministic tileset $\tau(s)$ such that the limit set Λ_s of

S.B. Cooper, A. Dawar, and B. Löwe (Eds.): CiE 2012, LNCS 7318, pp. 462–471, 2012.

the substitution is equal to the letter by letter projection $\pi\left(\mathcal{X}_{\tau(s)}\right)$ of the set of tilings $\mathcal{X}_{\tau(s)}$. Our approach is based on geometric constructions. After introducing the necessary definitions, the 104 tiles aperiodic tileset from [7] is extended as a strongly aperiodic tileset isomorphic to the tileset of Kari and Papasoglu [2] (section 2). Given a substitution, this tileset is then decorated to obtain a tileset deterministic in one direction that enforces the limit set (section 3). Finally, four copies of the previous tileset are synchronized to obtain the wanted strongly deterministic tileset (section 4). As a consequence, techniques from [7] can be combined to Lukkarila [4] to simplify the embedding of a Turing machine computation inside strongly deterministic tilesets to prove the undecidability of the Domino Problem and transfer more general results on tilings to the strongly deterministic case.

1   Definitions

A *tileset* is a triple $(\tau, \mathcal{H}, \mathcal{V})$ where τ is a finite alphabet of *tiles* and $\mathcal{H}, \mathcal{V} \subseteq \tau^2$ are finite sets of couples representing respectively the compatible horizontal and vertical neighbors. A *tiling* T is a map $T : \mathbb{Z}^2 \to \tau$ associating a tile to every cell of $\mathbb{Z}^2$ in such a way that every tile is compatible with its neighbors relatively to $\mathcal{H}$ and $\mathcal{V}$: $\forall(x,y) \in \mathbb{Z}^2$, $(T(x,y), T(x+1,y)) \in \mathcal{H}$ and $(T(x,y), T(x,y+1)) \in \mathcal{V}$. The compatibility rules represented by $\mathcal{H}$ and $\mathcal{V}$ are denoted as the *local rules* of the tileset. Implicitly associating the local rules to the tiles, we usually denote the tileset $(\tau, \mathcal{H}, \mathcal{V})$ simply as τ. The set of all tilings by a tileset τ is denoted as $\mathcal{X}_\tau \subseteq \tau^{\mathbb{Z}^2}$. A tiling $T \in \tau^{\mathbb{Z}^2}$ is *periodic* if there exists a translation vector $p \in \mathbb{Z}^2$ such that $\forall x \in \mathbb{Z}^2$, $T(x+p) = T(x)$. A tileset τ is *aperiodic* if it can tile the plane (i.e., $\mathcal{X}_\tau \neq \varnothing$) but never in a periodic way. We use the abbreviations *NE*, *NW*, *SE*, *SW* to denote the directions north-east, north-west, south-east, south-west respectively. A tileset τ is *NE-deterministic* if for all couples of tiles $(t_{\mathrm{W}}, t_{\mathrm{S}}) \in \tau^2$, there exists at most one tile $t_{\mathrm{NE}} \in \tau$ simultaneously compatible to the east with t_{W} and to the north with t_{S}: $(t_{\mathrm{W}}, t_{\mathrm{NE}}) \in \mathcal{H}$ and $(t_{\mathrm{S}}, t_{\mathrm{NE}}) \in \mathcal{V}$. $\{NW,SE,SW\}$-*determinism* is defined the same way. A tileset is *strongly deterministic* if it is simultaneously NE, NW, SE and SW-deterministic.

Given a finite alphabet Σ, a Σ-*coloring* c of the discrete plane $\mathbb{Z}^2$ by Σ is a map $c : \mathbb{Z}^2 \to \Sigma$. The *translation* of a coloring c by a vector $u \in \mathbb{Z}^2$ is the map $u \cdot c \in \Sigma^{\mathbb{Z}^2}$ verifying $\forall x \in \mathbb{Z}^2$, $u \cdot c(x) = c(x - u)$. Endowed with the product topology over $\mathbb{Z}^2$ of the discrete topology over Σ, $\Sigma^{\mathbb{Z}^2}$ is a compact topological space. A *subshift* $\mathcal{Y} \subseteq \Sigma^{\mathbb{Z}^2}$ is a topologically closed and translation invariant subset of $\Sigma^{\mathbb{Z}^2}$. E.g. the set of tilings $\mathcal{X}_\tau$ of a tileset τ is a subshift of $\tau^{\mathbb{Z}^2}$. A subshift $\mathcal{Y} \subseteq \Sigma^{\mathbb{Z}^2}$ is *sofic* if it can be recognized by local constraints in the following sense: there exist a tileset τ and an alphabetical projection $\pi : \tau \to \Sigma$, naturally extended to τ-colorings $\pi : \tau^{\mathbb{Z}^2} \to \Sigma^{\mathbb{Z}^2}$, such that $\pi(\mathcal{X}_\tau) = \mathcal{Y}$.

Let us now introduce a strengthened version of the soficity that we refer to as *directional soficity*. A subshift $\mathcal{Y} \subseteq \Sigma^{\mathbb{Z}^2}$ is *NE-sofic* if there exist a *NE-deterministic* tileset τ and an alphabetical projection $\pi : \tau \to \Sigma$ such that $\pi(\mathcal{X}_\tau) = \mathcal{Y}$. $\{NW,SW,SE\}$-*soficity* is defined the same way. A subshift $\mathcal{Y} \subseteq \Sigma^{\mathbb{Z}^2}$

464 B. Le Gloannec and N. Ollinger

is *4-way sofic* if there exist a *strongly deterministic* tileset τ and an alphabetical projection $\pi : \tau \to \Sigma$ such that $\pi(\mathcal{X}_\tau) = \mathcal{Y}$.

Let $\boxplus$ denote the finite set $\{0,1\} \times \{0,1\}$. A 2×2 *substitution* over an alphabet Σ is a map $s : \Sigma \to \Sigma^\boxplus$. s is naturally extended to its *global map* $S : \Sigma^{\mathbb{Z}^2} \to \Sigma^{\mathbb{Z}^2}$ verifying: $\forall c \in \Sigma^{\mathbb{Z}^2}$, $\forall x \in \mathbb{Z}^d$, $\forall u \in \boxplus$, $S(c)(2x + u) = s(c(x))(u)$. Following a dynamical systems point of view, we define the set of colorings of the plane generated by a substitution s as its *limit set* $\Lambda_s = \bigcap_{n \geq 0} \left\{ u \cdot S^n \left(\Sigma^{\mathbb{Z}^2} \right) \right\}_{u \in \mathbb{Z}^2}$ which is a subshift of $\Sigma^{\mathbb{Z}^2}$. Let us conclude by stating a useful characterization of the limit set for our construction. Given a substitution s over Σ, a Σ-coloring $c \in \Sigma^{\mathbb{Z}^2}$ admits a *history* if there exists a sequence $(u_n, c_n)_{n \geq 0}$, with, $\forall n \geq 0$, $u_n \in \boxplus$ and $c_n \in \Sigma^{\mathbb{Z}^2}$, such that $c_0 = c$ and $\forall n \geq 0$, $c_n = u_n \cdot S(c_{n+1})$. Then Λ_s is exactly the set of colorings admitting a history.

2   A Strongly Deterministic Aperiodic Tileset

In this section, we build a strongly deterministic aperiodic tileset by enriching the aperiodic tileset τ of 104 tiles introduced in [7].

The Aperiodic Tileset τ of 104 Tiles. The tileset τ can be defined as the smallest fixed point of a 2×2 *substitution scheme*. It is indeed *self-simulating* for a certain 2×2 substitution s on τ depicted on figure 1: both tilesets τ and $s(\tau)$ (which tiles are 2×2 *macro-tiles* on τ) are isomorphic and every tiling by τ can be uniquely decomposed into a tiling by $s(\tau)$. The substitution s is *non-ambiguous*: every coloring of its limit set Λ_s admits a unique pre-image by S. This forces in particular Λ_s to contain only non-periodic colorings. As $\mathcal{X}_\tau \subseteq \Lambda_s$, τ is aperiodic. For a detailed presentation of this construction, the reader is invited to refer to [7].

Let us now fix some useful notations. Let ⊞, ⊟, ⊡, ⊞ (i.e., SW, SE, NW, NE) denote the four colors of the layer 1 (parity), respectively represented in dark blue, blue, dark red, red on the figures; and X, H, V the three general types: *cross*, *horizontal bridge* and *vertical bridge* respectively, of decorations of the layer 2. Paths of information on the layer 2 are called *cables*. We finally use the notations ⌐, ⌐, ⌐, ⌐ to indicate the cables of positions SW, SE, NW, NE respectively of a tile with layer 2 of type cross X.

Remark 1. Observe that s forces the layer 2 (cables) of any tiling to describe an infinite stacking of *parity grids* (where square colors are alternating parity in both directions) which nth level is a grid of step $2^n + 1$. To establish the soficity of a limit set, one could code a coloring of the history on each level. Also note that s forces each level of the parity grid to be translated of $1/2$ step of the immediately inferior level grid in the SW direction.

The 4-Way Deterministic Enriched Tileset τ'. The tileset τ is not deterministic in any direction: the only reason is that is it not possible to determine

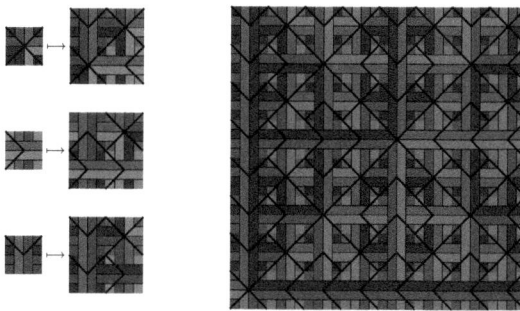

Fig. 1. Substitution s and pattern of $s^3(\tau)$

whether the type of layer 2 H or V must be chosen when producing a tile of parity ▦.

Let us define an alphabet of *labels* $\mathcal{L} = \{X, H, V\}$ and label the cables of the layer 2 by elements of ⊞ × $\mathcal{L}$ (instead of ⊞). The local rule is extended so that the labels of $\mathcal{L}$ are preserved (as the colors of ⊞) all along a same cable. Observe that any tile t of parity ▦ is directly encircled by a cable on the layer 2 of its eight neighbors. Let us enrich the local rules in respect to the four directly neighboring tiles so that this label corresponds to the type of the tile t: if t is a cross, the label must be X; if t is an horizontal bridge, the label must be H; if t is a vertical bridge, the label must be V. Let us observe that the substitutive structure of the tilings already forces every tile of parity ▦ directly encircled by a cable of color ▦, ▦, ▦ to be a horizontal bridge, vertical bridge, cross respectively. Hence the added labeling is only necessary for cables of color ▦ as the other couples color-label appearing in a tiling must always be (▦, H), (▦, V) and (▦, X).

Those new labels and local rules force the choice between types H and V on the layer 2 of tiles of parity ▦. However the production of the labels on cables of color ▦ is not deterministic: e.g., when the considered determinism direction is NE, the label on the cable at position ∟ of a tile of type X is not determined when this cable is of color ▦. That piece of information can only be obtained back in the history of the coloring coded by the tiling. For this purpose, we add some *wires* as depicted on the figure 2(a), carrying a label of $\mathcal{L}$. Those new wires must start on every corner of a square of color ▦ (figure 2(b)) and follow cables of color ▦ (figures 2(d) and 2(e)) until they reach an orthogonal cable of superior level (figures 2(c) and 2(e)). Tiles receiving those new decorations are hence perfectly identified. Associated local rules are: the label carried by a wire must be preserved all along; on both ends of a wire, the label must corresponds to the ones held by the cables to which it is connected.

The proof of the following result is mainly an exhaustive verification.

Theorem 1. τ' *is an aperiodic strongly deterministic tileset.*

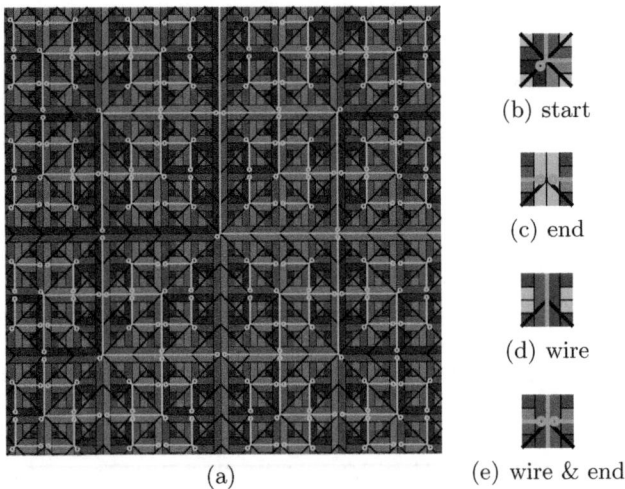

(b) start

(c) end

(d) wire

(e) wire & end

(a)

Fig. 2. Pattern of a tiling by τ' (a) and tiles holding new wires

3   NE-Soficity of Substitutions Limit Sets

Enforcing the Limit Set of a Substitution: The Enriched Tileset $\tau'(s')$.
Given a 2×2 substitution s' over an alphabet Σ, we use the quaternary tree
structure constituted by the cables of color ▦ and contained in every tiling, to
carry the history of a configuration of the limit set of s'. For that purpose, every
cable of color ▦ must carry a letter of Σ: those cables are hence labeled by
elements of ▦ $\times \mathcal{L} \times \Sigma$ (instead of ▦ $\times \mathcal{L}$). The added labels must verify certain
rules to enforce the tree to hold the hierarchy imposed by the substitution s':
on V tiles where two cables of color ▦ cross, the letters $a \in \Sigma$ carried by the
superior vertical cable and $b \in \Sigma$ carried by the inferior horizontal cable must
verify $b = s'(a)(x,y)$ where $x = 0$ (resp. $x = 1$) if the superior ▦ cable is on
right (resp. left) position and $y = 0$ (resp. $y = 1$) if the inferior ▦ cable is on
bottom (resp. top) position; symmetrically, on H tiles where two cables of color
▦ cross, the letters a carried by the superior horizontal cable and b carried by
the inferior vertical cable must verify $b = s'(a)(x,y)$ where $y = 0$ (resp. $y = 1$) if
the superior ▦ cable is on top (resp. bottom) position and $x = 0$ (resp. $x = 1$)
if the inferior ▦ cable is on left (resp. right) position; finally, on every cross X in
▦ position, the letters a carried by the ▦ cable of the layer 2 and b carried by
the layer 1 must verify $b = s'(a)(u)$ where $u = $ ▦ (resp. ▦,▦,▦) if the ▦ cable
is in position ∟ (resp. ⌐,⌐,⌐) in the cross. We also enrich the local rules to force
the four tiles of a same 2×2 parity block to share the same letter of Σ. For all
tiles $t \in \tau'(s')$ with layer 1 $(u, a) \in ▦ \times \Sigma$, let us define $\pi(t) = s'(a)(u)$.

Theorem 2 ([7]). $\pi\left(\mathcal{X}_{\tau'(s')}\right) = \Lambda_{s'}$, hence limit sets of 2×2 substitutions are
sofic.

So that in every level of a tiling all cables symmetrically carry the hierarchical information, we ultimately enrich $\tau'(s')$ by adding letters of Σ on cables of colors ⊞, ⊞ and ⊞ and forcing every ⊞ cable to share its letter with the three other cables of its parity block: on every tile (X included) where a cable of color ⊞ (resp. ⊞) appears at the left of a ⊞ (resp. ⊞) cable, they must share the same letter of Σ; similarly, on every tile where a cable of color ⊞ (resp. ⊞) appears at the top of a ⊞ (resp. ⊞) cable, they must share the same letter of Σ. We refer to these rules as *sharing rules*. That way, every level of the parity grid of every tiling t by $\tau'(s')$ codes a coloring of the history of $\pi(t)$ where each letter is carried by a parity block.

The NE-Deterministic Tileset $\tau_{\mathrm{NE}}(s')$. The tileset $\tau'(s')$ is not deterministic in any direction. The reason is that the letter of Σ carried by the corner of a cable on a X tile cannot always be deterministically determined when its position corresponds to the considered direction of determinism and no sharing rule force this letter: e.g., for the determinism direction NE, the letters of ⊞ cables of position ⌞ on X tiles are not determined. Note that this problem also arises in the particular case of the level 0: e.g., for the determinism direction NE, the letter of Σ held by the parity layer of any tile of parity ⊞ cannot be obtained deterministically. That piece of information can be obtained in the superior level of the history which appears, according to the remark 1, to be translated in the SW direction. In the following, we build a NE-deterministic tileset based on that observation.

For that purpose, we must add a color of ⊞ and a letter of Σ on some of the wires added in the previous section: those that link the SW corner of a ⊞ cable to the closest superior level cable in the south or west direction, e.g., the vertical wires running to the south depicted on the figure 3(a) (or we could have chosen equivalently the corresponding horizontal wires running to the west). Those wires are then labeled by elements of $\mathcal{L} \times \boxplus \times \Sigma$ (instead of $\mathcal{L}$). Tiles holding a wire are represented on figure 3. On tiles carrying the end (figure 3(c)) of such a wire, the color and letter of the wire must be the same as the ones of the cable to which it is connected. On tiles carrying the start (figure 3(b)) of such a wire, the color $u \in \boxplus$ and letter $a \in \Sigma$ carried by the wire and the letter $b \in \Sigma$ carried by the ⊞ cable must verify $b = s'(a)(u)$. Associated local rules are: the labels carried by a wire must be preserved all along its propagation.

Solving the same problem at level 0, i.e., predicting the letter of the parity layer of ⊞ tiles, is easier. Remember that every ⊞ tile is directly encircled by a cable on its eight neighbors. Then simply add the following local rules: the letters $b \in \Sigma$ of the parity layer of any ⊞ tile and $a \in \Sigma$ of the direct encirclement cable, of color $u \in \boxplus$, in any of its four direct neighbors, must verify $b = s'(a)(u)$.

$\tau_{\mathrm{NE}}(s')$ is a NE-deterministic tileset. The proof of this result is an exhaustive verification. It should be quite clear as the enrichments were done in purpose. The soficity result is obviously still valid (for a naturally extended projection π simply removing all additional decorations): we have $\pi\left(\mathcal{X}_{\tau_{\mathrm{NE}}(s')}\right) = \Lambda_{s'}$. This lead to the following result.

Theorem 3. *Limit sets of 2×2 substitutions are NE-sofic.*

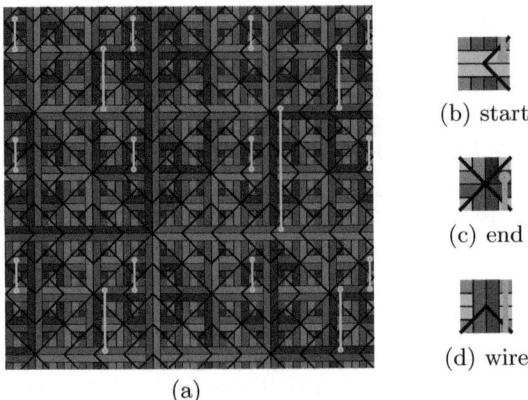

(a)

(b) start

(c) end

(d) wire

Fig. 3. Pattern of a tiling by $\tau_{\mathrm{NE}}(s')$ (a) and tiles holding wires

4   4-Way Soficity of Substitutions Limit Sets

In this section, we enrich the NE-deterministic tileset $\tau_{\mathrm{NE}}(s')$ into a strongly deterministic tileset $\tau_{4w}(s')$.

The 4-Way Deterministic Tileset $\tau_{4w}(s')$. We have underlined in remark 1 the fact that the substitution s translates superior levels of the parity grid towards the SW direction. Considering the three other symmetrical possible choices for this substitution represented on the figure 4, we can similarly build tilesets $\tau_{\mathrm{NW}}(s')$, $\tau_{\mathrm{SE}}(s')$ $\tau_{\mathrm{SW}}(s')$ that are NW, SE and SW-deterministic respectively. To build a strongly deterministic tileset $\tau_{4w}(s')$, let us consider the tileset constituted by the cartesian product $\tau_{\mathrm{NE}}(s') \times \tau_{\mathrm{NW}}(s') \times \tau_{\mathrm{SE}}(s') \times \tau_{\mathrm{SW}}(s')$ for which we require the local rules to be verified on each of the components of the product. For any tile $t = (t_1, t_2, t_3, t_4)$, we moreover require the four components $t_1, \ldots, t_4$ to share the same layer 1, i.e., same parity and same letter of Σ. That way, the coloring coded by each of the components of a tiling $T = (T_1, T_2, T_3, T_4)$ is the same: $\pi_1(T_1) = \pi_2(T_2) = \pi_3(T_3) = \pi_4(T_4)$ with π_i the associated projections. Each of the components of the product tileset is deterministic in one direction. The idea is to use this component to make the three others deterministic in its direction. We must then *synchronize* the histories coded by the four components of a tiling so that they code the same coloring at every level. The synchronized tileset should then be strongly deterministic.

Recall that the obstacle to determinism is that the letter of Σ carried by the corner of a cable on a tile X cannot always be deterministically determined when its position corresponds to the considered direction of determinism and no sharing rule force this letter.

Analyzing the Case of Two Opposite Directions. Let us examine the case of two components of opposite determinism directions. Without loss of generality, let us consider SW and NE directions and, in this paragraph only, the

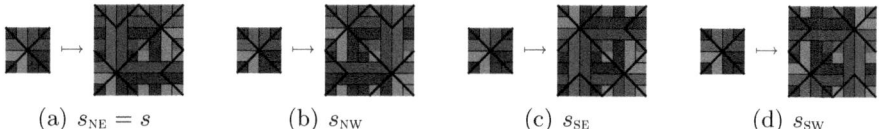

(a) $s_{\mathrm{NE}} = s$ (b) s_{NW} (c) s_{SE} (d) s_{SW}

Fig. 4. Four symmetrical substitutions

$\tau_{\mathrm{NE}}(s') \times \tau_{\mathrm{SW}}(s')$ associated product (i.e., components 1 and 4 of the previously introduced product) for which a tiling pattern is depicted of figure 5(a) (where on each tile, the first and second components are represented in the SW and NE corners respectively, and colors of the second components are dark green, green, orange, yellow for ▦,▦,▦,▦ respectively). In any tiling, only the parity layers of both components are actually synchronized so that they both code the same coloring at level 0. But let us assume one moment that we have synchronized both components on every level so that both components code exactly the same history of that coloring. Let us now pick the determinism direction SW, and analyze how the first component can be made SW-deterministic using the second component. By symmetry, the same analyze would make sense for the other component in the opposite determinism direction. In our choice, for the determinism direction SW on the first component, the letters of ▦ cables of position ⌐ on X tiles are not determined. As illustrated by the scattered representations of figures 5(c) and 5(d), we claim, when the tiling is fully synchronized, that the letters of these cables are precisely those carried by the corresponding ▦ cables (in yellow) of the second component and same hierarchical level that appear on neighboring tiles pointed by a vector $(-1, -1)$, i.e., one diagonal shift away in the SW direction. It is not convenient that this piece of information is available one shift away in the considered determinism direction: it precisely arrives one step later the position we need it. Nevertheless, observe that, at every level, that shift is constant $(-1, -1)$.

The case of two components of orthogonal determinism directions (e.g., $\tau_{\mathrm{NE}}(s') \times \tau_{\mathrm{NW}}(s')$ depicted on figure 5(b)) is similar (required information available one shift away).

Grouping Tiles. The analyses lead before show that the required pieces of information to synchronize the histories of the four components and by this way make the tileset strongly deterministic always are available one shift away, unfortunately in the considered determinism direction, of the tile to predict.

A simple solution to this constant shift problem is a 3×3 grouping of tiles: let τ_0 be the previously defined product tileset, let ▦ denote the set $\{0, 1, 2\}^2$, $\tau_1 \subseteq \tau_0^{▦}$ denote the set of valid 3×3 patterns (in respect to τ_0 local rules) over τ_0 and consider the grouped tileset $(\tau_1, \mathcal{H}_1, \mathcal{V}_1)$ verifying, $\forall (t, t') \in \tau_1^2$, $(t, t') \in \mathcal{H}_1 \Leftrightarrow \forall y \in \{0, 1, 2\}$, $t_{2,y} = t'_{0,y}$ and $(t, t') \in \mathcal{V}_1 \Leftrightarrow \forall x \in \{0, 1, 2\}$, $t_{x,2} = t'_{x,0}$. We then obtain $\tau_{4w}(s')$ from τ_1 by adding some synchronization requirements on the tiles. Let us get back to the example used in the case of two opposite

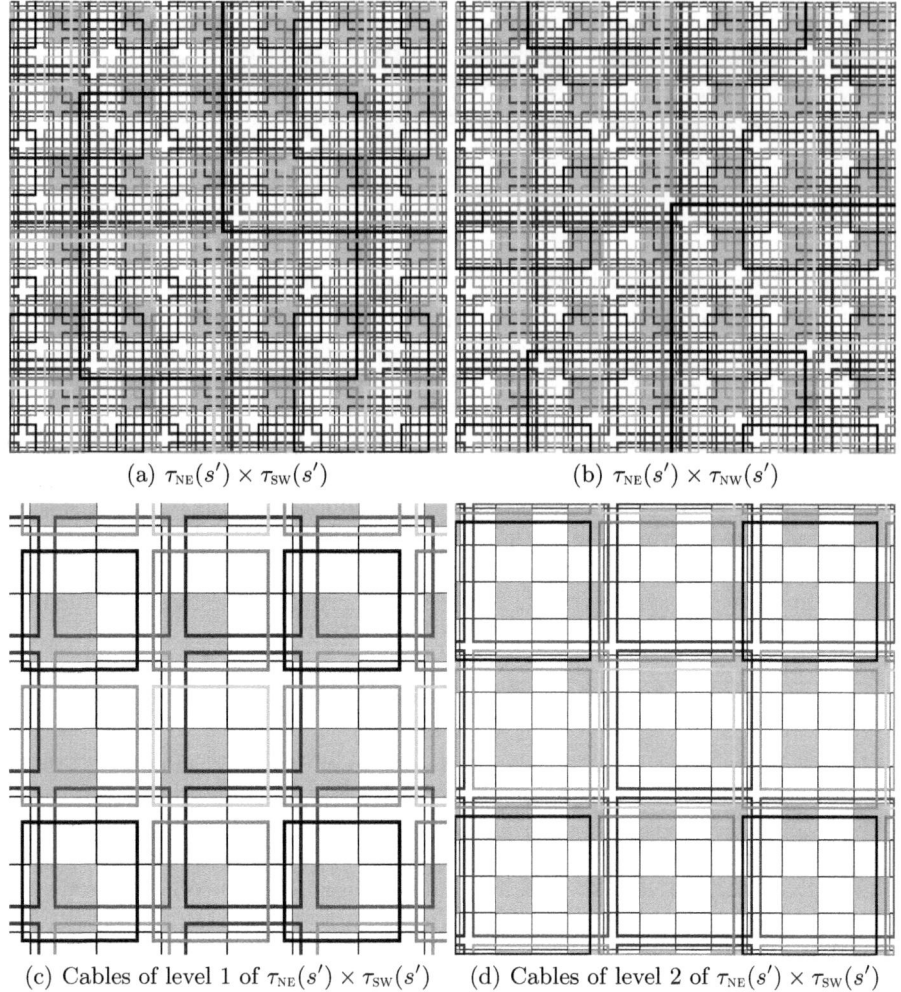

(a) $\tau_{\mathrm{NE}}(s') \times \tau_{\mathrm{SW}}(s')$ (b) $\tau_{\mathrm{NE}}(s') \times \tau_{\mathrm{NW}}(s')$

(c) Cables of level 1 of $\tau_{\mathrm{NE}}(s') \times \tau_{\mathrm{SW}}(s')$ (d) Cables of level 2 of $\tau_{\mathrm{NE}}(s') \times \tau_{\mathrm{SW}}(s')$

Fig. 5. Scattered view of the cables of the two-components tilesets and views level-by-level of the cables of $\tau_{\mathrm{NE}}(s') \times \tau_{\mathrm{SW}}(s')$

directions: on each tile t for which the central sub-tile $t_{1,1}$ has a first component of type X with a ▦ cable in ⌐ position labeled by a letter $a \in \Sigma$, there must be, as t forms a 3×3 valid pattern for τ_0, a yellow ▦ cable going along the fourth components of the sub-tiles $t_{0,2}, t_{0,1}, t_{0,0}, t_{1,0}, t_{2,0}$ and, still for validity reasons, carrying a same letter $b \in \Sigma$. We require in that case, based on the previous analysis, to have $a = b$. Considering all the symmetrical requirements and the analog requirements for the orthogonal case, we obtain the $\tau_{4w}(s')$ tileset.

Results. $\tau_{4w}(s')$ is a strongly deterministic tileset. The proof of this result is an exhaustive verification. The cases examined before should convince the reader of its correctness. Again, the soficity result is obviously still valid (for a naturally extended projection π simply removing all additional decorations): we still have $\pi\left(\mathcal{X}_{\tau_{4w}(s')}\right) = \Lambda_{s'}$. This lead to the final result.

Theorem 4. *Limit sets of 2×2 substitutions are 4-way sofic.*

References

1. Kari, J.: The nilpotency problem of one-dimensional cellular automata. SIAM J. Comput. 21(3), 571–586 (1992)
2. Kari, J., Papasoglu, P.: Deterministic aperiodic tile sets. Geometric and Functional Analysis 9, 353–369 (1999)
3. Robinson, R.: Undecidability and nonperiodicity for tilings of the plane. Inventiones Mathematicae 12(3), 177–209 (1971)
4. Lukkarila, V.: The 4-way deterministic tiling problem is undecidable. Theor. Comput. Sci. 410(16), 1516–1533 (2009)
5. Durand, B., Romashchenko, A., Shen, A.: Fixed Point and Aperiodic Tilings. In: Ito, M., Toyama, M. (eds.) DLT 2008. LNCS, vol. 5257, pp. 276–288. Springer, Heidelberg (2008)
6. Berger, R.: The undecidability of the domino problem. Memoirs American Mathematical Society 66 (1966)
7. Ollinger, N.: Two-by-Two Substitution Systems and the Undecidability of the Domino Problem. In: Beckmann, A., Dimitracopoulos, C., Löwe, B. (eds.) CiE 2008. LNCS, vol. 5028, pp. 476–485. Springer, Heidelberg (2008)
8. Mozes, S.: Tilings, substitution systems and dynamical systems generated by them. Journal d'Analyse Mathématique 53(1), 139–186 (1989)
9. Goodman-Strauss, C.: Matching rules and substitution tilings. Annals of Mathematics 147(1), 181–223 (1998)
10. Fernique, T., Ollinger, N., TUCS: Combinatorial Substitutions and Sofic Tilings. In: Proceedings of JAC 2010 Journées Automates Cellulaires 2010, Turku Finland, pp. 100–110 (2010)

The Computing Spacetime

Fotini Markopoulou

Perimeter Institute for Theoretical Physics, 31 Caroline St. N., Waterloo Ontario
N2L 2Y5, Canada
fotinimk@gmail.com

Abstract. The idea that the Universe is a program in a giant quantum
computer is both fascinating and suffers from various problems. Nonetheless, it can provide a unified picture of physics and this is very useful for
the problem of Quantum Gravity where such a unification is necessary.
We give an introduction to the idea of the universe as a quantum computation, the problem of Quantum Gravity, and Quantum Graphity, a
simple way to model a dynamical spacetime as a quantum computation.

1 Introduction

That the Universe can be thought of as a giant computation is a straightforward
corollary of the existence of a universal Turing machine. The basic idea (nicely
summarized by Deutsch in [9]) goes as follows. The laws of physics allow for
a machine, the universal Turing machine, such that its possible motions correspond to all possible motions of all possible physical objects. That is, a universal
quantum computer can simulate every physical entity and its behavior. This
means that physics, the study of all possible physical systems, is isomorphic to
the study of all programs that could run on a universal quantum computer. We
can think of our universe as software running on a universal computer.

Should this logical inference affect our understanding of physics, or even
change the way we do science? Several different lines of thought say yes, an
idea most concretely articulated in the field of cellular automata and quantum
information theory. In 1969, Konrad Zuse, in his book *Calculating Space*, proposed that the physical laws of the universe are fundamentally discrete, and
that the entire universe is the output of a deterministic computation on a giant
cellular automaton [30]. Cellular automata (CA) are regular grids of cells, each
cell in one of a finite number of states, usually just black/white. An initial state
of the CA is updated in global discrete time steps, in which each cell's new state
changes as a function of its old state and that of a small number of neighbors.
A concrete example of Zuse's vision is Conway's Game of Life. The rules are
simple: If a cell has 2 black neighbors, it stays the same; if it has 3 black neighbors, it becomes black; otherwise it becomes white. The result is remarkably rich
behavior on the border between randomness and order. A striking feature is the
occurrence of gliders, small groups of cells that move like independent emergent
entities. We can arrange the automaton so that the gliders interact to perform
computations, and it can be shown that the Game of Life is a universal Turing

S.B. Cooper, A. Dawar, and B. Löwe (Eds.): CiE 2012, LNCS 7318, pp. 472–484, 2012.
© Springer-Verlag Berlin Heidelberg 2012

machine [3]. It is simple to see how this evokes the possibility that we live in a giant CA [8]: In our CA Universe, what we think of as elementary particles may just be emergent gliders. Since CAs exist that are Turing machines, it is in principle possible to have any kind of glider behavior generated by a CA, including gliders observing the laws of elementary particle physics. We don't know how these are generated because we have no access to the microscopic cells, so we make physical theories about particle-like objects, but, in reality, we live in a CA.

Quantum information theory has given a new and interesting twist on the Universe as a Computation. Many practitioners in this field argue that everything fundamentally is information, an old idea that can be traced at least back to Wheeler's influential *it from bit* [29]. In that view, all interactions between physical systems in the universe are instances of information processing. The information involved in those processes is more primary than the physical systems themselves. Instead of thinking of particles as colliding, we should think of the information content of the particles being involved in a computation. By simple interpolation, the entire universe is nothing but a giant computation. As Lloyd puts it in [21], the universe computes "its own dynamical evolution; as the computation proceeds, reality unfolds".

These are fascinating ideas when loosely interpreted, but with obvious problems, including:

1. What does it mean that information is more fundamental than its physical instantiation?
2. Since any observation we can make, and anything physics describes, is just the program, there is no way to know the hardware that runs that program. The program can perhaps give us some hints as to what machine could efficiently run it, but at the end of the day this scenario assumes a fundamentally unknowable machine.
3. Is that machine running just one program, our universe? If yes, how is that "mother computation" chosen? If no, we need a meta-program that runs multiple programs, a computer version of the *multiverse* idea [27]. By one more iteration, multiple computers, each running multiple programs, are a logical possibility, leading to an extreme form of a multiverse. Or are we secretly assuming a Programmer?
4. The idea requires that all of physics is computable.
5. The CA Universe, in addition, advocates that the universe is fundamentally discrete. Fundamental discreteness is a very old and attractive idea but it remains to be seen whether it can be reconciled with observable physics, and, in particular, with quantum mechanics and Lorentz invariance. Quantum mechanics makes essential use of the complex numbers, a continuous field. It is, of course, logically possible to push fundamental discreteness to an extremely small scale, perhaps the Planck scale, and claim that the world appears continuous only by approximation, because we have no access to that small scale. This is where Lorentz invariance comes in. The Lorentz transformations form a non-compact group, meaning that by boosting an

observer sufficiently, we can blow up any tiny amount of discreteness to arbitrarily large size. Depending on the details of the physics, even scales smaller than the Planck scale can thus become observationally accessible. Reconciling observational constraints on Lorentz invariance violations and fundamental discreteness is a very active subject of research in quantum gravity and quantum gravity phenomenology.[1]

At the end of the day, the Universe as a Computation idea may just reflect the current way we understand and bring order to our surroundings. Maybe it just shifts us a little from the "Blind Watchmaker" to the "Blind Programmer". I find it very likely that the Universe as Computation is a culturally determined and temporary idea. Fun as it may be to speculate about the universe being a computer, it is rather sterile to do so in the abstract. The interesting question is whether this scenario can be put to good uses: Does it give us useful new tools and methods with which we can solve problems we couldn't solve before? Does it raise new interesting questions? The purpose of this article is to argue that it does, and that the relevant area of physics to use the idea of the Universe as a Computation is the field of Quantum Gravity and Quantum Cosmology. Quantum Gravity needs to unify quantum field theory, the physics of matter, with general relativity, the physics of spacetime, into a single consistent theory. The universe as a Computation suggests a new kind of unification: physical systems and their dynamics can be represented in terms of their information content and their dynamics is the processing of that information.

We shall illustrate this view with an example. In [16,12], we initiated a study of quantum gravity using spin systems as toy models for emergent geometry and gravity. These *quantum graphity models* are based on quantum networks with no a priori geometric notions. We have repeatedly found the quantum information perspective to be useful, both as a tool chest (for example, the Lieb-Robinson speed of information propagation can be used to derive the speed of light [20,23], or error correction to define conserved quantities [4]) and as an aid to conceptual clarity: the information theoretic language allows us to do physics without reference to a background geometry.

The purpose of this introductory article is to illustrate these ideas in a brief and self-contained format and invite discussion and exchange of ideas between the fields of quantum gravity and computer science. Technical details can be found in the suggested references. In the next Section, we state the problem of Quantum Gravity in terms of the breakdown of classical spacetime at Planck scale and the problem of time. In Section 3, we summarize the basics of Quantum Graphity, the representation of pre-geometry as qubits of adjacency, a sketch of the derivation of the speed of light from the fundamental dynamics, the toy trapped surfaces that arise in these systems, and the mechanism by which matter inside that world sees an emergent curved geometry. We briefly summarize our conclusions in Section 4.

---

[1] Cf., e.g., [14] and references therein.

2   The Problem of Quantum Gravity

The field of Quantum Gravity is the attempt to unify General Relativity and Quantum Theory. In spite of their impressive successes, the two theories leave us with a gap in situations in which the quantum effects of the gravitational field become important. This hampers our understanding of some of the most fascinating modern physics, such as the physics of black holes [26], Hawking radiation [28], and the very early universe,[2] or leads to absurdities such as the cosmological constant problem [7]. In all these situations, the curvature of spacetime is so high we are not confident in the reliability of quantum field theory.

The length scale where quantum gravitational effects become significant is, by dimensional analysis, the Planck length, $l_{\mathrm{Pl}} = \sqrt{\frac{G_N \hbar}{c^3}}$, the combination of Newton's constant G_N, Planck's constant $\hbar$, and the speed of light c. This is incredibly small, $l_{\mathrm{Pl}} \sim 10^{-35} m$, corresponding to energy scales of $10^{19} GeV$. At Planck scale, the concepts of size and distance break down. Any microscopic probe energetic enough to precisely measure a Planck-sized object needs to be so energetic (to measure l_{Pl}, its Compton wavelength must be $\sim l_{Pl}$) that it would completely distort the region of space it was supposed to measure. In this sense, the notion of a classical spacetime manifold breaks down at Planck scale. A quantum theory of gravity that reconciles general relativity and quantum theory, or replaces them, is required to understand physics, including spacetime, at that scale.

In spite of decades of research, a satisfactory quantum theory of gravity still eludes us. Much of the difficulty comes from the fundamentally different assumptions that these theories make on how the universe works. General relativity describes spacetime as a manifold $\mathcal{M}$ with a dynamical metric field $g_{\mu\nu}$, and gravity as the curvature that spacetime. Quantum field theory describes particle fields on a flat and fixed spacetime. Naive quantization of gravity, treating it as another quantum field, leads to nonsense as gravity is non-renormalizable. The difference between the two theories can be phrased in terms of the way each treats *time*. A fundamental lesson of general relativity is that there is no fixed spacetime background spacetime: *geometry tells matter where to go and matter tells geometry how to curve*. The spacetime geometry is a dynamical field. In addition, physical quantities are invariant under diffeomorphisms of $\mathcal{M}$. This means (roughly) that general relativity is a relational theory, i.e., the only physically relevant information is the relationship between events in spacetime [2]. On the other hand, quantum theory requires a fixed background spacetime, either a Newtonian one (quantum mechanics), or a fixed Minkowski spacetime (quantum field theory). Time in quantum theory is not a dynamical field, it is a background parameter.

Turning spacetime geometry into a quantum field is possible and the task of conservative approaches to quantum gravity such as Loop Quantum Gravity [1]. The result, however, of such quantizations is peculiar. We obtain a so-called *wavefunction of the universe* $|\Psi_U\rangle$, i.e., the diffeomorphism invariant quantization

---

[2] Cf., e.g., [5].

of the metric $g_{\mu\nu}$ projected on a spatial slice of $\mathcal{M}$. Instead of a Schrödiger equation, the evolution of $|\Psi_U\rangle$ is governed by the *Wheeler-deWitt equation*:

$$\widehat{H}|\Psi_U\rangle = 0, \tag{1}$$

where $\widehat{H}$ is the quantization of the "projection" of the Einstein equations in the direction normal to the spatial slice (for the actual details of this procedure, see, for example, [25]). The Wheeler-deWitt equation is peculiar on two (related) counts: it describes the evolution of the entire universe, not just a localized system as in the Schrödinger equation, and the right hand side is zero (not time evolution). That zero can be traced to the diffeomorphism invariance of general relativity and the fact that the Einstein equations describe the dynamics of the *entire* universe. The diffeomorphism symmetry gets mixed up with evolution in ways that are very difficult to untangle.[3]

Much more can be said about this, but the purpose of the present note is to point out that, since quantum gravity needs to unify quantum theory and general relativity, a unification of the corresponding descriptions of the physical world is required, and that quantum information theory can provide this. Reducing both quantum fields and differential manifolds to their information theoretic content can provide a common framework. The Universe as a Computation can, in that sense, be seen as a useful and practical tool to solve a long-standing problem. Note that we do not need to resolve whether information precedes its physical instantiation, or answer most of the problems listed above in order to put this notion to useful work. All we need is that an information theoretic description is possible, both for the physics of matter and for the physics of space-time. We have been pursuing this idea in the *Quantum Graphity models* for quantum gravity and we shall give a concrete example of such a model in the next Section.

3   Quantum Graphity

Quantum Graphity models [16,12] are toy models for emergent geometry and gravity. They are based on graphs whose adjacency is quantum and dynamical: the edges can be on (connected), off (disconnected), or in a superposition of on and off. We interpret the graph as pregeometry (the graph connectivity tells us who is neighbouring whom). A graphity model is given by such graph states evolving under a local Hamiltonian. The model of [12] which describe in the rest of this section, is a toy model for interacting matter and geometry, a Bose-Hubbard model where the interactions, or adjacencies, are quantum variables.

3.1   Qubits of Adjacency

Let us assume a universe consisting of N fundamental constituents, systems labeled by $i = 1, ..., N$. These are quantum mechanical, so we have $\{\mathcal{H}_i\}; i = 1, ..., N$

---

[3] For a classic review of the longstanding effort to find gravity's true degrees of freedom (metrics modulo diffeomorphisms), cf. [15].

Hibert spaces. Let K_N denote the complete graph that has these N systems as its vertices, a graph with $\frac{N(N-1)}{2}$ links $\mathbf{e} \equiv (i,j)$. To every such $\mathbf{e}$ we associate a Hilbert space $\mathcal{H}_{\mathbf{e}} \simeq \mathbf{C}^2$. Basis states on $\mathcal{H}_{\mathbf{e}}$ can be labeled by $|1\rangle, |0\rangle$, and we choose to interpret $|1\rangle$ as the link $\mathbf{e}$ being *on*, or present, and $|0\rangle$ as the link being *off*, or missing. The total Hilbert space corresponding to K_N then is $\mathcal{H}_{\text{graph}} = \bigotimes_{\mathbf{e}=1}^{N(N-1)/2} \mathcal{H}_{\mathbf{e}}$.

Our choice of basis in $\mathcal{H}_{\mathbf{e}}$ means that every basis state in $\mathcal{H}_{\text{graph}}$ corresponds to a subgraph of K_N. A generic state $|\Psi_{\text{graph}}\rangle \in \mathcal{H}_{\text{graph}}$ is a quantum superposition of subgraphs of K_N. For N very large, the state space contains superpositions of all possible finite graphs. By analogy with the adjacency matrix of a graph, we call $\mathcal{H}_{\mathbf{e}}$ a *qubit of adjacency*. States in $\mathcal{H}_{\text{graph}}$ then provide a simple discrete precursor to quantum geometry. Note, however, that since we cannot assume a pre-existing spacetime on which our N systems live, we cannot interpret the N vertices of K_N as points in that spacetime. That is, we do not have a discretization of a geometry, the geometry corresponding to a state is to be *inferred* from the behavior of matter interacting with $|\Psi_{\text{graph}}\rangle$.

To see how this works, let us next define a simple form of matter.

3.2   Interacting Matter and Geometry

We shall assign simple matter degrees of freedom to the vertices of K_N by assigning the Hilbert space $\mathcal{H}_i$ of a harmonic oscillator to each vertex i. We denote its creation and annihilation operators by $\hat{b}_i^\dagger, \hat{b}_i$ respectively, where $\hat{b}_i^\dagger |0\rangle_i = |1\rangle_i, \hat{b}_i |1\rangle_i = |0\rangle_i$, satisfying the usual bosonic relations, $[\hat{b}^\dagger, \hat{b}^\dagger] = 0 = [\hat{b}, \hat{b}], [\hat{b}, \hat{b}^\dagger] = 1$. Our N physical systems then are N bosonic particles and the total Hilbert space for these bosons is given by $\mathcal{H}_{\text{bosons}} = \bigotimes_{i=1}^{N} \mathcal{H}_i$.

The total Hilbert space of the theory is the state space of the combined matter and connectivity degrees of freedom, $\mathcal{H} = \mathcal{H}_{\text{bosons}} \otimes \mathcal{H}_{\text{graph}}$. A basis state in $\mathcal{H}$ has the form $|\Psi\rangle \equiv |\Psi_{\text{bosons}}\rangle \otimes |\Psi_{\text{graph}}\rangle \equiv |n_1, ..., n_N\rangle \otimes |e_1, ..., e_{\frac{N(N-1)}{2}}\rangle$. The first factor tells us how many bosons there are at every site i, while the second factor tells us which pairs $\mathbf{e}$ interact. This is an unusual bosonic system, as the structure of interactions is now promoted to a quantum degree of freedom.

This is interesting as a generic state can be a quantum superposition of "interactions". For example, for i and j in the state $|\phi_{ij}\rangle = (|10\rangle \otimes |1\rangle_{ij} + |10\rangle \otimes |0\rangle_{ij})/\sqrt{2}$, means a particle in i and no particle in j, and a quantum superposition between i and j interacting or not. The state, $|\phi_{ij}\rangle = (|00\rangle \otimes |1\rangle_{ij} + |11\rangle \otimes |0\rangle_{ij})/\sqrt{2}$, represents a different superposition, in which the bosonic degrees of freedom and the graph degrees of freedom are entangled. It is a significant feature of the model that matter can be entangled with geometry.

In [12], we proposed a simple Hamiltonian for the dynamics of the matter-geometry interaction. If the bosons are not interacting, their total Hamiltonian is trivial, $\hat{H}_v = \sum_{i=1}^{N} \hat{H}_i = -\sum_i \mu \hat{b}_i^\dagger \hat{b}_i$. An interesting interaction term is hopping, the physical process in which a quantum i is destroyed at i and one is created at j. We shall require that hopping is possible only if there is an *on* edge between i and j. Such dynamics is described by a Hamiltonian of the form

$$\hat{H}_{\text{hop}} = -E_{\text{hop}} \sum_{(i,j)} \hat{P}_{ij} \otimes (\hat{b}_i^\dagger \hat{b}_j + \hat{b}_i \hat{b}_j^\dagger), \tag{2}$$

where $\hat{P}_{ij} = |1\rangle\langle 1|_{(i,j)}$ is the projector on the edge (i,j) being in the *on* state. This projector is important, it means that it is the dynamics of the particles described by $\hat{H}_{\text{hop}}$ that gives to the link degrees of freedom the meaning of geometry: the state of the graph determines where the matter is allowed to go.

In the spirit of "geometry tells matter where to go and matter tells geometry how to curve", we need interacting graph and matter. To avoid interpretational problems, this interaction should be unitary. The simplest unitary interaction is

$$\hat{H}_{\text{ex}} = k \sum_{(i,j)} |0\rangle\langle 1|_{(i,j)} \otimes (\hat{b}_i^\dagger \hat{b}_j^\dagger)^R + |1\rangle\langle 0|_{(i,j)} \otimes (\hat{b}_i \hat{b}_j)^R. \tag{3}$$

This destroys an edge (i,j) and create R quanta at i and R quanta at j, or, vice-versa, destroys R quanta at i and R quanta at j to convert them into an edge. An example is shown in Figure 1. Of course, we need dynamics also for the graph degrees of freedom alone. The simplest choice is simply to assign some energy to every edge, $\hat{H}_{\text{link}} = -U \sum_{(i,j)} \sigma_{(i,j)}^z$.

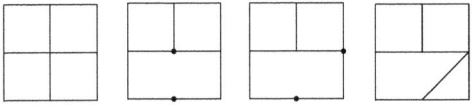

Fig. 1. Graph-matter dynamics: A link is is exchanged for two particles at its vertices; the particles hop on existing links; two particles are destroyed and a link is created between the corresponding vertices

This final step brings us to the total Hamiltonian for the model proposed in [12]:

$$\hat{H} = \hat{H}_{\text{link}} + \hat{H}_v + \hat{H}_{\text{ex}} + \hat{H}_{\text{hop}}. \tag{4}$$

It is possible to design such systems in the laboratory. For instance, one can use arrays of Josephson junctions whose interaction is mediated by a quantum dot with two levels.

This is a peculiar system in that who interacts with whom is a quantum degree of freedom, but otherwise it is an extremely simple system. Does it lead to any interesting behavior? Yes, more than one would expect, as we shall see next.

3.3   Calculating the Speed of Light as Propagation of Information

$\hat{H}_{\text{hop}}$ tells us that it takes a finite amount of time to go from i to j. If the graph was a chain, it would take a finite amount of time (modulo exponential decaying terms) for a particle to go from one end of the chain to another. This results to a "spacetime" picture (the evolution of the adjacency graph in time) with a finite lightcone structure. We can calculate this speed of light from the speed with which information propagates using methods from quantum information theory. From a local Hamiltonian, that is, a Hamiltonian that is the product of local terms, $H = \sum_{\langle ij \rangle} h_{ij}$, we can define the *Lieb-Robinson speed of information propagation* [20,23] as follows. Consider two points P and Q on a lattice, distance d_{PQ} apart. A disturbance at P is felt at Q a time t later with strength $\|[O_P(0), O_Q(t)]\|$, where $O_P(0)$ and $O_Q(t)$ are operators at P at time 0 and Q at time t respectively. It is shown in [20,23] that this signal strength is bound by

$$\|[O_P(0), O_Q(t)]\| \leq 2\|O_P\|\|O_Q\| \sum_n \frac{(2|t|h_{\max})^2}{n!} N_{PQ}(n), \tag{5}$$

where $h_{\max}$ is the maximum coupling strength in the Hamiltonian and N_{PQ} is the number of paths of length n in the lattice that connect P and Q. This can be rewritten as

$$\|[O_P(0), O_Q(t)]\| \leq 2\|O_P\|\|O_Q\| C \exp\left[-a\left(d_{PQ} - vt\right)\right]. \tag{6}$$

Saturating the above inequality defines the maximum speed v of information propagation in this system. Combining the two bounds allows us to calculate this speed in terms of the couplings in the Hamiltonian and the connectivity of the lattice.

In [11], we tested that v above can be the speed of light, by showing that in string net condensation, a spin system whose emergent excitations are photons [17,18,19], v agrees with the speed of light in the emergent Maxwell equations.[4] This is an interesting result as it allows us to reconcile emergent finite light cones and non-relativistic quantum mechanics. The underlying physics is, of course, quantum mechanics, but, within the bounds defined above, the system appears to have a finite light cone. A signal is possible outside the light cone, but it is exponentially suppressed. In fact, recent results show that in the continuum limit the finite light cone becomes exact as that signal vanishes in the continuum limit. Further work in [24] and results in [10] indicate that the emergence of a Minkowski metric is a behavior that can be extended to infinite-dimensional systems, i.e., the physics we are studying is not limited to spin systems.

Note that this speed v increases with the number N_{PQ} of paths connecting P and Q, and therefore it is a function of the connectivity of the lattice. Higher connectivity (vertex degree) means higher speed of light. This is used in what follows to derive the effective curved geometry matter sees.

3.4   Analogue Black Holes

One of the features of the above Hamiltonian acting on states which are not degree-regular graphs, observed in [12], was the trapping of bosons into regions

---

[4] Note, however, that this derivation does not assure us that this maximum speed is *universal* for all species of matter.

of high degree (cf. Fig. 2). The basic idea is the following: consider two sets of nodes, A and B, separated by a set of points C on the boundary, and let the vertices in A be of much higher degree than the vertices in B, $d_A \gg d_B$. If the number of edges departing from the set C and going to the set A is much higher than the number of edges going from C to B, then the hopping particles have a high probability of being "trapped" in the region A.

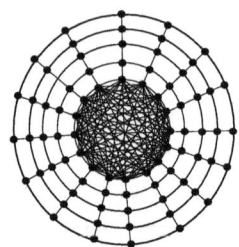

Fig. 2. Toy black hole configuration

A way to see the trapping is to study the Lieb-Robinson speed of particle propagation. Since the speed of propagation of particle probability is degree dependent, in the two regions we have two different speeds. Their ratio depends uniquely on the ratio of the degree of the two regions. Then, using the laws of optics, probability is reflected at the boundary due to Snell's law: for $\frac{d_B}{d_A} \sim \frac{1}{N} \rightarrow$ 0, and the region A acts as a trap.

This heuristic argument can be made precise. In [6], it was shown that matter propagating on the graph state shown in Fig. 2 has a unique ground state which is protected by a gap which increases linearly with the size N of the completely connected region. That is, high connectivity configurations are spin-system analogues of trapped surfaces.

3.5   Geometry: What the Matter Sees

While we can assign a metric to the graph itself, for the purposes of an analogue model for quantum gravity, the relevant geometry is the one the metric sees. This metric can be affected both by the graph state and by the matter Hamiltonian. Solving this problem is not easy. Our model may be simple conceptually, but it is a Hubbard model on a dynamical irregular lattice and, as is well-known, the Hubbard model beyond 1d quickly becomes intractable. Luckily, it turns out that a large and interesting sector of our model can be reduced to an effective 1d Hubbard model with modified couplings.

In order to do this, we restrict the time-dependent Schrödinger equation to the manifold formed by the classical states, i.e., single-particle states with a well-defined but unknown position. The equation of motion obtained corresponds to the equation of propagation of light in inhomogeneous media, similarly to black hole analogue systems. Once we have such a wave equation, we can extract the corresponding metric. This is a one-dimensional Hubbard model on a lattice with variable vertex degree and multiple edges between the same two vertices. The probability density for the matter obeys a (discrete) differential equation closed in the classical regime. This is a wave equation in which the vertex degree is related to the local speed of propagation of probability. This allows an interpretation of the probability density of particles similar to that in analogue gravity systems: *the matter sees a curved spacetime*. This establishes the desired relation between the connectivity of the graph and the curvature of its continuous analogue geometry.

4 Discussion

We argued that the idea of the Universe as a Computation is useful because it provides a route to a unification of matter and geometry. In the above, however, the Computation idea is a tool, not necessarily a fundamental ontology, and our results do not imply or require *it from bit*. It simply helps to talk about information as the primary object. It can be tempting to interpret this phrase as meaning that *bit* pre-exists *it*. However, I must admit that I do not understand what we can mean by information as decoupled from its physical realization. I prefer to simply take advantage of the fact that it allows us to study a system without having to specify the details of unknown physics.

In summary, it is *possible* to formulate Planck scale physics as a quantum information processing system, as we demonstrated in previous work (Quantum Causal Histories [22,13]). It is *useful* to formulate Planck scale physics as a quantum information processing system because: 1. Quantum information provides an unambiguous description of physics before geometry. 2. It is suitable for emergence problems, just as classical information is useful for statistical physics. 3. It provides a new toolbox well-adapted to background independent problems and lets us import methods from statistical physics to a background independent context. Information is useful because it allows us to study a system without committing to a particular ontology, necessary when the ontology is ambiguous, as is the case in emergence approaches to quantum gravity.

Even though I am not a believer in the full-blown Universe as a Computation philosophy, it is fun to explore some of the questions it raises from the rather concrete viewpoint described above. We shall end with a sample of such questions.

Information as unification. The old version of unification is the picture of group theory and symmetry breaking and the convergence of fundamental couplings. While this is now outdated, some level of unification is needed for quantum matter to interact with dynamical spacetime, as the language clash between differentiable manifolds and quantum fields on a fixed background has long been an obstacle to quantum gravity. The idea that information underlies everything allows a new path: express both gravity and matter in information theoretic terms. Quantum graphity models are a first step in that direction. It is a long way to go but we are catching a promising glimpse of a novel form of unification.

Why is the universe so stable? If the universe is a computer program, how come it doesn't crash, or at least it hasn't crashed yet? This sounds like a joke, but it is a relevant question in cosmology: is our universe a stable attractor, and if so why? It is interesting to look for potential commonalities between mechanisms for stability in computers and in physics. In computers, stability comes from some kind of built-in redundancy that provides error correction. In physics, certain symmetries can be seen as a kind of error correction. Elsewhere, we noted that the notion of decoherence-free subsystems used in quantum computing to protect against noise and errors is very similar to the notion of conserved quantities, something we used in [4] to find a large class of conserved quantities in Loop Quantum Gravity. I believe these results are only just scratching the surface.

How are the physics laws/computer program selected? Why our universe is what it is is a perennial problem in quantum gravity and cosmology. In the Universe as a Computation scenario it directly translates to the question of how the Program is selected, and this new viewpoint brings new possibilities. There are four commonly given answers: 1. Anthropic arguments: by construction, the universe we observe has to be compatible with the conscious life that observes it, hence it is unremarkable that the fundamental constants happen to fall within the narrow range that allows life. This is currently a very popular idea, supported by logic and possibly inflation and string theory, but also widely criticized as unscientific and non-explanatory. 2. Our laws have evolved through the history of the universe. This generally leads to meta-laws selecting the laws and a resulting circular argument. 3. Multiverse: our universe is one of many physically realized universes. The many can be arranged in various ways which have been thoroughly classified by Tegmark [27]. Unlike the anthropic argument, this scenario is wider, and, in some forms, in principle testable. But there is a huge proliferation of potential universes, not just those we can generate by allowing the fundamental constants to take other values (the usual multiverse), but also all possible laws or programs. 4. Ideas of self-organized criticality (SOC): our universe is a stable attractor. One may think that this should be the most promising direction, however, such ideas have hardly been explored. To a great extend, there is a serious technical obstacle. SOC is typically observed in non-equilibrium systems, while all of fundamental physics uses equilibrium quantum field theory. Properly introducing SOC ideas in cosmology requires a departure from the standard framework. Since many of the results in this area are already expressed in algorithmic terms, a description of the Universe as a computation can make it easier to introduce SOC ideas to a (quantum) cosmological setting. It will, of course, be necessary to study quantum systems that exhibit SOC first. This is a fascinating long-term direction for this kind of work.

Acknowledgements. This work was supported by the Alexander von Humboldt and the Templeton Foundations.

References

1. Ashtekar, A., Lewandowski, J.: Background independent quantum gravity: A status report, Class. Quant. Grav. 21, R53 (2004)
2. Ashtekar, A., Stachel, J.: Conceptual Problems of Quantum Gravity. Springer (1991)
3. Berlekamp, E.R., Conway, J.H., Guy, R.K.: Winning Ways for your Mathematical Plays. AK Peters Ltd. (2001)
4. Bilson-Thompson, S.O., Markopoulou, F., Smolin, L.: Quantum Gravity and the Standard Model. Class. Quant. Grav. 24, 3975 (2005)
5. Brandenberger, R.: Inflationary Cosmology: Progress and Problems, hep-ph/9910410
6. Caravelli, F., Hamma, A., Markopoulou, F., Riera, A.: Trapped surfaces and emergent curved space in the Bose-Hubbard model. Phys. Rev. D, arxiv:1108 (to appear, 2013)
7. Carroll, S.: The Cosmological Constant, astro-ph/0004075
8. Dennett, D.C.: Consciousness explained. Back Bay Books, Boston (2001)
9. Deutsch, D.: Physics, Philosophy and Quantum Technology. In: Shapiro, J.H., Hirota, O. (eds.) The Sixth International Conference on Quantum Communication, Measurement and Computing. Rinton Press, Princeton (2003)
10. Eisert, J., Osborne, T.J.: General Entanglement Scaling Laws from Time Evolution. Phys. Rev. Lett. 97, 150404 (2006)
11. Hamma, A., Markopoulou, F., Premont-Schwarz, I., Severini, S.: Lieb-Robinson bounds and the speed of light from topological order. Phys. Rev. Lett. 102, 017204, arXiv:0808.2495v2 [quant-ph]
12. Hamma, A., Markopoulou, F., Lloyd, S., Caravelli, F., Severini, S., Markstrom, K.: A quantum Bose-Hubbard model with evolving graph as toy model for emergent spacetime. Phys. Rev. D 81, 104032 (2010)
13. Hawkins, E., Markopoulou, F., Sahlmann, H.: Algebraic Causal Histories. Class. Q. Grav. 20, 3839 (2003)
14. Hossenfelder, S.: Experimental Search for Quantum Gravity, arxiv:1010.3420
15. Isham, C.J.: Prima Facie Questions in Quantum Gravity, gr-qc/9310031
16. Konopka, T., Markopoulou, F., Severini, S.: Quantum Graphity: a model of emergent locality. Phys. Rev. D 77, 104029 (2008)
17. Levin, M., Wen, X.G.: Fermions, strings, and gauge fields in lattice spin models. Phys. Rev. B 67, 245316 (2003)
18. Levin, M.A., Wen, X.G.: String-net condensation: A physical mechanism for topological phases. Phys. Rev. B 71, 045110 (2005)
19. Levin, M., Wen, X.G.: Quantum ether: Photons and electrons from a rotor model. arXiv:hep-th/0507118
20. Lieb, E.H., Robinson, D.W.: The finite group velocity of quantum spin systems. Commun. Math. Phys. 28, 251–257 (1972)
21. Lloyd, S.: Programming the Universe: A Quantum Computer Scientist Takes On the Cosmos, Knopf (2006)
22. Markopoulou, F.: Quantum Causal Histories. Class. Q. Grav. 17, 2059 (2000)
23. Nachtergaele, B., Sims, R.: Lieb-Robinson Bounds in Quantum Many-Body Physics. In: Sims, R., Ueltschi, D. (eds.) Entropy and the Quantum. Contemporary Mathematics, vol. 529, pp. 141–176. American Mathematical Society (2010)
24. Prémont-Schwarz, I., Hamma, A., Klich, I., Markopoulou-Kalamara, F.: Lieb-Robinson bounds for commutator-bounded operators, arXiv:0912.4544v1 [quant-ph]

25. Rovelli, C.: Quantum Gravity. Cambridge U. Press, New York (2004)
26. Susskind, L.: The Black Hole War, Little, Brown (2008)
27. Tegmark, M.: The Multiverse Hierarchy, arxiv:0905.1283
28. Wald, R.M.: The thermodynamics of black holes, gr-qc/9912119
29. Wheeler, J.A., Ford, K.: Geons, black holes and quantum foam: a life in physics. W.W. Norton Company, Inc., New York (1998)
30. Zuse, K.: Rechnender Raum. Elektronische Datenverarbeitung 8, 336–344 (1967)

Unifiability and Admissibility in Finite Algebras

George Metcalfe and Christoph Röthlisberger

Mathematics Institute, University of Bern, Sidlerstrasse 5, 3012 Bern, Switzerland
{george.metcalfe,christoph.roethlisberger}@math.unibe.ch

Abstract. Unifiability of finite sets of equations and admissibility of quasiequations in finite algebras are decidable problems, but a naive approach is computationally unfeasible even for small algebras. Algorithms are given here for obtaining more efficient proof systems deciding these problems, and some applications of the algorithms are described.

1 Introduction

Checking validity in finite algebras (similarly, derivability in finite-valued logics) has been studied extensively in the literature, and may be considered a "solved problem" in the sense that there exist both general methods for obtaining proof systems for checking validity (tableaux, resolution, multisequents, etc.) and standard optimization techniques for such systems (lemma generation, indexing, etc.) (cf., e.g., [14, 26, 3]). In this paper, we consider the related problems of checking the unifiability of finite sets of equations and admissibility of quasiequations in finite algebras. These problems are decidable (cf. [22]); however, a naive approach leads to computationally unfeasible procedures even for small algebras.

Let us fix a class of algebras $\mathcal{K}$ in the same language, noting that we will usually be concerned here with the case where $\mathcal{K}$ contains one finite algebra $\mathbf{A}$. A *unifier* in $\mathcal{K}$ for a finite set of equations Σ (in the same language) is a substitution σ such that for each $\varphi \approx \psi$ in Σ, the equation $\sigma(\varphi) \approx \sigma(\psi)$ holds in all members of $\mathcal{K}$. The (equational) unifiability problem for $\mathcal{K}$ consists of checking whether or not a given finite set of equations Σ has a unifier in $\mathcal{K}$. A more challenging problem, not considered in this work, is to find "most general bases" for unifiers in $\mathcal{K}$ (i.e., such that each unifier of Σ in $\mathcal{K}$ is obtained from a member of the basis by applying a further substitution), and to determine the unification type (unary, finitary, infinitary, or nullary) of $\mathcal{K}$ (cf., e.g., [2, 9, 10]). Unifiability allowing additional free constants may also be investigated (cf., e.g., [2, 24, 1]).

If the language of $\mathcal{K}$ contains constants, then checking unifiability in $\mathcal{K}$ amounts to checking satisfiability in the corresponding ground algebra: namely, $\mathbf{F}_{\mathcal{K}}(0)$, the free algebra of $\mathcal{K}$ on zero generators. If this algebra is finite (e.g., if $\mathcal{K}$ generates a locally finite variety), then it can be used to check unifiability, employing tools such as MUltlog/MUltseq [23, 12] or $_3T^4P$ [5] to obtain a corresponding proof system. If the language of $\mathcal{K}$ lacks a constant, then unifiability in $\mathcal{K}$ amounts to satisfiability in any subalgebra of $\mathbf{F}_{\mathcal{K}}(1)$, the free algebra of $\mathcal{K}$ on one generator. Again, if $\mathbf{F}_{\mathcal{K}}(1)$ is finite (guaranteed if $\mathcal{K}$ generates a locally finite variety), then

S.B. Cooper, A. Dawar, and B. Löwe (Eds.): CiE 2012, LNCS 7318, pp. 485–495, 2012.

using a tool such as the Algebra Workbench [25], the smallest such subalgebra can be found and a corresponding proof system obtained.

Checking the admissibility of a rule or quasiequation is more challenging. Intuitively, a rule is admissible in a logical system if it can be added to the system without producing any new theorems. Algebraically, a quasiequation $\Sigma \Rightarrow \varphi \approx \psi$ is admissible in $\mathcal{K}$ if every unifier of Σ in $\mathcal{K}$ is also a unifier of $\varphi \approx \psi$ in $\mathcal{K}$. Equivalently, $\Sigma \Rightarrow \varphi \approx \psi$ is admissible if and only if (henceforth iff) it holds in $\mathbf{F}_{\mathcal{K}}(\omega)$, the free algebra on countably infinitely many generators of $\mathcal{K}$. Note that Σ is unifiable in $\mathcal{K}$ iff $\Sigma \Rightarrow p \approx q$ is *not* admissible in $\mathcal{K}$ where p, q are any distinct variables not occurring in Σ. Admissibility (in tandem with unification) has been studied intensively in the context of intermediate and transitive modal logics and their algebras [22, 15, 9, 10, 17, 8], leading also to proof systems for checking admissibility [11, 16, 4], and certain many-valued logics and their algebras [22, 7, 18–20], but a general theory for this latter case has so far been lacking.

If $\mathcal{K}$ contains a single algebra $\mathbf{A}$ with n elements, then admissibility amounts to holding in the finite free algebra $\mathbf{F}_{\mathbf{A}}(n)$ and is decidable [22]. However, even for small n, the size of $\mathbf{F}_{\mathbf{A}}(n)$ may be prohibitively large. This is particularly striking when compared with the fact that derivability and admissibility in $\mathcal{K}$ may coincide; that is, $\mathcal{K}$ may be *structurally complete* (cf., e.g., [22, 7]). In other cases, admissibility of a quasiequation corresponds to holding in other, often quite small, algebras (cf., e.g., [20]). An algorithm is provided in this paper for discovering such algebras. Namely, we show that admissibility in $\mathbf{A}$ corresponds to holding in any subalgebra $\mathbf{B}$ of $\mathbf{F}_{\mathbf{A}}(n)$ where $\mathbf{A}$ is a homomorphic image of $\mathbf{B}$. Since tools such as the Algebra Workbench [25] allow us to calculate subalgebras and check homomorphic images, we obtain a procedure that is feasible even when the size of the free algebras is relatively large. We illustrate our procedure using a selection of well-known finite algebras, confirming some known results from the literature, and establishing new ones. We also investigate a situation where the smallest algebra for checking admissibility may not be the best choice, larger algebras with the structure of a product sometimes being more suitable for automated reasoning methods. Finally, we address, via an example, the issue that subalgebras of the free algebra $\mathbf{F}_{\mathbf{A}}(n)$ may not always give the best result, since there may exist smaller suitable subalgebras of a product of $\mathbf{F}_{\mathbf{A}}(n)$.

Unifiability and admissibility play a fundamental meta-level role in describing "hidden properties" of (classes of) algebras and logics. In particular, establishing the completeness of a logic or proof system with respect to some restricted class of algebras (perhaps just one standard algebra) often involves showing that a certain rule or quasiequation is admissible. Similarly, deciding unifiability of concepts can be a useful tool for database redundancy checking in description logics [1]. Finally, a longer term goal of the current work is to automatically obtain admissible rules for finite algebras and logics that can be used to simplify reasoning steps or to speed up derivations for checking validity.

2   Preliminaries

Let us first recall some ideas from universal algebra, referring to [6, 13] for further details. For a language $\mathcal{L}$, we denote the formula algebra over countably infinitely many variables by $\mathbf{Fm}_{\mathcal{L}}$ and let φ, ψ stand for arbitrary members of the universe $\mathrm{Fm}_{\mathcal{L}}$ called $\mathcal{L}$-*formulas*. An $\mathcal{L}$-*equation* is a pair of $\mathcal{L}$-formulas, written $\varphi \approx \psi$. An $\mathcal{L}$-*clause* is an ordered pair of finite sets of $\mathcal{L}$-equations, written $\Sigma \Rightarrow \Delta$, called an $\mathcal{L}$-*quasiequation* if $|\Delta| = 1$ and an $\mathcal{L}$-*negative clause* if $\Delta = \emptyset$. As usual, if the language is clear from the context we may omit the prefix $\mathcal{L}$.

Let us fix $\mathcal{K}$ to be a class of $\mathcal{L}$-algebras, noting that in this paper $\mathcal{K}$ will often consist of a single finite $\mathcal{L}$-algebra $\mathbf{A}$, which we typically write without brackets. A set Σ of $\mathcal{L}$-equations is *satisfiable* in an $\mathcal{L}$-algebra $\mathbf{A}$ if $\Sigma \subseteq \ker h$ for some homomorphism $h \colon \mathbf{Fm}_{\mathcal{L}} \to \mathbf{A}$. We write $\Sigma \models_{\mathcal{K}} \Delta$ and say, if $\Sigma \cup \Delta$ is finite, that the $\mathcal{L}$-clause $\Sigma \Rightarrow \Delta$ "holds in $\mathcal{K}$" if for every $\mathbf{A} \in \mathcal{K}$ and homomorphism $h \colon \mathbf{Fm}_{\mathcal{L}} \to \mathbf{A}$, $\Sigma \subseteq \ker h$ implies $\Delta \cap \ker h \neq \emptyset$.

$\mathcal{K}$ is said to be *axiomatized* by a set of $\mathcal{L}$-clauses Λ if $\mathcal{K}$ is the class of $\mathcal{L}$-algebras $\mathbf{A}$ such that all $\mathcal{L}$-clauses in Λ hold in $\mathbf{A}$. $\mathcal{K}$ is called a *variety, quasivariety,* or *antivariety* if it is axiomatized by a set of $\mathcal{L}$-equations, $\mathcal{L}$-quasiequations, or $\mathcal{L}$-negative clauses, respectively. The variety $\mathbb{V}(\mathcal{K})$, quasivariety $\mathbb{Q}(\mathcal{K})$, and antivariety $\mathbb{V}^{-}(\mathcal{K})$ *generated by* $\mathcal{K}$ are the smallest variety, quasivariety, and antivariety containing $\mathcal{K}$, respectively. Let $\mathbb{H}, \mathbb{I}, \mathbb{S}, \mathbb{P}, \mathbb{P}_U, \mathbb{P}_U^*$, and $\mathbb{H}^{-1}$ be, respectively, the class operators of taking homomorphic images, isomorphic images, subalgebras, products, ultraproducts, non-empty ultraproducts, and homomorphic preimages. Then $\mathbb{V}(\mathcal{K}) = \mathbb{HSP}(\mathcal{K})$, $\mathbb{Q}(\mathcal{K}) = \mathbb{ISPP}_U(\mathcal{K})$, and $\mathbb{V}^{-}(\mathcal{K}) = \mathbb{H}^{-1}\mathbb{SP}_U^*(\mathcal{K})$; for a finite algebra $\mathbf{A}$ these latter equivalences refine to $\mathbb{Q}(\mathbf{A}) = \mathbb{ISP}(\mathbf{A})$ and $\mathbb{V}^{-}(\mathbf{A}) = \mathbb{H}^{-1}\mathbb{S}(\mathbf{A})$ (cf. [6, 13] for details).

For a cardinal κ, let $\mathbf{F}_{\mathcal{K}}(\kappa)$ denote the free algebra of $\mathcal{K}$ with κ generators, recalling that $\mathbb{V}(\mathcal{K}) = \mathbb{V}(\mathbf{F}_{\mathcal{K}}(\omega))$ and that $\mathbb{V}(\mathcal{K}_1) = \mathbb{V}(\mathcal{K}_2)$ implies $\mathbf{F}_{\mathcal{K}_1}(\omega) = \mathbf{F}_{\mathcal{K}_2}(\omega)$ (cf. [6]). Note that when considering unifiability and admissibility, it can be helpful to view the elements of $\mathbf{F}_{\mathcal{K}}(\kappa)$ for $\kappa \leq \omega$ as equivalence classes $[\varphi]$ of formulas φ containing at most κ variables, defined with respect to the congruence relating φ and ψ whenever $\models_{\mathcal{K}} \varphi \approx \psi$.

3   Unifiability

As a warm-up for the harder case of admissibility, let us first consider the problem of checking whether a set of $\mathcal{L}$-equations Σ is *unifiable* in a class $\mathcal{K}$ of $\mathcal{L}$-algebras. Formally, a homomorphism (substitution) $\sigma \colon \mathbf{Fm}_{\mathcal{L}} \to \mathbf{Fm}_{\mathcal{L}}$ is called a $\mathcal{K}$-*unifier for* Σ if $\models_{\mathcal{K}} \sigma(\varphi) \approx \sigma(\psi)$ for every $(\varphi \approx \psi) \in \Sigma$. If such a homomorphism exists, then Σ is called $\mathcal{K}$-*unifiable*. In this paper, we shall be particularly interested in the case where Σ is finite and $\mathcal{K}$ consists of just one finite algebra (or is at least a locally finite variety). Our goal will be to find for an arbitrary finite algebra $\mathbf{A}$, another algebra $\mathbf{B}$ such that checking unifiability in $\mathbf{A}$ corresponds to checking satisfiability in $\mathbf{B}$. Standard tools for finite algebras and logics such as MUltlog/MUltseq [23, 12] or ${}_3T^AP$ [5] may then be used to automatically obtain a proof system for this latter task.

Let us fix $\mathcal{K}$ to be a class of $\mathcal{L}$-algebras and Σ to be a set of $\mathcal{L}$-equations. Observe first that Σ is $\mathcal{K}$-unifiable iff there is a $\mathcal{K}$-unifier for Σ mapping each formula to a formula with at most one variable. Simply note that if σ is a $\mathcal{K}$-unifier for Σ, then also $\tilde{\sigma} \circ \sigma$ is a $\mathcal{K}$-unifier for Σ where $\tilde{\sigma}$ maps every variable to a fixed variable p. In fact $\tilde{\sigma} \circ \sigma$ is a $\mathcal{K}$-unifier for Σ for *any* substitution $\tilde{\sigma}$. In particular, if $\mathcal{L}$ contains at least one constant, then we can map formulas to ground formulas. Building on this simple observation, the following lemmas establish that checking $\mathcal{K}$-unifiability amounts to checking satisfiability in any subalgebra of $\mathbf{F}_{\mathcal{K}}(1)$, and, moreover, that smallest finite subalgebras of $\mathbf{F}_{\mathcal{K}}(1)$ are the smallest algebras for which $\mathcal{K}$-unifiability corresponds to satisfiability.

Lemma 1. *The following are equivalent for any* $\mathbf{B} \in \mathbb{IS}(\mathbf{F}_{\mathcal{K}}(1))$:

(1) Σ *is* $\mathcal{K}$*-unifiable.*
(2) Σ *is satisfiable in* $\mathbf{F}_{\mathcal{K}}(1)$.
(3) Σ *is satisfiable in* $\mathbf{B}$.

Proof. $(1) \Rightarrow (2)$. If Σ is $\mathcal{K}$-unifiable, then there is a $\mathcal{K}$-unifier σ for Σ mapping each formula to a formula with at most one variable. It follows that Σ is satisfied in $\mathbf{F}_{\mathcal{K}}(1)$ by the unique homomorphism $h \colon \mathbf{Fm}_{\mathcal{L}} \to \mathbf{F}_{\mathcal{K}}(1)$ mapping each variable p to the equivalence class of $\sigma(p)$ in $\mathbf{F}_{\mathcal{K}}(1)$.

$(2) \Rightarrow (3)$. Suppose that $\Sigma \subseteq \ker h$ for some $h \colon \mathbf{Fm}_{\mathcal{L}} \to \mathbf{F}_{\mathcal{K}}(1)$. Define a homomorphism $k \colon \mathbf{F}_{\mathcal{K}}(1) \to \mathbf{B}$ by $k(x) = a$ for some $a \in \mathbf{B}$ where x is the single generator of $\mathbf{F}_{\mathcal{K}}(1)$. Then $\Sigma \subseteq \ker(k \circ h)$ as required.

$(3) \Rightarrow (1)$. If $\Sigma \subseteq \ker h$ for some $h \colon \mathbf{Fm}_{\mathcal{L}} \to \mathbf{B}$ and e embeds $\mathbf{B}$ into $\mathbf{F}_{\mathcal{K}}(1)$, then any substitution mapping p to a member of the equivalence class $e(h(p))$ is a $\mathcal{K}$-unifier of Σ.   □

Proposition 1. *The following are equivalent for any* $\mathbf{B} \in \mathcal{K}$:

(1) Σ *is* $\mathcal{K}$*-unifiable iff* Σ *is satisfiable in* $\mathbf{B}$.
(2) $\mathbb{V}^-(\mathbf{F}_{\mathcal{K}}(1)) = \mathbb{V}^-(\mathbf{B})$.

Moreover, if $\mathbf{C}$ *is a smallest finite subalgebra of* $\mathbf{F}_{\mathcal{K}}(1)$, *then* $|C| \leq |B|$ *for any* $\mathbf{B} \in \mathcal{K}$ *satisfying* (1) *and* (2).

Proof. Note first that $\mathbb{V}^-(\mathbf{F}_{\mathcal{K}}(1)) = \mathbb{V}^-(\mathbf{B})$ iff the same negative clauses $\Sigma \Rightarrow \emptyset$ hold in $\mathbf{F}_{\mathcal{K}}(1)$ and $\mathbf{B}$. But $\Sigma \Rightarrow \emptyset$ holds in $\mathbf{F}_{\mathcal{K}}(1)$ iff Σ is not satisfiable in $\mathbf{F}_{\mathcal{K}}(1)$ iff, by Lemma 1, Σ is not $\mathcal{K}$-unifiable. Moreover, $\Sigma \Rightarrow \emptyset$ holds in $\mathbf{B}$ iff Σ is not satisfiable in $\mathbf{B}$. Hence $\mathbb{V}^-(\mathbf{F}_{\mathcal{K}}(1)) = \mathbb{V}^-(\mathbf{B})$ iff whenever Σ is $\mathcal{K}$-unifiable, then Σ is satisfiable in $\mathbf{B}$, and vice versa. Suppose now that $\mathbf{C}$ is a smallest finite subalgebra of $\mathbf{F}_{\mathcal{K}}(1)$. By Lemma 1, $\mathbf{C}$ satisfies (1) and hence (2); i.e., $\mathbb{V}^-(\mathbf{F}_{\mathcal{K}}(1)) = \mathbb{V}^-(\mathbf{C})$. So for any $\mathbf{B} \in \mathcal{K}$ satisfying (1) and (2), $\mathbf{B} \in \mathbb{V}^-(\mathbf{F}_{\mathcal{K}}(1)) = \mathbb{V}^-(\mathbf{C}) = \mathbb{H}^{-1}\mathbb{S}(\mathbf{C}) = \mathbb{H}^{-1}(\mathbf{C})$, and it follows that $|C| \leq |B|$.   □

Hence we obtain an (in some sense) optimal proof system for checking unifiability in a given finite algebra by (i) calculating the finite algebra $\mathbf{F}_{\mathbf{A}}(1)$, and (ii) calculating a smallest subalgebra $\mathbf{B}$ of $\mathbf{F}_{\mathbf{A}}(1)$ (e.g., using the Algebra Workbench [25]), and then (iii) deriving a proof system for checking satisfiability in $\mathbf{B}$ (e.g., using MUltlog/MUltseq [23, 12] or ${}_3\mathcal{TAP}$ [5]).

Example 1. Consider the 4-element algebra $\mathbf{D_4} = \langle\{\bot, a, b, \top\}, \wedge, \vee, \neg, \bot, \top\rangle$ (noting that $\mathbb{Q}(\mathbf{D_4})$ is the variety of De Morgan algebras) consisting of a distributive bounded lattice with an involutive negation defined as shown below:

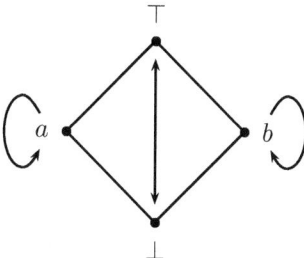

Since there are constants in the language of $\mathbf{D_4}$, the smallest algebra for checking unifiability is the ground algebra $\mathbf{F_{D_4}}(0)$ with elements associated to $\bot$ and $\top$. That is, checking unifiability amounts to checking classical satisfiability. E.g., $p \wedge \neg p \approx p \vee \neg p$ is not unifiable in $\mathbf{D_4}$, since $\top \wedge \neg\top \neq \top \vee \neg\top$ and $\bot \wedge \neg\bot \neq \bot \vee \neg\bot$. Moreover, although the constant-free case of $\mathbf{D_4^L} = \langle\{\bot, a, b, \top\}, \wedge, \vee, \neg\rangle$ (which generates the variety of De Morgan lattices) is not so immediate, there exists a smallest subalgebra of $\mathbf{F_{D_4^L}}(1)$ with elements corresponding to $p \wedge \neg p$ and $p \vee \neg p$. So checking unifiability in $\mathbf{D_4^L}$ amounts again to checking classical satisfiability.

$\mathcal{K}$ need not consist of just one or finitely many finite algebras. It is enough that the $\mathbf{F_{\mathcal{K}}}(0)$ algebra, if it exists, or otherwise the $\mathbf{F_{\mathcal{K}}}(1)$ algebra, is finite. This is the case for any locally finite (i.e., finitely generated algebras are finite) variety.

Example 2. Consider the (locally finite) variety of *bounded positive Sugihara monoids* studied in [21]: algebras $\langle A, \wedge, \vee, \cdot, \rightarrow, t, \bot, \top\rangle$ where $\langle A, \wedge, \vee, \bot, \top\rangle$ is a bounded lattice, $\langle A, \cdot, t\rangle$ is a commutative idempotent monoid, and $\rightarrow$ is the residual of $\cdot$ (i.e., $x \leq y \rightarrow z$ iff $x \cdot y \leq z$) satisfying $t \leq (x \rightarrow y) \vee (y \rightarrow x)$ and $[(((x \rightarrow y) \rightarrow y) \rightarrow x) \rightarrow z] \cdot [(((y \rightarrow x) \rightarrow x) \rightarrow y) \rightarrow z] \leq z$. The free algebra of this variety on zero generators has 6 elements: the equivalence classes of t, $\top$, $\bot$, $\top \rightarrow t$, $(\top \rightarrow t) \rightarrow \bot$, and $((\top \rightarrow t) \rightarrow \bot) \wedge t$. Hence unification in the variety can be decided using a proof system for satisfiability in this algebra.

Example 3. Even if the free algebra on zero generators is infinite, it may have a manageable structure. Consider the variety of bounded distributive lattices with a unary operator $\square$ satisfying $\square(x \wedge y) \approx \square x \wedge \square y$ and $\square\top \approx \top$. The free algebra on zero generators may be represented as a chain $\bot < \square\bot < \square\square\bot < \ldots < \top$ isomorphic to $\langle \mathbb{N} \cup \{\infty\}, \min, \max, s\rangle$ where $s(x) = x + 1$. Satisfiability in this algebra and hence also checking unifiability in the variety is decidable. Note, however, that adding Boolean negation to the language gives algebras for the modal logic K; decidability of unification in this variety is one of the most challenging open problems in the area (cf., e.g., [17, 1]).

4   Admissibility

An $\mathcal{L}$-quasiequation $\Sigma \Rightarrow \varphi \approx \psi$ is said to be *admissible* in a class of $\mathcal{L}$-algebras $\mathcal{K}$ if every $\mathcal{K}$-unifier of Σ is a $\mathcal{K}$-unifier of $\varphi \approx \psi$. $\mathcal{K}$ is called *structurally complete* if admissibility and validity coincide: i.e., $\Sigma \Rightarrow \varphi \approx \psi$ is admissible in $\mathcal{K}$ iff $\Sigma \models_{\mathcal{K}} \varphi \approx \psi$. Admissibility generalizes unifiability in the following sense: a finite set of $\mathcal{L}$-equations Σ is unifiable in a non-trivial class $\mathcal{K}$ of $\mathcal{L}$-algebras iff the quasiequation $\Sigma \Rightarrow p \approx q$ with p and q not occurring in Σ is *not* admissible.

Quasiequations admissible in $\mathcal{K}$ are, equivalently, quasiequations that hold in $\mathbf{F}_{\mathcal{K}}(\omega)$, the free algebra on countably infinitely many generators of $\mathcal{K}$.

Lemma 2 ([20]). *For a class of $\mathcal{L}$-algebras $\mathcal{K}$:*

(a) $\Sigma \Rightarrow \varphi \approx \psi$ *is admissible in $\mathcal{K}$ iff $\Sigma \models_{\mathbf{F}_{\mathcal{K}}(\omega)} \varphi \approx \psi$.*
(b) $\mathcal{K}$ *is structurally complete iff $\mathbb{Q}(\mathcal{K}) = \mathbb{Q}(\mathbf{F}_{\mathcal{K}}(\omega))$.*

Decidability of admissibility in finite algebras is guaranteed by the following:

Lemma 3 ([22]). *Let $\mathbf{A}$ be an n-element algebra. Then $\mathbf{F}_{\mathbf{A}}(m)$ is finite for all $m \in \mathbb{N}$ and $\mathbb{Q}(\mathbf{F}_{\mathbf{A}}(\omega)) = \mathbb{Q}(\mathbf{F}_{\mathbf{A}}(n))$.*

However, even free algebras on a small number of generators can be very large. E.g., the free algebra $\mathbf{F}_{\mathbf{D}_4}(2)$ (cf. Example 1) has 168 elements. We therefore seek smaller algebras that also generate $\mathbb{Q}(\mathbf{F}_{\mathbf{A}}(\omega))$ as a quasivariety.

Proposition 2. *Let $\mathbf{A}, \mathbf{B}$ be $\mathcal{L}$-algebras such that $\mathbf{B} \in \mathbb{Q}(\mathbf{F}_{\mathbf{A}}(\omega))$, and $\mathbf{A} \in \mathbb{V}(\mathbf{B})$ (in particular, if $\mathbf{B} \in \mathbb{S}(\mathbf{F}_{\mathbf{A}}(\omega))$ and $\mathbf{A} \in \mathbb{H}(\mathbf{B})$). Then*

(a) $\mathbb{Q}(\mathbf{B}) = \mathbb{Q}(\mathbf{F}_{\mathbf{A}}(\omega))$.
(b) $\Sigma \Rightarrow \varphi \approx \psi$ *is admissible in $\mathbf{A}$ iff $\Sigma \models_{\mathbf{B}} \varphi \approx \psi$.*

Proof. (a) Since $\mathbf{B} \in \mathbb{Q}(\mathbf{F}_{\mathbf{A}}(\omega))$, immediately $\mathbb{Q}(\mathbf{B}) \subseteq \mathbb{Q}(\mathbf{F}_{\mathbf{A}}(\omega))$. For the other direction, note that $\mathbb{V}(\mathbf{A}) \subseteq \mathbb{V}(\mathbf{B}) \subseteq \mathbb{V}(\mathbb{Q}(\mathbf{F}_{\mathbf{A}}(\omega))) = \mathbb{V}(\mathbf{F}_{\mathbf{A}}(\omega)) = \mathbb{V}(\mathbf{A})$; i.e. $\mathbb{V}(\mathbf{A}) = \mathbb{V}(\mathbf{B})$. Hence $\mathbf{F}_{\mathbf{A}}(\omega) = \mathbf{F}_{\mathbf{B}}(\omega) \in \mathbb{Q}(\mathbf{B})$ and $\mathbb{Q}(\mathbf{F}_{\mathbf{A}}(\omega)) \subseteq \mathbb{Q}(\mathbf{B})$.
(b) follows directly from (a) using Lemma 2.   □

This suggests the following procedure when $\mathbf{A}$ is finite:

(i) Find the smallest free algebra $\mathbf{F}_{\mathbf{A}}(m)$ ($m \leq |A|$) such that $\mathbf{A} \in \mathbb{H}(\mathbf{F}_{\mathbf{A}}(m))$.
(ii) Compute the set $\mathrm{Sub}(\mathbf{F}_{\mathbf{A}}(m))$ of subalgebras of $\mathbf{F}_{\mathbf{A}}(m)$.
(iii) Construct the set $\mathrm{Adm}(\mathbf{A})$ of all $\mathbf{B} \in \mathrm{Sub}(\mathbf{F}_{\mathbf{A}}(m))$ such that $\mathbf{A} \in \mathbb{H}(\mathbf{B})$.
(iv) Derive a proof system for checking satisfiability in a smallest $\mathbf{B} \in \mathrm{Adm}(\mathbf{A})$.

Steps (i)-(iii) of the procedure have been implemented using macros implemented for the Algebra Workbench [25]. Step (iv) can be implemented directly making use of a system such as MUltlog/MUltseq [23, 12] or $\mathcal{3T^AP}$ [5].

Example 4. Consider the algebra $\mathbf{S}_3^{\rightarrow\neg} = \langle \{-1, 0, 1\}, \rightarrow, \neg \rangle$ with operations

$\rightarrow$	-1	0	1
-1	1	1	1
0	-1	0	1
1	-1	-1	1

$\neg$	
-1	1
0	0
1	-1

noting that an equation of the form $\varphi \approx \varphi \rightarrow \varphi$ holds in $\mathbf{S}_3^{\rightarrow\neg}$ iff φ is a theorem of the $\{\rightarrow, \neg\}$-fragment of the logic RM. Now, following our procedure, we obtain:

(i) $\mathbf{S_3^{\rightarrow\neg}} \notin \mathbb{H}(\mathbf{F_{S_3^{\rightarrow\neg}}}(1))$, but $\mathbf{S_3^{\rightarrow\neg}} \in \mathbb{H}(\mathbf{F_{S_3^{\rightarrow\neg}}}(2))$, so $\mathbb{Q}(\mathbf{F_{S_3^{\rightarrow\neg}}}(\omega)) = \mathbb{Q}(\mathbf{F_{S_3^{\rightarrow\neg}}}(2))$.

(ii) $\mathbf{F_{S_3^{\rightarrow\neg}}}(2)$ has 264 elements and $\mathrm{Sub}(\mathbf{F_{S_3^{\rightarrow\neg}}}(2))$ contains its 5134 subalgebras.

(iii) $\mathrm{Adm}(\mathbf{S_3^{\rightarrow\neg}})$ contains 989 subalgebras $\mathbf{B}$ of $\mathbf{F_{S_3^{\rightarrow\neg}}}(2)$ such that $\mathbf{S_3^{\rightarrow\neg}} \in \mathbb{H}(\mathbf{B})$.

(iv) The two smallest algebras in $\mathrm{Adm}(\mathbf{S_3^{\rightarrow\neg}})$ have 6 elements; e.g.,

$$\mathbf{B} = \langle\{[p \rightarrow p], [\neg(p \rightarrow p)], [(p \rightarrow q) \rightarrow (p \rightarrow q)], [(q \rightarrow q) \rightarrow (p \rightarrow p)],$$
$$[\neg((p \rightarrow q) \rightarrow (p \rightarrow q))], [\neg((q \rightarrow q) \rightarrow (p \rightarrow p))]\}, \rightarrow, \neg\rangle.$$

Example 5. In some cases, it is possible to establish structural completeness results for the algebra $\mathbf{A}$ (equivalently, the quasivariety generated by $\mathbf{A}$) using the described procedure. The smallest $\mathbf{B} \in \mathrm{Adm}(\mathbf{A})$ may be an isomorphic copy of $\mathbf{A}$ itself; i.e., $\mathbf{A}$ can be embedded into $\mathbf{F_A}(m)$ for some m. In particular, known structural completeness results have been confirmed for

$\mathbf{L_3^{\rightarrow}} = \langle\{0, \frac{1}{2}, 1\}, \rightarrow_L\rangle$ the 3-element Komori C-algebra
$\mathbf{B_1} = \langle\{0, \frac{1}{2}, 1\}, \min, \max, \neg_G\rangle$ the 3-element Stone algebra
$\mathbf{G_3} = \langle\{0, \frac{1}{2}, 1\}, \min, \max, \rightarrow_G\rangle$ the 3-element positive Gödel algebra
$\mathbf{S_3^{\rightarrow}} = \langle\{-1, 0, 1\}, \rightarrow_S\rangle$ the 3-element implicational Sugihara monoid

where $x \rightarrow_L y = \min(1, 1 - x + y)$, $x \rightarrow_G y$ is y if $x > y$, otherwise 1, $\neg_G x = x \rightarrow_G 0$, and $\rightarrow_S$ is the operation $\rightarrow$ of $\mathbf{S_3^{\rightarrow\neg}}$ from Example 4. A new structural completeness result has also been established for the pseudocomplemented distributive lattice $\mathbf{B_2}$ obtained by adding a top element to the 4-element Boolean algebra. Note, however, that the procedure timed out for the case of the 9-element algebra $\mathbf{B_3}$.

Example 6. Recall from Example 1 that the algebras $\mathbf{D_4}$ and $\mathbf{D_4^L}$ generate the varieties of De Morgan algebras and De Morgan lattices, respectively. In both cases, the algebras $\mathbf{D_4}$ and $\mathbf{D_4^L}$ are homomorphic images of the corresponding free algebras on two generators (with 168 and 166 elements, respectively) but not on one generator. The smallest algebra found automatically in $\mathrm{Adm}(\mathbf{D_4})$ is isomorphic to the product $\mathbf{D_4} \times \mathbf{2}$ where $\mathbf{2}$ is the 2-element Boolean algebra, while the smallest algebra in $\mathrm{Adm}(\mathbf{D_4^L})$ is isomorphic to $\mathbf{D_4} \times \mathbf{2}$ with extra top and bottom elements. The fact that these smallest algebras are themselves structurally complete and generate the varieties of De Morgan algebras and lattices was established in [20]. Our procedure here confirms these results automatically.

Similar results were also obtained in [20] for Kleene algebras and lattices (subvarieties of De Morgan algebras and lattices) generated by the 3-element chains $\mathbf{C_3} = \langle\{\top, a, \bot\}, \wedge, \vee, \neg, \bot, \top\rangle$ and $\mathbf{C_3^L} = \langle\{\top, a, \bot\}, \wedge, \vee, \neg\rangle$ where $\neg$ swaps $\bot$ and $\top$ and leaves a fixed. In both cases the smallest algebra in $\mathrm{Adm}(\mathbf{C_3})$ and $\mathrm{Adm}(\mathbf{C_3^L})$, found automatically by our procedure, is a 4-element chain.

For certain classes of algebras $\mathcal{K}$, structural completeness fails but a quasiequation $\Sigma \Rightarrow \varphi \approx \psi$ is admissible iff $\Sigma \models_{\mathcal{K}} \varphi \approx \psi$ or Σ is not unifiable in $\mathcal{K}$. Moreover, following the previous section, the task of checking unifiability in $\mathcal{K}$ corresponds to checking satisfiability in some subalgebra of $\mathbf{F}_{\mathcal{K}}(1)$. Below we characterize this situation in the case where $\mathcal{K}$ contains just one algebra.

Lemma 4. *For an $\mathcal{L}$-algebra $\mathbf{A}$ and $\mathbf{A} \times \mathbf{B} \in \mathbb{Q}(\mathbf{A})$, the following are equivalent:*

(1) $\Sigma \models_{\mathbf{A} \times \mathbf{B}} \varphi \approx \psi$.
(2) $\Sigma \models_{\mathbf{A}} \varphi \approx \psi$ *or Σ is not satisfiable in $\mathbf{B}$.*

Proof. (1) $\Rightarrow$ (2). If Σ is satisfiable in $\mathbf{B}$, then there exists a homomorphism $h\colon \mathbf{Fm}_{\mathcal{L}} \to \mathbf{B}$ with $\Sigma \subseteq \ker h$. For any homomorphism $k\colon \mathbf{Fm}_{\mathcal{L}} \to \mathbf{A}$ with $\Sigma \subseteq \ker k$, define $e\colon \mathbf{Fm}_{\mathcal{L}} \to \mathbf{A} \times \mathbf{B}$ by $e(u) = (k(u), h(u))$. Then $\Sigma \subseteq \ker e$ so, using (1), $e(\varphi) = e(\psi)$ and $k(\varphi) = k(\psi)$. I.e., $\Sigma \models_{\mathbf{A}} \varphi \approx \psi$.

(2) $\Rightarrow$ (1). If $\Sigma \models_{\mathbf{A}} \varphi \approx \psi$, then $\Sigma \models_{\mathbf{A} \times \mathbf{B}} \varphi \approx \psi$, since $\mathbf{A} \times \mathbf{B} \in \mathbb{Q}(\mathbf{A})$. If Σ is not satisfiable in $\mathbf{B}$, then Σ is not satisfiable in $\mathbf{A} \times \mathbf{B}$, so $\Sigma \models_{\mathbf{A} \times \mathbf{B}} \varphi \approx \psi$.   □

Corollary 1. *If $\mathbf{A}$ is an $\mathcal{L}$-algebra and $\mathbb{Q}(\mathbf{F}_{\mathbf{A}}(\omega)) = \mathbb{Q}(\mathbf{A} \times \mathbf{B})$, then the following are equivalent:*

(1) $\Sigma \Rightarrow \varphi \approx \psi$ *is admissible in $\mathbf{A}$.*
(2) $\Sigma \models_{\mathbf{A}} \varphi \approx \psi$ *or Σ is not satisfiable in $\mathbf{B}$.*

Example 7. Consider the 3-valued Łukasiewicz algebra $\mathbf{L_3} = \langle \{0, \frac{1}{2}, 1\}, \to, \neg\rangle$ with $x \to y = \min(1, 1 - x + y)$ and $\neg x = 1 - x$. Following our procedure:

(i) $\mathbf{L_3} \notin \mathbb{H}(\mathbf{F}_{\mathbf{L_3}}(0))$, but $\mathbf{L_3} \in \mathbb{H}(\mathbf{F}_{\mathbf{L_3}}(1))$, so $\mathbb{Q}(\mathbf{F}_{\mathbf{L_3}}(\omega)) = \mathbb{Q}(\mathbf{F}_{\mathbf{L_3}}(1))$.
(ii) $\mathbf{F}_{\mathbf{L_3}}(1)$ has 12 elements and $\mathrm{Sub}(\mathbf{F}_{\mathbf{L_3}}(1))$ contains its 8 subalgebras.
(iii) $\mathrm{Adm}(\mathbf{L_3})$ contains just one proper subalgebra $\mathbf{B}$ of $\mathbf{F}_{\mathbf{L_3}}(1)$ with $\mathbf{L_3} \in \mathbb{H}(\mathbf{B})$; $\mathbf{B}$ is isomorphic to $\mathbf{L_3} \times \mathbf{2}$ (where $\mathbf{2} = \langle \{0,1\}, \to, \neg\rangle$), so by Corollary 1, $\Sigma \Rightarrow \varphi \approx \psi$ is admissible iff $\Sigma \models_{\mathbf{L_3}} \varphi \approx \psi$ or Σ is not classically satisfiable.

Note finally that, unlike the situation for unification, our procedure may not find the smallest algebra that generates $\mathbb{Q}(\mathbf{F}_{\mathbf{A}}(\omega))$. There may be a subalgebra of a product of copies of $\mathbf{F}_{\mathbf{A}}(m)$ which also generates this quasivariety, is not a subalgebra of $\mathbf{F}_{\mathbf{A}}(\omega)$, and has fewer elements than any member of $\mathrm{Adm}(\mathbf{A})$.

Example 8. Consider the algebra $\mathbf{P} = \langle \{a, b, c, d\}, \star\rangle$ where the unary operation $\star$ and the free algebras $\mathbf{F}_{\mathbf{P}}(n)$ are described by the following diagrams:

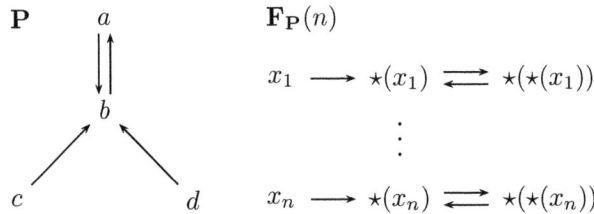

Following our procedure:

(i) $\mathbf{P} \notin \mathbb{H}(\mathbf{F}_{\mathbf{P}}(1))$, but $\mathbf{P} \in \mathbb{H}(\mathbf{F}_{\mathbf{P}}(2))$, so $\mathbb{Q}(\mathbf{F}_{\mathbf{P}}(\omega)) = \mathbb{Q}(\mathbf{F}_{\mathbf{P}}(2))$.
(ii) $\mathbf{F}_{\mathbf{P}}(2)$ has 6 elements and $\mathrm{Sub}(\mathbf{F}_{\mathbf{P}}(2))$ contains its 8 subalgebras.
(iii) $\mathrm{Adm}(\mathbf{P})$ contains just one algebra, $\mathbf{F}_{\mathbf{P}}(2)$ itself.

However, $\mathbf{P}$ can also be embedded into $\mathbf{F}_{\mathbf{P}}(1) \times \mathbf{F}_{\mathbf{P}}(1)$; that is, $\mathbf{P} \in \mathbb{Q}(\mathbf{F}_{\mathbf{P}}(1))$. Hence $\mathbb{Q}(\mathbf{P}) = \mathbb{Q}(\mathbf{F}_{\mathbf{P}}(\omega))$ and, by Lemma 2, $\mathbf{P}$ is structurally complete.

Table 1. Algebras for checking admissibility

A	|A|	Quasivariety $\mathbb{Q}(\mathbf{A})$	Free algebra	|Output Algebra|
$\mathbf{L_3}$	3	algebras for L_3 (Ex. 7)	$\|\mathbf{F_A}(1)\| = 12$	6
$\mathbf{B_1}$	3	Stone algebras (Ex. 5)	$\|\mathbf{F_A}(1)\| = 6$	3
$\mathbf{C_3}$	3	Kleene algebras (Ex. 6)	$\|\mathbf{F_A}(1)\| = 6$	4
$\mathbf{L_3^{\rightarrow}}$	3	algebras for $L_3^{\rightarrow}$ (Ex. 5)	$\|\mathbf{F_A}(2)\| = 40$	3
$\mathbf{C_3^L}$	3	Kleene lattices (Ex. 6)	$\|\mathbf{F_A}(2)\| = 82$	4
$\mathbf{S_3^{\rightarrow\neg}}$	3	algebras for $RM^{\rightarrow\neg}$ (Ex. 4)	$\|\mathbf{F_A}(2)\| = 264$	6
$\mathbf{S_3^{\rightarrow}}$	3	algebras for $RM^{\rightarrow}$ (Ex. 5)	$\|\mathbf{F_A}(2)\| = 60$	3
$\mathbf{G_3}$	3	algebras for G_3 (Ex. 5)	$\|\mathbf{F_A}(2)\| = 18$	3
$\mathbf{D_4^L}$	4	De Morgan lattices (Exs. 1,6)	$\|\mathbf{F_A}(2)\| = 166$	8
$\mathbf{D_4}$	4	De Morgan algebras (Exs. 1,6)	$\|\mathbf{F_A}(2)\| = 168$	10
$\mathbf{P}$	4	$\mathbb{Q}(\mathbf{P})$ (Ex. 8)	$\|\mathbf{F_A}(2)\| = 6$	6
$\mathbf{B_2}$	5	$\mathbb{Q}(\mathbf{B_2})$ (Ex. 5)	$\|\mathbf{F_A}(1)\| = 7$	5

A partial solution to this last issue when $\mathbb{V}(\mathbf{A}) = \mathbb{Q}(\mathbf{A})$ would be to represent $\mathbf{A}$ as a subdirect product of subdirectly irreducible algebras $\mathbf{A}_1, \ldots, \mathbf{A}_n$ in $\mathbb{V}(\mathbf{A})$. Then $\mathbb{V}(\mathbf{A}) = \mathbb{V}(\{\mathbf{A}_1, \ldots, \mathbf{A}_n\})$ and we can find $\mathbf{C}_1, \ldots, \mathbf{C}_n$ such that $\mathbf{C}_i$ is the smallest subalgebra of some $\mathbf{F_A}(m_i)$ with $\mathbf{A}_i \in \mathbb{H}(\mathbf{C}_i)$. In particular, if $\mathbb{V}(\mathbf{A})$ is structurally complete, then each $\mathbf{C}_i$ will be an isomorphic copy of $\mathbf{A}_i$.

5   Concluding Remarks

In this work we have introduced a basic framework for checking unifiability and admissibility in finite algebras, obtaining concrete algorithms that derive potentially efficient proof systems. Sample results on admissibility obtained by the procedure for 3, 4, and 5 element algebras have been described, and are summarized in Table 1. This approach may also be extended to cope with arbitrary finite-valued logics, making some minor modifications to deal with designated values. Note, however, that in order to improve the speed of the procedure and to cope with algebras of higher cardinality, which may (or may not) have very large free algebras even on small numbers of generators, more efficient strategies should be implemented. In particular, rather than calculating all the subalgebras of a free algebra and then checking which are homomorphic preimages of the original algebra, an interlaced procedure should be used which stores the current most suitable subalgebra and rules out larger candidates. For particularly large cases, we might also consider iteratively bounding the size of the subalgebras considered. Finally, we recall that although the procedure described in Section 3 obtains the smallest possible algebra for checking unifiability, this is not the case for the admissibility procedure described in Section 4 since the smallest suitable algebra could be a subalgebra of a product of free algebras. Tackling this issue, and the question of whether larger algebras (e.g., those with a product structure, cf. Corollary 1) might be better suited for checking admissibility will be the subject of future work.

Acknowledgements. Supported by Swiss National Science Foundation grant 20002_129507.

References

1. Baader, F., Morawska, B.: Unification in the description logic EL. Logical Methods in Computer Science 6(3) (2010)
2. Baader, F., Snyder, W.: Unification theory. In: Handbook of Automated Reasoning, vol. I, ch. 8, pp. 447–533. Elsevier (2001)
3. Baaz, M., Fermüller, C.G., Salzer, G.: Automated deduction for many-valued logics. In: Handbook of Automated Reasoning, vol. II, ch.20, pp. 1355–1402. Elsevier Science B.V. (2001)
4. Babenyshev, S., Rybakov, V., Schmidt, R.A., Tishkovsky, D.: A tableau method for checking rule admissibility in S4. In: Proceedings of UNIF 2009. ENTCS, vol. 262, pp. 17–32 (2010)
5. Beckert, B., Hähnle, R., Oel, P., Sulzmann, M.: The Tableau-Based Theorem Prover $_3T^A P$, Version 4.0. In: McRobbie, M.A., Slaney, J.K. (eds.) CADE 1996. LNCS, vol. 1104, pp. 303–307. Springer, Heidelberg (1996)
6. Burris, S., Sankappanavar, H.P.: A Course in Universal Algebra. Graduate Texts in Mathematics, vol. 78. Springer, New York (1981)
7. Cintula, P., Metcalfe, G.: Structural completeness in fuzzy logics. Notre Dame Journal of Formal Logic 50(2), 153–183 (2009)
8. Cintula, P., Metcalfe, G.: Admissible rules in the implication-negation fragment of intuitionistic logic. Annals of Pure and Applied Logic 162(10), 162–171 (2010)
9. Ghilardi, S.: Unification in intuitionistic logic. Journal of Symbolic Logic 64(2), 859–880 (1999)
10. Ghilardi, S.: Best solving modal equations. Annals of Pure and Applied Logic 102(3), 184–198 (2000)
11. Ghilardi, S.: A resolution/tableaux algorithm for projective approximations in IPC. Logic Journal of the IGPL 10(3), 227–241 (2002)
12. Gil, A.J., Salzer, G.: Homepage of MUltseq, http://www.logic.at/multseq
13. Gorbunov, V.A.: Algebraic Theory of Quasivarieties. Springer (1998)
14. Hähnle, R.: Automated Deduction in Multiple-Valued Logics. OUP (1993)
15. Iemhoff, R.: On the admissible rules of intuitionistic propositional logic. Journal of Symbolic Logic 66(1), 281–294 (2001)
16. Iemhoff, R., Metcalfe, G.: Proof theory for admissible rules. Annals of Pure and Applied Logic 159(1-2), 171–186 (2009)
17. Jeřábek, E.: Admissible rules of modal logics. Journal of Logic and Computation 15, 411–431 (2005)
18. Jeřábek, E.: Admissible rules of Łukasiewicz logic. Journal of Logic and Computation 20(2), 425–447 (2010)
19. Jeřábek, E.: Bases of admissible rules of Łukasiewicz logic. Journal of Logic and Computation 20(6), 1149–1163 (2010)
20. Metcalfe, G., Röthlisberger, C.: Admissibility in De Morgan algebras. Soft Computing (February 2012)

21. Olson, J.S., Raftery, J.G.: Positive Sugihara monoids. Algebra Universalis 57, 75–99 (2007)
22. Rybakov, V.: Admissibility of Logical Inference Rules. Studies in Logic and the Foundations of Mathematics, vol. 136. Elsevier, Amsterdam (1997)
23. Salzer, G.: Homepage of MUltlog, `http://www.logic.at/multlog`
24. Sofronie-Stokkermans, V.: On unification for bounded distributive lattices. ACM Transactions on Computational Logic 8(2) (2007)
25. Sprenger, M.: Algebra Workbench. Homepage,
    `http://www.algebraworkbench.net`
26. Zach, R.: Proof theory of finite-valued logics. Master's thesis, Technische Universität Wien (1993)

Natural Signs

Ruth Garrett Millikan

Department of Philosophy, University of Connecticut, Storrs,
CT 06269-2054, United States of America

Abstract. A description of natural signs and natural information is
proposed that interprets the presence of natural information as an affordance for the particular animal or species that would interpret it. This
solves the reference class problem that undercuts earlier correlational accounts. It explains how there can be natural signs of individuals and also
various non-correlational signs. The effect of superimposition of natural
signs is then described.

1 Introduction

A fundamental problem in the philosophy of information on which little recent
progress seems to have been made is the problem what it is for a signal or sign
to carry information with a certain content. It is my belief that this question can
be given a useful answer that is univocal across signs of all kinds, for example,
natural signs, conventional signs and the signals with which animals communicate. Today there is time to say some words only about natural information
and natural signs, but the description I will give of them can be generalized.
I believe, to all signs that genuinely carry information, for example, to animal
signals and natural language sentences that are non-accidentally true.

Natural information, as I will use the term, is that resource in the proximal
environment that can supply an animal with knowledge of its more distal environment if the animal possesses appropriate cognitive tools. Natural information
is carried by natural signs. Natural signs are what constitute the basic materials in nature that can be used to support perception and cognition in animals
including humans. If we want to understand how perception and cognition are
possible, we first need to understand what natural signs are and the various
forms they can take.

Here are some examples of natural signs. Despite first appearances, I will
argue, they are all fundamentally alike.

1. That the water is boiling is a sign that it has reached 212 degrees Fahrenheit.
2. Black clouds are often a sign of immanent rain; fever is often a sign of
   infection, sometimes a sign of measles, other times of flu, and so forth.
3. Given a wooden frame made of four straight boards paired in length and
   properly nailed end to end in a closed figure, that the frame has equal diagonals is a sign, often used by carpenters, that the sides are parallel and the
   corners at right angles.

S.B. Cooper, A. Dawar, and B. Löwe (Eds.): CiE 2012, LNCS 7318, pp. 496–506, 2012.
© Springer-Verlag Berlin Heidelberg 2012

4. That the head is like that of an elephant is likely to be a sign that the tail is like that of an elephant.
5. 'O say can you see' sung to a certain tune is likely to be a sign that 'by the dawn's early light' is coming next.
6. Bluster, who made the bomb I am to carry to Muggins, said that if by any chance I should hear it fizzing I should drop it dead and run, for that would be a sign that it was about to go off prematurely.
7. The direction of the North Star is a sign of the direction of geographic north; the pull on the southern pole of the magnetosome in a northern hemisphere marine bacterium is a sign of the direction of lesser oxygen.
8. That Jim has gone to the party is a good sign that that is where Jane is also.
9. This certain quality of voice is a sign for me that it is Aino, my daughter, who is speaking.
10. Traveling north from Rt. 89 on Wormwood Hill Road, the pond on the right is a sign that our house is coming up next.
11. Suzy's mitten lying on the walk to the side door, given that it wasn't there earlier today, is a sign that Suzy is already home from school.
12. The pointed tracks in the snow leading from the tattered azaleas on the north of our yard to the stripped rhododendrons on the east are a sign that it was the same hungry deer that nibbled on both.

The above examples are chosen in part to challenge assumptions that seem frequently to be made about the nature of natural signs.

One such assumption is that natural signs are always related by causal laws to what they signify. But in (3) the equal diagonals are not causes of the parallel sides and in (7) geographic north is not causally related to the direction of the North Star nor is the direction of lesser oxygen causally related to the pull of the magnetosome. In (10) the pond on Wormwood Hill Road is not causally connected with my house not a cause of it, not an effect of it, not an effect of a common cause. And in examples (9) through (12), the signs are signs of individuals, but there are no causal laws that concern individuals as such. In (12) the sign is of an identity relation, which is also an odd candidate for participation in causal laws.

A second suggestion has been that natural signs, signs carrying natural information, correspond to their signifieds with certainty, with 'a probability of 1' [1]. The idea is that a natural sign is always such qua being of a some uniform physical type and that this type must be such that every token of it necessarily corresponds to a real signified of a another uniform type. If that this water is boiling is a natural sign that it is at 212 degrees, then if any water boils (in this kind of situation) it must be at 212 degrees. If you know how to read a certain kind of natural sign then, unlike knowing how to read a conventional sign, you can't go wrong in taking every token just like it to signify something actual of the same kind as the other tokens do. What a wonderful support for knowledge that would be, as Dretske [1] tried to show.

Dretske then qualified his claim, however, with a reference to 'channel conditions,' conditions that had to be assumed to be in place, mediating between sign and signified, to make the probability be 1. But unless presence of the right channel conditions is to be considered part of the natural sign itself, this wholly undercuts the assumption of certainty. That the water is boiling will not be a sign that it has reached 212 if it is on top of Mt. Everest or on the moon. Knowing how to read a certain kind of sign by assuming that the channel conditions are right cannot yield certainty. Certainty about the channel conditions would have to be added.

2   Correlational Signs

Indeed, the kind of signs on which perception and cognition actually rest virtually never have forms/shapes/intrinsic properties that correspond unfailingly to the same sort of signified, nor are channel conditions that might produce unfailing correspondence (should they be known) univocally signed to senses like ours unless still further channel conditions which are not necessarily univocally signed are assumed. Actual animal cognition is not supported by infallible indicators. Said another way, tokens of signs that support actual perception/cognition do not carry their signifying types on their sleeves. A token fever might be a token of any of many signifying types; it might mean measles, or rather flu, or even allergy, and so forth. Physical twins of natural sign tokens that don't mean the same abound. Compare homonyms, which sound the same but are different words.

The capacities of humans and other animals to perceive and to know systematically to acquire accurate perceptions and true beliefs often appear to rest heavily on mere correlations in nature rather than necessities. In line with this, it has been suggested that we recognize a kind of natural information Nicholas Shea [7] calls it 'correlational information' that is produced when there is a non-accidentally continuing correlation (typically assumed to be underwritten by a causal connection, but we can admit other kinds of non-accidental correlation too) between one kind of thing and another, such that the probability of the one is raised given the other [2,6,7]. Lloyd [2], at least, is clear that where A types carry this kind of information about B types, an A token does not carry this information unless there actually exists a corresponding B token (p. 64). For example, no matter how high the non-accidental correlation between fever and measles is in the area, if Johnny's fever is caused by flu then his particular fever is not a correlational sign of measles. Considering correlational signs as candidates for a kind of natural sign signs that support perception and cognition we would need Lloyd's restriction. Signs carrying misinformation will not, as such, support cognition.

An obvious difficulty for the above description of correlational signs concerns the strength of correlation to be required. But more basic, I think, is that correlations and conditional probabilities are defined relative to reference classes. It could of course be that given all space-time as the reference class, the boiling of water does raise the probability that it is at 212 degrees. But boiling is likely to raise the probability more that water is at one or another of various other

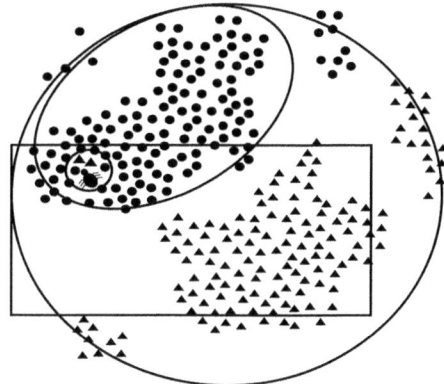

Fig. 1. The Reference Class Problem (With permission from Cambridge University Press) The diagram represents a space-time continuum collapsed to two dimensions. The dots are centers of strong positive correlation of A with B, fading in strength outwards. The triangles are centers of strong negative correlation of A with B. Outside the dots and triangles there is little or no correlation of A with B. Assume that the bug is currently encountering some specific token A that is coupled with B. Which circumscribed area determines the reference class relative to which this specific token A is or is not, objectively, a correlational sign of B? Suggestion: the relevant area would be one that precisely encloses the past and ongoing path of the bug (or its species).

temperatures, there being no reason to think that one earth-atmosphere is an especially common pressure.

Similarly, we have no evidence that a elephant-like head at one end raises the probability of a elephant-like tail at the other throughout all space-time, and it is quite certain that the direction of the North Star does not raise the probability of that direction being geographic north universe wide. If natural signs were correlational signs, clearly the reference classes for these correlations would have to be restricted in some way. How then would we decide, for any candidate natural sign token, what the boundaries should be for the restricted class by reference to which it would be required to be an instance of a correlation? How, for example, would we determine that, given this little bacterium right here now, the current direction of magnetic north is a natural sign of lesser oxygen, considering that the relevant correlation holds (probably) neither universe wide nor (certainly) earth wide?

Whispering in our ear may be the reply that the relevant reference class is the earth's northern oceans because, well, 'that's where the correlation is'. But the correlation is lots of other areas the bacterium is in as well myriad smaller areas in some of which it may be higher, and also myriad larger areas, ones, for example, that overlap with or include the northern hemisphere. And the correlation fails to hold in various other areas the bacterium is equally in, perhaps in some smaller areas such as the very square yard right around the bacterium which could be affected by a bar magnet someone has dropped underneath or by bubbles coming up from a diver. And it certainly fails in various larger areas that include where the bacterium is along with the southern seas. (For a visual image of this kind of problem, see figure 1) That there exists a correlation (of some positive strength or other) within some arbitrarily chosen reference class which includes this item surely cannot define for us what a natural sign is given our purposes. It is not facts of this anemic kind that underlie the possibility of perception and cognition. Besides, we should remember examples of signs such as the fizzing bomb in (6) and Suzy's mitten in (11). These are unique cases, the coinciding of sign with signified occurring only once.

3   A Reference Class for Determining Natural Correlational Signs

Clearly no cognizing creature samples space and time randomly, or samples very much of it at all. If correlations or conditional probabilities are relevant to understanding what in nature supports cognition, I suggest that the only relevant or non-arbitrary reference class is the class of (candidate) signs that occur within the path of the animal that would read them, or if it is a species that learns to read them, within the various paths of that species' members. The correlation must exist over the reference class that is this path, or over a portion or portions of this path that the animal must be able to distinguish, or over a long enough period of the animal's life for it to learn it. I am not suggesting that the reference class should merely include the animal's or animals' path(s) or relevant parts of it. That would raise exactly the same problem over again. The idea is that the relevant non-arbitrary reference class consists in the very samples sampled by the animal or species that uses a natural sign, or of such samples as are further discriminable by it in some way as relevant. Useful correlations have to occur within the experience of the animals using them, or within portions of this experience the animal can learn to discriminate as such. Or they have to occur among things prior signs of which occur within the experience of the animal using them, and so forth.

This changes the game quite radically, however, from the one played by traditional theorists of natural signs, who have hoped to find natural signs and natural information as an 'objective commodity' [1] out in the world, as structures that God could have seen long before he fashioned creatures to harvest them. Natural signs will instead be affordances like food or like shelter, things that are what they are only relative to an animal who would use them, in this case relative, in the first instance, to the actual location of the animal, not merely to its capacities. Nothing will be a natural sign of anything else absolutely. Rather, there will be things that can serve as signs or do serve as signs for a given animal or species. Notice, however, that from this perspective the question how strong a correlation must be to create a natural sign is not a problem. The correlation must be strong enough and of the right kind for the particular animal to detect it, and for it to be possibly useful, given the animal's purposes, to bet on it. Or, if the reference to possible use for a given animal is too vague, we can always fall back on the underlying idea that serving as a natural sign is what is basic rather than being a natural sign.

Strong correlations between things throughout all space and time are chiefly restricted to what follows from universal natural laws, these universal laws, as science knows, being hard to discern. But the space-time paths of individual animals and species often seem to be reference classes of very strong correlation between fairly numerous and fairly evident kinds of things. Suppose that we pause to wonder why this is so. Why are the correlations among samples an individual animal or a species happens to encounter along one section of its own idiosyncratic space-time path likely to continue to hold for the samples in another? Why are strong correlations in an animal's sampling past often good signs that these correlations will remain strong in its sampling future, regardless of the strength of these correlations universe wide? This, it seems to me, is the fundamental question we should be trying to answer about correlational natural signs, about how they can help to make cognition possible. Moreover I will argue that the answer to this question generalizes beyond correlational signs. It can foot an understanding of the nature of other natural signs that are not correlational, as in examples (6) and (10) through (12).

4   Answering the Fundamental Question

We need to superimpose two familiar facts. First, the path of an animal through space and time is continuous. It does not skip from one distant time and place to another. Similarly, a species spreads out over space and time on continuous individual paths that branch from one another, confined by natural selection within geographical areas with friendly properties. Second, the usual reason for the presence of local correlations is that things persist through time, staying in the same or connected locales, or that they self-maintain or cycle in the same or in connected locales or, in post-archeozoic times on earth, that they reproduce or are reproduced (artifacts, animal signals, linguistic forms) in the same or in connected locales. Whether conditions, events or entities, much that is near the animal today will be near it tomorrow, or something much like it will be. Channel conditions remain in place and events that manifest themselves through these channels recur. For these reasons, where there are good correlations in one portion of an animal's space-time path, frequently these correlations extend to other portions.

Mountains, valleys, rivers, oceans, the earth's atmosphere, persist. Individual trees, rocks, houses, roads, paths persist in definite places. Individual animals and plants and also species of these self-maintain hence persist, remaining in roughly the same or connected areas. The earth persists and rotates causing cyclical daily patterns, and it circles around the persisting sun causing cyclical yearly patterns, cyclical weather patterns and cyclical patterns in plant and animal behaviors. (It is not a natural law that black clouds tend to burst. Under different conditions they too would persist.) People cycle from home to work or school and back home again at fairly regular hours. Animal and plant species reproduce, stabilized by homeostasis in their gene pools and by natural selection in a continuing environment. Artifacts that work well or that please, songs and stories, red and green Christmas decorations and steeples on churches are all numerously reproduced. Where local sampling shows correlations, typically the causes of the correlations are being actively preserved or cycled or reproduced and will continue to be so for some time through neighboring times and places. An animal's path typically persists among, crosses or overlaps the single or branching paths of many such persisting, cycling or reproducing items.

Traveling north up Wormwood Hill Road, passing the pond on the right is repeatedly followed by passing our house because the relation between our house and the pond persists (10). In the circles I walk in the peculiar quality of my daughter Aino's voice is repeatedly reproduced by my daughter's larynx which persists in areas that frequently overlap mine (9). The correlation between lesser oxygen and the direction southern poles of magnetosomes of northern hemisphere bacteria point persists because the relation between magnetic north and the surface of the ocean persists as does the scarcity of other sources of magnetic fields on the earth (7). The correlation between a elephant-like head at one end with a elephant-like tail at the other persists on earth because elephants reproduce (4). The correlation between 'O say can you see' and 'by the dawn's early light' persists because the U.S. national anthem is copied over and over by people who teach it to one another (5). The correlation of fever with measles persists because the measles virus reproduces in people and people reproduce thus reproducing the conditions that lead from the measles virus to the fever (2). The correlation between where Jim is to be found in the evening and where Jane is to be found persists because Jim and Jane and a certain bond between them also persist often bringing them together in the evenings (8).

Understanding why a correlation persists, what kind of endurings or cyclings or reproducings are accounting for it, can of course be a huge help in trying to project the more exact path or paths that a correlation will take. Knowing how a disease spreads, for example, helps in deciding whether certain symptoms are likely to be signs of it or rather of something else. Knowing that Jim and Jane have split, I will no longer expect Jim's location to be a sign of Jane's. But it is not always necessary to know reasons. I understand that only edible mushrooms grow on dead hardwood trees, but I have no idea why. Most fundamental, however, is that many animals are born, live and die such that their life paths taken entire mark out reference classes in which many of the correlations they depend on actively persist. There is no need for them to recognize boundaries.

5 Generalizing to Non-correlational Signs

Viewing a correlation merely as a repeated pattern, and viewing natural sign reading simply as pattern completion, it will be possible to generalize this picture in several ways to include many more natural signs than those that depend upon correlations.

A continuing correlation can be viewed simply as a very small recurring pattern, a repeated pattern of an A state of affairs being so-related to a B state of affairs. What repeats is a black cloud being close overhead at a place and time with rain being at that place shortly after that time, or a elephant-like head being at one end of a thing and an elephant-like tail at the other, or child having a rash and fever when the child harbors the measles virus. What repeats in a sign-signed relation is a relation between a certain kind of sign and a certain kind of signed; reading the sign is completing the pattern.

In the simplest cases, the relation that determines the part of the pattern that is signed from the part that is the sign consists in the identity of an aspect of the state of affairs that is signed with an aspect of the state of affairs that is the sign. The same thing that has an elephant head at one end has an elephant tail at the other. The same child that has a rash and fever also harbors the measels virus. In the same place that the black clouds appear, the rain comes down soon after. In the same car and at the same time, that the gas guage needle is at the top of the dial, the gas tank is full. Or, generalizing this, In the same car and at the same time that the guage needle is a certain rough proportion of the way between the bottom and the top lines on the dial, the gas in the tank is that same rough proportion of the way from empty to full. But the relation between sign and signed may rest on a function other than identity. What is needed is only that the repeating pattern involve a determining relation between sign and signed, what is signed being determined as a function of the sign. Notice, for example, that in the case of black clouds and rain, the time of the rain is only a function of the time of the clouds, namely, just after. Similarly for lightning and thunder, of course.

What repeats in the case of basic natural laws often involves patterns of this more general kind. The temperature at which water boils is not 212 degrees but a function of the pressure; the basic pattern is a recurring relation between (the same portion of) water's temperature, its pressure and its boiling. For a very different kind of example, although American alligators range from nine inches long to more than fifteen feet long, the length of the alligator's snout is always the same rough proportion of the length of its tail. There is a recurring relation between snout length and tail length, a recurring pattern in which the one is determined as a function of the other. And so, of course, for every other pair of the alligator's parts. Relations between signs and what they sign of this kind may be called "projection relations" and we can call pattern repetitions of this more abstract sort repetitions by projection.

6   Natural Signs Based on a Single Repetition

Significant correlatons between one thing and another require many repetitions of the one given the other. But single recurrences of extremely complex patterns may also yield natural signs. When a very detailed complex pattern repeats even once in a nearby location, it is likely that the repetition is no accident, the chance of accident going down sharply with the detail and complexity of the pattern. Rather, the pattern has probably endured (it's the same individual again) or been repeated as a result of the persistence of conditions along with the cycling of events and/or some process involving reproduction or copying. If you encounter two detailed paintings that are exactly alike it is more than likely either that one has been reproduced from the other or both from some third or perhaps it is the same painting again. Moreover if you find that one half of a detailed painting is exactly like one half of another, thus making it likely that the one has been reproduced from the other or from some third, or that it is the same one over again, then it is very likely that the other halves match as well. Suddenly noticing the passing surroundings in my town from the back seat of the car and seeing that they exactly match the road and yard in front of Alice's house and that the house is alike as well, it is very likely that all the other features of Alice's house are to be found there also, for it is likely that it IS Alice's house, which has continued to endure since I last saw it.

Besides recurrence of the size relation between snout and tail of the alligator, the shape, composition and structure of the whole alligator, all its parts and all their relations to one another a concrete and very detailed pattern keeps recurring. Similarly, the patterns that are whole elephants and whole daisies repeat, and also whole symphonies and whole Protestant wedding ceremonies and whole Toyota Camrys and whole Gothic churches and whole chairs and whole books (copies of them) and whole McDonalds restaurants. Locally repeating patterns of these kinds tend to be much less abstract, much more detailed and richer, than are, say, universal lawlike patterns, and they are usually much easier to notice. Mentally completing complex patterns to match those that have been encountered in one's past in this way can often supply knowledge on a base of only one or a very few prior samples. Possibly I have only seen one elephant or one alligator before, or been at Alice's house only once before.

The notion of a correlation has dropped far into the background here, a much broader notion of local pattern repetition and pattern completion emerging as the base principle explaining natural signs and the reading of natural information. It seems best to claim then that the general form that natural signs take is that of patterns that repeat, either directly or by projection, within an animal's experience and that reading natural signs is just pattern completion. Natural sign reading may be based, at the one extreme, on observation of many repetitions of a very simple pattern, at the other extreme, on a single prior observation of a very complex pattern. Having had it rain many times when black clouds form I am prepared for rain the next time I see black clouds. Having seen only one whole elephant, I am prepared for the tail on the next.

It is often possible to recognize a complex repeated pattern from any of many different samplings of its various features. An elephant may be recognized from the front or the back or the side, by the trunk, by the head, by the legs, by the tusks, by the tail, by the trumpet. Each of these may serve as a sign of an elephant and as a sign of each other elephant feature. Similarly, of course, an automobile or a premises may be recognized from the front or the back or the side and so forth. A song may be recognized by hearing the beginning or the end or only a line in the middle. There are many ways to recognize a baseball game or a restaurant or a piano, also many ways to test for a chemical substance, and so forth. Thus the completion of complex patterns is a doubly versatile way that nature affords of reading natural signs. Surely it is not just correlations of simple pairs of features but repeated clumpings of multiple features that supply the bulk of natural signs supporting cognition.

7   Natural Signs of Individuals

An immediate and notable implication of this description of root signs is that there can be natural root signs of individual objects. There can be natural information that concerns individual objects. (This is a result that has not, to my knowledge, been obtained by a theory of information before) We are considering information to be a kind of affordance relative to an animal that can appreciate it, in particular, relative to the reference class of actual samplings of the animal over time. These samplings are taken along the animal's space-time path, which crosses the paths of other locally enduring individual objects, sometimes tracing them for a period of time, often re-crossing them at later times. Many of these objects will have simple features or combinations of features that remain unique in the experience of the animal. These features are natural signs for the animal, carrying natural information for it about individual objects. Nor is the animal dependent only on features of individuals that are unique in the animal's experience. I remarked earlier that although many of the correlations that animals depend on may actively persist throughout their lives, in other cases the animal may recognize rough boundaries for a domain of correlation, learning ways to recognize or track this domain. Individual objects persist through single, continuous space-time paths, different kinds of individuals displaying different characteristic patterns of spatial displacement over time. Big rocks, trees and houses usually say in the same place; squirrels have certain characteristic ways of moving about, automobiles another. Experience with these various patterns of rest or movement enables fallible tracking of the sign domains for various individuals of these kinds over longer or shorter periods of time.

8   Ovelapping Natural-Sign Patterns

I have presented natural signs as resulting from endurances and repetitions of patterns along paths that cris-cross and interweave with the paths of the animals for whom these signs are affordances. Such patterns often cris-cross or interweave with one another as well. Repeated patterns may be superimposed on one another. If part of one pattern is recognized as such, then recognized as also being a part of a second superimposed pattern, pattern completion may reveal relations between various other parts of these patterns as well. Think of employing two maps that overlap in content to determine relations between places shown just on one with those shown just on the other. Compare recognizing the repeated pattern that is small stray items lying outdoors where the people who own them have recently been, as superimposed over the repeated pattern that is Suzy, whose stray mitten one recognizes, entering from school by the side door about this time of day. This superpositioning yields Suzy having already returned from school (11). Bluster, who made the bomb I am to carry to Muggins, has mentally superimposed and chained together many small repeating patterns involving one simple kind of part that his bomb contains interacting with another simple part, which completed chain would determine it to go off as planned. But he knows of other sometimes repeated patterns that, superimposed, would include a hissing sound and immediate premature detonation, and it is this latter pattern that he is warning me about (6).

Acknowledgements. Portions of this paper are a revised form from [5], used with permission.

References

1. Dretske, F.: Knowledge and the Flow of Information. The MIT Press, Cambridge (1981)
2. Lloyd, D.: Simple Minds. The MIT Press, Cambridge (1989)
3. Millikan, R.G.: Varieties of Meaning. The MIT Press, Cambridge (2004)
4. Millikan, R.G.: On knowing the meaning; with a coda on swampman. Mind 119(473), 43–81 (2010)
5. Millikan, R.G.: Natural Information, Intentional Signs and Animal Communication. In: Stegmann, U. (ed.) Animal Communication Theory: Information and Influence. Cambridge University Press (forthcoming)
6. Price, C.: Functions in Mind; A theory of Intentional Content. Oxford University Press, Oxford (2001)
7. Shea, N.: Consumers need information: Supplementing teleosemantics with an input condition. Philosophy and Phenomenological Research 75(2), 404–435 (2007)

Characteristics of Minimal Effective Programming Systems

Samuel E. Moelius III

IDA Center for Computing Sciences, 17100 Science Drive, Bowie, MD 20715-4300,
United States of America
semoeli@super.org

Abstract. The Rogers semilattice of effective programming systems (epses) is the collection of all effective numberings of the partial computable functions ordered such that $\theta \leq \psi$ whenever θ-programs can be algorithmically translated into ψ-programs. Herein, it is shown that an eps ψ is minimal in this ordering if and only if, for each translation function t into ψ, there exists a computably enumerable equivalence relation (ceer) R such that (i) R is a subrelation of ψ's program equivalence relation, and (ii) R equates each ψ-program to some program in the range of t. It is also shown that there exists a minimal eps for which no *single* such R does the work for all such t. In fact, there exists a minimal eps ψ such that, for each ceer R, either R contradicts ψ's program equivalence relation, or there exists a translation function t into ψ such that the range of t *fails* to intersect *infinitely many* of R's equivalence classes.

1 Introduction

Let $\mathbb{N}$ be the set of natural numbers, i.e., $\{0, 1, 2, ...\}$. An effective programming systems (eps) is a partial computable function $\lambda p, x \, . \, \psi_p(x)$ mapping $\mathbb{N}^2$ to $\mathbb{N}$, and having the following property. For each partial computable function ζ mapping $\mathbb{N}$ to $\mathbb{N}$, there exists a p such that $\psi_p = \zeta$. Effective programming systems abstract the notion of *programming language* in the following sense. One can think of p as a *program*, and of ψ_p as the partial computable function denoted by p within some programming language corresponding to ψ.

Rogers [19] introduced the following ordering on epses. For epses θ and ψ, $\theta \leq \psi$ iff there exists a computable function $t : \mathbb{N} \to \mathbb{N}$ such that, for each p, $\theta_p = \psi_{t(p)}$. Intuitively, $\theta \leq \psi$ whenever θ-programs can be algorithmically translated into ψ-programs. Moreover, an eps ψ is *minimal* in this ordering iff having the ability to algorithmically translate θ-programs into ψ-programs implies having the ability to algorithmically translate ψ-programs into θ-programs, for each eps θ.

Arguably, the most well studied collection of minimal epses is that of the Friedberg numberings [3, 11]. Recall that a Friedberg numbering is an eps that is 1-1, i.e., for each p and q, $\psi_p = \psi_q$ implies $p = q$. Examples of works that make use of this concept include [12, 16, 18, 2, 22, 21, 10, 23, 5, 6, 7].

S.B. Cooper, A. Dawar, and B. Löwe (Eds.): CiE 2012, LNCS 7318, pp. 507–516, 2012.
© Springer-Verlag Berlin Heidelberg 2012

In [17], Pour-El asked whether every minimal eps is equivalent to some Friedberg numbering. Ershov [1, §5] showed that there exists a minimal effective numbering of the *computably enumerable sets* that is not equivalent to any 1-1 numbering. Shortly thereafter, his student, Khutoretskii, established the analogous result for the partial computable functions, thereby answering Pour-El's question.

Theorem 1 (Khutoretskii [8, Ex. 1 and Cor. 4]). There exists a minimal eps that is *not* equivalent to any Friedberg numbering.

For the purposes of this paper, Theorem 1 is best viewed through the following folklore theorem.

Theorem 2 (Folklore). For each eps ψ, ψ is equivalent to a Friedberg numbering iff ψ's program equivalence relation is computable.

Thus, Theorem 1 could be restated as: there exists a minimal eps whose program equivalence relation is *not* computable. On the other hand, as noted in the proof of Theorem 1, the constructed eps's program equivalence relation is computably enumerable. (In particular, exactly one such equivalence class is a simple set [20, §8.1], and all others a singletons.) Thus, one has the following.

Theorem 3 (Khutoretskii, corollary of Thm. 2 and proof of Thm. 1). There exists an eps whose program equivalence relation is computably enumerable, but *not* computable.

Subsequent to the above, Khutoretskii showed the following.

Theorem 4 (Khutoretskii, corollary of [9, Thm. 1]). There exists a minimal eps whose program equivalence relation is *not* computably enumerable.

Clearly, Theorems 3 and 4 can be viewed as a sharpening of Theorem 1. Herein, we sharpen Khutoretskii's results even further.

To facilitate the statement of our results, we first give a few definitions. Suppose that ψ is an eps. For each $t : \mathbb{N} \to \mathbb{N}$, we say that t is a *translation function into* ψ iff there exists an eps θ such that t witnesses $\theta \leq \psi$. The following definition is equivalent. For each $t : \mathbb{N} \to \mathbb{N}$, t is a translation function into ψ iff t is computable and the partial function $\lambda p, x \cdot \psi_{t(p)}(x)$ is an eps.

Definition 5. Suppose that ψ is an eps, and that t is a translation function into ψ. Then, for each equivalence relation R, (a) and (b) below.

(a) R *strongly ties* t *into* ψ iff R satisfies (i) and (ii) just below.[1]
    (i) R is a subrelation of ψ's program equivalence relation.
    (ii) The range of t intersects each of R's equivalence classes.
(b) R *weakly ties* t *into* ψ iff R satisfies (i) just above and (ii*) just below.[2]
    (ii*) The range of t intersects all but finitely many of R's equivalence classes.

---

[1] In some places, we omit the phrase "into ψ" when it is clear from context.
[2] See footnote 1.

Thus, if equivalence relation R strongly ties translation function t into eps ψ, then R equates each ψ-program to some program in the range of t. If R merely weakly ties t into ψ, then there may be infinitely many ψ-programs that R does *not* equate to any program in the range of t. However, those infinitely many such ψ-programs will form only finitely many equivalence classes.

Our first main result is that the minimal epses may be *characterized* as follows.

Theorem 6. For each eps ψ, (a)-(c) below are equivalent.

(a) ψ is minimal.
(b) For each translation function t into ψ, there exists a computably enumerable equivalence relation (ceer)[3] that strongly ties t into ψ.
(c) For each translation function t into ψ, there exists a ceer that weakly ties t into ψ.

Note that Theorem 4 is about a *single* equivalence relation, i.e., the program equivalence relation of a certain eps, whereas Theorem 6 is about one equivalence relation *per* translation function into any given eps. Thus, one might ask: if ψ is a minimal eps, then might there always exist a *single* ceer that strongly ties each translation function into ψ? The answer, as it turns out, is *no*. In fact, as Theorem 7 below states, there need not even exist a single ceer that *weakly* ties each translation function into ψ.

Theorem 7. There exists an eps ψ satisfying (a) and (b) below.

(a) ψ is minimal.
(b) For each ceer R, there exists a translation function t into ψ such that R does *not* weakly tie t into ψ.

Continuing with this line of thought, one finds that the strong and weak notions of Definition 5 separate when one considers single equivalence relations.

Theorem 8. There exists an eps ψ and a ceer R satisfying (a) and (b) below.

(a) For each translation function t into ψ, R weakly ties t into ψ.
(b) For each ceer R', there exists a translation function t into ψ such that R' does *not* strongly tie t into ψ.

Clearly, if ψ is an eps, and ψ's program equivalence relation is computably enumerable, then there exists a single ceer R that strongly ties each translation function into ψ, i.e., R is ψ's program equivalence relation. Thus, one might ask: does the converse hold? Theorem 9, just below, establishes that it does *not*.

Theorem 9. There exists an eps ψ and a ceer R satisfying (a) and (b) below.

---

[3] We pronounce ceer like the first syllable of "series". Computably enumerable equivalence relations are of interest in their own right. Gao and Gerdes [4] give an excellent survey.

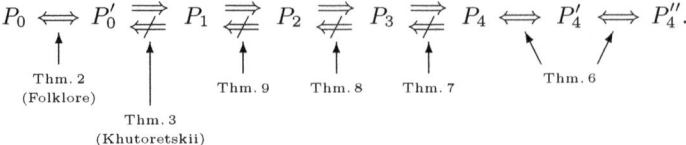

- $P_0(\psi) \Leftrightarrow \psi$ is equivalent to a Friedberg numbering.
- $P_0'(\psi) \Leftrightarrow \psi$'s program equivalence relation is computable.
- $P_1(\psi) \Leftrightarrow \psi$'s program equivalence relation is computably enumerable.
- $P_2(\psi) \Leftrightarrow$ there exists a **ceer** R that strongly ties each translation function into ψ.
- $P_3(\psi) \Leftrightarrow$ there exists a **ceer** R that weakly ties each translation function into ψ.
- $P_4(\psi) \Leftrightarrow$ for each translation function t into ψ, there exists a **ceer** that strongly ties t into ψ.
- $P_4'(\psi) \Leftrightarrow$ for each translation function t into ψ, there exists a **ceer** that weakly ties t into ψ.
- $P_4''(\psi) \Leftrightarrow \psi$ is minimal.

Fig. 1. A summary of the results mentioned in Section 1. In addition to the above: Mal'cev [13, 14] showed that $P_1 \Rightarrow P_4''$, and Khutoretskii [9] showed that $P_1 \not\Leftarrow P_4''$ (see Theorem 4).

(a) For each translation function t into ψ, R strongly ties t into ψ.
(b) ψ's program equivalence relation is *not* computably enumerable.

Figure 1 summarizes the results mentioned in this section. The remainder of this paper is organized as follows. Section 2 covers preliminaries. Section 3 gives the proof of Theorem 6, and a sketch of the proof of Theorem 7. Complete proofs of Theorems 7 through 9 can be found in the expanded version of this paper [15].

2   Preliminaries

Computability-theoretic concepts not covered below are treated in [20].

Lowercase math-italic letters (e.g., i, p, x), with or without decorations, range over elements of $\mathbb{N}$, unless stated otherwise. Uppercase math-italic letters (e.g., I, P, X), with or without decorations, range over subsets of $\mathbb{N}$, unless stated otherwise. For each *non*-empty X, $\min X$ denotes the minimum element of X. $\min \emptyset \overset{\text{def}}{=} \infty$. For each *non*-empty, finite X, $\max X$ denotes the maximum element of X. $\max \emptyset \overset{\text{def}}{=} -1$. $\mathcal{Fin}$ denotes the collection of all finite subsets of $\mathbb{N}$.

$\langle \cdot, \cdot \rangle$ denotes any fixed pairing function, i.e., a 1-1, onto, computable function of type $\mathbb{N}^2 \to \mathbb{N}$ [20, page 64]. For each x, y, and z, $\langle x, y, z \rangle \overset{\text{def}}{=} \langle x, \langle y, z \rangle \rangle$. For each X and Y, $X \times Y \overset{\text{def}}{=} \{\langle x, y \rangle \mid x \in X \ \wedge \ y \in Y\}$.

Every partial function considered herein maps $\mathbb{N}$ to $\mathbb{N}$, unless stated otherwise. For each partial function ζ, and each x, $\zeta(x){\downarrow}$ denotes that $\zeta(x)$ converges; whereas, $\zeta(x){\uparrow}$ denotes that $\zeta(x)$ diverges. We use $\uparrow$ to denote the value of a divergent computation. For the sake of some subsequent proofs, it is convenient to have the following notation. For each i and n,

$$i^{<n} \stackrel{\text{def}}{=} \lambda x. \begin{cases} i, & \text{if } x < n; \\ \uparrow, & \text{otherwise.} \end{cases} \tag{1}$$

Thus, $i^{<n}$ is the partial function that maps each value less than n to i, and that diverges everywhere else. For each partial function ζ, $\text{rng}(\zeta)$ denotes the range of ζ, i.e., $\text{rng}(\zeta) \stackrel{\text{def}}{=} \{y \mid (\exists x)[\zeta(x) = y]\}$. $\mathit{PartComp}$ denotes the set of all partial computable functions (mapping $\mathbb{N}$ to $\mathbb{N}$).

φ denotes any fixed acceptable (i.e., maximal) eps [19, 20, 16, 18, 21]. For each p, $W_p \stackrel{\text{def}}{=} \{x \mid \varphi_p(x)\downarrow\}$. For each p and s, the following.

$$\varphi_p^s \stackrel{\text{def}}{=} \lambda x. \begin{cases} \varphi_p(x), & \text{if } x < s \text{ and } \varphi_p(x) \text{ converges in fewer than } s \text{ steps;} \\ \uparrow, & \text{otherwise.} \end{cases} \tag{2}$$

$$W_p^s \stackrel{\text{def}}{=} \{x \mid \varphi_p^s(x)\downarrow\}. \tag{3}$$

For each eps ψ, $\text{Equiv}(\psi)$ denotes ψ's program equivalence relation, i.e.,

$$\text{Equiv}(\psi) \stackrel{\text{def}}{=} \{\langle p, q \rangle \mid \psi_p = \psi_q\}. \tag{4}$$

For each equivalence relation R, $\mathit{Classes}(R)$ denotes the set of R's equivalence classes, i.e., $\mathit{Classes}(R)$ is the set of exactly those E satisfying (a)-(c) below.

(a) $E \neq \emptyset$.
(b) $(\forall p, q \in E)[\langle p, q \rangle \in R]$.
(c) $(\forall p \in E)(\forall q \notin E)[\langle p, q \rangle \notin R]$.

3   Results

This section gives the proof of Theorem 6, and a sketch of the proof of Theorem 7. Complete proofs of Theorems 7 through 9 can be found in the expanded version of this paper [15].

Our first main result is that the minimal epses may be *characterized* as per Theorem 6, restated just below. Recall from Definition 5 that if equivalence relation R strongly ties translation function t into eps ψ, then (i) R is a subrelation of ψ's program equivalence relation, and (ii) the range of t intersects each of R's equivalence classes. On the other hand, if R merely weakly ties t into ψ, then the range of t need only intersect all but finitely many of R's equivalence classes.

Theorem 6. For each eps ψ, (a)-(c) below are equivalent.

(a) ψ is minimal.
(b) For each translation function t into ψ, there exists a ceer that strongly ties t into ψ.
(c) For each translation function t into ψ, there exists a ceer that weakly ties t into ψ.

Proof. Let ψ be given.

(a) $\Rightarrow$ (b): Suppose that ψ is minimal. Let t be any translation function into ψ, and let θ be such that t witnesses $\theta \leq \psi$. Since ψ is minimal, there exists a $t' : \mathbb{N} \to \mathbb{N}$ witnessing $\psi \leq \theta$. Let R be the reflexive, symmetric, transitive closure of

$$\{\langle p, (t \circ t')(p)\rangle \mid p \in \mathbb{N}\}. \tag{5}$$

Clearly, R is a ceer and $R \subseteq \text{Equiv}(\psi)$. It remains to show that, for each $E \in \textit{Classes}(R)$, $\text{rng}(t) \cap E \neq \emptyset$. So, let $E \in \textit{Classes}(R)$ be given, and let $p \in E$ be arbitrary. Then, clearly, $(t \circ t')(p) \in \text{rng}(t) \cap E$.

(b) $\Rightarrow$ (c): Immediate.

(c) $\Rightarrow$ (a): Suppose (c). Further suppose that θ is an eps, and that $t : \mathbb{N} \to \mathbb{N}$ witnesses $\theta \leq \psi$. Then, by (c), there exists a ceer $R \subseteq \text{Equiv}(\psi)$ such that, for all but finitely many $E \in \textit{Classes}(R)$, $\text{rng}(t) \cap E \neq \emptyset$. Let n be the number of elements of $\textit{Classes}(R)$ that do *not* intersect $\text{rng}(t)$, and let $E_0, ..., E_{n-1}$ be those elements. Choose $q_0, ..., q_{n-1}$ such that, for each $i < n$ and $p \in E_i$, $\theta_{q_i} = \psi_p$. Note that, for each p, either R equates p to some element of $\text{rng}(t)$, or $p \in E_i$, for some $i < n$. It follows that the function $t' : \mathbb{N} \to \mathbb{N}$, defined next, is computable.

$$t' = \lambda p \cdot \begin{cases} q, & \text{where } q \text{ is first found such that } \langle p, t(q)\rangle \in R, \\ & \text{if such a } q \text{ exists}; \\ q_i, & \text{otherwise, where } i \text{ is such that } p \in E_i. \end{cases} \tag{6}$$

It is straightforward to verify that t' witnesses $\psi \leq \theta$. $\square$ (**Theorem 6**)

Theorem 7, restated just below, is our second main result. It establishes that there there exists a minimal eps ψ such that, for each ceer R, either R contradicts ψ's program equivalence relation, or there exists a translation function t into ψ such that the range of t *fails* to intersect *infinitely many* of R's equivalence classes.

Theorem 7. There exists an eps ψ satisfying (a) and (b) below.

(a) ψ is minimal.
(b) For each ceer R, there exists a translation function t into ψ such that R does *not* weakly tie t into ψ.

Proof (Sketch). The eps ψ is constructed below, following some necessary definitions. Let $\mathcal{A}ux \subseteq \mathcal{P}art\mathcal{C}omp$ be such that

$$\mathcal{A}ux = \mathcal{P}art\mathcal{C}omp \setminus \{\langle i, j\rangle^{<k+1} \mid i, j \in \mathbb{N} \land k < 2^i\}. \tag{7}$$

It is straightforward to show that $\mathcal{A}ux$ is 1-1, computably enumerable. So, let $(\alpha_\ell)_{\ell \in \mathbb{N}}$ be a 1-1, effective numbering of $\mathcal{A}ux$.

As is common, ψ is constructed in stages, i.e., ψ is the union of $\psi^0 \subseteq \psi^1 \subseteq \cdots$. In conjunction with ψ, four computable predicates are constructed: $\lambda i, s.[i \in R\text{-flags}^s]$, $\lambda i, j, \ell, s.[\langle i, j, \ell\rangle \in t\text{-flags}^s]$, $\lambda \ell, s.[\ell \in \text{Src}^s]$, and $\lambda p, s.[p \in \text{Dst}^s]$. The purposes of these predicates are as follows.

- The R-flags predicate keeps track of which i are such that W_i contradicts ψ's program equivalence relation. More precisely, for each i, if there exists an s such that $i \in R\text{-flags}^s$, then $W_i \not\subseteq \text{Equiv}(\psi)$.
- The t-flags predicate helps to keep track of which ℓ *may be* such that φ_ℓ is a translation function into ψ. It will turn out that: if i and ℓ are such that $W_i \subseteq \text{Equiv}(\psi)$ and φ_ℓ is a translation function into ψ, then, for each j, and all but finitely many s, $\langle i, j, \ell \rangle \in t\text{-flags}^s$.
- The Src predicate keeps track of which ℓ are such that α_ℓ has not yet been assigned to any ψ-program. In particular, if ℓ and s are such that $\ell \in \text{Src}^s$ and $\alpha_\ell \neq \lambda x.\!\uparrow$, then, for each p, $\psi_p^s \neq \alpha_\ell$.
- The Dst predicate keeps track of which ψ-programs have not yet been used. More precisely, if p and s are such that $p \in \text{Dst}^s$, then $\psi_p^s = \lambda x.\!\uparrow$.

For each i and s, $i \in R\text{-flags}^{s+1}$ iff $i \in R\text{-flags}^s$, unless stated otherwise. Analogous statements apply to the t-flags, Src, and Dst predicates, as well. The following will be clear from the construction of ψ, for each s.

$$R\text{-flags}^s \subseteq R\text{-flags}^{s+1}. \qquad (8) \qquad\qquad \text{Src}^s \supseteq \text{Src}^{s+1}. \qquad (10)$$
$$t\text{-flags}^s \subseteq t\text{-flags}^{s+1}. \qquad (9) \qquad\qquad \text{Dst}^s \supseteq \text{Dst}^{s+1}. \qquad (11)$$

Let height $: \mathbb{N}^3 \to \mathbb{N}$ be such that, for each i, j, and s,

$$\text{height}_{i,j}^s = |\{\ell \mid \langle i, j, \ell \rangle \in t\text{-flags}^s\}|. \qquad (12)$$

It will be clear from the construction of ψ that, for each i, j, ℓ, and s,

$$\langle i, j, \ell \rangle \in t\text{-flags}^s \Rightarrow \ell < i. \qquad (13)$$

Thus, for each i, j, and s,

$$\text{height}_{i,j}^s \leq i. \qquad (14)$$

Let num $: \mathbb{N}^3 \to \mathbb{N}$ be such that, for each i, j, and s,

$$\text{num}_{i,j}^s = 2^{i-h}, \text{ where } h = \text{height}_{i,j}^s. \qquad (15)$$

Let $f : \mathbb{N}^3 \to \mathbb{N}$ be such that, for each i, j, and k,

$$f_{i,j}(k) = 2\langle i, j \cdot 2^{i+1} + k \rangle. \qquad (16)$$

For each i, j, s, and $k < \text{num}_{i,j}^s$, let $E_{i,j,k}^s \in \mathcal{F}in$ and $\bar{E}_{i,j,k}^s \in \mathcal{F}in$ be as follows, with $h = \text{height}_{i,j}^s$.

$$E_{i,j,k}^s = \{f_{i,j}(k \cdot 2^{h+1} \quad\quad), ..., f_{i,j}(\quad k \quad\quad \cdot 2^{h+1} + 2^h - 1)\}. \qquad (17)$$
$$\bar{E}_{i,j,k}^s = \{f_{i,j}(k \cdot 2^{h+1} + 2^h), ..., f_{i,j}((k+1) \cdot 2^{h+1} \quad\quad - 1)\}. \qquad (18)$$

Suppose that i, j, and s are such that $\text{height}_{i,j}^{s+1} = h + 1$, where $h = \text{height}_{i,j}^s$. Then, it can be shown that, for each $k < \text{num}_{i,j}^{s+1}$, the following.

$$E_{i,j,k}^{s+1} = E_{i,j,2k}^s \quad \cup \bar{E}_{i,j,2k}^s \quad . \qquad (19)$$
$$\bar{E}_{i,j,k}^{s+1} = E_{i,j,2k+1}^s \cup \bar{E}_{i,j,2k+1}^s. \qquad (20)$$

– STAGE $s = -1$. Do the following.
  - Set $R\text{-flags}^0 = \emptyset$.
  - Set $t\text{-flags}^0 = \emptyset$.
  - Set $\text{Src}^0 = \mathbb{N}$.
  - Set $\text{Dst}^0 = 2\mathbb{N} + 1$.
  - For each i, j, and $k < 2^i$, set $\psi^0_{f_{i,j}(2k)} = \psi^0_{f_{i,j}(2k+1)} = \langle i, j \rangle^{<k+1}$.
  - For each $p \in 2\mathbb{N} + 1$, set $\psi^0_p = \lambda x.\uparrow$.

– STAGE $s = \langle 0, \ell \rangle$. If $\ell \in \text{Src}^s$, then do the following.
  - Set $\text{Src}^{s+1} = \text{Src}^s \setminus \{\ell\}$.
  - Set $\text{Dst}^{s+1} = \text{Dst}^s \setminus \{\min \text{Dst}^s\}$.
  - Set $\psi^{s+1}_{\min \text{Dst}^s} = \alpha_\ell$.

– STAGE $s = \langle i+1, 0, - \rangle$. Determine whether there exist j and k satisfying conditions (a)-(c) just below.
  (a) $i \notin R\text{-flags}^s$.
  (b) $k < \text{num}^s_{i,j}$.
  (c) $W^s_i \cap (E^s_{i,j,k} \times \bar{E}^s_{i,j,k}) \neq \emptyset$.
  If such j and k exist, then do the following.
  - Set $R\text{-flags}^{s+1} = R\text{-flags}^s \cup \{i\}$.
  - Choose any $\ell, m \in \text{Src}^s$ such that $\ell \neq m$ and $\langle i, j \rangle^{<2^i} \subseteq \alpha_\ell \cap \alpha_m$.
  - Let $d : \mathbb{N} \to \mathbb{N}$ be any 1-1, computable function such that $\text{rng}(d)$ is computable, $\text{rng}(d) \subseteq \text{Dst}^s$, and $\text{Dst}^s \setminus \text{rng}(d)$ is infinite.
  - Set $\text{Src}^{s+1} = \text{Src}^s \setminus \{\ell, m\}$.
  - Set $\text{Dst}^{s+1} = \text{Dst}^s \setminus \text{rng}(d)$.
  - For each j, each $k < \text{num}^s_{i,j}$, and each $p \in E_{i,j,k}$, set $\psi^{s+1}_p = \alpha_\ell$.
  - For each j, each $k < \text{num}^s_{i,j}$, and each $q \in \bar{E}_{i,j,k}$, set $\psi^{s+1}_q = \alpha_m$.
  - For each j and $k < \text{num}^s_{i,j}$, set $\psi^{s+1}_{d(n+k)} = \langle i, j \rangle^{<(k+1)\cdot 2^h}$, where $n = \sum_{\hat{j}<j} \text{num}^s_{i,\hat{j}}$ and $h = \text{height}^s_{i,j}$.

– STAGE $s = \langle i+1, j+1, \ell, - \rangle$. Let $h = \text{height}^s_{i,j}$. Determine whether conditions (i)-(iv) just below are satisfied.
  (i) $\ell < i$.
  (ii) $i \notin R\text{-flags}^s$.
  (iii) $\langle i, j, \ell \rangle \notin t\text{-flags}^s$.
  (iv) For each $k < \text{num}^s_{i,j}$, $\text{rng}(\varphi^s_\ell) \cap (E^s_{i,j,k} \cup \bar{E}^s_{i,j,k}) \neq \emptyset$.
  If so, then do the following.
  - Set $t\text{-flags}^{s+1} = t\text{-flags}^s \cup \{\langle i, j, \ell \rangle\}$. (Note that this implies $\text{height}^{s+1}_{i,j} = \text{height}^s_{i,j} + 1$.)
  - Let $n = \text{num}^{s+1}_{i,j}$. (Note that, by the just previous step, $n = \text{num}^s_{i,j}/2$.)
  - Let $\{q_0 < q_1 < \cdots < q_{n-1}\}$ be the n least elements of Dst^s.
  - Set $\text{Dst}^{s+1} = \text{Dst}^s \setminus \{q_0, q_1, ..., q_{n-1}\}$.
  - For each $k < n$ and $p \in (E^{s+1}_{i,j,k} \cup \bar{E}^{s+1}_{i,j,k})$, set $\psi^{s+1}_p = \langle i, j \rangle^{(2k+2)\cdot 2^h}$.
  - For each $k < n$, set $\psi^{s+1}_{q_k} = \langle i, j \rangle^{<(2k+1)\cdot 2^h}$.

Fig. 2. The construction of ψ in the proof of Theorem 7. The symbols height, num, f, E, and $\bar{E}$ are defined in (12), (15), (16), (17), and (18), respectively.

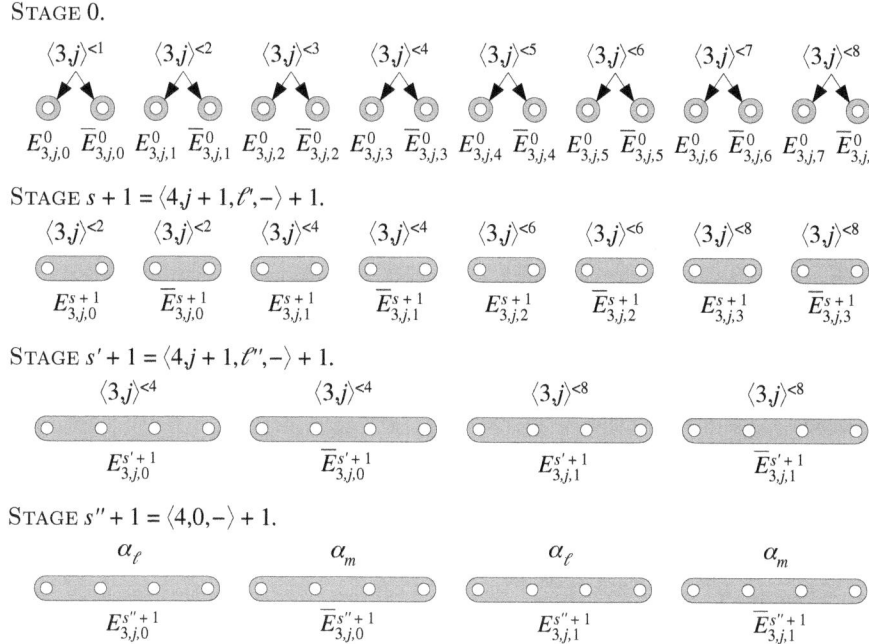

Fig. 3. A depiction of what *could* happen in the proof of Theorem 7 with respect to the ψ-programs of the form $f_{3,j}(k)$, where j is arbitrary and $k < 16$ (see text)

The partial function ψ is constructed in Figure 2. To help to give some of the intuition behind the construction, Figure 3 depicts what *could* happen with respect to the ψ-programs of the form $f_{3,j}(k)$, where j is arbitrary and $k < 16$. In stage 0, the programs will form *eight* pairs of equivalence classes, where the kth pair computes $\langle 3, j \rangle^{<k+1}$ (the first such pair being the 0th). If, subsequently, the conditions of some stage s of the form $\langle 4, j + 1, \ell', - \rangle$ are satisfied, then, in stage $s + 1$, the programs will form *four* pairs of equivalence classes, where the kth pair computes $\langle 3, j \rangle^{<2k+2}$. If, similarly, the conditions of some stage s' of the form $\langle 4, j + 1, \ell'', - \rangle$ are satisfied (where $\ell' \neq \ell''$), then, in stage $s' + 1$, the programs will form *two* pairs of equivalence classes, where the kth pair computes $\langle 3, j \rangle^{<4k+4}$. If, finally, the conditions of some stage s'' of the form $\langle 4, 0, - \rangle$ are satisfied, then, in stage $s'' + 1$, the equivalence classes will alternate in computing α_ℓ and α_m, for some *distinct* ℓ and m. $\approx \square$ (**Theorem 7**)

Acknowledgements. The author would like to thank James S. Royer for many helpful conversations relevant to this paper.

References

1. Ershov, Y.L.: On computable enumerations. Algebra and Logic 7(5), 330–346 (1968)
2. Freivalds, R., Kinber, E.B., Wiehagen, R.: Inductive inference and computable one-one numberings. Zeitschrift für Mathematische Logik und Grundlagen der Mathematik 28(27-32), 463–479 (1982)
3. Friedberg, R.M.: Three theorems on recursive enumeration. I. Decomposition. II. Maximal Set. III. Enumeration without duplication. Journal of Symbolic Logic 23(3), 309–316 (1958)
4. Gao, S., Gerdes, P.: Computably enumerable equivalence relations. Studia Logica 67(1), 27–59 (2001)
5. Goncharov, S., Yakhnis, A., Yakhnis, V.: Some effectively infinite classes of enumerations. Annals of Pure and Applied Logic 60(3), 207–235 (1993)
6. Herrmann, E., Kummer, M.: Diagonals and D-maximal sets. Journal of Symbolic Logic 59(1), 60–72 (1994)
7. Jain, S., Stephan, F., Teutsch, J.: Index sets and universal numberings. Journal of Computer and System Sciences 77(4), 760–773 (2011)
8. Khutoretskii, A.B.: On the reducibility of computable numerations. Algebra and Logic 8(2), 145–151 (1969)
9. Khutoretskii, A.B.: Two existence theorems for computable numerations. Algebra and Logic 8(4), 277–282 (1969)
10. Kummer, M.: A note on direct sums of Friedbergnumberings. Journal of Symbolic Logic 54(3), 1009–1010 (1989)
11. Kummer, M.: An easy priority-free proof of a theorem of Friedberg. Theoretical Computer Science 74(2), 249–251 (1990)
12. Lavrov, I.A.: Computable numberings. In: Logic, Foundations of Mathematics and Computability Theory, pp. 195–206 (1977)
13. Mal'cev, A.I.: Positive and negative numerations. Proceedings of the USSR Academy of Sciences 160(2), 278–280 (1965)
14. Mal'cev, A.I.: Positive and negative numberings. In: The Metamathematics of Algebraic Systems. Studies in Logic and the Foundations of Mathematics, vol. 66, pp. 379–383. Elsevier (1971); Translated by B.F. Wells III
15. Moelius III, S.E.: Characteristics of minimal effective programming systems. Submitted to arXiv (2012)
16. Machtey, M., Winklmann, K., Young, P.: Simple Gödel numberings, isomorphisms, and programming properties. SIAM Journal on Computing 7(1), 39–60 (1978)
17. Pour-El, M.B.: Gödel numberings versus Friedberg numberings. Proceedings of the American Mathematical Society 15(2), 252–256 (1964)
18. Riccardi, G.A.: The independence of control structures in abstract programming systems. Journal of Computer and System Sciences 22(2), 107–143 (1981)
19. Rogers Jr., H.: Gödel numberings of partial recursive functions. Journal of Symbolic Logic 23(3), 331–341 (1958)
20. Rogers Jr., H.: Theory of Recursive Functions and Effective Computability. McGraw Hill (1987); Reprinted. MIT Press (1967)
21. Royer, J.S.: A Connotational Theory of Program Structure. LNCS, vol. 273. Springer, Heidelberg (1987)
22. Schinzel, B.: On decomposition of Gödelnumberings into Friedbergnumberings. Journal of Symbolic Logic 47(2), 267–274 (1982)
23. Spreen, D.: Computable one-to-one enumerations of effective domains. Information and Computation 84(1), 26–46 (1990)

After Turing: Mathematical Modelling in the Biomedical and Social Sciences

From Animal Coat Patterns to Brain Tumours to Saving Marriages

James D. Murray

1 Centre for Mathematical Biology, Mathematical Institute, 24-29 St Giles', Oxford, OX1 3LB, England
2 Applied & Computational Mathematics, Ecology & Evolutionary Biology, Princeton University Princeton NJ 085440-1003, United States of America

Abstract. Turing's 1952 paper on reaction diffusion models for spatial pattern formation was important in the early development of the application of mathematical modelling in biology and medicine. We describe here three very different problems which have been studied in depth and which have proved informative and useful in understanding specific phenomena. We describe an early study of a reaction diffusion model which helped explain the diverse coat patterns observed on animal coats. We then describe a basic, but surprisingly informative and accurate model, currently used medically, for quantifying the growth of gliomablastoma brain tumours. It enhances imaging techniques beyond any brain scanning procedure currently available and is used to estimate patient life expectancy and explain some current patient brain tumour anomalies. Finally we describe a modelling example from the social sciences, which quantifies marital interaction which was used to predict divorce with surprising accuracy and has helped design a new scientific marital therapy which is currently used.

1 Introduction

The rediscovery of Turing's 1952 paper in the late 1960's had a major influence in the development of mathematical biology which grew rapidly from the mid-1970's. Mathematical modelling is now used in practically every field in the biomedical sciences with the involvement in the social sciences clearly an important growth field of the near future. Much of the research in the late 20th century was on biological pattern formation. With the explosion of genetic studies the belief that genetics would solve all these developmental problems has certainly not been borne out. The underlying mechanisms involved in developmental biology and medicine are still largely unknown and is a major interdisciplinary area of mathematics in the biological sciences.

A survey of just some of the early work in this important interdisciplinary field, now known as mathematical biology, or theoretical biology, or systems

S.B. Cooper, A. Dawar, and B. Löwe (Eds.): CiE 2012, LNCS 7318, pp. 517–527, 2012.

biology, and its remarkable growth since the early late 1960's and 1970's is given in an increasing number of books necessarily becoming ever more specialised. By the mid-1980's it was becoming more widely accepted that any real contribution to the biological sciences from modelling must be genuinely interdisciplinary and hence related to real biology.

2   How the Leopard Gets Its Spots

Turing's contribution to mathematical biology is *The Chemical Basis of Morphogenesis* [32] which has been seminal in several areas of spatial patterning modelling in development, ecology and other biological areas since its rediscovery in the late 1960's. He did not apply his model to any specific biological problem.

Turing showed how a system of reacting chemicals which can also diffuse, could generate a steady state heterogeneous spatial pattern of chemical concentrations. He called these chemicals *morphogens*. He hypothesized that these morphogenetic prepatterns could cue cell differentiation and result in observed spatial patterns. His model is encapsulated in the coupled system of reaction diffusion equations, of the general form,

$$\frac{\partial u}{\partial t} = \gamma f(u,v) + D_u \nabla^2 u \qquad \frac{\partial v}{\partial t} = \gamma g(u,v) + D_v \nabla^2 v \qquad (1)$$

where the functions $f(u,v)$ and $g(u,v)$ denote the reaction kinetics associated with the chemicals u, called the activator, and v, the inhibitor with D_u and D_v the diffusion coefficients of u and v respectively. The parameter, γ, arises when the system is written in this nondimensional form: it is a measure of domain scale. A stability analysis of the steady states of the kinetics shows that to generate spatial patterns in u and v it is necessary, among other things, that the inhibitor has a higher diffusion rate than the activator, that is $D_v > D_u$ (cf., e.g., [15, 16]). A review article by [11] is specifically devoted to Turing's theory.

A specific and the first experimental reaction diffusion mechanism [30] was used to study how animal coat patterns and butterfly wing patterns might be formed [12–14]. The [30] reaction terms used in (1) were

$$\begin{aligned} f(u,v) &= a - u - h(u,v) \\ g(u,v) &= \alpha(b-v) - h(u,v) \\ h(u,v) &= \frac{\rho uv}{1+u+Ku^2} \end{aligned} \qquad (2)$$

where a, b, α, ρ and K are constants the values of which were chosen which result in steady state spatially heterogeneous solutions. These parameters were kept fixed for all the calculations. *Only* the scale and geometry of the domain were varied in the analysis. For a given domain size and geometry each set of initial conditions gave a qualitatively similar but *unique* pattern, a fact reflected in nature.

It was shown [12, 14] that a single pre-patterning mechanism was capable of generating the typical geometry of mammalian coat patterns, from the mouse to the elephant and almost everything in between, with the final pattern governed simply by the size and shape of the embryo at the time the pattern formation process was initiated. In solving the reaction diffusion partial differential equation systems the domain size and shape is crucial in influencing the spatial patterns obtained. For a given mechanism if you try and simulate solutions in a very small domain it is not possible to obtain steady state spatial patterns. A minimum size is needed to drive any sustainable spatial pattern. As the size of the domain is increased, a series of increasingly complex spatial patterns emerge.

The concept behind the model is that the simulated spatial patterns solutions of a reaction diffusion mechanism reflect the final morphogen melanin patterning observed on animal coats. With this scenario the cells react to a given level in morphogen concentration, thus producing melanin (or rather becoming melanocytes - cells which produce melanin). In Figures 1(a) and 1(c) the black regions represent levels of melanin concentration higher than the uniform steady state. It should be emphasized that this model is a hypothetical one which has not been verified experimentally biologically but certainly circumstantially. Such an approach has also been used in many patterning processes such as in butterfly wing patterns [13, 15, 19] the last of which presents experimental confirmation of their theoretical predictions.

The solutions of the reaction diffusion system (1) and (2) in domains shown in Figure 1(a) were first computed as an example of how the geometry constrains the possible pattern modes. When the domain is very narrow, only simple, essentially one-dimensional, modes can exist. Two-dimensional patterns require the domain to have enough two dimensionality. With a tapering cylinder as in Figure 1(a) if the radius at one end is large enough, two-dimensional patterns can exist on the surface. So, a tapering cylinder can exhibit a gradation from a two-dimensional pattern to simple stripes as illustrated in Figure 1(a).

This shows that the conical domain mandates that it is not possible to have a tail with spots at its tip and stripes at its base, but only the converse: Figure 1(a) shows some examples of specific animal tails. This phenomenon is a genuine example of a *developmental constraint*. Cheetahs, of which a photo of one is shown in Figure 1(b), are prime examples of this, as well as other spotted animals such as genets. If the threshold level of morphogen is different, a different but related pattern can develop. In this way unique, but globally similar, patterns can be formed and could be the explanation for the different types of patterns on different species of the same animal genre, such as the giraffe. Because the initial conditions for an individual animal involve a limited randomness it implies the uniqueness of each animal pattern in the same species.

The interpretation of Figure 1(c) is that if the animal embryo is too small when the patterning mechanism is activated, as in the mouse, or too large, as in the hippopotamus and elephant for example, then no clear pattern will be observed and these animals are essentially uniform in color.

There have been numerous developments and an increased understanding of how coat patterns on animals, fish and butterflies, for example, are formed with the addition and combination of other pattern forming mechanisms, such as

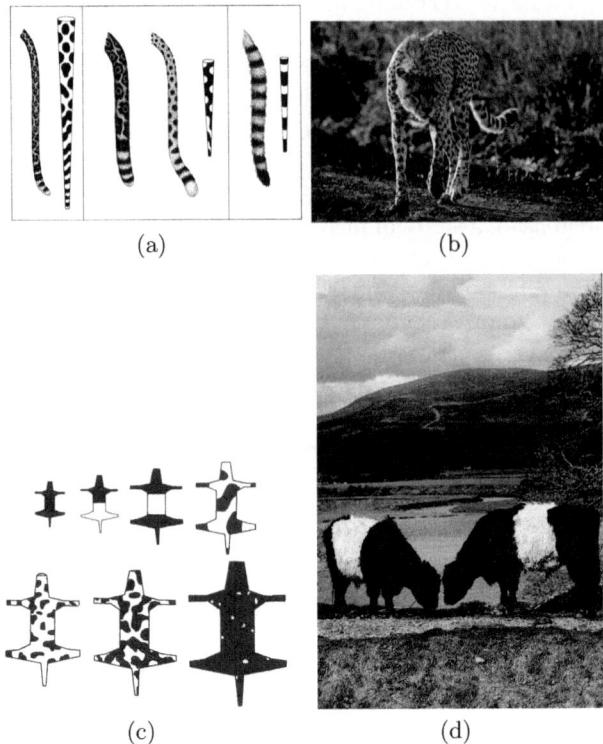

Fig. 1. (a) Examples of a *developmental constraint*. Spotted animals can have striped tails but not the other way round. From left to right are typical of the tail of the leopard, the cheetah and the genet together with the solutions from a reaction diffusion system which can generate steady state spatial patterns. The geometry and scale when the pattern mechanism is activated play crucial roles in the resulting coat patterns. Dark regions represent areas of high morphogen concentration. (Tail art work reproduced from [14] with permission of Patricia Wynne). (b) A cheetah (*Acinonyx jubatus*) which is an example of the developmental constraint described in (a). (Photograph courtesy of Professor Andrew Dobson). (c) Numerical simulations of the reaction diffusion model analysis for the generation of general coat markings on an illustrative domain. The model parameters were also the same; only the scale parameter, γ, was varied. The domain sizes have been reduced to fit in a single figure but in the simulations there was a scale factor of 1,000 between the smallest and the largest figure. (d) Belted Galloway cows are examples of the second bifurcation (Photograph courtesy of Allan Wright). Cf. [14–16] for numerous other examples.

chemotaxis whereby there is movement of cells up chemical gradients. There are numerous review articles and books, for example, [21] who consider fish stripes, [10, 25, 9] who discuss evolving fish patterns among other patterned species.

By the end of the 1970s it was becoming clear that a major problem with reaction diffusion models of pattern formation was the paucity of any real biological system which confirmed or otherwise the chemical basis of morphogenesis.

This gave rise in the early 1980s to the new experimentally verifiable approach to modeling morphogenesis, namely the mechanical theory in [20] and which is discussed at length with numerous applications, the predictions of which were confirmed by experiment (cf. [15, 16] for a detailed survey).

3 Brain Tumours: Enhancing Imaging Techniques, Quantifying Therapy Efficacy and Estimating Patient Life Expectancy

High grade glioblastoma brain tumours are the most aggressive brain tumours and are always fatal with an approximate median life expectancy of 9-12 months. In spite of increasing accuracy of imaging techniques they still cannot detect cancer cell densities sufficiently accurately. A practical model, which encompasses the two key elements in the growth of glioblastoma brain tumours, namely the invasive diffusive properties of the cancer cells and their growth rate, is given by

$$\frac{\partial c}{\partial t} = \nabla.D(\mathbf{x})\nabla c + \rho c \tag{3}$$

where $c(\mathbf{x}, t)$ is the tumour cell density, cells/mm^3, at position, $\mathbf{x}$, in the brain at time, t (months), and $D(\mathbf{x})$ is diffusion (mm^2/month) which quantifies the invasiveness of the cancer cells at position $\mathbf{x}$ in the brain. ρ is the proliferation rate (/month) of the cancer cells which gives the turnover time as $(\log 2)/\rho$ (months). Diffusion in grey matter is smaller than in white matter.

Solutions of (3) are unbounded in time because of the form of the growth term which implies exponential growth but in the time scale relevant to glioma growth and patient survival time it does not contribute significantly to the solutions relevant to patients.

In the original model [5] the brain was considered to be homogeneous matter bounded by the ventricles and skull. Even with such a simple anatomical model the tumour growth predictions of the analysis were broadly in line with patient observation of both low and high grade brain tumours. The limitations of current imaging techniques were clear. The model was used to mimic various accepted medical treatments, specifically radiation, surgical resection (removal) [33, 29] and chemotherapy [31, 26, 23]. A three dimensional model was proposed and studied in [1] who were the first to demonstrate that cancer cell diffusion, mainly ignored up to that time, is a major component of glioma growth. They showed that only those tumours with a low diffusion rate could benefit from wide surgical resection although eventually there will be multifocal recurrence as found clinically [24]: cf. [16] for a full discussion and review which encompasses anatomically correct brains.

4 Virtual Gliomas for Anatomically Correct Brains: Enhanced Imaging and Current Limitations

A major advance in the practical application of the model (3) was the availability of the brain web atlas [3]. This allowed the model to be applied to anatomically correct brains [27, 26, 29, 28, 16, 17]. Among other things it made it possible to refine the gross anatomic boundaries and to vary the degree of motility of glioma cells depending on whether it was in grey or white matter. With the BrainWeb Atlas it was possible to solve equation (3) in a three dimensional anatomically correct brain in which the grey and white matter are clearly delineated.

The patient specific procedure is to evaluate the tumour size from brain scans and, what is essential, estimate the parameter values for each patient [27] to obtain the average diffusion coefficient and the average growth rate. There is a threshold of detection of cancer cells with all imaging techniques, whether by imaging or microscopic studies. To use the predictive potential of the mathematical model (3) for preliminary predictions, serial imaging of the tumour was used to calculate its volume which was then taken as the volume of an equivalent sphere with radius r, namely $\frac{4\pi r^3}{3}$. We then consider the model to be radially symmetric with a constant diffusion coefficient, based on averaging the values from imaging. Equation (3) then becomes

$$\frac{\partial c}{\partial t} = D \left[\frac{\partial^2 c}{\partial r^2} + \frac{2\partial c}{r\partial r} \right] + \rho c \tag{4}$$

We consider that at time $t = 0$ there is a concentrated number of cancer cells, N cells/mm^3, at $r = 0$ in which case the analytical solution of (4) is given by

$$c(r,t) = \frac{N\exp(\rho t - \frac{r^2}{4Dt})}{8(\pi Dt)^{\frac{3}{2}}} \tag{5}$$

If the smallest level of image detection is denoted by c_1 cells/mm^3, then the radius, r, of the tumour for this cell density is obtained from (5) which asymptotically gives

$$r \sim 2t\sqrt{D\rho} \Rightarrow v = r/t \sim 2\sqrt{D\rho} \tag{6}$$

where v is the velocity of tumour growth. That is, the equivalent radial growth of the tumour is linear in time a finding confirmed by patient data: cf., e.g., [16].

5   Approximate *in vivo* Patient Survival Time

Based on accepted medical practice, we consider detection is when the spherical equivalent tumour volume is of radius 15mm and that death occurs when the radius is 30mm. The approximate survival time, t_{survival} (months), from detection, in the absence of any treatments, is, from (6),

$$t_{\text{survival}} = t_{r=30} - t_{r=15} = \frac{7.5}{\sqrt{D\rho}} \tag{7}$$

Typical growth rates vary widely, approximately from 1-5 /month and diffusion rates from 1-8 mm^2/month. The medians for 9 patients in the study in [23] are

$D = 0.9\text{mm}^2/\text{month}$ and $\rho = 1.16/\text{month}$ which gives a median survival time of 7.34 months.

Simulations for an anatomically correct brain highlights the problems with current imaging limitations. Figure 2 is a computed solution of (3) in a three dimensional anatomically correct brain with the diffusion coefficient spatially dependent on whether or not it is in grey or white matter. It shows the detectable tumour at death and the spread of the tumour cells beyond what can be detected by the most accurate current CT or MRI, or any other, imaging techniques.

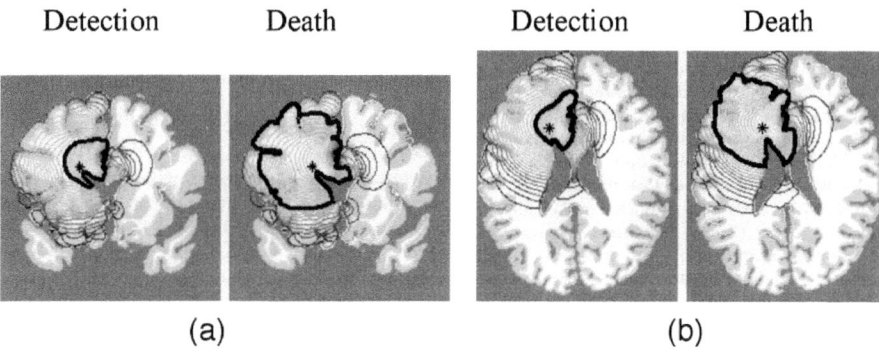

Detection   Death   Detection   Death

(a)   (b)

Fig. 2. Computed solutions of equation (3) in a three dimensional anatomically accurate brain with typical patient parameter values. These show the horizontal section of the virtual human brain through the site of the original tumour (+ in (a), * in (b)). The left image in each is the tumour at original detection while the right image is the same tumour at time of death. The thick black contour defines the edge of the tumour that can be detected by enhanced computer tomography (CT). The blue contours outside this black line represent the same cancer cell densities peripheral to the imaging limits. (a) Tumour in grey matter: the time from diagnosis to death is approximately 256 days. (b) Tumour in white matter: the time from diagnosis to death is approximately 158 days. (Figures extracted from [27]).

The model described here has been used to quantify the effect of different treatment efficacies for individual patients prior to their use, such as [31, 26, 33]. The last of these predicted patient survival rates which compared surprisingly accurately with the extant data at the time and recently published in [22]. A full discussion of how such treatments are incorporated in the model and their comparison with patient data is given in [16]. The model has been used to determine when tumours start [17, 18].

6   Marital Interaction and Divorce Prediction

The rise in divorce rates in developed countries is widespread, important, but a poorly understood phenomenon. What makes some marriages happy but others miserable? Previous attempts at predicting marriage dissolution tended to be based on mismatches in the couple's personality or areas of disagreement: these have not been too successful. The original model to quantify the interaction of a couple discussing a problem of contention, which they chose, was proposed in [4]: cf. also further developments in [6, 8, 7].

In [4] is developed, a simple, but surprisingly accurate predictive, mathematical model based on only a few variables descriptive of specific marital interaction patterns, such as the difference between positivity and negativity and how each partner is influenced by the other during a 15 minute (filmed) discussion of a subject of ongoing contention [4]. These topics could be on any topic, for example, money, in-laws, housing, sex, food or politics; the couple chose the topic they discussed. A major part of the longitudinal study on which the model was tested involved several hundred newly married couples. With the data, and the model analysis of it, it was possible to identify patterns predictive as to whether the couple would divorce or stay married happily or unhappily.

Based on the hypothesis that without a theoretical understanding of the processes related to marital, or relationship, stability and dissolution, it is difficult to design and evaluate effective relationship therapies. Using a psychological coding system to "score" each partner in a conflict conversation, the resultant time series can visually describe the ebb and flow of the interaction for a relationship conversation. The scoring system used is SPAFF (Specific Affect Scoring system); cf., e.g., a full description in [2]. The system consists of scoring each comment according to the specific scoring system. During the couple's discussion, of the topic of contention, an integer number between +4 and -4 is given according to the sentiment expressed. For example, positive numbers are given for affection (+4), humour (+4), interest (+2), while negative numbers are given for sadness (-1), anger (-1), contempt (-4), whining (-1), belligerence (-2) disgust (-3), stonewalling (-2) and so on. There is thus a continual scoring of the conversation as a function of time.

The model equations are:

$$W_{t+1} = a + r_1 W_t + I_{HW}(H_t) \qquad H_{t+1} = b + r_2 H_t + I_{WH}(W_t) \qquad (8)$$

where W_t and H_t are the scores of the wife and husband obtained from what they said at time t. Consider the first equation, the wife's. The influence function $I_{HW}(H_t)$ is the influence the husband has on the wife as a function of his score when he speaks at time t and it reflects the influence he has when the wife replies, namely at time $t + 1$. The parameter r_1, is the inertia parameter, which quantifies how easy, or willing, it is for the wife to change her attitude reflected in her score the last time she spoke, namely W_t, at time t. Here $0 \le r_1, r_2 < 1$. For example, if r_1 is small the wife is not fixed on what she last said since this term is small and is more willing to change her view. The parameter, a, reflects how the wife feels about the marriage when the husband is not influencing her, that is when $I_{HW}(H_t) = 0$.

Although each couple has a unique relationship quantified by their own characteristic model parameters, an analysis of the longitudinal study of couples only five dominant couple's styles were found [4]. These are primarily based on the influence functions, namely (i) validating, (ii) volatile, (iii) conflict-avoiding, (v) hostile and (v) hostile-detached. Only 2 of these styles, (i) and (iii), are stable, typically resulting in long-term happy relationships.

The ability to predict the longitudinal course of marital relationships using this modelling approach, with an average of 94% accuracy based on several hundred couples, has now been reported in the laboratories in four separate longitudinal studies: cf. the book [7]. Here the model has been extended to incorporate the effect of a baby on marriages and also to committed cohabiting gay male and lesbian relationships.

References

1. Burgess, P., Kulesa, P., Murray, J., Alvord, E.: The interaction of growth rates and diffusion coefficients in a three-dimensional mathematical model of gliomas. J. Neuropathol. Exp. Neurol. 56, 704–713 (1997)
2. Coan, J., Gottman, J.: The Specific Affect (SPAFF) coding system. In: Coan, J., Allen, J. (eds.) Handbook of Emotion Elicitation and Assessment, pp. 106–123. Oxford University Press, New York (2007)
3. Cocosco, C., Kollokian, V., Kwan, R.K.-S., Evans, A.: Brain Web: Online Interface to a 3D MRI Simulated Brain Database. In: Proceedings of 3rd International Conference on Functional Mapping of the Human Brain, vol. 5 (1997)
4. Cook, J., Tyson, R., White, K., Rushe, R., Gottman, J., Murray, J.: Mathematics of marital conflict: Qualitative dynamic mathematical modeling of marital interaction. J. Family Psychology 9, 110–130 (1995)
5. Cruywagen, G., Woodward, D., Tracqui, P., Bartoo, G., Murray, J., Alvord, E.: The modelling of diffusive tumors. J. Biol. Systems 3, 937–945 (1995)
6. Gottman, J., Guralnick, M., Wilson, B., Swanson, C., Swanson, K., Murray, J.: What should be the focus of emotion regulation in children? A nonlinear dynamic mathematical model of children's peer interaction in groups. Development & Psychopathology 9(2), 421–452 (1997)
7. Gottman, J., Murray, J., Swanson, C., Tyson, R., Swanson, K.: The Mathematics of Marriage: Dynamic Nonlinear Models. MIT Press, Cambridge (2002)
8. Gottman, J., Swanson, K., Murray, J.: The mathematics of marital conflict: dynamic mathematical nonlinear modeling of newlywed marital interaction. J. Family Psychol. 13, 1–17 (1999)
9. Kondo, S., Iwashita, M., Yamaguchi, M.: How animals get their skin patterns: fish pigment pattern as a live Turing wave. Inst. J. Dev. Biol. 53, 851–856 (2009)
10. Maini, P.: How the mouse got its stripes. Proc. Nat. Acad. Sci. 100, 9656–9657 (2003)
11. Maini, P.: Using mathematical models to help understand biological pattern formation. C. R. Biologies 327, 225–234 (2004)
12. Murray, J.: A pre-pattern formation mechanism for animal coat markings. J. Theor. Biol. 88, 161–199 (1981)
13. Murray, J.: On pattern formation mechanisms for lepidopteran wing patterns and mammalian coat markings. Phil. Trans. Roy. Soc. (Lond.) B 295, 473–496 (1981)

14. Murray, J.: Mammalian coat patterns: How the leopard gets its spots. Scientific American 256, 80–87 (1988)
15. Murray, J.: Mathematical Biology. Springer, Heidelberg (1989)
16. Murray, J.: Mathematical Biology: II. Spatial Models and Biomedical Applications, 3rd edn., vol. 2. Springer, New York (2003)
17. Murray, J.: On the Growth of Brain Tumours: enhancing imaging techniques, highlighting limitations of current imaging, quantifying therapy efficacy and estimating patient life expectancy. In: Lenaerts, T., Giacobini, M., Bersini, H., Bourgigne, P., Dorigo, M., Doursat, R. (eds.) Proceedings of the Eleventh European Conference on the Synthesis and Simulation of Living Systems, Advances in Artificial Life, ECAL 2011, pp. 23–26. MIT Press (2011)
18. Murray, J.: Glioblastoma brain tumours: Estimating the time from brain tumour initiation and resolution of a patient survival anomaly after similar treatment protocols. J. Biol. Dyn. (in press, 2012)
19. Nijhout, N., Maini, P., Madzvamuse, A., Wathen, A., Sekimura, T.: Pigmentation pattern formation in butterflies: experiments and models. C. R. Biologies 326, 717–727 (2003)
20. Oster, G., Murray, J., Harris, A.: Mechanical aspects of mesenchymal morphogenesis. J. Embryol. Exp. Morph. 78, 83–125 (1983)
21. Painter, K., Maini, P., Othmer, H.: Stripe formation in juvenile Pomacanthus explained by a generalized Turing mechanism with chemotaxis. Proc. Nat. Acad. Sci. 96, 5549–5554 (1999)
22. Ramakrishna, R., Barber, J., Kennedy, G., Win, R.R., Ojemann, G., Berger, M., Spence, A., Rostomily, R.: Imaging features of invasion and preoperative and postoperative tumor burden in previously untreated gliomablastomas: Correlation with survival. Surg. Neurol. Int. 1, 40–51 (2010)
23. Rockne, R., Rockhill, J., Mrugala, M., Spence, A., Kalet, I.K., Hendrickson, K., Cloughesy, A.L., Alvord, E., Swanson, K.: Prediciitng the efficacy of radiotherapy in individual glioblastoma patients in viv: a mathematical modelling approach. Phys. Med. Biol. 55, 3271–3285 (2010)
24. Silbergeld, D., Rostomily, R., Alvord, E.: The cause of death in patients with glioblastomas is multifocal: Clinical factors and autopsy findings in 117 cases of supratentorial glioblastomas in adults. J. Neuro-Oncol. 10, 179–185 (1991)
25. Suzuki, N., Hirata, M., Kondo, S.: Traveling stripes on the skin of a mutant mouse. Proc. Nat. Acad. Sci. USA 100, 9680–9685 (2003)
26. Swanson, K., Alvord, E., Murray, J.: Quantifying efficacy of chemotherapy of brain tumors (gliomas) with homogeneous and heterogeneous drug delivery therapy. Acta Biotheoretica 50(6), 223–237 (2002)
27. Swanson, K., Alvord, E., Murray, J.: Virtual brain tumours (gliomas) enhance the reality of medical imaging and highlight inadequacies of current therapy. British J. Cancer 86, 14–18 (2002); Abstracted and featured in the Year Book of the Institute of Oncology Elsevier Science (2003)
28. Swanson, K., Alvord, E., Murray, J.: Virtual and real brain tumors: using mathematical modelling to quantify glioma growth and invasion. J. Neurological Sciences 216(3), 1–10 (2003)
29. Swanson, K., Alvord, E., Murray, J.: Virtual resection of gliomas: effect of extent of resection on recurrence. Mathematical and Computer Modelling 37(11), 1177–1190 (2003)

30. Thomas, D.: Artificial enzyme membranes, transport, memory, and oscillatory phenomena. In: Thomas, D., Kernevez, J.P. (eds.) Analysis and Control of Immobilized Enzyme Systems, pp. 115–150. Springer, Heidelberg (1975)
31. Tracqui, P., Cruywagen, G., Woodward, D., Bartoo, G., Murray, J., Alvord, E.: A mathematical model of glioma growth: the effect of chemotherapy on spatial-temporal growth. Cell Prolif. 28, 17–31 (1995)
32. Turing, A.: The chemical basis of morphogenesis. Phil. Trans. Roy. Soc. B 237, 37–72 (1952)
33. Woodward, D., Cook, J., Tracqui, P., Cruywagen, G., Murray, J., Alvord, E.: A mathematical model of glioma growth: the effect of extent of surgical resection. Cell Prolif. 29, 269–288 (1996)

Existence of Faster than Light Signals Implies Hypercomputation already in Special Relativity

Péter Németi and Gergely Székely

Alfréd Rényi Institute of Mathematics, Hungarian Academy of Sciences, POB 127,
1364 Budapest, Hungary
{nemeti.peter,szekely.gergely}@renyi.mta.hu

Abstract. Within an axiomatic framework, we investigate the possibility of hypercomputation in special relativity via faster than light signals. We formally show that hypercomputation is theoretically possible in special relativity if and only if there are faster than light signals.

1 Introduction

The theory of relativistic hypercomputation (i.e., the investigation of relativity theory based physical computational scenarios which are able to solve non-Turing-computable problems) has an extensive literature and it is investigated by several researchers in the past decades, cf., e.g., [3, 4, 6, 7, 9, 12]. For an overview of different approaches to hypercomputation, cf., e.g., [24].

It is well-known that hypercomputation is not possible in special relativity in the usual sense (i.e., the sense of Malament–Hogarth spacetimes), cf., e.g., [9]. In this paper, we show that it is possible to perform relativistic hypercomputation via ordinary computers (Turing machines) in special relativity if there are faster than light (FTL) signals, e.g., particles. We shall also show that there have to be FTL signals if relativistic hypercomputation is possible in special relativity (via Turing machines), cf. Thm.2.

It is interesting in and of itself to investigate the (logical) consequences of the assumption that FTL objects exist, independently of the question whether they really exist or not in our actual physical universe. Logic based axiomatic investigations typically aim for describing all the theoretically possible universes and not just our actual one. Moreover, so far we have not excluded the possibility of the existence of FTL entities in our actual universe; and from time to time there appear theories and experimental results suggesting the existence of FTL objects. Recently, the OPERA experiment, cf. [17], raised the interest in the possibility of FTL particles.

Contrary to the common belief, the existence of FTL particles does not lead to a logical contradiction within special relativity. For a formal axiomatic proof of this fact, cf. [26]. However, it is interesting to note that, in contrast with this result, the impossibility of the existence of FTL inertial *observers* follows from special relativity, cf., e.g., [1].

The investigation of FTL motion in relativity theory goes back (at least) to Tolman, cf., e.g., [28, p.54-55]. Since then a great many works dealing with FTL

S.B. Cooper, A. Dawar, and B. Löwe (Eds.): CiE 2012, LNCS 7318, pp. 528–538, 2012.
© Springer-Verlag Berlin Heidelberg 2012

motion have appeared in the literature, cf., e.g., [13, 15, 19–21, 23, 29] to mention only a few.

2   Hypercomputation in SR

It is well-known that we can send information back to the past if there are FTL particles, cf., e.g., [26] and [28, p.54-55]. It is natural to try using this possibility to design computers with greater computational power. We shall show that uniformly accelerated relativistic computers can compute beyond the Church–Turing barrier via using FTL signals. In this section, we show this fact informally. In Sect.5, we reconstruct our informal ideas of this section within an axiomatic theory of special relativity extended with accelerated observers.

Our first observation is that if we can send out an FTL signal with a certain speed, we also have to be able to send out arbitrarily fast signals, by the principle of relativity. Prop.1 is a formal statement of this observation. To informally justify this statement, let us assume that we can send out an FTL signal by a certain experiment, say with speed 1.01c. According to special relativity, for any FTL speed, say 10^{10}c, there is a inertial reference frame (moving relative to our frame) according to which our signal moves with this speed. By the principle of relativity, inertial frames are experimentally indistinguishable, cf. [8, §5, pp.149-159] and [27, pp.176-178]. So the experiment which is configured in our reference frame as our original experiment is seen by this moving inertial frame as yielding an FTL signal moving with speed 10^{10}c in our frame. Therefore, in our (or any other inertial) reference frame, it is possible to send out an FTL signal with any speed.

Let us see the construction of our special relativistic hypercomputer. Let the computer be accelerated uniformly with respect to an inertial observer, cf. Fig.1.[1] There is an event O with the following property: any event E on the worldline of our uniformly accelerated computer is simultaneous with O, according to the inertial observer comoving with the computer at E, cf., e.g., [14, Fig.6.4, p.173], and [18, Fig.5.13, p.152].

Now let us show that this configuration can be used to decide non-Turing-computable questions if there are FTL signals. Let us set the computer to work on some recursively enumerable but non-Turing-computable problem, say the decision problem for the consistency of ZF set theory; the computer enumerates one by one all the consequences of ZF. Let us fix an event M on the worldline of the programmer which is later than O according to him. Now, if the computer finds a contradiction, let it send out a fast enough signal which reaches the programmer before event M. Such signal exists since, by our first observation, the computer can send out a signal which is arbitrarily fast with respect to his coordinate system (i.e., any half line in the "upper" half space determined by the comoving observer's simultaneity can be the worldline of the signal). Therefore, if the programmer receives a signal between events O and M, he knows that ZF

---

[1] In relativity theory, uniform acceleration means motion along a hyperbola (according to inertial observers), cf., e.g., [5, §3.8, pp.37-38], [14, §6], and [22, §12.4, pp.267-272].

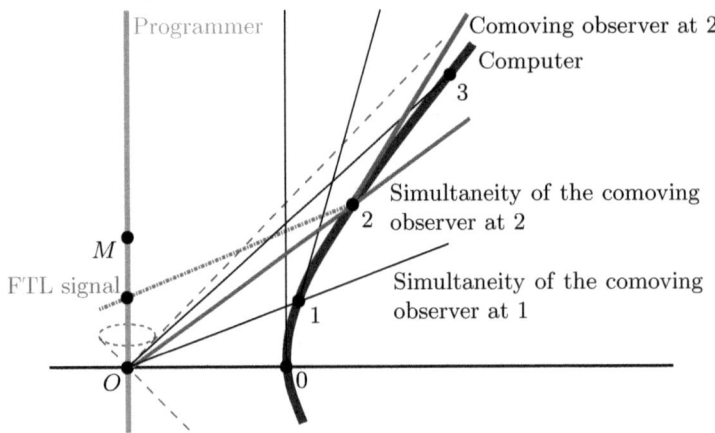

Fig. 1. Illustration of hypercomputation via FTL particles

is inconsistent; and if there is no signal between M and O, he knows that the computer has not found any contradiction, so after event M the programmer can conclude that there is no contradiction in ZF set theory. The same way, by this thought experiment using FTL signals, we can decide (experimentally) any recursively enumerable set of numbers.

If there are no FTL signals, then the whole computation has to happen in the causal past of the event when the programmer learns the result of computation. However, in special relativity, the computer remaining within the causal past of any event has only finite time to compute by the twin paradox theorem. That is why hypercomputation is not possible in special relativity without FTL signals. This argument is also the basis of proving that Minkowski spacetime is not a Malament–Hogarth spacetime.

3   The Language of Our Axiom Systems

To formalize the result of Sect.2, we need an axiomatic theory of special relativity extended with accelerated observers. Here we shall use the following two-sorted language of first-order logic parametrized by a natural number $d \geq 2$ representing the dimension of spacetime:

$$\{\, B, Q\,;\, \mathsf{Ob}, \mathsf{IOb}, \mathsf{Ph}, +, \cdot, \leq, \mathsf{W}\,\},$$

where B (bodies) and Q (quantities) are the two sorts, Ob (observers), IOb (inertial observers) and Ph (light signals) are one-place relation symbols of sort B, $+$ and $\cdot$ are two-place function symbols of sort Q, $\leq$ is a two-place relation symbol of sort Q, and W (the worldview relation) is a $d+2$-place relation symbol the first two arguments of which are of sort B and the rest are of sort Q.

Relations $\mathsf{Ob}(o)$, $\mathsf{IOb}(m)$ and $\mathsf{Ph}(p)$ are translated as "*o is an observer*," "*m is an inertial observer*," and "*p is a light signal*," respectively. To speak about co-ordinatization, we translate $\mathsf{W}(k, b, x_1, x_2, \ldots, x_d)$ as "*body k coordinatizes body b at space-time location* $\langle x_1, x_2, \ldots, x_d \rangle$," (i.e., at space location $\langle x_2, \ldots, x_d \rangle$ and instant x_1).

To make them easier to read, we omit the outermost universal quantifiers from the formalizations of our axioms, i.e., all the free variables are universally quantified.

We use the notation Q^d for the set of all d-tuples of elements of Q. If $\bar{x} \in Q^d$, we assume that $\bar{x} = \langle x_1, \ldots, x_d \rangle$, i.e., x_i denotes the i-th component of the d-tuple $\bar{x}$. Specially, we write $\mathsf{W}(m, b, \bar{x})$ in place of $\mathsf{W}(m, b, x_1, \ldots, x_d)$, and we write $\forall \bar{x}$ in place of $\forall x_1 \ldots \forall x_d$, etc.

4 Axioms of Special Relativity

Let us recall some of our axioms for special relativity. Our first axiom states some basic properties of addition, multiplication and ordering true for real numbers.

<u>AxOField:</u> The quantity part $\langle Q, +, \cdot, \leq \rangle$ is an ordered field.

In the next axiom, we shall use the concepts of time difference and spatial distance. The **time difference** of coordinate points $\bar{x}, \bar{y} \in Q^d$ is defined as:

$$\mathsf{time}(\bar{x}, \bar{y}) := x_1 - y_1.$$

To speak about the spatial distance of any two coordinate points, we have to use squared distance since it is possible that the distance of two points is not amongst the quantities, e.g., the distance of points $\langle 0, 0 \rangle$ and $\langle 1, 1 \rangle$ is $\sqrt{2}$. So in the field of rational numbers, $\langle 0, 0 \rangle$ and $\langle 1, 1 \rangle$ do not have distance just squared distance. Therefore, we define the **squared spatial distance** of $\bar{x}, \bar{y} \in Q^d$ as:

$$\mathsf{space}^2(\bar{x}, \bar{y}) := (x_2 - y_2)^2 + \ldots + (x_d - y_d)^2.$$

Our next axiom is the key axiom of our axiom system of special relativity. This axiom is the outcome of the Michelson–Morley experiment, and it has been continuously tested ever since then. Nowadays it is tested by GPS technology.

<u>AxPh :</u> For any inertial observer, the speed of light is the same everywhere and in every direction (and it is finite). Furthermore, it is possible to send out a light signal in any direction everywhere:

$$\mathsf{IOb}(m) \rightarrow \exists c_m \left(c_m > 0 \wedge \forall \bar{x} \bar{y} \Big[\mathsf{space}^2(\bar{x}, \bar{y}) = c_m^2 \cdot \mathsf{time}(\bar{x}, \bar{y})^2 \right.$$
$$\left. \leftrightarrow \exists p \big[\mathsf{Ph}(p) \wedge \mathsf{W}(m, p, \bar{x}) \wedge \mathsf{W}(m, p, \bar{y}) \big] \Big] \right).$$

Let us note here that AxPh does not require (by itself) that the speed of light is the same for every inertial observer. It requires only that the speed of light according to a fixed inertial observer is a positive quantity which does not depend on the direction or the location. However, by AxPh, we can define the **speed of light** according to inertial observer m as the following binary relation:

$$c(m, v) \overset{def}{\Longleftrightarrow} v > 0 \wedge \forall \bar{x}\bar{y}\Big[\exists p\big[\mathsf{Ph}(p) \wedge \mathsf{W}(m, p, \bar{x}) \wedge \mathsf{W}(m, p, \bar{y})\big]$$
$$\rightarrow \mathsf{space}^2(\bar{x}, \bar{y}) = v^2 \cdot \mathsf{time}(\bar{x}, \bar{y})^2\Big].$$

By AxPh, there is one and only one speed v for every inertial observer m such that $c(m, v)$ holds. From now on, we shall denote this unique speed by c_m.

Our next axiom connects the worldviews of different inertial observers by saying that they coordinatize the same "external" reality (the same set of events). By the **event** occurring for observer m at coordinate point $\bar{x}$, we mean the set of bodies m coordinatizes at $\bar{x}$:

$$\mathsf{ev}_m(\bar{x}) := \{b : \mathsf{W}(m, b, \bar{x})\}.$$

<u>AxEv:</u> All inertial observers coordinatize the same set of events:

$$\mathsf{IOb}(m) \wedge \mathsf{IOb}(k) \rightarrow \exists \bar{y}\, \forall b\big[\mathsf{W}(m, b, \bar{x}) \leftrightarrow \mathsf{W}(k, b, \bar{y})\big].$$

From now on, we shall abbreviate the subformula $\forall b\big[\mathsf{W}(m, b, \bar{x}) \leftrightarrow \mathsf{W}(k, b, \bar{y})\big]$ of AxEv to $\mathsf{ev}_m(\bar{x}) = \mathsf{ev}_k(\bar{y})$. The next two axioms are only simplifying ones.

<u>AxSelf:</u> Any inertial observer is stationary relative to himself:

$$\mathsf{IOb}(m) \rightarrow \forall \bar{x}\big[\mathsf{W}(m, m, \bar{x}) \leftrightarrow x_2 = \ldots = x_d = 0\big].$$

AxSymD: Any two inertial observers agree as to the spatial distance between two events if these two events are simultaneous for both of them. Furthermore, the speed of light is 1 for all observers:

$$\mathsf{IOb}(m) \wedge \mathsf{IOb}(k) \wedge x_1 = y_1 \wedge x_1' = y_1' \wedge \mathsf{ev}_m(\bar{x}) = \mathsf{ev}_k(\bar{x}')$$
$$\wedge\, \mathsf{ev}_m(\bar{y}) = \mathsf{ev}_k(\bar{y}') \rightarrow \mathsf{space}^2(\bar{x}, \bar{y}) = \mathsf{space}^2(\bar{x}', \bar{y}'), \text{ and}$$
$$\mathsf{IOb}(m) \rightarrow \exists p\big[\mathsf{Ph}(p) \wedge \mathsf{W}(m, p, 0, \ldots, 0) \wedge \mathsf{W}(m, p, 1, 1, 0, \ldots, 0)\big].$$

Our axiom system SpecRel is the collection of the five simple axioms above:

$$\mathsf{SpecRel} := \{\mathsf{AxOField}, \mathsf{AxPh}, \mathsf{AxEv}, \mathsf{AxSelf}, \mathsf{AxSymD}\}.$$

To show that SpecRel captures the kinematics of special relativity, let us introduce the **worldview transformation** between observers m and k (in symbols, w_{mk}) as the binary relation on Q^d connecting the coordinate points where m and k coordinatize the same (nonempty) events:

$$\mathsf{w}_{mk}(\bar{x}, \bar{y}) \overset{def}{\Longleftrightarrow} \mathsf{ev}_m(\bar{x}) = \mathsf{ev}_k(\bar{y}) \neq \emptyset.$$

Map $P : Q^d \to Q^d$ is called a Poincaré transformation iff it is an affine bijection such that, for all $\bar{x}, \bar{y}, \bar{x}', \bar{y}' \in Q^d$ for which $P(\bar{x}) = \bar{x}'$ and $P(\bar{y}) = \bar{y}'$,

$$\mathsf{time}(\bar{x}, \bar{y})^2 - \mathsf{space}^2(\bar{x}, \bar{y}) = \mathsf{time}(\bar{x}', \bar{y}')^2 - \mathsf{space}^2(\bar{x}', \bar{y}').$$

Thm.1 shows that our streamlined axiom system SpecRel perfectly captures the kinematics of special relativity since it implies that the worldview transformations between inertial observers are the same as in the standard non-axiomatic approaches. For the proof of Thm.1, cf. [2].

Theorem 1. *Let $d \geq 3$. Assume* SpecRel. *Then* w_{mk} *is a Poincaré transformation if m and k are inertial observers.*

The so-called **worldline** of body b according to observer m is defined as:

$$\mathsf{wl}_m(b) := \{\bar{x} : \mathsf{W}(m, b, \bar{x})\}.$$

Corollary 1. *Let $d \geq 3$. Assume* SpecRel. *The* $\mathsf{wl}_m(k)$ *is a straight line if m and k are inertial observers.*

To extend SpecRel to accelerated observers, we need further axioms. We connect the worldviews of accelerated and inertial observers by the next axiom.

AxCmv: At each moment of its world-line, each observer coordinatizes the nearby world for a short while as an inertial observer does.

Axiom AxCmv is captured by formalizing the following statement: at each point of the worldline of an observer there is an inertial comoving observer such that the derivative of the worldview transformation between them is the identity map, cf., e.g., [1] [25, §6] for details. We shall also use the generalized (localized) versions of axioms AxEv and AxSelf of SpecRel assumed for every observer.

AxEv⁻: Observers coordinatize all the events in which they participate:

$$\mathsf{Ob}(k) \wedge \mathsf{W}(m, k, \bar{x}) \to \exists \bar{y} \;\; \mathsf{ev}_m(\bar{x}) = \mathsf{ev}_k(\bar{y}).$$

AxSelf⁻: In his own worldview, the worldline of any observer is an interval of the time axis containing all the coordinate points of the time axis where the observer coordinatizes something:

$$\big[\mathsf{W}(m, m, \bar{x}) \to x_2 = \ldots = x_d = 0\big] \wedge$$
$$\big[\mathsf{W}(m, m, \bar{y}) \wedge \mathsf{W}(m, m, \bar{z}) \wedge y_1 < t < z_1 \to \mathsf{W}(m, m, t, 0, \ldots, 0)\big] \wedge$$
$$\exists b \big[\mathsf{W}(m, b, t, 0, \ldots, 0) \to \mathsf{W}(m, m, t, 0, \ldots, 0)\big].$$

Let us add these three axioms to SpecRel to get a theory of accelerated observers:

$$\mathsf{AccRel}_0 := \mathsf{SpecRel} \cup \{\mathsf{AxCmv}, \mathsf{AxEv}^-, \mathsf{AxSelf}^-\}.$$

Since AxCmv ties the behavior of accelerated observers to the inertial ones and SpecRel captures the kinematics of special relativity perfectly by Thm.1, it is quite natural to think that $\mathsf{AccRel_0}$ is a theory strong enough to prove the most fundamental theorems about accelerated observers. However, $\mathsf{AccRel_0}$ does not even imply the most basic predictions of relativity theory about accelerated observers, such as the twin paradox. Moreover, it can be proved that even if we add the whole first-order logic theory of real numbers to $\mathsf{AccRel_0}$ is not enough to get a theory that implies (predicts) the twin paradox, cf., e.g., [11] and [25, §7].

In the models of $\mathsf{AccRel_0}$ in which the twin paradox is not true, there are some definable gaps in W. Our next assumption excludes these gaps.

CONT: Every parametrically definable, bounded and nonempty subset of Q has a supremum (i.e., least upper bound) with respect to $\leq$.

In CONT, "definable" means "definable in the language of AccRel, parametrically." For a precise formulation of CONT, cf. [11, p.692] or [25, §10.1]. When Q is the ordered field of real numbers, CONT is automatically true.

Let us extend $\mathsf{AccRel_0}$ with axiom schema CONT:

$$\mathsf{AccRel} := \mathsf{AccRel_0} \cup \mathsf{CONT}.$$

It can be proved that AccRel implies the twin paradox, cf. [11] and [25, §7.2].

Let us now introduce some auxiliary axioms we shall use here but not listed so far. To do so, let us call a linear bijection of Q^d **trivial transformation** if leaves the time components (i.e., first coordinates) of coordinate points unchanged and it fixes the points of the time axis, i.e., the set of trivial transformation is:

$$Triv := \{\, T : T \text{ is a linear bijection of } Q^d,$$
$$T(\bar{y})_1 = \bar{y}_1 \text{ and } T(\bar{x}) = \bar{x} \text{ if } \bar{x}_s = \bar{o} \,\},$$

where $\bar{o}$ denotes the **origin**, i.e., coordinate point $\langle 0, \dots, 0 \rangle$.

AxThExp$^{\#}$: Inertial observers can move with any speed less than the speed of light and new inertial reference frames can be constructed from other inertial reference frames by transforming them by trivial transformations and translations along the time axis:

$$\exists h \, \mathsf{IOb}(h) \wedge \Big[\mathsf{IOb}(m) \wedge \mathsf{space}^2(\bar{x}, \bar{y}) < c_m^2 \cdot \mathsf{time}(\bar{x}, \bar{y})^2$$
$$\wedge \, T \in Triv \to \exists k m' \big[\mathsf{IOb}(k) \wedge \mathsf{IOb}(m') \wedge \mathsf{W}(m, k, \bar{x}) \wedge \mathsf{W}(m, k, \bar{y})$$
$$\wedge \, \mathsf{ev}_m(\bar{x}) = \mathsf{ev}_k(\bar{o}) \wedge \mathsf{w}_{mm'} = T\big]\Big].$$

The following axiom is a consequence of the principle of relativity. Cf. [10, 26] for a formalization of the principle of relativity in our first-order logic language.

AxVel: If one observer can send out a body with a certain speed in a certain direction, then any other inertial observer can send out a body with this speed in this direction.

$$\mathsf{IOb}(m) \wedge \mathsf{IOb}(k) \rightarrow$$
$$\Big[\exists b\big[\mathsf{W}(m,b,\bar{x}) \wedge \mathsf{W}(m,b,\bar{y})\big] \leftrightarrow \exists b\big[\mathsf{W}(k,b,\bar{x}) \wedge \mathsf{W}(k,b,\bar{y})\big]\Big].$$

We call body b **inertial body** iff there is an inertial observer m according to who b moves with uniform rectilinear motion:

$$\mathsf{IB}(b) \stackrel{def}{\Longleftrightarrow} \exists m\bar{x}\bar{y}\Big[\mathsf{IOb}(m) \wedge \bar{x} \neq \bar{y} \wedge \mathsf{W}(m,b,\bar{x}) \wedge \mathsf{W}(m,b,\bar{y})\wedge$$
$$\forall \bar{z}\big(\mathsf{W}(m,b,\bar{z}) \leftrightarrow \exists \lambda \big[\bar{z} = \bar{x} + \lambda(\bar{y}-\bar{x})\big]\big)\Big].$$

Let us now formulate the possibility of the existence of FTL inertial bodies.

∃FTLBody: There is an inertial observer who can send out an FTL inertial body:

$$\exists mb\bar{x}\bar{y}\big[\mathsf{IB}(b) \wedge \mathsf{IOb}(m) \wedge \mathsf{W}(m,b,\bar{x}) \wedge \mathsf{W}(m,b,\bar{y})\wedge$$
$$\mathsf{space}^2(\bar{x},\bar{y}) > \mathsf{c}_m^2 \cdot \mathsf{time}(\bar{x},\bar{y})^2\big].$$

∃FTLBody implies that inertial observers can send out a body with arbitrary large speed in any direction if SpecRel, AxThExp$^{\#}$, CONT and AxVel are assumed:

Proposition 1. *Let $d \geq 3$. Assume* SpecRel, AxThExp$^{\#}$, CONT, AxVel *and* ∃FTLBody. *Then any inertial observer can send out a body with any speed in any direction:*
$$\mathsf{IOb}(m) \rightarrow \exists b\big[\mathsf{W}(m,b,\bar{x}) \wedge \mathsf{W}(m,b,\bar{y})\big].$$

For the proof of Prop.1, cf. [16, §6].

5   Hypercomputation in AccRel

In this section, we formulate our statement on the logical equivalence between the existence of FTL signals and the possibility of hypercomputation in special relativity as a theorem in our first-order logic language. To formulate the possibility of hypercomputation as a formula of our first-order logic language, let us define the **life-curve** $\mathsf{lc}_m(k)$ of observer k according to observer m as the world-line of k according to m parametrized by the time measured by k, formally:

$$\mathsf{lc}_m(k) := \{\,\langle t,\bar{x}\rangle \in Q \times Q^d \ : \ \exists \bar{y} \ k \in \mathsf{ev}_k(\bar{y}) = \mathsf{ev}_m(\bar{x}) \wedge y_1 = t\,\}.$$

The **range** and **domain** of a binary relation R, is defined as:

$$Ran\,R := \{\,y : \exists x \ R(x,y)\,\} \quad \text{and} \quad Dom\,R := \{\,x : \exists y \ R(x,y)\,\}.$$

The following formula of our language captures the possibility of relativistic hypercomputation in the sense used in the theory of relativistic computation.

HypComp: There are two observers a programmer p and a computer c and an instant τ in the programmer's worldline such that the computer has infinite time to compute, and during its computation the computer can send a signal s_t to the programmer which reaches the programmer before the fixed instant:

$$\exists pc\tau \Big[\mathsf{Ob}(p) \wedge \mathsf{Ob}(c) \wedge \forall mx \Big(\mathsf{IOb}(m) \wedge x \geq 0 \rightarrow x \in Dom\, \mathsf{lc}_m(c) \wedge$$
$$\forall t \Big[t > 0 \rightarrow \exists t' s_t \big[0 < t' < \tau \wedge s_t \in \mathsf{ev}_m\big(\mathsf{lc}_m(c)(t)\big) \cap \mathsf{ev}_m\big(\mathsf{lc}_m(p)(t')\big) \big] \Big] \Big) \Big].$$

The following axiom ensures the existence of uniformly accelerated observers.

Ax∃UnifOb: It is possible to accelerate an observer uniformly:

$$\mathsf{IOb}(m) \rightarrow \exists k \Big[\mathsf{Ob}(k) \wedge Dom\, \mathsf{lc}_m(k) = Q$$
$$\wedge \forall \bar{x} \big[\bar{x} \in Ran\, \mathsf{lc}_m(k) \leftrightarrow x_2^2 - x_1^2 = a^2 \wedge x_3 = \ldots = x_d = 0 \big] \Big].$$

Now we can state our theorem on the logical equivalence between the existence of FTL signals and the possibility of hypercomputation in special relativity:

Theorem 2. *Let $d \geq 3$. Then*

$$\{\mathsf{AccRel}, \mathsf{AxThExp}^\#, \mathsf{Ax\exists Unifob}, \mathsf{AxVel}\} \models \exists \mathsf{FTLBody} \leftrightarrow \mathsf{HypComp}.$$

For the proof of Thm.2, cf. [16, §6].

6   Concluding Remarks

We have shown that, in special relativity, the possibility of hypercomputation is equivalent to the existence of FTL signals. A natural continuation is to investigate the question concerning the limits of the possibility of using FTL particles in hypercomputation in special and general relativity theories. For example, is there a natural assumption on spacetime which does not forbid the existence of FTL particles, but makes it impossible to use them for hypercomputation?

Of course our construction contains several engineering difficulties. For example, the larger the distance the more difficult to aim with a signal. Therefore, the computer has to calculate the speed of the FTL signal more and more accurately to ensure that the signal arrives to the programmer between events O and M, cf. Fig.1. Thus the computer has to be able to aim with the FTL signal with arbitrary precision.

Acknowledgments. This research is supported by the Hungarian Scientific Research Fund for basic research grants No. T81188 and No. PD84093.

References

1. Andréka, H., Madarász, J.X., Németi, I., Székely, G.: A logic road from special relativity to general relativity. Synthese, 1–17 (2011) (online-first)
2. Andréka, H., Madarász, J.X., Németi, I., Székely, G.: What are the numbers in which spacetime? (2012), arXiv:1204.1350
3. Andréka, H., Németi, I., Németi, P.: General relativistic hypercomputing and foundation of mathematics. Nat. Comput. 8(3), 499–516 (2009)
4. Dávid, G., Németi, I.: Relativistic computers and the Turing barrier. Appl. Math. Comput. 178(1), 118–142 (2006)
5. d'Inverno, R.: Introducing Einstein's relativity. Oxford University Press, New York (1992)
6. Earman, J., Norton, J.D.: Forever is a day: supertasks in Pitowsky and Malament–Hogarth spacetimes. Philos. Sci. 60(1), 22–42 (1993)
7. Etesi, G., Németi, I.: Non-Turing computations via Malament–Hogarth spacetimes. Internat. J. Theoret. Phys. 41(2), 341–370 (2002)
8. Friedman, M.: Foundations of Space-Time Theories. Relativistic Physics and Philosophy of Science. Princeton University Press, Princeton (1983)
9. Hogarth, M.L.: Does general relativity allow an observer to view an eternity in a finite time? Found. Phys. Lett. 5(2), 173–181 (1992)
10. Madarász, J.X.: Logic and Relativity (in the light of definability theory). Ph.D. thesis, Eötvös Loránd Univ., Budapest (2002), http://www.math-inst.hu/pub/algebraic-logic/Contents.html
11. Madarász, J.X., Németi, I., Székely, G.: Twin paradox and the logical foundation of relativity theory. Found. Phys. 36(5), 681–714 (2006)
12. Manchak, J.B.: On the possibility of supertasks in general relativity. Found. Phys. 40(3), 276–288 (2010)
13. Matolcsi, T., Rodrigues Jr., W.A.: The geometry of space-time with superluminal phenomena. Algebras Groups Geom. 14(1), 1–16 (1997)
14. Misner, C.W., Thorne, K.S., Wheeler, J.A.: Gravitation. W. H. Freeman and Co., San Francisco (1973)
15. Mittelstaedt, P.: What if there are superluminal signals? The European Physical Journal B - Condensed Matter and Complex Systems 13, 353–355 (2000)
16. Németi, P., Székely, G.: Special relativistic hypercomputation is possible if there are faster than light signals (2012) (preprint version); arXiv:1204.1773
17. OPERA collaboration: Measurement of the neutrino velocity with the OPERA detector in the CNGS beam (2011), arXiv:1109.4897
18. Petkov, V.: Relativity and the nature of spacetime, 2nd edn. Frontiers Collection. Springer, Berlin (2009)
19. Recami, E.: Tachyon kinematics and causality: a systematic thorough analysis of the tachyon causal paradoxes. Found. Phys. 17(3), 239–296 (1987)
20. Recami, E.: Superluminal motions? A bird's-eye view of the experimental situation. Found. Phys. 31, 1119–1135 (2001)
21. Recami, E., Fontana, F., Garavaglia, R.: Special relativity and superluminal motions: a discussion of some recent experiments. Internat. J. Modern Phys. A 15(18), 2793–2812 (2000)
22. Rindler, W.: Relativity. Special, general, and cosmological, 2nd edn. Oxford University Press, New York (2006)
23. Selleri, F.: Superluminal signals and the resolution of the causal paradox. Found. Phys. 36, 443–463 (2006)

24. Stannett, M.: The case for hypercomputation. Appl. Math. Comput. 178(1), 8–24 (2006)
25. Székely, G.: First-Order Logic Investigation of Relativity Theory with an Emphasis on Accelerated Observers. Ph.D. thesis, Eötvös Loránd Univ., Budapest (2009)
26. Székely, G.: The existence of superluminal particles is consistent with the kinematics of Einstein's special theory of relativity (2012), arXiv:1202.5790
27. Taylor, E.F., Wheeler, J.A.: Spacetime Physics. W. H. Freeman and Company, New York (1997)
28. Tolman, R.C.: The Theory of the Relativity of Motion. University of California, Berkely (1917)
29. Weinstein, S.: Super luminal signaling and relativity. Synthese 148, 381–399 (2006)

Turing Computable Embeddings
and Coding Families of Sets

Víctor A. Ocasio-González

Department of Mathematics, University of Notre Dame, Notre Dame IN 46617, USA
`vocasiog@nd.edu`

Abstract. In [7] the notion of Turing computable embeddings is introduced as an effective counterpart for Borel embeddings. The former
allows for the study of classes of structures with universe a subset of
ω. It also allows for finer distinctions, in particular, among classes with
$\aleph_0$ isomorphism types. The hierarchy of effective cardinalities that arises
from TC embeddings has been studied, among other places, in [7] and
[2]. In this work, we prove that the special class of 'daisy graphs', a subclass of undirected graphs used to code families of sets, has the same
effective cardinality as the class of archimedian real closed fields. As a
consequence, the class of abelian p-groups and the class of archimedian
real closed fields are TC incomparable.

1 Introduction

An interesting problem in mathematics is that of determining when two structures of the same class are isomorphic. A way to approach this problem, within
a class, is to determine what properties must a structure share with another so
that an isomorphism can be constructed. Once this has been determined for two
classes, a natural question to ask is how do the complexity of the 'isomorphism
problems' for the same classes of structures compare with each other. Various
notions of reducibility exist for comparing the classes of structures. Among them
is the notion of Turing computable embeddings, introduced in [7], which provides an effective version of a Borel embedding, see [3]. It is based on the Turing
reduction of atomic diagrams of structures, done in such a way as to preserve
isomorphism types.

Definition 1. *A Turing computable embedding (TC embedding) of a class of
structures K in another class K' is a Turing operator $\Phi = \varphi_e$ such that:*

- *for each $A \in K$, there is a $B \in K'$ such that $\varphi_e^{D(A)} = \chi_{D(B)}$,*
- *if A, A' correspond, respectively, to $B, B' \in K'$, then $A \cong A'$ iff $B \cong B'$*

We say that $K \leq_{\mathrm{tc}} K'$ iff there is a TC embedding from K into K'. Thus
$\leq_{\mathrm{tc}}$ provides a partial ordering on classes of structures. Also, we say K and K'
have the same *effective cardinality* if $K \equiv_{\mathrm{tc}} K'$, i.e if $K \leq_{\mathrm{tc}} K'$ and $K' \leq_{\mathrm{tc}} K$.
Note that Turing computable embeddings are *uniform* computable procedures.

S.B. Cooper, A. Dawar, and B. Löwe (Eds.): CiE 2012, LNCS 7318, pp. 539–548, 2012.
© Springer-Verlag Berlin Heidelberg 2012

A drawback of Borel embeddings is that they are unable to distinguish between classes with $\aleph_0$ isomorphism types. Furthermore, structures are required to have universe all of ω, while for TC embeddings we only require that the universe is a subset of ω. Exploiting this feature, the authors, in [7], were able to prove, among other things, that the class of finite prime fields lies strictly below (in the sense of TC-embeddings) the class of $\mathbb{Q}$ vector spaces. In those cases where we have a TC embedding from one class of structures to another, and both classes have no structure which is finite, then we can see that if $K \leq_{\text{tc}} K'$ then $K \leq_B K'$, so that $\leq_{\text{tc}}$ is a refinement of Borel embeddings on classes with computable languages and whose structures have infinite countable universe. More precisely, whenever we have a structure with infinite countable universe we can always modify it so that the universe is all of ω. So, that any TC embedding between two classes which have no finite structures can be thought of as a Borel embedding between said classes after we make the change in the universe of the structures. Evidence for this refinement is seen when comparing the work in [2] and work from Friedman and Stanley. In Borel embeddings, it was known that the class of abelian p-groups (A_pG) is not reducible to the class of families of 'daisy graphs' DG (which we can think of as families of subsets of ω) and vice versa, we say these classes are Borel incomparable. In [2], among other things, these results are translated to the realm of TC embeddings using (morally) the same ideas as Friedman and Stanley and it is proven that these classes are incomparable from the point of view of $\leq_{\text{tc}}$.

The main body of work in this paper consists in describing the Turing operators which provide the embedding of DG into ARCF, the class of archimedean real closed fields, and vice versa. From this it follows that ARCF and A_pG are TC incomparable. It is known, from work in [1], that incomparable classes of structures exist in the hierarchy of $\leq_{\text{tc}}$. This work provides an example of two 'natural' classes that are incomparable.

1.1   Facts about Theory of Real Closed Fields

Throughout this paper, when talking about a real closed field, we assume we are working in the language of ordered rings, $\mathbb{L} = \{+, \cdot, 0, 1, <\}$. It is known that the theory of real closed fields in this language is O-minimal, complete, model complete and decidable. Further, it admits quantifier elimination, and hence the definable sets are exactly the sets defined by (finite) boolean combinations of quantifier free formulas. These facts are freely used throughout this work. For detailed proofs and explanations the author recommends [9]. In proving Theorem 4 and Theorem 5, we shall use the following fact for O-minimal structures, whose proof can be found in [8]:

Theorem 1. *Suppose $\varphi(\bar{a}, \bar{x})$ is true of $\bar{b}$, a tuple algebraically independent over $\bar{a}$. There there is an open box, B, around $\bar{b}$ such that for all $\bar{x} \in B$, $\varphi(\bar{a}, \bar{x})$ holds.*

The statement above is not explicitly stated as a theorem in [8], but rather it is a result within the proof of one of the main theorems of the paper. We shall also need the following result of Van den Dries which can be found in [10] and [9]:

Theorem 2. *Let R be a real closed field, then R has definable Skolem functions.*

In fact, many schemes exists for skolemizing the structures in RCF but in this paper we use the one do to Van den Driess in [10]. A proof on how this skolemization works on subsets of higher dimensions can be found in [9]. The skolemization explicitly defines what is the value of the new function symbols introduced to the language. Because we can do this uniformly for all structures in RCF, we shall call RCF$^+$ the theory of real closed fields in the language of ordered rings augmented by the functions symbols that witness the skolemization. It follows that RCF$^+$ is also complete, decidable, O-minimal and admits quantifier elimination. It should also be noted that R regarded as a structure of RCF$^+$ has no new definable subsets. In Theorem 5, during the process of defining the algebraic closure of an element, we shall use the built-in definable Skolem functions of RCF$^+$. This way we do not have to worry too much about finding the roots of polynomials, a crucial step in enumerating the algebraic closure of elements, and as a consequence we might just end up with multiple names for the elements.

2 Results

Throughout the paper, all languages considered are computable and structures have universe a subset of ω.

2.1 Construction of Embeddings

We define a *daisy graph* as an undirected graph, G, with a distinguished vertex, say x_0 which we call a center, and a set of edges E, such that every other vertex in G is part of a unique loop containing x_0. Since any subset of ω can be coded as a daisy graph, without loss of generality, for $S \subset \omega$, we can think of a daisy graph as having a loop of size $2n+3$ if $n \in S$ or a loop of size $2n+4$ if $n \notin S$. The class DG is composed of all (countable) collections of daisy graphs, i.e coding (countable) families of subsets of ω, such that there is at most one daisy coding each set. It follows that structures $A, B \in$ DG are isomorphic iff they code the same subsets of ω. This characterization provides a means for embedding ARCF into DG. Recall that a real number r is uniquely determined by its Dedekind cut, i.e., $\{q \in \mathbb{Q} : q \leq r\}$. So, fix an enumeration of the rational numbers $(q_n)_{n<\omega}$. We can effectively construct a structure $A \in$ DG for each structure $\mathsf{R} \in$ ARCF by associating each $r \in \mathsf{R}$ with the daisy that codes the set $S_r = \{n : q_n < r\}$. The construction of the daisies in A is done in stages. Further, multiple daisies must be constructed simultaneously, since at any stage, we have seen only finitely many elements of R and know finitely much about each. Our main strategy, at any stage s, will consist of taking a real $r \in \mathsf{R}$, which we know it is truly a real number since the field is archimedian, and determining for $r < s$ whether $q_n < r$. If $q_n < r$, then we enumerate sentences into the diagram of A, enough to construct a loop of size $2n + 3$ if $q_n \leq r$. Otherwise, we construct a loop of size $2n + 4$. This procedure is uniform on structures in ARCF and it can be shown that isomorphic structures in ARCF give rise to isomorphic structures in DG. Hence, we have proved the following:

Theorem 3. ARCF $\leq_{\text{tc}}$ DG

Our goal now is to prove that the converse of Theorem 3 is also true. For this we need to construct a computable Turing operator Φ which uniformly in A (using as oracle $D(A)$) computes the atomic diagram of a structure R in ARCF. This is, in practice, more complex. Two ARCF's are isomorphic iff they have the same reals. Thus, doing the above construction 'backwards' fails to provide a Turing computable embedding, since by taking the algebraic closure of the reals defined by the Dedekind cuts we might end up with two non-isomorphic $A, A' \in$ DG mapping to the same $R \in$ ARCF.

Our strategy will be to construct Φ as a composition of Turing operators. First, we identify each node in $2^{<\omega}$ with a closed interval with rational endpoints $[q, q'] \subseteq [0, 1]$ so that every path through $2^{<\omega}$ corresponds to a countable intersection of (nested) closed intervals. We do this with a computable operator T in such a way, that paths through $2^{<\omega}$ represent algebraically independent reals. Thus, embedding any structure $A \in$ DG into $2^{<\omega}$ provides a way to transform information about whether or not a number n is in a given set, to information about whether or not a real is in a given interval. The latter is, of course, more useful when deciding facts about the ARCF we ultimately wish to construct. Finally, we use the image of A inside $2^{<\omega}$ to output the atomic diagram of a structure in ARCF.

In working towards constructing T, as described above, we make precise those properties it must have. The following notation will allow a better way to express the conditions we wish to impose to T.

Definition 2. *Suppose* $T : 2^{<\omega} \to \mathbb{Q} \times \mathbb{Q}$ *is a computable operator mapping* $\sigma \mapsto (q, q')$. *We define* $I_\sigma^T = [q, q']$ *to be the closed interval with rational endpoints* q *and* q'.

Once we have fixed such a T, we shall identify $\sigma \in 2^{<\omega}$ with I_σ^T. Furthermore, we shall write I_σ for I_σ^T when there is no ambiguity about the operator T. From now on, we shall identify an operator T defined as above with the tree it defines in 2^ω and we shall denote them both by T. Recall that a binary tree is *perfect* if for every σ in the tree there is a τ extending σ with $\tau^\wedge 0$ and $\tau^\wedge 1$ also in the tree. Our goal for now will be to prove the following theorem.

Theorem 4. *There is a perfect, computable, binary tree* T *whose continuum many paths represent algebraically independent reals over* $\mathbb{Q}$.

In order to prove the Theorem above we shall prove that there exists a computable operator $T : 2^{<\omega} \to \mathbb{Q} \times \mathbb{Q}$ satisfying the following conditions:

1. $T(\emptyset) = [0, 1]$
2. If $\sigma \preceq \tau$, then $I_\tau \subseteq I_\sigma$.
3. If $length(\sigma) = n$, then $diameter(I_\sigma) \leq 2^{-n}$, where $diameter(a, b)$, for an interval (a, b), is defined to be $b - a$.
4. If σ, τ are incomparable and both of length n, $I_\sigma \cap I_\tau = \emptyset$

5. For $f \in 2^\omega$, let r_f be the unique real in $\bigcap_{\sigma \subseteq f} I_\sigma$. Then for distinct $f_1, \ldots, f_n \in 2^\omega$, $r_{f_1}, \ldots, r_{f_n}$ are algebraically independent.

Observe that properties (1)-(4) are not hard to realize. The first one is a definition while the other three can be simultaneously made true by subdividing a given interval I into appropiate pieces. Further, note that by property (4) and the fact that T has $2^{<\omega}$ as its domain, the tree T will be perfect. Most of the work will be proving property (5). It is done by a classic diagonalization argument and recurrent applications of the following lemma whose proof we omit but can be seen to be true by the O-minimality of RCF.

Lemma 1. *Given disjoint intervals $I_1, \ldots, I_n \subset [0,1] \subset \mathbb{R}$ with known rational endpoints and given a non-trivial $p(x_1, \ldots, x_n) \in \mathbb{Z}[x_1, \ldots, x_n]$, we can effectively find disjoint intervals with rational endpoints $J_1, \ldots, J_n$ with $J_i \subseteq I_i$, such that for all reals $a_1 \in J_1, \ldots, a_n \in J_n$ $p(a_1, \ldots, a_n) \neq 0$.*

Our main strategy for proving Theorem 4 will be to apply the above lemma (several times if necessary) at each stage. Note that once we have finished searching for the rational endpoints of the J_i's, we can pick, for each i, $J_i' \subseteq J_i$, also with rational endpoints, in such a way, that properties (2)-(4) are satisfied. For example, if $J_i = [q_1, q_2]$, pick $J_i' = [q_1', q_2']$ such that $q_1' = \frac{1}{3}(q_1 + q_2)$ and $q_2' = \frac{2}{3}(q_1 + q_2)$. This will be enough to satisfy properties (2)-(4). In fact, there is no need to subdivide the intervals in pieces with the same diameter, nor is it important that there are 3 pieces. Any partition of J_i is acceptable as long as we satisfy the conditions. With this in mind let's prove Theorem 4.

Proof (4)
Fix an enumeration of all polynomials with integer coefficients in arbitrarily long (but finite) tuples of variables. Recall that n reals are algebraically independent (over $\mathbb{Q}$) if they satisfy no formula of the form $p(x_1, \ldots, x_n) = 0$, if $p(x_1, \ldots, x_n) \in \mathbb{Q}[x_1, \ldots, x_n]$ (equivalently $\mathbb{Z}[x_1 \ldots, x_n]$). So at each stage, we satisfy a requirement of the form:

$$R_{<\sigma_1, \ldots, \sigma_n, k>} \equiv \text{There exists } I_{\tau_1} \subseteq I_{\sigma_1}, \ldots, I_{\tau_n} \subseteq I_{\sigma_n} \text{ such that for all reals}$$
$$r_1 \in I_{\tau_1}, \ldots, r_n \in I_{\tau_n} \; p_k(r_1, \ldots, r_n) \neq 0$$

Each requirement is said to require attention when the image under T of all $\sigma_1, \ldots, \sigma_n$ has been defined. We construct $T : 2^{<\omega} \to \mathbb{Q} \times \mathbb{Q}$ in stages:
-At stage $s = 0$; $T_0(\emptyset) = [0, 1]$
-At stage $s + 1$;

Let T_s be the (partial) function at stage s and let $R_{\sigma_1, \ldots, \sigma_n, k}$ be the least requirement needing attention. Let End_{σ_i} be the set of all end points of T_s extending I_{σ_i}, i.e $I_\gamma \in End_{\sigma_i}$ iff $\sigma_i \prec \gamma$ and $T_s(\gamma^\wedge 0), T_s(\gamma^\wedge 1)$ are undefined. Note that for $i \neq j$, $End_{\sigma_i} \cap End_{\sigma_j}$ will not necessarily be empty. In fact, End_{σ_i} and End_{σ_j} could be equal. But since we are satisfying property (4) at each stage, then for $I_\gamma \in End_{\sigma_i}$ and $I_{\gamma'} \in End_{\sigma_j}$ with $\gamma \neq \gamma'$, hence incomparable since they are end points, $I_\gamma \cap I_{\gamma'} = \emptyset$. Further, if we take a tuple $I_{\gamma_1}, \ldots, I_{\gamma_n}$ with $I_{\gamma_i} \in End_{\sigma_i}$, and there is some $I_{\gamma_i} = I_{\gamma_j}$ then we do nothing since Lemma 1 does not apply. This, we argue, does not affect the conclusion of the Theorem since if two reals r_1, r_2 correspond to two distinct paths through $T(2^{<\omega})$, then they can be separated by disjoint intervals by property (4). So from now on in this proof we assume that whenever we take a tuple as above, it is a tuple with no repetitions.

Take a tuple $I_{\gamma_1}, \ldots, I_{\gamma_n}$ with $I_{\gamma_i} \in End_{\sigma_i}$ and apply Lemma 1 to get $J_1, \ldots, J_n$. By construction, $J_i \subseteq I_{\gamma_i}$ for all i. Pick another tuple as above (if any are available) and apply Lemma 1 again but this time for all I_{γ_i} that appear again we use the corresponding J_i instead of I_{γ_i} when applying the lemma. The idea is to refine each I_{γ_i} as much as possible so the $J_i' \subseteq I_{\gamma_i}$ that we end up with works with any other refinement of the I_{γ_j}, $j \neq i$. So, as before, pick tuples until all possibilities are exhausted, which they will since End_{σ_i} is finite for all i. Now, for all $J_j \subseteq I_{\gamma_j}$, $I_{\gamma_j} \in End_{\sigma_i}$, $i < n$, subdivide as in the comment below Lemma 1 J_j into J_j^0 and J_j^1. Define $T_{s+1} = T_s \cup \left\{ (\gamma_j^\wedge 0, J_j^0) : I_{\gamma_j} \in End_{\sigma_i} \text{ for some } i \right\} \cup \left\{ (\gamma_j^\wedge 1, J_j^1) : I_{\gamma_j} \in End_{\sigma_i} \text{ for some } i \right\}$.

Observe that the fact that T is perfect comes both from the fact that $2^{<\omega}$ is perfect and that for any σ we are mapping $\sigma^\wedge 0$ and $\sigma^\wedge 1$ to disjoint intervals.

Now that we have T, the next step towards proving the embeddability of the class DG into ARCF is to show that given $A \in$ DG, we can get, through a uniformly computable process, a subtree of $2^{<\omega}$, T_A, whose labeled paths represent the sets coded by daisies in A. To define the notion of labeled path we first need to say what is a label or mark on a tree. We introduce to the language (of trees) a relation M (for marks), and a set of constants $\{x_i : i < \omega\}$, and we say that a path $f \in [T_A]$ is a *labeled path* if there exists one (and only one) constant x_i such that for infinitely many $\sigma \prec f$, $(\sigma, x_i) \in M$. The tree $T_{A,s}$ will have the feature that every node will have attached to it countably (possibly finitely) many 'marks'. This type of bookkeeping is unnecesary for the construction of $T_{A,s}$ itself but attaching to the tree these marks will allow us, in the construction in Theorem 5, to guarantee that we can associate to each daisy in A a 'name' or constant, with only finitely much information, and allow it to keep the name. Further, it is possible that T_A contains paths that codes no daisy in A, hence labeling the paths permits us to identify which paths do. The tree T_A may be perfect, but in any case, its labeled paths will denote algebraically independent reals, in the sense of Theorem 4. There will be no terminal nodes. This follows from the fact that at any stage, say s, the terminal nodes of $T_{A,s}$ correspond to distinct daisies. How we extend these terminal nodes (in $T_{A,s}$), that is choosing left or right, will be determined by what information is coded in the corresponding daisy. Since daisies give both positive and negative information about a set, all terminal nodes are (eventually) extended to elements of $2^{<\omega}$.

Lemma 2. *There is a computable operator $H : \text{DG} \to 2^{<\omega}$ that, uniformly in A, computes a tree T_A whose labeled paths correspond to algebraically independent reals.*

Proof (2)
Let $D(A)$ be the atomic diagram of A and let T be as in Theorem 4. Let C be the set of all elements of A that are centers of daisies. Let $X = \{x_i : i < \omega\}$ of new constants and let $M \subseteq T \times X$ be a binary relation symbol. We construct T_A, M, and C in stages:
-At stage 0; $T_{A,0} = \emptyset$, i.e the tree consists of only the empty node; $C = \emptyset$; $M = \emptyset$.
-At stage $s + 1$;
At this stage, $T_{A,s}$ has at most s endpoints, $C = \{c_0, \ldots, c_k\}$ for $k \leq s$, and there are at most s tuples in M. In particular, any $\sigma \in T_{A,s}$ has at most $s < \infty$ marks. We allow ourselves to look only at the first s sentences of $D(A)$, call this set of sentences $D(A)_s$. Using these, identify elements in A that appear to be centers of daisy graphs. We can identify the centers since they are the only elements that are connected by an edge to more than 2 elements. Further, $D(A)_s$ tells us that they are distinct. If we see a new center, say c_{k+1}, then we put $c_{k+1} \in C$ and identify it with a $\sigma_{c_{k+1}}$ which starts at the empty node of T, as explained below. We introduce a set of tuples of the form (c_t, i), called *Check*, where $c_t \in C$ and $i < \omega$. This set keeps track of questions we have asked about centers of daisy graphs. Now, fix a $c \in C$ and verify if $D(A)_s$ has enough information to say the following for all $i \leq j \leq s$, where i is the largest such that $(c, i) \in Check$ (if no tuple involving c is in *Check* we let $i = 0$):

1. The element c is in a loop of size $2j + 3$.
2. The element c is in a loop of size $2j + 4$.

Of course, there is no sentence saying (1) or (2) above, since this is an existential statement, but we regard it as true if we have enough information in $D(A)_s$ to construct a loop of the appropiate size. Note that by our convention on daisy graphs only one must be true but we might at this stage not see which. Start by checking if (1) or (2) is true for $j = i$. If (1) is true then let $\tau \in T$ be such that $\sigma_c {}^\wedge 0 \preceq \tau$, $\tau {}^\wedge 0 \in T$, $\tau {}^\wedge 1 \in T$, and for all $\tau' \succ \sigma_c$, $\tau' \succ \tau$. That is, τ extends σ_c by going to the left and does so in such a way that τ stops at the next splitting. This can be done since T is perfect. If (2) is true then find $\tau \in T$ with same properties except this time τ extends σ_c from the right. Either way, add (c, j) to *Check*. If $j = c$ (recall the universe of A is a subset of ω) or $(\sigma_c, x_c) \in M$, add $(\tau, x_c) \in M$ and $(\sigma, x_c) \in M$ for all $\sigma \prec \tau$. The first condition guarantees all daisies (and hence paths) are eventually associated with a mark. The second provides continuity, so that once we have a mark we can follow that mark on the tree forever. Finally, let $\sigma_c = \tau$ and repeat the process for the remaining j. If at any point we get a false answer for both (1) and (2), or $j = s + 1$, then we stop asking questions about c and pick another $c' \in C$. Now define $T_{A,s+1} = T_s \cup \{\sigma_c : c \in C\}$. Finally, to conclude that $T_A = \bigcup_{s < \omega} T_{A,s}$ has the desired properties we still need to prove the following. If f is a labeled path through T_A, then there exists a unique x_c such that for infinitely many $\sigma \prec f$, $(\sigma, x_c) \in M$.

To see this is so, note first that labeled paths through T_A correspond in a one-to-one fashion with the centers of daisies in A. Furthermore, if f corresponds to a daisy whose center is c then it will be labeled, i.e., $(\sigma, x_c) \in M$ for some $\sigma \prec f$, when we can answer whether or not there is a loop of size $2c + 3$ or $2c + 4$ in said daisy. Once this initial labeling has been done, for every $\tau \succ \sigma$ there is $(\tau, x_c) \in M$. To see that x_c is unique note that if we think of daisies as subsets of ω, then for any two daisies, say S_1, S_2, there is a least n such that $n \in S_1$ and $n \notin S_2$. So that if f is the path associated, without loss of generality, with S_1 and σ is the initial segment of f we have at the stage where we witness there is a loop if size $2n + 3$ in S_1, then the τ extending σ will recieve a mark for S_1 and will never recieve a mark for S_2.

We are now ready to construct the second part of the computable operator Φ. The first part was given by Lemma 2, the second by the theorem below. Instead of working in RCF, the theory of real closed fields, we shall be working on the skolemization of RCF, RCF^+, as defined in Section 1.1. We also add constants naming all terms in the language of RCF^+, with extra constants $\{c_i : i < \omega\}$ for possible centers of daisies. This simplifies the process of finding the algebraic closure of an element. No longer do we need to define a constant to be (interpreted as) a root of a particular polynomial in the structure we are constructing. We know the constant is already 'picked out' by a Skolem function. Thus we shall eventually find the root of a polynomial if we check the constants one by one. This will all be made precise in the proof of Theorem 5.

Theorem 5. DG $\leq_{\text{tc}}$ ARCF

Proof (5)
We show how to convert a family of daisy graphs, A, to an ARCF, R. By Lemma 2 we freely identify A with T_A for the rest of the proof. The field R will contain one real for each path in T_A, all algebraically independent from each other. We shall enumerate the diagram of R, $D(R)$, in stages, assuming we have the atomic diagram of A, $D(A)$.

Let L^+ be the language of RCF^+ and we augment it with the following two sets of constants, $C = \{c_i : i \in \omega\}$ and $R = \{r_i : i < \omega; r_i = t(\bar{c})\}$, where $t(\bar{x})$ is a term in $L^+ \cup C$. The set C will serve as an auxiliary set since it will help us keep track of how many independent reals we have at any given stage. We identify each mark, x_{c_i}, of $T_{A,s}$, the tree at stage s, with the constant c_i to ensure we get the appropiate independent reals in R. By *identifying* a constant c with a mark we mean to add to $D(R)$ sentences saying that c lies in the rational interval at the node marked in $T_{A,s}$. We denote this interval as I_c. Note we shall use $D(T_A)$ to reconstruct T_A but we shall be more focused on finding the marks on T_A since these will allow us to say new facts about the constants. On the other hand, the set R, or in some cases a subset of R, will serve as the universe of the structure R we shall construct. We might worry that if we use only elements from the set R we might not be adding the reals coded by the daisies in A but notice that for every c_i there is some r_j such that $r_j = c_i$, just by how we defined the elements in R.

Now, recall that by quantifier elimination any first-order formula in $L^+ \cup R$ is equivalent to a finite boolean combination of equalities and inequalities of polynomials. Let $\{\alpha_n(\bar{z})\}_{n \in \omega}$ be an effective enumeration of all finite boolean combinations of equalities and inequalities in $L^+ \cup R$ with parameters $\bar{z}$ which are finite but arbitrary in length.

From the previous discussion it follows that we must satisfy the following requirement for all i: Decide whether $\alpha_i(\bar{z}) \in D(R)$ or $\neg\alpha_i(\bar{z}) \in D(R)$, where $\bar{z}$ is a tuple (of meaningful) elements of R. By a *meaningful element* of R we mean an element r whose definition as a term relies on constants of C that we have already identified with a mark of T_A.

Each stage will be divided in two steps. First we check if we can extend our knowledge of the constants c. Then we shall enumerate into $D(R)$ all facts we can decide with the amount of information available.

Construction

-At stage $s = -1$; $D(R) = \emptyset$, and no marks have been identified in T_A.

-At stage $s + 1$;

Let $T_{A,s}$ be the part of T_A we can construct using the first s sentences of $D(A)$. Here we are thinking of running the construction described in Lemma 2 as far as we can using the information given about A. So at this stage we have $k \leq s$ constants in C identified with marks of $T_{A,s}$. Now construct $T_{A,s+1}$ and do, of the following, what applies: If no new mark appears, do nothing. If a new mark x_c appears, that is, we see a new tuple $(\sigma, x_c) \in M$, identify the constant c with it. Note that if this is not the first identification of c, we could be adding a sentence to $D(R)$ that adds no new (useful) information about c. This might happen if we see $(\sigma, x_c) \in M$ now but we had already seen that $(\tau, x_c) \in M$ for some $\tau \succ \sigma$. Because of this, we define I_c to be the interval with smallest diameter where we know c lies.

Let R'_{s+1} be the first $s+1$ constants $r \in R$ which are meaningful (as described above). Then for each $i \leq s+1$ we make tuples of elements in R'_{s+1} of appropriate length, where appropriate is determined by the length of $\bar{z}$ in $\alpha_i(\bar{z})$. Notice that since every $r \in R'_{s+1}$ can be defined in terms of c's, then in fact, $\alpha_i(r)$ is equivalent to some $\beta(c_{i_1}, \ldots, c_{i_n})$, where $\beta(\bar{z})$ is a sentence in $L^+ \cup C$. This $\beta(\bar{z})$ can be found by just replacing each r by its definition in terms of the c's. Further, by Theorem 1, if $\beta(\bar{c})$ is true then there exists $J_1, \ldots, J_n$ with $c_{i_j} \in J_j$, $J_j \subseteq I_{c_{i_j}}$, such that $\forall a_1 \in J_1 \cdots \forall a_n \in J_n \, \beta(a_1, \ldots, a_n)$ is also true. Similarly, if $\neg\beta(\bar{c})$ is true.

So in order (according to your list), pick $i \leq s+1$ and an (appropriate) tuple $\bar{r}$. Find the corresponding $\beta(\bar{c})$ and ask which of the following is true:

$$\forall a_1 \in I_{c_{i_1}} \cdots \forall a_n \in I_{c_{i_n}} \, \beta(a_1, \ldots, a_n)$$
$$\forall a_1 \in I_{c_{i_1}} \cdots \forall a_n \in I_{c_{i_n}} \, \neg\beta(a_1, \ldots, a_n)$$

Recall that by quantifier elimination the above sentences are equivalent to quantifier free sentences and by decidability we can check their truth value. At this stage we might think both are false, but by completeness of RCF, and hence RCF^+, one must be true. So if we think both are false we do nothing. If we can tell which one is true then enumerate $\alpha_i(\bar{r})$ into $D(R)$ if the first is true or

$\neg \alpha_i(\bar{r})$ if the second is true. When we have gone through all of our list, of both tuples and α_i's, we pass to the next stage.

We claim that the above procedure suffices, so we need to prove that all sentences are decided at some stage. But this follows from Theorem 1 and the fact that a path through T_A corresponds to a nested sequence of intervals.

To conclude, notice that for any two non-isomorphic daisy graphs, A and B, there exists a daisy S coded in A which is not coded in B or vice versa. Thus the ARCF's, $\Phi(A)$ and $\Phi(B)$, to which they are mapped respectively will contain at least one real which does not appear in the other since the constants c in the construction are all algebraically independent.

References

1. Calvert, W., Cummins, D., Knight, J.F., Miller, S.: Comparison of classes of Finite structures. Algebra and Logic 43(6), 374–392 (2004)
2. Fokina, E., Knight, J.F., Melnikov, A., Quinn, S.M., Safranski, C.: Classes of ulm type and coding rank-homogeneous trees in other structures. J. Symbolic Logic 76(3), 846–869 (2011)
3. Friedman, H., Stanley, L.: A Borel reducibility theory for classes of countable structures. J. Symbolic Logic 54(3), 894–914 (1989)
4. Goncharov, S., Harizanov, V., Knight, J., McCoy, C., Miller, R., Solomon, R.: Enumerations in computable structure theory. Annals of Pure and Applied Logic 136, 219–246 (2005)
5. Chisholm, J., Fokina, E., Goncharov, S., Harizanov, V., Knight, J., Quinn, S.: Intrinsic bounds on complexity and definability at limit levels. J. Symbolic Logic 74(3), 1047–1060 (2009)
6. Hirschfeldt, D., Khoussainov, B., Slinko, A., Shore, R.: Degree spectra and computable dimensions in algebraic structures. Annals of Pure and Appl. Logic 115, 71–113 (2002)
7. Knight, J.F., Miller, S., Vanden Boom, M.: Turing computable embeddings. J. Symbolic Logic 72(3), 901–918 (2007)
8. Knight, J.F., Pillay, A., Steinhorn, C.: Definable sets in ordered structures II. Trans. Amer. Math. Soc. 295(2), 593–605 (1986)
9. Marker, D.: Model theory: an introduction. Graduate Texts in Mathematics, vol. 217. Springer (2000)
10. Van den Dries, L.: Algebraic theories with definable skolem functions. J. Symbolic Logic 49(2), 625–629 (1984)

On the Behavior of Tile Assembly System at High Temperatures

Shinnosuke Seki and Yasushi Okuno

[1] Department of Information and Computer Science, Aalto University,
P.O. Box 15400, 00076, Aalto, Finland
shinnosuke.seki@aalto.fi
[2] Department of Systems Bioscience for Drug Discovery, Kyoto University, 46-29,
Yoshida-Shimo-Adachi-cho, Sakyo-ku, Kyoto, 606-8501, Japan
okuno@pharm.kyoto-u.ac.jp

Abstract. Behaviors of Winfree's tile assembly systems (TASs) at high temperatures are investigated in combination with integer programming of a specific form called threshold programming. First, we propose a way to build bridges from the Boolean satisfiability problem (SAT) to threshold programming, and further to TAS's behavior, in order to prove the NP-hardness of optimizing temperatures of TASs that behave in a way given as input. These bridges will take us further to two important results on the behavior of TASs at high temperatures. The first says that arbitrarily high temperatures are required to assemble some shape by a TAS of "reasonable" size. The second is that for any temperature $\tau \geq 4$ given as a parameter, it is NP-hard to find the minimum size TAS that self-assembles a given shape and works at a temperature below τ.

1   Introduction

The abstract Tile Assembly Model (aTAM), which has been introduced by Winfree [13] based on a dynamic version of Wang tiling [12], is a model of "programmable crystal growth" with algorithmically-rich theoretical background and results. Self-assembling (square) tiles have been experimentally implemented in 1982 by Seeman [11] as monomers with sticky ends that are designed so ingeniously that they are guaranteed to bind deterministically into a single target shape (a commonly-used implementation of such tiles is based on DNA double-crossover molecules [7]). The last three decades have seen drastic advancements on the reliability of DNA tile assembly; in fact, the error rate of 10% per tile in 2004 was improved down to 0.13% in 2009 [8].

In the aTAM, sticky ends are abstracted to be a glue label, and their strengths are assigned by a strength function g. A square tile adheres stably to an aggregate of tiles whenever the sum of the strengths of neighboring sides with matching labels according to g exceeds a threshold τ, which is a system parameter called *temperature*. A certain "behavior" of tile assembly systems (TASs) was proved to require a very large gap between binding strengths as well as ones exponentially small in terms of the larger gaps [5] (for a formal definition of TAS's behavior as

S.B. Cooper, A. Dawar, and B. Löwe (Eds.): CiE 2012, LNCS 7318, pp. 549–559, 2012.

strength-free TAS, see Sect. 3). "Temperature" in aTAM is rather metaphorical than actually describing the physical temperature of the experimental environment. It nonetheless can be interpreted as a physical metric for the granularity with which different energy levels must be distinguished in order for TASs to behave as expected. One can quantitize the minimum distinction and rescale so that this quantity is normalized to be 1. As such, aTAM assumes that the glue strengths and temperature are integers.

The stability of tile attachment is determined by the strengths that at most 4 cooperative binding sites offer, relative to τ. This motivates us to study a system of inequalities whose left-hand side consists of at most 4 terms (non-negative integers or constants) and whose right-hand side is τ. We call such an inequality a τ-*inequality of at most 4 terms*; we use this term often with the replacement of τ by either $\geq_\tau$ (at least τ) or $<_\tau$ (strictly less than τ) to specify its sign. Then, we can say that τ-inequalities of at most 4 terms dominate the behavior of a TAS at the micro, or *local*, level, that is, every attachment event. Optimizing (minimizing) τ under τ-inequalities is a specific type of integer programming we call *threshold programming* (TP). We prove that TP is NP-hard even under the condition that all $\geq_\tau$-inequalities involved be of at most 4 terms and all $<_\tau$-inequalities involved be of at most 3 terms (Lemma 2). This condition enables us to reduce TP to the local behavior of a TAS, and this implies that optimizing the temperature of TASs that behave in a way specified as input cannot be done in a polynomial time, unless P = NP (Theorem 2).

The TP instances obtained in this reduction will lead us further to the study of terminal behaviors of TASs. For any temperature τ, one of the instances is converted into a shape such that TASs of "reasonable size" can self-assemble the shape only at the temperatures above τ (Theorem 3). They are also converted differently into a shape to prove that, for any $\tau \geq 4$, it is NP-hard to compute the minimum number of tile types required for TASs at a temperature at most τ to self-assemble the shape (Theorem 4).

Current laboratory techniques allow us to handle only at most 2 distinct energy levels, that is, temperature 2 (even making a distinction between two energy levels is difficult) [3,6]. Therefore, as of this date, we cannot help but interpret our results as computational infeasibility of determining whether behaviors of TAS can be implemented physically.

2   Abstract Tile Assembly Model

Let $\mathbb{N}_0$ denote the set of non-negative integers, and $\mathbb{N} = \mathbb{N}_0 \setminus \{0\}$. A *tile type* t is a unit square with four sides, each having a *glue label* (often represented as a finite string taken from a label set Λ). In this paper, we list glue labels in the counter-clockwise order starting from north (N) in order to specify a tile type. For each direction $d \in \{N, W, S, E\}$, let $t(d)$ be the glue label at the d side of t. We assume a finite set T of tile types, but an infinite number of copies of each tile type, each copy referred to as a *tile*. An *assembly* (a.k.a., *supertile*) is a positioning of tiles on (part of) the integer lattice $\mathbb{Z}^2$; i.e., a partial function

$\mathbb{Z}^2 \dashrightarrow T$. A *strength function* is a function $g : \Lambda \to \mathbb{N}_0$ indicating, for each glue label ℓ, the strength $g(\ell)$ with which it binds. Two adjacent tiles in an assembly *interact* if the glue labels on their abutting sides are the same and have positive strength according to g. Each assembly induces a *binding graph*, a grid graph whose vertices are tiles, with an edge between two tiles if they interact. The assembly is τ-*stable* if every cut of its binding graph has strength at least τ, where the weight of an edge is the strength of the glue it represents. That is, the assembly is τ-stable if at least energy τ is required to separate it into two parts.

A *tile assembly system* (TAS) is a quadruple $\mathcal{T} = (T, \sigma, g, \tau)$, where T and g are as stated above, $\sigma : \mathbb{Z}^2 \dashrightarrow T$ is a finite τ-stable *seed assembly*, and $\tau \in \mathbb{N}$ is a parameter called *temperature*. We assume that all seed assemblies σ consist of a single tile (i.e., $|\mathrm{dom}\,\sigma| = 1$). An assembly α is *producible* by $\mathcal{T}$ if either $\alpha = \sigma$, or β is a producible assembly and α can be obtained from β by placing a single tile on an empty space such that the resulting assembly α is τ-stable. An assembly is *terminal* if no tile can be stably attached to it. Let $\mathcal{A}[\mathcal{T}]$ be the set of producible assemblies of $\mathcal{T}$, and let $\mathcal{A}_\square[\mathcal{T}] \subseteq \mathcal{A}[\mathcal{T}]$ be the set of producible terminal assemblies of $\mathcal{T}$. A TAS $\mathcal{T}$ is *directed* if $|\mathcal{A}_\square[\mathcal{T}]| = 1$. Given a (connected) shape $S \subseteq \mathbb{Z}^2$, a TAS $\mathcal{T}$ *strictly self-assembles* S if every terminal assembly produced by $\mathcal{T}$ has shape S.

A directed TAS that strictly self-assembles a shape S can be regarded as a "program" to output S. Hence, a descriptional complexity of S that captures the notion of how concisely one can describe such a TAS has been investigated. Rothemund and Winfree introduced the *temperature-2 directed tile complexity* of S [9], which is defined as the minimum number of tile types required for a directed TAS at the temperature 2 to strictly self-assemble S. This complexity measure is naturally extended for an arbitrary temperature τ as *directed tile complexity of S at the temperature τ*; its definition must follow in a straightforward manner. We denote it by $\mathrm{C}^{\mathrm{dtilec}(\tau)}(S)$. Let $\mathrm{C}^{\mathrm{dtilec}(<\tau)}(S) = \min\{\mathrm{C}^{\mathrm{dtilec}(i)}(S) \mid 1 \le i < \tau\}$ and $\mathrm{C}^{\mathrm{dtilec}(\le\tau)}(S) = \min\{\mathrm{C}^{\mathrm{dtilec}(<\tau)}(S), \mathrm{C}^{\mathrm{dtilec}(\tau)}(S)\}$.

Proposition 1. *For any $\tau \in \mathbb{N}$, TASs at the temperature τ can be simulated at any temperature that is a multiple of τ.*

Proof. This simulation is simply done by multiplying the strengths and τ of a given TAS $\mathcal{T}_1 = (T, \sigma, g, \tau)$ by a proper constant c. The TAS thus obtained is $\mathcal{T}_2 = (T, \sigma, g', c\tau)$ with $g'(\ell) = cg(\ell)$ for each glue label ℓ in T.

Let us consider the $\mathcal{T}_1$ and $\mathcal{T}_2$ in this proof with the assumption that they are directed. It goes without saying that $\mathcal{T}_1$ and $\mathcal{T}_2$ reach the same terminal assembly (i.e., $\mathcal{A}_\square[\mathcal{T}_1] = \mathcal{A}_\square[\mathcal{T}_2]$), they exhibit exactly the same behaviors not only at the macro level as such but also at the micro, or local, level, that is, $\mathcal{A}[\mathcal{T}_1] = \mathcal{A}[\mathcal{T}_2]$. Let us formalize this behavioral equivalence at the local level next.

3   Behavioral Equivalences among TASs

The "behavior" of a TAS $\mathcal{T} = (T, \sigma, g, \tau)$ is determined fully by its strength function g and temperature τ. We can rephrase this as: g and τ determine the global behavior of $\mathcal{T}$ by specifying the local behavior of each tile type $t \in T$ as *cooperation set of t with respect to g and τ* [5], which is defined as:

$$\mathcal{D}_{g,\tau}(t) = \Big\{ D \subseteq \{\mathtt{N}, \mathtt{W}, \mathtt{S}, \mathtt{E}\} \;\Big|\; \textstyle\sum_{d \in D} g(t(d)) \geq \tau \Big\}.$$

This is the collection of subsets of sides of t whose glues have sufficient strengths to bind cooperatively. As such, a cooperation set is closed under superset operation. A tile type with empty cooperation set can be rid from T because a tile of that type never attaches. Then, for any $t \in T$, $\{\mathtt{N}, \mathtt{W}, \mathtt{S}, \mathtt{E}\} \in \mathcal{D}_{g,\tau}(t)$.

Cooperation sets provides a behavioral equivalence among TASs. Given TASs $\mathcal{T}_1 = (T, \sigma, g_1, \tau_1)$ and $\mathcal{T}_2 = (T, \sigma, g_2, \tau_2)$ that share the tile set T and seed σ, if $\mathcal{D}_{g_1,\tau_1}(t) = \mathcal{D}_{g_2,\tau_2}(t)$ for each tile type $t \in T$, then the local behaviors of $\mathcal{T}_1$ and $\mathcal{T}_2$ are exactly the same. Then these TASs are said to be *locally equivalent* (written as $\mathcal{T}_1 \sim \mathcal{T}_2$). Note that $\mathcal{T}_1 \sim \mathcal{T}_2$ implies $\mathcal{A}[\mathcal{T}_1] = \mathcal{A}[\mathcal{T}_2]$ and $\mathcal{A}_\square[\mathcal{T}_1] = \mathcal{A}_\square[\mathcal{T}_2]$. Thus, given locally equivalent TASs, one is directed and strictly self-assembles a shape if and only if so is and does the other.

By the local equivalence $\sim$, the set of all TASs is divided into the equivalence classes, and all the TASs in an equivalence class behave exactly in the same way locally. Hence, we can choose a TAS $\mathcal{T} = (T, \sigma, g, \tau)$ arbitrarily from the class, and let it describe the behavior of this class by the pair $(T, \{\mathcal{D}_{g,\tau}(t) \mid t \in T\})$. This suggests a way to define a variant of TAS by assigning each $t \in T$ with a set of subsets of $\{\mathtt{N}, \mathtt{W}, \mathtt{S}, \mathtt{E}\}$ as a cooperation set. This variant is free from strength function or temperature, and hence, called the *strength-free TAS* [5]. Formally, a *strength-free TAS* is a triple $(T, \sigma, \mathcal{D})$, where T and σ are the same as those for TAS, while $\mathcal{D} : T \to \mathcal{P}(\mathcal{P}(\{\mathtt{N}, \mathtt{W}, \mathtt{S}, \mathtt{E}\}))$ is a function from a tile type $t \in T$ to a set of subsets of $\{\mathtt{N}, \mathtt{W}, \mathtt{S}, \mathtt{E}\}$ that is closed under superset operation, where $\mathcal{P}$ means the power set. Given a strength-free TAS $\mathcal{T}_{\mathrm{sf}} = (T, \sigma, \mathcal{D})$ and a (conventional) TAS (T, σ, g, τ), they are *locally equivalent* if $\mathcal{D}(t) = \mathcal{D}_{g,\tau}(t)$ for each $t \in T$. For an equivalent class, if $\mathcal{T}_{\mathrm{sf}}$ is locally equivalent to its element, then so is it to all of its elements. Hence, we consider that $\mathcal{T}_{\mathrm{sf}}$ represents the local behavior of this class. Of particular note is that such a $\mathcal{T}_{\mathrm{sf}}$ is unique. It must be also noted that there exists a strength-free TAS that is locally equivalent to no TAS, and this implementability can be checked in a polynomial time [5].

The strength-free TAS was introduced as a technical tool to solve an open problem posed by Adleman et al. [2]. In the paper, they proposed an algorithm to find a minimum size directed TAS $\mathcal{T} = (T, \sigma, g, 2)$ that strictly self-assembles the $n \times n$ square Sq_n, subject to the constraint that the TAS's temperature is 2. This algorithm enumerates all temperature-2 TASs with at most $\mathrm{C}^{\mathrm{dtilec}(2)}(Sq_n)$ tile types, and checks whether each of them is directed and strictly self-assembles Sq_n (this is proved to be polynomial-time checkable in n). The temperature of a system to be checked need not be 2, but rather the temperature-2 restriction[1] is utilized to upper-bound the number of all candidates to be thus checked by a polynomial in n, and as a result, this algorithm runs in a polynomial time in n.

Chen, Doty, and Seki improved this algorithm by removing the temperature constraint [5]. Though being based on the above-mentioned idea by

---

[1] This can be replaced with the temperature-c restriction for any constant $c \geq 1$.

Adleman et al., their algorithm enumerates rather all strength-free TASs with at most $C^{\mathrm{dtilec}(2)}(Sq_n)$ tile types, and commits extra check for the implementability. A problem arises of whether this algorithm can be modified so as to output among all the minimum size directed TASs for Sq_n the one(s) working at the lowest temperature. This motivates us to study the following optimization:

FINDOPTIMALSTRENGTH
INPUT : a strength-free TAS $\mathcal{T}_{\mathrm{sf}}$
OUTPUT : a TAS of minimal temperature that is locally equivalent to $\mathcal{T}_{\mathrm{sf}}$, if any.

We shall prove that this is NP-hard (Theorem 2), which was left open in [5].

3.1   Threshold Programming

In order to prove the NP-hardness of FINDOPTIMALSTRENGTH, let us introduce a subclass of integer programming (IP) that aims at optimizing τ subject to only $\geq_\tau$-inequalities and $<_\tau$-inequalities. We call an IP a *threshold programming* (TP). This is formalized as: for given integer matrices C_1, C_2, minimize τ on condition that $C_1 \boldsymbol{x} \geq \tau \boldsymbol{1}$ and $C_2 \boldsymbol{x} < \tau \boldsymbol{1}$, where $\boldsymbol{x}$ is a vector of non-negative integer variables and $\boldsymbol{1} = (1, 1, \ldots, 1)$. FINDOPTIMALSTRENGTH is a TP any of whose constraints is either a $\geq_\tau$-inequality of at most 4 terms or a $<_\tau$-inequality of at most 3 terms. Such a TP is denoted by TP(4, 3). Its decision variant, denoted by τ-THRESHOLDPROGRAMMING(4, 3) or simply τ-TP(4, 3), is of interest in which τ is regarded rather as a constant, and one is asked to decide whether $\boldsymbol{x}$ exists.

Our arguments below mainly consist of designing systems of τ-inequalities for various purposes. As a tool, we introduce a sub-system that will be embedded into these systems and force their variables to assume at least or exactly some specific value. It is built on the following pair of τ-inequalities:

$$x_1 + x_a \geq \tau \text{ and } x_a < \tau. \tag{1}$$

This pair implies $x_1 \geq 1$. Once (1) being embedded into a τ-inequality system, the variable x_1 cannot help but assume a positive value itself (x_a is assumed to be an auxiliary variable occurring only in (1)). In the rest, when we say that a system has a positive variable x, we assume that its positiveness is thus forced.

Let us build a sub-system called 2^{i+1}-*adder (to the lower bound of a variable)*[2]. From a variable x with a lower bound n, i.e., $x \geq n$, it creates a variable with a lower bound $n + 2^{i+1}$. It has $5i+5$ positive integer variables $z_1, z_2, x_0, x_b, x_c, A_k, A'_k, A''_k, B'_k, B''_k (1 \leq k \leq i)$ and is designed as follows: for $2 \leq j \leq i$,

$$
\begin{array}{lll}
A'_1 + B'_1 + x_0, & A''_1 + B''_1 + x_0 & \geq \tau, \\
A_1 + B'_1 + x_0, & A'_1 + B''_1 + x_0 & < \tau, \\
A_{j-1} + B'_j + A'_j, A_{j-1} + B''_j + A''_j & \geq \tau, \\
A''_{j-1} + B'_j + A_j, A'_j + B''_j + A''_{j-1} & < \tau, \\
A''_i + x_b, & z_1 + x_c & < \tau, \\
z_1 + x_b, & z_2 + x_c & \geq \tau.
\end{array} \tag{2}
$$

---

[2]   This is a modification of a system of inequalities proposed in [5].

Solving (2) gives $A_i'' \geq A_i + 2^{i+1} - 2$ for $1 \leq k \leq i$. The four inequalities in the last two lines of (2) yield $z_1 \geq A_i'' + 1$ and $z_2 \geq z_1 + 1$ (in fact, these four inequalities implement a 2^1-adder). As a result, $z_2 \geq A_i + 2^{i+1}$.

Assume that $A_i \geq n$; then $z_2 \geq n + 2^{i+1}$. Note that under this assumption the new system has a solution for any $\tau \geq n + 2^{i+1} + 1$. With the fact that any positive number can be written as a sum of powers of 2, this property makes possible to combine multiple copies of 2^{i+1}-adders inductively in order to provide a variable with an arbitrarily large lower bound.

In Sect. 4, the 2^{i+1}-adder and what we shall build based on it in Sect. 3.2 will be transformed into shapes in order to prove two of our main results (Theorems 3 and 4). To this end, it must be noted that these systems of τ-inequalities are *quadripartite*, that is, we can quarter their variable set such that each inequality contains at most one variable from each of the four resulting variable subsets.

3.2   FindOptimalStrength is NP-hard

Let us prove the NP-hardness of FindOptimalStrength. For the reduction, we employ a variant of 3-Sat called (monotone) 1-in-3-Sat introduced by Schaefer [10], in which no literal is negated and one is required to find a truth assignment such that each clause has *exactly one* true literal. He proved its NP-completeness. We propose its restricted variant called *quadripartite* 1-in-3-Sat, whose instance consists of a variable set that is a union of four pairwise disjoint sets U_1, U_2, U_3, U_4 and clauses that contain at most one variable from each of these four subsets.

Lemma 1. *Quadripartite* 1-in-3-Sat *is NP-complete.*

Theorem 1. *For any $\tau \geq 4$, τ-ThresholdProgramming$(4,3)$ is NP-complete.*

Proof. A proof for $\tau = 4$ comes first. Due to Lemma 1, we convert an instance of quadripartite 1-in-3-Sat, given as a pair of a set of Boolean variables $U = \{u_1, \ldots, u_n\}$ and that of clauses $C = \{c_1, \ldots, c_m\}$, into an instance of τ-TP$(4,3)$. The quadripartite property is not needed here, but will be so in Sect. 4.

Let us convert this SAT instance into a system $\mathcal{S}$ of τ-inequalities with positive integer variables $v_1, v_2, \ldots, v_n$ (needless to say, (1) is used here for their positiveness), which correspond to the SAT variables in U, such that the SAT instance is satisfiable if and only if the system is solvable. In $\mathcal{S}$, the j-th clause of C, $c_j = \{u_{j_1}, u_{j_2}, u_{j_3}\}$ with $1 \leq j_1, j_2, j_3 \leq n$, is represented as

$$v_{j_1} + v_{j_2} + v_{j_3} \geq 4, v_{j_1} + v_{j_2} < 4, v_{j_1} + v_{j_3} < 4, \text{ and } v_{j_2} + v_{j_3} < 4, \qquad (3)$$

which is equivalent to the equation $v_{j_1} + v_{j_2} + v_{j_3} = 4$ due to the assumption that $v_{j_1}, v_{j_2}, v_{j_3} \geq 1$. Its solution must be that exactly one of the three variables is 2 and the others are 1. Therefore, if $\mathcal{S}$ is solvable, then by interpreting those in $v_1, \ldots, v_n$ with value 2 be positive and the other (that is, with value 1) negative, we can retrieve a way to satisfy the 1-in-3-Sat instance; and vice versa. Thus, 4-TP$(4,3)$ is NP-complete (in fact, we proved that even 4-TP$(3,2)$ is so).

Now the result is generalized for an arbitrary $\tau \geq 4$. As presented in Sect. 3.1, we can design τ-inequalities of at most 3 terms that provide auxiliary variables

$x_1, x_2, x_{\tau-4}$ with lower bounds 1, 2, and $\tau-4$, respectively. Combining them with an inequality $x_1 + x_2 + x_{\tau-4} < \tau$ implies that in any solution of $\mathcal{S}$, $x_{\tau-4}$ must assume the value $\tau-4$. We add this "constant" to the inequalities in (3) as

$$v_{j_1} + v_{j_2} + v_{j_3} + x_{\tau-4} \geq \tau, \qquad v_{j_1} + v_{j_2} + x_{\tau-4} < \tau \qquad (4)$$

and the analogs of this $<_\tau$-inequality for $v_{j_1}+v_{j_3}$ and $v_{j_2}+v_{j_3}$. Since $x_{\tau-4} = \tau-4$, these four τ-inequalities are equivalent to $v_{j_1} + v_{j_2} + v_{j_3} = 4$.

We conclude this proof by noting that any τ-inequality used is either a $\geq_\tau$-inequality of at most 4 terms or a $<_\tau$-inequality of at most 3 terms. □

Theorem 1 leads us to the NP-hardness of TP. Deleting the constant term $x_{\tau-4}$ from the τ-inequalities $x_1 + x_2 + x_{\tau-4} < \tau$ and (4) yields $x_1 + x_2 < \tau$ and

$$v_{j_1} + v_{j_2} + v_{j_3} \geq \tau, v_{j_1} + v_{j_2} < \tau, v_{j_1} + v_{j_3} < \tau, \text{ and } v_{j_2} + v_{j_3} < \tau, \qquad (5)$$

and consider optimizing τ. Due to $x_1 + x_2 < \tau$, the minimal possible value of τ is 4. Then its optimal value is 4 if and only if the 1-IN-3-SAT instance is satisfiable.

Lemma 2. THRESHOLDPROGRAMMING$(4, 3)$ *is* NP-*hard.*

Using this instance, now we can prove that FINDOPTIMALSTRENGTH is NP-hard. Making use of the fact that all of its τ-inequalities contain at most 4 terms, we transform them into cooperation sets of a strength-free TAS. The inequalities (5), which are for the clause c_j, are encoded as $t_{c_j} = (v_{j_1}, v_{j_2}, v_{j_3}, x_\tau)$ with $D(t_{c_j}) = \mathcal{P}(\{\mathbb{N}, \mathbb{W}, \mathbb{S}, \mathbb{E}\}) \setminus \{\{\mathbb{N}\}, \{\mathbb{W}\}, \{\mathbb{S}\}, \{\mathbb{N}, \mathbb{W}\}, \{\mathbb{N}, \mathbb{S}\}, \{\mathbb{W}, \mathbb{S}\}\}$, where an auxiliary variable x_τ is introduced whose glue is so strong that its attachment needs no cooperation. The other inequalities are encoded similarly.

Theorem 2. FINDOPTIMALSTRENGTH *is* NP-*hard.*

One observation of technical importance is that all the systems built in this section can be solved such that the values of all variables are strictly less than τ. Its check is left to the reader.

4 Tile Complexity at High Temperatures

In Sect. 3, bridges from 1-IN-3-SAT to TP(4, 3) and further to the *local*, or *micro*, behavior of a TAS have been established. Since the study of TAS aims at facilitating the design of nano-scale structures, we should shift our focus onto their *global* (or *macro*, *terminal*) behavior; how a given shape is built by TASs. Problems of interest include: for any temperature τ given as parameter,

1. Is there a shape S_τ that prefers the temperatures above τ to the lower ones in terms of tile complexity; that is, $\mathrm{C}^{\mathrm{dtilec}(<\tau)}(S_\tau) > \mathrm{C}^{\mathrm{dtilec}(\tau)}(S_\tau)$? If so, then can we design an algorithm to construct S_τ?
2. Can we compute $\mathrm{C}^{\mathrm{dtilec}(\leq\tau)}(S)$ of a shape S in a polynomial time?

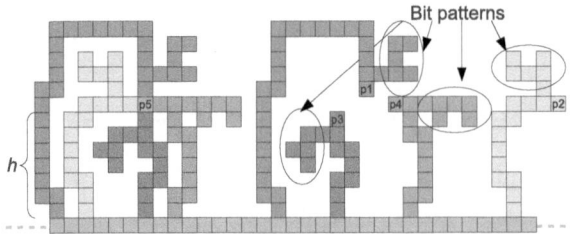

Fig. 1. A logical component for a $\geq_\tau$-inequality of 4 terms. Positions p_1, p_2, p_3, p_4 (specified by their subscripts) are at the tips of the four trees of color red, green, blue, and yellow from the left. These trees are associated with its four variables. The left is a gadget consisting of a single gray tile at the position p_5 supported by four pillars that are of the same shape as the four respective variable trees.

For $\tau = 2$, these problems have been studied intensively. Adleman et al. proved that $C^{\mathrm{dtilec}(2)}(Sq_n) = O(\frac{\log n}{\log \log n})$ for any $n \times n$ square Sq_n [1]. In contrast, $C^{\mathrm{dtilec}(1)}(Sq_n) \leq 2n - 1$ and conjectured to be tight [2], which is highly probable. Thus, $C^{\mathrm{dtilec}(<2)}(Sq_n) \gg C^{\mathrm{dtilec}(2)}(Sq_n)$, provided the conjecture is true. As for the second problem, computing $C^{\mathrm{dtilec}(\leq 2)}(S)$ of a given shape S is NP-hard [2].

We shall work on these problems without any constraint on temperature, and answer them as follows.

Theorem 3. *For any* $\tau \geq 2$, *there is a shape* S_τ *with* $C^{\mathrm{dtilec}(<\tau)}(S_\tau) > C^{\mathrm{dtilec}(\tau)}(S_\tau)$.

Theorem 4. *For any* $\tau \geq 4$, *it is* NP*-hard to compute the directed tile complexity of a shape at the temperatures below* τ.

These theorems will be proved by choosing a proper system of τ-inequalities among those built in Sect. 3 (recall that they are quadripartite) and transforming it into a shape S and a constant c. Let $V = \{v_1, v_2, \ldots, v_n\}$ be the set of all variables occurring in the system, which can be divided into four subsets V_N, V_W, V_S, V_E due to the quadripartite property. In addition, as mentioned at the end of Sect. 3, we can assume that $v_i < \tau$ for all $1 \leq i \leq n$; this is only for simplifying the design of S and our explanation below.

Consider a directed TAS $\mathcal{T} = (T, \sigma, g, \tau)$ that self-assembles S, whatever it is. First of all, we convert the variables in V into trees of height h (a parameter adjustable for our convenience) including a special position called *tip*. In Fig. 1, the positions $p_1, \ldots, p_4$ are the tip of trees including them respectively. Trees for the variables in V_N (V_W, V_S, and V_E) are designed so as to stick out their tip from north (resp., west, south, and east) as the red (yellow, green, and blue) tree in Fig. 1. We say that these trees are of north, west, south, and east type, respectively. Trees of the same directional type are of the identical shape except their identification "bit patterns" of length $\lceil \log n \rceil$. These n trees will be a part of S, and $\mathcal{T}$ actually needs $nh + \sum_{1 \leq i \leq n} c_i$ tile types (see [2] for details) no matter

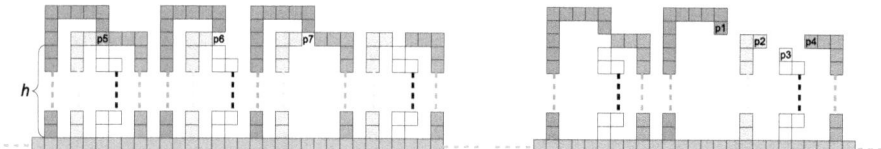

Fig. 2. A logical component for $<_\tau$-inequalities. The variable trees are written concisely, but only due to the space limit; they should be of shape as illustrated in Fig. 1.

what its temperature is (tree's tile complexity is temperature-independent); $h+c_i$ tile types are exclusively for the tree for v_i, where c_i is a constant. We adjust h such that $h \gg \max\{c_i \mid 1 \le i \le n\}$. By such an adjustment, one can also force $\mathcal{T}$ to put its seed somewhere on the scaffold (this technique is used in [2,4]).

Let us convert one $\ge_\tau$-inequality of 4 terms in the system into a shape. The system's quadripartite property guarantees that its four variables are represented as trees of four different directional types. Let $\mathcal{T}$ assemble the whole shape in Fig. 1. It must put tiles of distinct types t_1, t_2, t_3, t_4 at p_1, p_2, p_3, p_4, respectively. As seen in Fig. 1, these trees can be bundled closer into a gadget so that glues on their tips can cooperate to attach the grey tile at the position p_5. This actually suggests a way for $\mathcal{T}$ to accomplish the assembly process using strictly less than $5h$ tile types; that is, reusing the four variable trees for the gadget and letting the tip glues $t_1(\mathbf{S})$, $t_2(\mathbf{E})$, $t_3(\mathbf{N})$, and $t_4(\mathbf{W})$ cooperate to attach the grey tile. For this cooperation to happen, g must assign these glues with strengths such that

$$g(t_1(\mathbf{S})) + g(t_2(\mathbf{E})) + g(t_3(\mathbf{N})) + g(t_4(\mathbf{W})) \ge \tau. \tag{6}$$

We can interpret this as; $\mathcal{T}$ assigns the variables that are represented by the trees, or more precisely, by their tip glues with values so that their sum is at least τ. This means that posing a proper upper bound on the number of available tile types turns $\mathcal{T}$ into a mechanism to solve the given $\ge_\tau$-inequality.

There must be an objection that $\mathcal{T}$ could be designed so as to assemble the shape in a different manner. For example, another green tree could be made for the gadget from scratch and $\mathcal{T}$ lets it singly-support the grey tile, which in turn needs no interaction with the others any more. Then, the component gets irrelevant to the $\ge_\tau$-inequality (6). However, this "cheating" costs $\mathcal{T}$ extra $O(h)$ tile types for the new green tree. Without any proof, we claim that unless $5h$ tile types were available, $\mathcal{T}$ could not help but adopt our original way of assembly, and hence, it assigns the tip glues (variables) with strengths (values) so as to satisfy (6).

Due to the lack of space, a logical component for $<_\tau$-inequalities of at most 3 terms can be only illustrated in Fig. 2. One thing to be emphasized is that this cannot be done simply by modifying the design for $\ge_\tau$-inequalities by leaving the position p_5 empty. This is because no attachment may not mean the insufficient strength but label unmatching. We claim that unless $5h$ tile types are available for $\mathcal{T}$, the strengths assigned to the tip glues of red, yellow, and blue trees must be all positive and their sum must be strictly less than τ.

Every τ-inequality is thus converted into a component (note that the above-mentioned component designs can be easily modified for τ-inequalities of less terms). When a variable is common in inequalities, the resulting components should share the corresponding tree. The quadripartite property makes this sharing possible. Being mounted on a scaffold, these components amount to S.

The constant c should be set to $(nh + \sum_{1 \le i \le n} c_i) + h + c'$, where the extra h is for the auxiliary (white) tree in the component for $<_\tau$-inequalities, and c' is the number of tiles for the scaffold. Setting $h > \sum_{1 \le i \le n} c_i + c'$ secures $c < (n+2)h$. For any $k \ge 2$, now we can transform the system of τ-inequalities designed in Sect. 3.1 that forces τ be at least k into a shape S_k such that $C^{\mathrm{dtilec}(<k)}(S_k) > C^{\mathrm{dtilec}(k)}(S_k)$ in a straightforward way. This proves Theorem 3. Thus converting the TP instance built for Theorem 1 instead brings Theorem 4.

Acknowledgements. The authors thank anonymous referees for their valuable comments and suggestions on an earlier version and David Doty for fruitful and insightful discussions. This work was carried out under the financial support from Funding Program for Next Generation World-Leading Researchers (NEXT program) to Y. O. and from Kyoto University Start-up Grant-in-Aid for Young Scientists No. 021530 to S. S. Part of this work was also financially supported by the Department of Information and Computer Science of Aalto University.

References

1. Adleman, L.M., Cheng, Q., Goel, A., Huang, M.D.: Running time and program size for self-assembled squares. In: Proc. of STOC 2001, pp. 740–748. ACM (2001)
2. Adleman, L.M., Cheng, Q., Goel, A., Huang, M.D., Kempe, D., de Espanés, P.M., Rothemund, P.W.K.: Combinatorial optimization problems in self-assembly. In: Proc. of STOC 2002, pp. 23–32 (2002)
3. Barish, R.D., Schulman, R., Rothemund, P.W., Winfree, E.: An information-bearing seed for nucleating algorithmic self-assembly. Proceedings of the National Academy of Sciences 106(15), 6054–6059 (2009)
4. Bryans, N., Chiniforooshan, E., Doty, D., Kari, L., Seki, S.: The power of nondeterminism in self-assembly. In: Proc. of SODA 2011, pp. 590–602 (2011)
5. Chen, H.L., Doty, D., Seki, S.: Program Size and Temperature in Self-Assembly. In: Asano, T., Nakano, S.-I., Okamoto, Y., Watanabe, O. (eds.) ISAAC 2011. LNCS, vol. 7074, pp. 445–453. Springer, Heidelberg (2011)
6. Chen, H.L., Schulman, R., Goel, A., Winfree, E.: Reducing facet nucleation during algorithmic self-assembly. Nano Letters 7(9), 2913–2919 (2007)
7. Fu, T.J., Seeman, N.C.: DNA double-crossover molecules. Biochemistry 32, 3211–3220 (1993)
8. Fujibayashi, K., Hariadi, R., Park, S.H., Winfree, E., Murata, S.: Toward reliable algorithmic self-assembly of DNA tiles: A fixed-width cellular automaton pattern. Nano Letters 8(7), 1791–1797 (2009)
9. Rothemund, P.W.K., Winfree, E.: The program-size complexity of self-assembled squares (extended abstract). In: Proc. of STOC 2000, pp. 459–468 (2000)
10. Schaefer, T.J.: The complexity of satisfiability problems. In: Proc. of STOC 1978, pp. 216–226 (1978)

11. Seeman, N.C.: Nucleic-acid junctions and lattices. Journal of Theoretical Biology 99, 237–247 (1982)
12. Wang, H.: Proving theorems by pattern recognition - II. The Bell System Technical Journal XL (1), 1–41 (1961)
13. Winfree, E.: Algorithmic Self-Assembly of DNA. Ph.D. thesis, California Institute of Technology (June 1998)

Abstract Partial Cylindrical Algebraic Decomposition I: The Lifting Phase

Grant Olney Passmore[1,2] and Paul B. Jackson[2]

[1] Clare Hall, University of Cambridge, Herschel Road,
Cambridge CB3 9AL, United Kingdom
grant.passmore@cl.cam.ac.uk
[2] Laboratory for Foundations of Computer Science, School of Informatics,
University of Edinburgh, 10 Crichton Street, Edinburgh EH8 9AB, Scotland
pbj@inf.ed.ac.uk

Abstract. Though decidable, the theory of real closed fields (**RCF**) is fundamentally infeasible. This is unfortunate, as automatic proof methods for nonlinear real arithmetic are crucially needed in both formalised mathematics and the verification of real-world cyber-physical systems. Consequently, many researchers have proposed fast, sound but incomplete **RCF** proof procedures which are useful in various practical applications. We show how such practically useful, sound but incomplete **RCF** proof methods may be systematically utilised in the context of a complete **RCF** proof method without sacrificing its completeness. In particular, we present an extension of the **RCF** quantifier elimination method Partial CAD (P-CAD) which uses incomplete $\exists$ **RCF** proof procedures to "short-circuit" expensive computations during the lifting phase of P-CAD. We present the theoretical framework and preliminary experiments arising from an implementation in our **RCF** proof tool **RAHD**.

1 Introduction

Tarski's theorem that the elementary theory of real closed fields (**RCF**) admits effective elimination of quantifiers is one of the longstanding hallmarks of mathematical logic [13]. From this result, the decidability of elementary algebra and geometry readily follow, and a most tantalising situation arises: In principle, every elementary arithmetical conjecture over finite-dimensional real and complex spaces may be decided simply by formalising the conjecture and asking a computer of its truth. So why then do we still not know how many unit hyperspheres may kiss[1] in five dimensions? Is it 41? 42?

The issue is one of complexity. Though decidable, **RCF** is fundamentally infeasible. Due to Davenport-Heintz [5], it is known that there exist families of n-dimensional **RCF** formulas of length $O(n)$ whose only quantifier-free equivalences must contain polynomials of degree $2^{2^{\Omega(n)}}$ and of length $2^{2^{\Omega(n)}}$.

---

[1] See, e.g., [11] for background on the kissing problem for n-dimensional hyperspheres.

S.B. Cooper, A. Dawar, and B. Löwe (Eds.): CiE 2012, LNCS 7318, pp. 560–570, 2012.
© Springer-Verlag Berlin Heidelberg 2012

Nevertheless, there are countless examples of difficult, high-dimensional **RCF** problems solved in mathematical and engineering practice. What is the disconnect? (1) **RCF** problems solved in practice are most often solved using an ad hoc combination of methods, *not* by a general decision method. (2) **RCF** problems arising in practice commonly have structural properties dictated by the application domain from which they originated. Such structural properties can often be exploited making such problems more amenable to analysis and pushing them within the reaches of restricted, more efficient variants of known decision methods.

With this in mind, many researchers have proposed fast, sound but *incomplete* **RCF** proof procedures, many of them being of substantial practical use [1,7,14,10,12,6,4]. This is especially true for formal methods, where improved automated **RCF** proof methods are needed in the formal verification of cyberphysical systems. In these cases, as the **RCF** problems to be analysed are usually machine-generated (and incomprehensibly large), incomplete proof procedures can go a long way. For example, there is no denying the fact that applying a full quantifier elimination algorithm to decide the falsity of a formula such as $\exists x_1, \ldots, x_{100} \in \mathbb{R} \ (x_1 * x_1 + \ldots + x_{100} * x_{100} < 0)$ is an obvious misappropriation of resources. While such an example may seem contrived, consider the fact that when an **RCF** proof method is used in formal verification efforts, it is often fed huge collections of machine-generated formulas which may be (un)satisfiable for extremely simple reasons. Ideally, one would like to be able to use fast, sound but incomplete proof procedures as much as possible, falling back on the far more computationally expensive complete methods only when necessary. It would be desirable to have a principled manner in which incomplete proof methods could be used to improve the performance of a complete method without sacrificing its completeness.

We present *Abstract Partial Cylindrical Algebraic Decomposition* (AP-CAD), an extension of the **RCF** quantifier elimination procedure partial CAD. In AP-CAD, arbitrary sound but possibly incomplete $\exists$ **RCF** proof procedures may be used to "short-circuit" certain expensive computations during CAD construction. This is done in such a way that the completeness of the combined proof method is guaranteed. We restrict our AP-CAD presentation to the practically useful case of $\exists$ **RCF**. We have implemented AP-CAD within our **RCF** proof tool **RAHD** [9] for the case of full-dimensional cell decompositions and present experiments. **RAHD** contains many **RCF** proof methods and allows users to combine them into their own heuristic **RCF** proof procedures through a *proof strategy language*. This is ideal for AP-CAD, as the proof procedure parameters used by AP-CAD can be formally realised as **RAHD** proof strategies.

2   CAD Preliminaries

For a detailed account of CAD, we refer the reader to [2]. We present only the background on (P-)CAD required to understand AP-CAD for $\exists$ **RCF**. P-CAD is currently the most efficient known general quantifier elimination method for $\mathbf{RCF}^2$. An important fact is that the complexity of the (P-)CAD decision algorithm is doubly exponential in the dimension (number of variables) of its input formula. Generally, the most expensive phase of the (P-)CAD algorithm is the so-called "lifting phase." Let us fix some notation.

A *semialgebraic set* is a subset of $\mathbb{R}^n$ definable by a quantifier-free formula in the language of ordered rings. A *region* of $\mathbb{R}^n$ is a connected component of $\mathbb{R}^n$. An *algebraic decomposition* of $\mathbb{R}^n$ is a decomposition of $\mathbb{R}^n$ into finitely many semialgebraic regions. A *cylindrical algebraic decomposition* is a special type of algebraic decomposition whose regions are in a sense "well-behaved" with respect to projections onto lower dimensions. A *cell* is a region of a CAD.

Before delving into technical details, let us discuss how we can use a CAD to make $\exists \mathbf{RCF}$ decisions. By "the polynomials of (an $\exists \mathbf{RCF}$ formula) φ," we shall mean the collection of polynomials obtained by zeroing the RHS of every atom in φ through subtracting the RHS from both sides. We assume each such $\exists \mathbf{RCF}$ formula is in prenex normal form, so that it is an $\exists$-closed boolean combination of *sign conditions*, i.e., of atoms of the form $(p \odot 0)$ with $p \in \mathbb{Z}[x_1, \ldots, x_n]$, $\odot \in \{<, \leq, =, \geq, >\}$. We use $QF(\varphi)$ to mean the quantifier-free matrix of φ.

The key point is that if we have in hand a suitable CAD $C = \{c_1, \ldots, c_m\} \subset 2^{\mathbb{R}^n}$ derived from an $\exists \mathbf{RCF}$ formula φ, we can decide the truth of φ from the CAD directly. The reason is simple: C will have the property that every polynomial of φ has *constant sign* on each c_i, i.e., given p a polynomial of φ and a c_i a cell, it shall hold that $\forall \boldsymbol{r} \in c_i(p(\boldsymbol{r}) = 0) \vee \forall \boldsymbol{r} \in c_i(p(\boldsymbol{r}) > 0) \vee \forall \boldsymbol{r} \in c_i(p(\boldsymbol{r}) < 0)$. Consequently, $QF(\varphi)$ has constant truth value at every point in a given cell. Thus, to decide φ, we simply substitute a single sample point from each c_i into $QF(\varphi)$ and see if it ever evaluates to **true**. It will evaluate to **true** on at least one sample point if and only if φ is **true** over $\mathbb{R}^n$.

We shall define CAD by induction on dimension[3]. A CAD of $\mathbb{R}$ is a decomposition of $\mathbb{R}$ into finitely many cells $c_i \subseteq \mathbb{R}$ s.t. each c_i is of the form (i) $\{\alpha_1\}$, or (ii) $]\alpha_1, \alpha_2[$, or (iii) $]-\infty, \alpha_1[$ or $]\alpha_1, +\infty[$ for algebraic real numbers α_i. Let $\mathcal{A}$ be a region of $\mathbb{R}^i$. We call $\mathcal{A} \times \mathbb{R}$ the *cylinder over* $\mathcal{A}$ and denote it by $Z(\mathcal{A})$.

Definition 1 (Stack). *Let $f_1, \ldots, f_k \in \mathcal{C}(\mathcal{A}, \mathbb{R})$. That is, f_j is a continuous function from $\mathcal{A}$ to $\mathbb{R}$. Furthermore, suppose that the images of the f_j are ordered over $\mathcal{A}$ s.t. $\forall \alpha \in \mathcal{A}(f_j(\alpha) < f_{j+1}(\alpha))$. Then, $f_1, \ldots, f_k$ induce a stack $\mathfrak{S}$ over $\mathcal{A}$, where $\mathfrak{S}$ is a decomposition of $Z(\mathcal{A})$ into $2k+1$ regions of the following form:*

$- \; r_1 = \{\langle \alpha, x \rangle \mid \alpha \in \mathcal{A}, x < f_1(\alpha)\},$
$\quad r_3 = \{\langle \alpha, x \rangle \mid \alpha \in \mathcal{A}, f_1(\alpha) < x < f_2(\alpha)\},$

$\quad \vdots$

$\quad r_{2k-1} = \{\langle \alpha, x \rangle \mid \alpha \in \mathcal{A}, f_{k-1}(\alpha) < x < f_k(\alpha)\},$
$\quad r_{2k+1} = \{\langle \alpha, x \rangle \mid \alpha \in \mathcal{A}, f_k(\alpha) < x\},$

---

[2] See [8] for an explanation as to why P-CAD is also currently the best known general decision method for practical $\exists \mathbf{RCF}$ problems, despite the fact that $\exists \mathbf{RCF}$ has a theoretical exponential speed-up over $\mathbf{RCF}$.

[3] We shall speak freely of the symbolic manipulation and arithmetic of (irrational) real algebraic numbers. See, e.g., [2] for an algorithmic account.

$- r_2 = \{\langle \alpha, x \rangle \mid \alpha \in \mathcal{A}, x = f_1(\alpha)\},$

$\vdots$

$r_{2k} = \{\langle \alpha, x \rangle \mid \alpha \in \mathcal{A}, x = f_k(\alpha)\}.$

A CAD of $\mathbb{R}^{i+1}$ will be obtained from a CAD C of $\mathbb{R}^i$ by constructing a stack over every cell in C.

Definition 2 (CAD in $\mathbb{R}^{i+1}$). *An algebraic decomposition C_{i+1} of $\mathbb{R}^{i+1}$ is a CAD iff C_{i+1} is a union of stacks $C_{i+1} = \bigcup_{j=1}^{k} w_j$, s.t. the stack w_j is constructed over cell c_j in a CAD $C_i = \{c_1, \ldots, c_k\}$ of $\mathbb{R}^i$.*

The P-invariance property will allow us to use CADs to make $\exists \, \mathbf{RCF}$ decisions.

Definition 3 (P-invariance). *Let $P = \{p_1, \ldots, p_k\} \subset \mathbb{Z}[x_1, \ldots, x_n]$ and $\mathcal{A}$ be a region of $\mathbb{R}^n$. Then, we say $\mathcal{A}$ is P-invariant iff every member of P has constant sign on $\mathcal{A}$. That is given any $p_i \in P$,*

$$\forall r \in \mathcal{A}(p_i(r) = 0) \quad \vee \quad \forall r \in \mathcal{A}(p_i(r) > 0) \quad \vee \quad \forall r \in \mathcal{A}(p_i(r) < 0).$$

Given a CAD C, we say C is P-invariant iff every cell of C is P-invariant.

2.1   CAD Construction and Evaluation for $\exists$ RCF

The use of CADs for deciding $\exists \, \mathbf{RCF}$ sentences will take place in four steps. In what follows, φ is an $\exists \, \mathbf{RCF}$ sentence and $P = \{p_1, \ldots, p_k\} \subset \mathbb{Z}[x_1, \ldots, x_n]$ is the collection of polynomials of φ.

Projection. The *projection phase* will begin with P and iteratively apply a projection operator $Proj_i$ of the form $Proj_i : 2^{\mathbb{Z}[x_1, \ldots, x_{i+1}]} \to 2^{\mathbb{Z}[x_1, \ldots, x_i]}$ until a set of polynomials is obtained over $\mathbb{Z}[x_1]$. This process will consist of levels, one for each dimension, s.t. at each level i we shall have what is called a *level-i projection set*, $P_i \subset \mathbb{Z}[x_1, \ldots, x_i]$. These level-$i$ projection sets will have a special property: If we have a P_i-invariant CAD of $\mathbb{R}^i$, then we can use this CAD to construct a P_{i+1}-invariant CAD of $\mathbb{R}^{i+1}$.

Base. The *base phase* consists of computing a P_1-invariant CAD of $\mathbb{R}^1$, implicitly described as a sequence of sample points, one for each cell in the CAD. This can be done by *univariate real root isolation* and basic machinery for arithmetic with real algebraic numbers. Let us suppose we have done this and our sequence of sample points is $s_1 < s_2 < \ldots < s_{2m+1}$.

Lifting. The *lifting phase* will take an implicit description of a P_1-invariant CAD of $\mathbb{R}^1$ and progressively transform it into an implicit description of P_n-invariant CAD of $\mathbb{R}^n$. Let $C = \{c_1, \ldots, c_m\}$ be the P_i-invariant CAD for $\mathbb{R}^i$ which we shall lift to a P_{i+1}-invariant CAD of $\mathbb{R}^{i+1}$. Let $S = \{s_1, \ldots, s_m\}$ be our set of sample points, one from each cell in C. Then, for each cell c_j, we shall use the sample point $s_j \in c_j$ to construct a set of sample points in $\mathbb{R}^{i+1}$ corresponding to a *stack over c_j*:

1. As $s_j \in \mathbb{R}^i$, we have that $s_j = \langle r_1, \ldots, r_i \rangle$ for some $r_1, \ldots, r_i \in \mathbb{R}$.
2. Let $P_{i+1}[s_j]$ denote $P_{i+1}[x_1 \mapsto r_1, x_2 \mapsto r_2, \ldots, x_i \mapsto r_i]$. Then $P_{i+1}[s_j] \subset \mathbb{Z}[x_{i+1}]$ is a univariate family of polynomials.
3. Using the same process as we did in the base phase, compute a $P_{i+1}[s_j]$-invariant CAD of $\mathbb{R}^1$. Let this CAD be represented by a sequence of sample points $t_1 < t_2 < \ldots < t_{2v+1} \in \mathbb{R}$.
4. Then, the *stack over* c_j will be represented by the set of $2v + 1$ sample points obtained by appending each t_j to the lower-dimensional sample point s_j. That is, our stack over c_j will be represented by the following sequence of sample points $z_1, \ldots, z_{2v+1}$ in $\mathbb{R}^{i+1}$: $z_1 = \langle r_1, \ldots, r_i, t_1 \rangle$, $z_2 = \langle r_1, \ldots, r_i, t_2 \rangle$, $\ldots$, $z_{2v+1} = \langle r_1, \ldots, r_i, t_{2v+1} \rangle$.

In the above construction, we call the cell c_j (or the sample point representing it, s_j) the *parent* of the stack $\{z_1, \ldots, z_{2v+1}\}$.

Evaluation. Let $S = \{s_1, \ldots, s_m\} \subset \mathbb{R}^n$ be our final set of sample points. Return the boolean value $\bigvee_{r \in S} QF(\varphi)[r]$.

2.2 Partial CAD

Let us now sketch the idea of partial CAD, due to Collins and Hong [3]. As it stands, the CAD construction algorithm will build a P-invariant CAD induced by the polynomials P of an $\exists$ **RCF** formula φ without paying any attention to the logical content of the formula itself. But, when performing lifting, i.e., constructing a stack of regions of $\mathbb{R}^{i+1}$ over a lower-dimensional cell $c_j \subset \mathbb{R}^i$, we may be easily able to see — simply by substitution and evaluation — that the formula $QF(\varphi)$ could *never* be satisfied over c_j. For instance, let $QF(\varphi) = \left((x_4^4 + x_3 x_2^3 + 3x_1 > 2x_1^4) \wedge (x_1^2 > x_2 + x_3)\right)$. If c_j is a cell in a P_3-invariant CAD of $\mathbb{R}^3$ represented by the sample point $s_j = \langle 0, 1, 5 \rangle$, then we can see $QF(\varphi)$ will *never* be satisfied over a cell in a stack which is a child of c_j. Thus, we need not lift over c_j and can eliminate it.

This is the idea behind *partial CAD* when applied to $\exists$ **RCF** formulas: Before lifting over a cell in a CAD of $\mathbb{R}^i$, check if there are any atoms in your formula only involving the variables $x_1, \ldots, x_i$. If so, then perform *partial* evaluation of your formula by evaluating those atoms upon your sample point in $\mathbb{R}^i$, and then use simple propositional reasoning to try to deduce the truth of your formula. This can also allow us to find a *satisfying* assignment for the variables in $QF(\varphi)$ without constructing a whole CAD. For instance, let $QF(\varphi) = \left((x_4^4 + x_3 x_2^3 + 3x_1 > 2x_1^4) \vee (x_1^2 < x_2)\right)$. If c_j is a cell in a P_2-invariant CAD of $\mathbb{R}^2$ represented by the sample point $s_j = \langle -1, 2 \rangle$, then we can see immediately by substitution that $QF(\varphi)$ is satisfiable over $\mathbb{R}^4$. As a witness to this satisfiability, we may return $\langle -1, 2, r_3, r_4 \rangle$ where $r_3, r_4 \in \mathbb{R}$ are arbitrary.

3 Abstract Partial CAD

From a high level of abstraction, we can see partial CAD to be normal CAD augmented with three pieces of algorithmic data:

1. A strategy for selecting lower-dimensional cells to use for evaluating lower-dimensional atoms in our input formula,
2. An algorithm which when given a cell c_j will construct a formula $F(c_j)$ which, if it both has a truth value and is decided, can be used to tell (i) if the cell c_j can be thrown away (i.e., $F(c_j)$ is decided to be **false**), or (ii) if a satisfying assignment for our formula can be extracted already from a lower-dimensional cell (i.e., $F(c_j)$ is decided to be **true**),
3. A proof procedure which will be used to decide the formulas $F(c_j)$ generated by the algorithm above.

In fact, in their original paper on partial CAD, Collins and Hong make the point that different cell selection strategies could be used and even implement and experiment with a number of them[4]. For partial CAD restricted to $\exists$ **RCF**, these three pieces of algorithmic data described above would be:

1. Select cells $c_i \in C$ in some specified enumeration order (specified by s):
   $c_{s(1)}, c_{s(2)}, c_{s(3)}, \ldots$.
2. Given a cell c_j represented by a sample point $s_j = \langle r_1, \ldots, r_i \rangle \in \mathbb{R}^i$, the formula $F(c_j)$ will be constructed from our original $\exists$ **RCF** formula φ by the following process:
   (a) Let φ' be $QF(\varphi)$ augmented by instantiating x_1 with r_1, x_2 with r_2, ...
   (b) Evaluate all variable-free atoms in φ' to obtain a new formula φ''.
   (c) Replace all (unique) variable-containing atoms in φ'' with fresh propositional variables to obtain a new formula $F(c_j)$.
3. Use a propositional logic proof procedure to attempt to decide $F(c_j)$.

If $F(c_j)$ is **false** (i.e., unsatisfiable), cell c_j can be abandoned and we need not lift over it. If $F(c_j)$ is **true** (i.e., tautologous), then we can extract a witness to the truth of φ from the sample point s_j. Otherwise, we lift over c_j. These three pieces of data give us the widely-used partial CAD of Collins and Hong. But, from this point of view, we see that there are *many other choices* we could make.

3.1   Stages, Theatres and Lifting

The fundamental notion of AP-CAD will be that of a *stage*[5]. Let $\mathcal{L}_{\exists OR}$ be the fragment of the language of ordered rings consisting of purely $\exists$ prenex sentences.

Definition 4 (Stage). *A stage* $\mathfrak{A} = \langle \langle \mathbb{S}, w \rangle, \mathbb{F}, \mathbb{P} \rangle$ *will be given by three pieces of algorithmic data. We describe a stage by how it acts in the context of a fixed (but arbitrary) i-dimensional space* $\mathbb{R}^i$. *These data are as follows:*

---

[4] For Collins and Hong, a cell selection strategy selects *single* cells in some specified order. In Abstract Partial CAD, cell selection strategies will select *sets* of cells in some specified order and $\exists$ **RCF** proof procedures will be applied to see if every cell in a selected set of cells may be eliminated.
[5] The intended connotation is of a *stage* in a *theatre*.

1. A cell selection strategy *for selecting* subsets *of C_i for analysis (we call such a subset a "selection of cells"),*
2. A formula construction strategy *for constructing an* $\exists$ **RCF** *formula whose truth value will correspond to the relevance of a selection of cells (we call such a formula a "cell selection relevance formula"),*
3. An $\exists$ **RCF** *proof procedure used to (attempt to) decide the truth or falsity of a cell selection relevance formula.*

*In the context of CAD construction, sample points will be eliminated when their corresponding cells are deemed to be irrelevant to the $\exists$ **RCF** formula inducing the CAD. This removal might then result in a set of sample points for which the cell selection function behaves differently than it did initially. This motivates the* containment axiom *for covering width functions, so that these dynamics do not result in a non-terminating CAD-based decision algorithm employing the stage machinery. In what follows, let $R_i = \{s \subset \mathbb{R}^i \mid |s| < \omega\}$.*

1. A cell selection strategy *consists of two components:*
   (a) *A* covering width function $w : R_i \to \mathbb{N}$,
   (b) *A* cell selection function $\mathbb{S} : R_i \times \mathbb{N} \to R_i$ *obeying for all $S_i \in R_i$ and all $j \in \{1, \ldots, w(S_i)\}$ the* containment axiom*: $\mathbb{S}(S_i, j) \subset S_i$.*
2. A formula construction strategy *is a function* $\mathbb{F} : \mathcal{L}_{\exists OR} \times R_i \to \mathcal{L}_{\exists OR}$ *obeying certain* relevance judgment axioms*. To describe these axioms, we need the context of a fixed (but arbitrary) $\exists$ **RCF** formula and an associated P_i-invariant CAD of $\mathbb{R}^i$. Let φ be an $\exists$ **RCF** formula with polynomials $P \subset \mathbb{Z}[x_1, \ldots, x_n]$ and let $P_n, \ldots, P_1$ be a sequence of level-$(n, \ldots, 1)$ projection sets rooted in P (recall $P_n = P$). Let $C_i = \{c_1, \ldots, c_m\}$ be a P_i-invariant CAD of $\mathbb{R}^i$ with S_i a set of sample points drawn from a subset of the cells in C_i. If we are given a set of sample points $\{s_{a_1}, \ldots s_{a_v}\} \subseteq S_i$, then $\triangle(\{s_{a_1}, \ldots s_{a_v}\})$ will denote the set of cells from which the sample points s_{a_j} are drawn. Then, for each set of sample points S_i and each $j \in \{1, \ldots, w(S_i)\}$ the following* relevance judgment axioms *must hold: $[(\textbf{RCF} \models \neg \mathbb{F}(\varphi, \mathbb{S}(S_i, j))) \implies \mathcal{N}(\varphi, \mathbb{S}(S_i, j))]$, and $[(\textbf{RCF} \models \mathbb{F}(\varphi, \mathbb{S}(S_i, j))) \implies \textbf{RCF} \models \varphi]$, where $\mathcal{N}(\varphi, \{s_{a_1}, \ldots s_{a_v}\})$ means that no child (at any ancestral depth, i.e., in a P_{i+1}-invariant CAD of $\mathbb{R}^{i+1}$, in a P_{i+2}-invariant CAD of $\mathbb{R}^{i+2}, \ldots$, in a P_n-invariant CAD of $\mathbb{R}^n$) of any cell in the set $\triangle(\{s_{a_1}, \ldots, s_{a_v}\})$ will satisfy $QF(\varphi)$.*
3. An $\exists$ **RCF** *proof procedure is a function*

$$\mathbb{P} : \mathcal{L}_{\exists OR} \to \left(\{\textbf{true}, \textbf{false}, \textbf{unknown}\} \cup \bigcup_{j \in \mathbb{N}^+} \mathbb{R}^j \right)$$

obeying the soundness axioms*: $((\mathbb{P}(\psi) = \textbf{true}) \implies \textbf{RCF} \models \psi)$, $((\mathbb{P}(\psi) = \textbf{false}) \implies \textbf{RCF} \models \neg\psi)$, $((\mathbb{P}(\psi) \in \mathbb{R}^j) \implies \textbf{RCF} \models QF(\psi)[\mathbb{P}(\psi)])$ for arbitrary $\psi \in \mathcal{L}_{\exists OR}$ and with $QF(\psi)[\mathbb{P}(\psi)]$ in the final axiom being the substitution of the j-vector $\mathbb{P}(\psi)$ (or an arbitrary extension of it to the dimension of the polynomials appearing in ψ) into ψ, resulting in a variable-free formula. In this case, we call $\mathbb{P}(\psi)$ a* witness *to the truth of ψ.*

We shall want to have the freedom to give our AP-CAD algorithm a *sequence* of stages, one for each dimension $1, \ldots, n$. The intuition is as follows: Stages are introduced so that one can present a *strategy* to an underlying CAD decision algorithm which will prescribe a method for the algorithm to recognise when it can short-circuit certain expensive computations. If we can abandon a cell at a low dimension, for instance at the base phase or when beginning to lift over cells of $\mathbb{R}^2$, then this can potentially give us *hyper-exponential* savings down the line. Thus, it makes sense to arrange stages $\mathfrak{A}_1, \mathfrak{A}_2, \ldots \mathfrak{A}_n$ so that stage $\mathfrak{A}_1$ works *hardest* to make relevance judgments about cells. For if $\mathfrak{A}_1$ causes us to throw away cell $c_j \subset \mathbb{R}^1$, then this could lead to huge savings later. Then, $\mathfrak{A}_2$ might still work hard but a bit *less* hard, and so on. This collection of stages gives rise to the notion of an *n-theatre*. In what follows, let Θ be the set of all stages.

Definition 5 (Theatre). *An n-theatre $\mathbb{T}$ is a function $\mathbb{T} : \{1, \ldots, n\} \to \Theta$.*

Stage i in a theatre will be used to make judgments about cells in a P_i-invariant (partial) CAD of $\mathbb{R}^i$ (i.e., at level i). Let us describe a decision method we shall use for deciding $\exists\,\mathbf{RCF}$ sentences in the framework of AP-CAD.

Algorithm 1 (AP-CAD with Theatrical Lifting). *Suppose we are given an $\exists\,\mathbf{RCF}$ sentence φ with polynomials $P \subset \mathbb{Z}[x_1, \ldots, x_n]$, and an n-theatre $\mathbb{T}$.*

1. **Projection.** *As with standard P-CAD, compute a sequence of projection sets $P_1, \ldots, P_n$.*
2. **Base.** *As with standard P-CAD, compute a P_1-invariant CAD of $\mathbb{R}^1$, $C_1 = \{c_1, \ldots, c_{2m+1}\}$ represented by sample points $S_1 = \{s_1, \ldots, s_{2m+1}\}$. Set the current dimension $i := 1$.*
3. **Lifting.** *Let $\mathbb{T}(i) = \mathfrak{A}_i = \langle \langle \mathbb{S}_i, w_i \rangle, \mathbb{F}_i, \mathbb{P}_i \rangle$ be the stage for dimension i, and S_i the set of sample points for the P_i-invariant (partial) CAD of $\mathbb{R}^i$ over which we need to lift.*
   (a) *Let $U := w_i(S_i)$ and let $j := 1$.*
   (b) **While** *$j \leq U$* **do**
      i. *Let $\{s_{a_1}, \ldots, s_{a_v}\} := \mathbb{S}_i(S_i, j)$.*
      ii. *Let $\chi := \mathbb{P}_i(\mathbb{F}_i(\{s_{a_1}, \ldots, s_{a_v}\}))$.*
      iii. *If $\chi = \mathbf{true}$, then* **return true**.
      iv. *If $\chi = \langle x_1, \ldots, x_w \rangle \in \mathbb{R}^w$ for some $w \leq n$, then*
         A. *Fix an n-dim. extension of χ, e.g., $\mathbf{r} = \langle x_1, \ldots, x_w, \mathbf{0} \rangle \in \mathbb{R}^n$.*
         B. *Evaluate $QF(\psi)[\mathbf{r}]$ and set $R \in \{\mathbf{true}, \mathbf{false}\}$ to this result.*
         C. *If $R = \mathbf{true}$, then* **return** *$\mathbf{r}$ as a witness to the truth of φ.*
         D. *If $R = \mathbf{false}$, then* **return true**[6].
      v. *If $\chi = \mathbf{false}$, then set $S'_i := S_i \setminus \{s_{a_1}, \ldots, s_{a_v}\}$, else set $S'_i := S_i$.*

---

[6] This is perhaps the one counter-intuitive part of the algorithm. Note, however, that this is actually correct: By the combination of the second *relevance judgment axiom* for $\mathbb{F}_i$ and the *soundness axioms* for $\mathbb{P}_i$, the fact that $\mathbf{RCF} \models \mathbb{F}_i(\mathbb{S}_i(S_i, j))$ means that φ is **true**. It's just that the witness $\mathbb{P}_i$ computed for the truth of $\mathbb{F}_i(\mathbb{S}_i(S_i, j))$ might fail to be a witness for φ. In this case, we simply know φ is true without knowing a witness for it.

      *vi. If $S_i' = \emptyset$ then **return false**.*

    vii. If $S_i' = S_i$ then set $j := j + 1$.

  viii. If $S_i' \subset S_i$ then

      A. Set $S_i := S_i'$.

      B. Set $U := w_i(S_i)$.

      C. Set $j := 1$.

(c) Now, $S_i = \{t_1, \ldots, t_u\}$ contains sample points corresponding to the cells we have not deemed to be irrelevant. We need to lift over them.

    i. Let $S_{i+1} := \emptyset$.

    *ii. **For j from 1 to u do***

      A. Substitute the components of t_j in for the variables $x_1, \ldots, x_i$ in P_{i+1} to obtain a univariate family $P_{i+1}[t_j] \subset \mathbb{Z}[x_{i+1}]$.

      B. Compute a $P_{i+1}[t_j]$-invariant CAD of $\mathbb{R}^1$, represented by sample points K_j.

      C. Set $S_{i+1} := S_{i+1} \cup K_j$.

(d) Increase the current dimension by setting $i := i + 1$.

(e) If $i = n$ then lifting is done and we may proceed to the evaluation phase.

(f) If $i < n$ then we loop and begin the lifting process again, but now with the set of sample points S_{i+1}.

*4. **Evaluation.** Return the boolean value $\bigvee_{r \in S_n} QF(\varphi)[r]$.*

Theorem 2. *Algorithm 1 is a sound and complete $\exists$ **RCF** proof procedure.*

Proof. By induction on dimension. (See the extended version of this paper.)

4   Experimental Results

As an experiment (explicated in the extended version of this paper), we built a concrete AP-CAD theatre combining interval constraint reasoning with standard partial CAD [9]. As CAD performance is strongly dependent on the number of cells retained at each dimension, we compared this for three CAD variants: (i) CAD, (ii) standard P-CAD, and (iii) AP-CAD, w.r.t. an $\exists$ **RCF** sentence φ s.t.

$$QF(\varphi) = \begin{bmatrix} (x_1 x_4 + x_2 x_4 + x_3 x_2 < 0) \wedge (x_2 > 0) \wedge (x_3 > 0) \wedge (x_4 > 0) \\ \wedge (x_3 x_4 - x_4^2 + x_3^2 + 1 < 0) \end{bmatrix}.$$

As $QF(\varphi)$ involves only strict inequalities, we appeal to a theorem of McCallum allowing us to only consider full-dimensional cells (selecting rational sample points), and compare the methods w.r.t. this CAD variant [9]. We observe that the AP-CAD method retains less cells than the other methods. This is supported by experiments we have done with other $\exists$ **RCF** formulas. In all cases, the cost of AP-CAD theatre execution measured $< 0.01\%$ of the total CPU time, indicating that there is much positive impact to be made by using incomplete **RCF** proof procedures to enhance the performance of CAD-based decision methods. The cell retainment counts are as follows:

	CAD	P-CAD	AP-CAD
$\mathbb{Q}^1$	2	2	1
$\mathbb{Q}^2$	14	7	5
$\mathbb{Q}^3$	40	10	7
$\mathbb{Q}^4$	200	50	35

5 Conclusion

AP-CAD allows strategic algorithmic data to be used to "short-circuit" expensive computations during the *lifting phase* of a CAD-based decision algorithm. This provides a principled approach for utilising fast, sound but possibly incomplete $\exists$ **RCF** proof procedures to enhance a *complete* decision method without threatening its completeness. We see many ways this work might be extended. It would be very interesting to work out similar machinery to be used during the *projection phase* of P-CAD. For this work to bear serious practical fruit, many more AP-CAD stages should be constructed and experimented with heavily.

Acknowledgements. This paper reports on work presented in Chapters 7 and 8 of the first author's 2011 University of Edinburgh Ph.D. thesis [9], supervised by the second author. This research was supported by the EPSRC [grant numbers EP/I011005/1 and EP/I010335/1]. We thank the referees for their helpful comments and suggestions.

References

1. Avigad, J., Friedman, H.: Combining Decision Procedures for the Reals. In: Logical Methods in Computer Science (2006)
2. Basu, S., Pollack, R., Roy, M.F.: Algorithms in Real Algebraic Geometry. Springer, USA (2006)
3. Collins, G.E., Hong, H.: Partial Cylindrical Algebraic Decomposition for Quantifier Elimination. J. Sym. Comp. 12(3), 299–328 (1991)
4. Daumas, M., Lester, D., Muñoz, C.: Verified Real Number Calculations: A Library for Interval Arithmetic. IEEE Trans. Comp. 58(2), 226–237 (2009)
5. Davenport, J.H., Heintz, J.: Real Quantifier Elimination is Doubly Exponential. J. Symb. Comput. 5, 29–35 (1988), http://dx.doi.org/10.1016/S0747-71718880004-X
6. Gao, S., Ganai, M., Ivancic, F., Gupta, A., Sankaranarayanan, S., Clarke, E.: Integrating ICP and LRA solvers for deciding nonlinear real arithmetic problems. In: FMCAD 2010, pp. 81–89 (2010)
7. Harrison, J.: Verifying Nonlinear Real Formulas via Sums of Squares. In: Schneider, K., Brandt, J. (eds.) TPHOLs 2007. LNCS, vol. 4732, pp. 102–118. Springer, Heidelberg (2007), http://portal.acm.org/citation.cfm?id=1792233.1792242
8. Hong, H.: Comparison of Several Decision Algorithms for the Existential Theory of the Reals. Tech. rep., RISC (1991)
9. Passmore, G.O.: Combined Decision Procedures for Nonlinear Arithmetics, Real and Complex. Ph.D. thesis, University of Edinburgh (2011)

10. Passmore, G.O., Jackson, P.B.: Combined Decision Techniques for the Existential Theory of the Reals. In: Carette, J., Dixon, L., Coen, C.S., Watt, S.M. (eds.) MKM 2009, Held as Part of CICM 2009. LNCS (LNAI), vol. 5625, pp. 122–137. Springer, Heidelberg (2009)
11. Pfender, F., Ziegler, G.M.: Kissing Numbers, Sphere Packings, and Some Unexpected Proofs. Notices of the A.M.S. 51, 873–883 (2004)
12. Platzer, A., Quesel, J.-D., Rümmer, P.: Real World Verification. In: Schmidt, R.A. (ed.) CADE-22. LNCS, vol. 5663, pp. 485–501. Springer, Heidelberg (2009)
13. Tarski, A.: A Decision Method for Elementary Algebra and Geometry. RAND Corporation (1948)
14. Tiwari, A.: An Algebraic Approach for the Unsatisfiability of Nonlinear Constraints. In: Ong, L. (ed.) CSL 2005. LNCS, vol. 3634, pp. 248–262. Springer, Heidelberg (2005)

Multi-valued Functions in Computability Theory

Arno Pauly

Clare College, University of Cambridge, Trinity Lane,
Cambridge CB2 1TL, United Kingdom
Arno.Pauly@cl.cam.ac.uk

Abstract. Multi-valued functions are common in computable analysis (built upon the Type 2 Theory of Effectivity), and have made an appearance in complexity theory under the moniker *search problems* leading to complexity classes such as PPAD and PLS being studied. However, a systematic investigation of the resulting degree structures has only been initiated in the former situation so far (the Weihrauch-degrees).

A more general understanding is possible, if the category-theoretic properties of multi-valued functions are taken into account. In the present paper, the category-theoretic framework is established, and it is demonstrated that many-one degrees of multi-valued functions form a distributive lattice under very general conditions, regardless of the actual reducibility notions used (e.g., Cook, Karp, Weihrauch).

Beyond this, an abundance of open questions arises. Some classic results for reductions between functions carry over to multi-valued functions, but others do not. The basic theme here again depends on category-theoretic differences between functions and multi-valued functions.

Keywords: Multi-valued functions, many-one reduction, Weihrauch reducibility, category theory, degree structure.

1 Introduction

What Are Multi-valued Functions? A (partial) multi-valued function $f :\subseteq A \rightrightarrows B$ is just a set $f \subseteq A \times B$—i.e., a relation. However, the category of multi-valued functions is not the category of relations! We write $f(a)$ for $\{b \in B \mid (a, b) \in f\}$ and $\text{dom}(f) = \{a \in A \mid \exists b \in f(a)\}$. Then the composition of multi-valued functions $f :\subseteq A \rightrightarrows B$, $g :\subseteq B \rightrightarrows C$ is defined via $c \in (g \circ f)(a)$ iff $f(a) \subseteq \text{dom}(g)$ and $\exists b \in f(a)$ s.t. $c \in g(a)$. In the usual definition of the composition for relations, the former condition is absent!

The intended interpretation of a multi-valued function $f :\subseteq A \rightrightarrows B$ is that it links problem instances to solutions. This draws interest to the following partial order:

$$f \preceq g \ \Leftrightarrow \ \text{dom}(f) \subseteq \text{dom}(g) \wedge g_{|\text{dom}(f)} \subseteq f$$

We can read $f \preceq g$ as f *is easier as* g: There may be fewer instances for f than for g, and a solution to a problem instance in g is a solution for it in f, too, where applicable. This has the consequence that any procedure solving g also solves f.

S.B. Cooper, A. Dawar, and B. Löwe (Eds.): CiE 2012, LNCS 7318, pp. 571–580, 2012.
© Springer-Verlag Berlin Heidelberg 2012

For any two multi-valued functions $f, g :\subseteq A \rightrightarrows B$ there exists a hardest multi-valued function easier than both, i.e., there are binary infima w.r.t. $\preceq$. These are given by $f \wedge g = (f \cup g)_{|\,\mathrm{dom}(f) \cap \mathrm{dom}(g)}$.

Why Use Them? First, multi-valued functions are natural: From elimination orders on graphs over Nash equilibria in games to fixed points of continuous mappings, there are plenty of problems without a natural way to specify the desired solution uniquely. In fact, if one accepts their formulation as multi-valued functions, one can even prove that the latter two are non-equivalent to any function!

Then, they are well-behaved under realizability: It is a common situation in computability and complexity theory that we have an algorithmic notion for some functions on some special sets X, Y which we intend to lift to more general sets A, B. We do this by fixing surjective encodings $\delta_A :\subseteq X \to A$, $\delta_B :\subseteq Y \to B$, and then calling, e.g., a function $f : A \to B$ computable, iff there is a computable function $F :\subseteq X \to Y$ such that the following diagram commutes:

$$X \xrightarrow{\ F\ } Y$$
$$\downarrow{\delta_A} \qquad \downarrow{\delta_B}$$
$$A \xrightarrow{\ f\ } B$$

In general (depending on δ_A, δ_B), there will be algorithms (i.e., functions $F :\subseteq X \to Y$) which do not compute any function $f : A \to B$, which leads to the canonization problem: The desire to find an algorithm $C_A :\subseteq X \to X$ with the properties $C_A(x) = C_A(y)$ whenever $\delta_A(x) = \delta_A(y)$, and $\delta_A(C_A(x)) = \delta_A(x)$.

On the other hand, every algorithm computes a multi-valued function, hence, the canonization problem is relegated to a far less fundamental position.

Algorithms lacking semantics as a function can nonetheless be very meaningful. A common example for this is the multi-valued function $\chi : \mathbb{R} \to \{0,1\}$ with $0 \in \chi(x)$ iff $x \leq 1$ and $1 \in \chi(x)$ iff $x \geq 0$. χ is computable – but the only computable functions from $\mathbb{R}$ to $\{0,1\}$ are the constant ones! Hence, when working with real numbers, tests will have to be non-deterministic, i.e., multi-valued functions.

Finally, as will be demonstrated in this paper, the properties of multi-valued functions have a nice impact on the degree-structure of many-one reductions: One always obtains a distributive lattice here.

Due to lack of space, proofs are omitted here. They can be found in [17,18].

2   Background

Many-one reductions between multi-valued functions have been studied in complexity theory for several decades now, with the complexity classes PPAD [14] and PLS [12] garnering a lot of attention. Both have a number of very interesting complete problems, we just mention finding Nash equilibria in finite two player games with integer payoffs as a complete problem for PPAD [7].

There also are a several problems which are known to be in both PPAD and PLS, but where this is the best classification available. Deciding the winner in parity or discounted payoff games is a typical example here. Despite this strong motivation to study PPAD∩PLS, only recently it was noticed (in a publication) that this class actually has complete problems [8]—a fact that is an obvious consequence of the degree structure being a distributive lattice (which we show here). A systematic investigation of the degree structure seems to be missing so far.

In another setting for many-one reductions between multi-valued functions is the programme to classify the computational content of mathematical theorems initiated in [3,10]. Here a mathematical theorem of the form

$$\forall x \in X \ (x \in D \Rightarrow \exists y \in Y \ T(x,y))$$

is read as a multi-valued function $T :\subseteq X \rightrightarrows Y$ with $\mathrm{dom}(T) = D$ which has to find a witness $y \in Y$ given some $x \in X$. The tool for classification is Weihrauch-reducibility, a form of many-one reducibility introduced originally in [20,21].

Various theorems been classified in this framework: e.g., the Hahn-Banach theorem [10], Weak König's Lemma, the Intermediate Value theorem [4], Nash's theorem on the existence of equilibria [15], Bolzano-Weierstrass [5], Brouwer's Fixed Point theorem [6].

Accompanying the investigation of specific degrees, also the overall degree structure has been studied. The Weihrauch degrees form a distributive lattice [4,16], and can be turned into a Kleene algebra when equipped with additional natural operations ×, * [11]. While some additional results in this area do depend on specific properties of Weihrauch reducibility, the fundamental ones only use generic properties of many-one reductions and multi-valued functions—and as such would also apply to the study of PPAD, PLS, etc.!

In the present paper we outline how the notion of *generic properties of many-one reductions between multi-valued functions* can be formalized, and show the fundamental structural results derivable from them. Then we introduce some properties that do depend on specific reducibilities, and both present some results and open questions.

3   The Category of Multi-valued Functions

It is easy to see that composition of multi-valued functions is associative, so they form a category *Mult*. One can lift disjoint unions and cartesian products from sets to multi-valued functions in the straight-forward way, we shall denote the results by $f + g$ and $f \times g$. The disjoint union retains its rôle as the coproduct, however, the cartesian product is **not** the categorical product!

This situation is reminiscent of categories of partial functions, and indeed we can borrow the following:

Definition 1 (Robinson and Rosolini [19]). *A p-category is a category $\mathcal{C}$ together with a naturally associative and naturally commutative bifunctor $\times$: $\mathcal{C} \times \mathcal{C} \to \mathcal{C}$ (the (cartesian) product), a natural transformation Δ (the diagonal) between the identity functor and the derived functor $X \mapsto X \times X$, and two families of natural transformations $(\pi_1^A)_{A \in Ob(\mathcal{C})}$ and $(\pi_2^B)_{B \in Ob(\mathcal{C})}$ (the projections) where π_1^A is between the derived functor $X \mapsto X \times A$ and the identity, while π_2^B is between the derived functor $X \mapsto B \times X$ and the identity, such that the following properties are given:*

$$\pi_1^X(X) \circ \Delta(X) = \pi_2^X(X) \circ \Delta(X) = \mathrm{id}_X \quad (\pi_1^Y(X) \times \pi_2^X(Y)) \circ \Delta(X \times Y) = \mathrm{id}_{X \times Y}$$
$$\pi_1^Y(X) \circ (\mathrm{id}_X \times \pi_1^Z(Y)) = \pi_1^{(Y \times Z)}(X) \quad \pi_1^Z(X) \circ (\mathrm{id}_X \times \pi_2^Y(Z)) = \pi_1^{(Y \times Z)}(X)$$
$$\pi_2^X(Y) \circ (\pi_1^Y(X) \times \mathrm{id}_Z) = \pi_2^{(X \times Y)}(Z) \quad \pi_2^X(Z) \circ (\pi_2^X(Y) \times \mathrm{id}_Z) = \pi_2^{(X \times Y)}(Z)$$

For easier reading, we shall write $\pi_1^{X,Y}$ instead of $\pi_1^Y(X)$, $\pi_2^{X,Y}$ for $\pi_2^X(Y)$ and finally Δ_X in place of $\Delta(X)$. If the superscripts are obvious from the context, they may be dropped.

The treatment of partial maps in a categorical framework causes the concept of the domain of a map to split into two separate ones. With $\mathrm{Dom}(f)$ we denote the object A, if $f : A \to B$ is a morphism (and likewise CDom denotes B here). Following DiPaola and Heller [9], we write $\mathrm{dom}(f)$ for the morphism $\pi_1^{A,B} \circ (\mathrm{id}_A \times f) \circ \Delta_A$, where π_1 is the first projection of the product $X \times Y$. One can interpret $\mathrm{dom}(f) : A \to A$ as the partial identity on that part of A where the partial map f is actually defined. If $\mathrm{dom}(f) = \mathrm{id}_{\mathrm{CDom}(f)}$, we call f *total*.

Additionally we shall assume that the categories underlying our p-categories have coproducts, and that the functor $\times$ distributes over the coproducts.

We already mentioned the fundamental partial order $\preceq$. As it is compatible with the composition of multi-valued functions, as well as with the cartesian product and the coproduct, we introduce the notion of a *poset enriched p-category* for such structures. If also binary infima exists, and are compatible with composition, cartesian products and coproducts, we have a *meet-semilattice enriched p-category*. These concepts come with a natural concept of a substructure, which we shall use.

Moreover, we need two minor properties: A sub poset enriched p-category is called *wide*, if it contains all objects of the superstructure. A poset-enriched p-category is totally connected, if for any two objects A, B there is a morphisms $c_{A,B} : A \to B$. With these notions available, we can now provide the setting we need to introduce many-one reductions:

Definition 2. *A many-one category extension (moce) shall be a meet-semilattice enriched p-category $\mathcal{P}$ with coproducts, together with a wide and totally connected sub-poset enriched p-category $\mathcal{S}$, where $\times$ distributes over the coproducts.*

The intuition behind the preceding definition is that $\mathcal{P}$ is the category of problems one wishes to structure by reductions, whereas $\mathcal{S}$ is the subcategory of *simple* multi-valued functions that serve as reduction witnesses. Typical choices for $\mathcal{S}$ would be computable or polynomial-time computable functions (or multi-valued functions).

4   The Lattice of Many-One Degrees

There are two definitions of many-one reductions commonly found in the literature on (multi-valued) functions, which differ in the question whether the post-processing of the oracle answer still has access to the input. Forgetting the input leads to a simpler definition, and may make proofs of non-reducibility easier, while retaining it yields the nicer degree structure and allows to formulate stronger and more meaningful separation statements. We shall speak of strong many-one reductions if the original input is forgotten, and of many-one reductions otherwise.

Throughout this subsection, we assume that some moce $(\mathcal{P}, \mathcal{S}, \times, \preceq)$ is given, and refrain from mentioning it explicitly where this is unnecessary.

Definition 3 (Strong many-one reductions). *Let $f \leq_{sm} g$ hold for $f, g \in \mathcal{P}$, if there are $H, K \in \mathcal{S}$ with $f \preceq H \circ g \circ K$.*

Definition 4 (Many-one reductions). *Let $f \leq_m g$ hold for $f, g \in \mathcal{P}$, if there are $H, K \in \mathcal{S}$ with $f \preceq H \circ (\mathrm{id}_{\mathrm{Dom}(f)} \times (g \circ K)) \circ \Delta_{\mathrm{Dom}(f)}$.*

Proposition 1. *Both $\leq_{sm}$ and $\leq_m$ define preorders on $\mathcal{P}$. For any, $f, g \in \mathcal{P}$, $f \leq_{sm} g$ implies $f \leq_m g$.*

By $\mathfrak{D}$ we shall denote the partially ordered class of $\leq_m$-equivalence classes in $\mathcal{P}$. Both the coproduct $+$ and the cartesian product $\times$ in $\mathcal{P}$ can be lifted to operations on $\mathfrak{D}$, which we shall denote by $+, \times$ again. We need a third operation to be lifted from $\mathcal{P}$ to $\mathfrak{D}$. The coproduct injections shall be $\iota_1^{X,Y} : X \to X + Y$ and $\iota_2^{X,Y} : Y \to X + Y$, and we denote the infimum regarding $\preceq$ via $\wedge$. Now for $f : X \to Y, g : A \to B$ define $f \oplus g : X \times A \to Y + B$ via:

$$f \oplus g = [(\iota_1^{Y,B} \circ \pi_1^{Y,B}) \wedge (\iota_2^{Y,B} \circ \pi_2^{Y,B})] \circ (f \times g)$$

Informally, $f \oplus g$ receives a problem instance to each of f and g, and has to produce a solution to one of them. Unlike $+, \times$, it is clear that $\oplus$ lacks a counterpart for functions. Its degree-theoretic relevance follows from the following main result:

Theorem 1. $\mathfrak{D}$ *is a distributive lattice, with $\oplus$ as infimum and $+$ as supremum.*

The presence of certain distinguished objects in $\mathcal{P}$ translates into the existence of special degrees in $\mathfrak{D}$. As usual, we call an object $I \in Ob(\mathcal{P})$ *initial*, iff for any object $A \in Ob(\mathcal{P})$ there is exactly one morphism $f : I \to A$. The concept is extended to domains in p-categories by calling dom i initial, iff for any $A \in Ob(\mathcal{P})$ there is exactly one morphism f with $\mathrm{CDom}(f) = A$ and $f = f \circ \mathrm{dom}\, i$.

Our notion of emptiness for objects in p-categories does not amount to emptiness in the underlying category. Instead, we call $E \in Ob(\mathcal{P})$ *empty*, iff for any total morphism $g : A \to E$ we find A to be initial. Likewise, a domain dom e is empty, iff for any total morphism g with $(\mathrm{dom}\, e) \circ g = g$, we find $\mathrm{Dom}(g)$ to be initial. Note that empty implies initial. Objects are initial (empty), iff the corresponding identity is initial (empty) as a domain.

An object $F \in Ob(\mathcal{P})$ is called final, iff for any object $A \in Ob(\mathcal{P})$ there is exactly one total morphism $g : A \to F$. Likewise, a domain $\operatorname{dom} f$ is final, iff for any $A \in Ob(\mathcal{P})$ there is exactly one total morphism g with $\operatorname{Dom}(g) = A$ and $\operatorname{dom} f \circ g = g$. Objects are initial (empty, final), iff the corresponding identity is initial (empty, final) as a domain.

Proposition 2. *Let* $\operatorname{dom} i$ *be initial in both* $\mathcal{S}$ *and* $\mathcal{P}$. *Then its degree (denoted by 0) is the bottom element in* $\mathfrak{D}$.

In particular, this shows that all initial domains are equivalent w.r.t. $\leq_m$. The same holds for final domains, whose degree (if present) we denote by 1. Then we find:

Theorem 2. *If* $\mathcal{P}$ *and* $\mathcal{S}$ *share an empty domain and a final domain, then* $\mathfrak{D}$ *with the operations* $\times$, $+$ *and the constants* 0, 1 *is an idempotent commutative semiring, i.e., the following equations hold for all* $\mathbf{a}, \mathbf{b}, \mathbf{c} \in \mathfrak{D}$:

1. $\mathbf{a} + \mathbf{a} = \mathbf{a}$, $\mathbf{a} + \mathbf{b} = \mathbf{b} + \mathbf{a}$, $\mathbf{a} + (\mathbf{b} + \mathbf{c}) = (\mathbf{a} + \mathbf{b}) + \mathbf{c}$
2. $\mathbf{a} \times \mathbf{b} = \mathbf{b} \times \mathbf{a}$, $\mathbf{a} \times (\mathbf{b} \times \mathbf{c}) = (\mathbf{a} \times \mathbf{b}) \times \mathbf{c}$
3. $\mathbf{a} \times (\mathbf{b} + \mathbf{c}) = (\mathbf{a} \times \mathbf{b}) + (\mathbf{a} \times \mathbf{c})$
4. $0 + \mathbf{a} = \mathbf{a}$, $0 \times \mathbf{a} = 0$, $1 \times \mathbf{a} = \mathbf{a}$

We remark that in a similar fashion, an operation $*$ can be introduced (albeit requiring slightly more assumption) turning $\mathfrak{D}$ into a Kleene-algebra. As a consequence, a definition of wtt-reductions can be derived from our definition of many-one reductions.

5   Some Examples

In this section, we shall exhibit two basic examples for our framework, namely the adapted versions of Karp and Cook reducibilities to multi-valued functions. Another prime example is Weihrauch reducibility. The structural investigation of Weihrauch degrees served as inspiration for the present work, and we refer to the original literature for the details [4,16,2,11].

These examples do not exhaust the range of applicability, though: Medvedev-reducibility and many-one reductions between parameterized search problems are omitted due to limited space; resource-bounded variants of Weihrauch reducibility also satisfy the requirements. Beyond computability, also continuity may be used as the decisive property of reduction witnesses.

5.1   Computable Many-One Reductions

Here we consider the category of multi-valued functions from $\{0,1\}^*$ to $\{0,1\}^*$ in the rôle of $\mathcal{P}$, while the category of reduction witnesses $\mathcal{S}$ is given by the category $\mathcal{C}_1$ of restrictions of partial computable functions. These categories satisfy our conditions, with $f + g$, $f \times g$ defined via $(f + g)(0x) = 0f(x)$, $(f + g)(1x) = 1g(x)$, $(f \times g)(\langle x, y \rangle) = \langle f(x), g(x) \rangle$.

Definition 5 (special case of Definition 4). *For two multi-valued functions* $f, g :\subseteq \{0,1\}^* \rightrightarrows \{0,1\}^*$, *define* $f \leq_m g$, *if there are computable functions* $H, K :\subseteq \{0,1\}^* \rightarrow \{0,1\}^*$ *with* $H\langle x, y\rangle \in f(x)$ *whenever* $y \in g(K(x))$.

We use $\mathfrak{C}_1$ to denote the set of degrees in this setting.

Corollary 1 (of Theorem 1). $(\mathfrak{C}_1, \oplus, +)$ *is a distributive lattice.*

In $\mathcal{C}_1$, there exists both an empty domain and final domains, namely the nowhere defined multi-valued function $\emptyset \subset \{0,1\}^* \times \{0,1\}^*$ and any $\{(x,x)\} \subseteq \{0,1\}^* \times \{0,1\}^*$. The corresponding degrees shall be denoted by $0, 1 \in \mathfrak{C}_1$.

Proposition 3. 1 *is the least element in* $\mathfrak{C}_1 \setminus \{0\}$ *and contains exactly those multi-valued functions admitting a computable choice function.*

We do point out that decision problems cannot be considered as a special case of multi-valued functions in the straight-forward way, as our definition of many-one reductions allows modifications of the output; in particular, the characteristic function of a set is trivially equivalent to the characteristic function of its complement. However, many results proven for many-one reductions between search problems hold—with identical proofs—also for Turing reductions with the number of oracle queries limited to 1, which corresponds to the notion employed here.

For example, YATES' result [22] regarding the existence of minimal pairs applies here as follows:

Proposition 4 (Yates [22]). *There are* $\mathbf{a}, \mathbf{b} \in \mathfrak{C}_1 \setminus \{0,1\}$ *with total representatives such that for any* $\mathbf{c} \leq_m (\mathbf{a} \oplus \mathbf{b})$ *that has a representative* $f \in \mathbf{c}$ *of the type* $f : \{0,1\}^* \rightarrow \{0,1\}$, *we find* $\mathbf{c} = 1$.

However, the cumbersome restriction to degrees admitting a function representative is necessary, as minimal pairs for multi-valued functions do not exist in the computable case:

Proposition 5. *If* $\mathbf{a}, \mathbf{b} \in \mathfrak{C}_1$ *have total representatives, then* $\mathbf{a} \oplus \mathbf{b} = 1$ *implies* $\mathbf{a} = 1$ *or* $\mathbf{b} = 1$.

The proof of the preceding proposition is based on the following technical lemma:

Lemma 1. *There are Turing functionals* Ψ, Φ, *such that for all total multi-valued functions* $f, g : \{0,1\}^* \rightrightarrows \{0,1\}^*$ *and for any choice function* I *of* $(f \oplus g)$, *either* Ψ^I *is a choice function of* f *or* Φ^I *is a choice function of* g.

5.2 Polynomial-Time Many Reductions

Proceeding as above, but taking the category of polynomial-time computable functions as the category $\mathcal{S}$ of reduction witnesses, we again obtain a degree structure w.r.t. many-one reductions in the generic way, which we shall denote by $\mathfrak{P}_1$, and the reducibility by $\leq_m^p$.

Corollary 2 (of Theorem 1). $(\mathfrak{P}_1, \oplus, +)$ *is a distributive lattice.*

Again, 0 and 1 exist and are the bottom and second-least element respectively. $0 \in \mathfrak{P}_1$ contains only the no-where defined multi-valued function, whereas 1 contains exactly those multi-valued functions with non-empty domain admitting a polynomial-time computable choice function.

   Again, some results for functions or decision problems can be transferred. As a demonstration, we extend LADNER's density result [13, Theorem 2] to multi-valued functions. For this, note that two notions coinciding for single-valued functions differ for multi-valued functions, namely the existence of a computable choice function and the decidability of the graph. We call those multi-valued functions satisfying the former condition *computable*. The latter condition has the disadvantage of not being preserved downwards by many-one reductions. However, a decidable graph is the condition needed for the following theorem. Its proof closely resembles the one of [13, Theorem 2].

Theorem 3. *Let* $\mathbf{a}, \mathbf{b} \in \mathfrak{P}_1$ *admit representatives with decidable graphs and satisfy* $\mathbf{b} \not\leq_m^p \mathbf{a}$. *Then there are* $\mathbf{b}_0, \mathbf{b}_1 \in \mathfrak{P}_1$ *with* $\mathbf{b} = \mathbf{b}_0 + \mathbf{b}_1$, $\mathbf{b}_i \not\leq_m^p \mathbf{a}$ *and* $\mathbf{b} \not\leq_m^p \mathbf{a} + \mathbf{b}_i$ *for both* $i \in \{0, 1\}$.

Corollary 3. *The degrees in* $\mathcal{P}_1$ *admitting decidable graphs are dense (in themselves).*

A question that has received a lot of attention regarding (polynomial-time) many-one reductions between decision problems is about the existence and nature of minimal pairs. In terms of lattice theory, this asks whether the degree 1 is meet-irreducible, and if not, what kind of pairs can satisfy $\mathbf{a} \oplus \mathbf{b} = 1$. Following the initial result by LADNER that minimal pairs for polynomial-time many-one reductions between decision problems exist [13], AMBOS-SPIES could prove that every computable super-polynomial degree is part of a minimal pair [1].

   For search problems, however, the question remains open:

Question 1. Is $1 \in \mathcal{P}_1$ meet-irreducible?

The techniques used to construct a minimal pair in [13,1] diagonalize against pairs of reductions R_e, R_f trying to prevent $R_e(a) = R_f(b)$ for the constructed representatives a, b. If the equality cannot be prevented, then one can prove that the resulting set is already polynomial-time decidable using a constant prefix of b, hence, polynomial-time decidable. However, for search problems any pair of reductions to a pair of search problems produces a search problem, namely $R_e(a) \cup R_f(b)$.

   A non-computable minimal pair for Type-2 search problems was constructed in [11] by HIGUCHI and PAULY. There, the crucial part is the identifiability of hard and easy instances, which is not available in a Type-1 setting. The negative answer we obtained for computable many-one reductions in Subsection 5.1 relied on Lemma 1, which again cannot be transferred to the time-bounded case: There are polynomial-time decidable relations R such that neither R nor its inverse $\neg R^\dagger$ admit a polynomial-time choice function, even if $P = NP$ should hold.

6   Outlook

Hopefully we have made the case that investigating the degree structures of many-one reductions between multi-valued functions is both intrinsically and extrinsically interesting. The basic results follow from our generic results in Section 4, but beyond that the various kinds require specific attention. To some extent proof concepts can be extended from the traditional setting of many-one reductions between functions, but beyond that novel techniques are called for.

Acknowledgements. This paper is based on the second chapter of the authors thesis [18]. An extended version including proofs is available at the arXiv, see [17].

References

1. Ambos-Spies, K.: Minimal pairs for polynomial time reducibilities. In: Börger, E. (ed.) Computation Theory and Logic. LNCS, vol. 270, pp. 1–13. Springer, Heidelberg (1987)
2. Brattka, V., de Brecht, M., Pauly, A.: Closed choice and a uniform low basis theorem. Annals of Pure and Applied Logic (2012)
3. Brattka, V., Gherardi, G.: Effective choice and boundedness principles in computable analysis. Journal of Symbolic Logic 76, 143–176 (2011) arXiv:0905.4685
4. Brattka, V., Gherardi, G.: Weihrauch degrees, omniscience principles and weak computability. Bulletin of Symbolic Logic 17, 73–117 (2011) arXiv:0905.4679
5. Brattka, V., Gherardi, G., Marcone, A.: The Bolzano-Weierstrass Theorem is the jump of Weak König's Lemma. Annals of Pure and Applied Logic 163(6), 623–655 (2012)
6. Brattka, V., le Roux, S., Pauly, A.: On the Computational Content of the Brouwer Fixed Point Theorem. In: Cooper, S.B., Dawar, A., Löwe, B. (eds.) CiE 2012. LNCS, vol. 7318, pp. 57–68. Springer, Heidelberg (2012)
7. Chen, X., Deng, X.: Settling the complexity of 2-player Nash-equilibrium. Tech. Rep. 134, Electronic Colloquium on Computational Complexity (2005)
8. Daskalakis, C., Papadimitriou, C.: Continuous local search. In: Proceedings of SODA (2011)
9. Di Paola, R., Heller, A.: Dominical categories: Recursion theory without elements. Journal of Symbolic Logic 52, 594–635 (1987)
10. Gherardi, G., Marcone, A.: How incomputable is the separable Hahn-Banach theorem? Notre Dame Journal of Formal Logic 50(4), 393–425 (2009)
11. Higuchi, K., Pauly, A.: The degree-structure of Weihrauch-reducibility. arXiv 1101.0112 (2011)
12. Johnson, D.S., Papadimtriou, C.H., Yannakakis, M.: How easy is local search? Journal of Computer and System Sciences 37(1), 79–100 (1988)
13. Ladner, R.E.: On the structure of polynomial time reducibility. Journal of the ACM 22(1), 155–171 (1975)
14. Papadimitriou, C.H.: On the complexity of the parity argument and other inefficient proofs of existence. Journal of Computer and Systems Science 48(3), 498–532 (1994)

15. Pauly, A.: How incomputable is finding Nash equilibria? Journal of Universal Computer Science 16(18), 2686–2710 (2010)
16. Pauly, A.: On the (semi)lattices induced by continuous reducibilities. Mathematical Logic Quarterly 56(5), 488–502 (2010)
17. Pauly, A.: Many-one reductions between search problems. arXiv 1102.3151 (2011)
18. Pauly, A.: Computable Metamathematics and its Application to Game Theory. Ph.D. thesis, University of Cambridge (2012)
19. Robinson, E., Rosolini, G.: Categories of partial maps. Information and Computation 79(2), 95–130 (1988)
20. Weihrauch, K.: The degrees of discontinuity of some translators between representations of the real numbers. Informatik Berichte 129, FernUniversität Hagen, Hagen (July 1992)
21. Weihrauch, K.: The TTE-interpretation of three hierarchies of omniscience principles. Informatik Berichte 130, FernUniversität Hagen, Hagen (September 1992)
22. Yates, C.E.M.: A minimal pair of recursively enumerable degrees. The Journal of Symbolic Logic 31(2), 159–168 (1966)

Relative Randomness
for Martin-Löf Random Sets

NingNing Peng, Kojiro Higuchi, Takeshi Yamazaki, and Kazuyuki Tanaka

Mathematical Institute, Tohoku University, Sendai 980-8578, Japan
{sa8m42,sa7m24,yamazaki,tanaka}@math.tohoku.ac.jp

Abstract. Let Γ be a set of functions on the natural numbers. We introduce a new randomness notion called semi Γ-randomness, which is associated with a Γ-indexed test. Fix a computable sequence $\{G_n\}_{n\in\omega}$ of all c.e. open sets. For any $f \in \Gamma$, $\{G_{f(n)}\}_{n\in\omega}$ is called a Γ-indexed test if $\mu(G_{f(n)}) \leq 2^{-n}$ for all n. We prove that weak n-randomness is strictly stronger than semi Δ_n^0-randomness, for $n > 2$. Moreover, we investigate the relationships between various definitions of randomness.

1 Introduction

The present paper is concerned with the algorithmic notion of randomness such as the one originally introduced by P. Martin-Löf [6] in 1966. A *Martin-Löf test*, or *ML-test* for short, is a uniformly c.e. sequence $(G_m)_{m\in\omega}$ of open sets such that $\mu(G_m) \leq 2^{-m}$ for all $m \in \omega$. A set $Z \subseteq \omega$ *fails* the test if $Z \in \bigcap_m G_m$, otherwise Z *passes* the test. A set Z is *Martin-Löf random* if Z passes each ML-test.

There are many different approaches to randomness, most of which define a notion stronger than Martin-Löf randomness. One approach is to generalize the Martin-Löf test by specifying the m-th component (a c.e. set of measure at most 2^{-m}) via a function in some function class Γ. For example, the notion of Demuth randomness, introduced and studied by Demuth [1,2], is defined in this way. Thus, a Demuth test is a sequence $\{G_n\}_{n\in\omega}$ of c.e open sets such that $G_n = W_{f(n)}$ for all $n \in \omega$ and $\mu(G_n) \leq 2^{-n}$ where f is ω-c.e..

A main purpose of this paper is to give a general framework for such randomness notions. To do this, we introduce the notion of semi Γ-randomness. Note that our definition of passing a test is different from Demuth randomness, where a set Z fails a Demuth test if Z is in infinitely many G_m's, but fails a Γ-indexed test if it is in all of the G_m, see Section 3.

Γ-randomness was first studied by the first author [10], and is strongly connected with Yu's research [11]. The Γ-randomness notion could sometimes produce alternative proofs of existing results. For instance, some properties of $\varnothing'$-Schnorr randomness are proved more easily by the characterization due to $\mathbb{L}$-randomness than the usual methods. In Section 2, we prove a new result about $\mathbb{L}$-randomness. Another motivation of studying semi Γ-randomness is to consider weak versions of Γ-randomness. Last of all, the main results of this paper

S.B. Cooper, A. Dawar, and B. Löwe (Eds.): CiE 2012, LNCS 7318, pp. 581–588, 2012.
© Springer-Verlag Berlin Heidelberg 2012

appear in section 3. We show that for $n > 2$, weak n-randomness is strictly stronger than semi Δ_n^0-randomness from a general property of semi Γ- randomness. Moreover, we prove that weak 2-randomness, which is clearly equivalent to semi Δ_2^0-randomness, is strictly stronger than semi $\mathbb{L}$-randomness.

2   Preliminaries

The collection of binary strings is regarded as the set $2^{<\omega}$ of all functions from $\{0, \ldots, n-1\}$ to $\{0,1\}$ for some $n \in \omega$. We use $\sigma, \tau, \cdots$ to denote the elements of $2^{<\omega}$. Let 2^{ω} denote the set of infinite binary sequences. Subsets of ω can be identified with elements of 2^{ω}.

For $\sigma \in 2^{<\omega}$, we write $\mathrm{lh}(\sigma)$ for the length of σ. The empty string is denoted by λ. For strings σ and τ, let $\sigma \preceq \tau$ denote that σ is a prefix string of τ, i.e., $\mathrm{dom}(\sigma) \subseteq \mathrm{dom}(\tau)$ and $\sigma(m) = \tau(m)$ holds for each $m \in \mathrm{dom}(\sigma)$. The concatenation of two strings σ and τ is denoted by $\sigma\tau$. For a set A, $A \upharpoonright n$ is the prefix of A of length n. A topology on 2^{ω} is induced by the basic open sets $[\sigma] = \{X \in 2^{\omega} : X \succeq \sigma\}$ for all strings $\sigma \in 2^{<\omega}$. So each open set of 2^{ω} is generated by a subset of $2^{<\omega}$, that is $[S]^{\preceq} = \{X \in 2^{\omega} : \exists \sigma \in S \, (\sigma \preceq X)\}$. With this topology, 2^{ω} is called *Cantor space*.

The *Lebesgue measure* μ on 2^{ω} is induced by letting the measure be $\mu([\sigma]) := 2^{-|\sigma|}$ for each basic open set $[\sigma]$. If a class $G \subseteq 2^{\omega}$ is open then $\mu(G) = \sum_{\sigma \in B} 2^{-|\sigma|}$ where B is a prefix-free set of strings such that $G = \bigcup_{\sigma \in B}[\sigma]$. A class $\mathcal{C} \subseteq 2^{\omega}$ is called *null* if $\mu(\mathcal{C}) = 0$. If $2^{\omega} - \mathcal{C}$ is null, we say that $\mathcal{C}$ is *conull*.

Martin-Löf randomness is a central notion of algorithmic randomness for subsets of ω, which defined in the following way.

Definition 1 (Martin-Löf [6])

    (*i*) *A Martin-Löf test, or ML-test for short, is a uniformly c.e. sequence* $\{G_m\}_{m\in\omega}$ *of open sets such that* $\mu(G_m) \leq 2^{-m}$ *for all* $m \in \omega$.
    (*ii*) *A set* $Z \subseteq \omega$ *fails the test if* $Z \in \bigcap_m G_m$, *otherwise* Z *passes the test.*
    (*iii*) Z *is Martin-Löf random, or ML-random, if* Z *passes each ML-test.*

In this way, we are presenting the Martin-Löf random sets as the sets that pass all reasonable statistical tests in the form of effectively presented null sets. The randomness notions which are stronger than Martin-Löf randomness have been studied. 2-randomness is Martin-Löf randomness relative to $\varnothing'$. It was first studied by Kurtz [4] in 1981. He also considered weak 2-randomness, an interesting notion lying strictly between Martin-Löf randomness and 2-randomness. A set is weakly random (or Kurtz random) if it is not a member of any null Π_1^0 class. Similarly, it is weakly 2-random if it is not a member of any null Π_2^0 class.

Definition 2 (Kurtz [4])

    (*i*) *A generalized ML-test is a uniformly c.e. sequence* $\{G_m\}_{m\in\omega}$ *of open sets such that* $\mu(\bigcap_m G_m) = 0$.
    (*ii*) Z *is weakly 2-random if it passes every generalized ML-test.*

Fact 1. (*i*) *2-randomness* ⇒ *weak 2-randomness* ⇒ *Martin-Löf randomness.* (*ii*) *The reverse implication fails (Kurtz, Kautz).*

We recall some notions from [10]. Let ω^ω be the set of all functions from ω to ω.

Definition 3. *Let $\Gamma \subset \omega^\omega$. A set Z is Γ-random if Z is ML-random relative to f for all $f \in \Gamma$. Any ML-test relative to $f \in \Gamma$ is called a Γ-test.*

For $f \in \omega^\omega$, we say f-random and f-test instead of $\{f\}$-random and $\{f\}$-test, respectively.

Theorem 1 (Miller/Yu [7]). *Let $X, Y, Z \in 2^\omega$. If X is ML-random, Y is Z-random and $X \leq_T Y$, then X is Z-random.*

Proof. See [7, Theorem 4.3].

Corollary 1. *Let X be ML-random and $\Gamma \subset \omega^\omega$. If $X \leq_T Y$ and Y is Γ-random, then X is Γ-random.*

Let $\mathbb{L}$ denote the set of all low functions, where a function is *low* if its Turing jump is Turing reducible to the Turing jump of a computable function. We investigate some properties of $\mathbb{L}$-randomness. Note that there exists a low ML-random set. Clearly, any low set is not $\mathbb{L}$-random. Thus, $\mathbb{L}$-randomness is strictly stronger than ML-randomness. In fact, it turned out that $\mathbb{L}$-randomness is equivalent to $\varnothing'$-Schnorr randomness.

Theorem 2 (Yu [11]). $\mathbb{L}$*-randomness is equivalent to $\varnothing'$-Schnorr randomness.*

Proof. See [11, Theorem 4.1].

$\mathbb{L}$-randomness can be also given by subsets of $\mathbb{L}$. Let $\mathbb{G}$ denote the set of all 1-generic functions, where a function f is called a *1-generic* if for any c.e. set $W \subset \omega^{<\omega}$, there exists $\sigma \prec f$ such that $\sigma \in W$ or σ is incomparable with any element in W. Modifying the proof of Theorem 2, one can show that $\mathbb{L} \cap \mathbb{G}$-randomness is equivalent to $\varnothing'$-Schnorr randomness.

We would like to introduce another characterization of $\mathbb{L}$-randomness. Then, the next lemma is useful. Let $f, g \in \omega^\omega$. We say that f is *LR-reducible* to g, abbreviated $f \leq_{\mathrm{LR}} g$, if g-randomness implies f-randomness.

Lemma 1. *Let $\Gamma, \Gamma' \subset \omega^\omega$ such that for any $f \in \Gamma$ there is a function $g \in \Gamma \cap \Gamma'$ with $f \leq_{\mathrm{LR}} g$. Then, Γ-randomness is equivalent to $\Gamma \cap \Gamma'$-randomness.*

Proof. It suffices to show that every set which is not Γ-random is not $\Gamma \cap \Gamma'$-random. Fix $A \in 2^\omega$ such that A is not Γ-random. There exists $f \in \Gamma$ such that A is not f-random. By the assumption, there exists $g \in \Gamma \cap \Gamma'$ such that $f \leq_{\mathrm{LR}} g$. By the definition of LR-reducibility, A is not g-random. Thus A is not $\Gamma \cap \Gamma'$-random.

Let PA denote the set of all functions of PA degrees.

Proposition 1. $\mathbb{L}$-*randomness is equivalent to* $\mathbb{L} \cap \text{PA}$-*randomness.*

Proof. This is obtained from lemma 1, using the fact that for any low f, there exists a low g such that $f \leq_T g$ and g is of PA degree relative to f by Low Basis Theorem.

Let MLR denote the set of all ML-random sets.

Conjecture 1. $\mathbb{L}$-*randomness is equivalent to* $\mathbb{L} \cap \text{MLR}$-*randomness.*

The proof of Proposition 1 does not apply to this Conjecture since any c.e. set which is not low for random cannot be computed by a low random.

3   Semi Γ-Randomness

In this section, we investigate a new randomness notion weaker than Γ-random. We concentrate on an index function for the componets of a ML-test.

Definition 4. *Let* $\Gamma \subseteq \omega^\omega$.

(i) *We say that a sequence* $\{G_n\}_{n \in \omega}$ *of c.e open sets is a* Γ-*indexed test if there exists* $f \in \Gamma$ *such that* $G_n = W_{f(n)}$ *for all* $n \in \omega$ *and* $\mu(G_n) \leq 2^{-n}$.
(ii) *A set fails the test if it is in* $\bigcap_n G_n$, *otherwise the set passes the test.*
(iii) *A set is semi* Γ-*random if it passes every* Γ-*indexed test.*

Note that, there is no semi ω^ω-random set. It is straightforward from the definition that Martin-Löf randomness is equivalent to semi Δ_1^0-randomness. In [5], independently with us, Kučera and Nies define a semi Δ_2^0-random set nevertheless they call it limit randomness.

Theorem 3. *A set is semi* Δ_2^0-*random if and only if it is weakly 2-random.*

Proof. Let $\{W_{f(n)}\}_{n \in \omega}$ be such that $\mu(\bigcap_{n \in \omega} W_{f(n)}) = 0$, where f is computable. Since $\mu(\bigcap_{m < l} W_{f(m)}) \leq 2^{-n}$ is Π_1^0, there is a function $g \leq_T \varnothing'$ such that $\mu(\bigcap_{m < g(n)} W_{f(m)}) \leq 2^{-n}$ for all $n \in \omega$. Also, there is a function $h \leq_T \varnothing'$ such that $W_{h(n)} = \bigcap_{m < g(n)} W_{f(m)}$ for any n. So, for any generalized ML-test $\{W_{f(n)}\}$, there exists a semi Δ_2^0-test $\{W_{h(n)}\}$ that passes the same subset of 2^ω.

Let $\{W_{f(n)}\}_{n \in \omega}$ be a Δ_2^0-indexed test with $f \leq_T \varnothing'$. Then $Z \in \bigcap_{n \in \omega} W_{f(n)}$ is Π_2^0. Therefore there exists uniformly c.e. sequences $\{U_n\}_{n \in \omega}$ such that $\bigcap_{n \in \omega} U_n = \bigcap_{n \in \omega} W_{f(n)}$. Since $\mu(\bigcap_{n \in \omega} W_{f(n)}) = 0$, $\{U_n\}_{n \in \omega}$ is a generalized ML-test sequence which passes the same subset of 2^ω.

Proposition 2. *Every* $\mathbb{L}$-*random set is semi* $\mathbb{L}$-*random.*

Proof. This follows from a simple fact that every semi $\mathbb{L}$-test is a $\mathbb{L}$-test.

Definition 5. *For* $f, g \in \omega^\omega$, g *is said to be a tail of* f *if there exists a string* σ *such that* $\sigma g = f$.

Theorem 4. *If $\Gamma \subset \omega^\omega$ is closed under $\equiv_T$, then there is no universal Γ-indexed test unless ML-randomness is equivalent to semi Γ-randomness.*

Proof. Let Γ be a subset of ω^ω closed under $\equiv_T$. Choose a ML-random set X which is not semi Γ-random. Note that no tail of X is semi Γ-random by the assumption on Γ. Now, fix a Γ-indexed test $\{W_{g(i)}\}_{i \in \omega}$. Since the complement of $W_{g(1)}$ in 2^ω has at least measure $1/2$, there exists a tail Y of X such that $Y \notin W_{g(1)}$. Thus, $\{W_{g(i)}\}_{i \in \omega}$ is not a universal Γ-indexed test since Y is not semi Γ random and $Y \notin \bigcap_{i \in \omega} W_{g(i)}$.

The following lemma is used for separating between weak randomness and semi Γ-randomness.

Lemma 2. *For any $\Gamma, \Gamma' \subset \omega^\omega$ such that Γ-randomness is not equivalent to ML-randomness and Γ' is countable, there exists a semi Γ'-random set which is not Γ-random.*

Proof. Choose a ML-random set $X \in \bigcap_{i \in \omega} V_i$, where $\{V_i\}_{i \in \omega}$ is a universal f-test for some $f \in \Gamma$. Let $\{\{U_{g_i(j)}\}_{j \in \omega}\}_{i \in \omega}$ be a sequence of all Γ'-indexed tests. We construct a function h and a $\subset$-increasing sequence $\{\sigma_i\}_{i \in \omega}$ such that $[\sigma_i] \subset V_i$ and $(\lim_{j \to \infty} \sigma_j) \notin U_{g_i(h(i))}$ for any $i \in \omega$.

We shall describe the construction of h and $\{\sigma_i\}_{i \in \omega}$ at stage s. Let $\sigma = \bigcup_{i < s} \sigma_i$. As the inductive hypothesis, we may suppose that $[\sigma] \setminus \bigcup_{i < s} U_{g_i(h(i))}$ has positive measure. Define $h(s)$ by the least number x such that $[\sigma] \setminus (\bigcup_{i < s} U_{g_i(h(i))} \cup U_{g_s(x)})$ has positive measure. Choose a tail Y of X such that

$$\sigma Y \in [\sigma] \setminus \bigcup_{i \leq s} U_{g_i(h(i))}$$

Note that $\sigma Y \in \bigcap_{i \in \omega} V_i$ since $X \in \bigcap_{i \in \omega} V_i$. Thus, we can choose $\sigma_s \supsetneq \sigma$ such that $\sigma_s \subset \sigma Y$ and $[\sigma_s] \subset V_s$. This completes the construction at stage s.

It is clear that h and $\{\sigma_i\}_{i \in \omega}$ have the desired properties. Hence $\lim_{i \to \infty} \sigma_i$ is a semi Γ'-random set which is not Γ-random.

Theorem 5. *If $n > 2$, then weak n-randomness is strictly stronger than semi Δ_n^0-randomness.*

Proof. By Lemma 2, there exists a semi Δ_n^0-random set which is not $n - 1$-random. Since weak n-randomness implies $n - 1$-randomness and semi Δ_n^0-randomness, weak n-randomness is strictly stronger than semi Δ_n^0-randomness.

Finally, we consider the case $n = 2$.

Theorem 6. *Let Γ be the set of all functions of Turing degrees below $\deg_T(\varnothing')$. There exists a $\Pi_1^0(\varnothing')$ null set P containing a semi Γ-random set.*

Proof. Let f be a $\varnothing'$-computable increasing function such that $f(0) > 0$ and $\{x\}(x) \downarrow$ implies $\{x\}_{f(x)}(x) \downarrow$ for all $x \in \omega$. Note that any function which dominates f computes $\varnothing'$. Thus no element of Γ dominates f. Define $F \subset \omega^{<\omega}$, $H : \omega \to \omega$ and $M : \omega \to \omega$ by

$$F = \{\sigma \in \omega^{<\omega} \mid (\forall i < \mathrm{lh}(\sigma))[\sigma(i) < f(i)]\},$$

$$H(x) = (\text{least } y)[f(x) \cdot 2^{-y} \leq 2^{-1}]$$

and $M(0) = 0$ and

$$M(x+1) = M(x) + H(x) + 1$$

for all $x \in \omega$. Let $\mathcal{P}_{\mathrm{fin}}(\omega)$ be the set of all finite subsets of ω. Define $O : F \to \mathcal{P}_{\mathrm{fin}}(\omega)$ by $O(\lambda) = \varnothing$ and

$$O(\sigma i) = \begin{cases} O(\sigma) \cup \{i\} & \text{if } \mu(W_i) \leq 2^{-M(\mathrm{lh}(\sigma))-1}, \\ O(\sigma) & \text{otherwise.} \end{cases}$$

Define $T : F \to 2^{<\omega}$ by recursively as follows: let $T(\lambda) = \lambda$; and let $T(\sigma i)$ be a finite binary string τ of length $(\mathrm{lh}(T(\sigma)) + H(\mathrm{lh}(\sigma)))$ such that $\tau \succeq T(\sigma)$ and

$$\mu([\tau] \setminus \bigcup_{j \in O(\sigma i)} W_j) \geq 2^{-M(\mathrm{lh}(\sigma i))}.$$

We show that T is total by induction on $\sigma \in F$. Clearly, $T(\lambda)$ is defined and $\mu([T(\lambda)] \setminus \bigcup_{j \in O(\lambda)} W_j) \geq 2^{-M(\mathrm{lh}(\lambda))}$ holds. Let $\sigma i \in F$. Suppose that $\mu([T(\sigma)] \setminus \bigcup_{j \in O(\sigma)} W_j) \geq 2^{-M(\mathrm{lh}(\sigma))}$ holds as an inductive hypothesis. We only consider the case that $\mu(W_i) \leq 2^{-M(\mathrm{lh}(\sigma))-1}$. In this case we have that $O(\sigma i) = O(\sigma) \cup \{i\}$ and, therefore,

$$\mu([T(\sigma)] \setminus \bigcup_{j \in O(\sigma i)} W_j) \geq 2^{-M(\mathrm{lh}(\sigma))-1}.$$

Since $M(\mathrm{lh}(\sigma i)) = M(\mathrm{lh}(\sigma)) + H(\mathrm{lh}(\sigma)) + 1$ holds by the definition of M, $T(\sigma i)$ is defined and satisfies that $\mu([T(\sigma i)] \setminus \bigcup_{j \in O(\sigma i)} W_j) \geq 2^{-M(\mathrm{lh}(\sigma i))}$.

Let $P = \{X \in 2^\omega \mid (\forall x)(\exists \sigma \in F \cap \omega^x)[T(\sigma) \subset X]\}$. It is easy to see that P is $\Pi_1^0(\varnothing')$ since f, F, H, M, O and T are $\varnothing'$-computable.

We show that P is null. Every $\sigma \in F$ has exactly $f(\mathrm{lh}(\sigma))$-many immediate successors in F. Thus, by the definition of H and T, we have

$$\mu([\{T(\sigma) \mid \sigma \in F \cap \omega^{x+1}\}]^{\preceq}) \leq 2^{-1} \cdot \mu([\{T(\sigma) \mid \sigma \in F \cap \omega^{x+1}\}]^{\preceq})$$

for all $x \in \omega$. Notice that $T(\sigma)$ has length $\sum_{k < \mathrm{lh}(\sigma)} H(k)$ by the definition of T, and $\{T(\sigma) \mid \sigma \in F \cap \omega^x\}$ is the set of all extendible elements of P of length $\sum_{k < x} H(k)$ for all $x \in \omega$. Hence P is null.

Next, we show that P has a semi Γ-random set X. For a path p through F, we define $T(p) = \bigcup_{x \in \omega} T(p \upharpoonright x)$ and $O(p) = \bigcup_{x \in \omega} O(p \upharpoonright x)$. Note that $T(p) \notin \bigcup_{i \in O(p)} W_i$ by the definition of T. Let $\{\{W_{g_i(j)}\}_{j \in \omega}\}_{i \in \omega}$ be a sequence of all Γ-indexed tests. By Padding lemma, we can safely assume that g_i is strictly increasing.

We shall construct a strict $\subset$-increasing sequence $\{\sigma_i\}_{i\in\omega}$ of elements of F such that for any $i \in \omega$, there exists $j \in \omega$ with $g_i(j) \in O(\sigma_{i+1})$. Let $\sigma_0 = \lambda$. Fix $i \in \omega$. Suppose that for any $i' < i$, there exists $j \in \omega$ with $g_{i'}(j) \in O(\sigma_i)$. We want to find $\sigma_{i+1} \supset \sigma_i$ in F such that $g_i(j) \in O(\sigma_{i+1})$ for some $j \in \omega$.

To find such σ_i and j, we first construct g_i-computable function $q : \omega \to \omega$ recursively. $q(0) = \max\{f(\mathrm{lh}(\sigma_i)), g_i(M(\mathrm{lh}(\sigma)) + 1)\}$. Suppose that $q(x)$ is defined. Let $y_0 = (\text{least } y)[q(x) \cdot 2^{-y} \le 2^{-1}]$ and let $y_1 = q(x) + y_0 + 1$. Define $q(x + 1) = g_i(y_1 + 1)$. Since $q <_T \varnothing'$, we can choose the least natural number $x \in \omega$ such that $q(x + 1) < f(\mathrm{lh}(\sigma_i) + x + 1)$. Let $\sigma_{i+1} = \sigma_i 0^x q(x + 1)$. By the choice of x, one can easily prove that

$$(\forall y \le x)[f(\mathrm{lh}(\sigma_i) + y), M(\mathrm{lh}(\sigma_i) + y) \le q(y)]$$

by induction on y. Choose $y_1 \in \omega$ such that $q(x + 1) = g_i(y_1 + 1)$. Then $\mu(W_{g_i(y_1+1)}) = \mu(W_{g_i(y_1+1)}) \le 2^{-g_i(y_1+1)}$ and $g_i(y_1 + 1) = q(x+1) < f(\mathrm{lh}(\sigma_i) + x + 1)$ hold. Thus, we have $M(\mathrm{lh}(\sigma_i) + x) \le q(x) < y_1 < g_i(y_1 + 1) = q(x + 1)$ by the definition of $q(x+1)$. Hence $\mu(W_{q(x+1)}) \le 2^{-M(\mathrm{lh}(\sigma_i)+x)}$ holds. Therefore $g_i(y_1 + 1) = q(x + 1) \in O(\sigma_{i+1})$.

Now it is clear that $p = \bigcup_{i\in\omega} T(\sigma_i)$ is an element of P such that p is not Γ-random.

Corollary 2. *Weak 2-randomness is strictly stronger than semi* $\mathbb{L}$*-randomness.*

Proof. By Theorem 3, we know that weak 2-randomness is stronger than semi $\mathbb{L}$-randomness. By Theorem 6, there exists a semi $\mathbb{L}$-random set which is not $\varnothing'$-Kurtz random. This implies that weak 2-randomness is strictly stronger than semi $\mathbb{L}$-randomness since any weak 2-randomness is $\varnothing'$-Kurtz random.

Proposition 3. *2-randomness does not imply semi* Δ_3^0*-randomness.*

Proof. Since there is a Δ_3^0 2-random set, we show that there is no Δ_3^0 semi Δ_3^0-random set. Let X be a Δ_3^0 set, then there is a Δ_3^0 function f such that $W_{f(n)} = \{X \restriction n\}$. Note that $W_{f(n)}$ is a set containing only $X \restriction n$ and $\mu(W_{f(n)}) \le 2^{-n}$. So, $\{W_{f(n)}\}_{n\in\omega}$ is a semi Δ_3^0-test. But $A \in \bigcap_{n\in\omega} W_{f(n)}$. Hence, A is not semi Δ_3^0-random.

4   Conclusions and Future Research

In this paper, we introduced semi Γ-randomness for a set of number functions. We showed that weak n-randomness is strictly stronger than semi Δ_n^0-randomness, if $n > 2$, although weak 2-randomness is equivalent to semi Δ_2^0-randomness. On the other hand, we showed that weak 2-randomness is strictly stronger than semi $\mathbb{L}$-randomness where $\mathbb{L}$ is the set of low sets. In the future literature, we shall investigate a characterization of semi $\mathbb{L}$-randomness in terms of Kolmogorov complexity or martingales.

Acknowledgments. The first author was partially supported by a Global COE Program and scientific research plan project of the education department of Shaanxi province (Grant: 2010JK697). The second author was partially supported by JSPS Research Fellowship for Young Scientists. This research was partially supported by KAKENHI 23340020.

We would like to thank Takayuki Kihara, Masahiro Kumabe, Steve Simpson and ChenGuang Liu for their valuable comments and discussions. We also express our heartfelt thanks to the anonymous reviewers for their constructive comments.

References

1. Demuth, O.: Some classes of arithmetical real numbers. Commentations Mathematicae Universitatis Carolinae 23(3), 453–465 (1982) (Russian)
2. Demuth, O.: Remarks on the structure of tt-degrees based on constructive measure theory. Commentations Mathematicae Universitatis Carolinae 29(2), 233–247 (1988)
3. Downey, R.G., Hirshfeldt, D.R.: Hirshfeldt: Algorithmic Randomness and Complexity. Springer, Berlin (2010)
4. Kurtz, S.A.: Randomness and genericity in the degrees of unsolvability. Ph.D. Dissertation, University of Illinois, Urbana (1981)
5. Kučera, A., Nies, A.: Demuth randomness and computational complexity. Annals of Pure and Applied Logic 162(7), 504–513 (2011)
6. Martin-Löf, P.: The definition of random sequences. Information and Control 9(6), 602–619 (1966)
7. Miller, J.S., Yu, L.: On initial segment complexity and degrees of randomness. Transactions of the American Mathematical Society 360(6), 3193–3210 (2008)
8. Nies, A.: Computability and Randomness. Oxford University Press (2009)
9. Peng, N.N.: Algorithmic randomness and lowness notions. Master Thesis, Mathematical Institute, Tohoku University, Sendai, Japan (2010)
10. Peng, N.N.: The notions between Martin Löf randomness and 2-randomness. RIMS Kôkyûroku (1792), 117–122 (2010)
11. Yu, L.: Characterizing strong randomness via Martin-Löf randomness. Annals of Pure and Applied Logic 163(3), 214–224 (2012)

On the Tarski-Lindenbaum Algebra of the Class of all Strongly Constructivizable Prime Models

Mikhail G. Peretyat'kin

Institute of Mathematics, 125 Pushkin Street, 050010 Almaty, Kazakhstan
m.g.peretyatkin@predicate-logic.org

Abstract. We study the class $P_{s.c}$ of all strongly constructivizable prime models of a finite rich signature σ. It is proven that the Tarski-Lindenbaum algebra $\mathcal{L}(P_{s.c})$ considered together with a Gödel numbering γ of the sentences is a Boolean Π_4^0-algebra whose computable ultrafilters form a dense set in the set of all ultrafilters; moreover, the numerated Boolean algebra $(\mathcal{L}(P_{s.c}), \gamma)$ is universal relative to the class of all Boolean Σ_3^0-algebras. This gives an important characterization of the Tarski-Lindenbaum algebra $\mathcal{L}(P_{s.c})$ of the semantic class $P_{s.c.}$.

Theories in first-order predicate logic with equality are considered. General concepts of model theory, algorithm theory, Boolean algebras, and constructive models can be found in Hodges [5], Rogers [9], Goncharov and Ershov [2], and Goncharov [1].

1 Preliminaries

A finite signature is called *rich*, if it contains at least one n-ary predicate or function symbol for $n > 1$, or two unary function symbols. The following notations are used: $FL(\sigma)$ is the set of all formulas of signature σ, $SL(\sigma)$ is the set of all sentences (i.e., closed formulas) of signature σ. In the work, we use a finite rich signature σ, and consider a fixed Gödel numbering Φ_i, $i \in \mathbb{N}$, of the set $SL(\sigma)$. A theory F is called *finitely axiomatizable* if it is defined by a finite set of axioms and its signature is finite. Generally, *incomplete* theories are considered.

Let T be a theory of signature σ. On the set of sentences $SL(\sigma)$, an equivalence relation $\sim_T$ is defined by the rule $\Phi \sim_T \Psi \Leftrightarrow T \vdash (\Phi \leftrightarrow \Psi)$. The logical connectives $\vee$, $\wedge$, and $\neg$ generate Boolean operations $\cup, \cap$, and $-$ on the quotient set $SL(\sigma)/\sim_T$; One can easily check that, these operations are well-defined on the $\sim_T$-classes. Thereby, we obtain an algebra of the form

$$L(T) = \big(SL(\sigma)/\sim_T \; ; \; \cup, \cap, -, \mathbf{0}, \mathbf{1}\big),$$

that, in fact, is a Boolean algebra. It is called the *Tarski-Lindenbaum algebra* of the theory T. By $\mathcal{L}(T)$, we denote the Tarski-Lindenbaum algebra $L(T)$ considered together with a Gödel numbering γ; thereby, the concept of a computable isomorphism is applicable to such objects. For a class of models M, we write briefly $\mathcal{L}(M)$ instead of $\mathcal{L}\big(\mathrm{Th}(M)\big)$.

S.B. Cooper, A. Dawar, and B. Löwe (Eds.): CiE 2012, LNCS 7318, pp. 589–598, 2012.

The set of all finite tuples α of the form $\alpha = \langle \alpha_0, \alpha_1, \ldots, \alpha_n \rangle$, $\alpha_i \in \{0, 1\}$, is denoted by $2^{<\omega}$. The empty tuple is denoted by $\varnothing$. The *canonical (Gödel) index* of a finite tuple of zeros and ones of the form $\varepsilon = \langle \varepsilon_0, \varepsilon_1, \ldots, \varepsilon_{n-1} \rangle$, $\varepsilon_i \in \{0, 1\}$, is the number $\mathrm{Nom}(\varepsilon) = 2^n + \varepsilon_0 2^{n-1} + \varepsilon_1 2^{n-2} + \ldots + \varepsilon_{n-1} - 1$. We often write shortly $\langle \varepsilon \rangle$ instead of $\mathrm{Nom}(\varepsilon)$. We consider Boolean algebras in signature $\sigma_{BA} = \{\cup, \cap, -, \mathbf{0}, \mathbf{1}\}$. Besides, we consider two following binary relations, which are first-order definable in the theory of Boolean algebras by formulas $a \subseteq b \Leftrightarrow (a \cap b = a)$, $a \supseteq b \Leftrightarrow (a \cap b = b)$. Let $\mathcal{B}$ be a Boolean algebra, and $a \in \mathcal{B}$. By $\mathcal{B}[a]$, we denote the restriction of the Boolean algebra $\mathcal{B}$ on the set of all subelements of the element $a \in \mathcal{B}$ counting that $\mathbf{1} = a$ and $-x$ is defined as $a \setminus x$ in $\mathcal{B}[a]$. If b is an element of a Boolean algebra and $\alpha \in \{0, 1\}$, then b^α means b for $\alpha = 1$ and $-b$ for $\alpha = 0$. Similarly, if $\varPhi$ is a formula and $\alpha \in \{0, 1\}$, then $\varPhi^\alpha$ means $\varPhi$ for $\alpha = 1$ and $\neg \varPhi$ for $\alpha = 0$. We use notation $\mathcal{P}(A)$ for the power-set of A, and $\mathrm{Card}(A)$ for cardinality of the set A.

Definition of the concept of a binary tree can be found in [7, Sec, 2.1]. In the work, we use a more specialized term *compact binary trees* for them. An element n of a compact binary tree $\mathcal{D}$ such that $L(n) \not\subseteq \mathcal{D}$ is called a *dead end* of the tree $\mathcal{D}$. The set of all dead-end elements of a tree $\mathcal{D}$ is denoted by $\mathrm{Dend}(\mathcal{D})$. A tree is called *atomic* if above each of its elements, there is at least one dead-end element. A *chain* is a set $\pi \subseteq \mathbb{N}$ for which two following conditions are satisfied:

$$m \preccurlyeq n \wedge n \in \pi \Rightarrow m \in \pi, \ \textit{for all } m, n \in \mathbb{N},$$
$$m, n \in \pi \Rightarrow m \preccurlyeq n \vee n \preccurlyeq m, \ \textit{for all } m, n \in \mathbb{N}.$$

If $\mathcal{D}$ is a compact binary tree, we denote by $\varPi(\mathcal{D})$ the set of all maximal chains, while $\varPi^{\mathrm{fin}}(\mathcal{D})$ denotes the set of all maximal finite chains of the tree $\mathcal{D}$. We also use the following technical notations: by $\mathfrak{P}_n$, we denote the table condition with the Gödel number n, $n \in \mathbb{N}$; $A \models \mathfrak{P}_i$ means that the table condition is satisfied in the set A, $A \subseteq \mathbb{N}$;

$$\varOmega(m) = \{A \subseteq \mathbb{N} \mid (\forall i \in W_m) \, A \models \mathfrak{P}_i\},$$

is the parametric Stone space with index m; $\mathcal{D}_n$ is the closure of the set W_n up to a compact binary tree; and $\mathcal{D}_n^X$ is the closure of the set W_n^X up to a compact binary tree, where W_n denotes the computably enumerable set with c.e index n, while W_n^X denotes the computably enumerable set with c.e. index n relative to computability with an oracle $X \subseteq \mathbb{N}$, cf. [9]. It can be easily checked that the tree $\mathcal{D}_n$ is computably enumerable, while the tree $\mathcal{D}_n^X$ is computably enumerable with oracle X; moreover, each c.e. tree is presented in the sequence $\mathcal{D}_n$, $n \in \mathbb{N}$, and each c.e. tree in computation with oracle X is presented in the sequence $\mathcal{D}_n^X$, $n \in \mathbb{N}$. In accordance with [7, Sec.2.3], the number n plays the role of a c.e. index for the tree $\mathcal{D}_n$; furthermore, n plays the role of a c.e. index for the tree $\mathcal{D}_n^X$ in computability with an oracle X.

Finally, mention that, the main statement of the *canonical construction* of finitely axiomatizable theories is found in [7, Theorem 3.1.1.] (the old title "intermediate construction" is considered as obsolete).

2   Main Claim

Hereafter, we fix a finite rich signature σ. We denote by $P(\sigma)$ the class of all prime models of signature σ, and by $SC(\sigma)$, the class of all strongly constructivizable models of signature σ. Intersection of these classes $P_{s.c}(\sigma) = P(\sigma) \cap SC(\sigma)$ is the main object of our further study.

Theorem 1. *The following assertions hold:*

(a) $\mathcal{L}(P_{s.c}(\sigma))$ *is a Boolean* Π_4^0-*algebra,*
(b) computable ultrafilters of $\mathcal{L}(P_{s.c}(\sigma))$ *represent a dense set among arbitrary ultrafilters in the algebra,*
(c) for an arbitrary Boolean Σ_3^0-*algebra* $(\mathcal{B}, \nu)$ *there is a sentence* Φ *of signature* σ, *such that* $(\mathcal{B}, \nu) \cong (\mathcal{L}(\mathrm{Th}(\mathrm{Mod}(\Phi) \cap P_{s.c}(\sigma))), \gamma)$, *where* γ *is a Gödel numbering of the sentences of signature* σ.

Proof. By criterion of Goncharov and Harrington [3,4], a prime model $\mathfrak{N}$ of a complete decidable theory T is strongly constructivizable if and only if the family of principal types realized in $\mathfrak{N}$ is computable. From this we obtain that a sentence Ψ has a strongly constructivizable prime model if and only if there is a complete decidable theory T whose principal types are dense in the family of all types, and the family of its principal types is computable; moreover, $\Psi \in T$. An immediate calculation gives the prefix $\exists\forall\exists\forall$ for this condition. Finally, sentences Φ and Ψ are equivalent on the class $P_{s.c}(\sigma)$ of all strongly constructivizable prime models if and only if $(\Phi \wedge \neg\Psi) \vee (\Psi \wedge \neg\Phi)$ does not have a model in this class. This gives the required prefix $\forall\exists\forall\exists$ for (a).

Let T be an arbitrary complete theory extending $\mathrm{Th}(P_{s.c}(\sigma))$, and $\Psi \in T$. Obviously, Ψ has a model $\mathfrak{N} \in P_{s.c}(\sigma)$. From this we have that complete decidable theory $T' = \mathrm{Th}(\mathfrak{N})$ presenting a computable ultrafilter in $\mathrm{St}(\mathrm{Th}(P_{s.c}(\sigma)))$, is found in the neighborhood Ψ of the given arbitrary ultrafilter T of this Stone space. This gives the required density property posed in (b).

Proof of Part (c) is considered in Sections 3–4.

3   A Technical Statement

In this section, we use the concept of a compact binary tree (cf. preliminaries).

Lemma 2. *There exists a total computable functional* $\Theta : \mathcal{P}(\mathbb{N}) \to \mathcal{P}(\mathbb{N})$, *that satisfies the following properties:*
(a) $\mathcal{D} = \Theta(A)$ *is an atomic tree, for all* $A \subseteq \mathbb{N}$,
(b) $\mathcal{D} = \Theta(A)$ *is a computably enumerable tree, if the set* $A \subseteq \mathbb{N}$ *is computable,*
(c) for any computable $A \subseteq \mathbb{N}$, *in the tree* $\mathcal{D} = \Theta(A)$, *the family* $\Pi^{fin}(\mathcal{D})$ *is computable if and only if* $(\forall k)\left[k \in A \Rightarrow W_k \text{ is finite}\right]$.

Proof. We describe an effective process of constructing a computably enumerable tree $\mathcal{D}^{(A)}$, which depends on an input parameter $A \subseteq \mathbb{N}$. Following Rogers [9], we use notation W_n for nth computably enumerable set in Post's numbering of the family of all c.e. sets. Denote by W_n^t a finite part of the set W_n that can be computed in t steps. Introduce the following function dependent on A together with a pair of integer parameters:

$$h(A, k, t) = \text{Card}\big(\textstyle\bigcup_{i \in A \cap \{0,\dots,k\}} W_i^t \big).$$

It is obvious that for any $A \subseteq \mathbb{N}$ the function $h(A, k, t)$ is monotone in both k and t, and the following property holds for an arbitrary $A \subseteq \mathbb{N}$:

$$(\exists^\omega k)\big[\lim_{t \to \infty} h(A, k, t) = \infty \big] \Leftrightarrow (\exists j \in A)[\, W_j \text{ is infinite}\,]. \qquad (3.1)$$

There exists a strong sequence of finite sets $\pi_{n,i}^t$ such that
(1) $\pi_{n,i}^t$ is a chain for all n, i, t (cf. preliminaries),
(2) $\pi_{n,i}^t \subseteq \pi_{n,i}^{t+1}$ for all n, i, t,
(3) any computable family of chains coincides with one of the families of the form $\Pi_n = \{\pi_{n,i} | i \in \mathbb{N}\}$, $n \in \mathbb{N}$, where $\pi_{n,i} = \cup\{\pi_{n,i}^t | t \in \mathbb{N}\}$.

We now pass to construction of the tree $\mathcal{D}^{(A)}$. For the sake of brevity, we denote it by $\mathcal{D}$. By $\mathcal{D}^t$ we denote a part of $\mathcal{D}$, which is already constructed in the moment t. We shall use the markers $\square_n^k$, $n < k$. At some moment, the marker $\square_n^k$ may be placed on a chain of family Π_n passing through the element:

$$q_k = 2^{k+2} - 3 = L \underbrace{RR...R}_{k}(0), \qquad (3.2)$$

cf. [7, Fig.2.1.2(c)]. After that the marker is not moved. Some of the markers may be unused. At each step, we consider a pair (n, k), $n < k$; moreover, each such pair should be considered infinitely many times.

Construction.

Step $t = 0$. Let us set

$$\mathcal{D}^0 = \{0\} \cup \{1, 2, 5, 6, \dots, 2^{k+1} - 3, 2^{k+1} - 2, \dots; k \in \mathbb{N}\}.$$

The set $\mathcal{D}^0$ is an infinite tree having the last right chain, as in [7, Fig. 2.1.2(c)]. None of the markers are used at the initial moment.

Step $t > 0$. We consider a pair (n, k).

Case 1: the marker $\square_n^k$, is not yet placed. In this case, we verify whether there exists $j < t$ such that $q_k \in \pi_{n,j}^t$. In the case of a positive answer, we put the marker $\square_n^k$ on one of the chains $\pi_{n,j}$ subject to this condition. In addition, we set $\mathcal{D}^t = \mathcal{D}^{t-1}$.

Case 2: the marker $\square_n^k$ is already placed, and it marks the chain $\pi_{n,j}$. Let e be a maximal element of the set $\pi_{n,j}^t \cap \mathcal{D}^{t-1}$. We put

$$\mathcal{D}^t = \begin{cases} \mathcal{D}^{t-1} \cup \{L(e), R(e)\}, & \text{if } e < h(A, k, t), \\ \mathcal{D}^{t-1} & \text{otherwise.} \end{cases}$$

The step t is complete.

Assume that π is an infinite chain of the tree $\mathcal{D} = \cup\{\mathcal{D}^t | t \in \mathbb{N}\}$. We also assume that π is not the last right chain. Then, π passes through one of the elements (3.2). By construction, π must coincide with some chain $\pi_{n,j}$ marked by one of the markers $\square_n^k$. Moreover, it is necessary that the condition $\lim_{t\to\infty} h(A, k, t) = \infty$ is satisfied. The above reasoning implies that at most k different infinite chains can pass through the element q_k, but only in the case of growth of the function $h(A, k, t)$ as t increases. Moreover, all these chains are computably enumerable. Consequently, they are computable. Thus, in any case, the tree $\mathcal{D}$ is superatomic; this implies that the target tree $\mathcal{D}$ is atomic.

We now prove the statement in Part (c) of Lemma 2.

First, suppose that $(\forall j)[j \in A \Rightarrow W_j \text{ is finite}]$. By (3.1), we have

$$(\exists k_0)(\forall k > k_0) \lim_{t\to\infty} h(A, k, t) < \infty.$$

In this case, the tree $\mathcal{D}$ can contain only a finite set of infinite chains, and each of these chains is computable. Hence, the family $\varPi^{\text{fin}}(\mathcal{D})$ is computable.

Now, suppose that $(\exists j \in A)[W_j \text{ is infinite}]$. Consider a family $\varPi_n$ containing only finite chains. It is required to prove that $\varPi_n \neq \varPi^{\text{fin}}(\mathcal{D})$. To this end, we take $k > n$ such that

$$\lim_{t\to\infty} h(A, k, t) = \infty.$$

In the case when the marker $\square_n^k$ is not used, we obtain that none of the chains of the family $\varPi_n$ passes through q_k; therefore, $\varPi_n$ cannot coincide with the family of all finite chains of the tree $\mathcal{D}$. Consider another case when the marker $\square_n^k$ is placed on $\pi_{n,j} \in \varPi_n$ at some moment. Let e be its maximal element. Since the function $(\lambda t) h(A, k, t)$ increases unboundedly, the elements $L(e)$ and $R(e)$ will be included in $\mathcal{D}^t$. As a result, the chain $\pi_{n,j}$ is not a maximal chain of the tree $\mathcal{D} = \mathcal{D}^{(A)}$. In this case, we also obtain $\varPi_n \neq \varPi^{\text{fin}}(\mathcal{D})$.

These relations yield the necessary statement for Part (c) of Lemma 2

4   Proof to Part (c) of Theorem 1

Given a numerated Boolean Σ_3^0-algebra $(\mathcal{B}, \nu)$. By [8, Theorem 2.1], there is a numeration ν' of $\mathcal{B}$ such that $(\mathcal{B}, \nu')$ is a Boolean Σ_3^0-algebra whose computable ultrafilters represent a dense set in the set of all ultrafilters. For the sake of simplicity, we shall assume that the source numerated algebra $(\mathcal{B}, \nu)$ has such properties, that is:

$$\text{computable ultrafilters of } (\mathcal{B}, \nu) \text{ form a dense set in } \text{St}(\mathcal{B}). \qquad (4.0)$$

We assume, that $\mathcal{B}$ is a nontrivial algebra. By definition, signature operations $\cup$, $\cap$ and $-$ in $\mathcal{B}$ are presentable by computable functions on ν-numbers, and the equality relation is a Σ_3^0-relation in numeration ν:

$$\nu(x) = \nu(y) \iff H(x, y), \quad H \in \Sigma_3^0.$$

Consequently, there exists a unary relation H^* in this class Σ_3^0 such that for any finite tuple of zeros and ones $\alpha = \langle \alpha_0, \alpha_1, \ldots, \alpha_n \rangle$, we have

$$\nu(0)^{\alpha_0} \cap \nu(1)^{\alpha_1} \cap \ldots \cap \nu(n)^{\alpha_n} = \mathbf{0} \iff \langle \alpha_0, \alpha_1, \ldots, \alpha_n \rangle \in H^*, \quad H^* \in \Sigma_3^0.$$

For our construction, we shall use the following m-complete in class Σ_3^0 set, cf. [9, Sec. 14.8, Th. XV]:

$$E_3 = \{ n \mid (\exists k) \, [\, k \in W_n \wedge W_k \text{ is infinite} \,] \, \}. \tag{4.1}$$

Therefore, its complement is a set that is m-complete in class Π_3^0:

$$A_3 = \{ n \mid (\forall k) \, [\, k \in W_n \Rightarrow W_k \text{ is finite} \,] \, \}. \tag{4.2}$$

Particularly, we have $n \in A_3 \iff n \notin E_3$ for all $n \in \mathbb{N}$.

Since any Σ_3^0-set is m-reducible to E_3, there is a general computable function $f(x)$ such that for an arbitrary tuple $\alpha \in 2^{<\omega}$, $\alpha = \langle \alpha_0, \ldots, \alpha_n \rangle$, we have

$$\nu(0)^{\alpha_0} \cap \nu(1)^{\alpha_1} \cap \ldots \cap \nu(n)^{\alpha_n} = \mathbf{0} \iff (\exists k) \, [\, k \in W_{f(\alpha)} \wedge W_k \text{ is infinite} \,], \tag{4.3}$$

or equivalently,

$$\nu(0)^{\alpha_0} \cap \nu(1)^{\alpha_1} \cap \ldots \cap \nu(n)^{\alpha_n} \neq \mathbf{0} \iff (\forall k) \, [\, k \in W_{f(\alpha)} \Rightarrow W_k \text{ is finite} \,]. \tag{4.4}$$

Lemma 3. *The following assertions are true:*

(a) $f(\varnothing) \in A_3$,
(b) for any α in $2^{<\omega}$, $f(\alpha) \in A_3 \iff f(\alpha 0) \in A_3$ or $f(\alpha 1) \in A_3$,
(c) for any α in $2^{<\omega}$, $f(\alpha) \in E_3 \iff f(\alpha 0) \in E_3$ and $f(\alpha 1) \in E_3$.

Proof. The property (a) is a consequence of the fact that algebra $\mathcal{B}$ is nontrivial. The property (b) follows from the definition of the function $f(x)$ and relation $H^*(x)$, representing an ideal in the free Boolean algebra. The property (c) is a corollary from (b).

Now, our goal is to choose some pair (m, s) of integer parameters.

Choice of m. We choose m such that $\Omega(m) = \mathcal{P}(\mathbb{N})$ (cf. preliminaries). For this purpose, it is enough to get m such that $W_m = \varnothing$.

Choice of s. For this purpose, we describe a computable functional Ψ from $\mathcal{P}(\mathbb{N})$ to $\mathcal{P}(\mathbb{N})$, actually, yielding compact binary trees.

Given a set $A \subseteq \mathbb{N}$. Let $\alpha = \langle \alpha_0, \alpha_1, \ldots, \alpha_k, \ldots \rangle$ be the characteristic sequence for A, i.e., the following is satisfied:

$$\alpha_k = \begin{cases} 1, & \text{if } k \in A, \\ 0, & \text{if } k \notin A. \end{cases} \tag{4.5}$$

Taking A as an input parameter, let us construct the following set

$$Q_A = W_{f(\varnothing)} \cup W_{f(\langle \alpha_0 \rangle)} \cup \ldots \cup W_{f(\langle \alpha_0, \ldots, \alpha_n \rangle)} \cup \ldots. \tag{4.6}$$

It can easily be checked that, in the case when A is computable, the set Q_A is computably enumerable. Use this set Q_A as an input parameter for the construction $X \mapsto \Theta(X)$ described in Lemma 2. As a result, we obtain a subset $\mathcal{D}$ of $\mathbb{N}$, which actually is a tree by Lemma 2 (a). Thereby, the transformation $A \mapsto \Psi(A)$ is presented by the following rule:

$$A \mapsto Q_A \mapsto \Theta(Q_A) = \mathcal{D} = \Psi(A). \tag{4.7}$$

End of description of the operator Ψ.

It can easily be checked that the described transformation $A \mapsto \Psi(A)$ is realized by an algorithm using the set $A \subseteq \mathbb{N}$ as its input parameter. Thereby, the transformation $A \mapsto \Psi(A)$ can be considered as a computation by an algorithm $\mathcal{M}$ with an oracle A. Let s be a Gödel number of the algorithm $\mathcal{M}$. In accordance with the basic definitions of the theory of algorithms, we obtain the following form of the operator Ψ:

$$\Psi(A) = \mathcal{D}_s^A. \tag{4.8}$$

Choice of the pair of parameters (m, s) is finished.

Study main properties of the transformation $\Psi : A \mapsto \mathcal{D}_s^A$.

Lemma 4. *The following assertions hold:*

(a) For any $A \subseteq \mathbb{N}$, $\mathcal{D}_s^A$ is an atomic tree,
(b) For any computable $A \subseteq \mathbb{N}$, $\mathcal{D}_s^A$ is a computably enumerable tree.

Proof. Statement of Part (a) is provided by Lemma 2 (a), while Part (b) is provided by Lemma 2 (b).

Now we immediately pass to the proof of Part (c) of Theorem 1.

First of all, we have to point out a sentence Φ of the given finite rich signature σ, as it is stated in Part (c) of Theorem 1. For this, we use the canonical construction, cf. [7, Ch. 3, Th. 3.1.1]. Apply this construction to the pair (m, s) specifying also signature σ. As a result, we obtain a finitely axiomatizable theory $F = \mathbb{F}\mathbb{C}(m, s, \sigma)$ of signature σ. As Φ, we get a conjunction of axioms of the theory F. After that, our principal aim is to show that the sentence Φ satisfies all the requirements listed in Theorem 1 (c).

By the main statement of the canonical construction, there exists an effective sequence of sentences θ_n, $n \in \mathbb{N}$, of signature σ such that the family of extensions of F defined for each $A \subseteq \mathbb{N}$ by

$$F[A] = F \cup \{\theta_i | i \in A\} \cup \{\neg \theta_j | j \in \mathbb{N} \setminus A\}, \tag{4.9}$$

satisfies the following properties:

C1 *For any $A \subseteq \mathbb{N}$, the theory $F[A]$ is either complete or contradictory;*
C2 *The theory $F[A]$, $A \subseteq \mathbb{N}$, is consistent if and only if $A \in \Omega(m)$;*
C3 *For an arbitrary $A \in \Omega(m)$, the following statements are true:*

  (a) *theory $F[A]$ has a prime model if and only if the tree $\mathcal{D}_s^A$ is atomic,*

(b)   *a prime model of the theory $F[A]$, if it exists, is strongly constructivizable if and only if the set A is computable and the family of chains $\Pi^{fin}(\mathcal{D}_s^A)$ is computable.*

Consider an arbitrary finite tuple of zeros and ones $\alpha = \langle \alpha_0, ..., \alpha_k \rangle$. Construct an elementary intersection of the elements in $\mathcal{B}$ by the rule

$$b_\alpha = \nu(0)^{\alpha_0} \cap \nu(1)^{\alpha_1} \cap ... \cap \nu(k)^{\alpha_k}, \qquad (4.10)$$

as well as an elementary conjunction of corresponding sentences by the rule

$$\beta_\alpha = \theta_0^{\alpha_0} \wedge \theta_1^{\alpha_1} \wedge ... \wedge \theta_k^{\alpha_k} . \qquad (4.11)$$

The main idea of the construction is to provide the following relation:

Lemma 5. *For any tuple $\alpha \in 2^{<\omega}$, $b_\alpha \neq \mathbf{0}$ if and only if $\Phi \wedge \beta_\alpha$ has a strongly constructivizable prime model.*

Proof. Assume that $b_\alpha \neq \mathbf{0}$. By (4.0), computable ultrafilters form a dense set among arbitrary ultrafilters in the Boolean algebra $(\mathcal{B}, \nu)$. From this we obtain that there is an infinite sequence $\alpha^* = \langle \alpha_i \mid i < \omega \rangle$ extending α such that the set A related to α^* by (4.5) is computable, and

$$\nu(0)^{\alpha_0} \cap ... \cap \nu(i)^{\alpha_i} \neq \mathbf{0}, \quad \text{for all } i \in \mathbb{N}. \qquad (4.12)$$

By (4.4), we obtain that each set $W_{f(\langle \alpha_0, ..., \alpha_i \rangle)}$, $i \in \mathbb{N}$, contains only indices of finite sets; therefore, the set (4.6) also satisfies this property. Thereby, by Lemma 2, the tree $\Psi(A) = \mathcal{D}_s^A$ defined in (4.7) and (4.8) is an atomic computable tree with computable family $\Pi^{fin}(\mathcal{D}_s^A)$. By C1, C2, and C3(a,b), theory $F[A]$ is consistent, complete, and has a prime model $\mathfrak{N}$, which is strongly constructivizable. This ensures that the formula $\Phi \wedge \beta_\alpha$ is satisfied in the strongly constructivizable model $\mathfrak{N}$ since it is provable from $F[A]$.

Now, we assume that the sentence $\Phi \wedge \beta_\alpha$ has strongly constructivizable prime model $\mathfrak{N}$. Consider the set

$$A = \{\theta_i \mid \mathfrak{N} \models \theta_i\}, \qquad (4.13)$$

which is obviously computable. Build an infinite sequence $\alpha^* = \langle \alpha_i \mid i < \omega \rangle$ related to A by (4.5). Since $A \in \Omega(m)$, by C1 and C2, the theory $F[A]$ is consistent and complete. Moreover, this theory is decidable by Janiczak theorem since it is computably axiomatizable. By (4.13), all axioms of $F[A]$ are satisfied on the model $\mathfrak{N}$. Thereby, we have that A is computable and $F[A]$ has a strongly constructivizable prime model. By C3(a,b), we conclude that the tree $\mathcal{D}_s^A$ is atomic and the family $\Pi^{fin}(\mathcal{D}_s^A)$ is computable. In accordance with rules (4.7) and (4.8), we have $\mathcal{D}_s^A = \Theta(Q_A)$, where Q_A is defined by (4.6). By Lemma 2, Q_A must contain only indices of finite sets. Applying (4.4), we finally obtain $b_\alpha \neq \mathbf{0}$.

Lemma 5 is proven.

Let us map elements $\nu(i)$, $i \in \mathbb{N}$, of Boolean algebra $\mathcal{B}$ to sentences θ_i, $i \in \mathbb{N}$, of signature σ by the rule:

$$\lambda^*(\nu(k)) = \theta_k, \quad k \in \mathbb{N}. \tag{4.14}$$

Now, we shall extend the partial mapping (4.14) up to a computable isomorphism of the algebras under consideration. Define a mapping

$$\lambda : \mathcal{B} \to \mathcal{L}\big(\mathrm{Th}\big(\mathrm{Mod}(\Phi) \cap P_{s.c}(\sigma)\big)\big)$$

by the following rule: for an arbitrary finite sequence of finite binary tuples

$$\alpha_0, \alpha_1, \ldots \alpha_n \in 2^{<\omega},$$

we put

$$\lambda(b_{\alpha_0} \cup b_{\alpha_1} \cup \ldots \cup b_{\alpha_n}) = \beta_{\alpha_0} \cup \beta_{\alpha_1} \cup \ldots \cup \beta_{\alpha_n}. \tag{4.15}$$

The mapping λ is defined on all elements of $\mathcal{B}$ since the set of expressions involved in (4.15) embraces all elements of this algebra; moreover, μ is a homomorphic embedding of $\mathcal{B}$ into $\mathcal{L}\big(\mathrm{Th}(\mathrm{Mod}(\Phi) \cap P_{s.c}(\sigma))\big)$. Based on notation (4.9) together with property C1 for the canonical construction, by [7, Lem.0.3.2], we obtain that the set $\{\theta_i \mid i \in \mathbb{N}\}$ is generating in $\mathcal{L}(\Phi)$, and thus, in the Boolean algebra $\mathcal{L}\big(\mathrm{Th}(\mathrm{Mod}(\Phi) \cap P_{s.c}(\sigma))\big)$, ensuring that μ is a homomorphism "onto". Taking into consideration the principal property stated in Lemma 5, we obtain that μ is a one-to-one mapping. At last, Boolean operations above unions of elements b_α of the form (4.10) and disjunctions of elementary conjunctions β_α of the form (4.11), after omitting of zero terms, are produced by same rules. Therefore, λ is an isomorphism. Obviously, it is computable in numerations ν and γ obtaining finally the required computable isomorphism

$$\lambda : (\mathcal{B}, \nu) \to \Big(\mathcal{L}\big(\mathrm{Th}(\mathrm{Mod}(\Phi) \cap P_{s.c}(\sigma))\big), \gamma\Big).$$

Thereby, Part (a) of Theorem 1 is completely proven.

5   Final Remarks

The statement of Theorem 1 characterizes the Tarski-Lindenbaum algebra of the class of all strongly constructivizable prime models. It demonstrates a possibility to include "computable Brute Force" into a finitely axiomatizable theory. As a complement to Theorem 1, we can point out an earlier result, [7, Th.8.2.5(c)], that elementary theory of the class of all strongly constructivizable prime models is a Π_4^0-complete set. As it is proven in [6, Sec.2, Th.1, Th.2], any Boolean Π_4^0-algebra admits a Σ_3^0-numeration. Thus, Theorem 1 represents a significant characterization of the Tarski-Lindenbaum algebra $\mathcal{L}(P_{s.c.}(\sigma))$.

References

1. Goncharov, S.S.: Countable Boolean Algebras and Decidability. Plenum, New York (1997)
2. Goncharov, S.S., Ershov, Y.L.: Constructive models. Plenum, New York (1999)
3. Goncharov, S.S., Nurtazin, A.T.: Constructive models of complete decidable theories. Algebra Logika 12(2), 67–77 (1973)
4. Harrington, L.: Recursively presented prime models. J. Symbolic Logic 39(2), 305–309 (1974)
5. Hodges, W.: A shorter model theory. Cambridge University Press, Cambridge (1997)
6. Odintsov, S.P., Selivanov, V.L.: Arithmetical hierarchy and ideals of numerated Boolean algebras. Siberian Math. Journal 30(6), 140–149 (1989) (Russian)
7. Peretyat'kin, M.G.: Finitely axiomatizable theories. Plenum, New York (1997)
8. Peretyat'kin, M.G.: On the numerated Boolean algebras with a dense set of computable ultrafilters. Siberian Electronic Mathematical Reports (SEMR), 6 pp. (in publication, 2012)
9. Rogers, H.J.: Theory of Recursive Functions and Effective Computability. McGraw-Hill Book Co., New York (1967)

Lower Bound on Weights
of Large Degree Threshold Functions

Vladimir V. Podolskii

Steklov Mathematical Institute, Gubkina str. 8, 119991, Moscow, Russia
podolskii@mi.ras.ru

Abstract. An integer polynomial p of n variables is called a *threshold gate* for the Boolean function f of n variables if for all $x \in \{0,1\}^n$ $f(x) = 1$ if and only if $p(x) \geqslant 0$. The *weight* of a threshold gate is the sum of its absolute values.

In this paper we study how large weight might be needed if we fix some function and some threshold degree. We prove $2^{\Omega(2^{2n/5})}$ lower bound on this value. The best previous bound was $2^{\Omega(2^{n/8})}$ [12].

In addition we present substantially simpler proof of the weaker $2^{\Omega(2^{n/4})}$ lower bound. This proof is conceptually similar to other proofs of the bounds on weights of nonlinear threshold gates, but avoids a lot of technical details arising in other proofs. We hope that this proof will help to show the ideas behind the construction used to prove these lower bounds.

1   Introduction

Let $f\colon \{0,1\}^n \to \{0,1\}$ be a Boolean function. *A threshold gate* for the Boolean function f is an integer polynomial $p(x)$ of n variables $x = (x_1, \ldots, x_n)$ such that for any $x \in \{0,1\}^n$ we have $f(x) = 1$ if and only if $p(x) \geqslant 0$. In other words, for all $x \in \{0,1\}^n$ it is true that $f(x) = \operatorname{sgn} p(x)$, where we adopt the following definition of the sign function: $\operatorname{sgn}(t) = 1$ if $t \geqslant 0$ and $\operatorname{sgn}(t) = 0$ otherwise.

Thus, threshold gates are just representations of Boolean functions as the signs of polynomials. The formal study of such representations started in 1968 with the seminal monograph of Minsky and Papert [7]. Since then representations of this form found a lot of applications in circuit complexity, structural complexity, learning theory and communication complexity (see, for example [14,5,2,16]).

Two key complexity measures of threshold gates are the degree of the threshold gate and its weight. *The degree* $\deg p$ of a threshold gate p is just the degree of the polynomial. *The weight* $W(p)$ of a threshold gate p is the sum of absolute values of all its coefficients.

The complexity measures of a Boolean function f related to these complexity measures of threshold gates are the minimal threshold degree of a threshold gate for f which we denote by $\deg_{\pm} f$ and call *the threshold degree* and the minimal weight of a threshold gate for f. Both of these complexity measures play an important role in theoretical computer science (see the references above). In this

S.B. Cooper, A. Dawar, and B. Löwe (Eds.): CiE 2012, LNCS 7318, pp. 599–608, 2012.

paper we are interested in the minimal possible value of the weight of a threshold gate for some function f when the degree of the threshold gate is bounded. It is convenient to denote by $W(f,d)$ the minimal weight of a threshold gate of degree at most d for f. Note that this value is defined only if $d \geqslant \deg_\pm f$. It is also not hard to see that for all f we have $\deg_\pm f \leqslant n$ and $W(f,n) \leqslant 2^{O(n)}$ (just consider the polynomial p such that $p(x) = f(x)$ for all x).

The first results on the value of $W(f,d)$ were proven for $d = 1$. In [9] (see also [8] and [4]) it was proven that for all f with $\deg_\pm f = 1$ it is true that $W(f,1) = n^{O(n)}$. For a long time only lower bounds of the form $W(f,1) = 2^{\Omega(n)}$ were known (see [10] for one of early results). Tight lower bound was proven in [4], that is the function f with $\deg_\pm f = 1$ was constructed such that $W(f,1) = n^{\Omega(n)}$.

Concerning higher degree d, upper bound can be easily extended from the case $d = 1$. Namely for all f with $\deg_\pm f \leqslant d$ it is true (and easy to see) that $W(f,d) = n^{O(dn^d)}$ (see [15,3,12]). Note that this upper bound is much worse than for the case $d = 1$. As for the lower bound, the one which is better than for the linear case (and also tight for constant d) appeared in [12]. For any constant d the function f of threshold degree d was constructed there such that $W(f,d) = n^{\Omega(n^d)}$ (constant in Ω here depends on d). It is implicit in that paper though that the argument works for nonconstant d also and the resulting lower bound (with the dependence on d) is

$$\left(\frac{n}{d}\right)^{\frac{1}{2}(\frac{n}{2d})^d - o((\frac{n}{d})^d)}.\tag{1}$$

For this result another proof was given in [1].

Thus it turns out that the required weight grows with the growth of degree d. In this paper we are interested how large it might grow (note that for $d = n$ the weight is small again: $W(f,n) \leqslant 2^{O(n)}$). That is we study the value

$$W = \max_d \max_{f:\ \deg_\pm(f) \geqslant d} W(f,d).$$

The lower bound (1) works even for d linearly depending on n and so gives doubly exponential lower bound on this value. But it works only for $d \leqslant (n-c)/32$, where c is some constant, so the best lower bound we get from [12] is $2^{\Omega(2^{n/8})}$.

In this paper we prove the following bound.

Theorem 1. $W \geqslant 2^{\Omega(2^{2n/5})}$.

We note that the best upper bound known is simple $2^{O(n2^n)}$ (this can be deduced from the upper bound for the case $d = 1$ and the fact that there are at most 2^n monomials).

To prove our lower bound we adopt the strategy of [12] and provide a unified treatment of the argument of that paper. In short, the proof strategy is as follows. Starting from some function of threshold degree 1 (with some additional properties) that requires large weight when represented by degree-1 threshold gates we construct its "d-dimensional" generalization in a very specific way. For this generalization we are able to prove lower bounds for degree-d threshold gates and due to the specific features of our generalization we can prove strong lower bound.

In the paper [12] the construction of the function starts with the function constructed by Håstad in [4] to prove the optimal lower bound for the case $d = 1$. This helps to get n in the base of the exponent in the lower bound and thus to prove strong lower bound for the case of constant d. On the other hand, Håstad's function is very complicated and have desired properties only for large enough number of variables (16 variables). This does not allow us to prove lower bound for d close to n. In this paper we start with a much simpler functions having required properties starting from just 3 variables. With this function we cannot get n in the base of the exponent, but on the other hand we are able now to prove bounds for much larger d and thus get better lower bound on W.

We start exposition of our result by giving a simpler proof of the weaker bound of $2^{\Omega(2^{n/4})}$. In this proof we are able to avoid a lot of technical complications arising in the proof of [12] and make the function for which we prove the bound much simpler (here we use as a starting function of threshold degree 1 well known "greater than" function). We hope that this makes the proof easier to read and helps to show the ideas behind the construction which were not very clear in [12].

After that we define another starting function and explain how to change the proof to get $W \geqslant 2^{\Omega(2^{2n/5})}$ lower bound. The idea here is not only that we can prove the bound for larger d, but also that, roughly speaking, choosing the good function we can remove the constant 2 from the denominator of the term $\left(\frac{n}{2d}\right)$ in the exponent in the bound (1).

Besides representation of Boolean functions as $f \colon \{0,1\}^n \to \{0,1\}$, also representation of the form $f \colon \{-1,+1\}^n \to \{-1,+1\}$ turns out to be useful in complexity theory. For this representation we can also consider threshold gates and also define corresponding measures of the functions. Note that we can switch from one representation to another one by a simple linear transform. Thus the threshold degree of the function does not depend on the representation and the threshold weight may change only by 2^n multiplicative factor (see [6] for more information on relations between threshold weights in these two settings). Since this factor is very small compared to our lower bound, our result is true for both representations and in the proof we can choose the one of two presentations of Boolean functions which is more convenient to us. For the proof of weaker bound we shall use $\{0,1\}$ variables and for the stronger one — $\{-1,1\}$ variables.

We note in the conclusion that if we have the lower bound S on the minimal weight for the function of n variables and for degree d, it is easy to translate it to exactly the same bound S for $n' = n+c$ and $d' = d+c$ for any c, even depending on n (see, for example, [13], Corollary 1). This observation allows us to deduce strong lower bounds on weights of threshold gates of degree close to n.

Theorem 2. *For any $\epsilon > 0$ and $d \leqslant (1 - \epsilon)n$ there is an explicit function f such that $W(f,d) = 2^{2^{\Omega(n)}}$. For any $d \leqslant n - 2(1 + \epsilon) \log n$ there is an explicit function f such that $W(f,d) = 2^{\Omega(n^{1+\epsilon})}$.*

The rest of the paper is devoted to the formulation and the proof of our results. In Sections 2 and 3 we give a simple proof for the weaker bound: in the former we construct the function for which in the latter we prove the lower bound. In Section 4 we explain how to change the proof to give the stronger bound.

2   Construction of the Function

In this section we present the construction of the function for which we prove a weaker form of our bound. Our function is the generalization of GT function.

Definition 1. *For Boolean* $x, y \in \{0,1\}^n$ *let* $\mathrm{GT}(x, y) = 1$ *iff* $x \geqslant y$, *where* $x = (x_1, \ldots, x_n)$ *and* $y = (y_1, \ldots, y_n)$ *are considered as binary representations of integers with* x_n *and* y_n *being the most significant bits.*

Our function will depend on $m = 2n$ variables $(x, y) = (x_1, \ldots, x_n, y_1, \ldots, y_n) \in \{0,1\}^m$. Let us fix some $k_1, \ldots, k_d$ such that $\sum_{i=1}^{d} k_i = m$ and partition the input variables x and y in d groups of size $k_1, \ldots, k_d$, that is

$$(x, y) = (x^1, x^2, \ldots, x^d, y^1, y^2, \ldots, y^d),$$

where $x^i, y^i \in \{0,1\}^{k_i}$ for all i.

Let us denote by $[k]$ the set $\{1, \ldots, k\}$. Let us denote by $<_1$ the following ordering of the set $[k]$: $1, 2, 3, \ldots, k-1, k$, and by $<_0$ the reverse ordering: $k, k-1, k-2, \ldots, 2, 1$. We shall use these orders on sets $[k_1], \ldots, [k_d]$. It will always be clear from the context which set we consider.

Let us denote by $\mathrm{num}_i l$ the ordinal number of $l \in [k]$ w.r.t. the order $<_i$.

To define our function we need to define a specific order on the set $K = [k_1] \times \ldots \times [k_d]$. The construction below is essentially the same as in [12]. Our order will be similar to the lexicographic one, that is to compare two tuples from K we shall compare their components one by one until we find the difference. But as opposed to the lexicographic order, where each component of the tuples is compared w.r.t. the same ordering, in our order of the tuples components might be compared w.r.t. different orderings. Moreover, the ordering in which we compare the current component depends not only on the ordinal number of the component but on the values of previous components of the tuples.

Formally, suppose we want to compare tuples $\alpha = (\alpha_1, \ldots, \alpha_d) \in K$ and $\beta = (\beta_1, \ldots, \beta_d) \in K$. First we compare α_1 and β_1 w.r.t. the ordering $<_1$. If they are not equal then we have already compared the tuples: the larger the first component is the larger the tuple is. If they are equal we proceed to the second components. To compare them we use the following recursive rule to choose the next order.

Assume that the order $<_{i_l}$ to compare the lth components of the tuples is already determined and it happens that $\alpha_l = \beta_l$. The order to compare $(l+1)$st components is determined by the ordinal number of α_l (which coincides with β_l by the assumption) w.r.t. the order $<_{i_l}$. Namely,

$$i_{l+1} = \mathrm{num}_{i_l} \alpha_l \pmod 2. \tag{2}$$

In other words, we compare the $(l+1)$st coordinates w.r.t. the order $<_0$ if α_l has even ordinal number w.r.t. the order $<_{i_l}$ and we compare $(l+1)$st coordinates w.r.t. the order $<_1$ otherwise.

To say it the other way, we associate with coordinates of any tuple $\alpha = (\alpha_1, \ldots, \alpha_d) \in K$ orders $<_{i_1}, \ldots, <_{i_d}$ according to the rule (2). We use these orders to compare coordinates of α with coordinates of other tuples. Note that for two tuples $\alpha = (\alpha_1, \ldots, \alpha_d)$ and $\beta = (\beta_1, \ldots, \beta_d)$ the orders corresponding to their components coincides until we meet the first difference. After the first difference the orders corresponding to components might be different in α and β but we do not need to compare coordinates any further. Let us denote by $\text{num}_{\alpha,l}\alpha_l$ the ordinal number of the lth component of α w.r.t. the corresponding order.

Now we can define our function.

Definition 2. *For given* $(x,y) = (x^1, x^2, \ldots, x^d, y^1, y^2, \ldots, y^d)$, *where* $x^i = (x^i_1, \ldots, x^i_{k_i})$, $y^i = (y^i_1, \ldots, y^i_{k_i}) \in \{0,1\}^{k_i}$ *let* $\alpha = (\alpha_1, \ldots, \alpha_d) \in K$ *be the largest tuple w.r.t. the introduced order such that* $\prod_{i=1}^d (x^i_{\alpha_i} - y^i_{\alpha_i}) \neq 0$. *Then let*

$$f(x^1, \ldots, x^d, y^1, \ldots, y^d) = \text{sgn} \prod_{i=1}^d (x^i_{\alpha_i} - y^i_{\alpha_i}).$$

If there is no such α *let* $f(x^1, \ldots, x^d, y^1, \ldots, y^d) = 1$.

Note that if $d = 1$ our function is exactly the GT function.

3 $2^{\Omega(2^{n/4})}$ Lower Bound

First we note that our function is computable by a degree d threshold gate.

Lemma 1. $\deg_\pm(f) \leqslant d$.

The proof of this lemma is rather simple. One just have to order monomials $\prod_{i=1}^d (x^i_{\alpha_i} - y^i_{\alpha_i})$ for all $\alpha \in K$ in the increasing order and sum them up with suitable increasing coefficients. The details of the proof are omitted.

Now we proceed to the main result of this section.

Theorem 3. *Let* $k_i \geqslant 2$ *be even for* $i < d$ *and* $k_d \geqslant 3$. *Then*

$$\deg_\pm(f) = d$$

$$W(f,d) \geqslant 2^{(k_d-2)\prod_{i=1}^{d-1} k_i - d}$$

Remark 1. We can state analogous theorem for arbitrary $k_i \geqslant 2$ for $i < d$ and not only for even. However, with this assumption the proof and the bound are cleaner and at the same time the theorem still gives $2^{\Omega(2^{n/4})}$ bound.

The $2^{\Omega(2^{n/4})}$ lower bound follows easily from Theorem 3.

3.1   Proof of Theorem 3

Let us consider an arbitrary threshold gate p for f of degree at most d. That is, for any $x, y \in \{0, 1\}^n$ we have $f(x, y) = \text{sgn}(p(x, y))$.

It will be convenient for us to work in variables

$$u_j^i = x_j^i - y_j^i, \qquad v_j^i = x_j^i + y_j^i. \tag{3}$$

So after substituting $x_j^i = (u_j^i + v_j^i)/2$ and $y_j^i = (u_j^i - v_j^i)/2$ and multiplying the polynomial by 2^d to make the coefficients integer we obtain the polynomial p' in variables u_j^i, v_j^i that sign-represents f. That is

$$f(x, y) = \text{sgn}(p'(x - y, x + y)).$$

It is easy to see that the weight of the new polynomial is almost the same as the weight of p (compared to the value of our bound). Namely, we have the following bound.

Lemma 2. $W(p') \leqslant 2^d W(p)$.

The proof of this lemma is simple and is omitted.

Remark 2. A similar bound holds in the other direction too, but we do not need it.

Now we have to prove that $W(p') \geqslant 2^{(k_d - 2)} \prod_{i=1}^{d-1} k_i$.

First we shall prove that we can assume that p' has a nice structure. Lemmas similar to the next one appeared in [4,12] (see [13] for a more general version).

Lemma 3. *If we substitute by 0 all coefficients of the monomials of p' in which variables from one of the groups $u^1, \ldots, u^d$ are not presented, the resulting polynomial q will also sign-represent f.*

The proof of this lemma is based on the fact that the function f is anti-symmetric with respect to variables x^i and y^i for any i, that is if we permute variables x^i and y^i then the function will change the value (except for some singular inputs). From this it is easy to see that all monomials that are even in some variables u^i can be eliminated from p'. The details of the proof are omitted.

As a byproduct of the proof of this lemma we have the following corollary.

Corollary 1. $\deg_{\pm}(f) = d$.

Since $W(p') \geqslant W(q)$ it is enough to prove that $W(q) \geqslant 2^{(k_d - 2)} \prod_{i=1}^{d-1} k_i$.

Now we need a lemma concerning degree 1 threshold gates for GT. The argument is quite standard (see [10,11]) and is omitted.

Lemma 4. *Let $p = \sum_{i=1}^{k} w_i u_i$ be a degree 1 threshold gate for $\text{GT}(x, y)$ where $x, y \in \{0, 1\}^k$. Then for $j \geqslant 2$*

$$w_j \geqslant 2^{j-2} w_1 > 0 \tag{4}$$

and

$$w_j \geqslant w_{j-1}. \tag{5}$$

It will be convenient for us to consider two variants of the GT function: we denote by GT_1 the usual GT function and by GT_0 the analogous function but now on the reversed input. That is, $GT_0(x, y) = 1$ if and only if $x \geqslant y$, where $x = (x_1, \ldots, x_k), y = (y_1, \ldots, y_k)$ are considered as binary representations of integer numbers where the most significant bits are x_1, y_1. It is easy to see that if $p = \sum_{i=1}^{k} w_i u_i$ is a threshold gate for GT_0 then we have $w_{j-1} \geqslant w_j$ and $w_{n-j+1} \geqslant 2^{j-2} w_n > 0$, where $j = 2, \ldots n$.

Now we can prove the main lemma.

Lemma 5. *For all $l \leqslant d$ if $\alpha \in K$ is such that $\text{num}_{\alpha,i} \alpha_i = 1$ for all $i \geqslant l$ and $\beta = (\alpha_1, \ldots, k_l - \alpha_l + 1, \ldots, \alpha_d)$. Then*

$$w_\beta \geqslant w_\alpha 2^{(k_d - 2)} \prod_{i=l}^{d-1} k_i.$$

The idea is the following: we fix variables in all groups $u^1, \ldots, u^d$ except one. Then the function f becomes essentially the GT function and we can apply Lemma 4. Repeating this trick we can accumulate the large factor due to the specific construction of our order. More specifically the proof goes by induction on decreasing l. For the base of induction $l = d$ we fix all variables except u^d and applying inequality (4) immediately obtain the desired result. For the induction step we first apply the induction hypothesis to $l+1$ and then apply inequality (5) to the lth coordinate. Then we can again apply induction hypothesis to $l + 1$ and so forth. In this way we can apply induction hypothesis k_l times and obtain the desired result. The proof details are omitted.

It is easy to prove Theorem 3 now. Applying Lemma 5 with $l = 1$ we get

$$w_\beta \geqslant w_\alpha 2^{(k_d - 2)} \prod_{i=1}^{d-1} k_i.$$

Now, it is easy to see that $w_\alpha > 0$ (just substitute $u_{\alpha_1}^1 = -1$, $u_{\alpha_i}^i = 1$ for all $i \neq 1$ and $u_j^i = 0$ for all i and all $j \neq \alpha_i$). We conclude that $w_\alpha \geqslant 1$ and

$$w_\beta \geqslant 2^{(k_d - 2)} \prod_{i=1}^{d-1} k_i.$$

4   Improved Lower Bound

In this section we improve the argument of the previous sections to obtain the better lower bound. More precisely we prove

$$W \geqslant 2^{\Omega(2^{2n/5})}.$$

We shall work with Boolean variables $\{-1, +1\}$, so we change the definition of sgn-function: $\text{sgn}(x) = 1$ if $x \geqslant 0$ and $\text{sgn}(x) = -1$ otherwise.

The idea is to use another function instead of GT as a building block in our construction.

Definition 3. *For $x = (x_1, \ldots, x_k) \in \{-1, +1\}^k$ let $g(x_1, \ldots, x_k)$ be equal to $-x_k$ if bits $x_1, \ldots, x_k$ are not all equal and let it be x_k if they are all equal.*

This function can be easily represented by a linear threshold gate:

$$g(x_1, \ldots, x_k) = \operatorname{sgn}(\sum_{i=1}^{k-1} x_i - (k-2)x_k).$$

We need an alternative definition of g. Consider k linear form: $L_i(x) = x_i - x_{i+1}$ for $i = 1, \ldots, k-1$ and $L_0(x) = x_1 + x_k$ (L_i will play a role of u^i in the previous proof). The alternative (equivalent) definition of g is that $g(x)$ is equal to the sign of the last nonzero in the sequence $L_0(x), L_1(x), L_2(x), \ldots, L_{k-1}(x)$. Note that now it is more convenient for us to start numeration of L_i from 0. The reason for this is that the actual benefit we shall get only from linear forms $L_1, \ldots, L_{k-1}$. The form L_0 is needed only for technical reasons (the same role was previously played by variables v^i).

Note that in the proof of weaker bound we needed actually not one base function, but two of them. They were very similar though: GT_0 and GT_1. In case of stronger bound two functions will differ more substantially. Again, g_1 is just the function g we defined above. As for g_0 we let $g_0(x)$ be x_1 if not all bits of the input are equal and $g_0(x) = -x_1$ if all bits of input are equal. That is now we not only reverse the order of variables but also multiply the value of the function by -1. Note that $g_0(x)$ is equal to the sign of the last nonzero in the sequence $-L_0(x), L_{k-1}(x), L_{k-2}(x), \ldots, L_1(x)$.

Now we can apply the previous proof scheme with the functions g_1 and g_0 on first $d-1$ components and with GT on the last component. We denote the new function by f again. Below we state what changes in the proof.

For the new function the construction of the ordering is the same except that we use different orderings $<'_0$ and $<'_1$ on the first $d-1$ coordinates, namely we let $<'_1$ to be $0, 1, 2, \ldots, k-2, k-1$ and $<'_0$ to be $0, k-1, k-2, \ldots, 2, 1$, that is, 0 is always the smallest element. The orderings on the last coordinate remains the same as before (as well as the rule (2) defining the ordering on each next coordinate).

In the Definition 2 we now have only variables $x^1, x^2, \ldots, x^d, y^d$ and we let

$$f(x) = \operatorname{sgn}((-1)^{c_1+\ldots+c_{d-1}} L_{\alpha_1}(x^1) L_{\alpha_2}(x^2) \ldots L_{\alpha_{d-1}}(x^{d-1})(x^d_{\alpha_d} - y^d_{\alpha_d}))$$

for the largest α for which the expression is nonzero, where $c_i = 1$ if $\alpha_i = 0$ and the order corresponding to the i-th coordinate of α is $<'_0$ and $c_i = 0$ otherwise. If there is no such α (which can happen only if $x^d = y^d$) we let $f(x)$ to be 1. Note that the number of variables $n = \sum_{i=1}^{d-1} k_i + 2k_d$ is almost twice less than before.

The theorem we prove has the following form.

Theorem 4. *Let $k_i \geqslant 3$ be odd for $i < d$ and $k_d \geqslant 3$. Then*

$$\deg_\pm(f) = d$$

$$W(f, d) \geqslant 2^{(k_d-2)\prod_{i=1}^{d-1}(k_i-1)-d\log n}$$

Instead of the variables (3) for the first $d-1$ coordinates we now let $u^i_j = L_j(x^i)$ for $j = 0, \ldots, k_i - 1$. We do not have now variables v^i for $i \leqslant d-1$.

The proof of Lemma 2 remains the same, but now the factor appearing in it can be upper bounded by n^d. The proof strategy of Lemma 3 remains essentially the same, only this time instead of permutation of variables on first $d-1$ coordinates we multiply variables by -1. This shows the oddity of the polynomial in variables u^i for $i \leq d$ and the rest part is the same.

We present now the analog of Lemma 4.

Lemma 6. *Let $p = \sum_{i=0}^{k-1} w_i u_i$ be a degree 1 threshold gate for $g_1(x)$ where $x \in \{-1,+1\}^k$. Then for $j = 0,1,\ldots,k-1$ we have $w_j > 0$ and for $j = 2,\ldots,k-1$ we have $w_j > w_{j-1}$.*

For the function g_0 analogous statement is true.

Lemma 7. *Let $p = \sum_{i=0}^{k-1} w_i u_i$ be a degree 1 threshold gate for $g_0(x)$ where $x \in \{-1,+1\}^k$. Then we have $w_0 < 0$, for $j = 1,\ldots,k-1$ we have $w_j > 0$ and for $j = 2,\ldots,k-1$ we have $w_{j-1} > w_j$.*

Proofs of these lemmas are omitted.

The analog of Lemma 5 is very similar to the previous version, but becomes a little bit clumsy since we distinguish cases of $l = d$ and $l < d$.

Lemma 8. *For all $l \leq d$ if $\alpha \in K$ is such that $\text{num}_{\alpha,i}\alpha_i = 1$ for all $i \geq l$ and $\beta = (\alpha_1,\ldots,k_l-\alpha_l+\delta_{l,d},\ldots,\alpha_d)$, where $\delta_{l,d}$ is a Kronecker delta (that is $\delta_{ij} = 1$ if $i = j$ and $\delta_{ij} = 0$ otherwise) then*

$$w_\beta \geq w_\alpha 2^{(k_d-2)} \prod_{i=l}^{d-1}(k_i-1).$$

Concerning the proof of the lemma, the base of induction remains completely the same (note, that the statement is the same also). As for the induction step, it also remains the same but now we can apply the induction hypothesis $k_l - 1$ times instead of k_l times in the previous proof.

To conclude the proof of our lower bound we have to choose the values of k_i to maximize the lower bound we have. It is not hard to see that the optimal way is to take $k_d = 3$ as before and to take $k_1 = k_2 = \ldots = k_{d-1}$, let us denote the value of them by k. Then the exponent of our bound will be about $(k-1)^{n/k}$. Simple analysis shows that the maximum (over integers) is attained when $k = 5$. Thus we have a lower bound of $2^{2^{2(n-6)/5}-n}$.

To prove the first part of Theorem 2 we can just let $m = \frac{5}{4}\epsilon n$ and consider the function from the previous paragraph with m variables. Then we have lower bound $2^{\Omega(2^{2m/5})}$ for degree $m/5$ threshold gates and applying the observation preceding Theorem 2 we get the desired bound. For the second part of the theorem let $m = \frac{5}{2}(1+\epsilon)\log n$.

Finally we note that the result of [12] can also be reproved by the same argument and with better constants if we use Håstad's function in the last coordinate and g function in other coordinates.

Acknowledgements. The work is supported by the Russian Foundation for Basic Research and the programme "Leading Scientific Schools" (grant no. NSh-5593.2012.1).

References

1. Babai, L., Hansen, K.A., Podolskii, V.V., Sun, X.: Weights of exact threshold functions. In: Hliněný, P., Kučera, A. (eds.) MFCS 2010. LNCS, vol. 6281, pp. 66–77. Springer, Heidelberg (2010)
2. Beigel, R.: Perceptrons, PP, and the polynomial hierarchy. Computational Complexity 4, 339–349 (1994)
3. Buhrman, H., Vereshchagin, N.K., de Wolf, R.: On computation and communication with small bias. In: Proc. of the 22nd Conf. on Computational Complexity (CCC), pp. 24–32 (2007)
4. Håstad, J.: On the size of weights for threshold gates. SIAM J. Discret. Math. 7(3), 484–492 (1994)
5. Klivans, A.R., Servedio, R.A.: Learning DNF in time $2^{\tilde{O}(n^{1/3})}$. J. Comput. Syst. Sci. 68(2), 303–318 (2004)
6. Krause, M., Pudlák, P.: Computing Boolean functions by polynomials and threshold circuits. Comput. Complex. 7(4), 346–370 (1998)
7. Minsky, M.L., Papert, S.A.: Perceptrons: Expanded edition. MIT Press, Cambridge (1988)
8. Muroga, S.: Threshold logic and its applications. Wiley Interscience, Chichester (1971)
9. Muroga, S., Toda, I., Takasu, S.: Theory of majority decision elements. Journal of the Franklin Institute 271(5), 376–418 (1961)
10. Myhill, J., Kautz, W.H.: On the size of weights required for linear-input switching functions. IRE Trans. on Electronic Computers 10(2), 288–290 (1961)
11. Parberry, I.: Circuit complexity and neural networks. MIT Press, Cambridge (1994)
12. Podolskii, V.V.: Perceptrons of large weight. Probl. Inf. Transm. 45, 46–53 (2009)
13. Podolskii, V.V., Sherstov, A.A.: A small decrease in the degree of a polynomial with a given sign function can exponentially increase its weight and length. Mathematical Notes 87, 860–873 (2010)
14. Razborov, A.A.: On small depth threshold circuits. In: Proceedings of the Third Scandinavian Workshop on Algorithm Theory, pp. 42–52. Springer, London (1992)
15. Saks, M.E.: Slicing the hypercube. Surveys in Combinatorics, pp. 211–255 (1993)
16. Sherstov, A.A.: Communication lower bounds using dual polynomials. Bulletin of the EATCS 95, 59–93 (2008)

What Are Computers
(If They're not Thinking Things)?

John Preston

Department of Philosophy, The University of Reading, Whiteknights,
P.O. Box 217, Reading, Berkshire, RG6 6AH, United Kingdom

Many of us now imagine that in the future humans either will, or at least could, 'in theory', construct an electronic digital computer which would really be a thinking thing. Alan Turing was one of the first and surely the most notable exponent of this view, and a significant proportion of his published work was devoted to arguing for it.

However, even if one accepts this 'computationalist' view, the question 'What *are* computers?' is still worth asking, since almost no-one thinks of past and existing computers, or even of most foreseeable computers, as thinking things. We still need an account of what these devices that now surround us are, even if computationalists are right to think of them as proto-thinkers, as it were. In fact, I think that when one sees the answer to this question, the temptation to think that even more sophisticated computers really would be thinking things evaporates.

1 Our Technologies

The electronic digital computer is the most prominent among a proliferation of devices whose operations we regularly and quite naturally describe in some terms that can also be used to characterise human actions: mousetraps *catch* mice, washing machines *wash* clothes, dishwashers *wash* the crockery, thermostats *turn* the central heating on and off, pocket calculators *calculate*, guided missiles *seek* the exhaust heat of aircraft in order to destroy them, etc, etc.

Sometimes the underlying idea is still that these devices are things that we use in order to carry out these tasks, activities or functions. (*I* caught the mouse (using the mousetrap), *you* washed the clothes (by putting them in the washing machine), etc.). In the case of some of these devices, though, we have somehow learned to take seriously talk of the device *itself* carrying out the activity. Usually, no misunderstanding results. But the computer falls into this group, since we now think and talk of digital electronic computers as if *they* perform computations, calculate, search for data, store information, execute instructions, etc. This is entirely contingent, incidentally: when Turing was writing, the term 'computer' meant *person who performs computations*, and it could easily have kept that meaning. If someone had invented a handy term or acronym for what we now call computers (e.g., 'EDMs', for 'electronic digital machines'), 'computer' could have kept its original meaning, and we would easily and naturally have thought of EDMs as 'the devices that computers use' to do these things.

S.B. Cooper, A. Dawar, and B. Löwe (Eds.): CiE 2012, LNCS 7318, pp. 609–615, 2012.

When we attribute such achievements there are three aspects to the attribution: the task gets done, it gets done correctly, and it's the device that does it. To take all this at face value would be to conclude that these devices can correctly be said to be actually performing that function or carrying out that activity which would otherwise have to be performed or carried out by a person, using the activity-term in the same sense that it has when applied to humans. So we can speak of computers not only as 'searching for' data, 'storing' information, 'executing' instructions, but also as playing chess, scheduling tasks, controlling the operation of other machines, etc.

I'd like to suggest, though, that this way of talking when applied to computers encourages us to conceive of them in the wrong way, that it involves getting their entire *relation* to us wrong. In the case of computers, it's quite remarkable that hardly anyone stands up for this alternative and, I think, commonsense view of them and their activities. A relatively superficial feature of the language we use to talk about such devices is allowing computationalists to set the terms of the debate about 'machine intelligence'.[1]

2   Replacing-Technologies

We human beings aren't the only tool-using animals, but by any standard our use of tools is the most widespread and the most impressive. We use tools in situations where we want something done but don't want to use (only) our bodies to do it. Our reasons for not using only our bodies can be various: not wanting to get hurt, not being able to fit into or reach a particular space, not wanting to get dirty, performing some task (like driving in a screw, or cutting a piece of wood) for which no part of the human body will do, as well as considerations of speed, cost, efficiency, etc. Alongside using tools (in this core sense) that relieve us of tasks that are too dangerous, dirty, or physically unmanageable for our bodies, we now also use technologies that relieve us of tasks that are boring, complicated, time-consuming or expensive. This is often where computers come in. Of course, we also use computers to help us control machines of other kinds - cars, telephones, industrial machines, etc..

Computationalists are right to think that there's something importantly *different* about the computer. The computer is a new *kind* of technology. But it's a kind that's nevertheless contiguous with other kinds of technologies, and to understand it properly as a technology is to understand both this contiguity *and* the way in which it goes beyond previous technologies.

For each technology there's a crucial distinction between the process or activity involved and the product that results. What I call '*replacing-technologies*' are machines and devices that *replace* human activities. (Not all our machines

---

[1] One small group of philosophers who *do* resist the computationalist current is inspired by certain remarks that Ludwig Wittgenstein made about computation, partly in response to Turing, and I in turn have taken inspiration from them in this article. Their works in question here are [3], [4], and [6]. None of these people are to be blamed for any problematic developments of their ideas that I make here, though.

are replacing-technologies: no number of human beings, unaided by technology, could ever produce certain electromagnetic phenomena, or atomic power, or a nuclear explosion). We use replacing-technologies to ensure that the product or *end-state* of some activity that could only formerly be done by humans is brought about, even though the process or activity itself hasn't taken place. Of course, *some* process has taken place, but this process isn't the one that we formerly used, unaided, to produce that result. The device in question replaces a stretch of human activity, but does so in such a way that the arrangement which would otherwise have been the end-state of that activity does result.

Our *pre*-computational technology mostly involved tools and implements for the performance of tasks that used to require a significant amount of human physical effort. Computers, though, mostly relieve us of a different *kind* of activity, the kind of activity that used to require thinking, reasoning, inferring, calculating, remembering, and certain other psychological skills. Because the activities they replace are of this kind, intellectual rather than manual, and because these activities are often quite distinctive to human beings (they can't be performed by other creatures), computers are a very special, and especially impressive, kind of replacement-technology, a technology capable of being especially close to our minds (and thus perhaps to our hearts, as it were).

This distinctiveness is of course down to Alan Turing, who showed not only that a certain kind of mathematical function can be encoded in a way that makes it ideally suited to be mechanically implemented (in electrical circuitry), but also that what we now call 'Turing machines' are devices with a certain kind of *universality* (the kind specified in [7, §6]). Since almost all our programmed electronic devices can be thought of as (horrendously complicated) Turing machines, our stored-program digital computers are *general-purpose* technologies in a specific way, a way in which no previous technology has ever been. Their ever-increasing importance to us derives at least partly from this fact.

3   Intentional Actions and Non-intentional Operations

Because the main philosophical debate has been about machine *intelligence*, computationalists tend to think that if one could show that intelligent operations can be 'broken down into' unintelligent ones, and then show that Turing machines operate entirely on the basis of millions of operations of just this unintelligent kind, one would have successfully opened the door to the idea of an intelligent machine. There are problems with this strategy, but regardless of whether it works, I suggest instead that the key issue is really not intelligence but another, more fundamental feature of certain psychological concepts, the one that philosophers call *intentionality*. Intentionality is a feature of certain psychological phenomena crucially, the same psychological phenomena that computationalists are centrally concerned with, such as thoughts, beliefs, and desires. Roughly, a psychological verb ψ is *intentional* if it can be true to say of a person that he or she ψ's that p, on a particular occasion, despite the fact that p isn't the case on that occasion. So, for example, I can *want* Usain Bolt to set a new world record in the

100 metres at the London Olympics even if it turns out that he doesn't, I can *think* or *believe* that he's Trinidadian even though he isn't, I can even *regret* that he didn't win the 200 metres at the 2009 Berlin World Championships despite the fact that (unbeknownst to me at that time, of course) he did. Intentionality means that these psychological phenomena can still be correctly attributed to people even when the 'object' of those phenomena doesn't exist, or doesn't come to pass. Its exact technical definition and history aren't vital here but, the thing to note is that whether or not the high-level operations of computers are rightly described in 'intentional' terms (as computationalists would like, and as some of our ordinary *and* some of our technical talk about them suggests), those operations can always be broken down into sub-operations that are *non*-intentional. At the lowest level, as it were (think of this as the level of the Turing machine), the operations involved are clearly and *purely* mechanical. They are operations such as: scanning a square on the machine-tape, registering the contents of that square, erasing the symbol in the square, writing another such symbol, etc. That the operations of almost all our computers can be broken down into *such* steps is one of the things that Turing's 'On Computable Numbers' absolutely guarantees. But the fact that these operations are purely mechanical is a sure sign that they're *not* genuinely intentional.

For the fact is that neither human actions nor human psychological skills can be thus 'broken down'. Complex actions and skills can be thought of as composed of simple ones, of course, but these simple ones are *still* intentional.[2] This is a *logical* feature of intentionality, not merely a contingent one, and it marks a fundamental, *categorical* divide between the way in which computers work and the nature of human beings. However complex our computers become, they will only involve 'more of the same' (i.e., more such purely mechanical operations), and this can't yield thought or intelligence, just because it can't yield intentionality.

4   The Intentional Stance?

Isn't it the case, though, as Daniel Dennett has suggested,[3] that we take the *intentional stance* with respect to computers? That is, that we (at least on occasion) think of them in intentional terms, ascribing intentional properties to them?

On a very superficial level it's true that most of us do occasionally and casually say of a computer that it 'wants' or is 'waiting for' an instruction, that it's 'recalling' certain information, or even (when it's taking a long time processing) that it's 'thinking about' an issue. And when things go *wrong* we sometimes think of the computer we're using as frustrating our plans, perhaps even as conspiring to do so.

On a deeper level, as I've already stressed, although adult human beings don't often think of non-computational devices in such psychological terms, we *do*

---

[2] Here I am indebted to Tony Palmer's important but neglected paper [5].

[3] Cf. several of the essays reprinted in [1,2].

describe computers as if they achieve the psychological tasks they in fact replace (computing, calculating, storing information, executing instructions, etc.).

But to conclude from either of these habits that we do or can take the 'intentional stance' towards them would be a wild exaggeration. The range of intentional terms that we apply to computers is radically restricted and thus a *very* weak analogue of the full range of intentional terms that can be applied to humans. There's simply no application for most of the associated psychological terms we predicate of other humans, such as hoping, fearing, expecting, aspiring, anticipating, despairing, regretting, considering, reconsidering, contemplating, pondering, reflecting, etc. This is a consequence of what philosophers call the '*holism*' of the intentional: the serious and literal application of an intentional term brings with it the actuality, or at least the possibility, of applying *other* intentional terms to that same person. What's more, even when we consider those few intentional terms that we do apply to computers, we should remember that humans are of course quite capable of applying them to *non*-computational devices, and even to artifacts which aren't devices at all (such as dolls and puppets). So although computers (machines generally) can be described using a very weak *image* or *shadow* of one range of our psychological vocabulary, it does no harm whatsoever to think that our applications of such terms to machines, including computers, are always 'in scare-quotes', and not to be taken seriously.

Nevertheless, to talk of past, present or, I would argue, future computers as *achieving* the tasks they replace, or even occasionally as performing intelligent activities, such as thought, does at least have two very good rationales. The first is the fact that what their activities replace are indeed usually stretches of *thought* or intellectual activity. Where this clearly isn't the case, as for example in the case of a car's engine management system, which is also computational, there's very little temptation to think of the device in intentional terms. So where we humans, unaided, would have to think in order to attain some result, a temptation to conceive also of the computer which can be made to achieve that result (perhaps much faster and more reliably than we can) as a *thinking* thing is quite naturally felt.

The second rationale for talking of computers using a very limited range of intentional terms is that most of us simply don't know how computers operate. Their operation is *recondite*, that is, hidden from our inspection, and its full mechanical description is simply *unwieldy*. It would be quite wrong though, to suggest, as Dennett sometimes does, that the *only* way of coming to understand 'what the computer is doing' is from the 'intentional stance'. That is, we all know that there *is* a purely mechanical description of their activity, although most of us couldn't say what it is. Nevertheless, taking the 'intentional stance' towards computers (where this is understood in a *very* limited way, as a matter of predicting and explaining their operations and products in terms of goals, searches, decisions, etc.) is quite natural when one knows of no other way of explaining their operations, or even when this other way is simply too long-winded (which is almost always the case with computers, simply due to the number of operations they are capable of performing at great speed). This also explains why computer science texts usually explain the operation of computers in terms of analogies with human rule-following activities (such as games, recipes, knitting patterns, etc.).

Both these rationales, however, are very clearly *pragmatic* in nature. They can do nothing to support the serious philosophical conclusion that such machines are *really* performing intentional operations. It's surely ironic that serious thinkers who would otherwise scout such considerations have allowed them to prevail in this context, since elsewhere such rationales would normally be regarded as hopelessly insufficient for a philosophical conclusion.

5   Conclusion

In sum, computers are the latest kind of labour-saving technologies. The labour they save isn't that of the 'workers by hand', though, but that of the 'workers by brain'. *However* sophisticated, computers aren't things which compute, but things that we use to *replace* the human activity of computation. Perhaps they might thus be better described as things that people compute with, or things that people use to compute. But even these descriptions have to be taken with a pinch of salt, since in cases where I use a computational device to generate for me the result of a calculation, for example, no actual calculation (or computation) has taken place.

Among those working in the field of artificial intelligence, it is perhaps the designers of 'expert systems' who have come closest to realising the replacement status of computer technologies, since they're quite explicitly aware that they are (in their terms) trying to 'mechanise' the boring bits of experts' jobs. How ironic that the least interesting bit of artificial intelligence (from the point of view of philosophy) should have the clearest conception, not merely of the goals of the subject, but also of the nature of the devices it produces!

Where does all this leave Alan Turing himself? Turing manifestly still counts as a great thinker, one whose mathematical work made possible the development of devices that are effecting an enormous and unprecedented transformation in human work- (and leisure-) activity. He is thus the herald of a new machine-age (for better or worse, of course). But if we cease to think of him as the herald of thinking or intelligent machines, even though he would himself have protested and in some circles his reputation will be thought to suffer, his genuine achievements will in no way be underestimated.

References

1. Dennett, D.C.: Brainstorms: Philosophical Essays on Mind and Psychology. Harvester Press, Sussex (1981)
2. Dennett, D.C.: The Intentional Stance. MIT Press, Cambridge (1987)
3. Hacker, P.M.S.: Men, Minds and Machines. In: Wittgenstein: Meaning and Mind, pp. 147–170. Blackwell, Oxford (1990)

4. Hyman, J.: Introduction to Investigating Psychology: Sciences of the Mind after Wittgenstein, pp. 1–24. Routledge, London (1991)
5. Palmer, A.: The Limits of AI: Thought Experiments and Conceptual Investigations. In: Torrance, S. (ed.) The Mind and the Machine: Philosophical Aspects of Artificial Intelligence, pp. 43–50. Ellis Horwood, Chichester (1984)
6. Shanker, S.: Wittgenstein's Remarks on the Foundations of AI. Routledge, London (1998)
7. Turing, A.M.: On computable numbers, with an application to the Entscheidungsproblem. Proc. Lond. Math. Soc., II. Ser. 42, 230–265 (1936)

Compactness and the Effectivity
of Uniformization

Robert Rettinger

Fakultät für Mathematik und Informatik, FernUniversität in Hagen,
Universitätsstraße 1, 58097 Hagen, Germany

Abstract. We give new proofs of effective versions of the Riemann mapping theorem, its extension to multiply connected domains and the uniformization on Riemann surfaces. Astonishingly, in the presented proofs we need barely more than computational compactness and the classical results.

1   Introduction

The Riemann mapping theorem is probably one of the most fundamental theorems in complex analysis. Thus, not surprisingly, effectivization of this theorem was on the agenda ever since the theorem was proven. The first general constructive solution for Riemann mappings were given by P. Koebe in 1910 using the osculation method (cf. [5]). More constructivity results were added over the years based on different methods like potential theoretic methods, circle packings and random walks. Using the osculation method, P. Hertling [6] gave the first exact characterization of the effectiveness of the Riemann mapping based on the type-2-theory of effectivity. Similar results were recently given for multiply connected domains (cf. [7]), whereas effectivity results for the uniformization of Riemann surfaces were not known before.

Despite the belief that the classical proof of the Riemann mapping theorem is highly non-constructive, we will show that we have to add only a few things to make it constructive. More precisely, our main ingredient will be computational compactness of a certain class of formal power series. This property will even show effectiveness of other (non-constructively proven) results such as conformal mappings for multiply connected domains and the uniformization of Riemann surfaces. To be more precise, we list the classical results which we are going to effectivize in the next sections below.

Let $\mathbb{D}_\varepsilon(z_0)$ denote the disc $\{z \in \mathbb{C} \mid |z - z_0| < \varepsilon\}$. Furthermore we use $\mathbb{D}_\varepsilon$ and $\mathbb{D}$ instead of $\mathbb{D}_\varepsilon(0)$ and $\mathbb{D}_1(0)$, respectively. The classical Riemann theorem guarantees the existence of conformal mappings for simply connected domains (cf., e.g., [1]):

Theorem 1. *Let U be a proper and simply connected open subset of $\mathbb{C}$, $0 \in U$. Then there exists a unique bi-holomorphic mapping $f : U \to \mathbb{D}$ with $f(0) = 0$ and $f'(u) > 0$.*

S.B. Cooper, A. Dawar, and B. Löwe (Eds.): CiE 2012, LNCS 7318, pp. 616–625, 2012.

A similar result holds for multiply connected domains. There is a small technicality concerning this case: For $k > 1$ there does not exist a single k-connected
domain D so that all k-connected domains can be bi-holomorphically mapped
onto D, that is there does not exist a canonical k-connected domain. Here we
use circles with concentric circular slits but any of the families found in literature (cf., e.g., [9]) would work as well. To define the class $\mathcal{D}_k$ of k-connected
circles with concentric circular slits let $\iota : \partial\mathbb{D} \times [0; 1] \times [0; 1] \to 2^{\mathbb{D}}$ be the mapping $\iota(\gamma, a, b) = \{b \cdot \gamma' \mid \exists 0 \leq \alpha \leq a.\gamma' = \gamma \cdot e^{2\pi i\alpha}\}$. Then a set D is in
$\mathcal{D}_k$ iff $D = \mathbb{D} \setminus (\iota(\gamma_1, a_1, b_1) \cup ... \cup \iota(\gamma_{k-1}, a_{k-1}, b_{k-1}))$ for pairwise disjoint sets
$\iota(\gamma_1, a_1, b_1), ..., \iota(\gamma_{k-1}, a_{k-1}, b_{k-1})$ with $\gamma_i \in \partial\mathbb{D}$, $a_i \in [0; 1)$ and $b_i \in (0; 1)$.

Theorem 2. *Let $k > 0$ be a natural number, U be a k-connected, proper subset
of $\mathbb{C}$ and K be a connected component of the boundary ∂U of U, $0 \in U$. Then
there exists a unique domain $D \in \mathcal{D}_k$ and a unique bi-holomorphic mapping
$f : U \to D$ so that $f(z) = 0$, $f'(0) > 0$ and $f(K) \subseteq \partial\mathbb{D}$.*

A Riemann surface is a topological space T together with an atlas τ, i.e., τ is
a family $\tau = (\tau_i)_{i\in\mathbb{N}}$ of topological mappings $\tau_i : \mathbb{D} \to T$ so that $T = \bigcup_i \tau_i(\mathbb{D})$
and $\tau_i^{-1} \circ \tau_j$ is bi-holomorphic on $\tau_j^{-1}(\tau_i(\mathbb{D}))$ for all $i, j \in \mathbb{N}$. Thus the maps of
the atlas, i.e., the τ_i, define a complex structure on T. Furthermore, a mapping
$f : T \to T'$ of two Riemann surfaces (T, τ) and (T', τ') is called meromorphic iff
for all $i, j \in \mathbb{N}$ the mapping $\tau_j^{-1} \circ f \circ \tau_i$ is holomorphic.

Theorem 3. *Let (T, τ) be a simply connected Riemann surface. Then T can be
bi-meromorphically mapped onto exactly one of the following spaces*

   (1) $\mathbb{D}$ *(hyperbolic case)*,   (2) $\mathbb{C}$ *(spherical case) or* (3) $\mathbb{C}_\infty$ *(Euclidean case)*,

*where $\mathbb{C}_\infty$ denotes the complex sphere. More precisely, let $z_0, z_1, z_\infty \in T$ and R
either $\mathbb{D}$, $\mathbb{C}$, $\mathbb{C}_\infty$. Then there exists a unique bi-holomorphic mapping*

(1) $f_{z_0} : T \to \mathbb{D}$ with $f_{z_0}(z_0) = 0$ and $f'_{z_0}(z_0) > 0$ in the hyperbolic case,
(2) $f_{z_0} : T \to \mathbb{C}$ with $f_{z_0}(z_0) = 0$ and $f'_{z_0}(z_0) = 1$ in the spherical case and
*(3) $f_{z_0, z_1, z_\infty} : T \to \mathbb{C}_\infty$ with $f_{z_0, z_1, z_\infty}(z_0) = 0$, $f_{z_1, z_1, z_\infty}(z_1) = 1$ and
$f_{z_0, z_1, z_\infty}(z_\infty) = \infty$ in the Euclidean case.*

2   Type-2-Theory, Enumerability and Computational Compactness

Most of the following notions and more details can be found, e.g., in the book
[12]. We will start with some basic, well established concepts of type-2-theory:
relative computability and representations. Afterwards we will introduce two less
frequently used notions: r.e. open sets and computationally compact spaces.

   Let Σ^ω denote the set of infinite sequences $\alpha_0\alpha_1...$ of elements of the finite
alphabet Σ and Prefix(ω) denote the set of finite prefixes of the sequence $\omega \in \Sigma^\omega$.
To introduce computability on Σ^ω we use the classical Turing-machine: A partial

function $f :\subseteq \Sigma^\omega \to \Sigma^\omega$ is called computable iff there exists a computable function $f_e :\subseteq \Sigma^* \to \Sigma^*$ so that

$$\omega \in \mathrm{dom}(f) \text{ iff } \mathrm{Prefix}(\omega) \subseteq \mathrm{dom}(f_e) \text{ and } f_e(\mathrm{Prefix}(\omega)) = \mathrm{Prefix}(f(\omega)).$$

Given two representations ρ, ρ' of the sets M, M', i.e., surjective functions $\rho :\subseteq \Sigma^\omega \to M$ and $\rho' :\subseteq \Sigma^\omega \to M'$, we call a function $f :\subseteq M \to M'$ (ρ, ρ')-computable, iff there exists a computable realization g of f, i.e., a computable function $g :\subseteq \Sigma^\omega \to \Sigma^\omega$ so that $f \circ \rho(\omega) = \rho' \circ g(\omega)$ for all $\omega \in \mathrm{dom}(f \circ \rho)$. Thus computability on M and M' depends on the chosen representation ρ and ρ' up to equivalence, i.e., computable translations between these representations. In the same way we can introduce computability on finite products and sets given by numberings $\nu : \mathbb{N} \to M$. If the representations are obvious by the context or if we use so called standard representations of spaces then we will simply say that a function f is computable instead of f is (ρ, ρ')-computable. Notice that all the following (standard) representations are admissible with respect to the usual topology on these spaces.

Using a standard tupling function in a straight forward way, we can immediately define, for given numbering ν of M and representations ρ, ρ' of sets T, T', respectively, representations $\nu^*, \prod \nu, \nu \times \rho, \rho \times \rho', \prod \rho$ and ρ^* of $M^*, \prod_{i=0}^\infty M, M \times T, T \times T', \prod_{i=0}^\infty T$ and T^*, respectively, where $T^* = \bigcup_k T^k$, $\prod \nu(0^{i_0} 10^{i_1} 1 ...) = (\nu(i_0), \nu(i_1), ...)$, $\prod \rho(\alpha_0 \alpha_1 ...) = (\rho(\alpha_{\langle 0,0\rangle} \alpha_{\langle 0,1\rangle} ...), \rho(\alpha_{\langle 1,0\rangle} \alpha_{\langle 1,1\rangle} ...), ...)$ and the other products can be determined in the same way.

Let $\nu_\mathbb{Q}$ denote a standard numbering of $\mathbb{Q}$ then the standard representation $\rho_\mathbb{R}$ of $\mathbb{R}$ is defined to be the restriction of $\prod \nu_\mathbb{Q}$ to fast converging sequences, i.e., sequences $(q_0, q_1, ...)$ of rational numbers so that $|q_i - q_j| < 2^{-\min(i,j)}$. The standard representations of $\mathbb{C}$, $\mathbb{R}^n$ and $\mathbb{C}^n$ for arbitrary n is then defined by finite products of $\rho_\mathbb{R}$. Furthermore, the standard representation ρ_U of the open subsets of $\mathbb{C}$ is defined via the infinite product $\prod(\nu_\mathbb{Q}^2 \times \nu_\mathbb{Q})$, where a sequence $((q_0, q_0'), q_0''), ((q_1, q_1'), q_1''), ...$ represents the set $\bigcup_i \mathbb{D}_{q_i''}(q_i + i \cdot q_i')$.

Using the standard representation $\rho_\mathbb{C}$ of the complex numbers, we get a standard representation $\rho_F = \prod \rho_\mathbb{C}$ of the set $\mathcal{FP}$ of formal power series over $\mathbb{C}$. We will use the usual notation $f = \sum_i a_i \cdot z^i$ instead of $(a_0, a_1, ...)$ in the sequel. Furthermore, if A is some domain with $0 \in A$, a formal power series defines a holomorphic function on a neighborhood of 0 iff it converges on this neighborhood. We will denote this function by f, too.

Let (T, τ) be some Riemann surface. Then we identify T with the set of functions $\tau_i^{-1} \circ \tau_j$. We will do this as follows: Let ρ_R be the restriction of $\prod \rho_\mathbb{C} \times \prod \rho_U \times \prod \rho_F$ to those triples $((z_0, z_1, ...), (U_0, U_1, ...), (f_0, f_1, ...))$ so that $z_i \in \mathbb{D}$, $U_i \subseteq \mathbb{D}$ and there exists a Riemann surface (T, τ) so that $\mathrm{dom}(\tau_i^{-1} \circ \tau_j) = U_{\langle i,j\rangle}$ and $f_{\langle i,j\rangle} = \tau_i^{-1} \circ \tau_j$ on $U_{\langle i,j\rangle}$. Furthermore, we call a Riemann surface (T, τ) finitely generated iff there exist i so that $T = \bigcup_{j=0}^i \tau_j(\mathbb{D})$. In this case we can define a representation ρ_{FR} of these Riemann surfaces as a restriction of $\rho_\mathbb{C}^* \times \rho_U^* \times \rho_F^*$ in the same way. For $\mathbb{C}_\infty$ we fix the standard representations given by the maps τ_0, τ_1, so that $\tau_0(z) = 2 \cdot z$ and $\tau_1(z) = 1/z$ for all $z \in \mathbb{D}$.

Finally we use the following standard representation of the Sierpinski space $\mathbb{S} = \{0, 1\}$: Let $\rho_\mathbb{S}(0^\omega) = 0$ and $\rho_\mathbb{S}(w) = 1$ for all $w \in \{0, 1\}^\omega \setminus \{0^\omega\}$. Notice that this is a standard representation with respect to the usual topology $\emptyset, \{1\}, \{0, 1\}$ on $\mathbb{S}$. Using the Sierpinski space, we say that a subset O of a space A with a standard representation ν of A is *r.e. open* iff there exists a computable (total) function $f : A \to \mathbb{S}$ so that $O = f^{-1}(1)$. Similarly, for a family $g :\subseteq B \to 2^A$ with index set B we call g *uniformly r.e. open*, iff there exists a computable function $f : B \times A \to \mathbb{S}$ so that for all $b \in \mathrm{dom}(g)$ we have $g(b) = \{w \mid (b, w) \in f^{-t}(1)\}$. To simplify matters we will use in most cases non-uniform recursive enumerability. However, all the results mentioned below do indeed hold even uniformly for computable families g. Finally, we define *relative r.e. openness* for subsets $O \subset A$ of a represented space A as follows: O is r.e. open in $B \subseteq A$ iff there exists a computable function $f :\subseteq A \to \mathbb{S}$ so that $B \subseteq \mathrm{dom}(f)$ and $O = f^{-1}(1)$. Notice that A is always r.e. open in A.

For subsets of $\mathbb{C}$ r.e. openness is equivalent to the enumerablility of all rational closed balls inside the set ($\overline{A}$ denotes the topological closure of A):

Lemma 1. *Let O be a subset of $\mathbb{C}$. Then O is r.e. open iff there exists an r.e. set $R \subseteq \mathbb{Q} \times \mathbb{Q}[i]$ so that $O = \bigcup_{(q, q') \in R} \overline{\mathbb{D}_q(q')}$.*

Furthermore it is well known, that several constructions preserve r.e. openness.

Lemma 2. *Let T, T' be r.e. open sets and $f : T \to T'$ be computable. Then*

$$T \times T', \quad T \cap T', \quad T \cup T' \text{ and } f^{-1}(T')$$

are r.e. open.

Given two sequences $(q_i)_i$ and $(q'_i)_i$ and some j it is easy to decide whether $(q_i - q'_i) > 2^{-i+2}$, which means that the set $\{(x, x') \in \mathbb{R}^2 \mid x > x'\}$ is r.e. open and we get the following result by Lemma 2.

Corollary 1. *Let $f, g : U \to \mathbb{R}$ be computable functions and U be r.e. open. Then the sets*

$$\{z \mid f(z) > g(z)\} \text{ and } \{z \mid f(z) < g(z)\}$$

are r.e. open.

To define *computable compactness* let T be a separable T_0 space and $(B_i)_i$ be an enumeration of a basis of T. Then we call $K \subseteq T$ *computably compact* iff K is a compact Hausdorff space and there exists an r.e. set $R \subseteq \mathbb{N}^*$ so that the set $\{(B_{i_0}, ..., B_{i_n}) \mid (i_0, ..., i_n) \in R\}$ is the set of all finite coverings of K by basis elements. Similar to the r.e. open case, we define a uniform version of computable compactness and, although we give in most cases only the non-uniform results, all the results mentioned below hold even in the uniform setting. Let $g :\subseteq B \to 2^T$ be a family of closed subsets of T with index set B. We call g *uniformly computably compact* iff there exists a computable function $f : B \to (\mathbb{N}^*)^\omega$ so that, for all $b \in \mathrm{dom}(g)$, we have that $g(b)$ is a compact Hausdorff space and

$\{(B_{i_0}, ..., B_{i_n}) \mid (i_0, ..., i_n) \in f(b)\}$ is the set of all finite coverings of $g(b)$ by basis elements. For the spaces used in this paper we choose a canonical enumeration of basis elements (which are, anyway, already chosen by saying that our standard representations are admissible).

Simple examples of computationally compact spaces are the compact intervals $[p; q]$ for computable p and q or $\partial\mathbb{D}$. The following lemma gives us all the computationally compact spaces which we will need throughout this paper.

Lemma 3. *Let T, T' be computationally compact spaces and $f : T \to T'$ be computable. Then*

$$T \times T', \quad \prod_{i=0}^{\infty} T \text{ and } f(T)$$

are computationally compact.

The result on finite products is already shown in [13]. The infinite product, i.e., the effective version of the Theorem of Tychonoff, can be easily proven by the very same methods used to prove the finite case. An even more general effective version of Tychonoff's Theorem can be found in [11]. As an immediate consequence of Lemma 3 we get that the space of bounded formal power series are computationally compact:

Corollary 2. *Let $(r_i)_i$ be computable sequence of computable numbers. Then the class $\mathcal{FP}[(r_i)_i] := \{f = \sum_i a_i z^i \mid \forall i.|a_i| \leq r_i\}$ is computably compact.*

Having all these notions at hand the following "computability by compactness" argument is quite simple to prove. This principle is common place but nevertheless quite powerful as we will see in the next section. (For a different application of this principle, cf., e.g., [4].)

Let T now be some computationally compact space with an enumeration ν of a basis. Furthermore let $t \in T$ be some element so that $T \setminus \{t\}$ is r.e. open. Then t is computable with respect to the standard admissible representation: We have to enumerate all (indexes of) basis elements A so that $t \in A$. To do so, let $(B_i)_i$ and $(C_i)_i$ be enumerations of all finite coverings (finite set of indexes of basis elements) of T and an enumeration of all indexes of basis elements so that $\bigcup_i \nu(C_i) = T \setminus \{t\}$, respectively. We can easily enumerate all indexes i of basis elements of T so that $\{i\} \cup \bigcup_{j=0}^{n}\{C_j\} \supset B_j$ for some $i, j, n \in \mathbb{N}$. If $t \in \nu(i)$ then there exists some n and j so that the previous equality is fulfilled because $\nu(i) \cup (T \setminus \{t\}) = T$, T is compact, i.e., there exists some finite sub-covering of the open covering $\nu(i), \nu(C_0), \nu(C_1), ...$ of T and $(B_i)_i$ is an enumeration of all (finite sets of indexes of) finite coverings of T. If, on the other hand, $t \notin \nu(i)$ then $\nu(i) \cup (T \setminus \{t\}) = T \setminus \{t\}$ and the above equality is never fulfilled, as $(B_i)_i$ contains only sets of indexes of coverings of T. We can state this argument as follows:

Lemma 4. *Let T be a computationally compact space (with respect to an admissible representation) and $t \in T$ be given so that $T \setminus \{t\}$ is r.e. open. Then t is computable.*

3 Conformal Mappings for Simply and Multiply Connected Domains

In this section we will apply Lemma 4 to re-prove effective versions of Theorem 1 and Theorem 2. For simplicity reasons we will restrict ourselves to bounded domains and discuss the case of unbounded domains at the end of this section.

Given an open computable function $f : T \to T'$, r.e. openness of T does not necessarily imply r.e. openness of T'. For holomorphic functions, however, this can be easily shown (cf. [6]). Given an $\epsilon > 0$ and an r.e. open domain $A \subseteq \mathbb{C}$ we denote the set of formal power series which determine non-constant holomorphic functions f on A with $f(A) \subseteq \mathbb{D}_\varepsilon$ by $\mathcal{H}[\mathcal{A}, \mathbb{D}_\varepsilon]$. Notice that we will assume that A is a domain with $0 \in A$ without further mentioning. For non-connected A the following reformulation of the Effective Open Mapping Theorem in [6] hold, if one replaces A by the connected component of A which contains 0.

Lemma 5. *Let $\varepsilon > 0$ and A with $0 \in A$ be r.e. open. Then the family $h :\subseteq \mathcal{FP} \to 2^{\mathbb{C}}$, $h(f) = f(A)$ for all $f \in \mathcal{H}[\mathcal{A}, \mathbb{D}_\varepsilon]$, is r.e. open in $\mathcal{H}[\mathcal{A}, \mathbb{D}_\varepsilon]$.*

We will see below that the above lemma will help to eliminate all those holomorphic functions which cannot be Riemann mappings because their images contain points which they actually shouldn't. However, to do this, we have to eliminate all formal power series, which do not define a holomorphic function on A, first.

Lemma 6. *Let A be a r.e. open domain $A \subseteq \mathbb{C}$ and an $\varepsilon > 0$ be given. Then the set of all $f \in \mathcal{FP}$ which do not determine a function $f : A \to \overline{\mathbb{D}}_\varepsilon$ is r.e. open.*

Proof. Let $A_0, A_1, \dots$ be an enumeration of rational balls so that $A = \bigcup_i \overline{A_i}$ and let $f \in \mathcal{FP}$ be given. We would like to apply standard continuation techniques (cf., e.g., [8] or [10]) in connection with the *Kreiskettenverfahren*. However, to apply this techniques we need to guarantee that f is indeed a function in contrast to the fact that we want to enumerate all those f which are not functions.

The solution to this problem is quite simple. Let's assume that f is indeed a function $f : A \to \mathbb{D}_\varepsilon$, first. Using the Riemann inequality we get a bound on the coefficients of the Taylor series in each point of each A_i. On the other hand, it is easy to verify that a formal power series whose coefficients are bounded in this way do indeed define a function $f : A \to \mathbb{D}_{2\varepsilon}$. Thus we have simply to assume that f determines a function $f : A \to \mathbb{D}_{2\varepsilon}$ and apply the *Kreiskettenverfahren*. In this way we will eventually realize that f does not define a suitable function by testing the size of the coefficients and values. For simply connected domains A this is already enough to prove the statement of the lemma. For multiply connected domains, f need not be a function although the above construction would say that it is a function. This is because f can depend on the chosen path we use the *Kreiskettenverfahren* on, i.e., the path along which we continue f. Notice that we don't have to consider paths but only finite sequences of $A_{i_0} = A_0, A_{i_1}, \dots, A_{i_s}$ so that $A_{i_n} \cap A_{i_{n+1}} \neq \emptyset$, because such a sequence defines uniquely the continuation of the function. Using continuation along these paths we can enumerate those f which are not functions and prove the lemma even in the multiply connected case.

Now, given a computable, r.e. open family $g : B \dashrightarrow 2^A$, $A \subseteq \mathbb{C}$ and a computable sequence $(z_i)_i$ of points in $\mathbb{C}$, it is easy to see that the set $\{b \mid \exists i.z_i \in g(b)\}$ is r.e. open. Furthermore by testing whether $f(z) \neq f(z')$ for different $z, z' \in A$ one can guarantee that constant f will never be enumerated. Combining this with Lemma 5 and Lemma 6 we get the following corollary:

Corollary 3. *Let A, B be domains in $\mathbb{C}$ with $0 \in A, B$ so that*

(i) A is r.e. open and
(ii) there exists a computable sequence $(z_i)_i$ of computable complex numbers which is dense in the boundary ∂B of B.

Then the class of formal power series $f \in \mathcal{FP}$ so that f does not determine a holomorphic function $f : A \to B$, is r.e. open.

This is already enough to prove an effective version of Theorem 1. As we are only considering bounded domains, we can restrict ourselves to a set of suitably bounded formal power series which we will again denote by $\mathcal{FP}$ in the sequel.

Theorem 4. *Let U be some bounded, simply connected and r.e. open subset of $\mathbb{C}$, $0 \in U$, so that there exists a computable sequence of points on the boundary of U which is dense for this boundary. Then the Riemann mapping $f : U \to \mathbb{D}$ of U with $f'(0) > 0$ and $f(0) = 0$ is computable.*

Proof. Let NF and NF$^-$ denote the sets

$$\text{NF} = \{f \in \mathcal{FP} \mid f \notin \mathcal{H}(U, \mathbb{D})\} \text{ and } \text{NF}^- = \{f \in \mathcal{FP} \mid f \notin \mathcal{H}(\mathbb{D}, U)\},$$

respectively. By Corollary 3 these sets are r.e. open. By uniform versions of Corollary 1 the sets

$$\text{INV} = \{(f, g) \in \mathcal{FP}^2 \mid f \in \mathcal{H}(U, \mathbb{D}), g \in \mathcal{H}(\mathbb{D}, U), f \circ g \neq id\},$$

$$\text{INV}^- = \{(f, g) \in \mathcal{FP}^2 \mid f \in \mathcal{H}(U, \mathbb{D}), g \in \mathcal{H}(\mathbb{D}, U), g \circ f \neq id\},$$

$$\text{NN} = \{f \in \mathcal{FP} \mid f \in \mathcal{H}(U, \mathbb{D}), f(0) \neq 0 \vee f'(0) \not> 0\}$$

are r.e. open in $\mathcal{H}(U, \mathbb{D}) \times \mathcal{H}(\mathbb{D}, U)$ and $\mathcal{H}(U, \mathbb{D})$, respectively. Furthermore

$$\mathcal{FP}^2 \setminus ((\text{NF} \times \text{NF}^-) \cup \text{INV} \cup \text{INV}^- \cup (\text{NN} \times \mathcal{FP})) = \{(f, f^{-1})\}$$

where f is the uniquely determined function of Theorem 1. Thus the statement of the theorem follows by Lemma 4, where we get in addition the inverse of the Riemann mapping for free.

This theorem even characterize the class of domains for which the Riemann mapping is computable (cf. [6]).

To get the effective analogue of Theorem 2 we have to determine in addition the circular slit domain. This seems to be a problem, as the class of these domains is not even compact. However, relaxing the conditions on this class slightly gives us a computably compact space. The Riemann mapping will then take care of the right domain itself:

Theorem 5. *Let U be some bounded, k-connected and r.e. open subset of $\mathbb{C}$ with $0 \in U$ so that there exists a computable sequence of points on the boundary of U which is dense for this boundary. Furthermore let K be the connected component of ∂U determined by the unbounded connected component C of $\mathbb{C} \setminus U$.*

Then there exists a unique computable set $D \in \mathcal{D}_k$ and a computable bi-holomorphic mapping $f : U \to D$ with $f'(0) > 0$, $f(0) = 0$ and $f(K) = \partial \mathbb{D}$.

Proof. Similar to $\mathcal{D}_k$ we define $\hat{\mathcal{D}}_k$ to be the class of sets

$$D = \mathbb{D} \setminus (\iota(\gamma_1, a_1, b_1) \cup \ldots \cup \iota(\gamma_{k-1}, a_{k-1}, b_{k-1}))$$

with $\gamma_i \in \partial \mathbb{D}$, $a_i \in [0; 1]$ and $b_i \in [0; 1]$. Notice that $\mathcal{D}_k \subseteq \hat{\mathcal{D}}_k$ but there exists sets in $\hat{\mathcal{D}}_k$ with several connected components. However, if we restrict the above construction to k-connected sets in $\hat{\mathcal{D}}_k$ we get essentially $\mathcal{D}_k$. Furthermore, the class $(\partial \mathbb{D} \times [0; 1] \times [0; 1])^{k-1}$ is computably compact and r.e. open by Lemma 1.

By a uniform version of Corollary 3 and Lemma 3 the family

$$g_{\mathrm{NF}}(D) = \{(f, g) \mid f \notin \mathcal{H}(U, D) \vee g \notin \mathcal{H}(D, U)\}$$

is r.e. open. Similarly the families g_{INV}, g_{INV^-} and $g_{\mathrm{NN}}x$ defined by

$$g_{\mathrm{INV}}(D) = \{(f, g) \mid f \in \mathcal{H}(U, D) \vee g \in \mathcal{H}(D, U), f \circ g \neq id\},$$

$$g_{\mathrm{INV}^-}(D) = \{(f, g) \mid f \in \mathcal{H}(U, D) \vee g \in \mathcal{H}(D, U), g \circ f \neq id\},$$

$$g_{\mathrm{NN}}x(D) = \{(f, g) \mid f \in \mathcal{H}(U, \mathbb{D}), f(0) \neq 0 \vee f'(0) \not> 0\}$$

are open relative to a suitable superset and thus

$$C = \bigcup D \times (g_{\mathrm{NF}}(D) \cup g_{\mathrm{INV}}(D) \cup g_{\mathrm{INV}^-}(D) \cup g_{\mathrm{NN}}x(D))$$

is r.e. open as well. Then (D, f, f^{-1}) with

$$\{(D, f, f^{-1})\} = (\partial \mathbb{D} \times [0; 1] \times [0; 1])^{k-1} \times \mathcal{FP} \times \mathcal{FP} \setminus C$$

is computable which proves the theorem. Notice, that D is indeed a domain in $\mathcal{D}_k$ because f is a bi-holomorphic function which preserves k-connectivity.

Finally, the effective versions of Theorem 1 and Theorem 2 can be proven by well known reductions to the bounded cases given above by using the square root trick (cf., e.g., [5]). An alternative proof could follow our lines, however, it seems to be unavoidable to use some deeper results from complex analysis, e.g., Koebe's 1/4 Theorem, in this case. Theorem 5 gives in its uniform version essentially the same result as the main result in [7]. However, the above proof is slightly more general as we can adopt it to show that we actually do not need the exact value of the connectedness of the domain.

4   Uniformization

We will present effective versions of Theorem 3 in this section. We omit detailed proofs, which are essentially analog to the proofs of Corollary 3 and Theorem 4, Theorem 5 in the previous section.

Given a finite sequence $\tau_0, \ldots \tau_n$ of maps of a Riemann surface T and a formal power series f, we can proceed exactly as we have done in the prove of Lemma 6 to continue f to T and exclude the case, that f is not a bi-meromorphic map onto $\mathbb{D}$. Notice, that we get a different mapping f_i for each map i. The only thing we have to test in addition to Lemma 6 is that f indeed defines a mapping on T, i.e., we have to exclude those f so that f_i differs on some z from f_j at $\tau_j^{-1} \circ \tau_i(z)$. But this can be done exactly as we have done with the other properties. Thus, in the case that T is defined by finitely many maps and is equivalent to $\mathbb{D}$ the uniformization mapping of Theorem 3 is computable. In a similar way one can show computability in the other cases where T is defined by finitely many maps.

Theorem 6. *Let (T, τ) be a finitely generated, simply connected and computable Riemann surface. Then there exists computable bi-meromorphic uniformization functions $f : T \to \mathbb{D}$, $f : T \to \mathbb{C}$ or $f : T \to \mathbb{C}_\infty$.*

As a consequence we get the following general result for compact Riemann surfaces:

Corollary 4. *For any computable and compact, simply connected Riemann surface (T, τ) there exists a bi-holomorphic computable function $f : T \to \mathbb{C}$.*

Such generalizations of Theorem 6 are not true in the hyperbolic or Euclidean case. However, we can give a characterization similar to the one given in Theorem 4 and 5. To this end let d_T denote the hyperbolic metric on T.

Theorem 7. *Let (T, τ) be a computable, hyperbolic and simply connected Riemann surface, z_0 be a computable element in T and $(z_{i,j})_{i,j}$ be a computable sequence in T so that*

1. *$d(z_{i,j}, z_0) > j$ for all $i, j \in \mathbb{N}$ and*
2. *$\forall z \in T, n \in \mathbb{N}.d(z, z_0) > n \Rightarrow \exists i, j.j \geq n \wedge d(z, z_{i,j}) < 1/n$.*

Then there exists a computable bi-meromorphic mapping $f : T \to \mathbb{D}$.

5   Discussion

We have presented examples for the very powerful and simple principle given in Lemma 4. Although problems in complex analysis are well suited for this principle it is likely that this principle is of use even in other fields.

Furthermore the results given are purely computability results. Using the same method it seems, however, possible to achieve at least certain space bounds. As for all three effective versions of conformal mappings there is a lower complexity bound $\sharp P$ (cf., e.g., [2]), thus an upper bound $PSPACE$ would be indeed almost giving the exact complexity. Thus it seems to be worth to investigate even the complexity aspects of the principle given in Lemma 4.

Finally, most of the presented results can be given in a uniform way. This is obvious for simply and multiply connected sub-domains of $\mathbb{C}$ or, for Riemann surfaces, in the hyperbolic case. In the spherical and Euclidean case, however, this seems impossible with the information given by the representations of this paper.

References

1. Ahlfors, L.: Complex Analysis, 3rd edn. McGraw-Hill Science/Engineering/Math. (1979)
2. Binder, I., Braverman, M., Yampolsky, M.: On computational complexity of Riemann mapping. Arkiv for Matematik 45(2) (2007)
3. Brattka, V., Presser, G.: Computability on subsets of metric spaces. Theoretical Computer Science 305, 43–76 (2003)
4. Galatolo, S., Hoyrup, M., Rojas, C.: Dynamics and abstract computability:computing invariant measures. Discrete Contin. Dyn. Sys. 29(1), 193–212 (2011)
5. Henrici, P.: Applied and computational complex analysis. Wiley Classics Library Series, vol. 3 (1986)
6. Hertling, P.: An effective Riemann mapping theorem. Theoretical Computer Science 219(1-2), 225–265 (1999)
7. Andreev, V., McNicholl, T.: Computing Conformal Maps onto Canonical Slit Domains. In: 6th Int'l Conf. on Computability and Complexity in Analysis. Schloss Dagstuhl - Leibniz-Zentrum fuer Informatik, Germany (2009)
8. Müller, N.: Uniform Computational Complexity of Taylor series. In: Ottmann, T. (ed.) ICALP 1987. LNCS, vol. 267, pp. 435–444. Springer, Heidelberg (1987)
9. Nehari, Z.: Conformal mapping. McGraw-Hill, New York (1952)
10. Rettinger, R.: Lower Bounds on the Continuation of Holomorphic Functions. Electr. Notes Theor. Comput. Sci 221, 207–217 (2008)
11. Rettinger, R., Weihrauch, K.: Tychonoff for compact sets is computable (submitted)
12. Weihrauch, K.: Computable Analysis. Springer, Berlin (2000)
13. Weihrauch, K., Grubba, T.: Elementary Computational Topology. Journal of Universal Computer Science 15(6), 1381–1422 (2009)

On the Computability Power
of Membrane Systems with Controlled Mobility

Shankara Narayanan Krishna[1], Bogdan Aman[2,3], and Gabriel Ciobanu[2,3]

[1] Department of Computer Science and Engineering, IIT Bombay, Powai, Mumbai 400076, India
krishnas@cse.iitb.ac.in
[2] Institute of Computer Science, Romanian Academy, Str. T. Codrescu nr.2, cod 700481, Iasi, Romania
baman@iit.tuiasi.ro
[3] "A.I.Cuza" University of Iaşi, Bulevardul Carol I, Nr.11, 700506, Iasi, Romania
gabriel@info.uaic.ro

Abstract. In a previous paper we have shown that membrane systems with controlled mobility are able to solve a Π_2^P complete problem. Then, an enriched model with forced endocytosis and forced exocytosis enables us to move to the fourth level in the polynomial hierarchy, the model having $\Sigma_4^P \cup \Pi_4^P$ as lower bound. In this paper we study the computability power of this model (using forced endocytosis and forced exocytosis), and determine the border condition for achieving computational completeness: 4 membranes provide Turing completeness, while 3 membranes do not. Moreover, we show that the restricted division operation (which is crucial in achieving the $\Sigma_4^P \cup \Pi_4^P$ lower bound) does not provide computational completeness. However, Turing completeness can be achieved with pairs of operations (exocytosis, inhibitive endocytosis) and (inhibitive exocytosis, endocytosis) by using 4 membranes. Finally, we present some computability results expressing that membrane systems which use the operations of restricted division, restricted exocytosis and inhibitive endocytosis cannot yield computational completeness.

1 Introduction

The main data structure investigated in membrane computing [12] is the multiset, a set of objects with multiplicities associated with its elements. One of the basic operations on multisets, also corresponding to bio-chemical reactions taking place in a cell, is multiset rewriting. The evolution of P systems was defined mainly by multiset rewriting rules over the objects present in the compartments of the membrane structure, while the structure remained the same during the evolution/computation [12]. However, from a mathematical and biological point of view, this is a too restricted case. It is natural to also allow the membrane structure to evolve by division rules, endocytosis rules (taking a membrane that is outside a neighbouring membrane and moving it inside) and exocytosis rules (moving a membrane outside the membrane that contains it).

S.B. Cooper, A. Dawar, and B. Löwe (Eds.): CiE 2012, LNCS 7318, pp. 626–635, 2012.

When membrane systems are seen as computing devices, two main research directions are usually considered: computational power in comparison with the classical notion of Turing computability, and efficiency in algorithmically solving hard problems (e.g., NP-problems) in polynomial time [12]. In this respect, membrane systems define classes of computing devices which are both powerful, usually equivalent to Turing machines, and efficient.

Polynomial-time solutions to NP-complete problems using P systems with active membranes, through the generation of an exponential space, are presented comprehensively in [13]. Some solutions using systems of mobile membranes have been proposed recently in [1,8,5]. This paper is a continuation of the attempts made to answer essentially an open question posed in [12]: *Is it possible to efficiently solve* **PSPACE***-complete problems ... without using any of the operations of non-elementary membrane division ...?* In [8], we solved a Π_2^P complete problem using P systems with mobile membranes by using elementary membrane division, endocytosis, exocytosis and inhibitive endocytosis. Recently we enriched the model in [8], and showed that a restricted use of the operations *endocytosis* (endo), *exocytosis* (exo), *forced endocytosis* (fendo), *forced exocytosis* (fexo), *inhibitive endocytosis* (iendo) and *elementary membrane division* (div) can give a semi-uniform solution of a Π_4^P complete problem in polynomial time [5]. This is the first known result that connects the complexity class Π_4^P within the P area. It is not clear whether this model can efficiently solve **PSPACE**-complete problems, but the fact that it does not use any of the powerful operations mentioned in the open question, and is capable of reaching level 4 of the polynomial hierarchy seems encouraging.

Here, we analyse the computational power of the various operations used in our model. The interplay between endo, exo, fendo, fexo is quite powerful, and the computational power of a Turing machine is obtained using twelve membranes [6]. The computability aspects of the different forms of mobility in enhanced mobile membranes are investigated. Several combinations of operations that provide Turing completeness are obtained: ten membranes along with fendo, fexo; nine membranes with endo, exo, pendo; eight membranes with fendo, fexo, pendo and twelve membranes with endo, exo, fendo give computational completeness. A form of restricted mobility described by some new rules does not lead to computational completeness.

We conjecture that the operations used to reach level four of the polynomial hierarchy are not "powerful enough" to yield computational completeness. Other important results obtained in this paper characterize tight bounds for obtaining computational completeness. We observe that the results obtained in this paper use three, four or five membranes, a major improvement with respect to the results obtained in [7] where eight, nine or ten membranes are used. It is worth noting that three is the smallest number of membranes when effectively using the movement of membranes given by endocytosis and exocytosis.

2   Enhanced Mobile Membranes with Controlled Mobility

In what follows, the set of natural numbers is denoted by $\mathbf{N}$, and V denotes a finite alphabet. The free monoid generated by V under the operation of concatenation and the empty string denoted by λ, as unit element, is denoted by V^*.

We assume that the reader is familiar with membrane computing; for the state of the art, see [12]. The class of mobile membrane systems used here is an extension of *P systems with mobile membranes and controlled mobility* [8], which is a construct $\Pi = (V, H, \mu, w_1, \dots, w_n, R, I)$, where: $n \geq 1$ (the initial *degree* of the system); V is an alphabet (its elements are called *objects*); H is a finite set of *labels* for membranes; μ is a *membrane structure*, consisting of n membranes, labelled with elements of H; $w_1, w_2, \dots, w_n$ are strings over V, describing the initial *multisets of objects* placed in the n membranes of μ, I is the set of elementary membranes of μ, representing the output membranes of the system, and R is a finite set of *developmental rules* of the following forms, where membrane h is elementary and membrane m is not necessarily elementary:

(a) $[a]_h[\ ]_m \rightarrow [[w]_h]_m$, for $h, m \in H, a \in V, w \in V^*$ endocytosis
a membrane labelled h enters the adjacent membrane labelled m under the control of object a; the labels h and m remain unchanged during this process; however, the object a is modified to w during the operation.

(b) $[[a]_h]_m \rightarrow [w]_h[\ ]_m$, for $h, m \in H, a \in V, w \in V^*$ exocytosis
a membrane labelled h is sent out of a membrane labelled m under the control of object a; the labels of the two membranes remain unchanged; however the object a from membrane h is modified during this operation;

(c) $[\ ]_h[a]_m \rightarrow [[\ ]_h w]_m$, for $h, m \in H, a \in V, w \in V^*$ forced endocytosis
a membrane labelled h enters the adjacent membrane labelled m under the control of object a of m; the labels h and m remain unchanged during this process; however, the object a is modified to w during the operation.

(d) $[a[\ ]_h]_m \rightarrow [\ ]_h[w]_m$, for $h, m \in H, a \in V, w \in V^*$ forced exocytosis
a membrane labelled h is sent out of a membrane labelled m under the control of object a of m; the labels of the two membranes remain unchanged; however the object a of membrane m is modified to w during this operation.

(e) $[a]_h[]_{m/\neg S} \rightarrow [[a]_h]_m$, for $h, m \in H, a \in V$ inhibitive endocytosis
a membrane labelled h containing a can enter m provided m does not contain any object from S; the object a and the labels h and m of the membranes also remain unchanged. The objects of S are *inhibitors* that prevent membrane h from entering membrane m whenever h contains the object a.

(f) $[a[]]_{h\ m/\neg S} \rightarrow []_m[a]_h$, for $h, m \in H, a \in V$ inhibitive exocytosis
a membrane labelled h containing a can exit m provided m does not contain any object from S; the object a does not evolve in the process; the labels h and m of the membranes also remain unchanged.

(g) $[a]_h \rightarrow [u]_h[v]_h$, for $h \in H, a \in V, u, v \in V^*$ elementary division
in reaction with an object a, the membrane labelled h is divided into two membranes labelled h, with the object a replaced in the two new membranes by possibly new objects; the other objects remain unchanged.

The paper [8] did not consider the operations of forced endocytosis and forced exocytosis. The rules are applied according to the following principles:

1. All rules are applied in parallel, non-deterministically choosing the rules, membranes, and objects, but in such a way that the parallelism is maximal (in each step we apply a multiset of rules such that no further rule can be added, no further membranes and objects can evolve at the same time).
2. Membrane m is said to be *passive*, while membrane h is said to be *active*. In a computation step, any object or any active membrane can be involved in at most one rule, but the passive membranes are not considered involved (hence they can be used by several rules at the same time as passive membranes).
3. The evolution of objects and membranes takes place in a bottom-up manner. After having a (maximal) multiset of rules chosen, they are applied starting from the innermost membranes, level by level, up to the skin membrane (all these sub-steps form a unique evolution step, called a *transition* step).
4. When a membrane is moved across another membrane by rules (a)-(f), its whole contents (its objects) are moved.
5. All objects and membranes which do not evolve at a given step, are passed unchanged to the next configuration of the system.

By using the rules in this way, we get transitions among the configurations of the system. A computation is a sequence of transitions that starts from an initial configuration, such that each two transitions are connected by a common configuration. A computation is successful if it halts (it reaches a configuration where no rule can be applied).

At the end of a halting computation, the number of objects in the membranes I is considered to be the result of the computation. A non-halting computation provides no output. The family of all sets of numbers $\mathbf{N}(\Pi)$ which are obtained as a result of a halting computation by a P system Π with enhanced mobile membranes and controlled mobility of degree at most n using rules $\alpha \subseteq \{\mathsf{exo}, \mathsf{endo}, \mathsf{fendo}, \mathsf{fexo}, \mathsf{iendo}, \mathsf{rexo}, \mathsf{div}\}$, is denoted by $\mathbf{NEMCM}_n(\alpha)$. Here $\mathsf{iendo}, \mathsf{rexo}$ represent inhibitive endocytosis and inhibitive exocytosis, and div denotes division. When the number of membranes is finite, and not fixed as any particular n, we denote it as $*$.

An alternate notion of obtaining a result of a computation is to look at the language generated by the system. Given I, let $f_i, i \in I$ be the multiset content of membrane i at the end of a halting computation. Then $L(\Pi) = \{\uplus_{i \in I} f_i\}$ is the language associated with Π, namely the union of the multisets in all membranes. The family of all languages $L(\Pi)$ which are obtained as a result of a halting computation by a P system Π with enhanced mobile membranes and controlled mobility of degree at most n using rules $\alpha \subseteq \{\mathsf{exo}, \mathsf{endo}, \mathsf{fendo}, \mathsf{fexo}, \mathsf{iendo}, \mathsf{rexo}, div\}$, is denoted by $LEMCM_n(\alpha)$.

When $|w| = 1$ in rules (a)–(d), we call the operations "restricted", and use $\mathsf{rendo}, \mathsf{rexo}, \mathsf{rfendo}, \mathsf{rfexo}$ to denote them. Here, r stands for restricted. Likewise, when $|u| = |v| = 1$ in rule (g), we denote it by rdiv.

A special case of the use of iendo (rexo) rules is when the inhibitor set is $S = \varnothing$. Then, there are no symbols whose absence is mandatory for the mobility.

However, the thing to note here is that there is no evolution of objects. Thus, when $S = \varnothing$, an iendo rule is written as $[a]_i[\ ]_j \rightarrow [[a]_i]_j$. Similar for rexo.

3   Computability Power of Controlled Mobility

The language and automata approach is fundamental in introducing computing devices and investigating their computing power, compared with the power of Turing machines and other classical models of computation. **NRE**, **NCS**, **NO**$^\lambda$, **NMAT**, and **NFIN** denote the family of recursively enumerable sets, context-sensitive sets, ordered sets, sets of matrix grammars without appearance checking and finite sets of natural numbers, respectively. It is known that $\mathbf{NMAT}^\lambda, \mathbf{NMAT}_{ac}, \mathbf{NCS}, \mathbf{NO}^\lambda \subset \mathbf{NRE}$ [3].

In this section we explore the computability power of the class consisting of operations rendo, rexo, rfendo, rfexo, iendo, rexo, rdiv. The complexity of this class can be found in [5], where the $\Sigma_4^P \cup \Pi_4^P$ lower bound is proved.

The next result defines the border for computational completeness while using the operations endo, exo, fendo and fexo. Here, we show that four membranes are sufficient for computational completeness by using the fact that each recursively enumerable language can be generated by a matrix grammar in the strong binary normal form [4]. This result is tight from the known result in [6]: three membranes do not give computational completeness.

First we define the notions used in the proof: a *matrix grammar with appearance checking* is a tuple $G = (N, T, M, S, F)$ where N, T are sets of non-terminals and terminals respectively, S is the start symbol, M is a finite set of matrices of the form $(r_1, \ldots, r_m), m \geq 1$, with context-free rewriting rules $r_i : \alpha_i \rightarrow \beta_i$, $\alpha_i \in N$, $\beta_i \in (N \cup T)^*$, and F is a subset of the totality of occurrences of rules in the matrices of M. For any two strings x, y we say that $x \Rightarrow_{ac} y$ iff there are strings $x_0, \ldots, x_n$ and a matrix $(r_1, \ldots, r_m) \in M$ such that $x_0 = x, x_n = y$, and $x_{i-1} = x'_{i-1}\alpha_i x''_{i-1}, x_i = x'_{i-1}\beta_i x''_{i-1}$ for some $x'_{i-1}, x''_{i-1} \in (N \cup T)^*$, for all $1 \leq i \leq n - 1$, or else, the rule $\alpha_i \rightarrow \beta_i \in F$, α_i is not a subword of x_{i-1}, and $x_{i-1} = x_i$. In other words, a direct derivation in G corresponds to applying the rules of a matrix, in order, skipping the rules of F. We denote by $\mathrm{MAT}_{ac}^\lambda$ the families of languages generated. In case all rules are λ-free, we remove the superscript λ from the notation. G is in the *strong binary normal form*, if $N = N_1 \cup N_2 \cup \{S, \dagger\}$, with these three sets mutually disjoint, two distinguished symbols $B^{(1)}, B^{(2)} \in N_2$, and the matrices in M of one of the following forms:

1. $(S \rightarrow XA)$, with $X \in N_1, A \in N_2$,
2. $(X \rightarrow Y, A \rightarrow x)$, with $X, Y \in N_1, A \in N_2, x \in (N_2 \cup T)^*, |x| \leq 2$,
3. $(X \rightarrow Y, B^{(j)} \rightarrow \dagger)$, with $X, Y \in N_1, j = 1, 2$,
4. $(X \rightarrow \lambda, A \rightarrow x)$, with $X \in N_1, A \in N_2, x \in T^*, |x| \leq 2$.

There is only one matrix of type 1, and F consists of the rules $B^{(j)} \rightarrow \dagger$ with $j = 1, 2$ appearing in matrices of type 3 ($\dagger$ is a trap-symbol, once introduced it is never removed). A matrix of type 4 is used only once, in the last step of a derivation. It is proved in [4] that each recursively enumerable language can be generated by a matrix grammar in the strong binary normal form.

Theorem 1. $\mathrm{NEMCM}_4(\mathsf{endo},\mathsf{exo},\mathsf{fendo},\mathsf{fexo}) = \mathrm{NRE}$.

Proof. We simulate a matrix grammar with appearance checking $G = (N, T, M, S, F)$ in the strong binary normal form. We construct the P system $\Pi = (V, \{0,1,2,3\}, [\ [\ ]_1\ [\ ]_2\ [\ ]_3]_0, \varnothing, \{XA\}, \varnothing, \varnothing, R, \{1,2,3\})$. The symbols X, A correspond to the initial matrix $(S \to XA)$. Let there be n_1 matrices of types 2,4 labelled $1, \ldots, n_1$ and n_2 matrices of type-3 labelled $n_1 + 1, \ldots, n_1 + n_2$. We replace matrices of type-4 with $(X \to Z, A \to x)$ where Z is a new special symbol. $V = N \cup T \cup \{Y_j, A_j \mid Y \in N_1, A \in N_2, 1 \le j \le n_1 + n_2\} \cup \{Y', Y^{(1)}, Y^{(2)}, \overline{Y}_1, \overline{Y}_2 \mid Y \in N_1\} \cup \{Z, \dagger\} \cup \{\langle x\rangle \mid x \in (N_2 \cup T)^*, |x| \le 2\}$. The rules are:

Simulation of a type-2 matrix $m_i : (X \to Y, A \to x)$

1. $[X]_1[\ ]_2 \to [[Y_i]_1]_2, [[A]_1]_2 \to [A_j]_1[\ ]_2, X \in N_1, A \in N_2$ (endo, exo)
   provided matrices m_i, m_j have rules for X, A respectively
2. $[Y_k]_1[\ ]_3 \to [[Y_{k-1}]_1]_3, [[A_k]_1]_3 \to [A_{k-1}]_1[\ ]_3, k \ge 2$ (endo, exo)
3. $[Y_1]_1[\ ]_3 \to [[Y']_1]_3, [[A_1]_1]_3 \to [\langle x\rangle]_1[\ ]_3$ (endo, exo)
4. $[[\langle x\rangle]_1]_3 \to [\dagger]_1[\ ]_3, [[Y_j]_1]_3 \to [\dagger]_1[\ ]_3$ (exo, exo)
5. $[Y']_1[\ ]_3 \to [Y[\ ]_3]_1, [A_j[\ ]_3]_1 \to [\dagger]_1[\ ]_3, j \ge 1$ (fendo, fexo)
6. $[\langle x\rangle[\ ]_3]_1 \to [x]_1[\ ]_3$ (fexo)

Simulation of a type-3 matrix $m_i : (X \to Y, B^{(j)} \to \dagger)$

1. $[X]_1[\ ]_2 \to [Y^{(1)}[\ ]_2]_1$ if $(X \to Y, B^{(1)} \to \dagger)$ is a type-3 matrix   (fendo)
2. $[X]_1[\ ]_3 \to [Y^{(2)}[\ ]_3]_1$ if $(X \to Y, B^{(2)} \to \dagger)$ is a type-3 matrix   (fendo)
3. $[B^{(1)}[\ ]_2]_1 \to [\dagger]_1[\ ]_2, [B^{(2)}[\ ]_3]_1 \to [\dagger]_1[\ ]_3$ (fexo)
4. $[Y^{(1)}]_1[\ ]_3 \to [\overline{Y}_1\overline{Y}_2[\ ]_3]_1, [Y^{(2)}]_1[\ ]_2 \to [\overline{Y}_1\overline{Y}_2[\ ]_2]_1$ (fendo)
5. $[\overline{Y}_1[\ ]_2]_1 \to [\lambda]_1[\ ]_2, [\overline{Y}_2[\ ]_3]_1 \to [Y]_1[\ ]_3$ (fexo)

Termination

1. $[Z]_1[\ ]_2 \to [[\lambda]_1]_2, [A_j]_1[\ ]_3 \to [\dagger[\ ]_3]_1, A \in N_2$
2. $[\dagger]_1[\ ]_i \to [[\dagger]_1]_i, [[\dagger]_1]_i \to [\dagger]_1[\ ]_i, i = 2,3$ (endo, exo)

The initial configuration is $[[XA]_1[\ ]_2[\ ]_3]_0$. Assume there is a type-2 matrix $m_i : (X \to Y, A \to x)$. Then the following rules are used: membrane 1 enters 2 replacing X with Y_i, and membrane 1 comes out replacing some $A \in N_2$ with A_j. Membrane 1 keeps moving in and out of membrane 3 reducing the indices i, j. When we have Y_1, membrane 1 enters membrane 3, replacing Y_1 with Y'; if simultaneously we have A_1, it is replaced with $\langle x\rangle$. Consider the case $j < i$. We have $\langle x\rangle$ before obtaining Y_1. Then, $\langle x\rangle$ is replaced with $\dagger$ when membrane 1 comes out of membrane 3. Suppose $j > i$. We have $Y'A_l$ in membrane 1, $l \ge 1$. Then a fendo rule is used, by which membrane 3 enters membrane 1, replacing Y' with Y; A_l is replaced with $\dagger$. The case when X is replaced with X_i, and there are no symbols $A \in N_2$ is handled by the rule $[[Y_j]_1]_3 \to [\dagger]_1[\ ]_3$.

    The simulation of a type-3 matrix proceeds as follows: we have $[X]_1$ in membrane 1. If X corresponds to a type-3 matrix, $m_i : (X \to Y, B^{(1)} \to \dagger)$, then membrane 2 enters membrane 1 using a fendo rule, replacing X with $Y^{(1)}$. If there is a $B^{(1)}$ present in membrane 1, then membrane 2 replaces this with a † triggering an infinite computation; in parallel, membrane 3 enters membrane 1 replacing $Y^{(1)}$ with $\overline{Y}_1\overline{Y}_2$. Consider the two possibilities : (1) we have the configuration $[[\overline{Y}_1\overline{Y}_2[\ ]_2[\ ]_3]_1]_0$ (hence, $B^{(1)}$ is absent). In this case, membranes 2, 3 leave membrane 1: $\overline{Y}_1$ is erased, while $\overline{Y}_2$ is replaced with Y. (2) we have the configuration $[[\overline{Y}_1\overline{Y}_2†[\ ]_3]_1[\ ]_2]_0$ (hence, $B^{(1)}$ was present). In this case, membrane 3 leaves membrane 1 replacing $\overline{Y}_2$ with Y, but this does not matter, since we are anyway getting into an infinite computation. The case when X is part of the matrix $(X \to Y, B^{(2)} \to †)$ is symmetric; membranes 2, 3 exchange roles.

To terminate, we first simulate a type-4 matrix, at the end of which we obtain a Z in membrane 1. This Z is erased when membrane 1 enters membrane 2. Note that if there are symbols of N_2 present, then membrane 1 will come out of membrane 2 replacing $A \in N_2$ with A_j. In this case, membrane 3 enters membrane 1 replacing A_j with †. Note that this is the only applicable rule in this case. In the case when membrane 1 has no symbols of N_2, it remains trapped inside membrane 2, the contents of membrane 1 giving the output.   □

We shall now show that using inhibitive endocytosis and exocytosis, it is possible to obtain computational completeness with 4 membranes.

Theorem 2. *1.* $\mathrm{NEMCM}_4(\mathsf{iendo}, \mathsf{exo}) = \mathbf{NRE}$,
2. $\mathrm{NEMCM}_4(\mathsf{endo}, \mathsf{rexo}) = \mathbf{NRE}$.

Remark 1 Note that the semantics of the rules $\mathsf{iendo}, \mathsf{exo}$ allow a membrane i to enter membrane j using an iendo rule $[a]_i[\ ]_{j\backslash\neg S} \to [[a]_i]_j$, and then membrane i exits membrane j using a rule $[[b]_i]_j \to [w]_i[\ ]_j$, $b \neq a$.

In what follows we consider a restricted use of the rules iendo, exo: if a membrane i enters membrane j using an iendo rule $[a]_i[\ ]_{j\backslash\neg S} \to [[a]_i]_j$ in a step $k \geq 1$, then during step $k+1$, membrane i comes out of membrane j replacing the same symbol a with some w. We call this a *restricted semantics of* $\mathsf{iendo}, \mathsf{exo}$. It is clear that by enforcing this *restricted semantics*, Theorem 2 still remains valid.

We denote by $\mathrm{NEMCM}_n^r(\mathsf{iendo}, \mathsf{exo})$ the families of numbers $\mathbf{N}(\Pi)$ computed by systems Π under the restricted semantics of $\mathsf{iendo}, \mathsf{exo}$. The next theorem says that such systems with 3 membranes are strictly less than RE, by using ordered grammars that are known to be less than RE. An *ordered grammar* is a quadruple $G = (N, T, S, P)$ where N is finite set of non-terminals, T is a finite set of terminals, $S \in N$ is the start symbol and P is a finite partially ordered set of context-free production rules of the form $A \to x$, $A \in N, x \in (N \cup T)^*$. For $x, y \in (N \cup T)^*$, we say that x directly derives y, written $x \Rightarrow y$ iff there is a production $p : A \to w \in P$ such that $x = x'Ax''$, $y = x'wx''$, and there is no production $q : B \to v \in P$ such that $p \prec q$ and B occurs in x. The generated language is $L(G) = \{w \mid w \in T^*, S \Rightarrow^* w\}$. The family of languages generated is denoted O^λ. When λ-free rules are used, we remove λ from the notation.

Theorem 3. *1.* $\mathrm{NEMCM}_4^r(\mathsf{iendo}, \mathsf{exo}) = \mathbf{NRE}$,
2. $\mathrm{NEMCM}_3^r(\mathsf{iendo}, \mathsf{exo}) \subseteq \mathbf{NO}^\lambda \subset \mathbf{NRE}$.

Proof. 1. Follows from Theorem 2 and Remark 1 above.

2. Given a P system with 3 membranes $\Pi = (V, \{1, 2, 3\}, \mu, w_1, w_2, w_3, R, I)$, we construct an ordered grammar G that simulates Π such that $\mathbf{N}(\Pi) = \mathbf{N}(L(G))$. Without loss of generality, assume that $\mu = [\ [[\ ]_1]_2]_3$. Define $G = (N, T, S, P)$ with $N = V \cup V_1 \cup V_2 \cup V' \cup \overline{V}_1 \cup \overline{V}_2 \cup \{f_{r,a_i}^{in} \mid i = 1, 2, r \in R\} \cup \{\dagger\}$, $T = \overline{V}_1$. Here, $V_i = \{a_i \mid a \in V\}, \overline{V}_i = \{\overline{a}_i \mid a \in V\}$, $i = 1, 2$ and $V' = \{a' \mid a \in V\}$. Let the rules of R be labelled $r_1, \dots, r_n$. The production rules of P are as follows:

(a) $p_1 : S \rightarrow a'x_1w_2$, with $w_1 = ax_1$, and exists an exo rule $[[a]_1]_2 \rightarrow [y]_1[\ ]_2$

(b) $p_2 : a' \rightarrow y_1$

(c) $p_3 : S \rightarrow \overline{w}_1$ if none of the symbols of w_1 have exo rules in R

(d) $p_{r,a,i} : a_i \rightarrow f_{r,a_i}^{in}$, such that $r : [a]_i[\ ]_{j\backslash\neg X} \rightarrow [[a]_i]_j \in R$, $i \in \{1, 2\}$

(e) $p'_{b,r,j} : b_j \rightarrow \dagger$ for $b \in X$, given $r \in R$ as above

(f) $p''_{a,i} : f_{r,a_i}^{in} \rightarrow y_i$ if there is an exo rule $[[a]_i]_j \rightarrow [y]_i[\ ]_j$, $i, j \in \{1, 2\}$

(g) $p_{a,i} : a_i \rightarrow \overline{a}_i$ if there is no iendo rule $[a]_i[\ ]_{j\backslash\neg X} \rightarrow [[a]_i]_j \in R$, $i \in \{1, 2\}$.

(h) $p_{t,a,2} : \overline{a}_2 \rightarrow \lambda$

with order $p'_{b,r,j} \succ p_{r,a,i}, p_{r,a,i} \succ p_{a,i}, p''_{a,i} \succ p_{r,b,j}$ for $i, j \in \{1, 2\}$ and $a, b \in V$.

Let w_1, w_2, w_3 be the contents of membranes 1, 2 and 3. Since the initial configuration is $[[[w_1]_1w_2]_2w_3]_3$, the computation begins with an exo rule, simulated with the rules p_1, p_2. Any of the symbols having an exo rule are picked up in membrane 1, primed and replaced using p_2. Here on, the restrictive semantics comes into play. The configuration after the first exo rule is $[[y_1x_1]_1[w_2]_2w_3]_3$.

One of the membranes 1, 2 can enter the other using an iendo rule. If a membrane (assume membrane 1) enters the other (membrane 2) using an iendo rule r involving a, then in the next step, a is replaced using an exo rule due to the restrictive semantics of iendo, exo. Rule $p_{r,a,1}$ models the entry of membrane 1 into membrane 2. Note that if the inhibitor set X in rule r has symbols $b^1, \dots, b^k$, then the rules $p'_{b^l_2,r,2}$ all have a greater priority than $p_{r,a,1}$; $1 \le l \le k$. That is, if any of the symbols $b^1, \dots, b^k$ are present in membrane 2, then the iendo rule cannot be used. Note that when $p_{r,a,1}$ is used, then certainly none of the symbols $b^1, \dots, b^k$ are present in membrane 2, and a_1 is replaced with f_{r,a_1}^{in}. This denotes the fact that membrane 1 is "inside" membrane 2, using rule r, involving symbol a. The only next applicable rule is $p''_{a,1}$ simulating the exo rule replacing this symbol a. Note the order which enforces this: in the presence of a symbol f_{r,a_i}^{in}, no other f_{q,b_j}^{in} can be introduced before f_{r,a_i}^{in} is rewritten. This is continued until we reach a configuration which is the halting configuration.

Note that by the definition of restrictive semantics, if we are in a configuration $[[[y_i]_ix_j]_jw_3]_3$ that is not the initial configuration, then this cannot be halting (since we achieved this configuration using an iendo rule in the previous step). Thus, the halting configuration can only be $[[y_1]_1[x_2]_2w_3]_3$. To simulate halting, we use rules $p_{a,i}$ for $i = 1, 2$. Note the order $p_{r,a,i} \succ p_{a,i}$ prevents replacing symbols a_i with $\overline{a}_i$ if an iendo rule is applicable. Thus, we obtain a string over $\overline{V}_1 \cup \overline{V}_2$ only when we reach a halting configuration. This is followed by rule $p_{t,a,2}$ which erases all symbols of membrane 2 at the end of a halting configuration.

In case the initial configuration is $[\,[\,w_1]_1[\,w_2]_2 w_3\,]_3$, then the initial rule will be $S \rightarrow f^{in}_{r,a_1} x_1 w_2$ with $w_1 = a x_1$ or $S \rightarrow f^{in}_{r,a_2} x_2 w_1$ with $w_2 = a x_2$. The erasing rule $p_{t,a,2}$ will not be there in this case.   □

Open problem It is known that $\mathbf{NEMCM}_3^r(\mathsf{iendo}, \mathsf{exo}) \subseteq \mathbf{NEMCM}_3(\mathsf{iendo}, \mathsf{exo}) \subseteq \overline{\mathbf{NEMCM}_4^r}(\mathsf{iendo}, \mathsf{exo}) = \mathbf{NRE}$. Which inclusions are strict? What is the power of $\mathrm{EMCM}_3(\mathsf{iendo}, \mathsf{exo})$? An interesting class is $\mathrm{EMCM}_3(\mathsf{iendo}, \mathsf{exo})$—on the one hand, it seems very difficult to simulate a matrix grammar with appearance checking/ two counter machine, since we have just two membranes to play with; on the other hand, the inherent appearance checking in the iendo rules, coupled with the general (erasing) nature of the exo rules make it difficult to find mechanisms less than RE to simulate these.

A related question is the power of $\mathbf{NEMCM}_*(\mathsf{iendo}, \mathsf{exo}^{-\lambda})$ where $\mathsf{exo}^{-\lambda}$ stands for exocytosis rules where $a \in V$ is replaced with $w \in V^+$. If we assume the initial membrane structure $\mu = [\,[\ ]_1[\ ]_2 \cdots [\ ]_{n-1}]_n$, then it can be shown that $\mathbf{NEMCM}_*(\mathsf{iendo}, \mathsf{exo}^{-\lambda}) \subseteq \mathbf{NMAT}_{\mathsf{ac}} \subset \mathbf{NRE}$. The intuition behind this statement is: keep a flag to remember the current membrane structure, have a matrix of rules, using appearance checking to simulate iendo rules. No erasing is needed, since $I = \{1, \ldots, n-1\}$. A quotient of the computed language with respect to the flag completes the simulation. However, for a general μ, the computational power is not clear, since I could be a proper subset of $\{1, \ldots, n\}$, and the contents of membranes in $\{1, \ldots, n\} \backslash I$ have to be erased at the end.

We show that restrictive mobility is not powerful without rdiv (Theorem 4), while rdiv offers limited power in the absence of restrictive mobility (Theorem 5).

Theorem 4. $\mathbf{NEMCM}_*(\mathsf{rendo}, \mathsf{rexo}, \mathsf{rfendo}, \mathsf{rfexo}, \mathsf{iendo}, \mathsf{rexo}) \subseteq \mathbf{NFIN}$.

Theorem 5. $\mathbf{NEMCM}_*(\mathsf{rdiv}) \subseteq \mathbf{NCS}$.

The operations used to establish the $\Sigma_4^p \cup \Pi_4^p$ lower bound are $\{\mathsf{rdiv}, \mathsf{rendo}, \mathsf{rexo}, \mathsf{rfendo}, \mathsf{rfexo}, \mathsf{iendo}\}$. Using Theorems 4 and 5, we conjecture that $\mathbf{NEMCM}_*(\mathsf{rdiv}, \mathsf{rendo}, \mathsf{rexo}, \mathsf{rfendo}, \mathsf{rfexo}, \mathsf{iendo}, \mathsf{rexo}) \subset \mathbf{NRE}$. Looking at the proof of Theorem 2, one could think of replacing exo rules using rdiv, while considering matrix grammars (with or without appearance checking) to simulate the rules $A \rightarrow x$, ($|x| \leq 2$ helps). However, this creates extra copies of symbols, which have to be done away with, and the lack of erasing rules poses a problem. The next theorem puts together the restricted division with restricted mobility, and compare them with $\mathbf{NMAT}^\lambda$ and $ET0L$.

An *E0L system* is a context-free pure grammar with parallel derivations: $G = (V, T, \omega, R)$ where V is an alphabet, $T \subseteq V$ is a terminal alphabet, $\omega \in V^*$ is an axiom, and $R = \{a \rightarrow v \mid a \in V$ and $v \in V^*\}$ is a finite set of rules such that for each $a \in V$ there is at least one rule $a \rightarrow v$ in R. For $w_1, w_2 \in V^*$, we say that $w_1 \Rightarrow w_2$ if $w_1 = a_1 \ldots a_n$, $w_2 = v_1 \ldots v_n$ for $a_i \rightarrow v_i \in R$, $1 \leq i \leq n$. An *ET0L system* is a construct $G = (V, T, \omega, R_1, \ldots R_n)$ such that each quadruple (V, T, ω, R_i) is an E0L system. The generated language is defined as $L(G) = \{x \in T^* \mid \omega \Rightarrow_{R_{j_1}} w_1 \Rightarrow_{R_{j_2}} \cdots \Rightarrow_{R_{j_m}} w_m = x\}$, where $m \geq 0, 1 \leq j_i \leq n, 1 \leq i \leq m$.

Theorem 6. *1.* $\mathrm{NEMCM}_4(\mathsf{rendo}, \mathsf{rexo}, \mathsf{rdiv}) - \mathbf{NMAT}^\lambda \neq \varnothing$,

2. $\mathrm{LEMCM}_5(\mathsf{rendo}, \mathsf{rexo}, \mathsf{rdiv}) - ET0L \neq \varnothing$.

4   Conclusion

The results presented in this paper use a possibly minimal number of membranes (four) to achieve Turing completeness with pairs of operations (exocytosis, inhibitive endocytosis) and (inhibitive exocytosis, endocytosis). This is a major improvement with respect to the results obtained in [7] where nine membrane are used. We also present some computability results expressing that membrane systems using the operations of restricted division, restricted exocytosis and inhibitive endocytosis cannot yield computational completeness, defining the frontier between RE and non-RE for controlled mobility in membrane computing.

Acknowledgements. The work of Bogdan Aman and Gabriel Ciobanu was supported by a grant of the Romanian National Authority for Scientific Research, CNCS-UEFISCDI, project number PN-II-ID-PCE-2011-3-0919.

References

1. Aman, B., Ciobanu, G.: Solving a Weak NP-complete Problem in Polynomial Time by Using Mutual Mobile Membrane Systems. Acta Informatica 48(7-8), 409–415 (2011)
2. Aman, B., Ciobanu, G.: Mobility in Process Calculi and Natural Computing. Springer (2011)
3. Dassow, J., Păun, G.: Regulated Rewriting in Formal Language Theory (1989)
4. Freund, R., Păun, G.: On the Number of Non-terminal Symbols in Graph-Controlled, Programmed and Matrix Grammars. In: Margenstern, M., Rogozhin, Y. (eds.) MCU 2001. LNCS, vol. 2055, pp. 214–225. Springer, Heidelberg (2001)
5. Krishna, S.N., Aman, B., Ciobanu, G.: Solving the 4QBF Problem in Polynomial Time by Using the Biological-Inspired Mobility (preprint)
6. Krishna, S.N., Ciobanu, G.: On the Computational Power of Enhanced Mobile Membranes. In: Beckmann, A., Dimitracopoulos, C., Löwe, B. (eds.) CiE 2008. LNCS, vol. 5028, pp. 326–335. Springer, Heidelberg (2008)
7. Krishna, S.N., Ciobanu, G.: Computability Power of Mobility in Enhanced Mobile Membranes. In: Löwe, B., Normann, D., Soskov, I., Soskova, A. (eds.) CiE 2011. LNCS, vol. 6735, pp. 160–170. Springer, Heidelberg (2011)
8. Krishna, S.N., Ciobanu, G.: A $\Sigma_2^P \cup \Pi_2^P$ Lower Bound Using Mobile Membranes. In: Holzer, M. (ed.) DCFS 2011. LNCS, vol. 6808, pp. 275–288. Springer, Heidelberg (2011)
9. Krishna, S.N., Păun, G.: P Systems with Mobile Membranes. Natural Computing 4, 255–274 (2005)
10. Minsky, M.L.: Computation: Finite and Infinite Machines. Prentice-Hall (1967)
11. Păun, G.: P Systems with Active Membranes: Attacking NP-Complete Problems. Journal of Automata, Languages and Combinatorics 6(1), 75–90 (2001)
12. Păun, G., Rozenberg, G., Salomaa, A. (eds.): The Oxford Handbook of Membrane Computing. Oxford University Press (2010)
13. Pérez-Jiménez, M.J., Riscos-Núñez, A., Romero-Jiménez, A., Woods, D.: Complexity-Membrane Division, Membrane Creation. In: [12], pp. 302–336
14. Rozenberg, G., Salomaa, A. (eds.): The Mathematical Theory of L Systems. Academic Press (1980)

On Shift Spaces with Algebraic Structure

Ville Salo and Ilkka Törmä

Department of Mathematics, Turun Yliopisto, 20014 Turku, Finland
{vosalo,iatorm}@utu.fi

Abstract. We investigate subshifts with a general algebraic structure and cellular automata on them, with an emphasis on (order-theoretic) lattices. Our main results concern the characterization of Boolean algebraic subshifts, conditions for algebraic subshifts to be recoded into cellwise algebras and the limit dynamics of homomorphic cellular automata on lattice subshifts.

Keywords: subshifts, algebraic structure, lattices, cellular automata.

1 Introduction

There has been considerable interest in the connections between groups and symbolic dynamics in the literature. Linear cellular automata have been investigated extensively [5] [3] [10] [11] [6], and some authors [12] [7] [1] have looked into subshifts with a group structure. Since the study of linear automata and group subshifts has proven fruitful, it seems natural to consider whether other algebraic structures produce interesting behavior in subshifts and CA. This paper is, as far as we are aware, the first foray in this direction, with an emphasis on order-theoretic lattices.

In the first section, we look at the case of giving a subshift the structure of an order-theoretic lattice, with the operations defined cellwise. We characterize all such subshifts in terms of the local rules that define them. In particular, we find that a result of Kitchens [7] for group subshifts also holds for Boolean algebras, namely that they are, up to conjugacy, products of a full shift and a finite subshift. In our case, the conjugacy can even be made algebraic.

In the second section, we find necessary and sufficient conditions for when an algebra operation can be recoded to one defined cellwise. The conditions are stated in terms of the affine maps of the algebra, that is, functions built from the algebra operations. We prove that the recoding can be done when all the maps are 'local', in the sense that their radius is bounded as cellular automata. Many natural classes of algebras, in particular groups and Boolean algebras, always satisfy this condition.

In the third section, we consider cellular automata that are also algebra homomorphisms, in the case that the algebra has the so-called congruence-product property (lattices exhibit this property). We prove that all such CA are stable, and the limit set is conjugate to a product of full shifts and finite shifts.

S.B. Cooper, A. Dawar, and B. Löwe (Eds.): CiE 2012, LNCS 7318, pp. 636–645, 2012.
© Springer-Verlag Berlin Heidelberg 2012

2   Definitions

If S is a finite set, the *alphabet*, the space $S^{\mathbb{Z}}$ is called the *full shift over* S. If $x \in S^{\mathbb{Z}}$, we denote the ith coordinate of x with x_i, and abbreviate $x_i x_{i+1} \cdots x_{i+n-1}$ by $x_{[i,i+n-1]}$. For a word $w \in S^n$, we say that w *occurs in* x, if there exists i such that $w = x_{[i,i+n-1]}$. On $S^{\mathbb{Z}}$ we define the *shift map* σ_S (or simply σ, if S is clear from the context) by $\sigma_S(x)_i = x_{i+1}$ for all i.

We assume a basic acquaintance with symbolic dynamics, in particular the notions of subshift, forbidden word, SFT, sofic shift, mixingness, block map and cellular automaton. A clear exposition can be found in [9]. The set of words of length n occurring in a subshift X is denoted by $B_n(X)$.

Let T be a set of pairs (f, n), where f is a symbol and $n \in \mathbb{N}$, with the property that $(f, m), (f, n) \in T \Rightarrow m = n$. Here, n is called the *arity* of f. An *algebra of type* T is a pair (S, F), where S is a set and for each $(f, n) \in T$ we have a function $f' : S^n \to S$ in the set F. In this case, we identify f' with f, and the functions f are called *algebra operations*. We usually identify (S, F) with S, if F is clear from the context.

A *variety of type* T *defined by the set of identities* I is the class of algebras of type T that satisfy the identities in I. We do not define identities rigorously, but are satisfied with an intuitive presentation (see the examples below). If $\mathcal{F}$ is a variety, $(S, F) \in \mathcal{F}$ and $R \subset S$ is closed under the operations of S, then we call (R, F) a *subalgebra* of S. The set of subalgebras of S is denoted $\mathrm{Sub}(S)$. A function $g : S \to R$ between two algebras in $\mathcal{F}$ is called a *homomorphism* or an *$\mathcal{F}$-morphism* if $g(f(s_1, \ldots, s_n)) = f(g(s_1), \ldots, g(s_n))$ for each n-ary operation f. If g is bijective, it is called an *isomorphism*. The *direct product* of an indexed family $(S_i)_{i \in \mathcal{I}}$ of algebras is the algebra $\prod_{i \in \mathcal{I}} S_i$, where the operations are defined cellwise $(f(s^1, \ldots, s^n)_i = f(s_i^1, \ldots, s_i^n))$. An algebra is *directly indecomposable*, if it is not isomorphic to a product of two nontrivial algebras. All finite algebras are isomorphic to a finite product of directly indecomposable finite algebras. All varieties are closed under subalgebras, homomorphic images and products [2].

If $\sim \subset S \times S$ is an equivalence relation that satisfies

$$ s_1 \sim t_1, \ldots, s_n \sim t_n \Longrightarrow f(s_1, \ldots, s_n) \sim f(t_1, \ldots, t_n) $$

for all $s_i, t_i \in S$ and all n-ary operations f, we say that $\sim$ is a *congruence* on S. The set of congruences on an algebra S is denoted $\mathrm{Con}(S)$. A natural algebraic structure is induced by the algebra operations on the set of equivalence classes $S/\sim$, the kernel $\ker(g) = \{(s, t) \in S \times S \mid g(s) = g(t)\}$ of a homomorphism g is always a congruence, and $S/\ker(g)$ is isomorphic to $g(S)$ [2] (the last claim is known as the *Homomorphism Theorem*).

The variety of *lattices* has type $\{(\wedge, 2), (\vee, 2)\}$ and is defined by the identities

$$ x \wedge x \approx x, \; x \wedge y \approx y \wedge x \; , $$
$$ (x \wedge y) \wedge z \approx x \wedge (y \wedge z), \; x \wedge (x \vee y) \approx x \; , $$

and the same identities with $\wedge$ and $\vee$ interchanged (their *dual versions*). The operations $\wedge$ and $\vee$ are called *meet* and *join*, respectively. It is known that the variety of lattices coincides with the class of partially ordered sets where all pairs of elements have suprema and infima, where the correspondence is given by $x \wedge y = \inf\{x, y\}$ and $x \vee y = \sup\{x, y\}$. If S is a lattice and $a, b \in S$, then we denote $[a, b] = \{c \in S \mid a \leq c \leq b\}$, where $\leq$ is the partial order of S. A lattice S is *modular* if it satisfies the additional constraint that whenever $a, b, c \in S$ and $a \leq b$, then $a \vee (b \wedge c) = b \wedge (a \vee c)$. The variety of *distributive lattices* satisfies the lattice identities and the additional identity

$$x \vee (y \wedge z) = (x \vee y) \wedge (x \vee z)$$

and its dual version. All distributive lattices are modular [2]. Of particular interest is the binary lattice **2** containing the elements $\{0, 1\}$ with their usual numerical order.

The variety of *Boolean algebras* has type $\{(\wedge, 2), (\vee, 2), (\bar{\ }, 1), (1, 0), (0, 0)\}$. A Boolean algebra is a distributive lattice w.r.t. $\wedge$ and $\vee$, and also satisfies

$$x \wedge 0 \approx 0, \; x \vee 1 \approx 1 \; ,$$
$$x \wedge \bar{x} \approx 0, \; x \vee \bar{x} \approx 1 \; .$$

It is known that every finite Boolean algebra is isomorphic to the algebra of subsets 2^T of some set T where the ordering is given by set inclusion [2].

Let $\mathcal{F}$ be a variety of algebras. We call $X \subset S^{\mathbb{Z}}$ an $\mathcal{F}$-*subshift*, if it is a subshift and has an algebra structure in $\mathcal{F}$ whose operations are block maps. In particular, every finite $S \in \mathcal{F}$ induces a natural cellwise algebra structure on $S^{\mathbb{Z}}$ by $f(x^1, \ldots, x^n)_i = f(x_i^1, \ldots, x_i^n)$ for all i (that is, $S^{\mathbb{Z}}$ is taken to be a direct product), and in this case, if $X \subset S^{\mathbb{Z}}$ is a subshift in $\mathrm{Sub}(S^{\mathbb{Z}})$, it is called a *cellwise $\mathcal{F}$-subshift*. A conjugacy that is also an $\mathcal{F}$-morphism is called *algebraic*. A cellular automaton with state set S that is also an $\mathcal{F}$-morphism is called $\mathcal{F}$-*linear*.

3   Cellwise Lattice Subshifts

In this section, let S be a finite lattice ordered by $\leq$.

We begin by giving a characterization for the cellwise lattice subshifts.

Definition 1. *Let $X \subset S^{\mathbb{Z}}$ be a subshift and $a \in S$. For such X we define m^a and M^a by*

$$\forall i \in \mathbb{Z} : m_i^a = \bigwedge_{\substack{x \in X \\ x_0 \geq a}} x_i, \; M_i^a = \bigvee_{\substack{x \in X \\ x_0 \leq a}} x_i \; .$$

It is easy to see that $a \leq b$ implies $m^a \leq m^b$ and $M^a \leq M^b$.

Lemma 1. *Let $X \subset S^{\mathbb{Z}}$ be a subshift with $B_1(X) = S$. The following are equivalent:*

- *X is a cellwise lattice subshift.*
- *For all $x \in S^{\mathbb{Z}}$, we have $x \in X$ iff $x \geq \sigma^{-i}(m^{x_i})$ holds for all i.*
- *For all $x \in S^{\mathbb{Z}}$, we have $x \in X$ iff $x \leq \sigma^{-i}(M^{x_i})$ holds for all i.*

Proof. Assume first that X is a cellwise lattice subshift. Let $a \in S$, and for all $i \in \mathbb{Z}$, let $x^i \in X$ such that $x_i^i = m_i^a$. Such x^i exist, since $X \in \mathrm{Sub}(S^{\mathbb{Z}})$ and has alphabet S. Then the sequence $\bigwedge_{i=-n}^{n} x^i$ approaches m^a as n grows, so $m^a \in X$. The second condition is then necessary by the definition of the m^{x_i}, and it is sufficient since the sequence $\bigvee_{i=-n}^{n} \sigma^{-i}(m^{x_i})$ approaches x as n grows.

Assume then that $X \notin \mathrm{Sub}(S^{\mathbb{Z}})$. Suppose first that there exist $x, y \in X$ with $x \wedge y \notin X$, and let $a_i = x_i \wedge y_i$. Now, for all i, we have that $x \wedge y \geq \sigma^{-i}(m^{x_i}) \wedge \sigma^{-i}(m^{y_i}) \geq \sigma^{-i}(m^{a_i})$, so that the second condition does not hold.

Suppose then that $x \vee y \notin X$, and let $a_i = x_i \vee y_i$. Then we have $x \vee y \geq \sigma^{-i}(m^{x_i})$ and $x \vee y \geq \sigma^{-i}(m^{y_i})$, so that $x \vee y \geq \sigma^{-i}(m^{a_i})$, and again the second condition fails.

The third condition is dual to the second, and is proven similarly. $\square$

Remark 1. It can be proven that a cellwise lattice subshift X is sofic if and only if all the m^a are eventually periodic in both directions (and the dual claim for M^a holds as well).

Lemma 2. *A nontrivial subshift $X \subset 2^{\mathbb{Z}}$ is a cellwise lattice subshift if and only if one of the following holds:*

- $X = S^{\mathbb{Z}}$.
- $X = \{x \in S^{\mathbb{Z}} \mid x = \sigma^n(x)\}$ *for some n.*
- $X = \{x \in S^{\mathbb{Z}} \mid \forall i \in \mathbb{Z}, p \in P : x_i = 1 \Rightarrow x_{i+p} = 1\}$ *for a finite set $P \subset \mathbb{N}$.*
- $X = \{x \in S^{\mathbb{Z}} \mid \forall i \in \mathbb{Z}, p \in P : x_i = 1 \Rightarrow x_{i-p} = 1\}$ *for a finite set $P \subset \mathbb{N}$.*

Proof. Let X be a cellwise lattice subshift, and let $K = \{i \in \mathbb{Z} \mid m_i^1 = 1\}$. If $K = \{0\}$, then $X = S^{\mathbb{Z}}$, so suppose that $K \neq \{0\}$. From Lemma 1 and the fact that $m^1 \in X$ it follows that K is a subsemigroup of $\mathbb{Z}$, i.e.,

$$\{a_1 n_1 + \cdots + a_k n_k \mid a_i \in \mathbb{N}, n_i \in K\} \subset K .$$

A standard pigeonhole argument then shows that K is generated by some finite set $P \subset \mathbb{Z}$. The third and fourth items correspond to the cases $P \subset \mathbb{N}$ and $-P \subset \mathbb{N}$, respectively.

If P contains both positive and negative elements, let $n = \gcd P$. An application of Bezout's identity then shows that $K = \{pn \mid p \in \mathbb{Z}\}$, and the second case follows. $\square$

Corollary 1. *Every cellwise lattice subshift of $2^{\mathbb{Z}}$ is an SFT.*

Corollary 2. *If X is also closed under complementation in Lemma 2, only the first two cases can occur.*

Proof. Assume the contrary, say case 3. Then X contains the point $x = {}^{\infty}01^{\infty}$. But now $\bar{x} = {}^{\infty}10^{\infty} \in X$, and thus $m^1 = {}^{\infty}010^{\infty}$. $\square$

We have now completely characterized all binary Boolean algebraic subshifts.

Definition 2. *Let* $S = 2^T$ *be a finite Boolean algebra and* $X \subset S^{\mathbb{Z}}$ *a Boolean algebraic subshift. For each* $t \in T$, *define the block map* $\pi_t : X \to \{0,1\}^{\mathbb{Z}}$ *by*

$$\pi_t(x)_i = \begin{cases} 1, \text{ if } t \in x_i \\ 0, \text{ if } t \notin x_i \end{cases} .$$

If $R \subset T$, *we also define* $\pi_R : X \to (\{0,1\}^R)^{\mathbb{Z}}$ *by* $\pi_R(x) = (\pi_t(x))_{t \in R}$.

Note that each π_t and π_R above is also a homomorphism of Boolean algebras, and that π_T is an injective block map from X to $(\{0,1\}^T)^{\mathbb{Z}}$.

Definition 3. *Let* $S = 2^T$ *be a finite Boolean algebra. A subshift* $X \subset S^{\mathbb{Z}}$ *is called* simple *if there exists a set* $R \subset T$ *such that the following conditions hold.*

- *For every* $t \in T$, *we have either* $\pi_t(X) = \{x \in 2^{\mathbb{Z}} \mid x = \sigma^n(x)\}$ *for some* $n \in \mathbb{N}$, *or* $\pi_t(X) = 2^{\mathbb{Z}}$.
- *For all* $t \in T$, *there exists* $r \in R$ *and* $k \in \mathbb{Z}$ *with* $\pi_t = \sigma^k \circ \pi_r$.
- *The elements of* R *are independent in the sense that if* $r \in R$, $x \in \pi_r(X)$ *and* $y \in \pi_{R-\{r\}}(X)$, *then there exists* $z \in X$ *with* $\pi_r(z) = x$ *and* $\pi_{R-\{r\}}(z) = y$.

Informally, a subshift is simple if it is essentially a product of full shifts and periodic subshifts. Note that in particular, such a subshift is conjugate to a product of a full shift and a finite shift. Clearly, all simple shifts are cellwise Boolean algebraic subshifts. The converse also holds:

Theorem 1. *Let* $S = 2^T$ *be a finite Boolean algebra and* $X \subset S^{\mathbb{Z}}$ *a subshift with* $B_1(X) = S$. *If* $X \in \mathrm{Sub}(S^{\mathbb{Z}})$, *then* X *is simple.*

Proof. Since each $\pi_t(X)$ is a cellwise Boolean algebraic subshift over 2, the first condition follows from Corollary 2.

Define an equivalence relation $\sim$ on T by $t \sim s$ if $\pi_t = \sigma^k \circ \pi_s$ for some $k \in \mathbb{Z}$. Let R be a set of representatives for the equivalence classes of $\sim$. Then for this R, the second condition holds.

Consider then the configuration $m^{\{r\}}$ for some $r \in R$. If $t \in m_i^{\{r\}}$ for any $t \in T$ and $i \in \mathbb{Z}$, then $\pi_t(x) \geq \sigma^{-i}(\pi_r(x))$, and by complementing, $\pi_t(x) \leq \sigma^{-i}(\pi_r(x))$, holds for all $x \in X$. The converse is also true, which means that $R \cap m_i^{\{r\}} \subset \{r\}$ for all i. The third condition then follows from Lemma 1, since X is also a cellwise lattice subshift. $\square$

Example 1. Let S be a lattice with at least 3 elements, and let $a \neq 1$ be a least successor of 0. Consider the rule 'if $x_i = 1$, then $x_{i \pm 2^j} \geq a$ for all $j \in \mathbb{N}$'. By Lemma 1, the subshift generated by this rule forms a lattice. It is easy to see that this subshift is mixing, but is not even sofic.

4   General Algebraic Subshifts

In this section, $\mathcal{F}$ is a variety of algebras. We now consider $\mathcal{F}$-subshifts X that are not necessarily cellwise. One way to define such general shifts is to use a cellwise $\mathcal{F}$-subshift Y and a conjugacy $\phi : X \to Y$, and define

$$f(x_1, \ldots, x_n) = \phi^{-1}(f(\phi(x_1), \ldots, \phi(x_n)))$$

for all n-ary algebra operations f of $\mathcal{F}$. Clearly, ϕ now becomes an algebraic conjugacy. The main theorems in this section address the issue of deciding whether a given $\mathcal{F}$-subshift is algebraically conjugate to some cellwise $\mathcal{F}$-subshift.

Definition 4. *An* affine map *of an algebra S is inductively defined as either $t(\xi) = \xi$ (the identity map), $t(\xi) = a$ for some $a \in S$ (the constant map) or $t(\xi) = f(a_1, \ldots, a_n)$, where f is an n-ary operation, one of the a_i is an affine map and the rest are constants $a_j \in S$. Here, $\xi \notin S$ is used as a variable. To each affine map t we also associate a function $\mathrm{Eval}(t) : S \to S$ by replacing ξ with the function argument and evaluating the resulting expression. We may denote $\mathrm{Eval}(t)(a)$ by $t(a)$ if $a \in S$. The set of affine maps of S is denoted by $\mathrm{Af}(S)$.*

For example, in the ring $\mathbb{Z}$, the term $t(\xi) = 2 \cdot (3 + (\xi \cdot (-4)))$ is an affine map, and $\mathrm{Eval}(t)(i) = -8i + 6$ for all $i \in \mathbb{Z}$.

The following is a dynamical characterization of $\mathcal{F}$-subshifts that are cellwise up to algebraic conjugacy. The proof uses a common recoding technique found, for example, in the Recoding Construction 4.3.1 of [8].

Theorem 2. *Let $X \subset S^{\mathbb{Z}}$ be an $\mathcal{F}$-subshift. Then there exists a cellwise $\mathcal{F}$-subshift Y and an algebraic conjugacy $\phi : X \to Y$ if and only if there is an r such that for all $t \in \mathrm{Af}(X)$, the block map $\mathrm{Eval}(t)$ has radius at most r.*

Proof. Suppose first that such an r exists. Then, we can also meaningfully apply an affine map $t \in \mathrm{Af}(X)$ to a word $w \in B_{2r+1}(X)$ rooted at the origin by extending it arbitrarily to a configuration $x \in X$, and taking the center cell of $t(x)$. We define the following equivalence relation on $B_{2r+1}(X)$:

$$\forall v, w \in B_{2r+1}(X) : v \sim w \iff \forall t \in \mathrm{Af}(X) : t(v)_0 = t(w)_0 .$$

Note that, in particular, $v \sim w \implies v_0 = w_0$. We define an injective block map $\psi : X \to (B_{2r+1}(X)/\sim)^{\mathbb{Z}}$ by $\psi(x)_i = x_{[i-r,i+r]}/\sim$, and denote $Y = \psi(X)$. We denote the obtained conjugacy by $\phi : X \to Y$.

In order to make the algebra operations commute with ϕ, we define

$$f(y_1, \ldots, y_n) = \phi(f(\phi^{-1}(y_1), \ldots, \phi^{-1}(y_n)))$$

for all n-ary algebra operations f, which is obviously well-defined. Now ϕ extends to a bijection between $\mathrm{Af}(X)$ and $\mathrm{Af}(Y)$ in a natural way. Let us show that every algebra operation f is then defined cellwise in Y. Consider two points $y, y' \in Y$ with $y_0 = y_0'$. We need to show that $\phi(t)(y)_0 = \phi(t)(y')_0$ for all $\phi(t) \in \mathrm{Af}(Y)$. Assume the contrary, that $\phi(t)(y)_0 \neq \phi(t)(y')_0$ for some $\phi(t) \in \mathrm{Af}(Y)$. Let $x = \phi^{-1}(y)$ and $x' = \phi^{-1}(y')$. Then also $t(x)_{[-r,r]} = v \not\sim w = t(x')_{[-r,r]}$, and thus there exists $t' \in \mathrm{Af}(X)$ such that $t'(v) \neq t'(w)$. But by the assumption on affine maps, $t'' = t' \circ t$ has radius r. Now we have $t''(x)_0 \neq t''(x')_0$, which is a contradiction, since $x_{[-r,r]} \sim x'_{[-r,r]}$.

For the converse, note that if X is algebraically conjugate to a cellwise $\mathcal{F}$-subshift Y via the conjugacy ϕ, then the radius of every translate is at most the sum of the radii of ϕ and ϕ^{-1}. $\quad\square$

We also obtain a sufficient algebraic condition for algebraic conjugacy with a cellwise $\mathcal{F}$-subshift. In the special case of the full shift, this becomes a characterization. We start with a definition.

Definition 5. *We define the* depth *of an affine map as the number of nested algebra operations in it. We say an algebra S is k-shallow if for every $t \in \mathrm{Af}(S)$ there exists $t' \in \mathrm{Af}(S)$ of depth at most k such that $\mathrm{Eval}(t) = \mathrm{Eval}(t')$.*

Theorem 3. *Let $X \subset S^{\mathbb{Z}}$ be an $\mathcal{F}$-subshift. If X is k-shallow, then it is cellwise up to algebraic conjugacy. If X is algebraically conjugate to $R^{\mathbb{Z}}$ where $R \in \mathcal{F}$, then X is k-shallow for some k.*

Proof. If X is k-shallow, then clearly all affine maps have uniformly bounded radii, and Theorem 2 gives the result.

For the other claim, it suffices to show that $R^{\mathbb{Z}}$ is k-shallow for some k. Since R is finite, the set $\Gamma = \{\mathrm{Eval}(t) \mid t \in \mathrm{Af}(R)\}$ is finite. For an affine map $t \in \mathrm{Af}(R^{\mathbb{Z}})$ we denote by t_i the affine map in $\mathrm{Af}(R)$ that t computes in coordinate i. For each affine map $t \in \mathrm{Af}(R^{\mathbb{Z}})$ we define $\Delta_t = \{\mathrm{Eval}(t_i) \mid i \in \mathbb{Z}\}$. Let $n = |\Gamma|$ and note that since R^n is finite, it is k-shallow for some k.

Let $t \in \mathrm{Af}(R^{\mathbb{Z}})$ and let $j_1, \ldots, j_n$ be coordinates such that for all $h \in \Delta_t$ we have $\mathrm{Eval}(t_{j_i}) = h$ for some i. Construct an affine map $s \in \mathrm{Af}(R^n)$ by $s_i = t_{j_i}$. Since R^n is k-shallow we find some affine map $s' \in \mathrm{Af}(R^n)$ of depth at most k with $\mathrm{Eval}(s) = \mathrm{Eval}(s')$. We may now define an affine map $t' \in \mathrm{Af}(R^{\mathbb{Z}})$ by $t'_i = s'_{j'_i}$ for all i, where j'_i is such that $\mathrm{Eval}(t_i) = \mathrm{Eval}(s_{j'_i})$. Since $\mathrm{Eval}(s_j) = \mathrm{Eval}(s'_j)$ for all j, we have $\mathrm{Eval}(t) = \mathrm{Eval}(t')$. $\square$

Corollary 3. *Up to algebraic conjugacy, every distributive lattice, Boolean algebra, ring, semigroup, monoid and group subshift is defined cellwise.*

Proof. By finding suitable normal forms, one easily sees that distributive lattices, semigroups and monoids are 2-shallow, while Boolean algebras, groups and rings are 3-shallow. Note that affine maps have at most one unknown variable. $\square$

Corollary 4. *Every Boolean algebraic subshift is algebraically conjugate to a product of a full shift and a finite shift.*

Proof. This follows from the previous corollary and Theorem 1. $\square$

The previous corollary is an analogue of a result of Kitchens [7] for group subshifts.

Example 2. In the proof of Theorem 3, for the claim that algebraic conjugacy implies k-shallowness, the use of a full shift is crucial: the claim is false even for the mixing one-step SFT $X \subset \{0, 1, 2, \bot\}^{\mathbb{Z}}$ with the forbidden pairs $\{10, 11, 20, 21\}$ equipped with the cellwise binary operation

$\cdot$	0	1	2	$\bot$
0	0	0	0	$\bot$
1	1	2	1	$\bot$
2	2	2	2	$\bot$
$\bot$	$\bot$	$\bot$	$\bot$	$\bot$

Since lattices are not shallow in general, Theorem 2 and Theorem 3 do not tell us much about lattice subshifts. In fact, the following example gives a lattice subshift on a mixing SFT which is not cellwise up to algebraic conjugacy.

Example 3. Let $S = \{1_+, 1_-, 0_+, 0_-\}$ have the lattice structure $1_+ > 0_+ > 0_-$ and $1_+ > 1_- > 0_-$, and define $X \subset S^{\mathbb{Z}}$ as the SFT with forbidden words 0_-1_+ and $a_-b_-c_\delta$ for all $(a,b,c) \neq (0,0,0)$ and $\delta \in \{+,-\}$. Define X to have a lattice structure where $\vee$ is defined cellwise, and $\wedge$ is defined as first applying the cellwise meet of $S^{\mathbb{Z}}$, and then rewriting instances of 0_-1_+ to 0_-0_+ and instances of $a_-b_-c_\delta$ to $0_-0_-0_\delta$. An easy calculation confirms that the operations are well-defined and X indeed forms a lattice subshift.

We now show that X has affine maps of arbitrary radius by giving two right asymptotic points $x, y \in X$ with $x_0 \neq y_0$, and then constructing affine maps $t_k \in \mathrm{Af}(X)$ for all $k \in \mathbb{N}$ such that $t_k(x)_k \neq t_k(y)_k$. Let $x = {}^{\infty}1_+.0_+1_+{}^{\infty}$ and $y = {}^{\infty}1_+{}^{\infty}$. Define $z_k = \sigma^{-k}({}^{\infty}1_+.1_-1_+{}^{\infty})$ and $z' = {}^{\infty}0_+{}^{\infty}$, and let

$$t_k(\xi) = ((\cdots(((\xi \wedge z_0) \vee z') \wedge z_1) \vee z' \cdots) \wedge z_k) \vee z' .$$

It is easy to see that $t_k(x) = {}^{\infty}1_+.(0_+)^{k+2}1_+{}^{\infty}$, and on the other hand, $t_k(y) = y$.

We are not aware of a modular lattice which is not shallow. This raises the natural question whether every modular lattice subshift defined cellwise up to algebraic conjugacy.

5 $\mathcal{F}$-Linear Cellular Automata

Let $\mathcal{F}$ be again a variety, and S a finite member of $\mathcal{F}$. We only consider cellwise defined algebraic structures of $S^{\mathbb{Z}}$ in this section. We begin with the following useful lemma, characterizing all $\mathcal{F}$-linear cellular automata.

Lemma 3. *A CA G with neighborhood radius r is $\mathcal{F}$-linear if and only if its local rule g is an $\mathcal{F}$-morphism from S^{2r+1} to S.*

Definition 6. *The variety $\mathcal{F}$ has the congruence-product property, if for all finite families $(S_i)_{i\in[1,n]}$ of algebras in $\mathcal{F}$ we have that*

$$\mathrm{Con}(\prod_{i=1}^{n} S_i) = \prod_{i=1}^{n} \mathrm{Con}(S_i) .$$

A proof of the following can be found, for example, in [4].

Lemma 4. *The variety of lattices has the congruence-product property.*

In the remainder of this section, we show that if $\mathcal{F}$ has the congruence-product property, then the $\mathcal{F}$-linear cellular automata have very simple limit sets and limit dynamics. In particular, by the above lemma, our results hold for lattice-linear automata.

Lemma 5. *Let $\mathcal{F}$ be a variety with the congruence-product property and $S \in \mathcal{F}$ finite. Let G be an $\mathcal{F}$-linear CA on $S^{\mathbb{Z}}$ with radius r and local function g. Denote by $R \subset S$ the alphabet of the limit set of G (which is clearly a subalgebra), and denote by π'_k the canonical projections $R^{2r+1} \to R$. Let $\prod_{i=1}^m R_i$ be the decomposition of R into directly indecomposable algebras, and denote by π_i the canonical projections $R \to R_i$. Then for each $i \in [1, m]$ there exist j_i, k_i and a surjective $\mathcal{F}$-morphism $h_i : R_{j_i} \to R_i$ such that $\pi_i \circ g|_{R^{2r+1}} = h_i \circ \pi_{j_i} \circ \pi'_{k_i}$.*

Since the proof of Lemma 4 can also be carried out for Boolean algebras, we have the following.

Corollary 5. *If $S = 2^T$ is a Boolean algebra and G a Boolean-linear CA on $S^{\mathbb{Z}}$, then for each $t \in T$, either $\pi_t(G(S^{\mathbb{Z}}))$ is trivial, or we have $i \in \mathbb{Z}$ and $t' \in T$ such that $\pi_t \circ G = \pi_{t'} \circ \sigma^i$.*

Theorem 4. *Let $\mathcal{F}$ be a variety with the congruence-product property and $S \in \mathcal{F}$ finite. The limit set X of an $\mathcal{F}$-linear cellular automaton G on $S^{\mathbb{Z}}$ is algebraically conjugate to a product of full shifts, and G is stable. Furthermore, there exists $p \in \mathbb{N}$ such that $G^p|_X$ is a product of powers of shift maps.*

Proof. Let $\prod_{i=1}^m S_i$ be the decomposition of S into directly indecomposable algebras. Define $H = (\{S_1, \dots, S_m\}, E)$ as the directed graph where $(S_i, S_j) \in E$ iff the domain of the surjective map h_j given by Lemma 5 is S_i. Since each S_i has exactly one incoming arrow, every strongly connected component of H is a cycle or a single vertex. Let S_i be in a cycle, say $S_i \to S_{i_1} \to \cdots \to S_{i_{p'-1}} \to S_i$. Since S_i is finite, the map $f_i = h_i \circ h_{i_{p'-1}} \circ \cdots \circ h_{i_1}$ is an automorphism of S_i, and there exists $p_i \in \mathbb{N}$ such that $f_i^{p_i}$ is the identity map of S_i. This in turn implies that for all $x \in S^{\mathbb{Z}}$, we have

$$\pi_i(G^{p_i}(x)) = \pi_i(\sigma^{k_i}(x))$$

for some $k_i \in \mathbb{Z}$. That is, G^{p_i} simply shifts the S_i-components of points by a constant amount. Let $\mathcal{I}$ be the set of indices i such that S_i occurs in a cycle, and let $p = \text{lcm}_{i \in \mathcal{I}}(p_i)$. Clearly, G has a natural reversible restriction on the full shift $S_{\mathcal{I}}^{\mathbb{Z}}$, where $S_{\mathcal{I}} = \prod_{i \in \mathcal{I}} S_i$.

Consider then S_j for some $j \in \mathcal{J} = [1, m] - \mathcal{I}$. By following the incoming arrows we necessarily find an $i(j) \in \mathcal{I}$ and a path of the form

$$S_{i(j)} \to S_{i_1} \to \cdots \to S_{i_{p'-1}} \to S_{i(j)} \to S_{j_1} \to \cdots \to S_{j_{q'-1}} \to S_j \ ,$$

where $j_k \in \mathcal{J}$ for all k. Denote by $q(j)$ the length q' of the path from $S_{i(j)}$ to S_j, and let $q = \max_{j \in \mathcal{J}} q(j)$.

Clearly, if $y = G^q(x)$ and $j \in \mathcal{J}$, then $\pi_j(y)$ is a function of $\pi_{i(j)}(x)$, which in turn is a function of some $\pi_i(y)$ with $i \in \mathcal{I}$. But this means that the $\mathcal{J}$-components of y are uniquely determined by its $\mathcal{I}$-components. This and the fact that G is reversible on $S_{\mathcal{I}}^{\mathbb{Z}}$ imply that $X = G^q(S^{\mathbb{Z}})$, X is algebraically conjugate to $S_{\mathcal{I}}^{\mathbb{Z}}$, and $G^p|_X$ is a product of powers of shift maps.   □

6   Future Work

In this paper we have only considered limit sets in the case when the cellular automaton starts from the full shift. It would be interesting to study limit sets of $\mathcal{F}$-linear automata starting from more complicated shifts. We do not know if our approach generalizes to, say, mixing $\mathcal{F}$-subshifts of finite type, assuming the congruence-product property. Future work might also involve studying the connections between other properties of the variety $\mathcal{F}$ and the $\mathcal{F}$-linear CA.

Acknowledgements. This research was supported by the Academy of Finland Grant 131558.

References

1. Boyle, M., Schraudner, M.: $\mathbb{Z}^d$ group shifts and Bernoulli factors. Ergodic Theory Dynam. Systems 28(2), 367–387 (2008),
   http://dx.doi.org/10.1017/S0143385707000697
2. Burris, S., Sankappanavar, H.P.: A course in universal algebra. Graduate Texts in Mathematics, vol. 78. Springer, New York (1981)
3. Cattaneo, G., Formenti, E., Manzini, G., Margara, L.: Ergodicity, transitivity, and regularity for linear cellular automata over $\mathbb{Z}_m$. Theoret. Comput. Sci. 233(1-2), 147–164 (2000), http://dx.doi.org/10.1016/S0304-3975(98)00005-X
4. Grätzer, G.: Lattice theory. First concepts and distributive lattices. W. H. Freeman and Co., San Francisco (1971)
5. Itô, M., Ôsato, N., Nasu, M.: Linear cellular automata over $\mathbb{Z}_m$. J. Comput. System Sci. 27(1), 125–140 (1983), http://dx.doi.org/10.1016/0022-0000(83)90033-8
6. Kari, J.: Linear Cellular Automata with Multiple State Variables. In: Reichel, H., Tison, S. (eds.) STACS 2000. LNCS, vol. 1770, pp. 110–121. Springer, Heidelberg (2000)
7. Kitchens, B.P.: Expansive dynamics on zero-dimensional groups. Ergodic Theory Dyn. Syst. 7, 249–261 (1987)
8. Kitchens, B.P.: Symbolic dynamics – One-sided, two-sided and countable state Markov shifts. Universitext. Springer, Berlin (1998)
9. Lind, D., Marcus, B.: An introduction to symbolic dynamics and coding. Cambridge University Press, Cambridge (1995),
   http://dx.doi.org/10.1017/CBO9780511626302
10. Manzini, G., Margara, L.: Invertible linear cellular automata over $\mathbb{Z}_m$: algorithmic and dynamical aspects. J. Comput. System Sci. 56(1), 60–67 (1998),
    http://dx.doi.org/10.1006/jcss.1997.1535
11. Sato, T.: Ergodicity of linear cellular automata over $\mathbb{Z}_m$. Inform. Process. Lett. 61(3), 169–172 (1997),
    http://dx.doi.org/10.1016/S0020-0190(96)00206-2
12. Schmidt, K.: Dynamical systems of algebraic origin. Progress in Mathematics, vol. 128. Birkhäuser Verlag, Basel (1995)

Finite State Verifiers with Constant Randomness

A.C. Cem Say[1] and Abuzer Yakaryılmaz[2]

[1] Boğaziçi University, Department of Computer Engineering,
Bebek 34342 İstanbul, Turkey
say@boun.edu.tr
[2] University of Latvia, Faculty of Computing, Raina bulv. 19, Rīga, 1586, Latvia
abuzer@lu.lv

Abstract. We give a new characterization of NL as the class of languages whose members have certificates that can be verified with small error in polynomial time by finite state machines that use a constant number of random bits, as opposed to its conventional description in terms of deterministic logarithmic-space verifiers. It turns out that allowing two-way interaction with the prover does not change the class of verifiable languages, and that no polynomially bounded amount of randomness is useful for constant-memory computers when used as language recognizers, or public-coin verifiers.

1 Introduction

It is known that allowing constant-memory computers to use random bits and to commit small amounts of error increases their power, both as language recognizers [9], and as verifiers of membership proofs [6]. In this paper, we examine the effects of restricting such probabilistic machines (2pfa's) to use only a constant number of random bits, independent of the length of the input. We prove that such constant-randomness 2pfa's are able to verify membership in precisely the languages in NL. This is an interesting addition to the facts that NL has deterministic logspace verifiers, and NP is the class of languages that has logspace verifiers that use logarithmically many random bits [5]. We obtain this result by demonstrating that such verifiers are equivalent to multihead finite automata. Allowing these constant-coin verifiers to use logarithmic space, and to have two-way interaction with the prover, does not augment the class of verifiable languages. No nonregular language has such an interactive proof system if the verifier is restricted to use public coins. We also show that, when used as recognizers, no amount of polynomially-bounded randomness gives standard 2pfa's any power beyond their deterministic versions. The rest of this paper is structured as follows: Section 2 provides the necessary background. Our results on the new characterization of NL in terms of finite state verifiers, and the public-coin case, are presented in Section 3. Section 4 is a conclusion.

2 Preliminaries

For background on interactive proof systems with bounds on the usage of space and/or randomness, the reader is referred to [4].

S.B. Cooper, A. Dawar, and B. Löwe (Eds.): CiE 2012, LNCS 7318, pp. 646–654, 2012.
© Springer-Verlag Berlin Heidelberg 2012

The main model of verifier that we shall use is a Turing machine with a read-only input tape and a single read/write work tape. The input tape holds the input string between two occurrences of the end-marker symbol ¢, and we assume that the machine's transition function never attempts to move the input head beyond the end-markers. The input tape head is on the left end-marker at the start of the process. The verifier exchanges information with a prover by writing and reading one symbol at a time from the communication alphabet Γ in a communication cell. Using this information channel, the prover attempts to prove the membership of the input string in the language under consideration. Of course, one should not trust this blindly, and we even allow the possibility that the prover sends an infinite sequence of symbols, a contingency that could cause careless verifiers to run forever. The machine also has access to a source of random bits. The state set of the verifier TM is $Q = R \cup D \cup \{q_a, q_r\}$, where R is the set of coin-tossing states, D is the set of deterministic states, and q_a (accept) and q_r (reject) are the halting states. Associated with each state $q \in Q$, there is a communication symbol $\gamma_q \in \Gamma$. The special "null symbol" ϵ is guaranteed to be a member of Γ. One of the non-halting states is designated as the start state. When any state $q \in R \cup D$ with $\gamma_q \neq \epsilon$ is entered, γ_q is written in the communication cell. The prover can be modeled as a prover transition function ρ, which determines the symbol $\gamma \in \Gamma$ to be written in response, based on the input string and the entire communication that has taken place so far. Let $\Diamond = \{-1, 0, +1\}$ denote the set of possible head movement directions. When the verifier reads the response of the prover, it behaves according to the verifier transition function δ as follows: For $q \in R$, $\delta(q, \sigma, \theta, \gamma, b) = (q', \theta', d_i, d_w)$ indicates that the machine will switch to state q', write θ' on the work tape, move the input head in direction $d_i \in \Diamond$, and the work tape head in direction $d_w \in \Diamond$, if it is originally in state q, scanning the symbols σ, θ, and γ in the input and work tapes, and the communication cell, respectively, and seeing the random bit b as a result of the coin toss. For $q \in D$, $\delta(q, \sigma, \theta, \gamma) = (q', \theta', d_i, d_w)$ has a similar meaning, but without the randomness. If $\gamma_q = \epsilon$, the verifier transition function described above is applied directly, without any communication.

We say that language L has a *(private-coin) interactive proof system (IPS) with error probability* ε if there exists a prover P and a verifier V such that

1. for every $x \in L$, the interaction of P and V on input w results in acceptance with probability at least $1 - \varepsilon$, and,
2. for every $x \notin L$, and for any prover P^*, the interaction of P^* and V on input x results in rejection with probability at least $1 - \varepsilon$.

Interactive proof systems where the verifier accepts every member of the language with probability 1 are said to have perfect completeness.

$\mathsf{IP}(t, s, r)$ is the class of languages that have interactive proof systems with verifiers whose expected runtime is $t(n)$ steps, and which use $s(n)$ space (i.e., work tape cells) and $r(n)$ random bits on any input of length n. We use the notations cons, log, poly, and exp to stand for functions that are $O(1)$, $O(\log n)$, $O(n^k)$, and $2^{O(n^k)}$ for any constant k, respectively. For verifiers with no bound on a specific parameter, we shall use the symbol ∞ for that parameter.

Replacing condition 2 in the definition of interactive proof systems with the weaker condition

2'. for every $x \notin L$, and for any prover P^*, the interaction of P^* and V on input x results in acceptance with probability at most ε

leads to our definition of $\mathsf{IP}_w(t,s,r)$, the counterpart of $\mathsf{IP}(t,s,r)$ with these alternative kinds of verifiers that do not have to halt with high probability for all inputs. It should be noted that the time bound of t has a different meaning in the definition of IP_w, and indicates only that P is able to convince V on all inputs which are members of the language L with high probability in time t.

A *one-way interactive proof system* [3] is an IPS where the prover is restricted so that it maps the set of input strings to the set of sequences from the communication alphabet. For input string w, the prover writes the ith symbol of the corresponding sequence in the communication cell at the ith time the verifier enters a state q with $\gamma_q \neq \epsilon$. This ensures that the communication between the prover and the verifier is one-way. The corresponding language classes are named by prefixing the class names mentioned above with the designation "oneway-". Note that one can define a one-way IPS without mentioning a prover at all, as in the conventional "certificate" definitions of nondeterministic classes like NP.

The following equalities are trivial:

$$\mathsf{NP} = \mathsf{oneway\text{-}IP}_w(\mathsf{poly},\mathsf{poly},0) \tag{1}$$

$$\mathsf{NL} = \mathsf{oneway\text{-}IP}_w(\mathsf{poly},\mathsf{log},0) \tag{2}$$

Note that specifying 0 as the randomness complexity of the verifier is just a way of saying that it is deterministic.

Allowing logarithmic amounts of randomness yields the characterization [5]

$$\mathsf{NP} = \mathsf{oneway\text{-}IP}_w(\mathsf{poly},\mathsf{log},\mathsf{log}) = \mathsf{IP}_w(\mathsf{poly},\mathsf{log},\mathsf{log}), \tag{3}$$

with an improvement in the space bound. Relaxing the randomness bound further does not help on its own, since [3]

$$\mathsf{oneway\text{-}IP}_w(\mathsf{poly},\mathsf{log},\mathsf{poly}) = \mathsf{NP}, \tag{4}$$

but allowing interaction as well famously yields [2,16]

$$\mathsf{IP}_w(\mathsf{poly},\mathsf{log},\mathsf{poly}) = \mathsf{PSPACE}. \tag{5}$$

A *public-coin IPS*, also known as an *Arthur-Merlin game*, is an IPS where the verifier sends precisely its state, tape and head updates at every step to the prover, thereby ensuring that the prover always knows the verifier's configuration. The public-coin version of $\mathsf{IP}_w(t,s,r)$ is named $\mathsf{AM}_w(t,s,r)$. It is known [1,10,16] that

$$\mathsf{AM}_w(\mathsf{exp},\mathsf{log},\mathsf{exp}) = \mathsf{P}, \tag{6}$$

and

$$\mathsf{AM}_w(\mathsf{poly},\mathsf{poly},\mathsf{poly}) = \mathsf{PSPACE}. \tag{7}$$

The relationships in Equations 1-7 remain true for the strong definition of IPSs, since logarithmically bounded space is sufficient to cut off unacceptably long computational paths. When one considers finite state verifiers, [8] which use only a constant amount of cells on the work tape,[1] the difference between the weak and strong definitions of becomes evident.

With no limits on the runtime, or the number of random bits to be used, weak finite state verifiers are very powerful; oneway-$\mathsf{IP}_w(\infty,\mathsf{cons},\infty)$ contains every recursively enumerable language, whereas $\mathsf{IP}(\infty,\mathsf{cons},\infty)$ is contained in $\mathsf{SPACE}(2^{2^{O(n)}})$ [6]. It has been proven [8] that Arthur-Merlin games with finite state verifiers exist for languages outside the class 2PFA of languages recognizable by "stand-alone" 2pfa's, and that some languages have linear-time finite state verifiers only if the public-coin restriction is not enforced, in contrast to Equations 5 and 7.

We shall focus on verifiers which use a constant number of random bits for any input.

We start our examination of the effects of limiting the number of random bits by noting that machines that are not helped by a prover about their input are very weak when restricted to work with constant workspace, and polynomially bounded randomness.

Theorem 1. *For any polynomial p, every 2pfa whose expected number of coin tosses on halting computational branches is $O(p(n))$ for input strings of length n recognizes a regular language with bounded error.*

Proof. This is a straightforward modification of the proof (in [7]) of the following fact [7,14]:

> *For any polynomial p, 2pfa's with expected runtime $O(p(n))$ recognize only the regular languages with bounded error.*

We omit the details due to space constraints.   □

In the next section, we shall demonstrate an interesting relationship between constant-space, constant-randomness verifiers and multihead finite automata. A k-head finite automaton (2nfa(k)) is simply a nondeterministic finite-state machine with k two-way heads that it can direct on a read-only tape containing the input string, flanked by two end-markers. A configuration of a 2nfa(k) is a tuple consisting of its current state and head positions. The classes of languages recognized by these machine families will be denoted as 2NFA(k). Deterministic single-head versions of multihead automata are denoted as 2dfa's. Detailed information about 2nfa(k)'s can be found in [12]. We note the following important facts that will be used in our proofs.

---

[1] It is easy to see that such machines can be simulated by machines with longer programs which have no work tape at all, namely, two-way probabilistic finite automata (2pfa's) [9].

Fact 1. $\cup_{k \geq 1} 2\mathsf{NFA}(k) = \mathsf{NL}$ *[11]*.

Fact 2. *Every 2nfa(k) has an equivalent 2nfa(2k) that halts in $O(n^k)$ time on every computational branch.*

3   2pfa Verifiers with Constant Randomness and 2nfa(k)'s

Our new characterization of NL is demonstrated by the following two lemmas.

Lemma 1. $\mathsf{NL} \subseteq \text{oneway-IP}_w(\text{poly,cons,cons})$.

Proof. Let L be any language in NL. By Fact 1, L is recognized by a 2nfa(k) M. We show how to construct a weak one-way IPS with the required properties that recognizes L with perfect completeness for any desired error probability $\varepsilon < \frac{1}{2}$. As mentioned above, this is equivalent to demonstrating how every member of L has a membership certificate that can be checked with such a verifier. We start by building a verifier V that simulates one run of M, by consulting the certificate for choosing among the nondeterministic branches of M. V uses just $r = \lceil \log k \rceil$ random bits to branch to k computation paths (each path has probability at least 2^{-r}) while scanning the left input end-marker. Each such path will use its head to track the position of the corresponding head of M. For every step of the simulation of M, the certificate contains a symbol conveying the list of k symbols that would be scanned by M's heads at this step, together with an indication of which nondeterministic choice should now be taken by M to eventually reach the accept state. The ith path of V rejects immediately if it sees that the present certificate symbol is inconsistent with what the ith head is currently scanning, and updates its state and head position according to M's program and the information given by the certificate otherwise.

If the input string is accepted by M, the certificate will lead all paths of V to acceptance, by giving correct information about what the heads are seeing and the nondeterministic choice at every step, yielding a total acceptance probability of 1. Otherwise, any certificate must "lie" about at least one head in order to make some paths accept, causing the path responsible for that head to reject, so the acceptance probability in that case is at most $1 - 2^{-r}$. To reduce the unacceptably high error bound for nonmembers, we chain several copies of V to run one after another,[2] on a correspondingly long certificate, and accept if and only if all copies accept, rejecting otherwise. It is easy to see that a chain of m copies of V involves an error of $(1 - 2^{-r})^m$, and therefore $m \geq \frac{\log \varepsilon}{\log(1 - 2^{-r})}$ iterations are sufficient to obtain an error of ε, where the total number of random bits used by the resulting verifier would be $O(k \log k \log \frac{1}{\varepsilon})$. Note that a 2nfa($k$) with state set Q has at most $|Q|(n+2)^k$ distinct reachable configurations on any input of length n, and therefore V runs in polynomial time for correct proofs of membership.   □

---

[2] Note that the certificate guides the paths of V to position their heads back on the left end-marker and to start the next round of coin-flipping simultaneously.

The reason why our construction does not yield an IPS according to the strong definition is that an evil prover can supply an infinitely long fake certificate that makes some paths of the verifier enter infinite loops by lying[3] about a head that those paths cannot see, at the cost of being rejected by the path responsible for that head.

The proof of Lemma 1 shows that 2NFA(2), which contains nonregular languages, has verifiers with two random bits. This is the minimum number of useful random bits for 2pfa verifiers. A single coin toss would create just two computational paths with equal probability. Since a probabilistic machine that always responds correctly can be replaced by its deterministic counterpart, we must have the verifier err for at least one input string. But the probability of such an error is at least $\frac{1}{2}$ in a machine that tosses its coin only once, which would violate our bounded error condition.

The reader should also note that the IPSs of Lemma 1 are strictly stronger than 2pfa's unaided by a prover, even when the latter are allowed to use an unbounded number of fair coins, since it is known [13,15] that the class of languages recognizable by such stand-alone 2pfa's is properly contained in the class L.

We shall now show that two-way interaction with the prover does not augment the power of constant-randomness verifiers, even if they are allowed to use logarithmic space.

Lemma 2. $\mathsf{IP}_w(\text{poly},\text{log},\text{cons}) \subseteq \mathsf{NL}$.

Proof. Suppose that a language L has an IPS with prover P and logspace verifier V, that use the communication alphabet Γ. V always uses at most r random bits, and has state set $Q = C \cup N$, where $C = \{q \in Q \mid \gamma_q \neq \epsilon\}$ is the set of states which communicate with the prover, and N is the set of "noncommunicating" states. Let $t(n)$ denote the polynomial time bound of V. We start by transforming V to a set $\mathsf{S}=\{M_1, M_2, \ldots, M_{2^r}\}$ of deterministic logspace verifiers, each of which simulates a version of V with a different assignment to the r-bit random string.

We build a one-way IPS with a deterministic logspace verifier M for L. The purported membership certificate that M will check consists of 2^r tracks, each with alphabet $T = \Gamma \cup \{\bigcirc, \infty\}$. The ith track is supposed to contain a transcript of M_i's communications with the prover about the input x. The ith track square of the jth certificate symbol contains

- γ, if M_i receives the prover response γ in its jth communication step,
- $\bigcirc$, if M_i performs a halting computation with fewer than j communications, and,
- ∞ otherwise, that is, if M_i enters a nonhalting path of noncommunicating states after performing fewer than j communications.

To process the jth certificate symbol, M simulates all the M_is that are indicated to be on a halting path on the input x until they reach their jth communication

---

[3] We can assume that the simulated multihead automaton has the desirable property mentioned in Fact 2. Any prover that causes a long runtime must therefore be lying.

step, or terminate. M rejects if it detects a mismatch between the track content and the actual computation of M_i.

Note that some members of S can have the same communication transcript, that is, they can send precisely the same sequence of symbols to the prover, and therefore receive the same sequence of responses, during their execution on x. Partition S into blocks, each of which correspond to a different communication transcript. M discovers this partition as it goes through the certificate. At the start, it considers all the M_is as in the same block in the initial partition. Whenever it scans a new certificate symbol, M refines the partition to separate the M_is that send different symbols, or perform no communication, and rejects if the certificate is claiming that different prover messages are being received by two verifiers in the same block of the new partition. If any track contains a communication symbol after the appearance of a $\bigcirc$ on an ∞, M rejects. If it detects that it has been running for more than $n \cdot t(n)$ steps, M rejects. M accepts if the certificate survives these tests, and a majority of the M_is are verified to terminate with acceptance.

Clearly, a majority of the members of S accept as a result of their interaction with P on the input x if and only if $x \in L$. If the input is not in L, there is no prover that can fool V for more than half of its possible coin strings to cause acceptance together, and no certificate can make M accept this input. □

We have proven that

Theorem 2. oneway-IP$_w$(poly,cons,cons) = IP$_w$(poly,log,cons) = NL.

Let REG denote the set of regular languages. We also have the following to say about the public-coin versions of these verifiers.

Theorem 3. AM$_w$(∞,cons,cons) = REG.

Proof. One direction is obvious. For the other direction, we adapt the proof of Lemma 2 to this simpler case. The members of the set S of 2^r deterministic verifiers (the M_is) used in that proof are now 2dfa's. Since the prover is now free to send different messages to each of these verifiers, we do not have to worry about checking for consistency among the supplied communication transcripts of those machines. We can therefore simulate them sequentially, rather than in parallel, requiring the certificate to just present the transcripts of the M_is one after another. But such a certificate can be checked by a deterministic finite state verifier with just one head, meaning that the machine we construct is in fact a 2nfa, and therefore can recognize only regular languages. □

4   Open Questions

We have been able to represent the relationship between NL and NP in the form

$$\text{NL} = \text{oneway-IP}_w(\text{poly,cons,cons}) \subseteq \text{oneway-IP}_w(\text{poly,log,log}) = \text{NP}.$$

Further examination of other classes like oneway-IP$_w$(poly,cons,log) would be interesting.

Is the inclusion

$$\text{oneway-IP(poly,cons,cons)} \subseteq \text{oneway-IP}_w\text{(poly,cons,cons)}$$

proper or not? Every language that can be verified by a constant-randomness 2pfa that halts with probability 1 is recognized in linear time by a 2nfa(k) for some k. Does there exist a language in NL which cannot be recognized in linear time by any 2nfa(k)?

Acknowledgements. Say was partially supported by TÜBİTAK with grant 108E142; Yakaryılmaz was partially supported by TÜBİTAK with grant 108E142 and FP7 FET-Open project QCS.

We are grateful to Martin Kutrib, who helped us immensely with our questions about nfa(k)'s. We also thank Taylan Cemgil and Richard Lipton for their helpful answers.

References

1. Condon, A.: Computational Models of Games. MIT Press (1989)
2. Condon, A.: Space-bounded probabilistic game automata. Journal of the ACM 38(2), 472–494 (1991)
3. Condon, A.: The complexity of the max word problem and the power of one-way interactive proof systems. Computational Complexity 3(3), 292–305 (1993)
4. Condon, A.: The complexity of space bounded interactive proof systems. In: Complexity Theory: Current Research, pp. 147–190. Cambridge University Press (1993)
5. Condon, A., Ladner, R.: Interactive proof systems with polynomially bounded strategies. Journal of Computer and System Sciences 50(3), 506–518 (1995)
6. Condon, A., Lipton, R.J.: On the complexity of space bounded interactive proofs. In: Proceedings of the 30th Annual Symposium on Foundations of Computer Science. pp. 462–467 (1989),
http://portal.acm.org/citation.cfm?id=1398514.1398732
7. Dwork, C., Stockmeyer, L.: A time complexity gap for two-way probabilistic finite-state automata. SIAM Journal on Computing 19(6), 1011–1123 (1990)
8. Dwork, C., Stockmeyer, L.: Finite state verifiers I: The power of interaction. Journal of the ACM 39(4), 800–828 (1992)
9. Freivalds, R.: Probabilistic two-way machines. In: Proceedings of the International Symposium on Mathematical Foundations of Computer Science, pp. 33–45 (1981)
10. Goldwasser, S., Sipser, M.: Private coins versus public coins in interactive proof systems. In: Proceedings of the 18th Annual ACM Symposium on Theory of Computing (STOC 1986), pp. 59–68 (1986)
11. Hartmanis, J.: On non-determinancy in simple computing devices. Acta Informatica 1, 336–344 (1972)
12. Holzer, M., Kutrib, M., Malcher, A.: Complexity of multi-head finite automata: Origins and directions. Theoretical Computer Science 412, 83–96 (2011)

13. Kaņeps, J.: Stochasticity of the languages acceptable by two-way finite probabilistic automata. Diskretnaya Matematika 1, 63–67 (1989) (Russian)
14. Kaņeps, J., Freivalds, R.: Running Time to Recognize Nonregular Languages by 2-Way Probabilistic Automata. In: Leach Albert, J., Monien, B., Rodríguez-Artalejo, M. (eds.) ICALP 1991. LNCS, vol. 510, pp. 174–185. Springer, Heidelberg (1991)
15. Macarie, I.I.: Space-efficient deterministic simulation of probabilistic automata. SIAM Journal on Computing 27(2), 448–465 (1998)
16. Shamir, A.: IP = PSPACE. Journal of the ACM 39(4), 869–877 (1992)

Game Arguments in Computability Theory and Algorithmic Information Theory

Alexander Shen[1,2]

[1] Laboratoire d'Informatique, de Robotique et de Microélectronique de Montpellier (LIRMM), Université Montpellier 2, UMR 5506—CC477, 161 rue Ada, 34095 Montpellier Cedex 5, France
alexander.shen@lirmm.fr
[2] On leave from Institute for Information Transmission Problems, Russian Academy of Sciences, Bolshoy Karetny per. 19, Moscow, 127994, Russia
sasha.shen@gmail.com

Abstract. We provide some examples showing how game-theoretic arguments can be used in computability theory and algorithmic information theory: unique numbering theorem (Friedberg), the gap between conditional complexity and total conditional complexity, Epstein–Levin theorem and some (yet unpublished) result of Muchnik and Vyugin.

1 Friedberg's Unique Numbering

It often happens that some result in computability theory or algorithmic information theory is essentially about the existence of a winning strategy in some game. This statement can be elaborated in a more formal way [5,6,7], but in this paper we illustrate this approach by several examples.

Our first example is a classical result of R. Friedberg [4]: the existence of unique numberings.

Theorem 1 (Friedberg). *There exists a partial computable function $F(\cdot, \cdot)$ of two natural variables such that:*

1. *F is universal, i.e., every computable function $f(\cdot)$ of one variable appears among the functions $F_n : x \mapsto F(n, x)$;*
2. *all the functions F_n are different.*

Proof. The proof can be decomposed in two parts. First, we describe some game and explain why the existence of a (computable) winning strategy for one of the players makes the statement of Friedberg's theorem true.

In the second part of the proof we construct a winning strategy and therefore finish the proof.

Game

The game is infinite and is played on two boards. Each board is a table with an infinite number of columns (numbered $0, 1, 2 \ldots$ from left to right) and rows

S.B. Cooper, A. Dawar, and B. Löwe (Eds.): CiE 2012, LNCS 7318, pp. 655–666, 2012.

(numbered $0, 1, 2, \ldots$ starting from the top). Each player (we call them **A**lice and **B**ob, as usual) plays on its own board. The players alternate. At each move player can fill some (finitely many) cells at her/his choice with any natural numbers (s)he wishes. Once a cell is filled, it keeps this number forever (it cannot be erased).

The game is infinite, so in the limit we have two tables A (filled by Alice) and B (filled by Bob). Some cells in the limit tables may remain empty. The winner is determined by the following rule: Bob wins if

- for each row in A-table there exists an identical row in B-table;
- all the rows in B-table are different.

Lemma 1. *Assume that Bob has a computable winning strategy in this game. Then the statement of Theorem 1 is true.*

Proof. A table represents a partial function of two arguments in a natural way: the number in ith row and jth column is the value of the function on (i, j); if the cell is not filled, the value is undefined.

Let Alice fill A-table with the values of some universal function (so the jth cell in the ith row is the output of ith program on input j). Alice does this at her own pace simulating in parallel all the programs (and ignoring Bob's moves). Let Bob apply his computable winning strategy against the described strategy of Alice. Then his table also corresponds to some computable function B (since the entire process is algorithmic). This function satisfies both requirements of Theorem 1: since A-function is universal, every computable function appears in some row of A-table and therefore (due to the winning condition) also in some row of B-table. So B is universal. On the other hand, all B_n are different since the rows of B-table (containing B_n) are different.

Remark 1. If Alice had a computable winning strategy in our game, the statement of Theorem 1 would be false. Indeed, let Bob fill his table with the values of a universal function that satisfies the requirements of the theorem (ignoring Alice's moves). Then Alice fills her table in a computable way and wins. This means that some row of Alice's table does not appear in Bob's table (so his function is not universal) or two rows in Bob's table coincide (so his function does not satisfy the uniqueness requirement).

So we can try this game approach even not knowing for sure who wins in the game; finding out who wins in the game would tell us whether the statement of the theorem is true or false (assuming the strategy is computable).

Winning Strategy

Lemma 2. *Bob has a computable winning strategy in the game described.*

Proving this lemma we may completely forget about computability and just describe the winning strategy explicitly (this is the main advantage of the game approach). We do this in two steps: first we consider a simplified version of the game and explain how Bob can win. Then we explain what he should do in the full version of the game.

In the simplified version of the game Bob, except for filling B-table, could *kill* some rows in it. The rows that were killed are not taken into account when the winner is determined. So Bob wins if the final (limit) contents of the tables satisfies two requirements: (1) for each row in A-table there exists an identical valid (non-killed) row in B-table, and (2) all the valid rows in B-table are different. (According to this definition, after the row is killed its content does not matter.)

To win the game, Bob hires a countable number of assistants and makes ith assistant responsible for ith row in A-table. The assistants start their work one by one; let us agree that ith assistant starts working at move i, so at every moment only finitely many assistants are active. Assistant starts her work by *reserving* some row in B-table not reserved by other assistants, and then continues by copying the current contents of ith row of A-table (for which she is responsible) into this reserved row. Also at some point the assistant may decide to kill the current row reserved by her, reserve a new row, and start copying the current content of ith row into the new reserved row. Later in the game she may kill the reserved row again, etc.

The instructions for the assistant say when the reserved row is killed. They should guarantee that

- if ith row in the final (limit) state of A-table coincides with some previous row, then ith assistant kills her reserved row infinitely many times (so none of the reserved rows remain active);
- if it is not the case, i.e., if ith row is different from all previous rows in the final A-table, then ith assistant kills her row only finitely many times (and after that faithfully copies ith row of A-table into the reserved row).

If this is arranged, the valid rows of B-table correspond to the first occurences of rows with given contents in A-table, so they are all different, and contain all the rows of A-table.

The instruction for the assistant number i: keep track of the number of rows that you have already killed in some counter k; if in the current state of A-table the first k positions in ith row are identical to the first k positions of some previous row, kill the current reserved row in B-table (and increase the counter); if not, continue copying ith row into the current row.

Let us see why these instructions indeed have the required properties. Imagine that in the limit state of A-table the row i is the first row with given content, i.e., is different from all the previous rows. For each of the previous rows let us select and fix some position (column) where the rows differ, and consider the moment T when these positions reach the final state. Let N be the maximal number of the selected columns. After step T the ith row in A-table differs from all previous rows in one of the first N positions, so if the counter of killed rows exceeds N, no more killings are possible (for this assistant).

On the other hand, assume that ith assistant kills her row finitely many times and N is the maximal value of her counter. After N is reached, the contents of ith row in A-table is always different from the previous rows in one of the first

N positions, and the same is true in the limit (since this rectangle reaches its limit state at some moment).

So Bob can win in the simplified game, and to finish the proof of Lemma 2 we need to explain how Bob can refrain from killing and still win the game.

Let us say that a row is *odd* if it contains a finite odd number of non-empty cells. Bob will now ignore odd rows of A-table and at the same time guarantee that all possible odd rows (there are countably many possibilities) appear in B-table exactly once. We may assume now without loss of generality that odd rows never appear in A-table: if she adds some element in a row making this row odd, this element is ignored by Bob until Alice wants to add another element in this row, and then the pair is added. This makes the A-table that Bob sees slightly different from what Alice actually does, but all the rows in the limit A-table that are not odd (i.e., are infinite or have even number of filled cells) will get through — and Bob separately takes care of odd rows.

Now the instructions for assistants change: instead of killing some row, she should fill some cells in this row making it odd, and ensure that this odd row is new (different from all other odd rows of the current B-table). After that, this row is considered like if it were killed (no more changes). This guarantees that all non-odd rows of A-table appear in B-table exactly once.

Also Bob hires an additional assistant who ensures that all possible odd rows appear in B-table: she looks at all the possibilities one by one; if some odd row has not appeared yet, she reserves some row and puts the desired content there. (Unlike other assistants, she reserves more and more rows.) This behavior guarantees that all possible odd rows appear in B-table exactly once. (Recall that other assistants also avoid repetitions among odd rows.) Lemma 2 and Theorem 1 are proven.

2   Total Conditional Complexity

In this section we switch from the general computability theory to the algorithmic information theory and consider a simple example. We compare the conditional complexity $C(x|y)$ and the minimal length of the program of a *total* function that maps y to x. It turns out that the second quantity can be much bigger (see, e.g., [1].) But let us recall first the definitions.

The *conditional complexity* of a binary string x relative to a binary string y (a *condition*) is defined as the length of the shortest program that maps y to x. The definition depends on the choice of the programming language, and one should select an optimal one that makes the complexity minimal (up to $O(1)$ additive term). When the condition y is empty, we get (unconditional plain) complexity of x. See, e.g., [9] for more details. The conditional complexity of x relative to y is denoted by $C(x|y)$; the unconditional complexity of x is denoted by $C(x)$.

It is easy to see that $C(x|y)$ can also be defined as the minimal *complexity* of a program that maps y to x. (This definition coincides with the previous one up to $O(1)$ additive term.) But in some applications (e.g., in algorithmic statistics, see [10]) we are interested in *total* programs, i.e. programs that are guaranteed to terminate for every input. Let us define $\mathrm{CT}(x|y)$ as the minimal complexity of a *total* program that maps y to x. In general, this restriction could increase complexity, but how significant could be this increase? It turns out that these two quantities may differ drastically, as the following simple theorem shows:

Theorem 2. *For every n there exist two strings x_n and y_n of length n such that $C(x_n|y_n) = O(1)$ but $\mathrm{CT}(x_n|y_n) \geq n$.*

Proof. To prove this theorem, consider a game G_n (for each n). In this game Alice constructs a partial function A from $\mathbb{B}^n$ to $\mathbb{B}^n$, i.e., a function defined on (some) n-bit strings, whose values are also n-bit strings. Bob constructs a list $B_1, \ldots, B_k$ of total functions of type $\mathbb{B}^n \to \mathbb{B}^n$. (Here $\mathbb{B} = \{0, 1\}$.)

The players alternate; at each move Alice can add several strings to the domain of A and choose some values for A on these strings; the existing values cannot be changed. Bob can add some total functions to the list, but the total length of the list should remain less than 2^n. The players can also leave their data unchanged; the game, though infinite by definition, is essentially finite since only finite number of nontrivial moves is possible. The winner is determined as follows: Alice wins if there exists a n-bit string y such that $A(y)$ is defined and is different from $B_1(y), \ldots, B_k(y)$.

Lemma 3. *Alice has a computable (uniformly in n) winning strategy in this game.*[1]

Before proving this lemma, let us explain how it helps to prove Theorem 2. Let (for every n) Alice play against the following strategy of Bob: he just enumerates all the total functions of type $\mathbb{B}^n \to \mathbb{B}^n$ who have complexity less than n, and adds them to the list when they appear. (As in the previous section, Bob does not really care about Alice's moves.) The behavior of Alice is then also computable since she plays a computable strategy againt a computable opponent. Let y_n be the string where Alice wins, and let x_n be equal to $A(y_n)$ where A is the function constructed by Alice.

It is easy to see that $C(x_n|y_n) = O(1)$; indeed, knowing y_n, we know n, can simulate the game, and find x_n during this simulation. On the other hand, if there were a total function of complexity less than n that maps y_n to x_n, then this function would be in the list and Bob would win.

So it remains to prove the lemma by showing the strategy for Alice. This strategy is straightforward: first Alice selects some y and says that $A(y)$ is equal to some x. (This choice can be done in arbitrary way, if Bob has not selected any functions yet; we may always assume it is the case by postponing the first move of Bob; the timing is not important in this game.) Then Alice waits until one of Bob's functions maps y to x. This may never happen; in this case Alice does nothing else and wins with x and y. But if this happens, Alice selects another y and chooses x that is different from $B_1(y), \ldots, B_k(y)$ for all total

---

[1] Since the game is effectively finite, in fact the existence of a winning strategy implies the existence of a computable one. But it is easy to describe the computable strategy explicitly.

functions $B_1, \ldots, B_k$ that are currently in Bob's list. Since there are less than 2^n total functions in the list, it is always possible. Also, since Bob can make at most $2^n - 1$ nontrivial moves, Alice will not run out of strings y. Lemma 3 and theorem 2 are proven.

3   Epstein–Levin Theorem

In this section we discuss a game-theoretic interpretation of an important recent result of Epstein and Levin [3]. This result can be considered as an extension of some previous observations made by Vereshchagin (see [10]). Let us first recall some notions from the algorithmic information theory.

For a finite object x one may consider two quantities. The first one, the complexity of x, shows how many bits we need to describe x (using an optimal description method). The second one, a priori probability of x, measures how probable is the appearance of x in a (universal) algorithmic random process. The first approach goes back to Kolmogorov while the second one was suggested earlier by Solomonoff.[2] The relation between these two notions in a most clean form was established by Levin and later by Chaitin (see [2] for more details). For that purpose Levin modified the notion of complexity and introduced *prefix complexity* $\mathrm{K}(x)$ where programs (descriptions) satisfy an additional property: if p is a program that outputs x, then every extension of p (every string having prefix p) also outputs x. (Chaitin used another restriction: the set of programs should be prefix-free, i.e., none of the programs is a prefix of another one; though it is a significantly different restriction, it leads to the same notion of complexity up to $O(1)$ additive term.) The notion of a priori probability can be formally defined in the following way: consider a randomized algorithm M without input that outputs some natural number and then terminates. The output number depends on the outcomes of the internal random bit generator (fair coin), so for every x there is some probability m_x to get x as output. (The sum $\sum m_x$ does not exceed 1; it can be less if the machine M performs a non-terminating computation with positive probability.) There exists a universal machine M of this type, i.e., the machine for which function $x \mapsto m_x$ is maximal up to a constant factor. (For example, M can start by choosing a random machine in such a way that every choice has positive probability, and then simulate this machine.) We fix some universal machine M and call the probability m_x to get x on its output *a priori* probability of x.

The relation between prefix complexity and a priory probability is quite close: Levin and Chaitin have shown that $\mathrm{K}(x) = -\log_2 m_x + O(1)$. However, the situation changes if we extend prefix complexity and a priori probability to sets. Let X be a set of natural numbers. Then we can consider two quantities that measure the difficulty of a task "produce some element of X": *complexity* of X, defined as the minimal length of a program that produces some element in X, and *a priori probability* of X, the probability to get some element of X as

---

[2] Solomonoff also mentioned the optimal description method as a technical tool somewhere in his paper.

an output of the universal machine M. For an arbitrary set of integers these two quantities are not related in the same way as before: complexity can differ significantly from the minus logarithm of a priori probability. In other words, for an arbitrary set X the quantities

$$\max_{x \in X} m_x \qquad \text{and} \qquad \sum_{x \in X} m_x$$

(the first one corresponds to the complexity, the second one corresponds to a priori probability) could be very different. For example, if X is the set of strings of length n that have complexity close to n, the first quantity is rather small (since all m_x are close to 2^{-n}) while the second one is quite big (a string chosen randomly with respect to the uniform distribution on n-bit strings, has complexity close to n with high probability).

Epstein–Levin theorem says that such a big difference is *not* possible if the set X is *stochastic*. The notion of a stochastic object was introduced in the algorithmic statistics. A finite object X is called *stochastic* if there exist a finite distribution (a probability distribution with finite domain and rational probabilities) P that has small complexity, and the randomness deficiency of X with respect to P, defined as $-\log P(X) - \mathrm{K}(X|P)$, is small.

We do not go into the further details about Epstein–Levin theorem. Instead we just present a game that corresponds to it and prove that one of the players has a computable winning strategy; interested readers can check that indeed the existence of such a strategy implies the statement of Epstein–Levin theorem. (See also the comments after the definition of the game.)

The Epstein–Levin game happens on a finite bipartite graph $E \subset L \times R$ with left part L and right part R. A probability distribution ρ on R with rational values is fixed, together with three parameters: some natural number k, some natural number l and some positive rational number δ. After all these objects are fixed, we consider the following game.

Alice assigns some rational *weights* to vertices in L. Initially all the weights are zeros, but Alice can increase them during the game. The total weight of L (the sum of weights) should never exceed 1. Bob can *mark* some vertices on the left and some vertices on the right. After a vertex is marked, it remains marked forever. The restrictions for Bob: he can mark at most l vertices on the left, and the total ρ-probability of marked vertices on the right should be at most δ. The winner is determined as follows: Bob wins if every vertex y on the right for which the (limit) total weight of all its L-neighbors exceeds 2^{-k}, either is marked itself (at some point), or has a marked (at some point) neighbor.

Evidently, the task of Bob becomes harder if l or δ decrease (he has less freedom in marking vertices), and becomes easier if k decreases (he needs to take care of less vertices). So the greater k and the smaller δ is, the bigger l is needed by Bob to win. The following lemma gives a bound (with some absolute constant in $O(1)$):

Lemma 4. *For $l = 2^k(\log(1/\delta) + O(1))$ Bob has a computable winning strategy in the described game.*

Before we prove this lemma, let us explain the connection between this game and the statement of Epstein–Levin theorem. Vertices in R are finite sets of integers; vertices in L are integers, and the edges correspond to $\in$-relation. Alice weights are a priori probabilities of integers (more precisely, increasing approximations to them). The distribution P on R is a simple distribution (on a finite family of finite sets) that is assumed to make X (from Levin–Epstein theorem) stochastic. Bob can mark X making it non-random with respect to P (the marked vertices form a P-small subset and therefore all have big randomness deficiency), so Epstein and Levin do not need to care about it any more. If X is not marked and has big total weight (= the total a priori probability), X is guaranteed to have a marked neighbor. This means that some element of X is marked and therefore has small complexity (since there are only few marked elements); this is what Epstein–Levin theorem requires. (Of course, one needs to use some specific bounds instead of "small" and "large" etc., but that is all.)

Proof. To prove the existence of a winning strategy for Bob, we use the following (quite unusual — I've seen only one other example) type of argument: we exhibit a simple *probabilistic* strategy for Bob that guarantees some positive probability of winning against any strategy of Alice. Since the game is essentially a finite game with full information (see the comments at the end of the proof about how to make it really finite), either Alice or Bob have a winning strategy. And if Alice had one, no probabilistic strategy for Bob could have a positive probability of winning.

Let us describe this strategy for Bob. It is rather simple: if Alice increases weight of some vertex x in L by an additional $\varepsilon > 0$, Bob in his turn marks x with probability $c2^k\varepsilon$, while $c > 1$ is some constant to be chosen later. If $c2^k\varepsilon > 1$ (this always happens if ε is 2^{-k} or more), Bob marks this vertex for sure, and this will satisfy all its R-neighbors. Note that without loss of generality we may assume that Alice increases weights one at a time, since we can split her move into a sequence of moves.

We have explained how Bob marks L-vertices; if at some point this does not help, i.e., if there is a R-vertex that currently has total weight at least 2^{-k} but no marked neighbors, Bob marks this R-vertex.

The probabilistic strategy for Bob is described, and we need to show that the probability of winning the game for Bob (for suitable c, see below) is positive. It is true even for a stronger definition of winning: let us require that for each y in R whose total weight reaches the threshold after some Alice's move, the requirement is satisfied immediately (after the next move of Bob). Fix some y. Assume that the weights of neighbors of y were increased by $\varepsilon_1, \ldots, \varepsilon_u$ during the game, and $\sum \varepsilon_i \geq 2^{-k}$. After each increase the corresponding neighbor of y was marked with probability $c2^k\varepsilon_i$, so the probability that the requirement for y is still false (and y is marked), does not exceed

$$(1 - c2^k\varepsilon_1) \cdot \ldots \cdot (1 - c2^k\varepsilon_u) \leq e^{-c2^k(\varepsilon_1 + \ldots + \varepsilon_u)} \leq e^{-c}$$

(recall that $(1 - t) \leq e^{-t}$ and that $\sum \varepsilon_i \geq 2^{-k}$). Therefore the expected ρ-measure of marked vertices on the right (the weighted average of numbers not exceeding e^{-c}) does not exceed e^{-c}, and the probability to exceed $2e^{-c}$ is at most $1/2$. So it is enough to let c be $\log_2(1/\delta) + O(1)$.

It remains to make two comments. First, the estimate for probability should be done more accurately, since the values of $\varepsilon_1, \ldots, \varepsilon_u$ may depend on Bob's moves. However, the coin tossing used to decide whether to mark some vertex on the left, is independent of what happened before, and this makes the estimate valid. (Formally, we need a backward induction in the tree of possibilities.)

Second, to switch from a probabilistic strategy to a deterministic one, we need to make the game finite. One may assume that current weights of vertices on the left all have the form 2^{-m} for some integer m (replacing weights by approximations from below, we can compensate for an additional factor of 2 by changing k by 1). Still the game is not finite, since Alice can start with very small weights. However, this is not important: the graph is finite, and all very small weights can be replaced by some 2^{-m}. If $2^{-m} \cdot \#L < 1$, then the sum of weights still does not exceed 2, and this again is a constant factor.

4   Muchnik–Vyugin Theorem

Consider the following problem. Let m be some constant. Given a string x_0 and integer n, we want to find strings $x_1, \ldots, x_m$ such that $C(x_i|x_j) = n + O(1)$ for all pairs of different i, j in the range $0, \ldots, m$. (Note that both i and j can be equal to 0). This is possible only if x_0 has high enough complexity, at least n, since $C(x_0|x_j)$ is bounded by $C(x_0)$. It turns out that such $x_1, \ldots, x_m$ indeed exist if $C(x_0)$ is high enough (though the required complexity of x_0 is greater than n), and the constant hidden in $O(1)$-notation does not depend on n (but depends on m).

This statement is non-trivial even for $n = 1$: it says that for every n and for every string x of high enough complexity there exists a string y such that both $C(x|y)$ and $C(y|x)$ are equal to $n + O(1)$.

Here is the exact statement (it specifies also the dependence of $O(1)$-constant on m):

Theorem 3 (An.A. Muchnik, M. Vyugin). *For every m and n and for every binary string x_0 such that*

$$C(x_0) > n(m^2 + m + 1) + O(\log m)$$

there exist strings $x_1, \ldots, x_m$ such that

$$n \leq C(x_i|x_j) \leq n + O(\log m)$$

for every two different $i, j \in \{0, \ldots, n\}$.

Note that the high precision is what makes this theorem non-trivial (if an additional term $O(\log C(x_0))$ were allowed, one could take the shortest program for x_0 and replace first n bits in it by m independent random strings).

Let us explain the game that corresponds to this statement. It is played on graph with $(m+1)$ parts $X_0, \ldots, X_m$ (there are countably many vertices in each part X_i representing possible values of x_i). As usual, there are two players: Alice and Bob. Alice may connect vertices from different parts by *undirected* edges, while Bob can connect them by *directed* edges. Alice and Bob make alternating moves; at each move they can add any finite set of edges. Alice can also *mark* vertices x_0 in X_0. The restrictions are:

- Alice may mark at most $m2^{n+1+nm(m+1)}$ vertices (in X_0);
- for each vertex $x_i \in X_i$ and for each $j \neq i$, Alice may draw at most $m(m+1)2^n$ undirected edges connecting x_i with vertices in X_j;
- for each vertex $x_i \in X_i$ and for each $j \neq i$, Bob may draw at most 2^n outgoing edges from x_i to vertices in X_j.

The game is infinite. Alice wins if (in the limit) for every non-marked vertex $x_0 \in X_0$ there exist vertices $x_1, \ldots, x_m$ from $X_1, \ldots, X_m$ such that each two vertices x_i, x_j (where $i \neq j$) are connected by an undirected (Alice) edge, but not connected by a directed (Bob) edge.

Lemma 5. *Alice has a computable winning strategy in this game.*

It is easy to see how this lemma can be used to prove the statement. Imagine that Bob draws an edge $x_i \to x_j$ when he discovers that $C(x_j|x_i) < n$. Then he never violates the restriction. Alice can computably win against this strategy; every marked vertex then has small complexity, since a marked vertex can be described by its ordinal number in the enumeration order. For every non-marked vertex x_0 there exist $x_1, \ldots, x_m$ that satisfy the winning conditions. For them $C(x_j|x_i) \geq n$ (otherwise Bob would connect them by a directed edge), and $C(x_j|x_i) \leq n + O(\log m)$, since x_j can be obtained from x_i if we know i, j, and the ordinal number of undirected edge x_i-x_j (among all the edges that connect x_i to X_j).

So it remains to prove the lemma. Alice does the following. First of all, it groups all vertices into disjoint cliques of size $m+1$ that contain vertices from all $m+1$ parts; each vertex is included in one of the cliques. (Formally Alice cannot create infinitely many edges at a time, but this can be artificially postponed.) A clique becomes invalid when Bob creates an edge between two vertices in this clique. Then the clique is destroyed and in some cases a new one is created (see below). We may assume without loss of generality that Bob creates edges only between vertices of some currently active clique, and only one edge is created at each move (the creation of other edges can be postponed).

When a clique becomes invalid, all the vertices record this fact in their "internal memory". Each vertex remembers $m(m+1)$ non-negative integers corresponding to ordered pairs (i, j). This tuple is called an "index" of a vertex. When the clique becomes invalid because of Bob's edge going from X_i to X_j, the component of index corresponding to (i, j) is incremented by 1 in all the vertices of the clique, the clique is disbanded, and all its vertices become free.

Alice then tries to form a new clique involving x_0 from the disbanded one, in such a way that all vertices of this clique (from all $m+1$ parts) have equal indices (that coincide with the updated index of x_0), are not yet connected to each other by Bob's edges, and the other vertices of the clique (except x_0) are free, i.e., are not included in any current clique. If this is not possible, no clique is created, and the vertex x_0 is marked.

This strategy keeps the following invariant relations:

- for each clique, all its vertices have the same index;
- for every clique and for every pair i, j (where $i \neq j$) the $(i \to j)$th component of indices of clique vertices equals to the number of Bob's edges going from x_i to X_j, as well as the number of Bob's edges going from X_i to x_j (where x_i and x_j are clique vertices in X_i and X_j); in particular, all components of the index are less than 2^n (since they are equal to the number of outgoing edges from x_i to X_j); here we use that Bob at each step draws only one edge, and this edge is parallel to some edge of some active clique;
- at every step only finitely many vertices have non-zero indices, and for each value of the index we have the same number of vertices of that index in all the parts X_i.
- the number of free vertices in all the parts is the same; in X_0 free vertices are marked; the number of free vertices of a given index is also the same in all parts;

To finish the proof, it remains to prove the bound for the number of marked vertices. For that we estimate the number of marked vertices of each index (the number of possible indices is bounded by $2^{nm(m+1)}$ since its components are bounded by 2^n). The idea here is simple: if we have many (more than $2m2^n$) free vertices of some index, we can always find a clique (made of them) for every vertex $x_0 \in X_0$ of that index that has lost its old clique: indeed, we do it sequentially in $X_1, \ldots, X_m$, and at every step we can find a vertex that is not connected to already selected vertices, since the number of outgoing edges, as well as the number of incoming edges, is bounded by $n2^m$.

Acknowledgments. Supported in part by NAFIT EMER-008-01 grant and RFBR 09-01-00709-a grant. The author is grateful to Leonid Levin, Peter Gács, the participants of Kolmogorov seminar (Moscow) and his colleagues in LIRMM (Montpellier) and LIAFA (Paris). M. Vyugin kindly allowed me to include the unpublished result of An. Muchnik (1958–2007) and himself in Section 4.

References

1. Bauwens, B.: Computability in statistical hypotheses testing, and characterizations of independence and directed influences in time series using Kolmogorov complexity. PhD thesis, Ugent (May 2010)
2. Bienvenu, L., Shen, A.: Algorithmic information theory and martingales. arXiv:0906.2614v1 (2009) (preprint)
3. Epstein, S., Levin, L.A.: Sets Have Simple Members. arXiv:1107.1458v5 (2011) (preprint)

4. Friedberg, R.: Three theorems on recursive numberings. J. of Symbolic Logic 23, 309–316 (1958)
5. Muchnik, A.A.: On the basic structures of the descriptive theory of algorithms. Soviet Math. Dokl. 32, 671–674 (1985)
6. Muchnik, A.A., Mezhirov, I., Shen, A., Vereshchagin, N.: Game interpretation of Kolmogorov complexity. arXiv:1003.4712v1 (2010) (preprint)
7. Vereshchagin, N.: Kolmogorov complexity and Games. Bulletin of the European Association for Theoretical Computer Science 94, 51–83 (2008)
8. Muchnik, A.A., Vyugin, M.: Information distance for complex strings (2007) (unpublished work)
9. Shen, A.: Algorithmic Information Theory and Kolmogorov Complexity, Technical Report, Uppsala University, TR2000-034,
   http://www.it.uu.se/research/publications/reports/2000-034/
10. Vereshchagin, N.K., Vitanyi, P.M.B.: Rate Distortion and Denoising of Individual Data Using Kolmogorov Complexity. IEEE Transactions on Information Theory 56(7), 3438–3454 (2010)

Turing Patterns in Deserts

Jonathan A. Sherratt

Department of Mathematics and Maxwell Institute for Mathematical Sciences,
Heriot-Watt University, Edinburgh EH14 4AS, United Kingdom
jas@ma.hw.ac.uk

Abstract. Self-organised patterns of vegetation are a characteristic feature of many semi-arid regions. In particular, banded vegetation is typical on hillsides. Mathematical modelling is widely used to study these
banded patterns, because there are no laboratory replicates. I will describe the development of spatial patterns in an established model for
banded vegetation via a Turing bifurcation. I will discuss numerical simulations of the phenomenon, and I will summarise nonlinear analysis on
the existence and form of spatial patterns as a function of the model
parameter that corresponds to mean annual rainfall.

1 Introduction

Self-organised patterns of vegetation are a characteristic feature of semi-deserts.
The most striking and best studied example is striped patterns on gentle slopes
(see [1, 2] for review). These occur in many parts of the world, and are particularly well documented in Australia [3, 4], Mexico/South-Western USA [5, 6]
and sub-Saharan Africa [7–9] Bands of grass, shrubs or trees run along contours,
separated by bare ground; wavelengths of about 1km are typical for trees and
shrubs, with shorter wavelengths observed for grasses.

There are no laboratory replicates of banded vegetation, so that empirical
study is limited to observation of existing patterns. Because the timescale of
pattern evolution is very slow (decades), such observational data is ineffective
as a basis for assessing the implications of changes in environmental parameters
such as rainfall. Therefore theoretical models are an important and widely used
tool for studying these patterns [10]. This paper is concerned with pattern formation in one model for banded vegetation, due originally to Klausmeier [11].
It comprises coupled partial differential equations for plant and water densities,
and is the basic model for patterning due to water redistribution. Many extensions of the Klausmeier model have been proposed over the last decade. Most of
these involve separate variables for soil and surface water [12–16]. Some authors
have also incorporated features such as rainfall variability [17–19] and a herbi-
vore population [20]; see [21, 22] for other recent extensions. Note also that the
Klausmeier model and its extensions are not the only theoretical explanation for
vegetation stripes. Lejeune and coworkers [23–26] have studied in detail a model
based on the combination of short-range activation and long-range inhibition
between neighbouring plants. Here the activation is due to shading of one plant

S.B. Cooper, A. Dawar, and B. Löwe (Eds.): CiE 2012, LNCS 7318, pp. 667–674, 2012.

by another, while competition for water results in inhibition; the difference in length scales of these processes is due to the root system within the soil being much more extensive than the parts of the plants above ground. In this model, slope acts as a selector rather than an initiator of spatial patterning.

This paper is concerned with the original Klausmeier model [11]. I will present a detailed discussion of pattern solutions of this model, which arise via a Turing bifurcation in the model partial differential equations. I will show that studying these pattern solutions can provide valuable new ecological insights into the formation and maintainance of vegetation patterns in semi-deserts.

2   Model Equations

The dimensionless form of the Klausmeier model is:

$$\partial u/\partial t = \overbrace{wu^2}^{\substack{\text{plant} \\ \text{growth}}} - \overbrace{Bu}^{\substack{\text{plant} \\ \text{loss}}} + \overbrace{\partial^2 u/\partial x^2}^{\substack{\text{plant} \\ \text{dispersal}}} \tag{1a}$$

$$\partial w/\partial t = \underbrace{A}_{\substack{\text{rain-} \\ \text{fall}}} - \underbrace{w}_{\substack{\text{evap-} \\ \text{oration}}} - \underbrace{wu^2}_{\substack{\text{uptake} \\ \text{by plants}}} + \underbrace{\nu \partial w/\partial x}_{\substack{\text{flow} \\ \text{downhill}}}. \tag{1b}$$

Here $u(x,t)$ is plant density, $w(x,t)$ is water density, t is time and x is space in a one-dimensional domain of constant slope, with the positive direction being uphill. The (dimensionless) parameters A, B and ν reflect rainfall, plant loss and slope gradient respectively. For full details of the dimensional model and nondimensionalisation, see Klausmeier (1999), Sherratt (2005) or Sherratt & Lord (2007).

For all parameter values, (1) has a stable trivial steady state $u = 0$, $w = A$, corresponding to bare ground, without vegetation. When $A \geq 2B$, there are also two other homogeneous steady states which arise from a saddle node bifurcation:

$$u = u_1 \equiv \frac{2B}{A - \sqrt{A^2 - 4B^2}}, \quad w = w_1 \equiv \frac{A - \sqrt{A^2 - 4B^2}}{2} \tag{2}$$

$$\text{and} \quad u = u_2 \equiv \frac{2B}{A + \sqrt{A^2 - 4B^2}}, \quad w = w_2 \equiv \frac{A + \sqrt{A^2 - 4B^2}}{2}. \tag{3}$$

The first of these (2) is always unstable to homogeneous perturbations; the second is the key equilibrium from which patterns develop. This steady state is linearly stable to homogeneous perturbations whenever $B < 2$. For larger values of B and small A, (3) can become unstable, giving complicated local dynamics including a limit cycle, but realistic parameter values for semi-arid environments imply that $B < 2$.

For large values of the rainfall parameter A, (3) is also stable to inhomogeneous perturbations, so that the model predicts the spatially uniform vegetation that characterises temperate parts of the world. However as A is decreased a Turing bifurcation occurs: (3) becomes unstable to some spatially inhomogeneous perturbations, and spatial patterns develop. The patterns consist of periodically repeating peaks and troughs of vegetation (Figure 1), and as the rainfall parameter A is decreased further, these solutions gradually increase in amplitude, resembling more closely the empirically observed patterns. In the prototypical Turing system of two coupled reaction-diffusion equations, the patterns arising from a Turing bifurcation are stationary. However the advection term in (1a) causes the patterns to move, in the positive x direction (uphill). There has been a long-running debate in the ecological literature about this uphill migration, with some field studies reporting stationary patterns (e.g., [3]). However, the majority of data sets spanning a time period sufficient to address this issue do indicate uphill migration, with speeds in the range 0.2–$1\,\mathrm{m\,year}^{-1}$ (see Table 5 of [1]). A recent and very detailed study using photographic data from satellites [27, Chapter 10] confirms migration, with speeds in this range, for three out of six geographical locations. The ecological cause of uphill migration is that moisture levels are higher on the uphill edge of the bands than on their downhill edge, leading to reduced plant death and greater seedling density [28, 29].

3   Travelling Wave Solutions

Mathematically, patterns moving with constant shape and speed can be studied via the ansatz $u(x,t) = U(z)$ and $w(x,t) = W(z)$, where $z = x - ct$ with c being the migration speed. Substituting these solution forms into (1) gives the travelling wave equations

$$d^2U/dz^2 + c\,dU/dz + WU^2 - BU = 0 \tag{4a}$$
$$(\nu + c)dW/dz + A - W - WU^2 = 0. \tag{4b}$$

Patterned solutions correspond to periodic solutions of (4). In [30], Gabriel Lord and I used numerical bifurcation analysis to study these periodic solutions. We showed that for a given value of the migration speed c, patterns occur for a range of rainfall parameter values A. For most values of c, this range is bounded by a Hopf bifurcation point for (3,4) and a homoclinic solution of (4). However for some values of c there is a fold in the branch of periodic travelling wave solutions, and this then constitutes on end of the rainfall range for patterns [31, 32]. A typical result is illustrated in Figure 2, which shows the loci of the Hopf bifurcation point and the homoclinic solution in the A–c parameter plane, for fixed values of B and ν.

Analytical study of (1) is made more complicated by the advective term in the u-equation. For example, linear stability analysis of (3) to investigate the Turing bifurcation is significantly more complicated in (1) than for a system of two reaction-diffusion equations [34], and indeed one cannot obtain an exact closed-form expression for the value of A at which the bifurcation occurs. However, the slope parameter ν is much larger than A and B: Klausmeier [11] estimated $\nu = 182.5$, $A = 0.1$–3.0 and $B=0.05$–2.0. This large value is not due to the slope itself being steep: banded vegetation is restricted to slopes of a few percent, and

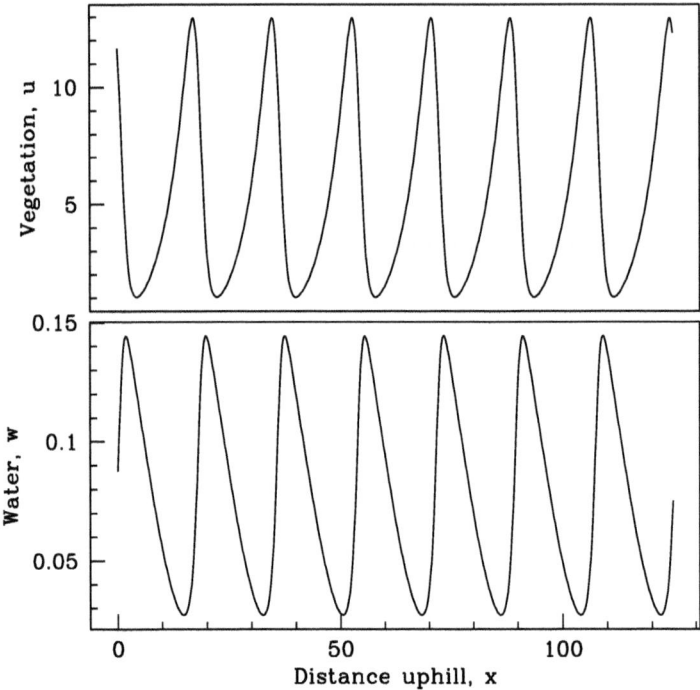

Fig. 1. An illustration of a typical vegetation pattern, as predicted by the Klausmeier model (1). There is a periodic pattern of peaks in vegetation density u, separated by regions in which vegetation is almost absent. The surface water density w also has a periodic form; it is largest on the uphill side of a vegetation stripe, and gradually decreases with distance uphill to the next stripe. The pattern moves slowly uphill; in this case the (dimensionless) migration speed is approximately 0.9. The parameter values are $A = 2.5$, $B = 0.45$, $\nu = 182.5$, which are in the range of Klausmeier's (1999) parameter estimates for grass. The equations were solved numerically using a finite difference scheme (see [30] for details) on the domain $0 < x < 125$ with periodic boundary conditions.

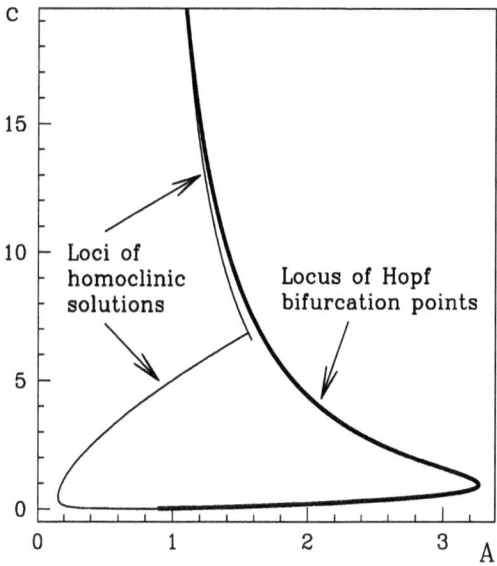

Fig. 2. A typical example of the part of the A–c parameter plane in which there are patterned solutions of (1), which corresponds to limit cycles in (4). I plot the loci of Hopf bifurcation points (━━━━) and homoclinic solutions (─────) in (4), which bound the pattern region. The other parameters are $B = 0.45$ and $\nu = 182.5$. The plot is truncated at $c \approx 20$: patterns actually exist for values of c up to about 50. The numerical solutions were performed using AUTO [33]. The loci of homoclinic orbits are approximations; they are in fact the loci of solutions of a fixed but very long wavelength (3000). Further details of the numerical continuation approach are given in [30].

on steeper slopes, different processes occur because rainwater generates gullies. Rather, ν is large because the plant diffusion coefficient is small compared to the advection rate of water, and it is the relative values of these quantities that determines the nondimensional parameter ν [11, 34]. By exploiting this large value of ν it is possible to obtain leading order approximations for various key points in the A–c parameter plane. In particular:

- The Turing bifurcation occurs at $A = (\sqrt{2}-1)^{1/2}\nu^{1/2}B^{5/4}$, $c = A^2/(2B^2\nu) + B^3\nu/(2A^2)$ [32].
- The maximum migration speed for patterns is $c = \nu B/(2 - B)$ [31].
- The base of the "tusk-shaped" region (see Figure 2) occurs at $c = 0.881B^{3/4}\nu^{1/2}$ [35, 36].
- Pattern solutions exist for arbitrarily small values of the migration speed c [32].

4 Conclusion

Ecological pattern formation at the level of whole ecosystems is a new, exciting, and rapidly growing research area, whose study is influenced strongly by Turing's ideas [37]. Vegetation patterns in semi-arid regions represent one example of such patterns, but there are many others, including regular isolated spots of trees and shrubs in savanna grasslands [38, 39], patterns of open-water pools in peatlands [40, 41], labyrinthine patterns in mussel beds [42, 43], striped patterns of tree lines ("ribbon forests") in the Rocky Mountains [44, 45]. Mathematical modelling is an important tool for the study of landscape patterns, and the Klausmeier model (1) is one of the most generic models: as well as semi-arid vegetation, it has been used to model fog-dependent plant ecosystems [46] and (with a slight modification) mussel beds in river estuaries [47]. I have outlined the mathematical analysis of pattern formation in the Klausmeier model, showing how such analysis can make clear and quantitative predictions concerning critical levels of the rainfall level and wave speed.

References

1. Valentin, C., d'Herbès, J.M., Poesen, J.: Soil and water components of banded vegetation patterns. Catena 37, 1–24 (1999)
2. Rietkerk, M., Dekker, S.C., de Ruiter, P.C., van de Koppel, J.: Self–organized patchiness and catastrophic shifts in ecosystems. Science 305, 1926–1929 (2004)
3. Dunkerley, D.L., Brown, K.J.: Oblique vegetation banding in the Australian arid zone: implications for theories of pattern evolution and maintenance. J. Arid Environ. 52, 163–181 (2002)
4. Berg, S.S., Dunkerley, D.L.: Patterned mulga near Alice Springs, central Australia, and the potential threat of firewood collection on this vegetation community. J. Arid Environ. 59, 313–350 (2004)
5. Montaña, C.: The colonization of bare areas in two–phase mosaics of an arid ecosystem. J. Ecol. 80, 315–327 (1992)
6. McDonald, A.K., Kinucan, R.J., Loomis, L.E.: Ecohydrological interactions within banded vegetation in the northeastern Chihuahuan Desert, USA. Ecohydrology 2, 66–71 (2009)
7. MacFadyen, W.: Vegetation patterns in the semi–desert plains of British Somaliland. Geographical J. 115, 199–211 (1950)
8. Valentin, C., d'Herbès, J.M.: Niger tiger bush as a natural water harvesting system. Catena 37, 231–256 (1999)
9. Couteron, P., Mahamane, A., Ouedraogo, P., Seghieri, J.: Differences between banded thickets (tiger bush) at two sites in West Africa. J. Veg. Sci. 11, 321–328 (2000)
10. Borgogno, F., D'Odorico, P., Laio, F., Ridolfi, L.: Mathematical models of vegetation pattern formation in ecohydrology. Rev. Geophys. 47, art. no. RG1005 (2009)
11. Klausmeier, C.A.: Regular and irregular patterns in semiarid vegetation. Science 284, 1826–1828 (1999)
12. HilleRisLambers, R., Rietkerk, M., van de Bosch, F., Prins, H.H.T., de Kroon, H.: Vegetation pattern formation in semi–arid grazing systems. Ecology 82, 50–61 (2001)

13. Rietkerk, M., Boerlijst, M.C., van Langevelde, F., HilleRisLambers, R., van de Koppel, J., Prins, H.H.T., de Roos, A.: Self–organisation of vegetation in arid ecosystems, Am. Am. Nat. 160, 524–530 (2002)

14. Gilad, E., von Hardenberg, J., Provenzale, A., Shachak, M., Meron, E.: A mathematical model of plants as ecosystem engineers, J. J. Theor. Biol. 244, 680–691 (2007)

15. Ursino, N.: Modeling banded vegetation patterns in semiarid regions: inter–dependence between biomass growth rate and relevant hydrological processes. Water Resour. Res. 43, W04412 (2007)

16. Ursino, N.: Above and below ground biomass patterns in arid lands. Ecological Modelling 220, 1411–1418 (2009)

17. Ursino, N., Contarini, S.: Stability of banded vegetation patterns under seasonal rainfall and limited soil moisture storage capacity. Adv. Water Resour. 29, 1556–1564 (2006)

18. Guttal, V., Jayaprakash, C.: Self–organisation and productivity in semi–arid ecosystems: implications of seasonality in rainfall. J. Theor Biol. 248, 290–500 (2007)

19. Kletter, A.Y., von Hardenberg, J., Meron, E., Provenzale, A.: Patterned vegetation and rainfall intermittency. J. Theor. Biol. 256, 574–583 (2009)

20. van de Koppel, J., Rietkerk, M., van Langevelde, F., Kumar, L., Klausmeier, C.A., Fryxell, J.M., Hearne, J.W., van Andel, J., de Ridder, N., Skidmore, M.A., Stroosnijder, L., Prins, H.H.T.: Spatial heterogeneity and irreversible vegetation change in semiarid grazing systems. Am. Nat. 159, 209–218 (2002)

21. Pueyo, Y., Kefi, S., Alados, C.L., Rietkerk, M.: Dispersal strategies and spatial organization of vegetation in arid ecosystems. Oikos 117, 1522–1532 (2008)

22. Kefi, S., Rietkerk, M., Katul, G.G.: Vegetation pattern shift as a result of rising atmospheric CO2 in arid ecosystems. Theor. Pop. Biol. 74, 332–344 (2008)

23. Lefever, R., Lejeune, O.: On the origin of tiger bush. Bull. Math. Biol. 59, 263–294 (1997)

24. Couteron, P., Lejeune, O.: Periodic spotted patterns in semi–arid vegetation explained by a propagation–inhibition model. J. Ecol. 89, 616–628 (2001)

25. Barbier, N., Couteron, P., Lefever, R., Deblauwe, V., Lejeune, O.: Spatial decoupling of facilitation, competition at the origin of gapped vegetation patterns. Ecology 89, 1521–1531 (2008)

26. Lefever, R., Barbier, N., Couteron, P., Lejeune, O.: Deeply gapped vegetation patterns: on crown/root allometry, criticality and desertification. J. Theor. Biol. 261, 194–209 (2009)

27. Deblauwe, V.: Modulation des structures de végétation auto–organisées en milieu aride / Self–organized vegetation pattern modulation in arid climates. PhD thesis, Université Libre de Bruxelles (2010)

28. Tongway, D.J., Ludwig, J.A.: Theories on the origins, maintainance, dynamics, and functioning of banded landscapes. In: Tongway, D.J., Valentin, C., Seghieri, J. (eds.) Banded Vegetation Patterning in Arid and Semi–Arid Environments, pp. 20–31. Springer, New York (2001)

29. Montaña, C., Seghieri, J., Cornet, A.: Vegetation dynamics: recruitment, regeneration in two–phase mosaics. In: Tongway, D.J., Valentin, C., Seghieri, J. (eds.) Banded Vegetation Patterning in Arid and Semi–Arid Environments, pp. 132–145. Springer, New York (2001)

30. Sherratt, J.A., Lord, G.J.: Nonlinear dynamics, pattern bifurcations in a model for vegetation stripes in semi–arid environments. Theor. Pop. Biol. 71, 1–11 (2007)

31. Sherratt, J.A.: Pattern solutions of the Klausmeier model for banded vegetation in semi–arid environments II: patterns with the largest possible propagation speeds. Proc. R. Soc. Lond. A 467, 3272–3294 (2011)

32. Sherratt, J.A.: Pattern solutions of the Klausmeier model for banded vegetation semi–arid environments IV: slowly moving patterns and their stability (submitted)

33. Doedel, E.J.: AUTO, a program for the automatic bifurcation analysis of autonomous systems. Cong. Numer. 30, 265–384 (1981)

34. Sherratt, J.A.: An analysis of vegetation stripe formation in semi–arid landscapes. J. Math. Biol. 51, 183–197 (2005)

35. Sherratt, J.A.: Pattern solutions of the Klausmeier model for banded vegetation in semi–arid environments I. Nonlinearity 23, 2657–2675 (2010)

36. Sherratt, J.A.: Pattern solutions of the Klausmeier model for banded vegetation in semi–arid environments III: the transition between homoclinic solutions (submitted)

37. Turing, A.M.: The chemical basis of morphogenesis. Phil. Trans. R. Soc. Lond. B 237, 37–72 (1952)

38. Lejeune, O., Tlidi, M., Couteron, P.: Localized vegetation patches: a self–organized response to resource scarcity. Phys. Rev. E 66, 010901 (2002)

39. Ben Wu, X., Archer, S.R.: Scale–dependent influence of topography–based hydrologic features on patterns of woody plant encroachment in savanna landscapes. Landscape Ecol. 20, 733–742 (2005)

40. Belyea, L.R.: Climatic and topographic limits to the abundance of bog pools. Hydrological Processes 21, 675–687 (2007)

41. Eppinga, M.B., de Ruiter, P.C., Wassen, M.J., Rietkerk, M.: Nutrients and hydrology indicate the driving mechanisms of peatland surface patterning. Am. Nat. 173, 803–818 (2009)

42. van de Koppel, J., Rietkerk, M., Dankers, N., Herman, P.M.J.: Scale–dependent feedback and regular spatial patterns in young mussel beds. Am. Nat. 165, E66–E77 (2005)

43. van de Koppel, J., Gascoigne, J.C., Theraulaz, G., Rietkerk, M., Mooij, W.M., Herman, P.M.J.: Experimental evidence for spatial self–organization and its emergent effects in mussel bed ecosystems. Science 322, 739–742 (2008)

44. Hiemstra, C.A., Liston, G.E., Reiners, W.A.: Observing, modelling, and validating snow redistribution by wind in a Wyoming upper treeline landscape. Ecol. Modelling 197, 35–51 (2006)

45. Bekker, M.F., Clark, J.T., Jackson, M.W.: Landscape metrics indicate differences in patterns and dominant controls of ribbon forests in the Rocky Mountains, USA. Applied Vegetation Science 12, 237–249 (2009)

46. Borthagaray, A.I., Fuentes, M.A., Marquet, P.A.: Vegetation pattern formation in a fog–dependent ecosystem. J. Theor. Biol. 265, 18–26 (2010)

47. Wang, R.H., Liu, Q.X., Sun, G.Q., Jin, Z., van de Koppel, J.: Nonlinear dynamic and pattern bifurcations in a model for spatial patterns in young mussel beds. J. R. Soc. Interface 6, 705–718 (2009)

Subsymbolic Computation Theory for the Human Intuitive Processor

Paul Smolensky

Department of Cognitive Science, Johns Hopkins University, 237 Krieger Hall, 3400 North Charles Street, Baltimore, Maryland 21218, United States of America
smolensky@jhu.edu

Abstract. The classic theory of computation initiated by Turing and his contemporaries provides a theory of effective procedures—algorithms that can be executed by the human mind, deploying cognitive processes constituting the *conscious rule interpreter*. The cognitive processes constituting the human *intuitive processor* potentially call for a different theory of computation. Assuming that important functions computed by the intuitive processor can be described abstractly as symbolic recursive functions and symbolic grammars, we ask which symbolic functions can be computed by the human intuitive processor, and how those functions are best specified—given that these functions must be computed using neural computation. Characterizing the automata of neural computation, we begin the construction of a class of recursive symbolic functions computable by these automata, and the construction of a class of neural networks that embody the grammars defining formal languages.

1 Effective Procedures vs. Intuitive Cognitive Processes

What is the set of functions computable by human intuition? Intuitive cognitive processes contrast with conscious rule interpretation [10], the mental processes enabling a clerk to take a list of English instructions and carry out a complex mathematical calculation by consciously interpreting those instructions one after another. Less mundanely, the *Entscheidungsproblem* also concerns the capabilities of conscious rule interpretation: Turing's 1936 paper [15] and other classic works provide a theory of the set of functions computable by human conscious rule interpretation, i.e., by effective procedures. Other notions of computation yielding different sets of computable functions derive from alternative conceptions of 'machine' in the sense of physical artifact (e.g., differential analyzer, optical or quantum computer).

But it is a natural biological system, the brain, that motivates the conception of computation of interest here. Within one stream of cognitive science [11], models of the brain and models of mental processes converged around 30 years ago upon a class of computational systems for modeling intuition, a class known variously as 'connectionist', 'parallel distributed processing', or 'abstract neural network' computational architectures [8].

S.B. Cooper, A. Dawar, and B. Löwe (Eds.): CiE 2012, LNCS 7318, pp. 675–685, 2012.

Since the dawn of modern cognitive science in the middle of the 20th century, many cognitive scientists have treated intuition as formally identical with conscious rule interpretation, but inaccessible to consciousness [11]. Within this conception, all higher mental processes, and not just conscious rule interpretation, can be modeled as effective procedures using classical computation. Newell [5], for example, famously identified the architecture of the human mind with that of a universal classical computer: what he called a 'physical symbol system'. This mainstream approach can be called the *symbolic paradigm*: such computation centers on symbols which individually serve syntactically as the tokens of manipulation and semantically as the elements interpretable in terms of the contents of consciousness.

The connectionist view, in contrast, models intuitive cognition within a different class of computational system, which we call *subsymbolic* [10] (after [1]). Section 2 characterizes subsymbolic computation via a class of automata. Sections 4 and 5 consider classes of subsymbolically-computable functions respectively derived from recursive equations and from formal languages.

The research program we consider (formally synopsized in [12]) is called *sub*symbolic rather than *non*symbolic because, unlike other research exploring connectionist computation (e.g. [3]), the working hypothesis is that—at a less fine-grained, more abstract level of description—the functions computed by intuitive processes are often well approximated by symbolic descriptions. The intuitive mental processes that actually *compute* these functions do not however admit a symbolic description: these processes require subsymbolic, connectionist descriptions. A central question then is: *which **symbolic** functions can the human intuitive processor compute?* The formal foundation for pursuing this question is laid in Section 3 and revisited in Section 6.

We return briefly in Section 7 to implications for cognitive science of some of the results we present, but until then we put aside the psychological motivation for connectionist computation—from theories of intuitive cognitive processes—and focus on the biological motivation. Specifically, we consider the formal capabilities of (a certain conception of) neural computation.

An important part of the mind-body problem is to connect the mental to the physical brain, and computational reduction offers the first prospect for carrying this out rigorously. This requires, however, that the computational architecture deployed has primitive operations, data, and combinators that a brain can provide. And this is what the connectionist architectures employed in the subsymbolic paradigm achieve, according to our current best understanding of the appropriate level of neural organization. To summarize the subsequent discussion: a mental concept is encoded in the activity of many neurons, not a single one; the activity of a neuron is a continuously varying quantity at the relevant level of analysis; combination of information from multiple neurons has an approximately linear character, while the consequences of this combination for neural activation has a non-linear character that imposes minimum and maximum levels. The primitives provided by the continuous architectures of the subsymbolic paradigm are within the capabilities of the brain—that the brain has *greater* complexity than assumed in these architectures does not compromise the subsymbolic paradigm's reduction of mental to neural computation.

While brain theory is far from achieving a settled state, this conception of continuous neural processing has generally displaced earlier notions according to which the relevant level of analysis was taken to be one where neural activations are binary (firing/not-firing), as assumed by the early discrete-computational network descriptions of the brain developed by Turing [16] as well as McCulloch and Pitts [4]. Similarly, early on, the search for the meaning of neural activation targeted individual cells, but recent years have seen an explosion of research in which neural meaning is sought by recording 'population codes' over hundreds of neurons, or patterns of aggregated activity over hundreds of thousands of neurons (using functional Magnetic Resonance Imaging).

2   Subsymbolic Automata

From an automata-theoretic perspective, computation is characterized by four fundamental properties [8, and references therein].

1. The tokens manipulated by primitive operations are 'subsymbolic': they reside at a level lower than that of the symbols of the symbolic paradigm. Each token is called the 'activation value' of a 'unit' (or 'neuron'). Tokens are related many-to-many to consciously accessible concepts: a single concept is realized by a particular combination of many subsymbolic tokens—a 'pattern of activity'— and a single token is simultaneously part of the realization of multiple concepts. This crucial property is called *distributed representation* of concepts.

2. The tokens are real numbers, of which a given automaton has a fixed number n: a state is an element of $\mathbb{R}^n$. The primitive operations are those of continuous numerical (as opposed to discrete symbolic) computation. These primitives include multiplication and exponentiation, but not determination of whether two tokens are of the same type, nor binding a value to a variable.

3. Each automaton is a continuously evolving dynamical system: differential equations play the role of algorithms. The primitive operations combine not sequentially, but in parallel, as all activation values simultaneously evolve in time according to the dynamical equations.

4. The dynamical equations are quasi-linear: if $a_k(t)$ denotes the kth activation value at time t, then the dynamics is given by $da_k/dt = \sigma([\mathbf{W} \cdot \mathbf{a}(t)]_k) - a_k(t)$ where $\mathbf{a}(t)$ is the collection of all activation values at time t, $\{a_k(t)\}_{k=1}^n$. The state space of all possible $\mathbf{a}$ is assumed to be vector space, and $\mathbf{W}$ is a matrix of values $\{W_{kj}\}_{j,k=1}^n$ called 'connection weights' (or 'synaptic strengths'): W_{kj} is the weight of the 'connection' from unit j to unit k. $\mathbf{W} \cdot \mathbf{a}$ is the matrix-vector product: $[\mathbf{W} \cdot \mathbf{a}(t)]_k \equiv \sum_j W_{kj} a_j(t)$ is called the 'net input to unit k'. The possibly non-linear function $\sigma : \mathbb{R} \to \mathbb{R}$ can serve to keep activation values within certain limits; a common choice is the logistic function $\sigma(z) = 1/[1+e^{-z}]$.

The space of dynamical equations available to subsymbolic models is of course greater than that given here; what is most crucial for our purposes is that the representational states of the computational system form a vector space.[1]

Such a subsymbolic system is often depicted as a network, a graph with the units as nodes (labelled by activation values), and the connections as links (labelled by weights).

An important special case is the *linear associator*. First, $\sigma(z) \equiv z$; then the equilibrium state satisfies $\mathbf{W} \cdot \mathbf{a} = \mathbf{a}$. Second, the network is *two-layer, feedforward*: the units split into a set of input units (with activation vector $\mathbf{i}$) and a set of output units ($\mathbf{o}$), with non-zero weights only from input to output units. Then the equilibrium state satisfies $\hat{\mathbf{W}} \cdot \mathbf{i} = \mathbf{o}$ and the network is equivalent to a linear transformation (given by the matrix $\hat{\mathbf{W}}$ of input-to-output weights) from the vector space of input vectors $\mathbf{i}$ to the vector space of output vectors $\mathbf{o}$.

The implications of distributed representation—of taking vectors in $\mathbb{R}^n$ as the basic data type of the computational architecture—are many [14]. The basic operations of continuous mathematics provided by vector space theory now do the fundamental computational work. Consider, for example, language processing, a system of intuitive processes that is critical for cognitive science. To preview the remainder of the article: Instead of stringing together two conceptual-level symbols to form $\text{Frodo}_{\text{subject}} \ \text{lives}_{\text{predicate}}$, we add together two vectors, the first encoding 'Frodo as subject' and the second 'lives as predicate'. The vector encoding 'Frodo as subject' results from taking a vector encoding Frodo and a vector encoding 'subject' and combining them with a vector operation, the tensor product. The basic mapping operation of vector space theory, linear transformation, provides the core of mental processing. Vector similarity, defined using the vector inner product, serves to model conceptual similarity, a central notion of psychological theory. And a notion from dynamical systems theory proves to have major implications as well: the differential equations governing many subsymbolic models have the property that as time progresses, a quantity called the *Harmony* steadily increases. Harmony can be interpreted as a numerical measure of the *well-formedness* of the network's activation vector with respect to the weights. The objective of network computation is to produce the representation (vector) that is maximally well-formed—*optimal*. And Harmony turns out to bear a deep relation to well-formedness as defined by a grammar.

We now examine some of the details behind this summary, and proceed to consider some of the implications for intuitive computation in the human mind.

3 From Symbolic to Vectorial Representations

At the foundation of the subsymbolic (but not the nonsymbolic) connectionist program is the reduction of symbolic conceptual-level mental representations to

---

[1] A fifth property—learning—is key to much connectionist modeling; the connection W_{kj} is typically 'learned' from statistical inference concerning the co-activation of units k and j during 'training', in which example input-output pairs of a target function are presented. However our concern here is not learnability but computability.

more fine-grained vectorial subconceptual representations. The vectorial embedding of a set of symbol structures S proceeds as follows [14, ch. 5]. Throughout, we use the example of binary trees with node labels from an alphabet A.

First, each symbol structure is mapped to a set of (symbolic) *filler/role bindings* via a *filler-role decomposition*. For example, a particular role r_x might be identified with a node position x (e.g., left child of right child of root, denoted by the bit string $x = 01$); among the bindings for a particular tree $s \in S$, this role would be bound to a particular symbol $f \in A$ just in case the symbol f labels the node position x in the tree s; this binding is written f/r_x and the set of all bindings of s is $\beta(s)$.

Next, a vector space V_F called the *filler (vector) space* is selected, along with an injection ψ_F from the set of symbolic fillers to V_F. Similarly, a *role (vector) space* V_R is selected, along with an injection ψ_R from the set of symbolic roles to V_R. Then V_F and V_R are combined by the tensor product to form the actual representational (vector) space for trees $V_S \equiv V_F \otimes V_R$.[2]

Finally, the filler-role decomposition and vectorial realizations are combined: the vector realization of s, $\psi_S(s) \in V_S$, is the *sum* of the vector realizations of the filler-role bindings of s, each of which uses the tensor product to bind together the encodings of the filler and the role:

$$\psi_S(s) = \sum_{f/r \in \beta(s)} \psi_F(f) \otimes \psi_R(r)$$

Henceforth, ψ_S will be abbreviated ψ.

For binary trees, we use a role realization obeying the recursive requirements

$$\psi_R(r_{0x}) = \psi_R(r_0) \otimes \psi_R(r_x), \quad \psi_R(r_{1x}) = \psi_R(r_1) \otimes \psi_R(r_x)$$

where $0x$ denotes the concatenation of the bit 0 with the bit-string x. Thus, e.g., $\psi_R(r_{010}) = \psi_R(r_0) \otimes \psi_R(r_1) \otimes \psi_R(r_0)$; ψ_R is entirely determined by $\psi_R(r_0)$ and $\psi_R(r_1)$, the vectors in $V_R^{(1)}$ realizing the roles 'left-' and 'right-child of root'. These two vectors need to be linearly independent, so $\dim(V_R^{(1)}) \geq 2$; for concreteness, we assume it equals 2. Roles for positions at tree depth d are realized with d-fold tensor products; the total vector space V_R is the direct sum of spaces $V_R^{(d)}$, $d = 0, \dots, \infty$ for all tree depths: $V_R = \bigoplus_{d=0}^{\infty} V_R^{(d)}$, with $\dim(V_R^{(d)}) = 2^d$. Likewise, $V_S = \bigoplus_{d=0}^{\infty} V_S^{(d)} = \bigoplus_{d=0}^{\infty} \left[V_F \otimes V_R^{(d)} \right] = V_F \otimes V_R$. In the case $d = 0$, $V_R^{(0)}$ is a 1-dimensional vector space so $V_S^{(0)} \equiv V_F \otimes V_R^{(0)}$ is isomorphic to V_F.

---

[2] Some definitions: Given bases $\{\hat{\mathbf{f}}_u\}_{u=1}^n$ for V_F and $\{\hat{\mathbf{r}}_v\}_{v=1}^m$ for V_R, $\{\hat{\mathbf{f}}_u \otimes \hat{\mathbf{r}}_v\}$ is a basis for the $n \times m$-dimensional space V_S. The tensor product of vectors $(\mathbf{f}, \mathbf{r}) \mapsto \mathbf{f} \otimes \mathbf{r}$ mapping $V_F \times V_R$ to V_S is bilinear, i.e., linear in each of $\mathbf{f}$ and $\mathbf{r}$ independently. So if $\{f_u\}$ and $\{r_v\}$ are respectively the elements of $\mathbf{f} \in V_F$ and $\mathbf{r} \in V_R$ with respect to bases $\{\hat{\mathbf{f}}_u\}$ and $\{\hat{\mathbf{r}}_u\}$, then the elements of $\mathbf{f} \otimes \mathbf{r} \in V_S$ with respect to the basis $\{\hat{\mathbf{f}}_u \otimes \hat{\mathbf{r}}_v\}$ are $\{f_u r_v\}$. The direct sum of vector spaces U and V, $U \oplus V$, is $U \times V$ with the obvious linear operations: $\alpha(\mathbf{u}, \mathbf{v}) + \beta(\mathbf{u}', \mathbf{v}') \equiv (\alpha \mathbf{u} + \beta \mathbf{u}', \alpha \mathbf{v} + \beta \mathbf{v}')$, $[\mathbf{A} \oplus \mathbf{B}] \cdot (\mathbf{u}, \mathbf{v}) \equiv (\mathbf{A} \cdot \mathbf{u}, \mathbf{B} \cdot \mathbf{v})$; $\dim(U \oplus V) = \dim(U) + \dim(V)$.

A tree s of depth $d = 0$ is simply a symbol in A; in this case $\texttt{atom}(s) = \mathsf{T}$ (otherwise $\texttt{atom}(s) = \mathsf{F}$). The identity transformations on infinite-dimensional V_S and V_R, 2-dimensional $V_R^{(1)}$, and 1-dimensional $V_R^{(0)}$ are respectively denoted $\mathbb{1}_S$, $\mathbb{1}_R$, $\mathbf{1}_R$ and $\mathbf{1}_R$. The identity transformation on finite-dimensional V_F is $\mathbf{1}_F$.

4   Linearly Computable Functions by Recursion

We shall say that a function $f : S \to S$ is *linearly computable* (i.e., computable by a linear associator network) if there is a linear transformation $\mathbb{W}_f : V_S \to V_S$ such that $\psi_S \circ f = \mathbb{W}_f \circ \psi_S$, i.e., the following diagram commutes:

$$
\begin{array}{ccc}
S & \xrightarrow{\;f\;} & S \\
{\scriptstyle\psi}\big\downarrow & & \big\downarrow{\scriptstyle\psi} \\
V_S & \xrightarrow{\;\mathbb{W}_f\;} & V_S
\end{array}
$$

This section discusses work that begins to characterize the set of linearly computable symbolic functions over binary trees (cf. [2]). We proceed from the base case of functions over depth-0 trees (the symbols in A), to the closure under composition of the primitive tree-role construction and decomposition functions, to functions defined by a kind of primitive recursion.

We start then with the set $\mathcal{B}$ of functions f_g that map every symbol $\mathsf{A} \in A$ in a tree into the symbol $g(\mathsf{A})$, without changing which node it labels, where $g : A \to A$ is some (partial) function.

The functions in $\mathcal{B}$ are linearly computable. In the case of interest, the vectors in V_F realizing the symbols $\{\psi(a)|a \in A\}$ are linearly independent, hence form a basis for V_F (adding additional vectors if necessary); let its dual basis be $\{\psi^+(a)|a \in A\}$ (i.e., $[\psi^+(a)]^T \psi(a')$ is 1 if $a = a'$ otherwise 0). Then $f_g \in \mathcal{B}$ is realized by $\mathbb{W}_f \equiv \sum_{a \in A} \psi(g(a))[\psi^+(a)]^T \otimes \mathbb{1}_R$.

We next form $\mathcal{F}$, the smallest set of functions that (i) includes the functions in $\mathcal{B}$, (ii) is closed under composition with functions in $\mathcal{B}$, and (iii) is closed under composition with the primitive tree functions: $\texttt{cons}$, which constructs a new binary tree $\texttt{cons}(p, q) \equiv s$ with left sub-tree p and right sub-tree q; $\texttt{ex}_0$ and $\texttt{ex}_1$, which extract the left sub-tree $\texttt{ex}_0(s) = p$ and right sub-tree $\texttt{ex}_1(s) = q$ from $s = \texttt{cons}(p, q)$.

The primitive tree functions are all linearly computable [14, ch. 8]. There exist linear transformations $\mathbb{W}_f$ on V_S satisfying, for $i = 0, 1$:

$$
\psi \circ \texttt{cons}(p, q) = \mathbb{W}_{\texttt{cons}_0} \cdot \psi(p) + \mathbb{W}_{\texttt{cons}_1} \cdot \psi(q) \qquad \psi \circ \texttt{ex}_i(s) = \mathbb{W}_{\texttt{ex}_i} \cdot \psi(s)
$$

In fact any function f in $\mathcal{F}$ is linearly computable, by a linear transformation of the form

$$
\mathbb{W}_f = \mathbb{1} \otimes \underline{\mathbf{W}}_f \qquad (\text{e.g, } \mathbb{W}_{\texttt{cons}_0} = \mathbb{1} \otimes \underline{\mathbf{W}}_{\texttt{cons}_0})
$$

where $\mathbb{1}$ satisfies the recursion relation $\mathbb{1} = 1_R + \mathbb{1} \otimes 1_R$ [14, ch. 8]. The solution is $\mathbb{1}_R$, which can be written (defining the sum-of-powers operator $\wp$):

$$\mathbb{1}_R = \wp(1_R) \equiv \bigoplus_{d=1}^{\infty} (1_R)^{\otimes d} \equiv 1_R \oplus 1_R \oplus (1_R \otimes 1_R) \oplus (1_R \otimes 1_R \otimes 1_R) \oplus \cdots$$

$\underline{\mathbf{W}}_f$ is a transformation dependent on f; the compositional map $f \mapsto \underline{\mathbf{W}}_f$ is recursively defined by requirements such as

$$f = \mathtt{cons}(f', f'') \Rightarrow \underline{\mathbf{W}}_f = \underline{\mathbf{W}}_{\mathtt{cons}_0} \cdot \underline{\mathbf{W}}_{f'} + \underline{\mathbf{W}}_{\mathtt{cons}_1} \cdot \underline{\mathbf{W}}_{f''}$$

An example of a simple function not in $\mathcal{F}$ is $\mathtt{reverse}$, which reverses the left and right children of every node. This function is defined by the recursion

$$\mathtt{reverse}(s) = \begin{cases} s & \text{if } \mathtt{atom}(s) \\ \mathtt{cons}(\mathtt{reverse}(\mathtt{ex}_1(s)), \mathtt{reverse}(\mathtt{ex}_0(s))) & \text{otherwise} \end{cases}$$

$\mathtt{reverse}$ is linearly computable by a transformation $\mathbb{1}_F \otimes \mathbb{P}$ where $\mathbb{P}$ satisfies the recursion

$$\mathbb{P} = 1_R \oplus \mathbb{W}_{\mathtt{cons}_0} \cdot \mathbb{P} \cdot \mathbb{W}_{\mathtt{ex}_1} + \mathbb{W}_{\mathtt{cons}_1} \cdot \mathbb{P} \cdot \mathbb{W}_{\mathtt{ex}_0}$$

A solution is

$$\mathbb{P} = \wp(\underline{\mathbf{W}}_{\mathtt{reverse}}) \quad \text{where} \quad \underline{\mathbf{W}}_{\mathtt{reverse}} \equiv \underline{\mathbf{W}}_{\mathtt{cons}_0} \cdot \underline{\mathbf{W}}_{\mathtt{ex}_1} + \underline{\mathbf{W}}_{\mathtt{cons}_1} \cdot \underline{\mathbf{W}}_{\mathtt{ex}_0}$$

More generally, consider the recursion

$$f(s) = \begin{cases} g(s) & \text{if } \mathtt{atom}(s) \\ h(f(\mathtt{ex}_0(s)), f(\mathtt{ex}_1(s))) & \text{otherwise} \end{cases}$$

Suppose g and h are linearly computable:

$$\psi \circ g(s) = \mathbf{G} \cdot \psi(s) \quad \text{and} \quad \psi \circ h(p, q) = \mathbb{H}_0 \cdot \psi(p) + \mathbb{H}_1 \cdot \psi(q)$$

Then f will be linearly computable by a linear transformation $\mathbb{W}_f$ if a solution exists to the following recursion:

$$\mathbb{W}_f = \mathbf{G} \oplus \mathbb{H}_0 \cdot \mathbb{W}_f \cdot \mathbb{W}_{\mathtt{ex}_0} + \mathbb{H}_1 \cdot \mathbb{W}_f \cdot \mathbb{W}_{\mathtt{ex}_1}$$

A solution is

$$\mathbb{W}_f = \mathbf{G} \circledast \sum_{d=0}^{\infty} \mathbb{M}^{\circledast d} \quad \text{where} \quad \mathbb{M} \equiv \mathbb{H}_0 \circledast \mathbb{W}_{\mathtt{ex}_0} + \mathbb{H}_1 \circledast \mathbb{W}_{\mathtt{ex}_1}$$

Here $\circledast$ is a contracted tensor product operator the definition of which requires more detail concerning the direct-sum structure of V_S than is possible here.

5   Grammars and Optimization

Alongside recursive function theory is another classic approach to specifying functions: via formal languages, defined by grammars consisting of string rewriting rules (or productions). This approach has achieved considerable success in characterizing intuitive mental processing of natural language (including that underlying human performance in Turing's Imitation Game [17]); this work culminates in the theory of 'universal grammar', which formally characterizes what the grammatical systems of human languages share and exactly how they may differ. (We shall discuss universal grammar in Section 7.)

The relation to subsymbolic computation arises because *grammatical* well-formedness can be realized as a kind of *connectionist* well-formedness called *Harmony* (or negative 'energy') [14, ch. 9, and references therein]. Harmony arises as a quantity that is steadily increased by the dynamical processing within an important class of network which includes those quasi-linear networks having both a symmetric connection matrix ($W_{kj} = W_{jk}$) and a monotonically increasing function σ.[3] Network dynamics can be interpreted as computing a maximal-Harmony representation—*locally* maximal, that is: at an equilibrium of the dynamics, no infinitesimal state change can produce higher Harmony. To compute *global* Harmony maxima, stochastic differential equations can be used to define the network dynamics (achieving probabilistic convergence as $t \to \infty$).

Can the notion of well-formedness provided by grammars that are defined by string rewriting rules be realized as network Harmony? To investigate this question, a mediating notion—*Harmonic Grammar*—proves useful [14, ch. 6]. A Harmonic Grammar $H_{\mathcal{G}}$ assigns to any tree s a numerical well-formedness value $H_{\mathcal{G}}(s)$; the language specified by $H_{\mathcal{G}}$ is the set of trees with maximal Harmony: $\mathcal{L}_{H_{\mathcal{G}}} \equiv \mathrm{argmax}_{s \in S} H_{\mathcal{G}}(s)$.

A Harmonic Grammar $H_{\mathcal{G}}$ is realized in a network $\mathcal{N}$ iff for all $s \in S$, $H_{\mathcal{G}}(s) = H_{\mathcal{N}}(\psi(s))$, where $H_{\mathcal{N}}$ is the connectionist Harmony of $\mathcal{N}$. It turns out that the language $\mathcal{L}_{\mathcal{G}}$ generated by any rewrite-rule grammar $\mathcal{G}$ can be specified by a Harmonic Grammar $H_{\mathcal{G}}$ (that is, $\mathcal{L}_{\mathcal{G}} = \mathcal{L}_{H_{\mathcal{G}}}$), and $H_{\mathcal{G}}$ can be constructed directly from the rules of $\mathcal{G}$ [14, ch. 10].

Further, when $\mathcal{G}$ is a context-free grammar in Chomsky Normal Form (hence with binary derivation trees), our vectorial realization of binary trees can be used to construct a network $\mathcal{N}_{\mathcal{G}}$ that realizes $H_{\mathcal{G}}$ [14, ch. 8]. This $H_{\mathcal{G}}$ has the form $H_{\mathcal{G}}(s) = \sum_{b,b' \in \beta(s)} H_{b,b'}$ where for each pair of filler/role bindings (b, b') of s, the number $H_{b,b'}$ encodes the grammatical well-formedness of b and b' co-existing in s. Typically in linguistics $H_{b,b'}$ is non-positive, with $|H_{b,b'}|$ interpreted as the strength in the grammar $H_{\mathcal{G}}$ of the *constraint*: 'do not combine b and b' in the same structure'.[4]

---

[3] The Harmony of the state vector $\mathbf{a}$ of a network $\mathcal{N}$ with weight matrix $\mathbf{W}$ is $H_{\mathcal{N}}(\mathbf{a}) \equiv \frac{1}{2}\sum_{kj} a_k W_{kj} a_j + \sum_k h(a_k)$, $h(a) \equiv -\int_0^a \sigma^{-1}(a')da'$.

[4] E.g., if b is a singular subject (like 'Frodo') and b' a plural verb form (like 'live'), $h_{\mathrm{agree}} = H_{b,b'} < 0$ is a grammatical penalty for a verb failing to agree in number with its subject. Adding $h_{\mathrm{agree}}\psi(b)[\psi(b')^+]^T$ to the weight matrix $\mathbf{W}$ causes the Harmony of any structure containing b and b' (e.g., 'Frodo live') to decrease by $|h_{\mathrm{agree}}|$.

When $\mathcal{L}_\mathcal{G} = \mathcal{L}_{H_\mathcal{G}}$, any tree s_0 generated by $\mathcal{G}$ is realized by a vector $\psi(s_0)$ that has maximal network Harmony $H_{\mathcal{N}_\mathcal{G}}(\psi(s_0))$ *among the vectors* $\{\psi(s)|s \in S\}$ *realizing trees* (this discrete subset of V_S, the image of S under ψ, will be called "the grid").

Within V_S there are, however, vectors $\mathbf{v}^*$ off the grid which have higher Harmony than those grid states realizing grammatical trees: such a vector $\mathbf{v}^*$ does not realize a tree; rather it realizes a pseudo-tree with weighted blends of symbols bound to tree positions (e.g., instead of $\mathtt{A}/r_x$ there might be $[0.5\mathtt{A} - 0.1\mathtt{B}]/r_x$). Such non-grid states are optimal within the vector space because conflicts between the constraints encoded in H entail that optima constitute compromises that interpolate between the alternative discrete states favored by the conflicting constraints [13].

6   Quantization

In order for the network to produce bona fide trees as output, another component can be added to the dynamics of $\mathcal{N}_\mathcal{G}$. This (deterministic) *quantization* dynamics produces an attractor at all and only the vectors on the grid. As the computation proceeds, this quantization dynamics comes to overpower the stochastic Harmony-optimization dynamics, so that the final output will be an attractor—a grid state [13]. In a range of (very simple) applications to modeling human language processing, simulations show these networks to be capable of reliably computing the vectorial realizations of symbolic states that are globally-optimal among grid states. Further, when forced to terminate too quickly, these simulations show that the probability of outputting some error E appropriately declines with its Harmony $H(E)$: not only do the global optima of the Harmonic Grammar model correct linguistic competence, the Harmonic Grammar's evaluation of sub-optimal states models the errors characteristic of human linguistic performance.

7   Optimization and Universal Grammar

The shift from a rewrite-rule grammar $\mathcal{G}$ to a Harmonic Grammar $H_\mathcal{G}$ for specifying a language may seem minor, but in fact it constitutes a fairly radical reconception: from grammars as generators to grammars as evaluators, from grammatical as generated to grammatical as optimal. Does this afford a better formal system for capturing the functions computed by human linguistic intuition? There is reason to believe that it does.

Soon after the appearance of Harmonic Grammar came the empirical discovery that in human grammars, it is often the case that the strength of each constraint exceeds the combined strengths of all weaker constraints. The constraints can then be considered to form a *strict-domination hierarchy*, with each constraint having absolute priority over all weaker constraints; while numerical strengths are needed for the reduction to neural computation, only the priority ranking is needed to determine optimality. This is *Optimality Theory* [6,7], which has proved to make a significant contribution to the theory of universal grammar via its central principle: the grammars of all human languages consist in the same constraints; grammars may differ only in how the constraints are ranked into a strict-domination hierarchy. This means that all languages share the same well-formedness criteria: they differ only in which criteria take priority in cases of conflict. The empirical successes are cases when what is universally shared by human languages are preferences that outputs of the grammar should respect, as opposed to processes that generate those outputs; since rewrite-rules characterize grammatical knowledge as generative procedures, for the purposes of universal grammar, it proves advantageous that Optimality Theory characterizes grammatical knowledge as preferences over outputs [9]. In the case of language, at least, specifying computations via optimization has been shown to provide insight into the nature of the functions computed by human intuition.

References

1. Hofstadter, D.R.: Waking up from the Boolean dream, or, subcognition as computation. In: Hofstadter, D.R. (ed.) Metamagical Themas: Questing for the Essence of Mind and Pattern, pp. 631–665. Bantam Books (1986)
2. Kimoto, M., Takahashi, M.: On Computable Tree Functions. In: He, J., Sato, M. (eds.) ASIAN 2000. LNCS, vol. 1961, pp. 273–289. Springer, Heidelberg (2000)
3. McClelland, J., Botvinick, M., Noelle, D., Plaut, D., Rogers, T., Seidenberg, M., Smith, L.: Letting structure emerge: Connectionist and dynamical systems approaches to cognition. Trends in Cognitive Sciences 14(8), 348–356 (2010)
4. McCulloch, W., Pitts, W.: A logical calculus of the ideas immanent in nervous activity. Bulletin of Mathematical Biology 5(4), 115–133 (1943)
5. Newell, A.: Physical symbol systems. Cognitive Science 4(2), 135–183 (1980)
6. Prince, A., Smolensky, P.: Optimality Theory: Constraint interaction in generative grammar. Blackwell (1993/2004)
7. Prince, A., Smolensky, P.: Optimality: From neural networks to universal grammar. Science 275(5306), 1604–1610 (1997)
8. Rumelhart, D., McClelland, J., The PDP Research Group: Parallel distributed processing: Explorations in the microstructure of cognition. Foundations, vol. 1. MIT Press, Cambridge (1986)
9. Rutgers Optimality Archive, http://roa.rutgers.edu
10. Smolensky, P.: On the proper treatment of connectionism. Behavioral and Brain Sciences 11(01), 1–23 (1988)
11. Smolensky, P.: Cognition: Discrete or continuous computation? In: Cooper, S., van Leeuwen, J. (eds.) Alan Turing — His Work and Impact. Elsevier (2012)
12. Smolensky, P.: Symbolic functions from neural computation. Philosophical Transactions of the Royal Society – A: Mathematical, Physical and Engineering Sciences (in press, 2012)
13. Smolensky, P., Goldrick, M., Mathis, D.: Optimization and quantization in gradient symbol systems: A framework for integrating the continuous and the discrete in cognition. Cognitive Science (in press, 2012)

14. Smolensky, P., Legendre, G.: The harmonic mind: From neural computation to Optimality-Theoretic grammar, vol. 1: Cognitive architecture, vol. 2: Linguistic and philosophical implications. MIT Press, Cambridge (2006)
15. Turing, A.M.: On computable numbers, with an application to the Entscheidungsproblem. Proceedings of the London Mathematical Society 42, 230–265 (1936)
16. Turing, A.M.: Intelligent machinery: A report by Turing, A.M. National Physical Laboratory (1948)
17. Turing, A.M.: Computing machinery and intelligence. Mind 59(236), 433–460 (1950)

A Correspondence Principle
for Exact Constructive Dimension

Ludwig Staiger

Institut für Informatik, Martin-Luther-Universität Halle-Wittenberg,
von-Seckendorff-Platz 1, D-06099 Halle, Germany
staiger@informatik.uni-halle.de

Abstract. Exact constructive dimension as a generalisation of Lutz's
[10,11] approach to constructive dimension was recently introduced in
[19]. It was shown that it is in the same way closely related to a priori
complexity, a variant of Kolmogorov complexity, of infinite sequences as
their constructive dimension is related to asymptotic Kolmogorov complexity.

The aim of the present paper is to extend this to the results of [8,9,18]
(see also [2, Section 13.6]) where it is shown that the asymptotic Kolmogorov complexity of infinite sequences in Σ_2^0-definable sets is bounded
by their Hausdorff dimension.

Using Hausdorff's original definition one obtains upper bounds on the
a priori complexity functions of infinite sequences in Σ_2^0-definable sets
via the exact dimension of the sets.

Lutz's [10,11] effectivisation of classical Hausdorff dimension led to the definition
of constructive and computable dimensions of sets of infinite sequences. He put
also the question of whether there is a correspondence principle stating that the
constructive (or computable) dimension of sufficiently simple sets coincides with
their Hausdorff dimension (cf. [9]). A first positive answer for classical dimensions
and sets definable by finite automata follows from the results of [17,12], and for
Σ_2^0-definable sets positive answers were given in [8,9] and [18].

In a recent paper [19] the above mentioned results by Lutz and results by
Ryabko were generalised from the case of 'usual' (classical) constructive and
Hausdorff dimension to the case of exact dimension [5,7]. This concerns Lutz's
martingale characterisation of Hausdorff dimension and Ryabko's [15] (see also
[1]) determining of the dimension of the level sets of the constructive dimension
(or asymptotic Kolmogorov complexity) of sets of infinite sequences.

Usually, the Hausdorff dimension (here also called classical Hausdorff dimension) of a set of reals is a real number α characterising a certain density or
measure property of this set (see the textbooks [3,4] or [12]). If one looks to
Hausdorff's original paper [7], however, one finds that he defined the Hausdorff
dimension to be a non-decreasing, right continuous function $h : (0, \infty) \to (0, \infty)$,
nowadays called a *gauge function* [5].

The paper [19] provided a generalisation of the martingale characterisation of
Hausdorff dimension and the determining of the dimension of the level sets to

S.B. Cooper, A. Dawar, and B. Löwe (Eds.): CiE 2012, LNCS 7318, pp. 686–695, 2012.
© Springer-Verlag Berlin Heidelberg 2012

the case of exact dimension and to Kolmogorov complexity functions of infinite sequences.

In the papers [8,9] and [18] (see also [2, Section 13.6]) a tight bound on the maximum asymptotic Kolmogorov complexity of sequences in Σ_2^0-sets by its 'usual' Hausdorff dimension was presented and computable martingales successful on Σ_2^0-sets with an exponent close to the Hausdorff dimension were constructed.

The purpose of the present paper is to generalise these results to a correspondence principle for the case of exact dimensions. This results also in a more precise bound on the maximum Kolmogorov complexity of sequences in Σ_2^0-sets than the mere asymptotics given in the above mentioned papers.

The paper is organised as follows. After introducing some notation and some preliminaries on gauge functions and Hausdorff's original approach we present in Section 2 necessary results, mainly from [19] on exact Hausdorff dimension, martingales and their effectivisation. Then Sections 3.1 and 3.2 show that the correspondence principles for constructive and computable dimensions hold for Σ_2^0-definable sets of sequences and gauge functions satisfying some computability constraints. The proofs follow mainly the line of the proofs given in [18]. Due to space limitations they had to be omitted in this paper.

1   Notation and Preliminaries

In this section we introduce the notation used throughout the paper. By $\mathbb{N} = \{0, 1, 2, \ldots\}$ we denote the set of natural numbers and by $\mathbb{Q}$ the set of rational numbers. Let $X = \{0, 1, \ldots, r - 1\}$ be an alphabet of cardinality $|X| = r \geq 2$. By X^* we denote the set of finite words on X, including the *empty word* e, and X^ω is the set of infinite strings (ω-words) over X. Subsets of X^* will be referred to as *languages* and subsets of X^ω as ω-*languages*.

For $w \in X^*$ and $\eta \in X^* \cup X^\omega$ let $w \cdot \eta$ be their *concatenation*. This concatenation product extends in an obvious way to subsets $W \subseteq X^*$ and $B \subseteq X^* \cup X^\omega$.

We denote by $|w|$ the *length* of the word $w \in X^*$ and $\mathbf{pref}(B)$ is the set of all finite prefixes of strings in $B \subseteq X^* \cup X^\omega$. We shall abbreviate $w \in \mathbf{pref}(\eta)$ ($\eta \in X^* \cup X^\omega$) by $w \sqsubseteq \eta$, and $\eta \upharpoonright n$ is the n-length prefix of η provided $|\eta| \geq n$. The δ-*limit* of a language $V \subseteq X^*$ is the ω-language $V^\delta := \{\xi : \xi \in X^\omega \wedge |\mathbf{pref}(\xi) \cap V| = \infty\}$. A language $W \subseteq X^*$ is referred to as *prefix-free* if $w \sqsubseteq v$ and $w, v \in W$ imply $w = v$.

For a computable domain $\mathcal{D}$, such as $\mathbb{N}$, $\mathbb{Q}$ or X^*, we refer to a function $f : \mathcal{D} \to \mathbb{R}$ as *left computable* (or *approximable from below*) provided the set $\{(d, q) : d \in \mathcal{D} \wedge q \in \mathbb{Q} \wedge q < f(d)\}$ is computably enumerable. Accordingly, a function $f : \mathcal{D} \to \mathbb{R}$ is called *right computable* (or *approximable from above*) if the set $\{(d, q) : d \in \mathcal{D} \wedge q \in \mathbb{Q} \wedge q > f(d)\}$ is computably enumerable, and f is *computable* if f is right and left computable. In contrast to this we refer to a function $f : \mathcal{D} \to \mathbb{Q}$ as *computable* provided f returns the exact value $f(d) \in \mathbb{Q}$. Accordingly, a real number $\alpha \in \mathbb{R}$ is left computable, right computable or computable provided the constant function $c_\alpha(t) = \alpha$ is left computable, right computable or computable, respectively.

A *super-martingale* is a function $\mathcal{V} : X^* \to [0, \infty)$ which satisfies $\mathcal{V}(e) \leq 1$ and the super-martingale inequality

$$r \cdot \mathcal{V}(w) \geq \sum_{x \in X} \mathcal{V}(wx) \text{ for all } w \in X^*. \tag{1}$$

If Eq. (1) is satisfied with equality $\mathcal{V}$ is called a *martingale*. Closely related with (super-)martingales are continuous (or cylindrical) (semi-)measures $\mu : X^* \to [0, 1]$ where $\mu(e) \leq 1$ and $\mu(w) \geq \sum_{x \in X} \mu(wx)$ for all $w \in X^*$.

1.1   Gauge Functions and Hausdorff's Original Approach

A function $h : (0, \infty) \to (0, \infty)$ is referred to as a *gauge function* provided h is right continuous and non-decreasing.[1] If not stated otherwise, we shall always assume that $\lim_{t \to 0} h(t) = 0$.

The h-dimensional outer measure of F on the space X^ω is given by

$$\mathcal{H}^h(F) := \lim_{n \to \infty} \inf \Big\{ \sum_{v \in V} h(r^{-|v|}) : V \subseteq X^* \wedge F \subseteq V \cdot X^\omega \wedge \min_{v \in V} |v| \geq n \Big\}. \tag{2}$$

If $\lim_{t \to 0} h(t) > 0$ then $\mathcal{H}^h(F) < \infty$ if and only if F is finite.

The usual α-dimensional Hausdorff measure $\mathcal{H}^\alpha$ is defined by gauge functions $h_\alpha(t) = t^\alpha, \alpha \in [0, 1]$, that is, $\mathcal{H}^\alpha = \mathcal{H}^{h_\alpha}$.

In this case the (usual or classical) Hausdorff dimension of a set $F \subseteq X^\omega$ is defined as

$$\dim_{\mathrm{H}} F := \sup\{\alpha : \alpha = 0 \vee \mathcal{H}^\alpha(F) = \infty\} = \inf\{\alpha : \alpha \geq 0 \wedge \mathcal{H}^\alpha(F) = 0\}. \tag{3}$$

As we see from Eq. (2) for our purposes the behaviour of gauge function is of interest only in a small vicinity of 0. Moreover, in many cases we are not interested in the exact value of $\mathcal{H}^h(F)$ when $0 < \mathcal{H}^h(F) < \infty$. Thus we can often make use of scaling a gauge function and altering it in a range (ε, ∞) apart from 0.

The following properties of gauge functions h and the related measure $\mathcal{H}^h$ are proved in the standard way.

Property 1. Let h, h' be gauge functions.

1. If $c \cdot h(r^{-n}) \leq h'(r^{-n})$ for some $c > 0$, then $c \cdot \mathcal{H}^h(F) \leq \mathcal{H}^{h'}(F)$.
2. If $\lim_{n \to \infty} \frac{h(r^{-n})}{h'(r^{-n})} = 0$ then $\mathcal{H}^{h'}(F) < \infty$ implies $\mathcal{H}^h(F) = 0$, and $\mathcal{H}^h(F) > 0$ implies $\mathcal{H}^{h'}(F) = \infty$.

Here the first property implies a certain equivalence of gauge functions. In fact, if $c \cdot h \leq h'$ and $c \cdot h' \leq h$ in the sense of Property 1.1 then for all $F \subseteq X^\omega$ the measures $\mathcal{H}^h(F)$ and $\mathcal{H}^{h'}(F)$ are both zero, finite or infinite.

In the same way the second property gives a partial pre-order of gauge functions (see [6, Section 9.1]). By analogy to the change-over-point $\dim_{\mathrm{H}} F$ for $\mathcal{H}^\alpha(F)$ this partial pre-order yields a suitable notion of Hausdorff dimension in the range of arbitrary gauge functions.

---

[1] In fact, since we are only interested in the values $h(r^{-n})$, $n \in \mathbb{N}$, the requirement of right continuity is just to conform with the usual meaning (cf. [5]).

Definition 1. We refer to a gauge function h as an *exact Hausdorff dimension function* for $F \subseteq X^\omega$ provided

$$\mathcal{H}^{h'}(F) = \begin{cases} \infty, & \text{if } \lim\limits_{n \to \infty} \frac{h(r^{-n})}{h'(r^{-n})} = 0 \text{ , and} \\ 0, & \text{if } \lim\limits_{n \to \infty} \frac{h'(r^{-n})}{h(r^{-n})} = 0. \end{cases}$$

In fact, Hausdorff [7] defined the dimension of a set F as an equivalence class of gauge functions $[h]$ such that $0 < \mathcal{H}^h(F) < \infty$. Property 1 shows that our definition covers this case.

Definition 1 is not as simple as the one of the classical Hausdorff dimension in Eq. (3), and it seems to be much more difficult to find the exact borderline, if it exists, between gauge functions with $\mathcal{H}^h(F) = 0$ and such with $\mathcal{H}^h(F) = \infty$.

2 Previous Results

2.1 Exact Hausdorff Dimension and Martingales

In this section we show a generalisation of Lutz's martingale characterisation of Hausdorff dimension to exact dimension.

Let $S_{c,h}[\mathcal{V}] := \{\xi : \xi \in X^\omega \wedge \limsup_{n \to \infty} \frac{\mathcal{V}(\xi[0..n])}{r^n \cdot h(r^{-n})} \geq c\}$, for a super-martingale $\mathcal{V} : X^* \to [0, \infty)$, a gauge function h and a value $c \in (0, \infty]$. In particular, $S_{\infty,h}[\mathcal{V}]$ is the set of all ω-words on which the super-martingale $\mathcal{V}$ is *successful* w.r.t. the order function $f(n) = r^n \cdot h(r^{-n})$ in the sense of Schnorr [16]. $S_{\infty,h}[\mathcal{V}]$ is also referred to as the *success set* of the super-martingale $\mathcal{V}$ w.r.t. the order function $f(n) = r^n \cdot h(r^{-n})$.

Observe that $S_{c,h}[\mathcal{V}] \subseteq S_{c',h'}[\mathcal{V}]$ whenever $c, c' \in (0, \infty]$ and $\lim\limits_{n \to \infty} \frac{h'(r^{-n})}{h(r^{-n})} = 0$.

Now we can generalise Lutz's result.

Theorem 1 ([19, Theorem 1]). *Let $F \subseteq X^\omega$. Then a gauge function h is an exact Hausdorff dimension function for F if and only if*

1. *for all gauge functions h' with $\lim\limits_{n \to \infty} \frac{h'(r^{-n})}{h(r^{-n})} = 0$ there is a super-martingale $\mathcal{V}$ such that $F \subseteq S_{\infty,h'}[\mathcal{V}]$, and*

2. *for all gauge functions h'' with $\lim\limits_{n \to \infty} \frac{h(r^{-n})}{h''(r^{-n})} = 0$ and all super-martingales $\mathcal{V}$ it holds $F \not\subseteq S_{\infty,h''}[\mathcal{V}]$.*

2.2 Effectivisation of Exact Hausdorff Dimension

The constructive dimension is a variant of dimension defined analogously to Theorem 1 using only left computable super-martingales. For the usual family of gauge functions $h_\alpha(t) = t^\alpha$ it was introduced by Lutz [10,11] and resulted, similarly to $\dim_H$ in a real number assigned to a subset $F \subseteq X^\omega$. In the case of left computable super-martingales the situation turned out to be even simpler than in the case of arbitrary super-martingales because the results of Levin [21] and

Schnorr [16] show that there is an optimal left computable super-martingale $\mathcal{U}$, that is, every other left computable super-martingale $\mathcal{V}$ satisfies $\mathcal{V}(w) \leq c_{\mathcal{V}} \cdot \mathcal{U}(w)$ for all $w \in X^*$ and some constant $c_{\mathcal{V}} > 0$ not depending on w. Thus we may define (cf. [19])

Definition 2. *Let $F \subseteq X^\omega$. We refer to $h : \mathbb{R} \to \mathbb{R}$ as an exact constructive dimension function for F provided $F \subseteq S_{\infty, h'}[\mathcal{U}]$ for all $h', \lim_{t \to 0} \frac{h(t)}{h'(t)} = 0$, and $F \not\subseteq S_{\infty, h''}[\mathcal{U}]$ for all $h'', \lim_{t \to 0} \frac{h''(t)}{h(t)} = 0$.*

Originally, Levin [21] showed that there is an optimal left computable continuous semi-measure $\mathbf{M}$ on X^*. As usual, we call a function $\mu : X^* \to [0, \infty)$ a *continuous* (or *cylindrical*) *semi-measure* on X^* provided $\mu(e) \leq 1$ and $\mu(w) \geq \sum_{x \in X} \mu(wx)$ for all $w \in X^*$. One easily verifies that μ is a continuous semi-measure if and only if $\mathcal{V}(w) := r^{|w|} \cdot \mu(w)$ is a super-martingale. Thus we might use $\mathcal{U}_{\mathbf{M}}$ with $\mathcal{U}_{\mathbf{M}}(w) := r^{|w|} \cdot \mathbf{M}(w)$ as our optimal left computable super-martingale.

Closely related to Levin's optimal left computable semi-measure is the *a priori entropy* (or *complexity*) $\mathrm{KA} : X^* \to \mathbb{N}$ defined by[2]

$$\mathrm{KA}(w) := \lfloor -\log_r \mathbf{M}(w) \rfloor \tag{4}$$

The requirement $\mathrm{KA}(w) \geq 0$ is one reason why we assumed $\mathbf{M}(e) \leq 1$.

The following theorem derives a bound for the set of sequences whose KA-complexity function is bounded.

Theorem 2 ([19, Theorem 4]). *Let $-\infty < c < \infty$ and let h be a gauge function. Then there is a $c' > 0$ such that*
$$\{\xi : \mathrm{KA}(\xi[0..n]) \leq_{\mathrm{i.o.}} -\log_r h(r^{-n}) + c\} \subseteq S_{c', h}[\mathcal{U}].$$

Conversely, if $\xi \in S_{c, h}[\mathcal{U}], c < \infty$, then from Eq. (4) one easily calculates $\mathrm{KA}(\xi[0..n]) \leq_{\mathrm{i.o.}} -\log_r h(r^{-n}) + c''$ for some $c'' \in (0, \infty)$. Thus we obtain a complexity characterisation of the success sets of the universal super-martingale $\mathcal{U}$.

$$\bigcup_{c > 0} \{\xi : \mathrm{KA}(\xi[0..n]) \leq_{\mathrm{i.o.}} -\log_r h(r^{-n}) + c\} = \bigcup_{c > 0} S_{c, h}[\mathcal{U}] \tag{5}$$

For gauge functions h' tending faster to 0 than h the following relations follow from $S_{c, h}[\mathcal{U}] \subseteq S_{\infty, h'}[\mathcal{U}]$.

Corollary 1. *Let h, h' be gauge functions such that $\lim_{t \to 0} \frac{h'(t)}{h(t)} = 0$. Then*

1. $\{\xi : \exists c(\mathrm{KA}(\xi[0..n]) \leq_{\mathrm{i.o.}} -\log_r h(r^{-n}) + c)\} \subseteq S_{\infty, h'}[\mathcal{U}]$, *and*

2. $\mathcal{H}^{h'}\left(\{\xi : \exists c(\mathrm{KA}(\xi[0..n]) \leq_{\mathrm{i.o.}} -\log_r h(r^{-n}) + c)\}\right) = 0.$

---

[2] Here we follow the notation of [20], in [2] a priori complexity was denoted by KM.

3   The Results

In [8,9,18] the correspondence principle could be stated for arbitrary (real) values of classical dimension. In the case of gauge functions the situation is more complicated. On the one hand because of the involved Definition 1, and, on the other hand, for the following reason (cf. also [19, Remark 2]). Unlike the classical case where the computable (even the rational) numbers are dense in the reals, for gauge functions it holds that, if $\alpha \in (0,1)$ is not a computable real, there is no computable function between $h_\alpha(t) = t^\alpha$ and $h_\alpha(t) = t^\alpha + \log_r \frac{1}{t}$.

First we mention the following general lower bound to the complexity function KA from [13] together with Eq. (5) yields a tight estimate for gauge functions satisfying $F \not\subseteq S_{c,h''}[\mathcal{U}]$ for arbitrary $F \subseteq X^\omega$ (cf. Definition 2).

Theorem 3 ([13]). *Let $F \subseteq X^\omega$, h be a gauge function and $\mathcal{H}^h(F) > 0$.*
Then for every $c > 0$ with $\mathcal{H}^h(F) > c \cdot \mathcal{U}(e)$ there is a $\xi \in F$ such that $\mathrm{KA}(\xi[0..n]) \geq_{\mathrm{a.e.}} -\log_r h(r^{-n}) - \log_r c$.

In order to obtain the announced upper bound, in view of Eq. (5) in the following two parts we show that for Σ_2^0-definable subsets $F \subseteq X^\omega$ and gauge functions h satisfying some computability constraints there are left-computable super-martingales or computable martingales $\mathcal{V}$, respectively, such that $F \subseteq S_{\infty,h'}[\mathcal{V}]$ whenever $\lim_{t\to 0} \frac{h'(t)}{h(t)} = 0$ and $\mathcal{H}^h(F) = 0$.

3.1   Constructive Dimension

As in [18] we ask now for an estimate of the condition $F \subseteq S_{c,h'}[\mathcal{U}]$ of Definition 2. The results use the following construction.

We start with an auxiliary lemma characterising subsets $F \subseteq X^\omega$ having null measure.

Lemma 1 ([14]). *Let $F \subseteq X^\omega$ and h be a gauge function. Then $\mathcal{H}^h(F) = 0$ if and only if there is a language $V \subseteq X^*$ such that $F \subseteq V^\delta$ and $\sum_{v \in V} h(r^{-|v|}) < \infty$.*

The following theorem gives a constructive version of Lemma 1.

Theorem 4. *If $F \subseteq X^\omega$ is a Σ_2-definable ω-language and h is a right computable gauge function such that $\mathcal{H}^h(F) = 0$ then there are a computable non-decreasing function $\bar{h} : \{r^{-i} : i \in \mathbb{N}\} \to \mathbb{Q}$ and a computable language $V \subseteq X^*$ satisfying*

1. *$\bar{h}(r^{-n}) \geq h(r^{-n})$ for all $n \in \mathbb{N}$,*
2. *$F \subseteq V^\delta$ and $\sum_{v \in V} \bar{h}(r^{-n}) < \infty$.*

Interpolating the computable function $\bar{h}$ we obtain the following consequence.

Corollary 2. *If $F \subseteq X^\omega$ is a Σ_2-definable ω-language and h is a right computable gauge function such that $\mathcal{H}^h(F) = 0$ then there is a computable non-decreasing function $\bar{h} : \mathbb{Q} \to \mathbb{Q}$ satisfying $\mathcal{H}^{\bar{h}}(F) = 0$ and $\bar{h}(t) \geq h(t)$ for $t \in \mathbb{Q} \cap (0,1)$.*

Our Theorem 4 yields the required upper bound for the prefix complexity KP, and hence also of the a priori complexity KA of an ω-word in F.

To this end we use the characterisation of KP via discrete semi-measures (cf. [2,20]).[3]

A mapping $\nu : X^* \to \mathbb{R}$ is referred to as a *discrete semi-measure* provided $\sum_{w \in X^*} \nu(w) < \infty$. It is known that there is an optimal left computable discrete semi-measure, that is, a left computable discrete semi-measure $\mathbf{m}$ such that for every left computable discrete semi-measure ν there is a constant c_ν such that $\forall w (w \in X^* \to \nu(w) \leq c_\nu \cdot \mathbf{m}(w))$. This measure $\mathbf{m}$ defines the prefix complexity (similarly as $\mathbf{M}$ defines the a priory complexity KA) $\mathrm{KP}(w) := \lfloor -\log_r \mathbf{m}(w) \rfloor$.

If $V \subseteq X^*$ is computably enumerable and $\bar{h} : \{r^{-n} > n \in \mathbb{N}\} \to \mathbb{R}$ is a left computable function such that $\sum_{v \in V} \bar{h}(r^{-|v|}) < \infty$ then

$$\nu(w) := \begin{cases} \bar{h}(r^{-|w|}), & \text{if } w \in V \text{, and} \\ 0, & \text{otherwise} \end{cases} \qquad (6)$$

defines a left computable discrete semi-measure. Thus Theorem 4 implies the following upper bound on the complexity functions of ω-words.

Lemma 2. *If $F \subseteq X^\omega$ is a Σ_2-definable ω-language and h is a right computable gauge function such that $\mathcal{H}^h(F) = 0$ then*

$$\mathrm{KP}(\xi[0..n]) \leq_{\text{i.o.}} -\log_r h(r^{-n}) + O(1) \text{ for all } \xi \in F, \text{ and}$$
$$\mathrm{KA}(\xi[0..n]) \leq_{\text{i.o.}} -\log_r h(r^{-n}) + O(1) \text{ for all } \xi \in F.$$

The latter inequality follows from the former and $\mathrm{KA}(w) \leq \mathrm{KP}(w) + O(1)$ (see, e.g., [2,20]).

Finally, Lemma 2, Eq. (5) and Corollary 1 prove the following.

Theorem 5. *If $F \subseteq X^\omega$ is a union of Σ_2-definable sets and h is a right computable gauge function such that $\mathcal{H}^h(F) = 0$ then $F \subseteq S_{\infty,h'}[\mathcal{U}]$ for every gauge function h' such that $\lim_{t \to 0} \frac{h'(t)}{h(t)} = 0$.*

3.2   Computable Dimension

Computable dimension is based on computable super-martingales as constructive dimension was based on left computable super-martingales. In contrast to the latter, for the former there is no universal computable super-martingale (cf. [2,16]). Thus we define analogously to Theorem 1

Definition 3. *We refer to a gauge function h as an exact computable dimension function for $F \subseteq X^\omega$ provided*

1. *for all gauge functions h' with $\lim\limits_{n \to \infty} \frac{h'(r^{-n})}{h(r^{-n})} = 0$ there is a computable super-martingale $\mathcal{V}$ such that $F \subseteq S_{\infty,h'}[\mathcal{V}]$, and*

2. *for all gauge functions h'' with $\lim\limits_{n \to \infty} \frac{h(r^{-n})}{h''(r^{-n})} = 0$ and all computable super-martingales $\mathcal{V}$ it holds $F \not\subseteq S_{\infty,h''}[\mathcal{V}]$.*

---

[3] Here we follow also the notation of [20], in [2] prefix complexity was denoted by K.

As for the constructive case the second item is fulfilled provided $\mathcal{H}^h(F) > 0$. For Item 1 we prove that for computable gauge functions h and Σ_2^0-definable sets $F \subseteq X^\omega$ with $\mathcal{H}^h(F) = 0$ there is a computable martingale $\mathcal{V}$ such that $F \subseteq \bigcup_{c>0} S_{c,h}[\mathcal{V}]$.

In order to achieve our goal we introduce families of covering codes as in [18]. For a prefix code $C \subseteq X^*$ we define its *minimal complementary code* as

$$\widehat{C} := (X \cup \mathbf{pref}(C) \cdot X) \setminus \mathbf{pref}(C) .$$

If $C = \varnothing$ we have $\widehat{C} = X$, and if $C \neq \varnothing$ the set $\widehat{C}$ consists of all words $w \cdot x \notin \mathbf{pref}(C)$ where $w \in \mathbf{pref}(C)$ and $x \in X$. It is readily seen that $C \cup \widehat{C}$ is a maximal prefix code, $C \cap \widehat{C} = \varnothing$, and $\mathbf{pref}(C \cup \widehat{C}) = \{e\} \cup \mathbf{pref}(C) \cup \widehat{C}$.

We call $\mathfrak{C} := (C_w)_{w \in X^*}$ a *family of covering codes* provided each C_w is a finite prefix code. Since then the set $C_w \cup \widehat{C}_w$ is a finite maximal prefix code, every word $u \in X^*$ has a uniquely specified $\mathfrak{C}$-factorisation $u = u_1 \cdots u_n \cdot u'$ where $u_{i+1} \in C_{u_1 \cdots u_i} \cup \widehat{C}_{u_1 \cdots u_i}$ for $i = 0, \ldots, n-1$ ($u_1 \cdots u_i = e$, if $i = 0$) and $u' \in \mathbf{pref}(C_{u_1 \cdots u_n} \cup \widehat{C}_{u_1 \cdots u_n})$. Analogously, every $\xi \in X^\omega$ has a uniquely specified $\mathfrak{C}$-factorisation $\xi = u_1 \cdots u_i \cdots$ where $u_{i+1} \in C_{u_1 \cdots u_i} \cup \widehat{C}_{u_1 \cdots u_i}$ for $i = 1, \ldots$.

In what follows we use martingales derived from prefix codes in the following manner.

Lemma 3. *Let $h : \mathbb{R} \to \mathbb{R}$ a gauge function and $\varnothing \neq C \subseteq X^*$ be a prefix code satisfying $\sum_{v \in C} h(r^{-|v|}) < \infty$. Then there is a martingale $V_C^{(h)} : X^* \to [0, \infty)$ such that*

$$V_C^{(h)}(w) = \begin{cases} \dfrac{r^{|w|} \cdot h(r^{(-|w|)})}{\sum_{v \in C} h(r^{(-|v|)}) + \sum_{u \in \widehat{C}} r^{-|u|}} & \text{, for } w \in C \text{, and} \\[3ex] \dfrac{1}{\sum_{v \in C} h(r^{(-|v|)}) + \sum_{u \in \widehat{C}} r^{-|u|}} & \text{, for } w \in \widehat{C}. \end{cases} \tag{7}$$

Remark 1. If C is a finite prefix code and $h : \mathbb{Q} \to \mathbb{Q}$ is computable then $V_C^{(h)}$ is a computable martingale.

For a gauge function $h : \mathbb{R} \to \mathbb{R}$ let $h_w(t) := \frac{h(r^{-|w|} \cdot t)}{h(t)}$ and let $\mathfrak{C} := (C_w)_{w \in X^*}$ be a family of covering codes.

Using the martingales $V_{C_w}^{(h_w)}$ we define a new martingale $V_{\mathfrak{C}}$ as follows: For $u \in X^*$ consider the $\mathfrak{C}$-factorisation $u_1 \cdots u_n \cdot u'$, and put

$$V_{\mathfrak{C}}^{(h)}(u) := \left(\prod_{i=0}^{n-1} V_{C_{u_1 \cdots u_i}}^{(h_{u_1 \cdots u_i})}(u_{i+1}) \right) \cdot V_{C_{u_1 \cdots u_n}}^{(h_{u_1 \cdots u_n})}(u') ,$$

that is, $V_{\mathfrak{C}}^{(h)}$ is in some sense the concatenation of the martingales $V_{C_w}^{(h_w)}$. Observe that $V_{\mathfrak{C}}^{(h)}$ is computable if only $h : \mathbb{R} \to \mathbb{R}$ is a computable function, the codes C_w are finite and the function which assigns to every w the corresponding code C_w is computable.

We have the following.

Lemma 4. *Let* $h : \mathbb{N} \to \mathbb{Q}$ *be a gauge function and let* $\mathfrak{C} = (C_w)_{w \in X^*}$ *be a family of covering codes such that* $\sum_{v \in C_w} \frac{h(r^{-|wv|})}{h(r^{-|v|})} \leq r^{-|w|}$ *for all* $w \in X^*$.

If the ω-*word* $\xi \in X^\omega$ *has a* $\mathfrak{C}$-*factorisation* $\xi = u_1 \cdots u_i \cdots$ *such that for some* $n_\xi \in \mathbb{N}$ *and all* $i \geq n_\xi$ *the factors* u_{i+1} *belong to* $C_{u_1 \cdots u_i}$. *Then there is a constant* $c_\xi > 0$ *not depending on* i *for which*

$$\mathcal{V}_{\mathfrak{C}}(u_1 \cdots u_i) \geq c_\xi \cdot r^{|u_1 \cdots u_i|} \cdot h(r^{-|u_1 \cdots u_i|}) \ .$$

Now we derive the announced result.

Theorem 6. *For every* Σ_2-*definable* ω-*language* $F \subseteq X^\omega$ *and every computable gauge function* $h : \mathbb{Q} \to \mathbb{R}$ *such that* $\mathcal{H}^h(F) = 0$ *there is a computable martingale* $\mathcal{V}$ *such that* $F \subseteq \bigcup_{c>0} S_{c,h}[\mathcal{V}]$.

References

1. Cai, J.Y., Hartmanis, J.: On Hausdorff and topological dimensions of the Kolmogorov complexity of the real line. J. Comput. System Sci. 49(3), 605–619 (1994)
2. Downey, R.G., Hirschfeldt, D.R.: Algorithmic Randomness and Complexity. Theory and Applications of Computability. Springer, New York (2010)
3. Edgar, G.: Measure, topology, and fractal geometry, 2nd edn. Undergraduate Texts in Mathematics. Springer, New York (2008)
4. Falconer, K.: Fractal geometry. John Wiley & Sons Ltd., Chichester (1990)
5. Graf, S., Mauldin, R.D., Williams, S.C.: The exact Hausdorff dimension in random recursive constructions. Mem. Amer. Math. Soc. 71(381), x+121 (1988)
6. Graham, R.L., Knuth, D.E., Patashnik, O.: Concrete mathematics, 2nd edn. Addison-Wesley Publishing Company, Reading (1994)
7. Hausdorff, F.: Dimension und äußeres Maß. Math. Ann. 79(1-2), 157–179 (1918)
8. Hitchcock, J.M.: Correspondence Principles for Effective Dimensions. In: Widmayer, P., Triguero, F., Morales, R., Hennessy, M., Eidenbenz, S., Conejo, R. (eds.) ICALP 2002. LNCS, vol. 2380, pp. 561–571. Springer, Heidelberg (2002)
9. Hitchcock, J.M.: Correspondence principles for effective dimensions. Theory Comput. Syst. 38(5), 559–571 (2005)
10. Lutz, J.H.: Gales and the Constructive Dimension of Individual Sequences. In: Welzl, E., Montanari, U., Rolim, J.D.P. (eds.) ICALP 2000. LNCS, vol. 1853, pp. 902–913. Springer, Heidelberg (2000)
11. Lutz, J.H.: The dimensions of individual strings and sequences. Inform. and Comput. 187(1), 49–79 (2003)
12. Merzenich, W., Staiger, L.: Fractals, dimension, and formal languages. RAIRO Inform. Théor. Appl. 28(3-4), 361–386 (1994)
13. Mielke, J.: Refined bounds on Kolmogorov complexity for ω-languages. Electr. Notes Theor. Comput. Sci. 221, 181–189 (2008)
14. Reimann, J.: Computability and Fractal Dimension. Ph.D. thesis, Ruprecht-Karls-Universität Heidelberg (2004)
15. Ryabko, B.Y.: Coding of combinatorial sources and Hausdorff dimension. Dokl. Akad. Nauk SSSR 277(5), 1066–1070 (1984)
16. Schnorr, C.P.: Zufälligkeit und Wahrscheinlichkeit. In: Eine Algorithmische Begründung der Wahrscheinlichkeitstheorie. Lecture Notes in Mathematics, vol. 218. Springer, Berlin (1971)

17. Staiger, L.: Kolmogorov complexity and Hausdorff dimension. Inform. and Comput. 103(2), 159–194 (1993)
18. Staiger, L.: A tight upper bound on Kolmogorov complexity and uniformly optimal prediction. Theory Comput. Syst. 31(3), 215–229 (1998)
19. Staiger, L.: Constructive Dimension and Hausdorff Dimension: The Case of Exact Dimension. In: Owe, O., Steffen, M., Telle, J.A. (eds.) FCT 2011. LNCS, vol. 6914, pp. 252–263. Springer, Heidelberg (2011)
20. Uspensky, V.A., Shen, A.: Relations between varieties of Kolmogorov complexities. Math. Systems Theory 29(3), 271–292 (1996)
21. Zvonkin, A.K., Levin, L.A.: The complexity of finite objects and the basing of the concepts of information and randomness on the theory of algorithms. Uspehi Mat. Nauk 25(6(156)), 85–127 (1970)

Low$_n$ Boolean Subalgebras

Rebecca M. Steiner

Graduate Center of the City University of New York, 365 Fifth Avenue, New York,
NY 10016, United States of America
rsteiner@gc.cuny.edu

Abstract. Every low$_n$ Boolean algebra, for $1 \leq n \leq 4$, is isomorphic to
a computable Boolean algebra. It is not yet known whether the same is
true for $n > 4$. However, it is known that there exists a low$_5$ subalgebra
of the computable atomless Boolean algebra which, when viewed as a
relation on the computable atomless Boolean algebra, does not have
a computable copy. We adapt the proof of this recent result to show
that there exists a low$_4$ subalgebra of the computable atomless Boolean
algebra which, when viewed as a relation on the computable atomless
Boolean algebra, has no computable copy. This result provides a sharp
contrast with the one which shows that every low$_4$ Boolean algebra has
a computable copy. That is, the spectrum of the subalgebra as a unary
relation can contain a low$_4$ degree without containing the degree $\mathbf{0}$, even
though no spectrum of a Boolean algebra (viewed as a structure) can do
the same.

1 Introduction

Researchers in computability theory have been coding (or attempting to code)
low$_n$ sets into Boolean algebras for at least the past twenty-five years. The
question which has been the focus of much of this work is: For which n is it
true that every low$_n$ Boolean algebra is isomorphic to a computable Boolean
algebra? The *spectrum of a structure* $\mathcal{A}$ is the set $\mathrm{Spec}(\mathcal{A})$ of Turing degrees of
structures isomorphic to $\mathcal{A}$: $\mathrm{Spec}(\mathcal{A}) = \{\deg(\mathcal{D}) : \mathcal{D} \cong \mathcal{A}\}$. With this definition,
we can rephrase the above question:

Question 1. For which n is it true that the spectrum of a low$_n$ Boolean algebra
must contain the degree $\mathbf{0}$?

Question 1 has been settled for $n = 1$, 2, 3, and 4, all in the affirmative [2,11,6].
That is, if the spectrum of a Boolean algebra contains a low, low$_2$, low$_3$, or low$_4$
degree, it must also contain the degree $\mathbf{0}$.

The goal of this chapter is not to contribute to the solution of Question 1, but
rather to contribute to the solution of a different (but related) question: If $\mathcal{B}$ is
the computable atomless Boolean algebra and $\mathcal{A}$ is a subalgebra which is low$_n$
within $\mathcal{B}$, must there be a computable subalgebra $\mathcal{D}$ such that $(\mathcal{B}, \mathcal{A}) \cong (\mathcal{B}, \mathcal{D})$?
The *spectrum of a relation* R *on a computable structure* $\mathcal{M}$ is the set $\mathrm{DgSp}_{\mathcal{M}}(R)$
of Turing degrees of all images of R under isomorphisms from $\mathcal{M}$ onto other

S.B. Cooper, A. Dawar, and B. Löwe (Eds.): CiE 2012, LNCS 7318, pp. 696–702, 2012.
© Springer-Verlag Berlin Heidelberg 2012

computable structures: $\mathrm{DgSp}_{\mathcal{M}}(R) = \{\deg(S) : (\exists \mathcal{B} \leq_T \emptyset)\,[(\mathcal{B}, S) \cong (\mathcal{M}, R)]\}$. With this definition, it is equivalent to ask:

Question 2. For which n is it true that the spectrum of a low$_n$ Boolean algebra, viewed as a unary relation on some fixed copy of $\mathcal{B}$, must contain the degree $\mathbf{0}$?

Question 2 has recently been answered in the negative for $n = 5$ [7], which is a case still unsettled for the former question.

A negative answer to Question 2 for $n = 1$, 2, 3, or 4 would be significant because it would provide us with an example of a spectrum of a subalgebra as a unary relation on the computable atomless Boolean algebra $\mathcal{B}$ which cannot possibly be the spectrum of a Boolean algebra viewed as a structure in its own right – the spectrum of the subalgebra would contain a low$_n$ degree for $n = 1$, 2, 3, or 4 without containing the degree $\mathbf{0}$, and we know that no spectrum of a Boolean algebra as a structure has this property [6].

In this article, we adapt the proof of the result that the spectrum of a low$_5$ subalgebra of $\mathcal{B}$, viewed as a relation on $\mathcal{B}$, need not contain the degree $\mathbf{0}$ to show that the spectrum of a low$_4$ subalgebra of $\mathcal{B}$, viewed as a relation on $\mathcal{B}$, need not contain the degree $\mathbf{0}$. Thus we provide a negative answer to the $n = 4$ case for the spectrum of a subalgebra. We also point out that this implies another spectral difference between Boolean algebras as structures and Boolean subalgebras as relations on $\mathcal{B}$. A theorem of Jockusch and Soare in [5] says that if a Boolean algebra has n-th jump degree $\mathbf{d}$ (defined below in Definition 4), where $n < \omega$, then $\mathbf{d} = \mathbf{0}^{(n)}$. In other words, no Boolean algebra as a structure can have n-th jump degree larger than $\mathbf{0}^{(n)}$. We show below in Corollary 11 that a Boolean subalgebra of $\mathcal{B}$, considered as a relation on $\mathcal{B}$, *can* have n-th jump degree strictly larger than $\mathbf{0}^{(n)}$.

2 The History

The first result on low$_n$ Boolean algebras and computable copies was the result of Downey and Jockusch for $n = 1$ which appeared in 1994.

Theorem 1. [2, Thm. 1] *Every low Boolean algebra is isomorphic to a computable one.*

They converted an argument about Boolean algebras to an argument about linear orderings, and then applied a theorem of Remmel [8, Thm. 2.1] which gives sufficient conditions for two linear orderings to have isomorphic interval algebras.

The result for $n = 2$ was the work of Thurber, and it appeared very soon after the $n = 1$ result. Thurber used the same sort of linear ordering argument Downey and Jockusch used.

Theorem 2. [11, Thm. 1] *Every low$_2$ Boolean algebra is isomorphic to a computable one.*

Six years after the original result, Knight and Stob came out with the results for $n = 3$ and $n = 4$.

Theorem 3. [6, Cor. 3.3] *Every low$_3$ Boolean algebra is isomorphic to a computable one.*

Theorem 4. [6, Cor. 5.3] *Every low$_4$ Boolean algebra is isomorphic to a computable one.*

Knight and Stob, unlike Downey and Jockusch and Thurber, argued in terms of Boolean algebras directly.

In an effort to attempt a result for $n = 5$, Miller, in 2011, instead ended up with a result just as interesting about low$_5$ subalgebras of the computable atomless Boolean algebra $\mathcal{B}$ when viewed as a relation on $\mathcal{B}$.

Definition 1. The *spectrum of a structure* $\mathcal{A}$ is the set $\mathrm{Spec}(\mathcal{A})$ of Turing degrees of structures isomorphic to $\mathcal{A}$:

$$\mathrm{Spec}(\mathcal{A}) = \{\deg(\mathcal{D}) : \mathcal{D} \cong \mathcal{A}\}.$$

The *spectrum of a relation R on a computable structure* $\mathcal{M}$ is the set $\mathrm{DgSp}_{\mathcal{M}}(R)$ of Turing degrees of all images of R under isomorphisms from $\mathcal{M}$ onto other computable structures:

$$\mathrm{DgSp}_{\mathcal{M}}(R) = \{\deg(S) : (\exists \mathcal{B} \leq_{\mathrm{T}} \emptyset)\,[(\mathcal{B}, S) \cong (\mathcal{M}, R)]\}.$$

Theorem 5. [7, Thm. 2.4] *Let $\mathbf{c}$ be any Turing degree which is not low$_4$. Then there exists a Boolean subalgebra $\mathcal{A}$ of the computable atomless Boolean algebra $\mathcal{B}$ for which $DgSp_{\mathcal{B}}(\mathcal{A})$ contains $\mathbf{c}$ but does not contain $\mathbf{0}$. In particular, there are low$_5$ degrees for which this works.*

Proof. We present a sketch of this proof.

Let $\mathcal{B}$ be a Boolean algebra and let $\mathcal{A}$ be a subalgebra of $\mathcal{B}$. An element $x \in \mathcal{B}$ is called an $\mathcal{A}$-*supremum* if x is the least upper bound in $\mathcal{B}$ of an infinite set of $\mathcal{A}$-atoms. Such an x is called a *single $\mathcal{A}$-supremum* (also called a *1-atom of $\mathcal{A}$* in the literature) if x is not the union of two disjoint $\mathcal{A}$-suprema, and a k-*fold $\mathcal{A}$-supremum* if x is the union of k disjoint single $\mathcal{A}$-suprema. The property of being a k-fold $\mathcal{A}$-supremum is $\Sigma_4^{\mathcal{A}}$ uniformly in k.

Given an arbitrary nonlow$_4$ degree $\mathbf{c}$ and an arbitrary $C \in \mathbf{c}$, Miller uses a C-oracle to construct a subalgebra $\mathcal{A}$ of $\mathcal{B}$ so that $\mathbf{c} \in \mathrm{DgSp}_{\mathcal{B}}(\mathcal{A})$. He builds $\mathcal{A}$ to satisfy the following:

$$n \in C^{(4)} \iff \exists x \in \mathcal{A}\,[\,x \text{ is a } 2^n\text{-fold } \mathcal{A}\text{-supremum}\,].$$

(For the detailed constructions of the subalgebra $\mathcal{A}$, see Miller's proof in [7].) The statement

$$\exists x \in \mathcal{A}\,[\,x \text{ is a } 2^n\text{-fold } \mathcal{A}\text{-supremum}\,]$$

is Σ_4^0 in $\mathcal{A}$, and hence computable in $\mathcal{A}^{(4)}$. And because this statement is equivalent to n being in $C^{(4)}$, we have $C^{(4)} \leq_{\mathrm{T}} \mathcal{A}^{(4)}$.

Now, suppose $\mathbf{0} \in \mathrm{DgSp}_{\mathcal{B}}(\mathcal{A})$. In other words, suppose there is a computable subalgebra $\mathcal{D}$ of $\mathcal{B}$ such that $(\mathcal{B}, \mathcal{D}) \cong (\mathcal{B}, \mathcal{A})$. Then the uniform $\Sigma_4^{\mathcal{A}}$ definition of k-fold $\mathcal{A}$-suprema would convert to a uniform $\Sigma_4^{\mathcal{D}}$ definition of k-fold $\mathcal{D}$-suprema on $(\mathcal{B}, \mathcal{D})$, which of course means just a Σ_4^0 definition, since $\mathcal{D}$ is computable. Thus it is Σ_4^0 whether $\mathcal{D}$ contains a 2^n-fold $\mathcal{D}$-supremum, and this in turn implies that $C^{(4)} \leq_{\mathrm{T}} \emptyset^{(4)}$. This is a contradiction because we chose C to be nonlow$_4$. $\qquad\square$

3   The Main Result

Theorem 6. *Let $\mathbf{c}$ be any Turing degree which is not low$_3$. Then there exists a Boolean subalgebra $\mathcal{A}$ of the computable atomless Boolean algebra $\mathcal{B}$ for which $\mathrm{DgSp}_{\mathcal{B}}(\mathcal{A})$ contains $\mathbf{c}$ but does not contain $\mathbf{0}$.*

Proof. The construction of a $\mathcal{D}$ for which $(\mathcal{B}, \mathcal{D}) \cong (\mathcal{B}, \mathcal{A})$ and $\deg(\mathcal{D}) = \mathbf{c}$ follows exactly as in the proof of Theorem 5 and [7], even though we are dealing with a nonlow$_3$ degree here, rather than a nonlow$_4$ degree. This is still a $C^{(4)}$-construction.

It is actually a small (yet powerful) fact from a classic result known as the "Jump Theorem" which allows us to use a low$_3$ argument with a low$_4$ construction; we shall use this fact to show that $\mathrm{DgSp}_{\mathcal{B}}(\mathcal{A})$ does not contain the degree $\mathbf{0}$.

Fact 7. *[9, "Jump Theorem" 3.4.3(v)] Let $A \subseteq \omega$ and $B \subseteq \omega$. Then $A \leq_{\mathrm{T}} B$ if and only if $A' \leq_1 B'$.*

The reason this works is that the Turing reduction $C^{(4)} \leq_{\mathrm{T}} \mathcal{A}^{(4)}$ is actually much stronger than just a Turing reduction. Recall that according to Miller's construction in [7], we have

$$n \in C^{(4)} \iff \exists x \in \mathcal{A}\,[\,x \text{ is a } 2^n\text{-fold } \mathcal{A}\text{-supremum}\,]. \tag{1}$$

It turns out that we actually have a 1-reduction from $C^{(4)}$ to $\mathcal{A}^{(4)}$: Recall that

$$\exists x \in \mathcal{A}\,[\,x \text{ is a } 2^n\text{-fold } \mathcal{A}\text{-supremum}\,]$$

is a $\Sigma_4^{\mathcal{A}}$ property, and since all $\Sigma_4^{\mathcal{A}}$ sets 1-reduce to $\mathcal{A}^{(4)}$, there is a total computable function f such that for all n,

$$f(n) \in \mathcal{A}^{(4)} \iff \mathcal{A} \text{ contains a } 2^n\text{-fold } \mathcal{A}\text{-supremum}. \tag{2}$$

Then by equivalences (1) and (2), $n \in C^{(4)}$ iff $f(n) \in \mathcal{A}^{(4)}$. Thus $C^{(4)} \leq_1 \mathcal{A}^{(4)}$ via this f.

For any $\mathcal{D}$ for which $(\mathcal{B}, \mathcal{D}) \cong (\mathcal{B}, \mathcal{A})$, if $\mathcal{A}$ has a k-fold $\mathcal{A}$-supremum, then $\mathcal{D}$ also has a k-fold $\mathcal{D}$-supremum: limits must map to limits under an isomorphism (same for limits of limits, etc.), and so k-fold suprema are isomorphism-invariant. Thus $n \in C^{(4)}$ iff $f(n) \in \mathcal{D}^{(4)}$. So $C^{(4)} \leq_1 \mathcal{D}^{(4)}$.

If there were a computable such $\mathcal{D}$, then we would have $C^{(4)} \leq_1 \emptyset^{(4)}$, and then by Fact 7, we would have $C^{(3)} \leq_T \emptyset^{(3)}$, which is a contradiction since we chose C to be nonlow$_3$. $\qquad\square$

Corollary 8. *There is a low$_4$ degree* **c** *and a Boolean subalgebra $\mathcal{A}$ of $\mathcal{B}$ for which $DgSp_{\mathcal{B}}(\mathcal{A})$ contains* **c** *but does not contain* **0**.

Proof. Choose **c** to be low$_4$ but not low$_3$, and then apply Theorem 6. $\qquad\square$

Corollary 9. *The spectrum of a Boolean subalgebra of $\mathcal{B}$ can fail to be the spectrum of any Boolean algebra (as a structure).*

Proof. The spectrum of the Boolean subalgebra of $\mathcal{B}$ which we constructed in Theorem 6 contains a low$_4$ degree but not the degree **0**. The spectrum of a Boolean algebra (as a structure in its own right) cannot contain a low$_4$ degree without containing the degree **0**, as we stated in Theorem 4. $\qquad\square$

Definition 2. A structure $\mathcal{D}$ is called *ultrahomogeneous* if every isomorphism between finitely-generated substructures of $\mathcal{D}$ extends to an automorphism of $\mathcal{D}$.

Definition 3. Let $\mathfrak{K}$ be a class of finitely-generated structures. A structure $\mathcal{D}$ is the *Fraïssé limit* of $\mathfrak{K}$ if $\mathcal{D}$ is countable, ultrahomogeneous, and $\mathfrak{K}$ is the class of all finitely-generated substructures of $\mathcal{D}$.

Theorem 10. *If $\mathcal{A}$ is a Boolean subalgebra of the computable atomless Boolean algebra $\mathcal{B}$ and $\mathcal{A}$ is not intrinsically computable, then $DgSp_{\mathcal{B}}(\mathcal{A})$ is closed upwards in the Turing degrees.*

Proof. In [1], the authors proved that if $\mathcal{F}$ is the Fraïssé limit of a class **K** of finite structures over a finite language $\mathcal{L}$ where $Th_{\mathcal{L}}(\mathbf{K})$ is computably axiomatizable and locally finite, then if R is a unary relation on $\mathcal{F}$ which is not intrinsically computable, $DgSp_{\mathcal{F}}(R)$ is upward closed under Turing reducibility [1, Cor. 5.2]. These hypotheses all hold for the atomless Boolean algebra $\mathcal{B}$, and so if $\mathcal{A}$ is the unary relation on $\mathcal{B}$, we have exactly the result we want. $\qquad\square$

When Miller wrote up his proof of Theorem 5, he neglected to describe the spectrum of the Boolean algebra he built. So we describe it here.

Corollary 11. *The spectrum of the Boolean subalgebra constructed in [7] and used here in Theorem 6 is the set of all Turing degrees* **d** *such that*

$$\mathbf{c}''' \leq \mathbf{d}'''$$

where **c** *is the nonlow$_4$ or nonlow$_3$ degree in the theorem.*

Proof. Let $\mathcal{D}$ be a copy of the $\mathcal{A}$ we built, where $\deg(\mathcal{D}) = \mathbf{d}$. We know from the construction of $\mathcal{A}$ that $n \in C^{(4)} \iff \exists x \in \mathcal{A} \,[\, x \text{ is a } 2^n\text{-fold } \mathcal{A}\text{-supremum }]$. Since $\mathcal{D} \cong \mathcal{A}$ and suprema are isomorphism-invariant, we have $n \in C^{(4)} \iff$

$\exists x \in \mathcal{D}$ [x is a 2^n-fold $\mathcal{D}$-supremum]. In the proof of Theorem 6 we defined f for which $f(n) \in \mathcal{A}^{(4)}$ means $\mathcal{A}$ contains a 2^n-fold $\mathcal{A}$-supremum. So $C^{(4)} \leq_1 \mathcal{D}^{(4)}$ via this f, and by Fact 7, we have $C''' \leq_T \mathcal{D}'''$.

Now let D be a set of degree $\mathbf{d}$, and suppose $C''' \leq_T D'''$. By Fact 7, we then have $C^{(4)} \leq_1 D^{(4)}$. Just as in Miller's construction in [7], we use a D-oracle to build a Boolean subalgebra $\mathcal{R}$ of $\mathcal{B}$ with $(\mathcal{B}, \mathcal{A}) \cong (\mathcal{B}, \mathcal{R})$. This gives us $\mathcal{R} \leq_T D$, and now that we have Theorem 10 to give us the upward closure of $\mathrm{DgSp}_{\mathcal{B}}(\mathcal{A})$, we can say that $\mathbf{d} \in \mathrm{DgSp}_{\mathcal{B}}(\mathcal{A})$.   □

Last but not least, we recall a well-known theorem of Jockusch and Soare about Boolean algebras as structures which we now know does not hold for Boolean subalgebras as relations on $\mathcal{B}$.

Definition 4. For an ordinal α, the *α-th jump degree of a structure* $\mathcal{M}$ is the least degree among the degrees of α-th jumps of structures isomorphic to $\mathcal{M}$, if such a least degree exists.

Theorem 12. (Jockusch & Soare) [5, Thm. 1.3] *Let $n < \omega$. If a Boolean algebra has nth-jump degree $\mathbf{d}$, then $\mathbf{d} = \mathbf{0}^{(n)}$.*

This theorem, along with Corollary 11, differentiates the jump degree spectrum of a Boolean algebra and the jump degree spectrum of a Boolean subalgebra of $\mathcal{B}$. Theorem 12 states that the n-th jump degree of a Boolean algebra (as a structure) can't be any larger than $\mathbf{0}^{(n)}$, while Corollary 11, via $n = 3$, shows that the n-th jump degree of a Boolean subalgebra of $\mathcal{B}$, considered as a relation on $\mathcal{B}$, *can* be larger than $\mathbf{0}^{(n)}$.

Acknowledgements. This research was partially supported by grant #DMS-1001306 from the National Science Foundation.

References

1. Csima, B.F., Harizanov, V.S., Miller, R.G., Montalbán, A.: Computability of Fraïssé Limits. Journal of Symbolic Logic 76, 66–93 (2011)
2. Downey, R., Jockusch, C.G.: Every low Boolean algebra is isomorphic to a recursive one. Proceedings of the American Mathematical Society 122, 871–880 (1994)
3. Frolov, A., Harizanov, V.S., Kalimullin, I., Kudinov, O., Miller, R.G.: Degree spectra of high$_n$ and nonlow$_n$ degrees. To appear in the Journal of Logic and Computation
4. Harizanov, V.S., Miller, R.G.: Spectra of structures and relations. Journal of Symbolic Logic 72, 324–348 (2007)
5. Jockusch, C.G., Soare, R.I.: Boolean algebras, stone spaces, and the iterated Turing jump. Journal of Symbolic Logic 59, 1121–1138 (1994)
6. Knight, J.F., Stob, M.: Computable Boolean algebras. Journal of Symbolic Logic 65, 1605–1623 (2000)
7. Miller, R.G.: Low$_5$ Boolean subalgebras and computable copies. Journal of Symbolic Logic 76, 1061–1074 (2011)

8. Remmel, J.B.: Recursive isomorphism types of recursive Boolean algebras. Journal of Symbolic Logic 46, 572–594 (1981)
9. Soare, R.I.: Recursively Enumerable Sets and Degrees. Springer, Berlin (1987)
10. Steiner, R.M.: Effective Algebraicity (submitted for publication)
11. Thurber, J.J.: Every low$_2$ Boolean algebra has a recursive copy. Proceedings of the American Mathematical Society 123, 3859–3866 (1995)

Bringing Up Turing's 'Child-Machine'

Susan G. Sterrett

Department of Philosophy, Carnegie Mellon University, Pittsburgh,
PA 15213, United States of America
susangsterrett@gmail.com

Abstract. Turing wrote that the "guiding principle" of his investigation into the possibility of intelligent machinery was "The analogy [of machinery that might be made to show intelligent behavior] with the human brain." [10] In his discussion of the investigations that Turing said were guided by this analogy, however, he employs a more far-reaching analogy: he eventually expands the analogy from the human brain out to "the human community as a whole." Along the way, he takes note of an obvious fact in the bigger scheme of things regarding human intelligence: grownups were once children; this leads him to imagine what a machine analogue of childhood might be. In this paper, I'll discuss Turing's child-machine, what he said about different ways of educating it, and what impact the "bringing up" of a child-machine has on its ability to behave in ways that might be taken for intelligent. I'll also discuss how some of the various games he suggested humans might play with machines are related to this approach.

1 A 'Guiding Principle'

In his writings on intelligence and machinery, Turing often employs analogies. One analogy he states explicitly and calls the "guiding principle" of his investigation into "possible ways in which machinery might be made to show intelligent behavior" is "the analogy with the human brain." [10]

The analogy that Turing employs in the discussions that follow is not a simple analogy between machine and brain; it's more specific, and less physically-oriented, than the brief description of an analogy between computing machinery and the human brain quoted above might at first suggest. Turing says his investigation is mainly concerned with the analogy between *the ways in which a human* [with a human brain] is *educated* such that the *potentialities for human intelligence are realized*, and "an analogous teaching process applied to machines." [10].

That is, his investigation concerns identifying and evaluating proposals for filling in the part of the analogy that answers: if we want a machine to fulfil *its* potentialities for intelligence, *how should it be "educated"*? In his 1950 "Computing Machinery and Intelligence", he spoke of dividing the problem of building a machine that can imitate the human mind into two parts: "The child program and the education process." He mentions yet a third component, on

S.B. Cooper, A. Dawar, and B. Löwe (Eds.): CiE 2012, LNCS 7318, pp. 703–713, 2012.

this approach: "Other experience, not to be described as education, to which it [the machine] has been subjected." That is, there is a distinction between "the education process" and "other experience." But what is it that distinguishes the education process?

2   Intelligent Behavior versus Completely Disciplined Behavior

This analogy—between a machine that has undergone an education process and a human student who has been educated by a teacher—provides Turing with the means to respond to one of the most common objections raised against the possibility that a machine could be regarded as exhibiting intelligent behavior. This objection (to the possibility of intelligent machinery) is, in Turing's words, the view that "[i]nsofar as a machine can show intelligence this is to be regarded as nothing but a reflection of the intelligence of its creator." That view, he says, is much like the view that "the credit for the discoveries of a pupil should be given to his teacher", which can be rebutted as follows:

> " In such a case the teacher would be pleased with the success of his methods of education, but would not claim the results themselves unless he had actually communicated them to his pupil. He would certainly have envisaged in very broad outline the sort of thing his pupil might be expected to do, but would not expect to forsee any sort of detail. " [10, p. 2]

Turing contrasts "intelligent behavior" of a machine with "completely disciplined behavior." Both are exceptional sorts of behavior for a machine; he says that "Most of the programmes which we can put into the machine will result in it doing something that we cannot make sense of at all, or which we regard as completely random behavior."

In both intelligent behavior and completely disciplined behavior, we are able to make sense of the machine's behavior. But the kind of sense we make of it differs. When a machine is carrying out computations, the machine's behavior is "completely disciplined" and what we strive for is to have "a clear mental picture of the state of the machine at each moment in the computation." [9, p. 459] When a teacher is educating a machine with an intent to produce an intelligent machine, the goal of the education process entails that some of the machine's rules of behavior will be undergoing change. The teacher will be able "to some extent to predict the pupil's behavior", but, in contrast to the case of programming it to carry out computations, won't have a clear picture of what is going on within the machine being educated. Intelligent behavior is not a large departure from completely disciplined behavior, but it does differ qualitatively from completely disciplined behavior: the sense we make of it is distinctively different. Intelligent behavior escapes the predictability of completely disciplined machine behavior without veering off into random behavior.

One qualitative difference between completely disciplined behavior and intelligent behavior is the presence of initiative. When describing a universal machine with no special programming but able to carry out whatever program is put into it, Turing remarks that after carrying out the actions specified by the program, it would sink into inactivity until another action is required. It would lack initiative. This is one reason that the universal computer, even if produced from a child-machine by some machine analogue of an education process to produce completely disciplined behavior, is not a good candidate for a machine analogue of a human — even though the actions that it does take would be faultless. Turing thinks that an intelligent machine will be fallible, and that the feature of fallibility can be (or might be an indication of something that is) an important advantage. He also thinks that it might be required to have some sort of random element in a machine in order to produce a machine that is amenable to undergoing the kind of process that is analogous to the education of a human.

So, randomness probably has a part to play in producing a machine that might possibly be said to exhibit intelligence. Yet, intelligent machine behavior is not random behavior. The part that randomness plays in intelligent machinery is in the generation of possibilities among which some search process is then employed. [9, p. 459] Turing writes of using a random element to generate forms of behavior at one point [9, p. 459]; at another point he speaks of using a random element to generate different child-machines among which one then selects the best ones. [9, p. 456]. However, in both places he indicates that random generation alone does not seem very efficient, and that he would expect to supplement the generation of alternatives or the search among alternatives with some more directed, more informed process. Of the process of "finding" an appropriate child-machine, he writes:

> "One may hope, however, that this process will be more expeditious than evolution. The survival of the fittest is a slow method for measuring advantages. The experimenter, by the exercise of intelligence, should be able to speed it up. Equally important is the fact that he is not restricted to random mutations. If he can trace a cause for some of the weakness he can probably think of the kind of mutation which will improve it." [9, p. 456]

And, of the education process, which aims to find the appropriate behavior:

> "The systematic method [of trying out different possibilities in the search for a solution] has the disadvantage that there may be an enormous block without any solutions in the region which has to be investigated first. Now the learning process may be regarded as a search for a form of behaviour which will satisfy the teacher (or some other criterion). Since there is probably a very large number of satisfactory solutions the random method seems to be better than the systematic." [9, p. 459]

The education process is a matter of "intervening" on the machine. Just as the behavior of the early machines could be changed by using a screwdriver to change the machine's physical configuration by physical means, so the behavior of digital computers can be changed by using communication with it to change its rules of operation in some way. These two kinds of intervention are referred to as "screwdriver intervention" and "paper intervention", respectively. In "Intelligent Machinery", the guiding principle (the analogy mentioned earlier) is employed here, too—with some qualifications. Turing notes that human life is such that "interference is the rule rather than the exception." He identifies which part of human life he means to compare to a machine that might be regarded as exhibiting intelligence:

> " [A human] is in frequent communication with other [humans], and is continually receiving visual and other stimuli which themselves consti-tute a form of interference. *It will only be when the [human] is 'concentrating' with a view to eliminating these stimuli or 'distractions' that he approx-imates a machine without interference.*" [10, p. 8; emphasis added]

The human behavior during a time period when the human approximates a machine without interference, though, "is largely determined by the way he has been conditioned by previous interference."

Since, as he says, humans are constantly undergoing interference, how is the analogy between humans and machines supposed to go here? What is the dif-ference between undergoing an education process and being intervened upon in other ways? Well, he seems to think of education as a special kind of interference: it involves a teacher who intentionally tries to affect the behavior of the machine. It's interference directed towards some goal. So, even though humans undergo interference as a rule as they go about their daily lives (except for the times when they withdraw and concentrate on something), we still want to distinguish the kind of interference that is education from other kinds of interference.

The analogy may not be precise, but I think it is pretty clear: humans undergo education processes for a portion of their lives (which Turing estimates at about the first twenty years of their lives), and their behavior after that is very much affected by the education they have received, even though they still receive other interference—most of the time, in fact. The point is to approximate the human process of education with some analogous process suitable for machines. The ma-jor points of his proposal are that, on analogy with a human's life, we plan for these three stages of a machine: first, there is the infant stage of a machine, which is a machine that has not been educated and is at least partly unorganized. It need not be a blank slate, but it is important that large amounts of its behavior are undetermined. This is followed by the child-machine stage, during which the machine is educated. The first stage of education is to get the machine to a point where "it could be relied on to produce definite reactions to certain commands." [10, p. 118] Education involves a teacher who is intentionally trying to teach or modify the machine's behavior to effect some specific kinds of behavior. The machine's behavior is in flux during this time. Even if the machine is given the means to educate itself using some kind of program during the child-machine stage, there is still oversight and monitoring by a teacher of sorts who checks up on its progress and intervenes if necessary. At some point the education can be ended, and the machine that results when education is ended is supposed to behave in a way that can be predicted "in very broad outline" by someone familiar with how it has been educated — but its behavior might not, in fact probably will not, be fully predictable. Finally, there is the adult-machine, which is still capable of learning, but is also capable of quite complex behavior without additional intervention.

What about a process that would start with an unorganized machine, which we would then 'organize' by suitable interference to be a universal machine (e.g., a digital computer capable of being programmed)? Turing says that researchers should be interested in understanding the process that begins with an unorganized machine and results in a universal machine, but he doesn't regard such a process as the appropriate "analogous process" of human education: a universal machine isn't really the behavioral analogue of an adult. One of the differences between a human adult and a universal machine is the point mentioned above regarding the lack of initiative. There are other reasons, too: such an adult-machine would "obey orders given in an appropriate language, even if they were very complicated; he would have no common sense, and would obey the most ridiculous orders unflinchingly." [10, p. 116]

Turing describes an experiment in "educating" machines he carried out. It involved a process meant to be analogous to administering punishments and rewards; of giving the machine something analogous to pain and pleasure. The machine to be educated in his experiment was one whose description was incomplete, as he put it, meaning that its actions were not yet fully specified; thus, the machine's operation would give rise to specific cases where the action called for is not determined. When such a specific case arises, the following is done: an action is selected randomly and applied tentatively, by making the appropriate entry in the machine's description. This is the point where the teacher "educates" the machine.

The general idea of employing pleasure and pain he has in mind is revealed in his discussion of "pleasure-pain systems." We can get the general idea without getting into the details too much. He is considering unorganized machines whose states are described using two expressions, one of which he calls "character": "Pleasure interference tends to fix the character, i.e., towards preventing it changing, whereas pain stimuli tend to disrupt the character, causing features which had become fixed to change, or to become again subject to random variation." When he describes the particular experiment he carried out, though, which he refers to as a "particular type of pain-pleasure system", the analogy seems to employ the brain-machine analogy quite directly: "When a pain stimulus occurs all tentative entries are cancelled, and when a pleasure stimulus occurs they are all made permanent." [10, p. 118] At the time, he found it took too much work to pursue this means of educating a machine much farther than the rather simplified version of it he had carried out.

As his friend and colleague Donald Michie put it, they were waiting for hardware. Michie recounts a story about one of Turing's plans to program the "Manchester Baby" to investigate what would happen when two different programs for playing chess were pit against each other: "[Turing] was thwarted (rightly) by . . . the guardian of its scarce resources, Tom Kilburn." According to Michie, in the years leading up to Turing's 1948 and 1950 papers on intelligent machinery, Michie, Turing, and Jack Good "formed a sort of discussion club focused around Turing's astonishing 'child machine' concept.[1] His proposal was to use our knowledge of how the brain acquires its intelligence as a model for designing a *teachable intelligent machine*." [1] The idea that the source of learning might be sought in some random elements of neural physiology was well-known in psychology; decades earlier, in his *Principles of Psychology*, William James had concluded a discussion on the formation of pathways in the brain: " All this is vague to the last degree, and amounts to little more than saying that a new path may be formed by the sort of *chances* that in nervous material are likely to occur." [3, p. 104] The discussion club worked on developing the analogy for machines; one might say that what they were doing in that discussion club was developing the basic ideas of what has since become known as reinforcement learning. Michie later showed that reinforcement learning could indeed be successfully carried out in machines, proving sceptics wrong. [4] In 1948, though, there was still a "wait for hardware", and having to wait for the hardware to be available to test their ideas must have been frustrating.

Turing also outlined other approaches he would have liked to try: one might program one's "teaching policies" into the machine, and let it run for awhile, modifying its own programs, and periodically check to see how much progress it has made in its education. He regarded the problem of building a machine that would display "initiative" as well as "discipline" (as he put it) as crucial. Achieving discipline in a machine: that we can see how to do. What initiative adds to discipline in a machine: this is a matter of comparing humans and machines with complete discipline, and asking what humans that are able to communicate had in addition to discipline. Then, one could address what it was that should be copied in the machine. A question remained as to what process to use that achieves ending up with a machine that had both. In particular, in what order should these two be instilled in the machine: first, discipline, then initiative, or somehow both together?

3   Teachers, Singular and Plural

In most of Turing's examples of a teacher educating a machine, it seems he is thinking of one or at most a few individual teachers. The kind of machine under consideration is a universal (i.e., programmable) machine, specifically, a digital computer, equipped with a means of "at most, organs of sight, speech,

---

[1] Turing actively sought out discussion with colleagues. Another such colleague was the philosopher Ludwig Wittgenstein, whose seminar he attended. For Turing's 'constructive uses' of their discussions, cf. [2].

and hearing." His investigations are, as a result, biased towards activities that "require little contact with the outside world." [10, p. 117] There are other obstacles, too: even if a machine were equipped with the ability to navigate physically, there are limitations on its abilities to be socialized. More than once, he mentions the advantage that human learners have in that they are able to benefit from interactions with other humans.[2]

It is interesting that a major piece of research in cultural anthropology on cross-cultural features of child-rearing appeals to neural processes very much in line with Turing's "pain-pleasure systems", except that it is evaluations of goodness and badness, rather than physical pleasure and physical pain, that are administered. What is interesting is that this work [5] provides a model of what the education process of a human child would be on the "it takes a village to raise a child" view: "Cultural models of child rearing, thus, exploit the neural capacities of the children so reared, to achieve a result, adulthood, that could not be accomplished by the human brain alone." One can see the kind of issues this might raise: what if different members of the community contradict each other? This is exactly the issue ("constancy of [the child's] experience") that the cultural anthropologists in [5] discovered was universally deemed important.

Turing does not talk about this kind of education—education by a community, or education constrained and informed by cultural norms—of a child-machine, but he does talk about the role of the human community in the intellectual activity of humans. It occurs in his discussion of initiative. In that discussion, by the time he got to talking about human community, he had already treated the issues of discipline and initiative separately, and the education of the child-machine had been limited to the instillation of discipline into the machine. Yet, given that he did, albeit very briefly, indicate that the analogy between brain and machine might involve how the human community is involved in the education process of a human, the question of an analogue for a community as teacher during the education process of a machine arises quite naturally.

We might ask, what kind of community? Turing considered digital computers that have "organs" for sight, speech, and hearing. The newest cellphones (e.g., equipped with Siri) have such "organs", and they incorporate some machine learning capabilities, including learning the preferences and habits of their owners. These kinds of machines (state of the art cellphones) generally interact with and "learn from" a single human owner. Yet, they interact with other virtual agents who communicate with them. Now that we have these possibilities not available to Turing, we might consider following through in more detail on remarks that Turing made about "intellectual search" by humans immersed in a human community, and consider what the analogous processes for a machine might be.

---

[2] Harry Collins' research on imitation games extends this observation about the value of interactions with others. He also ties the kind of knowledge gained in this way with the kind of knowledge that can be exhibited using imitation games. http://www.cardiff.ac.uk/socsi/contactsandpeople/harrycollins/expertise-project/imitationgameresearch.html

4   Imitation and Intelligence

Turing concludes his 1948 paper "Intelligent Machinery" with a section on the concept of intelligence, which ends in the description of an experiment. The experiment involves three people and the game of chess.

> "It is not difficult to devise a paper machine which will play a not very bad game of chess. Now get three [humans] as subjects for the experiment A, B, C. A and C are to be rather poor chess players, B is the operator who works the paper machine. . . . Two rooms are used with some arrangement for communicating moves, and a game is played between C and either A or the paper machine. C may find it difficult to tell which he is playing." [10, p. 127]

By 'paper machine', Turing means creating "the effect of a computing machine by writing down a set of rules of procedure and asking a man to carry them out. " [10, p. 113] So, the human who operates the paper machine, in conjunction with the written rules of procedure, is imitating a machine. Now, although it is meant to be straightforward to imitate a machine by this method, Turing does not consider it trivially easy, for he advises using someone who is both a mathematician and a chess player to work the paper machine.

Now, this experimental setup is most assuredly not intended to be an *objective* measure of intelligence of the paper machine. To prevent any charge of interpretive license, let me quote Turing from this last section of the paper, which bears the heading "Intelligence as an emotional concept": "The extent to which we regard something as behaving in an intelligent manner is determined as much by our own state of mind and training as by the properties of the object under consideration."

Upon what, then, does regarding something as intelligent depend? His answer is given in terms of what it is that would rule out regarding something as intelligent: "If we are able to explain and predict its behaviour or if there seems to be little underlying plan, we have little temptation to imagine intelligence." Different people bring different skills with respect to explaining and predicting the behavior of something: "With the same object therefore it is possible that one man would consider it as intelligent and another would not; the second man would have found out the rules of its behavior."

The experiment is set up as a comparison: between A, a "rather poor" chess player, and B, a paper machine. The experiment is not in terms of whether B can beat A at chess—the way the experiment is set up, B, which is a man imitating the behavior of a machine by implementing rules that could be carried out by a machine, but which are written by a human and intended to be read by a human, will likely win some rounds. The comparison is not between chess-playing abilities, but between how transparent it is to C that B's behavior is being produced by following a set of written rules capable of being carried out by a machine.

While "Intelligent Machinery" closed with the description of a three person game about telling the difference between a performance generated by 'rules of behavior' and one by a human, "Computing Machinery and Intelligence" opened with such a three person game. The three persons in the game (called an imitation game) were named A, B, and C, too, and C was to distinguish between A and B. There was a difference, though: the moves being communicated were not positions in a game of chess, but taking one's turn in a conversation. The distinction was not a matter of distinguishing between which player was a person and which a machine, but between which conversationalist was a man and which was a woman.

The experimental setup in Turing's 1950 paper that I dubbed "The Original Imitation Game Test" is very like a TV game show that premiered six years later, in 1956, called "To Tell The Truth." It was played as follows:

> " Three challengers are introduced, all claiming to be the central character.
> [. . .] the host reads aloud a signed affadavit about the central character.
> The panelists are each given a period of time to question the challengers. Questions are directed to the challengers by number (Number One, Number Two and Number Three), with the central character sworn to give truthful answers, and the impostors permitted to lie and pretend to be the central character.
> After questioning is complete, each member of the panel votes on which of the challengers they believe to be the central character, [. . .] Once the votes are cast, the host asks, "Will the real [person's name] please stand up?" The central character then stands, [. . .] Prize money is awarded to the challengers based on the number of incorrect votes the impostors draw." [11]

Being a convincing imposter can be difficult. In previously published work, I have argued that the OIG Test is a better game than the one currently referred to as "the Turing Test." [8,7,6]. One reason I gave for my view was that the task given the machine in the OIG Test is the same as the task set for the human in the OIG Test: to imitate something that it is not.[3] The concept of a machine being set the task of imitation should not seem at all foreign here—in fact, the term "imitation" is used by Turing in describing a universal machine; he speaks of the ability of a universal machine to imitate other machines. Isn't imitation a straightforward task for a computer, then, you may ask?

No, I don't think it is. For an uneducated machine (such as an uneducated universal machine) to imitate is one thing—it amounts to implementing a program. For an educated machine to imitate is quite another. In fact, I argued, what is called for is not really imitation, but figuring out and carrying out what

---

[3] In [8] I also show that the two tests in [9] give different *quantitative*, as well as qualitative, results. I consider it a major contribution of [8] to give what amounts to a proof that the two tests are unequivocably different on significant points, and that the OIG Test need not be set up around gender differences. Secondary literature citing [8] has not always recognized these two major points.

it takes to be a convincing imposter. While the central character in "To Tell the Truth" may give an answer to a question without any fear of being led to another question that he or she cannot answer, the imposter has to think how to keep the conversation from turning to topics that might present problems for an imposter. An imposter has to constantly be on guard to override tendencies to respond in ways that have by now become habitual, but which are inappropriate while posing as an imposter. Talk of overriding habits is, I believe, no longer fanciful talk, as reading about work done by robotics researchers on the difficulties faced in applying imitation learning in robotics will reveal. If IBM is looking for suggestions for its next Grand Challenge, let me suggest "To Tell the Truth."[4]

I shall not repeat all the points I made in those earlier works on Turing and tests for intelligence. Rather, my point in this talk about the OIG Test concerns a question germane to the education of Turing's Child-Machines. I suggest that reflecting on the question of how machines produced using different methods for educating machines fare on the OIG Test leads to useful ways of thinking about machine intelligence.

Acknowledgements. I should like to thank an anonymous referee for some helpful remarks, and the audience at the Fourth Regional Wittgenstein Workshop held at Washington and Lee University on March 11th, 2012 for discussion on this paper.

References

1. BCS Computer Conservation Society. Recollections of early AI in Britain: 1942 - 1965 (2002); (video for the BCS Computer Conservation Society's October 2002 Conference on the history of AI in Britain) transcript (downloaded on March 25, 2012)
2. Floyd, J.: Turing, Wittgenstein, and Types: Philosophical Aspects of Turing's 'The Reform of Mathematical Notation and Phraseology' (1944-1945). In: Barry Cooper, S., van Leeuwen, J. (eds.) Alan Turing - His Work and Impact. The Collected Works of A. M. Turing, revised edn. North-Holland, Elsevier (2001) (to appear)
3. James, W.: The Principles of Psychology, vol. I. Henry Holt and Company, New York (1890)
4. Michie, D.: Trial and Error. Science Survey, part 2, 129–145(1961)
5. Quinn, N.: Cultural Selves. Annals of the New York Academy of Sciences 1001, 145–176 (2003)
6. Sterrett, S.G.: Too Many Instincts: Contrasting Philosophical Views on Intelligence in Humans and Non-Humans. JETAI Journal of Experimental and Theoretical Artificial Intelligence 14(1), 39–60 (2002b); reprinted in: Ford, K., Glymour, C., Hayes, P. (eds.) Thinking About Android Epistemology. MIT Press (March 2006)

---

[4] IBM's Deep Blue took on the challenge of a machine playing chess at the Grandmaster level. IBM's Watson (with DeepQA technology) took on the challenge of competing in the game show *Jeopardy!*

7. Sterrett, S.G.: Nested Algorithms and The Original Imitation Game Test: A Reply to James Moor. Minds and Machines 12, 131–136 (2002a)
8. Sterrett, S.G.: Turing's Two Tests for Intelligence. Minds and Machines 10, 541–559 (2000); reprinted in: Moor, J.H. (ed.) The Turing Test: The Elusive Standard of Artificial Intelligence. Kluwer Academic (2003)
9. Turing, A.M.: Computing machinery and intelligence. Mind 59, 433–460 (1950)
10. Turing, A.M.: Intelligent Machinery. In: Turing, A.M., Ince, D.C. (ed.) Mechanical Intelligence, Collected Works, pp. 107–127. North Holland (1948/1992)
11. Wikipedia contributors. To Tell the Truth. Wikipedia, The Free Encyclopedia (January 27, 2012), Web (January 27, 2012)

Is Turing's Thesis the Consequence of a More General Physical Principle?

Matthew P. Szudzik

Carnegie Mellon University, Education City, PO Box 24866, Doha, Qatar
mszudzik@cmu.edu

Abstract. We discuss historical attempts to formulate a physical hypothesis from which Turing's thesis may be derived, and also discuss some related attempts to establish the computability of mathematical models in physics. We show that these attempts are all related to a single, unified hypothesis.

1   Introduction

Alan Turing [14] proposed the concept of a computer—that is, the concept of a mechanical device that can be programmed to perform any conceivable calculation—after studying the processes that humans use to perform calculations. In particular, he claimed that any function of non-negative integers which can be effectively calculated by humans is a function that can be calculated by a Turing machine. This claim, known as *Turing's thesis*, is an empirical principle that has withstood many tests to its validity.[1] But why does Turing's thesis seem to be true? Can it be derived, for example, from a principle of contemporary sociology, from a principle of human biology, or perhaps from a principle of fundamental physics?

We shall be concerned with the last of these questions, namely the question of whether Turing's thesis is the consequence of a physical principle. But before discussing some of the historically important attempts to answer this question, let us introduce the following formalism. Given a physical system, define a *deterministic physical model* for the system to be

1. a set S of *states*

together with

2. a set A of functions from S to the real numbers, and
3. for each non-negative real number t, a function Φ_t from S to S.

If the system begins in state s, then we understand $\Phi_t(s)$ to be the state of the system after t units of time. Furthermore, we identify each member α of

---

[1] Indeed, every time a software developer attempts to write a computer program to implement an explicitly-described procedure, Turing's thesis is tested.

S.B. Cooper, A. Dawar, and B. Löwe (Eds.): CiE 2012, LNCS 7318, pp. 714–722, 2012.

A with an *observable quantity* of the system,[2] and we consider $\alpha(s)$ to be the *value predicted* for the observable quantity when the system is in state *s*. For example, the following is a deterministic physical model for a particle that is moving uniformly along a straight line with a velocity of 3 meters per second.

Model 1. Let *S* be the set of all real numbers and define $\Phi_t(x) = x + 3t$ for all states *x* in *S*, where *t* is the time measured in seconds. The position of the particle on the line, measured in meters, is given by the function $\alpha(x) = x$.

We say that a deterministic physical model is *faithful* if and only if there is a state s_0 such that, for each time *t*, the actual values of the observable quantities at that time agree with the values that are predicted when the system is in state $\Phi_t(s_0)$. Deterministic physical models are commonly studied in the theory of dynamical systems.

Now, the first notable attempt to derive Turing's thesis from a principle of physics appears to have been made by Robert Rosen [12]. Rosen hypothesized[3] that every physical system has a faithful deterministic physical model where

1. the set *S* is the set of all non-negative integers,
2. each α in *A* is a total recursive function,[4] and
3. when restricted to non-negative integer times *t*, $\Phi_t(s)$ is a total recursive function of *s* and *t*.

Note that since recursive functions are functions from non-negative integers to non-negative integers, Condition 2 implies that the values of all observable quantities are non-negative integers. This hypothesis is justified by the fact that actual physical measurements have only finitely many digits of precision. For example, a distance measured with a meterstick is a non-negative integer multiple *m* of the length of the smallest division on the meterstick. Therefore, without loss of generality, we can consider *m* to be the value of the measurement. Nevertheless, Condition 2 does not forbid time from being measured with a non-negative real number, since time is not regarded as an observable quantity in deterministic physical models. For example, it is consistent with Rosen's hypotheses to specify that

$$\Phi_t(s) = \Phi_{\lfloor t \rfloor}(s) \tag{1}$$

for all non-negative real numbers *t* and for all states *s*, where $\lfloor t \rfloor$ denotes the largest integer less than or equal to *t*.

Now, Turing's thesis can be derived from Rosen's hypotheses as follows. First, note that by definition, if a function ψ is *effectively calculable* then there is a

---

[2] To ensure that the predictions of the model are unambiguous, it is common practice in physics to define each observable quantity operationally [1].

[3] Rosen did not fully formalize his hypotheses. The hypotheses given here are reasonable formal interpretations of Rosen's "Hypothesis I" [12].

[4] We shall use terminology from the theory of recursive functions [11] throughout this paper. It follows from Turing's work [15] that the set of all functions from non-negative integers to non-negative integers that are computable by Turing machines is identical to the set of recursive functions.

physical system that can reliably be used to calculate $\psi(x)$ for every non-negative integer x. This means that there must be a system where the input can be observed, where the output can be observed, and where it is possible to observe that the system is finished with the calculation. In the context of deterministic physical models this means that the system has observable quantities α, β, and γ, respectively, such that for each non-negative integer x

1. there exists a state s and a time u such that $\alpha(s) = x$ and $\gamma(\Phi_t(s)) = 1$ for all $t \geq u$, and
2. if $\alpha(s) = x$ for any state s then $\beta(\Phi_t(s)) = \psi(x)$ where t is any time such that $\gamma(\Phi_t(s)) = 1$.

It then immediately follows from Rosen's hypotheses that every effectively calculable function is recursive, whether it is calculated by a human being or by any other physical system. That is, Turing's thesis is a consequence of Rosen's hypotheses.

Informally, Rosen's hypotheses can be understood as stating that the universe is discrete, deterministic, and computable. It is important then to ask whether we are living in a discrete, deterministic, computable universe. Unfortunately, according to the currently-understood laws of physics, the answer appears to be "No." The universe, as we currently understand it, does not seem to satisfy Rosen's hypotheses. Nevertheless, it may be possible to reformulate the currently-understood physical laws so that Rosen's hypotheses are satisfied. This sort of reformulation was first attempted by Konrad Zuse [18,19] in the 1960's. Since then, increasingly sophisticated attempts have been made by Edward Fredkin [3] and Stephen Wolfram [17]. In contrast, Roger Penrose [9,10] has speculated that the universe might not be computable, but efforts to find experimental evidence for this assertion have not succeeded. There is, in fact, very little that can currently be said. The question of whether Turing's thesis is the consequence of a valid physical principle is too difficult to be answered conclusively at this time.

2   Physical Models

Although deterministic physical models are widely used in the study of dynamical systems and have important real-world applications, they are not the only sort of model that is useful in physics. The sorts of models used in quantum electrodynamics [2], for example, are necessarily non-deterministic. And it is difficult to reconcile the way that time is modeled in general relativity [4] with the way that it is treated as a linear quantity external to the states in a deterministic physical model. Anyone who has attempted to reformulate the laws of physics so that Rosen's hypotheses are satisfied has had to grapple with these obstacles, and no one has had complete success.

For this reason, we have proposed [13] a more general sort of model which we simply call a *physical model*. A physical model for a system is a set S of states together with a set A of functions from S to the real numbers. Each member α of A is identified with an observable quantity of the system,[5] and $\alpha(s)$ is the value predicted for that observable quantity when the system is in state s. Deterministic physical models are special sorts of physical models. For example, the deterministic physical model that was described in the introduction (Model 1) can be expressed as the following physical model.

Model 2. Let S be the set of all triples (x, x_0, t) of real numbers such that $t > 0$ and $x = x_0 + 3t$. The position of the particle on the line, measured in meters, is given by the function $\alpha(x, x_0, t) = x$. The initial position of the particle (for example, recorded in the observer's notebook) is given by the function $\beta(x, x_0, t) = x_0$. The time, measured in seconds, is given by the function $\gamma(x, x_0, t) = t$.

In contrast to Model 1, time is treated as an observable quantity in this model, as is the initial position. The inclusion of these observable quantities in the model can be justified physically by noting that if a researcher were to test Model 1 in a laboratory experiment, he would be required to measure the initial position x_0 of the particle and the position x of the particle at some later time t. That is, besides the position x, observations of both the time and the initial position are fundamental to the system.

Physical models can also be used to describe non-deterministic systems. For example, suppose that one atom of the radioactive isotope nitrogen-13 is placed inside a detector at time $t = 0$. We say that the detector has status 1 at time t if the decay of the isotope was detected at any earlier time, and we say that the detector has status 0 otherwise. We use a non-negative integer to represent the history of the detector. In particular, if b_i is the status of the detector at time $t = i$, then the history of the detector at time $t = n$ is $h = \sum_{i=1}^{n} b_i 2^{n-i}$. Note that when h is written as an n-bit binary number, the ith bit (counting from left to right) is b_i. For example, if the isotope decays sometime between $t = 2$ and $t = 3$, then the history of the detector at time $t = 5$ is 7 because the binary representation of 7 is $(00111)_2$. Now, if the history of the detector is displayed on a computer screen, then the following is a physical model for the history of the detector as observed by a researcher looking at the screen.

Model 3. Let S be the set of all triples $(t, 2^d - 1, k)$ where d, t, and k are non-negative integers such that $d \leq t$, $t \neq 0$, and $2k \leq 2^d - 1$. The history of the detector is given by the function $\alpha(t, h, k) = h$, and the time, measured in units of the half-life of the isotope, is given by the function $\beta(t, h, k) = t$.

In a deterministic model such as Model 2, the state of the system at a particular time is uniquely determined from its initial state. But this is not the case for Model 3, which is a non-deterministic physical model. For example, if we are given as an initial condition that the detector had status 0 at time $t = 1$, then there are four different states at time $t = 3$ which are consistent with that initial condition.

$$(3, (000)_2, 0) \qquad (3, (001)_2, 0) \qquad (3, (011)_2, 0) \qquad (3, (011)_2, 1)$$

---

[5] As before, we require that each observable quantity be defined operationally.

In addition, Model 3 has the property that if these states are considered to be equally likely, then the corresponding probability of observing a given history at time $t = 3$ matches the probability predicted by the conventional theory of radioactive decay. For example, since the history $(011)_2$ is observed in half of the four states, there is a $\frac{1}{2}$ probability that the detector's history will be $(011)_2$ at time $t = 3$ if the detector had status 0 at time $t = 1$. See [13] for a more complex example of a non-deterministic physical model that involves incompatible quantum measurements.

3   Computable Physical Models

Define a *computable physical model* to be a physical model where S is a recursive set of non-negative integers and where each α in A is a total recursive function.[6] Of course, deterministic physical models that satisfy Rosen's hypotheses can all be expressed as computable physical models. Non-deterministic models, such as the model for radioactive decay (Model 3), can also be expressed as computable physical models. Now, we assert the following hypothesis.

Computable Universe Hypothesis. The laws of physics can be expressed as a computable physical model.

To show that Turing's thesis is a consequence of the computable universe hypothesis, first recall that if a function ψ is *effectively calculable* then there is a physical system that can reliably be used to calculate $\psi(x)$ for every non-negative integer x. This means that in some state of the universe it must be possible to observe a record of the the function's input, to observe the function's output, and to observe that the calculation has finished. In the context of a physical model for the universe, this means that there are observable quantities α, β, and γ, respectively, such that for each non-negative integer x

1. there exists a state s such that $\alpha(s) = x$ and $\gamma(s) = 1$, and
2. if $\alpha(s) = x$ and $\gamma(s) = 1$ for any state s, then $\beta(s) = \psi(x)$.

It then immediately follows from the computable universe hypothesis that every effectively calculable function is recursive, whether it is calculated by a human being or by any other system that is governed by physical law. Hence, Turing's thesis is a consequence of the computable universe hypothesis.

Of course, as was the case with Rosen's hypotheses, it is not known whether the computable universe hypothesis is true. But since the computable universe hypothesis allows for non-determinism and for more complex temporal relationships, it may be somewhat easier to reformulate the currently-understood physical laws so as to satisfy this less-restrictive hypothesis.

We conclude this section by considering the following example that helps to clarify certain features of our definition of effective calculability in the context of non-deterministic computable physical models. First, let $\langle x, y \rangle$ denote a

---

[6] See the isomorphism theorems in [13] for other, equivalent definitions of a computable physical model.

non-negative integer that encodes the pair (x, y) of non-negative integers. For example, using Cantor's pairing function, we could define

$$\langle x, y \rangle = \frac{1}{2} \left(x^2 + 2xy + y^2 + 3x + y \right) \ . \tag{2}$$

Triples (x, y, z) of non-negative integers can then be encoded as $\langle \langle x, y \rangle, z \rangle$. We shall use $\langle x, y, z \rangle$ as an abbreviation for $\langle \langle x, y \rangle, z \rangle$. Also define $x \bmod 2$ to be the rightmost bit in the binary representation of the non-negative integer x. Now consider the following computable physical model for the history of a 'noisy' detector.

Model 4. Let S be the set of all pairs $\langle t, h \rangle$ where t and h are non-negative integers such that $t \neq 0$ and $h \leq 2^t - 1$. The time is given by the function $\alpha \langle t, h \rangle = t$, the current status of the detector is given by the function $\beta \langle t, h \rangle = h \bmod 2$, a trivial observable quantity is given by the function $\gamma \langle t, h \rangle = 1$, and the history of the detector is given by the function $\delta \langle t, h \rangle = h$.

Although superficially similar to Model 3, all histories are possible in Model 4. That is, there are no restrictions on the status of the detector. At any point in time and regardless of the detector's past history, the status of the detector can be 1 or 0. Now imagine forming a tree by taking each state $\langle t, h \rangle$ of Model 4 as a node of the tree, and by taking the children of this node to be those states of the form $\langle t + 1, h' \rangle$ where the first t bits of h' agree with the first t bits of h. We call this the *tree of histories* for Model 4, and every branch on this tree corresponds to an alternate sequence of histories for the detector. Moreover, we can think of the status of the detector as defining a function along each branch of the tree. For each state s on the branch, the input of the function is $\alpha (s)$, the corresponding output is $\beta (s)$, and the fact that the calculation has finished is indicated by $\gamma (s)$. But since all histories are possible for this detector, every possible function ψ from the positive integers to the set $\{0, 1\}$ is calculated by the detector along some branch, including functions ψ which are not recursive (that is, not computable by a Turing machine). Therefore, Model 4 is an example of a computable physical model where non-recursive functions may be calculated along certain branches of the tree of histories. It is important to note, though, that according to our definition of effective calculability, these non-recursive functions are not effectively calculable, since for each state s of Model 4 there is another state s' such that $\alpha (s) = \alpha (s')$ and $\gamma (s) = \gamma (s')$, but $\beta (s) \neq \beta (s')$. In other words, the detector cannot reliably be used to calculate ψ because the detector is behaving non-deterministically.

4   Continuous Models

Although Turing's thesis holds in all discrete, deterministic, computable universes, Turing himself did not believe that the universe is discrete. In particular, Turing [16] stated that digital computers ... may be classified amongst the 'discrete state machines'. These are the machines which move by sudden jumps or clicks from one quite definite state to another. These states are sufficiently different for the possibility of confusion between them to be ignored. Strictly speaking there are no such machines. Everything really moves continuously.

But discreteness is not a prerequisite for computability. In fact, Georg Kreisel [5] has hypothesized that the universe may be continuous and computable.

Before we turn to a discussion of Kreisel's hypothesis, define a *functional physical model* for a system to be a set D of finitely many real-valued functions, each of which takes k real numbers as input, for some non-negative integer k. We identify each member δ of D with an observable quantity of the system, and also identify each of the k inputs with an observable quantity. The observable quantities that are identified with the inputs are said to be the *given quantities* for the model,[7] and the observable quantities that are identified with the members of D are said to be the *predicted quantities* of the model. We say that a functional physical model is *faithful* if and only if the values of the predicted quantities in the model match the values that are actually measured whenever the values of the given quantities in the model match the values that are actually measured.[8]

For example, the deterministic physical model that was described in the introduction (Model 1) can be expressed as the following functional physical model.

Model 5. Let the position of the particle on the line, measured in meters, be given by the function $\delta(x_0, t) = x_0 + 3t$. The initial position of the particle is x_0. The time, measured in seconds, is t.

In this case, the position of the particle is the only predicted quantity. The particle's initial position and the time are the given quantities.

Now, note that every integer i can be encoded as a non-negative integer $\zeta(i)$, where $\zeta(i) = 2i$ if $i \geq 0$ and where $\zeta(i) = -2i - 1$ if $i < 0$. Each rational number $\frac{a}{b}$ in lowest-terms with $b > 0$ can be encoded as a non-negative integer $\rho\left(\frac{a}{b}\right)$ where

$$\rho\left(\frac{a}{b}\right) = \zeta\left((\operatorname{sgn} a)\, 2^{\zeta(a_1 - b_1)} 3^{\zeta(a_2 - b_2)} 5^{\zeta(a_3 - b_3)} 7^{\zeta(a_4 - b_4)} 11^{\zeta(a_5 - b_5)} \cdots\right) \quad (3)$$

and where

$$a = (\operatorname{sgn} a)\, 2^{a_1} 3^{a_2} 5^{a_3} 7^{a_4} 11^{a_5} \cdots \quad (4)$$

---

[7] To ensure that each function δ in D is defined for all real number inputs, the scale of each given quantity should be adjusted so that it ranges over all real numbers. For example, if a given quantity is a temperature and if δ is undefined for temperatures below 0 degrees Kelvin, then a logarithmic temperature scale should be used instead of the Kelvin scale.

[8] In practice, of course, measurement errors will prevent us from knowing these values exactly.

and
$$b = 2^{b_1} 3^{b_2} 5^{b_3} 7^{b_4} 11^{b_5} \cdots \tag{5}$$
are the prime factorizations of a and b, respectively. For each pair of rational numbers q and r, define $(q\,;r)$ to be the non-negative integer $\langle \rho\,(q)\,, \rho\,(r) \rangle$. We shall use $(q\,;r)$ to represent the open interval with endpoints q and r.

Next, let $\mathbb{N}$ denote the set of non-negative integers and let $\mathbb{R}$ denote the set of real numbers. Note that every real number x can be represented as a nested sequence of open intervals whose intersection is x. We say that a function $\phi : \mathbb{N} \to \mathbb{N}$ is a *nested oracle* for $x \in \mathbb{R}$ if and only if $\phi\,(0) = (a_0\,;b_0)$, $\phi\,(1) = (a_1\,;b_1)$, $\phi\,(2) = (a_2\,;b_2)$, ... is a sequence of nested intervals whose intersection is x. A function $\delta : \mathbb{R} \to \mathbb{R}$ is said to be *computable* (in the sense defined by Lacombe [6,7,8]) if and only if there is a total recursive function ξ such that if ϕ is a nested oracle for $x \in \mathbb{R}$, then $\lambda m\,[\xi\,(\phi\,(m))]$ is a nested oracle for $\delta(x)$. This definition can naturally be extended to functions from $\mathbb{R}^k$ to $\mathbb{R}$, for any $k \in \mathbb{N}$. Finally, we say that a functional physical model is *computable* if and only if every member of the set D is computable in the sense that we have just described.

Let us now return to a discussion of Kreisel's hypothesis. Kreisel hypothesized that every faithful functional physical model is computable. For example, Model 5 is computable because in that model the function $\delta : \mathbb{R}^2 \to \mathbb{R}$ is computed by the total recursive function ξ that is defined so that
$$\xi\,((a\,;b)\,,(c\,;d)) = (a + 3c\,;b + 3d) \tag{6}$$
for all rational numbers a, b, c, and d. Moreover, every computable functional physical model can be expressed as a computable physical model. For example, Model 5 can be expressed as the following computable physical model.

Model 6. Let S be the set of all triples $\langle (a\,;b)\,, (c\,;d)\,, (a + 3c\,;b + 3d) \rangle$ where a, b, c, and d are rational numbers such that $a < b$ and $c < d$. The initial position of the particle on the line, represented as a range of positions measured in meters, is given by the function $\alpha\,\langle (a\,;b)\,, (c\,;d)\,, (e\,;f) \rangle = (a\,;b)$. The time interval, measured in seconds, is given by the function $\beta\,\langle (a\,;b)\,, (c\,;d)\,, (e\,;f) \rangle = (c\,;d)$. The position of the particle, represented as a range of positions measured in meters, is given by the function $\gamma\,\langle (a\,;b)\,, (c\,;d)\,, (e\,;f) \rangle = (e\,;f)$.

Note that the states of Model 6 are all of the form
$$\langle (a\,;b)\,, (c\,;d)\,, \xi\,((a\,;b)\,, (c\,;d)) \rangle \ , \tag{7}$$
and this guarantees that if ϕ is a nested oracle for some initial position x_0 and if ψ is a nested oracle for some time t, then there is a unique sequence of states s_0, s_1, s_2, ... in S such that $\alpha\,(s_m) = \phi\,(m)$ and $\beta\,(s_m) = \psi\,(m)$ for all $m \in \mathbb{N}$. Therefore,
$$\gamma\,(s_m) = \xi\,(\phi\,(m)\,, \psi\,(m)) \tag{8}$$
for all $m \in \mathbb{N}$, and $\lambda m\,[\gamma\,(s_m)]$ is a nested oracle for the position $\delta\,(x_0, t)$ that is predicted by Model 5. Thus, there is a direct correspondence between the functional physical model (Model 5) and the computable physical model (Model 6). See [13] for more information regarding this correspondence.

In summary, the computable physical models comprise a very general class of models, capable of expressing discrete deterministic models such as those studied by Rosen, non-deterministic models such as the model for radioactive decay, and continuous models such as those studied by Kreisel. Furthermore, Turing's thesis is a consequence of the hypothesis that the laws of physics can be expressed as a computable physical model. It is very tempting, therefore, to wonder whether this hypothesis might be true.

References

1. Bridgman, P.W.: The Logic of Modern Physics. Macmillan (1927)
2. Feynman, R.P.: QED: The Strange Theory of Light and Matter. Princeton University Press (1985)
3. Fredkin, E.: Digital mechanics. Physica D 45, 254–270 (1990)
4. Gödel, K.: An example of a new type of cosmological solutions of Einstein's field equations of gravitation. Reviews of Modern Physics 21, 447–450 (1949)
5. Kreisel, G.: A notion of mechanistic theory. Synthese 29, 11–26 (1974)
6. Lacombe, D.: Extension de la notion de fonction récursive aux fonctions d'une ou plusieurs variables réelles I. Comptes Rendus Hebdomadaires des Séances de l'Académie des Sciences 240, 2478–2480 (1955)
7. Lacombe, D.: Extension de la notion de fonction récursive aux fonctions d'une ou plusieurs variables réelles II. Comptes Rendus Hebdomadaires des Séances de l'Académie des Sciences 241, 13–14 (1955)
8. Lacombe, D.: Extension de la notion de fonction récursive aux fonctions d'une ou plusieurs variables réelles III. Comptes Rendus Hebdomadaires des Séances de l'Académie des Sciences 241, 151–153 (1955)
9. Penrose, R.: The Emperor's New Mind. Oxford University Press (1989)
10. Penrose, R.: Shadows of the Mind. Oxford University Press (1994)
11. Rogers, Jr., H.: Theory of Recursive Functions and Effective Computability. McGraw-Hill (1967)
12. Rosen, R.: Church's thesis and its relation to the concept of realizability in biology and physics. Bulletin of Mathematical Biophysics 24, 375–393 (1962)
13. Szudzik, M.P.: The computable universe hypothesis. In: Zenil, H. (ed.) A Computable Universe. World Scientific (forthcoming 2012)
14. Turing, A.M.: On computable numbers, with an application to the Entscheidungsproblem. Proceedings of the London Mathematical Society 42, 230–265 (1937)
15. Turing, A.M.: Computability and λ-definability. The Journal of Symbolic Logic 2, 153–163 (1937)
16. Turing, A.M.: Computing machinery and intelligence. Mind 59, 433–460 (1950)
17. Wolfram, S.: A New Kind of Science. Wolfram Media (2002)
18. Zuse, K.: Rechnender raum. Elektronische Datenverarbeitung 8, 336–344 (1967)
19. Zuse, K.: Rechnender Raum. Friedrich Vieweg & Sohn (1969), see [20] for an English translation
20. Zuse, K.: Calculating space. Tech. Rep. AZT-70-164-GEMIT, Massachusetts Institute of Technology, Project MAC (1970)

Some Natural Zero One Laws
for Ordinals Below ε_0

Andreas Weiermann and Alan R. Woods

Vakgroep Wiskunde, Universiteit Gent, Krijgslaan 281 S22, 9000 Ghent, Belgium
Andreas.Weiermann@UGent.be

Abstract. We are going to prove that every ordinal α with $\varepsilon_0 > \alpha \geq \omega^\omega$ satisfies a natural zero one law in the following sense. For $\alpha < \varepsilon_0$ let $N\alpha$ be the number of occurences of ω in the Cantor normal form of α. ($N\alpha$ is then the number of edges in the unordered tree which can canonically be associated with α.) We prove that for any α with $\omega^\omega \leq \alpha < \varepsilon_0$ and any sentence φ in the language of linear orders the limit $\delta_\varphi(\alpha) = \lim_{n\to\infty} \frac{\#\{\beta<\alpha:(\beta,\in)\models\varphi \,\wedge\, N\beta=n\}}{\#\{\beta<\alpha:N\beta=n\}}$ exists and that $\delta_\varphi(\alpha) \in \{0,1\}$. We further show that for any such sentence φ the limit $\delta_\varphi(\varepsilon_0)$ exists although this limit is in general in between 0 and 1. We also investigate corresponding asymptotic densities for ordinals below ω^ω.

1 Introduction

This paper concerns logical limit laws for infinite ordinals. It is based on methods and techniques from the theory of logical limit laws for classes of finite structures and some ingredients from the theory of linear orders. We heavily use the machinery (which to a large extent goes back to pioneering work of Compton) developed in the book by Burris [2].

In 2001 the first author (after having read [2] and having recognized that ordinals below ε_0 form a natural additive number system) discussed the possibility of limit laws for ordinals with Compton at an AOFA-workshop in Tatihoo and Compton suggested among other things to contact the second author because of his results on limit laws for trees. This led to a fruitful interaction over the years. Tragically, the second author passed away at the age of 58 unexpectedly in December 2011 and the first author feels responsible to make publically available the beautiful results which so far have been achieved (partially funded by DFG and the John Templeton Foundation).

Technically this article is based on a mixture of results from [4] (which when compared with the results from this article have preliminary nature) and [7]. Also some basic techniques from [2,3] (and implicitly [1]) are used.

More elaborate results of the authors (concerning larger ordinal segments and more general languages, like the second order monadic ones) on limit laws for ordinals are sketched at the end and will hopefully be treated at later occasions.

S.B. Cooper, A. Dawar, and B. Löwe (Eds.): CiE 2012, LNCS 7318, pp. 723–732, 2012.
© Springer-Verlag Berlin Heidelberg 2012

2   Some Basic Results

For two linear orders A and B let $G_n(A, B)$ be the following two person game, also well known as Ehrenfeucht Fraïsse game, between, lets say, Bob and Alice. A play consists of an ordered sequence of n repetitions of the following: Bob chooses an element of A or B and Alice chooses an element of the other. The element of A selected at the t-th turn is denoted a_t and the element of B selected at the t-th turn is denoted b_t. We say that Alice has won the game if for each $s, t \leq n$ the assertion $a_s < a_t \pmod{A}$ is equivalent with $b_s < b_t \pmod{B}$. Otherwise we say that Bob has won. As usual we define what a winning strategy for Alice is and we say $A \sim_n B$ iff Alice has a winning strategy for $G_n(A, B)$. (We took this standard exposition from page 99 in Rosenstein's classical text book [3] on linear orders.) Let $\mathcal{L}$ be the usual first order language of linear orders. Then $\alpha \sim_n \beta$ yields that α and β model exactly the same $\mathcal{L}$-sentences φ of quantifier rank not exceeding n.

Lemma 1. *Fix a natural number $n \geq 1$. Let $(a_i)_{i \leq n}$ and $(a'_i)_{i \leq n}$ be sequences of natural numbers such that for all $i \leq n$: $(a_i \geq 2^n \iff a'_i \geq 2^n)$ and $(a_i < 2^n \Rightarrow a_i = a'_i)$.*
   Then we have

$$\omega^n \cdot a_n + \cdots + \omega^0 \cdot a_0 \sim_n \omega^n \cdot a'_n + \cdots + \omega^0 \cdot a'_0. \tag{1}$$

Moreover for such sequences $(a_i)_{i \leq n}$ and $(a'_i)_{i \leq n}$ and for any pair of non zero ordinals α and β we have

$$\omega^{n+1} \cdot \alpha + \omega^n \cdot a_n + \cdots + \omega^0 \cdot a_0 \sim_n \omega^{n+1} \cdot \beta + \omega^n \cdot a'_n + \cdots + \omega^0 \cdot a'_0. \tag{2}$$

Proof. We have only to show assertion (1) (but we still need assertion (2) for applying the induction hypothesis). For, we have $\omega^{n+1} \cdot \alpha \sim_{2n+2} \omega^{n+1}$ and $\omega^{n+1} \cdot \beta \sim_{2n+2} \omega^{n+1}$ by assertion (1) of Theorem 6.18 in [3]. Assertion (1) above and closure of $\sim_n$ under sums (i.e., assertion (1) of Lemma 6.5 of [3]) together with downward conservativity (cf., e.g., Lemma 6.4 of [3]) then yield $\omega^{n+1} \cdot \alpha + \omega^n \cdot a_n + \cdots + \omega^0 \cdot a_0 \sim_n \omega^{n+1} \cdot \beta + \omega^n \cdot a'_n + \cdots + \omega^0 \cdot a'_0$. For proving assertion (1) we proceed by induction on n. The induction starts with $n = 1$ for which the assertion holds. Now consider

$$\gamma = \omega^n \cdot a_n + \cdots + \omega^0 \cdot a_0$$

and

$$\delta = \omega^n \cdot a'_n + \cdots + \omega^0 \cdot a'_0.$$

We are going to apply Theorem 6.6 in [3] i.e., we verify the Fraïsse conditions for games on linear orderings. So we have to show that for every splitting of γ (δ) into an initial and final segment there is a splitting of δ (γ) into corresponding initial and final segments for which the second player wins the corresponding games on initial and final segments separately with $n - 1$ moves.

Assume without loss of generality that Player one picks $\xi < \gamma$. (In case that he picks $\xi < \delta$ the argument is symmetrical.) Assume that

$$\xi = \omega^n \cdot a_n + \cdots + \omega^{i+1} \cdot a_{i+1} + \omega^i \cdot b_i + \cdots + \omega^0 \cdot b_0$$

and $b_i < a_i$.

Assume first that $a_i \geq 2^n$.

In this case also $a_i' \geq 2^n$. If $b_i < 2^{n-1}$ let $b_i' := b_i$. If If $b_i \geq 2^{n-1}$ and $a_i - b_i < 2^{n-1}$ let $b_i' := a_i' - (a_i - b_i)$ ensuring $b_i' \geq 2^{n-1}$. If $b_i \geq 2^{n-1}$ and $a_i - b_i \geq 2^{n-1}$ then let $b_i' := 2^{n-1}$. Let

$$\xi' := \omega^n \cdot a_n' + \cdots + \omega^{i+1} \cdot a_{i+1}' + \omega^i \cdot b_i' + \cdots + \omega^0 \cdot b_0$$

and let Player two play ξ'. By induction hypothesis applied to (2) and assertion (1) of Lemma 6.5 in [3] we obtain $\xi \sim_{n-1} \xi'$. So the initial segments determined by ξ and ξ' are $\sim_n$ equivalent and it remains that the final segments in γ and δ are also $\sim_n$ equivalent. To prove this let

$$X := \{\eta : \xi < \eta < \gamma\}$$

and

$$Y := \{\eta : \xi' < \eta < \delta\}.$$

For a set X of ordinals let as usual $\mathrm{otype}(X)$ denote its order type, i.e., the ordinal which is order isomorphic to it. Then

$$\mathrm{otype}(X) = \omega^i \cdot (a_i - b_i) + \omega^{i-1} \cdot a_{i-1} + \cdots + \omega^0 \cdot a_0.$$

If $b_i < 2^{n-1}$ then

$$\mathrm{otype}(Y) = \omega^i \cdot (a_i' - b_i) + \omega^{i-1} \cdot a_{i-1}' + \cdots + \omega^0 \cdot a_0'$$

where $a_i - b_i \geq 2^{n-1}$ since $a_i' \geq 2^n$.

If $b_i \geq 2^{n-1}$ and $a_i - b_i < 2^{n-1}$ then

$$\mathrm{otype}(Y) = \omega^i \cdot (a_i' - (a_i' - (a_i - b_i))) + \omega^{i-1} \cdot a_{i-1}' + \cdots + \omega^0 \cdot a_0'$$
$$= \omega^i \cdot (a_i - b_i) + \omega^{i-1} \cdot a_{i-1}' + \cdots + \omega^0 \cdot a_0'.$$

If $b_i \geq 2^{n-1}$ and $a_i - b_i \geq 2^{n-1}$ then

$$\mathrm{otype}(Y) = \omega^i \cdot (a_i' - 2^{n-1}) + \omega^{i-1} \cdot a_{i-1}' + \cdots + \omega^0 \cdot a_0'$$

where $a_i' - 2^{n-1} \geq 2^{n-1}$ since $a_i' \geq 2^n$. By induction hypothesis applied to (2) and closure of $\sim_{n-1}$ under sum (assertion (1) of Lemma 6.5 in [3]) we obtain $X \sim_{n-1} Y$.

Assume for the second case that $a_i < 2^n$. In this case by assumption on the a_i and a_i' we have necessarily $a_i' = a_i$. Let

$$\xi' := \omega^n \cdot a_n' + \cdots + \omega^{i+1} \cdot a_{i+1}' + \omega^i \cdot b_i + \cdots + \omega^0 \cdot b_0$$

be the response of Player two. To apply Theorem 6.6 of [3] we split as in the first case γ and δ into initial and final segments. By induction hypothesis applied to (2) and assertion (1) of Lemma 6.5 in [3] we obtain $\xi \sim_{n-1} \xi'$. As before let

$$X := \{\eta : \xi < \eta < \gamma\}$$

and

$$Y := \{\eta : \xi' < \eta < \delta\}.$$

$$\mathrm{otype}(X) = \omega^i \cdot (a_i - b_i) + \omega^{i-1} \cdot a_{i-1} + \cdots + \omega^0 \cdot a_0$$

and

$$\mathrm{otype}(Y) = \omega^i \cdot (a_i' - b_i) + \omega^{i-1} \cdot a_{i-1}' + \cdots + \omega^0 \cdot a_0'.$$

By induction hypothesis applied to (2) and closure of $\sim_{n-1}$ under sum we obtain $X \sim_{n-1} Y$. The assertion now follows from Theorem 6.6 of [3].

Let us define for each $\alpha < \varepsilon_0$ the norm of α, $N\alpha$, as follows by recursion on α. $N0 := 0$ and $N\alpha := n + N\alpha_1 + \cdots + N\alpha_n$ if α has Cantor normal form $\alpha = \omega^{\alpha_1} + \cdots + \omega^{\alpha_n}$. Then $N\alpha$ is the number of occurrences of ω in the Cantor normal form of α. If we associate with α an unordered tree in the canonical way (see the proof of Theorem 2) then $N\alpha$ is the number of edges in the tree representation of α and so N is in fact a canonical norm function on the ordinals below ε_0. For a given $\alpha < \varepsilon_0$ and $n < \omega$ there will only be finitely many $\xi < \alpha$ with $N\xi = n$. For a finite set M we denote its cardinality by $\#M$. The ordinal ε_0 is as usual the least ordinal ξ such that $\xi = \omega^\xi$.

For $\alpha < \varepsilon_0$ we therefore may then define

$$c_\alpha(n) = \#\{\xi < \alpha : N\xi = n\}$$

and for an $\mathcal{L}$-sentence φ we may further define

$$c_\alpha^\varphi(n) = \#\{\xi < \alpha : \xi \models \varphi \wedge N\xi = n\}.$$

We further define

$$\delta_\varphi(\alpha) := \lim_{n \to \infty} \frac{c_\alpha^\varphi(n)}{c_\alpha(n)}$$

if this limit exists. We say that α satisfies an $\mathcal{L}$-limit law if $\delta_\varphi(\alpha)$ exists for all φ and we say that α satisfies an $\mathcal{L}$ zero one law if $\delta_\varphi(\alpha)$ exists for all φ and if $\delta_\varphi(\alpha)$ is either 0 or 1. In the sequel we write $\alpha \models \varphi$ as an abbreviation for $(\alpha, \in) \models \varphi$.

Theorem 1. *Let $\varepsilon_0 > \alpha \geq \omega$. Then ω^α satisfies an $\mathcal{L}$ zero one law.*

Proof. Let φ be a sentence of the language of linear orders. Assume that n is the quantifier rank of φ. Let I be the set $\{\{0\}, \{1\}, \ldots, \{2^n - 1\}, \{m : m \geq 2^n\}\}$. Let $A_0, \ldots, A_n$ be a sequence of elements of I. Let

$$P_\alpha(A_0, \ldots, A_n) := \{\omega^{n+1} \cdot \beta + \omega^n \cdot a_n + \cdots + \omega^0 \cdot a_0 < \omega^\alpha : \beta > 0 \wedge (\forall i \leq n)[a_i \in A_i]\}$$

and

$$Q_\alpha(A_0,\ldots,A_n) := \{\omega^n \cdot a_n + \cdots + \omega^0 \cdot a_0 < \omega^\alpha : (\forall i \le n)[a_i \in A_i]\}.$$

Then the union of P_α's and Q_α's taken over all (finitely many) sequences of elements in I gives the set of all ordinals below ω^α. By the Lemma 1 we have $\gamma \sim_n \delta$ for any pair of elements of each of the sets $P_\alpha(A_0,\ldots,A_n)$ and $Q_\alpha(A_0,\ldots,A_n)$. The finite collection of the $P_\alpha(A_0,\ldots,A_n)$ and $Q_\alpha(A_0,\ldots,A_n)$ yields an effective (finite) description of the finitely many equivalence classes for $\sim_n$. Then

$$\{\pi < \omega^\alpha : \pi \models \varphi\} = \bigcup_{A_0,\ldots,A_n \in I : (\exists i A_i \ne \{\emptyset\}) \wedge \exists \xi \in P_\alpha(A_0,\ldots,A_n):\xi \models \varphi} P_\alpha(A_0,\ldots,A_n) \cup$$
$$\bigcup_{A_0,\ldots,A_n \in I : (\exists i A_i \ne \{\emptyset\}) \wedge \exists \xi \in Q_\alpha(A_0,\ldots,A_n):\xi \models \varphi} Q_\alpha(A_0,\ldots,A_n)$$

and we only have to show that the asymptotic density of each $P_\alpha(A_0,\ldots,A_n)$ and $Q_\alpha(A_0,\ldots,A_n)$ is either 0 or 1. Of course the value 1 can only be distributed once. Consider a set $P_\alpha(A_0,\ldots,A_n)$ where some A_i is a singleton containing exactly a_i. Let $\mathcal{P}_{n+1} := \{\omega^\eta : \alpha > \eta \ge n+1\}$. Let $\mathcal{P}_i := \{\omega^i\}$ for $i \le n$. Then $P_\alpha(A_0,\ldots,A_n)$ is a partition set in the sense of Burris [2] and can be written as

$$P(A_0,\ldots,A_n) = (\ge 0)\mathcal{P}_{n+1} + \cdots + a_i \cdot \mathcal{P}_i \cdots.$$

This means that elements of $P(A_0,\ldots,A_n)$ can be written as sums of arbitrary many members of $\mathcal{P}_{n+1}$ and a_i members of each $\mathcal{P}_i$ for $i \le n$. We have that the local count function $c_{\omega^\alpha}(n)$ for ω^α is in RT_1 by [4] and according to Compton's Theorem 4.1 in [2] we obtain that the asymptotic density of $P_\alpha(A_0,\ldots,A_n)$ is zero since a_i is a small index (following the terminology of Definition 3.23 in [2]. If all A_i are non singleton elements then Theorem 4.1 in [2] shows that the asymptotic density of $P_\alpha(A_0,\ldots,A_n)$ is one. Finally since $Q(A_0,\ldots,A_n)$ can be written following Definition 3.25 of [2] as $0 \cdot \mathcal{P}_{n+1} + \cdots$ Theorem 4.1 in [2] yields that the asymptotic density of $Q(A_0,\ldots,A_n)$ is zero since 0 is a small index.

Theorem 2. *For any $\mathcal{L}$ sentence φ the limit $\delta_\varphi(\varepsilon_0)$ exists.*

Proof. The proof starts as the proof of Theorem 1 and continues with Woods' tree theorem.

Let φ be a sentence of the language of linear orders. Assume that n is the quantifier rank of φ. Let I be the set $\{\{0\},\{1\},\ldots,\{2^n-1\},\{m : m \ge 2^n\}\}$. Let $A_0,\ldots,A_n$ be a sequence of elements of I. Let

$$P(A_0,\ldots,A_n) := \{\omega^{n+1}\cdot\beta+\omega^n\cdot a_n+\cdots+\omega^0\cdot a_0 < \varepsilon_0 : \beta > 0 \wedge (\forall i \le n)[a_i \in A_i]\}$$

and

$$Q(A_0,\ldots,A_n) := \{\omega^n \cdot a_n + \cdots + \omega^0 \cdot a_0 < \varepsilon_0 : (\forall i \le n)[a_i \in A_i]\}.$$

Then the union of P's and Q's taken over all (finitely many) sequences of elements in I gives the set of all ordinals below ε_0. By Lemma 1 we have $\gamma \sim_n \delta$ for any pair of elements of fixed set P or a fixed set Q. Thus we have an effective description of the finitely many equivalence classes for $\sim_n$. Then

$$\{\pi < \varepsilon_0 : \pi \models \varphi\} = \bigcup_{A_0,\ldots,A_n \in I:(\exists i A_i \neq \{\emptyset\}) \wedge \exists \xi \in P(A_0,\ldots,A_n):\xi \models \varphi} P_\alpha(A_0,\ldots,A_n) \cup$$

$$\bigcup_{A_0,\ldots,A_n \in I:(\exists i A_i \neq \{\emptyset\}) \wedge \exists \xi \in Q(A_0,\ldots,A_n):\xi \models \varphi} Q(A_0,\ldots,A_n)$$

and we only have to show we only have to show that the asymptotic density of each $P(A_0,\ldots,A_n)$ and $Q(A_0,\ldots,A_n)$ exists. Now consider a fixed $P(A_0,\ldots,A_n)$ (or $Q(A_0,\ldots,A_n)$). To each ordinal α less than ε_0 we may associate canonically a finite rooted non planar tree $T(\alpha)$ as follows. (Non planar refers in contrast to planar to the property that there is no canonical ordering assumed for the immediate subtrees of a given tree.) To 0 we associate the tree consisting of a root. If α has a normal form $\omega^{\alpha_1} + \cdots + \omega^{\alpha_n}$ then we may assume that we have associated trees $T(\alpha_i)$ to the ordinals α_i for $1 \leq i \leq n$. Now let $T(\alpha)$ be the tree consisting of a root and immediate subtrees $T(\alpha_1),\ldots,T(\alpha_n)$. Then $|T(\alpha)|$, the number of nodes of $T(\alpha)$, is $1 + N(\alpha)$. By the finitary character of the description of $P(A_0,\ldots,A_n)$ we may find a sentence ψ in the language of trees such that $P(A_0,\ldots,A_n) = \{\alpha < \varepsilon_0 : T(\alpha) \models \psi\}$. This follows from Lemma 1 and the fact that one can describe trees representing ordinals of the form ω^k by a first order formula in the language of trees. Theorem 1.1 of the second author from [7] yields that for any monadic second order property ψ the limiting distribution probability of the fraction of unlabelled rooted trees with n vertices and which satisfy ψ exists. Therefore

$$\lim_{n \to \infty} \frac{\#\{T : T \models \psi \wedge |T| = n+1\}}{\#\{T : |T| = n+1\}}$$

exists. But this limit is equal to

$$\lim_{n \to \infty} \frac{\#\{\alpha < \varepsilon_0 : T(\alpha) \models \psi \wedge N\alpha = n\}}{\#\{\alpha < \varepsilon_0 : N\alpha = n\}}$$

which is the asymptotic density of $P(A_0,\ldots,A_n)$.

It is clear that we cannot expect a zero one law for ϵ_0. Being a successor is a first order property which has in this case a limiting distribution probability strictly inbetween 0 and 1 as shown in [4].

3   Refinements

We now investigate limit laws for not necessarily additive principal ordinals below ε_0.

Lemma 2. *Let* $\alpha = \omega^\beta \cdot m + \gamma$ *where* $m > 0$ *and* $\gamma < \omega^\beta < \varepsilon_0$. *Let* φ *be an* $\mathcal{L}$*-sentence. Assume that*

$$\delta_\varphi(\omega^\beta \cdot m) = \lim_{n \to \infty} \frac{c^\varphi_{\omega^\beta \cdot m}(n)}{c_{\omega^\beta \cdot m}(n)}$$

exists. Then

$$\lim_{n \to \infty} \frac{c^\varphi_\alpha(n)}{c_\alpha(n)} = \delta_\varphi(\omega^\beta \cdot m).$$

Proof. This Lemma is an easy corollary of the proof of Lemma 3.1 in [4].

$$\frac{c^\varphi_\alpha(n)}{c_\alpha(n)} \geq \frac{c^\varphi_{\omega^\beta \cdot m}(n)}{c_\alpha(n)}$$

$$= \frac{c_{\omega^\beta \cdot m}(n)}{c_\alpha(n)} \cdot \frac{c^\varphi_{\omega^\beta \cdot m}(n)}{c_{\omega^\beta \cdot m}(n)} \to_{n \to \infty} 1 \cdot \delta_\varphi(\omega^\beta \cdot m)$$

by equation (2) in the proof of Lemma 3.1 of [4]. Moreover

$$\frac{c^\varphi_\alpha(n)}{c_\alpha(n)} \leq \frac{c^\varphi_{\omega^\beta \cdot m}(n)}{c_{\omega^\beta \cdot m}(n)} + \frac{\#\{\xi : \omega^\beta \cdot m \leq \xi < \alpha \wedge N\xi = n\}}{c_{\omega^\beta \cdot m}(n)}$$

$$\to_{n \to \infty} \delta_\varphi(\omega^\beta \cdot m) + 0$$

by equation (2) in the proof of Lemma 3.1 of of [4].

Lemma 3. *Assume that* $\alpha < \omega^\omega$ *where* $\alpha = \omega^k \cdot m_k + \cdots + \omega^0 \cdot m_0$ *with* $m_k > 0$. *Then for each* $\ell \in \{0, \ldots, m_k\}$ *there is a sentence* $\varphi \in \mathcal{L}$ *such that*

$$\lim_{n \to \infty} \frac{c^\varphi_\alpha(n)}{c_\alpha(n)} = \frac{\ell}{m_k}.$$

Proof. Let φ describe that there exist exactly $(m_k - \ell)$ k-limit points. (Recall that 1-limit points are just limit points, i.e., ordinals of the form $\omega(1 + \beta)$ and recall that k-limit points are limits of $(k-1)$-limit points. So k-limit points have the form $\omega^k(1 + \beta)$. Obviously being a k-limit point is first order definable property of an ordinal under consideration.)

We are going to prove that

$$\lim_{n \to \infty} \frac{c^\varphi_{\omega^k \cdot m_k}(n)}{c_{\omega^k \cdot m_k}(n)} = \frac{\ell}{m_k}.$$

This yields the claim by Lemma 2. We first see that

$$\{\delta < \omega^k \cdot m_k : \delta \models \varphi \wedge N\delta = n\} =$$
$$\{\delta < \omega^k \cdot m_k : \delta \geq \omega^k \cdot (m_k - \ell) \wedge N\delta = n\} =$$
$$\{\xi < \omega^k \cdot l : N\xi = n - N(\omega^k \cdot (m_k - \ell))\}$$

Thus $c^\varphi_{\omega^k \cdot m_k}(n) = c_{\omega^k \cdot \ell}(n - N(\omega^k \cdot (m_k - \ell))) \sim c_{\omega^k}(n) \cdot \ell$ as shown in the last line of the proof of Lemma 3.1 of [4] of [4]. Moreover $c_{\omega^k \cdot m_k}(n) \sim m \cdot c_{\omega^k}(n)$ and the assertion follows.

Lemma 4. *If $\gamma \geq \omega > m > 0$ then*

$$\lim_{n\to\infty} \frac{c^{\varphi}_{\omega^{\gamma}\cdot m}(n)}{c_{\omega^{\gamma}\cdot m}(n)} = \lim_{n\to\infty} \frac{c^{\varphi}_{\omega^{\gamma}}(n)}{c_{\omega^{\gamma}}(n)} \in \{0,1\}.$$

Proof. Fix φ. Assume that the quantifier rank of φ does not exceed the natural number r. For $2 \leq i \leq m$ let

$$M_i^1(n) := \{\delta < \omega^{\gamma} \cdot i : \delta \models \varphi \wedge N\delta = n \wedge \delta \geq \omega^{\gamma} \cdot (i-1) + \omega^{r+1}\}$$

and

$$M_i^2(n) := \{\delta < \omega^{\gamma} \cdot i : \delta \models \varphi \wedge N\delta = n \wedge \omega^{\gamma} \cdot (i-1) \leq \delta < \omega^{\gamma} \cdot (i-1) + \omega^{r+1}\}$$

Then

$$c^{\varphi}_{\omega^{\gamma}\cdot m}(n) \geq$$
$$\#\{\delta < \omega^{\gamma} \cdot m : \delta \models \varphi \wedge N\delta = n \wedge \delta \geq \omega^r\} +$$
$$\#M_2^1(n) + \#M_2^2(n) + \ldots + \#M_m^1(n) + \#M_m^2(n).$$

By Assertion (1) of Theorem 6.18 in [3] we know that $\omega^r \sim_{2\cdot r} \omega^r \cdot \beta$ for all $\beta > 0$. If $\delta = \omega^{\gamma} \cdot i + \xi$ where $\xi < \omega^{\gamma}$ and $i > 0$ then, since $\gamma \geq \omega$, $\delta = \omega^r \cdot \omega^{\gamma} \cdot i + \xi$ and we have the equivalence: $\delta \models \varphi \iff \omega^r + \xi \models \varphi$. Thus $\delta \mapsto \omega^r + \xi$ gives a projection into ω^{γ} which preserves the validity (invalidity) of φ. Assume first that $\lim_{n\to\infty} \frac{c^{\varphi}_{\omega^{\gamma}}(n)}{c_{\omega^{\gamma}}(n)} = 1$. The remaining case $\lim_{n\to\infty} \frac{c^{\varphi}_{\omega^{\gamma}}(n)}{c_{\omega^{\gamma}}(n)} = 0$ can be treated similarly. Now we have

$$\#M_i^1(n) = \#\{\xi < \omega^{\gamma} : \xi \models \varphi \wedge N\xi = n - (N\gamma + 1) \cdot i\}$$

hence

$$\frac{\#M_i^1(n)}{m \cdot c_{\omega^{\gamma}}(n)}$$

converges to $\frac{1}{m}$. We have

$$\#M_i^2(n) = \#\{\xi < \omega^{r+1} : \xi \models \varphi \wedge N\xi = n - (N\gamma + 1) \cdot i + r + 1\}$$

hence

$$\frac{\#M_i^2(n)}{m \cdot c_{\omega^{\gamma}}(n)}$$

converges to 0. We see that $\liminf_{n\to\infty} \frac{c^{\varphi}_{\omega^{\gamma}\cdot m}(n)}{c_{\omega^{\gamma}\cdot m}(n)} \geq 1$ and the assertion follows.

Corollary 1. *If $\varepsilon_0 > \gamma \geq \omega^{\omega}$ then γ satisfies an $\mathcal{L}$ zero one law.*

Theorem 3. *Assume that $\alpha < \omega^{\omega}$ where $\alpha = \omega^k \cdot m_k + \cdots + \omega^0 \cdot m_0$ with $m_k > 0$. Then for each sentence $\varphi \in \mathcal{L}$ there is an $\ell \in \{0, \ldots, m_k\}$ such that*

$$\lim_{n\to\infty} \frac{c^{\varphi}_{\alpha}(n)}{c_{\alpha}(n)} = \frac{\ell}{m_k}. \tag{3}$$

Proof. By Lemma 2 we may assume without loss of generality that $\alpha = \omega^k \cdot m$. Let φ be given and assume that the rank of φ is r. Let $\mathcal{P}_i := \{\omega^i\}$ for $i \leq k$. Then $\{\delta < \alpha : \delta \models \varphi\}$ can be written as a disjoint union of partition sets (in the sense of Definition 3.25 of [2]) in the form $\bigcup_{A_0,\dots,A_k} A_k \mathcal{P}_k + \cdots + A_0 \mathcal{P}_0$ where A_i is an index in $\{0, 1, \dots, 2^r - 1, (\geq 2^r)\}$. Now A_k has to be finite by the choice of α. If one A_i is finite for some $i < k$ then $A_{k-1}\mathcal{P}_{k-1} + \dots + A_0 \mathcal{P}_0$ is a partition set in ω^k such that by Theorem 4.1 of [2] we have

$$\lim_{n \to \infty} \# \frac{\{\delta \in A_{k-1}\mathcal{P}_{k-1} + \dots + A_0 \mathcal{P}_0 : N\delta = n\}}{c_{\omega^k}(n)} = 0.$$

Thus we may concentrate on $A_k \mathcal{P}_k + (\geq 2^r)\mathcal{P}_{k-1} + \cdots + (\geq 2^r)\mathcal{P}_0$. This set has the same asymptotic density as $A_k \mathcal{P}_k$ and this set has asymptotic density in the desired set according to the shape of A_k. Moreover the resulting finite sums of these densities are also in the desired set of values.

One referee of this paper suggested to report about possible extensions of the results of this paper. We give a short list but proofs will be reported elsewhere.

1. Let $(p_i)_{i \geq 1}$ be an enumeration of the prime numbers and for $\alpha < \varepsilon_0$ with Cantor normal form $\alpha = \omega^\beta + \gamma$ let $G\alpha := p_{G\beta} \cdot G\gamma$ where $G0 := 1$. (This Gödel numbering shows up in Schütte's proof theory book and is also known as Matula-coding.) Let

$$\Delta_\varphi(\alpha) := \lim_{n \to \infty} \frac{\#\{\beta < \alpha : \beta \models \varphi \wedge G\beta \leq n\}}{\#\{\beta < \alpha : G\beta \leq n\}}.$$

Then $\Delta_\varphi(\alpha) \in \{0, 1\}$ for $\omega^\omega \leq \alpha < \varepsilon_0$ and $\varphi \in \mathcal{L}$. This follows by adapting the proofs of this paper to the new situation since we are working with multiplicative numbers systems (in the sense of [2]) and we know from [4] that we have for $\alpha < \varepsilon_0$ that

$$n \mapsto \#\{\beta < \alpha : G\beta \leq n\} \in RV_0.$$

2. Let $(p_i)_{i \geq 1}$ again be the enumeration of the prime numbers and for $\alpha < \varepsilon_0$ with Cantor normal form $\alpha = \omega^{\alpha_1} + \cdots + \omega^{\alpha_n}$ let $G'\alpha := p_1^{G'\alpha_1} \cdot \dots \cdot p_n^{G'\alpha_n}$ where $G'0 := 1$. (Such a Gödel numbering shows up in Gödel's work).) Let

$$\Delta'_\varphi(\alpha) := \lim_{n \to \infty} \frac{\#\{\beta < \alpha : \beta \models \varphi \wedge G'\beta \leq n\}}{\#\{\beta < \alpha : G'\beta \leq n\}}.$$

Then we conjecture that $\Delta'_\varphi(\alpha) \in \{0, 1\}$ for $\omega^\omega \leq \alpha < \varepsilon_0$ and $\varphi \in \mathcal{L}$. We expect that this follows by adapting the proofs of this paper to the new situation. To carry this out it seems reasonable to investigate whether we have for $\alpha < \varepsilon_0$ that

$$n \mapsto \#\{\beta < \alpha : G'\beta \leq n\} \in RV_0.$$

A remaining problem is that with respect to G' we do not have a multiplicative number system in the sense of Burris [2].

3. Let us now come back to the additive situation which is based on the norm function N. If φ is in the monadic second order language of linear orders then a limit law (but in general no zero one law) can be proved for all α with $\omega^\omega \leq \alpha \leq \varepsilon_0$. A proof for this result has been obtained by applying as new ingredient Shelah's 'additive colouring' technique.

4. If φ is in the weak monadic second order language of linear orders with $+$ then a limit law can be proved for ε_0 but no algorithm can separate formulas having limiting probability 0 from formulas having limiting probability 1.

5. If φ is in the weak monadic second order language of linear orders with $+$ and $\cdot$ then a limit law can be proved for the thinned out domain of structures $A = \{\omega^\alpha : \alpha < \varepsilon_0\}$.

6. One referee of this paper asked whether limit laws are affected by the choice of notation. Using a formula by Lagrange 1775 (cf. Lemma 3.3 in [5]) it can easily be shown that an $\mathcal{L}$ zero one law breaks down for many ordinals between ω^ω and ε_0 when they are represented by the lexicographic path order over a signature with a binary function symbol and a constant (cf., e.g., the definition of $<$ in [5] p.6). But the expectation is that for any system of ordinal notations published in the literature at least limit laws will hold with respect to $\mathcal{L}$ and canonically extended norm functions. For more expressive languages one would expect Cesaro limit laws to hold in many situations covered by 'natural wellorderings' (cf., e.g., [6]).

References

1. Bell, J.P., Burris, S.N.: Asymptotics for logical limit laws: when the growth of the components is in an RT class. Trans. Amer. Math. Soc. 355(9), 3777–3794 (2003)
2. Burris, S.N.: Number Theoretic Density and Logical Limit Laws. Mathematical Surveys and Monographs, vol. 86. American Mathematical Society (2001)
3. Rosenstein, J.: Linear Orderings. Academic Press (1982)
4. Weiermann, A.: A zero one law characterization of ϵ_0. Mathematics and Computer Science II. In: Chauvin, B., Flajolet, P., Gardy, D., Mokkadem, A. (eds.) Proceedings of the Colloquium on Algorithms, Trees, Combinatorics and Probabilities, pp. 527–539. Birkhäuser (2002)
5. Weiermann, A.: Phase transition thresholds for some Friedman-style independence results. Mathematical Logic Quarterly 53(1), 4–18 (2007)
6. Weiermann, A.: Analytic combinatorics, proof-theoretic ordinals, and phase transitions for independence results. Annals of Pure and Applied Logic 136(1-2), 189–218 (2005)
7. Woods, A.R.: Coloring rules for finite trees, and probabilities of monadic second order sentences. Random Structures Algorithms 10(4), 453–485 (1997)

On the Road to Thinking Machines: Insights and Ideas

Jiří Wiedermann

Institute of Computer Science, Academy of Sciences of the Czech Republic,
Pod Vodárenskou věží 2, 182 07 Prague 8, Czech Republic
jiri.wiedermann@cs.cas.cz

Abstract. The quest for understanding the working of artificial minds attaining a human-like cognition is culminating. While still inspired by the functionality of biological brains, the realization of thinking machines need not slavishly copy the principles used by their living pendants. Achieving a higher-level artificial intelligence no longer seems to be a matter of a fundamental scientific breakthrough but rather a matter of exploiting our best algorithmic theories of thinking machines supported by our most advanced robotic and real time data processing technologies. We review recent examples of such theories, ideas and machines which could pave the road towards building interesting artificial brains.

1 Introduction

Recent media coverage of the IBM Corp.'s computer named Watson defeating the two most successful contestants in the history of the game show "Jeopardy!" in a three day competition in February 2011 illustrates quite nicely the following fact. Nowadays, people not only accept that machines can think, people even seem to accept that machines can and will rival human intellectual abilities. Watson the Computer has in fact showcased the current state of the artificial general intelligence. In contrast to Watson the Computer whose ability to perform intelligent action is restricted to a (narrow) mental domain there exist less media celebrated anthropomorphic bi-pedal robots (mostly of Japanese and Korean provenance) displaying impressive intelligent action in the domain of their human-like body motion. Do thinking machines à la Watson the Computer on one hand, and walking robots on the other hand, represent two independent "evolutionary lines", or is there anything in common to both kind of machines?

Watson the Computer and various kinds of humanoid robots represent two approaches to building human-like thinking machines. While robots rely on their embodiment and build their (initially prevailing motor) intelligence so to speak in a bottom-up manner, a very restricted expert intelligence of Watson the Computer has been given to it in a top-down manner, as though via a Nuremberg funnel. (According to German literary tradition, the Nuremberg funnel method enables teaching everything to even the most stupid student without any effort on his or her side, as though by "pouring" the required knowledge into the student's head with the help of a funnel.)

S.B. Cooper, A. Dawar, and B. Löwe (Eds.): CiE 2012, LNCS 7318, pp. 733–744, 2012.
© Springer-Verlag Berlin Heidelberg 2012

We shall argue that the first, bottom-up approach starting at the embodiment level, is an approach leading to a development at the end of which we may expect a humanoid artifact with human-like intellectual abilities, language acquisition, understanding, thinking, and certain forms of consciousness included. This development can eventually be complemented by the Nuremberg funnel method. In this way intelligence operating over the entire knowledge base currently available to mankind will be achieved. Such intelligence can continuously enlarge itself by its own discoveries using similar principles as existing human-like intelligence. Eventually, the underlying robot or robots could again be de-embodied to give rise to a thinking, knowledgeable non-robotic artifact. This artifact will possess the highest degree od artificial general intelligence upper-bounded by known computability limits (cf. [32]). Such state can be viewed as the singularity point—i.e., a state when machine intelligence will reach the level of human intelligence, considered, e.g., in [16]. Note that this final state will still be like a state of knowledge or wisdom that in principle would also be attainable by a single human. This is because the underlying operating principles will be the same. The difference will be that this new artificial intelligence will exploit computational, perceptory and knowledge resources by many order of magnitudes exceeding those available to a single person.

In the main Section 2 we describe a series of the most important recent achievements and trends in the theory of artificial thinking machines that will, at least to our mind, shape the field in the near future. These achievements will be presented in an order enabling insights into the present ideas of how thinking machines should be constructed. The most important among these ideas is a departure from biologism, automatically built internal world models, use of mirror neurons forming the basis for imitation learning, development of habits, understanding, thinking, realization of both phenomenal and functional consciousness, global workspace theory, exploitation of episodic memories, and real time massive data processing.

2   The Principles

During the past two decades an interesting and promising body of new knowledge has accumulated in the theory. When properly screened, selected and ordered, this body of knowledge has a potential to offer more or less coherent ideas about algorithmic principles on which computational cognitive systems could be based. The respective knowledge has been scattered in the literature, the emerging trends are often not formulated explicitly and important contributions seem to penetrate only slowly into the general awareness of people working in the field.

Next, we shall highlight the main reasons for believing that we already have enough knowledge to glimpse the algorithmic principles behind the working of interesting artificial minds attaining a high-level cognition. In what follows we shall list and comment on the main ideas, trends and important theoretical achievements we have in mind.

2.1   Escaping the Turing Test

In theory, instead of considering thinking machines that can pass the Turing test, often a more general notion of *humanoid cognitive systems* is considered. Humanoid cognitive systems are cognitive systems endowed with human-like intelligence, not necessarily with the intelligence that would be indistinguishable by the Turing test [26]. Namely, the Turing test is explicitly, and unnecessarily, anthropomorphic. If our ultimate goal is to create machines that could help people in an intellectual domain, then it does not make sense to insist that the behavior of such machines must closely resemble that of people.

When speaking about humanoid cognitive systems, nowadays we usually have in mind *humanoid robotic systems.* This is a substantial deviation from Turing's original ideas [26] when he had in mind only "disembodied" computers, with no sensors and effectors, communicating with people only via a terminal.

In humanoid robotic systems, the adjective "humanoid" concerns both the form and the contents of such robotic systems. Physically, such systems should take the form of a human body, with as much sensors and actuators, mirroring perception and motion of the human body, as possible. As we shall see later, this is of utmost importance since practically all cognitive functions, higher-level function included, are derived from the sensorimotor interaction of a robot with its environment. Should the cognitive functions under development in a robot be of human-like nature, then the sensorimotor interaction of that robot, inclusively its environment, should be of similar nature as in the case of humans. We say that a humanoid robotic cognitive system should be *embodied* in a human-like body, and *situated*, via its sensors and effectors, in a human-like environment (cf. [21,22] for more details).

As far as the "content" of a humanoid robotic system, i.e., its control system is concerned, it appears there is no need to mirror the architecture of the human brain, only its functionality — see the next item.

2.2   Escaping Biologism

The next idea is escaping from anthropomorphism, or biologism in the design of control part of cognitive systems. It is amazing how many pictures and schemes of the human brain we see during a conference devoted to cognitive system design. Compare that to a similar situation in a conference devoted to the aircraft design, plagued by pictures of birds and their anatomy.

Nowadays, in addition to human brain, we see many examples of possible architectures for cognition in the nature. For instance, the organization of the nervous system of an octopus, which is known to be capable of performing extraordinary intellectual feats, is totally different from the organization supporting higher level cognitive activities in the vertebrate brain [17]. Therefore, when thinking about artificial minds it only seems natural to concentrate on solutions permitted, and enabled by our technologies while, of course, being inspired by nature.

2.3   Internal World Models

If a humanoid cognitive system has to communicate "intelligently" with people, it should obviously have information how the people's world looks like, what could be the abilities of people under various circumstance, etc. In short, such a system should possess a kind of internal model of the external world (inclusively that of the self), represented in whatever useful way.

Nowadays, it is generally believed that in order to open the road towards higher brain functions in humanoid cognitive systems we need automatic computational mechanisms that will augment the semantic knowledge acquired in the interaction of the system with its environment. These mechanisms often make use of internal world models. Presently, prevailing trends seem to prefer representations of the internal worlds in the form of neural nets (cf. the mirror net and formation of concepts described in Parts 2.4. and 2.5) rather than in the form of rule-based symbol manipulation systems. For an overview of the recent state-of-the-art and a discussion on internal world models, cf. [13] or [6]. A cognitive system architecture exploiting the idea of a world model can be found, e.g., in [31] or [33].

2.4   Mirror Neurons

Mirror neurons were discovered during the 1990s (cf. [24]). Roughly speaking, the mirror neurons are neurons that fire if their owner performs a certain action as well if their owner observes the same species performing the same action. This can be interpreted as mirror neurons being a mechanism for "mind reading" of other subjects. Far-reaching conjectures on the importance of mirror neurons for understanding the intentions of other people, empathy, imitation learning and even for language readiness (cf. [1,14,23]) have been developed since 1990s. In [29] it was shown that mirror neurons can serve as a mechanism synthesizing the *multimodal* (i.e., motor, perceptional and proprioceptive) *information* and completing it, if necessary, by virtue of associativity so that an agent can remain situated even when parts of the multimodal information are missing. Such a mechanism forms a basis on which plausible explanation of the development of a host of mental abilities has been founded. These abilities range from imitation learning, communication via a sign language up to the dawn of thinking and consciousness (cf. Part 2.7). The respective results have built a bridge between the theory of embodied cognition and mirror neurons. These results have also justified the above mentioned hopes laid on the discovery of mirror neurons, indeed. The basic model from [29] has later been elaborated in a series of subsequent papers (cf. [30,31,33]).

Note that the net of mirror neurons can be seen as a specific kind of an *internal static world model*. In this model, the world is represented in the way as it is cognized by an agent's sensory and motor actions, i.e., by an agent's interaction with its environment. It can be termed as a *sensorimotor model* describing the "syntax" of the world. In the mirror net, the combinations of the exteroceptory and proprioceptory inputs are stored jointly with motor actions fitting together, which "make sense" for the agent. Note that since the proprioceptory information is always a part of multimodal information, also elements of an agent's own model are in fact available in the mirror net.

2.5   Algebra of Thoughts

Each occurrence of multimodal information in the net of mirror neurons represents an *embodied concept*. Moreover, new, so-called *abstract concepts* are formed from the existing (mainly embodied) concepts in the *control unit* of a cognitive agent by the following four principles: *contiguity in space, contiguity in time, similarity, and abstraction*. (The first three principles have already been identified by the 18th century Scottish philosopher D. Hume [15].) The concepts are connected via associations of various strengths. The concepts and the associations among them form the agent's memory. At each time, some concepts in it are in active state. These concepts represent the current *"mental state"* of the agent. When new multimodal information enters the control unit it activates a new set of concepts. Based on the current mental state and the set of newly activated concepts, a new set of concepts is activated. This set represents the new mental state of the agent and determines the next motor action of the unit.

In [28,29,30,31,33] it has been shown that, based on the above mentioned principles, the control unit is capable of solving simple cognitive tasks: learning *simultaneous occurrence* of concepts (by contiguity in space), their sequence, so-called *simple conditioning* (by contiguity in time), *similarity based behavior* and computing their *abstractions*. In fact, these are the unit's basic operations.

The control unit represents a specific model of the world capturing the "semantics" of the world. In this model, the relations among concepts are stored. Obviously, these relations correspond to real relationships among real objects and phenomena observed or generated by the agent during its existence. Similar relations are also maintained among the representations of these objects and phenomena. All this information represents a kind of a *dynamic internal world model*. One can also see this model as a depository of the *"patterns of behavior which make sense in a given situation."*

2.6   Habits

If one wants to go further in the realization of the cognitive tasks, one should consider special concepts called *affects*. The affects come in two forms: positive and negative ones. The basic affects are activated directly from the sensors. The ones corresponding to the positive feelings are positive whereas the ones corresponding to the negative feelings are negative. The excitatory (inhibitory) associations arise among positive (negative) affects and concepts. The role of the affects is to modulate the excitation mechanism. With the help of affects, one can simulate reinforcement learning (also called operant conditioning) and delayed reinforcement learning. Pavlovian conditioning (cf. [27], p. 217), reinforcement learning and delayed reinforcement learning seem to represent a set of minimal tests, which a cognitive system aspiring to produce a non-trivial behavior should pass.

Affects by themselves do not correspond to what O'Regan [20] calls *"raw feelings"*, but they create an important ingredient of phenomenal consciousness under fulfillment of other requirements described in Part 2.10.

In a stimulating environment during an agent's interaction with its environment concepts within the control unit start to self–organize, via property of similarity, into *clusters* whose centers are formed by abstract concepts. Moreover, by properties of time contiguity, chains of concepts, called *habits*, linked by associations start to emerge. The habits correspond to often performed activities. The behavior of agents governed by habits starts to prevail. In most cases, such a behavior unfolds effortlessly since habits are triggered independently from the agent's goals or intentions, by environmental cues (cf. [18]). Only at the "crossings" of some habits an additional piece of multimodal information from the mirror net is required directing the subsequent behavior. For more details concerning the work and cognitive abilities of the control unit, see the author's earlier papers where the control unit under the name "cogitoid" has been described.

2.7   Understanding of Understanding

The mechanism of imitation learning is a starting point for higher mental abilities, cf. [1,14]. Imagine the following situation: agent A observes agent B performing a certain well distinguishable task. If A has in its repository of behavioral units (i.e., in its internal syntactic world model) multimodal information which matches well the situation mediated by its sensors (which, by itself, is a difficult robotic task), then A's mirror net will identify the entire corresponding multimodal information (by virtue of associativity). At the same time, it will complement it by the flag saying *"this is not my own experience"* (because the proprioceptive part of information must have been filled in when completing the entire multimodal information) and deliver it to the central unit where it will be processed adequately. This can be seen as a germ of the self concept.

At that moment, A has at its disposal information on what B is about to do, and hence, it can "forecast" the future actions of B. "Forecasting" is done by following the associations in the control unit starting in the current mental state. Agent A can even reconstruct the "feelings" of B (via affects) since they are parts of the retrieved multimodal information. This might be called *empathy* in our model. Moreover, if we endow our agent with the ability to memorize short recent sequences of its mental states, then A can repeat the observed actions of B. This, of course, is called *imitation*.

The same mechanism helps to form a more detailed *model of self*. Namely, observing the activities of a similar agent from a distance helps the observer to "fill in" the gaps in its own dynamic internal world model (i.e., in the control unit), since from the beginning an observer only knows "what it feels like" if it perceives its own part of the body while doing the actions at hand. At this stage, we are close to *primitive communication* done with the help of *gestures*.

Indicating some action via a characteristic gesture, an agent "broadcasts" visual information that is completed by the observer's associative memory mechanism to the complete multimodal information. That is, with the help of a single gesture complex information can be mediated. A gesture acts like an element of a higher-level (proto)language. For articulating agents, it is possible to complement and subsequently even substitute gestures by *articulated sounds.* It is the birth of a spoken language. At about this time the process of stratification of abstract concepts from the embodied ones begins thanks to the abstraction potential of the control unit (cf. Part 2.5). Namely, the agents "understand" their gestures (language), defined in terms of abstract concepts, via empathy in terms of their embodiment or grounding, in the same sensorimotorics and proprioception (i.e., in the same embodied concepts) [10,11], and in a more involved case, in the same patterns of behavior (habits). Without having a body an agent could not understand its communication [22].

2.8   Thinking

Having communication ability, an agent is close to thinking: *thinking means communication with oneself*, similarly as in cf. [7]. By communicating with oneself, an agent triggers the mechanism of discriminating between the external stimuli (I listen to what I am saying) and the internal ones. This mechanism may be termed as *self-awareness* in our model. By a small modification (from the viewpoint of the agent's designer), one can achieve that the still self-communication can be arranged without the involvement of speaking organs at all. In this case, the respective motor instructions issued by the control unit will not reach these organs. However, these instructions still proceed to the internal world model. Here they will invoke the same multimodal information as in the case when an agent directly hears spoken language or perceives its gestures via proprioception. Obviously, while thinking an agent "switches off" any interaction with the external world (i.e., both perception and motor actions). In such a case, the same processes go on, but this time they are only based on the virtual information stored in the internal world model. One can say that in the thinking mode an agent works "off-line", while in the "standard" mode it works "on-line". More details about the development of higher mental abilities can be found, e.g., in [33].

2.9   Global Workspace Theory

Global workspace theory (GWT) is a simplistic, very high-level cognitive architecture that has been developed by Baars by the end of the last century [2,3] to explain emergence of a conscious process from large sets of unconscious processes in the human brain. Central to the theory is a model of information flow in which multiple, parallel, asynchronous specialist processes (corresponding to unconscious processes) compete and co-operate in an arena for access to a global workspace. A winning process then corresponds to a conscious process. This process is promoted to the global workspace and is allowed to broadcast information back to the arena. Based on this, the specialist processes invoke another set of unconscious processes and the whole cycle repeats itself. Note that the processes in the global workspace appear in a serial manner, one after the other, while each of them is the integrated product of parallel processing. The GWT can successfully model a number of characteristics of consciousness, such as its role in handling novel situations, its limited capacity, its sequential nature, and its ability to trigger a vast range of unconscious brain processes. Unfortunately, the GWT neither does explain the mechanism how an originally unconscious process becomes a conscious one nor the mechanism of process competition.

The GWT has been incorporated into a number of computational models (cf. [9]). The operation of a cogitoid, mentioned in Part 2.6, can be seen as a specific implementation of the GWT. It is perhaps interesting to observe that one "question/answer processing cycle" (cf. [8]) of Watson the Computer works, in fact, according to the GWT.

2.10   (Dis)solving the Hard Problem of Consciousness

For the past decade or two, the modern theory of consciousness has been stigmatized by the dichotomy between so-called *functional (or access) consciousness* and the so-called *phenomenal consciousness (or qualia)*. These two notions were famously introduced by American philosopher of the mind, Ned Block [4] and subsequently also adopted by other important protagonists (e.g., David Chalmers [5]) in the field. Functional consciousness consists of that information globally available in the cognitive system for the purposes of reasoning, speech and high-level action control. Phenomenal consciousness consists of subjective phenomenal experience and feelings. Nowadays, we have relatively good ideas how to implement functional consciousness (cf. [31]). On the other hand, phenomenal consciousness seemed to present a nut hard to crack. The problem of explaining how and why we have qualitative phenomenal experiences (of form "what is it like") presents so-called *hard problem of consciousness* [5]. Nevertheless, recently theories pointing to a common evolutionary [12] and sensorimotor basis (cf. [20] and other works by this author) for both phenomenal and functional consciousness have appeared. According to O'Regan [20], in order to have a "raw feel" (qualia) it will suffice for a robot already possessing functional consciousness to engage in an embodied (sensorimotor) interaction with its environment. More specifically, qualia can be seen as an engagement in exercising a "fixed sensorimotor skill" accompanied by (functional) conscious attendance to that engagement and the skill's quality. The respective real-world interaction has to possess the properties of richness, bodiliness, insubordinateness and grabbiness. *Richness* characterizes abundance in details. *Bodiliness* or corporality requires that voluntary motions of a body systematically affect sensory inputs. *Insubordinateness* means that the world has its own dynamic that we can affect only partially (if at all) causing that bodiliness is never complete. And finally, *grabbiness* means that the perceptual stimuli have the alerting capacity—they can peremptorily interfere with cognitive processing (e.g., they can cause an interrupt).

2.11 Episodic Memory

Episodic memory is what people "remember", i.e., the contextualized information about autobiographical events (times, places, associated emotions), and other contextual knowledge that can be explicitly stated. Multimodal information (mentioned in Part 2.4) can serve as the simplest example of a "unitary" episodic memory. It is obvious that such memory is important for an agent to know about its past. Therefore, as noted in [19], it is surprising that the vast majority of integrated intelligent systems have ignored episodic memory, which often dooms them to what can be achieved by people with amnesia, which is demonstrably limited. Nowadays we frequently witness "add-ons" to the existing models of cognitive systems architecture to account for episodic memory (cf. [19]). In [19] it is argued that episodic memory systems can support a vast number of cognitive capabilities which are mostly based on inspecting memories from the past that are "similar" to the present situation. Among these capabilities there is noticing novel situations, detecting repetitions, virtual sensing (reminded by some recall), future action modeling, planing ahead (cf. [34]), environment modelling, predicting success/failure, managing long term goals, etc. Incorporating episodic memories and their efficient management and especially their retrieval is a non-trivial matter and it is here where current massive data processing technologies can find their good use (cf. Part 2.12). In the simplest case, in the case of simple agents (like animals), the retrieval from episodic memory can be based directly on the associative ability of the mirror net (Part 2.4). Phenomenal consciousness seems to play the role of a tagging system for episodic memories. More involved cases can be based on exploiting more complex tags attached to stored episodes, in analogy with retrieval from picture databases. Tags can be seen as short natural language expressions pertinent to the stored episode. Tagging (and storing) of current multimodal information is performed each time when the subject is conscious of that information. Thus, an ability to describe what an agent perceives at a given moment, and a conscious awareness of it, seems to be a prerequisite of efficient episodic memory management and their exploitation. (By the way, the role of language and consciousness in episodic memory management seems to have escaped the attention of researchers.)

2.12 Real Time Massive Data Processing

In a sense, Watson the Computer can be seen as a crippled cognitive system specialized in doing efficient contextual retrieval invoked by clues over its preprocessed episodic memory. Its success was possible thanks to technological progress enabling maintenance of supercritical volumes of data and their searching and retrieval by supercritical speed. Could this be the case that we are witnessing the birth of a new paradigm? This paradigm states that *intelligence is not only a matter of suitable algorithms, but also, and mainly so, of the ability to accumulate (e.g., via learning and episodic memories storing), organize, and exploit large data volumes representing knowledge, at a speed matching the timescale of the environmental requirements.* In the case of a robot, its sensorimotor interactions must also possess this quality, i.e., they must involve real time processing.

Watson the Computer seems to be the first case where the real time aspect has boldly entered the game, enforcing a massively parallel solution which has become the main factor in Watson's victory.

2.13   De-embodiment of Robotic Humanoid Cognitive Systems

As mentioned earlier, the right starting point for constructing human-like thinking machines has been development of systems possessing a "complete" body. Having a body has been a necessary condition for the evolution of higher mental abilities. However, once having such a robot, for practical purposes (let us say for conversation or expert advice) we might want to de-embody it, leaving it with its basic communication abilities only. The result would then be something similar to the brain in the vat. The resulting cognitive system will still be operational capable of thinking and of further mental development thanks to its elaborated internal world model and to model's update mechanisms. Such systems will be relatively easy to replicate and each one can later be specialized (via "Nuremberg funnel technique") to become an expert (or just an intellectual companion) in some specific domain according to the needs of its user.

3   Conclusion

Should a breakthrough occur in building thinking machines, then this breakthrough will be a kind of engineering breakthrough involving integration of our best theories of mind with the most advanced robotic and real time data processing technologies. The man/year effort spent in the case of Watson the Computer shows clearly that this is going to be a formidable engineering task since only a negligible part of the thinking machine research agenda has been verified so far. It is unlikely that thinking machines will be developed by purely academic research since it is beyond its power to concentrate the necessary amount of manpower and technology. This cannot be accomplished by large international research programs either since a dedicated long-term open-ended effort of many researchers concentrated on a single practically non-decomposable task is needed. It seems to be a unique strategic opportunity for giant IT corporations. The road towards thinking machines glimpses ahead of us and it only is a matter of money whether we set off for a journey along this road.

Acknowledgements. This research was partially supported by RVO 67985807 and GA ČR grant No. P202/10/1333.

References

1. Arbib, M.A.: The mirror system hypothesis: how did protolanguage evolve? In: Tallerman, M. (ed.) Language Origins: Perspectives on Evolution. Oxford University Press (2005)

2. Baars, B.J.: A cognitive theory of consciousness. Cambridge University Press, Cambridge (1988)
3. Baars, B.J.: In the theater of consciousness: The workspace of the mind. Oxford University Press, Oxford (1997)
4. Block, N.: On a Confusion About a Function of Consciousness. Brain and Behavioral Sciences 18, 227–247 (1995)
5. Chalmers, D.: Facing Up To the Problem of Consciousness. Journal of Consciousness Studies 2(3), 200–219 (1995)
6. Cruse, H.: The evolution of cognition—a hypothesis. Cognitive Science 27(1) (2003)
7. Dennett, D.: Consciousness Explained. The Penguin Press (1991)
8. Ferrucci, D., et al.: Building Watson: An Overview DeepQA Project. AI Magazine, 200–214 (Fall 2010)
9. Franklin, S.: IDA: A conscious artifact? J. of Consciousness Studies 10(4-5), 47–66 (2003)
10. Feldman, J.: From Molecule to Metaphor. MIT Press, Cambridge (2006)
11. Harnad, S.: The symbol grounding problem. Physica D (42), 335–346 (1990)
12. Harvey, I.: Evolving Robot Consciousness: The Easy Problems and the Rest. In: Fetzer, J.H. (ed.) Evolving Consciousness. Advances in Consciousness Research Series, pp. 205–219. John Benjamins, Amsterdam (2002)
13. Holland, O., Goodman, R.: Robots with internal models: a route to machine consciousness? Journal of Consciousness Studies 10(4-5) (2003)
14. Hurford, J.R.: Language beyond our grasp: what mirror neurons can, and cannot, do for language evolution. In: Kimbrough, O., Griebel, U., Plunkett, K. (eds.) The Evolution of Communication systems: A Comparative Approach. The Viennna Series in Theoretical Biology. MIT Press, Cambridge (2002)
15. Hume, D.: Enquiry concerning human understanding. In: Selby-Bigge, L.A. (ed.) Enquiries Concerning Human Understanding and Concerning the Principles of Morals, 3rd edn. Clarendon Press, Oxford (2003), revised by Nidditch, P.H.
16. Kurzweil, R.: The Singularity is Near, p. 652. Viking Books (2005)
17. Llinàs, R.: I of the Vortex: From Neurons to Self. MIT Press (2001)
18. Neal, D.: The Pull of the Past: When Do Habits Persist Despite Conflict With Motives? Pers. Soc. Psychol. Bull. 37, 1428–1437 (2011)
19. Nuxoll, A.M., Laird, J.E.: Extending Cognitive Architecture with Episodic Memory. In: Proceedings of the Twenty-Second Conference on Artificial Intelligence. AAAI Press, Vancouver (2007)
20. Kevin O'Regan, J.: How to Build Consciousness into a Robot: The Sensorimotor Approach. In: Lungarella, M., Iida, F., Bongard, J.C., Pfeifer, R. (eds.) 50 Years of Aritficial Intelligence. LNCS (LNAI), vol. 4850, pp. 332–346. Springer, Heidelberg (2007)
21. Pfeifer, R., Scheier, C.: Understanding Intelligence. The MIT Press, Cambridge (1999)
22. Pfeifer, R., Bongard, J.: How the body shapes the way we think: a new view of intelligence. MIT Press (2006)
23. Ramachandran, V.S.: Mirror neurons and imitation as the driving force behind 'the great leap forward' in human evolution. EDGE: The third culture (2000), http://www.edge.org/3rd_culture/ramachandran/ramachandran_p1.html
24. Rizzolatti, G., Fadiga, L., Gallese, V., Fogassi, I.: Premotor cortex and the recognition of motor actions. Cognitive Brain Research 3, 131–141 (1996)
25. Shanahan, M.P.: Consciousness, emotion, and imagination: a brain-inspired architecture for cognitive robotics. In: Proceedings AISB 2005 Symposium on Next Generation Approaches to Machine Consciousness, pp. 26–35 (2005)

26. Turing, A.: Computing Machinery and Intelligence. Mind 59(236), 433–460 (1950)
27. Valiant, L.G.: Circuits of the mind. Oxford University Press, New York (1994)
28. Wiedermann, J.: Towards Algorithmic Explanation of Mind Evolution and Functioning (Extended Abstract). In: Brim, L., Gruska, J., Zlatuška, J. (eds.) MFCS 1998. LNCS, vol. 1450, pp. 152–166. Springer, Heidelberg (1998)
29. Wiedermann, J.: Mirror neurons, embodied cognitive agents and imitation learning. Computing and Informatics 22(6), 545–559 (2003)
30. Wiedermann, J.: HUGO: A Cognitive Architecture with an Incorporated World Model. In: Proc. of the European Conference on Complex Systems ECCS 2006, Said Business School. Oxford University (2006)
31. Wiedermann, J.: A high level model of a conscious embodied agent. In: Proc. of the 8th IEEE International Conference on Cognitive Informatics, pp. 448–456 (2010); expanded version appeared in: International Journal of Software Science and Computational Intelligence (IJSSCI) 2(3), 62–78 (2010)
32. Wiedermann, J.: A Computability Argument Against Superintelligence. In: Cognitive Computation, February 18. Springer (2012)
33. Wiedermann, J.: Towards Computational Models of Artificial Cognitive Systems That Can, in Principle, Pass the Turing Test*. In: Bieliková, M., Friedrich, G., Gottlob, G., Katzenbeisser, S., Turán, G. (eds.) SOFSEM 2012. LNCS, vol. 7147, pp. 44–63. Springer, Heidelberg (2012)
34. Zimmer, C.: The Brain: Memories Are Crucial for Looking Into the Future. DISCOVER Magazine (April 2011)

Making Solomonoff Induction Effective
Or: You Can Learn What You Can Bound

Jörg Zimmermann and Armin B. Cremers

Institute of Computer Science, Rheinische Friedrich-Wilhelms-Universität Bonn,
Römerstr. 164, 53117 Bonn, Germany
{jz,abc}@iai.uni-bonn.de

Abstract. The notion of effective learnability is analyzed by relating it to the proof-theoretic strength of an axiom system which is used to derive totality proofs for recursive functions. The main result, the generator-predictor theorem, states that an infinite sequence of bits is learnable if the axiom system proves the totality of a recursive function which dominates the time function of the bit sequence generating process.

This result establishes a tight connection between learnability and provability, thus reducing the question of what can be effectively learned to the foundational questions of mathematics with regard to set existence axioms. Results of reverse mathematics are used to illustrate the implications of the generator-predictor theorem by connecting a hierarchy of axiom systems with increasing logical strength to fast growing functions.

Our results are discussed in the context of the probabilistic universal induction framework pioneered by Solomonoff, showing how the integration of a proof system into the learning process leads to naturally defined effective instances of Solomonoff induction. Finally, we analyze the problem of effective learning in a framework where the time scales of the generator and the predictor are coupled, leading to a surprising conclusion.

1   Introduction

An effective learning system is a system which can be fully specified by a program on a universal Turing machine. In its most general form, this program transforms a stream of percepts generated by an environment (later called the generator) into a stream of actions possibly changing this environment. Via this sensomotoric loop the system is embedded into its environment, about which a priori nothing is known [7]. A more specific notion of learning is defined as the process which translates the stream of percepts into predictions for future percepts. These predictions can then be used for choosing actions. The analysis of the "design space" for effective learning systems leads to three major questions:

1. How should a learning system represent and process uncertainty, or, what is the proper inductive logic?
2. What set of possible models of the environment should the system consider?
3. How to relate the explanatory power of a model to its complexity?

S.B. Cooper, A. Dawar, and B. Löwe (Eds.): CiE 2012, LNCS 7318, pp. 745–754, 2012.
© Springer-Verlag Berlin Heidelberg 2012

In the long run, the learning system should be able to detect as many regularities in its percept stream as possible, while dealing sensibly with the inherent uncertainty of predictions based on a finite amount of data. The first question regarding the representation and processing of uncertainty is in itself a current area of research, where a plethora of different approaches are discussed [5]. An attempt to find a unifying perspective on these approaches has been made in [15]. However, this question is not the focus of this contribution, so we shall discuss only a probabilistic learning framework, which is essentially a dynamic extension of Bayesian inference. Here we focus on the second and third question.

The definition of proof-driven learning systems, combining search in proof and program space, was inspired by an algorithm solving all well-defined problems as fast as possible (save for a factor of 5 and additive terms) introduced by M. Hutter in [6]. We apply this idea in the context of learning in the limit and discuss in more detail the dependence of concepts like well-definedness of programs or effective learnability on the proof-theoretic strength of the employed background theory, thus establishing an explicit link between effective learnability and the foundational mathematical questions treated in proof theory and ordinal analysis [10]. Additionally, our results emphasize the fact that provability is not an absolute concept. The theoretical limits of effective learnability and computable approximations to Solomonoff induction are intensively discussed in various publications investigating different aspects of this topic, see, for example, [8] and the references therein, but to our knowledge none has made the connection between proof-theoretic strength, time complexity and effective learnability as explicit as it is stated in our main result, the generator-predictor theorem. The first learning in the limit framework, wich does not consider the uncertainty of predictions, was introduced by E. M. Gold [3]. A discussion of our results in the context of this framework can be found in [16].

2   The Generator Space

A clarification of the concept of learnability has to consider the following question: What is the set of possible generators for observed events? To answer this question, we shall explore the notion of "all possible generators" from a mathematical and computational point of view, and discuss the question of effective learnability in the context of such generic model spaces. The last sentence uses intentionally the plural form of model space, because we shall see that the notion of "all possible models" cannot be defined in an absolute sense, but only with regard to a reference proof system. This dependence will lead to the establishment of a relationship between the *time* complexity of the percept-generating environment and the *logical* complexity of an effective learning system, thus shedding new light on the undecidability of a general approach to induction developed by Solomonoff in the 1960s [12,13,9].

2.1 Algorithmic Ontology: Programs as Generators

Ontology is a part of philosophy which tries to define what exists, or, more specifically, what *possibly* could exist. In the realm of mathematics, this question leads to the set existence problem, which is (partially) answered by various set theories, most commonly by using the axiom system ZFC, Zermelo-Fraenkel set theory with the axiom of choice. But in the realm of computer science, existence has to be *effective* existence, i.e., the domain of interest and its operations must have effective representations.

For this reason the objects we consider are programs executed by a fixed universal Turing machine U having a one-way read-only input tape, some work tapes, and a one-way write-only output tape (such Turing machines are called monotone), which will be the reference machine for all what follows. The choice of the specific universal Turing machine has only a constant factor as effect on the space complexity of a program and at most a logarithmic factor as effect on the time complexity [1]. These effects are small enough to be neglected in the following foundational considerations, but in the context of alternative models of computation, like cellular automata on infinite grids, the choice of the reference machine may become an important issue. The program strings are chosen to be prefix-free, i.e., no program string is the prefix of another program string. This is advantageous from a coding point of view (enabling, for example, the application of Kraft's inequality $\sum_p 2^{-\text{length}(p)} \leq 1$), and does not restrict universality [9].

A program p (represented as finite binary string) is a generator of a possible world, if it outputs an infinite stream of bits when executed by U. Unfortunately, it is not decidable whether a given program p has this well-definedness property. This is the reason why the general approach to induction introduced by Solomonoff is incomputable: the inference process uses the whole set of programs (program space) as possible generators, even the programs which are not well-defined in the above sense.

This results in the following dilemma: either one restricts the model space to a decidable set of well-defined programs, which leads to an effective inference process but ignores possibly meaningful programs, or one keeps all well-defined programs, but at the price of necessarily keeping ill-defined programs as well, risking the incomputability of the inference process. However, in the following we propose an approach which tries to mitigate this dilemma by reducing the question of learnability to the question of provability.

3 Learning Systems

Here we introduce the notion of a *probabilistic learning system*, which takes a finite string of observed bits (the percept string) as input and produces a probabilistic prediction for the next bit as output:

Definition 1. *A* probabilistic learning system *is a function*

$$\Lambda : \{0,1\}^* \times \{0,1\} \to [0,1]_{\mathbf{Q}}, \quad \text{with } \Lambda(x,0) + \Lambda(x,1) = 1 \text{ for all } x \in \{0,1\}^*.$$

Λ is an *effective* probabilistic learning system if Λ is a total recursive function. We use rational numbers as probability values, because real numbers cannot be used directly in a context of computability questions, but have to be dealt with by effective approximations [14]. This would increase the complexity of our definitions significantly, and is not necessary for a first understanding of the fundamental relationship between learnability and provability. Also we treat only deterministic generators leading to one definite observable bit sequence. A generalization of our results to real valued probabilities and probabilistic generators should be possible, but is open to future research.

Next we extend the prediction horizon of Λ by feeding it with its own predictions. This leads to a learning system $\Lambda^{(k)}$ which makes probabilistic predictions for the next k bits (xy is the concatenation of string x and y):

$$\Lambda^{(1)} = \Lambda,$$
$$\Lambda^{(k+1)}(x, y1) = \Lambda^{(k)}(x, y) \cdot \Lambda(xy, 1), \quad x \in \{0,1\}^*, y \in \{0,1\}^k,$$
$$\Lambda^{(k+1)}(x, y0) = \Lambda^{(k)}(x, y) \cdot \Lambda(xy, 0).$$

Finally, we define the learnability of an infinite bit sequence s ($s_{i:j}$ is the subsequence of s starting with bit i and ending with bit j):

Definition 2. *An infinite bit sequence s is learnable in the limit by the probabilistic learning system Λ, if for all $\epsilon > 0$ there is an n_0 so that for all $n \geq n_0$ and all $k \geq 1$:*

$$\Lambda^{(k)}(s_{1:n}, s_{n+1:n+k}) > 1 - \epsilon.$$

4 Σ-Driven Learning Systems

We now introduce the learning systems we shall use to investigate the notion of effective learnability. But first we need the following definitions specifying the nature of background theories available to these learning systems.

Definition 3. *A logic frame is a 5-tuple $\Sigma = (S, F, \models, \vdash, \Phi)$, where elements of S are called the structures of Σ, elements of F are called the sentences of Σ, the relation $\models \subseteq S \times F$ is called the satisfaction relation of Σ, $\vdash : 2^F \to 2^F$ is the deduction system of Σ, and $\Phi \subseteq F$ is the core axiom system of Σ.*

A deduction system is *sound* if for all $\Psi \subseteq F$ it holds: if $\Psi \vdash \psi$, then for all $M \in S$: if $M \models \Psi$, then $M \models \psi$. Next we define admissibility for logic frames:

Definition 4. *Let $\Sigma = (S, F, \models, \vdash, \Phi)$ be a logic frame. Σ is admissible if $\vdash$ is sound, Φ is enumerable, $\vdash$ effectively maps enumerable sets of axioms to enumerable sets of logical consequences, and for all recursive functions f there is a sentence $\phi_{\mathrm{tot}}(f) \in F$ with the following property: if $\Phi \vdash \phi_{\mathrm{tot}}(f)$, then f is a total recursive function.*

Note that this definition only fixes the meaning of $\phi_{\mathrm{tot}}(f)$ when it is derivable from Φ. A pathological definition like $\phi_{\mathrm{tot}}(f) = \mathsf{FALSE}$ for all f is consistent with the above definition of a logic frame, but would imply that such a logic frame could not prove the totality of any recursive function f.

Let $\phi_1, \ldots, \phi_n$ be the first n sentences enumerated by the deduction system of an admissible logic frame Σ, then the set $\mathrm{Tot}_n(\Sigma)$ is defined as follows:

$$f \in \mathrm{Tot}_n(\Sigma) \quad \text{iff} \quad \phi_{\mathrm{tot}}(f) \in \{\phi_1, \ldots, \phi_n\}.$$

So $\mathrm{Tot}_n(\Sigma)$ contains the recursive functions which admit a totality proof within the first n sentences enumerated by the deduction system of Σ.

Now let $g_n(m) = max(\{f(m)|f \in \mathrm{Tot}_n(\Sigma)\})$ and $g(n) = g_n(n)$. g is called the *guard function* w.r.t. Σ. The maximum over an empty set is defined as 0. Note that the maximum is taken over a finite set of total recursive functions, so the guard function is a total recursive function, too. The guard function g will play a central role as a scheduler ensuring the effectiveness of learning systems.

Let Σ be an admissible logic frame, then the Σ-*driven probabilistic learning system* $\Lambda(\Sigma)$ is defined as follows: $\Lambda(\Sigma)$ uses a dynamic model space and dynamic prior probabilities and considers at inference step n only the part of the program space given by the programs with length at most n. Because n grows unboundedly, eventually every program will be part of the model space, but at every instant of time the model space is finite. This assumption implies that the posterior distributions and the probabilistic predictions can be computed exactly and all involved probabilities are rational numbers. $\Lambda(\Sigma)$ also manages three labels for programs in addition to their current probabilities, in contrast to classical Baysian inference: "candidate", "suspended", and "discarded". A program which is added to the model space initially gets the label "candidate", all other programs keep their label from the previous inference step. After the nth bit has been observed, $\Lambda(\Sigma)$ executes the following steps:

Input: the nth observed bit.
Output: a probabilistic prediction for the next bit.

1. Derive ϕ_n using the deduction system of Σ.
2. Compute $g(n)$.
3. Initialize $\mu_* = 0$. This variable accumulates preliminary posterior probability mass and is needed to normalize the posterior distribution.
4. Start the enumeration of all programs p with $\mathrm{length}(p) \leq n$.
5. If a program p is labeled "discarded", it has already probability 0 and nothing has to be done.
6. If a program p is labeled "suspended" or "candidate", evaluate program p till it outputs n bits or its step function reaches $g(n)$.
7. If p has generated a bit and it is the observed one, then label p as "candidate". Assign p a preliminary posterior probability of $2^{-(\mathrm{length}(p)+\mathrm{switch}(p,n))}$, where $\mathrm{switch}(p, n)$ counts the number of switches p has experienced from "suspended" status back to "candidate" status up to now. A higher number of switches is translated retroactively into a lower prior probability. Add the preliminary posterior probability to μ_*
8. If the generated bit is not the observed one, label p as "discarded" and set its posterior probability to 0.

9. If p has reached the time limit specified by the guard function g, then label p as "suspended" and set its posterior probability to 0.
10. Continue the enumeration of programs.
11. If the enumeration of all programs p with length$(p) \leq n$ is completed and no program has the label "candidate", return $(\frac{1}{2},\frac{1}{2})$ as probabilistic prediction for the next bit. Exit this inference step.
12. Rescale the preliminary posterior probabilities of all models labeled "candidate" by $1/\mu_*$, resulting in a normalized posterior distribution on the currently considered model space. Use the guarded versions $p \upharpoonright g$ of the programs p labeled "candidate" for computing the probabilistic prediction for the next bit. ($p \upharpoonright g$, p guarded by g, is the time limited version of p: if the step function (the function which just counts the transitions made by p) of $p(m)$ exceeds $g(m)$, then the computation of p is terminated and 0 is returned, else $p(m)$ is returned).

This finishes our definition of Σ-driven probabilistic learning systems. Next we shall formulate and prove a theorem characterizing their learning capabilities.

5   Generator-Predictor Theorem

If the generator functions (see below) of all programs p generating a bit sequence s grow so fast that they are not dominated by a provably total recursive function, then it is indistinguishable, even in the limit, from a non-effective bit sequence. So the quest for making Solomonoff induction effective can be reduced to the concept of provably total recursive functions. But which functions are provably total recursive and which are not? The answer is: it depends. It depends on the background theory or logic frame which is accepted for the construction of totality proofs. At this point we have reduced the problem of universal induction to a logical parameter, the logic frame Σ to which the learning system can refer. The next step is to relate the proof strength of a logic frame Σ to the time complexity of a program p which generates the bit stream observed by our learning system. This will yield a natural characterization of the learnable bit sequences relative to Σ.

Before we can state this relationship as a theorem, we need the notion of the *generator time function*, generator function for short, of a program p:

Definition 5. *The generator time function $G_p^{(U)} : \mathbf{N} \rightarrow \mathbf{N} \cup \{\infty\}$ of a program p w.r.t. the universal reference machine U assigns every $n \in \mathbf{N}$ the number of transitions needed to generate the first n bits by the reference machine U executing p. If n_0 is the smallest number for which p does not generate a new bit, then $G_p^{(U)}(n) = \infty$ for all $n \geq n_0$.*

In the following we shall drop the superscript (U) because we are working only with one reference machine.

In general there are several programs generating the same observable bit sequence. So one can not hope to learn exactly the program p which generates the observed bit sequence, but only a program p' which is *observation equivalent*. The equivalence class of programs corresponding to an infinite bit sequence s we shall denote by $[s]$. Now we have introduced all the notions and concepts we need in order to state our main result:

Generator-Predictor Theorem: Let Σ be an admissible logic frame and s an infinite bit sequence. s is learnable by the effective probabilistic learning system $\Lambda(\Sigma)$, if Σ proves the totality of a recursive dominator of a generator function G_p for at least one program $p \in [s]$.

Proof: Let $L_\Sigma(s) = \{p | p \in [s]$ and Σ proves the totality of a recursive dominator of the generator function of $p\}$. So $L_\Sigma(s)$ contains the programs which can be eventually used by $\Lambda(\Sigma)$ as perfect predictors. If $L_\Sigma(s)$ is empty, the theorem makes no statement about the learnability of s, so lets assume that $L_\Sigma(s)$ contains at least one element. We shall show that the sum of the posterior probabilities of the programs in $L_\Sigma(s)$ will converge to 1 as the number of observed bits increases. This will be conducted in two steps. Let $\alpha(p, n)$ be the preliminary posterior probability (i.e., the posterior probability assigned to p before normalization) of program p in the nth inference step (programs not contained in the nth model space are considered as having a preliminary posterior probability of 0), $\alpha(n)$ be the sum of the preliminary posterior probabilities of all programs in $L_\Sigma(s)$ and $\bar{\alpha}(n)$ be the sum of the preliminary posterior probabilities of all programs not in $L_\Sigma(s)$. Note that $\alpha(n) + \bar{\alpha}(n) \neq 1$, because preliminary and not final posterior probabilities are considered. First we shall show the existence of a number n_1 so that $\alpha(n) \geq c$ for some constant $c > 0$ and for all $n \geq n_1$. And second, we shall see that $\bar{\alpha}(n)$ converges to 0 as the number of inference steps n goes to infinity. This implies that the normalized value of $\alpha(n)$ has to go to 1 as n goes to infinity:

$$\alpha^{\text{norm}}(n) = \frac{\alpha(n)}{\alpha(n) + \bar{\alpha}(n)} = \frac{1}{1 + \frac{\bar{\alpha}(n)}{\alpha(n)}} \xrightarrow[n \to \infty]{} 1$$

So both steps together will establish the generator-predictor theorem.

If $p \in L_\Sigma(s)$, then there is a number n_0 so that the guard function g majorizes G_p for all $n \geq n_0$. Let $n_1 = \max(n_0, \text{length}(p))$. Then p is part of the model space for all $n \geq n_1$ and it is not discarded and not suspended. Its preliminary posterior probability is $2^{-(\text{length}(p) + \text{switch}(p, n_1))}$. This number does not change anymore for $n \geq n_1$ (no new switches), thus $c = 2^{-(\text{length}(p) + \text{switch}(p, n_1))} > 0$ is a lower bound for $\alpha(n)$ for all $n \geq n_1$. This completes the first step.

Kraft's inequality for prefix codes implies $\sum_{p \notin L_\Sigma(s)} 2^{-\text{length}(p)} \leq 1$. Thus for all $\epsilon > 0$ there is a k_0 with $\sum_{p \notin L_\Sigma(s), \text{length}(p) > k} 2^{-\text{length}(p)} < \epsilon$ for all $k \geq k_0$, because for a convergent sum the partial sums $\sum_{p \notin L_\Sigma(s), \text{length}(p) \leq k} 2^{-\text{length}(p)}$ converge to the limit. Now choose n_2 so that $\sum_{p \notin L_\Sigma(s), \text{length}(p) > n} 2^{-\text{length}(p)} < \epsilon/2$ for all $n \geq n_2$, and n_3 so that $\sum_{p \notin L_\Sigma(s), \text{length}(p) \leq n_2} \alpha(p, n) < \epsilon/2$ for all $n \geq n_3$. n_3 exists, because for a fixed n_2 there are only finitely many summands $\alpha(p, n)$ contributing to the sum, each dropping to 0 (discarded or suspended

752   J. Zimmermann and A.B. Cremers

forever) or converging to 0 (number of switches increases unboundedly) as n goes to infinity. Then for all $n \geq n_3$ we have:

$$\bar{\alpha}(n) = \sum_{p \notin L_\Sigma(s)} \alpha(p,n) = \sum_{\substack{p \notin L_\Sigma(s) \\ \text{length}(p) \leq n_2}} \alpha(p,n) \quad + \sum_{\substack{p \notin L_\Sigma(s) \\ \text{length}(p) > n_2}} \alpha(p,n)$$

$$\leq \sum_{\substack{p \notin L_\Sigma(s) \\ \text{length}(p) \leq n_2}} \alpha(p,n) \quad + \sum_{\substack{p \notin L_\Sigma(s) \\ \text{length}(p) > n_2}} 2^{-\text{length}(p)} \quad < \quad \epsilon/2 + \epsilon/2 \quad = \quad \epsilon$$

This finishes the second step and thus the proof of the generator-predictor theorem.

$\square$

For example, if $\Sigma_{\text{PA}} = (\text{FOL}, \text{PA})$ (First order logic Peano Arithmetic), then every program with a generator function which is dominated by a provably total recursive function w.r.t. PA can be learned by $\Lambda(\Sigma_{\text{PA}})$. As the Ackermann-function is provably total in PA, this is already a pretty large set, and PA allows totality proofs of functions which grow much faster than the Ackermann-function.

The Generator-Predictor theorem states that for learning a bit sequence s it is enough to prove the totality of a recursive dominator of the generator function of a program $p \in [s]$, it is not necessary to prove the totality of p itself. However, one can show that whenever there is a $p \in [s]$ so that the logic frame Σ can prove the totality of a recursive dominator of G_p, then there is a $q \in [s]$ for which Σ can prove totality, provided the logic frame satisfies some weak closure conditions [16].

6 Learnability and Reverse Mathematics

Reverse mathematics is a program in mathematical logic that seeks to determine which set existence axioms are required to prove theorems of mathematics in order to classify these theorems according to the strength of existence axioms necessary to derive them. The program was founded in the 1970s by H. Friedman [2]. Maybe the most remarkable fact of reverse mathematics is that many theorems of classical mathematics fall into five large equivalence classes, consisting of provably equivalent theorems. This leads to five standard axiom systems, which are linearly ordered according to their proof-theoretic strength. S. G. Simpson has named them the "Big Five" [11], starting with a system called RCA_0 (Recursive Comprehension Axiom), followed by WKL_0 (Weak König's Lemma), ACA_0 (Arithmetical Comprehension Axiom), ATR_0 (Arithmetical Transfinite Recursion), and $\Pi^1_1\text{-CA}_0$ (Π^1_1-Comprehension Axiom).

One can show that the functions which admit totality proofs by RCA_0 are exactly the primitive recursive functions. This implies that a bit stream $s_{\text{Ackermann}}$ which has a generating program with a generator function growing like the Ackermann-function—a function known not to be primitive recursive—but no generating program with a generator function dominated by a primitive recursive function, can not be learned by $\Lambda(\mathsf{RCA_0})$. In terms of totality proofs $\mathsf{WKL_0}$ adds no additional proof-strength to $\mathsf{RCA_0}$, but the next system, $\mathsf{ACA_0}$, can prove the totality of the Ackermann-function, so $\Lambda(\mathsf{ACA_0})$ can learn $s_{\mathrm{Ackermann}}$ and is therefore a stronger learning system than $\Lambda(\mathsf{RCA_0})$. But there are total recursive functions which have no totality proof in $\mathsf{ACA_0}$ (which is equivalent to Peano Arithmetic in this regard), for example the Goodstein-function [4]. This can be continued even beyond $\Pi_1^1\text{-}\mathsf{CA_0}$, leading to the foundational questions within mathematics concerning the introduction of ever stronger set existence axioms. Here it suffices to note that these examples are a good illustration of the fact that increased proof-theoretic strength translates into stronger learning systems.

7   Synchronous Learning Frameworks

A closer look on real world incremental learning situations, where both, the environment and the learning system, are not suspended while the other one is performing its transitions resp. computations, leads to the following notion of *synchrony* of a bit sequence s:

Definition 6. *s is* synchronous *if* $\lim\sup\limits_{n\to\infty}\frac{G_p(n)}{n} < \infty$ *for at least one* $p \in [s]$.

Synchrony entails that the time scales of the learning system and the environment are coupled, that they cannot ultimately drift apart. As long as one not assumes a malicious environment, synchrony seems to be a natural property. Such a setting for learning could be called a synchronous learning framework, in contrast to the above considered learning frameworks, which could be classified as asynchronous.

In order to learn in the case of synchrony, it suffices to prove that n^2 is total, because n^2 is a dominator of every generator function satisfying the synchrony condition. Thus, a much weaker background theory than, e.g., Peano Arithmetic would suffice for an effective learning system to learn all synchronously generated bit sequences. In fact, because $\mathsf{RCA_0}$—the weakest of the five standard axiom systems considered in reverse mathematics—proves the totality of all primitive recursive functions, $\Lambda(\mathsf{RCA_0})$ is a perfect learning system in a synchronous world.

8   Conclusion

We argue that the generator-predictor theorem establishes a natural perspective on the effective core of Solomonoff induction by shedding new light on the cause of the incomputability of the non-relativized Solomonoff induction, instead of directly introducing specific resource constraints in order to achieve computability, like this is done, for example, in [7] for the AIξ learning system. This shifts the questions related to learnability to questions related to provability, and therefore into the realm of the foundations of mathematics.

The problem of universal induction in the synchronous learning framework, however, is intrinsically effective, and the focus of future research in a synchronous framework can be on *efficiency* questions. In fact, the source of incomputability in the asynchronous learning framework can be traced back to the fact that the learning system does not know how much time the generator process has "invested" in order to produce the next bit. An extension of a bit sequence s by inserting "clock signals" (coding a clock signal by "00" and output bits by "10" and "11") marking the passing of time would transform every sequence s into a synchronous one, thus eliminating the incomputability of Solomonoff induction. So the synchronous learning framework seems to be perfectly suited for studying the problem of universal induction from a computational point of view.

References

1. Arora, S., Barak, B.: Complexity Theory: A Modern Approach. Cambridge University Press (2009)
2. Friedman, H.: Some systems of second order arithmetic and their use. In: Proceedings of the International Congress of Mathematicians, Vancouver, B.C, vol. 1, pp. 235–242. Canad. Math. Congress, Montreal (1974)
3. Gold, E.M.: Language identification in the limit. Information and Control 10(5), 447–474 (1967)
4. Goodstein, R.: On the restricted ordinal theorem. Journal of Symbolic Logic 9, 33–41 (1944)
5. Huber, F., Schmidt-Petri, C. (eds.): Degrees of Belief. Springer (2009)
6. Hutter, M.: The fastest and shortest algorithm for all well-defined problems. International Journal of Foundations of Computer Science 13(3), 431–443 (2002)
7. Hutter, M.: Universal Artificial Intelligence: Sequential Decisions based on Algorithmic Probability. Springer (2005)
8. Legg, S.: Is There an Elegant Universal Theory of Prediction? In: Balcázar, J.L., Long, P.M., Stephan, F. (eds.) ALT 2006. LNCS (LNAI), vol. 4264, pp. 274–287. Springer, Heidelberg (2006)
9. Li, M., Vitányi, P.M.B.: An introduction to Kolmogorov complexity and its applications, 3rd edn. Graduate Texts in Computer Science. Springer (2008)
10. Rathjen, M.: The art of ordinal analysis. In: Proceedings of the International Congress of Mathematicians, pp. 45–69. Eur. Math. Soc. (2006)
11. Simpson, S.G.: Subsystems of Second Order Arithmetic, 2nd edn. Cambridge University Press (2009)
12. Solomonoff, R.: A formal theory of inductive inference, part I. Information and Control 7(1), 1–22 (1964)
13. Solomonoff, R.: A formal theory of inductive inference, part II. Information and Control 7(2), 224–254 (1964)
14. Weihrauch, K.: Computable analysis. Springer (2000)
15. Zimmermann, J., Cremers, A.B.: The Quest for Uncertainty. In: Calude, C.S., Rozenberg, G., Salomaa, A. (eds.) Maurer Festschrift. LNCS, vol. 6570, pp. 270–283. Springer, Heidelberg (2011)
16. Zimmermann, J., Cremers, A.B.: Proof-driven learning systems (2012) (preprint)

Author Index

Afshari, Bahareh 1
Allender, Eric 11
Allo, Patrick 17
Aman, Bogdan 626
Antunes, Luís 29

Becher, Verónica 35
Beeson, Michael 46
Botman, Daniel 355
Brattka, Vasco 56
Bridges, Douglas S. 68
Brumleve, Dan 78

Carlucci, Lorenzo 89
Case, John 96
Celaya, Marcel 107
Chen, Yijia 118
Cholak, Peter A. 129
Ciobanu, Gabriel 626
Cordón–Franco, Andrés 440
Cremers, Armin B. 745

Das, Anupam 139
Dassow, Jürgen 151
Downey, Rod 162
Dries, Roland 355
Dzhafarov, Damir D. 129

Edmonds, Bruce 182
Ehrig, Hartmut 193
Ermel, Claudia 193

Fang, Chengling 203
Fernández Duque, David 212
Ferreira, Fernando 222
Finlayson, Mark Alan 228
Flum, Jörg 118

Gershenson, Carlos 182
Golan, Amos 237
Grattan-Guinness, Ivor 245
Guillon, Pierre 253
Gurevich, Yuri 264

Hamkins, Joel David 78
Hartmanis, Juris 276

Hartung, Sepp 283
Havea, Robin S. 68
Hendtlass, Matthew 293
Higuchi, Kojiro 303, 581
Hirst, Jeffry L. 129
Hüffner, Falk 193
Huschenbett, Martin 313

Iliev, Petar 323

Jackson, Paul B. 560
Jain, Sanjay 96
Jeandel, Emmanuel 334
Johannson, Anders 344
Joosten, Joost J. 212

Kaandorp, Jaap A. 355
Kartzow, Alexander 363
Kihara, Takayuki 303, 384
Kjos-Hanssen, Bjørn 395
Koepke, Peter 405
Kondo, Shigeru 416
Kooi, Barteld 323
Krishna, Shankara Narayanan 626
Kristiansen, Lars 422
Kulikov, Alexander S. 432

Lara–Martín, F. Félix 440
Lassègue, Jean 450
Le Gloannec, Bastien 462
Le Roux, Stéphane 56
Liu, Jiamou 363
Lohrey, Markus 363
Longo, Giuseppe 450

Manea, Florin 151
Markopoulou, Fotini 472
Melanich, Olga 432
Mercaş, Robert 151
Metcalfe, George 485
Mihajlin, Ivan 432
Millikan, Ruth Garrett 496
Moelius III, Samuel E. 507
Müller, Moritz 118
Murray, James D. 517

Németi, Péter 528
Nichterlein, André 283
Niedermeier, Rolf 193

Ocasio-González, Víctor A. 539
Okuno, Yasushi 549
Ollinger, Nicolas 462

Passmore, Grant Olney 560
Pauly, Arno 56, 571
Peng, NingNing 581
Peretyat'kin, Mikhail G. 589
Podolskii, Vladimir V. 599
Preston, John 609

Rathjen, Michael 1
Rettinger, Robert 616
Röthlisberger, Christoph 485
Runge, Olga 193
Ruskey, Frank 107

Salo, Ville 636
Say, A.C. Cem 646
Schlicht, Philipp 78
Schuster, Peter 293
Seah, Samuel 96
Seki, Shinnosuke 549
Seyfferth, Benjamin 405
Shen, Alexander 655
Shenling, Wang 203
Sherratt, Jonathan A. 667

Smolensky, Paul 675
Souto, Andre 29
Staiger, Ludwig 686
Steiner, Rebecca M. 696
Stephan, Frank 96
Sterrett, Susan G. 703
Székely, Gergely 528
Szudzik, Matthew P. 714

Tamulonis, Carlos 355
Tanaka, Kazuyuki 374, 581
Taveneaux, Antoine 395
Teixeira, Andreia 29
Thapen, Neil 395
Törmä, Ilkka 636

van der Hoek, Wiebe 323

Weiermann, Andreas 723
Wiedermann, Jiří 733
Woods, Alan R. 723
Wu, Guohua 203

Yakaryılmaz, Abuzer 646
Yamazaki, Takeshi 581
Yoshii, Keisuke 374

Zdanowski, Konrad 89
Zimmermann, Jörg 745
Zinoviadis, Charalampos 253
Zou, James 344

—